DYNAMICS OF POLYMERIC LIQUIDS

Volume 1 Fluid Mechanics

Volume 2 Kinetic Theory

DYNAMICS OF POLYMERIC LIQUIDS

VOLUME 1
FLUID MECHANICS

SECOND EDITION

R. BYRON BIRD
Chemical Engineering Department
and Rheology Research Center
University of Wisconsin-Madison
Madison, Wisconsin

ROBERT C. ARMSTRONG
Chemical Engineering Department
Massachusetts Institute of Technology
Cambridge, Massachusetts

OLE HASSAGER
Instituttet for Kemiteknik
Danmarks tekniske Højskole
Lyngby, Danmark

A Wiley-Interscience Publication

JOHN WILEY & SONS

New York • Chichester • Brisbane • Toronto • Singapore

Library of Congress Cataloging-in-Publication Data:

Dynamics of polymeric liquids.

"A Wiley-Interscience publication"
Contents: v. 1. Fluid mechanics/R. Byron Bird, Robert C. Armstrong, Ole Hassager—v. 2. Kinetic theory/R. Byron Bird... [et al.]
1. Polymers and polymerization—Mechanical properties. 2. Polymers and polymerization. I. Bird, R. Byron (Robert Byron), 1924- .

TA357.D95 1987   620.1'92   86-13230
ISBN 0-471-80245-X (v. 1)

Printed in the United States of America

10 9 8 7 6 5

PREFACE

Liquid motion has fascinated many generations of scientists and engineers. Although many years of research have been devoted to the study of fluids of low molecular weight, which are well described by the Navier–Stokes equations, many challenging problems in both theory and applications remain. But even more challenging are polymeric liquids, whose motions cannot be described at all by the Navier–Stokes equations. In Volume 1, *Fluid Mechanics*, we summarize the key experiments that show how polymeric fluids differ from structurally simple fluids, and we then present, in rough historical order, various methods for solving polymer fluid dynamics problems; in Volume 2, *Kinetic Theory*, we use molecular models and the methods of statistical mechanics to obtain relations between bulk flow behavior and polymer structure. Table 1 shows how we have chosen to organize the material of the two volumes.

Understanding polymer fluid dynamics is important in connection with plastics manufacture, performance of lubricants, application of paints, processing of foodstuffs, and movement of biological fluids. Rapid advances in all these areas, as well as in the associated fundamental sciences, have made it desirable to have an introductory textbook for students and research workers. Examples and problems have been included to assist the beginner and almost all the material in both volumes has been classroom tested. These volumes should also be useful as an introduction to the advanced monographs and research literature on continuum physics, rheometry, molecular theory, and polymer processing. Literature citations have been supplied throughout the text for the interested reader. Although these volumes have been prepared with fourth or fifth year chemical engineering and chemistry students in mind, they may also be of interest to students or research workers in polymer engineering and science, molecular biology, mechanical engineering, materials science, and applied mechanics.

Both experiment and theory in this field have undergone rapid development in recent years. Except for a few papers of historical interest, the literature citations are from the period after 1950, and many are from the 1970s and 1980s. These volumes summarize the efforts of our many colleagues around the world; they also contain a number of developments and ideas heretofore unpublished. Unfortunately the following subjects have had to be omitted from the text: optical and electrical phenomena, computer simulation techniques, boundary-layer theory, two-phase systems, thermodynamics, and liquid crystals. Research is going on in all these areas, but we have had to rule them outside the scope of these volumes.

In these two books there is enough material for two one-semester courses: a fluid dynamics course (Chapters 1 to 10), followed by a kinetic theory course (Chapters 11 to 20). Needless to say, many points can be better understood by working through the details of

TABLE 1

Contents of Volumes 1 and 2

specific problems, and to this end many illustrative examples have been included. Of particular importance are the problems at the end of the chapters, which are divided into four classes:

Class A Numerical calculations using equations from the chapter.
Class B Analyses similar to those in the text.
Class C More difficult or lengthy analyses.
Class D Mathematical developments.

Many of these problems are used to augment the material in the main text, and quite a few of them are referred to later in the book. No effort has been spared to provide generous crossreferencing within each volume and between the two volumes, and appendices are included to assist the reader in problem solving. Everywhere in the text and problems we have used SI units; viscosities are given in Pa·s (1 poise = 0.1 Pa·s = 0.1 $\mathrm{N \cdot s/m^2}$) and molecular dimensions are given in nm (1Å = 0.1 nm).

Volumes 1 and 2 cut across several fields and it is impossible to find notation that pleases everyone. Our aim has been to follow the recommendations of the Society of Rheology for the notation for rheological quantities: $\eta^* = \eta' - i\eta''$ for the complex viscosity, η for the shear-rate-dependent viscosity, η_0 for the zero-shear-rate viscosity, Ψ_1 and Ψ_2 for the normal-stress coefficients, and $\dot{\gamma}$ for the shear rate. Like many workers in kinetic theory and transport phenomena we write the total stress tensor as $\boldsymbol{\pi} = p\boldsymbol{\delta} + \boldsymbol{\tau}$, where p is the pressure, $\boldsymbol{\delta}$ is the unit tensor, and $\boldsymbol{\tau}$ is that part of the stress tensor that vanishes at equilibrium; that is, both p and τ_{ii} are positive in compression, and furthermore $\boldsymbol{\tau}$ obeys the well-established sign convention that is used for heat and mass fluxes. Usually continuum mechanicists and mechanical engineers use a sign convention opposite to ours for $\boldsymbol{\pi}$ and $\boldsymbol{\tau}$. Bear in mind that misunderstandings can also arise because of differences in definitions of tensor operations; we use $(\nabla \boldsymbol{v})_{ij} = (\partial/\partial x_i)v_j$, and $[\nabla \cdot \boldsymbol{\tau}]_i = \sum_k (\partial/\partial x_k)\tau_{ki}$. In addition, we define the rate-of-strain tensor as $\dot{\boldsymbol{\gamma}} = \nabla \boldsymbol{v} + (\nabla \boldsymbol{v})^\dagger$ and the vorticity tensor as $\boldsymbol{\omega} = \nabla \boldsymbol{v} - (\nabla \boldsymbol{v})^\dagger$, with no factors of $\frac{1}{2}$ included. Lists of notation are included at the beginning of each volume; Appendices A and E, dealing with vector and tensor operations, should also provide assistance in interpreting the notation.

In preparing these volumes we have drawn extensively on the research literature of the recent decades; we acknowledge with deep appreciation our many colleagues who have been working at the frontiers of the subject. Several of the staff members of the Rheology Research Center of the University of Wisconsin (Professors A. S. Lodge, M. W. Johnson, Jr., J. D. Ferry, J. L. Schrag, S. Kim, D. S. Malkus, and M. Renardy) have provided us with considerable assistance, as did several of our colleagues at the Massachusetts Institute of Technology (Professors R. A. Brown, J. F. Brady, and H. Brenner). Also, we wish to thank Professor M. Tirrell (University of Minnesota), Professor D. G. Baird (Virginia Polytechnic Institute), and their students, who used these volumes in manuscript form and supplied us with extensive lists of improvements. Especially helpful have been the detailed criticisms provided by those who have used the manuscript: H. C. Öttinger, John M. Wiest, Steven R. Burdette, Lewis E. Wedgewood, Tony W. Liu, Jay D. Schieber, Xi-Jun Fan, and Hassane H. Saab from the University of Wisconsin, Miguel A. Bibbo, Lidia M. Quinzani, Ali Nadim, Anthony N. Beris, John V. Lawler, Susan J. Muller, Robert C. King, Minas R. Apelian, and William P. Raiford from the Massachussetts Institute of Technology, and Sue Hirz, Neil A. Dotson, Mike Kent, Ted Frick, Gibson Batch, Robert Secor, and Kurt Hermann from the University of Minnesota.

Very special thanks are due the various organizations without whose financial support our basic research would not have been possible: the National Science Foundation, the U.S. Office of Naval Research, the Petroleum Research Fund, and the Vilas Trust Fund of the University of Wisconsin. Undertaking the preparation of the manuscript for the second edition was made easier by the support provided by the Vilas Trust Fund and the John D. MacArthur Professorship made available to Dr. Bird; in addition Dr. Armstrong wishes to thank Exxon Research and Engineering for financial support during part of the sabbatical year used for writing these volumes, Dr. Curtiss acknowledges financial support provided by the Graduate School of the University of Wisconsin, and Dr. Hassager's stay at the University of Wisconsin for 11 months in 1984 was made possible by salary support provided by the Technical University of Denmark and the Mathematics Research Center at the University of Wisconsin and by a travel grant from the NATO Science Fellowship Program. Manuscript preparation was made easier by our various expert secretaries: Miss Jane A. Smith (University of Wisconsin), Mrs. Glorianne Collver-Jacobson (MIT), and Miss Helga Wolff (DTH).

R. Byron Bird
Madison, Wisconsin

Robert C. Armstrong
Cambridge, Massachusetts

Ole Hassager
Lyngby, Denmark

Charles F. Curtiss
Madison, Wisconsin

January 1987

Comments on Volume 1

During the decade that has elapsed since the first edition of Volume 1 was published, many advances have been made in polymer fluid dynamics; consequently the second edition of the book has been almost completely rewritten. Even the outline of the book has been altered in a major way by postponing the most difficult subjects to the later chapters so that more material will be easily accessible to those persons entering the field. Differences in content between the first and second editions are largely the result of our desire to devote proportionately more space to the solution of fluid dynamics problems and less to rheology and continuum mechanics.

In the first three chapters and in Appendix A we provide background material for the study of polymer fluid dynamics. Chapter 1 provides a review of some of the main ideas of fluid dynamics as well as some important results for Newtonian fluids that are needed later. Appendix A on vectors and tensors is included primarily as reference material. The subject of polymer fluid dynamics is introduced in Chapter 2 through a series of photographs and drawings to show what effects have been seen in the laboratory. Experimentally observed phenomena must be kept in mind at all times as we proceed to the theoretical discussions that follow. Despite the great advances made in recent decades in the theoretical aspects of the subject, polymer fluid dynamics is still very much a field that relies heavily on experimental observations. The experimentally measured rheological properties such as viscosity, normal stresses, stress relaxation, and recoil are discussed in Chapter 3, and frequent reference is made in later chapters to the data summarized here.

Once the background material of Chapters 1–3 has been presented, we are ready to go on to the study of polymer fluid dynamics calculations. The accompanying table lists types of flow problems that we show how to solve. Here we set forth the scope and aims of Chapters 4–9. Each chapter features a certain class of constitutive equations (i.e., stress tensor expressions) as well as the kinds of flows for which it is appropriate.

Many important industrial problems involve flows that are approximately steady-state shear flows, and for these problems we can use the relatively simple generalized Newtonian fluid of Chapter 4. Engineers have used models of this type for over 50 years to get approximate solutions to flow problems as well as heat- and mass-transfer problems. Many time-dependent problems, of interest to polymer chemists, are well described by the linear viscoelastic fluid, which is the subject of Chapter 5. Of course modeling of this sort is limited strictly to motions with vanishingly small displacement gradients. "Retarded motions"—those in which the velocity gradients and their time derivatives are small—can be studied in a satisfying way by the retarded-motion expansion as described in Chapter 6. Yielding as they do to analytical solutions, these retarded-motion flows provide insight into the onset of elastic effects and, in particular, into the elucidation of interesting secondary flow phenomena.

Of greatest current interest in polymer fluid dynamics, however, are the nonlinear differential models of Chapter 7 and the nonlinear integral models of Chapter 8; these chapters are designed to describe arbitrary flow fields and to some extent they include the

TABLE 1

Chapter	Fluid Model	Examples	Flows Described	Main Applications
4	Generalized Newtonian fluid	Power-Law Model Bingham Model Carreau–Yasuda Model	Steady shear flows (viscous effects only)	Volume flow rate versus pressure drop Torque versus angular velocity Rough engineering flow calculations Heat and mass transfer
5	General linear viscoelastic fluid	Maxwell Model Jeffreys Model Generalized Maxwell Model	Small displacement gradients	Structure–property relations Fluid characterization Quality control Wave transmission
6	Retarded motion expansion	2nd-Order Fluid 3rd-Order Fluid	Velocity gradients and their time derivatives are small	Motion of bubbles and drops Particle orientation in suspensions Onset of secondary flows Limiting cases of arbitrary flows
7	Differential models	Convected Maxwell Convected Jeffreys White–Metzner Oldroyd Giesekus	Arbitrary	Exploratory viscoelastic calculations Stability analyses Rough process modeling (e.g., fiber spinning) Simulation of viscoelastic flows
8	Integral models	Lodge rubberlike liquid Rivlin–Sawyers fluid K-BKZ fluid	Arbitrary	Interrelation of rheological data Quality control Simulation of viscoelastic flows
9	Memory-integral expansion Criminale–Ericksen–Filbey equation Reiner–Rivlin equation		Arbitrary Steady shear flows Steady homogeneous irrotational flows	Interrelation of constitutive equations Normal stress effects in steady shear flows

models of Chapters 4, 5, and 6 as special cases. For the most part the constitutive equations presented in these chapters are empirical nonlinear extensions of the linear viscoelastic models of Chapter 5, but the most successful ones are those that have at least some molecular basis. Just about all of the constitutive relations that we emphasize in these chapters have been extensively tested by comparison with rheological data, and many have been used for fluid dynamics calculations.

After completing Chapter 8 the reader will have had a fairly thorough introduction to polymer fluid dynamics and constitutive modeling. Moreover the reader can get through Chapter 8 with only a modest knowledge of differential equations, tensors, and matrices, since the continuum mechanics basis for the earlier chapters is not given until Chapter 9. Effectively what we have done is to make the study of polymer fluid dynamics somewhat more accessible by using the main results of Chapter 9 prior to giving the mathematical justification.

Students of polymer fluid dynamics will ultimately want to learn more about the fundamentals of continuum mechanics, and Chapter 9 provides a short introduction to this branch of classical mechanics. Giving a brief survey of the analysis of the rheometric experiments in Chapter 10 serves to introduce the reader to another very large subject that is very important in polymer fluid dynamics. Only after the reader is familiar with constitutive equations and continuum mechanics is it really possible to understand the fluid dynamics description of the key measuring instruments that provide us with the basic rheological data. Last, but not least, we point out that Appendices A, B, and C contain considerable tabular material that can facilitate the setting up and solving of fluid dynamics problems.

Deletion of all the material on the corotational description of continuum mechanics in Chapters 7 and 8 of the first edition represents a major change in viewpoint, which perhaps requires some comment. Relatively few researchers have chosen to use the corotational formalism, and furthermore molecular theories seem to favor the codeformational point of view, which leads to the convected derivatives and finite strain tensors that we use in Chapters 6–10 of the second edition. Of course, in the retarded motion expansion of Chapter 6 and in the differential models of Chapter 7, either system may be used equally well, and interrelating the two systems involves only notational changes. Yet situations may arise where the use of a corotational point of view may be advantageous, in which case the reader will have to consult the first edition of this book.

Despite extensive checking and rechecking by the authors, some errors will inevitably creep into the final published book; we hope that readers will assist us by letting us know of errors that they come across.

R.B.B.
R.C.A.
O.H.

CONTENTS
VOLUME 1

NOTES ON NOTATION
FOR VOLUME 1

Vector and Tensor Notation (see Appendix A)

Fluid Properties

μ	Viscosity of Newtonian fluids	Eq. 1.2-1
k	Thermal conductivity	Eq. 1.2-3
ρ	Density	
$\hat{C}_p$	Heat capacity at constant pressure (per mass unit)	

Fluid Dynamics

$\boldsymbol{v}$	Fluid velocity	
p	Pressure	Eq. 1.2-2
$\mathscr{P}$	Modified pressure	Table 1.2-1
$\boldsymbol{\pi}$	Total stress tensor $(= p\boldsymbol{\delta} + \boldsymbol{\tau})$	Eq. 1.2-1
$\boldsymbol{\tau}$	Stress tensor (vanishes at equilibrium)	Eq. 1.2-1
t	Time	
$\dfrac{D}{Dt}\boldsymbol{\tau}$	Substantial (material) time derivative $\left(= \dfrac{\partial}{\partial t}\boldsymbol{\tau} + \{\boldsymbol{v}\cdot\nabla\boldsymbol{\tau}\}\right)$	Table 1.2-1
$\boldsymbol{\tau}_{(1)}$	Convected time derivative of $\boldsymbol{\tau}$ $\left(= \dfrac{D}{Dt}\boldsymbol{\tau} - \{(\nabla\boldsymbol{v})^\dagger\cdot\boldsymbol{\tau} + \boldsymbol{\tau}\cdot\nabla\boldsymbol{v}\}\right)$	Eq. 7.1-1 Table 9.4-1
ψ	Stream function	Eqs. 1.4-2 and 3 Table 1.4-1

Kinematic Tensors

$\nabla\boldsymbol{v}$	Velocity gradient tensor, with ijth component $(\partial/\partial x_i)v_j$	Eq. 6.1-4
$\dot{\boldsymbol{\gamma}} = \boldsymbol{\gamma}_{(1)} = \boldsymbol{\gamma}^{(1)}$	Rate-of-strain tensor $(= \nabla\boldsymbol{v} + (\nabla\boldsymbol{v})^\dagger)$	Eq. 1.2-2 Eq. 5.1-2 Eq. 6.1-2
$\boldsymbol{\omega}$	Vorticity tensor $(= \nabla\boldsymbol{v} - (\nabla\boldsymbol{v})^\dagger)$	Eq. 6.1-3
$\boldsymbol{\gamma}_{(n)}, \boldsymbol{\gamma}^{(n)}$	nth rate-of-strain tensors	Eqs. 6.1-5 and 6 Table 9.3-1
$\boldsymbol{\gamma}_{[0]}, \boldsymbol{\gamma}^{[0]}$	Relative finite strain tensors	Eqs. 8.1-9 and 8 Table 9.3-1

$\boldsymbol{\gamma}$	Infinitesimal strain tensor	Eq. 5.1-5
B	Finger strain tensor ($=\boldsymbol{\delta} - \boldsymbol{\gamma}_{[0]}$)	Eq. 8.1-7
$\mathbf{B}^{-1}$	Cauchy strain tensor ($=\boldsymbol{\delta} + \boldsymbol{\gamma}^{[0]}$)	Eq. 8.1-6
Δ, E	Displacement gradient tensors	Eqs. 8.1-3 and 4

Other Kinematic Quantities

II, III	Invariants: $II = \text{tr}\,\dot{\boldsymbol{\gamma}}^2$, $III = \text{tr}\,\dot{\boldsymbol{\gamma}}^3$	Eqs. 4.1-6 and 7
I_1, I_2	Invariants: $I_1 = \text{tr}\,\mathbf{B}$, $I_2 = \text{tr}\,\mathbf{B}^{-1}$	Eq. 8.3-4
$\dot{\gamma}$	Magnitude of $\dot{\boldsymbol{\gamma}}(=\sqrt{\frac{1}{2}II})$	Eq. 4.1-8
$\dot{\gamma}$	Shear rate	Eq. 3.1-1
$\dot{\varepsilon}$	Elongation rate	Eq. 3.1-3
b	Shearfree flow parameter	Eq. 3.1-3

Quantities in Constitutive Equations (see Tables 7.3-2, 7.6-1, 8.3-2, and 8.3-3)

η_0	Zero-shear-rate viscosity	Eqs. 4.1-9, 5.2-2, and 5.2-9
η_∞	Infinite-shear-rate viscosity	Eq. 4.1-9
η_j	Constants with dimensions of viscosity	Eqs. 5.2-14 and 15
λ	Time constant	Eq. 4.1-9
λ_1	"Relaxation time"	Eqs. 5.2-2, 5.2-9, and 7.2-1
λ_2	"Retardation time"	Eqs. 5.2-9, 7.2-1
λ_j	Time constants	Eqs. 5.2-14 and 15; Eq. 7.3-2
m, n	Power-law constants for η	Eq. 4.1-10
m', n'	Power-law constants for Ψ_1	Eq. 4.2-68
α	Dimensionless constant	Eq. 7.3-3
b_1, b_2, b_{11}, etc.	Retarded-motion constants	Eq. 6.2-1
$G(t-t')$	Relaxation modulus	Eq. 5.2-18
$M(t-t')$	Memory function	Eqs. 5.2-19 and 8.3-11
$W(I_1, I_2)$	Potential function	Eq. 8.3-11
$\phi_1(I_1, I_2)$, $\phi_2(I_1, I_2)$	Damping functions	Eq. 8.3-12

Material Functions (see Tables 3.4-1 and 3.5-1)

$\eta(\dot{\gamma})$	(Non-Newtonian) viscosity	Eq. 3.3-1
$\Psi_1(\dot{\gamma})$	First normal stress coefficient	Eq. 3.3-2
$\Psi_2(\dot{\gamma})$	Second normal stress coefficient	Eq. 3.3-3
$[\eta]$	Intrinsic viscosity	Eq. 3.3-6
$\eta^*(\omega)$	Complex viscosity ($=\eta' - i\eta''$)	Eq. 3.4-3b
$G^*(\omega)$	Complex modulus ($=G' + iG''$)	Eq. 3.4-3a
$\bar{\eta}(\dot{\varepsilon})$	Elongational viscosity	Eq. 3.5-3
$\bar{\eta}_1(\dot{\varepsilon}, b)$, $\bar{\eta}_2(\dot{\varepsilon}, b)$	Shearfree flow viscosities	Eqs. 3.5-1 and 2
(subscript) 0	Quantity evaluated at zero shear rate or zero elongation rate	
(superscript) −	Quantity associated with stress relaxation after steady flow	
(superscript) +	Quantity associated with stress growth at inception of steady flow	

Dimensionless Groups

De	Deborah number	Eq. 2.8-1
We	Weissenberg number	Eq. 2.8-2

Re	Reynolds number	Tables 1.2-1 and 4.4-1
Ca	Capillary number	Eq. 6.5-57
Fr	Froude number	Tables 1.2-1 and 4.4-1
Br	Brinkman number	Tables 1.2-1 and 4.4-1
Na	Nahme–Griffith number	Table 4.4-1
Pé	Péclet number	Tables 1.2-1 and 4.4-1
Gz	Graetz number	Tables 4.4-4 and 5
St	Stanton number	Eq. 2.7-2
Nu	Nusselt number	Eqs. 4.4-27 and 35

General

k	Boltzmann's constant	
M	Molecular weight	
$\bar{M}_w$, $\bar{M}_n$, $\bar{M}_v$	Average molecular weights	Eqs. 2.1-1 and 2
$\tilde{N}$	Avogadro's number	
T	Absolute temperature	

Coordinates (see Appendix A)

x, y, z	Cartesian
r, θ, z	Cylindrical
r, θ, ϕ	Spherical
ξ, θ, z	Bipolar

Mathematical Symbols (see also Appendix A)

$\Gamma(x)$	Gamma function $\left(=\int_0^\infty t^{x-1}e^{-t}\,dt\right)$	
$\boldsymbol{\delta}$	Unit tensor	Eq. A.3-10
$\boldsymbol{\delta}_i$	Unit vector	Fig. A.2-1
δ_{ij}	Kronecker delta	Eqs. A.2-1 and 2
$\delta(x)$	Dirac delta function	Eq. 5.2-10a
$H(x)$	Heaviside unit step function	Problem 5B.6
$\mathrm{erf}(x)$	Error function $\left(=\frac{2}{\sqrt{\pi}}\int_0^x e^{-t^2}\,dt\right)$	
$\mathcal{Im}(z)$	Imaginary part of the comlex number z	
$\mathcal{Re}(z)$	Real part of the complex number z	
$\zeta(n)$	Riemann zeta function $\left(=\sum_{k=1}^{\infty} k^{-n}\right)$	
$\boldsymbol{n}$	Outwardly directed unit normal vector	
$[\ldots]^{(n)}$, $[\ldots]_{(n)}$	Convected time derivatives of $[\ldots]$	
$\overline{(\cdots)}$	Complex conjugate of $(\cdots)$	

PART I

NEWTONIAN VERSUS NON-NEWTONIAN BEHAVIOR

The first three chapters of this volume are intended to provide background material for the remainder of the book, which deals with the fluid dynamics of polymeric liquids. Some or all of these three chapters may be unnecessary, depending on the background of the reader.

Before embarking on a study of the flow of polymeric fluids, which are non-Newtonian, it is important that the reader have some minimal acquaintance with the equations of Newtonian fluid dynamics as well as some of the techniques for solving them. In Chapter 1 we give a very brief summary of the derivation of the basic equations of fluid dynamics, and then we specialize them for the Newtonian fluid. In the illustrative examples and in the problems at the end of the chapter we present a number of fluid dynamics problems that can be solved exactly or approximately. Although the equations of classical fluid dynamics have been known for a century and a half, there are very few nontrivial flow problems that admit analytic solutions. We must keep this in mind when we get to later chapters, where we find that only a small handful of nonlinear viscoelastic flow problems have been solved. Newtonian fluid dynamics has been found to be completely satisfactory for describing the flow phenomena of gases and of liquids containing small molecules—that is, molecules with a molecular weight of less than about 1000. In Chapter 1 we do not actually show any comparisons between theory and experiment, inasmuch as decades of experiments have demonstrated the dependability of the theory sufficiently well.

After this brief review of Newtonian fluid mechanics, we then turn our attention to the kinds of flow phenomena that have been observed in polymeric fluids—polymer solutions, polymer melts, and two-phase systems in which a polymeric fluid is the continuous phase. In Chapter 2 we go through a dozen or so experiments in which polymeric liquids behave quite differently from liquids made up of small molecules. Some of these experiments involve flow situations that one might encounter in industry and are therefore important from the standpoint of developing intuition about qualitative behavior. Others have no obvious connection with practical problems, but are useful in illustrating the weird world of polymer fluid dynamics. Chapter 2 is intended to be a rather light-hearted chapter, and its main purpose is to catalyze interest in a very challenging subject. By the time the reader has finished it, he should be thoroughly convinced that polymer liquids do not obey the laws of classical (i.e., Newtonian) fluid dynamics; he should also be aware of the fact that there are enormous qualitative differences between polymeric fluids and fluids made up of small molecules, not just minor quantitative differences.

The flow patterns in the experiments of Chapter 2 are rather complicated. For the most part, they are not suitable for making very carefully controlled and refined measurements. In Chapter 3, therefore, we turn to the study of extremely simple and well-defined flows in which there are only a few nonzero components of the stress tensor. By measuring one or more of these stress components, we can obtain experimental data on the viscosity and other rheological properties of polymeric liquids. These various properties are called "material functions", and these measured functions are important for quality control in industry as well as for the evaluation of rheological and molecular theories. In Chapter 3 we give a sampling of experimental data on material functions for various types of polymeric liquids.

Chapters 2 and 3, then, emphasize experimental facts about polymeric liquids. The field of polymer fluid dynamics is still very much an experimental subject. Many important industrial problems must be solved by use of flow visualization and other experimental techniques. As our fundamental knowledge of the subject increases, tedious experimental approaches will gradually give way to computer modeling. Even then there will always be a need for carefully measured material funtions and for measurements of velocity and stress fields. Although most of this book deals with the theoretical development of the subject, the formulation of analytical descriptions of non-Newtonian flow phenomena, and the molecular theories of rheology, it is extremely important that the reader recognize the limitations of the theoretical approach.

CHAPTER 1

REVIEW OF NEWTONIAN FLUID DYNAMICS

In this introductory chapter we begin by giving the equations that describe the flow of fluids. These are the "equations of change," which indicate how the mass, momentum, and energy of the fluid change with position and time. In §1.1 we present these equations in terms of the fluxes so that they are valid for any kind of fluid, and in §1.2 we specialize the equations for the "Newtonian fluid." In §1.3 we illustrate the use of the equations for solving a variety of isothermal flow problems. In §1.4 we introduce the stream-function method as a powerful analytic technique for the solution of some flow problems.

In §§1.1 and 1.2 we present all the equations that are needed for solving nonisothermal flow problems and heat conduction problems. However, no illustrative examples are given for heat transfer. Nonisothermal flows are discussed in Chapter 4 for generalized Newtonian fluids, and full use is made there of the energy equation derived in §§1.1 and 1.2.

At the end of this chapter there is a rather long set of problems. Many of these problems involve the solution of the hydrodynamic equations for Newtonian fluids. Almost all of these flow problems are encountered again in later chapters, where they are solved for polymeric fluids using "non-Newtonian" models. Then it will be very convenient to have the Newtonian fluid solutions available.

§1.1 THE EQUATIONS OF CHANGE IN TERMS OF THE FLUXES

The motion of any fluid is described by the equations of conservation of mass, momentum, and energy. These equations are shown in two forms in Table 1.1-1. Since these equations form the basis for all of the subject material of the book, it is reasonable to discuss their derivation briefly. A more elementary derivation can be found elsewhere.[1]

In the derivation we consider an arbitrary *fixed* region in space of volume V and surface S, as shown in Fig. 1.1-1; sometimes such a mathematical region is referred to as a "control volume." On every surface element dS there is an outwardly directed normal unit vector $\boldsymbol{n}$. We imagine that this fixed region is in the midst of a fluid flow field and that the fluid moves across the boundaries of the region. We now want to apply the laws of conservation of mass, momentum, and energy to the fluid contained within this fixed region. It is presumed that the reader is familiar with the material on vector and tensor notation in §§A.1-A.7.

[1] See, for example, R. B. Bird, W. E. Stewart, and E. N. Lightfoot, *Transport Phenomena*, Wiley, New York (1960), Chapts. 3 and 10.

TABLE 1.1-1

The Equations of Change in Terms of the Fluxes π and q

	In terms of $\frac{\partial}{\partial t}$	
Mass:	$\frac{\partial}{\partial t}\rho = -(\nabla \cdot \rho \boldsymbol{v})$	(A)
Momentum:	$\frac{\partial}{\partial t}\rho \boldsymbol{v} = -[\nabla \cdot \rho \boldsymbol{v}\boldsymbol{v}] - [\nabla \cdot \boldsymbol{\pi}] + \rho \boldsymbol{g}$	(B)
Energy:	$\frac{\partial}{\partial t}\rho \hat{U} = -(\nabla \cdot \rho \hat{U}\boldsymbol{v}) - (\nabla \cdot \boldsymbol{q}) - (\boldsymbol{\pi} : \nabla \boldsymbol{v})$	(C)
	In terms of $\frac{D}{Dt} = \frac{\partial}{\partial t} + (\boldsymbol{v} \cdot \nabla)$	
Mass:	$\frac{D\rho}{Dt} = -\rho(\nabla \cdot \boldsymbol{v})$	(D)
Momentum:	$\rho \frac{D\boldsymbol{v}}{Dt} = -[\nabla \cdot \boldsymbol{\pi}] + \rho \boldsymbol{g}$	(E)
Energy:	$\rho \frac{D\hat{U}}{Dt} = -(\nabla \cdot \boldsymbol{q}) - (\boldsymbol{\pi} : \nabla \boldsymbol{v})$	(F)

a. Conservation of Mass

Suppose that at the infinitesimal surface element dS the fluid is crossing the surface of V with a velocity $\boldsymbol{v}$. Then the local volume rate of flow of fluid across dS is $(\boldsymbol{n} \cdot \boldsymbol{v})dS$. If the flow is outward, then $(\boldsymbol{n} \cdot \boldsymbol{v})dS$ is positive, whereas if the flow is inward $(\boldsymbol{n} \cdot \boldsymbol{v})dS$ is negative.[2] The local mass rate of flow is then $(\boldsymbol{n} \cdot \rho\boldsymbol{v})dS$. Note that $\rho\boldsymbol{v}$ is the mass flux (i.e., mass per unit area per unit time).

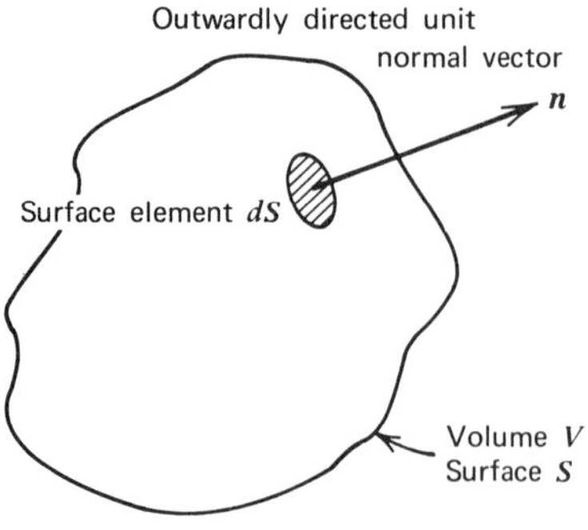

FIGURE 1.1-1. Arbitrary "control volume," fixed in space, over which mass, momentum, and energy balances are made.

[2] Recall that $(\boldsymbol{n} \cdot \boldsymbol{v})$ can be interpreted as the component of $\boldsymbol{v}$ in the direction of $\boldsymbol{n}$.

According to the law of conservation of mass, the total mass of fluid within V will increase only because of a net influx of fluid across the bounding surface S. Mathematically this is stated as

$$\frac{d}{dt}\int_V \rho\, dV = -\int_S (\boldsymbol{n}\cdot\rho\boldsymbol{v})dS \tag{1.1-1}$$

Rate of increase of mass of fluid within V	Rate of addition of mass across the surface S

When Gauss's divergence theorem (§A.5) is used, the surface integral can be transformed into a volume integral;

$$\frac{d}{dt}\int_V \rho\, dV = -\int_V (\boldsymbol{\nabla}\cdot\rho\boldsymbol{v})dV \tag{1.1-2}$$

The equation may be rearranged by bringing the time derivative inside the integral. This is permissible since the volume V is fixed. This gives:

$$\int_V \left[\frac{\partial\rho}{\partial t} + (\boldsymbol{\nabla}\cdot\rho\boldsymbol{v})\right]dV = 0 \tag{1.1-3}$$

We now have an integral over an arbitrary volume and this integral is equated to zero. Because of the arbitrariness of the volume, the integrand may now be set equal to zero. This gives:

$$\boxed{\frac{\partial\rho}{\partial t} = -(\boldsymbol{\nabla}\cdot\rho\boldsymbol{v})} \tag{1.1-4}$$

which is called the *equation of continuity*. If the fluid has constant density, then Eq. 1.1-4 simplifies to:

$$(\boldsymbol{\nabla}\cdot\boldsymbol{v}) = 0 \tag{1.1-5}$$

Use is often made of this relation for "incompressible fluids".

b. Conservation of Momentum

As pointed out above, the local volume rate of flow of a fluid across the surface element dS is $(\boldsymbol{n}\cdot\boldsymbol{v})dS$. If this is multiplied by the momentum per unit volume of the fluid, we then get $(\boldsymbol{n}\cdot\boldsymbol{v})\rho\boldsymbol{v}\, dS$; this is the rate at which momentum is carried across the element of surface dS just because the fluid itself flows across dS. This latter expression can be rearranged as $[\boldsymbol{n}\cdot\rho\boldsymbol{v}\boldsymbol{v}]dS$, and the quantity $\rho\boldsymbol{v}\boldsymbol{v}$ is the momentum flux (i.e., momentum per unit area per unit time) associated with the bulk flow of fluid. Sometimes this transport associated with bulk flow is referred to as "convective transport."

Note that there is a parallelism between the first paragraph of this subsection and the first paragraph of the preceding subsection, but that the tensorial order of the quantities involved is different. In the preceding discussion the entity being transported is mass (a scalar), and the mass flux is a vector ($\rho\boldsymbol{v}$). Here the entity being transported is momentum (a vector), and the momentum flux is a tensor ($\rho\boldsymbol{vv}$).

In addition to momentum transport by flow, there will also be momentum transferred by virtue of the molecular motions and interactions within the fluid. This additional momentum flux will be designated by the symbol $\boldsymbol{\pi}$, again a second-order tensor. We use the convention that the *ij*-component of this tensor π_{ij} represents the flux of positive *j*-momentum in the positive *i*-direction, associated with molecular processes. The rate of flow of momentum, resulting from molecular motions, across the element of surface dS with orientation $\boldsymbol{n}$ is then $[\boldsymbol{n}\cdot\boldsymbol{\pi}]dS$. It will be assumed throughout the entire book that $\boldsymbol{\pi}$ is a symmetric tensor (i.e., $\pi_{ij} = \pi_{ji}$). Thus far most of the kinetic theories for simple and macromolecular fluids yield symmetric momentum-flux tensors (for an exception see Example 18.4-2), and no experiments have been performed that enable one to measure any nonsymmetrical contributions.

We are now ready to write down the law of conservation of momentum. According to this law, the total momentum of the fluid within V will increase because of a net influx of momentum across the bounding surface—both by bulk flow and by molecular motions—and because of the external force of gravity acting on the fluid. When translated into mathematical terms this becomes

$$\frac{d}{dt}\int_V \rho\boldsymbol{v}\,dV = -\int_S [\boldsymbol{n}\cdot\rho\boldsymbol{vv}]dS - \int_S [\boldsymbol{n}\cdot\boldsymbol{\pi}]dS + \int_V \rho\boldsymbol{g}\,dV \tag{1.1-6}$$

Rate of increase of momentum of fluid within V	Rate of addition of momentum across S by bulk flow	Rate of addition of momentum across S by molecular motions and interactions	Force on fluid within V by gravity

where $\boldsymbol{g}$ is the force per unit mass due to gravity. Application of the Gauss divergence theorem enables us to rewrite the surface integrals as volume integrals:

$$\int_V \frac{\partial}{\partial t}\rho\boldsymbol{v}\,dV = -\int_V [\nabla\cdot\rho\boldsymbol{vv}]dV - \int_V [\nabla\cdot\boldsymbol{\pi}]dV + \int_V \rho\boldsymbol{g}\,dV \tag{1.1-7}$$

Then since the volume V is arbitrary, the integral signs may be removed to obtain

$$\boxed{\frac{\partial}{\partial t}\rho\boldsymbol{v} = -[\nabla\cdot\rho\boldsymbol{vv}] - [\nabla\cdot\boldsymbol{\pi}] + \rho\boldsymbol{g}} \tag{1.1-8}$$

and this is the *equation of motion.*

Before continuing it is appropriate to give an alternative interpretation of the tensor $\boldsymbol{\pi}$ and its components. In the derivation of Eq. 1.1-8 we could have used a somewhat different physical statement leading up to Eq. 1.1-6: the total momentum of the fluid within

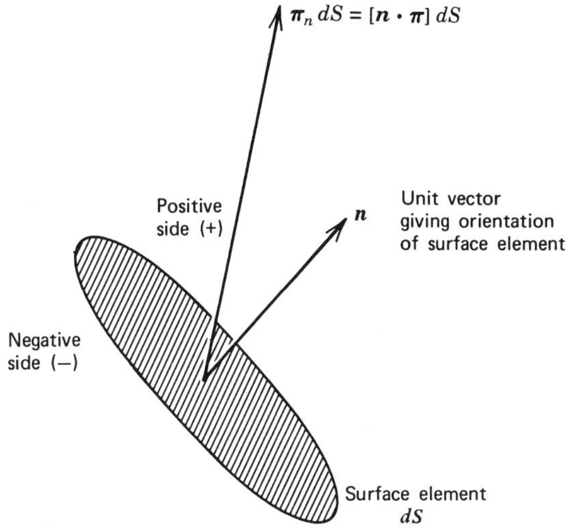

FIGURE 1.1-2. Element of surface dS across which a force $\pi_n\,dS$ is transmitted.

V will increase because of a net influx of momentum across the bounding surfaces by bulk flow, and because of the external forces acting on the fluid—both the surface force exerted by the surrounding fluid and the body force exerted by gravity. The surface force term in Eq. 1.1-6 would then have the form $-\int \pi_n\,dS$, where $\pi_n\,dS$ is a vector describing the force exerted by the fluid on the negative side of dS on the fluid on the positive side of dS (Fig. 1.1-2). Comparison of the above integral with the corresponding term in Eq. 1.1-6 shows that $\pi_n = [n \cdot \pi]$. That is, the force $\pi_n\,dS$ corresponding to any orientation n of dS can be obtained from the tensor π. When this interpretation is used, it is more natural to refer to π as the "stress tensor" (the term "pressure tensor" is also used). The component π_{ij} is the force per unit area acting in the positive j-direction on a surface perpendicular to the i-direction, the force being exerted by the negative material on the positive material (see Fig. 1.1-3).

If one uses this viewpoint, then the integral $-\int [n \cdot \pi]dS$ in Eq. 1.1-6 can be reinterpreted as "force of the fluid outside V acting on the fluid inside V across S." For some purposes it is useful to think of π as a momentum flux, whereas in some situations the concept of stress is more natural. We shall feel free to use both interpretations and the terms "momentum flux tensor" and "stress tensor" will be used interchangeably.[3]

[3] In most treatises on applied mechanics and mechanical engineering a different convention is used for the stress tensor. A stress tensor σ is defined by $\pi_{-n}\,dS = -\pi_n\,dS = [\sigma \cdot n]dS$; that is, one thinks about the fluid on the "positive side" exerting a force on the fluid on the "negative side." The tensors π and σ are related by $\pi = -\sigma^\dagger$, where † stands for "transpose"; thus they differ in sign and in the order of the indices. Since the stress tensor is usually assumed to be symmetric, the change in the order of the indices is not particularly worrisome, but the difference in sign convention is important. We have two reasons for preferring the convention adopted here: (i) In one-dimensional heat conduction described by Fourier's law $q_y = -k(dT/dy)$, it is customary to define q_y so that it is positive when heat is moving in the $+y$-direction, that is, when the temperature decreases with increasing y. Similarly in one-dimensional diffusion, described by Fick's law, $j_{Ay} = -\mathscr{D}_{AB}\,d\rho_A/dy$, the mass flux is defined as positive when species A is moving in the $+y$-direction, in the direction of decreasing concentration. Therefore in a shear flow described by Newton's law $\pi_{yx} = -\mu\,dv_x/dy$ it seems natural to define π_{yx} so that it is positive when x-momentum is moving in the $+y$-direction, that is, in the direction of decreasing velocity. Thus the linear laws for all three transport phenomena are formulated with the same sign convention. (ii) When the total stress tensor is broken down into two parts, a pressure contribution and a viscous contribution, as shown in Eq. 1.2-1, both parts have the same sign—that is, compression is positive in both terms in accordance with the sign convention normally used in thermodynamics.

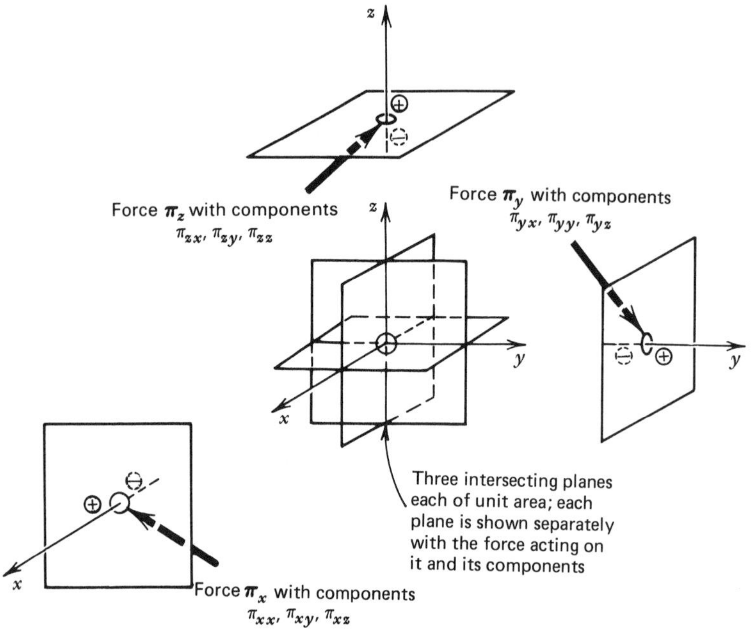

FIGURE 1.1-3. Sketch showing the sign convention and the index convention for the components of the stress tensor $\boldsymbol{\pi}$.

We conclude this subsection by giving two additional equations that may be obtained from Eq. 1.1-8. By forming the dot product of the local fluid velocity vector $\boldsymbol{v}$ with the entire equation of motion and using the equation of continuity we obtain an equation of change for kinetic energy:

$$\frac{\partial}{\partial t}(\tfrac{1}{2}\rho v^2) = -(\nabla \cdot \tfrac{1}{2}\rho v^2 \boldsymbol{v}) - (\boldsymbol{v} \cdot [\nabla \cdot \boldsymbol{\pi}]) + \rho(\boldsymbol{v} \cdot \boldsymbol{g}) \tag{1.1-9}$$

Furthermore, by forming the cross product of the position vector $\boldsymbol{r}$ with the equation of motion and assuming that the stress tensor $\boldsymbol{\pi}$ is symmetrical we obtain an equation of change for angular momentum:

$$\frac{\partial}{\partial t}(\rho[\boldsymbol{r} \times \boldsymbol{v}]) = -[\nabla \cdot \rho \boldsymbol{v}[\boldsymbol{r} \times \boldsymbol{v}]] - [\nabla \cdot \{\boldsymbol{r} \times \boldsymbol{\pi}\}^{\dagger}] + [\boldsymbol{r} \times \rho \boldsymbol{g}] \tag{1.1-10}$$

in which the symbol † stands for the transpose of a tensor. Note that the last two equations do not introduce additional unknowns, nor do they include new information.

c. Conservation of Energy

Once again we return to the point made earlier that the local volume rate of flow of a fluid across the surface element dS is $(\boldsymbol{n} \cdot \boldsymbol{v})dS$. If this is multiplied by $\frac{1}{2}\rho v^2$, which is the kinetic energy per unit volume, then we get $(\boldsymbol{n} \cdot \frac{1}{2}\rho v^2 \boldsymbol{v})dS$, which is the rate of convective flow of kinetic energy across the surface. Similarly, if $\hat{U}$ is the internal energy per unit mass, then $(\boldsymbol{n} \cdot \rho \hat{U} \boldsymbol{v})dS$ is the rate of convective flow of internal energy across the surface element.

In addition to the flow of kinetic and internal energy across dS because of the fluid flowing across dS, there may be energy transferred by molecular motions. This additional mode of transport is associated with heat conduction, and we designate this heat flux by the symbol $\boldsymbol{q}$. Then the heat flow across dS by heat conduction will be $(\boldsymbol{n} \cdot \boldsymbol{q})dS$.

As the fluid moves outwardly across the surface of V, it will do work on the fluid outside of V. It was mentioned earlier that $[\boldsymbol{n} \cdot \boldsymbol{\pi}]dS$ is the force exerted by the fluid *inside* V on the fluid *outside* V across dS. The rate of doing work[4] by the fluid inside V on the fluid outside V will be given by the scalar product of the force and the velocity. Hence the rate of doing work across dS is $([\boldsymbol{n} \cdot \boldsymbol{\pi}] \cdot \boldsymbol{v})dS$, which may also be written $(\boldsymbol{n} \cdot [\boldsymbol{\pi} \cdot \boldsymbol{v}])dS$.

We are now in a position to write down the law of conservation of energy (or the first law of thermodynamics) in mathematical terms. According to this law, the rate of increase of the sum of the kinetic and internal energies will equal the rate of energy addition (by flow and by heat conduction) minus the rate at which the fluid inside V is doing work (this includes work against the surrounding fluid across dS and work against the external force of gravity). When translated into mathematical symbols this becomes:

$$\frac{d}{dt}\int_V (\tfrac{1}{2}\rho v^2 + \rho\hat{U})dV = -\int_S (\boldsymbol{n} \cdot (\tfrac{1}{2}\rho v^2 + \rho\hat{U})\boldsymbol{v})dS$$

Rate of increase of kinetic and internal energy within V	Rate of addition of kinetic and internal energy across S by bulk flow

$$-\int_S (\boldsymbol{n} \cdot \boldsymbol{q})dS - \int_S (\boldsymbol{n} \cdot [\boldsymbol{\pi} \cdot \boldsymbol{v}])dS + \int_V (\boldsymbol{v} \cdot \rho\boldsymbol{g})dV \quad (1.1\text{-}11)$$

Rate of addition of energy across S by molecular motions	Rate at which fluid outside V does work against fluid inside V	Rate at which gravity does work on the fluid

Applying the divergence theorem and then removing the integral signs as before, we get

$$\frac{\partial}{\partial t}(\tfrac{1}{2}\rho v^2 + \rho\hat{U}) = -(\nabla \cdot (\tfrac{1}{2}\rho v^2 + \rho\hat{U})\boldsymbol{v}) - (\nabla \cdot \boldsymbol{q}) - (\nabla \cdot [\boldsymbol{\pi} \cdot \boldsymbol{v}]) + (\rho\boldsymbol{v} \cdot \boldsymbol{g}) \quad (1.1\text{-}12)$$

From this equation, which is an equation of change for the sum of the kinetic and internal energies, we can subtract the equation of change for kinetic energy alone, given in Eq. 1.1-9. This gives, if we make use of the assumption that $\boldsymbol{\pi}$ is symmetrical:

$$\boxed{\frac{\partial}{\partial t}\rho\hat{U} = -(\nabla \cdot \rho\hat{U}\boldsymbol{v}) - (\nabla \cdot \boldsymbol{q}) - (\boldsymbol{\pi} : \nabla\boldsymbol{v})} \quad (1.1\text{-}13)$$

which is the *internal energy equation.*

This completes the derivation of the three equations of change. In Table 1.1-1, where these equations are summarized, they are given in two forms: one in terms of $\partial/\partial t$, and the other in terms of D/Dt, which gives the time rate of change of a quantity as seen by an observer who is moving with the fluid. The operator $D/Dt = \partial/\partial t + (\boldsymbol{v} \cdot \nabla)$ is called the

[4] See Exercise 5 at the end of §A.1.

"substantial derivative" or the "material derivative," since it describes time changes taking place at a particular element of the "substance" or "material." It should be emphasized that there are *no* assumptions involved in going from Eqs. A, B, and C to Eqs. D, E, and F in Table 1.1-1.

The analytical solution to most of the problems in this book will begin with one of the three "boxed" equations above. Usually one will want them written out in component form in one of the standard orthogonal coordinate systems. For the reader's convenience, in Appendix B we give the equation of motion in rectangular, cylindrical, and spherical coordinates; in Appendix A many ∇-operations are tabulated in the same three coordinate systems and also in bipolar coordinates.

§1.2 THE EQUATIONS OF CHANGE IN TERMS OF THE TRANSPORT PROPERTIES

The equations in §1.1 are valid for any fluid. In this section we specialize these results for "Newtonian fluids" to obtain the equations of classical hydrodynamics. Then in the next section we give several examples of solutions of classical hydrodynamics problems. In doing so we select those problems that pertain to viscometry and to which we shall refer in subsequent chapters. Additional examples may be found in textbooks on transport phenomena.[1]

For structurally simple fluids such as gases, gaseous mixtures, and low-molecular-weight liquids and their mixtures, it has been established experimentally that in a simple shearing motion $v_x = v_x(y)$ the flux of x-momentum in the positive y-direction is given by "Newton's law of viscosity," $\pi_{yx} = -\mu \, dv_x/dy$, where μ is the *viscosity* of the fluid. The appropriate generalization for arbitrary, time-dependent flows is:[2,3]

$$\begin{aligned} \boldsymbol{\pi} &= p\boldsymbol{\delta} + \boldsymbol{\tau} \\ &= p\boldsymbol{\delta} - \mu[\nabla \boldsymbol{v} + (\nabla \boldsymbol{v})^\dagger] + (\tfrac{2}{3}\mu - \kappa)(\nabla \cdot \boldsymbol{v})\boldsymbol{\delta} \end{aligned} \tag{1.2-1}$$

where $(\nabla \boldsymbol{v})^\dagger$ is the transpose of the dyadic $\nabla \boldsymbol{v}$, and $\boldsymbol{\delta}$ is the unit tensor. This expression reduces to the hydrostatic pressure when there are no velocity gradients; it contains all possible combinations of first derivatives of velocity components that are allowed if one assumes that the fluid is isotropic and that the momentum flux tensor is symmetric.[2,3] The symbol p represents the thermodynamic pressure,[2] which is related to the density ρ and the temperature T through a "thermodynamic equation of state," $p = p(\rho, T)$; that is, this is taken to be the same function that one uses in thermal equilibrium.

The tensor $\boldsymbol{\tau}$ is the part of the momentum flux tensor or stress tensor that is associated with the viscosity of the fluid. We shall usually refer to it simply as the "momentum flux tensor" or "stress tensor," and use the terms "total momentum flux tensor" or "total stress tensor" for $\boldsymbol{\pi}$ when a distinction seems necessary. An equation that assigns a value to $\boldsymbol{\tau}$ is called a *constitutive equation.* Equation 1.2-1 is the constitutive equation for the Newtonian fluid.

Note that in generalizing Newton's law of viscosity to arbitrary flows an additional transport property κ, the *dilatational viscosity*, arises. The dilatational viscosity is identically

[1] See, for example, R. B. Bird, W. E. Stewart, and E. N. Lightfoot, *Transport Phenomena*, Wiley, New York (1960), Chapts. 3, 4, 10, 11.

[2] L. Landau and E. M. Lifshitz, *Fluid Mechanics*, Addison-Wesley, Reading, MA (1959), pp. 47–48, 187–188.

[3] G. K. Batchelor, *An Introduction to Fluid Dynamics*, Cambridge University Press, Cambridge (1967), Sects. 3.3 and 3.4.

zero for ideal, monatomic gases; for incompressible liquids $(\nabla \cdot \boldsymbol{v}) = 0$, and the term containing κ vanishes. Consequently the dilatational viscosity is of no importance in two key limiting cases, and no further mention will be made of it in this book.[4]

For all fluids the density ρ depends on the local thermodynamic state variables, such as pressure and temperature. However for liquids it is often a very good assumption to take the density to be constant. Such an idealized fluid is often called an "incompressible fluid", and the *momentum flux tensor* simplifies to

$$\boldsymbol{\pi} = p\boldsymbol{\delta} + \boldsymbol{\tau} = p\boldsymbol{\delta} - \mu\dot{\boldsymbol{\gamma}} \tag{1.2-2}$$

in which $\dot{\boldsymbol{\gamma}} = \nabla \boldsymbol{v} + (\nabla \boldsymbol{v})^\dagger$ is the *rate-of-strain tensor* or *rate-of-deformation tensor*. When the incompressibility assumption is made, a problem arises as to the meaning of p. For example, for a pure, incompressible fluid at constant temperature a plot of p vs. ρ is a vertical straight line; that is, the function $p(\rho)$ is many-valued. This poses no difficulty in solving hydrodynamic problems since only the gradient of p needs to be known. However, in connection with determining pressures at surfaces, an incompressible fluid theory can predict only pressure differences and not absolute values (unless, of course, the pressure on some bounding surface is specified through a boundary condition). For all discussions of Newtonian fluids in this book, Eq. 1.2-2 will be used for the momentum flux tensor; that is we will use the simple constitutive equation $\boldsymbol{\tau} = -\mu\dot{\boldsymbol{\gamma}}$ for incompressible Newtonian fluids.

The *heat flux* $\boldsymbol{q}$ for pure fluids and nondiffusing mixtures is given by "Fourier's law of heat conduction":

$$\boldsymbol{q} = -k\nabla T \tag{1.2-3}$$

in which k is the *thermal conductivity* and T is the temperature. For diffusing mixtures there are additional contributions to $\boldsymbol{q}$, but we do not discuss them here.[1,2]

Now that we have given the expressions in Eqs. 1.2-2 and 1.2-3 for the fluxes, let us turn to the equations of change, and particularly the *equations of change for incompressible Newtonian fluids*. These are listed in Table 1.2-1 and given in Appendix B in various coordinate systems. The equation of continuity was given earlier in Eq. 1.1-5. The equation of motion is obtained by substituting Eq. 1.2-2 into Eq. 1.1-8 and simplifying. The energy equation is obtained by first transforming Eq. 1.1-13 into an equation for temperature (by using standard thermodynamic transformations) and then inserting Fourier's law (Eq. 1.2-3) for $\boldsymbol{q}$. This process is outlined in Problem 1B.8; the final equation contains $\hat{C}_p$, which is the heat capacity at constant pressure per unit mass.

The equations of change in Table 1.2-1 are easy to interpret physically:

A and D: The *equation of continuity* states that within a small fixed volume there can be no net rate of addition of mass.

B and E: The *equation of motion* states that the mass-times-acceleration of a fluid element equals the sum of the pressure, viscous, and gravitational forces acting on the element.

C and F: The *energy equation* states that the temperature of a fluid element changes as it moves along with the fluid because of heat conduction (the k-term) and heat production by viscous heating (the μ-term).

[4] The dilatational viscosity of a liquid containing gas bubbles has been studied by G. K. Batchelor, *op. cit.*, pp. 253–255.

TABLE 1.2-1

Equations of Change for Newtonian Fluids with Constant ρ, μ, and k

	Dimensional Forms		Dimensionless Forms[a]	
Continuity	$(\nabla \cdot \boldsymbol{v}) = 0$	(A)	$(\nabla^* \cdot \boldsymbol{v}^*) = 0$	(D)
Motion[b,c]	$\rho \dfrac{D\boldsymbol{v}}{Dt} = -\nabla p + \mu\nabla^2\boldsymbol{v} + \rho\boldsymbol{g}$	(B)	$\mathrm{Re}\dfrac{D\boldsymbol{v}^*}{Dt^*} = -\nabla^* p^* + \nabla^{*2}\boldsymbol{v}^* + (\mathrm{Re}/\mathrm{Fr})\boldsymbol{g}/g$	(E)
Energy	$\rho\hat{C}_p \dfrac{DT}{Dt} = k\nabla^2 T + \frac{1}{2}\mu(\dot{\boldsymbol{\gamma}}:\dot{\boldsymbol{\gamma}})$	(C)	$\mathrm{P\acute{e}}\dfrac{DT^*}{Dt^*} = \nabla^{*2}T^* + \frac{1}{2}\,\mathrm{Br}(\dot{\boldsymbol{\gamma}}^*:\dot{\boldsymbol{\gamma}}^*)$	(F)

[a] The dimensionless forms are based on a reference length L, reference velocity V, a reference temperature T_0, and a reference temperature difference ΔT_0. In terms of these $\boldsymbol{v}^* = \boldsymbol{v}/V$, $\nabla^* = L\nabla$, $D/Dt^* = (L/V)D/Dt$, $p^* = (L/\mu V)p$, $T^* = (T - T_0)/\Delta T_0$ and $\dot{\boldsymbol{\gamma}}^* = (L/V)\dot{\boldsymbol{\gamma}}$. The Reynolds number $\mathrm{Re} = LV\rho/\mu$, the Froude number $\mathrm{Fr} = V^2/gL$, the Péclet number $\mathrm{P\acute{e}} = \rho\hat{C}_p LV/k$, and the Brinkman number $\mathrm{Br} = \mu V^2/k\Delta T_0$ are groups that appear as a result of writing the equations in dimensionless form. Other dimensionless groups may enter through the boundary conditions.

[b] For incompressible fluids we may combine the pressure and the gravity terms as $\nabla\mathscr{P} = \nabla p - \rho\boldsymbol{g}$ where $\mathscr{P}$ is called the "modified pressure." If the velocity is specified on the entire boundary, we can conclude that the gravitational acceleration has no effect on the velocity field. If forces are specified on part of the boundary, as in free surface flow, the modified pressure is not a useful concept. The nomenclature "modified pressure" was suggested by G. K. Batchelor, *An Introduction to Fluid Dynamics*, Cambridge University Press, Cambridge (1967), p. 176.

[c] The substantial derivative is defined as $D/Dt = \partial/\partial t + (\boldsymbol{v} \cdot \nabla)$.

Before ending this section we consider in Example 1.2-1 a result that is useful when the force on an object is desired in a fluid mechanical analysis.

EXAMPLE 1.2-1 Proof that Normal Stresses of Incompressible Newtonian Fluids Are Zero at Solid Surfaces

We consider a point P on a solid surface that is in contact with an incompressible Newtonian fluid. Use a rectangular coordinate system xyz whose origin is at P and whose z-axis is normal to the surface and points into the fluid, and show that $\tau_{zz}|_{z=0} = 0$.

SOLUTION The result follows from the definition of the normal stress component τ_{zz} and the mass conservation equation:

$$\begin{aligned}\tau_{zz}\Big|_{z=0} &= -2\mu\frac{\partial v_z}{\partial z}\Big|_{z=0} \\ &= 2\mu\left(\frac{\partial v_x}{\partial x} + \frac{\partial v_y}{\partial y}\right)\Big|_{z=0} = 0 \end{aligned} \qquad (1.2\text{-}4)$$

In the last step we have used the "no slip" condition on the solid surface.

Note: The result does not apply on surfaces with slip (see Eq. 1.4-43). Also in later chapters we shall find that for polymeric liquids normal stresses are *not* zero at solid surfaces.

§1.3 SOLUTION OF ISOTHERMAL FLOW PROBLEMS

In this section we illustrate the solution of fluid flow problems with the use of the equations for incompressible Newtonian fluids under isothermal conditions. Our starting equations are then Eqs. A and B in Table 1.2-1. These are four partial differential equations for the four unknowns: pressure and three components of velocity. Extensive experimental testing has shown that these equations describe the incompressible flow of Newtonian fluids exactly. Analytical solutions are, however, not always easy to obtain. In fact the equations are among the most challenging and extensively studied equations of mathematical physics. As a consequence we have available numerous treatises giving analytical solutions and solution procedures for Newtonian fluid mechanics.[1]

In connection with the examples we introduce two important approximate procedures: (i) in Examples 1.3-3 and 4 we use the lubrication approximation, in which the flow in a nearly constant cross section is approximated locally as flow in an equivalent constant-cross-section geometry; (ii) in Example 1.3-5 we introduce the quasi-steady-state approximation, in which an unsteady flow with small inertial effects is treated as a succession of steady-state flows.

EXAMPLE 1.3-1 Laminar Flow between Parallel Plane Surfaces

An incompressible Newtonian fluid is located in the space between two parallel plates that are separated by a distance B (see Fig. 1.3-1). The upper plate is moving in the $+x$-direction with a velocity V, thus contributing to the motion of the fluid. An additional contribution to the fluid motion is that due to a constant applied pressure gradient $\partial p/\partial x$. Find the velocity profile and the volume rate of flow. Assume that the flow is sufficiently slow that viscous heating is not important.

SOLUTION We postulate that in this system $v_x = v_x(y)$, $v_y = 0$, $v_z = 0$, $p = p(x, y)$, and $T =$ constant. We now apply these postulates to the equations of change in order to get the differential equations that describe the system. The equations of continuity and energy are clearly unimportant. The y-component of the equation of motion just gives the vertical pressure gradient, which is of no interest here. The x-component of the equation of motion becomes:

$$0 = -\frac{\partial p}{\partial x} + \mu \frac{d^2 v_x}{dy^2} \tag{1.3-1}$$

in which $\partial p/\partial x$ was stated to be a constant. This equation has to be integrated with respect to y with the boundary conditions:

$$\text{At } y = 0: \quad v_x = 0 \tag{1.3-2}$$

$$\text{At } y = B: \quad v_x = V \tag{1.3-3}$$

The result is the velocity distribution:

$$v_x = V\left(\frac{y}{B}\right) - \frac{B^2}{2\mu}\frac{\partial p}{\partial x}\left[\left(\frac{y}{B}\right) - \left(\frac{y}{B}\right)^2\right] \tag{1.3-4}$$

[1] For example, H. Lamb, *Hydrodynamics*, Dover, New York (1945); L. M. Milne-Thomson, *Theoretical Hydrodynamics*, 5th ed., Macmillan, New York (1967); H. L. Dryden, F. D. Murnaghan, and M. Bateman, *Hydrodynamics*, Dover, New York (1956); R. Berker, "Intégrations des équations du mouvement d'un fluide visqueux incompressible," *Handbuch der Physik*, Vol. VIII/2, Springer, Heidelberg (1963), pp. 1-384; H. Schlichting, *Boundary Layer Theory*, McGraw-Hill, New York, 4th ed. (1960); G. K. Batchelor, *An Introduction to Fluid Dynamics*, Cambridge Univ. Press (1967); L. D. Landau and E. M. Lifshitz, *Fluid Mechanics*, Pergamon, London (1959).

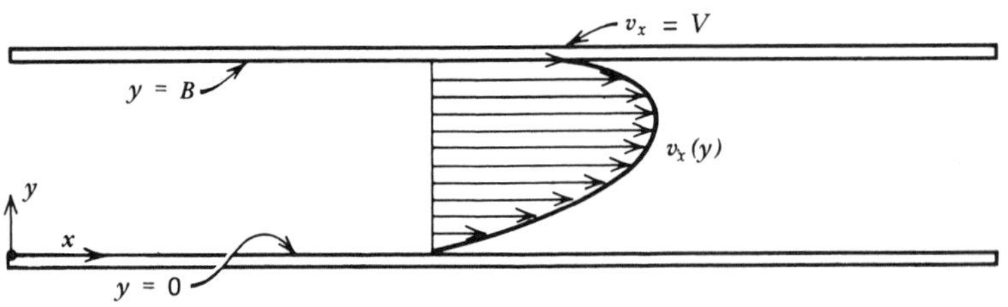

FIGURE 1.3-1. Flow between horizontal parallel planes with the upper plane moving and with an imposed pressure gradient in the flow direction.

The volume rate of flow Q for plates of width W is:

$$Q = WB\langle v_x \rangle = \int_0^W \int_0^B v_x \, dy \, dz$$
$$= WB \int_0^1 v_x d\left(\frac{y}{B}\right)$$
$$= \tfrac{1}{2}WBV - \frac{WB^3}{12\mu}\frac{\partial p}{\partial x} \tag{1.3-5}$$

Here the angular brackets $\langle\ \rangle$ indicate an average over the cross section. This result contains the solution for the problem where the pressure gradient and the wall motion both tend to drive the fluid in the same direction and also the problem where the pressure gradient and wall motion oppose one another. Furthermore, the solution in Eq. 1.3-5 is seen to be a sum of the solutions to the two separate problems of wall driven flow and pressure driven flow. This superposition of solutions results, of course, from the linearity of the governing equations and boundary conditions. (*Note*: It is not usually possible to perform this superposition for non-Newtonian flow problems where the governing equations are nonlinear.)

EXAMPLE 1.3-2 Laminar Flow in a Circular Tube

A fluid flows through a circular tube of radius R and length L. The tube makes an angle χ with the vertical direction. The modified pressures at the tube ends at $z = 0$ and $z = L$ (see Fig. 1.3-2) are $\mathscr{P}_0$ and $\mathscr{P}_L$, respectively. Find the steady-state velocity profile and the volume rate of flow, neglecting entrance and exit effects and assuming negligible viscous heating.

SOLUTION We postulate a solution of the form $v_z = v_z(r)$, $v_\theta = 0$, $v_r = 0$, and $p = p(r, \theta, z)$. The equation of continuity is satisfied identically, and only the z-component of the equation of motion is of interest:

$$0 = -\frac{\partial p}{\partial z} + \mu\frac{1}{r}\frac{d}{dr}\left(r\frac{dv_z}{dr}\right) + \rho g \cos\chi \tag{1.3-6}$$

This is to be solved with $v_z = 0$ at $r = R$ and v_z finite at $r = 0$.

Next we introduce[2] the "modified pressure" $\mathscr{P} = p - \rho g z \cos\chi$ so that

$$\frac{\partial \mathscr{P}}{\partial z} = \mu\frac{1}{r}\frac{d}{dr}\left(r\frac{dv_z}{dr}\right) \tag{1.3-7}$$

[2] In general, for incompressible fluids $\mathscr{P}$ is given by $\mathscr{P} = p + \rho g h$, where h is the distance *upward* (i.e., in the direction opposed to gravity) from some arbitrarily chosen reference plane. See also Table 1.2-1.

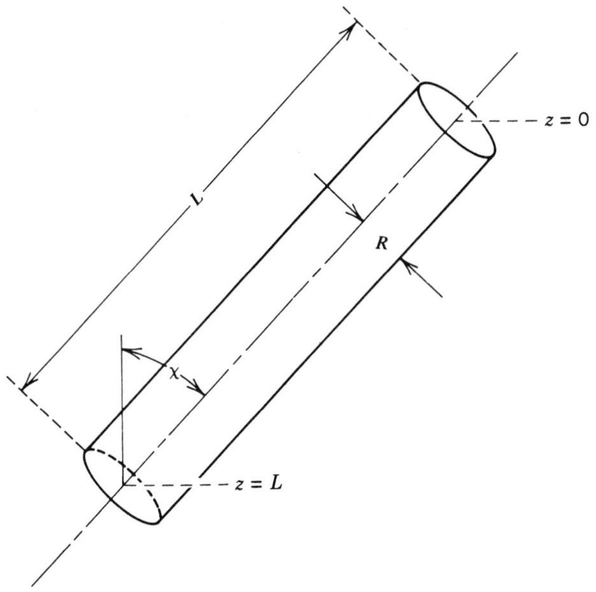

FIGURE 1.3-2. Flow through a circular tube that is inclined at an angle χ to the vertical.

The right side is a function of r alone; let us call it $F(r)$. Then

$$\mathscr{P} = F(r)z + G(r, \theta) \tag{1.3-8}$$

Application of the boundary conditions at $z = 0$ and $z = L$ gives

$$\mathscr{P} = -\frac{\mathscr{P}_0 - \mathscr{P}_L}{L}z + \mathscr{P}_0 \tag{1.3-9}$$

When this is substituted into Eq. 1.3-7, the latter can be integrated; use of the boundary conditions that $v_z = 0$ at $r = R$ and that v_z is finite at $r = 0$ then gives:

$$v_z = \frac{(\mathscr{P}_0 - \mathscr{P}_L)R^2}{4\mu L}\left[1 - \left(\frac{r}{R}\right)^2\right] \tag{1.3-10}$$

The volume rate of flow is then:

$$\begin{aligned} Q = \pi R^2 \langle v_z \rangle &= \int_0^{2\pi}\int_0^R v_z r\,dr\,d\theta \\ &= 2\pi R^2 \int_0^1 v_z \cdot \left(\frac{r}{R}\right) d\left(\frac{r}{R}\right) \\ &= \frac{\pi(\mathscr{P}_0 - \mathscr{P}_L)R^4}{8\mu L} \end{aligned} \tag{1.3-11}$$

which is the famous result of Hagen and Poiseuille.[3] This relation (accompanied by additional information about end corrections) is the basic equation needed to determine viscosity from tube flow data. It is valid for $\text{Re} < 2100$, where the Reynolds number $\text{Re} = 2R\langle v_z \rangle \rho/\mu = 2Q\rho/\pi\mu R$, the angular brackets indicating an average over the cross section. For $\text{Re} > 2100$, the flow will usually be turbulent.

[3] G. Hagen, *Ann. Phys. Chem.*, **46**, 423–442 (1839); J. L. Poiseuille, *Comptes Rendus*, **11**, 961, 1041 (1840); **12**, 112 (1841).

EXAMPLE 1.3-3 Flow in a Slightly Tapered Tube

An incompressible Newtonian fluid is flowing through the horizontal tapered tube shown in Fig. 1.3-3. Show that the analysis in Example 1.3-2 for a straight tube can be applied locally to this flow, and then use that result to obtain a relationship between the volume flow rate Q and the overall pressure drop $(\mathscr{P}_0 - \mathscr{P}_L) \equiv \Delta\mathscr{P}$.

SOLUTION The taper of the tube will require a flow in the radial direction and an acceleration in the axial direction, but it is reasonable to assume that the flow will maintain axial symmetry. Therefore we assume $v_z = v_z(r, z)$, $v_r = v_r(r, z)$, and $v_\theta = 0$. The continuity equation and the r- and z-components of the Navier-Stokes equations are then (cf. Eqs. B.2-4 and 6)

$$\frac{1}{r}\frac{\partial}{\partial r}(rv_r) + \frac{\partial v_z}{\partial z} = 0 \tag{1.3-12}$$

$$\rho\left(v_r \frac{\partial v_r}{\partial r} + v_z \frac{\partial v_r}{\partial z}\right) = \mu\left[\underline{\frac{\partial}{\partial r}\left(\frac{1}{r}\frac{\partial}{\partial r}(rv_r)\right)} + \underline{\frac{\partial^2 v_r}{\partial z^2}}\right] - \frac{\partial \mathscr{P}}{\partial r} \tag{1.3-13}$$

$$\rho\left(v_r \frac{\partial v_z}{\partial r} + v_z \frac{\partial v_z}{\partial z}\right) = \mu\left[\frac{1}{r}\frac{\partial}{\partial r}\left(r\frac{\partial v_z}{\partial r}\right) + \underline{\frac{\partial^2 v_z}{\partial z^2}}\right] - \frac{\partial \mathscr{P}}{\partial z} \tag{1.3-14}$$

The nonuniform geometry has changed the linear problem in Eq. 1.3-6 to a nonlinear one; Eqs. 1.3-12 through 14 are difficult to solve in general. However we can take advantage of the fact that the geometry changes slowly to show that these equations are dominated by only a few terms, and that if small terms are neglected, the problem is easily solved.

To see which terms must be kept and which can be neglected, we perform an *order of magnitude analysis* on Eqs. 1.3-12 through 14. First we estimate the sizes of the velocities. The axial velocity is determined by the volume flow rate and must be of order $(Q/\pi R_L^2) \equiv V$. We write this as

$$v_z \sim O(V) = O(Q/\pi R_L^2) \tag{1.3-15}$$

The size of v_r is dictated by the continuity equation. In this equation the $\partial v_z/\partial z$ term has magnitude

$$\begin{aligned}\frac{\partial v_z}{\partial z} &\sim O\left(\left(\frac{Q}{\pi R_L^2} - \frac{Q}{\pi R_0^2}\right)/L\right) \\ &= O(V(1 - (R_L/R_0)^2)/L)\end{aligned} \tag{1.3-16}$$

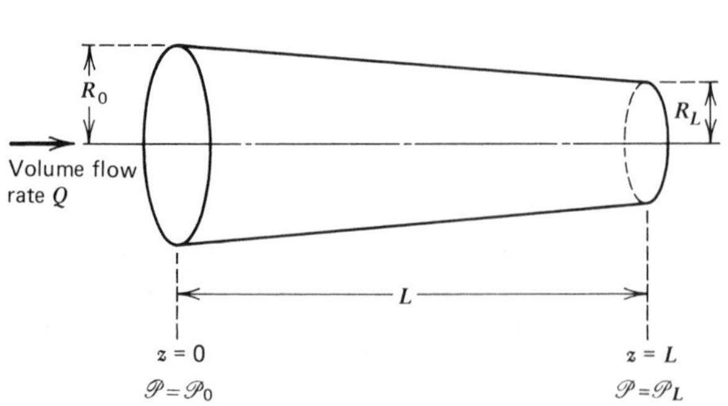

FIGURE 1.3-3. Tapered tube geometry analyzed in Example 1.3-3. The tube radius changes gradually from R_0 to R_L over a distance L. It is not necessary that dR/dz be constant as shown here, but it must be small for the analysis to hold.

Clearly we expect this derivative to be small since $(R_0 - R_L)/L \ll 1$. If we let U denote the order of v_r then the other contribution to the continuity equation is

$$\frac{1}{r}\frac{\partial}{\partial r}(rv_r) \sim O\left(\frac{U}{R_L}\right) \tag{1.3-17}$$

The radial derivative is approximated by estimating that v_r can vary from U to 0 over a distance of R_L. From the equation of continuity in Eq. 1.3-12 we may now find the relative sizes of the radial and axial velocities:

$$U = V\left(\frac{R_L}{L}\right)\left[1 - \left(\frac{R_L}{R_0}\right)^2\right] \tag{1.3-18}$$

We now look at the sizes of the contributions to the equation of motion. The viscous terms have orders:

$$\mu\frac{\partial}{\partial r}\left(\frac{1}{r}\frac{\partial}{\partial r}(rv_r)\right) \sim O\left(\mu\frac{V}{R_L L}\left[1 - \left(\frac{R_L}{R_0}\right)^2\right]\right) \tag{1.3-19}$$

$$\mu\frac{\partial^2 v_r}{\partial z^2} \sim O\left(\mu\frac{V}{L^3}R_L\left[1 - \left(\frac{R_L}{R_0}\right)^2\right]\right) \tag{1.3-20}$$

$$\mu\frac{1}{r}\frac{\partial}{\partial r}\left(r\frac{\partial v_z}{\partial r}\right) \sim O\left(\mu\frac{V}{R_L^2}\right) \tag{1.3-21}$$

$$\mu\frac{\partial^2 v_z}{\partial z^2} \sim O\left(\mu\frac{V}{L^2}\left[1 - \left(\frac{R_L}{R_0}\right)^2\right]\right) \tag{1.3-22}$$

From these results it is seen that the dashed underlined terms in Eqs. 1.3-13 and 14 are smaller than the unmarked viscous term by at least a factor of $(R_L/L)[1 - (R_L/R_0)^2] \ll 1$.

Similarly we estimate the sizes of the inertial terms:

$$\rho v_r\frac{\partial v_r}{\partial r} \sim O\left(\rho\frac{V^2 R_L}{L^2}\left[1 - \left(\frac{R_L}{R_0}\right)^2\right]^2\right) \tag{1.3-23}$$

$$\rho v_z\frac{\partial v_r}{\partial z} \sim O\left(\rho V^2\frac{R_L}{L^2}\left[1 - \left(\frac{R_L}{R_0}\right)^2\right]\right) \tag{1.3-24}$$

$$\rho v_r\frac{\partial v_z}{\partial r} \sim O\left(\rho\frac{V^2}{L}\left[1 - \left(\frac{R_L}{R_0}\right)^2\right]\right) \tag{1.3-25}$$

$$\rho v_z\frac{\partial v_z}{\partial z} \sim O\left(\rho\frac{V^2}{L}\left[1 - \left(\frac{R_L}{R_0}\right)^2\right]\right) \tag{1.3-26}$$

When the largest of these (Eqs. 1.3-25 and 26) are compared with the largest viscous term, Eq. 1.3-21, we see that all are negligible provided

$$\frac{\rho V R_L}{\mu}\left[\frac{R_L}{L}\left[1 - \left(\frac{R_L}{R_0}\right)^2\right]\right] = \tfrac{1}{2}\,\mathrm{Re}\left[\frac{R_L}{L}\left[1 - \left(\frac{R_L}{R_0}\right)^2\right]\right] \ll 1 \tag{1.3-27}$$

It is not necessary in this problem that the Reynolds number be small for the inertial terms to be negligible, since it is multiplied by a small geometric factor associated with the taper of the tube.

The pressure gradient terms are the last to be estimated. In order for the r and z components of the Navier-Stokes equations to be satisfied, the pressure gradient terms must be of the same size as the largest viscous term in each. Thus

$$\frac{\partial \mathscr{P}/\partial r}{\partial \mathscr{P}/\partial z} \sim O\left(\left(\frac{R_L}{L}\right)\left[1 - \left(\frac{R_L}{R_0}\right)^2\right]\right) \tag{1.3-28}$$

It can now be seen that if

$$\left(\frac{R_L}{L}\right)\left[1 - \left(\frac{R_L}{R_0}\right)^2\right] \ll 1 \tag{1.3-29}$$

the equation of motion is well approximated by Eq. 1.3-7, which describes flow in the straight tube. Integrating Eq. 1.3-7 with respect to r leads as before to

$$Q = \frac{\pi[R(z)]^4}{8\mu}\left(-\frac{d\mathscr{P}}{dz}\right) \tag{1.3-30}$$

Notice that we have had to hold z constant during this integration, because the boundary conditions vary with z. Although this result could easily have been adapted from Eq. 1.3-11, the above analysis serves to organize the approximation procedure and to document the limit of validity (Eq. 1.3-29) of the result. The process that we have illustrated here of adapting locally the results for a uniform geometry to a slowly varying geometry is known as the *lubrication approximation*[4]. It is a very powerful technique for estimating the flow rate vs. pressure drop relation in many complex flows.

Finally we obtain the relation between pressure drop $(\mathscr{P}_0 - \mathscr{P}_L)$ and volume flow rate. Rather than use z as the independent variable in Eq. 1.3-30, we can use $R = R_0 + (R_L - R_0)(z/L)$:

$$Q = \frac{\pi R^4}{8\mu}\left(-\frac{d\mathscr{P}}{dR}\right)\left(\frac{R_L - R_0}{L}\right) \tag{1.3-31}$$

But Q is constant for all z (and hence all R). Therefore the differential equation for $\mathscr{P}$ as a function of R can be integrated to give:

$$\begin{aligned} Q &= \frac{3\pi}{8\mu}\frac{(\mathscr{P}_0 - \mathscr{P}_L)}{L}\frac{(R_0 - R_L)}{(R_L^{-3} - R_0^{-3})} \\ &= \frac{\pi(\mathscr{P}_0 - \mathscr{P}_L)R_0^4}{8\mu L}\left[1 - \frac{1 + (R_L/R_0) + (R_L/R_0)^2 - 3(R_L/R_0)^3}{1 + (R_L/R_0) + (R_L/R_0)^2}\right] \end{aligned} \tag{1.3-32}$$

Hence the final result may be expressed as the Hagen-Poiseuille result multiplied by a correction factor.

EXAMPLE 1.3-4 The Cone-and-Plate Viscometer

The cone-and-plate geometry shown in Fig. 1.3-4 is a standard experimental arrangement for the measurement of viscosity (and other rheological properties, as we shall see later). Obtain the analytical relations needed to interpret the instrumental data:

[4] J. R. A. Pearson, *Mechanics of Polymer Processing*, Elsevier Applied Science Publishers, London (1985), pp. 165–177; R. I. Tanner, *Engineering Rheology*, Oxford Univ. Press (1985), pp. 228–236.

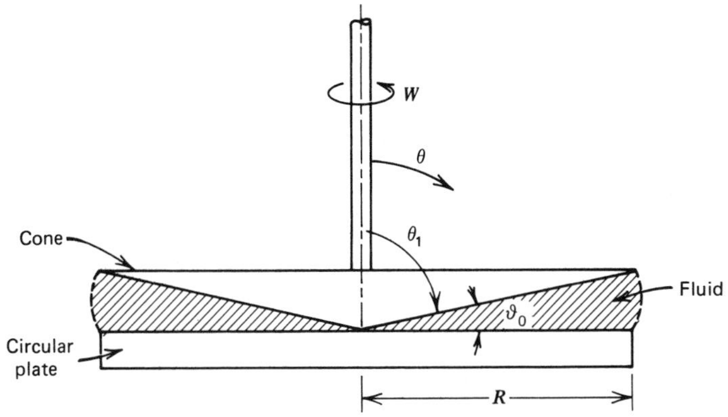

FIGURE 1.3-4. Cone-and-plate instrument; the angle ϑ_0 is usually between 0.5 and 8 degrees in commercial instruments.

a. A relation between the angular velocity W and the $\theta\phi$-component of the rate-of-strain tensor $\dot{\boldsymbol{\gamma}}$ in the gap.

b. A relationship between the torque $\mathscr{T}$ and the $\theta\phi$-component of the stress tensor $\boldsymbol{\tau}$ in the gap.

c. A relation giving the viscosity μ in terms of W and $\mathscr{T}$.

SOLUTION The simplest analysis of this experiment makes use of the fact that the angle ϑ_0 is so small that a lubrication approximation may be applied to the flow in the gap. In this example, however, we do not perform an order of magnitude analysis as in the previous example, but rather apply the lubrication approximation intuitively by regarding the flow to be locally that between parallel plates.

(a) The velocity v_ϕ at a radius r can be approximated by adapting Eq. 1.3-4 with $\partial p/\partial x = 0$ and with v_x replaced by v_ϕ. For the cone-and-plate system, at a distance r from the cone apex, the velocity of the cone will be Wr (this corresponds to V in Eq. 1.3-4), and the plate-cone separation will be given by $r \sin \vartheta_0 \doteq r\vartheta_0$ (this corresponds to B in Eq. 1.3-4). Hence the velocity profile will, to a good approximation, be

$$v_\phi = Wr\left(\frac{(\pi/2) - \theta}{(\pi/2) - \theta_1}\right) \tag{1.3-33}$$

The $\theta\phi$-component of the $\dot{\boldsymbol{\gamma}}$-tensor is then (cf. Eq. B.3-17):

$$\dot{\gamma}_{\theta\phi} = \frac{\sin\theta}{r}\frac{\partial}{\partial\theta}\left(\frac{v_\phi}{\sin\theta}\right) \doteq \frac{1}{r}\frac{\partial}{\partial\theta}v_\phi = -\frac{W}{\vartheta_0} \tag{1.3-34}$$

The approximation made here is that θ is so close to $\pi/2$ that $\sin\theta$ can be taken to be unity; this should be an excellent approximation. We see from Eq. 1.3-34 that $\dot{\gamma}_{\theta\phi}$ is constant throughout the cone-plate gap. This is one reason why this geometry is useful for macromolecular fluids where the viscosity depends on the shear rate.

(b) The torque required to maintain the motion will be obtained by integrating the product of the lever arm r and the force $\tau_{\theta\phi}|_{\theta=\pi/2} \cdot r dr d\phi$ over the surface of the plate

$$\mathscr{T} = \int_0^{2\pi}\int_0^R \tau_{\theta\phi}\bigg|_{\theta=\pi/2} r^2\, dr\, d\phi \tag{1.3-35}$$

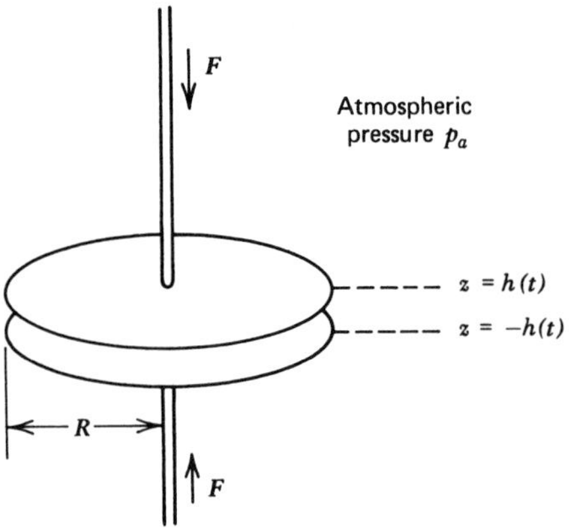

FIGURE 1.3-5. Squeezing flow between parallel disks each with radius R. The initial disk separation is $2h_0$.

Since $\dot{\gamma}_{\theta\phi}$ is constant throughout the gap, $\tau_{\theta\phi}$ will also be constant. Hence the integration is easily performed and one gets

$$\tau_{\theta\phi} = \frac{3\mathscr{T}}{2\pi R^3} \tag{1.3-36}$$

(c) Since $\boldsymbol{\tau} = -\mu\dot{\boldsymbol{\gamma}}$ for a Newtonian fluid, the $\theta\phi$-component of this equation, combined with Eqs. 1.3-34 and 36, gives

$$\mu = \frac{\tau_{\theta\phi}}{-\dot{\gamma}_{\theta\phi}} = \frac{3\mathscr{T}\vartheta_0}{2\pi R^3 W} \tag{1.3-37}$$

This gives the viscosity of the fluid in the gap in terms of the geometrical quantities R and ϑ_0 and the measured values of torque $\mathscr{T}$ and angular velocity W.

More complete analyses can be found elsewhere.[5] The treatment of this flow for non-Newtonian fluids is found in §10.2.

EXAMPLE 1.3-5 Squeezing Flow between Two Parallel Disks[6]

A fluid is placed in the gap between two parallel disks separated by a distance $2h_0$. The fluid completely fills the gap. A constant force F is applied to each disk as shown in Fig. 1.3-5. It is desired to obtain an expression for the change in gap separation with time. A "quasi-steady state" solution will be used; that is, at any time t the radial flow problem will be treated as a steady-state hydrodynamic problem. This means that the inertial terms in the equation of motion are neglected in the first approximation. In addition gravity is assumed negligible.

SOLUTION We introduce a cylindrical coordinate system with z-axis coinciding with the symmetry axis of the flow and with the two disks defined by $z = -h(t)$ and $z = h(t)$. Since $R \gg h$ the flow will primarily be in the r-direction so that $v_z \ll v_r$ and also $(\partial v_r/\partial r) \ll (\partial v_r/\partial z)$. Consistent with the

[5] R. B. Bird, W. E. Stewart, and E. N. Lightfoot, *op. cit.*, p. 119; S. Oka in F. R. Eirich, ed., *Rheology*, Vol. 3, Academic Press, New York (1960), Chapt. 2, pp. 61–62; K. Walters, *Rheometry*, Chapman and Hall, London (1975), Chapt. 4.

[6] J. Stefan, *Sitzungber. K. Akad. Wiss. Math. Natur. Wien*, **69**, Part 2, 713–735 (1874); see also L. Landau and E. M. Lifshitz, *op. cit.*, pp. 70–71.

quasi-steady-state approximation, we take $\rho \partial v_r/\partial t \ll \mu \partial^2 v_r/\partial z^2$. Consequently the equation of continuity and the r- and z- components of the equation of motion are well approximated by

$$\textit{Continuity:} \quad \frac{1}{r}\frac{\partial}{\partial r}(r v_r) + \frac{\partial v_z}{\partial z} = 0 \tag{1.3-38}$$

$$\textit{Motion } (r)\textit{:} \quad 0 = -\frac{\partial p}{\partial r} + \mu \frac{\partial^2 v_r}{\partial z^2} \tag{1.3-39}$$

$$\textit{Motion } (z)\textit{:} \quad 0 = -\frac{\partial p}{\partial z} \tag{1.3-40}$$

We seek a solution for the velocity field in the form $v_r = v_r(r, z, t)$ and $v_z = v_z(z, t)$. The continuity equation then demands that $v_r = rf(z, t)$. Furthermore, the equations of motion show that p must have the form $p = p_0 + p_2 r^2$, where p_0 and p_2 are constants to be determined. With these simplifications Eq. 1.3-40 is satisfied and Eqs. 1.3-38 and 39 give

$$2f + \frac{\partial v_z}{\partial z} = 0 \tag{1.3-41}$$

$$-2p_2 + \mu \frac{\partial^2 f}{\partial z^2} = 0 \tag{1.3-42}$$

The following boundary conditions are to be satisfied:

$$\partial f/\partial z = 0, \quad \text{at } z = 0; \quad f = 0, \quad \text{at } z = h \tag{1.3-43,44}$$

$$v_z = 0, \quad \text{at } z = 0; \quad v_z = \dot{h}, \quad \text{at } z = h \tag{1.3-45,46}$$

$$p = p_a, \quad \text{at } r = R \tag{1.3-47}$$

Here $\dot{h}$ stands for dh/dt. These 5 conditions suffice to determine p_0, p_2, and the 3 constants of integration of Eqs. 1.3-41 and 42. The results are

$$v_r = rf = \frac{3}{4}\frac{(-\dot{h})}{h} r \left[1 - \left(\frac{z}{h}\right)^2\right] \tag{1.3-48}$$

$$v_z = \frac{3}{2}\dot{h}\left[\left(\frac{z}{h}\right) - \frac{1}{3}\left(\frac{z}{h}\right)^3\right] \tag{1.3-49}$$

$$p - p_a = \frac{3(-\dot{h})\mu R^2}{4h^3}\left[1 - \left(\frac{r}{R}\right)^2\right] \tag{1.3-50}$$

To calculate the force on one plate all we need is the pressure distribution in Eq. 1.3-50, since we know from Example 1.2-1 that $\tau_{zz} = 0$ on the plates. Consequently we find

$$\begin{aligned} F &= \int_0^{2\pi}\int_0^R (p - p_a + \tau_{zz})\Big|_{z=h} r\,dr\,d\theta \\ &= 2\pi R^2 \int_0^1 (p - p_a)\left(\frac{r}{R}\right)d\left(\frac{r}{R}\right) \\ &= \frac{3\pi R^4 \mu(-\dot{h})}{8h^3} \end{aligned} \tag{1.3-51}$$

This is the *Stefan equation*, which shows how much force $F(t)$ must be applied in order to maintain the disk motion $h(t)$.

If we now ask what the disk motion will be for a *constant applied force* F, we have to solve the differential equation for $h(t)$ in Eq. 1.3-51 to give

$$\frac{1}{h^2} - \frac{1}{h_0^2} = \frac{16Ft}{3\pi R^4 \mu} \tag{1.3-52}$$

This gives the disk separation as a function of the elapsed time, when inertial effects may be neglected.

It is possible to correct Eq. 1.3-51 for small inertial effects by means of a perturbation approach. Then the creeping flow solution in Eqs. 1.3-48 and 49 is used to evaluate the inertial term $\rho[\partial v_r/\partial t + v_r(\partial v_r/\partial r) + v_z(\partial v_r/\partial z)]$ in Eq. 1.3-39. The steps described above are then repeated to find the perturbations in v_r, v_z and p. The final result for F is

$$F = \frac{3\pi R^4 \mu(-\dot{h})}{8h^3}\left[1 + \frac{5\rho h(-\dot{h})}{7\mu} + \frac{2\rho h^2 \ddot{h}}{5\mu\dot{h}}\right] \tag{1.3-53}$$

This result[7] can be used to estimate the importance of inertial effects in lubrication squeeze films and in parallel plate plastometers.

§1.4 SOLUTION OF ISOTHERMAL FLOW PROBLEMS BY USE OF THE STREAM FUNCTION

This section is devoted to the stream function as an important analytical tool for the solution of flow problems. For simplicity we restrict our attention to flows in which we may choose a coordinate system such that only two velocity components are nonzero. Following a brief introduction to the stream function we demonstrate its use in three illustrative examples in which we solve for the flow field around a translating sphere, a rising bubble, and a rotating sphere. The last example also introduces regular perturbation methods, which we will use in later chapters.

For illustrative purposes we start by considering two-dimensional "plane" flow, for which one may choose a rectangular coordinate system so that $v_x = v_x(x, y, t)$, $v_y = v_y(x, y, t)$, and $v_z = 0$. The continuity equation then reads

$$\frac{\partial v_x}{\partial x} + \frac{\partial v_y}{\partial y} = 0 \tag{1.4-1}$$

The form of this equation suggests the introduction of a function $\psi(x, y, t)$ called the *stream function* with the property[1] that

$$v_x = -\frac{\partial \psi}{\partial y} \tag{1.4-2}$$

$$v_y = \frac{\partial \psi}{\partial x} \tag{1.4-3}$$

[7] S. Ishizawa, *JSME Bull.*, **9**, 533–550 (1966); D. C. Kuzma, *Appl. Sci. Res.*, **A18**, 15–20 (1967); A. F. Jones and S. D. R. Wilson, *J. Lubr. Technol.*, **97**, 101–104 (1975); R. J. Grimm, *Appl. Sci. Res.*, **32**, 149–166 (1976).

[1] Some authors introduce a stream function equal to the negative of the one introduced here. Our sign convention agrees with that of H. Lamb, *Hydrodynamics*, 6th ed., Cambridge University Press, Cambridge (1932); L. M. Milne-Thomson, *Theoretical Hydrodynamics* Macmillan, New York (1955); and R. B. Bird, W. E. Stewart, and E. N. Lightfoot, *Transport Phenomena*, Wiley, New York (1960), §4.2.

The use of the variable ψ in place of v_x and v_y automatically ensures that the continuity equation is satisfied. When the expressions in Eqs. 1.4-2 and 3 are inserted into the equations of motion one obtains two equations for the stream function and pressure. It is customary to eliminate the pressure between these to obtain a single equation to be satisfied by ψ. In rectangular coordinates this is done by differentiating the x-component of the equations of motion with respect to y, differentiating the y-component with respect to x and subtracting the two resulting equations. In this way the pressure is eliminated, and one obtains a single fourth-order differential equation for ψ. Various forms of this equation are shown in Table 1.4-1. In the equation of motion we have allowed for the possibility that the body force per unit volume $\boldsymbol{f}$ could depend on position. The use of a position-dependent $\boldsymbol{f}$ is illustrated in Example 1.4-3, and it will be used in connection with perturbation solutions for non-Newtonian fluids in Chapter 6. If the body force is a gravitational force or any other force that does not depend on position, it will not enter in the equation for the stream function.

EXAMPLE 1.4-1 Flow around a Translating Sphere[2]

The low Reynolds number flow around a translating sphere is an important problem in classical fluid dynamics. We consider here the limit of "creeping flow" in which the inertial terms are neglected in the equation of motion. The sphere has radius R and is moving unidirectionally with constant velocity V in an incompressible Newtonian liquid of density ρ and viscosity μ. The fluid is at rest far from the sphere.

a. Find the velocity field for the flow around the sphere.

b. From the results in **(a)** obtain the expression for the drag force on the sphere (i.e., Stokes' law).

c. Find the disturbance in a quiescent fluid due to the movement of a sphere through the fluid, in terms of the force that the sphere exerts on the fluid.

SOLUTION **(a)** In order to simplify the calculations we represent the flow in a spherical coordinate system with origin at the center of the sphere, with the fluid approaching the sphere from the positive z-direction (see Fig. 1.4-1). In this coordinate system the flow is steady and the boundary conditions far from the sphere are

$$v_r \to -V \cos\theta, \qquad \text{for } r \to \infty \tag{1.4-4}$$

$$v_\theta \to V \sin\theta, \qquad \text{for } r \to \infty \tag{1.4-5}$$

At the surface of the sphere the conditions are

$$v_r = 0, \qquad \text{at } r = R \tag{1.4-6}$$

$$v_\theta = 0, \qquad \text{at } r = R \tag{1.4-7}$$

[2] H. Lamb, *op. cit.*, pp. 602 *et seq.*; see also L. Landau and E. M. Lifshitz, *Fluid Mechanics*, Addison-Wesley, Reading, MA (1959), pp. 63–71, 95–98; G. K. Batchelor, *An Introduction to Fluid Mechanics*, Cambridge University Press (1970), pp. 229–244; and H. Villat, *Leçons sur les fluides visqueux*, Gauthier-Villars, Paris (1943), Chapt. 7.

TABLE 1.4-1

Equations for the Stream Function[a]

Type of Motion	Coordinate System	Velocity Components	Differential Equations for ψ Which are Equivalent to[b,c] $\rho \frac{D}{Dt} \boldsymbol{v} = -\nabla p + \mu \nabla^2 \boldsymbol{v} + \boldsymbol{f}$ (A)	Expressions for Operators
Two-dimensional (planar)	Rectangular with $v_z = 0, f_z = 0$, and no z-dependence	$v_x = -\frac{\partial \psi}{\partial y}$ $v_y = +\frac{\partial \psi}{\partial x}$	$\rho\left(\frac{\partial}{\partial t}(\nabla^2\psi) + \frac{\partial(\psi, \nabla^2\psi)}{\partial(x, y)}\right) = \mu\nabla^4\psi + [\nabla \times \boldsymbol{f}]_z$ (B)	$\nabla^2 \equiv \frac{\partial^2}{\partial x^2} + \frac{\partial^2}{\partial y^2}$ $\nabla^4\psi \equiv \nabla^2(\nabla^2\psi) \equiv \left(\frac{\partial^4}{\partial x^4} + 2\frac{\partial^4}{\partial x^2 \partial y^2} + \frac{\partial^4}{\partial y^4}\right)\psi$ $[\nabla \times \boldsymbol{f}]_z = \frac{\partial}{\partial x} f_y - \frac{\partial}{\partial y} f_x$
	Cylindrical with $v_z = 0, f_z = 0$, and no z-dependence	$v_r = -\frac{1}{r}\frac{\partial \psi}{\partial \theta}$ $v_\theta = +\frac{\partial \psi}{\partial r}$	$\rho\left(\frac{\partial}{\partial t}(\nabla^2\psi) + \frac{1}{r}\frac{\partial(\psi, \nabla^2\psi)}{\partial(r, \theta)}\right) = \mu\nabla^4\psi + [\nabla \times \boldsymbol{f}]_z$ (C)	$\nabla^2 \equiv \frac{\partial^2}{\partial r^2} + \frac{1}{r}\frac{\partial}{\partial r} + \frac{1}{r^2}\frac{\partial^2}{\partial \theta^2}$ $[\nabla \times \boldsymbol{f}]_z = \frac{1}{r}\frac{\partial}{\partial r}(r f_\theta) - \frac{1}{r}\frac{\partial}{\partial \theta} f_r$

Axisymmetrical	Cylindrical with $v_\theta = 0$, $f_\theta = 0$, and no θ-dependence	$v_z = -\frac{1}{r}\frac{\partial \psi}{\partial r}$ $v_r = +\frac{1}{r}\frac{\partial \psi}{\partial z}$	$\rho\left(\frac{\partial}{\partial t}(E^2\psi) - \frac{1}{r}\frac{\partial(\psi, E^2\psi)}{\partial(r, z)} - \frac{2}{r^2}\frac{\partial \psi}{\partial z}E^2\psi\right)$ $= \mu E^4\psi + r[\nabla \times \boldsymbol{f}]_\theta$ (D)	$E^2 \equiv \frac{\partial^2}{\partial r^2} - \frac{1}{r}\frac{\partial}{\partial r} + \frac{\partial^2}{\partial z^2}$ $E^4\psi \equiv E^2(E^2\psi)$ $[\nabla \times \boldsymbol{f}]_\theta = \frac{\partial}{\partial z}f_r - \frac{\partial}{\partial r}f_z$
	Spherical with $v_\phi = 0$, $f_\phi = 0$, and no ϕ-dependence	$v_r = -\frac{1}{r^2 \sin\theta}\frac{\partial \psi}{\partial \theta}$ $v_\theta = +\frac{1}{r \sin\theta}\frac{\partial \psi}{\partial r}$	$\rho\left(\frac{\partial}{\partial t}(E^2\psi) + \frac{1}{r^2 \sin\theta}\frac{\partial(\psi, E^2\psi)}{\partial(r, \theta)} - \frac{2E^2\psi}{r^2 \sin^2\theta}\left(\frac{\partial \psi}{\partial r}\cos\theta - \frac{1}{r}\frac{\partial \psi}{\partial \theta}\sin\theta\right)\right)$ $= \mu E^4\psi + r \sin\theta[\nabla \times \boldsymbol{f}]_\phi$ (E)	$E^2 \equiv \frac{\partial^2}{\partial r^2} + \frac{\sin\theta}{r^2}\frac{\partial}{\partial \theta}\left(\frac{1}{\sin\theta}\frac{\partial}{\partial \theta}\right)$ $[\nabla \times \boldsymbol{f}]_\phi = \frac{1}{r}\frac{\partial}{\partial r}(rf_\theta) - \frac{1}{r}\frac{\partial}{\partial \theta}f_r$

[a] Adapted from R. B. Bird, W. E. Stewart, and E. N. Lightfoot, *Transport Phenomena*, Wiley, New York (1960), p. 131. Similar relations in general orthogonal coordinates may be found in S. Goldstein, *Modern Developments in Fluid Dynamics*, Oxford University Press, Oxford, England (1938), pp. 114–115; in this reference formulas are also given for axisymmetrical flows with a nonzero component of the velocity around the axis.

[b] Here the Jacobians are designated by $\frac{\partial(g, h)}{\partial(x, y)} = \begin{vmatrix} \partial g/\partial x & \partial g/\partial y \\ \partial h/\partial x & \partial h/\partial y \end{vmatrix}$.

[c] Here $\boldsymbol{f}$ denotes an arbitrary force per unit volume; its use is illustrated in Examples 1.4-3, 6.5-1, and 6.5-2.

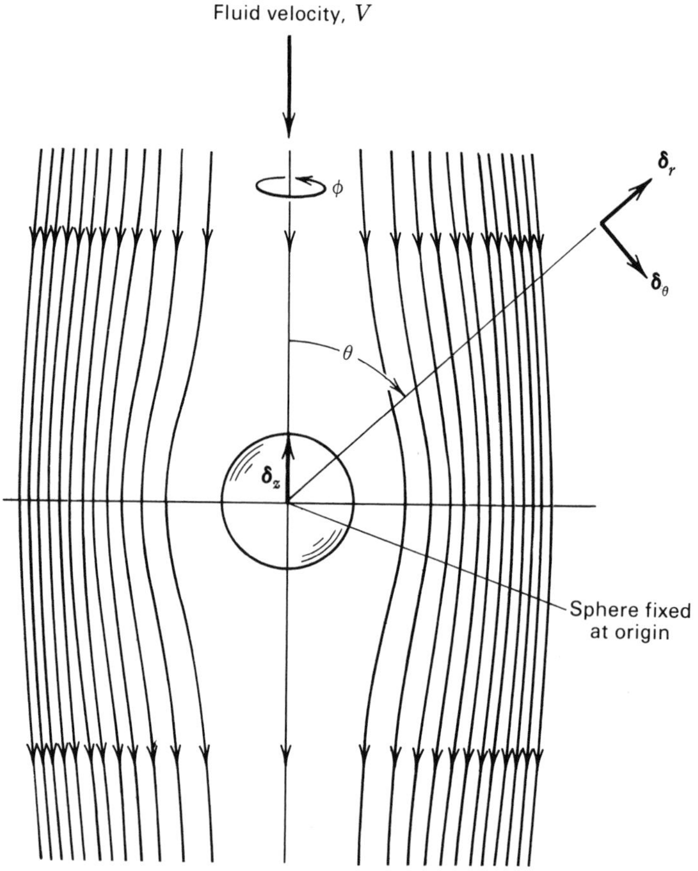

FIGURE 1.4-1. Flow around a sphere with fluid velocity $-V\boldsymbol{\delta}_z$ far from the sphere. Note that this corresponds to $v_r = (\boldsymbol{v} \cdot \boldsymbol{\delta}_r) = -V(\boldsymbol{\delta}_z \cdot \boldsymbol{\delta}_r) = -V\cos\theta$, and $v_\theta = (\boldsymbol{v} \cdot \boldsymbol{\delta}_\theta) = -V(\boldsymbol{\delta}_z \cdot \boldsymbol{\delta}_\theta) = -V(-\sin\theta)$; see Eqs. 1.4-4 and 5.

We solve the problem with the use of a stream function defined in spherical coordinates for axisymmetrical flow with no ϕ dependence. That is, according to Table 1.4-1 we use

$$v_r = -\frac{1}{r^2 \sin\theta} \frac{\partial\psi}{\partial\theta} \tag{1.4-8}$$

$$v_\theta = \frac{1}{r\sin\theta} \frac{\partial\psi}{\partial r} \tag{1.4-9}$$

Note that in terms of ψ the boundary conditions far from the sphere become

$$\psi \to \tfrac{1}{2}Vr^2 \sin^2\theta, \qquad \text{for } r \to \infty \tag{1.4-10}$$

In view of this condition we assume for ψ

$$\psi = f(r)\sin^2\theta \tag{1.4-11}$$

This expression also automatically ensures that $v_\theta = 0$ and $\partial v_r/\partial\theta = 0$ at $\theta = 0$ and π. The boundary conditions at $r = R$ then become

$$f(r) = 0, \qquad \text{at } r = R \tag{1.4-12}$$

$$f'(r) = 0, \qquad \text{at } r = R \tag{1.4-13}$$

For steady-state creeping flow the equation for ψ is (according to Eq. E of Table 1.4-1)

$$E^4\psi = 0 \tag{1.4-14}$$

where the operator E^2 is given in spherical coordinates in Table 1.4-1. If we substitute the expression for ψ in Eq. 1.4-11 into Eq. 1.4-14 we obtain

$$\left[\frac{\partial^2}{\partial r^2} + \frac{\sin\theta}{r^2}\frac{\partial}{\partial\theta}\left(\frac{1}{\sin\theta}\frac{\partial}{\partial\theta}\right)\right]\left[\sin^2\theta\left(\frac{d^2f}{dr^2} - \frac{2f}{r^2}\right)\right] = 0 \tag{1.4-15}$$

which may be rearranged to give

$$\left(\frac{d^2}{dr^2} - \frac{2}{r^2}\right)\left(\frac{d^2}{dr^2} - \frac{2}{r^2}\right)f = 0 \tag{1.4-16}$$

This is an "equidimensional" equation, known[3] to have solutions of the form $f = r^n$. If we substitute $f = r^n$ into Eq. 1.4-16, we see that n must satisfy the equation

$$[n(n-1) - 2][(n-2)(n-3) - 2] = 0 \tag{1.4-17}$$

so that $n = -1$, 1, 2, or 4. The general solution for f is then

$$f = A_1 r^{-1} + A_2 r + A_3 r^2 + A_4 r^4 \tag{1.4-18}$$

which has four constants as is appropriate for a fourth-order differential equation. The constants are determined from Eqs. 1.4-10, 12, and 13, and the stream function and velocity components become:

$$\psi = VR^2\left[\frac{1}{2}\left(\frac{r}{R}\right)^2 - \frac{3}{4}\left(\frac{r}{R}\right) + \frac{1}{4}\left(\frac{R}{r}\right)\right]\sin^2\theta \tag{1.4-19}$$

$$v_r = -V\left[1 - \frac{3}{2}\left(\frac{R}{r}\right) + \frac{1}{2}\left(\frac{R}{r}\right)^3\right]\cos\theta \tag{1.4-20}$$

$$v_\theta = V\left[1 - \frac{3}{4}\left(\frac{R}{r}\right) - \frac{1}{4}\left(\frac{R}{r}\right)^3\right]\sin\theta \tag{1.4-21}$$

(b) To find the drag force on the sphere we need the pressure $p(r, \theta)$. To obtain this we insert the above velocity field into the equations of motion; this leads to the following equations for p:

$$\frac{\partial p}{\partial r} = -3\mu VRr^{-3}\cos\theta - \rho g\cos\theta \tag{1.4-22}$$

$$\frac{1}{r}\frac{\partial p}{\partial\theta} = -\tfrac{3}{2}\mu VRr^{-3}\sin\theta + \rho g\sin\theta \tag{1.4-23}$$

where g is the gravitational acceleration acting in the negative z-direction. These equations are solved by

$$p = (\tfrac{3}{2}\mu VRr^{-2} - \rho gr)\cos\theta + p_0 \tag{1.4-24}$$

[3] C. M. Bender and S. A. Orzag, *Advanced Mathematical Methods for Scientists and Engineers*, McGraw-Hill, New York (1978), p. 12.

where p_0 is a constant. We now calculate the rr- and $r\theta$-components of the total stress tensor at the surface of the sphere

$$\pi_{rr}|_{r=R} = p|_{r=R} = (\tfrac{3}{2}\mu V R^{-1} - \rho g R)\cos\theta + p_0 \tag{1.4-25}$$

$$\pi_{r\theta}|_{r=R} = -\mu\left[r\frac{\partial}{\partial r}\left(\frac{v_\theta}{r}\right) + \frac{1}{r}\frac{\partial v_r}{\partial \theta}\right]\Bigg|_{r=R} = -\tfrac{3}{2}\mu V R^{-1}\sin\theta \tag{1.4-26}$$

Here we have used the result from Example 1.2-1 that $\tau_{rr} = 0$ at the solid surface. The force per unit area exerted by the fluid on the surface of the sphere is therefore $-[\boldsymbol{\delta}_r \cdot \boldsymbol{\pi}] = -[\pi_{rr}\boldsymbol{\delta}_r + \pi_{r\theta}\boldsymbol{\delta}_\theta]$. Certainly the net force will be in the z-direction and we therefore compute the z-component of this force:

$$-(\boldsymbol{\delta}_z \cdot [\pi_{rr}\boldsymbol{\delta}_r + \pi_{r\theta}\boldsymbol{\delta}_\theta]) = -\tfrac{3}{2}\mu V R^{-1} + \rho g R\cos^2\theta - p_0\cos\theta \tag{1.4-27}$$

where we have used $(\boldsymbol{\delta}_z \cdot \boldsymbol{\delta}_r) = \cos\theta$ and $(\boldsymbol{\delta}_z \cdot \boldsymbol{\delta}_\theta) = -\sin\theta$ (see Fig. 1.4-1). The total force in the positive z-direction on the sphere is then

$$\begin{aligned} F &= \int_0^{2\pi}\int_0^{\pi} (\boldsymbol{\delta}_z \cdot [-\boldsymbol{\delta}_r \cdot \boldsymbol{\pi}])R^2 \sin\theta\, d\theta\, d\phi \\ &= -6\pi\mu R V + \tfrac{4}{3}\pi R^3 \rho g \end{aligned} \tag{1.4-28}$$

The combination $6\pi\mu RV$ is called the drag force F_D and the combination $(4/3)\pi R^3\rho g$ is the buoyancy force. The result $F_D = 6\pi\mu RV$ is known as Stokes' law[2] and is a good approximation for Reynolds numbers $\text{Re} = 2RV\rho/\mu \leq 0.1$. The expression for the drag force has been extended[4] by a perturbation method to yield

$$F_D = 6\pi\mu RV[1 + \tfrac{3}{16}\,\text{Re} + \tfrac{9}{160}\,\text{Re}^2 \ln \text{Re} + O(\text{Re}^2)] \tag{1.4-29}$$

as an improved approximation for $\text{Re} < O(1)$. Experiments[5] show that an eddy forms at the back of the sphere at a Reynolds number of about 24. The eddy and wake flow begins to oscillate at $\text{Re} \doteq 130$ but remains laminar until $\text{Re} \doteq 200$ at which Reynolds number the flow gradually becomes turbulent.

(c) We now turn the problem around and let the fluid be at rest at infinity but let the sphere move with a constant velocity $\boldsymbol{v}_s = V\boldsymbol{\delta}_z$, with its center instantaneously at the origin (see Fig. 1.4-2). Then the velocity $\boldsymbol{v}$ of the fluid is given according to Eqs. 1.4-20 and 21 by

$$\begin{aligned} \boldsymbol{v} &= \frac{3}{4}\frac{R}{r}V(2\boldsymbol{\delta}_r\cos\theta - \boldsymbol{\delta}_\theta\sin\theta) + O\left(\frac{R^3}{r^3}\right) \\ &= \frac{3}{4}\frac{R}{r}V(\boldsymbol{\delta}_z + \boldsymbol{\delta}_r\cos\theta) + O\left(\frac{R^3}{r^3}\right) \end{aligned} \tag{1.4-30}$$

Now the force $\boldsymbol{F}_s$ exerted by the sphere on the fluid by virtue of its motion is

$$\boldsymbol{F}_s = 6\pi\mu RV\boldsymbol{\delta}_z \tag{1.4-31}$$

It is possible to express the velocity disturbance $\boldsymbol{v}$ of Eq. 1.4-30 directly in terms of $\boldsymbol{F}_s$ as follows:

$$\boldsymbol{v} = [\boldsymbol{\Omega} \cdot \boldsymbol{F}_s] + O\left(\frac{R^3}{r^3}\right) \tag{1.4-32}$$

[4] I. Proudman and J. R. A. Pearson, *J. Fluid Mech.*, **2**, 237–262 (1957).
[5] S. Taneda, *J. Phys. Soc. Jpn.*, **11**, 1104–1108 (1956).

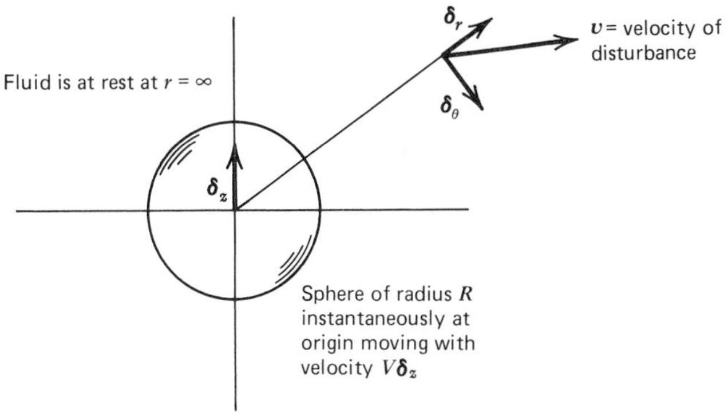

FIGURE 1.4-2. Velocity disturbance caused by a sphere moving linearly through a fluid with a constant velocity $V\boldsymbol{\delta}_z$ at that instant at which the center of the sphere is at the origin.

Here the tensor,

$$\boldsymbol{\Omega} = \frac{1}{8\pi\mu r}(\boldsymbol{\delta} + \boldsymbol{\delta}_r\boldsymbol{\delta}_r) \tag{1.4-33}$$

is called the *Oseen–Burgers tensor*[6] or *hydrodynamic interaction tensor*. It is shown in §§13.6, 14.6, and 15.4 how this tensor is used to describe the perturbation in the solvent velocity field near one part of a macromolecule because of the motion of another part of the macromolecule.

It is instructive to use the Oseen-Burgers tensor to attempt to calculate the total volume flow rate of liquid Q through the xy-plane ($\theta = \pi/2$) at the instant the sphere is at the origin. The fluid velocity at $z = 0$ and at a distance r from the origin is according to the Oseen-Burgers expression $\boldsymbol{v} = (F_s/8\pi\mu r)\boldsymbol{\delta}_z$. The volume flow rate is then

$$Q = \int_0^{2\pi}\int_R^{\infty} v_z r\,dr\,d\phi = 2\pi\left(\frac{F_s}{8\pi\mu}\right)\int_R^{\infty} dr \tag{1.4-34}$$

We see however that the integral is not convergent. This means that the Oseen tensor expression predicts that a finite force $F_s\boldsymbol{\delta}_z$ induces an infinite volume flux in the fluid. This is possible, of course, because the assumption of creeping flow is tantamount to assuming that the density of the fluid is exactly zero. The introduction of any finite fluid density, no matter how small, dramatically changes the velocity disturbance far from the sphere.[6] The slow, $O(r^{-1})$ decay of the velocity as $r \to \infty$ is then restricted to a "wake" region behind the sphere, and the Oseen-Burgers expression ceases to be a valid first approximation for the velocity disturbance far from the sphere.

EXAMPLE 1.4-2 Flow around a Rising Bubble[7]

A small spherical bubble of radius R is rising slowly in an incompressible Newtonian fluid of density ρ and viscosity μ. The fluid is at rest far from the bubble.

a. Find the velocity field for the flow around the bubble.

b. Develop a relationship between the gravitational acceleration g and the rise velocity of the bubble, V.

[6] C. W. Oseen, *Neuere Methoden und Ergebnisse in der Hydrodynamik*, Akad. Verlag, Leipzig (1927); J. M. Burgers, *Second Report on Viscosity and Plasticity*, Amsterdam Academy of Sciences, Amsterdam (1938), Chapt. 3.

[7] H. Lamb, *op. cit.*, pp. 600–601, and G. K. Batchelor, *op. cit.*, pp. 235–238, develop the solution for the flow around a translating spherical drop. In the limit as the viscosity of the drop approaches zero, these solutions then simplify to the solution given here.

SOLUTION **(a)** This flow problem is closely related to the flow around a solid sphere and we therefore adopt the same coordinate system and solution procedure illustrated in the previous example. The difference between the two situations is just in the boundary conditions at the surface of the spherical object; in this example we have the conditions

$$v_r = 0 \qquad \text{at } r = R \tag{1.4-35}$$

$$\tau_{r\theta} = -\mu\left[r\frac{\partial}{\partial r}\left(\frac{v_\theta}{r}\right) + \frac{1}{r}\frac{\partial v_r}{\partial \theta}\right] = 0 \qquad \text{at } r = R \tag{1.4-36}$$

This means that the boundary conditions for the function f of Eq. 1.4-18 are

$$f = 0 \qquad \text{at } r = R \tag{1.4-37}$$

$$-2f' + Rf'' = 0 \qquad \text{at } r = R \tag{1.4-38}$$

The four constants are now determined from Eq. 1.4-10 for $r \to \infty$ and the above equations for $r = R$ to yield

$$\psi = V(\tfrac{1}{2}r^2 - \tfrac{1}{2}Rr)\sin^2\theta \tag{1.4-39}$$

$$v_r = -V(1 - (R/r))\cos\theta \tag{1.4-40}$$

$$v_\theta = V(1 - \tfrac{1}{2}(R/r))\sin\theta \tag{1.4-41}$$

(b) To find the rise velocity we must know the pressure p. Here again we parallel the previous example to find

$$p = (\mu V R r^{-2} - \rho g r)\cos\theta + p_0 \tag{1.4-42}$$

where p_0 is a constant. The rr-component of the total stress tensor at the surface of the bubble is therefore

$$\begin{aligned}\pi_{rr}|_{r=R} &= p|_{r=R} - 2\mu\left.\frac{\partial v_r}{\partial r}\right|_{r=R} \\ &= (3\mu V R^{-1} - \rho g R)\cos\theta + p_0\end{aligned} \tag{1.4-43}$$

Notice that τ_{rr} does not vanish on the surface of the bubble as it does on the surface of the sphere. On the other hand $\pi_{r\theta}$ does vanish on the surface of the bubble. The net force per unit area in the z-direction is therefore

$$-(\boldsymbol{\delta}_z \cdot [\pi_{rr}\boldsymbol{\delta}_r]) = (-3\mu V R^{-1} + \rho g R)\cos^2\theta - p_0\cos\theta \tag{1.4-44}$$

where we have used $(\boldsymbol{\delta}_z \cdot \boldsymbol{\delta}_r) = \cos\theta$. The total force on the bubble is then

$$\begin{aligned}F &= \int_0^{2\pi}\int_0^{\pi} (\boldsymbol{\delta}_z \cdot [-\boldsymbol{\delta}_r \cdot \boldsymbol{\pi}])R^2 \sin\theta\, d\theta\, d\phi \\ &= -4\pi\mu R V + \tfrac{4}{3}\pi R^3 \rho g\end{aligned} \tag{1.4-45}$$

The first term in this expression describes the viscous drag on the bubble and the second term the buoyancy force on the bubble. The velocity of rise is found by using the condition that there is no net force on the bubble so that $V = (1/3)\rho g R^2/\mu$. Note that this gives $\pi_{rr} = p_0$ at the surface of the bubble independent of θ. This means that the bubble will remain spherical with no tendency to deform from that shape even with negligible surface tension.

For small but finite values of the Reynolds number, inertial forces will tend to deform bubbles, and the bubble shape will be given as a balance among viscous, inertial, and surface tension forces. Taylor and Acrivos[8] have derived the following expression to describe small deviations from the spherical shape:

$$R_s(\theta) = R[1 - \tfrac{5}{96}\,\mathrm{Re}\,\mathrm{Ca}\,(3\cos^2\theta - 1)] \tag{1.4-46}$$

Here $R_s(\theta)$ is the radius of the slightly deformed bubble as a function of θ, and R is the radius of an undeformed sphere of the same volume. The surface tension σ enters in the *capillary number* $\mathrm{Ca} = \mu V/\sigma$. The expression is valid when both $\mathrm{Re} = 2RV\rho/\mu \ll 1$ and $\mathrm{ReCa} \ll 1$, and it shows that inertial effects will deform the bubble into the shape of an ellipsoid flattened at the poles.

EXAMPLE 1.4-3 Rotating Sphere in a Viscous Liquid

A sphere of radius R is made to rotate with an angular velocity W in a Newtonian liquid of infinite extent, which is quiescent far from the sphere (see Fig. 1.4-3).

a. Find the velocity distribution in the surrounding liquid when the rotation is so slow that the inertial effects may be neglected.

b. Find the secondary flow due to small inertial effects.

c. Find the torque $\mathscr{T}$ which has to be applied to the sphere in order to maintain the rotation.

SOLUTION **(a)** We describe the flow in a spherical coordinate system with the origin at the center of the sphere and with the z-axis along the axis of rotation. The boundary conditions on the velocity field are then

$$v_r = 0,\ v_\theta = 0,\ v_\phi = RW\sin\theta, \qquad \text{at } r = R \tag{1.4-47}$$

$$v_r \to 0,\ v_\theta \to 0,\ v_\phi \to 0, \qquad \text{for } r \to \infty \tag{1.4-48}$$

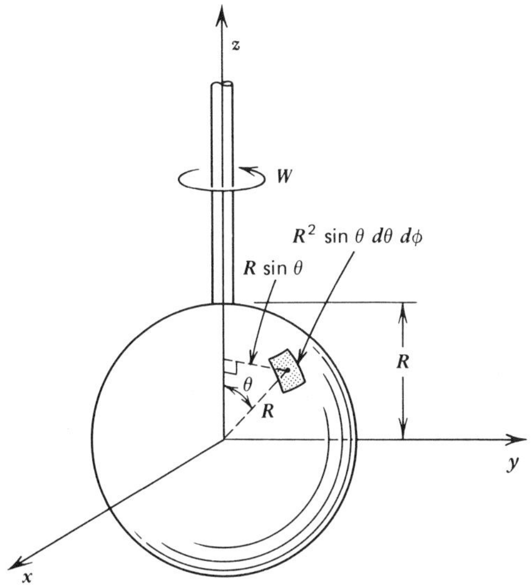

FIGURE 1.4-3. A sphere of radius R rotates with constant angular velocity W about the z-axis in an incompressible Newtonian fluid. The fluid is at rest far from the sphere.

[8] T. D. Taylor and A. Acrivos, *J. Fluid Mech.*, **18**, 466–476 (1964).

In view of these conditions we look for a solution to the creeping flow problem in the form $v_r = 0$, $v_\theta = 0$, $v_\phi = v(r) \sin \theta$ and with the modified pressure $\mathscr{P}$ being constant. The ϕ-component of the equation of motion with the inertial terms neglected then shows that $v(r)$ must satisfy

$$\frac{d}{dr}\left(r^2 \frac{dv}{dr}\right) - 2v = 0 \tag{1.4-49}$$

This is an equidimensional equation, which may be solved by the same technique that was used in Examples 1.4-1 and 2. The result is the following expression for the creeping flow around the rotating sphere

$$v_{0r} = 0; \quad v_{0\theta} = 0; \quad v_{0\phi} = \frac{R^3}{r^2} W \sin \theta \tag{1.4-50}$$

A subscript zero has been added to the velocity components to indicate that these correspond to very slow flow.

(b) To find the secondary flow due to small inertial effects we use a perturbation method. The perturbation parameter that describes the importance of the inertial effects is a Reynolds number $\text{Re} = R^2 W\rho/\mu$. For $\text{Re} \ll 1$ we assume then that the velocity field $\boldsymbol{v}$ and the pressure field p are

$$\boldsymbol{v} = \boldsymbol{v}_0 + \text{Re}\, \boldsymbol{v}_1 + \cdots \tag{1.4-51}$$

$$p = p_0 + \text{Re}\, p_1 + \cdots \tag{1.4-52}$$

Here $\boldsymbol{v}_0(r, \theta)$ is the velocity field given in component form in Eq. 1.4-50, and p_0 *is* $-\rho g r \cos \theta$. Since $\boldsymbol{v}_0$ satisfies the boundary conditions at the surface of the sphere and at infinity we see that $\boldsymbol{v}_1$ must satisfy

$$\boldsymbol{v}_1 = \boldsymbol{0}, \qquad \text{at } r = R \tag{1.4-53}$$

$$\boldsymbol{v}_1 \to \boldsymbol{0}, \qquad \text{for } r \to \infty \tag{1.4-54}$$

When the expansions are inserted into the steady state equation of motion and terms of first order in Re are equated, we find that $\boldsymbol{v}_1$ and p_1 must satisfy

$$-\boldsymbol{\nabla} p_1 + \mu \boldsymbol{\nabla}^2 \boldsymbol{v}_1 + \boldsymbol{f} = \boldsymbol{0} \tag{1.4-55}$$

where

$$\boldsymbol{f} = -\frac{\mu}{WR^2}[\boldsymbol{v}_0 \cdot \boldsymbol{\nabla} \boldsymbol{v}_0] = \mu R^4 W r^{-5} \sin \theta\, (\sin \theta\, \boldsymbol{\delta}_r + \cos \theta\, \boldsymbol{\delta}_\theta) \tag{1.4-56}$$

In addition $\boldsymbol{v}_1$ must satisfy the equation of continuity. The $\boldsymbol{\delta}_r$ and $\boldsymbol{\delta}_\theta$ are unit vectors associated with the spherical coordinate system, and we have used Eq. 1.4-50 for $\boldsymbol{v}_0$. We see that Eq. 1.4-55 has the same form as Eq. A of Table 1.4-1 with $\rho = 0$. The boundary conditions that $\boldsymbol{v}_1$ must be zero at the surface of the sphere and at infinity introduce no velocity in the ϕ-direction or any ϕ-dependence. Since furthermore $f_\phi = 0$ and $\boldsymbol{f}$ has no ϕ-dependence, we expect $\boldsymbol{v}_1$ to have the same properties. We may therefore derive $\boldsymbol{v}_1$ from a stream function $\psi(r, \theta)$ for axisymmetrical spherical flow. From Eq. E of Table 1.4-1 we see that ψ must satisfy

$$E^4 \psi = 6WR^4 r^{-5} \sin^2 \theta \cos \theta \tag{1.4-57}$$

We assume that ψ has the form

$$\psi = f(r) \sin^2 \theta \cos \theta \tag{1.4-58}$$

Then $f(r)$ must satisfy

$$\left(\frac{d^2}{dr^2} - \frac{6}{r^2}\right)\left(\frac{d^2}{dr^2} - \frac{6}{r^2}\right)f = 6WR^4r^{-5} \tag{1.4-59}$$

This equidimensional equation is solved by the method outlined in Example 1.4-1. The general solution is

$$f = -\tfrac{1}{4}WR^4r^{-1} + C_1r^{-2} + C_2 + C_3r^3 + C_4r^5 \tag{1.4-60}$$

The boundary condition at infinity in Eq. 1.4-54 implies that C_3 and C_4 must be zero. The boundary condition at the surface of the sphere in Eq. 1.4-53 corresponds to

$$f = 0, f' = 0, \qquad \text{at } r = R \tag{1.4-61}$$

These conditions determine C_1 and C_2. The final results for the components of the secondary flow field $\boldsymbol{v}_1$ are then

$$v_{1r} = -\tfrac{1}{8}WR^3\left(1 - \frac{R}{r}\right)^2 r^{-2}(3\cos^2\theta - 1) \tag{1.4-62}$$

$$v_{1\theta} = \tfrac{1}{4}WR^4\left(1 - \frac{R}{r}\right)r^{-3}\sin\theta\cos\theta \tag{1.4-63}$$

$$v_{1\phi} = 0 \tag{1.4-64}$$

These expressions show that inertial effects make the fluid move away from the sphere in the equatorial plane and towards the sphere along the axis in agreement with observations and physical intuition.

The rather straightforward perturbation solution given in this example is called a "regular perturbation" solution. Not all perturbation problems are equally straightforward. In particular the solutions from which the expressions in Eqs. 1.4-29 and 46 are obtained are examples of "singular perturbation" solutions, and these involve more elaborate techniques.

(c) We next find the torque $\mathscr{T}$ which must be applied to the sphere in order to maintain the rotation. The torque acting on a small elemental area $R^2 \sin\theta\, d\theta\, d\phi$ is the product of the shear force $\tau_{r\phi}|_{r=R}\, R^2 \sin\theta\, d\theta\, d\phi$ and the lever arm $R\sin\theta$ measured from the axis of rotation. Integration over the sphere surface then gives:

$$\begin{aligned}\mathscr{T} &= \int_0^{2\pi}\int_0^{\pi} \tau_{r\phi}\Big|_{r=R} (R\sin\theta)R^2\sin\theta\, d\theta\, d\phi \\ &= -2\pi\mu R^3 \int_0^{\pi} r\frac{\partial}{\partial r}\left(\frac{v_\phi}{r}\right)\Big|_{r=R} \sin^2\theta\, d\theta \\ &= +6\pi\mu R^3 W \int_0^{\pi} \sin^3\theta\, d\theta \\ &= 8\pi\mu R^3 W \end{aligned} \tag{1.4-65}$$

Note that the first-order term $\boldsymbol{v}_1$ in the expansion in Eq. 1.4-51 does not contribute to the torque. For other methods of finding the torque see Problem 1B.12.

PROBLEMS

1A.1 Volume Rate of Flow through a Circular Tube

The chlorinated biphenyl used by Sukanek and Laurence[1] in their research on viscous heating has the following properties at 313 K:

$$\mu = 400 \text{ poise} = 40 \text{ Pa} \cdot \text{s}$$

$$\rho = 1.6 \text{ g/cm}^3 = 1.6 \times 10^3 \text{ kg/m}^3$$

$$k = 9.4 \times 10^3 \text{ g} \cdot \text{cm/s}^3 \cdot \text{K} = 9.4 \times 10^{-2} \text{ W/m} \cdot \text{K}$$

$$\hat{C}_p = 0.235 \text{ cal/g} \cdot \text{C} = 983 \text{ J/kg} \cdot \text{K}$$

a. What will the volume rate of flow be for the flow of this fluid through a horizontal capillary tube of radius 3.2 mm and length 10.3 cm when the pressure drop is 1.67×10^5 Pa?

b. Compute the Reynolds number for the flow in order to verify that the flow is laminar.

c. What will happen if the radius of the tube is doubled?

Answers: **a.** 1.67×10^{-6} m^3/s
b. 1.33×10^{-2}

1A.2 Terminal Velocity of a Falling Sphere

a. By making a force balance on a sphere falling in a liquid and using the result in Eq. 1.4-28 show that the terminal velocity (the velocity attained at steady state) is

$$v_t = 2R^2(\rho_s - \rho)g/9\mu \tag{1A.2-1}$$

where ρ and ρ_s, are the densities of the liquid and the sphere, respectively.

b. What will be the terminal velocity of a steel sphere of radius 0.25 cm with density 7850 kg/m^3 falling in the liquid in Problem 1A.1?

c. Verify, by computing the Reynolds number for the flow, that it was permissible to use Eq. 1A.2-1.

d. What is the maximum radius of steel spheres that can be used so that Stokes' law can still be used for describing the viscous drag on the sphere falling in the liquid of Problem 1A.1?

Answers: **b.** 2.13×10^{-3} m/s
c. 4.26×10^{-4}
d. 1.54 cm

1A.3 Measurement of Viscosity in a Cone-and-Plate Viscometer

The system in Fig. 1.3-4 has the following geometrical measurements:

$$R = 5.2 \text{ cm} = 0.052 \text{ m}$$

$$\vartheta_0 = 0.35^\circ = 0.00611 \text{ rad}$$

[1] P. C. Sukanek and R. L. Laurence, *AIChE J.*, **20**, 474–484 (1974).

With a fluid completely filling the gap, a torque of 2.47 N·m is required to maintain an angular velocity of 1.28 rad/s.

a. What is the shear stess $\tau_{\theta\phi}$ in the gap?
b. What is $\dot{\gamma}_{\theta\phi}$ in the gap?
c. What is the viscosity of the fluid?

Answers: **a.** 8400 Pa
b. $-209\ s^{-1}$
c. 40 Pa · s

1A.4 Squeezing Flow Experiment

a. From Eq. 1.3-52 obtain a formula for the time required to squeeze out half of the material that is initially in the gap; call this quantity $t_{1/2}$.

b. For a silicone oil with a viscosity of 106 Pa·s Leider[2] found experimentally that $t_{1/2}$ was 499s for a squeezing flow system in which the disk radius was 2.54 cm, the initial disk separation (i.e., $2h_0$) was 0.01209 cm, and a mass of 4.07 kg was placed on the upper disk. What value of $t_{1/2}$ is calculated from the formula in (**a**)?

Answer: **b.** 535 s

1B.1 Flow in a Tube with a Sinusoidally Varying Cross-Section

Figure 1B.1 shows a tube with a radius of the form $R(z) = R_0[1 + \alpha \sin(\pi z/l)]$ where $|\alpha| < 1$ and R_0 is the mean radius. The pressure drop corresponding to the length of one full cycle, $2l$, is denoted by $\mathscr{P}_0 - \mathscr{P}_{2l}$.

a. When $l \gg R_0$ use a lubrication approximation to develop an expression for the volume rate of flow Q for a given average pressure gradient $(\mathscr{P}_0 - \mathscr{P}_{2l})/2l$.

b. When $|\alpha| \ll 1$ show that the expression becomes approximately

$$Q = \frac{\pi R_0^4(\mathscr{P}_0 - \mathscr{P}_{2l})}{16\mu l[1 + 5\alpha^2]} \tag{1B.1-1}$$

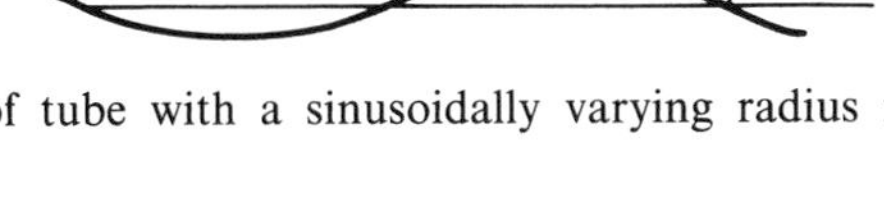

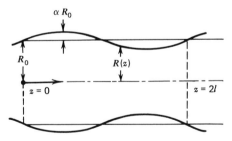

FIGURE 1B.1. Section of tube with a sinusoidally varying radius in a cylindrical coordinate system.

1B.2 Flow Between Coaxial Cylinders and Spheres

a. The space between two coaxial cylinders is filled with an incompressible, Newtonian fluid at constant temperature. The radii of the inner and outer wetted surfaces are κR and R, respectively. The angular velocities of rotation of the inner and outer cylinders are W_i and W_o. Determine the velocity distribution in the fluid, and the torques on the two cylinders needed to maintain the motion.

[2] P. J. Leider, *Ind. Eng. Chem. Fundam.*, **13**, 342–346 (1974).

b. Repeat part (**a**) for two concentric spheres. Assume that the flow is sufficiently slow that no secondary flows occur.

c. Find the torque per unit length $\mathscr{T}/L$ on a cylinder of radius R rotating slowly with angular velocity W in an infinite sea of liquid at rest at infinity.

d. Find the torque $\mathscr{T}$ on a sphere rotating slowly in an infinite sea of liquid at rest at infinity.

Answers: **a.** $$v_\theta = \frac{\kappa R}{1-\kappa^2}\left[(W_o - \kappa^2 W_i)\frac{r}{\kappa R} + (W_i - W_o)\frac{\kappa R}{r}\right] \qquad (1B.2\text{-}1)$$

b. $$v_\phi = \frac{\kappa R \sin\theta}{1-\kappa^3}\left[(W_o - \kappa^3 W_i)\frac{r}{\kappa R} + (W_i - W_o)\left(\frac{\kappa R}{r}\right)^2\right] \qquad (1B.2\text{-}2)$$

c. $$\mathscr{T}/L = 4\pi\mu W R^2 \qquad (1B.2\text{-}3)$$

d. $$\mathscr{T} = 8\pi\mu W R^3 \qquad (1B.2\text{-}4)$$

1B.3 Laminar Newtonian Flow in a Triangular Duct[3]

A straight duct has a length L and a cross section of triangular shape, bounded by the plane surfaces $y = H$, $y = \sqrt{3}x$, and $y = -\sqrt{3}x$. Verify that the velocity profile for the laminar flow of a Newtonian fluid in a duct of this type is

$$v_z = \frac{(\mathscr{P}_0 - \mathscr{P}_L)}{4\mu L H}(y - H)(3x^2 - y^2) \qquad (1B.3\text{-}1)$$

Then obtain the volume rate of flow

$$Q = \frac{\sqrt{3}(\mathscr{P}_0 - \mathscr{P}_L)H^4}{180\mu L} \qquad (1B.3\text{-}2)$$

1B.4 Contraction of a Newtonian Jet at Large Reynolds Number

The behavior of Newtonian jets issuing from circular tubes has been studied experimentally by Middleman and Gavis.[4] They found that the jets swell for Re < 16 and contract for Re > 16, where the Reynolds number Re is defined as $\text{Re} = 2R\langle v_z \rangle \rho/\mu$ (the angular brackets indicate an average over the tube cross section).

We assume that the flow is isothermal, and consider the equation of continuity (Eq. 1.1-4) and the equation of motion (Eq. 1.1-8) written for steady-state, incompressible flow.

a. First show that these two equations can be integrated over a volume V enclosed by a surface S to obtain

$$\int_S (\boldsymbol{n} \cdot \boldsymbol{v})\,dS = 0 \qquad (1B.4\text{-}1)$$

$$\int_S [\boldsymbol{n} \cdot \rho\boldsymbol{v}\boldsymbol{v}]\,dS + \int_S [\boldsymbol{n} \cdot \boldsymbol{\pi}]\,dS - \int_V \rho\boldsymbol{g}\,dV = 0 \qquad (1B.4\text{-}2)$$

[3] L. D. Landau and E. M. Lifshitz, *Fluid Mechanics*, Addison-Wesley, Reading, MA (1959), p. 58. See also R. Berker, *Handbuch der Physik*, Vol. VIII/2, Springer, Berlin (1963), pp. 67–77, for a summary of formulas for flow in conduits of various cross sections.

[4] S. Middleman and J. Gavis, *Phys. Fluids*, **4**, 355–359 (1961).

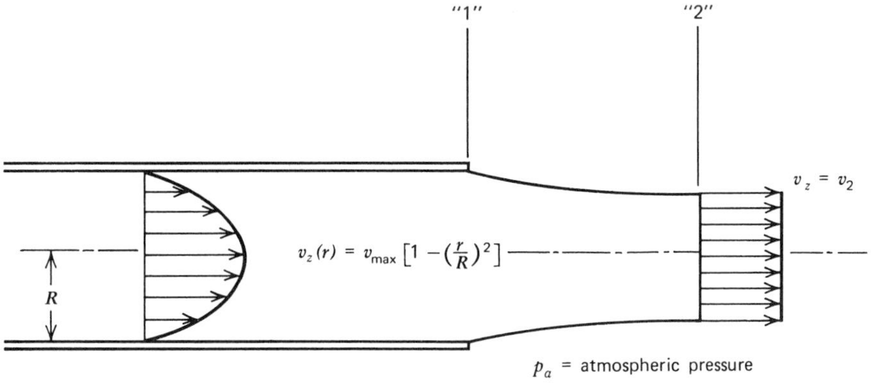

FIGURE 1B.4. Contraction of a Newtonian jet emerging from a circular tube. Within the tube the velocity profile is parabolic. At plane "2" the velocity is uniform. The cross sectional areas at "1" and "2" are S_1 and S_2, respectively. Newtonian jets have been found to contract for $\text{Re} = 2R\langle v_z \rangle \rho/\mu > 16$ and to swell for $\text{Re} < 16$. For $\text{Re} > 100$, it is found experimentally that $S_2/S_1 = \frac{3}{4}$.

When these equations are applied to the fluid contained in the region between planes "1" and "2" in Fig. 1B.4 they become

$$\langle v_z \rangle_1 S_1 - v_2 S_2 = 0 \tag{1B.4-3}$$

$$\rho \langle v_z^2 \rangle_1 S_1 - \rho v_2^2 S_2 + \langle \pi_{zz} \rangle_1 S_1 - p_a S_2 + p_a(S_2 - S_1) = 0 \tag{1B.4-4}$$

where p_a is the atmospheric pressure.

b. It is desired to use the equations developed in (a) to estimate S_2/S_1 for the jet in the limit of high Reynolds number laminar flow. We do this by assuming that in this limit:

1. The flow remains parabolic up to plane "1," so that $\langle v_z^2 \rangle_1 / \langle v_z \rangle_1^2 = 4/3$. (Verify this numerical value.)
2. $\langle \pi_{zz} \rangle_1$ can be approximated as p_a.

Neither of these approximations has been justified. If now one solves Eqs. 1B.4-3 and 4 simultaneously then one obtains $S_2/S_1 = 3/4$; verify this. This result was obtained by Harmon.[5] The value of 3/4 agrees with the experimental results of Middleman and Gavis for Reynolds numbers of more than 100. However at $\text{Re} = 16$, it was found experimentally that $S_2/S_1 = 1$, and as the Reynolds number approaches zero, S_2/S_1 becomes about 1.13.[6]

1B.5 Parallel-Disk Viscometer

A schematic diagram is given in Fig. 1B.5 of a parallel-disk viscometer. The fluid is placed in the gap of thickness B between the two circular disks. Develop a formula for deducing the fluid viscosity from the measurement of the torque $\mathscr{T}$ required to turn the upper disk at an angular velocity W.

a. Postulate that for small values of W the velocity and pressure profiles have the form: $v_r = 0$, $v_z = 0$, and $v_\theta = zf(r)$; and $p = p(r, z)$. Write down the resulting simplified equations of continuity and motion.

[5] D. B. Harmon, *J. Franklin Inst.*, **259**, 519–522 (1955); S. Middleman and J. Gavis, *Phys. Fluids*, **4**, 355–359 (1961).
[6] R. I. Tanner in J. R. A. Pearson and S. M. Richardson, eds., *Numerical Analysis of Polymer Processing*, Applied Science Publishers, London (1983); see also R. I. Tanner, *Engineering Rheology*, Elsevier Applied Science Publishers, London (1985), pp. 322–325.

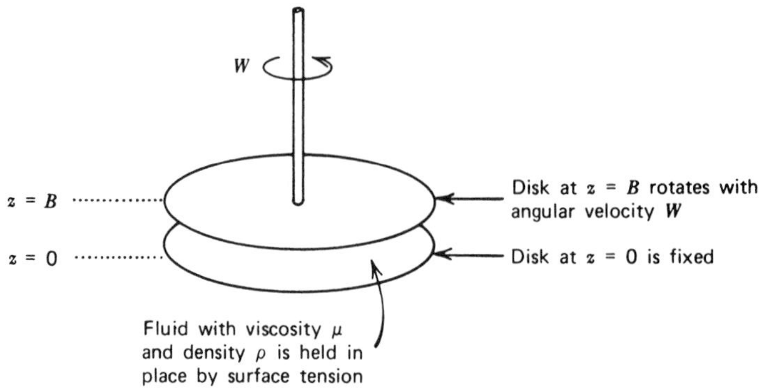

FIGURE 1B.5. Parallel-disk viscometer with two disks of radius R separated by a distance B, with dimensions such that $R \gg B$.

b. From the θ-component of the equation of motion, obtain a differential equation for $f(r)$. Solve this equation and evaluate the constants of integration. Finally obtain the velocity distribution $v_\theta = Wrz/B$.

c. Show that the torque required to turn the upper disk is $\mathscr{T} = \pi\mu WR^4/2B$.

1B.6 The Rayleigh Problem[7]

An incompressible Newtonian fluid occupies the half-space, from $y = 0$ to $y = \infty$, and is bounded at $y = 0$ by a solid plane surface. Before $t = 0$, the fluid is at rest. Then at time $t = 0$ the solid surface at $y = 0$ is made to move with constant velocity V in the positive x-direction. It is desired to find the velocity distribution $v_x = v_x(y, t)$.

a. Introduce a dimensionless velocity $\phi = v_x/V$. Inasmuch as there is no natural unit of length for the problem, and since the initial condition at $t = 0$ and the boundary condition at $y = \infty$ are the same, it seems reasonable to try a solution of the form

$$\phi = \phi(\eta) \quad \text{where} \quad \eta = \sqrt{\rho y^2/4\mu t} \tag{1B.6-1}$$

Show that with these variables the equation of motion for this flow becomes

$$\phi'' + 2\eta\phi' = 0 \tag{1B.6-2}$$

in which primes denote differentiation with respect to η. What are the boundary conditions that go with this equation?

b. Show that this equation can be solved to give

$$\phi = 1 - \frac{2}{\sqrt{\pi}}\int_0^\eta \exp(-x^2)\,dx = 1 - \operatorname{erf}\sqrt{\rho y^2/4\mu t} \tag{1B.6-3}$$

in which erf η is the error function of η.

1B.7 Steady Simple Elongational Flow and Elongational Viscosity

Here we study the type of flow that occurs when a cylindrical filament of fluid is stretched slowly in the absence of external body forces (see Fig. 1B.7). For steady-state simple elongational flow

$$v_z = \dot{\varepsilon}z \quad v_r = -\tfrac{1}{2}\dot{\varepsilon}r \quad v_\theta = 0 \tag{1B.7-1}$$

in which $\dot{\varepsilon}$ is the (constant) elongation rate.

[7] Lord Rayleigh (John William Strutt), *Phil. Mag.*, **21** (6), 697–711 (1911). See also R. B. Bird, W. E. Stewart, and E. N. Lightfoot, *op. cit.*, pp. 124–126.

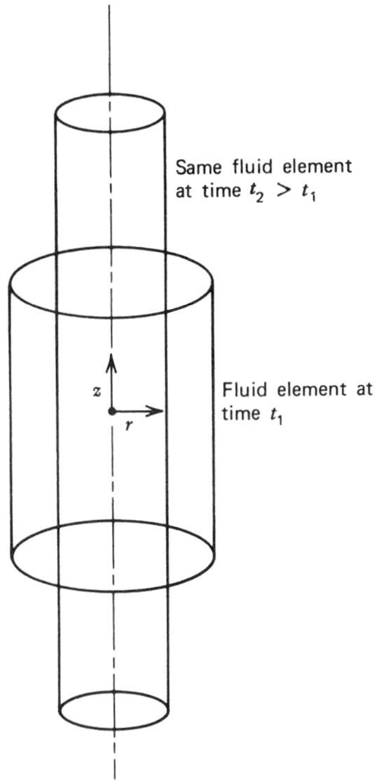

FIGURE 1B.7. Steady simple elongational flow, showing how a cylindrical portion of fluid deforms with time. The velocity components are $v_z = \dot{\varepsilon}z$, $v_r = -\frac{1}{2}\dot{\varepsilon}r$, $v_\theta = 0$, where $\dot{\varepsilon}$ is the (constant) elongation rate.

a. Verify that this velocity distribution satisfies the equation of continuity identically. Show further that for an incompressible Newtonian fluid in steady simple elongational flow

$$\tau_{rr} = \tau_{\theta\theta} = \mu\dot{\varepsilon}; \quad \tau_{zz} = -2\mu\dot{\varepsilon} \tag{1B.7-2}$$

b. Next use the equation of motion to gain further information about the components of the stress tensor. For steady-state flow with negligible inertial terms ($[\nabla \cdot \rho \boldsymbol{v}\boldsymbol{v}] = \mathbf{0}$) and no external forces ($\boldsymbol{g} = \mathbf{0}$), show that the components of $\boldsymbol{\pi}$ are constant throughout the fluid, and that specifically $p + \tau_{rr} = p_0$, where p_0 is the ambient pressure outside of the fluid fiber.

c. Show that the results of (**b**) lead to the conclusion that $\pi_{zz} - p_0 = \tau_{zz} - \tau_{rr} = -3\mu(dv_z/dz)$. The coefficient of proportionality between the normal stress difference $\tau_{zz} - \tau_{rr}$ and the negative of the elongation rate is called $\bar{\eta}$, the *elongational viscosity*[8] or *Trouton viscosity*.[9] Thus for Newtonian fluids we have the well-known result that

$$\bar{\eta} = 3\mu \tag{1B.7-3}$$

This result has been used for calculating the shape of a freely falling jet of liquid.[10]

[8] A. S. Lodge, *Elastic Liquids*, Academic Press, New York (1964), pp. 97–98, 114–118; see also §3.5.
[9] F. T. Trouton, *Proc. Roy. Soc.*, **A77**, 426 (1906).
[10] See Problem 1C.5.

1B.8 Alternative Form of the Energy Equation

It is desired to show how the energy equation in terms of the internal energy (Eq. C of Table 1.1-1) can be transformed into the energy equation in terms of the temperature (Eq. C of Table 1.2-1).

a. First show that $(\boldsymbol{\pi}:\nabla\boldsymbol{v}) = p(\nabla\cdot\boldsymbol{v}) + (\boldsymbol{\tau}:\nabla\boldsymbol{v})$ by using Eq. 1.2-1 and the definitions of the ∇-operations.

b. Next replace $\hat{U}$ by $\hat{H} - p\hat{V}$, where $\hat{H}$ is the enthalpy per unit mass. Show that

$$\frac{D}{Dt}p\hat{V} = \frac{p}{\rho}(\nabla\cdot\boldsymbol{v}) + \frac{1}{\rho}\frac{Dp}{Dt} \tag{1B.8-1}$$

by using the equation of continuity (Eq. D of Table 1.1-1). This enables us to rewrite the energy equation as an equation of change for enthalpy:

$$\rho\frac{D\hat{H}}{Dt} = -(\nabla\cdot\boldsymbol{q}) - (\boldsymbol{\tau}:\nabla\boldsymbol{v}) + \frac{Dp}{Dt} \tag{1B.8-2}$$

c. Next we asume that $\hat{H}$ can be expressed in terms of two state variables p and T, so that we can use the thermodynamic relation

$$\begin{aligned} d\hat{H} &= \left(\frac{\partial\hat{H}}{\partial T}\right)_p dT + \left(\frac{\partial\hat{H}}{\partial p}\right)_T dp \\ &= \hat{C}_p\,dT + \left[\hat{V} - T\left(\frac{\partial\hat{V}}{\partial T}\right)_p\right]dp \end{aligned} \tag{1B.8-3}$$

Use of this relation enables us to rewrite Eq. 1B.8-2 as

$$\rho\hat{C}_p\frac{DT}{Dt} = -(\nabla\cdot\boldsymbol{q}) + \left(\frac{\partial\ln\hat{V}}{\partial\ln T}\right)_p\frac{Dp}{Dt} - (\boldsymbol{\tau}:\nabla\boldsymbol{v}) \tag{1B.8-4}$$

Note that up to this point the only assumption that has been made is that $\hat{H}$ depends on only two state variables, and not on the state of strain in the system.

d. When the density is assumed constant, the $(\partial\ln\hat{V}/\partial\ln T)_p$ term in Eq. 1B.8-4 can be omitted. Next show that $(-\boldsymbol{\tau}:\nabla\boldsymbol{v})$ can be written as $+\frac{1}{2}\mu(\dot{\boldsymbol{\gamma}}:\dot{\boldsymbol{\gamma}})$ by using the expression in Eq. 1.2-2 for incompressible fluids. Then, show that introduction of Fourier's law and assumption of constant thermal conductivity leads finally to Eq. C of Table 1.2-1.

1B.9 Low Reynolds Number Flow around a Cylinder[11]

An incompressible Newtonian liquid approaches a cylinder with a constant velocity $\boldsymbol{V}$ in the positive x-direction (see Fig. 1B.9). When the equations of change are solved in the low Reynolds number regime, the following expressions are found for the pressure $p(r, \theta)$ and the fluid velocity $\boldsymbol{v}(r, \theta)$ for the region near the cylinder (they are *not* valid for large r/R):

$$p = p^0 - C\mu\frac{(\boldsymbol{V}\cdot\boldsymbol{r})}{r^2} + \rho(\boldsymbol{g}\cdot\boldsymbol{r}) \tag{1B.9-1}$$

$$\begin{aligned} \boldsymbol{v} &= C\boldsymbol{V}\left[\frac{1}{2}\ln\left(\frac{r}{R}\right) + \frac{1}{4} - \frac{1}{4}\left(\frac{R}{r}\right)^2\right] \\ &\quad - C\boldsymbol{r}\frac{(\boldsymbol{V}\cdot\boldsymbol{r})}{2r^2}\left[1 - \left(\frac{R}{r}\right)^2\right] \end{aligned} \tag{1B.9-2}$$

[11] See G. K. Batchelor, *An Introduction to Fluid Dynamics*, Cambridge University Press, Cambridge (1967), pp. 244–246, 261.

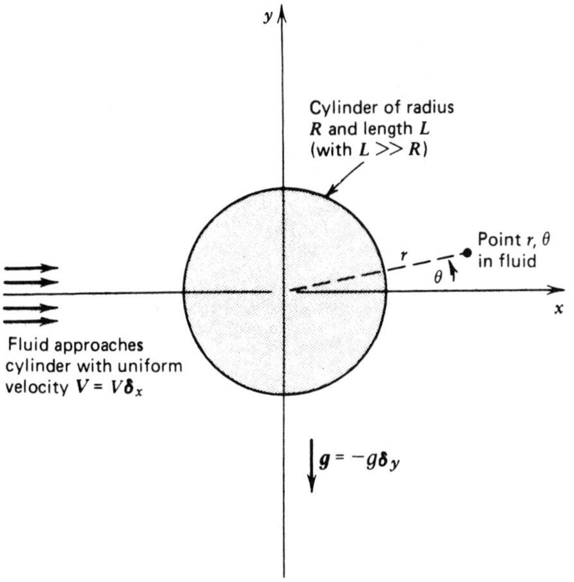

FIGURE 1B.9. Transverse flow around a cylinder, with the fluid approach velocity V and gravitational force g.

In these equations the constant C is given by $C = 2/\ln(7.4/\mathrm{Re})$, in which $\mathrm{Re} = 2RV\rho/\mu$ is the Reynolds number.

a. Show that, in the absence of flow, Eq. 1B.9-1 satisfies the equation of motion.

b. Write out $p(r, \theta)$, $v_r(r, \theta)$, and $v_\theta(r, \theta)$ explicitly in terms of the scalars V and g; that is, no vectors should appear in the expressions.

c. Use the results of (**b**) to obtain p, τ_{rr}, and $\tau_{r\theta}$ at $r = R$.

d. Next show that the x-component of the stress exerted by the fluid on the solid cylinder at any point on the surface is $([-\boldsymbol{\delta}_r \cdot \boldsymbol{\pi}] \cdot \boldsymbol{\delta}_x)|_{r=R}$, and that this quantity can be simplified to

$$-p|_{r=R} \cos\theta + \tau_{r\theta}|_{r=R} \sin\theta \tag{1B.9-3}$$

e. Use the results of (**c**) and (**d**) to get the force F_x in the x-direction exerted by the fluid on a length L of the cylinder

$$F_x = 2C\pi L\mu V \tag{1B.9-4}$$

1B.10 Radial Flow between Porous Cylinders

An incompressible Newtonian liquid is made to flow radially (in the $-r$-direction in cylindrical coordinates) from a porous cylinder of radius R to another porous cylinder of radius κR (see Fig. 1B.10).

a. Write down the equations of continuity and motion for the postulates that v_r and $\mathcal{P}$ are functions of r alone.

b. Integrate the equation of continuity to get v_r as a function of r. Determine the constant of integration by writing it in terms of Q, the volume rate of flow through a length L of the outer (or inner) surface.

c. Integrate the equation of motion to get the pressure difference required to maintain a flow rate Q. Express the pressure difference in terms of ρ, μ, Q, κ, R, and L.

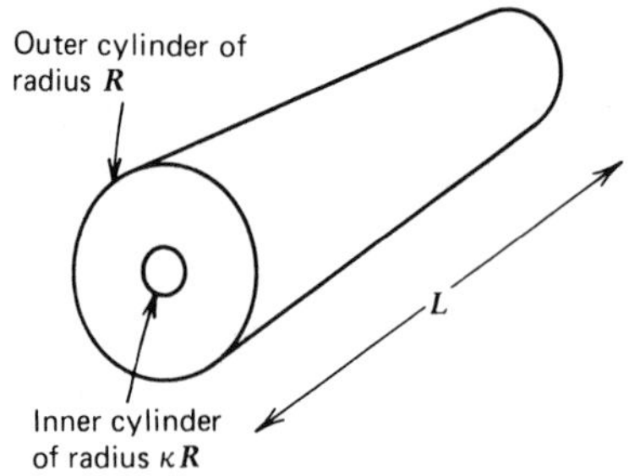

FIGURE 1B.10. Two porous cylinders, with purely radial flow from the outer cylinder to the inner cylinder.

1B.11 Flow into a Thin Slit[12]

A fluid in the region $x \geqslant 0$ flows into the slit in Fig. 1B.11, and then goes down the slit with volume flow rate Q. It is postulated that $v_\theta = 0$, $v_z = 0$, $v_r = v_r(r, \theta)$, and $\mathscr{P} = \mathscr{P}(r, \theta)$. The fluid is Newtonian and incompressible. The flow is symmetrical about the xz-plane.

a. Show that the equation of continuity leads to

$$v_r = \frac{1}{r} f(\theta) \tag{1B.11-1}$$

where f is a function of θ, with $df/d\theta = 0$ at $\theta = 0$, and $f = 0$ at $\theta = \pi/2$.

b. Next write down the r- and θ-components of the equation of motion in the "creeping-flow limit" (i.e., omit the $\rho D\mathbf{v}/Dt$ term), using the tables in Appendix B.

c. Substitute v_r from (**a**) into the equations in (**b**). Then differentiate the r-component of the equation of motion with respect to θ, and the θ-component with respect to r. Show that this leads to

$$d^3f/d\theta^3 + 4df/d\theta = 0 \tag{1B.11-2}$$

d. Solve the equation in (**c**) to obtain an expression for $f(\theta)$ containing three constants of integration.

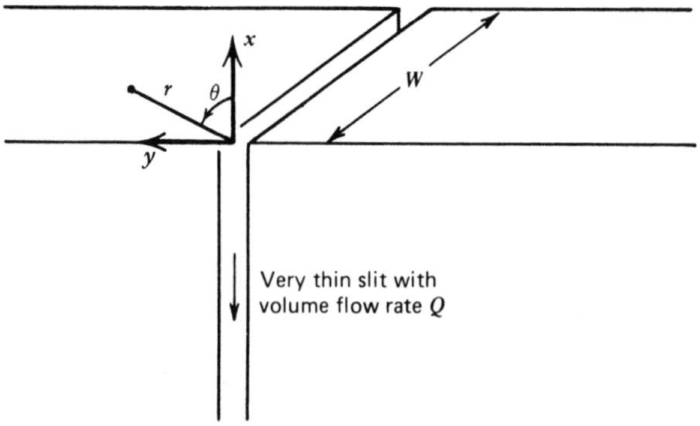

FIGURE 1B.11. A thin slit into which a fluid flows from the region $x \geq 0$. The z-direction is perpendicular to the plane of the paper.

[12] The corresponding viscoelastic problem is discussed in Problem 8C.1.

e. Determine the integration constants by using the two boundary conditions in (**a**) and the fact that the total volume flow rate through a cylindrical surface at any r must equal Q. This gives finally

$$v_r = -(2Q/\pi W r)\cos^2\theta \tag{1B.11-3}$$

f. Next obtain $\mathscr{P}(r, \theta)$ from the equations of motion in (**b**)

$$\mathscr{P}(r, \theta) = \mathscr{P}_\infty - \frac{2\mu Q}{\pi W r^2}\cos 2\theta \tag{1B.11-4}$$

What is the meaning of $\mathscr{P}_\infty$?

g. Next verify that the rate-of-deformation tensor is

$$\dot{\boldsymbol{\gamma}} = (2Q/\pi W r^2)[2(\boldsymbol{\delta}_r\boldsymbol{\delta}_r - \boldsymbol{\delta}_\theta\boldsymbol{\delta}_\theta)\cos^2\theta + (\boldsymbol{\delta}_r\boldsymbol{\delta}_\theta + \boldsymbol{\delta}_\theta\boldsymbol{\delta}_r)\sin 2\theta] \tag{1B.11-5}$$

h. Then obtain the total normal stress on the solid surface at $\theta = \pi/2$:

$$(p + \tau_{\theta\theta})|_{\theta=\pi/2} = p_\infty + (2\mu Q/\pi W r^2) \tag{1B.11-6}$$

i. What is $\tau_{\theta r}|_{\theta=\pi/2}$? Is this a surprising result?

j. Use the expression in Eq. 1B.11-3 to obtain $v_x(x, y)$ and $v_y(x, y)$ for this flow. Then obtain $\partial v_y/\partial x$ for $x = 0$. How does this tie in with (**i**)?

k. Find the volume rate of flow through the plane $x = 1$ by evaluating the integral $\int(\boldsymbol{n}\cdot\boldsymbol{v})\,dS$ for this surface.

1B.12 Torque on a Rotating Sphere

In Part (**c**) of Example 1.4-3, the torque required for the slow rotation of a sphere in a Newtonian fluid was found. Here we obtain the same result by two different methods.

a. Verify that the velocity distribution in Eq. 1.4-50 can be written in vector form:

$$\boldsymbol{v} = (R/r)^3[\boldsymbol{W} \times \boldsymbol{r}] \tag{1B.12-1}$$

and that for this velocity distribution

$$\nabla\boldsymbol{v} + (\nabla\boldsymbol{v})^\dagger = -\frac{3R^3}{r^5}\{\boldsymbol{r}[\boldsymbol{W} \times \boldsymbol{r}] + [\boldsymbol{W} \times \boldsymbol{r}]\boldsymbol{r}\} \tag{1B.12-2}$$

Next show that the torque exerted by a fluid on any solid body is

$$\mathscr{T} = \int [[\boldsymbol{n}\cdot\boldsymbol{\pi}] \times \boldsymbol{r}]dS \tag{1B.12-3}$$

where $\boldsymbol{n}$ is the outwardly directed unit normal at the solid surface. Next show that the above results may be used to get for the rotating sphere:

$$\begin{aligned}\mathscr{T} &= 3\mu R^3 \int_0^{2\pi}\int_0^{\pi}(\boldsymbol{W} - [\boldsymbol{W}\cdot\boldsymbol{nn}])\sin\theta\, d\theta\, d\phi \\ &= -8\pi\mu R^3 \boldsymbol{W}\end{aligned} \tag{1B.12-4}$$

which can be compared with Eq. 1.4-65.

b. The rate of work done on the rotating sphere is $\mathscr{T}W$. This has to be equal to the rate of energy dissipation in the surrounding fluid (in $R \leqslant r < \infty$):

$$\mathscr{T}W = \iiint (-\boldsymbol{\tau} : \nabla \boldsymbol{v}) r^2 \, dr \sin\theta \, d\theta \, d\phi \tag{1B.12-5}$$

Use the velocity distribution in Eq. 1B.12-1 to obtain the torque by doing the integral in Eq. 1B.12-5.

1C.1 Circulating Flow in an Annulus

A rod (radius κR) moves upward with constant velocity V through a cylindrical container (radius R) containing an incompressible viscous fluid. The fluid circulates in the cylinder, moving upward along the moving central core and moving back downward along the fixed wall. It is desired to find the velocity distribution in the annular region, away from the end disturbances (see Fig. 1C.1).

a. First consider the problem when the annular region is quite narrow—that is, κ is just slightly less than unity. In that case the annulus may be approximated by a thin plane slit, and its

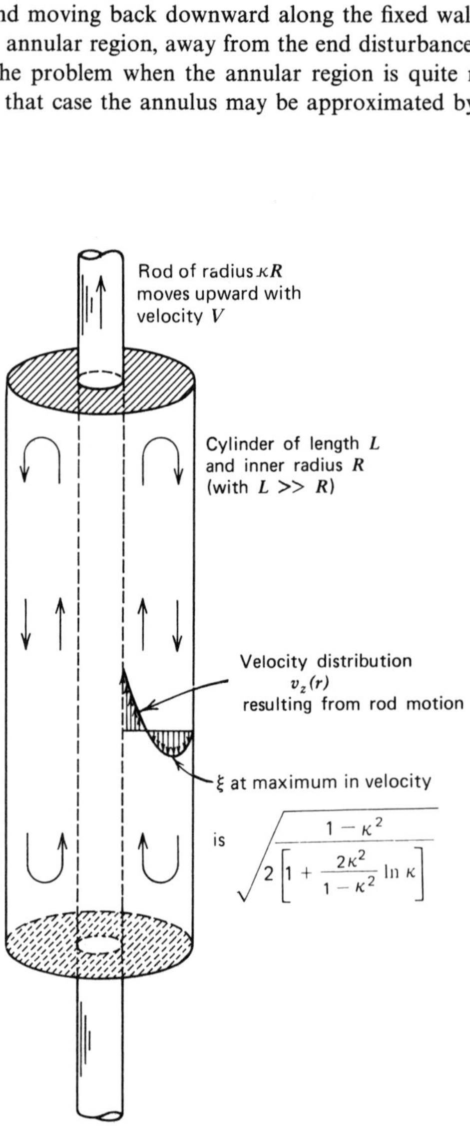

FIGURE 1C.1. Steady-state circulating flow in an annulus with the interior surface moving axially.

curvature can be neglected. Show that in this limit, the velocity distribution is given by

$$\frac{v_z}{V} = 3\zeta^2 - 4\zeta + 1 \tag{1C.1-1}$$

where $\zeta = (\xi - \kappa)/(1 - \kappa)$, and $\xi = r/R$.

b. Next, work the problem without the thin-slit assumption. Show that the velocity distribution is given by:

$$\frac{v_z}{V} = \frac{(1 - \xi^2)[1 - (2\kappa^2/(1 - \kappa^2)) \ln (1/\kappa)] - (1 - \kappa^2) \ln (1/\xi)}{(1 - \kappa^2) - (1 + \kappa^2) \ln (1/\kappa)} \tag{1C.1-2}$$

1C.2 Analysis of a Falling Cylinder Viscometer[13]

A falling cylinder viscometer consists of a long vertical cylinder (radius R) and a cylindrical slug (radius κR and height H) as shown in Fig. 1C.2. The slug is equipped with very thin fins so that its axis is maintained coincident with the axis of the tube.

The rate of descent of the cylindrical slug in the cylinder can be observed when the latter is filled with an incompressible fluid. It is desired to obtain an equation that gives the viscosity of the fluid in terms of the speed of fall V and the various geometric quantities. It is, of course, assumed that the measurement is made when a constant velocity of descent has been attained.

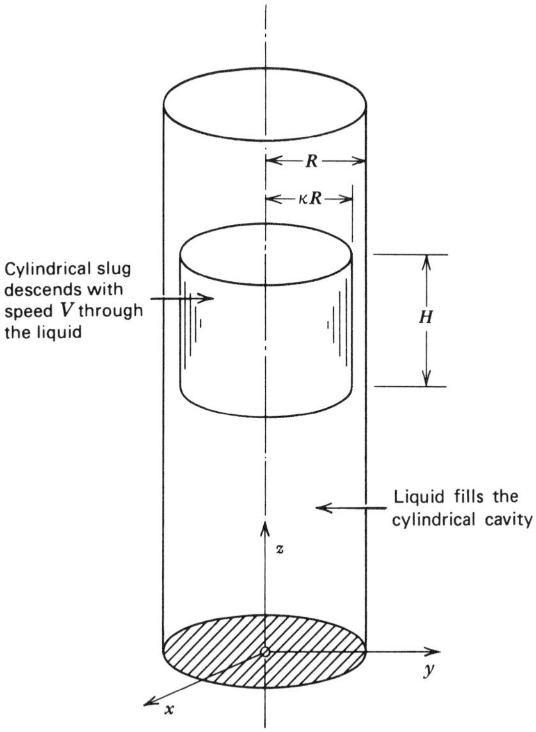

FIGURE 1C.2. Falling-cylinder viscometer, consisting of a tightly fitting cylindrical slug that descends through the liquid, which is forced upwards through the annular space.

[13] J. Lohrenz, G. W. Swift, and F. Kurata, *AIChE J.*, **6**, 547–550 (1960); *ibid.*, **7**, 6S (1961); E. Ashare, R. B. Bird, and J. A. Lescarboura, *AIChE J.*, **11**, 910–916 (1965).

a. First show that the velocity distribution in the annular slit is given by

$$\frac{v_z}{V} = -\frac{(1-\xi^2) - (1+\kappa^2)\ln(1/\xi)}{(1-\kappa^2) - (1+\kappa^2)\ln(1/\kappa)} \qquad (1C.2\text{-}1)$$

in which $\xi = r/R$ is a dimensionless radial coordinate.

b. Make a force balance on the cylindrical slug and obtain finally

$$\mu = \frac{(\rho_0 - \rho)g(\kappa R)^2}{2V}\left(\ln\frac{1}{\kappa} - \frac{1-\kappa^2}{1+\kappa^2}\right) \qquad (1C.2\text{-}2)$$

in which ρ is the density of the fluid and ρ_0 is the density of the slug.

c. Verify that for small slit width the result in (**b**) may be expanded in powers of $\varepsilon = 1 - \kappa$ to give

$$\mu = \frac{(\rho_0 - \rho)gR^2\varepsilon^3}{6V}\left(1 - \tfrac{1}{2}\varepsilon - \tfrac{13}{20}\varepsilon^2 + \cdots\right) \qquad (1C.2\text{-}3)$$

d. Draw a sketch showing the pressure distribution in the system.

e. Which of the two forces—pressure or viscous drag—is primarily responsible for balancing the gravitational force on the slug?

1C.3 Flow Near a Sharp Corner[14]

Consider the two dimensional plane creeping flow of an incompressible Newtonian fluid near a sharp corner. Since the equations for the velocity field are linear we may consider separately the antisymmetrical and symmetrical flow shown in Fig. 1C.3. Assume that the stream function may be expanded as

$$\psi = \sum_{n=1}^{\infty} \mathcal{Re}\{A_n r^{\lambda_n} f_n(\theta)\} \qquad (1C.3\text{-}1)$$

where the terms are ordered so that

$$1 < \mathcal{Re}\{\lambda_1\} < \mathcal{Re}\{\lambda_2\} < \cdots \qquad (1C.3\text{-}2)$$

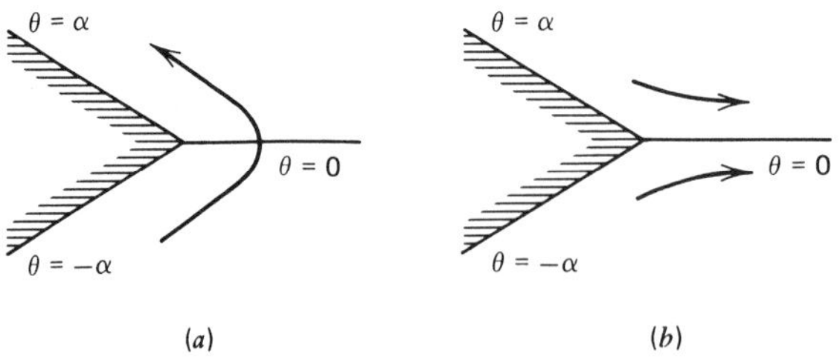

FIGURE 1C.3. Flow near a sharp corner (*a*) Antisymmetrical flow in the region $-\alpha < \theta < \alpha$; (*b*) symmetrical flow in the region $-\alpha < \theta < \alpha$. Note that situation (*b*) is also the flow in the region $0 < \theta < \alpha$ with a solid surface at $\theta = \alpha$ and a free surface at $\theta = 0$.

[14] H. K. Moffatt, *J. Fluid Mech.*, **18**, 1–18 (1964). This reference includes a careful discussion of the infinite sequence of eddies that occurs close to the corner when $2\alpha < 2\alpha_c$ (cf. Table 1C.3). Our interest here is in the flow when $2\alpha > 2\alpha_c$.

TABLE 1C.3

Values of λ_1 in Eqs. 1C.3-6 and 8 Corresponding to Sample Values of 2α.[a]

	Flow	
2α	Antisymmetrical ($2\alpha_c \doteq 146°$)	Symmetrical ($2\alpha_c \doteq 156°$)
$\pi + \varepsilon$	$2 - (2\varepsilon/\pi) + O(\varepsilon^2)$	$3 - (4\varepsilon/\pi) + O(\varepsilon^2)$
$3\pi/2$	1.544...	1.909...
$2\pi + \varepsilon$	$(3/2) + O(\varepsilon^2)$	$(3/2) - (\varepsilon/\pi) + O(\varepsilon^2)$

[a] When $2\alpha < 2\alpha_c$ the λ_1 are complex and the flow consists of a series of eddies.[14] In this table ε is a small, arbitrary number ($\varepsilon \ll 1$).

where $\mathscr{Re}\{z\}$ indicates the real part of the complex number z. Then the flow close to the corner will be dominated by the first nonzero term in the series.

a. Show that the general solution for f_n is

$$f_n = c_1 \cos \lambda_n\theta + c_2 \cos (\lambda_n - 2)\theta + c_3 \sin \lambda_n\theta + c_4 \sin (\lambda_n - 2)\theta \tag{1C.3-3}$$

where the first two terms are associated with antisymmetrical flow and the last two terms with symmetrical flow. In general the c_k are complex.

b. Show that the boundary conditions on f_n are

$$\text{At } \theta = \alpha: \quad f_n = 0; \quad f'_n = 0 \tag{1C.3-4}$$

$$\text{At } \theta = -\alpha: \quad f_n = 0; \quad f'_n = 0 \tag{1C.3-5}$$

c. When $2\alpha > 2\alpha_c$ it can be shown that λ_1 is real; therefore it is permissible to omit the real operator $\mathscr{Re}\{\ \}$ from the first term in Eq. 1C.3-1. Show that the stream function has the following asymptotic behavior near the corner for *antisymmetrical* flow:

$$\psi \sim A_1 r^{\lambda_1}[\cos(\lambda_1 - 2)\alpha \cos \lambda_1\theta - \cos \lambda_1\alpha \cos(\lambda_1 - 2)\theta] \quad (r \to 0) \tag{1C.3-6}$$

where λ_1 is the solution to

$$\sin 2(\lambda_1 - 1)\alpha = -(\lambda_1 - 1) \sin 2\alpha \tag{1C.3-7}$$

subject to the condition in Eq. 1C.3-2. Then for *symmetrical* flow show that

$$\psi \sim A_1 r^{\lambda_1}[\sin((\lambda_1 - 2)\alpha) \sin(\lambda_1\theta) - \sin(\lambda_1\alpha) \sin((\lambda_1 - 2)\theta)] \quad (r \to 0) \tag{1C.3-8}$$

where λ_1 is the solution to

$$\sin 2(\lambda_1 - 1)\alpha = +(\lambda_1 - 1) \sin 2\alpha \tag{1C.3-9}$$

d. Verify sample entries for λ_1 in Table 1C.3 and identify the entries corresponding to (1) simple shear flow, (2) flow around a 90° corner with fixed walls, and (3) flow of a jet from a slit into air with no or little change in cross section.

e. Verify that near the corner ($r \to 0$), for antisymmetrical flow

$$v_r \sim A_1 r^{\lambda_1 - 1}[\lambda_1 \cos(\lambda_1 - 2)\alpha \sin \lambda_1\theta - (\lambda_1 - 2) \cos \lambda_1\alpha \sin(\lambda_1 - 2)\theta] \tag{1C.3-10}$$

$$v_\theta \sim A_1 r^{\lambda_1 - 1}[\lambda_1 \cos(\lambda_1 - 2)\alpha \cos \lambda_1\theta - \lambda_1 \cos \lambda_1\alpha \cos(\lambda_1 - 2)\theta] \tag{1C.3-11}$$

$$p \sim -4\mu A_1 r^{\lambda_1 - 2}(\lambda_1 - 1) \cos \lambda_1\alpha \sin (\lambda_1 - 2)\theta + p_0 \tag{1C.3-12}$$

and for symmetrical flow

$$v_r \sim A_1 r^{\lambda_1 - 1}[-\lambda_1 \sin(\lambda_1 - 2)\alpha \cos \lambda_1\theta + (\lambda_1 - 2) \sin \lambda_1\alpha \cos(\lambda_1 - 2)\theta] \tag{1C.3-13}$$

$$v_\theta \sim A_1 r^{\lambda_1 - 1}[\lambda_1 \sin(\lambda_1 - 2)\alpha \sin \lambda_1\theta - \lambda_1 \sin \lambda_1\alpha \sin(\lambda_1 - 2)\theta] \tag{1C.3-14}$$

$$p \sim 4\mu A_1 r^{\lambda_1 - 2}(\lambda_1 - 1) \sin \lambda_1\alpha \cos(\lambda_1 - 2)\theta + p_0 \tag{1C.3-15}$$

where μ is the viscosity and p_0 is a constant.

f. Finally by substitution into the equation of motion show that the inertial terms are negligible compared to the viscous terms near the corner provided $\mathcal{R}e\{\lambda_1\} > 0$.

1C.4 Journal-Bearing Problem[15]

A journal-bearing system is shown in Fig. 1C.4. The journal of radius r_1 rotates with angular velocity W_1 in a stationary outer bearing of radius r_2. The journal and bearing are of length L in the z-direction, and the eccentric annular gap is completely filled with a Newtonian liquid. The origins of the rectangular and cylindrical coordinate systems are taken to be on the journal axis. The axes of the journal and bearings are separated by a distance a, and the difference in radii, $r_2 - r_1$, is called c, with $c \ll r_1$. It is desired to find the steady-state velocity distribution and pressure distribution in the system. In addition, we wish to know the torque and the force exerted by the fluid on the rotating journal.

a. We assume that the tangential flow in the annular gap can be approximated locally as that in a plane slit with one wall moving and with an imposed pressure gradient (cf. Example 1.3-1). Then at each θ we introduce a local XYZ-coordinate system with X pointing in the flow direction, Y pointing radially, and Z coinciding with z. Thus $Y = r - r_1$, and the surface at $Y = 0$ is moving with the linear velocity $W_1 r_1$, whereas the surface at $Y = B$ is fixed. Show that the local velocity profile will be

$$v_X = W_1 r_1\left(1 - \frac{Y}{B}\right) - \frac{B^2}{2\mu}\frac{dp}{dX}\frac{Y}{B}\left(1 - \frac{Y}{B}\right) \tag{1C.4-1}$$

and that the volume rate of flow will be

$$Q = \tfrac{1}{2}BLW_1 r_1 - \frac{B^3 L}{12\mu}\frac{dp}{dX} \tag{1C.4-2}$$

[15] This problem is patterned very closely after the discussion by A. Sommerfield, *Vorlesungen über Theoretische Physik*, Vol. 2, Dietrich, Wiesbaden (1947), Sect. 36. For a comprehensive theoretical treatment on lubrication see N. Tipei, *Theory of Lubrication*, Stanford University Press, Stanford, CA (1962). For an analytical solution in bipolar coordinates, see G. J. Farns, cited by R. Ehrlich and J. C. Slattery, *Ind. Eng. Chem. Fundam.*, 7, 239–246 (1968), Appendix B; and G. H. Wannier, *Quart. J. Appl. Math.*, **8**, 1–32 (1950).

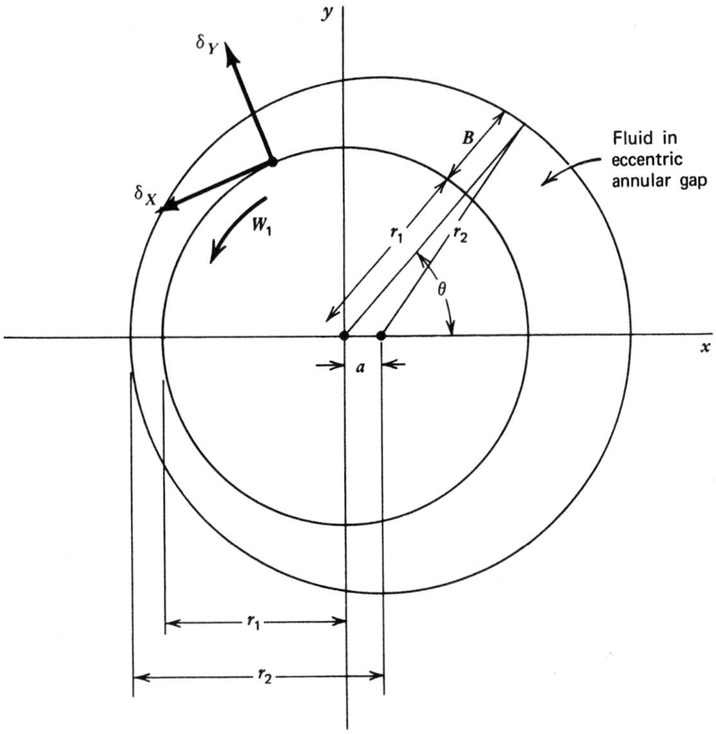

FIGURE 1C.4. Journal-bearing arrangement: cylinder of radius r_1 rotating with angular velocity W_1 in a cylindrical cavity of radius r_2. The gap width $B(\theta)$ is given approximately as $B \doteq c + a\cos\theta$, where $c = r_2 - r_1$.

In the journal-bearing problem we do not know what Q is; it is convenient to replace the unknown constant Q by an unknown constant B_0 defined by $Q = \frac{1}{2}B_0 L W_1 r_1$. Then show that

$$\frac{dp}{dX} = 6\mu W_1 r_1\left(\frac{1}{B^2} - \frac{B_0}{B^3}\right) \tag{1C.4-3}$$

$$v_X = W_1 r_1\left[1 - \left(\frac{4}{B} - \frac{3B_0}{B^2}\right)Y + 3\left(\frac{1}{B^2} - \frac{B_0}{B^3}\right)Y^2\right] \tag{1C.4-4}$$

$$\tau_{YX}\bigg|_{Y=0} = -\mu\frac{\partial v_X}{\partial Y}\bigg|_{Y=0} = \mu W_1 r_1\left(\frac{4}{B} - \frac{3B_0}{B^2}\right) \tag{1C.4-5}$$

b. Next apply Eqs. 1C.4-3 and 5 to the system in Fig. 1C.4 to obtain

$$\frac{1}{r_1}\frac{dp}{d\theta} = 6\mu W_1 r_1\left(\frac{1}{B^2} - \frac{B_0}{B^3}\right) \tag{1C.4-6}$$

$$\tau_{r\theta}\bigg|_{r=r_1} = \mu W_1 r_1\left(\frac{4}{B} - \frac{3B_0}{B^2}\right) \tag{1C.4-7}$$

in which it is understood that $B = B(\theta)$ as described in the figure caption for Fig. 1C.4.

c. Next, in preparation for (**d**), we need to know how to evaluate the following integrals

$$J_n = \int_0^{2\pi} \frac{d\theta}{(c + a\cos\theta)^n} \tag{1C.4-8}$$

$$K_n = \int_0^{2\pi} \frac{\cos\theta\, d\theta}{(c + a\cos\theta)^n} \tag{1C.4-9}$$

Show that

$$J_1 = 2\pi(c^2 - a^2)^{-1/2} \tag{1C.4-10}$$

$$J_2 = -\frac{\partial J_1}{\partial c} = 2\pi c(c^2 - a^2)^{-3/2} \tag{1C.4-11}$$

$$J_3 = -\frac{1}{2}\frac{\partial J_2}{\partial c} = 2\pi(c^2 + \tfrac{1}{2}a^2)(c^2 - a^2)^{-5/2} \tag{1C.4-12}$$

$$J_4 = -\frac{1}{3}\frac{\partial J_3}{\partial c} = 2\pi c(c^2 + \tfrac{3}{2}a^2)(c^2 - a^2)^{-7/2} \tag{1C.4-13}$$

$$K_2 = \left(\frac{1}{a}\right)(J_1 - cJ_2) = -2\pi a(c^2 - a^2)^{-3/2} \tag{1C.4-14}$$

$$K_3 = \left(\frac{1}{a}\right)(J_2 - cJ_3) = -2\pi a(\tfrac{3}{2}c)(c^2 - a^2)^{-5/2} \tag{1C.4-15}$$

$$K_4 = \left(\frac{1}{a}\right)(J_3 - cJ_4) = -2\pi a(2c^2 + \tfrac{1}{2}a^2)(c^2 - a^2)^{-7/2} \tag{1C.4-16}$$

The expressions for J_4 and K_4 are not needed here but are given for completeness inasmuch as they will be needed in Example 6.4-2.

d. Integrate Eq. 1C.4-6 between $\theta = 0$ and $\theta = 2\pi$. Then use the fact that p has to be the same at $\theta = 0$ and $\theta = 2\pi$ in Fig. 1C.4 to obtain

$$B_0 = \frac{J_2}{J_3} = c\,\frac{c^2 - a^2}{c^2 + \tfrac{1}{2}a^2} \tag{1C.4-17}$$

e. Now obtain the torque that the fluid exerts on the rotating journal:

$$\begin{aligned} \mathscr{T} &= L\int_0^{2\pi} (-\tau_{r\theta})|_{r=r_1} r_1^2\, d\theta \\ &= -\mu L W_1 r_1^3(4J_1 - 3B_0 J_2) \\ &= -\frac{2\pi\mu L W_1 r_1^3}{\sqrt{c^2 - a^2}}\frac{c^2 + 2a^2}{c^2 + \tfrac{1}{2}a^2} \end{aligned} \tag{1C.4-18}$$

f. Next we want to get the components of the force that the fluid exerts on the rotating journal. We first consider the force in the positive y-direction:

$$\begin{aligned} F_y &= L\int_0^{2\pi} [-(p + \tau_{rr})\sin\theta - \tau_{r\theta}\cos\theta]|_{r=r_1} r_1\, d\theta \\ &= L\int_0^{2\pi} \left(-\frac{dp}{d\theta} - \tau_{r\theta}\right)\Big|_{r=r_1} \cos\theta\, r_1\, d\theta \\ &= -6\mu L W_1 r_1^3(K_2 - B_0 K_3) + \text{higher-order terms from } \tau_{r\theta} \\ &\doteq -\frac{6\pi\mu L W_1 r_1^3 a}{\sqrt{c^2 - a^2}(c^2 + \tfrac{1}{2}a^2)} \end{aligned} \tag{1C.4-19}$$

In going from the first to the second line we omit the τ_{rr} contribution, since it is zero. We also perform an integration by parts. In going from the second to the third line, we omit the $\tau_{r\theta}$ contribution, since from Eqs. 1C.4-6, 7, and 17 we see that $dp/d\theta$ is of order $(r_1/c)^2$, whereas $\tau_{r\theta}$ is of order (r_1/c).

Next show that for the force in the positive x-direction:

$$F_x = L \int_0^{2\pi} [-(p + \tau_{rr}) \cos \theta + \tau_{r\theta} \sin \theta]|_{r=r_1} r_1 \, d\theta = 0 \tag{1C.4-20}$$

Interpret these results.

g. Finally, integrate Eq. 1C.4-6 to obtain the pressure distribution in the system

$$\begin{aligned} p &= p_0 + 6\mu W_1 r_1^2 \left[\int \frac{d\theta}{(c + a\cos\theta)^2} - B_0 \int \frac{d\theta}{(c + a\cos\theta)^3} \right] \\ &= p_0 + \frac{6\mu W_1 r_1^2 a \sin\theta\,(c + \frac{1}{2}a\cos\theta)}{(c^2 + \frac{1}{2}a^2)(c + a\cos\theta)^2} \end{aligned} \tag{1C.4-21}$$

where p_0 is an arbitary constant pressure. Both integrals can be written as $\int (c + a\cos\theta)^{-1}\,d\theta +$ additional terms, and when the indicated subtraction in [ ] is performed, it is found that the coefficient of $\int (c + a\cos\theta)^{-1}\,d\theta$ is zero so that this integral does not have to be evaluated.

1C.5 Thin Filament Equation for a Vertically Falling Stream of a Newtonian Liquid[16]

A liquid flowing axially in a vertical circular tube with a volume rate of flow Q emerges from the tube and falls downward. Its radius is $R(z)$, where z is the distance downward from the end of the tube. As an approximation we assume that the axial fluid velocity v is a function of z alone, so that the velocity v and the radius of the jet R are related by $Q = \pi[R(z)]^2 v(z)$. We further assume that the radius of the filament changes gradually with axial position, so that $dR/dz \ll 1$.

a. Write a momentum balance over the fixed control volume shown in Fig. 1C.5. Include the effects of surface tension with surface tension σ and the ambient atmospheric pressure p_{atm}. Show then that

$$\rho v v' = -\frac{d}{dz}\pi_{zz} + \frac{2R'}{R}\left(\frac{\sigma}{R} - \pi_{zz}\right) + \rho g + \frac{2R'}{R} p_{atm} \tag{1C.5-1}$$

where $v' = dv/dz$ and $R' = dR/dz$.

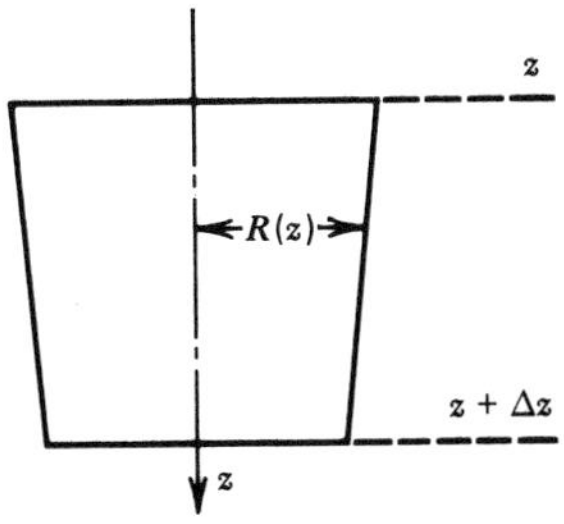

FIGURE 1C.5. Small section of liquid filament of linearly changing radius $R(z)$ within the small distance Δz.

[16] M. A. Matovich and J. R. A. Pearson, *Ind. Eng. Chem. Fundam.*, **8**, 512–520 (1969). See also C. J. S. Petrie, *Elongational Flows*, Pitman, London (1979) for an extensive discussion including bibliography. Numerical solutions of this equation have been reported and compared with data by A. Kaye and D. G. Vale, *Rheol. Acta*, **8**, 1–5 (1969), and R. J. Fisher, M. M. Denn, and R. I. Tanner, *Ind. Eng. Chem. Fundam.*, **19**, 195–197 (1980). See Example 7.4-3.

b. Introduce the boundary condition at the surface of the filament and rewrite Eq. 1C.5-1 as

$$\rho v v' = \left[-\frac{d}{dz}(\tau_{zz} - \tau_{rr}) + \frac{v'}{v}(\tau_{zz} - \tau_{rr}) \right] - \frac{\sigma}{2}\sqrt{\frac{\pi}{Qv}}\, v' + \rho g \tag{1C.5-2}$$

c. Specialize the above "thin filament equation" to Newtonian fluids to obtain

$$\rho v v' = 3\mu[v'' - (v')^2/v] - \frac{\sigma}{2}\sqrt{\frac{\pi}{Qv}}\, v' + \rho g \tag{1C.5-3}$$

d. Show that if all but viscous terms are neglected a general solution to Eq. 1C.5-3 is

$$v = c_1 \exp(-z/c_2) \tag{1C.5-4}$$

e. Show that if all but viscous and gravity terms are neglected, a solution to Eq. 1C.5-3 is:

$$v = (2\rho g/3\mu c_1)\sinh^2[\tfrac{1}{2}c_1^{1/2}(z + c_2)] \tag{1C.5-5}$$

In Eqs. 1C.5-4 and 5 the constants c_1 and c_2 are to be determined from boundary conditions.

1D.1 Oscillatory Flow of a Viscous Fluid

A viscous fluid occupies the region $y > 0$, and is bounded at the plane $y = 0$ by a solid surface that executes an oscillatory motion in the x-direction with a frequency ω and velocity amplitude V. Find the velocity distribution in the system after the initial transients fade away.

It is thus desired to solve the differential equation

$$\frac{\partial v_x}{\partial t} = \nu \frac{\partial^2 v_x}{\partial y^2} \tag{1D.1-1}$$

in which $\nu = \mu/\rho$, with the boundary conditions

$$\text{At } y = 0: \quad v_x = V\cos\omega t = V\,\mathcal{Re}\,\{e^{i\omega t}\} \tag{1D.1-2}$$

$$\text{At } y = \infty: \quad v_x = 0 \tag{1D.1-3}$$

a. Postulate a solution of the form

$$v_x(y, t) = \mathcal{Re}\,\{v^0(y)e^{i\omega t}\} \tag{1D.1-4}$$

in which $v^0(y)$ is in general complex. Show that the substitution of this function into Eq. 1D.1-1 leads to a differential equation for $v^0(y)$

$$\frac{d^2 v^0}{dy^2} - \left(\frac{i\omega}{\nu}\right)v^0 = 0 \tag{1D.1-5}$$

b. Show that this equation can be solved with the appropriate boundary conditions to give

$$v^0(y) = Ve^{-(1+i)\sqrt{\omega/2\nu}\,y} \tag{1D.1-6}$$

c. Show that substitution of this result into Eq. 1D.1-4 gives for the velocity distribution

$$v_x(y, t) = Ve^{-\sqrt{\omega/2\nu}\,y} \cos(\omega t - \sqrt{\omega/2\nu}\,y) \tag{1D.1-7}$$

Sketch the result.

d. Rework the problem if the fluid, instead of extending to $y = \infty$, is bounded[17] by a fixed plane at $y = B$. Show that the velocity profiles will be very nearly linear if $\sqrt{\rho\omega/2\mu}\,B \ll 1$.

1D.2 Viscous Heating in Oscillatory Flow[18]

A Newtonian fluid of viscosity μ and thermal conductivity k is located in the region between two parallel plates separated by a distance b. Both plates are maintained at a temperature T_0. The lower plate (at $x = 0$) is made to oscillate sinusoidally in the z-direction with a velocity amplitude V and a frequency ω. The upper plate (at $x = b$) is held fixed. Estimate the temperature rise that results from viscous heating. Consider the high-frequency limit only.

Answer: $T - T_0 = (\mu V^2/4k)[(1 - e^{-2a\xi}) - (1 - e^{-2a})\xi]$
where $\xi = x/b$ and $a = b\sqrt{\rho\omega/2\mu}$

[17] See, for example, L. D. Landau and E. M. Lifshitz, *Fluid Mechanics*, 2nd ed., Pergamon Press, Oxford (1987), p. 88.

[18] R. B. Bird, Chem. Eng. Prog., Symposium Series Number 58, **61**, 1–15 (1965); see Illustrative Example 1.

CHAPTER 2

FLOW PHENOMENA IN POLYMERIC LIQUIDS

A fluid that's macromolecular
Is really quite weird—in particular
The big normal stresses
The fluid possesses
Give rise to effects quite spectacular.

The purpose of this chapter is to demonstrate the striking qualitative difference between the behavior of Newtonian liquids and polymeric liquids. An appreciation of the qualitative behavior of polymeric liquids is important as background information for the quantitative treatments to follow in the remainder of the volumes.[1]

We begin the chapter with a brief review of the chemical constitution of polymeric liquids. It is of course this chemical composition that is responsible for the flow behavior of polymeric liquids. The treatment here is intended just to give the reader a minimum of understanding of the fluids encountered in later chapters and to provide a qualitative appreciation of the structural factors responsible for the flow properties of polymeric liquids. After this introductory section follows a sequence of six sections in which we compare and contrast the flow of polymeric liquids with Newtonian liquids. Some of the experiments presented have the character of "fun experiments." It is important to emphasize that they do not represent anomalous behavior shown by a few "strange" fluids, but are rather typical for fluids containing very large molecules. The classification of the experiments in six separate sections is to a large extent arbitrary and is done primarily for convenience. Indeed for any one polymeric liquid the behavior in the different experiments is of course related, since it is given by the underlying fundamental constitutive equation of the liquid. This connection will be explained in later chapters and some of the experiments presented here will then serve as useful reference experiments. It is also important to emphasize that the list of qualitative experiments is not intended to be exhaustive. Many other interesting flow phenomena have been observed, and still more remain to be discovered. In the final section we will briefly discuss the role of dimensionless groups in the fluid dynamics of polymeric liquids.

[1] Two interesting movies are available in which these differences are illustrated particularly well: H. Markovitz, *Rheological Behavior of Fluids*, Education Services, Inc., Watertown, MA (1965); and K. Walters and J. M. Broadbent, *Non-Newtonian Fluids*, Department of Applied Mathematics, University College of Wales, Aberystwyth, UK (1980) (the latter is available both as film and as videotape).

Polymeric fluids are sometimes called *viscoelastic fluids*. This means that the fluids have both viscous and "elastic" properties. The use of the word "elastic" to characterize a property of a fluid may seem a bit contradictory. By "elasticity" one usually means the ability of a material to return to some unique, original shape; on the other hand, by a "fluid" one means a material that will take the shape of any container in which it is left, and therefore does not possess a unique, original shape. Despite this apparent contradiction we will in this chapter show several flow situations in which polymeric fluids exhibit properties that one would describe as "elastic". In this chapter we shall use the word "elastic" to describe loosely effects associated with nonlinear or time-dependent properties of the constitutive equation, other than those associated with the non-Newtonian viscosity to be described in §2.2. Another concept that is closely tied to that of elasticity is the concept of "memory". Indeed a material that has no memory cannot be elastic, since it has no way of remembering a unique, original shape. Hence fluids exhibiting elastic properties are also often referred to as *memory fluids*.

§2.1 THE CHEMICAL NATURE OF POLYMERIC LIQUIDS[1]

A *macromolecule* (or *polymer*)[2] is a large molecule composed of many small simple chemical units, generally called *structural units*. In some polymers each structural unit is connected to precisely two other structural units, and the resulting chain structure is called a *linear* macromolecule. In other polymers most structural units are connected to two other units, although some structural units connect three or more units, and we talk of *branched* molecules. Where the chains terminate, special units called *end groups* are found. Figure 2.1-1 shows symbolic representations of linear and branched macromolecules. For the sake of completeness we mention also that in some macromolecular materials all structural units are interconnected resulting in a three-dimensional *cross-linked* or *network structure* rather than in separate molecules. Such materials, however, generally possess no fluid phase and are therefore outside the scope of this book.

It is sometimes useful to distinguish between synthetic and natural (biological) macromolecules. Many synthetic polymers are built from a single structural unit, and the polymer is then referred to as a *homopolymer*. Typical examples of synthetic homopolymers are polyethylene, polystyrene, and polyvinylchloride (see Table 2.1-1). In contrast *copolymers* are built from two or more different structural units. According to the manner in which the structural units combine, copolymers are further classified as random copolymers, block copolymers, or graft copolymers, as illustrated in Fig. 2.1-2. The motivation for producing copolymers is to obtain materials with a wider range of mechanical properties than is possible with the homopolymers alone.

Biological macromolecules, in contrast with synthetic macromolecules, generally contain a large number of different structural units. The polypeptide chains that make up proteins, for instance, consist of about 20 different structural units. Among other examples of biological macromolecules are the viruses and the interesting DNA molecules that carry

[1] Much more comprehensive treatments of the chemistry of macromolecules may be found in many standard references such as F. W. Billmeyer, Jr., *Textbook of Polymer Science*, 3rd ed., Wiley, New York (1984); P. J. Flory, *Principles of Polymer Chemistry*, Cornell University Press, Ithaca, NY (1953); and C. Tanford, *Physical Chemistry of Macromolecules*, Wiley, New York (1961).

[2] Some scientists are careful to distinguish between macromolecules (very large molecules) and polymers (macromolecules made up of repeating structural units). This fine distinction seems to be generally ignored, and we shall use the two terms synonymously in this book.

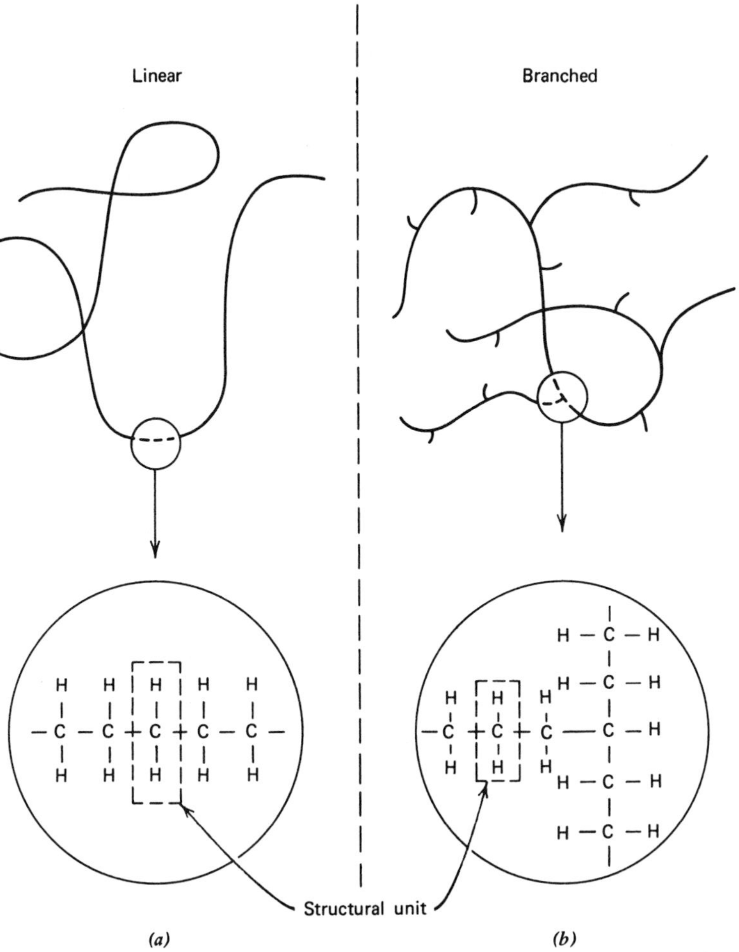

FIGURE 2.1-1. Symbolic representations of linear and branched macromolecules (high-density polyethylene and low-density polyethylene, respectively): (*a*) Linear, (*b*) branched.

in their structure the key to the inherited characteristics of organisms. More on biological polymers can be found elsewhere.[3]

Aside from unimportant corrections from end groups and branch points, the molecular weight of a macromolecule is the product of the molecular weight of a structural unit and the number of structural units in the molecule. Typical synthetic polymer molecules may have molecular weights between 10,000 and 1,000,000 g/mol. Biological macromolecules may have even larger molecular weights; for example, the molecular weight of tobacco mosaic virus is about 40,000,000 g/mol.

Naturally one can conceive of a polymer sample in which the molecular weight of all macromolecules is the same. Such a sample is called *monodisperse*. Indeed, some biological polymers are monodisperse. Synthetic monodisperse or "almost monodisperse"

[3] See, for example, C. Tanford, *Physical Chemistry of Macromolecules*, Wiley, New York (1961); H. Morawetz, *Macromolecules in Solution*, Vol. 21, High Polymers Series, Wiley, New York (1975); T. E. Creighton, *Proteins: Structures and Molecular Principles*, Freeman, New York (1983); J. King, *Protein and Nucleic Acid Structure and Dynamics*, Benjamin-Cummings, Menlo Park, CA (1985).

TABLE 2.1-1

Some Synthetic Polymers, Their Monomers and Their Structural Units

Polymer	Monomer(s)	Structural Unit
Polyethylene	$CH_2{=}CH_2$	$-CH_2-$
Polyvinylchloride	$CH_2{=}CHCl$	$-CH_2-CHCl-$
Polystyrene	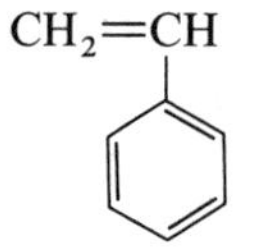	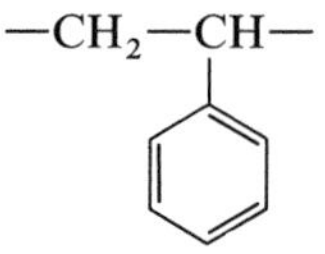
Polyacrylamide	$CH_2{=}\underset{\underset{CONH_2}{\mid}}{CH}$	$-CH_2-\underset{\underset{CONH_2}{\mid}}{CH}-$
Polyisobutylene	$CH_2{=}\overset{\overset{CH_3}{\mid}}{\underset{\underset{CH_3}{\mid}}{C}}$	$-CH_2-\overset{\overset{CH_3}{\mid}}{\underset{\underset{CH_3}{\mid}}{C}}-$
Polyisoprene (natural rubber)	$CH_2{=}\overset{\overset{CH_3}{\mid}}{C}-CH{=}CH_2$	$-CH_2-\overset{\overset{CH_3}{\mid}}{C}{=}CH-CH_2-$
Polydimethylsiloxane	$HO-\overset{\overset{CH_3}{\mid}}{\underset{\underset{CH_3}{\mid}}{Si}}-OH$	$-\overset{\overset{CH_3}{\mid}}{\underset{\underset{CH_3}{\mid}}{Si}}-O-$
Polyethyleneoxide (Polyox)	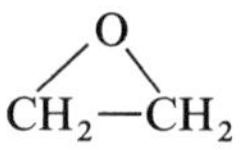	$-O-CH_2-CH_2-$
Polyhexamethylene adipamide (Nylon 66)	$NH_2-(CH_2)_6-NH_2$ and $HO-\overset{\overset{O}{\|}}{C}-(CH_2)_4-\overset{\overset{O}{\|}}{C}-OH$	$-NH-(CH_2)_6-NH-\overset{\overset{O}{\|}}{C}-(CH_2)_4-\overset{\overset{O}{\|}}{C}-$
Polyethylene terephthalate (polyester)	$HO-CH_2-CH_2-OH$ and 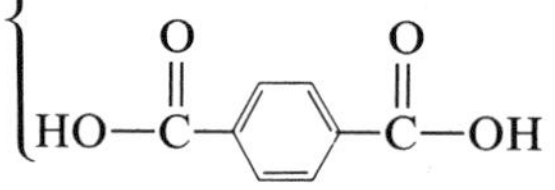	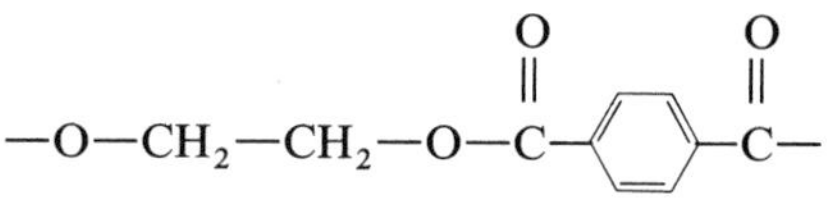

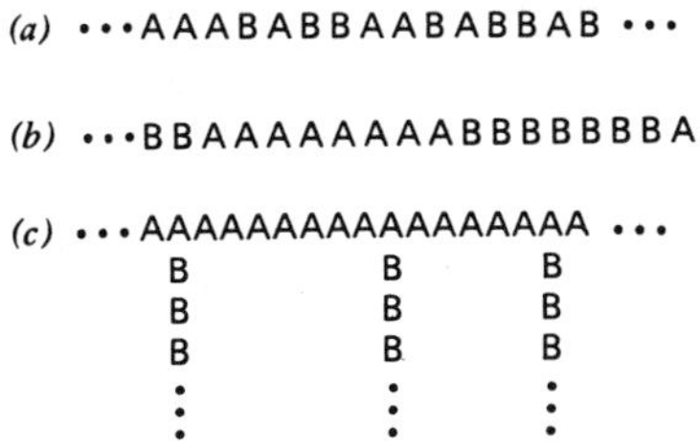

FIGURE 2.1-2. Schematic representations of (*a*) random, (*b*) block, and (*c*) graft copolymers. A and B represent two different kinds of structural units.

polymers may be prepared by special techniques, but are seldom used commercially. Most commercial polymers by contrast are *polydisperse*, that is, they contain molecules of many different molecular weights. Thus we may talk of a distribution of molecular weights.

In order to describe molecular weight distributions in simple quantitative terms, we introduce various molecular weight averages. Let us assume that we are given a polydisperse macromolecular sample, composed of a number of monodisperse fractions. Specifically let us say that fraction "1" contains N_1 moles of molecular weight M_1, fraction "2" contains N_2 moles of molecular weight M_2 and so forth. We may introduce an average molecular weight by multiplying the molecular weight of each fraction by the number of moles in that fraction, summing and dividing by the total number of moles,

$$\bar{M}_n = \frac{\sum_i N_i M_i}{\sum_i N_i} \tag{2.1-1}$$

This is called the *number-average molecular weight*; it is particularly sensitive to additions of small amounts of low molecular weight fractions. Since the mass of the ith fraction w_i is $w_i = N_i M_i$, we may alternatively form an average by weighting the M_i with respect to the mass of the fractions,

$$\bar{M}_w = \frac{\sum_i w_i M_i}{\sum_i w_i} = \frac{\sum_i N_i M_i^2}{\sum_i N_i M_i} \tag{2.1-2}$$

This average is called the *weight-average molecular weight*; it is more sensitive to the high molecular weight fractions than is $\bar{M}_n$. One may define further molecular weight averages by taking ratios of higher moments of the molecular weight distribution

$$\bar{M}_{z+j} = \frac{\sum_i N_i M_i^{3+j}}{\sum_i N_i M_i^{2+j}} \tag{2.1-3}$$

When $j = 0$ this defines the *z-average molecular weight*, and for $j > 0$ this defines the *$z + j$-average molecular weight*. For monodisperse samples all these averages are equal, but for polydisperse samples $\bar{M}_n < \bar{M}_w < \bar{M}_z < \bar{M}_{z+1}$ etc. The ratio $\bar{M}_w/\bar{M}_n$, known as the *heterogeneity index*, is often taken as a simple measure of the polydispersity of a sample (Problem 2B.1). In addition to the average molecular weights defined above, there is also the *viscosity-average molecular weight* $\bar{M}_v$ defined in Eq. 3.6-14. The average $\bar{M}_v$ lies between $\bar{M}_n$ and $\bar{M}_w$.

The chemical formulas in Table 2.1-1 serve as a useful symbolic notation for the chemist, but they do not convey any information about the extent to which the molecules are able to change their configurations. In fact most polymers are capable of assuming a huge number of configurations by rotations around chemical bonds. Furthermore the molecules will be continually changing their configurations due to thermal motions. Such configurational changes may be either local rearrangements of the backbone or there may be large overall changes in configuration. There is consequently an entire spectrum of time constants associated with the rates at which such thermally induced configurational changes take place. We call these the *time constants for the fluid*, and we say that the fluid has a *spectrum of relaxation times*. It is these time constants that give polymeric fluids at least a partial memory. Since polymeric fluids are not chemically cross-linked they will not have a permanent memory, but there will be a finite longest relaxation time. We say that the relaxation spectrum gives the fluids a *fading memory* of duration of the longest relaxation time. It is shown in §2.8 that when the longest relaxation time is equal to or greater than the characteristic time for the macroscopic flow system, marked deviations from Newtonian behavior are observed.

In closing we add that polymers are not the only liquids that exhibit departure from Newtonian behavior. For example, detergent solutions or fine clay suspensions will form large-scale structures due to hydrogen bonding, dipole interaction, and other intermolecular forces. The time constants for the rearrangement of these structures are comparable to typical times of flow and as a result such fluids are also non-Newtonian.[4]

§2.2 NON-NEWTONIAN VISCOSITY

Probably the single most important characteristic of polymeric liquids is the fact that they have a "shear-rate dependent" or "non-Newtonian" viscosity. A quantitative definition of this property is postponed until §3.3, but a simple qualitative experiment can be performed that illustrates the property. In this first experiment, we consider two identical, vertical tubes, the bottoms of which are covered by a flat plate. The two tubes are filled with fluids, one Newtonian and the other polymeric, chosen in such a way as to have the same viscosity in an experiment involving very low shear rates. This criterion is satisfied, for example, if two identical small spheres fall through each sample at the same rate[1] (Fig. 2.2-1*a*). In addition the density of the sphere should be much larger than the densities of the fluids, so that the difference in density of the fluids may be neglected. A particularly good Newtonian liquid to use is an aqueous glycerin solution. By varying the glycerin concentration, the viscosity of the mixture can be varied from 0.001002 to 1.490 Pa·s at 293 K. In Fig. 2.1-1*b* we see that when the plate is removed from the bottoms of the tubes and the fluids are allowed to flow out by gravity, the polymeric fluid drains much more quickly than the Newtonian fluid.

We know that if both fluids were Newtonian,[1] then the fact that the two spheres drop at the same rate in experiment (*a*) would mean that the fluids have the same viscosity. This in turn means that the fluids would drain at the same rate in experiment (*b*). We might then explain the results in Fig. 2.2-1 by saying that the polymer liquid has a lower viscosity in the high shear rate part of experiment (*b*) than in experiment (*a*). The decrease in viscosity with increasing shear rate is referred to as "shear thinning," and the fluid is said to be *shear thinning* or *pseudoplastic*. This effect can be quite dramatic, with the viscosity decreasing by a factor of as much as 10^3 or 10^4. Examples of pseudoplastic fluids are molten polyethylene

[4] See W. B. Russel, W. R. Schowalter, and D. A. Saville, *Colloidal Dispersions*, Cambridge University Press (1987).

[1] For Newtonian fluids, the flow around a translating sphere is analyzed in Example 1.4-1, and flow in a circular tube is analyzed in Example 1.3-2.

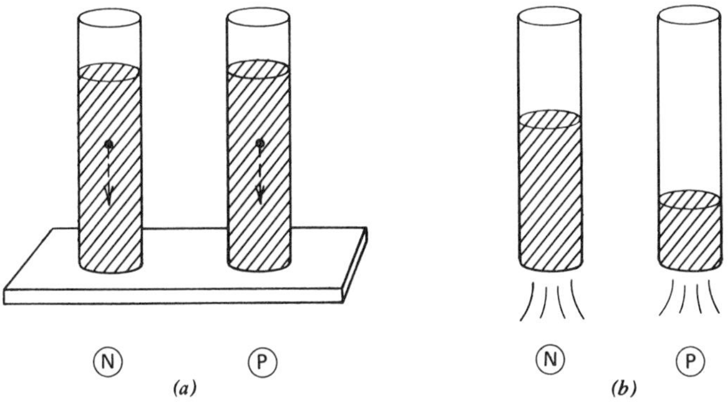

FIGURE 2.2-1. Tube flow and "shear thinning." In each part, the Newtonian behavior is shown on the left Ⓝ; the behavior of a polymer on the right Ⓟ. (*a*) A tiny sphere falls at the same rate through each; (*b*) the polymer flows out faster than the Newtonian fluid.

and polypropylene, and solutions of carboxymethylcellulose (CMC) in water, polyacrylamide in water and glycerin, and aluminum laurate in decalin and *m*-cresol. In fact almost all polymer solutions and melts that exhibit a shear-rate dependent viscosity are shear thinning.

A few fluids behave oppositely to what we have shown here, that is, they flow out of the tube in Fig. 2.2-1*b* more slowly than the corresponding Newtonian liquid. A fluid whose viscosity increases with shear rate is called *shear thickening* or *dilatant*.[2,3] This behavior is exhibited by fairly concentrated suspensions of very small particles; two reported examples are suspensions of titanium dioxide in a sucrose solution[4] and corn starch in an ethylene-glycol–water mixture.[5] As far as we know, relatively few polymer solutions have been shown to be dilatant.

Still different behavior is shown by fluids that will not flow unless acted on by at least some critical shear stress, called the *yield stress*. We call these *viscoplastic fluids*. Certain types of paints, greases, and pastes are examples of viscoplastic fluids. Carboxypolymethylene (pH 7) in water has also been reported to be viscoplastic.[6] Although the concept of yield stress has been very useful, its existence has been challenged[7] on the ground that all materials will flow provided that one waits long enough. For example, a paint film applied on a vertical wall will certainly flow, but for a good paint this process is slower than solvent evaporation and possibly chemical changes. Then the yield stress becomes a useful model property to describe the shear stress below which the flow of the paint becomes negligible. More will be said about the details of the dependence of viscosity on shear rate in the next chapter, where we discuss its measurement in polymeric fluids.

[2] A review of dilatant behavior is given by W. H. Bauer and E. A. Collins in F. R. Eirich, ed., *Rheology*, Vol. 4, Academic Press, New York (1967), Chapt. 8, pp. 423–459. See especially their Appendix II, p. 459, which contains a listing of dilatant systems.

[3] M. Reiner, *Deformation, Strain and Flow*, Interscience, New York (1960), pp. 306–309, and others use the term "dilatancy" to describe the change in volume of granular masses necessitated by a distortion. The standard example is the apparent drying of wet sand when stepped on.

[4] A. B. Metzner and M. Whitlock, *Trans. Soc. Rheol.*, **2**, 239–254 (1958).

[5] R. G. Green and R. G. Griskey, *Trans. Soc. Rheol.*, **12**, 13–25 (1968).

[6] A. G. Fredrickson, *Principles and Applications of Rheology*, Prentice-Hall, Englewood Cliffs, NJ (1964), p. 178; R. B. Bird, G. Dai, and B. J. Yarusso, *Revs. Chem. Engr.*, **1**, 1–70 (1982).

[7] H. A. Barnes and K. Walters, *Rheol. Acta*, **24**, 323–326 (1985).

§2.3 NORMAL STRESS EFFECTS

A number of important effects in the flow of polymeric liquids may be attributed to the fact that polymeric liquids exhibit normal stress differences in "shear flows" such as the flow in Fig. 1.3-1. Let us first establish some labeling conventions for referring to normal stress differences. If the fluid moves along one coordinate direction only and its velocity varies only in one other coordinate direction, then we call the direction of fluid velocity the "1" direction; the direction of velocity variation, the "2" direction; and the remaining neutral direction, the "3" direction. Then we call $\tau_{11} - \tau_{22}$ the *first normal stress difference.* Likewise, we call $\tau_{22} - \tau_{33}$ the *second normal stress difference.* In Fig. 1.3-1 the x, y, and z correspond to 1, 2, and 3 respectively. Thus the first normal stress difference is $\tau_{xx} - \tau_{yy}$ and the second normal stress difference is $\tau_{yy} - \tau_{zz}$. For Newtonian fluids the normal stress differences are exactly zero in shearing flow.

For polymeric fluids the first normal stress difference is practically always negative and numerically much larger than the second normal stress difference. This means that to a first approximation polymeric fluids exhibit in addition to the shear stresses an extra tension along the streamlines, that is, in the "1" direction. It was shown by Weissenberg[1] that the simple notion of an extra tension along streamlines may be used to obtain qualitative explanations of a large number of experiments. In terms of chemical structure, the extra tension along the streamlines in polymeric fluids arises from the stretching and alignment of the polymer molecules along the streamlines. The thermal motions make the polymer molecules act as small "rubber bands" wanting to snap back, and it is in this way that the extra tension arises. Thus there is also a structural background for Weissenberg's fundamental proposition, and we shall use it extensively in this and following sections.

The second normal stress difference has been found experimentally to be positive, but usually much smaller than the magnitude of the first normal stress difference. This means that in a shear flow the fluid exhibits a small extra tension in the "3" direction. A simple structural explanation for this extra tension is lacking, and the simplest kinetic theories of polymeric fluids are not capable of describing this effect; more elaborate theories are successful, however. We emphasize that the second normal stress difference is quite small, and it is normally observable only in situations where the first normal stress difference, for geometrical reasons, has no effect.

We now consider several experiments that enable us to see how normal stress effects manifest themselves.

a. Rod-Climbing

In this experiment we insert rotating rods into two beakers, one containing a Newtonian liquid and the other a polymer solution. In Fig. 2.3-1 we see that the Newtonian liquid near the rotating rod is pushed outward by the centrifugal force, and a dip in the liquid surface near the center of the beaker results. This is typical of the flow near the rotating shaft of a stirrer. The contrasting behavior of the polymer solution is striking. The polymer solution moves in the opposite direction, toward the center of the beaker and climbs up the rod. Moreover, for comparable rotational speeds, the response of the polymer solution can be far more dramatic than that of the Newtonian liquid, as seen in Fig. 2.3-1.

[1] K. Weissenberg, *Nature*, **159**, 310–311 (1947). In this pioneering analysis Weissenberg further proposed that the second normal stress difference should be precisely zero. This additional proposition has become known as the "Weissenberg hypothesis" and is now known not to be correct, albeit a good first approximation for some problems.

Ⓝ Ⓟ

FIGURE 2.3-1. Fixed cylinder with rotating rod. Ⓝ The Newtonian liquid, glycerin, shows a vortex; Ⓟ the polymer solution, polyacrylamide in glycerin, climbs the rod. The rod is rotated much faster in the glycerin than in the polyacrylamide solution. At comparable low rates of rotation of the shaft, the polymer will climb whereas the free surface of the Newtonian liquid will remain flat. [Photographs courtesy of Dr. F. Nazem, Rheology Research Center, University of Wisconsin-Madison.]

This phenomenon was first described by Garner and Nissan[2] and by Russel[3] but seems to have been known in the paint industry prior to its description in the scientific literature. The phenomenon may be interpreted in a rather simple fashion with the use of the notion[1] of an extra tension along the streamlines. In the rod-climbing experiment the streamlines are closed circles and the extra tension along these lines "strangulates" the fluid and forces it inwards against the centrifugal force and upwards against the gravitational force.

The above simple argument takes no account of the second normal stress difference. In Example 2.3-1 we present an analysis of the experiment based on the equations of motion, which shows that both normal stress differences actually contribute to the effect.

Finally we mention the interesting *Quelleffect*[4]. If instead of having a rotating rod in

[2] F. H. Garner and A. H. Nissan, *Nature*, **158**, 634–635 (1946).

[3] R. J. Russel, Ph.D. Thesis, Imperial College, University of London (1946) (unpublished), p. 58.

[4] K. Kirschke, *Polymères et Lubrification*, Colloques Internationaux du CNRS, No. 233, Éditions du CNRS, Paris (1974), pp. 137–144; G. Böhme, *Strömungsmechanik nicht-Newtonscher Fluide*, Teubner, Stuttgart (1981), pp. 127–128; G. R. Böhme and W. Warnecke, *Rheol. Acta*, **24**, 22–34 (1985).

the cylindrical container one rotates the bottom of the cylinder around its axis one obtains a bulge in the free surface around the axis. We expect that the extra tension along the streamlines is responsible also for this effect, which one may think of as "rod-climbing without a rod."

EXAMPLE 2.3-1 Interpretation of Free Surface Shapes in the Rod-Climbing Experiment[5]

Consider a fluid contained in the annular region between two vertical coaxial cylinders where the inner cylinder is rotating with a constant angular velocity. The fluid is bounded at the top by a "lubricated lid" that keeps the top surface flat by supporting a normal pressure distribution but no shear stress. Also the container is so deep that the influence from the bottom of the container is negligible near the lid. Use the equations of change to analyze the distribution of the normal pressure exerted by the fluid on the lubricated lid. Then by means of a simple argument explain what would cause the fluid to climb the rotating rod if the lubricated lid were replaced by a free surface. *Note*: In this experiment $1 = \theta$, $2 = r$, $3 = z$.

SOLUTION We begin by introducing a cylindrical coordinate system with z-axis coincident with the cylinder axis. Away from the bottom the only nonzero component of velocity is then v_θ. Furthermore this velocity component as well as the stress components and the pressure will be functions of r only. Hence the r- and θ-components of the equation of motion may be written, respectively,

Motion (r):
$$-\rho\frac{v_\theta^2}{r} = -\frac{1}{r}\frac{d}{dr}(r\tau_{rr}) + \frac{\tau_{\theta\theta}}{r} - \frac{dp}{dr} \tag{2.3-1}$$

Motion (θ):
$$0 = -\frac{1}{r^2}\frac{d}{dr}(r^2\tau_{r\theta}) \tag{2.3-2}$$

The normal pressure exerted on the lubricated lid is given by $p + \tau_{zz}$. An expression for the derivative of this quantity may be obtained by adding and subtracting $d\tau_{zz}/dr$ on the right side of Eq. 2.3-1, which may then be rearranged to:

$$\frac{d}{dr}(\tau_{zz} + p) = \frac{d}{dr}(\tau_{zz} - \tau_{rr}) + \frac{\tau_{\theta\theta} - \tau_{rr}}{r} + \rho\frac{v_\theta^2}{r} \tag{2.3-3}$$

We now wish to reformulate the first term on the right side of this equation so that a derivative $(d/d\tau_{r\theta})$ appears in place of (d/dr). To do this, note from Eq. 2.3-2 that

$$\frac{d\tau_{r\theta}}{dr} = -\frac{2}{r}\tau_{r\theta} \tag{2.3-4}$$

Then Eq. 2.3-3 may be formulated as

$$\begin{aligned}\frac{d(\tau_{zz} + p)}{d\ln r} &= 2\tau_{r\theta}\frac{d}{d\tau_{r\theta}}(\tau_{rr} - \tau_{zz}) + (\tau_{\theta\theta} - \tau_{rr}) + \rho v_\theta^2 \\ &= 2\tau_{21}\frac{d}{d\tau_{21}}(\tau_{22} - \tau_{33}) + (\tau_{11} - \tau_{22}) + \rho v_1^2\end{aligned} \tag{2.3-5}$$

[5] A. S. Lodge, *Elastic Liquids*, Academic Press, New York (1964), pp. 192–194; D. D. Joseph and R. L. Fosdick, *Arch. Rat. Mech. Anal.*, **49**, 321–401 (1973); D. D. Joseph, G. S. Beavers, A. Cers, C. Dewald, A. Hoger, and P. T. Than, *J. Rheol.*, **28**, 325–345 (1984).

In the second line on the right side, θ, r and z have been replaced by 1, 2 and 3 respectively. This has been done in agreement with the convention introduced at the beginning of this section with the assumption that the flow is locally a shearing flow.

For Newtonian fluids the first and second normal stress differences, $\tau_{11} - \tau_{22}$ and $\tau_{22} - \tau_{33}$, are both zero in shear flow, and Eq. 2.3-5 shows that the normal pressure exerted on the lubricated lid increases with the radius. Thus if the lid were removed we would expect that the fluid would rise near the cylinder wall and dip near the rod.

For polymeric fluids we have already indicated that $\tau_{11} - \tau_{22}$ is practically always negative with a numerical value much larger than that of $\tau_{22} - \tau_{33}$. We see that the normal stresses may cause the total normal pressure to decrease in the radial direction, and this is then consistent with the rod-climbing effect.

b. Convex Surface in a Tilted Trough

In this experiment we consider the low Reynolds number flow down an open, slightly inclined channel. The flow is driven by gravity and we observe the shape of the free surface of the fluid as shown in Fig. 2.3-2. The Newtonian fluid is seen to have a flat free surface, aside from the meniscus effect, whereas that of a typical non-Newtonian substance is slightly convex. The effect is small but reproducible. The experiment was first performed by Tanner[6] following a suggestion by Wineman and Pipkin.[7] *Note*: In this flow $1 = z$, $2 = x$, $3 = y$ (see Fig. 2.3-3).

This effect may be explained qualitatively by noting that there will be an extra tension in the y-direction. This tension will be the greatest near the vertical walls of the channel, since the velocity gradient dv_z/dx has its maximum value at the walls. Therefore the surface of the fluid will be "pulled down" near the walls and the fluid will bulge in the central region. In the following example we draw the same conclusion by using the equations of change.

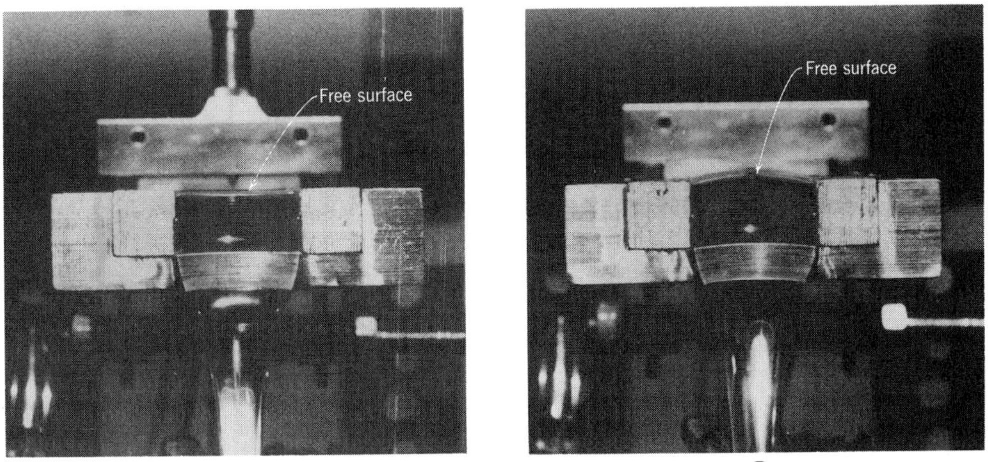

(N) (P)

FIGURE 2.3-2. Tanner's tilted trough. A Newtonian liquid flowing down an inclined channel exhibits a flat free surface (N); the free surface of a polyisobutylene solution is convex (P). [Reproduced from R. I. Tanner, *Trans. Soc. Rheol.*, **14**, 483–507 (1970).]

[6] R. I. Tanner, *Trans. Soc. Rheol.*, **14**, 483–507 (1970); R. I. Tanner, *Engineering Rheology*, Oxford University Press (1985), pp. 102–105.

[7] A. S. Wineman and A. C. Pipkin, *Acta Mech.*, **2**, 104–115 (1966).

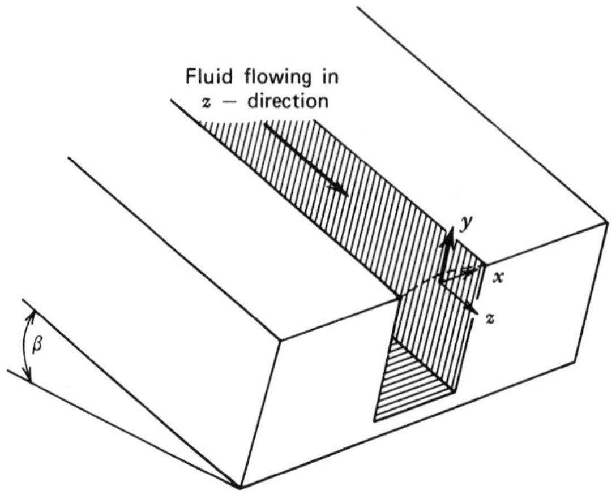

FIGURE 2.3-3. Coordinates for analysis of the tilted trough experiment. The trough is inclined at an angle β with respect to the horizontal.

EXAMPLE 2.3-2. Interpretation of Free-Surface Shapes in the Tilted Trough Experiment

For the flow in Fig. 2.3-2 explain why the free surface is flat for Newtonian liquids and convex for macromolecular fluids. To simplify the analysis, assume that the trough is very deep so that the presence of the bottom may be ignored. The trough is inclined at an angle β relative to the horizontal and has width W.

SOLUTION If the trough is very deep, we anticipate a velocity field of the form $v_z = v_z(x)$, $v_x = 0$, and $v_y = 0$ where the coordinate system is chosen as shown in Fig. 2.3-3. We further expect the stress components τ_{ij} to depend on x alone, and $p = p(x, y, z)$ since the surface is not necessarily flat. No information is gained from the continuity equation for the assumed velocity profile.

The equation of motion for this system gives

x-component:
$$0 = -\frac{\partial p}{\partial x} - \frac{d\tau_{xx}}{dx} \tag{2.3-6}$$

y-component:
$$0 = -\frac{\partial p}{\partial y} - \underline{\frac{d\tau_{xy}}{dx}} - \rho g \cos\beta \tag{2.3-7}$$

z-component:
$$0 = -\frac{\partial p}{\partial z} - \frac{d\tau_{xz}}{dx} + \rho g \sin\beta \tag{2.3-8}$$

The dashed-underlined term can be neglected if we assume that the free surface is not distorted enough to alter the stress field from that in steady shear flow between two infinite parallel plates. From these equations we can conclude that

$$\frac{\partial p}{\partial z} = 0 \tag{2.3-9}$$

and hence that

$$\tau_{xz} = (\rho g \sin\beta)x \tag{2.3-10}$$

Equation 2.3-9 is found by taking the partial derivative with respect to z of all three components of the equation of motion. From these results it can be seen that $\partial p/\partial z$ is a constant; and since $\partial p/\partial z$ is zero along the free surface, it must be zero everywhere.

Next, since π_{yy} is an analytic function of position, we may write:

$$d\pi_{yy} = \frac{\partial \pi_{yy}}{\partial x} dx + \frac{\partial \pi_{yy}}{\partial y} dy + \frac{\partial \pi_{yy}}{\partial z} dz = \frac{\partial \pi_{yy}}{\partial x} dx + \frac{\partial p}{\partial y} dy \tag{2.3-11}$$

By using Eqs. 2.3-6 and 7 together with Eq. 2.3-11, we find after integration that:

$$\pi_{yy}(x, y) = -(\tau_{xx} - \tau_{yy})|_x - \rho g(\cos \beta)y + C \tag{2.3-12}$$

where C is a constant of integration.

The total outward normal force per unit area exerted by the fluid at the free surface must be equal to the atmospheric pressure p_a. Provided the surface is not too curved, this requires that $\pi_{yy} = p_a$ and, thus, the free surface is given by the equation

$$y = \frac{-(\tau_{xx} - \tau_{yy})|_x + C - p_a}{\rho g \cos \beta} \tag{2.3-13}$$

At $x = 0$ all velocity gradients vanish so that the components of the stress tensor are all zero. For $x > 0$, if $(\tau_{xx} - \tau_{yy})|_x$ is positive, then the free surface will be lower than at $x = 0$. Thus the convex free surface shown in Fig. 2.3-2 corresponds to a positive second normal stress difference. For Newtonian fluids $(\tau_{xx} - \tau_{yy}) = 0$ for this flow, and Eq. 2.3-13 predicts a flat liquid surface as observed.

We can regard the second normal stress difference to be a function of the shear stress τ_{xz} instead of x. From Eq. 2.3-13 the total height of the fluid bulge h is given by

$$h = \frac{(\tau_{xx} - \tau_{yy})|_{\tau_{xz} = \tau_w}}{\rho g \cos \beta} \tag{2.3-14}$$

where $\tau_w = \frac{1}{2}\rho g W \sin \beta$ is the wall shear stress. Thus by measuring the size of the bulge in the fluid surface for very small values of the tilt angle β, we can measure the second normal stress difference at vanishingly small shear stresses. Since this method is restricted to low shear stress, any conclusions we have drawn about the second normal stress difference from this experiment are also limited. For larger values of shear stress, a more detailed analysis[8] has been performed to give information about the complete shape of the free surface and the onset of secondary flows.

Note that we have neglected surface tension in this analysis. If R denotes the radius (or half-width) of the channel, then the surface tension σ will not be important provided $(\rho g R^2 \cos \beta)/2\sigma \gg 1$. One of the nice results of this experiment is that it clearly establishes that the second normal stress difference is positive, at least at low shear rates.

c. Hole Pressure Effect

We now discuss measurement errors for polymeric fluids inherent in pressure transducers that are not flush-mounted.[9] To illustrate this error we consider the flow of fluids in a channel with a deep transverse slot as shown in Fig. 2.3-4. A fluid flows from left to right, driven by a pressure gradient in the x-direction. Flush-mounted pressure transducers are installed at the wall opposite the slot and at the bottom of the slot; the transducers show readings $(p + \tau_{yy})_1$ and $(p + \tau_{yy})_2$ respectively. It may be shown from the

[8] L. Sturges and D. D. Joseph, *Arch. Rat. Mech. Anal.*, **59**, 359–387 (1975).

[9] This measurement error was first confirmed by an extensive set of experiments described in J. M. Broadbent, A. Kaye, A. S. Lodge, and D. G. Vale, *Nature*, **217**, 55–56 (1968); A. Kaye, A. S. Lodge, and D. G. Vale, *Rheol. Acta*, **7**, 368–379 (1968); J. M. Broadbent and A. S. Lodge, *ibid.*, **10**, 557–573 (1972).

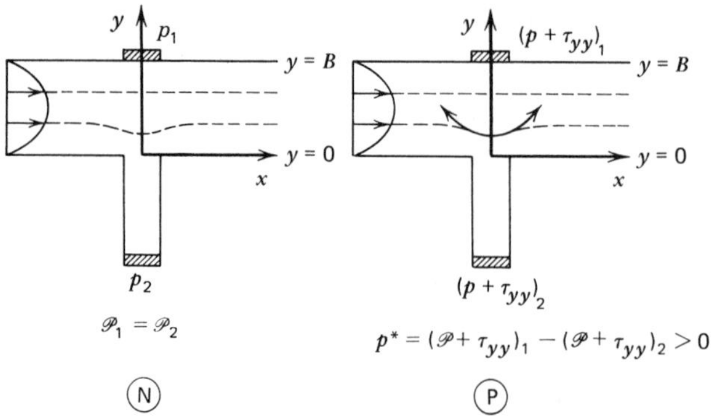

FIGURE 2.3-4. A fluid is flowing from left to right between two parallel plates across a deep transverse slot. "Pressures" are measured by flush-mounted transducer "1" and recessed transducer "2." Ⓝ For the Newtonian fluid $\mathscr{P}_1 = \mathscr{P}_2$. Ⓟ For polymeric fluids $(\mathscr{P} + \tau_{yy})_1 > (\mathscr{P} + \tau_{yy})_2$. The arrows tangent to the streamline indicate how the extra tension along a streamline tends to "lift" the fluid out of the hole, so that the recessed transducer gives a reading that is lower than that of the flush-mounted transducer.

equation of motion for Newtonian fluids that the centerline, $x = 0$, is a line of symmetry for the velocity field, if the flow is sufficiently slow that inertial effects may be neglected. This in turn means that the modified pressure $\mathscr{P}$ on the line of symmetry is independent of y. Thus, since τ_{yy} is zero at solid surfaces for Newtonian fluids, the modified pressures $\mathscr{P}_1$ and $\mathscr{P}_2$ are equal for Newtonian fluids, when inertial effects are negligible. On the other hand, for polymeric liquids we find[10] that generally $(\mathscr{P} + \tau_{yy})_1 > (\mathscr{P} + \tau_{yy})_2$. The difference $p^* = (\mathscr{P} + \tau_{yy})_1 - (\mathscr{P} + \tau_{yy})_2$, called the *hole pressure*, is related to the normal stresses in shear flow. For the transverse slot shown in Fig. 2.3-4 the following simple relation has been found[11] experimentally to hold

Transverse slot:
$$p^* = -\frac{1}{2}\int_0^{\tau_w} \frac{\tau_{11} - \tau_{22}}{\tau_{21}}\, d\tau_{21} \tag{2.3-15}$$

Here τ_w is the wall shear stress in the channel far from the slot. In this expression the 1, 2, and 3 directions have the meaning given at the beginning of this section, that is, in Fig. 2.3-4 we have $1 = x$, $2 = y$, $3 = z$. The stress tensor components to be used in the integral refer to the undisturbed shearing flow. The first normal stress difference is taken to be a function of shear stress. This is permissible, since we will see in Chapter 3 that both $\tau_{11} - \tau_{22}$ and τ_{21} are single-valued functions of the shear rate. Equation 2.3-15 was derived initially from theoretical arguments; however, the main justification for its continued use rests on the agreement with experiments.

Hole pressure effects occur whenever a polymeric liquid flows over a depression in a conduit wall. Two other geometries that have been studied are circular holes and long narrow slits aligned with the flow.[11] For these geometries the predicted relations are:

Circular hole:
$$p^* = -\frac{1}{3}\int_0^{\tau_w} \frac{(\tau_{11} - \tau_{22}) - (\tau_{22} - \tau_{33})}{\tau_{21}}\, d\tau_{21} \tag{2.3-16}$$

[10] D. Pike and D. G. Baird, *J. Rheol.*, **28**, 439–447 (1984); A. S. Lodge and L. de Vargas, *Rheol. Acta*, **22**, 151–170 (1983); A. S. Lodge, *Polym. News*, **9**, 242–246 (1984).

[11] K. Higashitani and W. G. Pritchard, *Trans. Soc. Rheol.*, **16**, 687–696 (1972).

Longitudinal slot: $$p^* = \int_0^{\tau_w} \frac{(\tau_{22} - \tau_{33})}{\tau_{21}} d\tau_{21} \tag{2.3-17}$$

This last relation has been subject to much less experimental investigation than the other two formulas.

§2.4 SECONDARY FLOWS

We now turn to the first of three experiments in which a flow field may be divided into a strong primary flow field and a weak secondary flow field. Roughly speaking the primary flows are associated with the viscous properties of the fluids, and the secondary flows with inertial effects and "elastic effects." We shall find in all three experiments that the secondary flows caused by elastic effects are opposite to those caused by inertial effects. Although there is no general principle that inertial effects and elastic effects produce secondary flows in opposite directions, that does seem to be a good rule of thumb.

a. Rotating Sphere

In this experiment we consider a sphere rotating in a large sea of fluid. This situation was analyzed in Example 1.4-3 for a Newtonian fluid. The steady primary flow field (in Eq. 1.4-50) is one in which fluid particles are carried in circles concentric with the axis of rotation of the sphere. Superimposed on this is a secondary flow driven by inertial effects (given in Eqs. 1.4-62, 63, and 64) in which the fluid moves towards the sphere near the axis of rotation and away from the sphere in the equatorial plane. We understand this intuitively since the centrifugal force is largest near the equator of the sphere, thus driving the fluid away from the sphere here.

The same experiment performed with a sphere in a 5% solution of polyacrylamide was performed by Giesekus[1] as shown in Fig. 2.4-1. The photograph clearly shows a rotating primary flow but a secondary flow toward the sphere in the equatorial plane and away from the sphere near the axis of rotation. We shall return to this experiment in Example 6.5-1 where a quantitative analysis is given on the basis of a simple constitutive equation that contains both viscous and "elastic" contributions to the stress tensor.

b. Cylindrical Tank with Rotating Lid

In this experiment a flow field is produced by placing a rotating disk on top of a beaker filled with fluid. Photographs of secondary flows in this system are shown[2] in Fig. 2.4-2. The primary fluid motion is in the tangential direction because of the rotation of the disk. Moreover, the fluid at the top of the container will be rotating with a larger angular velocity than that at the bottom; consequently, the fluid near the disk will experience a large outward centrifugal force. For Newtonian fluids there are no forces to counter this, and we are not surprised to find that there is a weak secondary flow, everywhere perpendicular to the primary flow, which is directed radially outward near the disk, down the side of the beaker, inward along the bottom, and finally back up near the center, as shown in Fig. 2.4-2 Ⓝ. The magnitude of the velocity in the secondary flow is roughly 10% of that in the primary flow.

[1] H. Giesekus in E. H. Lee, ed., *Proceedings of the Fourth International Congress on Rheology*, Wiley-Interscience, New York (1965), Part 1, pp. 249–266.

[2] C. T. Hill, *Trans. Soc. Rheol.*, **16**, 213–245 (1972). See also C. T. Hill, J. D. Huppler, and R. B. Bird, *Chem. Eng. Sci.*, **21**, 815–817 (1966). Theoretical analyses have been given by J. M. Kramer and M. W. Johnson, Jr., *Trans. Soc. Rheol.*, **16**, 197–212 (1972), and J. P. Nirschl and W. E. Stewart, *J. Non-Newtonian Fluid Mech.*, **16**, 233–250 (1984).

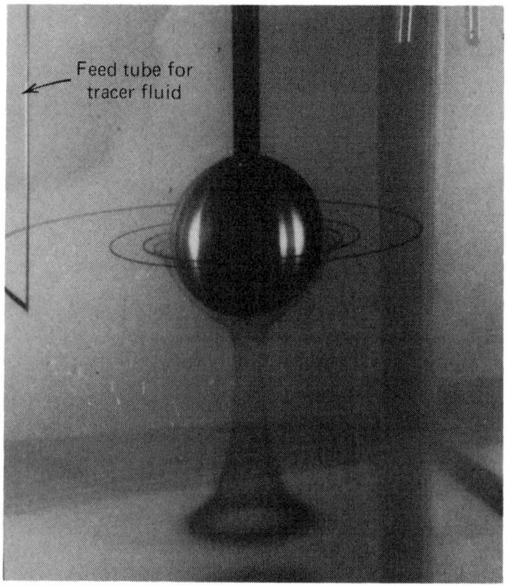

FIGURE 2.4-1. Secondary flow around a rotating sphere in a polyacrylamide solution. [Reproduced from H. Giesekus in E. H. Lee, ed., *Proceedings of the Fourth International Congress on Rheology*, Wiley-Interscience, New York (1965), Part 1, pp. 249–266.]

A polymer fluid placed in the disk and cylinder apparatus exhibits a secondary flow in the opposite direction. A typical flow pattern is shown in Fig. 2.4-2 Ⓟ for a solution of polyacrylamide in a solvent of glycerol and water. Qualitatively we expect that the normal stresses associated with the primary flow that act to produce the rod climbing are at work here. Moreover, the normal stresses do not have to be very large in an absolute sense in order to cause a reversal in the direction of secondary flow. At polymer concentrations so low that normal stresses could no longer be detected by standard techniques, the reverse secondary flow was still clearly visible at small disk rotation speeds.

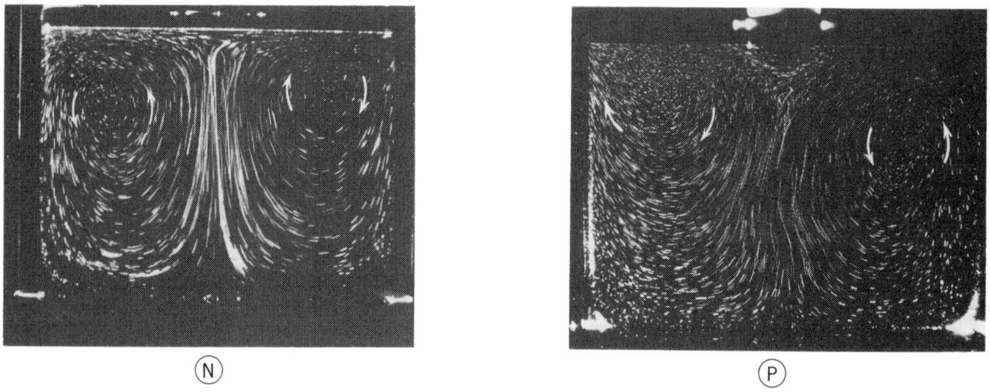

FIGURE 2.4-2. Secondary flows in the disk-cylinder system. Ⓝ The Newtonian fluid moves up at the center, whereas Ⓟ the viscoelastic fluid, polyacrylamide (Separan 30)–glycerol–water, moves down at the center. [Reproduced from C. T. Hill, *Trans. Soc. Rheol.*, **16**, 213–245 (1972).]

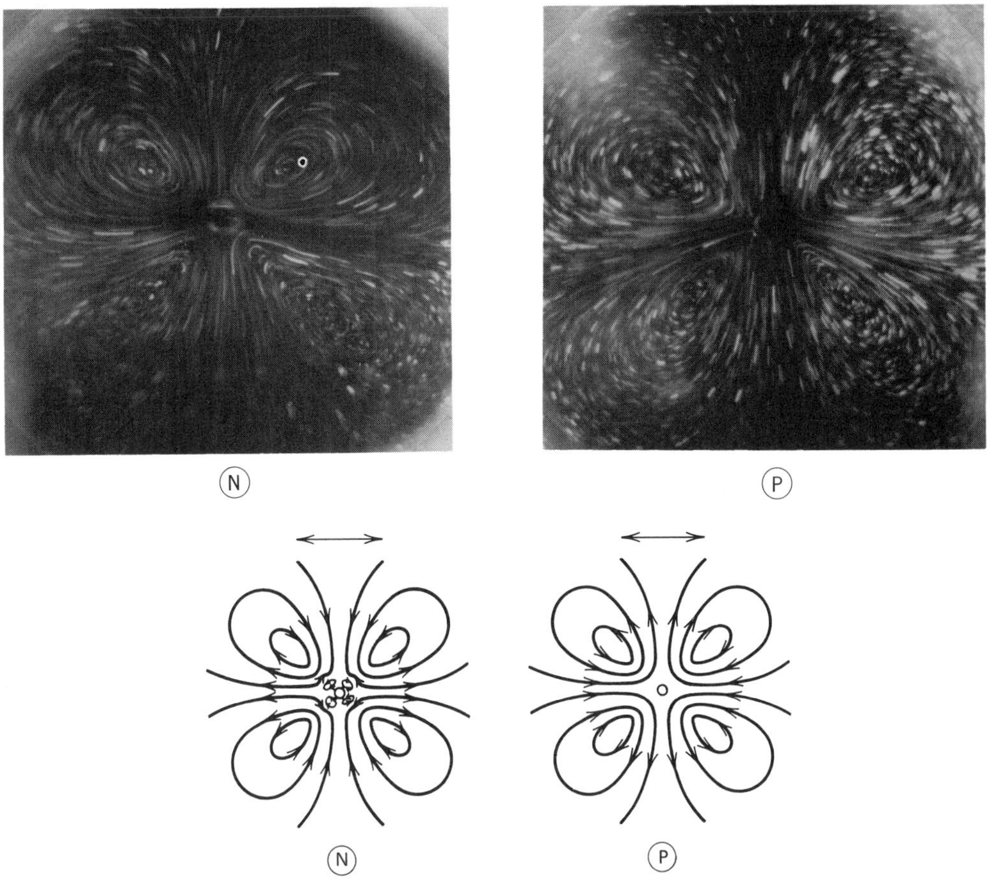

FIGURE 2.4-3. Steady streaming motion produced by a long cylinder oscillating normal to its axis. The cylinder is viewed on end and the direction of oscillation is shown by the double arrow. The photographs do not show streamlines but mean particle pathlines made visible by illuminating tiny spheres with a stroboscope synchronized with the cylinder frequency. Ⓝ A water–glycerin mixture: Ⓟ 100 ppm polyacrylamide in water–glycerin. [Reproduced from C. T. Chang and W. R. Schowalter, *Nature*, **252**, 686–688 (1974).]

c. Streaming Motion Produced by an Oscillating Cylinder

A long cylinder of radius R is made to oscillate with a frequency ω and amplitude δ normal to its axis. For small values of δ this produces a primary flow field of order δ with frequency ω which has mean value zero at any fixed position. In addition fluid particles will move with a nonzero time average velocity in what is known as *steady streaming motion* or, sometimes, *acoustic streaming*. Without going into detail we can describe the two kinds of contributions to this motion that may arise at order δ^2. First, since the flow field is inhomogeneous, the time average velocity of a given fluid particle will not be the same as the time average velocity at a given fixed position. Thus the primary flow field causes the fluid particles to move with a mean velocity of order δ^2. Second, nonlinear terms in the equation of motion or possibly in the constitutive equation produce a secondary flow field of order δ^2 with frequency 2ω and a nonzero mean value. This secondary flow field then directly gives a particle motion of order δ^2. The resulting steady streaming motion may be made visible by a flow visualization technique as shown in Fig. 2.4-3 for a Newtonian fluid Ⓝ and a polyacrylamide solution Ⓟ. For the Newtonian fluid the streaming motion is directly toward the oscillating cylinder along the axis of oscillation in an inner region, and opposite in an outer region. For the polymer solution, by contrast, there is only one region and the direction is toward the cylinder along the entire axis of oscillation.[3] Thus there is a flow reversal in most of the fluid, but we see also that the rule about the competing effects of inertia and elasticity is no more than a rule of thumb.

§2.5 OTHER ELASTIC EFFECTS

In this section we consider a further number of effects that are all in one way or another manifestations of elastic properties of polymeric fluids. However the experiments are more difficult to analyze than those encountered in the previous two sections. In addition the experiments form a more heterogeneous group in the sense that there is not one well-defined property, such as normal stresses, that is simply related to them all.

a. Extrudate Swell (also called "die swell")

In this experiment we consider a fluid that exits from a capillary of diameter D into air forming a jet of diameter D_e. For polymeric fluids it is customary to refer to the material in the jet as the *extrudate*. For Newtonian fluids D_e will be about 13% larger than D in the limit of small Reynolds number[1] and about 13% smaller than D in the limit of large Reynolds number laminar flow.[2] Hence a Newtonian fluid leaves the capillary without any dramatic change in diameter, as evidenced in Fig. 2.5-1 Ⓝ. The contrasting behavior of a

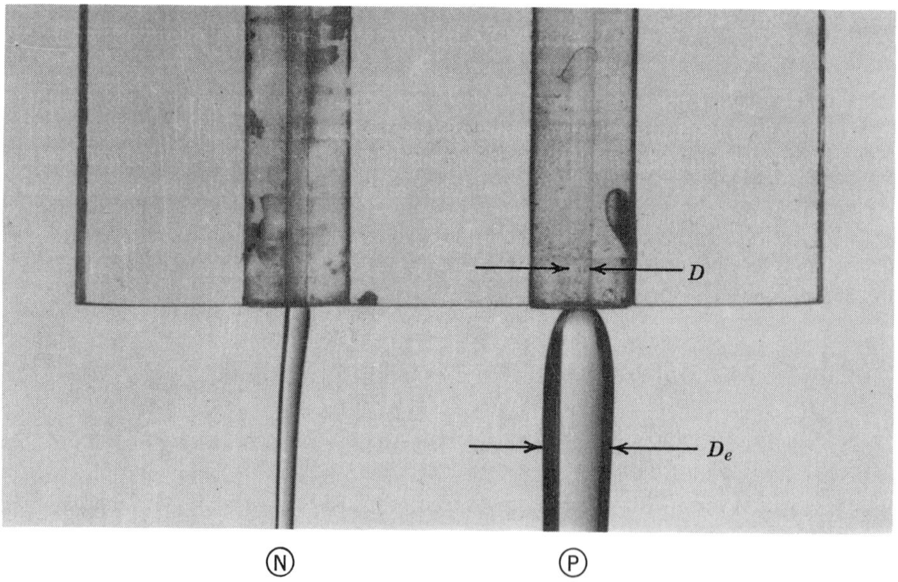

FIGURE 2.5-1. Behavior of fluids issuing from orifices. Ⓝ A stream of Newtonian fluid (silicone fluid) shows no diameter increase upon emergence from the capillary tube; Ⓟ a solution of 2.44 g of polymethylmethacrylate ($\bar{M}_n = 10^6$ g/mol) in 100 cm^3 of dimethylphthalate shows an increase by a factor of 3 in diameter as it flows downward out of the tube. [Reproduced from A. S. Lodge, *Elastic Liquids*, Academic Press, New York (1964), p. 242.]

[3] A detailed analysis is given by C. F. Chang and W. R. Schowalter, *J. Non-Newtonian Fluid Mech.*, **6**, 47–67 (1979).
[1] R. I. Tanner in J. R. A. Pearson and S. M. Richardson, eds., *Computational Analysis of Polymer Processing*, Applied Science, London (1983), p. 66.
[2] See Problem 1B.4.

polymeric fluid is shown in Fig. 2.5-1 Ⓟ, where D_e increases to about 300% of D. Extrudate diameters of two, three, or even four times the capillary diameter are encountered with polymers, and the phenomenon is referred to as *extrudate swell.*

A very oversimplified interpretation of extrudate swell may be obtained by using again the simple proposition of Weissenberg that flowing polymeric fluids have an extra tension along the streamlines. Once the fluid is outside the capillary this extra tension cannot be supported, and the fluid will contract axially and expand radially. Using this "elastic recoil" mechanism Tanner[3] proposed the following simple expression for extrudate swell:

$$\frac{D_e}{D} = 0.1 + \left[1 + \frac{1}{2}\left(\frac{\tau_{11} - \tau_{22}}{2\tau_{21}}\right)_w^2\right]^{1/6} \tag{2.5-1}$$

where the subscript w indicates that the stresses in steady tube flow $\tau_{11} - \tau_{22}$ and τ_{21} are to be evaluated at the wall. The simplicity of Eq. 2.5-1 and its success[4] in describing data on extrudate swell recommend its use for estimation purposes. Note, however, that the analysis in Example 2.3-1 of the rod-climbing experiment shows that one must be careful with simple arguments of this kind, and it may be that the second normal stress difference $\tau_{22} - \tau_{33}$ influences the swell ratio somewhat.

In addition to the relaxation of the normal stresses at the end of the capillary, there may be other phenomena that contribute to the extrudate swell. For example, viscous heating produces a temperature gradient across the tube cross-section. It has been demonstrated[5] that for Newtonian fluids with temperature-dependent viscosity, the viscosity near the tube wall is higher than at the center, where the temperature has a maximum, and extrudate swell results; this effect could also contribute to the extrudate swell in polymeric liquids. Still another suggestion[6] is that polymer molecules near the tube wall undergo more stretching than those near the center, and that in the extrudate the springing back of the molecules near the liquid-gas interface contributes to the jet diameter increase. It should be evident that extrudate swell is a complicated phenomenon for which no simple analysis can be entirely satisfactory.

We conclude this discussion with a few comments regarding the role of the length-to-diameter (L/D) ratio of the die. First consider a die with a very small L/D. A typical fluid element moving along the centerline begins as a short, fat cylinder in the reservoir, and then is squeezed into a long, thin cylinder within the die. Upon emerging from the die, the fluid element, "remembering" its former shape, tries to return to become short and fat again. This contribution to the extrudate swell is thus ascribed to the fluid memory of past kinematic events.

Next we consider a die with large L/D and the experiment by Lodge[7], which shows that the "memory effect" described in the last paragraph cannot give a complete explanation. He filled a tube with a large L/D with a sample of silicone "silly putty" and then allowed it to remain in the tube for a time longer than the memory of the fluid. The "silly putty" was then forced out of the tube with a plunger. As the material emerged from the tube extrudate swell was observed.

In some experiments with a die having $L/D = 150/8$, Giesekus[8] showed that with increased flow rate "delayed swell" occurs. That is, the extrudate has about the same

[3] R. I. Tanner, *J. Polym. Sci.*, **A8**, 2067–2078 (1970); R. I. Tanner, *Engineering Rheology*, Oxford University Press (1985), pp. 321–329, 338–343.
[4] J. Vlachopoulos, M. Horie, and S. Lidorikis, *Trans. Soc. Rheol.*, **16**, 669–685 (1972).
[5] H. B. Phuoc and R. I. Tanner, *J. Fluid Mech.*, **98**, 253–271 (1980).
[6] R. I. Tanner, *J. Non-Newtonian Fluid Mech.*, **6**, 289–302 (1980).
[7] A. S. Lodge, *Elastic Liquids*, Academic Press, New York (1964), pp. 243–244.
[8] H. Giesekus, *Rheol. Acta*, **8**, 411–421 (1969).

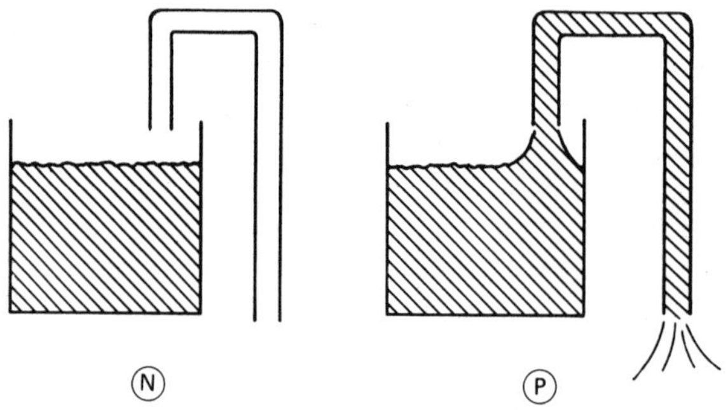

FIGURE 2.5-2. The tubeless siphon. (N) When the siphon tube is lifted out of the fluid, the Newtonian liquid stops flowing; (P) the macromolecular fluid continues to be siphoned.

diameter as the capillary for a distance several diameters downstream of the exit, and then at this point shows pronounced swelling.

Another interesting point is that for noncircular dies, the shape of the cross section of the extrudate is not the same as that of the die.

b. The Tubeless Siphon[9]

We turn now to an experiment involving the siphoning of Newtonian and non-Newtonian fluids. Imagine two identical experiments in which fluid is being siphoned out of a container—one experiment using a Newtonian fluid, the other a polymeric fluid (see Fig. 2.5-2). Now suppose that the tube is lifted up out of the fluid in the container. We immediately hear a slurping sound from the siphon that was in the Newtonian fluid as the liquid empties out of the tube and the siphoning stops. By contrast the non-Newtonian fluid continues to flow up to and through the siphon![10]

Photographs of fluid columns in working tubeless siphons are shown in Figs. 2.5-3 and 4. It is believed that it is the orientation and elongation of the polymer molecules along the streamlines that are responsible for the large axial stresses that make the siphon work. Note in Fig. 2.5-4 that these same stresses also prevent a bubble from rising in the fluid column.

c. The Uebler Effect

This effect was discovered in connection with the use of tiny gas bubbles in a flow visualization experiment for contraction flow. It was observed by Uebler[11] that when the diameter of the bubbles is equal to or larger than about $\frac{1}{6}$ to $\frac{1}{8}$ of the small tube diameter, a bubble moving along the centerline comes to a sudden stop right at the entrance to the small tube. Such a bubble can then remain stationary for a minute or more before finally proceeding on its journey down the tube.

[9] Tacitus relates that this effect has been utilized in connection with the harvesting of bitumen on the Dead Sea around AD 70. See *The Complete Works of Tacitus* (transl. by A. J. Church and W. J. Brodribb), The Modern Library, New York (1942), Book V, The History, p. 661. Tacitus was born AD 55. This reference, which seems to be the earliest description of a non-Newtonian flow phenomenon, was supplied by Professor Moshe Gottlieb.

[10] D. F. James, *Nature*, **212**, 754–756 (1966).

[11] A. B. Metzner, *AIChE J*, **13**, 316–318 (1967); E. A. Uebler, Ph.D. Thesis, University of Delaware, Newark (1966).

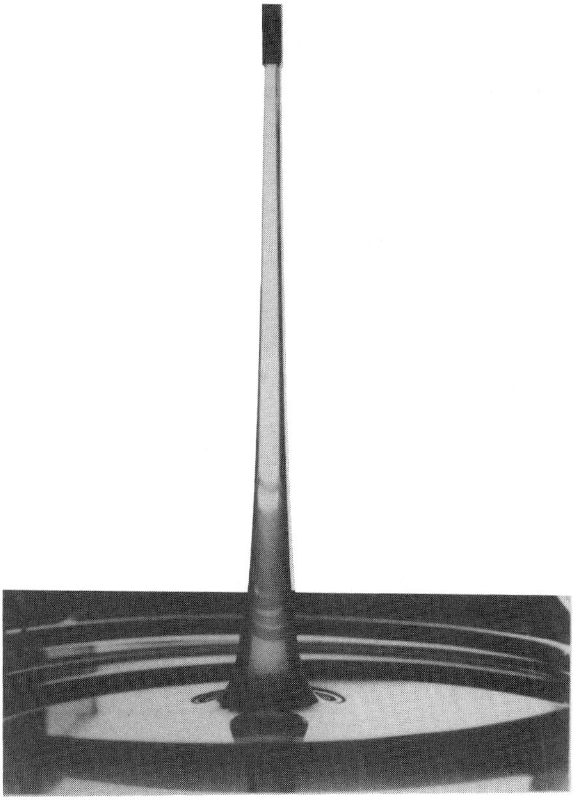

FIGURE 2.5-3. Fluid column in tubeless siphon experiment with a high molecular weight hydrocarbon polymer, AM-1 in JP-8 aviation fuel. [Reproduced from S. T. J. Peng and R. F. Landel, *J. Appl. Phys.* **47**, 4255 (1976)].

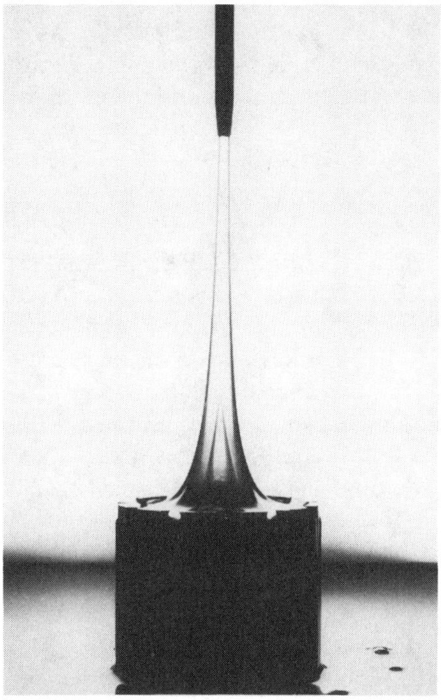

FIGURE 2.5-4. Fluid column in tubeless siphon experiment with the same polymer as in Fig. 2.5-3. Note the stable trapped bubble at the bottom of the column. [Reproduced from S. T. J. Peng and R. F. Landel in G. Astarita, G. Marrucci, and L. Nicolais, eds., *Rheology*, Vol. 2, Plenum Press, New York (1980), p. 388.]

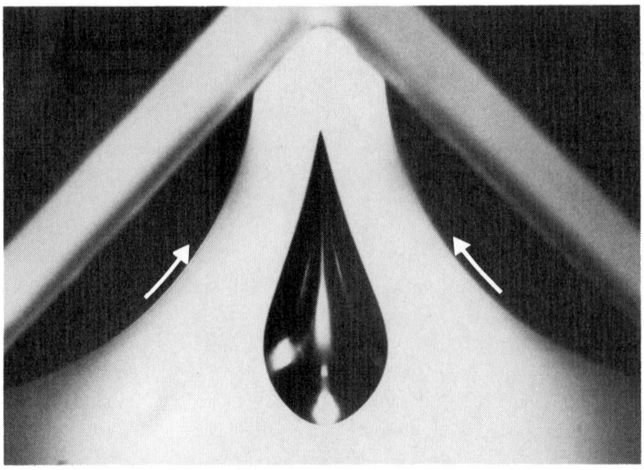

FIGURE 2.5-5. Gas bubble trapped in the flow field created by two parallel, horizontal, counter-rotating cylinders. The sense of rotation is indicated by arrows on the cylinders. In the absence of rotation the bubble passes between the cylinders. [Photograph by J. Michele in H. Giesekus, *Z. Angew. Math. Mech.*, **58**, T26–T37 (1978).]

A demonstration of this effect has been given by Giesekus,[12] who studied the flow field created by two counter-rotating cylinders immersed in a polymeric fluid as shown in Fig. 2.5-5. The axes of the cylinders are parallel and horizontal, and a flow field is generated in which fluid moves up in the region between the cylinders and down outside the cylinders. In this situation Giesekus showed that a large gas bubble may remain stationary on the upstream side of the contraction, even though common sense would suggest that both buoyancy and the flow field would make the bubble rise. As soon as the rotation of the cylinders is stopped, however, the bubble rises and escapes through the contraction. Note the similarity in the shape of this bubble and that in Fig. 2.5-4.

d. Contraction Flow

We now turn our attention to the details of the flow pattern in the neighborhood of a sudden contraction. Figure 2.5-6 shows two streak photographs taken by Giesekus[13] contrasting the velocity fields of glycerin and aqueous polyacrylamide as they flow from a large reservoir into a small tube. The Reynolds number in both photographs is very low. The behavior of the two fluids is not even qualitatively similar. A typical streamline in the Newtonian fluid is a straight line heading directly toward the entrance to the small tube; glycerin approaches the exit from a full 90 degrees in any direction about the centerline. However, for a polymer solution, only the fluid in a small conical region about the centerline moves linearly toward the entrance to the small tube. A significant portion of the polymer solution is trapped in a large circulation pattern and does not enter the small tube at all.

[12] H. Giesekus, *Z. Angew. Math. Mech.*, **58**, T26–T36 (1978).

[13] A. B. Metzner, E. A. Uebler, and C. F. Chan Man Fong, *AIChE J*, **15**, 750–758 (1969). For other photographs of converging flows of polymer solutions, see H. Giesekus, *Rheol. Acta*, **7**, 127–138 (1968); P. J. Cable and D. V. Boger, *AIChE J.*, **24**, 869–879, 992–999 (1978); **25**, 152–159 (1979); D. V. Boger and H. Nguyen, *Polym. Eng. Sci.*, **18**, 1037–1043 (1978).

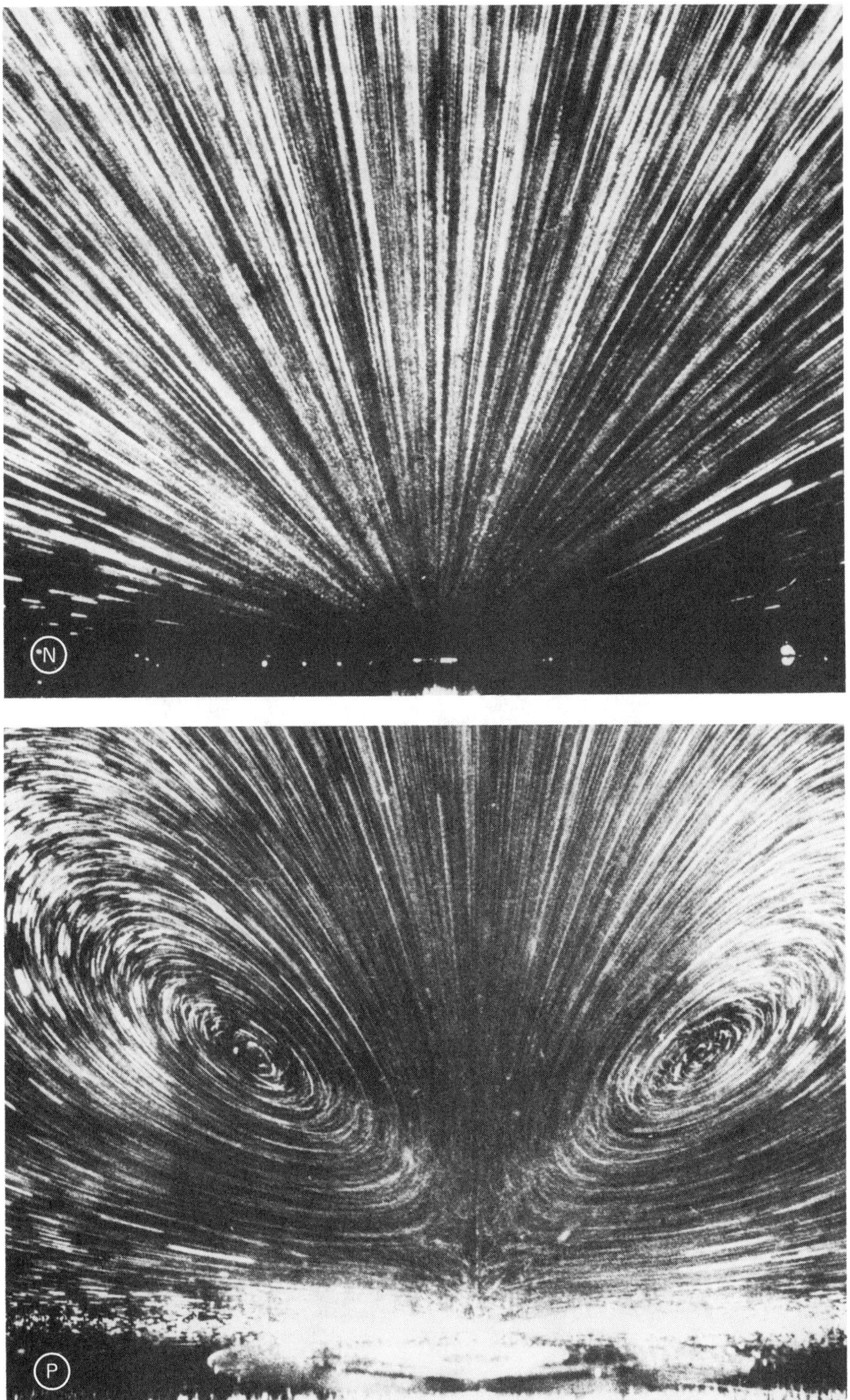

FIGURE 2.5-6. Velocity fields near a sudden contraction. The fluid moves from top to bottom in the photographs from a large reservoir into a small circular tube. Ⓝ The streamlines in glycerin are straight and all directed toward the exit; Ⓟ a 1.67% aqueous polyacrylamide solution shows a large toroidal vortex. [Photographs by H. Giesekus, given in A. B. Metzner, E. A. Uebler, and C. F. Chan Man Fong, *AIChE J.*, **15**, 750–758 (1969).]

Recirculation zones are generally considered detrimental in equipment for polymer processing. For this reason it is important to understand and be able to predict the conditions under which recirculation zones appear.

e. Elastic Recoil

We have referred earlier to an extra tension along the streamlines in polymeric fluids, caused by alignment and stretching of the polymer molecules. It was suggested that the tendency for the molecules to snap back when the external forces are removed produces "elastic recoil" of the fluid that could be at least partially responsible for extrudate swelling. We now turn to a number of experiments that we interpret as demonstrations of the elastic recoil effect.

The first experiment was performed by Kapoor[14] as shown in Fig. 2.5-7. In his experiment, a streak of charcoal slurry was first introduced into a 2% (weight) solution of carboxymethylcellulose in water by means of a syringe. A pressure gradient was then applied to the solution, and the deformation of the trace line was recorded in a sequence of

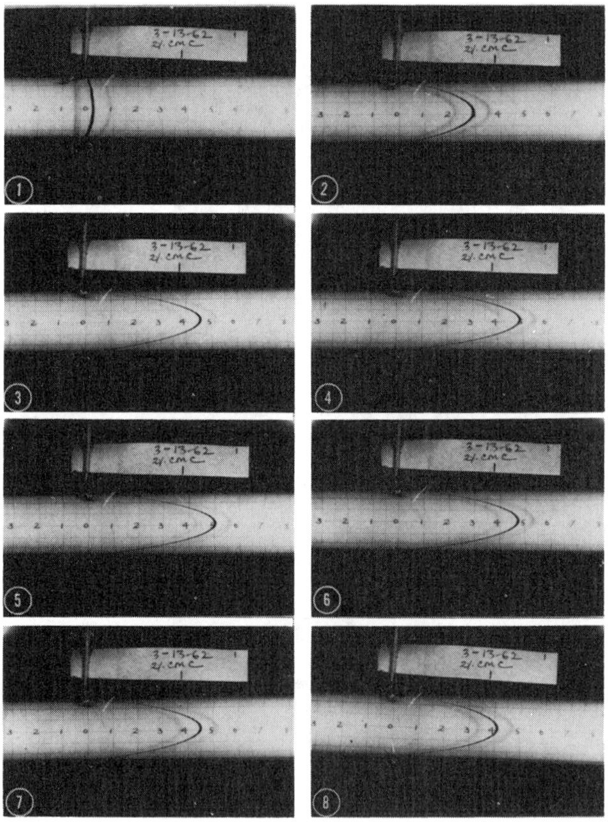

Ⓟ

FIGURE 2.5-7. A solution of 2% carboxymethylcellulose (CMC 70H) in water is made to flow under a pressure gradient that is turned off just before frame 5. The flow and subsequent recoil are shown by the charcoal tracer line. [Reproduced from A. G. Fredrickson, *Principles and Applications of Rheology*, © Prentice-Hall, Englewood Cliffs, NJ (1964), p. 120.]

[14] N. N. Kapoor, M.S. Thesis, University of Minnesota, Minneapolis (1964); A. G. Fredrickson, *Principles and Applications of Rheology*, Prentice-Hall, Englewood Cliffs, NJ (1964), p. 120.

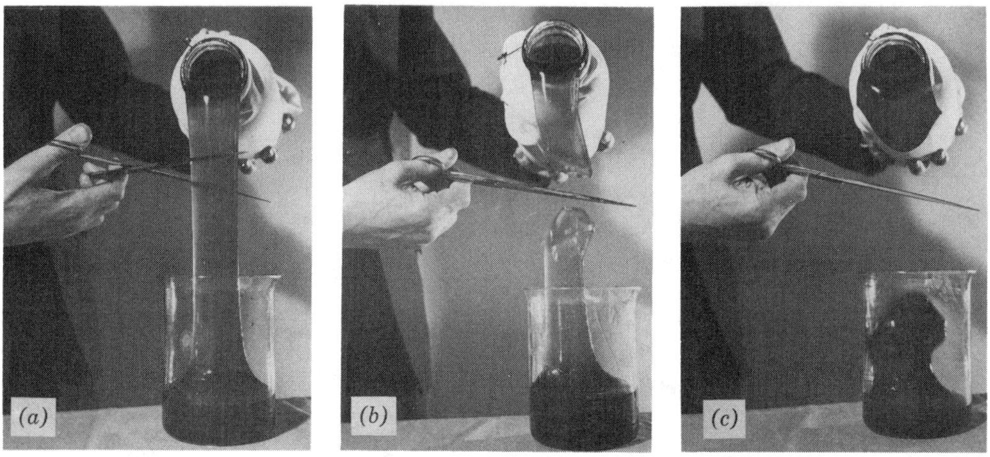

FIGURE 2.5-8. An aluminum soap solution, made of aluminum dilaurate in decalin and *m*-cresol, is (*a*) poured from a beaker and (*b*) cut in midstream. In (*c*), note that the liquid above the cut springs back to the beaker and only the fluid below the cut falls to the container. [Reproduced from A. S. Lodge, *Elastic Liquids*, Academic Press, New York (1964), p. 238. For a further discussion of aluminum soap solutions see N. Weber and W. H. Bauer, *J. Phys. Chem.*, **60**, 270–273 (1956).]

photographs. After a short time the pressure gradient was removed; the pictures taken after this time show the charcoal trace line receding as the fluid recoils. Recoil is not observed for Newtonian liquids.

In another experiment due to Lodge,[15] an aluminum soap solution is poured from a bottle into a beaker, as shown in Fig. 2.5-8 (*a*). The fluid is "pulled" out of the bottle by the weight of the entire fluid column and the tubeless siphon effect. When the fluid column is cut in two as shown in part (*b*) the top part of the column snaps back into the bottle by elastic recoil, as shown in part (*c*).

In quantitative experiments, Meissner[16] found that a filament of low-density polyethylene at 423 K which is stretched rapidly from 1 to 30 cm in length, at which point it is suddenly set free, will recover to a length of just 3 cm. The fact that the filament recovers to a length of 3 cm rather than to 1 cm shows that the "memory" of the original shape is not perfect. If the filament is held at a length of 30 cm for some time before it is set free, it will recover to some length larger than 3 cm. Finally, if the filament is held at a length of 30 cm for a long time before it is set free, there will be little or no recovery, thus demonstrating the "fading memory" of the fluid. In closing, we remark that a recovery by a factor of ten (from 30 cm to 3 cm) is very large. Typical cross-linked rubber bands cannot even be stretched by a factor of ten without breaking.

f. Filament stability

The ability of polymeric fluids to form stable filaments has been demonstrated in connection with the tubeless siphon experiment. In this section we show two further experiments that demonstrate the remarkable filament stability of polymeric fluids. Both

[15] A. S. Lodge, *Elastic Liquids*, Academic Press, New York (1964), p. 236.
[16] J. Meissner, *Rheol. Acta*, **10**, 230–242 (1971).

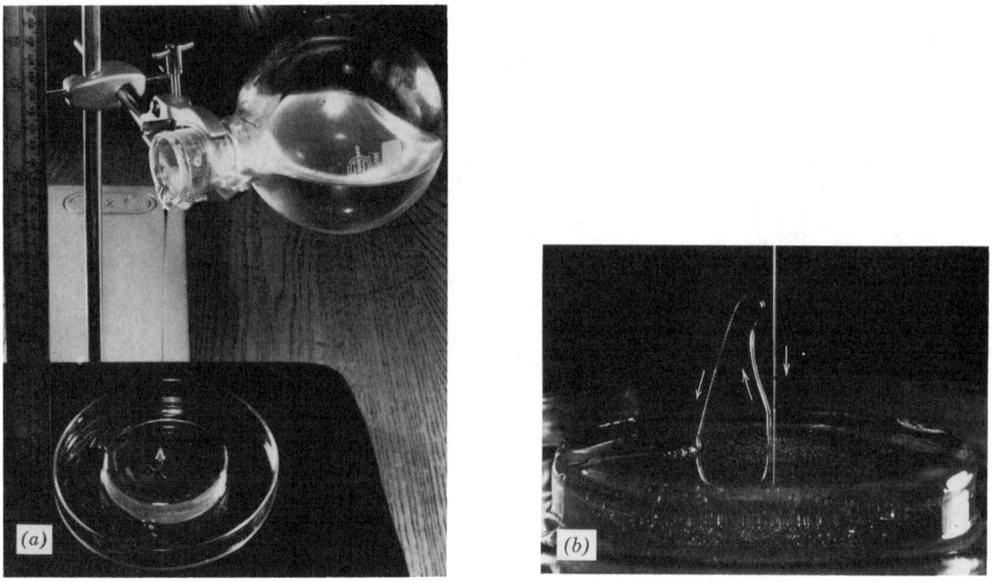

FIGURE 2.5-9. (*a*) A solution of 5.76 g of polyisobutylene ($\bar{M}_n \doteq 10^6$) in 100 cm^3 of decalin is poured onto a shallow pool of the same liquid. (*b*) A filament of the liquid stream rebounds from the pool if the stream being poured has a sufficiently small diameter. [Reproduced from A. S. Lodge, *Elastic Liquids*, Academic Press, New York (1964), p. 251.]

experiments involve very thin filaments, and surface forces are likely to contribute to the effects.

Kaye[17] has observed that when a solution of polyisobutylene in decalin is poured in a thin stream onto a shallow pool of the same liquid, the stream of liquid will bounce off the pool as shown in Fig. 2.5-9. Furthermore, the bouncing liquid stream forms a fine, continuous element. This behavior, which lasts for a period of about 1 s, alternates on a regular basis with a period of similar length, during which there is no bouncing stream. In the off-period, a small hump of liquid can be seen forming where the stream being poured strikes the surface of the pool.

Barnard and Pritchard[18] have observed that if the wall of a glass beaker, which has contained a solution of a high molecular weight polyisobutylene in low molecular weight polyisobutylene, is cleaned by tearing the polymer from the wall while wrapping it onto a glass rod, an interesting phenomenon is observed. Several very thin streams of fluid are found to emanate from irregularities on the surface of the liquid (see Fig. 2.5-10). The streams increase in length at a rate of several centimeters per second. It is believed that this is an electrically driven phenomenon. However, in Newtonian liquids of low conductivity similarly to that of the polyisobutylene, the streaming jet breaks up into a series of droplets, instead of remaining as a continuous filament such as the polymer shown here.[19]

[17] A. Kaye, *Nature*, **197**, 1001–1002 (1963).

[18] B. J. S. Barnard and W. G. Pritchard, *Nature*, **250**, 215–216 (1974).

[19] G. I. Taylor, *Proc. Roy. Soc., London*, **A313**, 453–475 (1969).

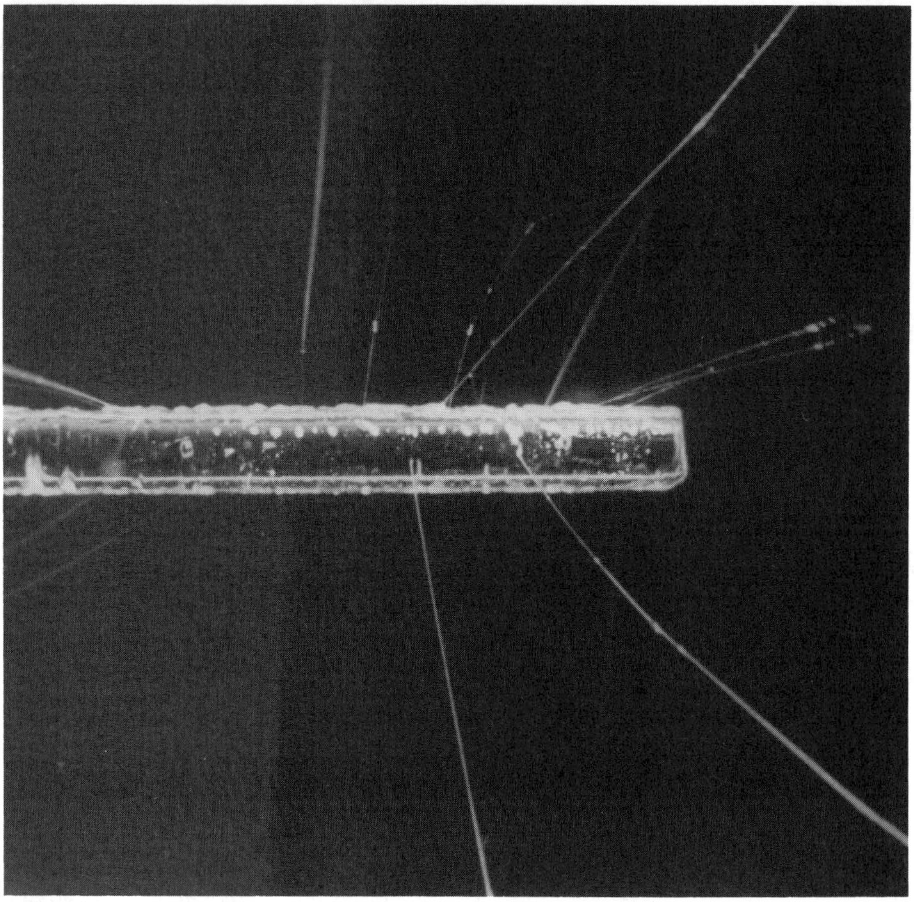

FIGURE 2.5-10. A solution of high molecular weight polyisobutylene in low molecular weight polyisobutylene forms fine, continuous streams emanating from a 6-mm glass rod. [Reproduced from B. J. S. Barnard and W. G. Pritchard, *Nature*, **250**, 215–216 (1974).]

g. Vortex Inhibition

The final experiment in this section is that of vortex inhibition.[20] The experiment is done by first filling a large tank with water, stirring the water to generate a circulation in the tank, and finally removing a plug from the center of the bottom to allow the water to drain. As the water empties from the tank, a very stable air core reaching all the way to the outlet forms, accompanied by a pronounced slurping sound (see Fig. 2.5-11 Ⓝ). Now, if a very small amount of certain polymers is added to the draining water, the air core suddenly disappears and the noise that goes with it ceases (see Fig. 2.5-11 Ⓟ). Moreover, the volume flow rate out of the tank nearly doubles after the air core is eliminated, provided that the level in the tank is kept constant.[21]

[20] R. J. Gordon and C. Balakrishnan, *J. Appl. Polym. Sci.*, **16**, 1629–1639 (1972).
[21] S. Ishikawa, Sc.D. Thesis, Massachusetts Institute of Technology, Cambridge (1979).

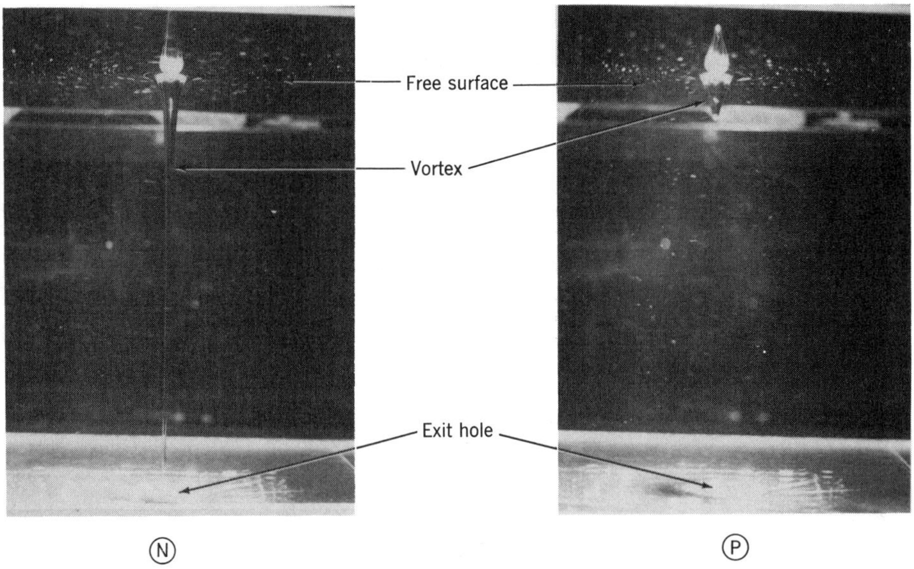

FIGURE 2.5-11. Vortex inhibition. Ⓝ shows the vortex that is formed as water drains form a tank, and Ⓟ shows the corresponding result after addition of a small amount of polymer to the water. A reflection of the vortex in the free surface of the liquid can be seen just above the vortex in both photographs. [Reproduced from photographs courtesy of Professor R. J. Gordon, Department of Chemical Engineering, University of Florida.]

Among the best polymers for vortex inhibition are polyethyleneoxide, effective at 7.5 ppm, and polyacrylamide, effective at only 3 ppm. These solutions are exceedingly dilute. It is even more interesting that the polymers that are good at vortex inhibition are generally also good in drag reduction—a dilute solution, turbulent flow phenomenon described in §2.7a. Furthermore, the same ordering in terms of effective concentrations seems to hold for both phenomena. For example, the two polymers mentioned above show significant drag reduction at concentrations of 9 and 5 ppm, respectively.

The mechanism for vortex inhibition is more amenable to understanding than that for drag reduction because the former involves laminar flow and the latter involves turbulent flow. The bulk of the flow out of the vortex tank occurs through a thin boundary layer[22] along the bottom surface of the container. As polymer molecules flow from this thin layer into the exit tube they experience very large rates of strain which are sufficient to stretch them nearly to their fully extended configuration.[21] In the fully extended configuration, the polymer molecules cause sufficient increase in the tension along streamlines near the exit hole that a larger fraction of the total flow passes through the boundary layer. This results in a slight increase in the size of the viscous core around the axis of rotation and the disappearance of the air core. This understanding of the mechanism for vortex inhibition and the close connection between effectiveness of polymers in vortex inhibition and drag

[22] The physical origin of this boundary layer is the no-slip condition on v_θ at the bottom of the container. In the bulk of the flow, an (inward) radial pressure gradient exists to counter the (outward) centrifugal force associated with the swirling flow ($v_\theta \propto 1/r$). Near the bottom wall v_θ goes to zero, and the unopposed radial pressure gradient causes an inward flow. Taylor has illustrated this boundary layer by stirring a cup of tea which contains tea leaves. When stirring is stopped, the inward flow in the boundary layer at the bottom of the cup causes the tea leaves to collect in a pile at the center. E. S. Taylor, *Illustrated Experiments in Fluid Mechanics, The NCFMF Book of Film Notes*, MIT Press, Cambridge, MA (1972), pp. 97–104.

reduction lends credence to the currently accepted belief that molecular extension is also responsible for drag reduction.

§2.6 BUBBLES AND PARTICLES

The fluid dynamics of systems with bubbles or particles is a vast subject even for Newtonian fluids.[1] When the bubbles or particles move in non-Newtonian fluids one encounters striking, new phenomena in practically every situation,[2] as exemplified by the following experiments.

a. Bubble Shapes

We now consider the buoyancy-driven rise of a gas bubble in a large container of fluid. If the fluid is Newtonian and the motion is so slow that inertial effects may be neglected then the bubble is spherical as shown in Example 1.4-2. Bubbles that rise very slowly in non-Newtonian fluids will also be spherical. However to observe spherical bubbles requires either very tiny bubbles or a very high viscosity of the fluid. Usually the bubbles will be deformed in some fashion from spherical shape, either because of inertial or elastic effects.

When bubbles in Newtonian fluids are deformed slightly due to inertial effects, they assume the shape of an ellipsoid of revolution flattened at the poles (an oblate ellipsoid). The actual deformation depends on the Reynolds number and the capillary number and is given to a first approximation in Eq. 1.4-46. By contrast bubbles in polymeric fluids that are deformed slightly due to elastic effects assume the shape of an ellipsoid of revolution that is elongated at the poles (a prolate ellipsoid) as shown in Fig. 2.6-1*a*. Qualitatively we can explain this by postulating that there is an extra tension along the streamlines that tends to squeeze the bubble at its equator and hence elongate it at the poles as shown in Fig. 2.6-1*b*. However such an extra tension could not be the result of a steady shear flow but the result of a transient stretching of the fluid elements. In Example 6.5-2 we present a quantitative computation of the deformation.

Bubbles with the shape shown in Fig. 2.6-1 are observed only when the motion is so slow that the deformation from the spherical shape is small. Larger or faster bubbles deform into a variety of shapes,[3] often with a cusp at the bottom end. An example of such a deformed shape is shown in Fig. 2.6-2. This bubble is not even rotationally symmetric but has a bottom end in the form of a knife edge, as evidenced by the two photographs of the same bubble taken at right angles. In fact the knife edge appears to continue into a thin sheet of air that eventually dissolves in the liquid. Thus one may say that the bubble in Fig. 2.6-1 is a "closed shape," whereas that in Fig. 2.6-2 is an "open shape." An important difference between bubbles with closed shapes and open shapes is that surface active impurities in the fluid may accumulate on the surface of the former but not on the latter type of bubbles. Surface-active impurities tend to immobilize bubble surfaces and hence retard the motion of gas bubbles. Thus a discontinuous change in bubble shape from a closed bubble (Fig. 2.6-1) to an open shape (Fig. 2.6-2) may be responsible for a discontinuous

[1] See J. F. Harper, *Adv. Appl. Mech.*, **12**, 59–129 (1972) and R. Clift, J. R. Grace, and M. E. Weber, *Bubbles, Drops, and Particles*, Academic Press, New York (1978).

[2] For extensive reviews see H. Giesekus, *Z. Angew. Math. Mech.*, **58**, T26–T36 (1978), and P. O. Brunn, *J. Non-Newtonian Fluid Mech.*, **7**, 271–288 (1980).

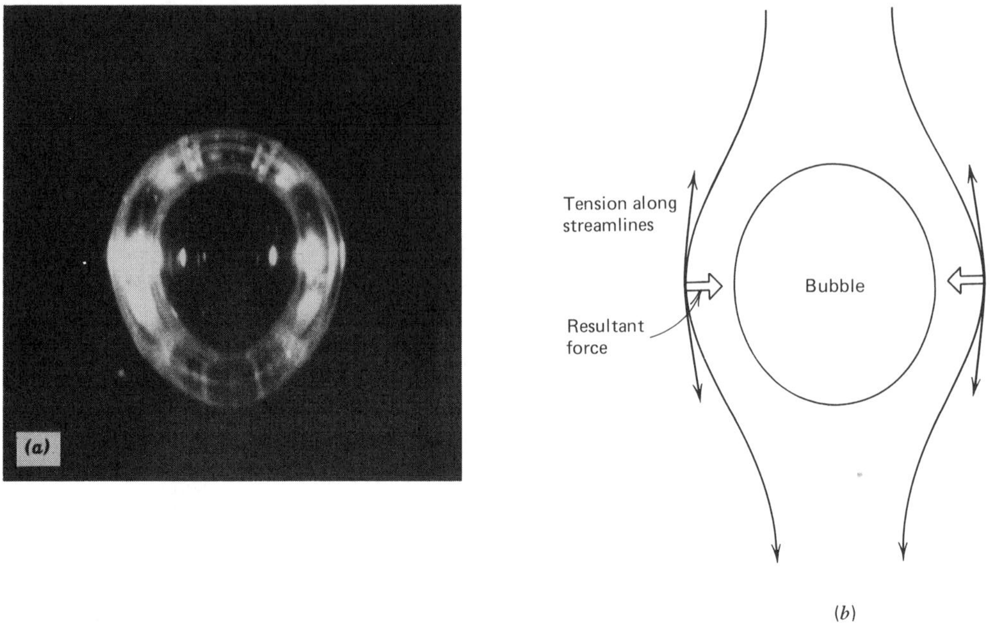

FIGURE 2.6-1. (*a*) Slightly deformed gas bubble of volume approximately 24 mm^3 rising with velocity 0.2 mm/s in a 1% solution of polyacrylamide in glycerin. [Photograph by O. Persson, Instituttet for Kemiteknik, Danmarks tekniske Højskole, Danmark]; (*b*) qualitative explanation of deformation in terms of an extra tension along the streamlines tending to squeeze the bubble at its equator. The streamlines are seen by an observer moving with the bubble.

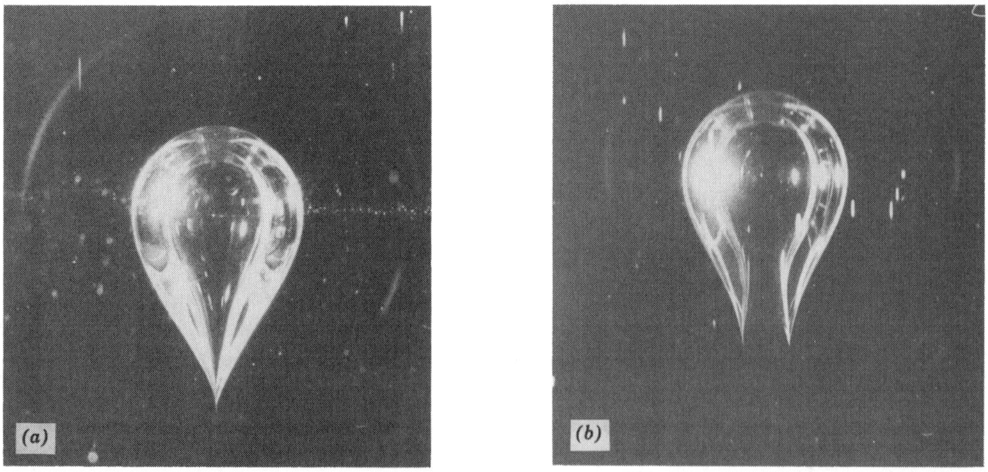

FIGURE 2.6-2. Highly distorted gas bubble of volume approximately 2100 mm^3 rising with velocity 10 mm/s in a polyacrylamide solution. The bubble is seen from two mutually perpendicular directions. Note that the bottom end is not axisymmetric. [Photographs by O. Persson, Instituttet for Kemiteknik, Danmarks tekniske Højskole, Danmark.]

increase in the rise velocity that is often encountered[3] when bubbles exceed a certain critical size.

b. Negative Wake

We now compare measurements of fluid velocity in the wakes behind gas bubbles rising in Newtonian and non-Newtonian fluids. The measurements reveal the unexpected phenomenon that the fluid velocity behind the bubbles in polymeric liquids is in the downward direction away from the rising bubbles (the velocities are referred to an observer at rest with respect to the liquid far from the bubbles). Thus the liquid velocity is opposite to the velocity in the usual wake behind objects moving in Newtonian fluids, and the phenomenon has been termed "negative wake."[4]

Consider in Fig. 2.6-3 Ⓝ the fluid velocity on the axis of a gas bubble rising in a viscous Newtonian liquid (a 99% glycerol-1% water mixture at 295 K). The position and velocity of the bubble is indicated in Fig. 2.6-3 Ⓝ by the width and height of the rectangle. The figure shows that while some fluid is pushed in front of the bubble, more fluid is pulled along behind the bubble. Consider now in Fig. 2.6-3 Ⓟ the fluid velocity on the axis of a gas bubble in a 1% solution of polyacrylamide in glycerol. The figure clearly shows that the fluid behind the bubble instead of being pulled along in the usual wake pattern is in fact "recoiling" away from the bubble. Negative wakes can be observed behind solid spheres as well as behind bubbles.[5]

c. Particle Interaction in Homogeneous Flow

A particularly simple example of a particle interaction phenomenon in non-Newtonian fluids occurs when two spheres of equal size and mass slowly settle along their line of centers in a fluid at rest far from the spheres. It is known[6] that in Newtonian fluids the two spheres will fall with the same speed so that the distance between them is constant, provided the Reynolds number is low. However in non-Newtonian fluids the two spheres will not fall with the same speed. Riddle, Narvaez, and Bird[7] discovered that if the initial separation is larger than a certain critical separation the spheres will diverge, whereas if it is smaller than this separation they will converge. The diverging of the spheres is probably related to a negative wake behind the leading sphere.

An even more spectacular interaction effect occurs in homogeneous shear flow of polymeric liquids with suspended spheres. During the shear flow, the suspended spheres rearrange to form chain structures. This effect seems to have been first described by Highgate and Whorlow.[8,9] In an investigation of the effect Michele, Pätzold, and Donis[10] used a suspension of glass spheres (60–70 μm in diameter) in a 0.5% solution of

[3] G. Astarita and G. Apuzzo, *AIChE J.*, **11**, 815–820 (1965); P. H. Calderbank, D. S. L. Johnson, and J. Loudon, *Chem. Eng. Sci.*, **25**, 235–256 (1970), L. G. Leal, J. Skoog, and A. Acrivos, *Can. J. Chem. Eng.*, **49**, 569–575 (1971), M. Coutanceau and M. Hajjam, *Appl. Sci. Res.*, **38**, 199–207 (1982).

[4] O. Hassager, *Nature*, **279**, 402–403 (1979).

[5] D. Sigli and M. Coutanceau, *J. Non-Newtonian Fluid Mech.*, **2**, 1–22 (1977); C. Bisgaard and O. Hassager, *Rheol. Acta*, **21**, 537–539 (1982); C. Bisgaard, *J. Non-Newtonian Fluid Mech.*, **12**, 283–302 (1983).

[6] J. Happel and H. Brenner, *Low Reynolds Number Hydrodynamics*, 2nd ed., Noordhoff, Leiden (1973).

[7] M. J. Riddle, C. Narvaez, and R. B. Bird, *J. Non-Newtonian Fluid Mech.*, **2**, 23–25 (1977).

[8] D. J. Highgate, *Nature*, **211**, 1390–1391 (1966); D. J. Highgate and R. W. Whorlow in R. E. Wetton and R. W. Whorlow, eds., *Polymer Systems: Deformation and Flow*, Macmillan, London (1968), pp. 251–261.

[9] D. J. Highgate and R. W. Whorlow, *Rheol. Acta*, **8**, 142–151 (1969).

[10] J. Michele, R. Pätzold, and R. Donis, *Rheol. Acta*, **16**, 317–321 (1977).

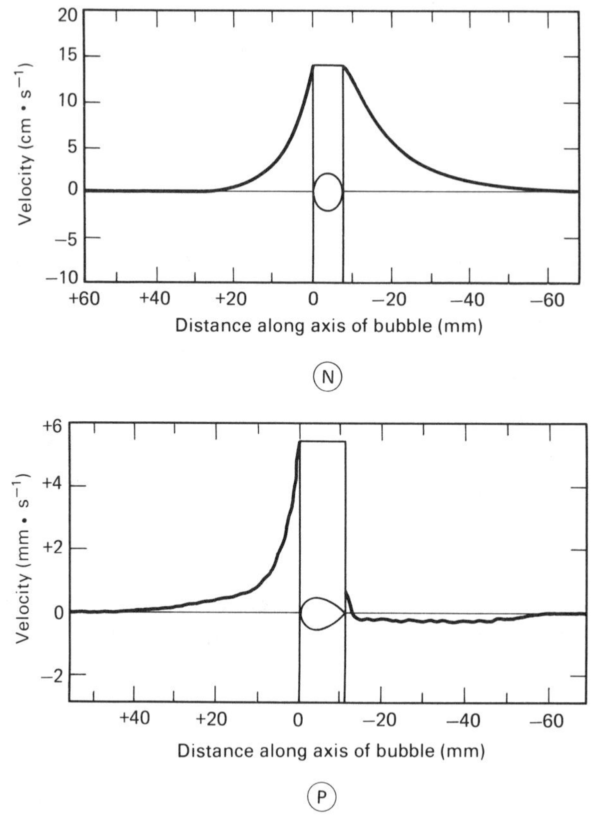

FIGURE 2.6-3. (N) Axial component of liquid velocity on the axis of a gas bubble of volume 400 mm^3 rising in a viscous Newtonian liquid (a 99% glycerol–1% water mixture at 295 K). The position of the bubble is indicated by the tall rectangle, the height of which indicates the rise velocity of the bubble (14.2 cm/s). The data are obtained from an experiment in which a bubble rising at constant speed along the axis of a glass cylinder passes a test section in which two photocells measure the rise velocity of the bubble and a laser-Doppler anemometer measures the axial component of the liquid velocity at a fixed point on the axis of the cylinder as a function of time. A third photocell in line with one of the anemometer beams indicates exactly when the bubble enters into and exits from the point of the velocity measurement. The constant rise velocity of the bubble allows a transformation of the velocity versus time record into the velocity versus space representation shown. (P) Representation of the axial component of fluid velocity on the axis of a gas bubble of volume 460 mm^3 rising in a solution of polyacrylamide in glycerol at 295 K. [Reprinted by permission from O. Hassager, *Nature*, **279**, 402–403 (1979). Copyright © 1979 Macmillan Journals Limited.]

polyacrylamide in water. A droplet of the suspension was placed between two glass plates which were squeezed together to a separation of approximately 100 μm. Figure 2.6-4*a* shows the more or less random distribution of the spheres after this process. Then the top plate was moved sideways, thus creating a shear flow between the plates. Figure 2.6-4*b* shows the distribution of the spheres after a movement of about 3 cm. Then the top plate was moved back and forth a number of times, and this gave the distribution in Fig. 2.6-4*c*. Further movement resulted in the distribution with just two chains of spheres in Fig. 2.6-4*d*. Michele, Pätzold, and Donis found that this striking alignment effect is strongly dependent on polymer concentration, being even more pronounced at higher polymer concentrations. They further concluded from, say, Fig. 2.6-4*b* that long range hydrodynamic forces must be

(a)

(b)

(c)

(d)

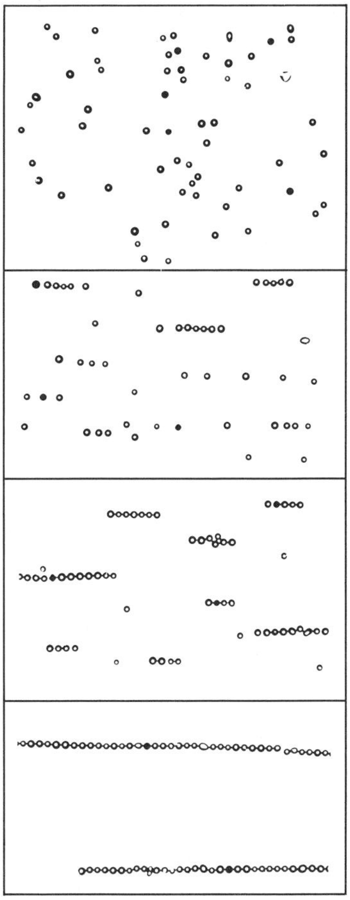

FIGURE 2.6-4. A 1.2% suspension of glass spheres (60–70 μm) in a 0.5% solution of polyacrylamide in water: (*a*) initial, random configuration; (*b*) after shearing movement of top plate of about 3 cm; (*c*) after top plate was moved back and forth several times; (*d*) and after further and faster movement of top plate. [Reproduced from J. Michele, R. Pätzold, and R. Donis, *Rheol. Acta*, **16**, 317–321 (1977).]

more important than short-range electrostatic forces between the spheres. An interpretation of the effect in terms of normal stresses has been given by Giesekus.[2]

§2.7. EFFECTS OF POLYMER ADDITIVES IN TURBULENT FLOW

Except for the discussion on vortex inhibition, we have been dealing with unusual phenomena in polymer melts and polymer solutions with concentrations roughly of the order of 0.1% and higher. Since the viscosity of such systems is usually very high, turbulence is seldom encountered. In this section we are going to see some non-Newtonian effects in solutions with polymer concentrations of the order of 10–500 ppm. With such minute amounts of polymer added to a Newtonian fluid the viscosity of the fluid is not appreciably changed when measured in laminar flow. However when the fluids undergo turbulent motion some dramatic effects are encountered, as illustrated in the following.

a. Drag Reduction

In 1948 Toms[1] discovered that if he added a very small amount of polymethylmethacrylate, approximately 10 ppm by weight, to monochlorobenzene undergoing turbulent tube flow, a substantial reduction in pressure drop at the given flow rate resulted. This reduction in pressure drop in the polymer solution relative to the pure solvent alone at the same flow rate is defined as *drag reduction.* Since then any number of polymer-solvent pairs have been found to show drag reduction. Three additional examples give an idea of the variety of possible systems: polyisobutylene in decalin, carboxymethylcellulose in water, and polyethylene oxide in water.

Drag reduction results may be presented conveniently in terms of the Fanning friction factor f:

$$f = \frac{1}{4}\left(\frac{D}{L}\right)\frac{\mathscr{P}_0 - \mathscr{P}_L}{\frac{1}{2}\rho\langle v\rangle^2} \tag{2.7-1}$$

where $\mathscr{P}_0 - \mathscr{P}_L$ is the modified pressure drop over a length L of the tube, D is the tube diameter, and $\langle v\rangle$ is the average velocity over a cross section of the tube. The friction factor is essentially a dimensionless pressure gradient, and it is a function only of the Reynolds

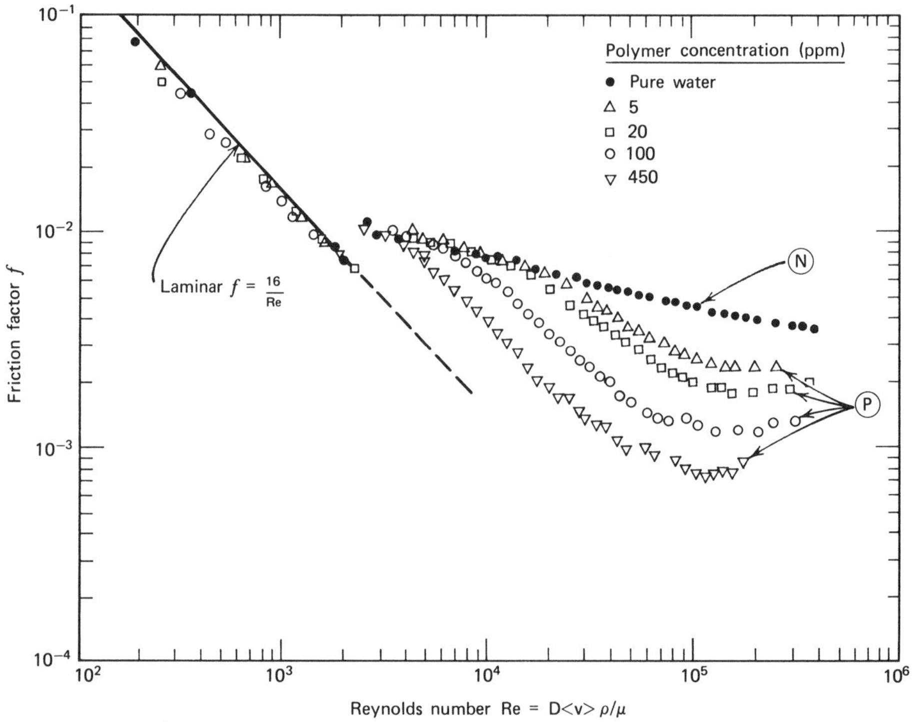

FIGURE 2.7-1. Friction factor for dilute aqueous solutions of polyethylene oxide, $\bar{M}_v = 6.1 \times 10^6$. In the turbulent regime, the curves for the polymer solutions lie below that of the solvent and illustrate drag reduction. [Data replotted from P. S. Virk, Sc.D. Thesis, Massachusetts Institute of Technology, Cambridge, MA (1966). See also C. B. Wang, *Ind. Eng. Chem. Fundam.*, **11**, 546–551 (1972).]

[1] B. A. Toms in *Proceedings of the International Congress on Rheology* (Holland, 1948), North-Holland, Amsterdam (1949), pp. II.135–II.141; *Phys. Fluids*, **20**, S3–S5 (1977). The latter reference is in a special number of the journal devoted entirely to turbulence and drag reduction. See also P. S. Virk, *AIChE J.*, **21**, 625–656 (1975); W. D. McComb and L. H. Rabie, *AIChE J.*, **28**, 547–557, 558–565 (1982). For a study of the drag-reducing properties of fish mucus, see T. L. Daniel, *Biol. Bull.*, **160**, 376–382 (1981).

number $\mathrm{Re} = D\langle v \rangle \rho/\mu$ for fully developed flow of Newtonian fluids. At the very tiny polymer concentrations of interest in drag reduction, the viscosity and density of the polymer solution differ only slightly from those of the pure solvent. Nonetheless, the effect of the polymer additive is to lower the value of the friction factor at a given Reynolds number. The amount by which the friction factor is lowered is a measure of the amount of drag reduction.[2]

Figure 2.7-1 shows the friction factor for water with and without a small amount of polyethylene oxide. It is clear from this figure that the drag reduction is strictly a turbulent flow phenomenon; it does not result from maintaining laminar flow past the usual transition region. Note also that whereas the addition of only 5 ppm of polyethylene oxide to water gives a 40% reduction in f at $\mathrm{Re} = 1.0 \times 10^5$, the viscosity of the solution is only 1.0% greater than the viscosity of the water alone.

Although the mechanism producing drag reduction is not yet known, a number of polymer characteristics making for good drag reducers have been determined. A long-chain backbone and flexibility are important characteristics of good drag-reducing agents. For instance, of two polymers with the same molecular weight and same structural units, a linear one will be more effective than a highly branched one. Also, for two different polymers of similar configuration and the same molecular weight, the one with the lower molecular weight monomer will have the greater drag-reducing effect if both are utilized at the same weight concentration.

b. Effect of Polymer Additives on Heat Transfer in Turbulent Flow

Consider a fluid flowing in a pipe of diameter D at a sufficiently high Reynolds number so that the flow is turbulent. The tube is jacketed by a heater so as to provide a constant heat flux q_0 to the fluid in the region of interest. As is customary, the heat transferred from the wall to the fluid may be expressed in terms of a local Stanton number St

$$St = \frac{q_0}{(T_w - T_b)\hat{C}_p G} = \frac{h}{\hat{C}_p G} \tag{2.7-2}$$

where $(T_w - T_b)$ represents the local difference between the wall temperature and the bulk temperature of the fluid, $\hat{C}_p$ is the fluid heat capacity per unit mass, $G = \langle \rho v \rangle$ is the mass velocity, and h is the local heat transfer coefficient.[3] On the order of 25 diameters down the tube from the entrance to the heated section, the local Stanton number becomes independent of position, and it is this measure of heat transfer that is of interest here.[4]

In Fig. 2.7-2 we present some data[5] taken on a polymer solution consisting of 50 ppm of a high molecular weight polyethylene oxide in water. Both friction factor and

[2] Provided the polymer additive has negligible effect on the viscosity and density of the solvent, then the drop in f at fixed Re will agree with the definition of the amount of drag reduction given before.

[3] For more on the various ways of presenting heat-transfer results see R. B. Bird, W. E. Stewart, and E. N. Lightfoot, *Transport Phenomena*, Wiley, New York (1960), Chapt. 13.

[4] See M. K. Gupta, A. B. Metzner, and J. P. Hartnett, *Int. J. Heat Mass Transfer*, **10**, 1211–1224 (1967).

[5] These data are given by P. M. Debrule and R. H. Sabersky, *Int. J. Heat Mass Transfer*, **17**, 529–540 (1974). Other data on this problem are given among others by K. A. Smith, G. H. Keurohlian, P. S. Virk, and E. W. Merrill, *AIChE J.*, **15**, 294–297 (1969); M. Poreh and U. Paz, *Int. J. Heat Mass Transfer*, **11**, 805–818 (1968); Y. Dimant and M. Poreh, *Adv. Heat Transfer*, **12**, 77–103 (1976); C. S. Wells, Jr., *AIChE J.*, **14**, 406–410 (1968); G. Marrucci and G. Astarita, *Ind. Eng. Chem. Fundam.*, **6**, 470–471 (1967); M. K. Gupta *et al.*, *loc. cit.*, Y. I. Cho, K. S. Ng, and J. P. Hartnett, *Lett. Heat Mass Transfer*, **7**, 347–351 (1980); E. F. Mathys and R. H. Sabersky, *Int. J. Heat Mass Transfer*, **25**, 1343–1351 (1982); and E. Y. Kwack, J. P. Hartnett, and Y. I. Cho, *Wärme- und Stoffübergang*, **16**, 35–44 (1982). Data on the analogous mass-transfer effect is given by Y. I. Cho and J. P. Hartnett, *Int. J. Heat Mass Transfer*, **24**, 945–951 (1981).

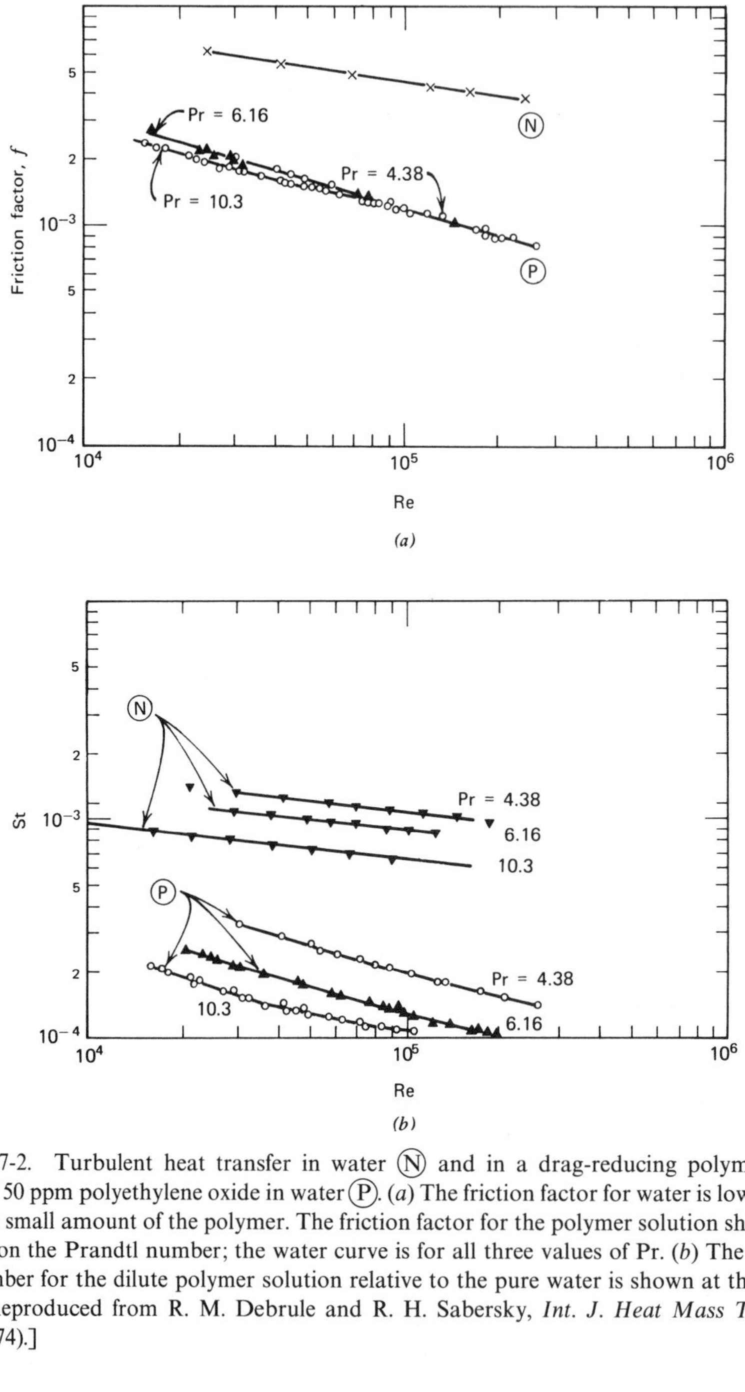

FIGURE 2.7-2. Turbulent heat transfer in water (N) and in a drag-reducing polymer solution consisting of 50 ppm polyethylene oxide in water (P). (*a*) The friction factor for water is lowered by the addition of a small amount of the polymer. The friction factor for the polymer solution shows a small dependence on the Prandtl number; the water curve is for all three values of Pr. (*b*) The decrease in Stanton number for the dilute polymer solution relative to the pure water is shown at three Prandtl numbers. [Reproduced from R. M. Debrule and R. H. Sabersky, *Int. J. Heat Mass Transfer*, **17**, 529–540 (1974).]

heat transfer results are included for comparison. The friction factor data show that drag reduction is indeed being realized. Note that varying the Prandtl number $\text{Pr} = \hat{C}_p\mu/k$, where k is the thermal conductivity of the fluid, has little effect on f, as expected. The results for the Stanton number show a corresponding decrease after the addition of 50 ppm of the polymer. Even though the magnitude of the Stanton number shows a noticeable dependence on Pr, the percent decrease in St is roughly constant at a given Reynolds number. For example, at $Re = 10^5$ the addition of 50 ppm polyethylene oxide reduces the Stanton number approximately 82% at all three Prandtl numbers shown. The decrease in St appears to be slightly more pronounced than the drop in friction factor (74% at $Re = 10^5$) for the drag-reducing system presented here. The results in Fig. 2.7-2 are for flow in a smooth tube. For a rough tube, the polymer additives produce an even larger drop in both f and St, with the Stanton number being lowered by nearly a factor of 10 by 50 ppm of polymer.

In nucleate boiling of dilute polymer solutions it has been found that polymer additives give rise to enhancement of heat transfer.[6] Photographic studies reveal differences in bubble size and bubble dynamics between pure fluids and those with polymeric additives.

c. Effect of Polymer Additives on Jet Breakup

In this final illustration we consider some high-speed photographs showing the break-up of jets.[7] In Fig. 2.7-3 we show two jets photographed 1 m from the orifice; the Newtonian jet is pure water, and the polymeric jet consists of water with 200-ppm polyethylene oxide added. Both jets are in the process of breakup, but they show a remarkable difference: around the pure water jet a number of satellite droplets appear, but

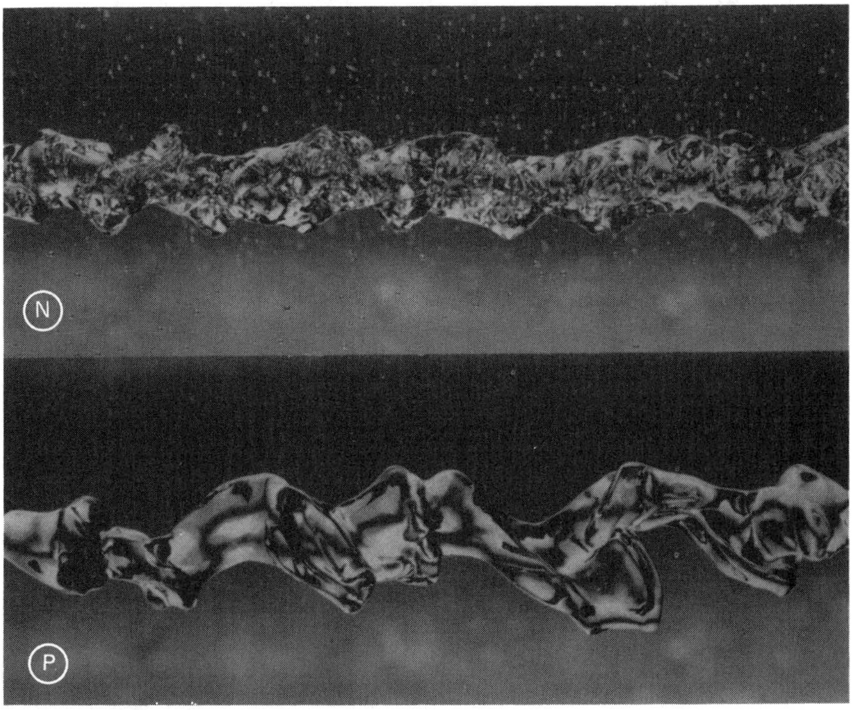

FIGURE 2.7-3. High-speed photographs of a jet 1 m from an orifice: Ⓝ pure water and Ⓟ 200 ppm solution of polyethyleneoxide in water. [Photographs courtesy of J. J. Taylor, Independent Consultant, Santa Barbara, CA, and J. W. Hoyt, Department of Mechanical Engineering, San Diego State University.]

[6] P. Kotchaphakdee and M. C. Williams, *Int. J. Heat Mass Transfer*, **13**, 835–848 (1970); H. J. Gannett, Jr., and M. C. Williams, *ibid.*, **14**, 1001–1005 (1971); D. D. Paul and S. I. Abdel-Khalik, *J. Rheol.*, **27**, 59–76 (1983).
[7] J. W. Hoyt and J. J. Taylor, *Phys. Fluids*, **20**, S253–S257 (1977).

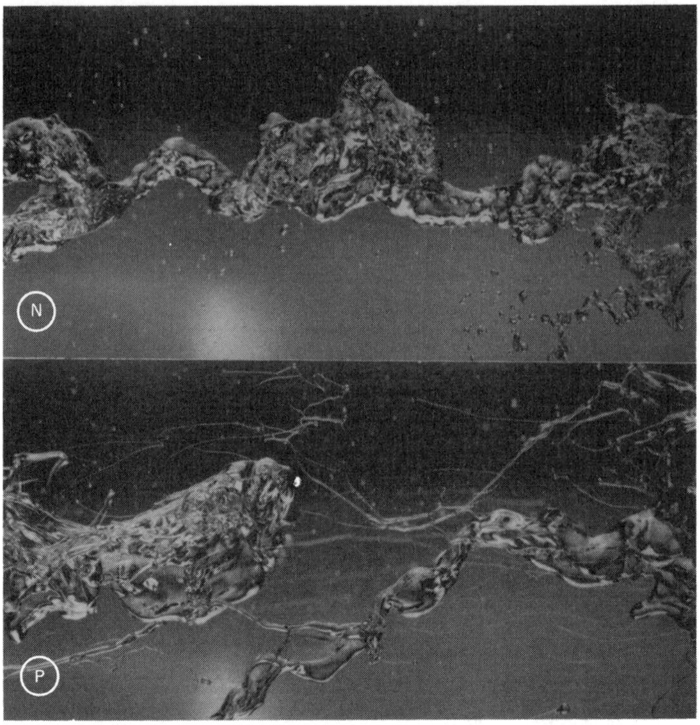

FIGURE 2.7-4. High-speed photographs taken 2 m from the orifice of the same two jets shown in Fig. 2.7-3. [Reproduced from photographs courtesy of J. J. Taylor, Independent Consultant, Santa Barbara, CA, and J. W. Hoyt, Department of Mechanical Engineering, San Diego State University.]

there are no satellite droplets near the jet with polymer additive. In Fig. 2.7-4 the same jets are shown 2 m from the orifice. Here both jets have disintegrated further. At this point some satellite droplets have also appeared in the jet with polymer additive. What is more striking, however, is that during breakup the polymer jet is pulled out into thin threads or filaments. Presumably it is the resistance of the polymer molecules to elongation that is responsible for this kind of jet breakup.

§2.8 DIMENSIONLESS GROUPS IN NON-NEWTONIAN FLUID MECHANICS

In Newtonian fluid mechanics the Reynolds number appears as a dimensionless group that may be interpreted roughly as a ratio of the magnitude of inertial forces to that of viscous forces. In any given flow situation other dimensionless numbers may arise (for example, geometric ratios), but the Reynolds number is generally the most important dimensionless group. Dimensionless groups are particularly useful for scaling arguments and also for cataloging flow regimes. For example, in tangential annular flow with the inner cylinder rotating, one can make visual observations and then determine the ranges of Reynolds numbers and radius ratios for which one has laminar flow, Taylor vortices, undulating vortices, and turbulence.

For viscoelastic fluids the key dimensionless group is the *Deborah number*, introduced by Reiner.[1] This number may be interpreted as the ratio of the magnitude of the

[1] M. Reiner, *Phys. Today*, **17**, 62 (Jan. 1964). Similar ideas (supported by experimental data on specific flow systems) were independently proposed by R. B. Bird, *Can. J. Chem. Eng.*, **43**, 161–167 (1965).

elastic forces to that of the viscous forces. It is defined as the ratio of a characteristic time (or "time scale") of the fluid, λ, to a characteristic time (or "time scale") of the flow system, t_{flow}

$$\text{De} = \lambda / t_{\text{flow}} \tag{2.8-1}$$

The characteristic time of the fluid is often taken to be the largest time constant describing the slowest molecular motions, or else some mean time constant determined by linear viscoelasticity (see Table 5.3-1); sometimes the characteristic time is chosen to be a time constant in a constitutive equation. The characteristic time for the flow is usually taken to be the time interval during which a typical fluid element experiences a significant sequence of kinematic events; sometimes it is taken to be the duration of an experiment or experimental observation. If the flow following a material particle is steady, then the characteristic time is taken to be the reciprocal of a characteristic strain rate. In Fig. 2.8-1 we show four steady-state flows and suggest useful definitions of the Deborah number.

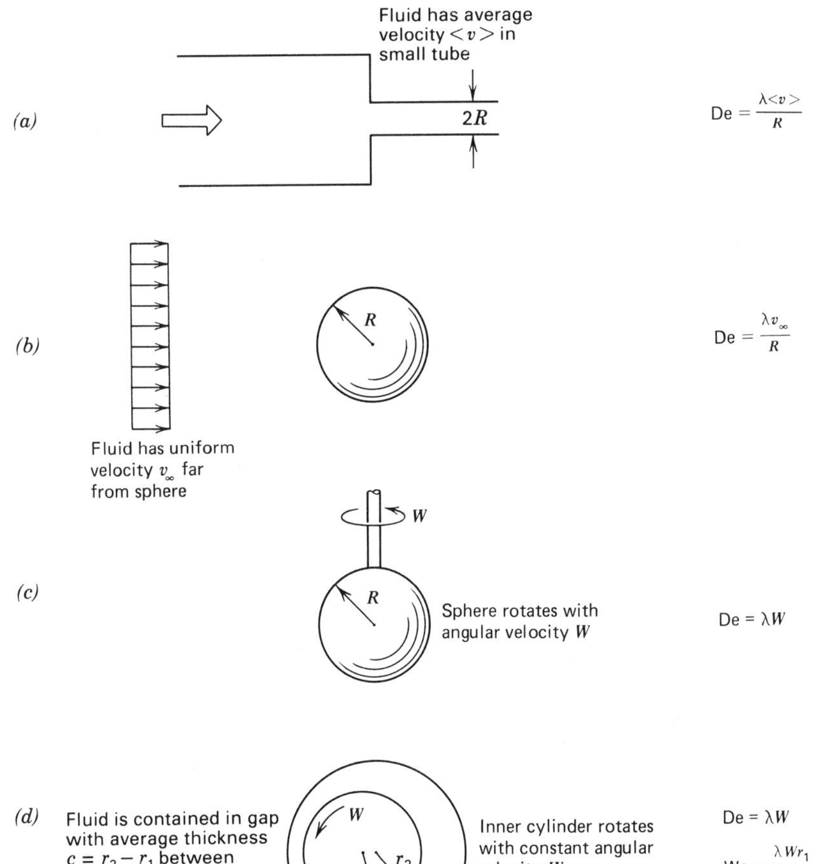

FIGURE 2.8-1. Four flow systems for illustrating the definition of De (Eq. 2.8-1) and We (Eq. 2.8-2): (*a*) flow through a sudden contraction; (*b*) flow around a sphere; (*c*) flow near a rotating sphere; and (*d*) flow between eccentric cylinders (journal bearing). Since two time scales for the flow can be identified for the eccentric cylinder flow, we use the characteristic shear rate to construct We, and the time required to execute one revolution to construct De.

For some problems there is more than one characteristic time for the flow that can be identified. For example in the eccentric cylinder flow in Fig. 2.8-1, one time scale is W^{-1} which gives the order of magnitude of the time required to move once around the cylinders; a second time scale for this problem is c/Wr_1 which is the reciprocal of the mean shear rate between the cylinders. A second dimensionless group, the *Weissenberg number*, that is sometimes used in polymer fluid dynamics involves a ratio of λ to this second characteristic time. The Weissenberg number We is defined by:

$$\mathrm{We} = \lambda\kappa \tag{2.8-2}$$

where κ is a characteristic strain rate in the flow. For many problems, however, there is only one characteristic time that can be identified, and for these problems we choose to use the Deborah number as the dimensionless group.

Two limiting values of the Deborah number can be identified with classical mechanics. If the Deborah number is small, then thermal motions keep the polymer molecules more or less in their equilibrium configurations, and the polymeric fluid shows only minor qualitative differences from a Newtonian fluid. We say that Newtonian fluid behavior is obtained in the limit $\mathrm{De} \to 0$. Conversely, if the Deborah number is large, polymer molecules that are distorted by the flow will not have time to relax during the time scale of the process or experiment. In the limit $\mathrm{De} \to \infty$ the experiment happens so fast that the polymer molecules have no time to change configuration, and the fluid behaves more or less as a Hookean elastic solid. For many polymeric liquids, λ lies between 10^{-3} s for dilute solutions and 10^3 s for concentrated solutions and melts. Since the characteristic times for flow systems are often in the same range, a wide spectrum of De values is easily obtainable. In the figures given earlier in this chapter the labeling Ⓝ and Ⓟ could just as well be replaced by the statements $\mathrm{De} < \mathrm{De}_{\mathrm{crit}}$ and $\mathrm{De} > \mathrm{De}_{\mathrm{crit}}$, where the actual value of the

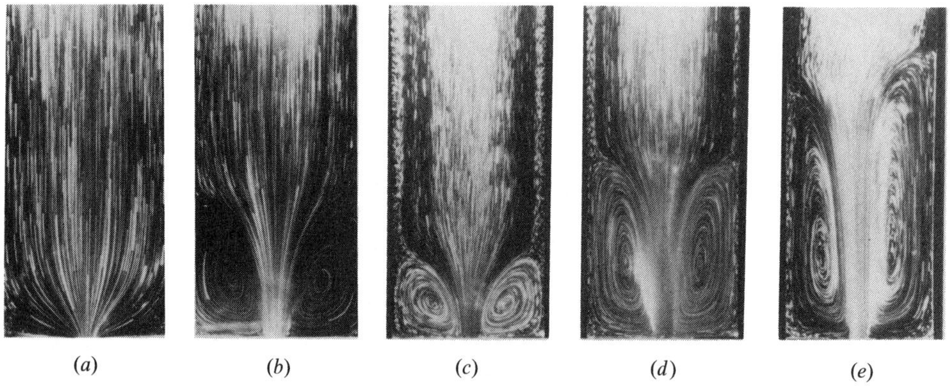

FIGURE 2.8-2. Streak photographs showing the streamlines for the flow downward through an axisymmetric sudden contraction with contraction ratio 7.675 to 1 as a function of De. (*a*) De = 0 for a Newtonian glucose syrup. (*b–e*) De = 0.2, 1, 3, and 8 respectively for a 0.057% polyacrylamide glucose solution. From De = 1 to 3, the non-Newtonian vortex grows, but the flow remains two-dimensional and axisymmetric. For De > 3, the vortex becomes asymmetric and a swirling component to the flow is observed. At Deborah numbers higher than 8, the flow becomes very erratic. The Deborah number is defined by $\lambda\langle v\rangle/R$ where λ is estimated as the longest relaxation time for the polyacrylamide solution and R is the small tube radius. [Reproduced from D. V. Boger and H. Nguyen, *Polym. Eng. Sci.* **18**, 1038–1043 (1978).]

critical Deborah number depends on the specific choices for λ and t_{flow}. For many systems De_{crit} has been found to be about unity.

Notice from the above discussion that there is in fact no such thing as a non-Newtonian fluid in and of itself. The value of the Deborah number which describes the extent of the non-Newtonian behavior depends on both λ and t_{flow}. It is in fact precisely this double dependence on the respective times of the fluid and the flow that motivated Reiner to name the ratio after the prophetess Deborah[1,2] who said: "The mountains flowed before the Lord." Indeed the time constants for mountain flow are so large that within the time span of human life they will behave as solids. However, the Lord has an infinite time available for observation and in this time span the mountains will flow as Newtonian fluids with De = 0.

We can illustrate the significance of De in two simple experiments with "silly putty." This unvulcanized silicone rubber has a time constant of the order of seconds, so that it behaves as a solid in experiments of duration much less than one second, and as a fluid in experiments lasting many minutes. For example, if we roll silly putty into a ball and bounce it on a hard surface, it will bounce like a solid rubber ball. However, if the silly putty is left in a container for an hour, it will flow so as to fill the container the way fluids do.

As an illustration of the classification of flow regimes in a complex flow by using De, we direct the reader's attention to Fig. 2.8-2. Here streak photographs by Boger and Nguyen[3] are shown for flow of a 0.057% polyacrylamide glucose solution through a sudden contraction for a series of different Deborah numbers. Since the fluid is fixed in these experiments, the time constant λ is also fixed. The Deborah number ($\lambda\langle v\rangle/R$; cf. Fig. 2.8-1) is thus varied by increasing the volume flow rate through the contraction. Note that when De is of order unity, the first very large deviations from Newtonian behavior are observed.

In a given flow problem we may define a Reynolds number in addition to the Deborah number and possibly the Weissenberg number. Then any combination of these numbers, such as their product or their ratio, is also a dimensionless number, and several such have in fact been named.[4] We make no use of these other combinations in this text.

PROBLEMS

2A.1 Calculation of Molecular Weight Averages

A polymer blend is being prepared by mixing 80 kg of one polymer sample A (number average molecular weight $\bar{M}_{n,A} = 10{,}000$ g/mol and weight average molecular weight $\bar{M}_{w,A} = 15{,}000$ g/mol) with 20 kg of another sample B ($\bar{M}_{n,B} = 20{,}000$ g/mol and $\bar{M}_{w,B} = 50{,}000$ g/mol). It is desired to

[2] From Deborah's Song, Judges, 5:5.

[3] D. V. Boger and H. Nguyen, *Polym. Eng. Sci.*, **18**, 1038–1043 (1978). More recent experiments by S. J. Muller, Sc.D. Thesis, Massachusetts Institute of Technology, Cambridge (1986) have shown that the flow transitions with increasing De are even more complex than those shown in Fig. 2.8-2. Prior to the development of the large vortex, as De is increased from small values a transition to a three-dimensional, time-periodic flow is found at De approximately equal to 0.6. The fluid is observed to undergo torsional oscillations as it moves through the contraction. For De approximately equal to unity, the flow again becomes steady state and two-dimensional. Once the flow has returned to steady state, the non-Newtonian vortex is observed. See also, J. V. Lawler, S. J. Muller, R. A. Brown, and R. C. Armstrong, *J. Non-Newtonian Fluid Mech.*, **20**, 51–93 (1986).

[4] For more on dimensionless groups, see G. Astarita and G. Marrucci, *Principles of Non-Newtonian Fluid Mechanics*, McGraw-Hill, New York (1974); J. R. A. Pearson, *Mechanics of Polymer Processing*, Elsevier, London (1985), Chapter 6; R. I. Tanner, *Engineering Rheology*, Oxford University Press (1985), pp. 190–192.

calculate the number average and weight average molecular weights of the mixture, $\bar{M}_{n,\text{mix}}$ and $\bar{M}_{w,\text{mix}}$, respectively.

Answer: $\bar{M}_{n,\text{mix}} = 1.11 \times 10^4$ g/mol

$\bar{M}_{w,\text{mix}} = 2.20 \times 10^4$ g/mol

2A.2 Comparison of Viscosity Determinations

The characteristics of two fluids, *A* and *B*, have been investigated by (1) measuring the terminal velocity of a sphere falling in a large bath of each fluid and by (2) measuring the volume flow rate of each at a specified pressure gradient. In (1) a 0.5 cm diameter sphere weighing 0.104 g fell with a terminal velocity of 0.75 cm/s in both fluids. However, in (2), fluids *A* and *B* had volumetric flow rates of 3.061×10^{-2} and 0.3455 cm^3/s, respectively, at an imposed pressure gradient of 6.857×10^4 Pa/m in a 0.210 cm diameter tube. The density of *A* is 1.0 g/cm^3, and the density of *B* is 1.014 g/cm^3. Based on these results which of the two fluids is definitely *not* Newtonian?

2B.1 The Heterogeneity Index

a. From the definitions, Eqs. 2.1-1 and 2 show that:

$$\frac{\bar{M}_w}{\bar{M}_n} - 1 = \frac{\sum_i \sum_j N_i N_j (M_i - M_j)^2}{2(\sum_i N_i M_i)^2} \tag{2B.1-1}$$

b. What does Eq. 2B.1-1 imply about the relative sizes of $\bar{M}_n$ and $\bar{M}_w$?

2B.2 Interpretation of the First Normal Stress Difference in Tangential Annular Flow[1]

a. In the limit that inertial forces are negligible, show that the equation of motion for tangential annular flow leads to:

$$\frac{d}{dr}(p + \tau_{rr}) = \frac{\tau_{\theta\theta} - \tau_{rr}}{r} \tag{2B.2-1}$$

b. Now consider the static forces that are present when a thin string wrapped around a cylinder of radius *R* is under tension *T* (see Fig. 2B.2). By making a force balance on a very short isolated segment of string, show that the reaction of the cylinder is a radial force of magnitude F_r, given by

$$\left(\frac{F_r}{l}\right) = \frac{T}{R} \qquad \left(\frac{l}{R} \ll 1\right) \tag{2B.2-2}$$

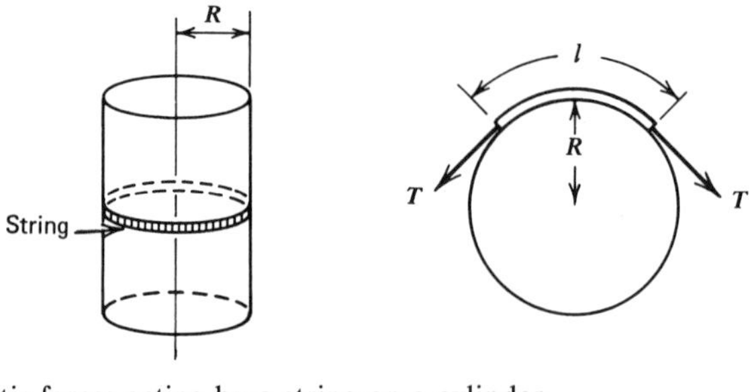

FIGURE 2B.2. Static forces acting by a string on a cylinder.

[1] This problem was suggested by A. S. Lodge, *Elastic Liquids*, Academic Press, New York (1964), p. 191.

c. Use the results from (**a**) and (**b**) to interpret physically the normal stresses in tangential annular flow. How does this relate to the rod-climbing experiment?

2B.3 Simplified Results for Hole Pressure

For many polymeric liquids the normal stress differences can be expressed as powers of the shear stress:[2]

$$\tau_{11} - \tau_{22} = -\kappa_1 |\tau_{21}|^{n_1} \tag{2B.3-1}$$

$$\tau_{22} - \tau_{33} = \kappa_2 |\tau_{21}|^{n_2} \tag{2B.3-2}$$

in which κ_1, κ_2, n_1, and n_2 are all constants. Show that for fluids in which these relations hold, the hole pressure results, Eqs. 2.3-15 to 17, simplify to

Circular hole:
$$p^* = -\frac{1}{3}\left(\frac{\tau_{11} - \tau_{22}}{n_1} - \frac{\tau_{22} - \tau_{33}}{n_2} \cdot\right)_w \tag{2B.3-3}$$

Transverse slot:
$$p^* = -\frac{1}{2n_1}(\tau_{11} - \tau_{22})_w \tag{2B.3-4}$$

Axial slot:
$$p^* = +\frac{1}{n_2}(\tau_{22} - \tau_{33})_w \tag{2B.3-5}$$

where the subscript w means "evaluated at the wall." One fluid model found in the literature, the second-order fluid, gives a value of 2 for both n_1 and n_2. The results above agree with derivations of the hole pressure made with the second-order fluid model assumed from the beginning[3] (Problem 6B.15).

2B.4 Free Surface Shape in a Tilted Trough[4]

It is desired to obtain an expression for the free surface shape of a fluid flowing down a tilted trough of semicircular cross section that is inclined at an angle β to the horizontal. To do this, parallel the development in Example 2.3-2.

a. First, consider steady flow in a tube of radius R inclined at an angle β. From the symmetry of the velocity field it can be shown that $\tau_{z\theta} = \tau_{r\theta} = 0$ for this flow (see §3.2). Combine this information with the equations of motion to show that the variation of $\pi_{\theta\theta}$ within the tube is given by

$$d\pi_{\theta\theta} = -\rho g \cos\beta \cos\theta \, r d\theta - \left(\frac{d}{dr}N_2 + \frac{N_2}{r} + \rho g \cos\beta \sin\theta\right) dr + \left(\frac{\partial p}{\partial z}\right) dz \tag{2B.4-1}$$

where $N_2 = \tau_{rr} - \tau_{\theta\theta}$ is the second normal stress difference and we have taken $\theta = 0$ to be horizontal.

b. Now imagine that the upper half of the tube and fluid is removed so that we are left with flow down a semicircular trough. The velocity profile in the trough will be identical to that in the lower half of the tube provided the normal stress distribution given in Eq. 2B.4-1 is maintained on the surface $\theta = 0$. Explain.

[2] K. Higashitani and W. G. Pritchard, *Trans. Soc. Rheol.*, **16**, 687–696 (1972).
[3] R. I. Tanner and A. C. Pipkin, *Trans. Soc. Rheol.*, **13**, 471–484 (1969), treat the transverse slot problem and E. A. Kearsley, *Trans. Soc. Rheol.*, **14**, 419–424 (1970) analyzes flow along an axial slot.
[4] A. C. Pipkin and R. I. Tanner, *Mech. Today*, **1**, 262–321 (1972); R. I. Tanner, *Trans. Soc. Rheol.*, 483–507 (1970). L. Sturges and D. D. Joseph, *Arch. Rat. Mech. Anal.*, **59**, 359–387 (1976), include surface tension in their analysis and also consider the onset of secondary flow with increasing β.

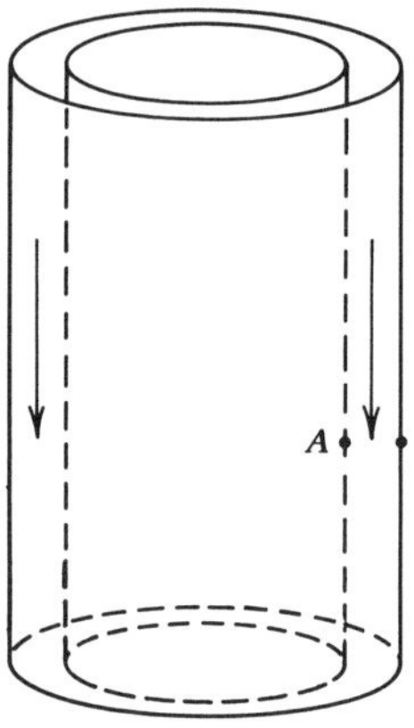

FIGURE 2B.5. Pressure measurements in axial annular flow. For a Newtonian fluid $p_A = p_B$, whereas for a polymeric fluid $\pi_{rr,A} > \pi_{rr,B}$.

c. In order to describe the free surface it is convenient to consider a Cartesian coordinate system oriented similarly to the one in Fig. 2.3-3: the z-axis is coincident with the axis of the trough, x lies in the horizontal plane, and y is positive in the upward direction. Show that the normal stress π_{yy} is given by

$$\pi_{yy}(x, y) = -\rho g y \cos\beta + \pi_{yy}(x, 0) + C$$

$$\pi_{yy}(x, 0) = -\int_0^x \frac{N_2}{x}\,dx - N_2(x) \tag{2B.4-2}$$

where C is an integration constant.

d. Show that the total bulge in the free surface of the fluid is

$$h = -\frac{\pi_{yy}(R, 0)}{\rho g \cos\beta} \tag{2B.4-3}$$

2B.5 Axial Annular Flow[5]

A fluid is pumped axially through the annular space between two concentric cylinders; the Reynolds number is small enough so that the flow is laminar. Small, flush-mounted pressure transducers (labeled A and B in Fig. 2B.5) are located on the cylinder walls opposite one another. For a Newtonian fluid, the pressure readings at A and B are found to be the same, so that $p_A = p_B$. For a viscoelastic fluid, it is found that $\pi_{rr,A}$ is slightly larger than $\pi_{rr,B}$. Use the r-component of the equation of motion to show that this difference in behavior between the Newtonian and polymeric fluids is consistent with a small, positive second normal stress difference. Assume that the two transducers are located at the same vertical position.

[5] J. W. Hayes and R. I. Tanner in E. H. Lee, ed., *Proceedings of the Fourth International Congress on Rheology*, Wiley-Interscience, New York (1965), Part 3, pp. 389–399; J. D. Huppler, *Trans. Soc. Rheol.*, **9:2**, 273–286 (1965); R. I. Tanner, *Trans. Soc. Rheol.*, **11**, 347–360 (1967). These measurements were made with hole-mounted transducers and actually showed $\pi_{rr,A} < \pi_{rr,B}$. When correction is made for the hole pressure, it is found that $\pi_{rr,A} > \pi_{rr,B}$ [A. C. Pipkin and R. I. Tanner, *Mech. Today*, **1**, 262–321 (1972)].

CHAPTER 3

MATERIAL FUNCTIONS FOR POLYMERIC LIQUIDS

Chapter 2 has demonstrated rather dramatically that Newton's law of viscosity is wholly inadequate for the description of macromolecular liquids. In this chapter we begin the task of discovering the proper way to describe these non-Newtonian fluids. The first step is the quantitative experimental characterization of their flow behavior. Recall that incompressible Newtonian fluids at constant temperature can be characterized by just two *material constants*: the density ρ and the viscosity μ. Once these quantities have been measured, the governing equations for the velocity and stress distributions in the fluid are fixed for any flow system. There are many steady-state and unsteady-state experiments from which μ can be determined.[1]

The experimental description of incompressible non-Newtonian fluids, on the other hand, is much more complicated. We can, of course, measure the density, but since we have no equation for $\boldsymbol{\tau}$ analogous to Eq. 1.2-2, we do not know what other property or properties need to be measured. From the "fun experiments" that were presented in the last chapter, we know a little about what to expect. For example, if we try to measure a "viscosity" using a viscometer, this "viscosity" will not necessarily be a constant. Furthermore, we know from analyses of rod climbing and the tilted trough that normal stresses should also be measured. The recoil experiment in §2.5 indicates that unsteady-state experiments might also lead to additional information.

Whereas different isothermal experiments on a Newtonian fluid yield a single *material constant*, namely the viscosity, a variety of experiments performed on a polymeric liquid will yield a host of *material functions* that depend on shear rate, frequency, time, and so on. These material functions serve to classify fluids, and they can be used to determine constants in specific non-Newtonian constitutive equations. In this chapter, then, we shall present the standard flow patterns used in characterizing polymeric liquids and discuss the material functions that can be obtained from each. Representative fluid behavior will also be shown by means of sample experimental data. Once the reader understands the kind of fluid responses characteristic of polymeric materials in various experiments, the constitutive equations developed and used in the rest of the book will be more meaningful.

We begin this chapter with a description of the two standard kinds of flows, shear and shearfree, that are used to characterize polymeric liquids; the velocity and stress fields for these flows are described in §§3.1 and 3.2. Then in §§3.3 and 3.4 and in §3.5 we give the most common material functions that arise in these two types of flows, respectively. In §3.6

[1] A rather complete listing of standard techniques is given in J. R. Van Wazer, J. W. Lyons, K. Y. Kim, and R. E. Colwell, *Viscosity and Flow Measurement*, Wiley, New York (1963).

we present some of the most important experimental findings about the concentation, temperature, and molecular weight dependence of the material functions. Finally in §3.7 we return to the subject of kinematics and give a more detailed discussion of the identification of shear and shearfree flows.[2]

Whereas this chapter deals with the *description* of the nature and diversity of material response to simple shearing and shearfree flows, the analysis of *experimental methods* for measuring these quantities is not dealt with until Chapter 10. This discussion is deferred because constitutive equations are either helpful or necessary in the analyses.

§3.1 SHEAR AND SHEARFREE FLOWS

The two types of flows most often used to characterize polymeric liquids are *shear* flow and *shearfree* flow. We give examples of each and show that the relative motion of material particles is very different in these two types of flow. It is thus not surprising that the material information learned from the two types of flow is quite different. In this section we restrict our attention to *homogeneous* (or uniform) deformations[1] in which the velocity gradient is independent of position; in §3.7 nonhomogeneous flows are discussed.

a. Shear Flow

A *simple shear flow* is given by the velocity field

$$v_x = \dot{\gamma}_{yx} y; \quad v_y = 0; \quad v_z = 0 \tag{3.1-1}$$

in which the velocity gradient $\dot{\gamma}_{yx}$ can be a function of time. The absolute value of $\dot{\gamma}_{yx}$ is called the *shear rate*[2] $\dot{\gamma}$. For steady shear flow (sometimes called a *viscometric*[3] flow) the shear rate is independent of time; it is presumed that the shear rate has been constant for such a long time that all the stresses in the fluid are time-independent. A simple shear flow is easily generated between parallel plates as illustrated in Fig. 3.1-1. A characteristic of steady shear flows is that the distance l between two neighboring fluid particles, which are initially on the y-axis and separated by a distance l_0, is

$$l = l_0\sqrt{1 + (\dot{\gamma}\Delta t)^2} \sim l_0 \dot{\gamma} \Delta t \qquad (\dot{\gamma} = \text{constant}) \tag{3.1-2}$$

after a time interval Δt; the second (approximate) expression is applicable when $\Delta t \to \infty$. The particles are seen to separate linearly in Δt for large Δt. A second distinguishing feature

[2] This section can be omitted on first reading.

[1] An extensive treatment of homogeneous deformation and flow is found in A. S. Lodge, *Elastic Liquids*, Academic Press, New York (1964); our description of kinematics has been strongly influenced by this book.

[2] The definition of shear rate $\dot{\gamma}$ given here is a special case of a more general formula that is given in Eqs. 3.7-3 and 4.1-8 and used in more complex shearing flows later in the book. The general formula gives $\dot{\gamma}$ as the magnitude of the rate-of-strain tensor (cf. Eq. A.3-21):

$$\dot{\gamma} = +\sqrt{(1/2)(\dot{\boldsymbol{\gamma}}:\dot{\boldsymbol{\gamma}})} \tag{3.1-1a}$$

so that $\dot{\gamma}$ is always a positive quantity.

[3] Viscometric flow is not restricted to $\dot{\gamma}_{yx}$ independent of position; see §3.7.

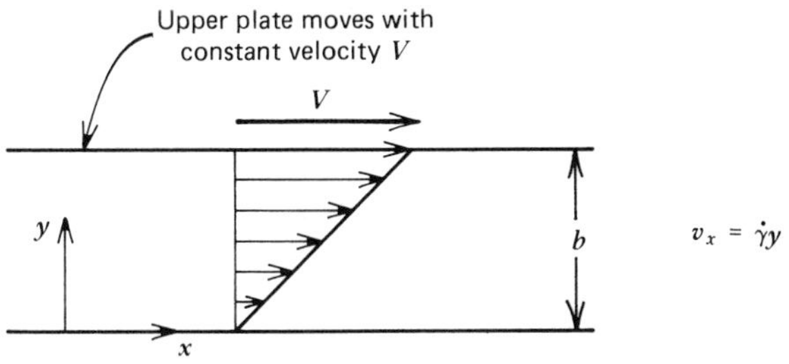

FIGURE 3.1-1. Steady simple shear flow with shear rate $\dot{\gamma} = V/b$. This flow is characterized by a one-parameter family of material surfaces, namely y = constant, which slide relative to one another.

of shear flows is that it is possible to identify a one-parameter family of material surfaces that slide relative to one another without stretching. For example, the flow in Fig. 3.1-1 is like the shearing of a deck of cards, with the family of material surfaces corresponding to the cards. Shearing flows are found in many polymer processing operations, for example, in injection molding and extrusion, and in many rheometer flows. In addition polymer melts and solutions are subjected to shear flow in transportation through ducts and in lubrication applications.

b. Shearfree Flow

Simple shearfree flows are given by the velocity field

$$\begin{aligned} v_x &= -\frac{1}{2}\dot{\varepsilon}(1+b)x \\ v_y &= -\frac{1}{2}\dot{\varepsilon}(1-b)y \\ v_z &= +\dot{\varepsilon}z \end{aligned} \tag{3.1-3}$$

where $0 \leq b \leq 1$ and $\dot{\varepsilon}$ is the *elongation rate*[4] which can depend on time. Several special shearfree flows are obtained for particular choices of the parameter b:

Elongational flow:	$(b = 0, \dot{\varepsilon} > 0)$
Biaxial stretching flow:	$(b = 0, \dot{\varepsilon} < 0)$
Planar elongational flow:	$(b = 1)$

The streamlines for elongational flow ($b = 0$) are illustrated in Fig. 3.1-2; the choice of b will affect the way the streamlines change with rotation about the z-axis. The effects of the three kinds of shearfree flow on a cube of material are illustrated for steady state in Fig. 3.1-3, where these deformations are compared with steady shearing. By steady shearfree flow we mean that $\dot{\varepsilon}$ is independent of time; it is presumed that the elongation rate has been constant for such a long time that all the stresses in the fluid are time-independent.

[4] Note that the magnitude of the rate-of-strain tensor $\dot{\gamma}$ given in Eq. 3.1-1a is equal to $\sqrt{3+b^2}|\dot{\varepsilon}|$ for simple shearfree flow.

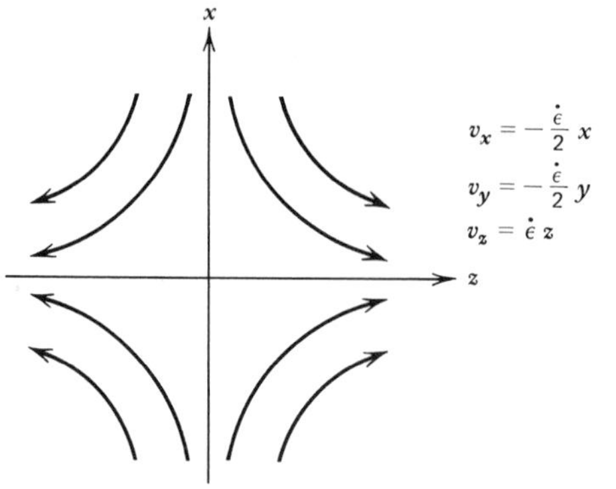

FIGURE 3.1-2. Steady elongational flow (shearfree flow with $b = 0$).

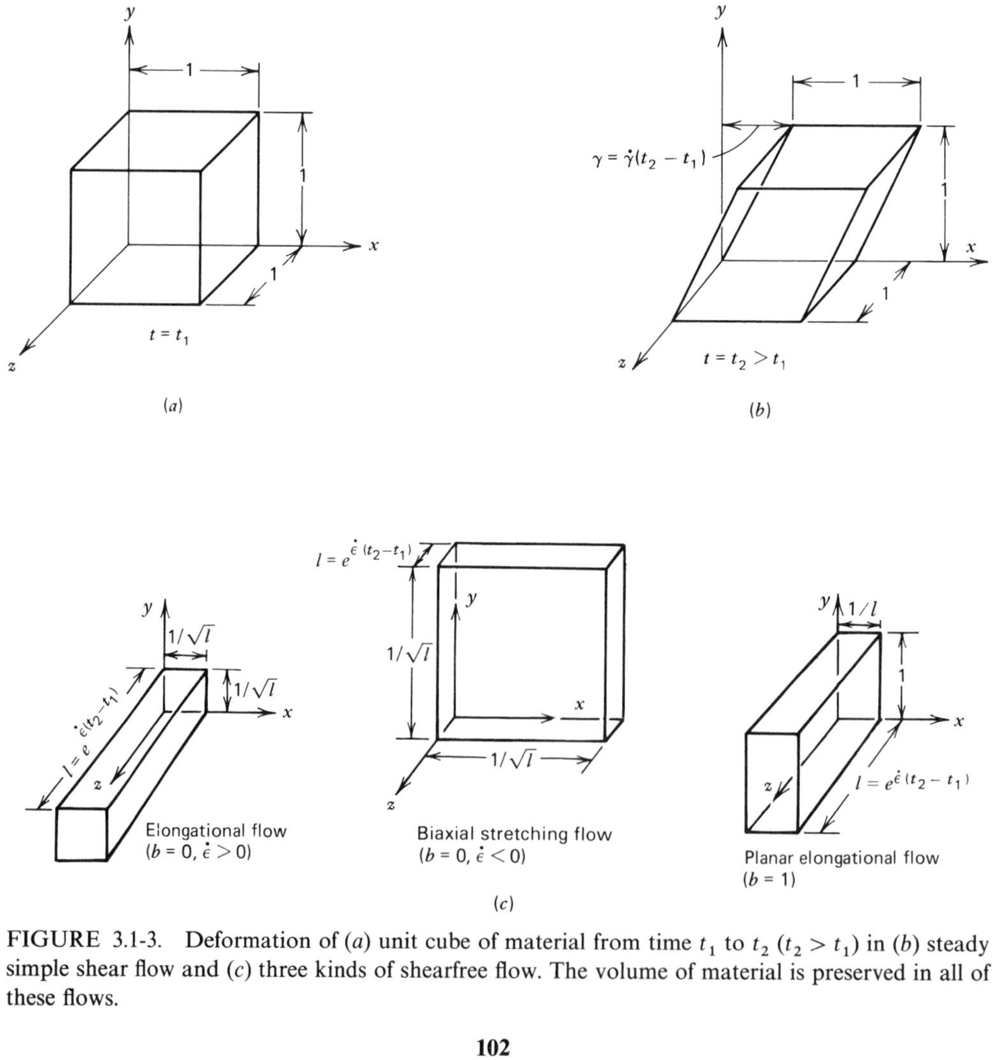

FIGURE 3.1-3. Deformation of (*a*) unit cube of material from time t_1 to t_2 ($t_2 > t_1$) in (*b*) steady simple shear flow and (*c*) three kinds of shearfree flow. The volume of material is preserved in all of these flows.

A distinguishing feature of steady shearfree flows that can be seen from Eq. 3.1-3 is that neighboring fluid elements move relative to one another at an exponential rate. For example two fluid particles on the z-axis which are initially separated by a distance l_0 will be separated by distance l after time Δt:

$$l = l_0 e^{\dot{\varepsilon}\Delta t} \qquad (\dot{\varepsilon} = \text{constant}) \tag{3.1-4}$$

This exponential rate of separation of fluid elements is much more rapid than the linear rate of separation in steady shear flow.

Shearfree flows are found in many polymer processing operations, for example, fiber spinning, film blowing, vacuum thermoforming, and sheet stretching; polymer foaming involves shearfree flow around bubbles. In addition many other flows, such as flows near stagnation points, flow through contractions, and flow around corners have a strong elongational character. Because of the shearing that must occur near the solid boundaries in these latter examples, they are not purely shearfree flows, however.

§3.2 THE STRESS TENSOR FOR SHEAR AND SHEARFREE FLOWS

Material functions relate the kinematics of a flow to the stress field needed to sustain the motion, so we must describe the stresses required for simple shear or shearfree motions. In this section we indicate how the stress tensor can be simplified for these two types of flows, and in addition we deduce which combinations of stress tensor components are experimentally accessible.

a. Shear Flows

For Newtonian liquids we know that in the shear flow of Eq. 3.1-1 only the shear stress τ_{yx} is nonzero. For non-Newtonian fluids we must assume, in the absence of a constitutive equation, that in any flow all six independent components of the stress tensor may be nonzero.[1] However, for simple shearing flows of incompressible liquids, it is possible to show that at most three independent combinations of components of the stress tensor can be measured (see Problem 3B.3). This conclusion is based on the invariance of the shear flow in Eq. 3.1-1 with respect to a 180° rotation about the z-axis and the assumptions that the fluid is isotropic, so that it has no preferred direction other than one introduced by the flow itself, and that the stress in the fluid depends only on the flow field. The most general form that the total stress tensor can have for a simple shearing flow is

$$\boldsymbol{\pi} = p\boldsymbol{\delta} + \boldsymbol{\tau} = \begin{pmatrix} p + \tau_{xx} & \tau_{yx} & 0 \\ \tau_{yx} & p + \tau_{yy} & 0 \\ 0 & 0 & p + \tau_{zz} \end{pmatrix} \tag{3.2-1}$$

[1] We will always take the stress tensor to be symmetrical. The theoretical possibility of a nonsymmetrical stress tensor is discussed by J. S. Dahler and L. E. Scriven, *Nature*, **192**, 36–37 (1961). To date no experiments have shown asymmetry in the stress tensor for amorphous liquids. Furthermore, almost all of the molecular theories for amorphous liquids give expressions for the stress tensor that are symmetrical (see §13.3 and other discussions in Volume 2).

However for incompressible fluids we cannot separate the pressure and normal stress contributions in normal force measurements on surfaces (see below Eq. 1.2-2). Hence, the only quantities of experimental interest are the shear stress and two normal stress differences. The stresses that are customarily used in conjunction with shear flow are:

Shear stress:	τ_{yx}	
First normal stress difference:	$\tau_{xx} - \tau_{yy}$	(3.2-2)[2]
Second normal stress difference:	$\tau_{yy} - \tau_{zz}$	

Hence there are only three independent, experimentally accessible quantities in simple shear flow.

b. Shearfree Flows

The shearfree flow given in Eq. 3.1-3 shows even more symmetry than does simple shear flow. From Fig. 3.1-3 it is clear that the description of the deformation is unaffected by a 180° rotation about the x-, y-, or z-axis. This symmetry coupled with the assumed fluid isotropy reduces the most general form of the stress tensor for shearfree flow to

$$\boldsymbol{\pi} = p\boldsymbol{\delta} + \boldsymbol{\tau} = \begin{pmatrix} p + \tau_{xx} & 0 & 0 \\ 0 & p + \tau_{yy} & 0 \\ 0 & 0 & p + \tau_{zz} \end{pmatrix} \tag{3.2-3}$$

When we restrict attention to incompressible fluids, there are only two normal stress differences of experimental interest:

$$\begin{gathered} \tau_{zz} - \tau_{xx} \\ \tau_{yy} - \tau_{xx} \end{gathered} \tag{3.2-4}$$

For the elongational and biaxial stretching flows for which $b = 0$ in Eq. 3.1-3, the x- and y-directions are indistinguishable so that $\tau_{xx} - \tau_{yy} = 0$ and there is only one normal stress difference to be determined for these flows.

We turn now in the next three sections to defining and describing material functions that give the three stress combinations in shear flow and the two normal stress differences in shearfree flow. A long list of material functions has been defined for these two types of flow fields, corresponding to the large variety of time-dependent $\dot{\gamma}_{yx}$ and $\dot{\varepsilon}$ that can be produced experimentally.

§3.3 STEADY SHEAR FLOW MATERIAL FUNCTIONS

As a consequence of the assumption made above that the stress tensor depends only on the flow field, we can say that the steady-state stresses in steady simple shear flow are functions only of the shear rate $\dot{\gamma}$. The *viscosity* η (also called the *non-Newtonian viscosity* or

[2] Some rheologists use N_1 to denote $-(\tau_{xx} - \tau_{yy})$ and N_2 to denote $-(\tau_{yy} - \tau_{zz})$.

shear-rate dependent viscosity) is defined analogously to the viscosity for Newtonian fluids (cf. Eq. 1.2-1):

$$\boxed{\tau_{yx} = -\eta(\dot{\gamma})\dot{\gamma}_{yx}} \tag{3.3-1}$$

Likewise, we can define normal stress coefficients Ψ_1 and Ψ_2 as follows:

$$\tau_{xx} - \tau_{yy} = -\Psi_1(\dot{\gamma})\dot{\gamma}_{yx}^2 \tag{3.3-2}$$

$$\tau_{yy} - \tau_{zz} = -\Psi_2(\dot{\gamma})\dot{\gamma}_{yx}^2 \tag{3.3-3}$$

The functions Ψ_1 and Ψ_2 are known as the *first* and *second normal stress coefficients*, respectively; η, Ψ_1, and Ψ_2 are sometimes collectively referred to as the *viscometric functions.* In Eq. 3.3-1 changing the sign of $\dot{\gamma}_{yx}$ will change the sign of the shear stress and therefore η must be an even function of $\dot{\gamma}_{yx}$. Similarly, in Eqs. 3.3-2 and 3 the normal stress differences will not change sign if $\dot{\gamma}_{yx}$ changes sign, so that Ψ_1 and Ψ_2 are also even functions of $\dot{\gamma}_{yx}$. This explains why η, Ψ_1, and Ψ_2 are written as function of $\dot{\gamma} = |\dot{\gamma}_{yx}|$ rather than $\dot{\gamma}_{yx}$. Sometimes it is more convenient to regard the viscometric functions as functions of the absolute value of the shear stress $\tau = |\tau_{yx}|$.

Experimentally, the viscosity is the best known of the viscometric functions. Some typical plots of $\eta(\dot{\gamma})$ are given in Figs. 3.3-1 to 4 for a polymer melt, concentrated solutions, and dilute solutions. At low shear rates, the shear stress is proportional to $\dot{\gamma}$, and the viscosity approaches a constant value η_0, the *zero-shear-rate viscosity.* At higher shear rates the viscosity of most polymeric liquids decreases with increasing shear rate.[1] For many

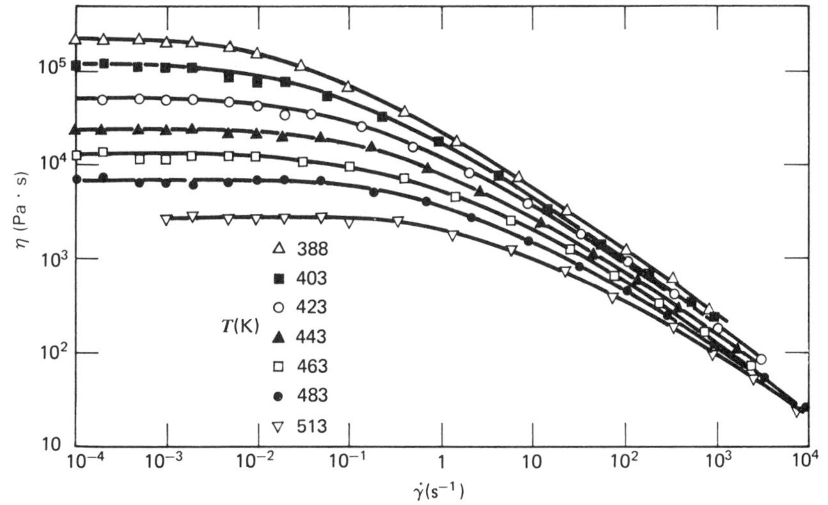

FIGURE 3.3-1. Non-Newtonian viscosity η of a low-density polyethylene melt (Melt I) at several different temperatures. The data taken at shear rates below 5×10^{-2} s^{-1} were obtained with a Weissenberg Rheogoniometer; the viscosity at higher shear rates was determined using a capillary viscometer. [J. Meissner, *Kunststoffe*, **61**, 576–582 (1971).]

[1] Almost all macromolecular fluids show this *shear thinning* or *pseudoplastic* behavior. A few fluids, called *shear thickening* or *dilatant fluids* are discussed by S. Burow, A. Peterlin, and D. T. Turner, *Polym. Lett.*, **2**, 67–70 (1964). See §2.2. For other "shear-thickening" data see M.-N. Layec-Raphalen and C. Wolff, *J. Non-Newtonian Fluid Mech.*, **1**, 159–173 (1976); they studied tube flow of dilute solutions of high molecular weight polystyrene in decalin.

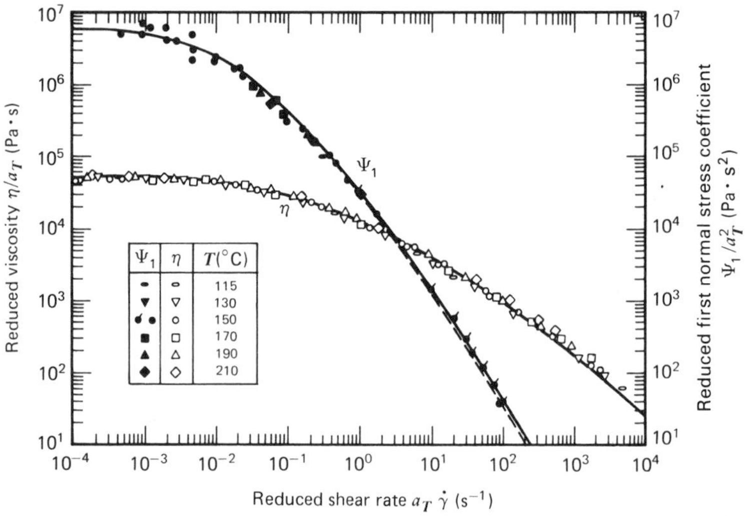

FIGURE 3.3-2. Master curves for the viscosity and first normal stress coefficient as functions of shear rate for the low-density polyethylene melt (Melt I) shown in Fig. 3.3-1. Data taken at the indicated temperatures were shifted to a reference temperature $T_0 = 423$ K. The shift factor a_T used here is given by $\eta_0(T)/\eta_0(T_0)$. The solid and dashed curves were calculated with the Wagner model described in §8.3, the solid line with a single term exponential "damping function" and the dashed line with two terms. [H. M. Laun, *Rheol. Acta*, **17**, 1–15 (1978).]

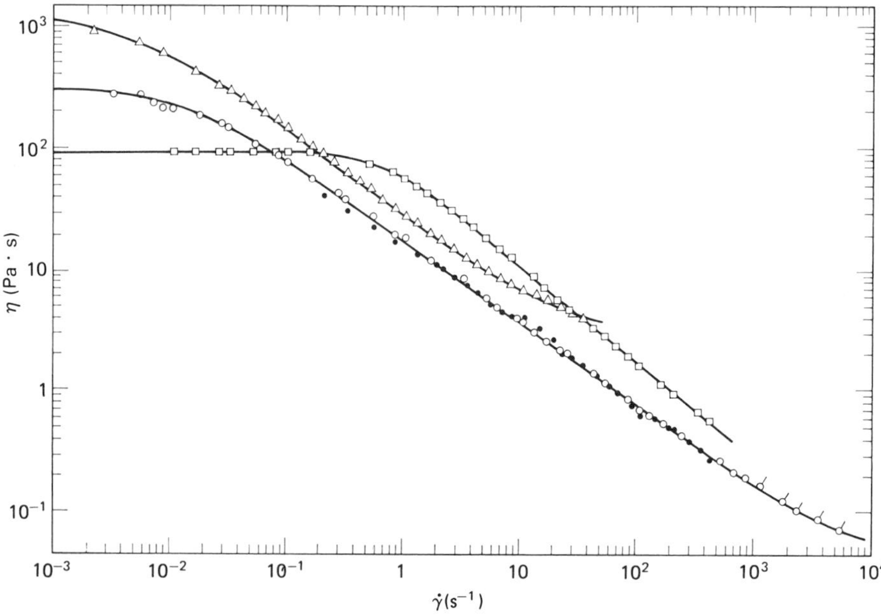

FIGURE 3.3-3. Dependence of viscosity on shear rate for two polymer solutions and an aluminum soap solution: ○ 1.5% polyacrylamide (Separan AP30) in a 50/50 mixture by weight of water and glycerin; △ 2.0% polyisobutylene in Primol; and □ 7% aluminum laurate in a mixture of decalin and *m*-cresol. The data shown by ○ were taken on a Ferranti–Shirley viscometer; all others are from a Weissenberg Rheogoniometer. All data were taken at 298 K. [Data of J. D. Huppler, E. Ashare, and L. A. Holmes, *Trans. Soc. Rheol.*, **11**, 159–179 (1967).]

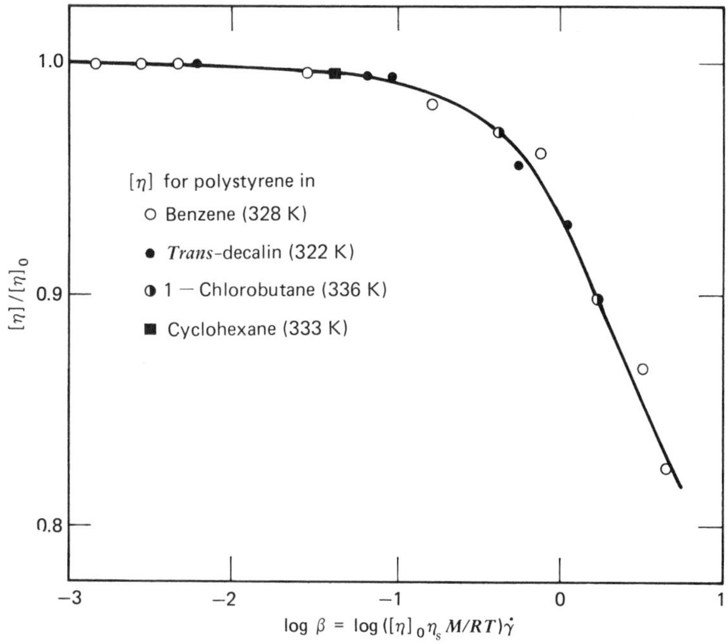

FIGURE 3.3-4. Intrinsic viscosity $[\eta]$ (see Eq. 3.3-6) of polystyrene solutions, with various solvents, as a function of reduced shear rate β. [Data of T. Kotaka, H. Suzuki, and H. Inagaki, *J. Chem. Phys.*, **45**, 2770–2773 (1966).]

engineering applications, this is the most important property of polymeric fluids. When plotted as $\log \eta$ versus $\log \dot{\gamma}$, the viscosity vs. shear rate curve exhibits a pronounced linear region at high shear rates that can persist over several decades of decreasing viscosity. The slope of the linear section, or "power-law" region, is found experimentally to be between -0.4 and -0.9 for typical polymeric liquids. The range of shear rates over which the transition from η_0 to the power law region occurs is fairly narrow for narrow molecular weight distributions. As the molecular weight distribution of the polymer is broadened, the transition region is also broadened and shifted to lower shear rates.[2] Finally, at very high rates of shear the viscosity may again become independent of shear rate and approach η_∞, the *infinite-shear-rate viscosity*. For concentrated solutions and melts, η_∞ is not usually measurable since polymer degradation becomes a serious problem before sufficiently high shear rates can be obtained.

When viscosity data are available for a wide variety of $\dot{\gamma}$ and T, it is convenient to display the data as "master curves." This is illustrated by taking the data of Fig. 3.3-1 for $\eta(\dot{\gamma}, T)$ and replotting them in Fig. 3.3-2 as though they were all taken at a single, reference temperature. In Fig. 3.3-1, it is seen that changing the temperature does not affect the functional dependence of η on $\dot{\gamma}$; it merely alters the zero-shear-rate viscosity and the shear rate at which the transition from constant viscosity to power-law behavior occurs. As temperature is increased, η_0 decreases and the transition shear rate increases. Data obtained at various temperatures can be superposed onto a single "master curve" by plotting $\eta(T)\eta_0(T_0)/\eta_0(T)$ versus $a_T\dot{\gamma}$ where $a_T = a_T(T_0)$ is a "shift factor."[3] This means that viscosity

[2] W. W. Graessley, *Adv. Polym. Sci.*, **16**, 1–179 (1974).

[3] In $a_T(T_0)$ the subscript T denotes the temperature at which data were actually taken; and the argument T_0, the reference temperature. The T_0 argument is normally suppressed.

measured at a temperature T and a shear rate $\dot{\gamma}$ is equivalent, after correction for the temperature dependence of η_0, to viscosity measured at temperature T_0 and shear rate $a_T\dot{\gamma}$. The success of this method is shown in the resulting master curve for a polyethylene melt in Fig. 3.3-2. The value of a_T is determined primarily by $\eta_0(T)$; for example the master curves in Fig. 3.3-2 were obtained by taking $a_T = \eta_0(T)/\eta_0(T_0)$. This means that $a_T < 1$ for $T > T_0$ and $a_T > 1$ for $T < T_0$. The technique we have illustrated here is known as *time–temperature superposition* or the *method of reduced variables.* We discuss the method more fully in §3.6.

In dilute solutions, the viscosity is dominated by the solvent, so that it is difficult to discern the effect of the polymer on $\eta(\dot{\gamma})$. The polymer contribution to the viscosity can be obtained by expanding the *relative viscosity* η_{rel}, defined as the ratio of solution viscosity to solvent viscosity η_s

$$\eta_{\text{rel}} = \frac{\eta}{\eta_s} \tag{3.3-4}$$

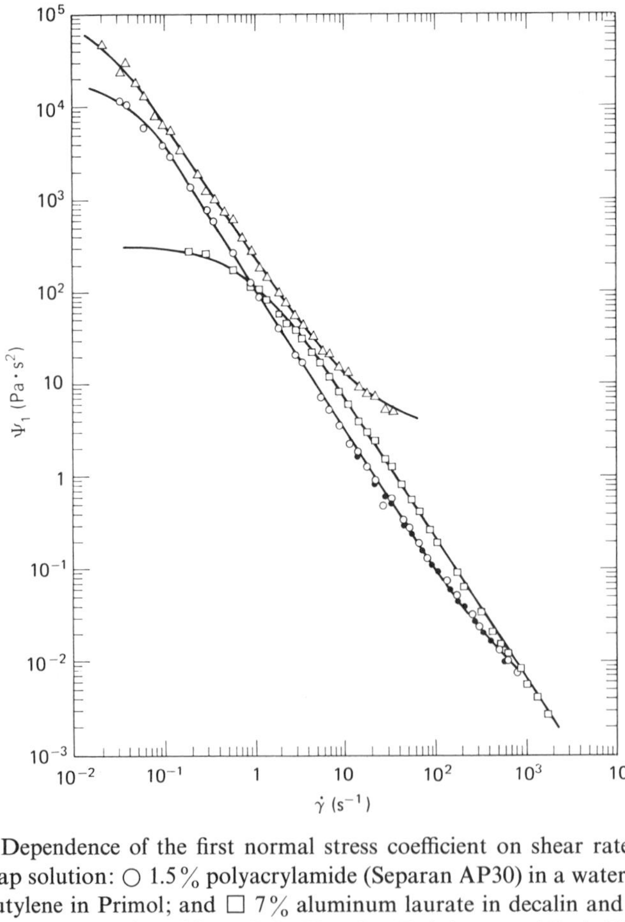

FIGURE 3.3-5. Dependence of the first normal stress coefficient on shear rate for two polymer solutions and a soap solution: ○ 1.5% polyacrylamide (Separan AP30) in a water–glycerin mixture; △ 2.0% polyisobutylene in Primol; and □ 7% aluminum laurate in decalin and *m*-cresol. All data were taken at 298 K. [Data replotted from J. D. Huppler, E. Ashare, and L. A. Holmes, *Trans. Soc. Rheol.*, **11**, 159–179 (1967).]

in a Taylor series in the mass concentration c

$$\eta_{\text{rel}} = 1 + [\eta]c + k'[\eta]^2c^2 + \cdots \tag{3.3-5}$$

in which $[\eta]$ and k' are independent of concentration. The coefficient $[\eta]$ is the *intrinsic viscosity* of the solution, and k' is the *Huggins coefficient*; it follows from Eq. 3.3-5 that

$$\boxed{[\eta] = \lim_{c \to 0} \left(\frac{\eta - \eta_s}{c\eta_s} \right)} \tag{3.3-6}$$

Note that the intrinsic viscosity has dimensions of reciprocal concentration. As shown in Fig. 3.3-4, the intrinsic viscosity depends on shear rate in much the same way as the viscosity of the polymer melt and polymer solutions.

The first normal stress coefficient is shown in Figs. 3.3-2 and 5. The polymer melt data in Fig. 3.3-2 are presented in terms of reduced variables in the same way as the viscosity (see §3.6 for more detail); the shift factors a_T are the same in both plots. It is seen that Ψ_1 is positive[4] and that it has a large power law region in which it decreases by as much as a factor of 10^6. Most often, the rate of decline of Ψ_1 with $\dot{\gamma}$ is greater than that of η with $\dot{\gamma}$. At low rates of shear the first normal stress difference is proportional to $\dot{\gamma}^2$, so that Ψ_1 tends to a constant $\Psi_{1,0}$, the zero-shear-rate first normal stress coefficient, as $\dot{\gamma}$ approaches zero. There does not seem to be a limiting value of Ψ_1 at high shear rates to correspond to η_∞. Some data do show, however, a levelling-off trend at the high shear rates.

The second normal stress coefficient is not nearly as well studied experimentally as η and Ψ_1. The most important points to note about Ψ_2 are that its magnitude is much smaller than Ψ_1, usually about 10% of Ψ_1, and that it is negative.[5] It was thought for some time that $\tau_{yy} = \tau_{zz}$ (or $\Psi_2 = 0$); this relation, called the "Weissenberg hypothesis," is now known not to be correct. In Fig. 3.3-6 we present data for Ψ_2. There it can be seen that Ψ_2 exhibits a large power-law region such as those for $\eta(\dot{\gamma})$ and $\Psi_1(\dot{\gamma})$. Values for $\Psi_{2,0}$ or $\Psi_{2,\infty}$, the zero- and infinite-shear-rate values of Ψ_2, are not observed in these data. In Fig. 3.3-7, values of the ratio $-\Psi_2/\Psi_1$ are seen to range from 0.01 to 0.2 for a polyacrylamide solution and a polyethylene oxide solution. Our knowledge about Ψ_2 is still incomplete. Furthermore, the general statements above regarding the sign and magnitude of Ψ_2 are based on data for moderately concentrated solutions of relatively few different kinds of polymers. Some data[6] suggest that Ψ_2 may even change sign and become positive at large shear rates.

[4] G. Kiss, *Phys. Today*, **37**, 15, 121 (1984), has pointed out that negative first normal stress coefficients have been observed in some liquid crystalline solutions and other systems. See G. Kiss and R. S. Porter, *J. Polym. Sci. Polym. Symp.*, **65**, 193–211 (1978); *J. Polym. Sci. Polym. Phys. Ed.*, **18**, 361–388 (1980).

[5] The first study in which Ψ_2 was reported to be negative is that of R. F. Ginn and A. B. Metzner in E. H. Lee, ed., Wiley-Interscience, *Proceedings of the Fourth International Congress on Rheology*, New York (1965), Part 2, pp. 583–601; *Trans. Soc. Rheol.*, **13**, 429–453 (1969). The uncertainty in the sign of Ψ_2 was resolved by the discovery of the hole pressure effect (see §2.3); see J. M. Broadbent, A. Kaye, A. S. Lodge, and D. G. Vale, *Nature*, **277**, 55–56 (1968); A. Kaye, A. S. Lodge, and D. G. Vale, *Rheol. Acta*, **7**, 368–379 (1968); J. M. Broadbent and A. S. Lodge, *Rheol. Acta*, **10**, 557–573 (1972). See also M. Keentok, A. G. Georgescu, A. A. Sherwood, and R. I. Tanner, *J. Non-Newtonian Fluid Mech.*, **6**, 303–324 (1980); H. W. Gao, S. Ramachandran, and E. B. Christiansen, *J. Rheol.*, **25**, 213–235 (1981); K. Walters, *IUPAC Macro 83* (Bucharest, Romania), Part 2, 227–237 (1983); S. Ramachandran, H. W. Gao, and E. B. Christiansen, *Macromolecules*, **18**, 695–699 (1985); W.-M. Kulicke and U. Wallbaum, *Chem. Eng. Sci.*, **40**, 961–972 (1985).

[6] W. G. Pritchard, *Phil. Trans. Roy. Soc. London*, **A270**, 507–556 (1971).

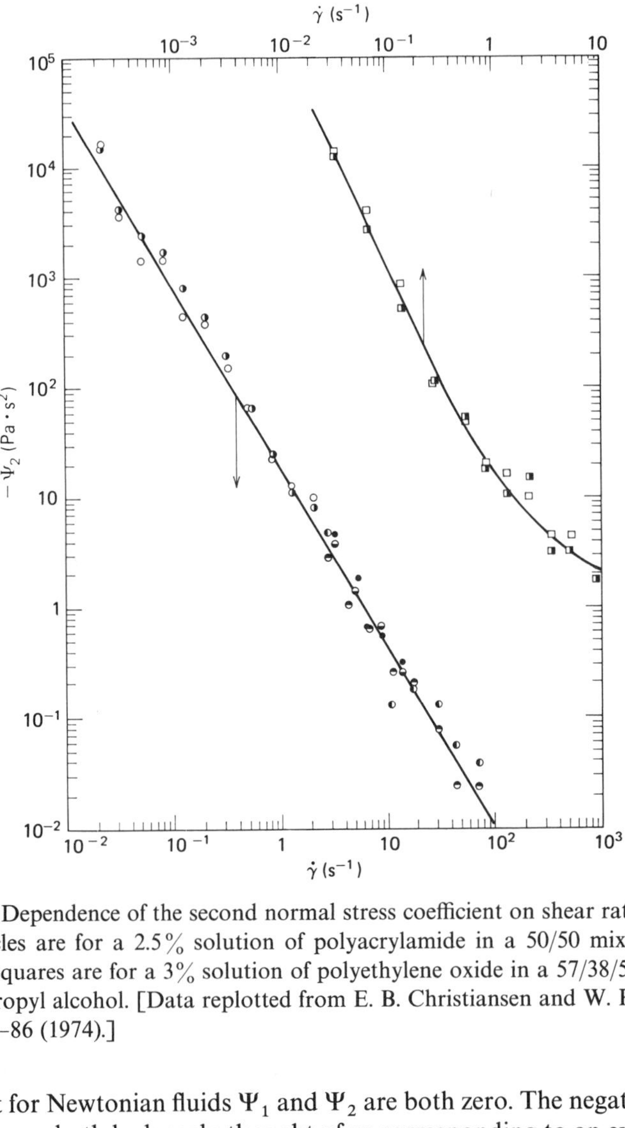

FIGURE 3.3-6. Dependence of the second normal stress coefficient on shear rate for two polymer solutions: the circles are for a 2.5% solution of polyacrylamide in a 50/50 mixture of water and glycerin, and the squares are for a 3% solution of polyethylene oxide in a 57/38/5 mixture of water, glycerin, and isopropyl alcohol. [Data replotted from E. B. Christiansen and W. R. Leppard, *Trans. Soc. Rheol.*, **18**, 65–86 (1974).]

Note that for Newtonian fluids Ψ_1 and Ψ_2 are both zero. The negative $\tau_{xx} - \tau_{yy}$ and positive $\tau_{yy} - \tau_{zz}$ can both be loosely thought of as corresponding to an extra compression in the y-direction. Consequently, in order to maintain steady shear flow between two parallel plates, a normal force must be applied to the plates to prevent them from separating when the fluid is polymeric. Only a shear stress is needed to maintain steady shear flow for a Newtonian fluid.

A measure of the 'elasticity' in the fluid is the *stress ratio*, $(\tau_{xx} - \tau_{yy})/\tau_{yx}$. This ratio is zero for Newtonian fluids, and also for non-Newtonian fluids in the limit of small shear rates. Figure 3.3-8 shows the stress ratio for a low density polyethylene melt and a polyacrylamide solution. The stress ratio for both is a monotone increasing function of shear rate with values as high as 2 to 3 being attained for the polyethylene and 20 to 30 for the polyacrylamide. Many polymers do not have stress ratios this large.

Now that we have desçribed the steady shear flow material functions, we point out briefly one common method for measuring them by using a cone-and-plate apparatus (see Fig. 1.3-4 and Example 1.3-4). As was seen in Example 1.3-4, by using a very small cone angle ϑ_0 a very nearly uniform shear rate of W/ϑ_0 can be achieved throughout the sample

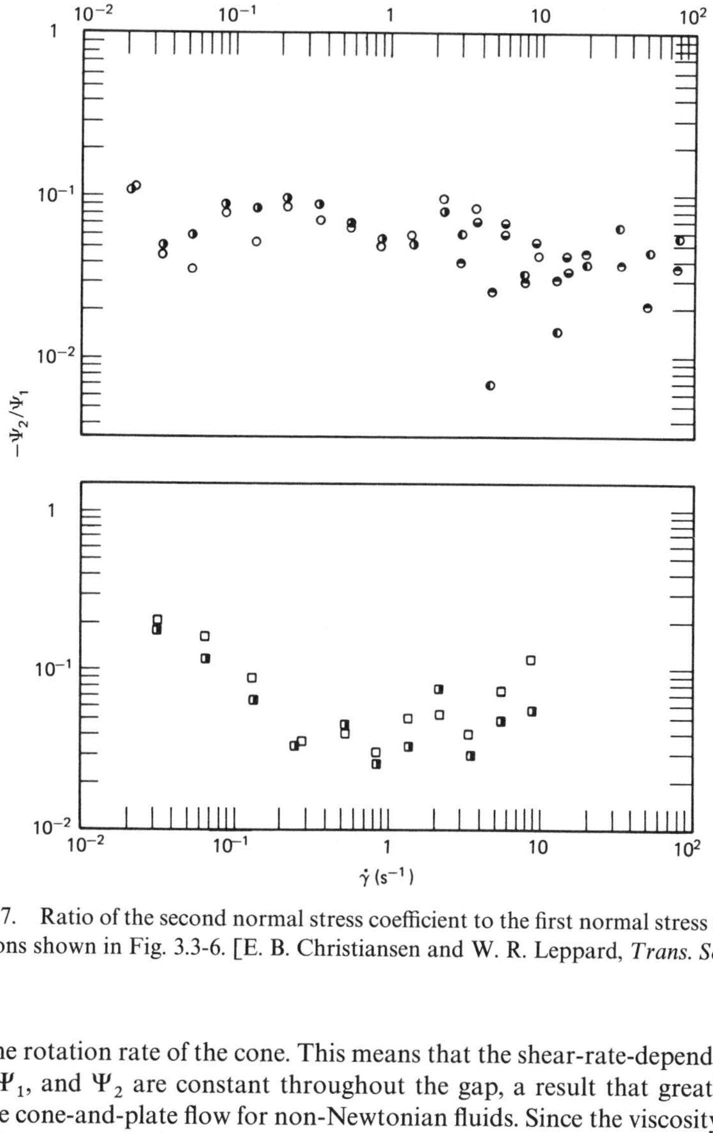

FIGURE 3.3-7. Ratio of the second normal stress coefficient to the first normal stress coefficient for the two solutions shown in Fig. 3.3-6. [E. B. Christiansen and W. R. Leppard, *Trans. Soc. Rheol.*, **18**, 65–86 (1974).]

where W is the rotation rate of the cone. This means that the shear-rate-dependent material functions η, Ψ_1, and Ψ_2 are constant throughout the gap, a result that greatly simplifies analysis of the cone-and-plate flow for non-Newtonian fluids. Since the viscosity is constant in this flow, the analysis of the shear stress from Example 1.3-4 can be taken over directly to give

$$\eta(\dot{\gamma}) = \frac{3\mathscr{T}\vartheta_0}{2\pi R^3 W} \tag{3.3-7}$$

$$\dot{\gamma} = W/\vartheta_0 \tag{3.3-8}$$

When used for polymer characterization, the cone-and-plate rheometer is normally instrumented to allow measurement of the total thrust $\mathscr{F}$ downward on the bottom plate in addition to the torque $\mathscr{T}$. From this total thrust measurement the first normal stress coefficient can be calculated by

$$\Psi_1 = \frac{2\mathscr{F}\vartheta_0^2}{\pi R^2 W^2} \tag{3.3-9}$$

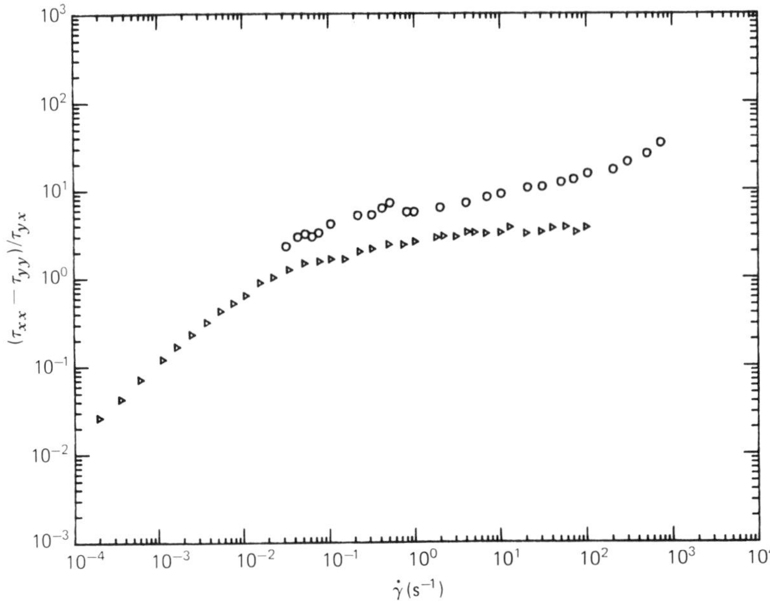

FIGURE 3.3-8. The stress ratio $(\tau_{xx} - \tau_{yy})/\tau_{yx}$ for the low-density polyethylene melt of Fig. 3.3-2 and the polyacrylamide solution of Figs. 3.3-3 and 5: ○ polyacrylamide; ▷ LDPE.

If in addition we can measure the local stress distribution across the plate surface, $\pi_{\theta\theta}(r)|_{\theta=\pi/2}$, then the second normal stress coefficient can be determined from

$$\Psi_1 + 2\Psi_2 = -\frac{1}{\dot{\gamma}^2} \frac{\partial \pi_{\theta\theta}}{\partial \ln r}\bigg|_{\theta=\pi/2} \tag{3.3-10}$$

in conjunction with the previous result for Ψ_1. In addition the second normal stress coefficient can be determined directly from the local stress $\pi_{\theta\theta}(R)|_{\theta=\pi/2}$ at the plate edge

$$\Psi_2 = \frac{p_a - \pi_{\theta\theta}(R)}{\dot{\gamma}^2} \tag{3.3-11}$$

where p_a is atmospheric pressure. A derivation of these results and a discussion of alternative methods for measuring the viscometric functions are given in Example 10.2-1.

§3.4 UNSTEADY SHEAR FLOW MATERIAL FUNCTIONS

We now consider the time-dependent behavior of polymeric liquids in unsteady simple shearing flow. A summary of the most commonly used shear flow experiments is given in Fig. 3.4-1. As in steady shear flow there are only three measurable stress quantities: the shear stress and two normal stress differences. These are represented by material functions defined analogously to η, Ψ_1, and Ψ_2 except that in the time-dependent shearing flows they can depend on time (or frequency) as well as the shear rate. A summary of the time-dependent material functions corresponding to the experiments shown in Fig. 3.4-1 is given in Table 3.4-1.

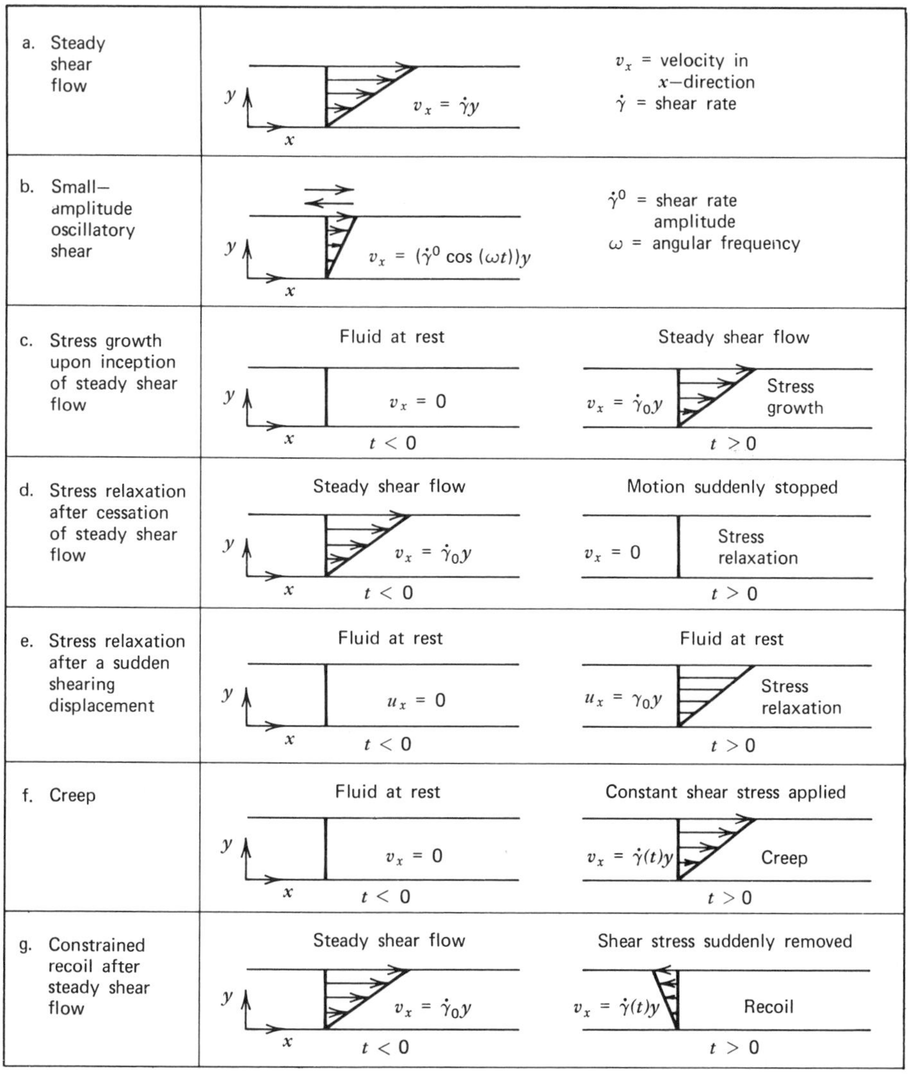

FIGURE 3.4-1. Various types of simple shear flow experiments used in rheology. For the shearing displacement experiment in (*e*), u_x is the displacement of a particle in the x-direction measured from its position just before $t = 0$; also the "magnitude of shear" γ_0 is the shear-displacement gradient and is $\gamma_0 = \int_0^t \dot{\gamma}(t')\,dt'$.

Experiment b: Small-Amplitude Oscillatory Shear Flow

The *small-amplitude oscillatory shear experiment* (Fig. 3.4-1b) involves measurement of the unsteady response of a sample that is contained between two parallel plates, the upper one of which undergoes small-amplitude sinusoidal oscillations in its own plane with frequency ω. The instantaneous velocity profile will be very nearly linear in y if $\omega\rho h^2/2\eta_0 \ll 1$, where h is the distance between the plates (see Problem 1D.1). For a linear velocity profile, the shear strain between times 0 and t, defined by $\gamma_{yx}(0, t) = \int_0^t \dot{\gamma}_{yx}(t')\,dt'$,

Table 3.4-1

Material Functions in Simple Shearing Flows $v_x = \dot{\gamma}_{yx}\, y,\ v_y = v_z = 0$

Flow	Material Function	Defining Equation
a. Steady shear flow $\dot{\gamma}_{yx} = \dot{\gamma} = \text{constant}$	$\eta(\dot{\gamma})$ $\Psi_1(\dot{\gamma})$ $\Psi_2(\dot{\gamma})$	$\tau_{yx} = -\eta\dot{\gamma}_{yx}$ $\tau_{xx} - \tau_{yy} = -\Psi_1\dot{\gamma}_{yx}^2$ $\tau_{yy} - \tau_{zz} = -\Psi_2\dot{\gamma}_{yx}^2$
b. Small-amplitude oscillatory shear $\dot{\gamma}_{yx} = \dot{\gamma}^0 \cos \omega t$ $= \gamma^0\omega \cos \omega t$	$\eta'(\omega)$ $\eta''(\omega)$ $G'(\omega) = \eta''\omega$ $G''(\omega) = \eta'\omega$	$\tau_{yx} = -\eta'\dot{\gamma}^0 \cos \omega t$ $-\eta''\dot{\gamma}^0 \sin \omega t$ $\tau_{yx} = -G'\gamma^0 \sin \omega t$ $-G''\gamma^0 \cos \omega t$
c. Stress growth upon inception of steady shear flow $\dot{\gamma}_{yx} = \begin{cases} 0 & t < 0 \\ \dot{\gamma}_0 & t \geq 0 \end{cases}$	$\eta^+(t, \dot{\gamma}_0)$ $\Psi_1^+(t, \dot{\gamma}_0)$ $\Psi_2^+(t, \dot{\gamma}_0)$	$\tau_{yx} = -\eta^+\dot{\gamma}_0$ $\tau_{xx} - \tau_{yy} = -\Psi_1^+\dot{\gamma}_0^2$ $\tau_{yy} - \tau_{zz} = -\Psi_2^+\dot{\gamma}_0^2$
d. Stress relaxation after cessation of steady shear flow $\dot{\gamma}_{yx} = \begin{cases} \dot{\gamma}_0 & t < 0 \\ 0 & t \geq 0 \end{cases}$	$\eta^-(t, \dot{\gamma}_0)$ $\Psi_1^-(t, \dot{\gamma}_0)$ $\Psi_2^-(t, \dot{\gamma}_0)$	$\tau_{yx} = -\eta^-\dot{\gamma}_0$ $\tau_{xx} - \tau_{yy} = -\Psi_1^-\dot{\gamma}_0^2$ $\tau_{yy} - \tau_{zz} = -\Psi_2^-\dot{\gamma}_0^2$
e. Stress relaxation after a sudden shearing displacement $\dot{\gamma}_{yx} = \gamma_0\delta(t)$	$G(t, \gamma_0)$ $G_{\Psi 1}(t, \gamma_0)$	$\tau_{yx} = -G\gamma_0$ $\tau_{xx} - \tau_{yy} = -G_{\Psi 1}\gamma_0^2$
f. Creep $\tau_{yx} = \begin{cases} 0 & t < 0 \\ \tau_0 & t \geq 0 \end{cases}$	$J(t, \tau_0)$	$\gamma_{yx}(0, t) = -J\tau_0$
g. Constrained recoil after steady shear flow $\tau_{yx} = \begin{cases} \tau_0 & t < 0 \\ 0 & t \geq 0 \end{cases}$	$\gamma_r(0, t, \tau_0)$ $\gamma_\infty(\tau_0)$ $J_e^0(\tau_0)$	$\gamma_r = \int_0^t \dot{\gamma}_{yx}(t')\, dt'$ $\gamma_\infty = \lim_{t\to\infty} \gamma_r$ $\gamma_\infty = J_e^0(\tau_0)\tau_0$

and the shear rate at time t in the fluid will be independent of position and are given respectively by

$$\gamma_{yx}(0, t) = \gamma^0 \sin \omega t \tag{3.4-1a}$$

$$\dot{\gamma}_{yx}(t) = \gamma^0\omega \cos \omega t = \dot{\gamma}^0 \cos \omega t \tag{3.4-1b}$$

where γ^0 and $\dot{\gamma}^0$ are the (positive) amplitudes of the shear strain and shear rate oscillations. For Newtonian fluids the shear stress τ_{yx} is in phase with the shear rate $\dot{\gamma}_{yx}$ and there are no normal stresses. For polymeric materials, the response is shown qualitatively in Fig. 3.4-2,

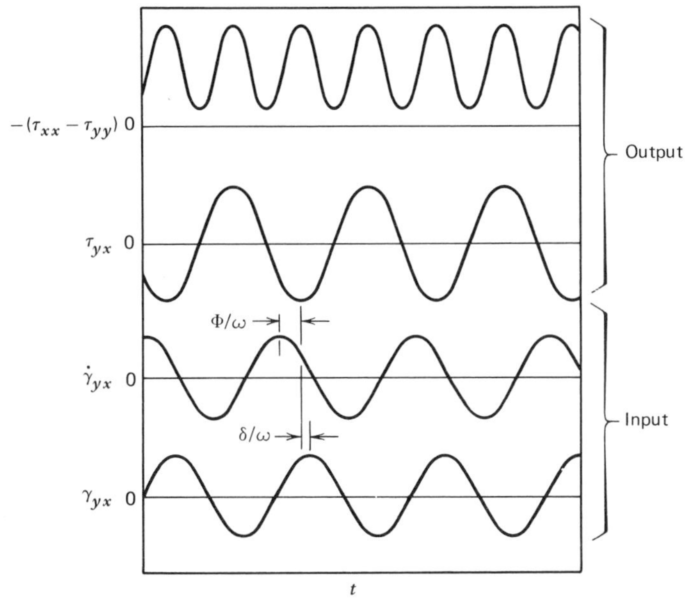

FIGURE 3.4-2. Oscillatory shear strain, shear rate, shear stress, and first normal stress difference in small-amplitude oscillatory shear flow. [After A. S. Lodge. *Elastic Liquids*, Academic Press, New York (1964), p. 113.]

where it is seen that the shear stress oscillates with frequency ω, but is not in phase with either the shear strain or shear rate, and that the normal stresses oscillate with a frequency 2ω about a nonzero mean value (see Problem 3C.1).

For shear stress one can measure the amplitude and phase shift as a function of the frequency ω. We will assume that the shear strain amplitude γ^0 is sufficiently small so that the shear stress is linear in strain or strain rate, and we write (corresponding to the two forms of Eq. 3.4-1):

$$\tau_{yx} = -A(\omega)\gamma^0 \sin(\omega t + \delta) \qquad (0 \leq \delta \leq \pi/2) \tag{3.4-2a}$$

$$\tau_{yx} = -B(\omega)\dot{\gamma}^0 \cos(\omega t - \Phi) \qquad (0 \leq \Phi \leq \pi/2) \tag{3.4-2b}$$

where $\Phi = (\pi/2) - \delta$. It is customary to rewrite Eqs. 3.4-2 to display the in-phase and out-of-phase parts of the shear stress. In this way we define the two equivalent sets of linear viscoelastic material functions G', G'' and η', η'':

$$\boxed{\begin{aligned} \tau_{yx} &= -G'(\omega)\gamma^0 \sin \omega t - G''(\omega)\gamma^0 \cos \omega t \\ \tau_{yx} &= -\eta'(\omega)\dot{\gamma}^0 \cos \omega t - \eta''(\omega)\dot{\gamma}^0 \sin \omega t \end{aligned}} \qquad \begin{aligned} &(3.4\text{-}3a) \\ &(3.4\text{-}3b) \end{aligned}$$

It is easy to see that G', G'' are related to A, δ and that η', η'' are related to B, Φ by

$$A(\omega) = \sqrt{G'^2 + G''^2} = |G^*|, \quad \tan \delta = G''/G' \tag{3.4-4a}$$

$$B(\omega) = \sqrt{\eta'^2 + \eta''^2} = |\eta^*|, \quad \tan \Phi = \eta''/\eta' \tag{3.4-4b}$$

in which we have introduced the notation $|G^*|$ and $|\eta^*|$ for the magnitudes of the *complex modulus* G^* and *complex viscosity* η^*.[1]

Because G' and G'', or alternatively η' and η'', determine the shear stress that is linear in the strain these material functions are often called linear viscoelastic properties. They are important in characterizing the behavior of a material in small deformations (see Chapter 5). To understand the information that these material functions contain, it is useful to recall that for a perfectly elastic solid G' is equal to the constant shear modulus G, and G'' is zero, whereas for a Newtonian fluid, η' is equal to the viscosity μ, and η'' is zero. For this reason G' $(=\eta''\omega)$ is known as the *storage modulus*; it gives information about the elastic character of the fluid or the energy storage that takes place during the deformation. On the other hand, G'' $(=\eta'\omega)$ is known as the *loss modulus*; it tells about the viscous character of the fluid, or the energy dissipation that occurs in flow. The quantity η' is called the *dynamic viscosity*; and the phase angle δ between stress and strain is normally given by the *loss tangent*, tan δ. For more detail on these and other linear viscoelastic properties, standard references should be consulted.[2] We shall use both G^* and η^* in specifying the linear viscoelastic properties.

Sample data showing the complex modulus for a polymer melt, the complex viscosity for three polymer solutions, and the intrinsic complex viscosity for a dilute solution are shown in Figs. 3.4-3 to 6. The dynamic viscosity is found to approach the zero-shear-rate viscosity at low frequency; this is to be expected. Correspondingly the loss modulus is asymptotic to $\eta_0\omega$ as $\omega \to 0$. On the other hand the out-of-phase part of the complex viscosity associated with energy storage is found to approach zero linearly in ω as $\omega \to 0$; likewise the storage modulus is proportional to ω^2. Thus there is a nonzero limiting value of $\eta''/\omega = G'/\omega^2$ at low frequency. At intermediate frequencies, η' and η''/ω both show large power-law regions similar to those we have previously encountered for the viscometric functions. Finally at high frequencies, η' may approach a limiting value η'_∞. As with η_∞, this quantity is usually observed only for dilute solutions where it is found to be slightly larger

[1] The complex modulus and viscosity arise naturally in the following alternative formulas for the stress given in Eq. 3.4-3. Let us use the complex notation

$$\gamma_{yx}(0, t) = -\gamma^0 \mathscr{Re} \{ie^{i\omega t}\} \tag{3.4-1a'}$$

$$\dot{\gamma}_{yx}(t) = \dot{\gamma}^0 \mathscr{Re} \{e^{i\omega t}\} \tag{3.4-1b'}$$

where γ^0 and $\dot{\gamma}^0$ are the (real, positive) amplitudes of the oscillatory shear strain and shear rate. For small deformations the shear stress is assumed to oscillate with the same frequency, but not necessarily in phase: $\tau_{yx} = \mathscr{Re} \{\tau_{yx}^0 e^{i\omega t}\}$ where τ_{yx}^0 is in general complex. Then if the complex modulus G^* and the complex viscosity η^* are defined by:

$$\tau_{yx}^0 = iG^*\gamma^0 \tag{3.4-3a'}$$

$$\tau_{yx}^0 = -\eta^*\dot{\gamma}^0 \tag{3.4-3b'}$$

and we take

$$G^* = G' + iG'' \tag{3.4-3c}$$

$$\eta^* = \eta' - i\eta'' \tag{3.4-3d}$$

then the above formulas are equivalent to Eqs. 3.4-3a and b. Also we have $G^* = i\omega\eta^*$.

[2] J. D. Ferry, *Viscoelastic Properties of Polymers*, 3rd ed., Wiley, New York (1980); H. Leaderman in F. R. Eirich, ed., *Rheology*, Vol. 2, Academic Press, New York (1958), Chapt. 1, pp. 1–61.; A. V. Tobolsky, *op. cit.*, pp. 63–81; T. Alfrey, Jr. and E. F. Gurnee in F. R. Eirich, ed., *Rheology*, Vol. 3, Academic Press, New York (1956), Chapt. 11, pp. 387–429.

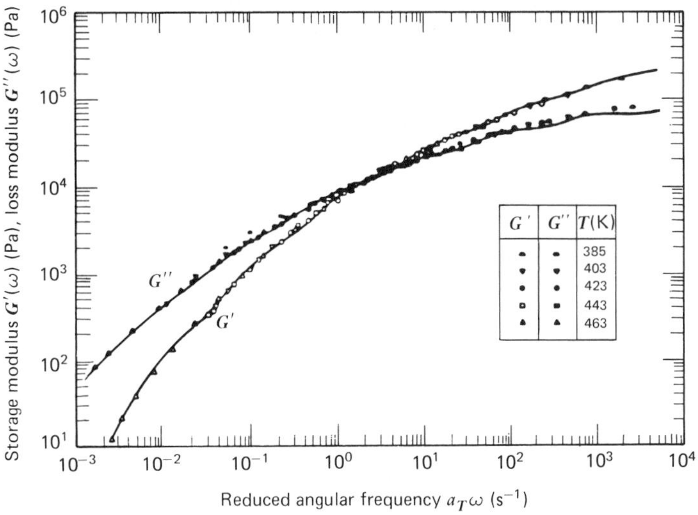

FIGURE 3.4-3. Storage and loss moduli, G' and G'', as functions of frequency ω at a reference temperature of $T_0 = 423$ K for the low-density polyethylene melt (Melt I) shown in Fig. 3.3-1. The shift factors a_T used to shift the data from the indicated temperatures to T_0 are the same as those used for the viscosity and first normal stress coefficient in Fig. 3.3-2. The solid curves are calculated from the generalized Maxwell model, Eqs. 5.2-13 through 15, with the constants tabulated in Table 5.3-2. [Data of A. Zosel (1972) as reported in H. M. Laun, *Rheol. Acta*, **17**, 1–15 (1978).]

than the solvent viscosity.[3] Also, at high frequencies η''/ω may become proportional to ω^{-2}. Since this corresponds to a storage modulus G' that is constant, the fluid becomes like a perfectly elastic solid, which indicates that there is not sufficient time for molecular rearrangements at the high frequencies.

Data for the dilute solution[4] in Fig. 3.4-6 are presented in terms of the intrinsic quantities $[\eta']$ and $[\eta'']/\omega$. These are obtained by extrapolating data for η' and η'' obtained at different polymer concentrations to zero concentration; they are thus infinite dilution properties and are dynamic analogs of the intrinsic viscosity discussed in §3.3. They are defined by:[5]

$$[\eta'] = \lim_{c \to 0} \left(\frac{\eta' - \eta_s}{c\eta_s} \right) \tag{3.4-5}$$

$$[\eta'']/\omega = \lim_{c \to 0} \left(\frac{\eta''/\omega}{c\eta_s} \right) \tag{3.4-6}$$

As seen in Fig. 3.4-2, the normal stresses oscillate about non-zero values. There are three material functions that are needed to describe each of the oscillatory normal stress differences $\tau_{xx} - \tau_{yy}$ and $\tau_{yy} - \tau_{zz}$. Not many experimental data are available for these material functions.[6]

[3] Values of η'_∞ less than η_s have also been reported for polystyrene, polyisoprene, and polybutadiene in a variety of solvents over a wide range of concentrations. See the Ph.D. Theses (Department of Chemistry, University of Wisconsin-Madison) of C. J. T. Landry (1985), D. R. Radtke (1986), T. M. Stokich (1987), and P. A. Merchak (1987); see also T. M. Stokich and J. L. Schrag, *Macromolecules*, **20** (1987).
[4] For a review of dilute solution viscoelastic properties see K. Osaki, *Adv. Polym. Sci.*, **12**, 1–64 (1973).
[5] The intrinsic quantities $[G']$ and $[G'']$ are defined similarly to $[\eta'']$ and $[\eta']$, respectively, except that the values to be extrapolated to infinite dilution are divided by c rather than $c\eta_s$. Thus $[G'] = \omega\eta_s[\eta'']$ and $[G''] = \omega\eta_s[\eta']$.
[6] See E. B. Christiansen and W. R. Leppard, *Trans. Soc. Rheol.*, **18**, 65–86 (1974).

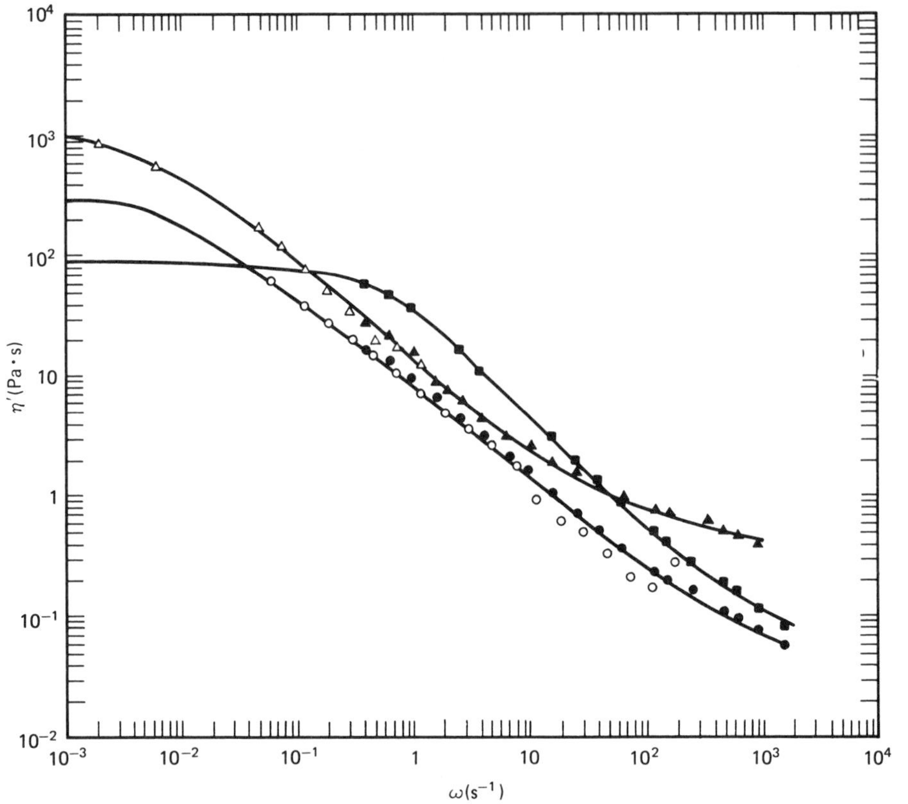

FIGURE 3.4-4. Dynamic viscosity η' as a function of frequency ω for two polymer solutions and a soap solution: ○ 1.5% polyacrylamide in a water–glycerin mixture; △ 2.0% polyisobutylene in Primol; and □ 7.0% aluminum laurate in a mixture of decalin and *m*-cresol. The hollow symbols represent data taken on a Weissenberg Rheogoniometer; the solid symbols, data taken on the Birnboim apparatus. All data were taken at 298 K. [Data of J. D. Huppler, E. Ashare, and L. A. Holmes, *Trans. Soc. Rheol.*, **11**, 159–179 (1967).]

Experiment c: Stress Growth upon the Inception of Steady Shear Flow

In a *stress growth experiment* (Fig. 3.4-1c), the fluid sample is presumed to be at rest for all times previous to $t = 0$; all components of stress are thus zero when the steady shearing is begun at time $t = 0$. For times $t \geq 0$ we denote the constant velocity gradient as $\dot{\gamma}_0$. The object of this experiment is to observe the approach of the stresses to their steady shear flow values. Material functions $\eta^+(t, \dot{\gamma}_0)$, $\Psi_1^+(t, \dot{\gamma}_0)$, and $\Psi_2^+(t, \dot{\gamma}_0)$ are defined (see Table 3.4-1) analogously to η, Ψ_1, and Ψ_2 to describe the transient shear stress and normal stress differences. The plus sign superscript emphasizes that a steady shear rate is applied for positive times. These transient properties can be measured in a cone-and-plate instrument.

Qualitatively it is found that only for vanishingly small shear rates does the shear stress approach its steady-state value monotonically. This monotone-increasing, low-shear-rate-limiting η^+ curve forms an envelope below which the η^+ curves at higher shear rates fall; it contains the same information as the linear viscoelastic properties η^* or G^*. For larger shear rates η^+ departs from the linear viscoelastic envelope, goes through a maximum, and then approaches the steady-state value, possibly after one or more oscillations about $\eta(\dot{\gamma}_0)$. The time at which η^+ departs from the linear envelope decreases as

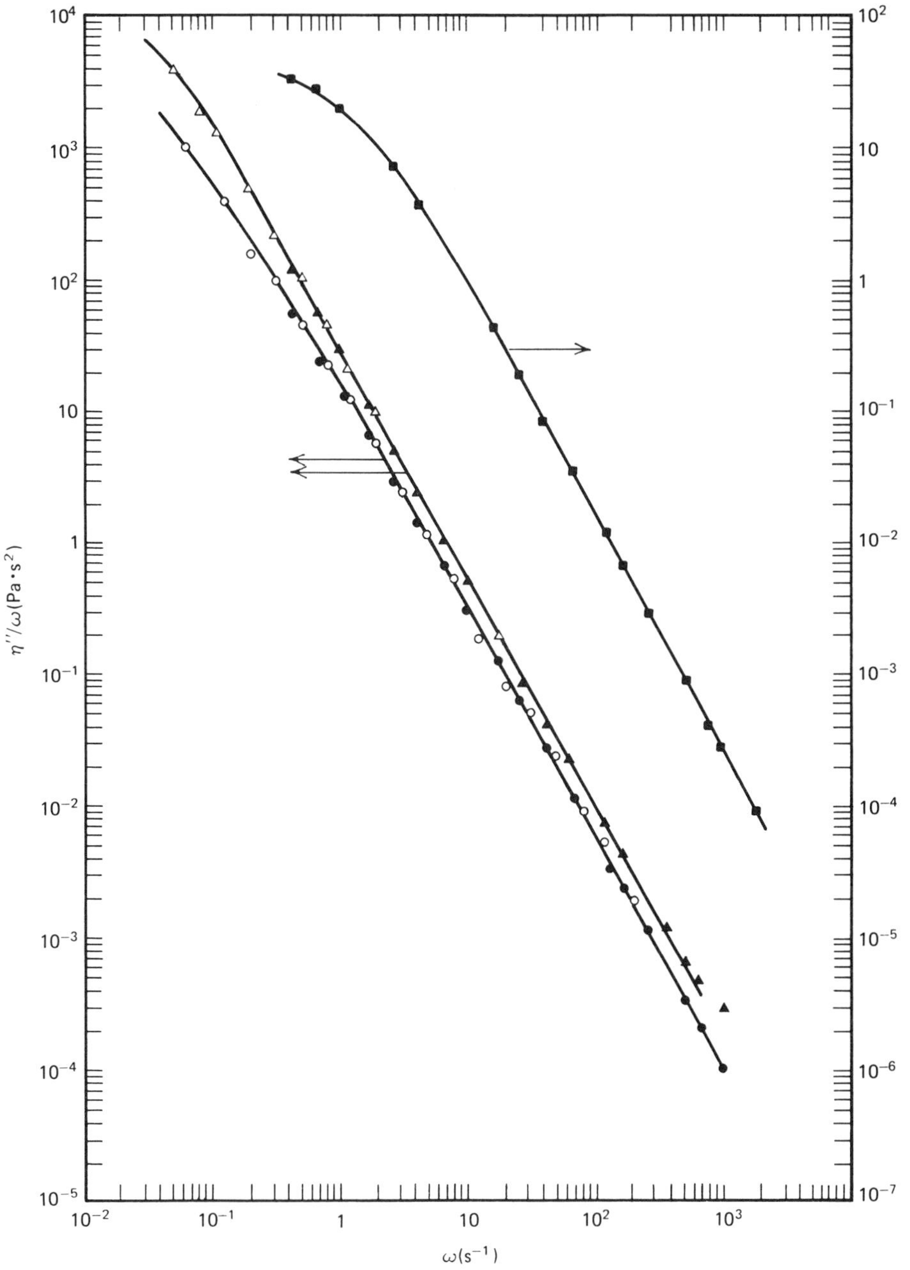

FIGURE 3.4-5. Plots of η''/ω as a function of frequency for the two polymer solutions and the aluminum soap solution shown in Fig. 3.4-4. [Data of J. D. Huppler, E. Ashare, and L. A. Holmes, *Trans. Soc. Rheol.*, **11**, 159–179 (1967).]

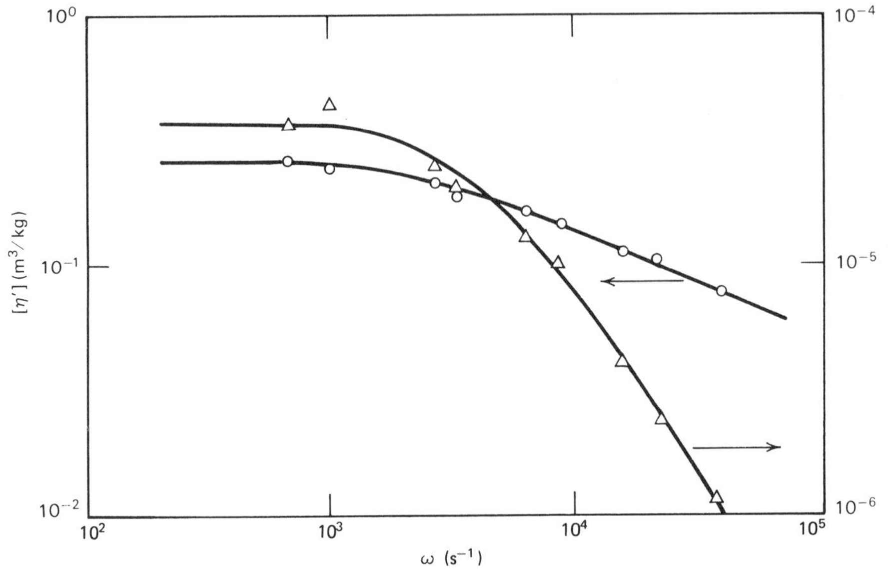

FIGURE 3.4-6. Real and imaginary parts of the intrinsic complex viscosity plotted as $[\eta']$ (circles) and $[\eta'']/\omega$ (triangles) versus frequency ω for poly-α-methylstyrene, with a narrow-distribution molecular weight of 1.43×10^6, in α-chloronaphthalene. [Data replotted from K. Osaki, J. L. Schrag, and J. D. Ferry, *Macromolecules*, **5**, 144–147 (1972).]

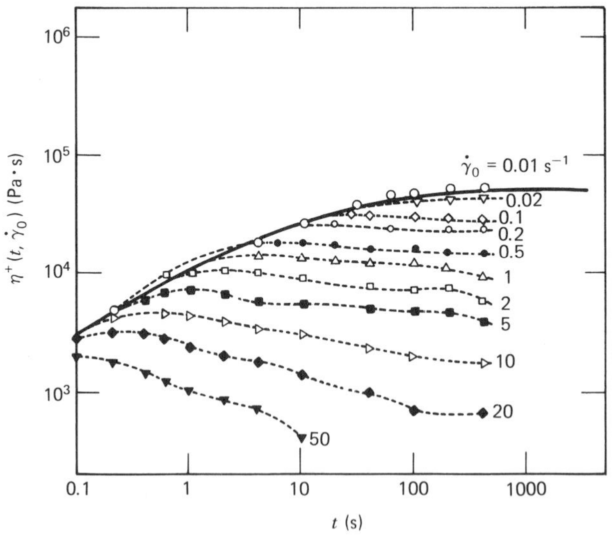

FIGURE 3.4-7. Shear stress growth function $\eta^+(t,\dot{\gamma}_0)$ data for a low-density polyethylene melt (Melt I). The maximum in η^+ occurs at smaller times as $\dot{\gamma}_0$ is increased. Note that all of the η^+ curves lie below the solid curve envelope, which is the value of η^+ in the linear viscoelastic limit. The solid curve is calculated from Eq. 5.3-25 with the spectrum in Table 5.3-2. The experimental data for small $\dot{\gamma}$ differ from the linear viscoelastic prediction at short times because of instrumental problems involved in the start-up of steady shear flow. [Reprinted with permission from M. H. Wagner and J. Meissner, *Macromol. Chem.*, **181**, 1533–1550 (1980), Hüthig and Wepf Verlag, Basel.]

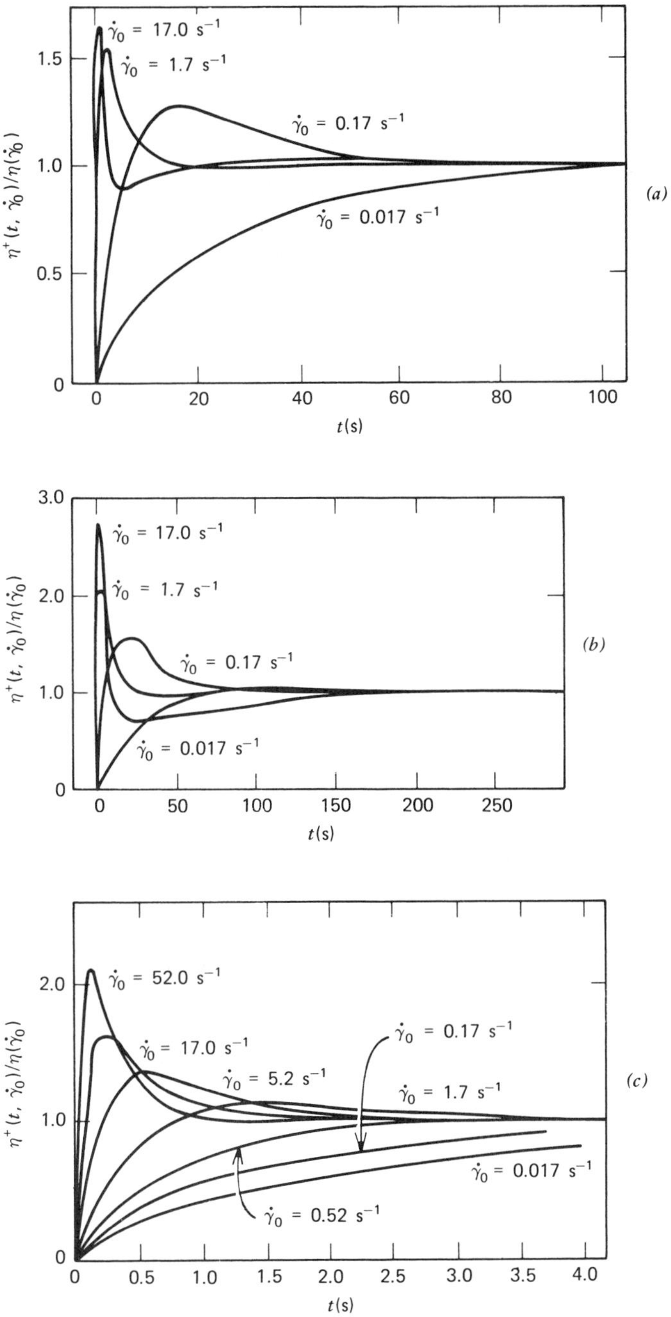

FIGURE 3.4-8. Shear stress growth function $\eta^+(t,\dot{\gamma}_0)/\eta(\dot{\gamma}_0)$ for two polymer solutions and an aluminum soap solution: (a) 1.5% polyacrylamide (Separan AP30) in a 50/50 mixture by weight of water and glycerin; (b) 2.0% polyisobutylene in Primol; and (c) 7% aluminum laurate in a mixture of decalin and m-cresol. All data were taken at 298 K. Note that the data are reduced with respect to $\eta(\dot{\gamma}_0)$ so that they all approach a common asymptote; all the data still lie within the linear viscoelastic envelope if plotted as $\eta^+(t, \dot{\gamma}_0)$. [Data of J. D. Huppler, I. F. Macdonald, E. Ashare, T. W. Spriggs, R. B. Bird, and L. A. Holmes, *Trans. Soc. Rheol.*, **11**, 181–204 (1967).]

$\dot{\gamma}_0$ is increased in such a way that the shear strain where non-linear effects are first detected is constant. For example, this critical strain for η^+ is about 1.5 for the low-density polyethylene melt in Fig. 3.4-7. For polymer melts in general, it appears that the maximum in η^+ also always occurs at about the same value of the shear strain $\gamma_{max} = t_{max}\dot{\gamma}_0$ for a given polymer,[7] and that γ_{max} is on the order of 2 to 3. Since the viscosity decreases with increasing shear rate, η^+ approaches successively lower steady-state asymptotes as $\dot{\gamma}_0$ is raised. The size of the shear-stress overshoot increases with increasing shear rate; this is reflected in plots of $\eta^+(t, \dot{\gamma}_0)/\eta(\dot{\gamma}_0)$, which are presented for three moderately concentrated solutions in Fig. 3.4-8.

The growing first normal stress difference shows the same qualitative dependence on $\dot{\gamma}_0$ as the shear stress (Figs. 3.4-9 and 10). The shear strains at which departure from the low-shear-rate envelope and at which the primary maximum occur for the first-normal-stress growth coefficient again appear to be independent of the imposed shear rate for melts, and they are larger than for the shear stress growth. For example, the deviation from the envelope occurs at about $\gamma = 2$ for Ψ_1^+ for the low-density polyethylene melt. The size of the maximum overshoot is larger for Ψ_1^+/Ψ_1 than η^+/η. Limited data[8] on $\Psi_2^+(t, \dot{\gamma}_0)$ indicate that $\Psi_2^+(t, \dot{\gamma}_0)/\Psi_1^+(t, \dot{\gamma}_0) = \Psi_2/\Psi_1$ for all $\dot{\gamma}_0$ and t.

Experiment d: Stress Relaxation after Cessation of Steady Shear Flow

In the *stress relaxation experiment* shown in Fig. 3.4-1d, the motion of a fluid that is undergoing steady shear flow with shear rate $\dot{\gamma}_0$ is suddenly stopped at say time $t = 0$ so that

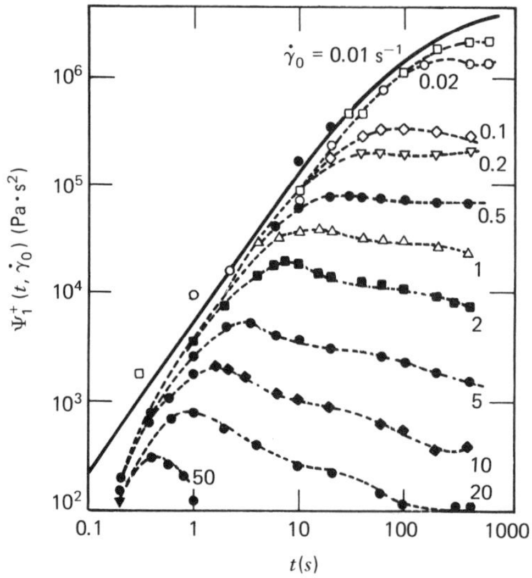

FIGURE 3.4-9. First normal stress growth function $\Psi_1^+(t, \dot{\gamma}_0)$ for the low-density polyethylene melt (Melt I) in Fig. 3.4-7. All Ψ_1^+ curves lie below the solid curve envelope which is approached for small shear rates; the solid curve is calculated from linear viscoelastic data (see Table 5.3-2) with the Lodge rubberlike liquid model (cf. Eqs. 8.2-1 and 4). [Reprinted with permission from M. H. Wagner and J. Meissner, *Macromol. Chem.*, **181**, 1533–1550 (1980), Hüthig and Wepf Verlag, Basel.]

[7] W. W. Graessley, *Adv. Polym. Sci.*, **16**, 1–179 (1974).

[8] W. R. Leppard and E. B. Christiansen, *AIChE J.*, **21**, 999–1006 (1975).

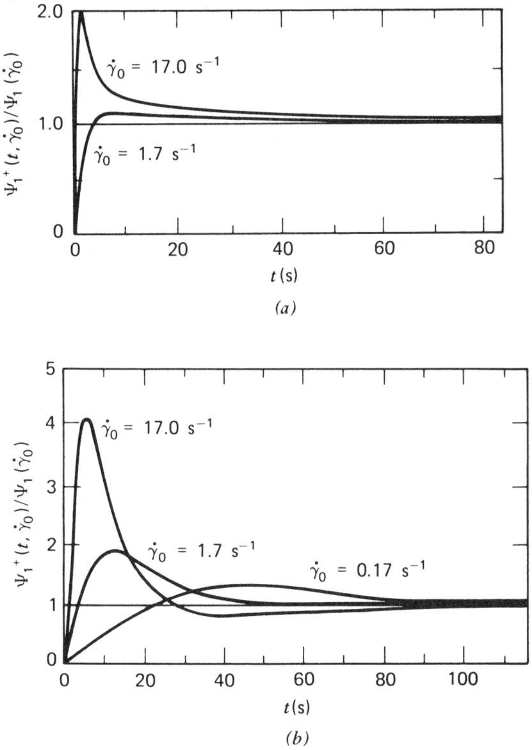

FIGURE 3.4-10. First normal stress growth function $\Psi_1^+(t, \dot{\gamma}_0)/\Psi_1(\dot{\gamma}_0)$ for (*a*) 1.5% polyacrylamide (Separan AP30) in a 50/50 mixture by weight of water and glycerin and (*b*) 2.0% polyisobutylene in Primol. All data were taken at 298 K. Note that Ψ_1^+ is reduced by the steady shear flow value Ψ_1 so that these curves show the relative sizes of the overshoot at different shear rates. [Data of J. D. Huppler, I. F. Macdonald, E. Ashare, T. W. Spriggs, R. B. Bird, and L. A. Holmes, *Trans. Soc. Rheol.*, **11**, 181–204 (1967).]

$\dot{\gamma} = 0$ for $t \geq 0$. The decay of the steady shear flow stresses to zero is then observed.[9] We can describe the relaxing stresses by the stress relaxation material functions $\eta^-(t, \dot{\gamma}_0)$, $\Psi_1^-(t, \dot{\gamma}_0)$, and $\Psi_2^-(t, \dot{\gamma}_0)$ (see Table 3.4-1) which are defined analogously to the viscometric functions. The superscripted minus sign is a reminder that the steady shear flow occurred for negative times. As in the definition of the stress growth functions, the $\dot{\gamma}_0$ in the argument of each material function indicates the parametric dependence on shear rate. For Newtonian fluids, $\eta^-(t, \dot{\gamma}_0) = \mu(1 - H(t))$, where $H(t)$ is the Heaviside unit step function, and $\Psi_1^-(t, \dot{\gamma}_0)$ and $\Psi_2^-(t, \dot{\gamma}_0)$ are both zero.

Experiments show that the stresses relax monotonically to zero and that they relax more rapidly as the shear rate $\dot{\gamma}_0$ in the preceding steady shear flow is increased. All the relaxation curves for η^- lie below the linear viscoelastic envelope approached in the limit $\dot{\gamma}_0 \to 0$. Furthermore it is found that the shear stress relaxes more rapidly than the first normal stress difference, which is similar to the slower response of the normal stresses in the stress growth experiment. Sample data for η^- and Ψ_1^- are shown in Figs. 3.4-11 to 14. Limited data on Ψ_2^- indicate[8] that $\Psi_2^-(t, \dot{\gamma}_0)/\Psi_1^-(t, \dot{\gamma}_0) = \Psi_2/\Psi_1$ for all $\dot{\gamma}_0$ and all t.

[9] For materials with a yield stress, the shear stress will not relax to zero but rather to the yield value. This method has been used to measure the yield stress of suspensions by N. Q. Dzuy and D. V. Boger, *J. Rheol.*, **27**, 321–350 (1983), and of foams by S. A. Khan, Sc.D. Thesis, Massachusetts Institute of Technology, Cambridge (1985).

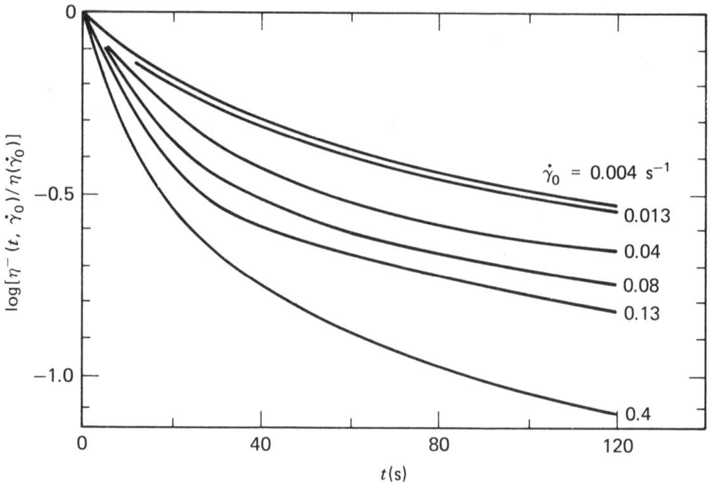

FIGURE 3.4-11. Shear stress relaxation function $\eta^-(t,\dot{\gamma}_0)/\eta(\dot{\gamma}_0)$ for a low molecular weight polyisobutylene melt ($\bar{M}_w \doteq 10^4$). [E. Mustafayev, A. Ya. Malkin, Ye. P. Plotnikova, and G. V. Vinogradov, *Vysokomol. Soedin.*, **6**, 1515–1521 (1964).]

Experiment e: Stress Relaxation after a Sudden Shearing Displacement

Another experiment in which we can observe relaxing stresses is *stress relaxation following a sudden shearing displacement* (Fig. 3.4-1e); this is sometimes referred to as *step-strain stress relaxation.* The shear strain γ_0 can be induced by applying a large, constant shear rate $\dot{\gamma}_0$ for a short time interval Δt, so that $\dot{\gamma}_0 \Delta t = \gamma_0$. As defined in Table 3.4-1, the time decay of the shear stress is described by the *relaxation modulus* $G(t, \gamma_0)$ and the relaxation of the first normal stress difference, by the function $G_{\Psi 1}(t, \gamma_0)$ for a step strain at $t = 0$.

Data for a low-density polyethylene melt and a polystyrene solution are shown in Figs. 3.4-15 and 16. For small shear strains, the relaxation modulus is found to be independent of γ_0

$$\lim_{\gamma_0 \to 0} G(t, \gamma_0) = G(t) \tag{3.4-7}$$

In this limit, the shear stress is linear in strain, so that the relaxation modulus is a function of time alone and contains the same linear viscoelastic information as G' and G'' (see Chapter 5). The effect of increasing strain on G and $G_{\Psi 1}$ is to decrease them, but not to alter their time dependence, except at very short times. Thus the curves for G and $G_{\Psi 1}$ in Fig. 3.4-15 and for G in Fig. 3.4-16 for different γ_0 are parallel and can be superposed by vertical shifting as demonstrated in Fig. 3.4-16. It is further observed that

$$\frac{G(t, \gamma_0)}{G_{\Psi 1}(t, \gamma_0)} = 1 \tag{3.4-8}$$

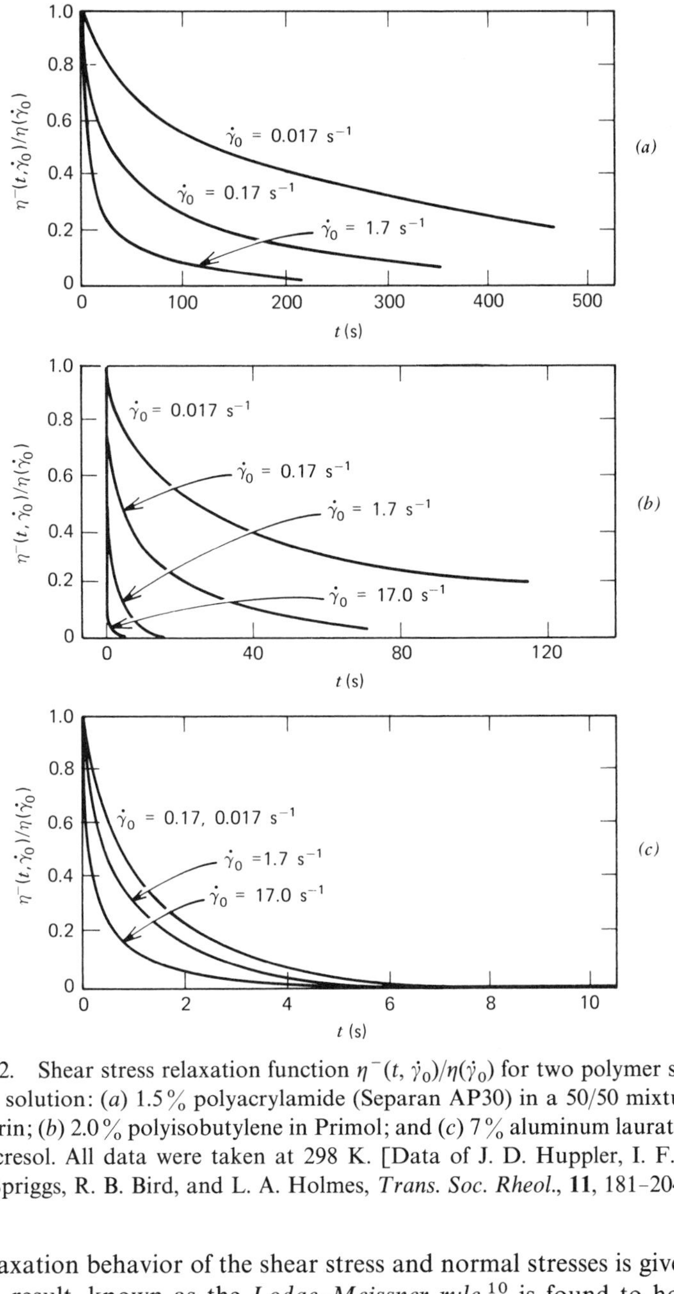

FIGURE 3.4-12. Shear stress relaxation function $\eta^-(t, \dot{\gamma}_0)/\eta(\dot{\gamma}_0)$ for two polymer solutions and an aluminum soap solution: (*a*) 1.5% polyacrylamide (Separan AP30) in a 50/50 mixture by weight of water and glycerin; (*b*) 2.0% polyisobutylene in Primol; and (*c*) 7% aluminum laurate in a mixture of decalin and *m*-cresol. All data were taken at 298 K. [Data of J. D. Huppler, I. F. Macdonald, E. Ashare, T. W. Spriggs, R. B. Bird, and L. A. Holmes, *Trans. Soc. Rheol.*, **11**, 181–204 (1967).]

that is, the relaxation behavior of the shear stress and normal stresses is given by the same function. This result, known as the *Lodge–Meissner rule*,[10] is found to hold for the few concentrated polymer solutions and melts for which it has been tested.

[10] A. S. Lodge and J. Meissner, *Rheol. Acta*, **11**, 351–352 (1972); see also A. S. Lodge, *Rheol. Acta*, **14**, 664–665 (1975), and O. Hassager and S. Pedersen, *J. Non-Newtonian Fluid Mech.*, **4**, 261–268 (1978). K. Osaki, S. Kimura, and M. Kurata, *J. Polym. Sci. Polym. Phys. Ed.*, **19**, 517–527 (1981), found that for concentrated solutions of polystyrene in Arochlor 1248 the ratio $-(\tau_{yy} - \tau_{zz})/(\tau_{xx} - \tau_{yy})$ is independent of time in stress relaxation following a step shear strain; that is, the two normal-stress differences relax with the same time dependence. However, the value of this ratio is found to depend on the size of the applied strain. For small strains this ratio is a constant, 0.17; for large strains, the ratio decreases with increasing strain.

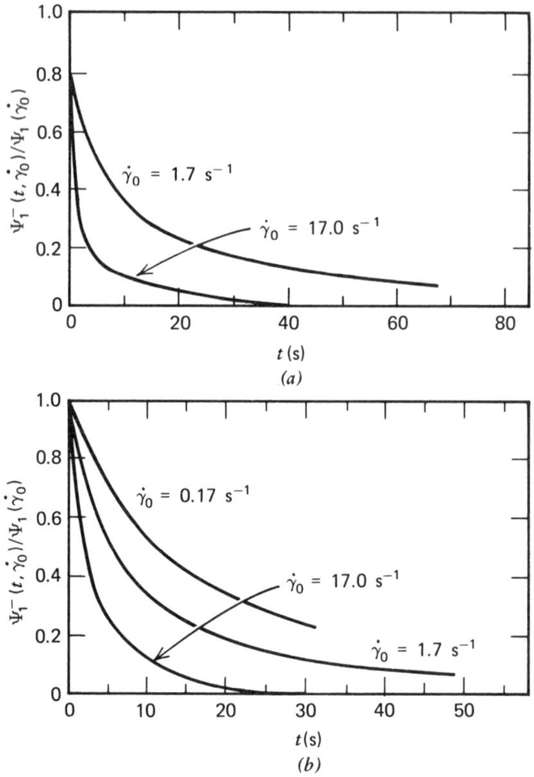

FIGURE 3.4-13. First normal stress relaxation function $\Psi_1^-(t, \dot\gamma_0)/\Psi_1(\dot\gamma_0)$ for (*a*) 1.5% polyacrylamide in water and glycerin and (*b*) 2.0% polyisobutylene in Primol. All data were taken at 298 K. [Data of J. D. Huppler, I. F. Macdonald, E. Ashare, T. W. Spriggs, R. B. Bird, and L. A. Holmes, *Trans. Soc. Rheol.*, **11**, 181–204 (1967).]

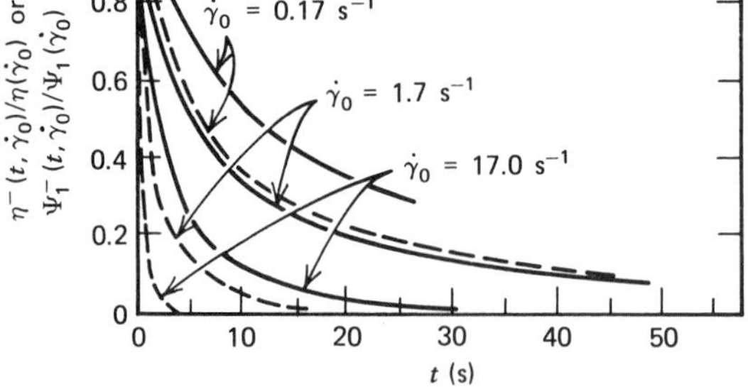

FIGURE 3.4-14. Comparison of first normal stress relaxation function $\Psi_1^-(t, \dot\gamma_0)/\Psi_1(\dot\gamma_0)$ and shear stress relaxation function $\eta^-(t, \dot\gamma_0)/\eta(\dot\gamma_0)$ for 2.0% polyisobutylene in Primol at 298 K. Note that the first normal stress difference (——) relaxes more slowly than the shear stress (----). [Data of J. D. Huppler, I. F. Macdonald, E. Ashare, T. W. Spriggs, R. B. Bird, and L. A. Holmes, *Trans Soc. Rheol.*, **11**, 181–204 (1967).]

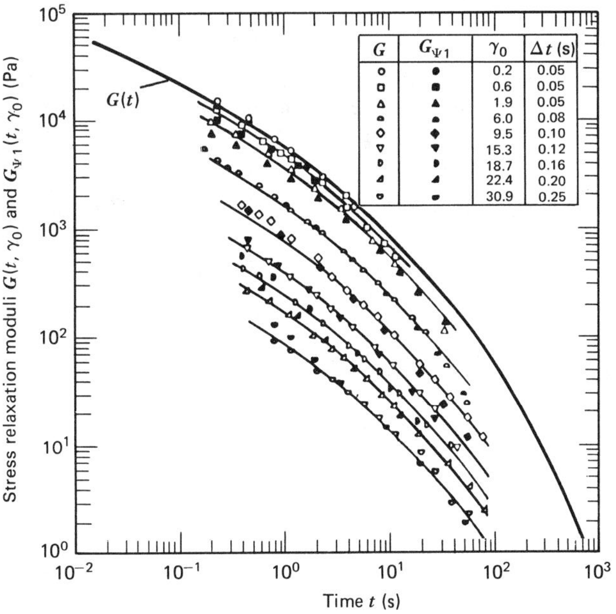

FIGURE 3.4-15. The relaxation modulus $G(t,\gamma_0)$ (open symbols) and normal stress relaxation function $G_{\Psi 1}(t, \gamma_0)$ (solid symbols) for a low-density polyethylene melt (Melt I). The indicated strains were obtained by applying a large, constant shear rate for time interval Δt; all data were taken at $T = 423$ K. The linear relaxation modulus curve $G(t)$ was calculated from the dynamic data in Fig. 3.4-3 (see Example 5.3-7). The solid curves connecting the sets of data points were obtained by shifting $G(t)$ vertically [H. M. Laun, *Rheol. Acta*, **17**, 1–15 (1978).]

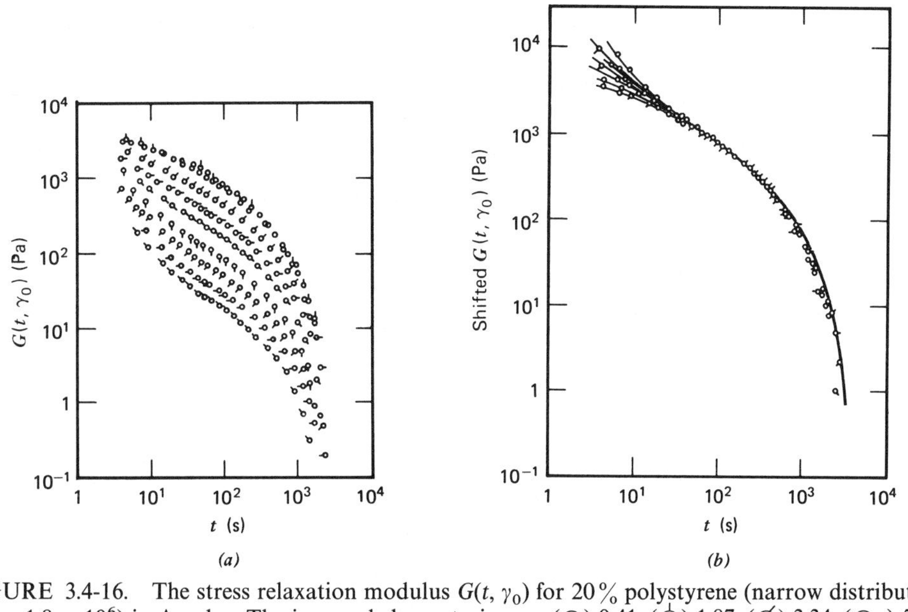

(*a*) (*b*)

FIGURE 3.4-16. The stress relaxation modulus $G(t, \gamma_0)$ for 20% polystyrene (narrow distribution $\bar{M}_w = 1.8 \times 10^6$) in Aroclor. The imposed shear strains are (○) 0.41, (○̍) 1.87, (○́) 3.34, (○-) 5.22, (Q) 6.68, (○̦) 10.0, (○̦) 13.4, (-○) 18.7, and (○̀) 25.4. Part (*a*) shows how $G(t, \gamma_0)$ varies with shear strain; note that the data for γ_0 equal to 0.41 and 1.87 are in the linear regime. In (*b*) the data are superposed by vertical shifting to show the similarity in $G(t, \gamma_0)$ at large times regardless of the imposed shear strain. [Y. Einaga, K. Osaki, M. Kurata, S. Kimura, and M. Tamura, *Polym. J. Jpn.*, **2**, 550–552 (1971).]

Experiment f: Creep

Instead of controlling the shear rate and measuring the resulting stresses, we can apply a prescribed time-dependent shear stress and measure the resulting shear strain. In the *creep experiment* (Fig. 3.4-1f), a constant shear stress $\tau_{yx} = \tau_0$ is applied at $t = 0$ to a fluid that was previously at rest and stress-free. It is convenient to measure the shear strain in the material relative to the application of the shear stress at $t = 0$

$$\gamma_{yx}(0, t) = \int_0^t \dot{\gamma}_{yx}(t')dt' \tag{3.4-9}$$

The "give" of the sample in response to the applied stress is sometimes described by the *creep compliance* $J(t, \tau_0)$ as defined in Table 3.4-1. For small values of τ_0 the strain is proportional to stress so that

$$\lim_{\tau_0 \to 0} J(t, \tau_0) = J(t) \tag{3.4-10}$$

The linear viscoelastic function $J(t)$ contains the same information as $G(t)$, G', and G''.[11]

Creep data for a low-density polyethylene melt are shown in Fig. 3.4-17. By a reduced time of 200 s the strain is proportional to time, so that steady state has been reached. Comparison of this time with the corresponding time of 500 to 600 s to reach steady state in start-up of steady shear flow with constant shear rate of 1 s^{-1} (this gives the same steady state in the two experiments since $\eta(1\ s^{-1}) = 10^4$ Pa·s) in Fig. 3.4-7, indicates that the time constant for approach to steady state is different for the two different experiments. At least for the low-density polyethylene shown here steady state is achieved

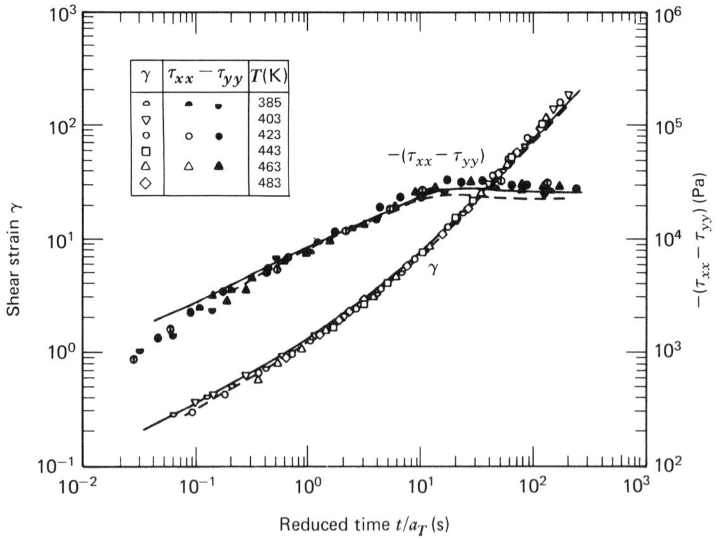

FIGURE 3.4-17. The shear strain and first normal stress difference for a low-density polyethylene melt (Melt I) as functions of reduced time t/a_T in a creep experiment. The data are all taken for an applied shear stress of 10^4 Pa, and shift factors a_T are the same as found in Fig. 3.3-2. The reference temperature is $T_0 = 423$ K. The solid and the dashed curves were calculated with the Wagner model (cf. Fig. 3.3-2). [M. H. Wagner and H. M. Laun, *Rheol. Acta*, **17**, 138–148 (1978).]

[11] J. D. Ferry, *Viscoelastic Properties of Polymers*, 3rd ed., Wiley, New York (1980), Chapt. 3.

faster in creep than in the constant shear rate test. Also shown in Fig. 3.4-17 is the transient first normal stress difference; note that it appears to show a small overshoot under these experimental conditions.

Experiment g: Constrained Recoil after Steady Shear Flow

If the shear stress is suddenly removed from a viscoelastic fluid undergoing steady shear flow (Fig. 3.4-1g) and if the separation between the plates is held constant, the fluid will recoil to some position it previously occupied. We have seen an example of this *constrained recoil* in tube flow of an aqueous carboxymethylcellulose solution in the sequence of photographs in §2.5. The recovery of the fluid is described by the shear strain $\gamma_r(0, t, \dot{\gamma}_0)$ during recoil measured relative to the time when the shear stress is removed; the time evolutions of the shear stress and shear strain are illustrated in Fig. 3.4-18.

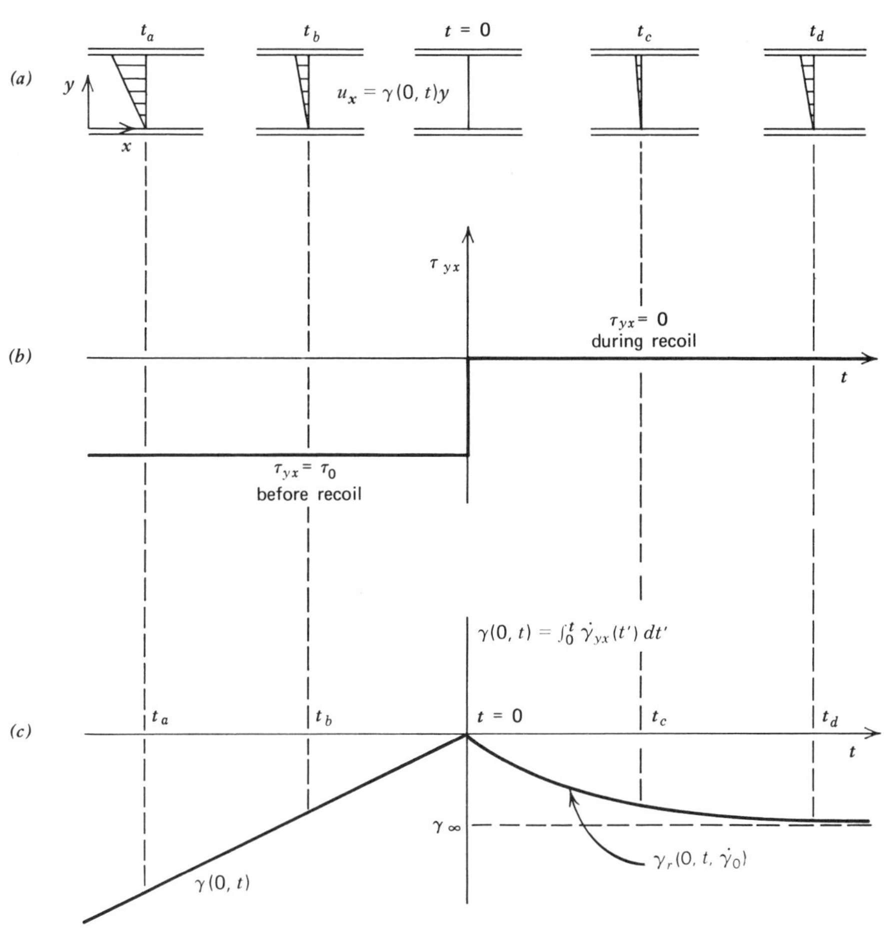

FIGURE 3.4-18. Constrained recoil after cessation of steady shear flow. (*a*) Diagram showing the fluid deformation before and after recoil begins. (*b*) Sketch of shear stress as a function of time. (*c*) Shear strain measured relative to the configuration at the time of shear stress removal $\gamma(0, t) = \int_0^t \dot{\gamma}_{yx}(t')\,dt'$; strain during recoil is denoted by γ_r and the ultimate recoil by γ_∞. Note that γ_r depends parametrically on the shear rate $\dot{\gamma}_0$ prior to the recoil experiment.

At long times the recoil approaches the *ultimate recoil* or *recoverable shear* $\gamma_\infty(\dot{\gamma}_0)$, which depends on the value of shear rate (or shear stress) in the steady shear flow preceding the recoil. The steady state compliance, J_e^0, is defined in terms of the ultimate recoil by

$$\begin{aligned}\gamma_\infty &= J_e^0(\tau_0)\tau_0 \\ &= -J_e^0(\dot{\gamma}_0)\eta(\dot{\gamma}_0)\dot{\gamma}_0\end{aligned} \tag{3.4-11}$$

In the limit of small τ_0 or $\dot{\gamma}_0$, the steady state compliance is a linear viscoelastic property equal to (see Eq. 5.3-33 and Table 5.3-1)

$$J_e^0 = \lim_{\omega \to 0} \frac{G'}{\omega^2 \eta_0^2} = \frac{\Psi_{1,0}}{2\eta_0^2} \tag{3.4-12}$$

Note that in the linear limit the (steady state) recoverable shear is

$$\boxed{\gamma_\infty = -\frac{\Psi_{1,0}\dot{\gamma}_0}{2\eta_0}} \tag{3.4-13}$$

This formula for recoverable shear is often (incorrectly) used outside of the linear viscoelastic regime with $\Psi_{1,0}$ and η_0 replaced by $\Psi_1(\dot{\gamma}_0)$ and $\eta(\dot{\gamma}_0)$. Although this is conventional, the connection formula between J_e^0 and $\Psi_1/2\eta^2$ in Eq. 3.4-12 does not hold outside of the zero-shear-rate regime.

A consequence of constraining the fluid recovery by maintaining a fixed gap between the plates is that the normal stresses do not go to zero when the shear stress is made zero; rather, they relax over a period of time comparable to that required for the recoil. If all

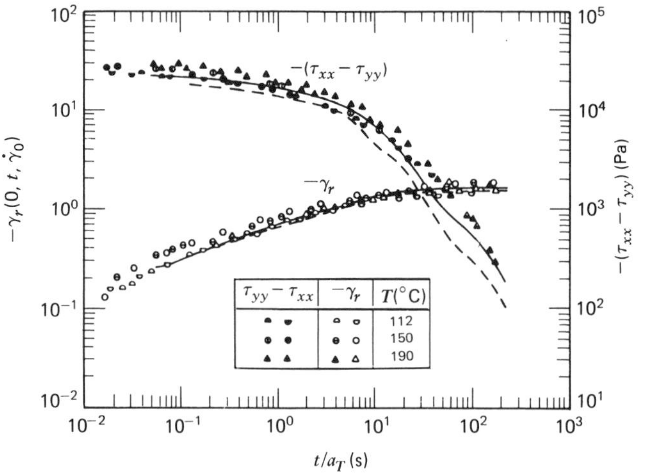

FIGURE 3.4-19. Constrained recoil after steady shear flow is achieved in the creep tests shown in Fig. 3.4-17. The shear strain is measured relative to the time at which the shear stress is set to zero; it is negative because recoil occurs in the direction opposite to the preceding shear flow. The data from three different temperatures are superposed on a single curve at reference temperature $T_0 = 150$°C by use of the reduced time scale t/a_T. The relaxing first normal stress difference shows that $\tau_{xx} - \tau_{yy}$ does not become zero immediately when τ_{yx} is set to zero. The solid and dashed curves are calculated with the Wagner model (cf. Fig. 3.3-2). [M. H. Wagner and H. M. Laun, *Rheol. Acta*, **17**, 138–148 (1978).]

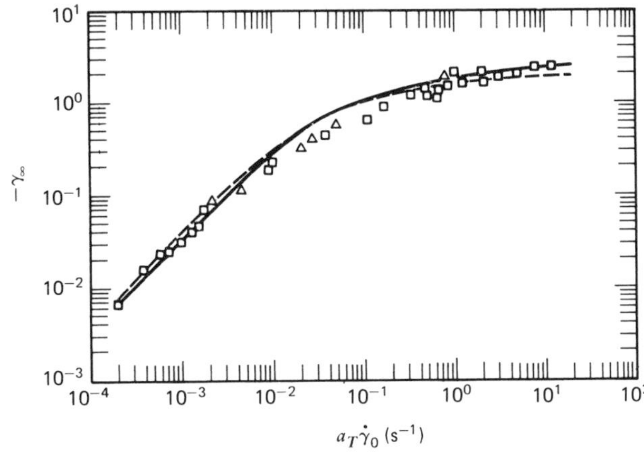

FIGURE 3.4-20. Dependence of the ultimate recoil γ_∞ on the shear rate $\dot{\gamma}_0$ of the steady shear flow preceding the recoil experiment for the low-density polyethylene melt shown in Fig. 3.4-19. Recoil data were taken at $T = 423$ K following steady shear flow obtained from both constant shear rate (□) and constant shear stress (○) experiments. The (○) data are from Fig. 3.4-19 and the (□) data from Fig. 3.4-21. The solid and dashed curves are calculated with the Wagner model (cf. Fig. 3.3-2). [M. H. Wagner and H. M. Laun, *Rheol. Acta*, **17**, 138–148 (1978).]

stresses were removed from the sample at $t = 0$, the fluid might expand laterally during the recoil.[12] Unconstrained or free recovery is thus not necessarily a shear flow. Since constrained recoil can be pictured as the opposite to a creep experiment, it is sometimes called "creep recovery".

Constrained recoil data for a low-density polyethylene melt are shown in Figs. 3.4-19 through 21. In Fig. 3.4-19 the recoil from the steady shear flow achieved after the creep

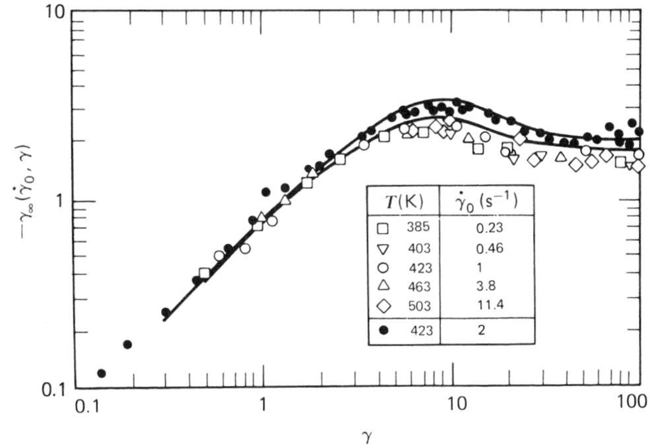

FIGURE 3.4-21. Ultimate recoil as a function of shear strain following start-up of steady shear flow; the shear stress was set to zero after the indicated strains in the start up of steady shear flow. The open symbols correspond to measurements made at the indicated temperatures and shear rates, and all correspond to a reduced shear rate of $a_T\dot{\gamma}_0 = 1\ \mathrm{s}^{-1}$ at a reference temperature of $T_0 = 423$ K. The solid symbols were taken at $\dot{\gamma}_0 = 2\ \mathrm{s}^{-1}$ and $T = 423$ K. The solid curves were calculated with the Wagner model (cf. Fig. 3.3-2). [M. H. Wagner and H. M. Laun, *Rheol. Acta*, **17**, 138–148 (1978).]

[12] A. S. Lodge, *Elastic Liquids*, Academic Press, New York (1964), pp. 131–138, 239–242.

experiments in Fig. 3.4-17 is shown. The time required to attain the ultimate recoil is approximately the same as the time to reach steady state in the creep experiments. The ultimate recoil increases with increasing value of the steady shear rate $\dot{\gamma}_0$ preceding the recovery as shown in Fig. 3.4-20; the larger the shear rate, the more strain is imposed on the sample within the memory span of the fluid. Recoil experiments can also be done on fluid that is not undergoing steady shear flow. For example start-up of steady shear flow experiments can be stopped short of steady state at various shear strains $\gamma(0, t) = \dot{\gamma}_0 t$ by setting τ_{yx} to zero and the ultimate[13] recoil $\gamma_\infty(\dot{\gamma}_0, \gamma)$ observed. These data are illustrated in Fig. 3.4-21 for two different shear rates where it is seen that a maximum value of $\gamma_\infty(\dot{\gamma}_0, \gamma)$ occurs at strains less than that required to attain steady shear flow.

§3.5 SHEARFREE FLOW MATERIAL FUNCTIONS

Since we assume that the fluid is isotropic, the stresses and material functions in simple shearfree flows depend only on $\dot{\varepsilon}(t)$ and the parameter b that defines the type of flow. For steady simple shearfree flows we define two viscosity functions $\bar{\eta}_1$ and $\bar{\eta}_2$ to describe the two normal stress differences introduced in §3.2

$$\tau_{zz} - \tau_{xx} = -\bar{\eta}_1(\dot{\varepsilon}, b)\dot{\varepsilon} \tag{3.5-1}$$

$$\tau_{yy} - \tau_{xx} = -\bar{\eta}_2(\dot{\varepsilon}, b)\dot{\varepsilon} \tag{3.5-2}$$

For the special steady-state shearfree flow where $b = 0$, $\bar{\eta}_2 = 0$ and $\bar{\eta}_1$ is equal to the *elongational viscosity* $\bar{\eta}$:

$$\bar{\eta}(\dot{\varepsilon}) = \bar{\eta}_1(\dot{\varepsilon}, 0); \qquad \bar{\eta}_2(\dot{\varepsilon}, 0) = 0 \tag{3.5-3}$$

For $\dot{\varepsilon} > 0$, $\bar{\eta}$ describes elongational flow, and for $\dot{\varepsilon} < 0$, it describes biaxial stretching. The elongational viscosity is sometimes called the "Trouton viscosity" or the "extensional viscosity."

Sample data for $\bar{\eta}$ are shown in Fig. 3.5-1 for a polystyrene melt.[1] Also shown for comparison is the viscosity at the same temperature. At low elongation rates the elongational viscosity approaches a constant value known as the *zero-elongation-rate elongational viscosity* $\bar{\eta}_0$, which is three times the zero-shear-rate viscosity. Recall that $\bar{\eta} = 3\mu$ for Newtonian fluids (cf. Eq. 1B.7-3). As the elongation rate is increased, $\bar{\eta}(\dot{\varepsilon})$ is seen to increase somewhat, and then decrease at still higher elongation rates. This is contrasted in Fig. 3.5-1 with the viscosity $\eta(\dot{\gamma})$, which decreases with increasing shear rate. The shear stress at which η begins to "shear thin" is approximately the same as the stress difference $(\tau_{zz} - \tau_{xx})$ at which $\bar{\eta}$ begins to deviate from $\bar{\eta}_0$. For other polymer melts, different behavior for $\bar{\eta}(\dot{\varepsilon})$ is found. For example, for a polyisobutylene–isoprene copolymer[2] $\bar{\eta}$ appears to be independent of elongation rate and no maximum is observed; for high-density polyethylene,[3] $\bar{\eta}$ is observed to decrease with increasing $\dot{\varepsilon}$ without ever going through a

[13] We use the same terminology, ultimate recoil, for both $\gamma_\infty(\dot{\gamma}_0)$ and $\gamma_\infty(\dot{\gamma}_0, \gamma)$. The number of arguments indicates whether the recoil is measured following steady shear flow.

[1] H. Münstedt, *J. Rheol.*, **24**, 847–867 (1980).

[2] J. F. Stevenson, *AIChE J.*, **18**, 540–547 (1972).

[3] H. M. Laun in G. Astarita, G. Marrucci, and L. Nicolais, eds., *Rheology, Vol. 2: Fluids*, Plenum Press, New York (1980), pp. 419–424.

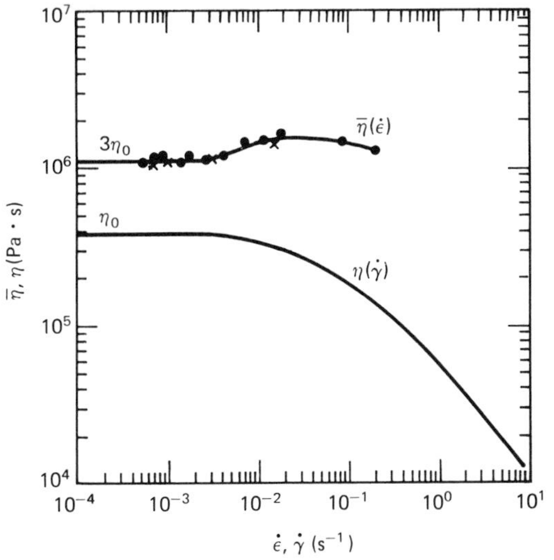

FIGURE 3.5-1. Elongational viscosity $\bar{\eta}$ and viscosity η for a polystyrene melt ($\bar{M}_w = 2.19 \times 10^5$, $\bar{M}_w/\bar{M}_n = 2.3$; sample denoted as PS IV in Fig. 3.5-2) as functions of elongation rate and shear rate, respectively. The experimentally determined zero-strain-rate values of $\bar{\eta}_0 = 1.1 \times 10^6$ Pa·s and $\eta_0 = 3.7 \times 10^5$ Pa·s agree closely with the relation $\bar{\eta}_0 = 3\eta_0$, as they must. The temperature is $T = 433$ K. [Data replotted from H. Münstedt, *J. Rheol.*, **24**, 847–867 (1980).]

TABLE 3.5-1

Definitions of Material Functions in Shearfree Flows

$$v_x = -\tfrac{1}{2}\dot{\varepsilon}(1+b)x, \quad v_y = -\tfrac{1}{2}\dot{\varepsilon}(1-b)y, \quad v_z = \dot{\varepsilon}z$$

Flow	Material Function	Defining Equation
a. Steady shearfree flow $\dot{\varepsilon} = \text{constant}$	$\bar{\eta}_1(\dot{\varepsilon})$ $\bar{\eta}_2(\dot{\varepsilon})$	$\tau_{zz} - \tau_{xx} = -\bar{\eta}_1\dot{\varepsilon}$ $\tau_{yy} - \tau_{xx} = -\bar{\eta}_2\dot{\varepsilon}$
b. Stress growth on inception of steady shearfree flow $\dot{\varepsilon} = \begin{cases} 0 & t < 0 \\ \dot{\varepsilon}_0 & t \geq 0 \end{cases}$	$\bar{\eta}_1^+(t, \dot{\varepsilon}_0)$ $\bar{\eta}_2^+(t, \dot{\varepsilon}_0)$	$\tau_{zz} - \tau_{xx} = -\bar{\eta}_1^+\dot{\varepsilon}_0$ $\tau_{yy} - \tau_{xx} = -\bar{\eta}_2^+\dot{\varepsilon}_0$
c. Elongational creep $(b = 0)$ $(\tau_{zz} - \tau_{xx}) = \begin{cases} 0 & t < 0 \\ \sigma_0 & t \geq 0 \end{cases}$	$D(t, \sigma_0)$	$\varepsilon(0, t) = -D\sigma_0$
d. Free recovery after steady elongational flow $(b = 0)$ $(\tau_{zz} - \tau_{xx}) = \begin{cases} \sigma_0 & t < 0 \\ 0 & t \geq 0 \end{cases}$	$\varepsilon_r(0, t, \sigma_0)$ $\varepsilon_\infty(\sigma_0)$	$\varepsilon_r = \int_0^t \dot{\varepsilon}(t')\, dt'$ $\varepsilon_\infty = \lim_{t\to\infty} \varepsilon_r$

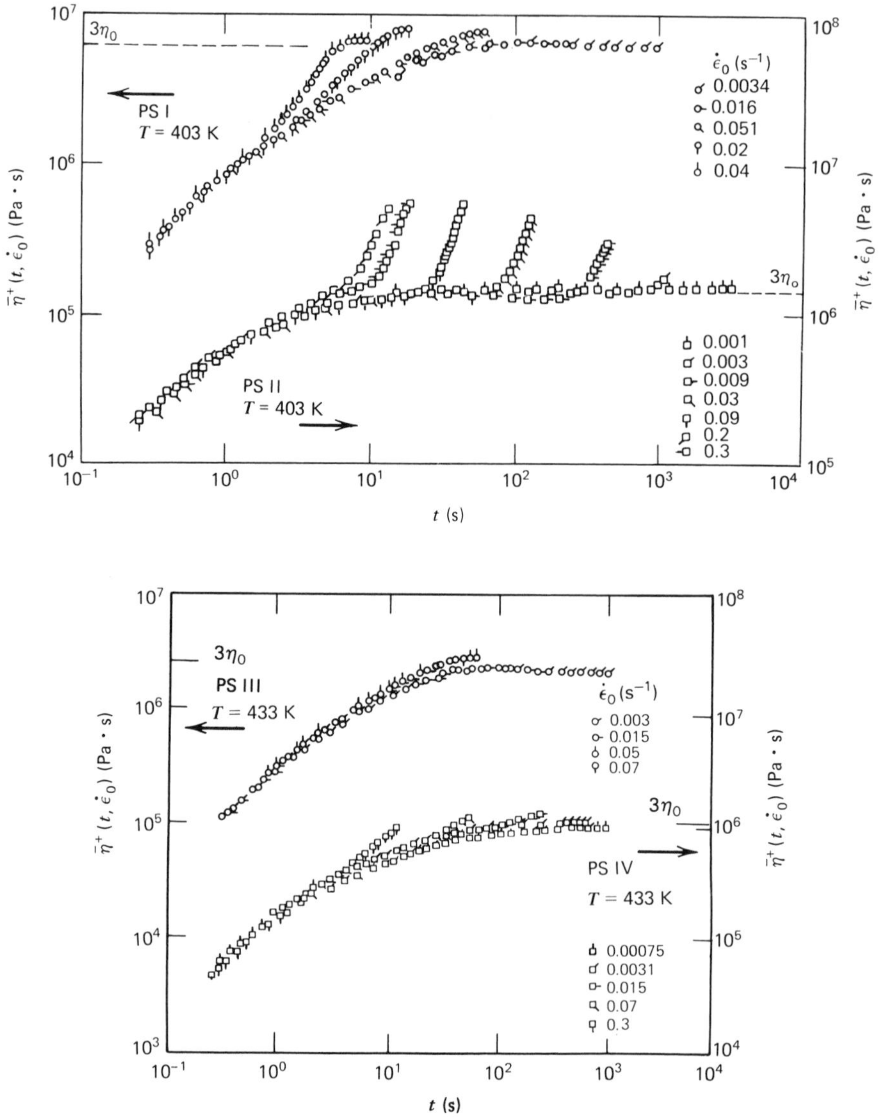

FIGURE 3.5-2. Time dependence of the elongational stress growth viscosity $\bar{\eta}^+$ for four polystyrene melts. The constant elongation rate $\dot{\varepsilon}_0$ is applied to the samples for $t \geq 0$. The number average and weight average molecular weights of the samples are as follows:

	$\bar{M}_w$	$\bar{M}_w/\bar{M}_n$
PS I	7.4×10^4	1.2
PS II	3.9×10^4	1.1
PS III	2.53×10^5	1.9
PS IV	2.19×10^5	2.3

[Data from H. Münstedt, *J. Rheol.*, **24**, 847–867 (1980).]

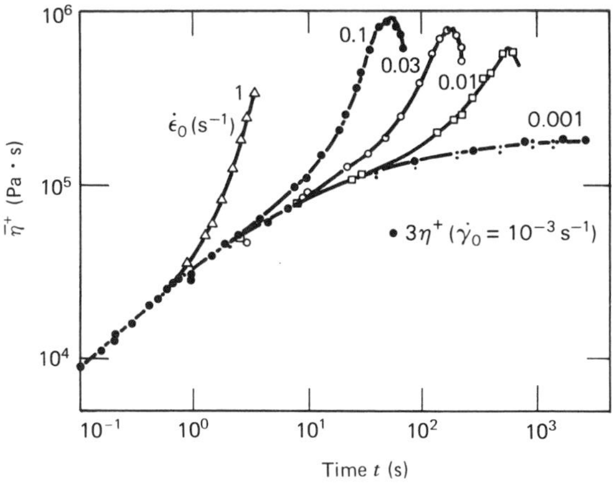

FIGURE 3.5-3. Elongational stress growth viscosity $\bar{\eta}^+$ as a function of time for a low-density polyethylene melt. Note the agreement between $\bar{\eta}^+$ and $3\eta^+$ at low $\dot{\varepsilon}_0$ and $\dot{\gamma}_0$. All data were taken at $T = 423$ K. [J. Meissner, *Chem. Engr. Commun.*, **33**, 159–180 (1985).]

maximum. For other polymer melts, e.g, low-density polyethylene, and polymer solutions it has been impossible thus far to achieve steady elongational flow and measure $\bar{\eta}$.

Because of the difficulty in reaching steady state in many shearfree flows of polymeric liquids, transient shearfree flows are very important. Several of the most commonly used of these are summarized in Table 3.5-1. For start-up of steady elongational flow the transient stress is described by the *elongational stress growth function* $\bar{\eta}^+$. Data for this material function are illustrated in Fig. 3.5-2 for several polystyrene melts including the one shown in Fig. 3.5-1. At low strain rates the transient viscosity $\bar{\eta}^+$ is seen to approach the zero-elongation-rate elongational viscosity exponentially. The limiting shape of the $\bar{\eta}^+$ curve at small $\dot{\varepsilon}_0$ is independent of $\dot{\varepsilon}_0$ and is equal to $3\eta^+$ in the limit of zero shear rate; $\bar{\eta}^+(t, 0)$ can be predicted from linear viscoelasticity. (see Chapter 5).

The most striking feature of the $\bar{\eta}^+$ curves is the abrupt upturn, or "strain hardening," that occurs at a roughly constant value of *Hencky strain* $\varepsilon(0, t) = \dot{\varepsilon}_0 t$, the value of which varies from fluid to fluid. Although steady state is approached in most of the tests shown, for the PS II sample there is no indication of a steady state even at the lowest applied strain rates. Note that the strains applied to the samples are quite large; the largest value of ε achieved of about 3.5 corresponds to a ratio of final to initial length of the sample of about $e^{3.5} = 33$ (see Eq. 3.1-4).

The unattainability of a steady state in $\bar{\eta}^+$ has been observed in other melts. The most striking example is the study of a low-density polyethylene melt[4] for which data are shown in Fig. 3.5-3. There $\bar{\eta}^+$ at $\dot{\varepsilon}_0 = 10^{-3}\text{s}^{-1}$ is seen to agree well with $3\eta^+$ measured at a shear rate of $\dot{\gamma} = 10^{-3}$ s^{-1}. For higher elongation rates the characteristic strain hardening is observed followed by a *maximum* in $\bar{\eta}^+$. No steady state is observed, and $\bar{\eta}^+$ is still decreasing at the highest values of strain. In these experiments strains of $\varepsilon = 7$ were achieved which correspond to increasing the sample length by a factor of 1100! It is not clear that $\bar{\eta}$ exists for some materials and conditions.

Measurements have also been made on the stress growth material functions for other types of shearfree flows. These are illustrated in Figs. 3.5-4 and 5 for a polyisobutylene. Figure 3.5-4 contains data for both elongational flow and biaxial elongational flow. The

[4] T. Raible, A. Demarmels, and J. Meissner, *Polym. Bull.*, **1**, 397–402 (1979); T. Raible, Ph.D. Dissertation No. 6751, ETH Zürich (1981).

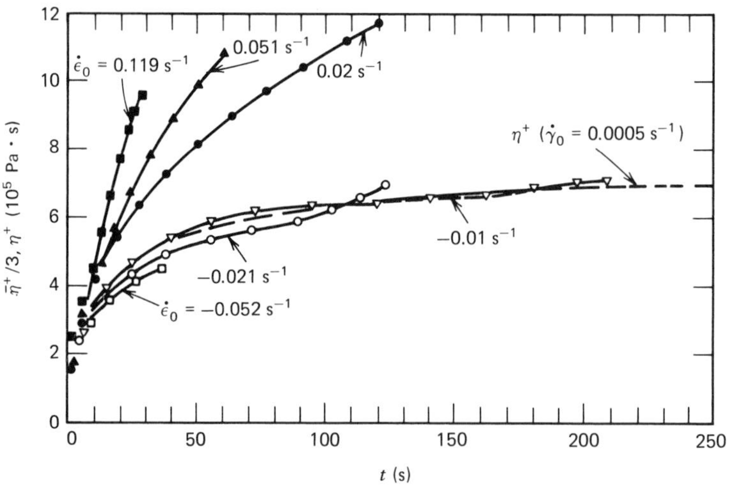

FIGURE 3.5-4. Elongational stress growth viscosity $\bar{\eta}^+$ for start-up of elongational ($\dot{\varepsilon}_0 > 0$) and biaxial elongational ($\dot{\varepsilon}_0 < 0$) flows. Data for η^+ at a very low shear rate are shown for comparison. The data for $\bar{\eta}^+$ are normalized so that they are equal to η^+ in the limit as $\dot{\varepsilon} \to 0$ and $\dot{\gamma} \to 0$. Data are for a polyisobutylene at $T = 296$ K. [J. Meissner, *Pure Appl. Chem.*, **56**, 369–384 (1984).]

data are normalized so that both sets should equal η^+ in the linear viscoelastic limit. Whereas $\bar{\eta}^+(\dot{\varepsilon}_0 > 0)$ rises above this limit as seen in the previous figures, $\bar{\eta}^+(\dot{\varepsilon}_0 < 0)$ falls below η^+ at small strains and then appears to exceed η^+ at large strains. For the planar elongational flow ($b = 1$) in Fig. 3.5-5, $\bar{\eta}_1^+/4$ at large $\dot{\varepsilon}_0$ exceeds η^+ at moderate strains but may fall below η^+ at larger strains. In contrast $\bar{\eta}_2^+/2$ appears to decrease from the linear viscoelastic envelope and remain below it. A few measurements have also been made on this polyisobutylene for shearfree flows with b between 0 and 1.[5]

Thus far we have considered the transient shearfree flows in which $\dot{\varepsilon}$ is varied and the stresses are measured. Next we consider two experiments in which the stress is varied and

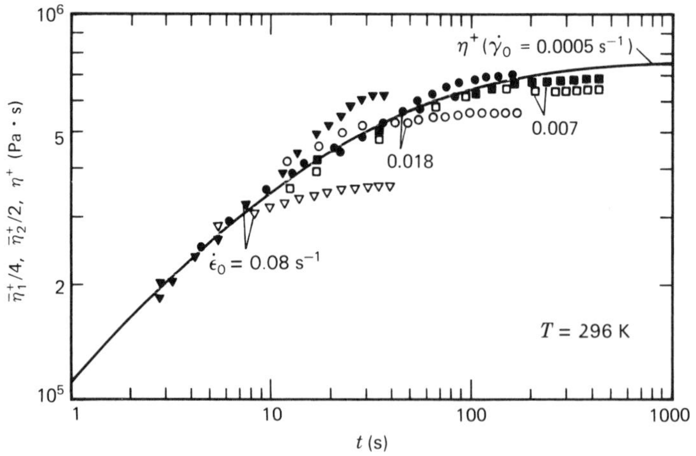

FIGURE 3.5-5. Stress growth viscosities $\bar{\eta}_1^+$ (filled symbols) and $\bar{\eta}_2^+$ (hollow symbols) for start-up of planar elongational flow ($b = 1$). Both $\bar{\eta}_1^+$ and $\bar{\eta}_2^+$ are normalized so they are equal to η^+ in the limit $\dot{\varepsilon}_0 \to 0$ and $\dot{\gamma}_0 \to 0$. Data are for the same material and temperature as Fig. 3.5-4. [J. Meissner, *Pure Appl. Chem.*, **56**, 369–384 (1984).]

[5] A. Demarmels, Ph.D. Dissertation No. 7345, ETH Zürich (1983).

the strain or strain rate is measured. We restrict attention to elongational flow, with $b = 0$, and we denote the normal stress difference $(\tau_{zz} - \tau_{xx})$ by σ. For a cylindrical sample being pulled in the z direction with its sides exposed to atmospheric pressure, σ is equal to the stress applied to the sample by the mechanical clamping arrangement. If σ is suddenly fixed at σ_0 for $t \geq 0$ on a sample that was at equilibrium prior to $t = 0$, then we have an *elongational creep* experiment. If after some amount of deformation ε in the creep experiment the stress σ is suddenly set to 0, then we can observe *free recovery* following elongational flow.

Data for the elongational creep compliance $D(t, \sigma_0)$ defined in Table 3.5-1 are shown in Fig. 3.5-6 for the four polystyrenes presented in Fig. 3.5-2. Steady state is achieved when the slope of log D vs log t is $+1$. As in the data presented in Fig. 3.5-2, no steady state is observed for the PS II sample except at the smallest value of σ.

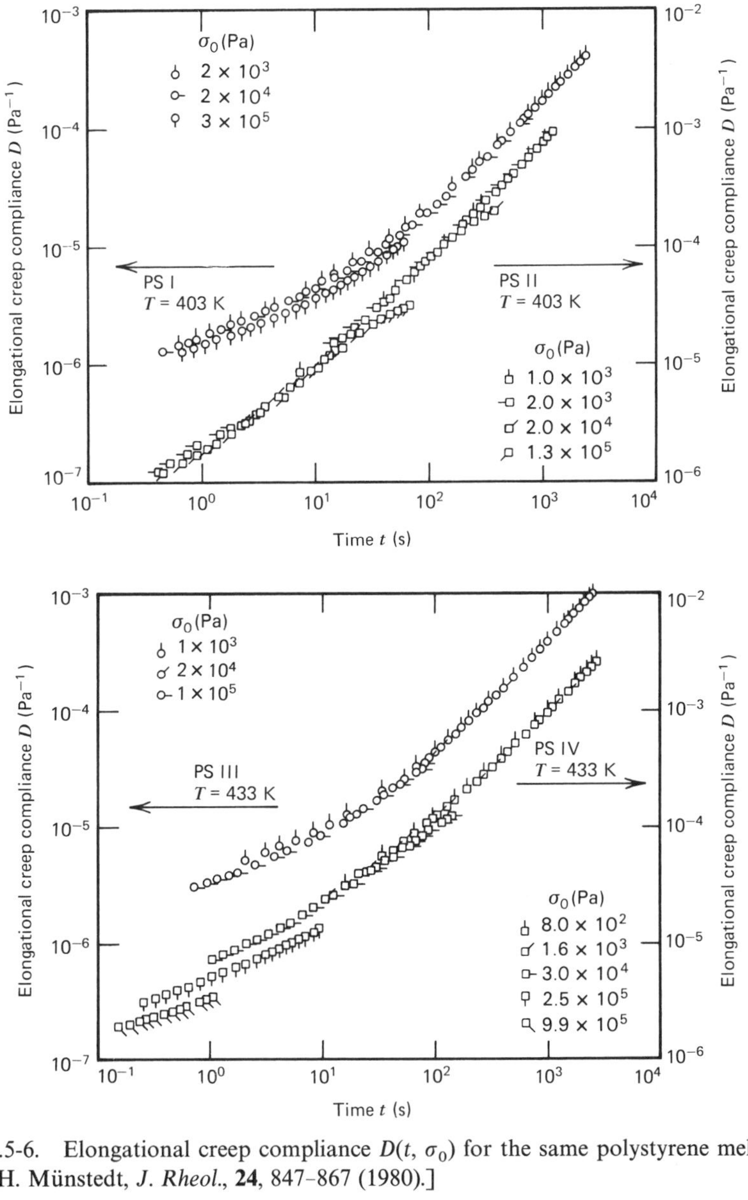

FIGURE 3.5-6. Elongational creep compliance $D(t, \sigma_0)$ for the same polystyrene melts shown in Fig. 3.5-2. [H. Münstedt, *J. Rheol.*, **24**, 847–867 (1980).]

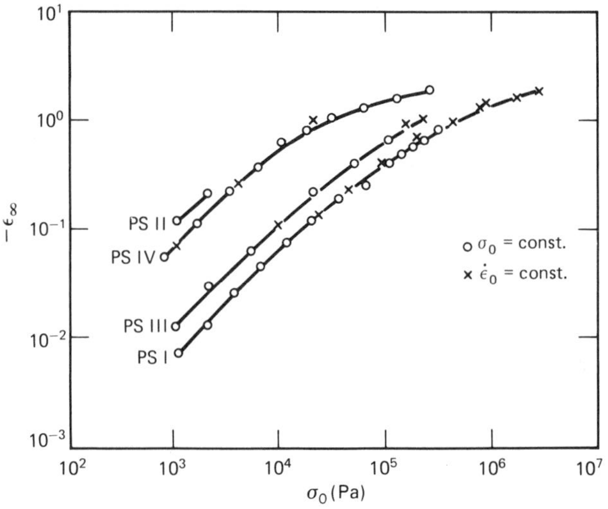

FIGURE 3.5-7. Ultimate recoil ε_∞ following steady elongational flow for the four polystyrene melts and conditions shown in Fig. 3.5-2. [H. Münstedt, *J. Rheol.*, **24**, 847–867 (1980).]

The *ultimate recoil* ε_∞ following steady elongational flow is shown in Fig. 3.5-7. Limited data on PS II is plotted since its steady state was generally not attained. Note that some of the recovery experiments were performed after start-up of steady elongational flow and some after elongational creep. Since steady state was reached prior to the recovery experiment, the two sets of data agree closely. The negative of the ultimate recoil increases monotonically with increasing applied strain rate or tensile stress. For large values of σ_0 the ultimate recoil approaches -2 for the PS I and PS IV samples. This means that the length of a specimen can decrease by more than seven fold following elongational flow! This is an amazing demonstration of "memory" in a liquid system.

Because it is difficult to achieve steady elongational flow, recovery experiments are often performed as a function of applied strain following the initiation of creep or constant

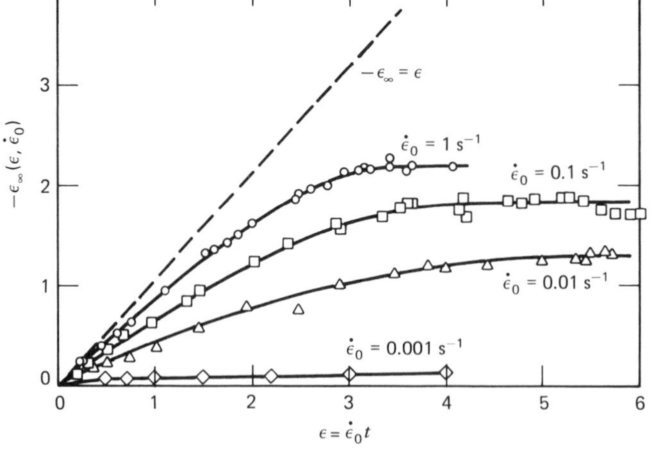

FIGURE 3.5-8. Ultimate recoil ε_∞ $(\varepsilon, \dot{\varepsilon}_0)$ after a strain of ε is applied in start-up of elongational flow with elongation rate $\dot{\varepsilon}_0$ for a low-density polyethylene melt (IUPAC A). All data are taken at T = 423 K. [Data from H. Münstedt and H. M. Laun, *Rheol. Acta*, **18**, 492–504 (1979).]

strain rate flow. In these experiments ultimate recoil results are presented as $\varepsilon_\infty(\varepsilon, \sigma_0)$ or $\varepsilon_\infty(\varepsilon, \dot{\varepsilon}_0)$, respectively. Here ε is the strain accrued from the beginning of the elongational flow to the point that σ_0 or $\dot{\varepsilon}_0$ is set to zero. Data for a low-density polyethylene melt for $\varepsilon_\infty(\varepsilon, \dot{\varepsilon}_0)$ are shown in Fig. 3.5-8. The highest amounts of recovery are around $-\varepsilon_\infty = 2.3$ which corresponds to a length contraction of a factor of 10! In this figure the $\varepsilon_\infty = -\varepsilon$ curve represents total recovery of the deformation. For high strain rates and short times, the actual recovery is seen to approach this limiting curve.

§3.6 USEFUL CORRELATIONS FOR MATERIAL FUNCTIONS[1–4]

In §§3.3 to 3.5 we described how the material functions depend on the kinematics, or fluid motion. These material properties also depend on the chemical constitution of the polymeric fluid—for example, solvent, type of polymer, molecular weight, and molecular weight distribution—and on the physical state of the fluid as measured by polymer concentration, temperature, and pressure. In this final section we wish to illustrate how these other variables affect the material functions.

a. Temperature Effects[5–7]

The material functions are strong functions of temperature. In Fig. 3.3-1, the influence of temperature on the viscosity of a low-density polyethylene melt is shown. There it can be seen that the zero-shear-rate viscosity decreases by two orders of magnitude as the temperature is raised from 388 to 513 K. More importantly, the viscosity-shear-rate curves evidently have similar shapes at the different temperatures. This similarity provides the basis for an important empirical method, known as the "method of reduced variables", for combining data taken at several different temperatures into one master curve for the sample. This method was described briefly in connection with Fig. 3.3-2.

In order to obtain a master curve for the viscosity function at an arbitrary reference temperature T_0 from plots of $\log \eta$ versus $\log \dot{\gamma}$ for a variety of temperatures T we follow a two step procedure: (1) the curve at temperature T is first shifted vertically upward by an amount $\log[\eta_0(T_0)/\eta_0(T)]$ and (2) the resulting curve is then shifted horizontally in such a way that any overlapping regions of the T_0-curve and shifted T-curve superpose. The amount by which $\eta(\dot{\gamma}, T)$ must be translated to the right in order to achieve superposition is defined as $\log a_T$.

Often it is found that the shift factor is given by[8]

$$a_T = \frac{\eta_0(T)T_0\rho_0}{\eta_0(T_0)T\rho} \tag{3.6-1}$$

[1] W. W. Graessley, *Adv. Polym. Sci.*, **16**, 1–179 (1974).
[2] S. Middleman, *The Flow of High Polymers*, Wiley-Interscience, New York (1968), Chapt. 5.
[3] A. Peterlin, *Adv. Macromol. Chem.*, **1**, 225–281 (1968).
[4] G. C. Berry and T. G. Fox, *Adv. Polym. Sci.*, **5**, 261–357 (1968).
[5] J. D. Ferry, *Viscoelastic Properties of Polymers*, 3rd ed., Wiley, New York (1980).
[6] G. V. Vinogradov and A. Ya. Malkin, *J. Polym. Sci.* Part A, **21**, 2357–2372 (1964); G. V. Vinogradov and A. Ya. Malkin, *Rheology of Polymers*, Mir Publishers, Moscow (1980), §§2.2, 3.2, 4.6.
[7] R. C. Armstrong and H. H. Winter in E. U. Schlünder, ed., *Heat Exchange Design Handbook*, Hemisphere, Washington, D.C. (1983), Chapt. 2.5.12.
[8] A theoretical basis for the method of reduced variables is provided by molecular theory (see Chapter 15). For dilute solutions Eq. 15.3-31 gives the same expression as Eq. 3.6-1 except that η_0 is replaced by $\eta_0 - \eta_s$ where η_s is the solvent viscosity.

where ρ is the density at temperature T, and ρ_0, the density at T_0. Thus the method of reduced variables predicts that a single *master curve* can be obtained by plotting reduced viscosity η_r vs. reduced shear rate $\dot{\gamma}_r$ where these are defined by

$$\eta_r = \eta(\dot{\gamma}, T)\frac{\eta_0(T_0)}{\eta_0(T)}$$

$$\doteq \frac{\eta(\dot{\gamma}, T)T_0}{a_T T} \quad (3.6\text{-}2)$$

$$\dot{\gamma}_r = a_T\dot{\gamma} \quad (3.6\text{-}3)$$

In the second line of Eq. 3.6-2 we have neglected the temperature dependence of density. Moreover the ratio $T_0\rho_0/T\rho$ is about unity and changes very little over ordinary temperature ranges. For example this ratio is 0.92 for low-density polyethylene at 150°C relative to 200°C, whereas the complete shift factor a_T has a value of 0.32 for the same temperature change.

If zero-shear-rate viscosity data are not available, then Eq. 3.6-2 cannot be used to calculate the reduced viscosity and empirical shifting along the viscosity and shear rate axes must be done simultaneously in order to determine a_T. A better procedure is to use the fact that the shear stress can be shifted without knowing η_0 (or a_T). From Eqs. 3.6-2 and 3 we define the reduced shear stress τ_r as

$$\tau_r(\dot{\gamma}, T_0) = \tau_{yx}(\dot{\gamma}, T)\frac{T_0\rho_0}{T\rho} \quad (3.6\text{-}4)$$

Thus τ_{yx} is fairly insensitive to temperature and can be shifted without knowing a_T. Horizontal shifting of different $\tau_r(\dot{\gamma}, T_0)$ curves will give a master curve of $\tau_r(\dot{\gamma}_r, T_0)$; the amount of shifting along the shear rate axis at each temperature gives a_T. Once this master curve is available it can be used to construct $\eta_r = -\tau_r/\dot{\gamma}_r$ vs. $\dot{\gamma}_r$.

The temperature dependence of a_T is illustrated in Figs. 3.6-1 and 2 for several polymer melts. Because of Eq. 3.6-1 these figures also give the temperature dependence of η_0. Two types of exponential functions have been used for describing the temperature dependence. Often an "Arrhenius dependence" of the form

$$a_T = \exp\left[\frac{\Delta\tilde{H}}{R}\left(\frac{1}{T} - \frac{1}{T_0}\right)\right] \quad (3.6\text{-}5)$$

is found to be appropriate (see Fig. 3.6-1). Here $\Delta\tilde{H}$ is known as the "activation energy for flow". This behavior is observed with low-molecular-weight fluids and with molten polymers 100 K or more above their glass transition temperatures. Typical values of $\Delta\tilde{H}/R$ are 4.5×10^3 K (low-density polyethylene), 2.83×10^3 K (high-density polyethylene), and 5.14×10^3 K (polypropylene). These values, of course, depend on molecular parameters of the polymer and therefore may change considerably within each class of polymer.

For temperatures between the glass-transition temperature T_g and $T_g + 100$, the *WLF equation* for a_T has been found to hold for a wide variety of polymers[9]

$$\log a_T = \frac{-c_1^0(T - T_0)}{c_2^0 + (T - T_0)} \quad (3.6\text{-}6)$$

[9] M. L. Williams, R. F. Landel, and J. D. Ferry, *J. Am. Chem. Soc.*, **77**, 3701–3707 (1955).

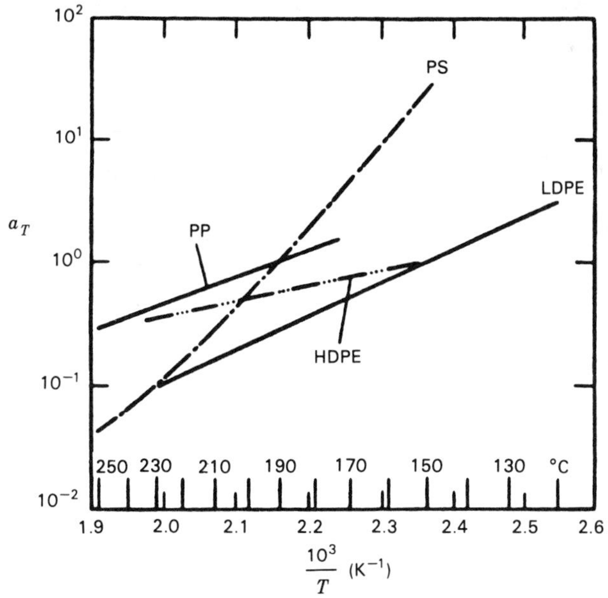

FIGURE 3.6-1. Arrhenius plot of the shift factors a_T for several molten polymers. The reference temperature T_0 is 423 K for low-density polyethylene (LDPE) and high-density polyethylene (HDPE) and 190°C for polypropylene (PP) and polystyrene (PS). [Reproduced with permission from H. Münstedt, *Kunststoffe*, **68**, 94 (1978), Hanser Publishers, Munich, GFR.]

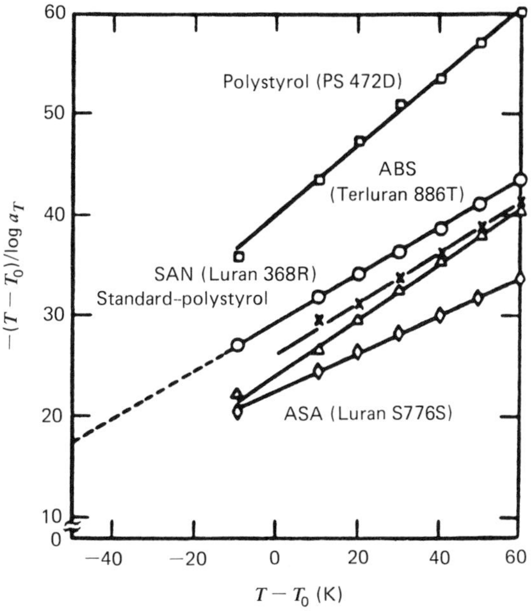

FIGURE 3.6-2. WLF plot of the shift factor a_T for molten polystyrenes. The polymers shown are polystyrene (PS 472D — □), acryonitrile-butadiene-styrene terpolymer (ABS — ○), styrene-acrylonitrile copolymer (SAN — ×), polystyrene (△), and acrylonitrile-styrene-acrylester terpolymer (ASA — ◇). The reference temperature T_0 is 463 K. [Reproduced with permission from H. Münstedt, *Kunststoffe*, **68**, 95 (1978), Hanser Publishers, Munich, GFR.]

If T_0 is taken to be the glass-transition temperature, then $c_1^0 = 17.44$ and $c_2^0 = 51.6$ K. These constants are useful when specific data on a polymer are not available. When some data are available, it is preferable to take $c_1^0 = 8.86$ and $c_2^0 = 101.6$ K, and then to choose T_0 to give a best fit of the data.

Styrene polymers are usually well described by the WLF equation. The agreement of the WLF equation is shown best by plotting $(T - T_0)/\log a_T$ vs. $(T - T_0)$ as shown in Fig. 3.6-2 for a variety of polystyrenes.

Master curves for other material functions are obtained similarly to η. To decide on a proper form for the reduced material functions we assume that all components of the stress are reduced in the same manner as the shear stress. For example, we assume the first normal stress difference $N_1 = +\Psi_1\dot{\gamma}^2$ shifts in the same way as τ_{yx}, so that the reduced first normal stress difference is given by

$$N_{1,r}(\dot{\gamma}, T_0) = N_1(\dot{\gamma}, T)\frac{T_0}{T}\frac{\rho_0}{\rho} \tag{3.6-7}$$

Thus the first normal stress coefficient is reduced as

$$\Psi_{1,r}(\dot{\gamma}, T_0) = \Psi_1 \frac{T_0}{a_T^2 T} \tag{3.6-8}$$

where we have taken $\rho_0/\rho = 1$.

For the linear viscoelastic properties (for which the method of reduced variables was originally developed), the reduced moduli are defined by

$$G_r'(\omega, T_0) = G'(\omega, T)\frac{T_0\rho_0}{T\rho} \tag{3.6-9}$$

$$G_r''(\omega, T_0) = G''(\omega, T)\frac{T_0\rho_0}{T\rho} \tag{3.6-10}$$

Note that the moduli are reduced in exactly the same way as stress. Corresponding to Eqs. 3.6-9 and 10 we have (cf. Eq. 3.6-2)

$$\eta_r'(\omega, T_0) = \eta'(\omega, T)\frac{T_0}{a_T T} \tag{3.6-11}$$

$$\eta_r''(\omega, T_0) = \eta''(\omega, T)\frac{T_0}{a_T T} \tag{3.6-12}$$

where we have set $\rho/\rho_0 = 1$.

The success of the method of reduced variables in producing master curves is illustrated in Figs. 3.3-2, 3.4-3, 17, 19 and 20 for low-density polyethylene. A successful master curve of η_r and $N_{1,r}$ for a polystyrene solution is shown in Fig. 3.6-3.

The principal use for the method of reduced variables is that it allows us to extend the effective shear-rate or frequency range of an experimental geometry. Because of the factor a_T that multiplies $\dot{\gamma}$ in the master curve, varying temperature at a fixed shear rate is equivalent to varying shear rate at a fixed temperature. Thus, by measuring viscosity over two or three decades of shear rate at many different temperatures and then combining the

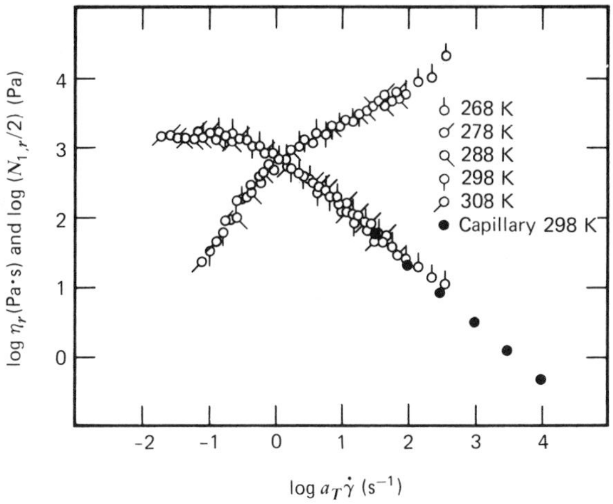

FIGURE 3.6-3. Master curves for viscosity and first normal stress difference for a solution of 2.0×10^6 molecular weight, monodisperse, linear polystyrene in 1-chloronaphthalene. The reduced quantities η_r and $N_{1,r}$ are defined in Eqs. 3.6-2 and 7. The concentration of polymer is 0.150 g/cm^3 and the reference temperature is $T_0 = 423$ K. High shear rate data taken in the capillary viscometer at 423 K were not shifted. [K. Yasuda, R. C. Armstrong, and R. E. Cohen, *Rheol. Acta*, **20**, 163–178 (1981).]

results from the separate tests into a composite curve, it is possible to obtain the viscosity function over six or more decades of shear rate. To obtain reliable results by this method, it is essential that substantial portions of the superposed segments overlap. Even then we must be careful in using results obtained this way, since the cumulative error in estimating several shifts may be large when a wide range of shear rates is used. Finally, when the method of reduced variables is used for dilute solutions, it is more convenient to work in terms of $\eta - \eta_s$ and $\eta_0 - \eta_s$ in place of η and η_0.

b. Effects of Concentration and Molecular Weight[10]

We have already been introduced to the effects of concentration on the viscosity function in §3.3 where we used the intrinsic viscosity to describe the initial rate of increase of η with increasing mass concentration c. In this subsection we shall extend that discussion to cover a full range of concentrations, from infinite dilution to the undiluted state. It is found that for homologous series of fractionated linear polymers, the relation between $[\eta]_0$ and molecular weight can be expressed as:[11]

$$[\eta]_0 = K'M^a \tag{3.6-13}$$

where the value of K' depends on the particular polymer–solvent pair. The parameter a, which is known as the *Mark–Houwink exponent*, lies in the range 0.5–0.8. Originally it was thought that $[\eta]_0$ was proportional to M, this linear relation being known as *Staudinger's rule*.[12]

[10] W. W. Graessley, *op cit.*, §§5 and 8. We have drawn heavily on Graessley's review article for this subsection.
[11] P. J. Flory, *Principles of Polymer Chemistry*, Cornell University Press, Ithaca, NY (1953), pp. 24, 310–314; C. Tanford, *Physical Chemistry of Macromolecules*, Wiley, New York (1961), pp. 407–412.
[12] H. Staudinger and W. Heuer, *Ber. Deutsch. Chem. Ges.*, **63**, 222–234 (1930); H. Staudinger, *Die hochmolekularen organischen Verbindungen*, Springer, Berlin (1932).

Equation 3.6-13 provides a convenient method for determining molecular weight of a polydisperse sample of a linear polymer.[11] For polydisperse samples the *viscosity average molecular weight* $\bar{M}_v$ is given by

$$[\eta]_0 = K'\bar{M}_v^a \tag{3.6-14}$$

where K' and a have the same values as for the monodisperse polymer.

Experimentally, it is found that at low concentrations (defined roughly as $c[\eta]_0 <$ 1 to 5) the material functions correlate well with $c[\eta]_0$ and at high concentrations they correlate with cM. Since it is found that $[\eta]_0$ is proportional to M^a, then we can combine the above and say that cM^x should be a good correlating parameter with x approximately equal to 0.68 at low concentration and x equal to unity at high concentrations. Because c and M occur together in both of these regimes, we shall discuss molecular weight and concentration effects together. The quantity $c[\eta]_0$, which gives a rough estimate of the degree to which the domains of different molecules are likely to overlap, is sometimes called the "coil overlap" parameter. Similarly cM gives an approximate measure of the number of intermolecular contacts per molecule. The fact that each is dominant in one concentration regime suggests two different kinds of interaction between molecules at high and low concentrations.

From the discussion in §3.3 we know that the non-Newtonian viscosity can be described to a very good approximation by giving three quantities: the zero-shear-rate viscosity η_0, the shear rate $\dot{\gamma}_0$ at which η begins to decrease from η_0, and the slope $n - 1$ (see §4.1) of the power-law region of a log η versus log $\dot{\gamma}$ curve. A convenient way to present solution data is to plot $(\eta - \eta_s)/(\eta_0 - \eta_s)$ as a function of a reduced shear rate $\lambda\dot{\gamma}$ where λ is some characteristic time constant for the polymer solution. The form of λ suggested by molecular theory (see Chapters 13 to 15) leads to a reduced shear rate $\beta \equiv \lambda\dot{\gamma} = (\eta_0 - \eta_s)M\dot{\gamma}/cRT$. Note that in the limit as c approaches zero this definition becomes $\beta = [\eta]_0\eta_s M\dot{\gamma}/RT$. This latter quantity was used in constructing Fig. 3.3-4 in which intrinsic viscosity data were presented for a variety of polystyrene solutions at different temperatures. We shall consider, then, the influence of concentration and molecular weight on η_0, $\dot{\gamma}_0$(or β_0), and $(n - 1)$.

The behavior of η_0 as a function of c and M is fairly well understood. At low concentrations an expression of the form $\eta_0 = \eta_0(c[\eta]_0)$ may be used to consolidate data for a given polymer–solvent system over a wide range of both concentration and molecular weight of the polymer. An example is the Martin equation:[13]

$$\eta_0 - \eta_s = \eta_s c[\eta]_0 e^{k''c[\eta]_0} \tag{3.6-15}$$

in which k'' is an arbitrary constant. Actually, a slightly better fit with data is obtained by replacing $[\eta]_0$ with $M^{a'}$ and choosing a' to fit data. It is found that a' is close to the Mark–Houwink exponent.

At high concentrations η_0 is governed by the product cM. The most striking feature of $\eta_0(cM)$ is illustrated in Fig. 3.6-4 for a variety of undiluted polymers. It is seen that η_0 goes from a linear to a 3.4 power dependence on M at some critical molecular weight M_c

$$\begin{aligned} \eta_0 &\propto M \qquad (M < M_c) \\ \eta_0 &\propto M^{3.4} \qquad (M > M_c) \end{aligned} \tag{3.6-16}$$

[13] L. Utracki and R. Simha, *J. Polym. Sci. Part A*, **1**, 1089–1098 (1963).

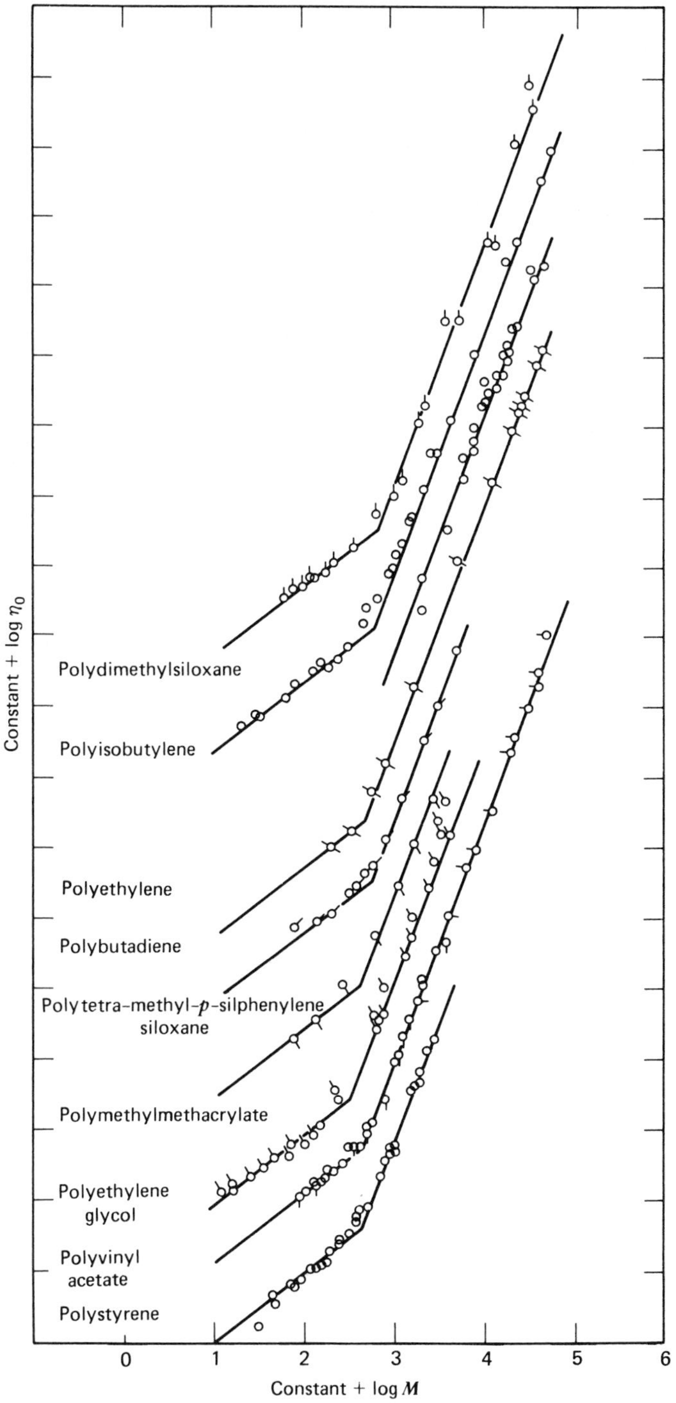

FIGURE 3.6-4. Plots of constant + log η_0 vs. constant + log M for nine different polymers. The two constants are different for each of the polymers, and the one appearing in the abscissa is proportional to concentration, which is constant for a given undiluted polymer. For each polymer the slopes of the left and right straight line regions are 1.0 and 3.4, respectively. [G. C. Berry and T. G. Fox, *Adv. Polym. Sci.*, **5**, 261–357 (1968).]

Actually there is no discontinuity in the slope of $\eta_0(cM)$; the transition is smooth and appears sharp because of the way in which the figure is plotted. The $M^{3.4}$ dependence is followed by an amazingly wide variety of linear polymers. For concentrated solutions, similar behavior is found with $(M_c)_{\text{solution}} = (\rho/c)M_c$ where ρ is the density of the solution. The $M^{3.4}$ behavior of melts and concentrated polymer solutions is thought by many polymer chemists to be due to the existence of "entanglements", or temporary, physical junctions between different polymer molecules (see Chapter 20). It is believed that M_c is a rough measure (within a factor of 2 or 3) of the molecular weight of polymer between two entanglement points. The molecular weight dependence of η_0 can also be interpreted in terms of the "reptation theory" for polymer melts (see Chapter 19).

Next we consider the variation[14] of the critical shear rate β_0 (or $\dot{\gamma}_0$) with c and M. For the sake of definiteness, β_0 is arbitrarily taken to be the value of the dimensionless shear rate at which $(\eta - \eta_s)/(\eta_0 - \eta_s) = 0.8$. At low concentrations β_0 is a constant between 1.5 and 2.0; this value persists down to infinite dilution. At the high concentrations, β_0 increases with cM; β_0 begins this increase at a value of cM slightly greater than ρM_c.

Finally, the power-law slope $(n - 1)$ is approximately -0.1 at infinite dilution. It decreases with $c[\eta]_0$ and approaches $(n - 1) = -0.8$ for $c[\eta]_0 > 20$. The fact that the power-law slope approaches a constant value at high concentrations and molecular weights suggests that for a given polymer–solvent system in the concentrated regime or for a given polymer melt, data taken at varying concentrations, molecular weights, and temperatures should all lie on a single curve when plotted as η/η_0 vs. $\dot{\gamma}/\dot{\gamma}_0$. This prediction is impressively confirmed in Fig. 3.6-5.

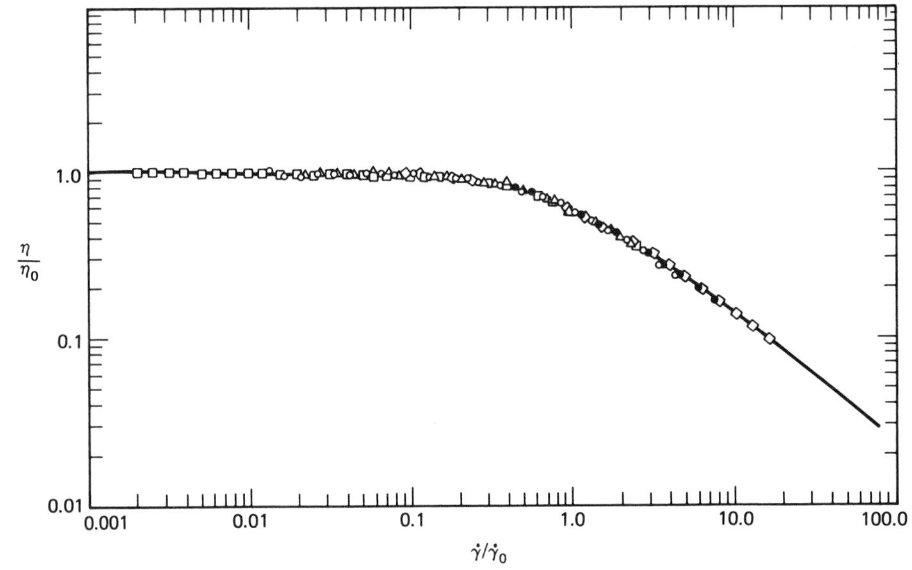

FIGURE 3.6-5. Composite plot of dimensionless viscosity η/η_0 versus dimensionless shear rate $\dot{\gamma}/\dot{\gamma}_0$ for several different concentrated polystyrene–n-butyl benzene solutions. Molecular weights varied from 1.6×10^5 to 2.4×10^6, concentrations from 0.255 to 0.55 g/cm^3, and temperatures from 303 to 333 K. [W. W. Graessley, *Adv. Polym. Sci.*, **16**, 1–179 (1974).]

[14] These results as well as those for the power-law slope are based on data for narrow molecular weight distribution polystyrene and poly-α-methylstyrene solutions and melts. See W. W. Graessley, *op. cit.*, and K. Yasuda, R. C. Armstrong, and R. E. Cohen, *Rheol. Acta*, **20**, 163–178 (1981).

Most of the above statements are for polymers with narrow molecular weight distributions. Roughly speaking, broadening the distribution lowers $\dot{\gamma}_0$ and broadens the range of shear rates over which the transition from η_0 to a power-law behavior occurs. Some work has been done on constructing phenomenological blending laws from which the properties of polymers of broad molecular weight distributions can be predicted from monodisperse polymer behavior. As an example of effort along these lines we mention the study of Harris on polystyrene solutions.[15] It was found that η_0 for a two-component blend could be related to η_0 of the separate components by

$$\log(\eta_0)_{\text{blend}} = w_1 \log(\eta_0)_1 + w_2 \log(\eta_0)_2 \tag{3.6-17}$$

where w_1 and w_2 are the weight fractions of the components of the blend. This relation does best when the difference between molecular weights of the components is not too great. The molecular weight distribution does not appear to affect either the power-law slope or $\dot{\gamma}_0$ inasmuch as $\dot{\gamma}_0$ is the same function of c and $\bar{M}_w$, and $(n - 1)$ is the same function of $c\bar{M}_n$ as for narrow distribution polymers. This study included molecular weights in the range $3.94 \times 10^5 < \bar{M}_w < 1.9 \times 10^6$ and concentrations of polystyrene in Aroclor between 0.0357 and 0.1502 g/cm^3. The kinetic theory of polydisperse polymer melts is discussed in Chapter 19.

The dynamic viscosity $\eta'(\omega)$ can be described by η_0, a characteristic frequency ω_0 at which $\eta'(\omega)$ begins to decrease, and a power-law slope for high frequencies. The dynamic viscosity η' is often presented as $(\eta' - \eta_s)/(\eta_0 - \eta_s)$ versus $\lambda\omega$ where λ is given by molecular theory as before (see text above Eq. 3.6-15). Experimentally, it is found that the characteristic frequency ω_0 is practically the same as $\dot{\gamma}_0$ for *all* polymeric fluids.[16] Hence the only aspect of η' that we need to describe is the high frequency behavior.[17] For molecules that are not of too high a molecular weight, η' is proportional to $\omega^{-1/3}$ at low concentrations and high frequencies. At $c[\eta]_0$ approximately equal to 3, the dependence on ω begins to change until for high concentrations and undiluted polymers, η' varies as $\omega^{-1/2}$. This sequence is sometimes referred to as a transition from "Zimmlike" to "Rouselike" behavior, as the Zimm and Rouse molecular theories account for the respective dependences on ω (see Chapter 15). As one proceeds from low to high concentrations, the high frequency behavior of η''/ω goes from $\omega^{-4/3}$ to $\omega^{-3/2}$.

Master curves for the dynamic moduli for a series of linear, narrow-distribution polystyrenes having different molecular weights are shown in Fig. 3.6-6. The transition region between zero-frequency behavior ($G' \propto \omega^2$) and high frequency behavior ($G' \propto \omega^{1/2}$) becomes broader as molecular weight increases. There is a pronounced common region in which the curves for different molecular weights overlap and in which the storage modulus is constant. This value is known as the *plateau modulus* G_{eN}^0. Clearly this modulus is independent of the overall molecular weight of the polymer. It is believed to be proportional to the molecular weight of polymer segments between 'entanglements' or between points where the motion of the polymer is severely constrained by neighboring molecules[18] (see Chapters 19 and 20).

The molecular weight dependence of several other material functions is now briefly described. In Fig. 3.6-7 the molecular weight dependence of the zero-shear-rate first normal stress coefficient and the steady-state compliance is contrasted with that of η_0 for

[15] E. K. Harris, Jr., *J. Appl. Polym. Sci.*, **17**, 1679–1692 (1973).
[16] W. W. Graessley, *op. cit.*, p. 126.
[17] J. D. Ferry, *op. cit.*, Chapt. 11.
[18] J. D. Ferry, *op. cit.*, Chapt. 13.

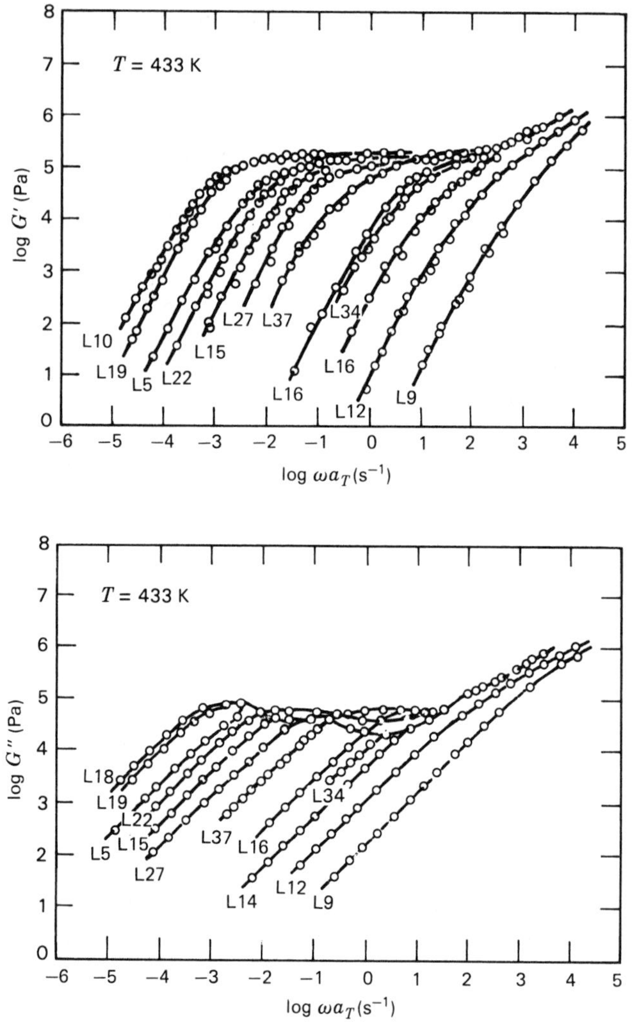

FIGURE 3.6-6. Master curves of G' and G'' for a series of narrow-distribution, linear polystyrenes. The value of G' in the frequency regime where it is constant is known as the plateau modulus. The viscosity average molecular weights of the samples are: L18-580,000; L19-510,000; L5-350,000; L22-275,000; L15-215,000; L27-167,000; L37-113,000; L16-58,700; L34-46,900; L14-28,900; L12-14,800; L9-8,900. The reference temperature is $T = 160°C$. [Reprinted with permission from S. Onogi, T. Masuda, and K. Kitagawa, *Macromolecules*, **3**, 109–116 (1970). Copyright (1970) American Chemical Society.]

polyamide-6 polymers. It is interesting that J_e^0 is independent of molecular weight and that $\Psi_{1,0}$ follows

$$\begin{aligned}\Psi_{1,0} &\propto \bar{M}_w^{7.0} \\ &\propto \eta_0^2 \end{aligned} \tag{3.6-18}$$

For polydisperse polymers $\Psi_{1,0}$ is more sensitive to high molecular weight fractions than either η_0 or J_e^0.

Since $\bar{\eta}_0 = 3\eta_0$, the molecular weight dependence of $\bar{\eta}_0$ should be the same as that for η_0. This is confirmed in Fig. 3.6-8 for the four polystyrene melts shown in Fig. 3.5-2. The

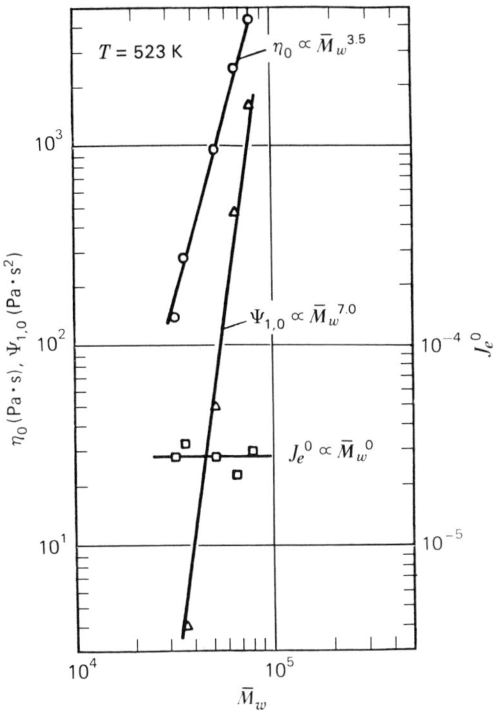

FIGURE 3.6-7. Molecular weight dependence of J_e^0, η_0, and $\Psi_{1,0}$ for polyamide-6 melts with Schultz–Flory molecular weight distributions. [Data from H. M. Laun, *J. Rheol.*, **30**, 459–501 (1986).]

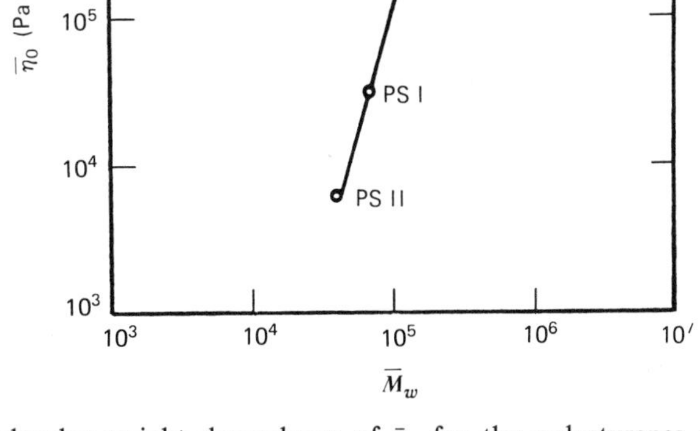

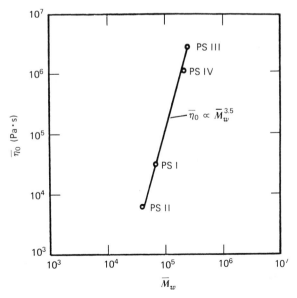

FIGURE 3.6-8. Molecular weight dependence of $\bar{\eta}_0$ for the polystyrenes shown in Fig. 3.5-2. $T = 433$ K. [H. Münstedt, *J. Rheol.*, **24**, 847–867 (1980).]

sensitivity of $\bar{\eta}^+$, D, and ε_∞ to molecular weight distribution has already been seen in Figs. 3.5-2, 6, and 7, respectively.

c. Relations between Linear Viscoelastic Properties and Viscometric Functions

From the sample viscosity and dynamic viscosity data presented in §§3.3 and 3.4 it is evident that $\eta(\dot{\gamma})$ and $\eta'(\omega)$ are similar functions of their arguments. In fact, we know that both approach η_0 as their arguments go to zero and also that both begin to decrease at comparable values of $\dot{\gamma}$ and ω. The main difference between these functions is their behavior at large shear rates and frequencies; it is found that η' decreases more rapidly with ω than η does with $\dot{\gamma}$. The *Cox-Merz rule*[19] has been suggested as a way of obtaining an improved relation between the linear viscoelastic properties and the viscosity. This empiricism predicts that the magnitude of the complex viscosity is equal to the viscosity at corresponding values of frequency and shear rate

$$\eta(\dot{\gamma}) = |\eta^*(\omega)|\Big|_{\omega=\dot{\gamma}} = \eta'(\omega)\left[1 + \left(\frac{\eta''}{\eta'}\right)^2\right]^{0.5}\Bigg|_{\omega=\dot{\gamma}} \tag{3.6-19}$$

The Cox-Merz rule has proven very useful in predicting $\eta(\dot{\gamma})$ when only linear viscoelastic data are available. In Fig. 3.6-9 we show an experimental test of Eq. 3.6-19 for three different polystyrene solutions. Also shown are data for the dynamic viscosity. It is clear that $|\eta^*(\omega)|$ follows $\eta(\dot{\gamma})$ more closely than $\eta'(\omega)$ does, and that the Cox-Merz rule is a good approximation. The agreement between $|\eta^*|$ and η is within experimental error for the linear polystyrene sample.

An interesting alternative to the Cox-Merz rule for predicting η from linear viscoelastic data is *Gleissle's mirror relation*:[20]

$$\eta(\dot{\gamma}) = \eta^+(t)|_{t=1/\dot{\gamma}} \tag{3.6-20}$$

in which $\eta^+(t)$ is the limiting curve of $\eta^+(t, \dot{\gamma})$ as $\dot{\gamma} \to 0$. Successful fits of the viscosity from the viscous stress growth coefficient have been constructed for a variety of polymers, including silicone oils, polyisobutylene, and polyethylene. The Cox-Merz and the Gleissele mirror relation are compared in Fig. 3.6-10 for low-density polyethylene.

It is desirable to be able to predict the normal stresses in steady shear flow from more easily made linear viscoelastic measurements. It may be recalled from §§3.3 and 3.4 that Ψ_1 and $2\eta''/\omega$ have the same values at low shear rates and frequencies and that the shapes of the curves are similar. This similarity is emphasized in Fig. 3.6-9 where $N_1/2$ and G' are compared; here $N_1 = +\Psi_1\dot{\gamma}^2$ is the negative of the first normal stress difference. The discrepancy between $N_1/2$ and G' is more pronounced than that between η and η'. A useful empiricism, analogous to the Cox-Merz rule, is *Laun's rule*[21]

$$\Psi_1(\dot{\gamma}) = \frac{2\eta''(\omega)}{\omega}\left[1 + \left(\frac{\eta''}{\eta'}\right)^2\right]^{0.7}\Bigg|_{\omega=\dot{\gamma}} \tag{3.6-21}$$

[19] W. P. Cox and E. H. Merz, *J. Polym. Sci.*, **28**, 619–622 (1958).

[20] W. Gleissle in G. Astarita, G. Marrucci, and L. Nicolais, eds., *Rheology*, *Vol.* 2: *Fluids*, Plenum Press, New York (1980), pp. 457–462.

[21] H. M. Laun, *J. Rheol.*, **30**, 459–501 (1986).

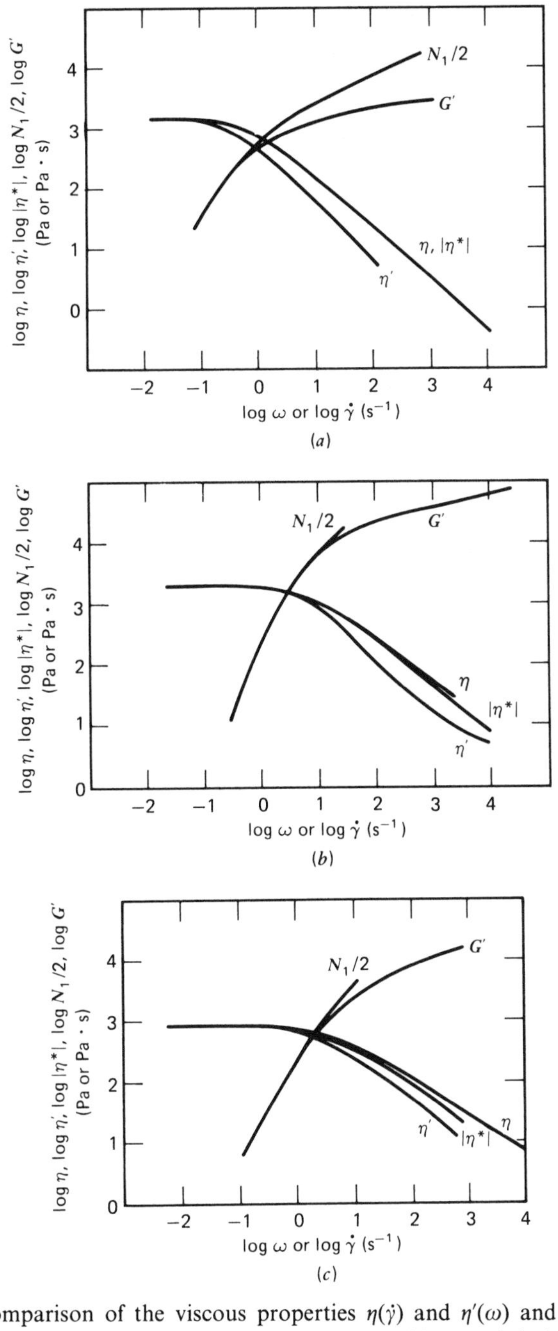

FIGURE 3.6-9. Comparison of the viscous properties $\eta(\dot{\gamma})$ and $\eta'(\omega)$ and the elastic properties $N_1(\dot{\gamma})/2$ and $G'(\omega)$ for several solutions of polystyrene in 1-chloronaphthalene. Also shown is the magnitude of the complex viscosity $|\eta^*|$ which should equal η according to the Cox–Merz rule. The solutions are (*a*) 0.15 g/ml of 2×10^6 $\bar{M}_w$ narrow distribution, linear polystyrene; (*b*) 0.45 g/ml of 7.9×10^5 $\bar{M}_w$ narrow distribution, star-branched polystyrene; and (*c*) 0.42 g/ml of 3.69×10^5 $\bar{M}_w$ broad distribution, linear polystyrene (Styron). [Data from K. Yasuda, R. C. Armstrong, and R. E. Cohen, *Rheol. Acta.*, **20**, 163–178 (1981).]

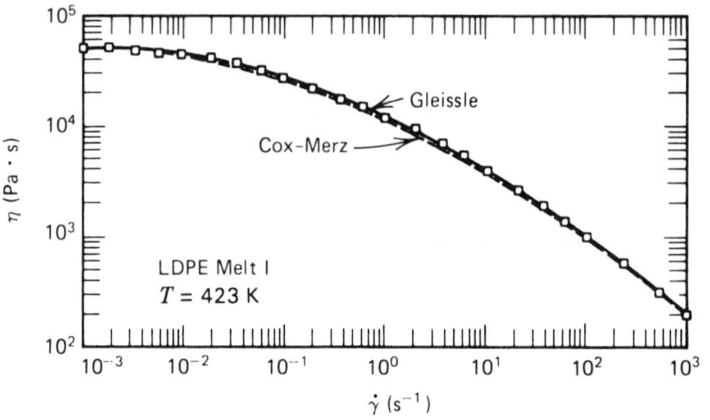

FIGURE 3.6-10. Comparison of the Cox-Merz rule (Eq. 3.6-19) and the Gleissle mirror rule (Eq. 3.6-20) for a low-density polyethylene melt (Melt I). The symbols are data; the dashed curve, Eq. 3.6-19; and the solid curve, Eq. 3.6-20. [H. M. Laun, *J. Rheol.*, **30**, 459-501 (1986).]

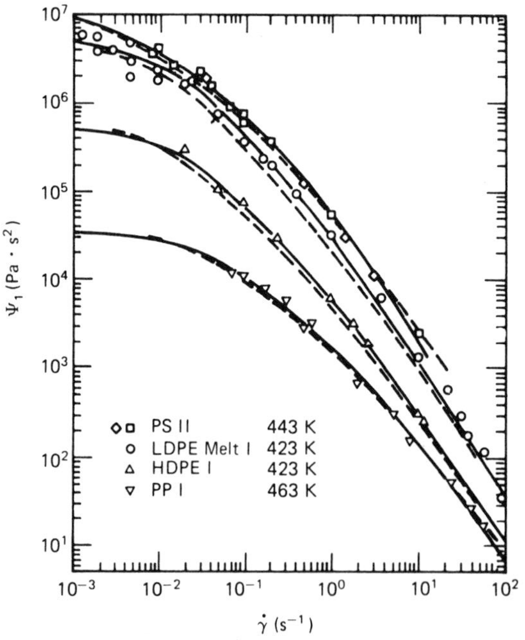

FIGURE 3.6-11. First normal stress coefficient vs. shear rate for polystyrene (PS), low-density polyethylene (LDPE), high-density polyethylene (HDPE), and polypropylene (PP). The symbols are measured values and the ---- curves are calculated from linear viscoelastic data by Eq. 3.6-21. The solid curves are calculated with the Wagner model with a single exponential damping function (see §8.3). [H. M. Laun, *J. Rheol.*, **30**, 459-501 (1986).]

Figure 3.6-11 shows a comparison between measured Ψ_1 and values computed from linear viscoelastic data according to Eq. 3.6-21. The agreement is excellent.

Gleissle[22] has also proposed a 'mirror relation' for Ψ_1 similar to Eq. 3.6-20:

$$\Psi_1(\dot{\gamma}) = \Psi_1^+(t)|_{t=k/\dot{\gamma}} \tag{3.6-22}$$

where k is an empirical constant which seems to be in the range of $2.5 \leq k \leq 3$ for many polymeric liquids. For silicone oils and low-density polyethylene a value of $k = 3$ gives a quantitative connection between Ψ_1 and Ψ_1^+.

Laun[23] has also reported an empiricism relating the recoverable shear γ_∞ to the complex viscosity

$$\begin{aligned} -\gamma_\infty(\dot{\gamma}) &= \frac{\eta''}{\eta'}\left[1 + \left(\frac{\eta''}{\eta'}\right)^2\right]^{1.5}\Bigg|_{\omega=\dot{\gamma}} \\ &= \frac{\Psi_1\dot{\gamma}}{2\eta}\left[1 + \left(\frac{\eta''}{\eta'}\right)^2\right]^{1.3}\Bigg|_{\omega=\dot{\gamma}} \end{aligned} \tag{3.6-23}$$

The second line of Eq. 3.6-23 is obtained by using Eqs. 3.6-19 and 21. Note the similarity between Eqs. 3.6-23 and 3.4-13.

§3.7 KINEMATICS AND CLASSIFICATION OF SHEAR AND SHEARFREE FLOWS

In §3.1 we introduced *simple* shear and shearfree flows; these flows were sufficient for defining the material functions in §§3.3-5. There are many other *non-simple* shear and shearfree flows for which these same material functions apply. The purpose of this section is to give a careful classification of these. This is important for two reasons: first, in later chapters we find constitutive equations that are valid for restricted types of flows, and we need to be able to identify those flows for which the constitutive equation can be legitimately applied; second, since the simple flows of §3.1 are not the most convenient for measuring the material functions, we need to be able to find alternative flows.

a. Shear Flows

In Table 3.7-1 we list all of the various categories of shear flows and give an example of each. This table should give some perspective to the definitions that follow. We begin by returning to the steady simple shear flow of Eq. 3.1-1 with $\dot{\gamma}$ independent of time. A number of characteristics[1] of this flow are:

[22] W. Gleissle, *op. cit.*

[23] H. M. Laun, *op. cit.*

[1] A thorough discussion of shear and other flows in terms of the response of material particles, lines, and planes is given in A. S. Lodge, *Elastic Liquids*, Academic Press, New York (1964); A. S. Lodge, *Body Tensor Fields in Continuum Mechanics*, Academic Press, New York (1974). See also R. I. Tanner, *Engineering Rheology*, Oxford University Press (1985), Chapter 3.

TABLE 3.7-1

Classification and Examples of Shear Flows

		Rheologically Unsteady		Rheologically Steady
General shear flow	N[a]	Tangential annular flow with the inner cylinder vibrating axially.		
	H[b]	Eccentric-disk (orthogonal) rheometer flow. See Problem 3B.1.		
Unidirectional shear flow	N	Oscillatory tangential annular flow.	N	Helical flow. See Table 3.7-2e.[c]
	H	Oscillatory cone-and-plate flow (approximate for small cone angles and high viscosity fluids). See Table 3.7-2d.	H	Tangential annular flow (approximate for narrow gaps and high viscosity fluids).[c]
Rectilinear shear flow	N	Pulsatile pressure-driven flow in tubes or slits.	N	Pressure-driven slit flow.[c]
	H	Wall-driven oscillatory flow between parallel plates (approximate for narrow gaps and high viscosity fluids).[d]	H	Wall-driven flow between parallel plates.[c,d]

[a] N = nonhomogeneous.
[b] H = homogeneous.
[c] These are called "viscometric flows."
[d] These are called "simple shear flows."

(a) Fluid planes that are parallel to the bounding plates move rigidly; since these planes consist of the same set of fluid particles at all times, we call them material planes or surfaces. These form a one-parameter family of material surfaces, y = constant, and each member is denoted as a *shearing plane*. Note that the distance between any two particles in a shearing plane is constant.

(b) The volume of every material element remains constant throughout the flow. The constant volume condition is a requirement for a shear flow independent of the assumption of incompressibility used in this book.

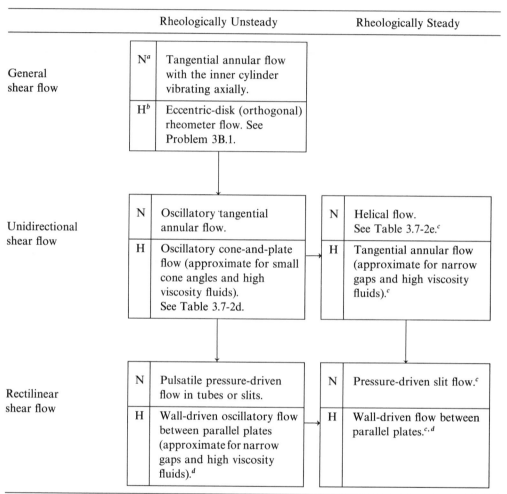

(c) The direction of relative motion of the shearing planes is denoted by the unit vector $\boldsymbol{\delta}_x$ along the x-coordinate axis. If at any instant we trace out a curve to which $\boldsymbol{\delta}_x$ is everywhere tangent, we generate a straight line that we call a *line of shear*. In the present flow all of the lines of shear are parallel to the x-axis, and they are coincident with fluid particle pathlines. This latter property does not hold in all shear flows as we shall see in Example 3.7-1. In simple shear flow, the lines of shear are material lines.

(d) The shear rate $\dot{\gamma}$, which is equal to the relative sliding velocity of any two shearing planes divided by the distance between them, is a constant.

(e) Finally, in Cartesian coordinates the velocity gradient tensor has the form

$$\nabla \boldsymbol{v} = \dot{\gamma}_{yx} \begin{pmatrix} 0 & 0 & 0 \\ 1 & 0 & 0 \\ 0 & 0 & 0 \end{pmatrix} = \boldsymbol{\delta}_y \boldsymbol{\delta}_x \dot{\gamma}_{yx} \tag{3.7-1}$$

and hence,

$$\dot{\boldsymbol{\gamma}} = \dot{\gamma}_{yx} \begin{pmatrix} 0 & 1 & 0 \\ 1 & 0 & 0 \\ 0 & 0 & 0 \end{pmatrix} = (\boldsymbol{\delta}_x \boldsymbol{\delta}_y + \boldsymbol{\delta}_y \boldsymbol{\delta}_x)\dot{\gamma}_{yx} \tag{3.7-2}$$

By using Eq. 3.7-2, it is not difficult to show that the shear rate $\dot{\gamma} = |\dot{\gamma}_{yx}|$ is related to the second invariant of $\dot{\boldsymbol{\gamma}}$ (see Eq. A.3-21):

$$\dot{\gamma} = \sqrt{\tfrac{1}{2}II} = \sqrt{\tfrac{1}{2}(\dot{\boldsymbol{\gamma}}:\dot{\boldsymbol{\gamma}})} \tag{3.7-3}$$

One can take as a definition for steady simple shear flow, either properties **(a)** to **(d)** or property **(e)**.

Now that we have discussed steady simple shear flow in some detail we are ready to give the definition of general shear flow: one that is not necessarily steady-state, rectilinear, or homogeneous and that does not have plane shearing surfaces. To this general definition, we then add a sequence of further requirements that define special kinds of shear flow; this specialization process will end at our original example, steady simple shear flow.

We define a *shear flow* to be a flow in which:

i. There is a one-parameter family of material surfaces, the *shearing surfaces*, which move isometrically, that is, the distance between any two neighboring particles in the surface is constant; and

ii. The volume of every fluid element is constant.

As an alternative to **(i)** and **(ii)** one could use **(i)** and

ii′. The separation of any two neighboring shearing surfaces is constant.

At any instant, a family of curves can be drawn on each shearing surface so that they are tangent to the direction of relative sliding motion at each particle. These curves are known as *lines of shear*. For an arbitrary shear flow defined by **(i)** and **(ii)** above, the lines of shear are not necessarily material lines, that is, they do not necessarily consist of the same set of material particles at every instant. From the material point of view, then, the lines of shear are in general functions of time, since they must be drawn differently on the shearing surfaces at different times.

We now define a type of shear flow for which the lines of shear are "constant." These are *unidirectional shear flows*, and they have the following property in addition to (i) and (ii):

iii. The lines of shear are material lines.

Unidirectional shear flow is the most frequently encountered kind of shear flow. The lines of shear are constant in the sense that if a line of shear is drawn on a shearing surface at any particular time, then the same material particles that lie along the curve at that time also fall on the same line of shear at any other time.

In order to describe the kinematics of shear flows mathematically, it is useful to introduce a coordinate system based on the shapes of the shearing surfaces. We construct an orthogonal curvilinear coordinate system with unit vectors $\hat{\boldsymbol{\delta}}_1$, $\hat{\boldsymbol{\delta}}_2$, and $\hat{\boldsymbol{\delta}}_3$ at every particle so that at any time $\hat{\boldsymbol{\delta}}_1$ and $\hat{\boldsymbol{\delta}}_3$ are tangent to a shearing surface and $\hat{\boldsymbol{\delta}}_2$ is normal to the shearing surface. Furthermore we take the grid of $\hat{x}_1$- and $\hat{x}_3$-coordinate curves to be imprinted on the shearing surfaces so that these curves are made up of material particles. We shall refer to the $\hat{\boldsymbol{\delta}}_i$ set of axes as "shear axes"; notice that from a spatial point of view the shear axes may be functions of time as well as position. Since the lines of shear are material lines for unidirectional shear flow, we may choose the $\hat{x}_1$-curves to coincide with the lines of shear for this kind of shearing motion. We then call the direction denoted by $\hat{\boldsymbol{\delta}}_1$ "the direction of shear." Referred to the $\hat{\boldsymbol{\delta}}_1\hat{\boldsymbol{\delta}}_2\hat{\boldsymbol{\delta}}_3$-axes, the velocity gradient tensor at every particle has the following form in unidirectional shear flow:[2]

$$\boxed{\nabla\hat{\boldsymbol{v}} = \hat{\gamma}_{21}(t)\begin{pmatrix} 0 & 0 & 0 \\ 1 & 0 & 0 \\ 0 & 0 & 0 \end{pmatrix} = \hat{\boldsymbol{\delta}}_2\hat{\boldsymbol{\delta}}_1\hat{\gamma}_{21}(t)} \qquad (3.7\text{-}4)$$

and the rate-of-strain tensor can be written

$$\boxed{\dot{\boldsymbol{\gamma}} = \hat{\gamma}_{21}(t)\begin{pmatrix} 0 & 1 & 0 \\ 1 & 0 & 0 \\ 0 & 0 & 0 \end{pmatrix} = (\hat{\boldsymbol{\delta}}_1\hat{\boldsymbol{\delta}}_2 + \hat{\boldsymbol{\delta}}_2\hat{\boldsymbol{\delta}}_1)\hat{\gamma}_{21}(t)} \qquad (3.7\text{-}5)$$

(See Problem 3C.2 for the corresponding form for $\dot{\boldsymbol{\gamma}}$ for arbitrary shear flows.) Here $\hat{\boldsymbol{v}}$ is the velocity as seen in the shear axes frame of reference. Since $\boldsymbol{v}$ and $\hat{\boldsymbol{v}}$ differ by at most a rigid rotation and rigid translation, $\nabla\boldsymbol{v}$ and $\nabla\hat{\boldsymbol{v}}$ may differ by a rigid rotation, but $\dot{\boldsymbol{\gamma}}$ and $\hat{\dot{\boldsymbol{\gamma}}}$ must be the same. Of course, the components $\dot{\gamma}_{ij}(t)$ and $\hat{\gamma}_{ij}(t)$ will be different unless the $\boldsymbol{\delta}_i$ and $\hat{\boldsymbol{\delta}}_i$ are coincident at time t. Note that $\dot{\gamma} = |\hat{\gamma}_{21}(t)|$ is related to the second invariant of the rate-of-strain tensor just as in steady simple shear flow; it can thus be computed using the definition of $\dot{\gamma}$ in Eq. 3.7-3 provided that $\dot{\boldsymbol{\gamma}}$ is known in *any* coordinate system. By comparing Eqs. 3.7-4 and 3.7-5 with Eqs. 3.7-1 and 3.7-2, we see that at every particle unidirectional shear flow is a simple shear flow when viewed from the particle shear axes.

The motion of a typical fluid element in unidirectional shear flow is depicted in Fig. 3.7-1. Note that the same fluid particles lie along $\hat{\boldsymbol{\delta}}_1$ and $\hat{\boldsymbol{\delta}}_3$ at times t and t'; whereas different fluid particles lie along $\hat{\boldsymbol{\delta}}_2$ at the two different times. From a particle point of view, different unidirectional shear flows are obtained by changing the time dependence of $\hat{\gamma}_{21}$ in Eq. 3.7-4. The various time-dependent unidirectional shear flow experiments used in rheology are illustrated in Fig. 3.4-1 for simple shear flow.

[2] For the use of matrices to display the components of a tensor, see §A.9.

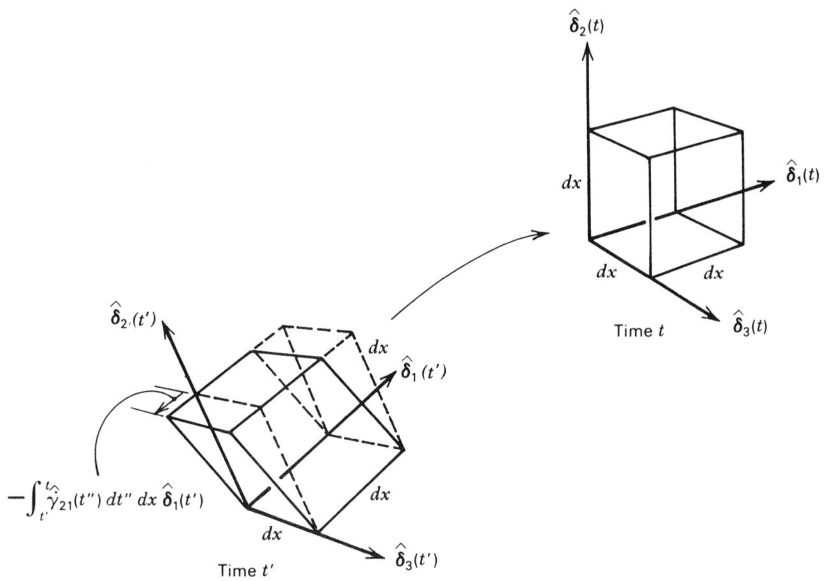

FIGURE 3.7-1. Motion of a typical fluid element of volume $(dx)^3$ undergoing unidirectional shear flow. The deformation that takes the element from its shape at some past time t' to its shape at the present time t is a simple shear flow with shear rate $\hat{\dot{\gamma}}_{21}(t)$. The unit vectors $\hat{\boldsymbol{\delta}}_1$, $\hat{\boldsymbol{\delta}}_2$, $\hat{\boldsymbol{\delta}}_3$ specify the coordinate system from which the flow is seen as simple shear. The same fluid particles lie along $\hat{\boldsymbol{\delta}}_1$ and $\hat{\boldsymbol{\delta}}_3$ at times t' and t, but different fluid particles mark $\hat{\boldsymbol{\delta}}_2$ at the two times. The top and bottom faces of the cube are parts of adjacent shearing surfaces.

Next, we define *rheologically steady shear flow* or *viscometric flow*[3] as a unidirectional shear flow for which:

iv. The velocity gradient $\hat{\dot{\gamma}}_{21}$ is independent of time at a given particle.

In view of Eq. 3.7-4, an alternative way[4] of defining viscometric flow is that the flow history at any particle be one of steady simple shear as seen from the shear axes at the particle.[5] Note that this definition requires the flow to be steady following a particle. In rheologically

[3] This definition of viscometric flow is not universal. It is the same definition used by B. D. Coleman, H. Markovitz, and W. Noll, *Viscometric Flows of Non-Newtonian Fluids*, Springer, New York (1966), and by Pipkin and Tanner (see footnote 5). Some rheologists use the word viscometric to stand for a unidirectional shear flow.

[4] For future reference the finite strain tensor $\boldsymbol{\gamma}^{[0]}(t, t')$, as defined in §8.1, has the form

$$\boldsymbol{\gamma}^{[0]} = \begin{pmatrix} 0 & -(t-t')\hat{\dot{\gamma}}_{21} & 0 \\ -(t-t')\hat{\dot{\gamma}}_{21} & (t-t')^2\hat{\dot{\gamma}}_{21}^2 & 0 \\ 0 & 0 & 0 \end{pmatrix} \tag{3.7-5a}$$

when referred to the shear axes. Requiring $\boldsymbol{\gamma}^{[0]}$ to be given in this way is equivalent to both definitions of viscometric flow given in the text. The rate-of-strain tensor may be calculated from $\boldsymbol{\gamma}^{[0]}$ using the relation given in Table 9.3-1 that

$$\dot{\boldsymbol{\gamma}}(t) = \left[\frac{\partial}{\partial t'}\boldsymbol{\gamma}^{[0]}(t, t')\right]\bigg|_{t'=t} \tag{3.7-5b}$$

[5] This is the approach taken in the review article by A. C. Pipkin and R. I. Tanner, *Mech. Today*, **1**, 262–321 (1972). We have drawn on this review in Example 3.7-1.

complex fluids that exhibit memory effects (cf. §2.5), this definition of steady flow is preferable to the usual hydrodynamic one. We refer, then, to flows steady at a fluid particle as "rheologically steady." The velocity gradient $\hat{\dot{\gamma}}_{21}$ at a particle is constant; however, $\hat{\dot{\gamma}}_{21}$ can vary from particle to particle. If $\hat{\dot{\gamma}}_{21}$ does not vary from particle to particle, then the flow is homogeneous as well. An example of a flow that is steady in the hydrodynamical sense that $\partial \boldsymbol{v}(\boldsymbol{r}, t)/\partial t = 0$, but that is rheologically unsteady is eccentric-disk rheometer flow (see Problem 3B.1). Some examples of rheologically steady shear flows are steady-state tube flow, steady tangential annular flow, and steady axial annular flow.

TABLE 3.7-2

Examples of Viscometric Flow

Flow	Velocity Field	Shear Axes Shear Rate	Shearing Surfaces Lines of Shear
a. Steady tube flow 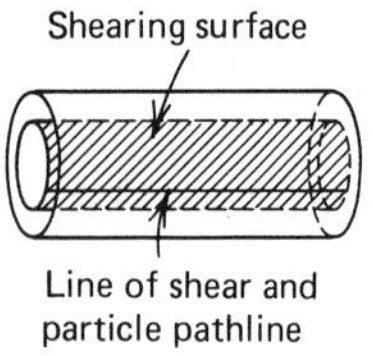 (*a*)	$v_z = v_z(r)$ $v_r = 0$ $v_\theta = 0$	$\hat{\boldsymbol{\delta}}_1 = \boldsymbol{\delta}_z$ $\hat{\boldsymbol{\delta}}_2 = \boldsymbol{\delta}_r$ $\hat{\boldsymbol{\delta}}_3 = \boldsymbol{\delta}_\theta$ $\dot{\gamma} = -\dfrac{d}{dr} v_z$	Concentric cylinders Straight lines parallel to the tube axis
b. Steady tangential annular flow 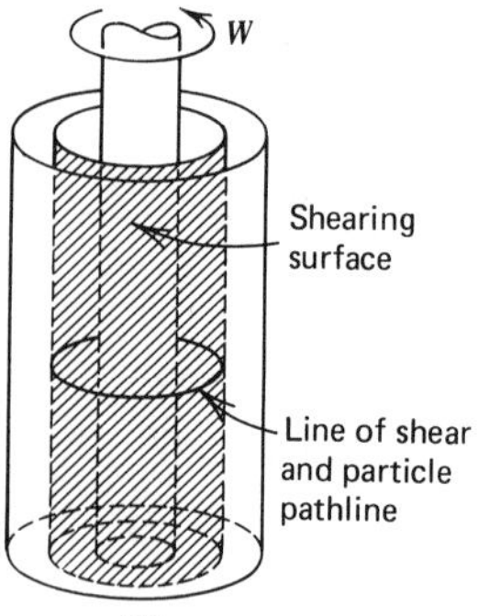 (*b*)	$v_\theta = v_\theta(r)$ $v_r = 0$ $v_z = 0$	$\hat{\boldsymbol{\delta}}_1 = \boldsymbol{\delta}_\theta$ $\hat{\boldsymbol{\delta}}_2 = \boldsymbol{\delta}_r$ $\hat{\boldsymbol{\delta}}_3 = -\boldsymbol{\delta}_z$ $\dot{\gamma} = -r \dfrac{d}{dr}\left(\dfrac{v_\theta}{r}\right)$	Concentric cylinders Circles of constant r and z
c. Steady torsional flow 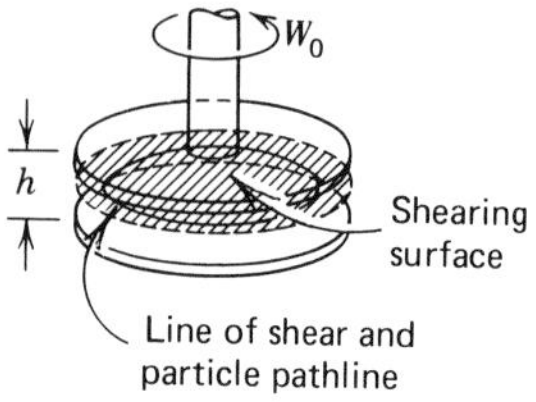 (*c*)	$v_\theta = \dfrac{rzW_0}{h}$ $v_r = 0$ $v_z = 0$	$\hat{\boldsymbol{\delta}}_1 = \boldsymbol{\delta}_\theta$ $\hat{\boldsymbol{\delta}}_2 = \boldsymbol{\delta}_z$ $\hat{\boldsymbol{\delta}}_3 = \boldsymbol{\delta}_r$ $\dot{\gamma} = rW_0/h$	Parallel disks Circles of constant r and z

TABLE 3.7-2 *(continued)*

d. Steady cone-and-plate flow (small cone angle)	$v_\phi = rf(\theta)$ $v_\theta = 0$ $v_r = 0$	$\hat{\boldsymbol{\delta}}_1 = \boldsymbol{\delta}_\phi$ $\hat{\boldsymbol{\delta}}_2 = -\boldsymbol{\delta}_\theta$ $\hat{\boldsymbol{\delta}}_3 = \boldsymbol{\delta}_r$ $\dot{\gamma} = -\frac{1}{r}\frac{\partial v_\phi}{\partial \theta}$	Cones of constant θ Circles of constant r and z

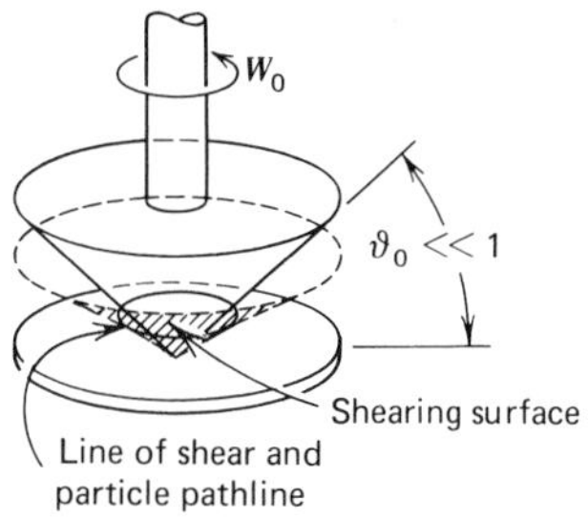

(*d*)

e. Steady helical flow	$v_\theta = v_\theta(r)$ $v_z = v_z(r)$ $v_r = 0$	$\hat{\boldsymbol{\delta}}_1 = \frac{1}{\dot{\gamma}}\left[\boldsymbol{\delta}_\theta r \frac{d}{dr}\left(\frac{v_\theta}{r}\right) + \boldsymbol{\delta}_z \frac{dv_z}{dr}\right]$ $\hat{\boldsymbol{\delta}}_2 = \boldsymbol{\delta}_r$ $\hat{\boldsymbol{\delta}}_3 = \frac{1}{\dot{\gamma}}\left[\boldsymbol{\delta}_\theta \frac{dv_z}{dr} - \boldsymbol{\delta}_z r \frac{d}{dr}\left(\frac{v_\theta}{r}\right)\right]$ $\dot{\gamma} = +\sqrt{\left(r\frac{d}{dr}\left(\frac{v_\theta}{r}\right)\right)^2 + \left(\frac{dv_z}{dr}\right)^2}$	Concentric cylinders Helices

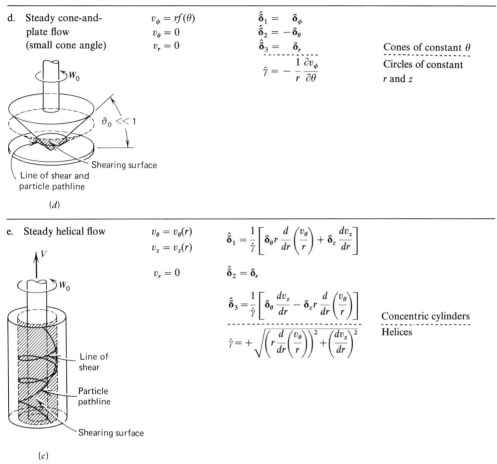

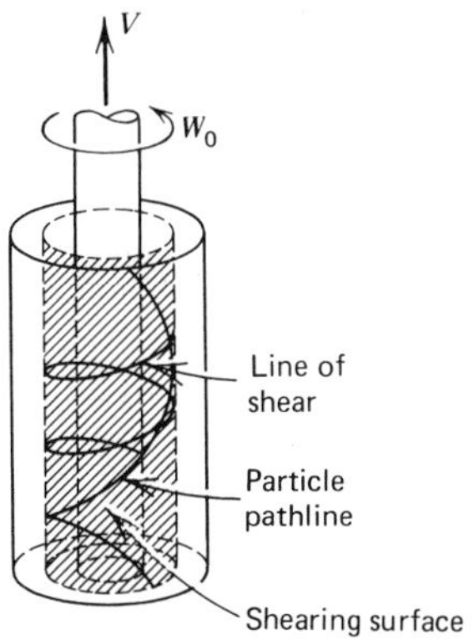

(*c*)

The viscometric flows in Table 3.7-1 are especially marked. It is important to be able to identify these flows, as we shall later obtain a constitutive equation that describes viscometric flows for a wide class of polymeric materials (see §9.6). In Example 3.7-1 we show that steady tube flow, tangential annular flow, and helical flow are viscometric flows. These and some other viscometric flows are listed in Table 3.7-2 along with some of their identifying characteristics.

Before proceeding to Example 3.7-1, we point out that *rectilinear shear flows* are defined by (i), (ii), (iii), and

v. The fluid particle pathlines are straight lines.

Finally, *steady simple shear flow* is a viscometric flow for which, in addition to (i), (ii), (iii), and (iv), we require

vi. The shearing surfaces are planes, that is, the shear axes $\hat{\boldsymbol{\delta}}_1$, $\hat{\boldsymbol{\delta}}_2$, $\hat{\boldsymbol{\delta}}_3$ are everywhere identical to the rectangular Cartesian axes $\boldsymbol{\delta}_x$, $\boldsymbol{\delta}_y$, $\boldsymbol{\delta}_z$.

vii. The flow is homogeneous, that is, the velocity gradient $\hat{\dot{\gamma}}_{21} = \dot{\gamma}_{yx}$ is independent of position.

EXAMPLE 3.7-1 Kinematics of Steady Tube Flow, Steady Tangential Annular Flow, and Steady Helical Flow

Consider the three fluid flows: **(a)** steady axial tube flow under a constant pressure gradient, **(b)** steady tangential annular flow between two concentric cylinders in relative rotation, and **(c)** steady helical flow between two coaxial cylinders, the inner one of which is rotating and is translating in the axial direction. In **(a)** the motion of the liquid will be such that $v_z = v_z(r)$, $v_r = 0$, and $v_\theta = 0$; in **(b)** the velocity field will have the form $v_\theta = v_\theta(r)$, $v_z = 0$, and $v_r = 0$; and in **(c)**, $v_\theta = v_\theta(r)$, $v_z = v_z(r)$, and $v_r = 0$. Show that all of these flows are viscometric according to the definition given in the text.

SOLUTION **(a)** Steady Tube Flow

Since the fluid velocity is in the z-direction only and varies in the r-direction only, the flow can be pictured as the relative axial sliding motion of concentric cylinders of fluid. These rigid cylindrical material surfaces are then the shearing surfaces and perform a telescoping motion. The shear axes correspond to the cylindrical coordinate axes as follows: $\hat{\boldsymbol{\delta}}_1 = \boldsymbol{\delta}_z$, $\hat{\boldsymbol{\delta}}_2 = \boldsymbol{\delta}_r$ and $\hat{\boldsymbol{\delta}}_3 = \boldsymbol{\delta}_\theta$. It is easy to see that the lines of shear are parallel to the z-axis and are also fluid pathlines. Referred to the shear axes, the velocity gradient tensor is

$$\nabla\hat{\boldsymbol{v}} = \hat{\boldsymbol{\delta}}_2\hat{\boldsymbol{\delta}}_1\hat{\gamma}_{21}(r) \tag{3.7-6}$$

where $\hat{\gamma}_{21}(r) = dv_z(r)/dr = -\dot{\gamma}$ is constant for each material particle since each particle moves at constant r. Thus the flow history at every particle is a steady simple shearing moton and the flow is viscometric.

(b) Steady Tangential Annular Flow

Once again the shearing surfaces are concentric cylinders. At any time t, the shear axes at a particle P are related to the cylindrical coordinate axes at the position of the particle as follows: $\hat{\boldsymbol{\delta}}_1 = \boldsymbol{\delta}_\theta$, $\hat{\boldsymbol{\delta}}_2 = \boldsymbol{\delta}_r$, and $\hat{\boldsymbol{\delta}}_3 = -\boldsymbol{\delta}_z$. In order to compute the velocity gradient tensor relative to the shear axes at P, we note that a point fixed relative to the shear axes has velocity $(v_\theta/r)_P r$ referred to the space-fixed cylindrical coordinate system. Here $(v_\theta/r)_P$ is the constant angular velocity of P as it moves in its circular trajectory. Thus when viewed from the shear axes, the velocity field is $\hat{v}_1 = v_\theta - (v_\theta/r)_P r$, $\hat{v}_2 = \hat{v}_3 = 0$. Since at any instant the shear coordinate system is coincident with the cylindrical one, we can use the relations for the components of $\nabla\boldsymbol{v}$ given in Table A.7-2 to find

$$\nabla\hat{\boldsymbol{v}} = \hat{\boldsymbol{\delta}}_2\hat{\boldsymbol{\delta}}_1\frac{d\hat{v}_1}{dr} - \hat{\boldsymbol{\delta}}_1\hat{\boldsymbol{\delta}}_2\frac{\hat{v}_1}{r} \tag{3.7-7}$$

But

$$\frac{d\hat{v}_1}{dr} = \frac{d}{dr}\left[v_\theta - \left(\frac{v_\theta}{r}\right)_P r\right] = \frac{dv_\theta}{dr} - \left(\frac{v_\theta}{r}\right)_P = r\frac{d}{dr}\left(\frac{v_\theta}{r}\right) + \frac{v_\theta}{r} - \left(\frac{v_\theta}{r}\right)_P \tag{3.7-8}$$

When Eq. 3.7-8 is substituted into Eq. 3.7-7 and the velocity gradient tensor evaluated at the particle P, we find

$$(\nabla\hat{\boldsymbol{v}})\bigg|_P = \hat{\boldsymbol{\delta}}_2\hat{\boldsymbol{\delta}}_1\hat{\gamma}_{21}(r) \tag{3.7-9}$$

where $\hat{\gamma}_{21} = rd(v_\theta/r)/dr$ is the constant shear rate at the particle and is equal to $-\dot{\gamma}$ if the inner cylinder is rotating and the outer one fixed. The flow at any particle P is thus steady simple shear as seen from the shear axes; according to the definition, then, tangential annular flow is a viscometric flow.

(c) Steady Helical Flow

In this flow the direction of shear is not so obvious as in the two preceding parts. Hence, in this part we show how the shear axes may be determined in the process of writing the velocity gradient tensor in the form appropriate to a steady simple shear flow. Again we compute the velocity gradient with respect to shear axes at an arbitrary particle *P*.

Because the shearing surfaces are concentric cylinders, at any time *t* the unit vector $\hat{\boldsymbol{\delta}}_2$ at *P* is coincident with $\boldsymbol{\delta}_r$. Furthermore, the unit vectors $\hat{\boldsymbol{\delta}}_1$ and $\hat{\boldsymbol{\delta}}_3$ are tangent to the cylindrical surfaces so that at any time $\boldsymbol{\delta}_\theta$ and $\boldsymbol{\delta}_z$ can be expressed in terms of them:

$$\begin{aligned} \boldsymbol{\delta}_\theta &= \hat{\boldsymbol{\delta}}_1 a + \hat{\boldsymbol{\delta}}_3 b \\ \boldsymbol{\delta}_z &= \hat{\boldsymbol{\delta}}_1 b - \hat{\boldsymbol{\delta}}_3 a \end{aligned} \tag{3.7-10}$$

where a and b depend on the details of the flow field and $\sqrt{a^2 + b^2} = 1$.

We now note that a point, which is stationary as seen from the shear axes at *P*, has velocity $\boldsymbol{v} = \boldsymbol{\delta}_\theta(v_\theta/r)_P r + \boldsymbol{\delta}_z(v_z)_P$ relative to the space-fixed cylindrical coordinate system. Thus the velocity $\hat{\boldsymbol{v}}$ of any fluid element relative to the shear axes at *P* may be written in terms of the cylindrical coordinate unit vectors:

$$\hat{\boldsymbol{v}} = \boldsymbol{\delta}_\theta\left[v_\theta - \left(\frac{v_\theta}{r}\right)_P r\right] + \boldsymbol{\delta}_z\left[v_z - (v_z)_P\right] \tag{3.7-11}$$

The velocity gradient relative to the shear axes is readily computed in the cylindrical coordinate system by using Table A.7-2:

$$\nabla\hat{\boldsymbol{v}} = \boldsymbol{\delta}_r\boldsymbol{\delta}_\theta \frac{d}{dr}\left[v_\theta - \left(\frac{v_\theta}{r}\right)_P r\right] + \boldsymbol{\delta}_r\boldsymbol{\delta}_z \frac{d}{dr}\left[v_z - (v_z)_P\right] + \boldsymbol{\delta}_\theta\boldsymbol{\delta}_r\left[-\frac{v_\theta}{r} + \left(\frac{v_\theta}{r}\right)_P\right] \tag{3.7-12}$$

When this is evaluated at the particle *P* we find

$$(\nabla\hat{\boldsymbol{v}})\Big|_P = \boldsymbol{\delta}_r\boldsymbol{\delta}_\theta r\frac{d}{dr}\left(\frac{v_\theta}{r}\right) + \boldsymbol{\delta}_r\boldsymbol{\delta}_z\frac{dv_z}{dr} = \hat{\boldsymbol{\delta}}_2\hat{\boldsymbol{\delta}}_1\hat{\dot{\gamma}}_{21} \tag{3.7-13}$$

where the last equality above emphasizes that when the velocity gradient is written in terms of the $\hat{\boldsymbol{\delta}}_i$, it must have the simple shear flow form. By substituting Eqs. 3.7-10 into Eq. 3.7-13 we can make the following identifications:

$$\hat{\boldsymbol{\delta}}_1 = \frac{1}{\dot{\gamma}}\left[\boldsymbol{\delta}_\theta r\frac{d}{dr}\left(\frac{v_\theta}{r}\right) + \boldsymbol{\delta}_z\frac{dv_z}{dr}\right] \tag{3.7-14}$$

$$\hat{\boldsymbol{\delta}}_3 = \frac{1}{\dot{\gamma}}\left[\boldsymbol{\delta}_\theta\frac{dv_z}{dr} - \boldsymbol{\delta}_z r\frac{d}{dr}\left(\frac{v_\theta}{r}\right)\right] \tag{3.7-15}$$

$$\dot{\gamma} = \sqrt{\left(r\frac{d}{dr}\left(\frac{v_\theta}{r}\right)\right)^2 + \left(\frac{dv_z}{dr}\right)^2} \tag{3.7-16}$$

Thus steady helical flow is a steady simple shear flow if seen from the shear axes given in Eqs. 3.7-14 and 3.7-15, and it is thus a viscometric flow. Note that Eq. 3.7-16 can also be obtained from the postulated velocity field, the definition of $\dot{\gamma}$ in Eq. 3.7-3, and Eqs. B.3-7 to 12. Equation 3.7-16 serves to emphasize that the terms "shear rate" and "velocity gradient" are not synonymous.

In parts **(a)** and **(b)** of this example, the lines of shear and the particle pathlines are identical, since $\hat{\boldsymbol{\delta}}_1$ is everywhere parallel to the local fluid velocity vector (see Table 3.7-2a and b). Steady helical flow, however, is a flow in which the particle pathlines and the lines of shear are different. The pathline is the curve traced out by a given particle as it moves about in space. Since the flow considered here is steady, the velocity vector at any position is constant, and thus the pathline is the curve that is everywhere tangent to the velocity vector $\boldsymbol{v} = \boldsymbol{\delta}_\theta v_\theta + \boldsymbol{\delta}_z v_z$. On the other hand the line of shear is defined as the curve that, at any instant, is everywhere tangent to $\hat{\boldsymbol{\delta}}_1$. Since the flow is steady, the line of shear is independent of time. From Eq. 3.7-14 we see that $\hat{\boldsymbol{\delta}}_1$ is not in general parallel to $\boldsymbol{v}$ so that the line of shear is different than the particle pathline (see Table 3.7-2e).

b. Shearfree Flows

A *shearfree flow*[6] is defined as a flow for which it is possible to select for every fluid element an orthogonal set of unit vectors $\hat{\boldsymbol{\delta}}_i$ fixed in the element so that referred to these axes the rate-of-strain tensor has a diagonal form:

$$\dot{\boldsymbol{\gamma}} = \begin{pmatrix} \hat{\dot{\gamma}}_{11} & 0 & 0 \\ 0 & \hat{\dot{\gamma}}_{22} & 0 \\ 0 & 0 & \hat{\dot{\gamma}}_{33} \end{pmatrix} = \hat{\boldsymbol{\delta}}_1\hat{\boldsymbol{\delta}}_1\hat{\dot{\gamma}}_{11} + \hat{\boldsymbol{\delta}}_2\hat{\boldsymbol{\delta}}_2\hat{\dot{\gamma}}_{22} + \hat{\boldsymbol{\delta}}_3\hat{\boldsymbol{\delta}}_3\hat{\dot{\gamma}}_{33} \tag{3.7-17}$$

In addition we require that the volume remain constant in a shearfree flow so that $\hat{\dot{\gamma}}_{11} + \hat{\dot{\gamma}}_{22} + \hat{\dot{\gamma}}_{33} = 0$. By "fixed in a fluid element" we mean that at every time t the same fluid particles lie along each of the $\hat{\boldsymbol{\delta}}_i$'s.[7] The particle-fixed axes $\hat{\boldsymbol{\delta}}_i$ used in Eq. 3.7-17 are known as the *principal axes of rate of strain.* At any instant, the rate-of-strain tensor has the same form given above when referred to an orthogonal space-fixed coordinate system which has the same orientation everywhere as the principal axes of rate of strain.

A shearfree flow is *homogeneous* if the $\hat{\dot{\gamma}}_{ij}$ are independent of position. A homogeneous, shearfree flow for which the ratios of the $\hat{\dot{\gamma}}_{ij}$ (i.e., $\hat{\dot{\gamma}}_{11}/\hat{\dot{\gamma}}_{22}$, $\hat{\dot{\gamma}}_{22}/\hat{\dot{\gamma}}_{33}$) are independent of time is said to be a *simple shearfree flow.*

EXAMPLE 3.7-2. Kinematics of Flow into a Line Sink

A fluid is flowing into a line sink located at $r = 0$ in a cylindrical coordinate system. The strength of the sink is Q where Q is the volume rate of flow into the sink per unit length. Show that this flow is a shearfree flow.

SOLUTION The fluid velocity field is of the form $v_r = v_r(r)$, $v_\theta = v_z = 0$. From Table B.3 we have the rate-of-strain tensor in cylindrical coordinates:

$$\dot{\boldsymbol{\gamma}} = \begin{pmatrix} 2\dfrac{dv_r}{dr} & 0 & 0 \\ 0 & 2\dfrac{v_r}{r} & 0 \\ 0 & 0 & 0 \end{pmatrix} \tag{3.7-18}$$

[6] See also A. S. Lodge, *Elastic Liquids*, Academic Press, New York (1964), pp. 35–38; A. S. Lodge, *Body Tensor Fields in Continuum Mechanics*, Academic Press, New York (1974), pp. 81–83, 175–176; B. D. Coleman, *Proc. Roy. Soc.*, **A306**, 449–476 (1968). Coleman calls these flows "extensions." Thorough reviews of theoretical and experimental aspects of shearfree flows are given by J. M. Dealy, *Polym. Eng. Sci.*, **11**, 433–445 (1971); K. Walters, *Rheometry*, Chapman and Hall, London (1975), Chapt. 7; and C. J. S. Petrie, *Elongational Flows*, Pitman, London (1979).

[7] Since $\dot{\boldsymbol{\gamma}}$ is symmetric, it is always possible at any time t to select an orthogonal set of unit vectors at a given fluid element such that $\dot{\boldsymbol{\gamma}}$ is diagonal when referred to these axes. However, in general, these axes will not be tangent to curves defined by the same set of material particles at different times, and they are therefore not fixed in the element in the sense used for the $\hat{\boldsymbol{\delta}}_i$.

where the components are given in the order r, θ, z. Since $\dot{\boldsymbol{\gamma}}$ is diagonal, the cylindrical coordinate unit vectors give the principal axes of rate-of-strain.

To show that this is a shearfree flow, it is necessary to be certain that the curves defined by the same material particles are tangent to the $\boldsymbol{\delta}_r$, $\boldsymbol{\delta}_\theta$, and $\boldsymbol{\delta}_z$ unit vectors for all times. This is clearly true of the given velocity field, since for example the fluid particles that lie on the θ-coordinate curve at some position r, θ, z will all have the same radial velocity v_r and will thus always lie on θ-coordinate curves.

PROBLEMS

3A.1 Relation between Viscosity and Complex Viscosity

Use the Cox-Merz rule to construct plots of $\eta(\dot{\gamma})$ for the polyacrylamide and aluminium soap solutions shown in Fig. 3.3-3 and Figs. 3.4-4 and 5. Also on the same plots, graph $\eta'(\omega)$ and some of the actual $\eta(\dot{\gamma})$ data. How well is the viscosity described by $|\eta^*|$? η'?

3A.2 Rod-Climbing for a Polyacrylamide Solution

In §2.3 the phenomenon of rod-climbing is described. Equation 2.3-5 was derived to predict whether or not a polymeric fluid would climb a rotating shaft. In that equation if $d\pi_{zz}/dr > 0$, then the fluid will not climb the rod; and if $d\pi_{zz}/dr < 0$, then the fluid will climb the rod. Use the data for the aqueous polyacrylamide solution in Figs. 3.3-3 and 5 to decide what the free surface for this fluid would look like in the rod-climbing experiment. Assume that inertial effects are negligible. Since Ψ_2 data are not available for this solution, assume that $-\Psi_2(\dot{\gamma})/\Psi_1(\dot{\gamma})$ is a constant and is the same as for the polyacrylamide solution in Figs. 3.3-6 and 7. Compare your conclusion with Fig. 2.3-1.

3B.1 Eccentric Disk Rheometer

The eccentric disk (or Maxwell orthogonal) rheometer[1] consists of two parallel disks, both of which rotate with a constant angular velocity W; the fluid to be tested is placed in the region between the two disks. The axes of rotation of the disks are separated by a distance a, and the gap between the disks is b (Fig. 3B.1). Fluid disks of constant z are assumed to rotate rigidly about points on the line connecting the centers of the bounding disks.

a. Is this flow a shear flow?
b. Show that the velocity field corresponding to the motion described above is

$$v_x = -W(y - Az)$$
$$v_y = Wx$$
$$v_z = 0 \tag{3B.1-1}$$

where $A = a/b$, and the origin of the Cartesian coordinate system is located at the center of the bottom disk.

In Example 10.1-2 it is shown that η' and η'' can be obtained from measurements of this instrument. For more on the kinematics of the eccentric disk flow see Problem 3C.2.

3B.2 Complex Representation of Oscillatory Quantities

Let $z = a + ib$ be a complex quantity whose complex conjugate is $\bar{z} = a - ib$ (here a and b are real).
a. Verify that the magnitude of z is

$$|z| \equiv \sqrt{a^2 + b^2} = (z\bar{z})^{1/2} \tag{3B.2-1}$$

[1] A. N. Gent, *Br. J. Appl. Phys.*, **11**, 165–167 (1960); B. Maxwell and R. P. Chartoff, *Trans. Soc. Rheol.*, **9**, 41–52 (1965); K. Walters, *Rheometry*, Chapman and Hall, London (1975), pp. 168–200, discusses other off-center instruments as well.

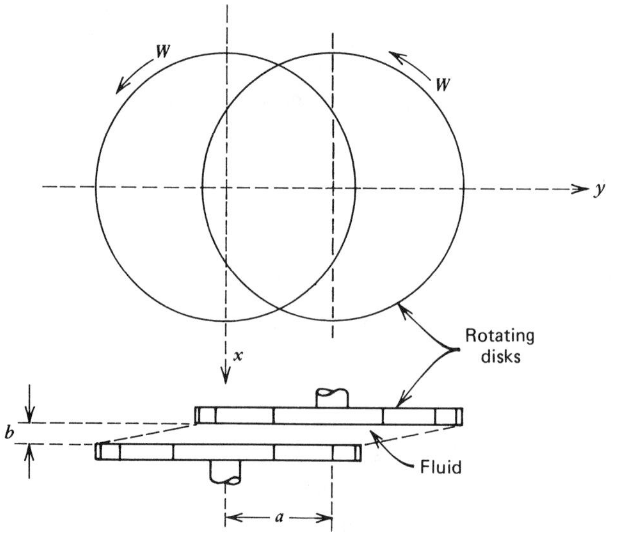

FIGURE 3B.1. Schematic diagram of the eccentric-disk (Maxwell orthogonal) rheometer; both circular disks rotate with the same constant angular velocity W. The origin of coordinates is at the center of the lower disk, and the z-axis points upward.

b. Show that z may be written as

$$z = |z| e^{i\theta} \tag{3B.2-2}$$

where $\theta = \arctan(b/a)$.

c. In some oscillatory rheological experiment, a part of the apparatus is forced to undergo sinusoidal oscillations with frequency ω which we represent as a cosine function $\cos \omega t$. The response of some other part of the equipment is observed to oscillate with frequency ω and amplitude R, but it oscillates out of phase with the driving function. Thus we represent the response function by $r = R\cos(\omega t + \delta)$; here δ is the angle by which the response lags the input. Show that r can also be represented by

$$r = \mathscr{Re}\{r^0 e^{i\omega t}\} \tag{3B.2-3}$$

where r^0 is a complex amplitude with

$$\mathscr{Re}\{r^0\} = R\cos\delta \tag{3B.2-4a}$$

$$\mathscr{Im}\{r^0\} = R\sin\delta \tag{3B.2-4b}$$

Sketch the driving and response functions and label the angle δ.

3B.3 Form of the Stress Tensor for Unidirectional Shear Flows[2]

For unidirectional shearing flows the velocity field near a particle can be written in the form (cf. Eq. 3.7-4)

$$\hat{v}_1 = \hat{\dot{\gamma}}_{21}\hat{x}_2, \hat{v}_2 = \hat{v}_3 = 0 \tag{3B.3-1}$$

[2] A. S. Lodge, *Elastic Liquids*, Academic Press, New York (1964).

where $\hat{\gamma}_{21}$ is the (time-dependent) velocity gradient. We want to use the symmetry of the velocity field to show that the stress tensor for an isotropic liquid undergoing unidirectional shear flow is of the form given in Eq. 3.2-1.

a. By considering the change of variables $\hat{x} \rightarrow \bar{x}$

$$\bar{x}_1 = -\hat{x}_1; \bar{x}_2 = -\hat{x}_2; \bar{x}_3 = \hat{x}_3 \tag{3B.3-2}$$

show that the kinematics are unaffected by a 180° rotation about the $\hat{x}_3$-axis.

b. For an isotropic fluid in which the stress in a fluid element depends only on the kinematic history of that element, the stress field must have the same symmetry as the velocity field. This allows us to conclude that the stress components in the $\hat{x}$ and $\bar{x}$ coordinate systems are the same:

$$\bar{\tau}_{ij} = \hat{\tau}_{ij} \tag{3B.3-3}$$

Calculate the force normal to the $\hat{x}_3$-direction, $[\hat{\boldsymbol{\delta}}_3 \cdot \boldsymbol{\tau}]$, in the two coordinate systems and use these expressions to conclude

$$\hat{\tau}_{31} = \hat{\tau}_{13} = \hat{\tau}_{32} = \hat{\tau}_{23} = 0 \tag{3B.3-4}$$

3B.4 Form of the Stress Tensor for Shearfree Flows

In a shearfree flow, the velocity field near an arbitrary fluid particle is given relative to the principal axes of rate of strain at that particle by (cf. Eq. 3.7-17)

$$\hat{\mathbf{v}} = \hat{\boldsymbol{\delta}}_1 \frac{1}{2} \hat{\gamma}_{11} \hat{x}_1 + \hat{\boldsymbol{\delta}}_2 \frac{1}{2} \hat{\gamma}_{22} \hat{x}_2 + \hat{\boldsymbol{\delta}}_3 \frac{1}{2} \hat{\gamma}_{33} \hat{x}_3 \tag{3B.4-1}$$

where the $\hat{\gamma}_{ii}$ may be functions of time at the given particle. By considering a 180° rotation about the $\hat{x}_1$-axis show that

$$\begin{aligned} \hat{\tau}_{12} &= \hat{\tau}_{21} = 0 \\ \hat{\tau}_{13} &= \hat{\tau}_{31} = 0 \end{aligned} \tag{3B.4-2}$$

Use the discussion in Problem 3B.3 as a guide. Similarly show that

$$\hat{\tau}_{23} = \hat{\tau}_{32} = 0 \tag{3B.4-3}$$

Thus the stress tensor must have the form shown in Eq. 3.2-3 for shearfree flows.

3B.5 Steady Radial Creeping Flow Between Two Circular Disks

A fluid flows radially between two circular disks (see Fig. 4B.3). The velocity distribution is $v_r = v_r(r, z)$, $v_\theta = 0$, and $v_z = 0$. Is this a shear flow?

3C.1 Functional Forms for the Viscometric Functions

Consider a steady simple shear flow with $v_x = \dot{\gamma}_{yx} y$, $v_y = v_z = 0$. By utilizing a rotation of the coordinate system through 180° about the y-axis and arguments similar to those in Problem 3B.3, show that η, Ψ_1, and Ψ_2 are all even functions of the velocity gradient $\dot{\gamma}_{yx}$. How can arguments such as those proposed here be used to explain the shapes of the normal-stress and shear-stress responses in the small-amplitude oscillatory shear experiment?

3C.2 Eccentric-Disk-Rheometer Flow and Kinematics of General Shear Flow[3]

In Eq. 3.7-5 we give the rate-of-strain tensor referred to the shear axes for a unidirectional shear flow. Here we give the corresponding result for general shear flows and then ask how this result applies to the eccentric-disk flow of Problem 3B.1.

We consider the same shear axes defined above Eq. 3.7-4 with $\hat{x}_1$- and $\hat{x}_3$-coordinate curves embedded in the shearing surfaces. It can be shown for a general shear flow that the rate-of-strain tensor is always of the form

$$\dot{\boldsymbol{\gamma}} = \begin{pmatrix} 0 & \hat{\dot{\gamma}}_{21} & 0 \\ \hat{\dot{\gamma}}_{21} & 0 & \hat{\dot{\gamma}}_{32} \\ 0 & \hat{\dot{\gamma}}_{32} & 0 \end{pmatrix}$$
$$= (\hat{\boldsymbol{\delta}}_1\hat{\boldsymbol{\delta}}_2 + \hat{\boldsymbol{\delta}}_2\hat{\boldsymbol{\delta}}_1)\hat{\dot{\gamma}}_{21} + (\hat{\boldsymbol{\delta}}_3\hat{\boldsymbol{\delta}}_2 + \hat{\boldsymbol{\delta}}_2\hat{\boldsymbol{\delta}}_3)\hat{\dot{\gamma}}_{32} \tag{3C.2-1}$$

when referred to this coordinate system. Here $\hat{\dot{\gamma}}_{21}$ and $\hat{\dot{\gamma}}_{32}$ can both be functions of time and can be positive or negative. Furthermore it is found that a line of shear makes an angle $\zeta(t)$ with the $\hat{x}_1$-curve where

$$\zeta(t) = \tan^{-1}(\hat{\dot{\gamma}}_{32}/\hat{\dot{\gamma}}_{21}) \tag{3C.2-2}$$

and that the shear rate is

$$\dot{\gamma} = \sqrt{\tfrac{1}{2}II} = \sqrt{\hat{\dot{\gamma}}_{21}^2 + \hat{\dot{\gamma}}_{32}^2} \tag{3C.2-3}$$

Note that if ζ is a function of time then the lines of shear cannot be material lines and thus the flow cannot be a unidirectional shear flow.

Now consider the eccentric-disk flow described by Eq. 3B.1-1.

a. What are the shearing surfaces for this flow?

b. Next examine the flow as seen from the shear axes at an arbitrary particle P. For convenience take $\hat{\boldsymbol{\delta}}_1 = \boldsymbol{\delta}_x$ and $\hat{\boldsymbol{\delta}}_3 = -\boldsymbol{\delta}_y$ at $t = 0$. Obtain relations between $\hat{\boldsymbol{\delta}}_1$, $\hat{\boldsymbol{\delta}}_3$ and $\boldsymbol{\delta}_x$, $\boldsymbol{\delta}_y$ at any time t.

c. Determine the velocity field near P relative to the shear axes and show that

$$\nabla\hat{\mathbf{v}} = \hat{\boldsymbol{\delta}}_2\hat{\boldsymbol{\delta}}_1 AW \cos Wt + \hat{\boldsymbol{\delta}}_2\hat{\boldsymbol{\delta}}_3 AW \sin Wt \tag{3C.2-4}$$

d. Show further that the shear rate $\dot{\gamma}$ is AW (independent of position and time) and that the direction of shear is always parallel to the x-axis.

e. Is this flow a unidirectional shear flow? Explain.

[3] A. S. Lodge, *Body Tensor Fields in Continuum Mechanics*, Academic Press, New York (1974), pp. 60–72, 76–78.

PART II
ELEMENTARY CONSTITUTIVE EQUATIONS AND THEIR USE IN SOLVING FLUID DYNAMICS PROBLEMS

We now come to the first chapters that deal with the solution of polymer fluid dynamics problems. In these chapters we begin facing the main challenge of the subject: given a fluid dynamics problem, obtain the velocity profiles by solving the equations of change together with a suitable "constitutive equation" (i.e., an expression for the stress tensor in terms of some kinematic tensors); then use the velocity distribution so obtained to get other information of engineering interest, such as volume rates of flow, pressure drops, torques, forces on solid surfaces, etc. The two chapters in this part of the book deal with the two "classical" approaches to the subject: the solution of steady-state shear flow problems by use of various "generalized Newtonian fluid" constitutive equations, and the description of fluid motions with very small displacement gradients by use of various "linear viscoelastic fluid" constitutive equations.

Chapter 4 is devoted to the generalized Newtonian fluid models. In many industrial problems the most important feature of polymeric liquids is the fact that their viscosities decrease markedly as the shear rate increases. This is the only rheological property of these fluids that is incorporated into the generalized Newtonian constitutive equations. In these constitutive equations one allows the viscosity to be a function of the shear rate, and many empirical expressions have been proposed for describing this "shear thinning," as it is often called. Engineers have relied on these simple empirical models for describing flows which are steady-state (or almost so) and which are shear flows (or very nearly so), and engineering designs based on these models have been by and large successful. Although the method is presented in Chapter 4 as a completely empirical approach, it will be found in Chapter 9 that the method can in fact be justified on the basis of more modern theories. The generalized Newtonian fluid models have been around for a long time—over a half century—and we feel that engineers should be familiar with them.

Chapter 5 deals with the linear viscoelastic fluid models. These are intended to be used only for flows in which the displacement gradients are exceedingly small; any kind of time-dependent behavior is allowed, however. The general linear viscoelastic fluid contains a "relaxation modulus" and many empirical expressions have been suggested for this quantity. The constitutive equations resulting from these various empiricisms have been widely used for interpreting time-dependent experiments, primarily by polymer chemists, who want to use these experiments to perform a sort of "mechanical spectroscopy" in order to draw inferences about the molecular architecture of polymers. In addition, the measurement of various linear viscoelastic material functions provides a basis for quality control in the plastics industry.

The two chapters in this part of the book are nonoverlapping as far as the types of flows considered: Chapter 4 deals with steady-state shear flows, whereas Chapter 5 deals with unsteady-state, small-amplitude motions. Given a sample of a polymeric liquid, an engineer measures the viscosity and then uses a generalized Newtonian fluid model to describe flows of interest to him, whereas a polymer chemist determines the relaxation modulus experimentally and then uses a linear viscoelastic fluid model to interpret other linear experiments. There is, at this stage of the development of the subject, virtually no communication between the engineer and the polymer chemist inasmuch as they measure different material functions and use totally different constitutive equations.

The two types of constitutive equations introduced in Chapters 4 and 5 are still in common use. Until about 1950 practitioners in the two fields mentioned above regarded their fields as being disjoint, inasmuch as there was no theoretical framework available for providing a connecting link. From 1950 on, however, considerable progress was made in formulating theories of continuum mechanics that made it possible to study steady-state shear flows and unsteady-state, small-amplitude motions as special cases in a more general theoretical framework. Later, in Chapter 9, we see how the constitutive equations of Chapters 4 and 5 may be obtained by simplifying a very general constitutive equation.

From a strictly logical point of view we should start with the general formulation in Chapter 9 and then teach the entire subject by considering various special cases. We feel, however, that following a more or less historical sequence enables the beginning student to proceed from the less mathematical to the more mathematical material, from the old-fashioned tried-and-tested material to the somewhat more speculative and untested, and from the study of restricted sets of flow problems to the study of much more general flows. It will be evident to the reader that when we adopt this sequence of study, the earlier chapters enable us to solve a wide variety of flow and heat-transfer problems using rather elementary mathematics, whereas in the later chapters the number of solvable problems becomes rather small, just because of the complexity of the constitutive equations.

CHAPTER 4

THE GENERALIZED NEWTONIAN FLUID

This chapter is devoted to the generalized Newtonian fluid, which results from a minor modification of the Newtonian fluid constitutive equation given in Eq. 1.2-2. This equation incorporates the idea of a shear-rate-dependent viscosity, and hence can describe the non-Newtonian viscosity curves shown in Figs. 3.3-1 to 4. It cannot, however, describe normal stress effects or time-dependent elastic effects. There are many industrial flow problems in which the non-Newtonian viscosity effects are of paramount importance, and hence the generalized Newtonian fluid is useful; in addition it is a relatively simple constitutive equation, and many problems have been solved using it.

In §4.1 we introduce the model and give several useful empiricisms for the non-Newtonian viscosity, in particular, the "power-law model." Then in §4.2 we show how to use the power-law model by going through a series of illustrative examples; the problems in these examples are sufficiently simple that analytical solutions may be obtained. For more complicated problems a variational principle is available, and it is discussed in §4.3. Up to this point it is presumed that the flow system is isothermal. Many industrial non-Newtonian problems are nonisothermal, and §4.4 provides a brief introduction to this rather large subject; particular attention is paid to viscous heating, because this effect is generally present in plastics processing operations and in lubrication problems. In §§4.2 to 4.4 all the illustrative examples are worked out for the power-law model; in §4.5 we discuss several other models and illustrate their use. Finally, in §4.6 we summarize the limitations on the generalized Newtonian fluid model and discuss the extent to which its use can be legitimized.

§4.1 THE GENERALIZED NEWTONIAN FLUID AND USEFUL EMPIRICISMS FOR THE NON-NEWTONIAN VISCOSITY

For the industrial chemical engineer, the most important property of macromolecular fluids discussed in Chapter 3 is the non-Newtonian viscosity—that is, the fact that the viscosity of the fluid changes with the shear rate. Since for some fluids the viscosity can change by a factor of 10, 100, or even 1000, it is evident that such an enormous change cannot be ignored in pipe flow calculations, lubrication problems, design of on-line viscometers, extruder operation, and polymer processing calculations. Therefore it is not surprising that one of the earliest empiricisms to be introduced was a modification of

Newton's law of viscosity in which the viscosity is allowed to vary with the shear rate. That is, for the elementary flow $v_x = v_x(y)$, $v_y = 0$, $v_z = 0$, the early rheologists replaced

Newtonian fluid: $$\tau_{yx} = -\mu \frac{dv_x}{dy}, \quad \mu = \text{a constant for a given temperature, pressure, and composition} \tag{4.1-1}$$

by the empiricism:[1]

Generalized Newtonian fluid: $$\tau_{yx} = -\eta \frac{dv_x}{dy}, \quad \eta = \text{a function of } |dv_x/dy| \tag{4.1-2}$$

Absolute value signs have been used at the right of Eq. 4.1-2, since one would expect the change in viscosity to depend on the magnitude but not on the sign of the velocity gradient. Having written Eq. 4.1-2 one can then introduce various empiricisms to describe the experimental non-Newtonian viscosity curves.

We now wish to extend the ideas above to an arbitrary flow. First, for an incompressible Newtonian fluid, for any flow field $\boldsymbol{v} = \boldsymbol{v}(x, y, z, t)$ we have:

Incompressible Newtonian fluid: $$\boxed{\boldsymbol{\tau} = -\mu\dot{\boldsymbol{\gamma}}} \quad \mu = \text{a constant for a given temperature, pressure, and composition} \tag{4.1-3}$$

in which $\dot{\boldsymbol{\gamma}}$ is the rate-of-strain tensor $\nabla\boldsymbol{v} + (\nabla\boldsymbol{v})^\dagger$ (cf. Eq. 1.2-2). To include the idea of a non-Newtonian viscosity, we write[1]

Incompressible generalized Newtonian fluid: $$\boxed{\boldsymbol{\tau} = -\eta\dot{\boldsymbol{\gamma}}} \quad \eta = \text{a function of the scalar invariants of } \dot{\boldsymbol{\gamma}} \tag{4.1-4}$$

If the non-Newtonian viscosity, a scalar, is to depend on the tensor $\dot{\boldsymbol{\gamma}}$, then it must depend only on those particular combinations of components of the tensor that are not dependent on the coordinate system.[2] As described in §A.3 we may select as three independent combinations:

$$I = \sum_i \dot{\gamma}_{ii} \tag{4.1-5}$$

$$II = \sum_i \sum_j \dot{\gamma}_{ij}\dot{\gamma}_{ji} \tag{4.1-6}$$

$$III = \sum_i \sum_j \sum_k \dot{\gamma}_{ij}\dot{\gamma}_{jk}\dot{\gamma}_{ki} \tag{4.1-7}$$

For an incompressible fluid $I = 2(\nabla \cdot \boldsymbol{v}) = 0$. For shearing flows III turns out to be zero (see Problem 4B.8); since, as we will point out later, Eq. 4.1-4 should be used only for shearing

[1] M. Reiner, *Deformation, Strain, and Flow*, Interscience, New York (1960).

[2] K. Hohenemser and W. Prager, **12**, 216–226 (1932); J. G. Oldroyd, *Proc. Camb. Phil. Soc.*, **45**, 595–611 (1949); **47**, 410–418 (1950).

flows, or at least flows that are very nearly shearing, omitting III from any further consideration is not a serious restriction. Hence η is taken to depend only on II. Actually we shall prefer to use $\dot{\gamma}$, the magnitude of the rate-of-strain tensor $\dot{\boldsymbol{\gamma}}$, instead of II

$$\dot{\gamma} = \sqrt{\frac{1}{2}\sum_i \sum_j \dot{\gamma}_{ij}\dot{\gamma}_{ji}} = \sqrt{\frac{1}{2} II} \tag{4.1-8}$$

In taking the square root one must affix the proper sign so that $\dot{\gamma}$ will be positive. In shearing flows $\dot{\gamma}$ is called the "shear rate."

Equation 4.1-4 with $\eta = \eta(\dot{\gamma})$ is then the starting point for all of the calculations in this chapter. Its principal usefulness is for calculating flow rates and shearing forces in steady-state shear flows, such as:

a. Tube flow.
b. Axial annular flow.
c. Tangential annular flow.
d. Helical annular flow.
e. Flow between parallel planes.
f. Flow between rotating disks.
g. Cone-and-plate flow.

Although Eq. 4.1-4 gives correct results for flow rates and shearing forces in steady shear flows, we hasten to add that engineers have not hesitated to apply this equation to somewhat more complicated flows and systems slowly varying with time. A thorough assessment of the errors inherent in such calculations is not available, but one feels intuitively that such a practice probably represents good engineering empiricism. We give a few examples of this presently, and then in §4.6 we discuss the limits on the use of the model.

Let us now turn to the empiricisms for $\eta(\dot{\gamma})$. From Chapter 3 we know the appearance of the η vs. $\dot{\gamma}$ curves. Although for some problems one can use the raw data for $\eta(\dot{\gamma})$, it is often useful to make calculations and derivations with simple empirical equations for $\eta(\dot{\gamma})$ that are known to describe the experimental data with sufficient accuracy. Many such empiricisms are available, and we make no attempt at completeness. We cite only two here and mention some others in Table 4.5-1.

a. The Carreau–Yasuda Model (Parameters: η_0, η_∞, λ, n, a)

This five-parameter model has sufficient flexibility to fit a wide variety of experimental $\eta(\dot{\gamma})$ curves; it has proven to be useful for numerical calculations in which one needs an analytical expression for the non-Newtonian viscosity curve. The model is[3,4]

$$\boxed{\frac{\eta - \eta_\infty}{\eta_0 - \eta_\infty} = [1 + (\lambda\dot{\gamma})^a]^{(n-1)/a}} \tag{4.1-9}$$

Here η_0 is the zero-shear-rate viscosity, η_∞ is the infinite-shear-rate viscosity, λ is a time constant, n is the "power-law exponent" (since it describes the slope of $(\eta - \eta_\infty)/(\eta_0 - \eta_\infty)$ in the "power-law region"), and a is a dimensionless parameter that describes the transition region between the zero-shear-rate region and the power-law region. Some examples of

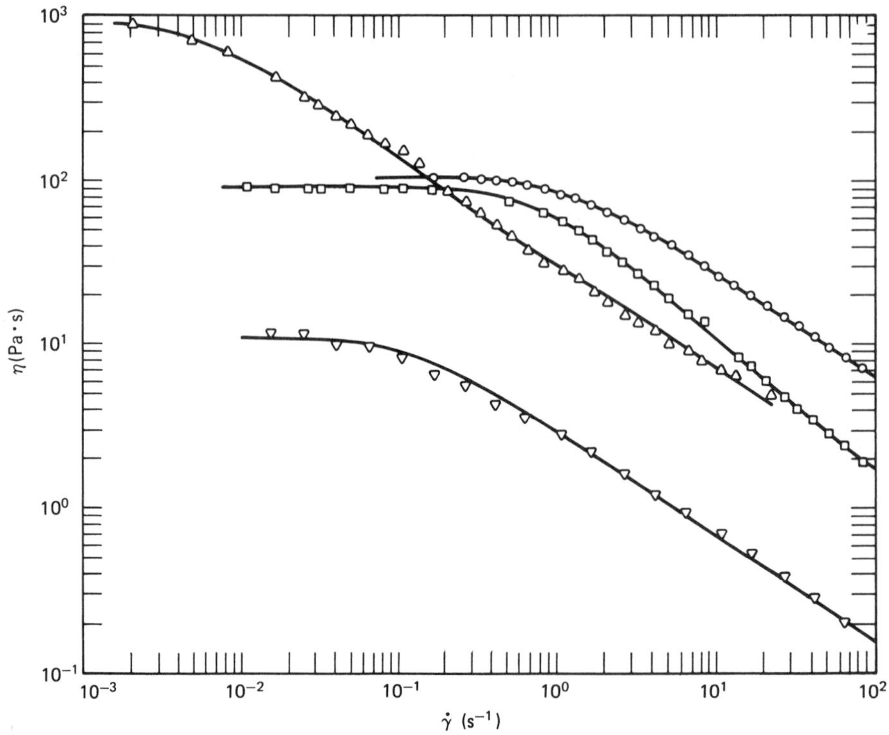

FIGURE 4.1-1. Non-Newtonian viscosity of three polymer solutions and a soap solution as fitted by the Carreau viscosity equation (Eq. 4.1-9, with $a = 2$). [R. B. Bird, O. Hassager, and S. I. Abdel-Khalik, *AIChE J.*, **20**, 1041–1066 (1974).] △ 2% polyisobutylene in Primol 355; data of J. D. Huppler, E. Ashare, and L. A. Holmes, *Trans. Soc. Rheol.*, **11**, 159–179 (1968): $\eta_0 = 9.23 \times 10^2$ Pa • s, $\eta_\infty = 1.50 \times 10^{-1}$ Pa • s, $\lambda = 191$ s, $n = 0.358$. ○ 5% polystyrene in Aroclor 1242; data of E. Ashare, Ph.D. Thesis, University of Wisconsin, Madison (1968): $\eta_0 = 1.01 \times 10^2$ Pa • s, $\eta_\infty = 5.9 \times 10^{-2}$ Pa • s, $\lambda = 0.84$ s, $n = 0.364$. ▽ 0.75% polyacrylamide (Separan-30) in a 95/5 mixture by weight of water and glycerin; data of B. D. Marsh (1967), as cited by P. J. Carreau, I. F. Macdonald, and R. B. Bird, *Chem. Eng. Sci.*, **23**, 901–911 (1968): $\eta_0 = 10.6$ Pa • s, $\eta_\infty = 10^{-2}$ Pa • s, $\lambda = 8.04$ s, $n = 0.364$. □ 7% aluminum soap in decalin and *m*-cresol; data of J. D. Huppler, E. Ashare, and L. A. Holmes, *loc. cit.*: $\eta_0 = 89.6$ Pa • s, $\eta_\infty = 10^{-2}$ Pa • s, $\lambda = 1.41$ s, $n = 0.200$.

curve-fitting of experimental data are given in Figs. 4.1-1, 2, and 3, and sample values of the parameters in the Carreau–Yasuda model are given in Table 4.1-1 for polystyrene solutions. For many concentrated polymer solutions and melts, good fits can be obtained for $a = 2$ and $\eta_\infty = 0$; then only three parameters η_0, λ, and n need to be determined. Equation 4.1-9, with $a = 2$, is usually referred to as the Carreau equation,[3] since the parameter a was added later by Yasuda.[4]

b. The "Power-Law" Model[5] of Ostwald and de Waele (Parameters: m and n)

In almost all industrial problems the descending linear region (the "power-law region") of the plot of $\log \eta$ vs. $\log \dot{\gamma}$, seen in Figs. 4.1-1, 2, and 3, is the most important

[3] P. J. Carreau, Ph.D. Thesis, University of Wisconsin, Madison (1968).

[4] K. Yasuda, Ph.D. Thesis, Massachusetts Institute of Technology, Cambridge (1979); K. Yasuda, R. C. Armstrong, and R. E. Cohen, *Rheol. Acta*, **20**, 163–178 (1981).

[5] W. Ostwald, *Kolloid-Z.*, **36**, 99–117 (1925); A. de Waele, *Oil Color Chem. Assoc. J.* **6**, 33–88 (1923).

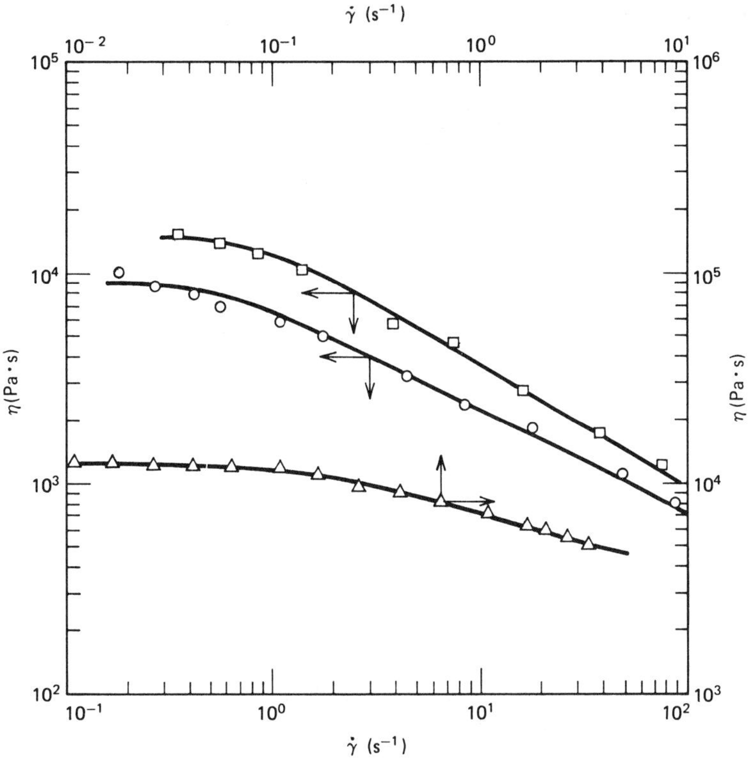

FIGURE 4.1-2. Non-Newtonian viscosity of three polymer melts as described by the Carreau viscosity equation (Eq. 4.1-9, with $a = 2$). [S. I. Abdel-Khalik, O. Hassager, and R. B. Bird, *Polym. Eng. Sci.*, **14**, 859–867 (1974).] □ Polystyrene at 453 K; data of T. F. Ballenger, I.-J. Chen, J. W. Crowder, G. E. Hagler, D. C. Bogue, and J. L. White, *Trans. Soc. Rheol.*, **15**, 195–215 (1971): $\eta_0 = 1.48 \times 10^4$ Pa·s, $\eta_\infty = 0$, $\lambda = 1.04$ s, $n = 0.398$. ○ High-density polyethylene at 443 K; data of Ballenger, *et al.*, *loc. cit.*: $\eta_0 = 8.92 \times 10^3$ Pa·s, $\eta_\infty = 0$, $\lambda = 1.58$ s, $n = 0.496$. △ Phenoxy-A at 485 K; data of B. D. Marsh as cited by P. J. Carreau, I. F. Macdonald, and R. B. Bird, *Chem. Eng. Sci.*, **23**, 901–911 (1968): $\eta_0 = 1.24 \times 10^4$ Pa·s, $\eta_\infty = 0$, $\lambda = 7.44$ s, $n = 0.728$.

region. In fact, for many inexpensive viscometers and for many fluids, it is almost impossible to obtain data for the horizontal region of the $\eta(\dot{\gamma})$ curve. The tilted straight line can be described by a simple "power-law" expression:

$$\boxed{\eta = m\dot{\gamma}^{n-1}} \tag{4.1-10}$$

which contains two parameters: m (with units of Pa·s^n), and n (dimensionless). Equation 4.1-10 may also be regarded as the limiting expression for high shear rates obtained from Eq. 4.1-9 (with $\eta_\infty = 0$); it is then evident that the exponent n in the power-law model has the same meaning as the n in the Carreau–Yasuda equation, and that the m of the power law (sometimes referred to as the "consistency index") is $\eta_0 \lambda^{n-1}$. When $n = 1$ and $m = \mu$ the Newtonian fluid is recovered. If $n < 1$, the fluid is said to be "pseudoplastic" or "shear thinning," and if $n > 1$, the fluid is called "dilatant" or "shear thickening."[6] Some sample

[6] M. Reiner, *op cit.*, pp. 306–308, and others use the term "dilatancy" to describe the change in volume of granular masses necessitated by a distortion. The standard example is the apparent drying of wet sand when stepped on.

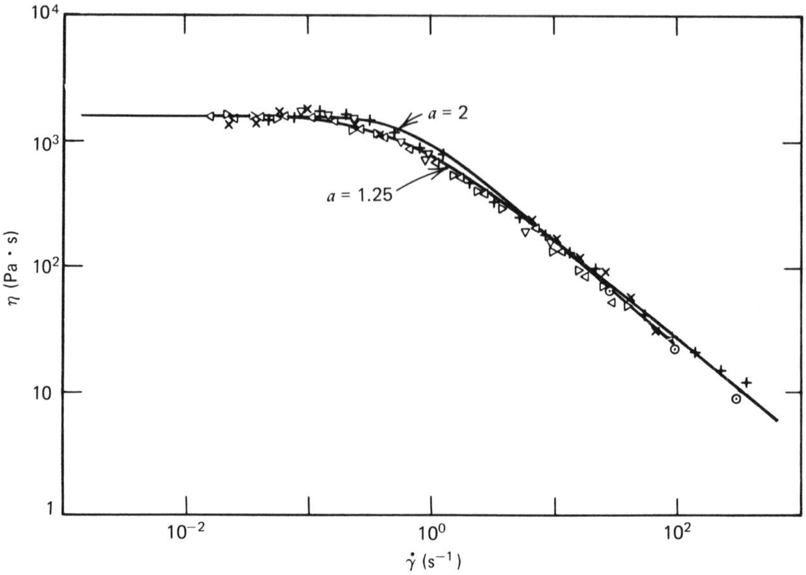

FIGURE 4.1-3. Non-Newtonian viscosity of a solution of linear, monodisperse polystyrene in 1-chloronaphthalene as fitted by the Carreau–Yasuda viscosity equation in Eq. 4.1-9. The concentration of polymer is 0.15 g/ml, and $\bar{M}_w = 2 \times 10^6$. Fits are shown for two different choices of the parameter a. For $a = 1.25$, the other model parameters are $\eta_0 = 1400$ Pa·s, $\lambda = 1.60$ s, $n = 0.2$, and $\eta_\infty = 0$. [K. Yasuda, Ph.D. Thesis, Massachusetts Institute of Technology, Cambridge (1979).]

TABLE 4.1-1

Parameters in Carreau–Yasuda Model for Some Solutions of Linear Polystyrene in 1-Chloronaphthalene[a]

Properties of Solution		Parameters in Eq. 4.1-9 (η_∞ is taken to be zero)			
$\bar{M}_w$ (g/mol)	c (g/ml)	η_0 (Pa·s)	λ (s)	n (---)	a (---)
2×10^6	0.15	1400	1.60	0.200	1.25
2×10^6	0.088	90	3.79×10^{-1}	0.265	0.98
3.9×10^5	0.45	8080	1.109	0.304	2
3.9×10^5	0.30	135	3.61×10^{-2}	0.305	2
1.1×10^5	0.52	1180	9.24×10^{-2}	0.441	2
1.1×10^5	0.45	166	1.73×10^{-2}	0.538	2
3.7×10^4	0.62	3930	1×10^{-1}	0.217	2

[a] Values of the parameters are taken from K. Yasuda, R. C. Armstrong, and R. E. Cohen, *Rheol. Acta*, **20**, 163–178 (1981).

TABLE 4.1-2

Power-Law Parameters for Aqueous Solutions[a]

Solution	Temperature (K)	m (Pa·s^n)	n (—)
0.5% Hydroxyethylcellulose	293	0.84	0.509
	313	0.30	0.595
	333	0.136	0.645
2.0% Hydroxyethylcellulose	293	93.5	0.189
	313	59.7	0.223
	333	38.5	0.254
1.0% Polyethylene oxide	293	0.994	0.532
	313	0.706	0.544
	333	0.486	0.599

[a] R. M. Turian, Ph.D. Thesis, University of Wisconsin, Madison (1964), pp. 142–148.

numerical values of m and n are given in Table 4.1-2; note that m and n are both temperature dependent, the parameter m decreasing rapidly with increasing temperature.

The power-law model for $\eta(\dot{\gamma})$ is the most well-known and widely-used empiricism in engineering work, because a wide variety of flow problems have been solved analytically for it. One can often get a rough estimate of the effect of the non-Newtonian viscosity by making a calculation based on the power-law model. However its shortcomings must not be overlooked: (i) it cannot describe the viscosity for very small shear rates, and in some problems this can lead to large errors, (ii) a characteristic time and a characteristic viscosity cannot be constructed from the parameters m and n alone, and this can be awkward in pursuing dimensional-analysis arguments (see §2.8), and (iii) there is no way to relate the parameters m and n to molecular weight and concentration, since the standard correlations are in terms of η_0 and $\eta^*(\omega)$ (see §3.6). Keep in mind that Eqs. 4.1-9 and 10 are just empirical curve fits of the experimental $\eta(\dot{\gamma})$ curves. Because of the widespread use of the power-law model we illustrate its application in the next section.

§4.2 ISOTHERMAL FLOW PROBLEMS

The procedure for the solution of elementary flow problems for generalized Newtonian fluids is exactly the same as for Newtonian fluids, except that the mathematics is more awkward because of the additional complexity introduced by the non-Newtonian viscosity function. In this section we give a series of examples to illustrate the solution procedure for power-law fluids. A summary of power-law solutions is given in Table 4.2-1.

EXAMPLE 4.2-1 Flow of a Power-Law Fluid in a Straight Circular Tube and in a Slightly Tapered Tube

Rework Examples 1.3-2 and 1.3-3 for the power-law model.

SOLUTION For a straight tube of uniform cross section we postulate a solution of the form $v_z = v_z(r)$, $v_\theta = 0$, $v_r = 0$, and $\mathscr{P} = \mathscr{P}(z)$. The z-component of the equation of motion, in terms of $\boldsymbol{\tau}$, is

$$0 = -\frac{d\mathscr{P}}{dz} - \frac{1}{r}\frac{d}{dr}(r\tau_{rz}) \qquad (4.2\text{-}1)$$

TABLE 4.2-1

Solutions to Flow Problems for Power-Law Model

Problem	Solution	
Axial flow through a slit of width W, thickness $2B$, and length L under a pressure drop $\mathscr{P}_0 - \mathscr{P}_L$ (with $B \ll W \ll L$)	$Q = \dfrac{2WB^2}{(1/n) + 2}\left(\dfrac{(\mathscr{P}_0 - \mathscr{P}_L)B}{mL}\right)^{1/n}$	(A)
Axial flow through a circular tube of radius R and length L under a pressure drop $\mathscr{P}_0 - \mathscr{P}_L$ (with $R \ll L$)	$Q = \dfrac{\pi R^3}{(1/n) + 3}\left(\dfrac{(\mathscr{P}_0 - \mathscr{P}_L)R}{2mL}\right)^{1/n}$	(B)
Axial flow through an annulus with inner and outer radii κR and R, and length L, under a pressure drop $\mathscr{P}_0 - \mathscr{P}_L$ ($r = \beta R$ is the location of the maximum in the velocity profile—see Table 4.2-2)	$Q = \dfrac{\pi R^3}{(1/n) + 3}\left(\dfrac{(\mathscr{P}_0 - \mathscr{P}_L)}{2mL}\right)^{1/n} \cdot [(1 - \beta^2)^{1+(1/n)} - \kappa^{1-(1/n)}(\beta^2 - \kappa^2)^{1+(1/n)}]$	(C)
Axial flow through an annulus with no axial pressure drop; inner cylinder moves with velocity V (κ, R, L have same meanings as in Eq. C)	$Q = \pi R^2 V \dfrac{(3 - (1/n))(1 - \kappa^2) - 2(1 - \kappa^{3-(1/n)})}{(3 - (1/n))(1 - \kappa^{1-(1/n)})}$	(D)
Applied torque $\mathscr{T}$ in tangential annular flow between two coaxial cylinders the inner one of which is being made to rotate at an angular velocity W (κ, R, L have the same meanings as in Eq. C)	$\mathscr{T} = 2\pi(\kappa R)^2 mL\left(\dfrac{2W/n}{1 - \kappa^{2/n}}\right)^n$	(E)
Radial flow between two parallel disks separated by $2B$, with pressure drop $\mathscr{P}_1 - \mathscr{P}_2$ over the distance from $r = R_1$ to $r = R_2$	$Q = \dfrac{4\pi B^2}{(1/n) + 2}\left(\dfrac{(\mathscr{P}_1 - \mathscr{P}_2)B(1 - n)}{m(R_2^{1-n} - R_1^{1-n})}\right)^{1/n}$	(F)[a]
Squeezing flow between two circular disks of radius R, with applied force $F(t)$ and instantaneous disk separation $2h$ (the instantaneous plate velocity is $\dot{h}$)	$F(t) = \dfrac{(-\dot{h})^n}{h^{2n+1}}\left(\dfrac{2n + 1}{2n}\right)^n\left(\dfrac{\pi m R^{n+3}}{n + 3}\right)$	(G)[b]

[a] Lubrication-approximation expression obtained by applying Eq. A locally in the space between disks.
[b] Approximate expression obtained by applying Eq. A locally and using a quasi-steady-state assumption.

or using the arguments given in Example 1.3-2

$$\frac{1}{r}\frac{d}{dr}(r\tau_{rz}) = \frac{\mathscr{P}_0 - \mathscr{P}_L}{L} \tag{4.2-2}$$

This may be integrated to give:

$$\tau_{rz} = \frac{(\mathscr{P}_0 - \mathscr{P}_L)r}{2L} + \frac{C_1}{r} \tag{4.2-3}$$

The constant C_1 has to be zero since one does not expect to have an infinite stress at the tube axis. Equation 4.2-3 then is the result of the application of the equation of motion (i.e., conservation of momentum). Equation 4.2-3 may be written in the alternative form

$$\tau_{rz} = \tau_R \cdot \frac{r}{R} \tag{4.2-4}$$

where τ_R is the wall shear stress; that is $\tau_{rz} = \tau_R$ at $r = R$.

Next we have to use the power-law equation for the stress, as given by Eqs. 4.1-4 and 4.1-10. In the latter equation $\dot{\gamma}$, which must be a positive quantity, is given by $(-dv_z/dr)$. Then

$$\tau_{rz} = -\eta \frac{dv_z}{dr} = -m\dot{\gamma}^{n-1}\frac{dv_z}{dr} = m\left(-\frac{dv_z}{dr}\right)^n \tag{4.2-5}$$

Combination of Eqs. 4.2-4 and 4.2-5 then gives the differential equation for v_z:

$$m\left(-\frac{dv_z}{dr}\right)^n = \tau_R \cdot \frac{r}{R} \tag{4.2-6}$$

Taking the nth root of both sides and integrating the first-order, separable equation gives

$$v_z = -\left(\frac{\tau_R}{mR}\right)^{1/n} \frac{r^{(1/n)+1}}{(1/n)+1} + C_2 \tag{4.2-7}$$

The constant C_2 is evaluated by requiring that v_z be zero at $r = R$. One then gets finally

$$v_z = \left(\frac{\tau_R}{m}\right)^{1/n} \frac{R}{(1/n)+1}\left[1 - \left(\frac{r}{R}\right)^{(1/n)+1}\right] \tag{4.2-8}$$

For $n < 1$ this gives a velocity profile that is flatter than the parabolic profile in Eq. 1.3-10 for Newtonian fluids (see Fig. 4.2-1).

It is then easy to get the volume rate of flow Q:

$$\begin{aligned} Q &= \int_0^{2\pi}\int_0^R v_z r\, dr\, d\theta \\ &= 2\pi R^2 \int_0^1 v_z \cdot \frac{r}{R}\, d\left(\frac{r}{R}\right) \\ &= \frac{\pi R^3}{(1/n)+3}\left(\frac{\tau_R}{m}\right)^{1/n} \\ &= \frac{\pi R^3}{(1/n)+3}\left[\frac{(\mathscr{P}_0 - \mathscr{P}_L)R}{2mL}\right]^{1/n} \end{aligned} \tag{4.2-9}$$

For $n = 1$ and $m = \mu$ this reduces to the Hagen–Poiseuille equation for Newtonian fluids.[1]

[1] Laminar-turbulent transition has been studied by D. W. Dodge and A. B. Metzner, *AIChE J.*, **5**, 189–204 (1959); they found that the laminar-turbulent transition occurred in the modified Reynolds number range $2100 < \text{Re}_n < 3100$, where $\text{Re}_n = (D^n\langle v_z\rangle^{2-n}\rho/m)(3 + (1/n))^{-n}2^{3-n}$ for power-law fluids. For other studies see N. W. Ryan and M. M. Johnson, *AIChE J.*, **5**, 433–435 (1959); R. W. Hanks, *ibid.*, **9**, 306–309 (1963); D. M. Meter, *ibid.*, **10**, 881–884 (1964).

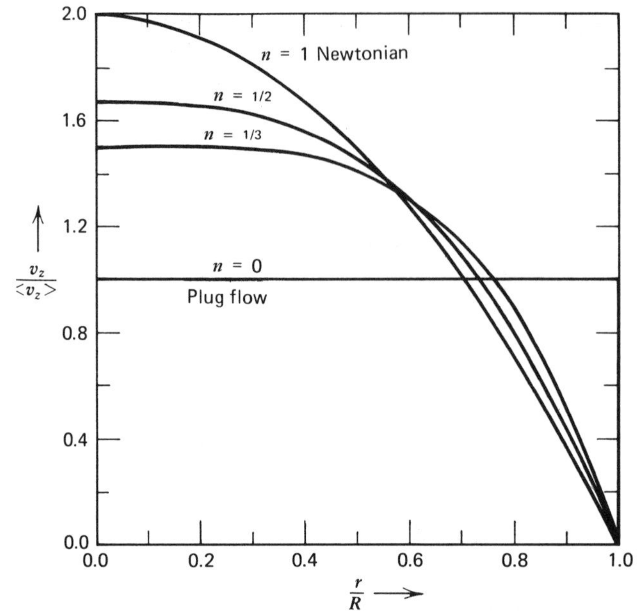

FIGURE 4.2-1. Tube flow velocity profiles for a power-law fluid from Eq. 4.2-8. Note that the profiles become increasingly flatter as n decreases; $n = 0$ corresponds to plug flow. The Newtonian (parabolic) profile is shown as $n = 1$.

The shear rate at the tube wall, $\dot{\gamma}_R = (-dv_z/dr)|_{r=R}$ is a quantity that is sometimes of interest in viscometry. An expression for this quantity can be obtained from Eqs. 4.2-6 and 9; it is found that $\dot{\gamma}_R = [(3n + 1)/4n]\dot{\gamma}_a$, where $\dot{\gamma}_a = 4\langle v_z \rangle/R$ is the "apparent shear rate" (note that for a Newtonian fluid $\dot{\gamma}_a$ is identical to $\dot{\gamma}_R$).

As already pointed out, the power-law model gives an unreasonably high value for the viscosity $\eta(\dot{\gamma})$ for small values of the shear rate. It is therefore appropriate to ask what limitations have to be placed on Eq. 4.2-9 because of this defect of the power law. It is shown in Problem 4B.13 that the power-law expression for Q can be expected to be reliable when $\tau_R \gg \eta_0 \dot{\gamma}_0$, where η_0 is the zero-shear-rate viscosity and $\dot{\gamma}_0$ is the value of the shear rate at which shear thinning begins (see §3.6).

For a tapered tube whose radius decreases linearly from R_0 at $z = 0$ to R_L at $z = L$, we get by applying Eq. 4.2-9 locally for a small segment dz of the tube

$$\mathscr{P}_0 - \mathscr{P}_L = \frac{2mL}{3n}\left[\frac{Q}{\pi}\left(\frac{1}{n} + 3\right)\right]^n \left(\frac{R_L^{-3n} - R_0^{-3n}}{R_0 - R_L}\right) \tag{4.2-10}$$

This gives the relation between pressure drop and volume rate of flow[2].

Sutterby[3] found that the use of a generalized Newtonian fluid model described adequately the Q vs. $\mathscr{P}_0 - \mathscr{P}_L$ relationship for the slow flow of polymer solutions in a converging tube. For very fast flow, on the other hand, the data were well described by the results of an ideal (inviscid) fluid calculation; this is perhaps not too surprising since the inviscid fluid corresponds to Re $\to \infty$.

The flow of non-Newtonian fluids in tapered tubes has also been studied by Oka and Murata.[4]

[2] J. M. McKelvey, V. Maire, and F. Haupt, *Chem. Eng.*, **83**, 94–102 (1976).

[3] J. L. Sutterby, *Trans. Soc. Rheol.*, **9**:2, 227–241 (1965).

[4] S. Oka, *Zairyō*, **14**, 241–244 (1965); S. Oka and T. Murata, *Jpn. J. Appl. Phys.*, **8**, 5–8 (1969); S. Oka, *Biorheology*, **10**, 207–212 (1973).

EXAMPLE 4.2-2 Thickness of a Film of Polymer Solution Flowing Down an Inclined Plate

Obtain an expression for the thickness of a film of polymer solution as it flows down an inclined plate making an angle α with the vertical. Use the power-law fluid model for the derivation. Take the origin of coordinates to be such that $x = 0$ at the film surface and $x = \delta$ at the plate; the film extends along the plate from $z = 0$ to $z = L$.

SOLUTION From the equation of motion we get

$$0 = -\frac{d\tau_{xz}}{dx} + \rho g \cos\alpha \tag{4.2-11}$$

When this is integrated, using the boundary condition that $\tau_{xz} = 0$ at the liquid–air interface at $x = 0$, we get

$$\tau_{xz} = \rho g x \cos\alpha \tag{4.2-12}$$

We know that v_z decreases with increasing x, so that $\dot{\gamma}$ will be taken to be $(-dv_z/dx)$ in order to ensure that $\dot{\gamma}$ be a positive quantity. Then the power law gives the following expression for τ_{xz}:

$$\begin{aligned}\tau_{xz} &= -\eta\frac{dv_z}{dx} = -m\left(-\frac{dv_z}{dx}\right)^{n-1}\frac{dv_z}{dx} \\ &= +m\left(-\frac{dv_z}{dx}\right)^{n}\end{aligned} \tag{4.2-13}$$

Combination of Eqs. 4.2-12 and 4.2-13 then gives

$$-\frac{dv_z}{dx} = \left(\frac{\rho g x}{m}\cos\alpha\right)^{1/n} \tag{4.2-14}$$

Integration with the boundary condition that $v_z = 0$ at $x = \delta$ gives

$$v_z = \left(\frac{\rho g \delta}{m}\cos\alpha\right)^{1/n}\frac{\delta}{(1/n)+1}\left[1 - \left(\frac{x}{\delta}\right)^{(1/n)+1}\right] \tag{4.2-15}$$

for the velocity distribution.

Integration over the cross section of flow (thickness δ and width W) gives for the volume rate of flow

$$Q = \frac{W\delta^2}{(1/n)+2}\left(\frac{\rho g \delta}{m}\cos\alpha\right)^{1/n} \tag{4.2-16}$$

Solution for the film thickness then gives

$$\delta = \left(\frac{m}{\rho g \cos\alpha}\right)^{1/(2n+1)}\left(\frac{Q[(1/n)+2]}{W}\right)^{n/(2n+1)} \tag{4.2-17}$$

This shows how the film thickness depends on the flow rate Q, the fluid properties m and n, and the angle of inclination α.

EXAMPLE 4.2-3 Plane Couette Flow[5]

A macromolecular fluid is confined to the space between two horizontal planes ($x = 0$ and $x = B$) the upper one of which is moving in the positive z-direction with a constant speed V. In addition there is a pressure gradient in the z-direction, the pressure at $z = 0$ being p_0 and that at $z = L$ being p_L. Obtain an expression for the volume rate of flow in the z-direction in the slit that results from the combined effects of the motion of the upper plate and the pressure gradient. Ignore the effect of gravity.

Problems of this type arise in diverse processing operations, such as in certain types of extruders, and in various lubrication problems.

SOLUTION For this flow the equation of motion is

$$0 = -\frac{dp}{dz} - \frac{d\tau_{xz}}{dx} \tag{4.2-18}$$

which may be integrated to give

$$\tau_{xz} = -\left[\frac{(p_0 - p_L)B}{L}\right](\beta - \xi) \tag{4.2-19}$$

where β is a constant of integration, and $\xi = x/B$.

Equation 4.1-4 gives

$$\tau_{xz} = -\eta \frac{dv_z}{dx} \tag{4.2-20}$$

By way of illustration, we use the power law for η. Two cases have to be distinguished here:

Case I: There is no maximum in the velocity profile $v_z(x)$

Case II: There is a maximum in the velocity profile $v_z(x)$

We consider only Case I, although the final results for Case II will be given.

For Case I, $\eta = m(dv_z/dx)^{n-1}$ and hence Eq. 4.2-20 becomes

$$\tau_{xz} = -m\left(\frac{dv_z}{dx}\right)^n \tag{4.2-21}$$

Combination of Eqs. 4.2-19 and 4.2-21 gives an equation for v_z as a function of x. Integration of this equation and application of the boundary conditions $v_z = 0$ at $x = 0$ and $v_z = V$ at $x = B$ gives

$$\phi(\xi) = \frac{v_z}{V} = \frac{\beta^{s+1} - (\beta - \xi)^{s+1}}{\beta^{s+1} - (\beta - 1)^{s+1}} \tag{4.2-22}$$

where $\xi = x/B$, $s = 1/n$, and Λ is a dimensionless parameter given by

$$\Lambda \equiv \frac{(p_0 - p_L)B}{mL}\left(\frac{B}{V}\right)^{1/s}$$

$$= \left[\frac{s+1}{\beta^{s+1} - (\beta - 1)^{s+1}}\right]^{1/s} \quad \textit{Case I}: \Lambda \leq (s+1)^{1/s} \tag{4.2-23}$$

[5] R. W. Flumerfelt, M. W. Pierick, S. L. Cooper, and R. B. Bird, *Ind. Eng. Chem. Fundam.*, **8**, 354–357 (1969); earlier work was done on this problem by F. W. Kroesser and S. Middleman, *Polym. Eng. Sci.*, **5**, 230–234 (1965) and Z. Tadmor, *Polym. Eng. Sci.*, **6**, 203–212 (1966). Annular Couette flow has been studied by S. H. Lin and C. C. Hsu, *Ind. Eng. Chem. Fundam.*, **19**, 421–424 (1980).

Hence, knowing $\beta = \beta(\Lambda, s)$ from Eq. 4.2-23, one can obtain the velocity profile from Eq. 4.2-22.

The volume rate of flow between two planes of width W is then found to be

$$\frac{Q}{WBV} = \int_0^1 \phi \, d\xi = \phi\xi \Big|_0^1 - \int_0^1 \frac{d\phi}{d\xi} \xi \, d\xi$$

$$= -(\beta - 1) + \left(\frac{s+1}{s+2}\right) \frac{\beta^{s+2} - (\beta - 1)^{s+2}}{\beta^{s+1} - (\beta - 1)^{s+1}} \qquad \textit{Case I}: \Lambda \leq (s+1)^{1/s} \qquad (4.2\text{-}24)$$

with β determined from Eq. 4.2-23.

The results analogous to Eqs. 4.2-23 and 4.2-24 for Case II can be shown to be

$$\Lambda = \left[\frac{s+1}{\beta^{s+1} - (1-\beta)^{s+1}}\right]^{1/s} \qquad \textit{Case II}: \Lambda \geq (s+1)^{1/s} \qquad (4.2\text{-}25)$$

$$\frac{Q}{WBV} = (1 - \beta) + \left(\frac{s+1}{s+2}\right) \frac{\beta^{s+2} + (1-\beta)^{s+2}}{\beta^{s+1} - (1-\beta)^{s+1}} \qquad \textit{Case II}: \Lambda \geq (s+1)^{1/s} \qquad (4.2\text{-}26)$$

Hence, the choice of Case I or Case II formulas depends on whether Λ is larger or smaller than $(s+1)^{1/s}$. A table of $\beta = \beta(\Lambda, s)$ has been prepared by Flumerfelt *et al.*[5] The dimensionless flow rate $\Omega = Q/WBV$ is shown in Fig. 4.2-2; this chart is so constructed that it includes the case $p_0 < p_L$ as well as $p_0 > p_L$.

In Case II there is a maximum or a minimum in the velocity profile. Near the maximum or minimum in the velocity the power law will overestimate the viscosity by a large amount. Errors in the volume flow rate Q may be large unless the bigger of $|\tau_{xz}(x = 0)|$ and $|\tau_{xz}(x = B)|$ is much greater than $\eta_0 \dot{\gamma}_0$, where $\dot{\gamma}_0$ is that shear rate at which shear thinning of the viscosity begins.

EXAMPLE 4.2-4 Axial Annular Flow[6–8]

Obtain the relation between the pressure drop and volume rate of flow for the pressure-driven flow of a power-law fluid through the annular gap between two coaxial cylinders of radii κR and R (with $\kappa < 1$). Let the maximum in the velocity distribution be located at $r = \beta R$, where β is a constant that has to be determined later.

SOLUTION We postulate that $v_z = v_z(r)$, $v_r = 0$, $v_\theta = 0$, and $\mathscr{P} = \mathscr{P}(z)$. Then the differential equation for τ_{rz}, obtained from the z-component of the equation of motion, is found to be Eq. 4.2-1; this can be integrated to give Eq. 4.2-3, but the constant C_1 cannot now be set equal to zero, because for this problem $\kappa R \leq r \leq R$. However we can rewrite Eq. 4.2-3 so that the constant β appears rather than C_1:

$$\tau_{rz} = \frac{(\mathscr{P}_0 - \mathscr{P}_L)R}{2L}\left(\frac{r}{R} - \beta^2 \frac{R}{r}\right) \qquad (4.2\text{-}27)$$

[6] A. G. Fredrickson and R. B. Bird, *Ind. Eng. Chem.*, **50**, 347–352 (1958); erratum, *Ind. Eng. Chem. Fundam.*, **3**, 383 (1964).

[7] R. W. Hanks and K. M. Larsen, *Ind. Eng. Chem. Fundam.*, **18**, 33–35 (1979).

[8] For flow in converging annular regions see J. Parnaby and R. A. Worth, *Proc. Inst. Mech. Eng.*, **188**, 357–364 (1974), and J. F. Dijksman and E. P. W. Savenije, *Rheol. Acta*, **24**, 105–118 (1985); in the latter a special toroidal coordinate system is developed and used.

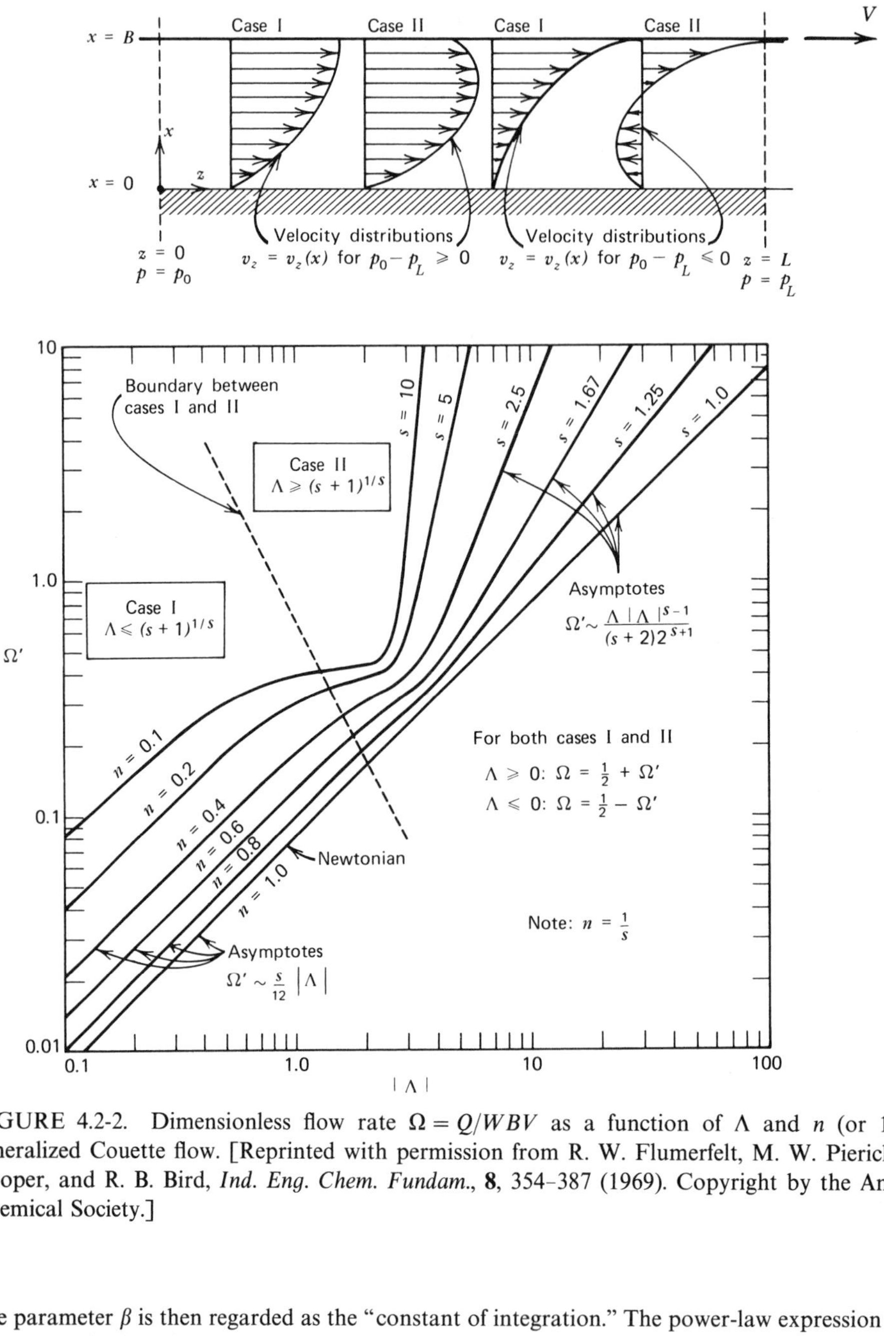

FIGURE 4.2-2. Dimensionless flow rate $\Omega = Q/WBV$ as a function of Λ and n (or $1/s$) for generalized Couette flow. [Reprinted with permission from R. W. Flumerfelt, M. W. Pierick, S. L. Cooper, and R. B. Bird, *Ind. Eng. Chem. Fundam.*, **8**, 354–387 (1969). Copyright by the American Chemical Society.]

The parameter β is then regarded as the "constant of integration." The power-law expression for the shear stress is given by

$$\tau_{rz} = -m\left(\frac{dv_z}{dr}\right)^n, \qquad \kappa R \leq r \leq \beta R \tag{4.2-28}$$

$$\tau_{rz} = m\left(-\frac{dv_z}{dr}\right)^n, \qquad \beta R \leq r \leq R \tag{4.2-29}$$

Substitution of these expressions into Eq. 4.2-27 leads then to differential equations for the velocity distribution in the two regions. These equations may now be integrated and the boundary conditions used: $v_z = 0$ at $r = \kappa R$ and at $r = R$. This leads to

$$v_z = R\left[\frac{(\mathscr{P}_0 - \mathscr{P}_L)R}{2mL}\right]^s \int_\kappa^\xi \left(\frac{\beta^2}{\xi'} - \xi'\right)^s d\xi', \qquad \kappa \le \xi \le \beta \tag{4.2-30}$$

$$v_z = R\left[\frac{(\mathscr{P}_0 - \mathscr{P}_L)R}{2mL}\right]^s \int_\xi^1 \left(\xi' - \frac{\beta^2}{\xi'}\right)^s d\xi', \qquad \beta \le \xi \le 1 \tag{4.2-31}$$

in which $\xi = r/R$ and $s = 1/n$.

Next the constant β is determined by requiring that Eqs. 4.2-30 and 31 match at the location of the velocity maximum; this gives at once

$$\int_\kappa^\beta \left(\frac{\beta^2}{\xi} - \xi\right)^s d\xi = \int_\beta^1 \left(\xi - \frac{\beta^2}{\xi}\right)^s d\xi \tag{4.2-32}$$

This relation gives β as a function of the geometrical quantity κ and the power-law exponent n (see Table 4.2-2).[7]

The volume rate of flow in the annulus is then[7]

$$Q = 2\pi \int_{\kappa R}^R v_z r\, dr = \pi R^3 \left[\frac{(\mathscr{P}_0 - \mathscr{P}_L)R}{2mL}\right]^s \int_\kappa^1 |\beta^2 - \xi^2|^{s+1} \xi^{-s}\, d\xi$$

$$= \frac{\pi R^{3+(1/n)}}{(1/n) + 3}\left(\frac{\mathscr{P}_0 - \mathscr{P}_L}{2mL}\right)^{1/n} [(1 - \beta^2)^{1+(1/n)} - \kappa^{1-(1/n)}(\beta^2 - \kappa^2)^{1+(1/n)}] \tag{4.2-33}$$

This result follows from substituting $v_z(r)$ from Eqs. 4.2-30 and 31 into the integral for Q, and then interchanging the order of integration and performing the inner integrals.

TABLE 4.2-2

Values of $\beta(\kappa, n)$ Computed from Eq. 4.2-32[a]

	κ								
n	0.10	0.20	0.30	0.40	0.50	0.60	0.70	0.80	0.90
0.10	0.3442	0.4687	0.5632	0.6431	0.7140	0.7788	0.8389	0.8954	0.9489
0.20	0.3682	0.4856	0.5749	0.6509	0.7191	0.7818	0.8404	0.8960	0.9491
0.30	0.3884	0.4991	0.5840	0.6570	0.7229	0.7840	0.8416	0.8965	0.9492
0.40	0.4052	0.5100	0.5912	0.6617	0.7259	0.7858	0.8426	0.8969	0.9493
0.50	0.4193	0.5189	0.5970	0.6655	0.7283	0.7872	0.8433	0.8972	0.9493
0.60	0.4312	0.5262	0.6018	0.6686	0.7303	0.7884	0.8439	0.8975	0.9494
0.70	0.4412	0.5324	0.6059	0.6713	0.7319	0.7893	0.8444	0.8977	0.9495
0.80	0.4498	0.5377	0.6093	0.6735	0.7333	0.7902	0.8449	0.8979	0.9495
0.90	0.4572	0.5422	0.6122	0.6754	0.7345	0.7909	0.8452	0.8980	0.9495
1.00	0.4637	0.5461	0.6147	0.6770	0.7355	0.7915	0.8455	0.8981	0.9496

[a] This table is abstracted from the more complete Table I in R. W. Hanks and K. M. Larsen, *Ind. Eng. Chem. Fundam.*, **18**, 33–35 (1979).

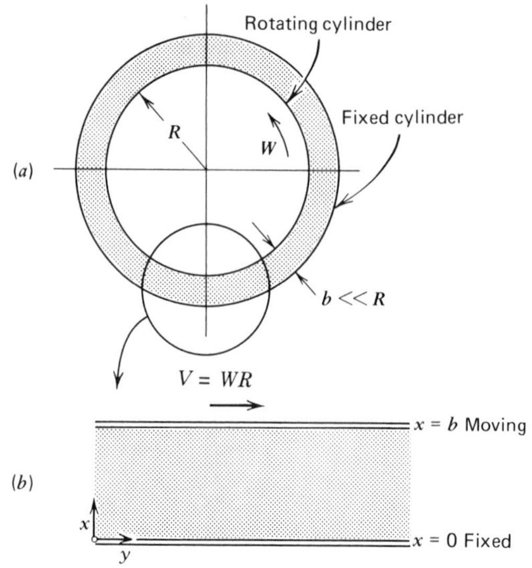

FIGURE 4.2-3. (*a*) Helical flow in a thin annulus; the fluid flows axially because of a pressure gradient and tangentially because of the rotating inner cylinder. (*b*) Coordinate system to be used for the equivalent problem neglecting curvature.

In this flow an appreciable portion of the fluid is in a low shear-rate region, namely that which passes through a washer-shaped region about $r = \beta R$. The width of this washer-shaped ring increases as the pressure drop decreases, and in that region the viscosity from the power law will be much greater than the experimental value; as a consequence volume flow rates from Eq. 4.2-33 can be expected to be lower than the experimental values. The errors introduced by overestimating the viscosity near $r = \beta R$ will be negligible if the wall shear stresses at the inner and outer cylinders, $\tau_{\kappa R}$ and τ_R, are both much greater than $\eta_0 \dot{\gamma}_0$. This point is pursued further in Example 4.5-2.

EXAMPLE 4.2-5 Enhancement of Axial Annular Flow by Rotating Inner Cylinder (Helical Flow of a Power-Law Fluid)[9]

Here we consider the axial flow in a very thin annulus where the inner cylinder radius is R and the gap width is b, which is much smaller than R. The flow is driven by a pressure gradient. We want to investigate the way in which the flow rate changes if the inner cylinder is made to rotate. In this illustrative example we have more than one nonvanishing component of $\dot{\boldsymbol{\gamma}}$, so that the flows in the two directions are coupled through the dependence of the non-Newtonian viscosity on the magnitude of $\dot{\boldsymbol{\gamma}}$ defined in Eq. 4.1-8.

SOLUTION The system is sketched in Fig. 4.2-3*a*. Because of the thinness of the slit, curvature can be neglected and the original problem becomes equivalent to the plane-slit problem shown in Fig. 4.2-3*b*; in this figure we also show the coordinate system we use. We postulate that $v_z = v_z(x)$, $v_y = v_y(x)$, $v_x = 0$, and $\mathscr{P} = \mathscr{P}(z)$.

[9] This problem has been studied experimentally and theoretically by A. C. Dierckes, Jr. and W. R. Schowalter, *Ind. Eng. Chem. Fundam.*, **5**, 263–271 (1966); numerical calculations for the Oldroyd model viscosity function (Eq. A of Table 7.3-3) have been made by J. G. Savins and G. C. Wallick, *AIChE J.*, **12**, 357–363 (1966). See also B. D. Coleman, H. Markovitz, and W. Noll, *Viscometric Flows of Non-Newtonian Fluids*, Springer, New York, (1966), pp. 37–41.

For these postulates the equations of motion become:

y-component
$$0 = -\frac{d}{dx}\tau_{xy} \tag{4.2-34}$$

z-component
$$0 = \frac{\mathscr{P}_0 - \mathscr{P}_L}{L} - \frac{d}{dx}\tau_{xz} \tag{4.2-35}$$

where $\mathscr{P}_0 - \mathscr{P}_L$ is the modified pressure drop between $z = 0$ and $z = L$. The components of the stress tensor for the power-law fluid are

$$\tau_{xy} = -\eta(\dot{\gamma})\dot{\gamma}_{xy} = -m\dot{\gamma}^{n-1}\dot{\gamma}_{xy} \tag{4.2-36}$$

$$\tau_{xz} = -\eta(\dot{\gamma})\dot{\gamma}_{xz} = -m\dot{\gamma}^{n-1}\dot{\gamma}_{xz} \tag{4.2-37}$$

in which the magnitude of the rate-of-strain tensor is

$$\dot{\gamma} = \sqrt{\left(\frac{dv_y}{dx}\right)^2 + \left(\frac{dv_z}{dx}\right)^2} \tag{4.2-38}$$

It is convenient to use dimensionless quantities

$$\bar{x} = \frac{x}{b}; \quad \bar{v}_i = \frac{v_i}{V}; \quad a = \frac{(\mathscr{P}_0 - \mathscr{P}_L)}{mL}\frac{b^{n+1}}{V^n} \tag{4.2-39}$$

and further to abbreviate $d\bar{v}_y/d\bar{x}$ by Y and $d\bar{v}_z/d\bar{x}$ by Z. Then the equations of motion, in terms of the velocity gradients, become

$$\frac{d}{d\bar{x}}[(Y^2 + Z^2)^{(n-1)/2}Y] = 0 \tag{4.2-40}$$

$$\frac{d}{d\bar{x}}[(Y^2 + Z^2)^{(n-1)/2}Z] = -a \tag{4.2-41}$$

These equations may be integrated at once, and we use the symbols C_1 and C_2 for the constants of integration that appear on the right side. When these equations are then solved for Y and Z, we get

$$Y = \frac{d\bar{v}_y}{d\bar{x}} = C_1[C_1^2 + (C_2 - a\bar{x})^2]^{(1-n)/2n} \tag{4.2-42}$$

$$Z = \frac{d\bar{v}_z}{d\bar{x}} = (C_2 - a\bar{x})[C_1^2 + (C_2 - a\bar{x})^2]^{(1-n)/2n} \tag{4.2-43}$$

From this point on we specialize to $n = 1/3$, since for this choice an analytical solution can be obtained. The data in §4.1 suggest that $n = 1/3$ is very nearly appropriate for some polymer solutions.

Integration of Eqs. 4.2-42 and 4.2-43 with $n = 1/3$ gives

$$\bar{v}_y = \int_0^{\bar{x}} C_1[C_1^2 + (C_2 - a\bar{x})^2]\, d\bar{x} \tag{4.2-44}$$

$$\bar{v}_z = \int_0^{\bar{x}} (C_2 - a\bar{x})[C_1^2 + (C_2 - a\bar{x})^2]\, d\bar{x} \tag{4.2-45}$$

in which the integration constants have been set equal to zero since both velocity components are zero at $\bar{x} = 0$. The boundary condition that $d\bar{v}_z/d\bar{x} = 0$ at $\bar{x} = 1/2$ then leads to

$$C_2 = \frac{a}{2} \tag{4.2-46}$$

The boundary condition that $\bar{v}_y = 1$ at $\bar{x} = 1$ leads to the cubic equation $C_1^3 + \frac{1}{12}a^2 C_1 - 1 = 0$, which has only one real root, according to Descartes' rule of signs. That root is

$$C_1 = A_+ + A_- \tag{4.2-47}$$

in which

$$A_\pm = \sqrt[3]{\frac{1}{2} \pm \sqrt{\frac{1}{4} + \frac{1}{27}\left(\frac{a^2}{12}\right)^3}} \tag{4.2-48}$$

For large values of a (i.e., large pressure drop or small angular velocity of the inner cylinder) this last expression can be expanded as

$$C_1 = (12/a^2) - (12/a^2)^4 + \cdots \tag{4.2-49}$$

This expansion is used presently.

The integrals in Eqs. 4.2-44 and 45 are now performed to give the velocity profiles (see Fig. 4.2-4):

$$\bar{v}_y = C_1\left[C_1^2 \bar{x} + \frac{a^2}{12}(3\bar{x} - 6\bar{x}^2 + 4\bar{x}^3)\right] \tag{4.2-50}$$

$$\bar{v}_z = \frac{a}{2}\left[\left(C_1^2 + \frac{a^2}{4}\right)\bar{x} - \frac{1}{2}a^2\bar{x}^2 + \frac{1}{3}a^2\bar{x}^3\right] - a\left[\left(C_1^2 + \frac{a^2}{4}\right)\frac{\bar{x}^2}{2} - \frac{1}{3}a^2\bar{x}^3 + \frac{1}{4}a^2\bar{x}^4\right] \tag{4.2-51}$$

and the axial volume rate of flow through the annulus is

$$Q = 2\pi R \int_0^b v_z\, dx = 2\pi R^2 b W \int_0^1 \bar{v}_z\, d\bar{x}$$

$$= \frac{\pi R^2 b W a^3}{40}\left[1 + \frac{20}{3}\left(\frac{C_1}{a}\right)^2\right] \tag{4.2-52}$$

in which $a = (b\Delta\mathscr{P}/mL)(b/WR)^{1/3}$. Use of Eq. 4.2-49 then leads to

$$Q = \frac{\pi R^2 b W a^3}{40}\left[1 + \frac{960}{a^6} + \cdots\right]$$

$$= \frac{\pi R b^2}{40}\left(\frac{b\Delta\mathscr{P}}{mL}\right)^3\left[1 + 960\left(\frac{WR}{b}\right)^2\left(\frac{mL}{b\Delta\mathscr{P}}\right)^6 + \cdots\right] \tag{4.2-53}$$

in which $\Delta\mathscr{P} = \mathscr{P}_0 - \mathscr{P}_L$. This shows that the flow in the axial direction is enhanced because of the imposed shearing in the tangential direction, since this additional shearing causes the viscosity to be lowered. Note that the correction term is very sensitive to the slit width, which enters as the inverse eighth power, and the pressure gradient, which appears to the minus sixth power. This is a good

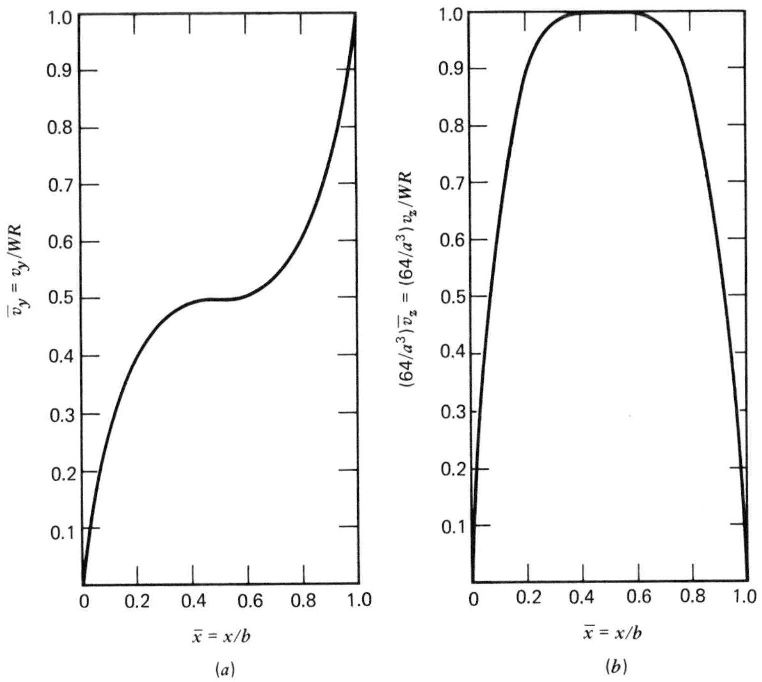

FIGURE 4.2-4. Velocity profiles for helical flow in a thin annular gap. (*a*) The dimensionless tangential velocity $\bar{v}_y = v_y/WR$, given by Eq. 4.2-50, and (*b*) the dimensionless axial velocity $\bar{v}_z = v_z/WR$, given by Eq. 4.2-51. In these velocity expressions C_1 has been taken to be $12/a^2$ (see Eq. 4.2-49). If there were no axial flow, $\bar{v}_y$ would be the linear function $\bar{v}_y = \bar{x}$; the deviation from linearity results from the change in viscosity across the cross section, brought about by the axial flow.

illustration of how the power-law model can be used to ascertain the sensitivity of a quantity (e.g., Q) to the key parameters in the system (e.g., b and $\Delta\mathscr{P}$).

EXAMPLE 4.2-6 Flow Enhancement Produced by a Pulsatile Pressure Drop in a Circular Tube (Quasi-Steady-State Approximation)[10]

A polymer is flowing axially in a horizontal circular tube of radius R and length L as a result of a sinusoidally varying pressure drop

$$\Delta p = \Delta p_0 \cdot (1 + \varepsilon \mathscr{R}e\{e^{i\omega t}\}) \tag{4.2-54}$$

in which $\Delta p = p_0 - p_L$, and Δp_0 is the corresponding quantity for $\varepsilon = 0$. The parameter ε is presumed to be small with respect to unity.

a. Find the volume rate of flow $Q(t)$ by applying Eq. B of Table 4.2-1 instantaneously. Then find $\langle Q \rangle$, the average value of Q over one cycle of oscillation; also find the flow-rate enhancement $(\langle Q \rangle - Q_0)/Q_0$, where Q_0 is the flow rate with $\varepsilon = 0$.

b. Find the power requirement for pumping the material through the circular tube with a pulsatile pressure gradient. What conclusions can you draw?

[10] H. A. Barnes, P. Townsend, and K. Walters, *Rheol. Acta*, **10**, 517–527 (1971); *Nature*, **224**, 585–587 (1969). See also J. M. Davies, S. Bhumiratana, and R. B. Bird, *J. Non-Newtonian Fluid Mech.*, **3**, 237–259 (1977/1978), and N. Phan-Thien and J. Dudek, *J. Non-Newtonian Fluid Mech.*, **11**, 147–161 (1982). Viscoelastic effects are discussed in Example 7.4-2.

SOLUTION (**a**) In the quasi-steady-state approximation

$$Q(t) = \frac{\pi R^3}{(1/n) + 3}\left(\frac{\Delta p_0 R}{2mL}\right)^{1/n}(1 + \varepsilon \mathscr{R}e\{e^{i\omega t}\})^{1/n} \tag{4.2-55}$$

for the power-law fluid model.

Then the time average volume flow rate is

$$\begin{aligned}\frac{\langle Q\rangle}{Q_0} &= \langle (1 + \varepsilon \mathscr{R}e\{e^{i\omega t}\})^{1/n}\rangle \\ &= 1 + \varepsilon \frac{1}{n}\langle \mathscr{R}e\{e^{i\omega t}\}\rangle + \varepsilon^2 \frac{1}{2}\left(\frac{1}{n}\right)\left(\frac{1}{n} - 1\right)\langle [\mathscr{R}e\{e^{i\omega t}\}]^2\rangle + \cdots\end{aligned} \tag{4.2-56}$$

in which an expansion for small ε has been made. The time average of $\mathscr{R}e\{e^{i\omega t}\}$ is zero, and the time average of its square is[11]

$$\langle [\mathscr{R}e\{e^{i\omega t}\}]^2\rangle = \tfrac{1}{2}\langle \mathscr{R}e\{e^{2i\omega t}\} + 1\rangle = \tfrac{1}{2} \tag{4.2-57}$$

The flow-rate enhancement is then given by:

$$\frac{\langle Q\rangle - Q_0}{Q_0} = \frac{1}{4}\left(\frac{1-n}{n^2}\right)\varepsilon^2 + O(\varepsilon^4) \tag{4.2-58}$$

The power law thus predicts that the enhancement increases as n decreases from unity (i.e., as the fluid becomes more non-Newtonian), and that it is independent of the frequency.

(**b**) The time-averaged power $\langle P\rangle$ required to pump the fluid is:[12]

$$\begin{aligned}\langle P\rangle &= \left\langle \iint v_z \Delta p \, r dr d\theta \right\rangle \\ &= \langle Q\Delta p\rangle \\ &= Q_0 \Delta p_0 \langle (1 + \varepsilon \mathscr{R}e\{e^{i\omega t}\})^{(1/n)+1}\rangle \\ &= Q_0 \Delta p_0 \left(1 + \varepsilon^2 \frac{1}{2}\left(\frac{1}{n} + 1\right)\left(\frac{1}{n}\right)\langle [\mathscr{R}e\{e^{i\omega t}\}]^2\rangle + \cdots\right)\end{aligned} \tag{4.2-59}$$

Hence the fractional increase in power needed is

$$\frac{\langle P\rangle - P_0}{P_0} = \frac{1}{4}\left(\frac{1+n}{n^2}\right)\varepsilon^2 + \ldots \tag{4.2-60}$$

[11] The following relation is useful:

$$\mathscr{R}e\{w_1\}\mathscr{R}e\{w_2\} = \tfrac{1}{2}[\mathscr{R}e\{w_1 w_2\} + \mathscr{R}e\{w_1 \bar{w}_2\}] \tag{4.2-56a}$$

in which w_1 and w_2 are two complex quantities and the overbar indicates a complex conjugate.

[12] S. Middleman, *Fundamentals of Polymer Processing*, McGraw-Hill, New York (1977), p. 114.

Comparison of Eqs. 4.2-58 and 60 shows that the flow-rate enhancement increases less rapidly than the power requirement. Hence there is no energetic advantage to pumping with a sinusoidal pressure gradient.

Flow enhancement under sinusoidal pumping has been observed experimentally; data comparisons are given in Example 7.4-2, where this problem is solved again using a constitutive equation that describes elastic effects, and without the quasi-steady-state assumption. It is found there that the enhancement in Eq. 4.2-58 is exact through order ε^2 for the power-law viscosity function.

EXAMPLE 4.2-7 Squeezing Flow between Two Parallel Circular Disks (Lubrication Approximation and Quasi-Steady-State Approximation)

Analyze the flow of a power-law fluid in the gap between two circular disks that approach one another according to some prescribed velocity (see Fig. 4.2-5). The velocity of the upper plate is given by $\dot{h} = dh/dt$. Use a lubrication approximation and a quasi-steady-state assumption; that is, assume that the instantaneous volume rate of flow $Q(r)$ across the cylindrical surface at r is that for flow through a slit of thickness $2h$ and width $2\pi r$. Equate this $Q(r)$ to the volumetric flow rate obtained by a conservation-of-mass statement.

Obtain the time required to squeeze out half of the liquid in the gap by the application of a constant force F on the disks.

SOLUTION Conservation of mass states that, for an incompressible fluid, the volume rate at which fluid crosses the cylindrical surface at r should equal the rate at which the volume between the two plates within the cylindrical surface at r decreases:

$$Q(r) = 2\pi r^2 \, (-\dot{h}) \tag{4.2-61}$$

To apply Eq. A of Table 4.2-1 to the flow between the disks in the region between r and $r + dr$ we make the following changes of notation:

$$W \to 2\pi r$$
$$B \to h$$
$$(\mathscr{P}_0 - \mathscr{P}_L)/L \to -dp/dr$$
$$Q \to Q(r)$$

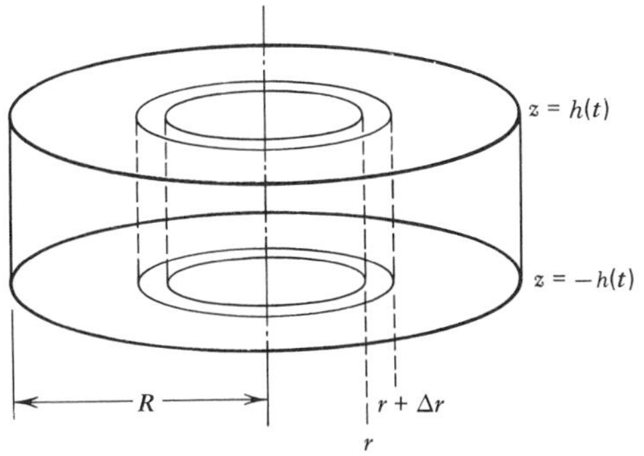

FIGURE 4.2-5. Squeezing flow between two circular disks of radius R. The instantaneous disk separation is $2h(t)$.

Then we get for the volume rate of flow:

$$Q(r) = \frac{2 \cdot 2\pi r \cdot h^2}{(1/n) + 2}\left(-\frac{h}{m}\frac{dp}{dr}\right)^{1/n} \tag{4.2-62}$$

When Eqs. 4.2-61 and 62 are equated, we get an equation for p as a function of r. When this is integrated with respect to r, with the boundary condition $p = p_a$ at $r = R$, we get

$$p - p_a = m\frac{(-\dot{h})^n}{h^{2n+1}}\left(\frac{2n+1}{2n}\right)^n \frac{R^{n+1}}{n+1}\left[1 - \left(\frac{r}{R}\right)^{n+1}\right] \tag{4.2-63}$$

where p_a is the atmospheric pressure.

The force on the upper plate required to move the plate at a speed $\dot{h}$ is then:

$$F(t) = \int_0^{2\pi}\int_0^R (p - p_a + \tau_{zz})\Big|_{z=h} r\,dr\,d\theta \tag{4.2-64}$$

The normal stress τ_{zz} at the upper plate is zero by the same arguments used for Newtonian fluids (see Example 1.2-1). When $p - p_a$ from Eq. 4.2-63 is substituted into the integral for $F(t)$ we get

$$F(t) = \frac{(-\dot{h})^n}{h^{2n+1}}\left(\frac{2n+1}{2n}\right)^n \frac{\pi m R^{n+3}}{n+3} \tag{4.2-65}$$

This is the *Scott equation*,[13,14] which was first developed for measuring the non-Newtonian viscosity of unvulcanized rubber stocks.

For a constant force F, Eq. 4.2-65 is an ordinary differential equation for $h(t)$. When this equation is integrated from $t = 0$ (when $h = h_0$) to $t = t_{1/2}$ (when $h = h_0/2$), the following equation is obtained for $t_{1/2}$:

$$\frac{t_{1/2}}{n} = K_n\left(\frac{\pi R^2 m}{F}\right)^{1/n}\left(\frac{R}{h_0}\right)^{1+(1/n)} \tag{4.2-66}$$

in which K_n is a constant that depends only on n

$$K_n = \left(\frac{2^{1+(1/n)} - 1}{2n}\right)\left(\frac{2n+1}{n+1}\right)\left(\frac{1}{n+3}\right)^{1/n} \tag{4.2-67}$$

Equation 4.2-66 has been tested by Leider[15] who measured $t_{1/2}$ and F in 181 experimental runs. His data are compared with the power-law result in Fig. 4.2-6, where suitably chosen dimensionless groups are used. The characteristic time constant λ for the fluid is defined as

$$\lambda = (m'/2m)^{1/(n'-n)} \tag{4.2-68}$$

[13] J. R. Scott, *Trans. Inst. Rubber Ind.*, **7**, 169–186 (1931); **10**, 481–493 (1935).

[14] S. Oka in F. R. Eirich, ed., *Rheology*, Vol. 3, Academic Press, New York (1960), Chapt. 2, pp. 73–75; A. Cameron, *The Principles of Lubrication*, Longmans, Green, and Co., London (1966), pp. 389–392; D. F. Moore, *The Friction and Lubrication of Elastomers*, Pergamon, Elmsford, NY (1972); A. B. Metzner, *Rheol. Acta*, **10**, 434–444 (1971); M. L. DeMartine and E. L. Cussler, *J. Pharm. Sci.*, **64**, 976–982 (1975); G. Brindley, J. M. Davies, and K. Walters, *J. Non-Newtonian Fluid Mech.*, **1**, 19–37 (1976).

[15] P. J. Leider, *Rheology Research Center Report No. 22*, Nov. 1973, University of Wisconsin, Madison; *Ind. Eng. Chem. Fundam.*, **13**, 342–346 (1974). Additional experimental testing of the squeeze-flow equations and further examination of the theory have been carried out by R. J. Grimm, *AIChE J.*, **24**, 427–439 (1978).

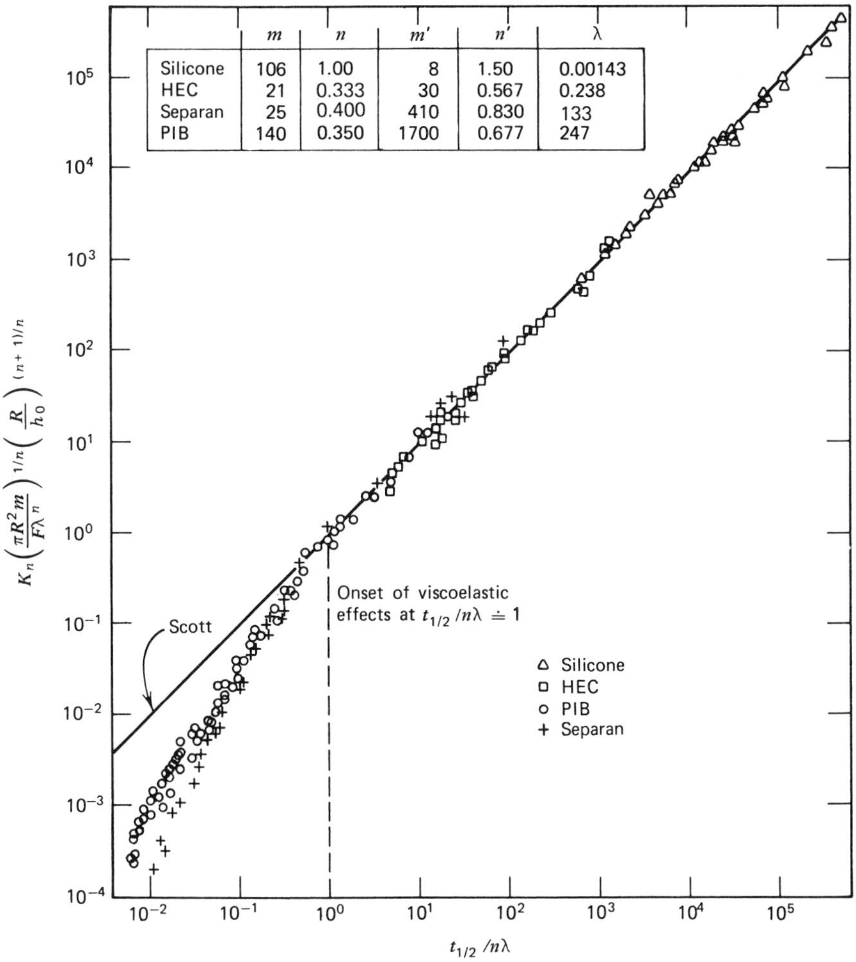

FIGURE 4.2-6. Squeeze flow data of P. J. Leider (University of Wisconsin Rheology Research Center Rept. No. 22, Nov. 1973) along with the Scott equation (Eq. 4.2-66). The fluid parameters for Eq. 4.2-68 are shown in the upper left-hand corner; m has units of $\text{Pa}\cdot\text{s}^n$, m' has units of $\text{Pa}\cdot\text{s}^{n'}$, and λ is given in seconds. [Reprinted with permission from P. J. Leider, *Ind. Eng. Chem. Fundam.*, **13**, 342–346 (1974). Copyright by the American Chemical Society.]

in which m and n are the constants in the power law ($\eta = m\dot{\gamma}^{n-1}$) and m' and n' are constants describing the power-law region of the first normal stress coefficient ($\Psi_1 = m'\dot{\gamma}^{n'-2}$). It is seen that for the fluids tested the power-law model describes the data down to $t_{1/2}/n\lambda = 1$, but below that value there are marked deviations from the power law because of elastic effects. This point is discussed further in §4.6. The success of the Scott equation indicates that the squeeze-flow experiment may be useful for determining fluid parameters from measurements of F and $t_{1/2}$. From Eq. 4.2-66 we see that a log–log plot of (Fh_0/R^3) vs. $(R/h_0 t_{1/2})$ will, for large $t_{1/2}$, give n from the slope of the straight line and m from an intercept

$$\log\left(\frac{Fh_0}{R^3}\right) = \log\left[\pi m(nK_n)^n\right] + n\log\left(\frac{R}{h_0 t_{1/2}}\right) \tag{4.2-69}$$

From a log–log plot of $(R^2/F)^{1/n}(R/h_0)^{1+(1/n)}$ vs. $t_{1/2}/n$ one can get λ from the break in the curve at $t_{1/2}/n = \lambda$. Explicit illustrations have been given by Leider.[15]

This example has illustrated the use of a generalized Newtonian fluid for a flow that is not a steady-state shear flow. In other words, we have violated the restriction placed on the model. Nonetheless, in the limit of slow squeezing flows, experiments show that we can apply the model. Experimental data were then used to define the limits of applicability of the model in terms of a characteristic time λ for each fluid. Hence the generalized Newtonian fluid model, together with dimensional analysis, has suggested a suitable form for the correlation, and the experimental data have established the correlation quantitatively.

§4.3 ISOTHERMAL FLOW PROBLEMS BY THE CALCULUS OF VARIATIONS AND WEIGHTED RESIDUAL METHODS[1]

The problems discussed in §4.2 involve flows in simple geometries. To solve the equations of motion for problems of somewhat greater complexity one generally turns to numerical methods. For generalized Newtonian fluids a variational principle is available, and this principle can serve as a basis for the numerical procedures.[2] In addition, the variational principle can be used to obtain approximate analytical solutions; the latter are often useful in those engineering design problems where quick and approximate solutions are needed. A closely related, alternative procedure for obtaining numerical solutions is the method of weighted residuals. We begin this section with a brief introduction to the calculus of variations,[3] illustrating the method for a system with one independent and one dependent variable. Then we give a similar introduction to the method of weighted residuals, followed by a description of the connection between the two methods. We conclude by giving a general variational statement for three-dimensional, creeping flows of generalized Newtonian fluids.

a. Introduction to the Variational Method

A functional is a mathematical expression that assigns numbers to functions. As an example of a functional, consider the expression for the distance between two points x_1 and x_2 along the curve defined by a function $y(x)$. For fixed endpoints x_1, y_1 and x_2, y_2 the distance depends on the particular function y chosen and is given by the functional J

$$J\{y(x)\} = \int_{x_1}^{x_2} \sqrt{1 + y'^2}\, dx \tag{4.3-1}$$

Corresponding to every curve $y(x)$ we can assign a number J, the distance between the end-points along the curve. A variational problem arises when we want to find the function y that gives the shortest distance between the two points x_1, y_1 and x_2, y_2 in the xy-plane—that is, the smallest value of J.

[1] This section may be omitted in an introductory course; it is not needed for reading §§4.4 to 4.6.

[2] M. J. Crochet, A. R. Davies, and K. Walters, *Numerical Simulation of Non-Newtonian Flow*, Elsevier, Amsterdam (1984).

[3] Very good short introductions to variational methods may be found in I. S. Sokolnikoff and R. M. Redheffer, *Mathematics of Physics and Modern Engineering*, 2nd ed., McGraw-Hill, New York (1966), Chapt. 5, §15; H. Margenau and G. M. Murphy, *The Mathematics of Physics and Chemistry*, 2nd ed., Van Nostrand, Princeton, NJ (1956), Chapt. 6; F. B. Hildebrand, *Methods of Applied Mathematics*, Prentice-Hall, Englewood Cliffs, NJ (1952), Chapt. 2; and V. S. Arpaci, *Conduction Heat Transfer*, Addison-Wesley, Reading, MA (1966), Chapt. 8. For applications to fluid dynamics and transport-phenomena problems see R. S. Schechter, *The Variational Method in Engineering*, McGraw-Hill, New York (1967).

Variational problems such as that of finding the minimum value of J in Eq. 4.3-1 can be shown to have equivalent statements as differential equations. To demonstrate this let us consider a more general functional J defined by

$$J\{y(x)\} = \int_{x_1}^{x_2} F(x, y, y')\, dx + G(x, y)\Big|_{x_1}^{x_2} \tag{4.3-2}$$

in which y is a function of x, and the derivative dy/dx is called y'; both F and G are presumed to be known functions of their arguments. For each function $y(x)$ a value of J can be calculated. We now ask: for what function $y(x)$ is J stationary (i.e., a maximum, a minimum, or locally constant)?

Let $y(x)$ be the function that makes J stationary, and let $\bar{y}(x) = y(x) + \varepsilon\eta(x)$ be some "neighboring" function for which J is not stationary; here $\eta(x)$ is an arbitrary function (it is presumed that η, η', and η'' are continuous) and ε is a small quantity. Then we require that the function

$$J(\varepsilon) = \int_{x_1}^{x_2} F(x, y + \varepsilon\eta, y' + \varepsilon\eta')\, dx + G(x, y + \varepsilon\eta)\Big|_{x_1}^{x_2} \tag{4.3-3}$$

be stationary at $\varepsilon = 0$. In other words, we require that $dJ/d\varepsilon$ be zero at $\varepsilon = 0$. To do this we expand the functions F and G in Taylor series about $\varepsilon = 0$ in the parameter ε:

$$J(\varepsilon) = \int_{x_1}^{x_2} \left[F(x, y, y') + \varepsilon\eta \frac{\partial F}{\partial y} + \varepsilon\eta' \frac{\partial F}{\partial y'} + \cdots \right] dx + \left[G(x, y) + \varepsilon\eta \frac{\partial G}{\partial y} + \cdots \right]_{x_1}^{x_2} \tag{4.3-4}$$

in which we display only terms through ε to the first power. Then

$$\left.\frac{dJ}{d\varepsilon}\right|_{\varepsilon=0} = \int_{x_1}^{x_2} \left[\eta \frac{\partial F}{\partial y} + \eta' \frac{\partial F}{\partial y'} \right] dx + \eta \frac{\partial G}{\partial y}\Big|_{x_1}^{x_2} = 0 \tag{4.3-5}$$

When the second term in the integral is integrated by parts, we obtain

$$\int_{x_1}^{x_2} \eta \left[\frac{\partial F}{\partial y} - \frac{d}{dx}\frac{\partial F}{\partial y'} \right] dx + \left[\eta \left(\frac{\partial F}{\partial y'} + \frac{\partial G}{\partial y} \right) \right]_{x_1}^{x_2} = 0 \tag{4.3-6}$$

Therefore, since $\eta(x)$ is an arbitrary function we can conclude that over the range $x_1 < x < x_2$:

$$\boxed{\frac{d}{dx}\frac{\partial F}{\partial y'} - \frac{\partial F}{\partial y} = 0} \tag{4.3-7}$$

and at the end points, $x = x_1$ and $x = x_2$:

$$\boxed{\left.\left(\frac{\partial F}{\partial y'} + \frac{\partial G}{\partial y} \right)\right|_{x_i} = 0} \qquad i = 1, 2 \tag{4.3-8}$$

for the known functions F and G. Conversely if a function $y(x)$ satisfies the differential equation in Eq. 4.3-7 with the boundary conditions in Eq. 4.3-8 then $y(x)$ has the property of making J in Eq. 4.3-2 stationary. Equation 4.3-7 is called the *Euler–Lagrange equation*, which y must satisfy over the interval $x_1 < x < x_2$ if J is to be stationary, and Eqs. 4.3-8 are called the *natural boundary conditions* for the equation.

So far we have not considered any restrictions on the functions $y(x)$ from which we wish to select the one that makes J stationary. Now we restrict our attention to functions that have a known value y_i at one (or both) of the endpoints, so that

$$\boxed{y(x_i) = y_i} \tag{4.3-9}$$

for $i = 1$ or 2 (or both). We then ask: for what function $y(x)$ that satisfies Eq. 4.3-9 is J stationary?

To answer this question we again introduce a function $\bar{y}(x) = y(x) + \varepsilon\eta(x)$ that now must satisfy Eq. 4.3-9. This means that $\eta(x)$ is no longer completely arbitrary since we must have

$$\eta(x_i) = 0 \tag{4.3-10}$$

for those end points where y is given by Eq. 4.3-9. We may then repeat the developments leading to Eq. 4.3-6 and we again arrive at the Euler–Lagrange equation in Eq. 4.3-7, which has to be satisfied by $y(x)$ in the range $x_1 < x < x_2$. At the endpoints $x = x_1$ and $x = x_2$, however, the natural boundary conditions in Eq. 4.3-8 apply only at those points where $\eta(x) \neq 0$. At the endpoints where $\eta(x) = 0$ the natural boundary conditions are replaced by the *essential boundary conditions* in Eq. 4.3-9. In many problems it is necessary to have at least one essential boundary condition in order to specify the function y completely. In Fig. 4.3-1 the specification of boundary conditions is illustrated in terms of the choice of the function $\bar{y}(x)$.

The variational method can be used in problem solving in two ways: (1) If one formulates a problem as the finding of a stationary value of a functional, then one may convert the problem to that of solving a differential equation with boundary conditions using standard procedures. (2) If one is faced with a differential equation and boundary conditions for which a variational functional J is known, one may obtain an *approximate* solution by substituting a trial function $y(x)$ containing some arbitrary parameters $a_1, \ldots a_n$ into the functional J and then setting $\partial J/\partial a_j = 0$ for $j = 1, 2, \ldots n$. In this way the n parameters a_j are determined, and the optimal function of the chosen form for $y(x)$ is found. Note that the trial function must be chosen to satisfy the essential boundary conditions for all values of the parameters. This means that the approximate solution will satisfy the essential boundary conditions exactly, whereas the differential equation and the natural boundary conditions will in general be satisfied only approximately. It is this second use of the variational method that is of interest here, since it allows us to find approximate solutions to difficult boundary value problems.

To illustrate the use of a variational principle in finding an approximate solution to a differential equation we consider a simple one-dimensional heat-transfer problem.[4] Consider a rod of length L (see Fig. 4.3-2) that is attached to a wall which is maintained at constant temperature T_0; the rod is surrounded by air at temperature T_a. The loss of heat from the rod to its surroundings is described by a heat-transfer coefficient h, and we assume the tip of the rod is insulated, so that heat transfer occurs only across the sides.

[4] For the derivation of Eq. 4.3-11 see R. B. Bird, W. E. Stewart, and E. N. Lightfoot, *Transport Phenomena*, Wiley, New York (1960), §9.7.

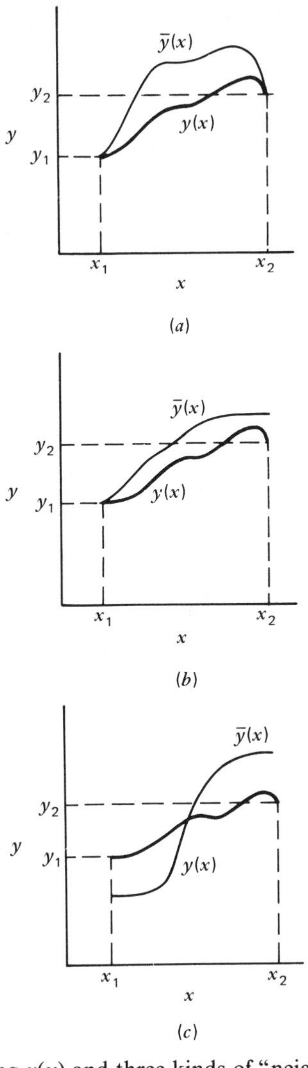

FIGURE 4.3-1. Sketches showing $y(x)$ and three kinds of "neighboring" functions $\bar{y}(x)$. In (*a*) the function $y(x)$ is given at both $x = x_1$ and $x = x_2$, whereas in (*b*) $y(x)$ is prescribed at $x = x_1$, and in (*c*) no boundary values are given.

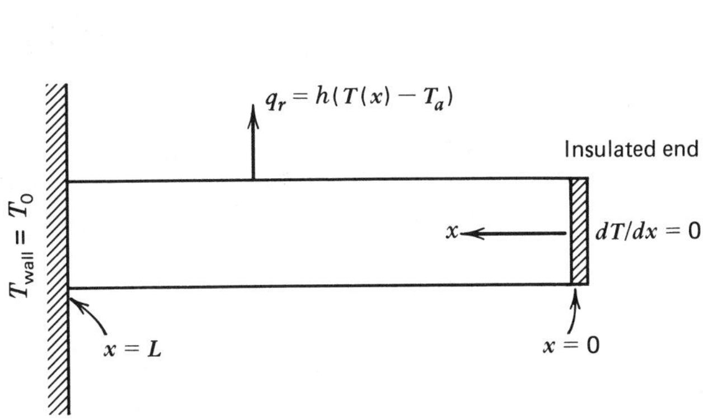

FIGURE 4.3-2. Rod attached to wall at constant temperature T_0. Heat loss to the surrounding air is given in terms of a heat-transfer coefficient h.

An energy balance on the rod gives the following differential equation for the temperature profile $T(x)$:

$$\frac{d^2T}{dx^2} - \left(\frac{hP}{kA}\right)(T - T_a) = 0 \tag{4.3-11}$$

where P and A are the perimeter and cross-sectional area of the rod, and k is its thermal conductivity. If we introduce dimensionless variables

$$\Theta = \frac{T - T_a}{T_0 - T_a}; \quad \xi = \frac{x}{L}; \quad N^2 = \frac{hPL^2}{kA} \tag{4.3-12}$$

then the dimensionless form of the differential equation is

$$\Theta'' - N^2\Theta = 0 \tag{4.3-13}$$

where $\Theta'' = d^2\Theta/d\xi^2$. The boundary conditions become:

$$\text{at} \quad \xi = 0 \qquad \Theta' = 0 \tag{4.3-14}$$

$$\text{at} \quad \xi = 1 \qquad \Theta = 1 \tag{4.3-15}$$

A variational statement that is equivalent to the problem in Eqs. 4.3-13 through 15 is the following: Find the function $\Theta(\xi)$ that satisfies Eq. 4.3-15 and makes the following integral stationary:

$$J\{\Theta(\xi)\} = \int_0^1 [(\Theta')^2 + N^2\Theta^2]\, d\xi \tag{4.3-16}$$

To see this we note that J is of the form of Eq. 4.3-2 with

$$F(\xi, \Theta, \Theta') = (\Theta')^2 + N^2\Theta^2 \tag{4.3-17}$$

and $G = 0$. With these forms for F and G it is then straightforward to show that the Euler–Lagrange equation (Eq. 4.3-7) is precisely twice Eq. 4.3-13 and that the natural boundary condition (Eq. 4.3-8) is just twice Eq. 4.3-14. It may in addition be shown that the stationary point is a minimum (Problem 4B.14).

We now illustrate how to use the variational principle to get an approximate solution to Eqs. 4.3-13 through 15. The first thing we have to do is to select a trial function for the dimensionless temperature profile $\Theta(\xi)$. Instead of picking just one trial function, however, we construct a whole set of trial functions of the form

$$\Theta^{(n)} = 1 + (1 - \xi) \sum_{j=1}^{n} a_j \xi^{j-1} \qquad (n = 1, 2, 3, \ldots) \tag{4.3-18}$$

That is, depending on whether n is 1, 2, 3... we can use a linear, quadratic, cubic, or higher-order trial function. In any event, the trial function satisfies the essential boundary condition in Eq. 4.3-15, but has no other restrictions.

Let us begin by selecting the linear trial function ($n = 1$). When this function is inserted into the functional J in Eq. 4.3-16 we find[5]:

$$J\{\Theta^{(1)}\} = a_1^2 + N^2(1 + a_1 + \tfrac{1}{3}a_1^2) \tag{4.3-19}$$

Of all possible linear trial functions the "best" is then the one that minimizes $J\{\Theta^{(1)}\}$. This function is obtained by solving the equation $dJ/da_1 = 0$ to find the optimal value of a_1

$$a_1 = -\frac{3N^2}{2(3 + N^2)} \tag{4.3-20}$$

When this value of a_1 is substituted into Eq. 4.3-18 with $n = 1$, the best approximate solution has been obtained that one can get with a linear trial function.

The above process can be repeated for a quadratic trial function by using Eq. 4.3-18 with $n = 2$. In this way we get

$$J\{\Theta^{(2)}\} = a_1^2 + \tfrac{1}{3}a_2^2 + N^2(1 + a_1 + \tfrac{1}{3}a_1^2 + \tfrac{1}{6}a_1a_2 + \tfrac{1}{3}a_2 + \tfrac{1}{30}a_2^2) \tag{4.3-21}$$

The "best" quadratic trial function is obtained by solving the simultaneous equations $\partial J/\partial a_1 = 0$ and $\partial J/\partial a_2 = 0$ to obtain

$$a_1 = -\frac{120N^2 + 2N^4}{240 + 104N^2 + 3N^4}; \qquad a_2 = -\frac{120N^2 + 10N^4}{240 + 104N^2 + 3N^4} \tag{4.3-22}$$

When these values are substituted into Eq. 4.3-18 with $n = 2$, we have the best approximate solution that is possible with a quadratic trial function. In the same way one can use cubic or higher polynomial trial functions. Of course, functions other than polynomials can also be used.

Now that two approximate solutions have been obtained, it seems reasonable to inquire how "good" the approximations are. For comparison we note that for this problem the exact solution is

$$\Theta = \frac{\cosh N\xi}{\cosh N} \tag{4.3-23}$$

[5] The following integral is useful for polynomial trial functions of the form assumed in Eq. 4.3-18:

$$\int_0^1 x^n(1 - x)^m\, dx = \frac{n!\, m!}{(n + m + 1)!} \tag{4.3-18a}$$

where n and m are non-negative integers. For even polynomial trial functions these integrals are useful:

$$\int_0^1 (1 - x^2)^m\, dx = \frac{(2m)!!}{(2m + 1)!!} \tag{4.3-18b}$$

$$\int_0^1 x^{2(m+1)}(1 - x^2)^n\, dx = \frac{(2m + 1)!!\,(2n)!!}{(2m + 2n + 3)!!} \tag{4.3-18c}$$

in which $m!! = m(m - 2)(m - 4)\cdots 2$ for m even and $m!! = m(m - 2)(m - 4)\cdots 1$ for m odd.

TABLE 4.3-1

Minimum Values of the Variational Functional in Eq. 4.3-16 for Linear and Quadratic Trial Functions as well as for the Exact Solution

	$N = \frac{1}{2}$	$N = 1$	$N = 2$	$N = 3$
$J_{\min}^{(1)}$	0.2355768	0.8125000	2.673469	3.937501
$J_{\min}^{(2)}$	0.2310597	0.7617675	1.939394	3.063425
$J_{\min}$	0.2310585	0.7615941	1.928055	2.985165

and the corresponding value of the functional is obtained from Eq. 4.3-16

$$J_{\min} = \frac{N \sinh 2N}{2 \cosh^2 N} \tag{4.3-24}$$

In general it is not a simple matter to make an unambiguous evaluation of different approximate solutions, even when the exact analytical solution is known. For example one could think of comparing the average value of an approximation with the average value of the exact solution. This is not necessarily a meaningful comparison, however, since functions can be constructed that have the same average value as the exact solution, but which otherwise bear little resemblance to the exact solution. With the variational functional available, however, it *is* possible to make an unambiguous evaluation of various approximations. This is because of our knowledge that the exact solution is the *only* function that will give the value of J in Eq. 4.3-24 and that all approximations will give higher values of J. Thus for any fixed value of N we may define the "best" of two approximations as the one that gives the lowest value of J. In Table 4.3-1 we show the values of J obtained for the best linear approximation $J_{\min}^{(1)}$ and for the best quadratic approximation $J_{\min}^{(2)}$. To examine the accuracy of the approximations further we compare in Table 4.3-2 the values of $\Theta^{(1)}(0)$ and $\Theta^{(2)}(0)$ with the exact values of $\Theta(0)$. We make the comparison at $\xi = 0$ because Θ is not specified here.

Several comments can be made in connection with the trial functions used above. (i) It is interesting that a linear trial function can even be used to obtain an approximate solution of a second-order differential equation. Note however that only first-order derivatives appear in the functional J. This reduction of order is characteristic for

TABLE 4.3-2

Values of Dimensionless Temperature at $\xi = 0$ for the Linear and Quadratic Trial Functions as well as for the Exact Solution

	$N = \frac{1}{2}$	$N = 1$	$N = 2$	$N = 3$
$\Theta^{(1)}(0)$	0.884615	0.625000	0.142857	−0.125000
$\Theta^{(2)}(0)$	0.886828	0.648415	0.272727	0.124735
$\Theta(0)$	0.886819	0.648052	0.265802	0.099328

variational principles and also for certain weighted residual methods as explained later. (ii) We see that for small values of N the coefficients a_1 and a_2 in the second-order solution are nearly equal. Indeed for $a_2 = a_1$ the natural boundary condition in Eq. 4.3-14 would be satisfied exactly. It is possible to force the quadratic trial function to obey the natural boundary condition exactly by imposing $a_1 = a_2$, but this increases the minimum value of J and is therefore not the best quadratic approximation as defined by the variational statement. (iii) Finally we see that both approximations deteriorate as N increases. In fact for the linear approximation we see that $\Theta^{(1)}(0)$ is negative for $N = 3$, indicating that the approximation has lost physical significance. Roughly speaking, there are two alternative ways in which further improved approximations can be constructed. One is to keep the polynomial trial function in Eq. 4.3-18, but to use higher values of n. Another is to replace the trial function in Eq. 4.3-18 by one which is only piecewise polynomial. This latter approach is taken in the method of finite elements.

b. The Method of Weighted Residuals and Its Relation to Variational Principles

We now turn to another method for obtaining approximate solutions to differential equations. The development of this method does not require that the differential equation have an equivalent variational functional. Rather than give a formal development,[6] however, we just use the simple one-dimensional heat transfer problem, described by Eqs. 4.3-13 through 15, to illustrate the method. Once again we use a trial function $\Theta^{(n)}$ that depends on a total of n parameters a_j. For the differential equation, Eq. 4.3-13, we use the function in Eq. 4.3-18 which automatically satisfies the boundary condition at $\xi = 1$. This time, however, the trial function is introduced directly into the differential equation and we then obtain a *residual*

$$R^{(n)}(\xi) = \frac{d^2\Theta^{(n)}}{d\xi^2} - N^2\Theta^{(n)} \tag{4.3-25}$$

since the trial function does not in general satisfy the differential equation. The parameters a_j are then chosen to make the residual "small." To do this we introduce a set of n weighting functions $w_i^{(n)}(\xi)$ where $i = 1, 2, \ldots n$. The a_j are then obtained by requiring that the *weighted residuals*, formed by multiplying the residual with the weighting functions and integrating over the domain of the problem, be zero

$$\int_0^1 \left(\frac{d^2\Theta^{(n)}}{d\xi^2} - N^2\Theta^{(n)}\right) w_i^{(n)}\, d\xi = 0, \qquad i = 1, 2, \ldots n \tag{4.3-26}$$

When the $w_i^{(n)}$ are chosen to be linearly independent this gives n equations that may be solved to obtain the parameters a_j. This is the *method of weighted residuals* for obtaining an approximate solution to a differential equation. A widely used example of this method is the *collocation method*[7] which arises from Eq. 4.3-26 when the weighting functions are chosen to

[6] More extensive descriptions of the method of weighted residuals may be found in B. A. Finlayson, *The Method of Weighted Residuals and Variational Principles*, Academic Press, New York (1972), and J. Villadsen and M. L. Michelsen, *Solution of Differential Equation Models by Polynomial Approximation*, Prentice-Hall, Englewood Cliffs, NJ (1978).

[7] J. Villadsen and W. E. Stewart, *Chem. Eng. Sci.*, **22**, 1483–1501 (1967); J. P. Nirschl and W. E. Stewart, *J. Non-Newtonian Fluid Mech.*, **16**, 233–250 (1984).

be Dirac delta functions. This forces the approximation $\Theta^{(n)}$ to satisfy the differential equation at selected points, the collocation points. Note however that the collocation method can not be used to obtain a useful approximation $\Theta^{(1)}$ of the form given by Eq. 4.3-18, since $d^2\Theta^{(1)}/d\xi^2$ vanishes identically.

We now consider a different weighted residual method which is closely related to the variational method. By integrating the first term in Eq. 4.3-26 by parts, the following alternative weighted residual formulation is obtained for the heat-conduction problem:

$$\left(\frac{d\Theta^{(n)}}{d\xi}\right)w_i^{(n)}\Bigg|_0^1 - \int_0^1 \left(\frac{d\Theta^{(n)}}{d\xi}\right)\left(\frac{dw_i^{(n)}}{d\xi}\right) d\xi - N^2 \int_0^1 \Theta^{(n)} w_i^{(n)}\, d\xi = 0, \quad i = 1, 2, \ldots n \tag{4.3-27}$$

Notice that only first-order derivatives appear in this equation; that is, there is the same reduction of order as in the variational formulation of the problem. Despite the fact that we have used Eq. 4.3-26 to derive Eq. 4.3-27, the latter has more general applicability than the former, since lower-order trial functions can be used in Eq. 4.3-27 than in Eq. 4.3-26.

Next we consider a special choice for the weighting function, which leads to *Galerkin's method*:[6]

$$w_i^{(n)} = \frac{\partial}{\partial a_i}\Theta^{(n)} \tag{4.3-28}$$

We illustrate the use of this method in connection with Eq. 4.3-27 to obtain the first-order approximation $\Theta^{(1)}$ in Eq. 4.3-18. According to Galerkin's method one obtains the single weighting function $w_1^{(1)} = (1 - \xi)$. Now $w_1^{(1)} = 0$ at $\xi = 1$ and hence the term evaluated at $\xi = 1$ in Eq. 4.3-27 drops out. Also, in the term at $\xi = 0$ we insert the natural boundary condition in Eq. 4.3-14 and hence the term at $\xi = 0$ drops out also. When the remaining integrals in Eq. 4.3-27 are performed we obtain

$$-a_1 - N^2(\tfrac{1}{2} + \tfrac{1}{3}a_1) = 0 \tag{4.3-29}$$

Solving this equation for a_1 gives precisely the same result as in Eq. 4.3-20 obtained by the variational principle. In the same fashion we could go on and derive the coefficients in the second-order approximation and the same coefficients obtained with the variational principle would be found.

For any problem for which there exists a variational principle, it is possible to select a Galerkin formulation that will be equivalent to it.[8] This is illustrated for the heat-conduction problem by substituting Eq. 4.3-28 into Eq. 4.3-27 to obtain

$$\frac{\partial}{\partial a_1}\int_0^1 \left[\left(\frac{d\Theta^{(n)}}{d\xi}\right)^2 + N^2\Theta^{(n)2}\right] d\xi = 0 \tag{4.3-30}$$

where the terms at $\xi = 0$ and $\xi = 1$ have been dropped due to the boundary conditions. Comparison of this equation with the minimization conditions for the variational method ($\partial/\partial a_i$ of Eq. 4.3-16) shows that the formulations are identical.

Finally we remark that Galerkin formulations as described above have even wider applicability than variational methods, since they can be applied to obtain approximate solutions to many differential equations that have no equivalent variational functional. For

[8] B. A. Finlayson, *op. cit.*, p. 229.

steady creeping flow of generalized Newtonian fluids, however, there is a variational principle, and we proceed now to present the principle and demonstrate its applicability.

c. Variational Principle for Three-Dimensional Steady Creeping Flow of Generalized Newtonian Fluids

We consider the flow of a generalized Newtonian fluid within a region V of three-dimensional space; this region is bounded by a surface S, which is divided into two non-overlapping regions S_v and S_f. On the boundary S_v the velocity $\boldsymbol{v}$ is specified, and on S_f a known external force per unit area $\boldsymbol{f}$ is acting on the fluid. In the variational principle the boundary conditions on S_v are treated as essential boundary conditions and the conditions on S_f as natural boundary conditions. We define a variational functional by

$$J = \int_V \left[\int_0^{\dot{\gamma}} \eta(\dot{\gamma})\dot{\gamma}\, d\dot{\gamma} - \tfrac{1}{2} p \operatorname{tr} \dot{\boldsymbol{\gamma}} - \rho(\boldsymbol{v} \cdot \boldsymbol{g}) \right] dV - \int_{S_f} (\boldsymbol{f} \cdot \boldsymbol{v}) dS \tag{4.3-31}$$

Here $\eta(\dot{\gamma})$ is the non-Newtonian viscosity and $\dot{\gamma}$ is the invariant defined by Eq. 4.1-8. For flow in conduits of uniform cross section A, we can take S_v to be the conduit wall and S_f to consist of an upstream plane "1" and a downstream plane "2" both perpendicular to the axis of the conduit. The surface integral in J then becomes $-(p_1 - p_2) \int_A v dS$ where v is the axial velocity and $(p_1 - p_2)$ is the pressure drop from the upstream to the downstream plane. The variational principle now states the following:[9]

> Of all the velocity fields $\boldsymbol{v}$ (that satisfy the boundary conditions on S_v) and pressure fields p, the particular pair $(\boldsymbol{v}, p)$ that causes J in Eq. 4.3-31 to have a *stationary* value is also the solution to the incompressible creeping flow equations of continuity and motion with the stated boundary conditions on S_v and S_f.

If the velocity fields are restricted a priori to satisfy the incompressibility condition, then the term with p in Eq. 4.3-31 drops out. It may then be shown that of all the velocity fields $\boldsymbol{v}$ (consistent with $(\nabla \cdot \boldsymbol{v}) = 0$ and the conditions on S_v) the one that satisfies the creeping flow equation of motion with the stated boundary conditions minimizes J in Eq. 4.3-31.

The variational principle was developed by Pawlowski,[10] Prager,[11] and others.[12-15] The analogous principle for Newtonian fluids was developed by Helmholtz and by Korteweg.[16]

[9] A proof of this principle is outlined in Problems 4D.3 and 4D.4.

[10] J. Pawlowski, *Kolloid-Z.*, **138**, 6–11 (1954).

[11] W. Prager (in *Studies in Mathematics and Mechanics*, R. von Mises Presentation Volume, Academic Press, New York (1954), pp. 208–216) developed a similar variational principle for Bingham fluids.

[12] Y. Tomita, *Bull. Jpn. Soc. Mech. Eng.*, **2**, 469–474 (1959); *Reorojii: Hisenkei Ryūtai no Rikigaku*, Corona, Tokyo (1975), pp. 273–282.

[13] R. B. Bird, *Phys. Fluids*, **3**, 539–541 (1960); comment: **5**, 502 (1962).

[14] M. W. Johnson, Jr., *Phys. Fluids*, **3**, 871–878 (1960); *Trans. Soc. Rheol.*, **5**, 9–21 (1961), has given both minimum and maximum principles. See also J. C. Slattery, *Chem. Eng. Sci.*, **19**, 801–806 (1964); R. W. Flumerfelt and J. C. Slattery, *ibid.*, **20**, 157–163 (1965).

[15] R. S. Schechter, *AIChE J.*, **7**, 445–448 (1961).

[16] H. Lamb, *Hydrodynamics*, 6th ed., Cambridge University Press (1932), §344.

We now present two examples to illustrate the use of this variational principle. We remind the reader that either of these problems could be solved equivalently by forming a weighted residual directly from the governing differential equation. Whether it is easier for a given problem to proceed by forming the weighted residual or by inserting the trial function into the general variational functional depends on the particular problem and also, to some extent, on personal taste.

EXAMPLE 4.3-1 Axial Annular Flow of a Power-Law Fluid

As in Example 4.2-4 we consider the axial flow of a fluid in a horizontal annulus of length L formed by cylindrical surfaces at $r = \kappa R$ and $r = R$. Here we study the flow of a power-law fluid for which an analytical solution is known, so that we can have an exact result with which we can compare our approximate variational result. It is desired to find the volume rate of flow Q for the fluid. Neglect the effect of gravity so that $\mathscr{P}_0 - \mathscr{P}_L = p_0 - p_L = \Delta p$.

SOLUTION A very simple approximate form for the velocity profile that satisfies the boundary conditions is the expression (see Fig. 4.3-3):

$$v_z = a(1 - |\sigma|^{(1/n)+1}) \tag{4.3-32}$$

where

$$\sigma = \frac{2\xi - (1 + \kappa)}{(1 - \kappa)} \tag{4.3-33}$$

This trial velocity distribution contains a as the variational parameter; $\xi = r/R$ is the dimensionless radial coordinate. The quantity $\dot{\gamma}$ is then

$$\dot{\gamma} = \pm \frac{dv_z}{dr} = \pm \frac{1}{R}\frac{dv_z}{d\xi} = \pm \left(\frac{2/R}{1 - \kappa}\right)\frac{dv_z}{d\sigma} = \mp \frac{2a\left(\dfrac{1}{n} + 1\right)}{R(1 - \kappa)}|\sigma|^{1/n} \tag{4.3-34}$$

where the upper sign is for $\sigma < 0$ and the lower for $\sigma > 0$.

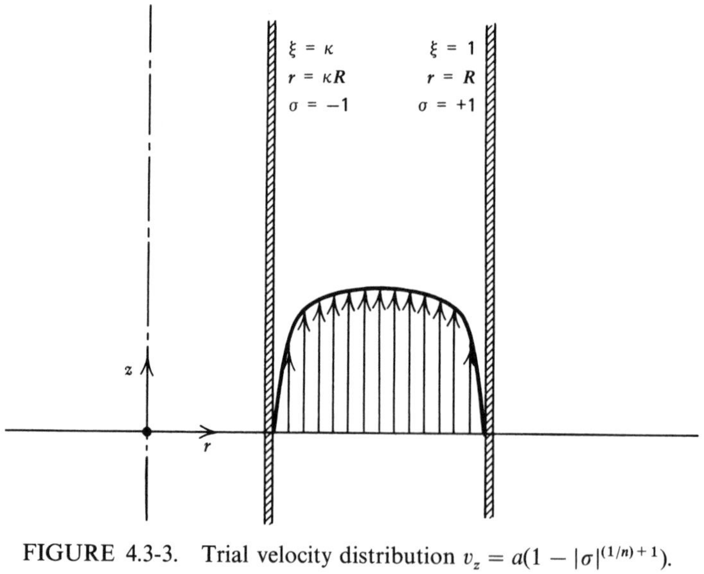

FIGURE 4.3-3. Trial velocity distribution $v_z = a(1 - |\sigma|^{(1/n)+1})$.

The variational functional J for this problem becomes

$$\begin{aligned}
J &= 2\pi L \int_{\kappa R}^{R} \frac{m}{n+1} \dot{\gamma}^{n+1} r\, dr - 2\pi\, \Delta p \int_{\kappa R}^{R} v_z r\, dr \\
&= \frac{2\pi L R^2 m}{n+1} \int_{-1}^{+1} \dot{\gamma}^{n+1} \left[\underline{\left(\frac{1-\kappa}{2}\right)\sigma} + \left(\frac{1+\kappa}{2}\right) \right] \left(\frac{1-\kappa}{2}\right) d\sigma \\
&\quad - 2\pi \Delta p R^2 \int_{-1}^{+1} v_z \left[\underline{\left(\frac{1-\kappa}{2}\right)\sigma} + \left(\frac{1+\kappa}{2}\right) \right] \left(\frac{1-\kappa}{2}\right) d\sigma \\
&= \frac{\pi L R^2 m}{n+1} (1-\kappa^2) \left(\frac{2a\left(\frac{1}{n}+1\right)}{R(1-\kappa)} \right)^{n+1} \int_0^1 \sigma^{(1/n)+1}\, d\sigma \\
&\quad - \pi \Delta p R^2 (1-\kappa^2) a \int_0^1 \left(1 - \sigma^{(1/n)+1}\right) d\sigma \\
&= \frac{\pi L R^2 m (1-\kappa^2)}{(n+1)\left(\frac{1}{n}+2\right)} \left(\frac{2a\left(\frac{1}{n}+1\right)}{R(1-\kappa)} \right)^{n+1} - \frac{\left(\frac{1}{n}+1\right)}{\left(\frac{1}{n}+2\right)} \pi \Delta p R^2 (1-\kappa^2) a
\end{aligned} \tag{4.3-35}$$

The dashed-underlined terms do not contribute since they are odd functions over the interval $(-1, 1)$.

When we set $\partial J/\partial a = 0$ and solve for a we get

$$a = \left(\frac{\Delta p R}{2mL}\right)^{1/n} \frac{R(1-\kappa)^{(1/n)+1}}{2\left(\frac{1}{n}+1\right)} \tag{4.3-36}$$

Then the volumetric flow rate Q is

$$Q = \pi R^3 \left(\frac{\Delta p R}{2mL}\right)^{1/n} \frac{(1-\kappa)^{(1/n)+2}}{\left(\frac{1}{n}+2\right)} \left(\frac{1+\kappa}{2}\right) \tag{4.3-37}$$

Fredrickson and Bird[17] have given a table of exact dimensionless Q values; when we compare Eq. 4.3-37 with their values we find

		$Q(\pi R^3)^{-1}\left(\frac{\Delta p R}{2mL}\right)^{-1/n} \frac{((1/n)+2)}{(1-\kappa)^{(1/n)+2}}$	
		Exact	Variational
$n = 0.5$	$\kappa = 0.5$	0.761	0.750
$n = 0.333$	$\kappa = 0.5$	0.765	0.750
$n = 0.333$	$\kappa = 0.2$	0.661	0.600

[17] A. G. Fredrickson and R. B. Bird, *Ind. Eng. Chem.*, **50**, 347–352 (1958); erratum: *Ind. Eng. Chem. Fundam.*, **3**, 383 (1964).

The variational results are within about 2% of the exact results for $\kappa > 0.5$ and $n > 0.25$, but become worse for smaller κ and n values. Better results could be obtained by using a trial function with two constants a_1 and a_2 instead of the one-constant function in Eq. 4.3-32.

A rather detailed variational calculation for the annular flow of a fluid described by the Sutterby model[18] has been carried out by Mitsuishi, Aoyagi, and Soeda.[19] Extensive comparisons between variational calculations and experimental data have also been made for ducts of elliptical, rectangular, triangular, and eccentric annular cross section.[20] A simple empirical technique for estimating flow rates in channels of unusual cross section has been suggested by Miller[21]. The use of variational methods for flow of various generalized Newtonian fluids around a sphere has been studied by Slattery and co-workers.[22]

EXAMPLE 4.3-2 Estimation of Velocity Distribution for Axial Eccentric Annular Flow (e.g., in a Wire-Coating Device)

A wire of radius R_1 is moving axially with speed V inside a cylindrical cavity of radius R_2 in which a molten polymer is flowing. The wire axis is displaced from the cavity axis by a distance δ. It is desired to estimate the velocity distribution and in particular to determine under what conditions the velocity will be significantly dependent on the variable in the tangential direction (i.e., around the wire). This problem was suggested by a recent study on stability of the wire-coating operation.[23]

SOLUTION It is appropriate to use bipolar coordinates (see Appendix A). In order to use them it is necessary to relate the geometric quantities R_1, R_2, and δ to the quantities a, ξ_1, and ξ_2 used in bipolar coordinates. This is done in the caption for Fig. A.7-1. The quantities ξ_1 and ξ_2 specify the locations of the bounding surfaces.

We now want to estimate the velocity profile. We assume a velocity distribution of the form:

$$\frac{v_z}{V} = \left(\frac{\xi - \xi_2}{\xi_1 - \xi_2}\right)[1 - A(\xi_1 - \xi)\cos\theta] \tag{4.3-38}$$

in which A will be the variational parameter. This function satisfies the boundary conditions at $\xi = \xi_1$ and $\xi = \xi_2$. For a concentric arrangement A would be zero, but in an eccentric arrangement it gives the importance of the θ-dependence of v_z. From Fig. A.7-1 it is evident that v_z should depend on a function of θ that is even, and the cosine function satisfies this requirement.

We now wish to determine the optimal value of A by minimizing J; for this problem we are concerned only with the non-Newtonian viscosity term in Eq. 4.3-31. For the flow under consideration (see Table A.7-4):

$$\dot{\gamma} = \sqrt{\tfrac{1}{2}(\dot{\boldsymbol{\gamma}}:\dot{\boldsymbol{\gamma}})} = \sqrt{\left(\frac{X}{a}\frac{\partial v_z}{\partial \xi}\right)^2 + \left(\frac{X}{a}\frac{\partial v_z}{\partial \theta}\right)^2} \tag{4.3-39}$$

[18] J. L. Sutterby, *Trans. Soc. Rheol.*, **9**:2, 227–241 (1965); *AIChE J.*, **12**, 63–68 (1966). See also Table 4.5-1.

[19] N. Mitsuishi, Y. Aoyagi, and H. Soeda, *Kagaku Kōgaku*, **36**, 186–192 (1972).

[20] N. Mitsuishi, Y. Kitayama, and Y. Aoyagi, *Kagaku Kōgaku*, **31**, 570–577 (1967); N. Mitsuishi and Y. Aoyagi, *Memoirs of the Faculty of Engineering*, Kyūshu University, Vol. 28, No. 3, pp. 223–241 (1969); T. Mizushina, N. Mitsuishi, R. Nakamura, *Kagaku Kōgaku*, **28**, 648–652 (1964); T. L. Guckes, *Trans. ASME, J. Eng. for Industry*, **97B**, 498–506 (1975).

[21] C. Miller, *Ind. Eng. Chem. Fundam.*, **11**, 524–528 (1972); R. W. Hanks, *Ind. Eng. Chem. Fundam.*, **13**, 62–66 (1974).

[22] J. C. Slattery, *AIChE J.*, **8**, 663–667 (1962); M. L. Wasserman and J. C. Slattery, *ibid.*, **10**, 383–388 (1964); S. W. Hopke and J. C. Slattery, *ibid.*, **16**, 224–229, 317–318 (1970).

[23] Z. Tadmor and R. B. Bird, *Polym. Eng. Sci.*, **14**, 124–136 (1974).

in which $X = \cosh \xi + \cos \theta$. For a very large wire speed V, the first term under the square-root sign will surely be much larger than the second term. Hence we write:

$$\dot{\gamma} \doteq +\frac{X}{a}\frac{\partial v_z}{\partial \xi} \tag{4.3-40}$$

The plus sign is chosen since ξ increases from the outer cylinder to the inner cylinder, as does the axial velocity of the fluid. We shall assume the polymer melt viscosity to be described by a power law so that

$$\begin{aligned}\int_0^{\dot{\gamma}} \eta\dot{\gamma}\, d\dot{\gamma} &= \frac{m}{n+1}\dot{\gamma}^{n+1} \\ &= \frac{m}{n+1}\left(\frac{X}{a}\frac{\partial v_z}{\partial \xi}\right)^{n+1} \\ &= \frac{m}{n+1}\left[\frac{X}{a}\frac{V}{\xi_1 - \xi_2}(1 - A(\xi_1 + \xi_2 - 2\xi)\cos\theta)\right]^{n+1}\end{aligned} \tag{4.3-41}$$

Hence the variational integral becomes for a tube of length L

$$\begin{aligned}J &= L\int_0^{2\pi}\int_{\xi_2}^{\xi_1} \frac{m}{n+1}\left(\frac{X}{a}\frac{\partial v_z}{\partial \xi}\right)^{n+1}\left(\frac{a}{X}\right)^2 d\xi\, d\theta \\ &= \frac{LmV}{(n+1)(\xi_1 - \xi_2)}\int_0^{2\pi}\int_{\xi_2}^{\xi_1}\left(\frac{a}{X}\right)^{n-1}[1 - A(\xi_1 + \xi_2 - 2\xi)\cos\theta]^{n-1}\, d\xi\, d\theta\end{aligned} \tag{4.3-42}$$

When $\partial J/\partial A$ is set equal to zero, it is found that[23] for $R_1 = 0.317$ mm and $R_2 = 0.635$ mm:

δ(mm)	ξ_1	ξ_2	a(mm)	$A(n = 0.5)$	$A(n = 0.75)$	$A(n = 1)$
0.0237	3.6895	2.9982	6.35	0.025	0.008	0
0.1407	1.8184	1.1948	0.953	0.175	0.060	0
0.261	0.8814	0.4812	0.318	0.73	0.25	0

The values of A were obtained by numerical integration. It is thus seen that the θ-dependence becomes more important as the eccentricity increases and as the index n decreases.

§4.4 NONISOTHERMAL FLOW PROBLEMS[1]

Up to this point we have given examples of flow problems in which it is assumed that the temperature is constant. In most polymer processing applications and in lubrication systems the changes of temperature with position and time are significant and cannot be ignored. In the manufacture of plastic objects one usually starts by melting plastic pellets and then performing a sequence of processing operations on the molten material; finally the

[1] For additional information on heat and mass transfer, we refer the reader to: A. B. Metzner, *Adv. Heat Transfer*, **2**, 357–397 (1965); H. H. Winter, *Adv. Heat Transfer*, **13**, 205–267 (1977); S. Middleman, *Fundamentals of Polymer Processing*, Wiley, New York (1977), Chapt. 13; Z. Tadmor and C. G. Gogos, *Principles of Polymer Processing*, Wiley, New York (1979), Chapt. 9; R. C. Armstrong and H. H. Winter in E. U. Schlünder, ed., *Heat Exchanger Design Handbook*, Hemisphere, New York (1983), §2.5.12; J. R. A. Pearson, *Mechanics of Polymer Processing*, Elsevier, London (1984), Parts III and IV and Appendix 4; J. Laven, Doctoral Dissertation, Delft University Press (1985); J. van Dam and H. Janeschitz-Kriegl, *Int. J. Heat Mass Transfer*, **28**, 395–406 (1985).

material must be cooled in order to obtain the finished product. Clearly heat transfer and phase change play important roles. In high speed processing operations, such as extrusion, and in lubrication problems the temperature rise by viscous heating is appreciable (because of the high viscosity of the polymeric liquids and because of the large velocity gradients); consequently the viscous-heating term must be included in the equation of change for temperature. Moreover, because of the low thermal conductivity of polymers, temperature increases due to viscous heating can be considerable and very nonuniform. The reliable estimation of viscous heating effects and local temperatures is of particular interest in polymer flow problems because of the thermal instability of polymeric liquids; chemical degradation can occur if hot spots develop in the processing line. The description of nonisothermal flows not only requires the simultaneous solution of three equations of change (continuity, motion, and energy) but in addition the temperature dependence of all physical properties (viscosity, thermal conductivity, density, and heat capacity) must in general be taken into account; moreover, for polymeric liquids the shear-rate dependence of the viscosity cannot be ignored.

The equations that will form the starting point for the heat-transfer and fluid-flow discussions here are very similar to those in Table 1.2-1:

Continuity:

$$(\nabla \cdot \boldsymbol{v}) = 0 \tag{4.4-1}$$

No mass sources within the fluid

Motion:

$$\rho \frac{D\boldsymbol{v}}{Dt} = -\nabla p + [\nabla \cdot \eta \dot{\boldsymbol{\gamma}}] + \rho \boldsymbol{g} \tag{4.4-2}$$

$\rho \frac{D\boldsymbol{v}}{Dt}$	$-\nabla p$	$[\nabla \cdot \eta \dot{\boldsymbol{\gamma}}]$	$\rho \boldsymbol{g}$
Mass per unit volume times acceleration	Pressure force per unit volume	Viscous force per unit volume	Gravitational or external force per unit volume

Energy:

$$\rho \hat{C}_p \frac{DT}{Dt} = (\nabla \cdot k \nabla T) + \tfrac{1}{2}\eta(\dot{\boldsymbol{\gamma}}:\dot{\boldsymbol{\gamma}}) \tag{4.4-3}$$

$\rho \hat{C}_p \frac{DT}{Dt}$	$(\nabla \cdot k \nabla T)$	$\frac{1}{2}\eta(\dot{\boldsymbol{\gamma}}:\dot{\boldsymbol{\gamma}})$
Rate of increase of internal energy per unit volume	Rate of addition of energy by heat conduction per unit volume	Rate of conversion of mechanical to thermal energy per unit volume

In obtaining Eq. 4.4-3 from Eq. 1.1-13 it has been assumed that $\hat{U} = \hat{U}(p, T)$; that is, we assume that $\hat{U}$ does not depend explicitly on any kinematic quantities such as strain and rate of strain. This has been a standard assumption in all heat-transfer calculations; although the assumption is generally regarded as reasonable, we know of no thorough experimental study of it.[2]

[2] The review by G. Astarita and G. C. Sarti in J. F. Hutton, J. R. A. Pearson, and K. Walters, eds., *Theoretical Rheology*, Wiley, New York (1974), pp. 123–127, contains a theoretical discussion of the validity of Eq. 4.4-3 for viscoelastic fluids.

TABLE 4.4-1

Scaled Dimensionless Forms of the Equations of Change for the Generalized Newtonian Fluid with Constant ρ and k[a–d]

$(\nabla^* \cdot \boldsymbol{v}^*) = 0$		(A)
$\mathrm{Re} \dfrac{D\boldsymbol{v}^*}{Dt^*} = -\nabla^* p^* + \left[\nabla^* \cdot \dfrac{\eta}{\eta^0} \nabla^* \boldsymbol{v}^*\right] + \dfrac{\mathrm{Re}}{\mathrm{Fr}} \dfrac{\boldsymbol{g}}{g}$		(B)
$\mathrm{P\acute{e}} \dfrac{DT^*}{Dt^*} = \nabla^{*2} T^* + \dfrac{1}{2} \mathrm{Gn} \left(\dfrac{\eta}{\eta^0}\right)(\dot{\boldsymbol{\gamma}}^* : \dot{\boldsymbol{\gamma}}^*)$		(C)

Dimensionless groups:	$\mathrm{Re} = HV\rho/\eta^0$	(D)	$\mathrm{Fr} = V^2/gH$	(E)
	$\mathrm{P\acute{e}} = \rho\hat{C}_p HV/k$	(F)	$\mathrm{Gn} = \eta^0 V^2/k\Delta T_0$	(G)

[a] The dimensionless quantities are defined by

$$\boldsymbol{v}^* = \boldsymbol{v}/V, \quad \nabla^* = H\nabla, \quad \frac{D}{Dt^*} = \left(\frac{H}{V}\right)\frac{D}{Dt},$$

$$p^* = \frac{Hp}{\eta^0 V}, \quad T^* = \frac{T - T_0}{\Delta T_0}, \quad \dot{\boldsymbol{\gamma}}^* = \left(\frac{V}{H}\right)\dot{\boldsymbol{\gamma}}$$

[b] To scale length, a characteristic dimension H is used; for time, (H/V) is used where V is a characteristic velocity. A characteristic shear rate and temperature are (V/H) and T_0; the characteristic viscosity η^0 is the viscosity evaluated at $\dot{\gamma} = (V/H)$ and $T = T_0$. There are several possibilities for ΔT_0:

$$\Delta T_{\mathrm{process}} = |T_w - T_i| \qquad \Delta T_{\mathrm{gen}} = \eta^0 V^2/k$$

$$\Delta T_{\mathrm{adiabatic}} = \Delta p/\rho\hat{C}_p \qquad \Delta T_{\mathrm{rheol}} = \left|\frac{\eta}{\partial\eta/\partial T}\right|_{\substack{T=T_0 \\ \dot{\gamma}=V/H}}$$

where T_i and T_w are two temperatures specified in boundary conditions, say an inlet and a wall temperature. $\Delta T_{\mathrm{adiabatic}}$ is a characteristic adiabatic temperature rise for the process; ΔT_{gen}, a characteristic temperature change associated with a balance between viscous heating and heat conduction; and $\Delta T_{\mathrm{rheol}}$ is a characteristic temperature rise necessary to change the viscosity substantially.

[c] If $\Delta T_0 = \Delta T_{\mathrm{process}}$ then,

$$\mathrm{Gn} = \mathrm{Br} = \eta^0 V^2/k|T_w - T_i|$$

where Br is the Brinkman number. If $\Delta T_0 = \Delta T_{\mathrm{rheol}}$ then,

$$\mathrm{Gn} = \mathrm{Na} = V^2|\partial\eta/\partial T|/k$$

where Na is the Nahme-Griffith number.

[d] A thorough discussion of scaling the energy equation is given by J. R. A. Pearson, *Polym. Eng. Sci.*, **18**, 222–229 (1978); see also J. R. A. Pearson, *Mechanics of Polymer Processing*, Elsevier, New York (1985).

The scaled, dimensionless forms of Eqs. 4.4-1 through 3 are listed in Table 4.4-1. There the convection term in the energy equation is seen to be scaled by the Péclet number, $\text{Pé} = \rho \hat{C}_p HV/k$. Because of the low thermal conductivity of polymer melts and concentrated solutions, $\text{Pé} \gg 1$ in typical nonisothermal flow problems, meaning that heat is convected much more rapidly than it is conducted. The viscous heating term is scaled by a heat generation number Gn, the form of which depends on the choice of a characteristic temperature difference ΔT_0 for a problem. If $\Delta T_{\text{process}} \ll \Delta T_{\text{rheol}}$ then it is reasonable to choose the characteristic ΔT_0 to be $\Delta T_{\text{process}}$. In this case the Gn number is the Brinkman number introduced in Chapter 1, except that here μ is replaced by η^0, the viscosity evaluated at the reference temperature and shear rate. If $\Delta T_{\text{process}}$ is not small compared with ΔT_{rheol} it is better to use the latter for ΔT_0. Then Gn becomes the Nahme-Griffith[3] number Na.

Let us now examine the fluid properties that appear in the equations of change and their dependence on temperature. In Table 4.4-2 some sample values of thermal conductivity k, heat capacity per unit mass $\hat{C}_p$, and density ρ, are given for a number of molten commercial polymers; it should be noted that the k values all lie within a rather narrow range (0.1–0.3 W/m·K). Then in Table 4.4-3 information is given on the temperature dependence of k, $\hat{C}_p$, and ρ. In most flow and heat-transfer calculations we assume that the thermal conductivity, heat capacity, and density of the fluid do not change appreciably with temperature or pressure. The assumption of constant thermal conductivity is not serious since k does not change much with temperature, nor does it appear to change very much with velocity gradient;[4] in addition there appears to be no need to replace the scalar thermal conductivity by a second-order tensor to account for nonisotropic heat conduction. The assumption of constant density means that our discussion will be restricted to forced convection and that free convection will be omitted.[5]

Although it is reasonable in many calculations to assume that k, $\hat{C}_p$, and ρ do not vary with temperature, the same cannot be said of the parameters in the generalized

TABLE 4.4-2

Thermal Conductivity and Heat Capacity of Some Molten Polymers at 150°C[a]

Polymer		k(W/m·K)	$\hat{C}_p$ (kJ/kg·K)	ρ (10^{-3} kg/m^3)
LDPE	Low-density polyethylene	0.241	2.57	0.782
HDPE	High-density polyethylene	0.255	2.65	0.782
PP	Polypropylene		2.80	
PVC	Polyvinyl chloride	0.166	1.53	1.31
PS	Polystyrene	0.167	2.04	0.997
PMMA	Polymethylmethacrylate	0.195		1.11

[a] Taken from H. H. Winter, *Adv. Heat Transfer*, **13**, 205–267 (1977).

[3] R. M. Griffith, *Ind. Eng. Chem. Fundam.*, **1**, 180–187 (1962); R. Nahme, *Ingenieur-Archiv.*, **11**, 191–209 (1940).

[4] A. A. Cocci, Jr. and J. J. C. Picot [*Polym. Eng. Sci.* **13**, 337–341 (1973)] have found that the thermal conductivity of polydimethylsiloxane samples increased by about 5% as the shear rate went from 0 to 300 s^{-1}.

[5] A theoretical analysis of free-convection heat transfer to non-Newtonian fluids was given by A. Acrivos, *AIChE J.*, **6**, 584–590 (1960). Experimental data were obtained for free convection near a heated vertical plate in an aqueous carboxy-polymethylene solution by I. G. Reilly, C. Tien, and M. Adelman, *Can. J. Chem. Eng.*, **43**, 157–160 (1965). Numerical calculations for the power-law and Ellis models have been made by H. Ozoe and S. W. Churchill, *AIChE J.*, **18**, 1196–1207 (1972).

TABLE 4.4-3

Temperature Dependence of Physical Properties of Polymers[a,b]

Property	Polymer	Temperature Range (°C)	Coefficients in Polynomial Representation					
			A	B	C	D	E	F
k (W/m·K)	HDPE	10–143	0.453	-8.59×10^{-4}	-5.29×10^{-6}	4.12×10^{-8}	-1.98×10^{-8}	
		143–200	0.26					
	LDPE	10–126	0.365	-4.07×10^{-4}	-7.34×10^{-6}	8.28×10^{-8}	-5.53×10^{-8}	
		126–200	0.223					
	PVC	0–200	0.168					
$\hat{C}_p$ (kJ/kg·K)	HDPE	10–88	1.597	3.61×10^{-3}	5.96×10^{-5}	-3.44×10^{-8}	9.77×10^{-9}	
		88–121	-1.983×10^{2}	6.17	-6.34×10^{-2}	2.19×10^{-4}		
		121–130	-2.837×10^{2}	2.41				
		130–133	1.208×10^{3}	-9.07				
		133–200	1.984	3.88×10^{-3}				
	LDPE	10–90	1.943	5.39×10^{-2}	2.56×10^{-2}	-3.23×10^{-6}	3.53×10^{-8}	
		90–105	8.497×10^{1}	-1.84	1.04×10^{-2}			
		105–110	-1.29×10^{2}	1.3				
		110–113.5	3.786×10^{2}	-3.31				
		113.5–200	1.98	3.70×10^{-3}				
	PVC	10–67	0.75	4.66×10^{-3}				
		67–96	1.361×10^{2}	-6.64	1.21×10^{-1}	-9.71×10^{-4}	2.90×10^{-6}	
		96–200	1.208	2.96×10^{-3}				
ρ^{-1} (cm^3/g)	HDPE	10–133	1.033	17.87×10^{-4}	-7.19×10^{-5}	16.11×10^{-7}	-15.45×10^{-9}	5.58×10^{-11}
		133–200	1.158	8.09×10^{-4}				
	LDPE	10–113.5	1.078	1.24×10^{-4}	2.68×10^{-5}	-3.95×10^{-7}	2.35×10^{-9}	
		113.5–200	1.158	8.09×10^{-4}				
	PVC	10–110	0.7154	1.02×10^{-4}	0.0781×10^{-5}	-0.0167×10^{-7}	0.0524×10^{-9}	
		110–200	0.6791	5.67×10^{-4}				

[a] All properties are given as polynomials of the form $A + BT + CT^2 + DT^3 + \cdots$ in which T is the temperature in *degrees centigrade*.
[b] The coefficients in this table were taken from U. Kleindienst, Doctoral Dissertation, Universität Stuttgart (1976)..

Newtonian model for η. For example, useful empiricisms for the temperature dependence of the power-law parameters m and n are[6]

$$m = m^0 e^{-A(T - T_0)} \tag{4.4-4}$$

$$n = n^0 + B(T - T_0) \tag{4.4-5}$$

in which T_0 is a reference temperature, and m^0 and n^0 are the values of the parameters at that temperature; the constants A and B are reciprocal characteristic temperature differences and are determined for each fluid from experimental data. As may be seen in Table 4.1-2, the parameter m is a strong function of temperature. Often B is found to be quite small; hence assuming n to be a constant is appropriate.

When the power-law model of Eq. 4.1-10, with the temperature-dependent parameters m and n (given, for example, by Eqs. 4.4-4 and 5), is inserted into Eqs. 4.4-2 and 3, then the equations of motion and energy are strongly coupled. In general, then, the solution of the set of equations of change can not be obtained analytically. This means that virtually all realistic nonisothermal polymer flow problems have to be done numerically.

On the other hand a number of analytical solutions to nonisothermal flow problems, with the assumption of constant physical properties, have been given in the engineering literature. These are useful for order-of-magnitude estimates and for checking computer programs in the limit that physical properties do not depend on temperature. We give a few analytical solutions at the end of this section. In addition we give two tables[7] of calculated Nusselt numbers, one for tubes (Table 4.4-4) and one for thin slits (Table 4.4-5). The "constant-wall-temperature" entries for circular tubes are derived in Examples 4.4-1 and 2.

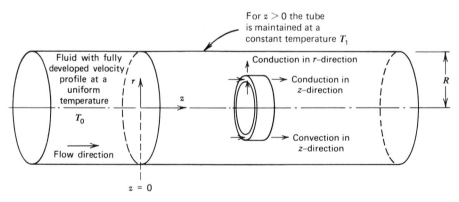

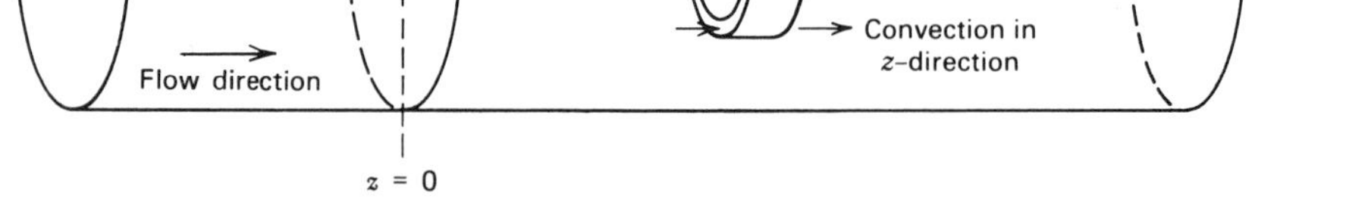

FIGURE 4.4-1. Axial flow in a circular tube with heat transport by conduction and convection. The wall temperature changes from T_0 to T_1 at $z = 0$.

[6] R. M. Turian, Ph.D. Thesis, University of Wisconsin, Madison (1964), suggested similar expressions for the power-law, p. 139; see also pp. 161–162 for the corresponding Ellis model results. For some purposes it is convenient to use

$$m^0/m = [1 + C(T - T_0)/T_0]^n \tag{4.4-4a}$$

as suggested by T. Mizushina, R. Itō, Y. Kuriwaki, and K. Yahikozawa, *Kagaku Kōgaku*, **31**, 250–255 (1967); here C is a constant, and n is the (constant) exponent in the power-law model.

[7] Tables 4.4-4 and 5 are based on the summary articles (in Dutch) by W. J. Beek and R. Eggink, *De Ingenieur*, **74**, Ch. 81–89 (1962), and J. M. Valstar and W. J. Beek, *ibid.*, **75**, Ch. 1–7 (1963).

The solution techniques needed to solve the "constant-wall-heat-flux" problems are given elsewhere.[8]

EXAMPLE 4.4-1 Flow in Tubes with Constant Wall Temperature[9–11] (Asymptotic Solution for Small z)

A polymeric fluid is flowing axially in a circular tube of radius R (see Fig. 4.4-1). Before the fluid arrives at the plane $z = 0$, it is at a uniform temperature T_0 and has the fully developed velocity distribution:

$$v_z = v_{\max}\left[1 - \left(\frac{r}{R}\right)^{s+1}\right] \tag{4.4-6}$$

where $s = 1/n$, and $v_{\max}$ is given in Eq. 4.2-8. Then for $z > 0$ the wall is maintained at a constant temperature T_1; if $T_0 - T_1$ is positive, heat is leaving the tube and the fluid is cooled, whereas if $T_0 - T_1$ is negative, heat is being added as would be the case if the tube were being heated with steam. Obtain an asymptotic expression for the temperature profile in the fluid for small z by using the "method of combination of variables." Then obtain the expression for the local heat-transfer coefficient and the Nusselt number.

SOLUTION To obtain an approximate description of this heating or cooling problem, we assume that the power-law constants m and n are not dependent on the temperature; also the temperature dependence of ρ, k, and $\hat{C}_p$ is ignored. The energy equation (Eq. 4.4-3) then becomes, for low flow rates where viscous heating is unimportant:

$$\underbrace{\rho\hat{C}_p v_{\max}\left[1 - \left(\frac{r}{R}\right)^{s+1}\right]\frac{\partial T}{\partial z}}_{\text{Heat convection in the } z\text{-direction}} = k\left[\underbrace{\frac{1}{r}\frac{\partial}{\partial r}\left(r\frac{\partial T}{\partial r}\right)}_{\text{Heat conduction in the } r\text{-direction}} + \underbrace{\frac{\partial^2 T}{\partial z^2}}_{\text{Heat conduction in the } z\text{-direction}}\right] \tag{4.4-7}$$

with boundary conditions:

$$B.C.\ 1:\quad \text{at } r = 0 \qquad T = \text{finite} \tag{4.4-8}$$

$$B.C.\ 2:\quad \text{at } r = R \qquad T = T_1 \tag{4.4-9}$$

$$B.C.\ 3:\quad \text{at } z \to -\infty \qquad T \to T_0 \tag{4.4-10}$$

$$B.C.\ 4:\quad \text{at } z \to +\infty \qquad T \to T_1 \tag{4.4-11}$$

[8] See, for example, pp. 363–364 and §9.8 of R. B. Bird, W. E. Stewart, and E. N. Lightfoot, *Transport Phenomena*, Wiley, New York (1960). The power-law solution was first given by U. Grigull, *Chem.-Ingen. Techn.*, **28**, 553–556 (1956); see also R. B. Bird, *ibid.*, **31**, 569–572 (1959). The problem has also been solved with n = constant and m varying according to Eq. 4.4-4a by T. Mizushina, R. Itō, Y. Kuriwaki, and K. Yahikozawa, *loc. cit.*; see also T. Mizushina and Y. Kuriwaki, *Mem. Fac. Eng. Kyōto Univ.*, **30**, 511–524 (1968). Further work on the problem has been done by N. Mitsuishi and O. Miyatake, *Kagaku Kōgaku*, **32**, 1222–1227 (1968); *Mem. Fac. Eng., Kyūshū Univ.*, **28**(2), 91–107 (1968). The last four papers listed include experimental data.

[9] R. B. Bird, W. E. Stewart, and E. N. Lightfoot, *Transport Phenomena*, Wiley, New York (1960), pp. 307–308, 349–350.

[10] R. L. Pigford, *CEP Symp. Ser.*, No. 17, **51**, 79–92 (1955).

[11] The analogous mass-transfer problem was discussed by H. Kramers and P. J. Kreyger, *Chem. Eng. Sci.*, **6**, 42–48 (1956).

TABLE 4.4-4

Asymptotic Results for Nusselt Numbers (Tube Flow)[a,b]; Nu = hD/k

All Values are *Local* Nu Numbers	Constant wall temperature 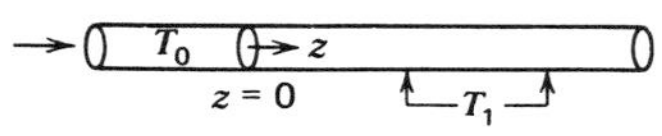		Constant wall heat flux 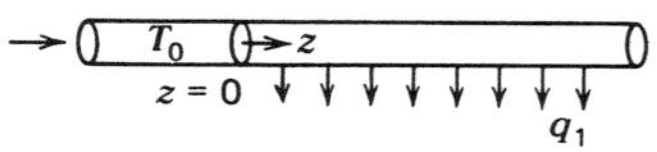	
Thermal entrance region[c] $\frac{\langle v_z \rangle D^2}{\alpha z} \gg 1$	Plug flow	$\mathrm{Nu} = \frac{1}{\sqrt{\pi}} \left(\frac{\langle v_z \rangle D^2}{\alpha z} \right)^{1/2}$ (A)	Plug flow	$\mathrm{Nu} = \frac{\sqrt{\pi}}{2} \left(\frac{\langle v_z \rangle D^2}{\alpha z} \right)^{1/2}$ (G)
	Laminar non-Newtonian flow	$\mathrm{Nu} = \frac{2}{9^{1/3}\Gamma(\frac{4}{3})} \left[\frac{\langle v_z \rangle D^2}{\alpha z} \left(-\frac{1}{4} \frac{d\phi}{d\xi}\bigg|_{\xi=1} \right) \right]^{1/3}$ (B)	Laminar non-Newtonian flow	$\mathrm{Nu} = \frac{2\Gamma(\frac{2}{3})}{9^{1/3}} \left[\frac{\langle v_z \rangle D^2}{\alpha z} \left(-\frac{1}{4} \frac{d\phi}{d\xi}\bigg|_{\xi=1} \right) \right]^{1/3}$ (H)
	Laminar Newtonian flow	$\mathrm{Nu} = \frac{2}{9^{1/3}\Gamma(\frac{4}{3})} \left(\frac{\langle v_z \rangle D^2}{\alpha z} \right)^{1/3}$ (C)	Laminar Newtonian flow	$\mathrm{Nu} = \frac{2\Gamma(\frac{2}{3})}{9^{1/3}} \left(\frac{\langle v_z \rangle D^2}{\alpha z} \right)^{1/3}$ (I)
Thermally fully developed flow $\frac{\langle v_z \rangle D^2}{\alpha z} \ll 1$	Plug flow	$\mathrm{Nu} = 5.772$ (D)	Plug flow	$\mathrm{Nu} = 8$ (J)
	Laminar non-Newtonian flow	$\mathrm{Nu} = \beta_1^2$, where β_1 is the *lowest* eigenvalue of $\frac{1}{\xi} \frac{d}{d\xi} \left(\xi \frac{dX_n}{d\xi} \right) + \beta_n^2 \phi(\xi) X_n = 0$; $X_n'(0) = 0,\ X_n(1) = 0$ (E)	Laminar non-Newtonian flow	$\mathrm{Nu} = \left[2 \int_0^1 \frac{1}{\xi} \left[\int_0^{\xi} \xi' \phi(\xi') d\xi' \right]^2 d\xi \right]^{-1}$ (K)
	Laminar Newtonian flow	$\mathrm{Nu} = 3.657$ (F)	Laminar Newtonian flow	$\mathrm{Nu} = \frac{48}{11}$ (L)

[a] Note: $\phi(\xi) = v_z/\langle v_z \rangle$ where $\xi = r/R$ and $R = D/2$; for Newtonian fluids $\langle v_z \rangle D^2/\alpha z = \mathrm{RePr}(D/z)$ with $\mathrm{Re} = D\langle v_z \rangle \rho/\mu$. Here $\alpha = k/\rho\hat{C}_p$.

[b] W. J. Beek and R. Eggink, *De Ingenieur*, **74**, Ch. 85–89 (1962).

[c] The grouping $\langle v_z \rangle D^2/\alpha z$ is sometimes written as $\mathrm{Gz} \cdot (L/z)$ where $\mathrm{Gz} = \langle v_z \rangle D^2/\alpha L$ is called the Graetz number; here L is the length of the pipe past $z = 0$. Thus the thermal entry region corresponds to large Graetz number.

TABLE 4.4-5

Asymptotic Results for Nusselt Numbers (Thin-Slit Flow)[a,b]; $\mathrm{Nu} = 4hB/k$

All Values are *Local* Nu Numbers	Constant wall temperature 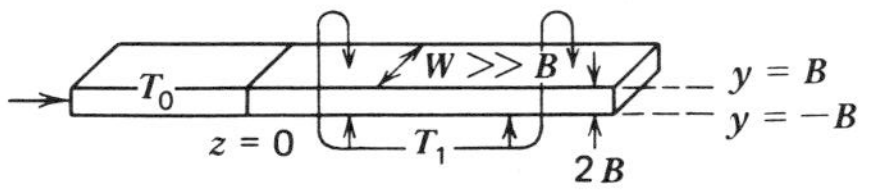		Constant wall heat flux 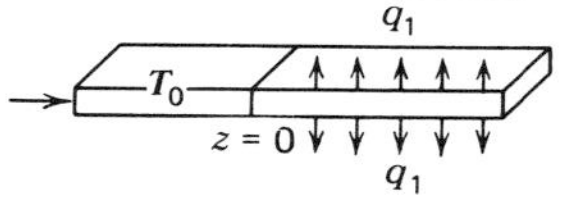	
Thermal entrance region[c] $\dfrac{\langle v_z\rangle B^2}{\alpha z} \gg 1$	Plug flow	$\mathrm{Nu} = \dfrac{4}{\sqrt{\pi}}\left(\dfrac{\langle v_z\rangle B^2}{\alpha z}\right)^{1/2}$ (A)	Plug flow	$\mathrm{Nu} = 2\sqrt{\pi}\left(\dfrac{\langle v_z\rangle B^2}{\alpha z}\right)^{1/2}$ (G)
	Laminar non-Newtonian flow	$\mathrm{Nu} = \dfrac{4}{9^{1/3}\Gamma(\frac{4}{3})}\left[\dfrac{\langle v_z\rangle B^2}{\alpha z}\left(-\left.\dfrac{d\phi}{d\sigma}\right\|_{\sigma=1}\right)\right]^{1/3}$ (B)	Laminar non-Newtonian flow	$\mathrm{Nu} = \dfrac{4\Gamma(\frac{2}{3})}{9^{1/3}}\left[\dfrac{\langle v_z\rangle B^2}{\alpha z}\left(-\left.\dfrac{d\phi}{d\sigma}\right\|_{\sigma=1}\right)\right]^{1/3}$ (H)
	Laminar Newtonian flow	$\mathrm{Nu} = \dfrac{4}{3^{1/3}\Gamma(\frac{4}{3})}\left(\dfrac{\langle v_z\rangle B^2}{\alpha z}\right)^{1/3}$ (C)	Laminar Newtonian flow	$\mathrm{Nu} = \dfrac{4\Gamma(\frac{2}{3})}{3^{1/3}}\left(\dfrac{\langle v_z\rangle B^2}{\alpha z}\right)^{1/3}$ (I)
Thermally fully developed flow $\dfrac{\langle v_z\rangle B^2}{\alpha z} \ll 1$	Plug flow	$\mathrm{Nu} = \pi^2$ (D)	Plug flow	$\mathrm{Nu} = 12$ (J)
	Laminar non-Newtonian flow	$\mathrm{Nu} = 4\beta_1^2$, where β_1 is the *lowest* eigenvalue of $\dfrac{d^2X_n}{d\sigma^2} + \beta_n^2\phi(\sigma)X_n = 0;\quad X_n(\pm 1) = 0$ (E)	Laminar non-Newtonian flow	$\mathrm{Nu} = \left[\dfrac{1}{4}\int_0^1\left[\int_0^\sigma \phi(\sigma')d\sigma'\right]^2 d\sigma\right]^{-1}$ (K)
	Laminar Newtonian flow	$\mathrm{Nu} = 7.54$ (F)	Laminar Newtonian flow	$\mathrm{Nu} = \dfrac{140}{17}$ (L)

[a] Note: $\phi(\sigma) = v_z/\langle v_z\rangle$ where $\sigma = y/B$; for Newtonian fluids $\langle v_z\rangle D^2/\alpha z = 4\,\mathrm{RePr}(B/z)$ with $\mathrm{Re} = 4B\langle v_z\rangle\rho/\mu$. Here $\alpha = k/\rho\hat{C}_p$.

[b] J. M. Valstar and W. J. Beek, *De Ingenieur*, **75**, Ch. 1-7 (1963).

[c] The grouping $\langle v_z\rangle B^2/\alpha z$ is sometimes written as $\mathrm{Gz}\cdot(L/z)$ where $\mathrm{Gz} = \langle v_z\rangle B^2/\alpha L$ is the Graetz number; here L is the length of the slit past $z = 0$. Thus the thermal entry region corresponds to large Graetz number.

The three terms in Eq. 4.4-7 are represented by arrows in Fig. 4.4-1. Except in unusual problems—mainly in liquid metals—the heat conduction in the z-direction is much smaller than the heat convection in the z-direction (the heat transport by fluid flow), so that the axial convection term should be balanced against the radial conduction term. Balancing these two terms allows us to choose an appropriate length scale L for transport in the z-direction. Equating the order of magnitude of these two terms gives

$$\rho\hat{C}_p\langle v_z\rangle\Delta T_0/L = k\Delta T_0/R^2 \tag{4.4-12}$$

where we have used R as a length scale for the radial conduction, the average velocity $\langle v_z\rangle$ as a scale for axial velocity, and $\Delta T_0 = T_0 - T_1$ as a characteristic temperature difference. From Eq. 4.4-12 we see that $L = \langle v_z\rangle R^2/\alpha$ and thus we are led to introduce the following dimensionless quantities:

$$\Theta = \frac{T - T_1}{T_0 - T_1}; \quad \xi = \frac{r}{R}; \quad \zeta = \frac{\alpha z}{\langle v_z\rangle R^2}; \quad \text{Pé} = \frac{\rho\hat{C}_p\langle v_z\rangle R}{k} \tag{4.4-13}$$

in which $\alpha = k/\rho\hat{C}_p$. We also use a dimensionless velocity profile defined as follows:

$$\phi(\xi) = \frac{v_z}{\langle v_z\rangle} = \frac{v_z}{Q/\pi R^2} = \left(\frac{s+3}{s+1}\right)(1 - \xi^{s+1}) \tag{4.4-14}$$

which follows from Eqs. 4.2-8 and 9. Then the dimensionless form of the differential equation is

$$\phi(\xi)\frac{\partial\Theta}{\partial\zeta} = \frac{1}{\xi}\frac{\partial}{\partial\xi}\left(\xi\frac{\partial\Theta}{\partial\xi}\right) + \underline{\frac{1}{\text{Pé}^2}\frac{\partial^2\Theta}{\partial\zeta^2}} \tag{4.4-15}$$

For large Péclet numbers,[12] typical of polymers because of their low thermal conductivities, the axial conduction (dashed underlined) term can be neglected, and one of the boundary conditions in the z-direction can be dropped. The dimensionless boundary conditions to be used with Eq. 4.4-15 (with the dashed underlined term omitted) are

$$B.C.\ 1: \quad \text{at } \zeta = 0 \quad \Theta = 1 \tag{4.4-16}$$

$$B.C.\ 2: \quad \text{at } \xi = 1 \quad \Theta = 0 \tag{4.4-17a}$$

$$B.C.\ 3: \quad \text{at } \xi = 0 \quad \Theta = \text{finite} \tag{4.4-17b}$$

For small z the heat removal (if $T_0 - T_1 > 0$) or the heat penetration (if $T_0 - T_1 < 0$) is restricted to a very thin cylindrical shell near the wall. In this situation the following three approximations lead to results that are accurate in the limit as $z \to 0$:

- **i.** Curvature effects may be ignored and the problem treated as though the wall were flat; the distance from the wall will be designated by $y = R - r$.
- **ii.** The fluid may be regarded as extending from the (flat) heat-transfer surface, $y = 0$, to $y = \infty$.
- **iii.** The velocity profile may be regarded as linear, with a slope given by the velocity gradient at the wall.

[12] See Problem 4D.2 for a discussion of small and moderate Pé.

When we introduce the symbol $\sigma = y/R = 1 - \xi$ for the dimensionless distance from the wall, and when the three mathematical approximations above are used, the problem statement becomes

$$\left(\frac{d\phi}{d\sigma}\right)\bigg|_{\sigma=0} \sigma \frac{\partial \Theta}{\partial \zeta} = \frac{\partial^2 \Theta}{\partial \sigma^2} \tag{4.4-18}$$

$$\textit{B.C. 1:} \quad \text{at } \zeta = 0 \qquad \Theta = 1 \tag{4.4-19}$$

$$\textit{B.C. 2:} \quad \text{at } \sigma = 0 \qquad \Theta = 0 \tag{4.4-20}$$

$$\textit{B.C. 3:} \quad \text{at } \sigma = \infty \qquad \Theta = 1 \tag{4.4-21}$$

For the power-law fluid $(d\phi/d\sigma)_{\sigma=0}$ is just $(s + 3)$, as may be seen from Eq. 4.4-14. We now postulate that a solution can be found of the form $\Theta = \Theta(\chi)$, where χ is a new variable formed from a combination of the independent variables:

$$\chi = \frac{\sigma}{\sqrt[3]{9\zeta/(s + 3)}} \tag{4.4-22}$$

Then the problem becomes

$$\frac{d^2\Theta}{d\chi^2} + 3\chi^2 \frac{d\Theta}{d\chi} = 0 \tag{4.4-23}$$

with $\Theta(0) = 0$ and $\Theta(\infty) = 1$. The solution is

$$\Theta = \frac{1}{\Gamma(\frac{4}{3})} \int_0^{\chi} e^{-\bar{\chi}^3}\, d\bar{\chi} \tag{4.4-24}$$

This is the temperature profile in dimensionless form.

For calculating the heat transfer to or from the tube it is convenient to work in terms of the local heat-transfer coefficient h, defined by

$$q_w = h(T_b - T_w) \tag{4.4-25}$$

in which q_w is the radial heat flux at the wall, T_w is the local wall temperature, and T_b is the bulk temperature, defined by

$$T_b(z) = \frac{\int_0^{2\pi} \int_0^R v_z(r) T(r, z) r\, dr\, d\theta}{\int_0^{2\pi} \int_0^R v_z(r) r\, dr\, d\theta} \tag{4.4-26}$$

Usually the dimensionless heat-transfer coefficient, or Nusselt number, $\text{Nu} = 2hR/k$, is tabulated. For this problem $T_b \doteq T_0$ and $T_w = T_1$, so that the Nusselt number is

$$\text{Nu} = \frac{2q_w R}{k(T_0 - T_1)} = \frac{2R}{k} \frac{(+k\, \partial T/\partial y)|_{y=0}}{(T_0 - T_1)} = +2 \frac{\partial \Theta}{\partial \sigma}\bigg|_{\sigma=0} \tag{4.4-27}$$

When Eq. 4.4-24 is substituted into Eq. 4.4-27 we then get:

$$\text{Nu} = 2 \frac{d\Theta}{d\chi}\bigg|_{\chi=0} \left(\frac{\partial \chi}{\partial \sigma}\right) = \frac{2}{\Gamma(\frac{4}{3})} \sqrt[3]{\frac{(s + 3)}{9\zeta}} = \frac{2}{\Gamma(\frac{4}{3})} \sqrt[3]{\frac{(s + 3)\langle v_z \rangle R^2}{9\alpha z}} \tag{4.4-28}$$

This is just a special case of the result given in Eq. B of Table 4.4-4.

EXAMPLE 4.4-2. Flow in Tubes with Constant Wall Temperature (Asymptotic Solution for Large z)[8,13]

Rework Example 4.4-1 by the "method of separation of variables" and obtain a formal expression for the Nusselt number valid for any value of z. Show that a particularly simple asymptotic result can be obtained for large z values.

SOLUTION In the method of separation of variables we seek a solution to Eq. 4.4-15 of the form of a product

$$\Theta(\xi, \zeta) = X(\xi)Z(\zeta) \tag{4.4-29}$$

and insert it into the partial differential equation. Then division by XZ gives

$$\frac{1}{Z}\frac{dZ}{d\zeta} = \frac{1}{\phi X \xi}\frac{d}{d\xi}\left(\xi \frac{dX}{d\xi}\right) \tag{4.4-30}$$

Both sides are equal to a constant, which we denote by $-\beta^2$. Hence we get two ordinary differential equations:

$$\frac{dZ}{d\zeta} = -\beta^2 Z \tag{4.4-31}$$

$$\frac{1}{\xi}\frac{d}{d\xi}\left(\xi \frac{dX}{d\xi}\right) + \beta^2 \phi X = 0 \tag{4.4-32}$$

The second of these has boundary conditions: $X'(0) = 0$ and $X(1) = 0$. The X-equation is a Sturm–Liouville problem, and there is a set of eigenfunctions X_i corresponding to the set of eigenvalues β_i. The complete solution must then be a linear combination of products of the form given in Eq. 4.4-29:

$$\Theta(\xi, \zeta) = \sum_{i=1}^{\infty} A_i X_i(\xi) e^{-\beta_i^2 \zeta} \tag{4.4-33}$$

The A_i are determined by the requirement that $\Theta = 1$ at $\zeta = 0$. After Eq. 4.4-33 is written for $\zeta = 0$, we multiply both sides by $X_j(\xi)\phi(\xi)\xi\, d\xi$ and integrate from $\xi = 0$ to $\xi = 1$. Then when use is made of the fact that the $X_i(\xi)$ are orthogonal over the range $\xi = 0$ to $\xi = 1$ with respect to the weighting function $\phi(\xi)\xi$ we obtain finally

$$A_i = \frac{\int_0^1 X_i \phi \xi\, d\xi}{\int_0^1 X_i^2 \phi \xi\, d\xi} \tag{4.4-34}$$

Equations 4.4-33 and 34 constitute a formal solution to the problem.

We next get an expression for the Nusselt number, which in this case is given by

$$\mathrm{Nu} = \frac{2hR}{k} = \frac{2R(-k\partial T/\partial r)|_{r=R}}{k(T_b - T_w)} = -2\frac{(\partial\Theta/\partial\xi)|_{\xi=1}}{\Theta_b} \tag{4.4-35}$$

[13] B. C. Lyche and R. B. Bird, *Chem. Eng. Sci.*, **6**, 35–41 (1956), gave the solution for constant m and n. The extension to temperature-dependent m and an associated experimental study were given by E. B. Christiansen and S. E. Craig, Jr., *AIChE J.*, **8**, 154–160 (1962). See also E. B. Christiansen, G. E. Jensen, and F. S. Tao, *ibid.*, **12**, 1196–1202 (1966) and E. B. Christiansen and G. E. Jensen, *ibid.*, **15**, 504–507 (1969).

in which $T_w = T_1$ and $\Theta_b = (T_b - T_1)/(T_0 - T_1)$. From Eq. 4.4-33 we find that

$$\left.\frac{\partial \Theta}{\partial \xi}\right|_{\xi=1} = \sum_{i=1}^{\infty} A_i e^{-\beta_i^2 \zeta} X_i'(1) \tag{4.4-36}$$

$$\begin{aligned}
\Theta_b &= \frac{\int_0^1 \phi(\xi)\Theta(\xi,\zeta)\xi\,d\xi}{\int_0^1 \phi(\xi)\xi\,d\xi} = 2\int_0^1 \phi(\xi)\Theta(\xi,\zeta)\xi\,d\xi \\
&= 2\sum_{i=1}^{\infty} A_i e^{-\beta_i^2\zeta} \int_0^1 \phi(\xi)X_i(\xi)\xi\,d\xi \\
&= -2\sum_{i=1}^{\infty} A_i e^{-\beta_i^2\zeta} \int_0^1 \frac{1}{\beta_i^2}\frac{d}{d\xi}\left(\xi\frac{dX_i}{d\xi}\right)d\xi \\
&= -2\sum_{i=1}^{\infty} A_i e^{-\beta_i^2\zeta} \frac{1}{\beta_i^2}\,\xi\left.\frac{dX_i}{d\xi}\right|_0^1 = -2\sum_{i=1}^{\infty}\frac{A_i}{\beta_i^2}e^{-\beta_i^2\zeta}X_i'(1)
\end{aligned} \tag{4.4-37}$$

In going from the second to the third line, use has been made of Eq. 4.4-32. The Nusselt number is then given by

$$\mathrm{Nu} = \frac{\sum_{i=1}^{\infty} A_i e^{-\beta_i^2\zeta} X_i'(1)}{\sum_{i=1}^{\infty} (A_i/\beta_i^2) e^{-\beta_i^2\zeta} X_i'(1)} \tag{4.4-38}$$

This expression is the local Nusselt number valid for all values of z. However, in order to use this result all the eigenvalues and eigenfunctions must be known.

For large values of z (or ζ) Eq. 4.4-38 can be simplified considerably inasmuch as only the first term in each sum is needed:

$$\lim_{\zeta\to\infty} \mathrm{Nu} = \beta_1^2 \tag{4.4-39}$$

This is just the result given in Eq. E of Table 4.4-4; according to this result we need calculate only the first eigenvalue for the boundary-value problem in Eq. 4.4-32. There are several ways of doing this; we illustrate here the use of the method of Stodola and Vianello[14] for the particular case of $n = \frac{1}{2}$ (or $s = 2$).

For $s = 2$ we have $\phi(\xi) = \frac{5}{3}(1 - \xi^3)$. As a first guess for the lowest eigenfunction we try $X_1(\xi) = 1 - \xi^3$. Then Eq. 4.4-32 becomes

$$\frac{d}{d\xi}\left(\xi\frac{dX}{d\xi}\right) = -\frac{5}{3}\beta_1^2\xi(1-\xi^3)^2 \tag{4.4-40}$$

[14] See, for example, F. B. Hildebrand, *Advanced Calculus for Applications*, Prentice-Hall, Englewood Cliffs, NJ (1962), pp. 200–206. To solve the differential equation

$$\frac{d}{dx}\left[p(x)\frac{dy}{dx}\right] + \lambda r(x)y = 0 \tag{4.4-39a}$$

with homogeneous boundary conditions at $x = a$ and $x = b$, do the following: (i) in the λ-term, replace $y(x)$ by a reasonable first guess $y_1(x)$; (ii) solve the resulting differential equation and obtain the solution $y(x) = \lambda f_1(x)$; then repeat (i) with a second guess $y_2(x) = f_1(x)$; solve the resulting differential equation to obtain the solution $y(x) = \lambda f_2(x)$, and continue the process as long as desired. At the nth stage in this process the nth approximation to the lowest eigenvalue λ_1 can be obtained from

$$\lambda_1^{(n)} = \frac{\int_a^b r(x)f_n(x)y_n(x)\,dx}{\int_a^b r(x)[f_n(x)]^2\,dx} \tag{4.4-39b}$$

This may be integrated to give

$$X(\xi) = \frac{5}{3}\beta_1^2\left(\frac{297}{64\cdot 25} - \frac{1}{4}\xi^2 + \frac{2}{25}\xi^5 - \frac{1}{64}\xi^8\right) \qquad (4.4\text{-}41)$$
$$\equiv \beta_1^2 f_1(\xi)$$

Then, using Eq. 4.4-39b we find

$$[\beta_1^2]^{(1)} = \frac{\int_0^1 \xi(1-\xi^3)f_1(\xi)(1-\xi^3)\,d\xi}{\int_0^1 \xi(1-\xi^3)[f_1(\xi)]^2\,d\xi} \qquad (4.4\text{-}42)$$
$$= 3.97$$

This approximate result can be compared with the value Nu $= 3.95$ obtained by a complete numerical solution to the eigenvalue problem. It should be evident that the method of Stodola and Vianello can provide relatively quick estimates of the Nusselt numbers for large z for problems of this type.

EXAMPLE 4.4-3. Flow in a Circular Die with Viscous Heating[15,16]

A molten plastic enters a horizontal circular die of radius R and length L at a temperature T_0. The fluid velocity is sufficiently high that viscous heating is known to be important. Find the temperature distribution in the die assuming (i) the die wall is maintained at temperature T_0; (ii) the fluid is described adequately by a power-law viscosity with m and n independent of temperature; (iii) the fluid has a fully developed velocity profile at the die entrance; and (iv) density, heat capacity, and thermal conductivity do not change with temperature or pressure.

What is the maximum temperature predicted for a polyethylene melt that enters the die at 463 K when the wall shear stress is 2×10^5 Pa? The melt has the following physical properties: $\rho\hat{C}_p = 1.80 \times 10^6$ J/m^{3}K; $k = 0.25$ W/m·K; $n = 0.50$; $m = 6.9 \times 10^3$ Pa·s$^{1/2}$. The radius and length of the die are 0.0394 and 1.28 cm, respectively.

SOLUTION We postulate that $v_z = v_z(r)$, $p = p(z)$, and $T = T(r, z)$. Then the equation of continuity is satisfied identically and the equation of motion (Eq. 4.4-2) and the equation of energy (Eq. 4.4-3) are

Motion:
$$0 = -\frac{dp}{dz} + \frac{1}{r}\frac{d}{dr}\left(\eta r \frac{dv_z}{dr}\right) \qquad (4.4\text{-}43)$$

Energy:
$$\rho\hat{C}_p v_z \frac{\partial T}{\partial z} = k\frac{1}{r}\frac{\partial}{\partial r}\left(r\frac{\partial T}{\partial r}\right) + \eta\left(\frac{dv_z}{dr}\right)^2 \qquad (4.4\text{-}44)$$

[15] Flow of a Newtonian fluid in a circular tube with viscous heating was first solved according to the method given here by H. C. Brinkman, *Appl. Sci. Res.*, **A2**, 120–124 (1951). The extension to power-law fluids was worked out by R. B. Bird, *J. Soc. Plast. Eng.*, **11**, 35–40 (1955), with later developments by N. Galili, R. Takserman-Krozer, and Z. Rigbi, *Rheol. Acta*, **14**, 550–567 (1975); N. Galili, Z. Rigbi, and R. Takserman-Krozer, *Rheol. Acta*, **14**, 816–831 (1975); and C. A. Hieber, *Rheol. Acta.* **16**, 553–567 (1977). Temperature-dependent viscosity effects were considered by J. B. Lyon and R. E. Gee, *Ind. Eng. Chem.*, **49**, 956–960 (1957), by E. A. Kearsley, *Trans. Soc. Rheol.*, **6**, 253–261 (1962), and by P. C. Sukanek, *Chem. Eng. Sci.*, **26**, 1775–1776 (1971). In the last two publications cited, analytical solutions are given for large z for the Newtonian and power-law fluids.

[16] See also S. M. Dinh and R. C. Armstrong, *AIChE J.*, **28**, 294–301 (1982).

Axial heat conduction has been neglected in the energy equation for reasons discussed in Example 4.4-1. For the power-law fluid we have to insert $\eta = m(-dv_z/dr)^{n-1}$, and then according to Example 4.2-1

$$v_z = v_{max}\left[1 - \left(\frac{r}{R}\right)^{s+1}\right] \tag{4.4-45}$$

where $v_{max} = [(p_0 - p_L)R/2mL]^s[R/(s + 1)]$ and $s = 1/n$.

It is convenient to rewrite Eq. 4.4-44 in dimensionless form by introducing the following dimensionless quantities:

$$\xi = \frac{r}{R}; \quad \zeta = \frac{\alpha z}{v_{max}R^2}; \quad \Theta = \frac{T - T_0}{\Delta T_0}; \quad \phi = \frac{v_z}{v_{max}} \tag{4.4-46}$$

where the scaling for z is selected as in Example 4.4-1 to correspond to large Péclet number. Here we have chosen to use v_{max} as a characteristic velocity. Since in this problem there is no temperature change imposed by the boundary conditions and the viscosity is assumed to be temperature independent, a reasonable choice for ΔT_0 to scale temperature changes is ΔT_{gen}, the temperature rise associated with a balance between viscous heating and heat conduction. From Table 4.4-1 (footnote *b*) this is

$$\Delta T_{gen} = \eta^0 V^2/k = \frac{mR^2}{k}\left(\frac{v_{max}}{R}\right)^{n+1} \tag{4.4-47}$$

where η^0 is the viscosity evaluated at a characteristic shear rate (v_{max}/R). Then the dimensionless temperature is

$$\Theta = \frac{(T - T_0)k}{mR^2(v_{max}/R)^{n+1}} \tag{4.4-48}$$

and the mathematical problem to be solved is

$$(1 - \xi^{s+1})\frac{\partial\Theta}{\partial\zeta} = \frac{1}{\xi}\frac{\partial}{\partial\xi}\left(\xi\frac{\partial\Theta}{\partial\xi}\right) + (s + 1)^{(1/s)+1}\xi^{s+1} \tag{4.4-49}$$

with boundary conditions

$$\textit{B.C. 1:} \quad \text{at } \zeta = 0 \quad \Theta = 0 \tag{4.4-50}$$
$$\textit{B.C. 2:} \quad \text{at } \xi = 0 \quad \partial\Theta/\partial\xi = 0 \tag{4.4-51}$$
$$\textit{B.C. 3:} \quad \text{at } \xi = 1 \quad \Theta = 0 \tag{4.4-52}$$

Because of the heat-production term, Eq. 4.4-49 cannot be directly solved by the method of separation of variables. Intuitively we know that for large distances down the tube a temperature distribution should be attained that depends only on the radial coordinate. Therefore it seems reasonable to postulate a solution of the form

$$\Theta(\xi, \zeta) = \Theta_1(\xi) - \Theta_2(\xi, \zeta) \tag{4.4-53}$$

in which Θ_1 is the solution for very large ζ, and Θ_2 is a function that just cancels Θ_1 at $\zeta = 0$ and which tends to zero as ζ becomes very large. The Θ_1-part is easily obtained by setting the left side of

Eq. 4.4-49 equal to zero and solving the remaining ordinary differential equation using the boundary conditions at $\xi = 0$ and $\xi = 1$. This gives

$$\Theta_1(\xi) = \frac{(s+1)^{(1/s)+1}}{(s+3)^2}(1 - \xi^{s+3}) \tag{4.4-54}$$

Now the results in Eqs. 4.4-53 and 54 can be inserted into Eq. 4.4-49, and in this way an equation is obtained for Θ_2

$$(1 - \xi^{s+1})\frac{\partial \Theta_2}{\partial \zeta} = \frac{1}{\xi}\frac{\partial}{\partial \xi}\left(\xi \frac{\partial \Theta_2}{\partial \xi}\right) \tag{4.4-55}$$

This equation can be solved by the method of separation of variables. We postulate that $\Theta_2(\xi, \zeta) = X(\xi)Z(\zeta)$ and this leads to

$$\frac{1}{Z}\frac{dZ}{d\zeta} = \frac{1}{X}\frac{1}{(1 - \xi^{s+1})}\frac{1}{\xi}\frac{d}{d\xi}\left(\xi \frac{dX}{d\xi}\right) = -a \tag{4.4-56}$$

in which a is the separation constant. This procedure thus leads to two separate ordinary differential equations for $X(\xi)$ and $Z(\zeta)$. The X-equation has to be solved with the boundary conditions that $X(\xi) = 0$ at $\xi = 1$, and $X'(\xi) = 0$ at $\xi = 0$. This boundary-value problem has an infinite set of solutions $X_k(\xi)$ corresponding to the infinite set of eigenvalues a_k. It may be shown directly from the differential equation that the $X_k(\xi)$ satisfy the orthogonality conditions

$$\int_0^1 (1 - \xi^{s+1})\xi X_k(\xi) X_l(\xi)\, d\xi = 0 \qquad (k \neq l) \tag{4.4-57}$$

Because of the linearity of the original partial differential equation, the expression for $\Theta_2(\xi, \zeta)$ must be given by a superposition of products of functions of ξ and functions of ζ. The final expression for Θ must therefore have the form

$$\Theta(\xi, \zeta) = \frac{(s+1)^{(1/s)+1}}{(s+3)^2}(1 - \xi^{s+3}) - \sum_{k=1}^{\infty} B_k X_k(\xi) e^{-a_k \zeta} \tag{4.4-58}$$

The B_k must be determined from the boundary condition at $\zeta = 0$. By multiplying Eq. 4.4-58 (with $\zeta = 0$ and $\Theta = 0$) by $\xi(1 - \xi^{s+1})X_l$ and integrating over ξ from 0 to 1, we get

$$B_l = \frac{(s+1)^{(1/s)+1}}{(s+3)^2} \frac{\int_0^1 X_l(1 - \xi^{s+3})(1 - \xi^{s+1})\xi\, d\xi}{\int_0^1 X_l^2(1 - \xi^{s+1})\xi\, d\xi} \tag{4.4-59}$$

The problem is thus solved once the eigenfunctions $X_k(\xi)$ and the eigenvalues a_k are known.

We illustrate here a method that may be used in the instances when $s = 1/n$ is an integer;[17] in particular we let $s = 1/n = 2$. We seek a solution to the $X(\xi)$-equation in Eq. 4.4-56 in the form of a

[17] For arbitrary values of n see C. A. Hieber, *Rheol. Acta*, **16**, 553–567 (1977). A numerical solution for arbitrary n of the corresponding eigenvalue problem for slit flow is given by R. M. Ybarra and R. E. Eckert, *AIChE J.*, **26**, 751–762 (1980); see also R. A. Mashelkar, V. V. Charan, and N. G. Karanth, *Chem. Eng. J.*, **6**, 75–77 (1973).

TABLE 4.4-6[a]

Constants for Use in Eq. 4.4-58 When $n = 1/2$

i	a_i	B_i
1	6.582	+0.2904
2	39.10	−0.1217
3	99.50	+0.06065
4	187.8	−0.03525

[a] This table is based on C. A. Hieber, *Rheol. Acta*, **16**, 553–567 (1977), Table 4 (with $a_i = [(s + 3)/(s + 1)]\,\omega_i$ and $B_i = -4^{n-1}[(s+1)/(s+3)]^{n+1}\,\tau_i$ where ω_i and τ_i are the eigenvalues and expansion coefficients in Hieber's notation). This kind of tabulation was first given by R. B. Bird, *SPE J.*, **11**, 35–40 (1955).

power series; the ith eigenfunction will thus have the form

$$X_i(\xi) = \sum_{k=0}^{\infty} b_{ik}\xi^k \tag{4.4-60}$$

in which b_{i0} is arbitrarily chosen to be unity, and the b_{ik} are zero for $k < 0$. The coefficient b_{i1} must be zero to satisfy the boundary condition at $\xi = 0$. Substitution of the above postulated solution into the differential equation leads to the following recursion formula: $b_{ik} = -(a_i/k^2)(b_{i,k-2} - b_{i,k-5})$. Hence all of the b_{ik} can be expressed in terms of the eigenvalues a_i, which are at this point still not known

$$\begin{aligned} b_{i2} &= -\tfrac{1}{4}a_i & b_{i5} &= +\tfrac{1}{25}a_i \\ b_{i3} &= 0 & b_{i6} &= -\tfrac{1}{2304}a_i^3 \\ b_{i4} &= +\tfrac{1}{64}a_i^2 & b_{i7} &= -\tfrac{29}{4900}a_i^2, \text{ etc.} \end{aligned} \tag{4.4-61}$$

The boundary condition at $\xi = 1$ then requires that $\sum_{k=0}^{\infty} b_{ik} = 0$. When the b's from Eq. 4.4-61 are substituted into this, we get an algebraic equation for a_i; since this equation is of infinite order, an infinite number of eigenvalues $a_1, a_2, a_3, \ldots$ will be obtained. The first few of these values are shown in Table 4.4-6. Since the X-equation is a second-order differential equation, there must be a second solution in addition to that in Eq. 4.4-60, with the b's provided by Eq. 4.4-61. It can be shown, however, that the second solution becomes infinite at $\xi = 0$ and is hence unacceptable here.

Once we have the eigenvalues a_i and the eigenfunctions X_i, it is a straightforward matter to get the B_i from Eq. 4.4-59. The first few B_i are shown in Table 4.4-6 along with the corresponding eigenvalues.[18] The temperature profiles computed from Eq. 4.4-58 are shown in Fig. 4.4-2. There it can be seen that, at short distances, there is a peak in the temperature profile near the wall where the velocity gradient and also the viscous heating are large. The temperature profiles for a thermally insulated wall are also shown in Fig. 4.4-2 for comparison.

[18] Note that the X-equations of Eqs. 4.4-32 and 4.4-56 are nearly the same. Since $\phi = v_z/\langle v_z \rangle$ (rather than $v_z/v_{z,\max}$) in Eq. 4.4-32, the eigenvalues in the two examples bear the relation $a_j = (v_{z,\max}/\langle v_z \rangle)\beta_j^2$. Hence the $a_1 = 6.582$ in Table 4.4-6 corresponds to $\beta_1^2 = 3.95$ cited just after Eq. 4.4-42.

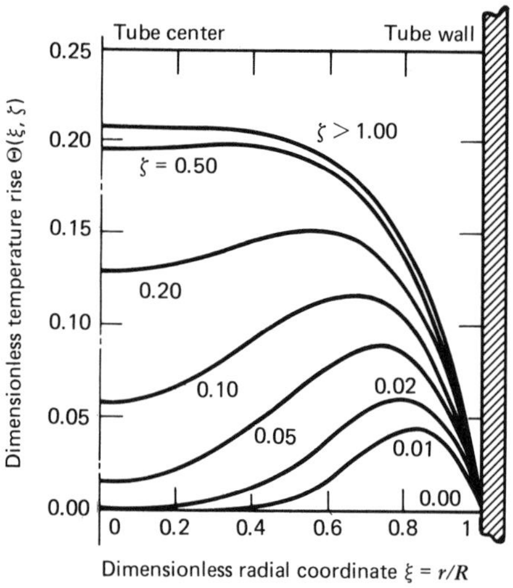

(*a*)

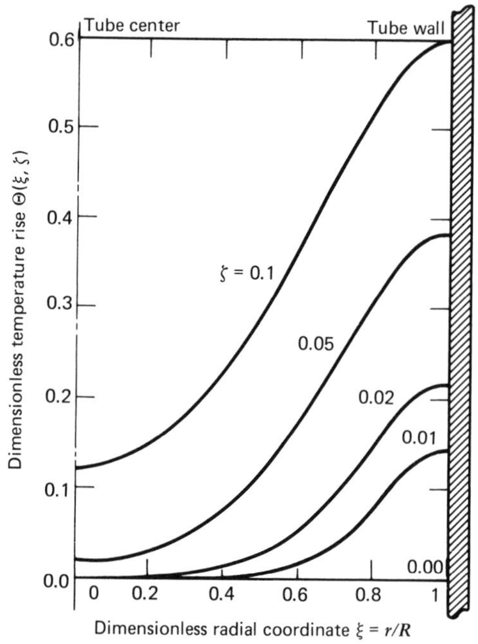

(*b*)

FIGURE 4.4-2. Temperature profiles for tube flow with viscous heating, based on the power-law model with $n = 1/2$, and with m and n taken to be constant. (*a*) Constant wall temperature; (*b*) zero wall heat flux. [After R. B. Bird, *SPE J.*, **11**, 35–40 (1955).]

For the particular experimental data given in the problem statement, the maximum temperature will occur at the tube exit, that is at $z = 0.0128$ m. The dimensionless axial distance ζ divided by Péclet number at the exit is:

$$\zeta = \frac{kz}{\rho \hat{C}_p v_{\max} R^2} = \frac{k(s+1)m^s z}{\rho \hat{C}_p \tau_R^s R^3}$$

$$= \frac{(0.25)(3)(6.9 \times 10^3)^2(0.0128)}{(1.80 \times 10^6)(2.0 \times 10^5)^2(3.94 \times 10^{-4})^3}$$

$$= 0.104 \tag{4.4-62}$$

From Fig. 4.4-2 we find by interpolation that the maximum dimensionless temperature Θ_{max} is about 0.12. Hence from Eq. 4.4-48 we get

$$\begin{aligned}(T - T_0)_{max} &= \frac{mR^2(\tau_R/m)^{(s+1)}}{(s+1)^{(1/s)+1}k}\Theta_{max} \\ &= \frac{R^2\tau_R^3\Theta_{max}}{3^{3/2}km^2} \\ &= \frac{(3.94 \times 10^{-4})^2(2.0 \times 10^5)^3(0.12)}{(6.9 \times 10^3)^2(0.25)(3^{3/2})} \\ &= 2.4 \text{ K} \end{aligned} \tag{4.4-63}$$

For the insulated wall problem, the temperature rise would have been about 12 K. It is thus evident that viscous heating can produce nonignorable temperature rises in extrusion operations.

We emphasize that the development here has been given for temperature-independent physical properties. The temperature dependence of m, however, cannot really be ignored if the temperature rise is more than a few degrees. Hence the use of the result in Fig. 4.4-2 should really be restricted to estimating flow speeds or pressure drops at which viscous heating just begins to be important.

We conclude with a few comments about the calculation of the eigenvalues a_i. Obtaining the a_i for $i \geq 4$ is tedious by the method outlined above. For $i \geq 4$ the WKBJ (Wentzel–Kramers–Brillouin–Jeffreys) method, widely used in quantum mechanics, is useful for getting both the eigenvalues and eigenfunctions. Applications to non-Newtonian heat-transfer problems have been given by Whiteman and Drake,[19] by Ziegenhagen,[20] and by Dinh and Armstrong.[21] For the problem at hand the application of the WKBJ method gives for any velocity profile v_z

$$a_i = \left[\frac{(i - \frac{1}{3})\pi}{\int_0^1 \sqrt{v_z(\xi)/v_{max}}\, d\xi}\right]^2 \tag{4.4-64}$$

Note that this result is not restricted to integer values of $s = 1/n$. For the power law with $s = 2$ (or $n = 0.5$) we find

$$a_i = \left[\frac{(i - \frac{1}{3})\pi}{\int_0^1 \sqrt{1 - \xi^3}\, d\xi}\right]^2 = \left[\frac{(i - \frac{1}{3})\pi}{\Gamma(\frac{3}{2})\Gamma(\frac{4}{3})/\Gamma(\frac{11}{6})}\right]^2 = \left[\frac{(i - \frac{1}{3})\pi}{0.841}\right]^2 \tag{4.4-65}$$

This gives $a_1 = 6.202$, $a_2 = 38.8$, $a_3 = 99.2$, and $a_4 = 187.6$; by comparing with the exact values in Table 4.4-6 we see that for small i, for which the method is not intended, the results are not too good, but that the error is less than 0.5% for $i = 3$.

EXAMPLE 4.4-4 Viscous Heating in a Cone-and-Plate Viscometer[22]

Obtain the temperature profile for the flow of a power-law fluid in the parallel-plate system of Fig. 1.3-1 for the special case that there is no superposed pressure gradient. First consider m, n, and k all to be constants. Then let m vary with temperature according to Eq. 4.4-4. Finally show how the results can be adapted for use in estimating viscous-heating effects in the cone-and-plate instrument.

[19] I. R. Whiteman and W. B. Drake, *Trans. ASME*, **80**, 728–732 (1958).

[20] A. J. Ziegenhagen, *Int. J. Heat Mass Transfer*, **8**, 499–505 (1965).

[21] S. M. Dinh and R. C. Armstrong, *AIChE J.*, **28**, 294–301 (1982), have obtained the eigenvalues and uniformly valid eigenfunctions for both the tube and slit problems and for wall boundary conditions ranging from insulated to constant temperature.

[22] The perturbation solution presented here is patterned after R. M. Turian, *Chem. Eng. Sci.*, **20**, 771–781 (1965); a closed-form solution was given by J. Gavis and R. L. Laurence, *Ind. Eng. Chem. Fundam.*, **7**, 525–527 (1968). The equation in Turian's paper which corresponds to Eq. 4.4-93 erroneously contains (1/20)Na/n rather than the correct value (1/20)Na; note that Turian's BrA corresponds to Na in this example.

SOLUTION **(a)** Flow between Parallel Plates with the Assumption of Constant Physical Properties.

We consider the flow in the x-direction of a fluid between two parallel plates located at $y = 0$ and $y = B$ (cf. Fig. 1.3-1); the lower plate is fixed and the upper one is moving in the positive x-direction with a constant velocity V and $\partial p/\partial x = 0$. For a generalized Newtonian fluid the equations of motion and energy that we need for describing the flow are the following simplifications of Eqs. 4.4-2 and 3:

$$0 = \frac{d}{dy}\left(\eta \frac{dv_x}{dy}\right) \tag{4.4-66}$$

$$0 = \frac{d}{dy}\left(k \frac{dT}{dy}\right) + \eta\left(\frac{dv_x}{dy}\right)^2 \tag{4.4-67}$$

into which, for this problem, we substitute $\eta = m(dv_x/dy)^{n-1}$.

We first work through the simplified problem in which m and n, as well as the thermal conductivity, are not varying with temperature. The problem can be written in dimensionless form as

$$0 = \frac{d}{d\xi}\left(\frac{d\phi}{d\xi}\right)^n \tag{4.4-68}$$

$$0 = \frac{d^2\Theta}{d\xi^2} + \left(\frac{d\phi}{d\xi}\right)^{n+1} \tag{4.4-69}$$

subject to the boundary conditions

$$\text{at } \xi = 0 \quad \phi = 0, \quad \Theta = 0 \tag{4.4-70}$$

$$\text{at } \xi = 1 \quad \phi = 1, \quad \Theta = 0 \tag{4.4-71}$$

The dimensionless variables are defined by

$$\phi = \frac{v_x}{V}; \quad \Theta = \frac{(T - T_0)}{\Delta T_0}; \quad \xi = \frac{y}{B} \tag{4.4-72}$$

in which ΔT_0 is a characteristic temperature difference in the problem. Since no reference ΔT can be formed from the boundary conditions or the temperature dependence of the viscosity, we choose ΔT_0 to be the temperature rise necessary for heat conduction to balance heat production by viscous heating. A balance between the order of magnitude of the viscous heating and heat conduction terms in Eq. 4.4-67 gives

$$k\frac{\Delta T_0}{B^2} = \eta\bigg|_{\dot{\gamma}=V/B}\left(\frac{V}{B}\right)^2 = m\left(\frac{V}{B}\right)^{n+1} \tag{4.4-73}$$

so that

$$\Delta T_0 = mV^{n+1}/B^{n-1}k \tag{4.4-74}$$

Equations 4.4-68 through 71 are easily solved to give

$$\phi = \xi \tag{4.4-75}$$

$$\Theta = \tfrac{1}{2}(\xi - \xi^2) \tag{4.4-76}$$

The maximum temperature occurs at the midplane and is given by

$$\Theta_{\max} = \tfrac{1}{8} \tag{4.4-77}$$

This simple result can be used for making rough estimates of the temperature rise that can be expected in the gap between two moving surfaces in the absence of a pressure gradient.

(b) Flow Between Parallel Plates with m Varying with the Temperature, but with n and k Constant. We let m vary with temperature according to Eq. 4.4-4 and again introduce dimensionless variables defined by Eq. 4.4-72. Now, however, we scale temperature by a characteristic temperature difference required to change the viscosity substantially, since the effects of temperature dependent viscosity are of primary interest here. This ΔT_0 can be calculated from Eq. 4.4-4 as

$$\Delta T_0 = \left|\frac{\eta}{\partial\eta/\partial T}\right|_{\substack{\dot\gamma = V/B \\ T = T_0}} = 1/A \tag{4.4-78}$$

and we thus take

$$\Theta = A(T - T_0) \tag{4.4-79}$$

The equations of motion and energy (Eqs. 4.4-66 and 67) become

$$e^{-\Theta}\left(\frac{d\phi}{d\xi}\right)^n = C \tag{4.4-80}$$

$$\frac{d^2\Theta}{d\xi^2} + \mathrm{Na}\, e^{-\Theta}\left(\frac{d\phi}{d\xi}\right)^{n+1} = 0 \tag{4.4-81}$$

in which $\mathrm{Na} = m^0 V^{n+1}A/B^{n-1}k$ is the Nahme-Griffith number for the power-law fluid, m^0 is the value of m at T_0 in Eq. 4.4-4, and C is an integration constant. The Nahme-Griffith number is a ratio of the temperature rise characteristic of the viscous heating in the problem (Eq. 4.4-74) to the temperature change necessary to alter the viscosity (Eq. 4.4-78). Thus for small values of Na the solution of part (a) is expected to be accurate.

From Eq. 4.4-80

$$\frac{d\phi}{d\xi} = C^s e^{s\Theta} = C^s(1 + a_1\Theta + a_2\Theta^2 + \cdots) \tag{4.4-82}$$

in which $s = 1/n$ and $a_j = s^j/j!$. This can be inserted into the energy equation Eq. 4.4-81 to give

$$\frac{d^2\Theta}{d\xi^2} = -\mathrm{Na}\, C^{s+1}e^{s\Theta} = -\mathrm{Na}\, C^{s+1}(1 + a_1\Theta + a_2\Theta^2 + \cdots) \tag{4.4-83}$$

Equations 4.4-82 and 83 are solved by a perturbation procedure, using the Nahme-Griffith number as the perturbation parameter. We know that the unperturbed problem discussed in (a) has the solution $\phi = \xi$ and $\Theta = (1/2)\mathrm{Na}(\xi - \xi^2)$. This suggests that we seek a solution of the form:

$$\phi = \xi + \mathrm{Na}\,\phi_1(\xi) + \mathrm{Na}^2\phi_2(\xi) + \cdots \tag{4.4-84}$$

$$\Theta = \tfrac{1}{2}\mathrm{Na}\,(\xi - \xi^2) + \mathrm{Na}^2\Theta_2(\xi) + \cdots \tag{4.4-85}$$

It is also necessary to expand the integration constant in a similar series:

$$C = C_0 + \text{Na}C_1 + \text{Na}^2C_2 + \cdots \tag{4.4-86}$$

$$C^s = C_0^s + \text{Na}\,(sC_0^{s-1}C_1) + \text{Na}^2\,(sC_0^{s-1}C_2 + \tfrac{1}{2}s(s-1)C_0^{s-2}C_1^2) + \cdots \tag{4.4-87}$$

When these expansions are substituted into Eqs. 4.4-82 and 83, sets of differential equations are obtained by equating the coefficients of equal powers of Na. The resulting differential equations are solved with the boundary conditions that Θ_j and ϕ_j are both zero at $\xi = 0, 1$. In this way, we obtain

$$\phi = \xi - \frac{1}{12n}\,\text{Na}\,(\xi - 3\xi^2 + 2\xi^3) + \cdots \tag{4.4-88}$$

$$\Theta = \frac{1}{2}\,\text{Na}(\xi - \xi^2) - \frac{1}{24n}\,\text{Na}^2(n\xi - (n+1)\xi^2 + 2\xi^3 - \xi^4) + \cdots \tag{4.4-89}$$

It is not difficult to show that $\phi(\xi)$ and $\Theta(\xi)$ are symmetric about the plane $\xi = 1/2$.

(c) Application to the Cone-and-Plate Viscometer

When viscous heating is appreciable, viscometers will give incorrect readings because of the temperature dependence of the rheological properties. We now adapt the above results in an approximate way to the cone-and-plate instrument in order to estimate the influence of viscous heating on the measurement of viscosity.

We can make a table of equivalences between the notation used for the parallel-plate system and that used in Example 1.3-4 for the cone-and-plate system:

Parallel-Plate System	Cone-and-Plate System
V	Wr
B	$\vartheta_0 r$
$\xi = \dfrac{y}{B}$	$\bar{\xi} = \dfrac{(\pi/2) - \theta}{\vartheta_0}$
$\phi = \dfrac{v_x}{V}$	$\bar{\phi} = \dfrac{v_\phi}{Wr}$
$\text{Na} = \dfrac{m^0 V^{n+1} A}{kB^{n-1}}$	$\overline{\text{Na}} = \widetilde{\text{Na}}\left(\dfrac{r}{R}\right)^2 = \dfrac{m^0 W^{n+1} R^2 A}{k\vartheta_0^{n-1}}\left(\dfrac{r}{R}\right)^2$

Recall that the torque required to maintain the rotary motion of the cone is, according to Eq. 1.3-35,

$$\mathscr{T} = \int_0^{2\pi}\int_0^R \tau_{\theta\phi}\Big|_{\theta=\pi/2} r^2\,dr\,d\phi \doteq 2\pi\int_0^R \left(-\eta\,\frac{1}{r}\frac{\partial v_\phi}{\partial\theta}\right)\Big|_{\theta=\pi/2} r^2\,dr \tag{4.4-90}$$

Here an approximate form of $\dot{\gamma}_{\theta\phi}$ is used since $\theta \doteq \pi/2$ throughout the gap and $\sin\theta \doteq 1$.

Now, using the power law gives

$$\begin{aligned}\mathscr{T} &= 2\pi R^3 \int_0^1 \left[m\left(-\frac{1}{r}\frac{\partial v_\phi}{\partial\theta}\right)^n\right]_{\theta=\pi/2} \left(\frac{r}{R}\right)^2 d\left(\frac{r}{R}\right) \\ &= 2\pi m^0 R^3 \int_0^1 \left[\left(\frac{d\bar{\phi}}{d\bar{\xi}}\frac{W}{\vartheta_0}\right)^n\right]_{\bar{\xi}=0} \left(\frac{r}{R}\right)^2 d\left(\frac{r}{R}\right)\end{aligned} \tag{4.4-91}$$

We now assume that $\bar{\phi}$ will be the same function of $\bar{\xi}$ and $\overline{\text{Na}}$ that ϕ is of ξ and Na. Then

$$\mathscr{T} = \frac{2\pi m^0 W^n R^3}{\vartheta_0^n} \int_0^1 \left[1 - \frac{1}{12n}\widetilde{\text{Na}}\left(\frac{r}{R}\right)^2 + \cdots\right]^n \left(\frac{r}{R}\right)^2 d\left(\frac{r}{R}\right) \tag{4.4-92}$$

Use of the binomial expansion and integration then finally gives

$$\mathscr{T} = \frac{2\pi m^0 W^n R^3}{3\vartheta_0^n}\left(1 - \tfrac{1}{20}\widetilde{\text{Na}} + \cdots\right) \tag{4.4-93}$$

This shows how viscous heating reduces the torque below that which one would expect for a power-law fluid with parameters m and n independent of temperature (i.e., $A = 0$ and $B = 0$ in Eqs. 4.4-4 and 5). Viscous heating can be a serious problem in interpreting cone-and-plate viscometer data (see §10.2).

§4.5 OTHER EMPIRICAL NON-NEWTONIAN VISCOSITY FUNCTIONS FOR USE IN THE GENERALIZED NEWTONIAN FLUID MODEL

In §4.1 we gave two empirical functions $\eta(\dot{\gamma})$: the 5-constant Carreau–Yasuda function, which can fit rather accurately many experimental viscosity curves over a wide range of shear rate; and the 2-constant power-law function, which can fit only a linear region of a log η vs. log $\dot{\gamma}$ plot if such exists. Because of the ease with which analytical solutions can be obtained for the power-law function, we used it extensively in §§4.2–4.4. In the engineering literature many other empirical functions have been used, and in this section we introduce a few of them and illustrate their use.

In Table 4.5-1 we give a half dozen expressions for $\eta(\dot{\gamma})$. Each of these models has a characteristic time. The three-parameter ($\eta_0, \dot{\gamma}_0, n$) *truncated power law* is a discontinuous function that has a zero-shear-rate viscosity up to a critical shear rate $\dot{\gamma}_0$, and a power-law region above $\dot{\gamma}_0$. The two-parameter (λ_0, τ_0) *Eyring model* was derived from a molecular theory of liquids,[1] and it has been improved upon by inclusion of additional empirical constants. The three-parameter ($\eta_0, \tau_{1/2}, \alpha$) *Ellis model* is an example of a model that gives η in terms of stress rather than velocity gradients; for very high shear stress the model exhibits a power-law region, so that the Ellis-model parameters and the power-law parameters are related by $m = \eta_0^{1/\alpha}\tau_{1/2}^{1-(1/\alpha)}$ and $n = 1/\alpha$. Finally the two-parameter (μ_0, τ_0) *Bingham model* is an illustration of a constitutive equation for a "viscoplastic" material,[2] that is, one with a yield stress, τ_0, below which there is no flow. The viscoplastic models are particularly useful for describing liquids with large amounts of suspended solids.

Because of the importance of plane-slit flow and cylindrical-tube flow in engineering applications, we have summarized the results for these two flows for five empirical $\eta(\dot{\gamma})$-functions in Table 4.5-2. These results are obtained by following the procedure in Example 4.2-1. In the next two examples we illustrate some other aspects of solving flow problems with generalized Newtonian fluids. Example 4.5-1 shows how to cope with the yield stress in

[1] See, for example, R. B. Bird, W. E. Stewart, and E. N. Lightfoot, *Transport Phenomena*, Wiley, New York (1960), pp. 26–29.

[2] For a literature survey on viscoplastic models (including the Bingham, Casson, and Herschel–Bulkley equations), experimental tests, and solution of flow problems, see R. B. Bird, G. C. Dai, and B. J. Yarusso, *Revs. Chem. Eng.*, **1**, 1–70 (1982). For plastic viscoelastic constitutive equations see J. L. White, *J. Non-Newtonian Fluid Mech.*, **8**, 195–202 (1981).

TABLE 4.5-1

Empirical η Functions for Use in the Generalized Newtonian Fluid Model[a]

Model	Equation	Time Constant	Comments
Spriggs' truncated power law[b]	$\eta = \eta_0 \quad (\dot{\gamma} \leq \dot{\gamma}_0) \quad$ (A) $\eta = \eta_0(\dot{\gamma}/\dot{\gamma}_0)^{n-1} \quad (\dot{\gamma} \geq \dot{\gamma}_0) \quad$ (B)	$1/\dot{\gamma}_0$	$\dot{\gamma}_0$ is the value of $\dot{\gamma}$ at which "shear thinning" begins
Eyring[c] ($\eta_1 = 0, \alpha = 1$); Powell–Eyring[d] ($\alpha = 1$); Sutterby[e] ($\eta_1 = 0$)	$\eta = \lambda_0 \tau_0 \left(\dfrac{\operatorname{arcsinh} \lambda_0 \dot{\gamma}}{\lambda_0 \dot{\gamma}}\right)^{\alpha} + \eta_1 \quad$ (C)	λ_0	The Eyring equation was the first $\eta(\dot{\gamma})$ expression obtained by a molecular theory
Ellis[f]	$\dfrac{\eta_0}{\eta} = 1 + \left(\dfrac{\tau}{\tau_{1/2}}\right)^{\alpha - 1} \quad$ (D)	$\eta_0/\tau_{1/2}$	$\tau_{1/2}$ is the value of $\tau = \sqrt{(\boldsymbol{\tau}:\boldsymbol{\tau})/2}$ at which $\eta = \eta_0/2$
Bingham[g]	$\eta = \infty \quad (\tau \leq \tau_0) \quad$ (E) $\eta = \mu_0 + \dfrac{\tau_0}{\dot{\gamma}} \quad (\tau \geq \tau_0) \quad$ (F)	μ_0/τ_0	τ_0 is the "yield stress" and $\tau = \sqrt{(\boldsymbol{\tau}:\boldsymbol{\tau})/2}$; Oldroyd[h] has done the most extensive theoretical work for this model

[a] In addition to the Carreau–Yasuda model (Eq. 4.1-9) and the power-law model (Eq. 4.1-10).
[b] T. W. Spriggs University of Wisconsin-Madison, unpublished (1965).
[c] J. F. Kincaid, H. Eyring, and A. E. Stearn, *Chem. Rev.*, **28**, 301–365 (1941); F. H. Ree, T. Ree, and H. Eyring, *Ind. Eng. Chem.*, **50**, 1036–1040 (1958).
[d] R. E. Powell and H. Eyring, *Nature*, **154**, 427–428 (1944); D. L. Salt, N. W. Ryan, and E. B. Christiansen, *J. Colloid Sci.*, **6**, 146–154 (1951). A temperature-dependent Powell–Eyring model has been used by E. B. Christiansen and S. J. Kelsey, *Chem. Eng. Sci.*, **28**, 1099–1113 (1973).
[e] J. L. Sutterby, *Trans. Soc. Rheol.*, **9**, 227–241 (1965); *AIChE J.*, **12**, 63–68 (1966).
[f] S. B. Ellis [see M. Reiner, *Deformation, Strain, and Flow*, Interscience, New York (1960), p. 246]. S. Matsuhisa and R. B. Bird, *AIChE J.*, **11**, 588–595 (1965).
[g] E. C. Bingham, *Fluidity and Plasticity*, McGraw-Hill, New York (1922), pp. 215–218; *U.S. Bur. Stand. Bull.*, **13**, 309–353 (1916). See also W. Prager, *Introduction to Mechanics of Continua*, Ginn, Boston (1961), Chapt. VII.
[h] J. G. Oldroyd, *Proc. Camb. Phil. Soc.*, **43**, 100–105, 383–395, 396–405, 521–532 (1947); **44**, 200–213, 214–228 (1948); **45**, 595–611 (1949); **47**, 410–418 (1951).

the Bingham model; Example 4.5-2 demonstrates how to adapt a slit-flow result to an axial annular flow problem in an approximate procedure, and it further illustrates how quantitative limits can be placed on the range of validity of the power-law model.

EXAMPLE 4.5-1 Tangential Flow of a Bingham Fluid in an Annulus

Determine the velocity distribution for the flow of a Bingham fluid in a coaxial annular region bounded by cylinders of radii κR and R (with $\kappa < 1$). The inner cylinder is kept fixed and the outer cylinder has an angular velocity W. As a result the fluid flows tangentially in the annular region. Obtain the relation between the angular velocity of the outer cylinder and the torque exerted on the outer cylinder.

TABLE 4.5-2

Flow Rates through Slits and Tubes

Model (Constants)	Volume Rate of Flow Q in Thin Slit (Width W, Thickness $2B$, Length L) $\tau_{xz}\vert_{x=B} \equiv \tau_B = (\mathscr{P}_0 - \mathscr{P}_L)B/L$		Volume Rate of Flow Q in Circular Tube (Radius R, Length L) $\tau_{rz}\vert_{r=R} \equiv \tau_R = (\mathscr{P}_0 - \mathscr{P}_L)R/2L$	
Power law (m, n)	$\dfrac{2WB^2}{(1/n)+2}\left(\dfrac{\tau_B}{m}\right)^{1/n}$	(A)	$\dfrac{\pi R^3}{(1/n)+3}\left(\dfrac{\tau_R}{m}\right)^{1/n}$	(F)
Ellis (η_0, $\tau_{1/2}$, α)	$\dfrac{2WB^2\tau_B}{3\eta_0}\left[1+\dfrac{3}{\alpha+2}\left(\dfrac{\tau_B}{\tau_{1/2}}\right)^{\alpha-1}\right]$	(B)[a]	$\dfrac{\pi R^3\tau_R}{4\eta_0}\left[1+\dfrac{4}{\alpha+3}\left(\dfrac{\tau_R}{\tau_{1/2}}\right)^{\alpha-1}\right]$	(G)[a]
Truncated power law (η_0, $\dot{\gamma}_0$, n)	$\dfrac{2WB^2\tau_B}{[(1/n)+2]\eta_0}\left[\left(\dfrac{\eta_0\dot{\gamma}_0}{\tau_B}\right)^{1-(1/n)}+\dfrac{1}{3}\left(\dfrac{1}{n}-1\right)\left(\dfrac{\eta_0\dot{\gamma}_0}{\tau_B}\right)^3\right]$ (when $\tau_B \gg \eta_0\dot{\gamma}_0$)	(C)	$\dfrac{\pi R^3\tau_R}{[(1/n)+3]\eta_0}\left[\left(\dfrac{\eta_0\dot{\gamma}_0}{\tau_R}\right)^{1-(1/n)}+\dfrac{1}{4}\left(\dfrac{1}{n}-1\right)\left(\dfrac{\eta_0\dot{\gamma}_0}{\tau_R}\right)^4\right]$ (when $\tau_R \gg \eta_0\dot{\gamma}_0$)	(H)
Eyring (τ_0, λ_0)	$\dfrac{2WB^2}{\lambda_0}\left(\dfrac{\tau_0}{\tau_B}\right)^2\left[\dfrac{\tau_B}{\tau_0}\cosh\dfrac{\tau_B}{\tau_0}-\sinh\dfrac{\tau_B}{\tau_0}\right]$	(D)	$\dfrac{2\pi R^3}{\lambda_0}\left(\dfrac{\tau_0}{\tau_R}\right)^3\left[\left\{\dfrac{1}{2}\left(\dfrac{\tau_R}{\tau_0}\right)^2+1\right\}\cosh\dfrac{\tau_R}{\tau_0}-\dfrac{\tau_R}{\tau_0}\sinh\dfrac{\tau_R}{\tau_0}-1\right]$	(I)
Bingham (μ_0, τ_0)	$\dfrac{2WB^2\tau_B}{3\mu_0}\left[1-\dfrac{3}{2}\left(\dfrac{\tau_0}{\tau_B}\right)+\dfrac{1}{2}\left(\dfrac{\tau_0}{\tau_B}\right)^3\right]$ (when $\tau_B \geq \tau_0$)	(E)	$\dfrac{\pi R^3\tau_R}{4\mu_0}\left[1-\dfrac{4}{3}\left(\dfrac{\tau_0}{\tau_R}\right)+\dfrac{1}{3}\left(\dfrac{\tau_0}{\tau_R}\right)^4\right]$ (when $\tau_R \geq \tau_0$)	(J)[b]

[a] See Problem 4B.9.

[b] Eq. J is called the *Buckingham-Reiner equation*; see E. Buckingham, *Proc. ASTM*, **117**, 1154–1161 (1921), and R. B. Bird, W. E. Stewart, and E. N. Lightfoot, *Transport Phenomena*, Wiley, New York (1960), pp. 48–50.

SOLUTION For this system $v_r = v_z = 0$ and $v_\theta = v_\theta(r)$. Hence the only nonvanishing components of $\boldsymbol{\tau}$ are $\tau_{r\theta}$ and $\tau_{\theta r}$, and the θ-equation of motion for steady state is in cylindrical coordinates:

$$0 = \frac{1}{r^2}\frac{d}{dr}(r^2\tau_{r\theta}) \tag{4.5-1}$$

This may be integrated to give

$$\tau_{r\theta} = \frac{C_1}{r^2} \tag{4.5-2}$$

If the torque at the outer cylinder is known to be $\mathscr{T}$, then

$$\mathscr{T} = -\tau_{r\theta}|_{r=R} \cdot 2\pi RL \cdot R \tag{4.5-3}$$

where the minus sign is chosen because the θ-momentum flux is in the r-direction. Hence, $C_1 = -\mathscr{T}/2\pi L$ and

$$\tau_{r\theta} = -\frac{\mathscr{T}}{2\pi L r^2} \tag{4.5-4}$$

This result, which is valid for any kind of fluid, can also be obtained by recognizing that angular momentum must be transmitted undiminished from the outer to the inner cylinder.

For the Bingham model, the analytical expression to be used depends on the value of $\tau = \sqrt{(\boldsymbol{\tau}:\boldsymbol{\tau})/2} = \sqrt{(\Sigma_i \Sigma_j \tau_{ij}^2)/2}$. Because $\tau_{r\theta}$ and $\tau_{\theta r}$ are the only nonvanishing components of $\boldsymbol{\tau}$, we have

$$\tau = \sqrt{(\boldsymbol{\tau}:\boldsymbol{\tau})/2} = \sqrt{\tau_{r\theta}^2} = -\tau_{r\theta} \tag{4.5-5}$$

Hence we use Eq. F of Table 4.5-1 when $\tau \geq \tau_0$ (i.e., when the critical shear stress is exceeded) and Eq. E when $\tau \leq \tau_0$. From Eq. 4.5-4, we find that we can define a quantity r_0, which is the value of r for which $\tau = \tau_0$

$$r_0 = \left(\frac{\mathscr{T}}{2\pi\tau_0 L}\right)^{1/2} \tag{4.5-6}$$

We can then discuss three situations:

a. If $r_0 \leq \kappa R$, then there will be no fluid motion at all.
b. If $\kappa R < r_0 \leq R$, then there will be viscous flow in the region $\kappa R < r < r_0$ and "plug flow" for $r_0 \leq r \leq R$.
c. If $r_0 \geq R$, then there is flow throughout.

Next, we write Eq. F of Table 4.5-1 for the system at hand. Inasmuch as

$$\dot{\gamma} = \sqrt{(\dot{\boldsymbol{\gamma}}:\dot{\boldsymbol{\gamma}})/2} = r\frac{d}{dr}\left(\frac{v_\theta}{r}\right) \tag{4.5-7}$$

we get from Eq. 4.1-4

$$\tau_{r\theta} = -\left\{\mu_0 + \frac{\tau_0}{r\dfrac{d}{dr}\left(\dfrac{v_\theta}{r}\right)}\right\} r\frac{d}{dr}\left(\frac{v_\theta}{r}\right) = -\tau_0 - \mu_0 r\frac{d}{dr}\left(\frac{v_\theta}{r}\right) \tag{4.5-8}$$

Since v_θ/r does not decrease with increasing r, we see that $\tau_{r\theta}$ is negative as expected.

By substituting Eq. 4.5-8 into Eq. 4.5-4 and integrating, we get for case (**b**)

$$\frac{v_\theta}{r} = W + \frac{\mathscr{T}}{4\pi L\mu_0 r_0^2}\left[1 - \left(\frac{r_0}{r}\right)^2\right] - \frac{\tau_0}{\mu_0}\ln\frac{r}{r_0} \qquad (\text{for } \kappa R \leq r \leq r_0) \tag{4.5-9}$$

$$\frac{v_\theta}{r} = W \qquad (\text{for } r_0 \leq r \leq R) \tag{4.5-10}$$

in which the boundary condition $v_\theta/r = W$ at $r = r_0$ has been used. For case (**c**), we use the boundary condition that $v_\theta/r = W$ at $r = R$ and get

$$\frac{v_\theta}{r} = W + \frac{\mathscr{T}}{4\pi L\mu_0 R^2}\left[1 - \left(\frac{R}{r}\right)^2\right] - \frac{\tau_0}{\mu_0}\ln\frac{r}{R}, \qquad (\text{for } \kappa R \leq r \leq R) \tag{4.5-11}$$

Finally, if in this last equation we set $v_\theta = 0$ at $r = \kappa R$, we can solve for W to get

$$(\text{case (c) only}) \qquad W = \frac{\mathscr{T}}{4\pi L\mu_0 R^2}\left(\frac{1}{\kappa^2} - 1\right) + \frac{\tau_0}{\mu_0}\ln\kappa \tag{4.5-12}$$

This relation between W and $\mathscr{T}$ (known as the *Reiner–Riwlin equation*[3]) gives a means for determining μ_0 and τ_0 from a concentric cylinder apparatus.

EXAMPLE 4.5-2 An Approximate Solution for Axial Annular Flow to Account for the Zero-Shear-Rate Region of the Non-Newtonian Viscosity[4]

At the end of Example 4.2-4 it was pointed out that for slow flow rates the power-law fluid solution for the axial flow in an annulus can be expected to be in error because the power law does not describe the $\eta(\dot{\gamma})$-function faithfully for small shear rates. In this example we pursue this point quantitatively and give a comparison with experimental data. We use the Ellis model in this analysis.

Since a completely analytical solution for the Ellis model is not possible, use the following approximate procedure: **a.** Write the expression for volume rate of flow Q in an annulus for a Newtonian fluid as the expression for Q for plane slit flow multiplied by a "curvature correction factor;" **b.** Multiply the plane slit flow expression for Q of an Ellis fluid by the same curvature correction factor as was found in (**a**).

SOLUTION (**a**) Curvature Correction Factor for Newtonian Fluid
For the Newtonian fluid the volume rate of flow in a coaxial annulus is given by[5]

$$Q_{\text{ann.}} = \frac{\pi(\mathscr{P}_0 - \mathscr{P}_L)R^4}{8\mu L}\left[(1 - \kappa^4) - \frac{(1 - \kappa^2)^2}{\ln(1/\kappa)}\right] \tag{4.5-13}$$

where $\mathscr{P}_0 - \mathscr{P}_L$ is the modified pressure drop over a length L of conduit. For the flow in a slit of width W, thickness $2B$, and length L ($B \ll W \ll L$) the volume rate of flow is easily found to be (see Eq. 1.3-5)

$$Q_{\text{slit}} = \frac{2}{3}\frac{(\mathscr{P}_0 - \mathscr{P}_L)WB^3}{\mu L} \tag{4.5-14}$$

[3] M. Reiner and R. Riwlin, *Kolloid-Z.*, **43**, 1–5 (1927).
[4] E. Ashare, R. B. Bird, and J. A. Lescarboura, *AIChE J.*, **11**, 910–916 (1965).
[5] See, for example, R. B. Bird, W. E. Stewart, and E. N. Lightfoot, *Transport Phenomena*, Wiley, New York (1960), pp. 51–54.

If in Eq. 4.5-14 we replace W by $2\pi R$ and $2B$ by $R(1-\kappa)$, we get

$$Q_{\text{ann.}}\,(\text{approximate}) = \frac{\pi(\mathscr{P}_0 - \mathscr{P}_L)R^4}{6\mu L}(1-\kappa)^3 \tag{4.5-15}$$

In both Eqs. 4.5-13 and 4.5-15 we set $1-\kappa$ equal to the small quantity ε. Then we divide one by the other to get the Newtonian curvature correction factor C_N

$$\begin{aligned} C_N &= Q_{\text{ann.}}/Q_{\text{ann.}}\,(\text{approximate}) \\ &= \frac{3}{4\varepsilon^3}\left\{[1-(1-\varepsilon)^4] - \frac{\varepsilon^2(2-\varepsilon)^2}{\varepsilon + \frac{1}{2}\varepsilon^2 + \frac{1}{3}\varepsilon^3 + \cdots}\right\} \\ &= 1 - \frac{1}{2}\varepsilon + \frac{1}{60}\varepsilon^2 + \frac{1}{120}\varepsilon^3 + \cdots \end{aligned} \tag{4.5-16}$$

(b) Approximate Formula for Axial Annular Flow of Ellis Fluid

We now take the slit flow formula for an Ellis fluid from Eq. B of Table 4.5-2 and replace the slit width

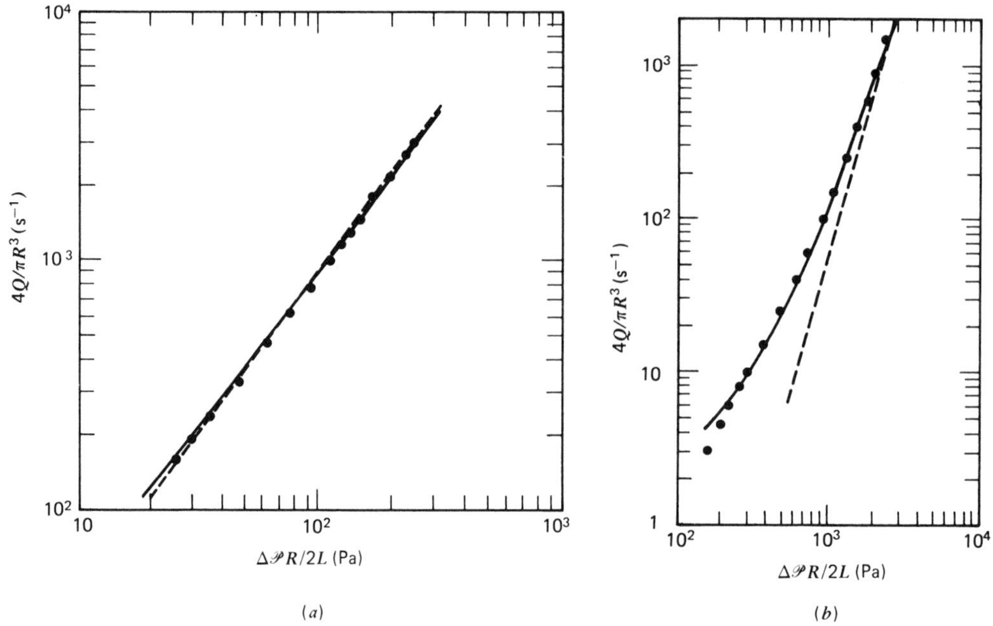

FIGURE 4.5-1. Illustration of the inadequacy of the power-law model in slow annular flow, and the improved description afforded by an $\eta(\dot{\gamma})$ expression that contains a zero-shear-rate viscosity (the Ellis model of Table 4.5-1). [E. Ashare, R. B. Bird, and J. A. Lescarboura, *AIChE J.*, **11**, 910–916 (1965).] (*a*) Volume flow rate of 1.0% Natrosol-G in annulus ($\kappa = 0.5043$). Solid curve: approximate equation obtained for the Ellis model (Eq. 4.5-17) ($\alpha = 1.5$, $\eta_0 = 0.031$ Pa·s, $\tau_{1/2} = 26$ Pa). Dashed curve: exact solution for the power-law model (Eq. C of Table 4.2-1) ($n = 0.763$, $m = 0.893$ Pa·s^n). Data points are taken from D. W. McEachern, Ph.D. Thesis, University of Wisconsin, Madison (1963). (*b*) Volume flow rate of 3.5% CMC-70-medium in annulus ($\kappa = 0.624$). Solid curve: Ellis model (Eq. 4.5-17) approximate ($\alpha = 3.0$, $\eta_0 = 2.27$ Pa·s, $\tau_{1/2} = 152$ Pa). Dashed curve: power-law (Eq. C of Table 4.2-1) exact ($n = 0.280$, $m = 60.6$ Pa·s^n). Data points are taken from A. G. Fredrickson, Ph.D. Thesis, University of Wisconsin, Madison (1959).

W by $2\pi R$ and the slit thickness $2B$ by $R(1-\kappa)$ as we did for the Newtonian fluid in (**a**); we then multiply by C_N:

$$Q_{\text{ann.}} \doteq \frac{\pi(\mathscr{P}_0 - \mathscr{P}_L)R^4}{6\eta_0 L}(1-\kappa)^3\left[1 + \frac{3}{\alpha+2}\left(\frac{(\mathscr{P}_0 - \mathscr{P}_L)R(1-\kappa)}{2\tau_{1/2}L}\right)^{\alpha-1}\right]\cdot C_N \qquad (4.5\text{-}17)$$

This is an approximate result because the Newtonian curvature correction C_N has been applied. It is believed that this approximate procedure is valid[6] for κ greater than about 0.6; this conclusion was drawn by comparing the exact solution for a power-law fluid[7] with the slit approximation for a power-law fluid multiplied by C_N.

In Fig. 4.5-1 we show annular flow experimental data for aqueous solutions of two different polymers.[7] It can be seen that Eq. 4.5-17 describes the data quite well. The power-law and Ellis models are seen to agree for $[3/(\alpha+2)][\Delta\mathscr{P}R(1-\kappa)/2\tau_{1/2}L]^{\alpha-1} \gg 1$. Otherwise the Ellis model is appreciably better than the power-law model. This inadequacy has also been pointed out by other authors.[8]

§4.6 LIMITATIONS OF GENERALIZED NEWTONIAN FLUID MODELS AND RECOMMENDATIONS FOR THEIR USE

In §4.1 the generalized Newtonian fluid was introduced as a simple empirical modification of the Newtonian fluid that accounts for the experimentally observed dependence of viscosity on shear rate. Because of the empirical nature of the model, the question then arises as to the limits of validity of the formulas developed in this chapter for solving flow problems. To determine these limits we can get some help from the subject of continuum mechanics and some from dimensional analysis, but the ultimate test is, of course, the comparison with experimental data.

In Example 9.6-3, in the chapter on continuum mechanics, it is shown that in steady-state shear flow an extremely wide class of viscoelastic constitutive equations simplifies to the Criminale-Ericksen-Filbey (CEF) equation. The first term of the CEF equation for $\boldsymbol{\tau}$ is just $-\eta(\dot{\gamma})\dot{\boldsymbol{\gamma}}$; the other two terms, containing Ψ_1 and Ψ_2, describe the elastic effects associated with the normal stresses. The latter turn out to be unimportant in getting volume-flow-rate versus pressure drop relations in flow in straight, uniform conduits and also in getting torque versus angular velocity relations in tangential or helical annular flows for annuli. In other words for steady-state shear flows the formulas obtained in this chapter for Q versus $\Delta\mathscr{P}$ and $\mathscr{T}$ versus W are exactly the same as would be obtained for much wider classes of nonlinear viscoelastic constitutive equations. Comparisons with experimental data have furthermore substantiated the conclusion that the methods given in this chapter for Q versus $\Delta\mathscr{P}$ and $\mathscr{T}$ versus W in steady shear flows are virtually exact, the only limitation resulting from the possible inadequacy of the particular form chosen for $\eta(\dot{\gamma})$; an example of the latter is the failure of the power-law model for the slow axial flow in an annulus (see Example 4.5-2).

In some problems in this chapter we have used the generalized Newtonian fluid to solve flow problems that are not shear flows and/or steady-state flows. In Example 4.2-7 (squeezing flow between parallel disks) we assumed that the flow is locally a *shear* flow—though it is not; in Example 4.2-6 (pulsatile flow in a circular tube) we assumed that the velocity distribution is instantaneously the same as that for a *steady* flow—though it is

[6] S. Matsuhisa and R. B. Bird, *AIChE J.*, **11**, 588–595 (1965).

[7] A. G. Fredrickson and R. B. Bird, *Ind. Eng. Chem.*, **50**, 347–352 (1958); *erratum*: *Ind. Eng. Chem. Fundam.*, **3**, 383 (1964); R. W. Hanks and K. M. Larsen, *Ind. Eng. Chem. Fundam.*, **18**, 33–35 (1979).

[8] R. D. Vaughn and P. D. Bergman, *Ind. Eng. Chem. Proc. Des. Dev.*, **55**, 44–47 (1966).

not. When is it permissible to make such assumptions? It is permissible if "elastic effects" are not important, and as pointed out in §2.8, this means that the Deborah number[1] in the flow problem must be smaller than some critical value.

We are now faced with the problem of specifying the time constant of the fluid. There are several possibilities: (a) use the time constant in the $\eta(\dot{\gamma})$ expression (i.e., the λ in the Carreau–Yasuda equation or the time constants in Table 4.5-1); (b) use a time constant constructed from the m and n of the power-law viscosity and the m' and n' of the power-law expression for Ψ_1 (i.e., the λ in Eq. 4.2-68); or (c) use the time constant provided by linear viscoelasticity (i.e., the time constant from Table 5.3-1). It can be argued that choice (a) is inappropriate, since elastic effects may be unrelated to the viscosity measured in steady shear flow; on the other hand, experimental data, many molecular theories, and many nonlinear viscoelastic constitutive equations do show a connection between the time constant determined from (a) and that determined from (c). We emphasize, however, that the time constants obtained by methods (a), (b), and (c) will not be exactly the same.

Once a choice has been made for the fluid time constant λ, then a comparison of experimental data with the generalized Newtonian fluid flow solution enables one to establish the critical Deborah number above which the generalized Newtonian fluid is inapplicable.[2] For example, in Fig. 4.2-6 it is seen that above $\text{De} \equiv \lambda/t_{1/2}$ of about three elastic effects are important, and the Scott equation, based on the power-law model, is not valid; note further that for very fast squeezing flows the Scott equation may be in error by more than a factor of 10 for the data shown. This should serve as a warning that the appearance of elastic effects can severely limit the usefulness of the generalized Newtonian fluid.

In Example 4.2-6 (pulsatile tube flow) a suitable characteristic time for the flow process would be $1/\omega$, where ω is the frequency of the sinusoidally varying pressure gradient. Then an appropriate Deborah number would be $\text{De} = \lambda\omega$. As long as De is small, we can expect that the flow-enhancement formula in Eq. 4.2-58 will be trustworthy for fluids whose viscosity is well described by the power-law model, but that when De is much greater than the critical value it will not be able to describe the data.

We have emphasized that the critical Deborah number, which specifies the limit on the range of validity of the generalized Newtonian fluid, has to be determined experimentally. As we shall see in Chapters 6–8, it is also possible to establish the critical Deborah number by using nonlinear viscoelastic models, provided that the model describes experimental material functions well and that the flow problem can be solved for the flow system concerned. In either case experimental data are needed.

In this chapter we have given a sampling of analytical solutions to relatively simple flow problems for the power-law generalized Newtonian fluid. Further examples of analytical solutions can be found in older books on rheology[3–5] and in engineering journals. In some of these references generalized Newtonian fluids are referred to as "purely viscous fluids" or "inelastic fluids." Many other analytical and numerical solutions to

[1] M. Reiner, *Phys. Today*, **17**, 62 (Jan. 1964); R. R. Huilgol, *Continuum Mechanics of Viscoelastic Liquids*, Wiley, New York (1975), pp. 226–228.

[2] The role of the ratio of characteristic times was illustrated by R. B. Bird, *Can. J. Chem. Eng.*, **43**, 161–168 (1965), who documented the importance of this ratio by means of experimental data on flow through porous media, around spheres, in a converging section, and in the oscillating manometer.

[3] M. Reiner, *Deformation, Strain, and Flow*, Wiley-Interscience, New York (1960), Chaps. VIII, XVI, XVII, XVIII, XIX.

[4] W. L. Wilkinson, *Non-Newtonian Fluids: Fluid Mechanics, Mixing and Heat Transfer*, Pergamon Press, New York (1960).

[5] S. Middleman, *The Flow of High Polymers*, Wiley-Interscience, New York (1968), Chapt. 3.

generalized Newtonian flow problems along with applications can be found in books on polymer processing.[6–10]

The engineering and applied science literature abounds with examples of the application of generalized Newtonian fluid models to problems, which are far from being steady-state shear flow problems: boundary-layer flows,[11] complex flows at the entry to tubes,[12] instability problems,[13] turbulent flows,[14] agitation and mixing,[15] fiber spinning,[16] mold filling,[17] films and coatings,[18] flow with pressure-dependent viscosity,[19] and flow around spheres.[20] In some instances there have been no (or very limited) comparisons with experimental data, so that the range of applications of the theoretical calculations is not known. Until such time as these types of problems have been studied with nonlinear viscoelastic models (see Chapters 7 and 8), generalized Newtonian models will continue to be used.

In conclusion, then, we recommend that the generalized Newtonian model be used for calculating Q vs. $\Delta\mathscr{P}$ for steady flow in rectilinear conduits of constant cross section, and for getting $\mathscr{T}$ vs. W for tangential and helical flow in annuli. Such calculations can be expected to be reliable, provided that the $\eta(\dot{\gamma})$ function describes the viscosity data well. The power-law viscosity function can be useful in many problems, but caution must be exercised if the flow field puts considerable emphasis on the zero-shear-rate region of the $\eta(\dot{\gamma})$ curve, where the power law is inadequate. The generalized Newtonian model can also be expected to give good results in systems that deviate somewhat from steady-state shear flows (i.e., they may have some time dependence or some nonshearing velocity gradients), provided that the Deborah number is much less than $\mathrm{De}_{\mathrm{crit}}$; generally this critical Deborah number

[6] J. M. McKelvey, *Polymer Processing*, Wiley, New York (1962).

[7] S. Middleman, *Fundamentals of Polymer Processing*, McGraw-Hill, New York (1977).

[8] Z. Tadmor and C. G. Gogos, *Principles of Polymer Processing*, Wiley, New York (1979).

[9] R. I. Tanner, *Engineering Rheology*, Oxford University Press (1985).

[10] J. R. A. Pearson, *Mechanics of Polymer Processing*, Elsevier Applied Science, London (1984).

[11] J. G. Oldroyd, *Proc. Camb. Phil. Soc.*, **43**, 383–395 (1947); W. R. Schowalter, *AIChE J.*, **6**, 24–28 (1960), *erratum*: **10**, 597 (1964); A. Acrivos, M. J. Shah, and E. E. Petersen, *ibid.*, **6**, 312–317 (1960); J. L. White and A. B. Metzner, *ibid.*, **11**, 324–330 (1965); S. Y. Lee and W. F. Ames, *ibid.*, **12**, 700–708 (1966); V. G. Fox, L. E. Erickson, and L. T. Fan, *ibid.*, **15**, 327–333 (1969); W. R. Schowalter, *Mechanics of Non-Newtonian Fluids*, Pergamon Press, New York (1978), pp. 205 *et seq.*

[12] S. S. Chen, L. T. Fan, C. L. Hwang, *AIChE J.*, **16**, 293–299 (1970); N. D. Sylvester and S. L. Rosen, *ibid.*, **16**, 964–972 (1970).

[13] J. P. Tordella in F. R. Eirich, ed., *Rheology*, Vol. 5, Academic Press, New York (1969) Chapt. 2; J. R. A. Pearson, *Ann. Rev. Fluid Mech.*, **8**, 163–181 (1976).

[14] A. B. Metzner and J. C. Reed, *AIChE J.*, **1**, 434–440 (1955); R. G. Shaver and E. W. Merrill, *ibid.*, **5**, 181–188 (1959); D. W. Dodge and A. B. Metzner, *ibid.*, **5**, 189–204 (1959); D. W. McEachern, *ibid.*, **15**, 885–889 (1969); H. Rubin and C. Elata, *ibid.*, **17**, 990–996 (1971).

[15] A. B. Metzner and R. E. Otto, *AIChE J.*, **3**, 3–10 (1957); E. S. Godleski and J. C. Smith, *ibid.*, **8**, 617–620 (1961); D. W. Hubbard and F. F. Calvetti, *ibid.*, **18**, 663–665 (1972).

[16] J. R. A. Pearson and Y. T. Shah, *Ind. Eng. Chem. Fundam.*, **13**, 134–138 (1974); J. R. A. Pearson, Y. T. Shah, and R. D. Mhaskar, *ibid.*, **15**, 31–37 (1976); C. J. S. Petrie, *Elongational Flows*, Pitman, London (1979).

[17] J. L. Berger and C. G. Gogos, *Polym. Eng. Sci.*, **13**, 102–112 (1973); P.-C. Wu, C. F. Huang, and C. G. Gogos, *ibid.*, **14**, 223–230 (1974).

[18] C. Gutfinger and J. A. Tallmadge, *AIChE J.*, **11**, 403–413 (1965); J. A. Tallmadge, *ibid.*, **16**, 925–930 (1970); S. A. Jenekhe, *Ind. Eng. Chem. Fundam.*, **23**, 425–432, 432–436 (1984); S. A. Jenekhe and S. B. Schuldt, *Chem. Eng. Commun.*, **33**, 135–147 (1985).

[19] N. Galili, R. Takserman-Krozer, and Z. Rigbi, *Rheol. Acta*, **14**, 550–567 (1975).

[20] A. N. Beris, J. A. Tsamopoulos, R. C. Armstrong, and R. A. Brown, *J. Fluid Mech.*, **158**, 219–244 (1985), (Bingham fluid); K. Adachi, N. Yoshioka, and K. Sakai, *J. Non-Newtonian Fluid Mech.*, **3**, 107–125 (1977/1978), ("extended Williamson model," i.e., Eq. 4.1-9 with $n = 1 - a$); M. B. Bush and N. Phan-Thien, *J. Non-Newtonian Fluid Mech.*, **16**, 303–313 (1984), (Carreau model, i.e., Eq. 4.1-9 with $a = 2$); Y. I. Cho and J. P. Hartnett, *J. Non-Newtonian Fluid Mech.*, **12**, 243–247 (1983), (power-law model).

must be determined experimentally. In other situations—elongational flows, flows rapidly changing in time, flows with several nonzero velocity components, etc.—the generalized Newtonian fluid should not be used, except as a last resort; good experimental measurements are definitely preferred.

PROBLEMS

4A.1 Pipe Flow of a Polyisoprene Solution

A 13.5% (by weight) solution of polyisoprene in isopentane has the following power-law parameters at 323 K: $m = 5 \times 10^3$ Pa · s^n, and $n = 0.2$. It is being pumped through a pipe that is 10.2 m long and has an internal diameter of 1.3 cm; the flow is known to be laminar. It is desired to build another pipe with a length of 20.4 m with the same volume rate of flow and the same pressure drop. What should its radius be?

Answer: 1 cm

4A.2 Flow of a Carboxymethylcellulose Soluton in an Annulus

A 3.5% aqueous carboxymethylcellulose solution has Ellis-model parameters as follows: $\eta_0 = 2.27$ Pa · s, $\tau_{1/2} = 152$ Pa, and $\alpha = 3.0$. Find the flow rate Q in a horizontal annulus with $R = 2$ cm and $\kappa = 0.624$ when $\Delta pR/2L = 500$ Pa. Use the approximate relation given in Eq. 4.5-17.

Answer: 154 cm^3/s

4A.3 Pipe Flow of a Polypropylene Melt (Ellis Model)

a. A commerical sample of polypropylene at 403 K has the following Ellis-model parameters:[1] $\eta_0 = 1.24 \times 10^4$ Pa · s, $\tau_{1/2} = 6.90 \times 10^3$ Pa, $\alpha = 2.82$. Calculate the volume rate of flow Q for a horizontal pipe, 5 cm internal diameter and length 17 m, when the pressure drop is 4.5×10^6 Pa.

b. What would be the power-law constants for the fluid? Repeat the calculations in (**a**) for the power law. Why are the results in (**a**) and (**b**) different? Under what conditions would they be expected to be the same?

Answers: **a.** 3.87×10^{-6} m^3/s
b. 5.92×10^{-7} m^3/s

4A.4 Pumping a Nylon Melt through a Capillary (Truncated Power Law)[2]

A nylon-6 melt is to be pumped through a capillary of radius 5×10^{-4} m and length 2.5×10^{-2} m with pressure drop of 2.67×10^7 Pa. What volume rate of flow is expected (neglect viscous dissipation as well as entrance and exit effects)? The viscosity data have been fit with a truncated power law, with the following constants: $\eta_0 = 2.50 \times 10^2$ Pa · s, $\dot{\gamma}_0 = 800$ s^{-1}, and $n = 0.5$.

Answer: 1.18×10^{-7} m^3/s

4A.5 Squeezing Flow of Polymer Solutions (Power Law)

One of the polymer solutions studied by Leider[3] in his squeezing-flow experiments was a 0.5% solution of polyacrylamide in glycerin for which the power-law parameters had been measured and

[1] N. Yamada, N. Kishi, and H. Iizuka, *Kōbunshi Kagaku*, **22**, 513–519 (1965).
[2] This problem was taken from W. J. Beek and D. B. Holmes, *Fysisch Technologische Aspecten van de Polymeerverwerking*, Laboratorium voor Technische Natuurkunde, Technische Hogeschool, Delft (1965).
[3] P. J. Leider, *Ind. Eng. Chem. Fundam.*, **13**, 342–346 (1974).

found to be: $m = 25$ Pa·s^n, $n = 1/3$; in addition the time constant λ, defined in Eq. 4.2-68, was found to be 129 s.

This fluid was placed between two circular disks with radius 2.54 cm, with an initial disk separation of $2h_0 = 4.98 \times 10^{-3}$ cm. A 4.07 kg mass was placed on the upper disk and it was found that the time for half the material to be squeezed out was 540 s. Compare this value with the value of $t_{1/2}$ computed from the Scott equation. Verify that the experiment described is indeed within the range of applicability of the Scott equation.

4A.6 Volume Rate of Flow in a Tube-Disk-Slit Assembly

a. The top curve in Fig. 4.1-2 is for a polystyrene melt at 453 K. Using the graph obtain the m and n of the power law appropriate for the power-law region of the data; show all work. Then check the values by obtaining the power-law constants from the Carreau-model parameters given in the figure caption.

b. Derive an equation (based on the power-law model) from which the volume rate of flow Q can be obtained for the apparatus shown in Fig. 4A.6, when the pressure drop $p_0 - p_3$ is given. Use the results of Table 4.2-1 and Problem 4B.3 for deriving the expression for Q.

c. List the assumptions that are implied in the result obtained in (b).

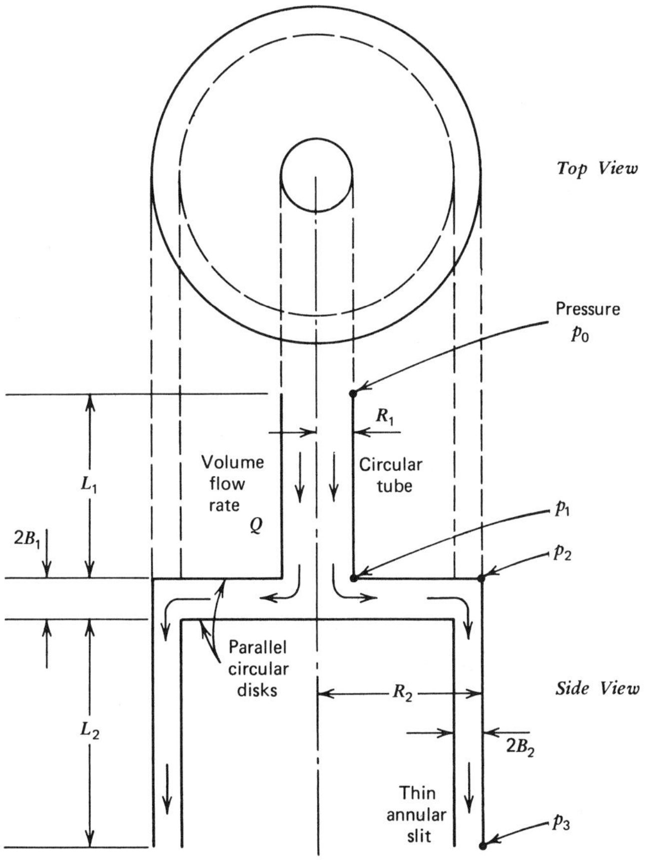

FIGURE 4A.6. Flow downward through a tube, radially between two parallel surfaces and then downward through an annular slit.

d. Calculate the volume flow rate in cm^3/s for a pressure drop $p_0 - p_3 = 2.1 \times 10^7$ Pa (which is approximately 210 atm) for the polystyrene melt cited in (a). The dimensions of the device are

$$R_1 = 0.5 \text{ cm} \qquad L_1 = 11 \text{ cm} \qquad B_1 = 0.2 \text{ cm}$$
$$R_2 = 7.0 \text{ cm} \qquad L_2 = 9 \text{ cm} \qquad B_2 = 0.15 \text{ cm}$$

The fluid density is 1 g/cm^3.

Answer: 75 cm^3/s (approximately)

4A.7 Tube Flow of a Polystyrene Melt

Viscosity data for a polystyrene melt at 453 K are shown in Fig. 4.1-2. It is desired to pump this melt at 453 K through a tube 8 cm long and with 0.5 cm radius. If a pressure drop of 5.6×10^6 Pa is applied, what is the volume flow rate if the fluid is assumed to be described by:

a. the power-law model

b. the truncated power-law model (cf. Eq. H in Table 4.5-2).

Justify the agreement (or lack of agreement) by comparing the wall shear stress in the tube τ_R with the critical shear stress $\eta_0 \dot{\gamma}_0$ where shear thinning begins in the truncated power law.

Answer: **a.** 36.1 cm^3/s

4A.8 Heat Transfer in Mold Filling of a Polystyrene Melt

A molten polystyrene is to be formed into a rectangular shape by injecting it into a cold mold that is 30 cm long, 7.5 cm wide, and 0.2 cm thick.[4] The fluid flow is parallel to the longest dimension of the mold. The time required to fill the mold is 2.5 s with a pressure gradient of 1.33×10^6 Pa/m. The temperature of the molten polymer entering the mold is 250°C and the mold walls are at 50°C. For polystyrene,[5] $k = 1.295 \times 10^{-1}$ J/K·m·s and $\alpha = 7.05 \times 10^{-8}$ m^2/s. Estimate the local Nusselt number to describe the heat transfer between the polymer and the mold.

4A.9 Pipe Flow of a Polymer Solution

A 1% aqueous solution of polyethylene oxide at 333 K has these power-law parameters: $m = 0.50$ Pa·s^n and $n = 0.60$. The solution is being pumped between two tanks, with Tank 1 at pressure p_1, and Tank 2 at pressure p_2; the pipe carrying the solution has an internal diameter of 27 cm, and a length of 14.7 m.

It has been decided to replace the single pipe by a pair of pipes of the same length, but with smaller diameter. What diameter should these pipes have in order for the volume flow rate to be the same as that in the single pipe?

4B.1 Axial Annular Flow with Inner Cylinder Moving Axially (Power-Law)

In the system shown in Fig. 4B.1 it is desired to find the velocity distribution in the annular region between A and B for a polymeric fluid. Use the power-law viscosity equation.

a. Combine the equation of motion and the power-law equation to obtain a differential equation for $v_z(r)$, taking the z-axis to be coincident with the axis of the moving rod.

b. Integrate the equation for $v_z(r)$ to get the velocity distribution in terms of two constants C_1, C_2.

c. What boundary conditions do you use to determine C_1, C_2?

[4] J. L. S. Wales, J. van Leeuwen, and R. van der Vijgh, *Polym. Eng. Sci.*, **12**, 359–363 (1972).

[5] J. Brandrup and E. H. Immergut, *Polymer Handbook*, Wiley Interscience, New York (1966).

d. Use those boundary conditions to find the result

$$\frac{v_z(r)}{V} = \frac{\xi^{1-s} - 1}{\kappa^{1-s} - 1} \tag{4B.1-1}$$

where $\xi = r/R$ and $s = 1/n$.

e. Show how this simplifies, for the Newtonian fluid, to

$$\frac{v_z(r)}{V} = \frac{\ln \xi}{\ln \kappa} \tag{4B.1-2}$$

f. What is the force acting on the wire in the region between A and B?
g. Is this a "shear flow" according to the definition in §3.7?
h. Find the volume flow rate Q of the fluid through the annular region.
i. Show how the expression for Q in (h) simplifies for a Newtonian fluid.

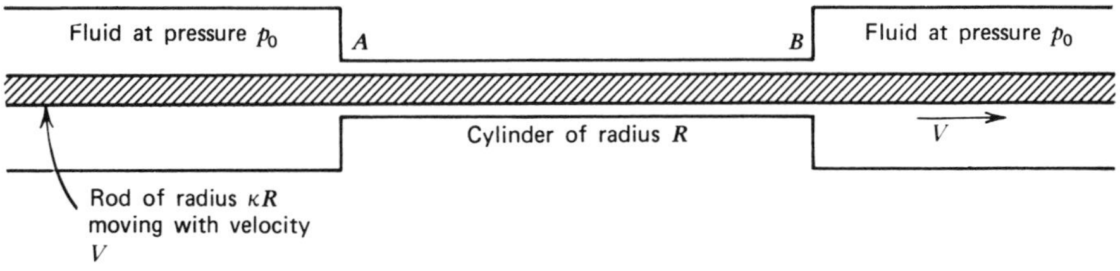

FIGURE 4B.1. Axial annular flow with inner cylinder moving axially.

4B.2 Tangential Annular Flow for a Polymer (Power Law)

A polymeric liquid is being sheared in the annular region between two cylindrical surfaces of length L and of radii κR and R (with $\kappa < 1$). The inner cylinder is rotating with an angular velocity W, and the outer cylinder is fixed.

a. Show that for this system the θ-component of the equation of motion simplifies to

$$0 = -\frac{1}{r^2}\frac{d}{dr}(r^2\tau_{r\theta}) \tag{4B.2-1}$$

and the $r\theta$-component of $\boldsymbol{\tau}$ for the power-law is

$$\tau_{r\theta} = m\left[-r\frac{d}{dr}\left(\frac{v_\theta}{r}\right)\right]^n \tag{4B.2-2}$$

b. Show that the equations in (a) along with the appropriate boundary conditions can be solved to give for the velocity distribution

$$\frac{v_\theta}{Wr} = \frac{(R/r)^{2/n} - 1}{(1/\kappa)^{2/n} - 1} \tag{4B.2-3}$$

c. Obtain an expression for the torque required to turn the inner cylinder.

Answer: **c.** $\mathscr{T} = 2\pi(\kappa R)^2 mL\left(\frac{2W/n}{1 - \kappa^{2/n}}\right)^n$

4B.3 Radial Flow between Parallel Disks (Power Law)[6]

a. First solve the problem of the power-law fluid flow through a slit of width W and thickness $2B$, and verify the result given in Eq. A of Table 4.2-1.

b. Then solve the problem of slow steady-state radial flow between two fixed parallel disks (see Fig. 4B.3), which are separated by a distance $2B$. Let the inner and outer radii of the disks be R_1 and R_2. Obtain an expression for the volume rate of flow by applying the result of (a) locally in the region between the two disks:

$$Q = \frac{4\pi B^2}{(1/n) + 2}\left(\frac{(p_1 - p_2)B(1 - n)}{m(R_2^{1-n} - R_1^{1-n})}\right)^{1/n} \tag{4B.3-1}$$

This equation[7] has been used by several groups of investigators[8–10] to describe radial flow data for some moderately viscoelastic polymer solutions.

c. Is this a "shear flow" according to the definition in §3.7?

d. How does Eq. 4B.3-1 simplify for Newtonian fluids?

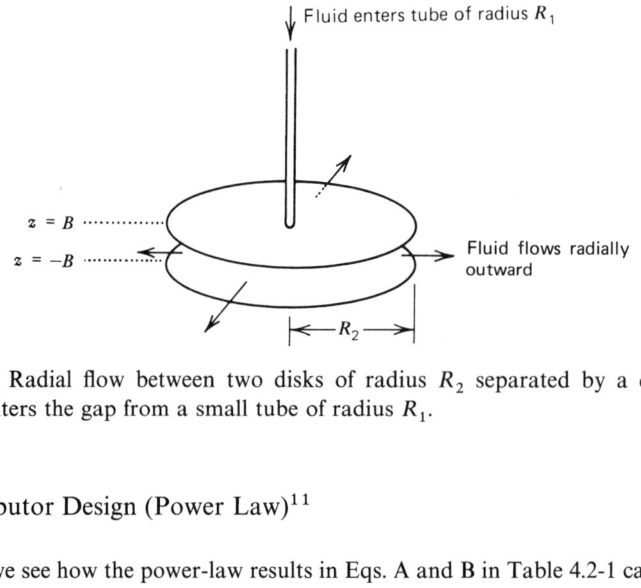

FIGURE 4B.3. Radial flow between two disks of radius R_2 separated by a distance $2B$, with $B \ll R_2$. Fluid enters the gap from a small tube of radius R_1.

4B.4 Distributor Design (Power Law)[11]

In this problem we see how the power-law results in Eqs. A and B in Table 4.2-1 can be combined to get an approximate solution to a rather complicated flow problem. In Fig. 4B.4 we see a distributor system that is supposed to deliver a polymeric fluid with a uniform efflux velocity as the fluid leaves the thin slit between parallel plates. The circular pipe and parallel plates are horizontal, and the fluid leaves the pipe and enters the parallel-plate system through a thin slit in the tube. The flow between the plates is in the x-direction because of the presence of vertical dividers that also serve to maintain the spacing between the plates.

[6] This problem has been solved numerically for the Carreau model by A. Co, *Ind. Eng. Chem. Fundam.*, **20**, 340–348 (1981), for the third-order fluid by A. Co and R. B. Bird, *Appl. Sci. Res.*, **33**, 385–404 (1977), and for a nonlinear differential model by A. Co and W. E. Stewart, *AIChE J.*, **28**, 644–655 (1982).

[7] T. Y. Na and A. G. Hansen, *Int. J. Nonlinear Mech.*, **2**, 261–273 (1967).

[8] B. R. Laurencena and M. C. Williams, *Trans. Soc. Rheol.*, **18**, 331–355 (1974).

[9] H. Amadou, P. M. Adler, and J.-M. Piau, *J. Non-Newtonian Fluid Mech.*, **4**, 229–237 (1978).

[10] L. R. Schmidt, *J. Rheol.*, **22**, 571–588 (1978).

[11] This problem was suggested by a conversation with F. H. Ancker, Union Carbide Co., Bound Brook, NJ, who solved a somewhat similar problem by the method used here.

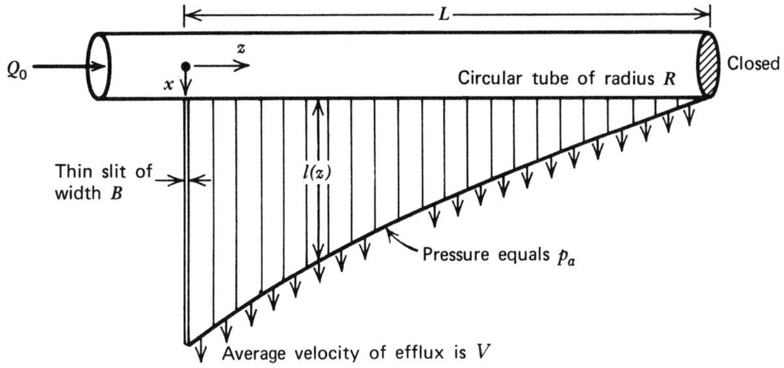

FIGURE 4B.4. Tube of radius R, with slit of width B attached, functioning as a distributor.

a. First consider the flow in the cylindrical tube. Verify that if Q_0 is the volume rate of flow entering the tube, then the volume rate of flow across any plane of constant z will be

$$Q(z) = Q_0\left(1 - \frac{z}{L}\right) \tag{4B.4-1}$$

Then apply the power-law result for a circular tube (Table 4.2-1) locally to obtain the following differential equation for the pressure:

$$Q_0\left(1 - \frac{z}{L}\right) = \frac{\pi R^3}{(1/n) + 3}\left(\frac{R}{2m}\right)^{1/n}\left(-\frac{dp}{dz}\right)^{1/n} \tag{4B.4-2}$$

Integrate the differential equation for $p(z)$ from an arbitrary distance z to the end of the tube $z = L$, and obtain

$$p - p_a = \frac{2mL}{R}\left(\frac{(1/n) + 3}{\pi R^3}\right)^n \frac{Q_0^n}{n+1}(1 - \zeta)^{n+1} \tag{4B.4-3}$$

where $\zeta = z/L$.

b. Next adapt the slit flow result in Table 4.2-1 to the problem at hand to obtain

$$V = \frac{B/2}{(1/n) + 2}\left[\frac{(p - p_a)B}{2ml(\zeta)}\right]^{1/n} \tag{4B.4-4}$$

c. From (a) and (b) find the equation of the curve $l(\zeta)$ that will ensure uniform efflux:

$$l(\zeta) = \frac{BL}{R(n+1)}\left(\frac{B}{2\pi R^3}\frac{(1/n) + 3}{(1/n) + 2}\right)^n\left(\frac{Q_0}{V}\right)^n(1 - \zeta)^{n+1} \tag{4B.4-5}$$

Note that m does not appear in the result!

4B.5 Flow in a Tapered Tube (Power Law)

Derive the expression for the relation between Q and $p_0 - p_L$ given in Eq. 4.2-10 for a tapered tube. Do this by writing Eq. 4.2-9 over a differential segment of the tube of length dz. Show how this adaptation of Eq. 4.2-9 can be integrated from $z = 0$ to $z = L$ to give Eq. 4.2-10. What limitations have to be placed on the result?

4B.6 Adjacent Flow of Two Immiscible Fluids (Power Law)[12]

a. Two immiscible polymeric fluids are contained in the space between two parallel planes, located at $x = 0$ and $x = B$, the upper one being in motion in the z-direction with a constant speed V, as shown in Fig. 4B.6. Both fluids are describable in terms of power-law parameters, m_I and n_I for fluid "I", and m_{II} and n_{II} for fluid "II." The interface is located at $x = \lambda B$, and there is no pressure gradient in the z-direction. Fluid I is more dense than Fluid II, and it is to be presumed that the flow rate is such that the interface between the fluids remains virtually planar. Find the velocity distribution.

b. (Lengthy!) Repeat the problem in part (a) when the upper plate is fixed, but there is a pressure gradient in the z-direction.

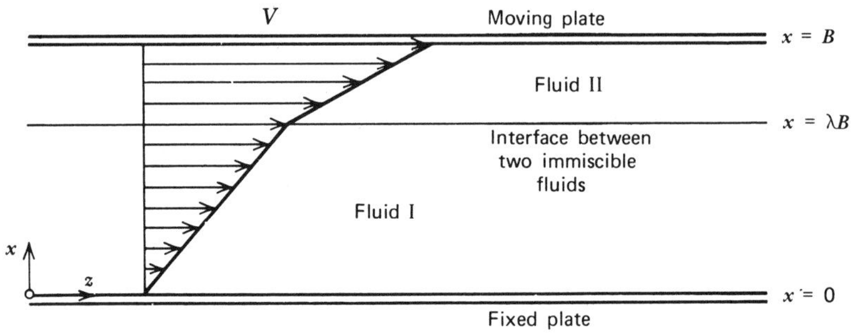

FIGURE 4B.6. Two immiscible power-law fluids being sheared in the region between two parallel plates, one of which is in motion.

4B.7 Flow in Circular Tubes and Slits (Any Generalized Newtonian Fluid)

In Example 4.2-1 it is shown how to obtain the volume rate of flow for a power-law fluid through circular tubes and the corresponding flow-rate formula for plane slits is given in Eq. A of Table 4.2-1. In Table 4.5-2 results are given for the flow of other kinds of "model" fluids in tubes and thin slits. Actually it is not particularly difficult to obtain formal expressions for the flow rates in tubes and slits for any kind of shear-rate dependence of the viscosity.

a. For circular tubes show that the general expression for volume rate of flow can be integrated by parts twice to obtain

$$Q = 2\pi \int_0^R v_z r\,dr$$

$$= \frac{\pi R^3 \dot{\gamma}_R}{3} - \frac{\pi}{3}\int_0^{\dot{\gamma}_R} r^3\,d\dot{\gamma} \qquad (4B.7\text{-}1)$$

in which $\dot{\gamma} = -dv_z/dr$ and $\dot{\gamma}_R = -dv_z/dr|_{r=R}$. Use of Eq. 4.2-4 and the definition of η in Eq. 4.1-4 enables us to make a change of variable:

$$r = (R/\tau_R)\eta\dot{\gamma} \qquad (4B.7\text{-}2)$$

[12] For discussions of the problems of multilayered flows with applications to manufacture of laminated sheets and bicomponent filaments and to mixing processes, see W. J. Schrenk, *Plast. Eng.*, **30**, 65–68 (1974); W. J. Schrenk and T. Alfrey, Jr., *SPE J.*, **29**, 38–42, 43–47 (1973); T. Alfrey, Jr., *Polym. Eng. Sci.*, **9**, 400–404 (1969); W. J. Schrenk and T. Alfrey, Jr., *Polym. Eng. Sci.*, **9**, 393–399 (1969); A. E. Everage, Jr., *Trans. Soc. Rheol.*, **17**, 629–646 (1973). Most of the calculations in these papers were made for Newtonian fluids; in the last reference a variational method was employed.

Show that this leads to

$$Q = \frac{\pi R^3 \dot{\gamma}_R}{3} - \frac{\pi}{3}\left(\frac{R}{\tau_R}\right)^3 \int_0^{\dot{\gamma}_R} \eta^3 \dot{\gamma}^3 \, d\dot{\gamma} \tag{4B.7-3}$$

where $\dot{\gamma}_R$ is obtained by solving the equation

$$\tau_R = \eta(\dot{\gamma}_R)\dot{\gamma}_R \tag{4B.7-4}$$

Hence, if one knows $\eta(\dot{\gamma})$ from a cone-and-plate viscometer, for example, then $\dot{\gamma}_R$ can be obtained from Eq. 4B.7-4 in terms of τ_R. When that result is inserted into Eq. 4B.7-3, the integration can be performed to get Q as a function of τ_R. The integrals will have to be performed numerically. Verify Eqs. 4B.7-3 and 4 by inserting the power law and showing that Eq. 4.2-9 is recovered.

b. Repeat the procedure outlined in (a), this time for the flow through a plane slit of width W and thickness $2B$. Show that the results corresponding to Eqs. 4B.7-3 and 4 are

$$Q = WB^2\dot{\gamma}_B - W\left(\frac{B}{\tau_B}\right)^2 \int_0^{\dot{\gamma}_B} \eta^2 \dot{\gamma}^2 \, d\dot{\gamma} \tag{4B.7-5}$$

$$\tau_B = \eta(\dot{\gamma}_B)\dot{\gamma}_B \tag{4B.7-6}$$

Check these results by inserting the power law and showing that Eq. A in Table 4.5-2 is recovered.

4B.8 Third Invariant of Rate-of-Strain Tensor for Shear Flow

Just after Eq. 4.1-7 it was stated that for shear flows III is identically zero. Verify this comment by evaluating III for the flow $v_x = \dot{\gamma}(t)y$, $v_y = 0$, $v_z = 0$.

4B.9 Flow in Circular Tubes and Slits (Ellis model)[13]

a. Verify Eqs. B and G of Table 4.5-2.

b. Verify that the Ellis and power-law model parameters are related by

$$m = \eta_0^{1/\alpha}\tau_{1/2}^{1-(1/\alpha)}, \quad n = 1/\alpha \tag{4B.9-1}$$

c. How can the Ellis model be used to determine the limits of validity of the power-law model?

4B.10 Flow of Blood in Tubes (Casson Equation)

To describe the flow of pigment-oil suspensions, Casson[14] proposed the following rheological equations for shearing flows:

$$\sqrt{\pm\tau_{yx}} = \sqrt{\tau_0} + \sqrt{\mu_0}\sqrt{\mp dv_x/dy} \qquad \text{for} \quad \tau_{yx} > \tau_0 \tag{4B.10-1}$$

$$\dot{\gamma}_{yx} = 0 \qquad \text{for} \quad \tau_{yx} < \tau_0 \tag{4B.10-2}$$

in which μ_0 and τ_0 are constants; the upper signs in Eq. 4B.10-1 are valid for positive momentum flux, and the lower signs for negative momentum flux.

[13] S. Matsuhisa and R. B. Bird, *AIChE J.*, **11**, 588–595 (1965).

[14] N. Casson in C. C. Mill, ed., *Rheology of Disperse Systems*, Pergamon Press, New York (1959), p. 84.

The Casson equation has proven useful for the description of the flow of blood on both glass and fibrin surfaces.[15,16] Some information is available on the relation between the model constants and the chemical composition of the blood.[17,18] In addition the Casson equation has been modified for suspensions of spherical particles in polymer solutions.[19]

a. What is the physical significance of the two constants in the equation? To what extent is this equation similar to the Bingham fluid?

b. Show that for the Casson equation the volume flow rate through a circular tube is given by[16]

$$Q = \frac{\pi R^4(\mathscr{P}_0 - \mathscr{P}_L)}{8\mu_0 L}\left(1 - \frac{16}{7}\sqrt{\xi} + \frac{4}{3}\xi - \frac{1}{21}\xi^4\right) \tag{4B.10-3}$$

in which the dimensionless parameter ξ is

$$\xi = \frac{\tau_0}{(\mathscr{P}_0 - \mathscr{P}_L)R/2L} \tag{4B.10-4}$$

4B.11 Flow of a Bingham Fluid in a Tube[20]

a. Show that when a Bingham fluid flows in a tube it will have a "plug flow" region in the center with the radius $r_0 = 2\tau_0 L/(\mathscr{P}_0 - \mathscr{P}_L)$.

b. Verify that the velocity distribution is:

$$v_z^> = \frac{(\mathscr{P}_0 - \mathscr{P}_L)R^2}{4\mu_0 L}\left[1 - \left(\frac{r}{R}\right)^2\right] - \frac{\tau_0 R}{\mu_0}\left[1 - \left(\frac{r}{R}\right)\right] \qquad r \geq r_0 \tag{4B.11-1}$$

$$v_z^< = \frac{(\mathscr{P}_0 - \mathscr{P}_L)R^2}{4\mu_0 L}\left(1 - \frac{r_0}{R}\right)^2 \qquad r \leq r_0 \tag{4B.11-2}$$

c. Obtain the expression for the volume flow rate given in Eq. J of Table 4.5-2.

4B.12 Temperature and Shear-Rate Dependence of Viscosity

Derive the *Bestul-Belcher equation*[21] relating various measurable derivatives:

$$\left(\frac{\partial \eta}{\partial T}\right)_{\dot{\gamma}} \Big/ \left(\frac{\partial \eta}{\partial T}\right)_{\tau_{yx}} = 1 + \left(\frac{\partial \ln \eta}{\partial \ln \dot{\gamma}}\right)_T \tag{4B.12-1}$$

in a steady shear flow $v_x = \dot{\gamma}y$, $v_y = 0$, $v_z = 0$. This equation has been used by Meissner.[22]

[15] V. L. Shah, *Adv. Transport Phenomena*, **1**, 1–57 (1980).
[16] S. Oka in A. L. Copley, ed., *Proceedings of the Fourth International Congress on Rheology*, Wiley, New York (1965), Part 4, pp. 81–92.
[17] E. W. Merrill, W. G. Margetts, G. R. Cokelet, and E. R. Gilliland in A. L. Copley, ed., *Proceedings of the Fourth International Congress on Rheology*, Wiley, New York (1965), Part 4, pp. 135–143.
[18] E. N. Lightfoot, *Transport Phenomena and Living Systems*, Wiley, New York (1974), pp. 35–38, 430, 438, 440; M. M. Lih, *Transport Phenomena in Medicine and Biology*, Wiley, New York (1975), pp. 378–386.
[19] S. Onogi, T. Matsumoto, and Y. Warashina, *Trans. Soc. Rheol.*, **17**, 175–190 (1973).
[20] For other Bingham flow problems see R. B. Bird, G. C. Dai, and B. J. Yarusso, *Rev. Chem. Eng.*, **1**, 1–70 (1982).
[21] A. B. Bestul and H. V. Belcher, *J. Appl. Phys.*, **24**, 696–702 (1953).
[22] J. Meissner in E. H. Lee, ed., *Proceedings of the Fourth International Congress on Rheology*, Wiley, New York (1965), Part 3, pp. 437–453.

4B.13 Tube Flow for the Truncated Power-Law Fluid

One way to estimate the importance of the zero-shear-rate viscosity on the volume flow rate in a tube is by using the simple truncated power-law model given in Table 4.5-1, Eqs. A and B. Let r_0 denote the radial position where $\dot{\gamma} = \dot{\gamma}_0$, that is where the viscosity changes from the zero-shear-rate value to the power-law function.

a. How must r_0 be restricted in order to use the truncated power-law model? Show that this corresponds to $(\eta_0\dot{\gamma}_0/\tau_R) < 1$ where τ_R is the wall shear stress in the tube.

b. Solve for the velocity field for $r \le r_0$ and $r \ge r_0$. Match the results at $r = r_0$ in order to eliminate the last integration constant.

c. Show that the volume flow rate is given by Eq. H in Table 4.5-2.

d. Show that the result in part (c) reduces to the power-law result given in Eq. B of Table 4.2-1 or Eq. F of Table 4.5-2 when $(\eta_0\dot{\gamma}_0/\tau_R) \ll 1$. Note that the power-law consistency index m is related to the truncated power-law constants by $m = \eta_0\dot{\gamma}_0^{1-n}$. Where does this come from?

4B.14 Variational Functional for Rod

Consider the functional in Eq. 4.3-16 and parallel the development in Eq. 4.3-3 *et seq.* In particular let $\overline{\Theta}(\xi) = \Theta(\xi) + \varepsilon\eta(\xi)$, where $\Theta(\xi)$ is the solution to Eqs. 4.3-13, 14 and 15, and $\eta(\xi)$ is an arbitrary function with $\eta(1) = 0$.

a. Show that when terms through second order in ε are retained one finds:

$$J\{\overline{\Theta}(\xi)\} = J\{\Theta(\xi)\} + \varepsilon^2 J\{\eta(\xi)\} \tag{4B.14-1}$$

b. Use the above to show that $\Theta(\xi)$ makes J a minimum.

4B.15 Development of Design Equation for Manifold of a "Coat-Hanger" Die (Power Law)[23]

Plastic sheeting can be made by extruding the molten polymer through a "coat-hanger" die made up of an entry tube, two manifolds, and a slit (see Fig. 4B.15). The manifold is a tube of circular cross section, whose radius $\bar{R}$ varies in the direction $\bar{z}$ of the manifold axis. Our object is to design the manifold (i.e., find $\bar{R}(\bar{z})$) so that the flow through the slit will be uniform; that is the volume flow rate of the slit must not vary in the x-direction.

The slit has a total width $2W$ and has a thickness $2B$. The volume rate of flow into the entry tube is $2Q_0$, with half of the fluid going into the left manifold and half into the right manifold.

a. Consider a width Δx of the slit. What is the volume flow rate through this portion of the slit?

b. Make a mass balance over a length $\Delta\bar{z}$ of the manifold tube and then let $\Delta\bar{z}$ go to zero to get the differential equation:

$$-\frac{d\bar{Q}}{d\bar{z}} = \frac{Q_0}{W}\cos\alpha \tag{4B.15-1}$$

where $\bar{Q}$ is the volume flow rate at $\bar{z}$. Draw a carefully labelled diagram to show how you derive this relation.

c. Let $\bar{p}(\bar{z})$ be the pressure as a function of $\bar{z}$ in the manifold. Let $p(z)$ be the pressure in the slit. Why is $p(-L(x)) \doteq \bar{p}(\bar{z})$? How are x and $\bar{z}$ related? How are W and $\bar{L}$ related?

[23] J. R. A. Pearson, *Trans. J. Plast. Inst.*, **32**, 239 (1964); J. R. A. Pearson, *Mechanics of Polymer Processing*, Elsevier Applied Science, New York (1985), §10.2. See also Z. Tadmor and C. G. Gogos, *Principles of Polymer Processing*, Wiley, New York (1979), pp. 545–551.

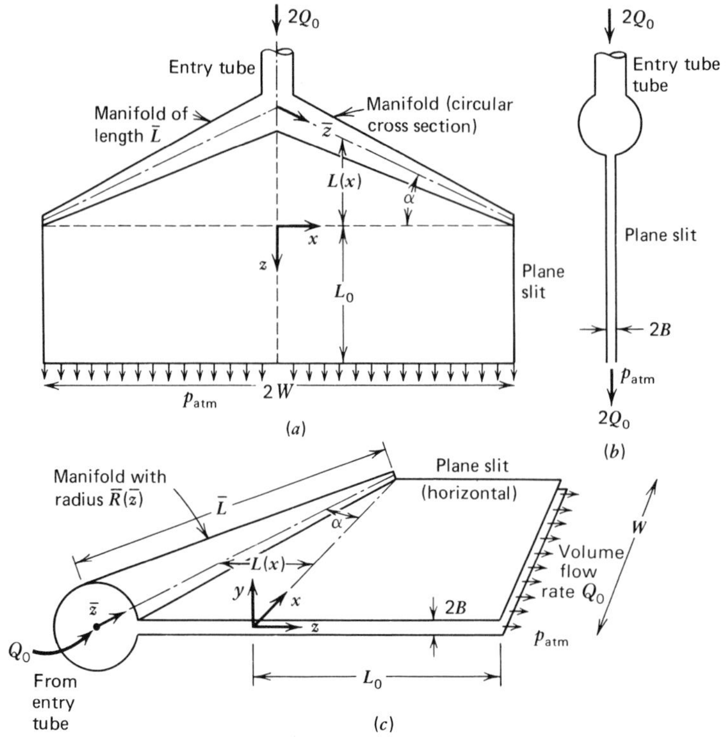

FIGURE 4B.15. The "coat-hanger" die, so named because the entry tube and manifold resemble a coat hanger. (*a*) Top view; (*b*) side view (looking in *x*-direction at a cut through the apparatus in the *yz*-plane); (*c*) one manifold and half of slit.

d. Adapt the power-law-fluid plane-slit-flow formula to a region of width Δx. Then use the results in **(c)** to obtain, finally,

$$-\frac{d\bar{p}}{d\bar{z}} = \left[\frac{Q_0}{W}\frac{[(1/n) + 2]}{2B^2}\right]^n \frac{m}{B} \sin\alpha \tag{4B.15-2}$$

e. Write the power-law analog of the Hagen–Poiseuille formula locally over a segment $d\bar{z}$ of the tube. This should contain $d\bar{p}/d\bar{z}$.

f. Combine the results of **(b)**, **(d)**, and **(e)** to get a differential equation for $\bar{R}$ as a function of $\bar{z}$. This should be of the form

$$\bar{R}^{(1/n)+2}\, d\bar{R}/d\bar{z} = \text{const} \tag{4B.15-3}$$

g. Integrate the differential equation to get $\bar{R}(\bar{z})$, the shape of the manifold. Take $\bar{R}$ to be zero at $\bar{z} = \bar{L}$ (just to make the problem a little simpler). Find ultimately that the desired design formula is

$$[\bar{R}(\bar{z})]^{3n+1} = \left[\frac{2}{\pi}\frac{[(1/n) + 3]}{[(1/n) + 2]}\right]^n \frac{2B^{2n+1}}{\sin\alpha}(W - x)^n$$

Show that if $n = 1/2$, then $\bar{R} \propto (W - x)^{1/5}$.

4B.16 Tube Flow with Slip at the Wall (Power Law)

Obtain a modification of Eq. B of Table 4.2-1 by using the Navier slip boundary condition at the liquid–solid interface:

$$v_z = \beta \tau_{rz} \quad \text{at} \quad r = R \tag{4B.16-1}$$

in which β is the *slip coefficient*. What is the significance of $\beta = 0$? The boundary condition in Eq. 4B.16-1 may be useful in the description of the "apparent slip" associated with the existence of a non-homogeneous region of fluid near the wall.[24] Instead of using a slip coefficient one can also model the system by postulating a thin film of fluid near the wall with different rheological properties.

4C.1 Flow out of a Tank with an Exit Tube (Power Law)

The tank-and-tube assembly in Fig. 4C.1 is initially filled with an incompressible non-Newtonian fluid of density ρ; the viscosity of the fluid is given by the power-law expression. During the draining

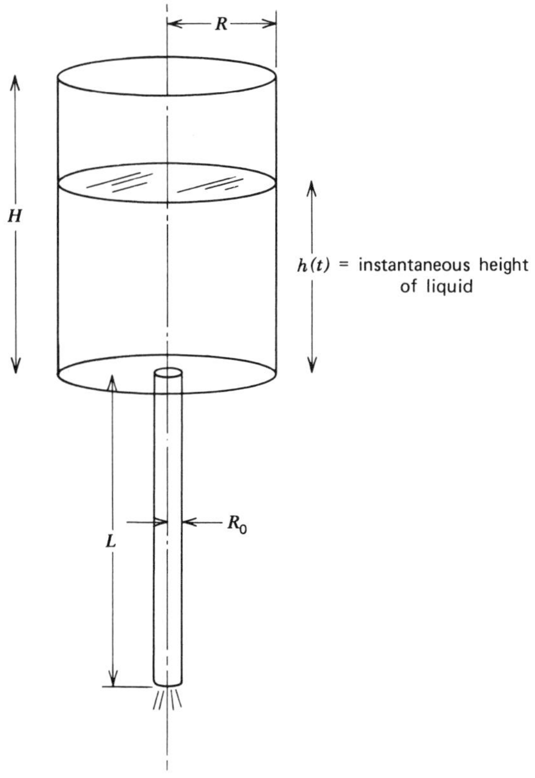

FIGURE 4C.1. Tank with a long tube attached; both the fluid surface and the tube exit are open to the atmosphere.

[24] For some experimental data and further literature references see Y. Cohen and A. B. Metzner, *J. Rheol.*, **29**, 67–102 (1985); see also A. M. Kraynik and W. R. Schowalter, *J. Rheol.*, **25**, 95–114 (1981), and P. J. Carreau, Q. H. Bui, and P. Leroux, *Rheol. Acta*, **18**, 600–608 (1979).

process it is assumed that the flow in the tube is laminar. Show that the time required to drain the tank (but not the pipe), t_{efflux}, is given by

$$t_{\text{efflux}} = \left(\frac{2mL}{\rho g R_0}\right)^{1/n}\left(\frac{1}{n} + 3\right)\frac{R^2}{R_0^3}\left[\frac{(H + L)^{1-(1/n)} - L^{1-(1/n)}}{1 - (1/n)}\right] \tag{4C.1-1}$$

4C.2 Falling Cylinder Viscometer (Power Law)

Rework Problem 1C.2 for a power-law fluid. In order to simplify the development, consider only the situation in which the clearance between the cylinder and tube is extremely small so that curvature effects can be completely neglected.

4C.3 The Rayleigh Problem for a Power-Law Fluid[25]

It is desired to solve the problem given in Problem 1B.6 for a power-law fluid. Use the same dimensionless velocity as before, but define a combined, independent variable appropriate to the power-law fluid

$$\phi_n = \frac{v_x}{V} \tag{4C.3-1}$$

$$r = (n + 1)^{-1} y(\rho/mtV^{n-1})^{1/(n+1)} \tag{4C.3-2}$$

and show that the differential equation for the dimensionless velocity distribution is

$$\phi_n''(-\phi_n')^{n-1} + (n + 1)^n n^{-1} r\phi_n' = 0 \tag{4C.3-3}$$

in which primes denote differentiation with respect to r. Obtain a formal solution to the problem for a pseudoplastic fluid ($n < 1$). Some sample velocity profiles are given in Fig. 4C.3.

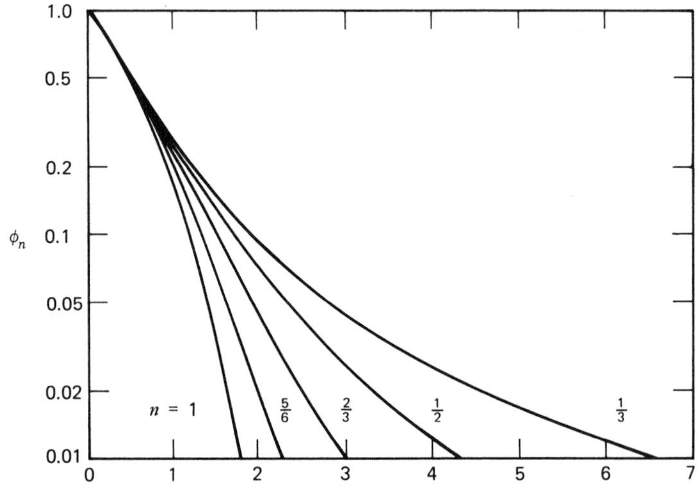

FIGURE 4C.3. Dimensionless velocity profiles for flow of a power-law fluid near a flat surface suddenly set in motion with a constant velocity. [R. B. Bird, *AIChE J.*, **5**, 565–566 (1959).]

[25] R. B. Bird, *AIChE J.*, **5**, 565–566 (1959).

It should be borne in mind that in unsteady-state problems of this type, viscoelastic effects will probably be important, and these are not described by the simple power-law model. See §7.4 for the solution to the Rayleigh problem for a viscoelastic fluid.

$$\text{Answer:} \quad \phi_n = \left[\frac{(1+n)^n(1-n)}{2n}\right]^{-1/(1-n)} \int_r^{\infty} (B_n + r^2)^{-1/(1-n)}\, dr$$

where B_n is a constant determined by $\phi_n = 1$ at $r = 0$.

4C.4 Axial Annular Flow (Bingham Fluid)

Repeat Example 4.2-4 for the Bingham fluid defined in Table 4.5-1. Use the following dimensionless quantities suggested by Fredrickson and Bird:[26]

$$T = \frac{2\tau_{rz}L}{(\mathscr{P}_0 - \mathscr{P}_L)R}; \quad T_0 = \frac{2\tau_0 L}{(\mathscr{P}_0 - \mathscr{P}_L)R} \tag{4C.4-1}$$

$$\phi = \left(\frac{2\mu_0 L}{(\mathscr{P}_0 - \mathscr{P}_L)R^2}\right)v_z; \quad \xi = \frac{r}{R} \tag{4C.4-2}$$

a. Show that the momentum flux distribution and the constitutive equation may be written

$$T = \xi - \beta^2\xi^{-1}; \quad T = \pm T_0 - \frac{d\phi}{d\xi} \tag{4C.4-3}$$

in which the $+$ sign is used when momentum is being transported in the $+r$-direction and the $-$ sign is used when transport is in the $-r$-direction. Show further that the bounds on the plug-flow region β_+ and β_- are given by

$$\pm T_0 = \beta_\pm - \left(\frac{\beta^2}{\beta_\pm}\right) \tag{4C.4-4}$$

where β is a constant of integration.

b. Obtain the velocity distribution appropriate for each of the three regions:

$$\phi_- = -T_0(\xi - \kappa) - \frac{1}{2}(\xi^2 - \kappa^2) + \beta^2 \ln\frac{\xi}{\kappa}, \qquad \kappa \le \xi \le \beta_- \tag{4C.4-5}$$

$$\phi_0 = \phi_-(\beta_-) = \phi_+(\beta_+), \qquad \beta_- \le \xi \le \beta_+ \tag{4C.4-6}$$

$$\phi_+ = -T_0(1 - \xi) + \frac{1}{2}(1 - \xi^2) + \beta^2 \ln \xi, \qquad \beta_+ \le \xi \le 1 \tag{4C.4-7}$$

c. Verify that the determining equation for β_+ is

$$2\beta_+(\beta_+ - T_0)\ln\frac{\beta_+ - T_0}{\beta_+\kappa} - 1 + (T_0 + \kappa)^2 + 2T_0(1 - \beta_+) = 0 \tag{4C.4-8}$$

[26] A Slibar and P. R. Paslay, *Z. Angew. Math. Mech.*, **37**, 441–449 (1957); A. G. Fredrickson and R. B. Bird, *Ind. Eng. Chem.*, **50**, 347–352 (1958); see also B. E. Anshus, *Ind. Eng. Chem. Fundam.*, **13**, 162–164 (1974). Helical flow of a Bingham fluid in an annulus has been solved approximately by P. R. Paslay and A. Slibar, *Petr. Trans. AIME*, **210**, 310–317 (1957).

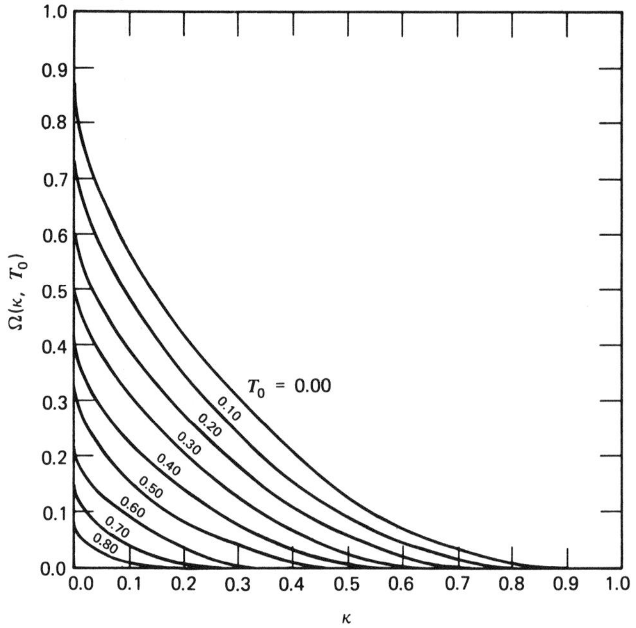

FIGURE 4C.4. The function $\Omega(\kappa, T_0)$ to be used with Eq. 4C.4-9 for computing Bingham flow in an annulus. A numerical table corresponding to this graph was given by Fredrickson and Bird. [Reprinted with permission from A. G. Fredrickson and R. B. Bird, *Ind. Eng. Chem.*, **50**, 347–352 (1958). Copyright by the American Chemical Society.]

d. Show that the volume rate of flow is

$$Q = \frac{\pi R^4(\mathscr{P}_0 - \mathscr{P}_L)}{8\mu_0 L}\Big[(1-\kappa^4) - 2\beta_+(\beta_+ - T_0)(1-\kappa^2) - \frac{4}{3}(1+\kappa^3)T_0 + \frac{1}{3}(2\beta_+ - T_0)^3 T_0\Big] \qquad (4C.4\text{-}9)$$

$$= \frac{\pi R^4(\mathscr{P}_0 - \mathscr{P}_L)}{8\mu_0 L}\,\Omega(\kappa, T_0)$$

In Fig. 4C.4 the function Ω is shown as a function of κ and T_0.

4C.5 An Empiricism for $\eta(II, III)$

a. Show that for the elongational flow $v_z = \dot{\varepsilon}z$, $v_x = -(\dot{\varepsilon}/2)x$, $v_y = -(\dot{\varepsilon}/2)y$, the invariants of Eqs. 4.1-5 to 7 are

$$I = 0; \qquad II = 6\dot{\varepsilon}^2; \quad III = 6\dot{\varepsilon}^3 \qquad (4C.5\text{-}1)$$

b. Now let us require that $\eta(II, III)$ satisfy the following two limiting expressions:

$$\eta(II, III) = m\left(\sqrt{II/2}\right)^{n-1}, \qquad \text{in steady shear flow} \qquad (4C.5\text{-}2)$$

$$\eta(II, III) = \eta_0, \qquad \text{in steady elongational flow} \qquad (4C.5\text{-}3)$$

Are these reasonable requirements? Show that the following function has the desired limits:

$$\eta(II, III) = \eta_0\left(\frac{III}{6(II/6)^{3/2}}\right) + m\left(\sqrt{(II - 6(III/6)^{2/3})/2}\right)^{n-1} \tag{4C.5-4}$$

c. What are the dimensions of $(m/\eta_0)^{1/(n-1)}$?

4D.1 Flow of a Polymeric Fluid through a Square Tube[27,28] (Power Law)

Obtain an approximate expression for the volume rate of flow of a power-law fluid through a horizontal tube of square cross section. Let the square be described by the lines $x = \pm B$, $y = \pm B$, and let the pressure drop over the tube of length L be $p_0 - p_L$. Use a variational approach and take the trial function to be

$$v_z = A\left[1 - \left(\frac{x}{B}\right)^2\right]\left[1 - \left(\frac{y}{B}\right)^2\right] \tag{4D.1-1}$$

where A is the variational parameter.

4D.2 Effect of Axial Conduction on Forced Convection Heat Transfer

In Examples 4.4-1 through 3 the axial conduction term in the energy equation was neglected. In this problem we consider its importance on the temperature profiles. Consider the flow of a polymeric fluid between parallel plates separated by a gap $2B$. Far upstream the fluid has uniform temperature T_0, and the walls are also at T_0; then at $z = 0$, the wall temperature is raised to T_1 and held at that value for all $z > 0$. For simplicity assume that the fluid is incompressible and that its physical properties do not depend on temperature. Further assume that the power-law index $n = 0$ so that the velocity profile is flat with $v_z = V$. This latter assumption is clearly a crude approximation for polymeric liquids; it is a good approximation for slurries, foam , and concentrated suspensions.

a. Solve for the temperature profiles in the regions up and downstream of $z = 0$. Show that

$z > 0$:
$$\Theta = \sum_{n=0}^{\infty} C_n \cos \lambda_n \eta \exp[(\text{Pé} - \sqrt{\text{Pé}^2 + 4\lambda_n^2})\zeta/2] \tag{4D.2-1}$$

$z < 0$:
$$\Theta = 1 + \sum_{n=0}^{\infty} D_n \cos \lambda_n \eta \exp[(\text{Pé} + \sqrt{\text{Pé}^2 + 4\lambda_n^2})\zeta/2] \tag{4D.2-2}$$

[27] For a thorough calculation of this type see R. S. Schechter, *AIChE J.*, **7**, 445–448 (1961). For other noncircular conduit calculations see N. Mitsuishi and Y. Aoyagi, *J. Chem. Eng. Jpn.*, **6**, 402–408 (1973), (eccentric annulus); *Chem. Eng. Sci.*, **24**, 309–319 (1969), (rectangular and isosceles triangular ducts); N. Mitsuishi, Y. Kitayama, and Y. Aoyagi, *Kagaku Kōgaku*, **31**, 570–577 (1967), (rectangular and isosceles triangular ducts); N. Mitsuishi, Y. Aoyagi, and H. Soeda, *Kagaku Kōgaku*, **36**, 182–192 (1972), (concentric annulus). In connection with flow through noncircular tubes it must be kept in mind that $\Psi_2 \neq 0$ is a necessary (but not sufficient) condition for secondary flows to exist [J. G. Oldroyd, *Proc. Roy. Soc.*, **A283**, 115–133 (1965)]; these secondary flows are so weak, however, that they do not appreciably influence the Q vs. $\Delta\mathscr{P}$ relation. For application of variational methods to flow around a sphere see A. J. Ziegenhagen, R. B. Bird, and M. W. Johnson, Jr., *Trans. Soc. Rheol.*, **5**, 47–49 (1961); A. J. Ziegenhagen, *Appl. Sci. Res.*, **A14**, 43–56 (1961); S. W. Hopke and J. C. Slattery, *AIChE J.*, **16**, 224–229 (1970); *ibid.*, **16**, 317–318 (1970).

[28] See also T.-J. Liu, *Ind. Eng. Chem. Fundam.*, **22**, 183–186 (1983), for power-law flow in ducts of arbitrary cross section by means of the Galerkin method.

in which $\Theta = (T_1 - T)/(T_1 - T_0)$ is the dimensionless temperature, $\text{Pé} = \rho \hat{C}_p VB/k$ is the Péclet number, $\eta = y/B$ and $\zeta = z/B$ are dimensionless position variables, and $\lambda_n = (n + (1/2))\pi$ for $n = 0, 1, 2 \ldots$ are eigenvalues for the problem.

b. Apply appropriate matching conditions at $z = 0$ to the two solutions in order to determine the C_n and D_n constants.

c. How do these results simplify in the limit of large Pé? Is this consistent with the treatment given in Example 4.4-2?

4D.3. Variational Principle for Several Dependent Variables in a Three-Dimensional Space

Consider n dependent variables y_j for $j = 1, 2, \ldots n$ defined in a three-dimensional volume V bounded by a surface S. The y_j are functions of the rectangular coordinates x_i for $i = 1, 2, 3$. The surface S is divided into two nonoverlapping regions S_v and S_f. On S_v the y_j satisfy essential boundary conditions

$$y_j = y_{j,c} \tag{4D.3-1}$$

where the $y_{j,c}$ are known functions defined on S_v. On S_f the y_j satisfy natural boundary conditions. Define a variational functional J by

$$J = \int_V F_v(x_i, y_j, y_{j,i})\, dV + \int_{S_f} F_s(x_i, y_j) dS \tag{4D.3-2}$$

where F_v and F_s are known functions of the arguments listed, and $\partial y_j/\partial x_i$ is denoted $y_{j,i}$ for brevity.

Prove the following variational principle: The set of functions y_j that satisfy the essential boundary conditions in Eq. 4D.3-1 and make J stationary are also the solutions to the partial differential equations

$$\frac{\partial F_v}{\partial y_j} - \sum_{i=1}^{3} \frac{\partial}{\partial x_i} \frac{\partial F_v}{\partial y_{j,i}} = 0, \qquad \text{for } j = 1, 2, \ldots n \tag{4D.3-3}$$

in the volume V with the natural boundary conditions

$$\frac{\partial F_s}{\partial y_j} + \sum_{i=1}^{3} n_i \frac{\partial F_v}{\partial y_{j,i}} = 0, \qquad \text{for } j = 1, 2, \ldots n \tag{4D.3-4}$$

and the essential boundary conditions in Eq. 4D.3-1. In Eq. 4D.3-4 the n_i are the rectangular components of the outwardly directed unit normal to the surface S_f.

It is suggested to follow the same outline as in §4.3 part a, and introduce functions $\bar{y}_j(x) = y_j(x) + \varepsilon\eta_j(x)$. Instead of the integration by parts use the Gauss–Ostrogradskii divergence theorem.

4D.4. Variational Principle for Generalized Newtonian Fluids

It is desired to use the general principle derived in Problem 4D.3 to prove the variational principle stated below Eq. 4.3-31. Use rectangular coordinates x_i with velocity components v_i for $i = 1, 2, 3$ and the following outline.

a. Show that

$$\frac{\partial \dot{\gamma}}{\partial v_{j,i}} = \frac{1}{\dot{\gamma}} \dot{\gamma}_{ij} \tag{4D.4-1}$$

where $v_{j,i} = \partial v_j/\partial x_i$.

b. Show that the condition that J in Eq. 4.3-31 is stationary is equivalent to the following differential equation:

$$\sum_{i=1}^{3} \frac{\partial}{\partial x_i}(\eta\dot{\gamma}_{ij}) - \frac{\partial}{\partial x_j} p + \rho g_j = 0, \qquad \text{for } j = 1, 2, 3 \tag{4D.4-2}$$

with the boundary condition

$$f_j + \sum_{i=1}^{3} n_i \eta \dot{\gamma}_{ij} + p n_j = 0, \qquad \text{for } j = 1, 2, 3 \tag{4D.4-3}$$

on S_f and essential boundary conditions on S_v.

c. Draw a picture of the control volume V and explain why the above equations complete the proof. Pay particular attention to Eq. 4D.4-3.

CHAPTER 5
THE GENERAL LINEAR VISCOELASTIC FLUID

In Chapter 4 we discussed an expression for the stress tensor that is particularly useful for engineers who must solve problems involving large-deformation flows, both without and with heat transfer. In such problems, as we have seen, the predominant feature of the rheological behavior of the macromolecular fluids is their shear-rate-dependent viscosity.

Although the generalized Newtonian fluid has proven to be of great value in solving problems of engineering interest, its use is strictly speaking limited to steady-state shearing flows. It is generally inappropriate for the description of unsteady flow phenomena, where the elastic response of the polymeric fluid becomes important. In this chapter we introduce a constitutive equation that can describe some of the time-dependent motions of macromolecular fluids, albeit only the restricted class of flows with very small displacement gradients.

Why do we spend a whole chapter on such a restricted class of flows? There are several very good reasons for studying this subject, known as "linear viscoelasticity": (1) polymer chemists have evolved several experiments that have enabled them to interrelate structure with the linear mechanical responses; (2) the material functions measured in these experiments have proven useful for characterization and quality control; and (3) some background in linear viscoelasticity is helpful to proceed to the subject of "nonlinear viscoelasticity", which is treated in Chapters 6 through 9 of this volume. It is this last reason that we shall consider the principal motivation here. For the reader interested in the experiments of linear viscoelasticity, their analysis and molecular interpretations, we recommend the outstanding treatise of Ferry,[1] where a wealth of information and an extensive bibliography are to be found.

We begin in §5.1 by comparing and contrasting Newton's "law" of viscosity and Hooke's "law" of elasticity, the two limiting idealizations for viscous liquids and elastic solids. Then in §5.2 we show how Maxwell combined the ideas of viscosity and elasticity to arrive at a simple equation for a "viscoelastic fluid"; after that we show how Maxwell's idea can be extended to obtain the constitutive equation for the *general linear viscoelastic fluid*, which has been widely used for many years to characterize the small-displacement behavior of polymeric liquids. It is this constitutive equation that is the primary subject of study in this chapter.

In §5.3 we use the constitutive equation to obtain expressions for some of the time-dependent material functions defined in Chapter 3. In the highly idealized flows discussed in this section, the velocity distribution is prescribed, and the shear stress is obtained directly from the constitutive equation. By contrast, in §5.4 we solve some linear viscoelastic flow problems that require the simultaneous consideration of the equations of change and the

[1] J. D. Ferry, *Viscoelastic Properties of Polymers*, 3rd ed., Wiley, New York (1980).

constitutive equation. In the final section, §5.5, we point out the limitations of the constitutive equation studied in this chapter.

§5.1 NEWTONIAN FLUIDS AND HOOKEAN SOLIDS

In Chapter 1 a discussion of the constitutive equation for the Newtonian fluid was given. Here we introduce the constitutive equation for the Hookean solid and point out some similarities and differences between these two "classical" constitutive equations. We do this by considering the two idealized experiments shown in Fig. 5.1-1.

The first experiment (Fig. 5.1-1*a*) is the *shearing motion of a Newtonian fluid* between two planes, the upper one of which moves with a velocity $V(t)$. We assume that the viscosity μ is so large and that the interplane distance B is so small that the velocity distribution $v_x(y, t)$ is a linear function of y. Then the velocity distribution is:

$$v_x(y, t) = \frac{V(t)}{B} y = \dot{\gamma}_{yx}(t) y \tag{5.1-1}$$

in which $\dot{\gamma}_{yx}$ is the yx-component of the *rate-of-strain tensor*

$$\dot{\boldsymbol{\gamma}} = \nabla \boldsymbol{v} + (\nabla \boldsymbol{v})^\dagger \qquad \left(\text{or } \dot{\gamma}_{ij} = \frac{\partial}{\partial x_i} v_j + \frac{\partial}{\partial x_j} v_i\right) \tag{5.1-2}$$

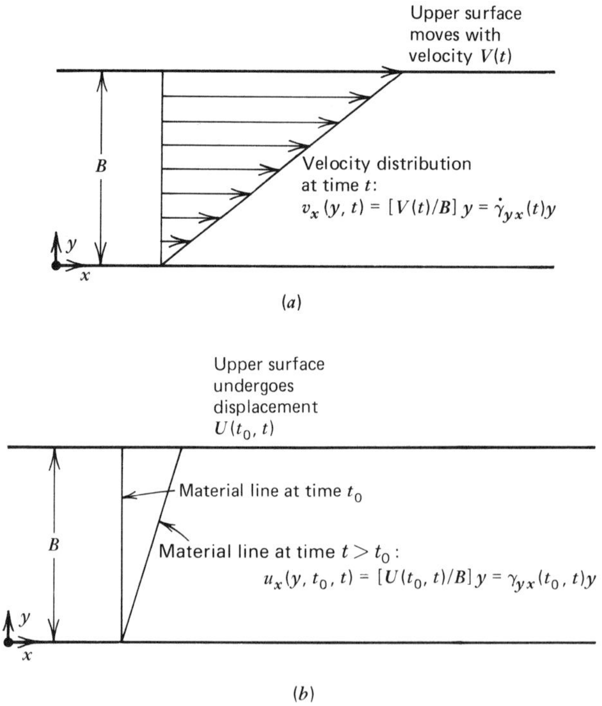

FIGURE 5.1-1. Material being sheared between two parallel planes, the upper one of which moves as a function of time. (*a*) The *velocity* profile for unsteady shear flow of a Newtonian fluid. (*b*) The *displacement* profile for the unsteady shearing motion of a Hookean solid. Note: at time t_0 the solid is at rest with no shear stress.

The shear stress is then given for a Newtonian fluid by

$$\tau_{yx}(t) = -\mu \frac{\partial v_x}{\partial y} = -\mu \dot{\gamma}_{yx}(t) \tag{5.1-3}$$

That is, the stress at time t is proportional to the velocity gradient *at the same time t*.

The second experiment (Fig 5.1-1*b*) is the *shearing motion of a Hookean solid* between two parallel planes. At some time t_0 the solid is in an isotropic stress state with no imposed external stresses other than atmospheric pressure. Then the upper plane undergoes an infinitesimal displacement $U(t_0, t)$. We assume that the displacement of the material in the gap is a linear function of the distance y above the lower plane, so that the displacement at any position is:

$$u_x(y, t_0, t) = \frac{U(t_0, t)}{B} y = \gamma_{yx}(t_0, t)y \tag{5.1-4}$$

Here $\gamma_{yx}(t_0, t)$ is the yx-component of the *infinitesimal strain tensor*, which is defined in terms of the *displacement gradient tensor* $\nabla \boldsymbol{u}$ as follows

$$\boldsymbol{\gamma} = \nabla \boldsymbol{u} + (\nabla \boldsymbol{u})^\dagger \qquad \left(\text{or } \gamma_{ij} = \frac{\partial}{\partial x_i} u_j + \frac{\partial}{\partial x_j} u_i\right) \tag{5.1-5}$$

Note that the strain tensor depends on two times and that $\gamma_{yx}(t_0, t_0) = 0$. Then the shear stress for a Hookean solid is given by

$$\tau_{yx}(t) = -G \frac{\partial u_x}{\partial y} = -G\gamma_{yx}(t_0, t) \tag{5.1-6}$$

where G is the elastic modulus; that is, the stress at time t is proportional to the strain at time t, referred to the isotropic stress state at time t_0. The Hookean solid "remembers" where it was at time t_0, in contrast to the Newtonian fluid which has no memory of past events.

The two discussions above are parallel except for two points: (i) in the discussion of solids we must specify a reference time with respect to which strain is measured; strain tensors always depend on two times, and we establish the convention that the first variable listed denotes the reference state; (ii) the stress relation, Eq. 5.1-6, is restricted to infinitesimally small displacement gradients $\partial u_j/\partial x_i$; for large displacement gradients Eq. 5.1-6 ceases to be valid, whereas Eq. 5.1-3 is valid for flow with arbitrarily large displacement gradients. The restriction that is placed on Eq. 5.1-6 has some important implications to be discussed in §5.5.

The kinematic quantities in the two experiments are related to one another. The velocities and displacements are related by:

$$\begin{cases} v_x(y, t) = \dfrac{\partial}{\partial t} u_x(y, t_0, t) & \text{(5.1-7a)} \\[2ex] u_x(y, t_0, t) = \displaystyle\int_{t_0}^{t} v_x(y, t')dt' & \text{(5.1-7b)} \end{cases}$$

Similarly the components of the rate-of-strain and infinitesimal strain tensors are related by:

$$\begin{cases} \dot{\gamma}_{yx}(t) = \dfrac{\partial}{\partial t} \gamma_{yx}(t_0, t) & \text{(5.1-8a)} \\[2ex] \gamma_{yx}(t_0, t) = \displaystyle\int_{t_0}^{t} \dot{\gamma}_{yx}(t') dt' & \text{(5.1-8b)} \end{cases}$$

We also point out that for the infinitesimal strain tensor

$$\gamma_{yx}(t_0, t) = \gamma_{yx}(t_0, t^*) + \gamma_{yx}(t^*, t) \tag{5.1-9}$$

That is, the infinitesimal strains are additive. These relations are referred to later.

§5.2 LINEAR VISCOELASTIC FLUIDS

In Chapter 2 it was found that *polymeric liquids are viscoelastic*. For example, the photographs of §2.5e show how a polymeric liquid recoils because of its elastic properties. Furthermore in §§3.4 and 3.5 several kinds of transient-response experiments were described in which elastic behavior was evident. For example, delayed stress relaxation after cessation of steady flow is an indication of fluid elasticity as is the large recovery following elongational flow. The main thrust of this chapter is to show how the ideas of viscosity and elasticity may be combined into a single constitutive equation that can describe these various elastic effects.

a. The Maxwell Model

The first attempt to obtain a viscoelastic constitutive equation appears to have been that of Maxwell,[1] who over a century ago developed a theory for viscoelasticity, because he thought that gases might be viscoelastic. He proposed that fluids with both viscosity and elasticity could be described by:

$$\tau_{yx} + \frac{\mu}{G} \frac{\partial \tau_{yx}}{\partial t} = -\mu \dot{\gamma}_{yx} \tag{5.2-1}$$

For steady-state motions this equation simplifies to the Newtonian fluid with viscosity μ. For sudden changes in stress, the time derivative term dominates the left side of the equation, and then integration with respect to time (with the help of Eq. 5.1-8b) gives the Hookean solid with elastic modulus G. This is the simplest expression for the shear stress for a fluid that is both viscous and elastic.

Since Hooke's law is valid only for infinitesimal displacement gradients, it seems reasonable to expect that Maxwell's equation would be subject to the same restriction. We must keep in mind that Newton's and Hooke's "laws" were both proposed empirically; these linear relations were put forward as modest but reasonable suggestions. Subsequent experimentation, coupled with calculations based on the equations of change and the

[1] J. C. Maxwell, *Phil. Trans. Roy. Soc.*, **A157**, 49–88 (1867).

proposed constitutive equations, showed that the linear relations of Newton and Hooke are indeed useful for a wide range of materials. However, it has long been recognized that these "classical" constitutive equations do have their limitations. Many materials require more complex mathematical descriptions.

Maxwell's proposal was also empirical. Although it turns out that its range of validity is somewhat limited, certain nonlinear generalizations of the Maxwell constitutive equation have proven to be very useful in polymer fluid dynamics, and we will eventually present and evaluate these generalizations in Chapters 7 and 8.

We now look at some alternative forms of the Maxwell model. First we generalize Eq. 5.2-1 to arbitrary, small displacement flows by putting the equation in tensor form. In addition, we adopt new symbols for the constants, since this notation will be more in keeping with subsequent developments: we replace μ by η_0 (the zero-shear-rate viscosity) and μ/G by λ_1 (a time constant, often called the "relaxation time"). Then the *Maxwell model* is:

$$\boxed{\boldsymbol{\tau} + \lambda_1 \frac{\partial}{\partial t} \boldsymbol{\tau} = -\eta_0 \dot{\boldsymbol{\gamma}}} \tag{5.2-2}$$

This is more complicated than Newton's and Hooke's equations since this is a differential equation for $\boldsymbol{\tau}$.

For many purposes it would be preferable to solve Eq. 5.2-2 for the stress tensor. That is easily done by recognizing that Eq. 5.2-2 is a first-order, linear equation[2] for $\boldsymbol{\tau}$ as a function of t, which can be integrated at once to give:

$$\boldsymbol{\tau}(t) = e^{-t/\lambda_1}\left[\int \left(-\frac{\eta_0}{\lambda_1}\dot{\boldsymbol{\gamma}}(t)\right)e^{t/\lambda_1}\,dt + \boldsymbol{\kappa}\right] \tag{5.2-3}$$

We now affix limits on the integral and write:

$$\boldsymbol{\tau}(t) = -\frac{\int_{-\infty}^{t} (\eta_0/\lambda_1)\dot{\boldsymbol{\gamma}}(t')e^{t'/\lambda_1}\,dt'}{e^{t/\lambda_1}} + \boldsymbol{\kappa}e^{-t/\lambda_1} \tag{5.2-4}$$

where primes have been added to t's in the integrand in order to avoid confusion; the choice of $-\infty$ as the lower limit is arbitrary—some other choice would result in a different value for the integration constant $\boldsymbol{\kappa}$. If we prescribe that the stress in the fluid is finite at $t = -\infty$, we must choose $\boldsymbol{\kappa}$ to be zero. But we must also check the first term on the right side of Eq. 5.2-4, since both numerator and denominator tend to zero as t goes to $-\infty$. When we use L'Hôpital's rule we get:

$$\lim_{t\to-\infty} \boldsymbol{\tau}(t) = \lim_{t\to-\infty} -\frac{(\eta_0/\lambda_1)\dot{\boldsymbol{\gamma}}(t)e^{t/\lambda_1}}{(1/\lambda_1)e^{t/\lambda_1}} = -\eta_0\dot{\boldsymbol{\gamma}}(-\infty) \tag{5.2-5}$$

[2] Recall that the differential equation $dy/dx + P(x)y = Q(x)$ has the solution:

$$y = e^{-\int P(x)dx}\left[\int Q(x)e^{+\int P(x)dx}\,dx + K\right] \tag{5.2-2a}$$

where K is a constant of integration.

Therefore, if $\dot{\gamma}(-\infty)$ is finite, the stress is also finite at $t = -\infty$. Consequently, the Maxwell constitutive equation can be written in the form

$$\boxed{\tau(t) = -\int_{-\infty}^{t} \left\{ \frac{\eta_0}{\lambda_1} e^{-(t-t')/\lambda_1} \right\} \dot{\gamma}(t')dt'} \tag{5.2-6}$$

The quantity within the braces is called the *relaxation modulus* for the Maxwell fluid. When written in this form, the Maxwell model says that the stress at the present time t depends on the rate of strain at time t as well as on the rate of strain at all past times t', with a weighting factor (the relaxation modulus) that decays exponentially as one goes backwards in time. Thus we see that this form of the Maxwell equation contains the notion of a "fading memory". The fluid remembers very well what it has experienced in the very recent past but has only a hazy recollection of events in the distant past. Sometimes we say that the stress at time t depends on the "history" of the rate of strain for all past times $-\infty < t' \leq t$.

Next, we want to put the Maxwell model into still another form by doing an integration by parts using Eq. 5.1-8b. However, in using the latter equation we have to specify a reference time (there called t_0). For fluids there is no unique reference state t_0 to use in describing strains, so that *for fluids it is customary to measure the strain at a past time t' relative to the configuration of the fluid at the present time t*. Hence, for fluids we generalize Eq. 5.1-8 thus:

$$\frac{\partial \gamma}{\partial t'} = \dot{\gamma}(t') \quad \text{and} \quad \gamma(t, t') = \int_{t}^{t'} \dot{\gamma}(t'')dt'' \tag{5.2-7}$$

These relations are valid for any flow pattern as long as the displacement gradients are infinitesimally small. Now integration by parts of Eq. 5.2-6 gives

$$\boxed{\tau(t) = +\int_{-\infty}^{t} \left\{ \frac{\eta_0}{\lambda_1^2} e^{-(t-t')/\lambda_1} \right\} \gamma(t, t')dt'} \tag{5.2-8}$$

The quantity within the braces is called the *memory function* for the Maxwell fluid. In this form the Maxwell model states that the stress at the present time t depends on the history of the strain for all past times $-\infty < t' \leq t$. The exponential factor in the integrand describes the fading memory.

All three forms of Maxwell's constitutive equation—Eqs. 5.2-2, 6, and 8—are equivalent provided that $\dot{\gamma}$ is finite at $t = -\infty$ and the displacement gradients are infinitesimally small. Equation 5.2-6 looks like a modified Newton's law and Eq. 5.2-8 like a modified Hooke's law. Maxwell's equation played a key role in the development of linear viscoelasticity, and as we shall see in Chapters 7 and 8, it has also been taken as the starting point for the development of nonlinear viscoelastic models. It is therefore important that the material presented in this section be thoroughly understood before continuing.

The two-constant Maxwell model was found to be inadequate for describing linear viscoelastic data. Therefore, through the years more elaborate equations were suggested; let us look at a few of these.

b. The Jeffreys Model

The Maxwell equation in Eq. 5.2-2 is a linear relation between τ and $\dot{\gamma}$. But one can easily invent other linear relations. For example, we can include the time derivative of $\dot{\gamma}$ and get the constitutive equation

$$\tau + \lambda_1 \frac{\partial \tau}{\partial t} = -\eta_0 \left(\dot{\gamma} + \lambda_2 \frac{\partial \dot{\gamma}}{\partial t} \right) \tag{5.2-9}$$

which is known as the *Jeffreys model.* This equation, containing two time constants λ_1 and λ_2 (the "relaxation time" and the "retardation time", respectively), was proposed for the study of wave propagation in the earth's mantle.[3]

The Jeffreys model can also be put into integral form. When Eq. 5.2-9 is integrated as a first-order differential equation, using the initial condition that τ be finite at $t = -\infty$, we find that (if $\dot{\gamma}$ and $\partial\dot{\gamma}/\partial t$ are both finite at $t = -\infty$)

$$\tau(t) = -\int_{-\infty}^{t} \frac{\eta_0}{\lambda_1} \left(1 - \frac{\lambda_2}{\lambda_1} \right) e^{-(t-t')/\lambda_1} \dot{\gamma}(t')dt' - \frac{\eta_0 \lambda_2}{\lambda_1} \dot{\gamma}(t) \tag{5.2-10}$$

From this form of the constitutive equation it can be seen that $\lambda_2 < \lambda_1$; otherwise, in stress relaxation after cessation of steady shear flow τ_{yx} would have the wrong sign.

We would like to put Eq. 5.2-10 into a form that is the same as Eq. 5.2-6, so that we can identify the relaxation modulus. This is done by using the Dirac delta function:[4]

$$\tau(t) = -\int_{-\infty}^{t} \left\{ \frac{\eta_0}{\lambda_1} \left(1 - \frac{\lambda_2}{\lambda_1} \right) e^{-(t-t')/\lambda_1} + 2 \frac{\eta_0 \lambda_2}{\lambda_1} \delta(t - t') \right\} \dot{\gamma}(t')dt' \tag{5.2-11}$$

[3] H. Jeffreys, *The Earth*, Cambridge University Press (1929), p. 265.

[4] P. A. M. Dirac, *The Principles of Quantum Mechanics*, 3rd ed., Oxford University Press (1947), pp. 58–61; M. J. Lighthill, *Fourier Analysis and Generalised Functions*, Cambridge University Press (1964), p. 17. Here we use the definition that:

$$\delta(x) = \lim_{n \to \infty} \sqrt{\frac{n}{\pi}}\, e^{-nx^2} \tag{5.2-10a}$$

From this it follows that:

$$\int_{-a}^{a} f(x)\delta(x)dx = 2\int_{0}^{a} f(x)\delta(x)dx = f(0) \tag{5.2-10b}$$

$$\int_{-a}^{a} f(x)\delta'(x)dx = -f'(0) \tag{5.2-10c}$$

in which $a > 0$, and the prime denotes differentiation with respect to x. Note particularly the occurrence of the factor of 2 in Eq. 5.2-10b when the integral is over the region from 0 to a. This explains the occurrence of the factors of 2 in the δ-function terms in Eqs. 5.2-11 and 12.

The quantity enclosed in the braces is the relaxation modulus for the Jeffreys model. Integration by parts then gives:

$$\tau(t) = +\int_{-\infty}^{t} \left\{\frac{\eta_0}{\lambda_1^2}\left(1 - \frac{\lambda_2}{\lambda_1}\right)e^{-(t-t')/\lambda_1} + \frac{2\eta_0\lambda_2}{\lambda_1}\frac{\partial}{\partial t'}\delta(t - t')\right\}\gamma(t, t')dt' \qquad (5.2\text{-}12)$$

The quantity in braces here is the memory function. The Jeffreys model is also used in Chapters 7 and 8 as a point of departure for proposing nonlinear viscoelastic models. Clearly more time derivatives could be added to the left and right sides of Eq. 5.2-9 thereby generating many more linear relations between τ and $\dot{\gamma}$ (see Problem 5C.1).

c. The Generalized Maxwell Model

Another way to generalize the Maxwell model is to construct a "superposition" of Maxwell models. We can write Eq. 5.2-2 for the kth partial stress τ_k using constants λ_k and η_k; then we get the total stress by summing the partial stresses. This constitutive equation can be integrated to give equations similar in form to Eqs. 5.2-6 and 5.2-8. To summarize, we have for the *generalized Maxwell model*:

$$\tau(t) = \sum_{k=1}^{\infty} \tau_k(t); \qquad \tau_k + \lambda_k \frac{\partial}{\partial t}\tau_k = -\eta_k \dot{\gamma} \qquad (5.2\text{-}13)$$

$$\tau(t) = -\int_{-\infty}^{t} \left\{\sum_{k=1}^{\infty} \frac{\eta_k}{\lambda_k} e^{-(t-t')/\lambda_k}\right\}\dot{\gamma}(t')dt' \qquad (5.2\text{-}14)$$

$$\tau(t) = +\int_{-\infty}^{t} \left\{\sum_{k=1}^{\infty} \frac{\eta_k}{\lambda_k^2} e^{-(t-t')/\lambda_k}\right\}\gamma(t, t')dt' \qquad (5.2\text{-}15)$$

We adopt the convention that $\lambda_1 > \lambda_2 > \lambda_3 \ldots$. This model contains infinitely many constants λ_k and η_k, thus allowing for a spectrum of relaxation times and viscosities. (Of course, one can set $\lambda_k = 0$ and $\eta_k = 0$ for k greater than some finite number K.) For some purposes it may be desirable to reduce the total number of parameters to three by use of the following empiricisms:[5]

$$\eta_k = \eta_0 \frac{\lambda_k}{\sum_k \lambda_k}; \quad \lambda_k = \frac{\lambda}{k^\alpha} \qquad (5.2\text{-}16,17)$$

where η_0 is the zero-shear-rate viscosity, λ is a time constant, and α is a dimensionless quantity, which describes the slope of η' versus ω on a log–log plot at high ω (for large ω, $\eta' \propto \omega^{(1/\alpha)-1}$). The relations above are not entirely empirical. The *Rouse molecular theory for dilute polymer solutions*[6] gives very nearly Eqs. 5.2-16,17 with $\alpha = 2$; in this theory the polymer molecules are modeled as freely jointed chains made up of beads connected linearly by Hookean springs. Experimental data for concentrated polymer situations and polymer melts seem to be portrayed reasonably well by Eqs. 5.2-16,17 with α in the range 2 to 4. On the other hand, the *Doi and Edwards molecular theory for polymer melts*[7] suggests an

[5] T. W. Spriggs, *Chem. Eng. Sci.*, **20**, 931–940 (1965).

[6] P. E. Rouse, Jr., *J. Chem. Phys.*, **21**, 1272–1280 (1953); see also Chapter 15.

[7] M. Doi and S. F. Edwards, *J. Chem. Soc. Faraday Trans. II*, **74**, 1789–1832 (1978); **75**, 38–54 (1979); see also Chapter 19.

empiricism with λ_k replaced by λ_k^2 in Eq. 5.2-16, and k taking on odd values only (see Problem 5B.8). Still other possibilities for describing $G(s)$ are the three-parameter empiricism in Problem 5B.2 and the finite sum of exponentials used in Example 5.3-7.

d. The General Linear Viscoelastic Model

When we compare Eq. 5.2-6 (for the Maxwell model), Eq. 5.2-11 (for the Jeffreys model), and Eq. 5.2-14 (for the generalized Maxwell model) we find that they are all of the same form: an integral over all past times of a relaxation modulus multiplied by a rate-of-deformation tensor. The only physical ideas that have been included in these models of varying degrees of complexity are those of viscosity and elasticity. An equation that includes all of these models, and many more of course, is the *general linear viscoelastic model*, which may be written in either of two equivalent forms:

$$\boldsymbol{\tau} = -\int_{-\infty}^{t} G(t - t')\dot{\boldsymbol{\gamma}}(t')dt' \tag{5.2-18}$$

$$\boldsymbol{\tau} = +\int_{-\infty}^{t} M(t - t')\boldsymbol{\gamma}(t, t')dt' \tag{5.2-19}$$

in which $G(t - t')$ is the *relaxation modulus* and $M(t - t') = \partial G(t - t')/\partial t'$ is the *memory function*. This model is the starting point for most of the development in this chapter and will be referred to in connection with nonlinear constitutive equations and molecular theories. Of course, for getting quantitative answers to problems we shall have to use some specific expression for $G(t - t')$, such as the expression between braces in Eqs. 5.2-6, 11, or 14. It should be kept in mind that Eqs. 5.2-18 and 19 are equivalent only if $\dot{\boldsymbol{\gamma}}$ is finite at $t = -\infty$ and if the displacement gradients are infinitesimally small.

Equations 5.2-18 and 19 have an important feature: the integrands consist of the product of two functions, *the first depending on the nature of the fluid* (since material parameters, such as η_k and λ_k, appear in $G(t - t')$ or $M(t - t')$) and *the second depending on the nature of the flow* (since the kinematics is described by $\dot{\boldsymbol{\gamma}}(t')$ or $\boldsymbol{\gamma}(t, t')$). In Chapter 8 we discuss nonlinear viscoelastic constitutive equations that have this same "factorized" structure.

In Eqs. 5.2-18 and 19 the functions $G(s)$ and $M(s)$ are positive functions which decrease monotonically to zero as $s = t - t'$ goes to infinity. Such viscoelastic fluids are often said to have "fading memory". If $G(s)$ has the form given by the generalized Maxwell model, the duration of the memory is governed by the largest relaxation time, λ_{max}. Although Eqs. 5.2-18 and 19 are strictly applicable only to flows with infinitesimally small displacement gradients, we can in some instances apply them outside this region because of the rapidly fading memory. For example, in steady shear flow with shear rate $\dot{\gamma} \ll 1/\lambda_{max}$, the memory of the large strains is negligible and does not contribute to the stress.

The general linear viscoelastic model in Eq. 5.2-18 may also be derived by more formal arguments.[8] One assumes that the stress at time t resulting from a step strain at time t' is linear in the strain and multiplied by a decaying function of the elapsed time $t - t'$. An actual flow history may then be regarded as made up of a number of small step strains. By *Boltzmann's superposition principle*[9] it is assumed that the stress contributions from the

[8] A. C. Pipkin, *Lectures on Viscoelasticity Theory*, Springer, New York (1972), pp. 7–12.
[9] L. Boltzmann, *Pogg. Ann. Phys.*, 7, 624–654 (1876).

individual small step strains at past times t' may be added to give the stress at time t. This results in the integral in Eq. 5.2-18. In the linear limit considered in this chapter the Boltzmann superposition principle seems reasonable inasmuch as coupling effects between two past deformations must be of second order in the applied deformations and hence negligible.

Equation 5.2-18 (or Eq. 5.2-19) is generally accepted as the correct starting point for the description of the rheology of incompressible viscoelastic fluids for small-displacement-gradient motions. In Chapters 7, 8, and 9 large-displacement-gradient flows are considered and equations more general than Eq. 5.2-18 are given.

§5.3 LINEAR VISCOELASTIC RHEOLOGICAL PROPERTIES

In the foregoing section it was shown that the elementary concepts of viscosity and elasticity can be combined in a number of ways to develop equations of increasing complexity, culminating with Eq. 5.2-18, which is the most general equation of linear viscoelasticity. In this section we illustrate the use of this equation by obtaining expressions for the time-dependent material functions in terms of the relaxation modulus $G(t - t')$.

Before discussing the transient-response phenomena, we mention briefly how Eq. 5.2-18 simplifies for steady-state shear flow. When the fluid has been flowing between parallel plates for a long time with constant velocity gradient $\dot{\gamma}_{yx}$ (where $\lambda_{\max}|\dot{\gamma}_{yx}| \ll 1$), then Eq. 5.2-18 becomes

$$\begin{aligned}\tau_{yx} &= -\int_{-\infty}^{t} G(t - t')\dot{\gamma}_{yx}\,dt' \\ &= -\left[\int_{0}^{\infty} G(s)ds\right]\dot{\gamma}_{yx} \\ &\equiv -\eta_0\dot{\gamma}_{yx} \end{aligned} \tag{5.3-1}$$

Here the change of variables $s = t - t'$ has been made. We see that the viscosity is just equal to the integral over the relaxation modulus. The subscript "0" on η_0 indicates that this is the zero-shear-rate viscosity. In the theory of linear viscoelasticity we are able to obtain the viscosity only in this limit of vanishingly small velocity gradients.

EXAMPLE 5.3-1 Small-Amplitude Oscillatory Motion

A polymeric fluid is located in the space between two parallel plates, the upper one of which is made to oscillate with frequency ω in its own plane in the x-direction. As shown in Figure 3.4-1b the velocity profile is assumed to be instantaneously linear,[1] which is a good assumption for highly viscous materials in very narrow slits (cf. Problem 1D.1). Therefore the velocity gradient is changing with time in the following way:

$$\dot{\gamma}_{yx}(t) = \dot{\gamma}_{yx}^{0} \cos \omega t \tag{5.3-2}$$

in which we take $\dot{\gamma}_{yx}^{0}$ to be real and positive. In order to satisfy the small-displacement-gradient restriction on Eq. 5.2-18, we require that $\dot{\gamma}_{yx}^{0}/\omega \ll 1$. Find the time-dependent shear stress τ_{yx} that is needed to maintain this oscillatory motion, and obtain expressions for the real and imaginary parts of the complex viscosity.

[1] Inertial effects have been considered by K. Walters and R. A. Kemp in R. E. Whetton and R. W. Whorlow, eds., *Polymer Systems*, Macmillan, London (1968), pp. 237–250, and *Rheol. Acta*, 7, 1–8 (1968).

SOLUTION Substitution of the velocity gradient of Eq. 5.3-2 into the constitutive equation in Eq. 5.2-18 gives:

$$\begin{aligned}\tau_{yx} &= -\int_{-\infty}^{t} G(t-t')\dot{\gamma}_{yx}^{0} \cos \omega t' \, dt' \\ &= -\dot{\gamma}_{yx}^{0} \int_{0}^{\infty} G(s) \cos \omega(t-s) \, ds \\ &= -\left[\int_{0}^{\infty} G(s) \cos \omega s \, ds\right] \dot{\gamma}_{yx}^{0} \cos \omega t \\ &\quad -\left[\int_{0}^{\infty} G(s) \sin \omega s \, ds\right] \dot{\gamma}_{yx}^{0} \sin \omega t\end{aligned} \tag{5.3-3}$$

Comparison of this result with the definitions of $\eta'(\omega)$ and $\eta''(\omega)$ in Eq. 3.4-3b shows that:

$$\eta'(\omega) = \int_{0}^{\infty} G(s) \cos \omega s \, ds \tag{5.3-4}$$

$$\eta''(\omega) = \int_{0}^{\infty} G(s) \sin \omega s \, ds \tag{5.3-5}$$

or, alternatively, we may write the results in terms of the "complex viscosity"

$$\eta^* = \eta' - i\eta'' = \int_{0}^{\infty} G(s) e^{-i\omega s} \, ds \tag{5.3-6}$$

The relaxation modulus $G(s)$ can be eliminated between Eqs. 5.3-4 and 5 to give the Kramers–Kronig relations (see Problem 5D.2), which interrelate $\eta'(\omega)$ and $\eta''(\omega)$. Finally, we note for future reference that:

$$\lim_{\omega \to 0} \frac{\eta''(\omega)/\omega}{\eta'(\omega)} = \frac{\int_0^\infty G(s) s \, ds}{\int_0^\infty G(s) \, ds} \tag{5.3-7}$$

We shall find as we go through this section that several other limiting quantities are also equal to the same ratio of integrals.

Let us now see what the expressions for $\eta'(\omega)$ and $\eta''(\omega)$ look like for the particular choice of relaxation modulus given by the generalized Maxwell model (quantity in braces in Eq. 5.2-14)

$$\eta'(\omega) = \sum_{k=1}^{\infty} \frac{\eta_k}{1 + (\lambda_k \omega)^2} \tag{5.3-8}$$

$$\frac{\eta''(\omega)}{\omega} = \sum_{k=1}^{\infty} \frac{\eta_k \lambda_k}{1 + (\lambda_k \omega)^2} \tag{5.3-9}$$

If, in addition, we introduce the expressions for η_k and λ_k given in Eqs. 5.2-16 and 17, these results become

$$\frac{\eta'}{\eta_0} = \frac{1}{\zeta(\alpha)} \sum_{k=1}^{\infty} \frac{k^{\alpha}}{k^{2\alpha} + (\lambda\omega)^2} \tag{5.3-10}$$

$$\frac{\eta''}{\eta_0} = \frac{\lambda\omega}{\zeta(\alpha)} \sum_{k=1}^{\infty} \frac{1}{k^{2\alpha} + (\lambda\omega)^2} \tag{5.3-11}$$

in which $\zeta(\alpha)$ is the Riemann zeta function.[2] These expressions are not particularly appropriate for computation of the functions $\eta'(\omega)$ and $\eta''(\omega)$. Instead we have available[3] an alternative pair of expressions useful for low frequencies ($\omega \ll \lambda^{-1}$)

$$\frac{\eta'}{\eta_0} = 1 - \left[\frac{(\lambda\omega)^2}{\zeta(\alpha)} \sum_{k=1}^{\infty} \frac{1}{k^{\alpha}(k^{2\alpha} + (\lambda\omega)^2)}\right] \tag{5.3-12}$$

$$\frac{\eta''}{\eta_0} = \lambda\omega\left[\frac{\zeta(2\alpha)}{\zeta(\alpha)} - \frac{(\lambda\omega)^2}{\zeta(\alpha)} \sum_{k=1}^{\infty} \frac{1}{k^{2\alpha}(k^{2\alpha} + (\lambda\omega)^2)}\right] \tag{5.3-13}$$

and another pair of asymptotic expressions that is excellent for high frequencies ($\omega \gg \lambda^{-1}$)

$$\frac{\eta'}{\eta_0} \sim \frac{1}{\zeta(\alpha)}\left[\frac{\pi(\lambda\omega)^{(1/\alpha)-1}}{2\alpha \sin((\alpha + 1)\pi/2\alpha)}\right] \tag{5.3-14}$$

$$\frac{\eta''}{\eta_0} \sim \frac{1}{\zeta(\alpha)}\left[\frac{\pi(\lambda\omega)^{(1/\alpha)-1}}{2\alpha \sin(\pi/2\alpha)} - \frac{(\lambda\omega)^{-1}}{2}\right] \tag{5.3-15}$$

These large-frequency expressions are obtained by using the Euler–Maclaurin expansion to convert the sums into integrals (see Problem 5B.5). Equations 5.3-10 and 11 (and their low- and high-frequency equivalents) are useful for curve-fitting data for the small-amplitude oscillatory experiment. In addition, in Example 5.3-7 it is shown how η_k and λ_k can be chosen so that Eqs. 5.3-8 and 9 accurately fit experimental data for η' and η''.

We conclude this illustrative example by reminding the reader that for a Newtonian fluid η' is just a constant (the viscosity), and that η'' is zero; that is, for the Newtonian fluid the shear stress is in phase with the velocity gradient.

EXAMPLE 5.3-2 Stress Relaxation after a Sudden Shearing Displacement

The purpose of this example is to show why the function $G(t - t')$ is called the "relaxation modulus". A polymeric liquid is at rest in the region between two parallel plates for time $t < t_0$. At time $t = t_0$ the upper plate is instantaneously moved slightly in the x-direction, as shown in Fig. 3.4-1e. Find the expression for the shear stress at time $t > t_0$, for a sudden shear strain $\gamma_0 \ll 1$.

SOLUTION To solve this problem we first imagine that the displacement actually occurs during the finite time interval from $t_0 - \varepsilon$ to t_0, and then later we let ε go to zero. That is, the yx-components of the infinitesimal strain tensor and of the rate-of-strain tensor are as shown in Fig. 5.3-1.

For the displacement occurring in the finite time interval ε, Eq. 5.2-18 becomes (for $t > t_0$)

$$\begin{aligned}\tau_{yx}(t) &= -\int_{-\infty}^{t_0-\varepsilon} G(t - t')\dot{\gamma}_{yx}(t')dt' - \int_{t_0-\varepsilon}^{t_0} G(t - t')\dot{\gamma}_{yx}(t')dt' - \int_{t_0}^{t} G(t - t')\dot{\gamma}_{yx}(t')dt' \\ &= -\frac{\gamma_0}{\varepsilon}\int_{t_0-\varepsilon}^{t_0} G(t - t')dt' \end{aligned} \tag{5.3-16}$$

[2] The Riemann zeta function is defined as:

$$\zeta(\alpha) = \sum_{k=1}^{\infty} k^{-\alpha}, \qquad \alpha > 1 \tag{5.3-11a}$$

a few sample values being $\zeta(2) = \pi^2/6$, $\zeta(4) = \pi^4/90$, $\zeta(6) = \pi^6/945$ (see M. Abramowitz and I. A. Stegun, eds., *Handbook of Mathematical Functions*, Nat. Bur. Stds. Appl. Math. Series No. 55, U.S. Govt. Ptg. Off., Washington, D.C. (1964), p. 807).

[3] T. W. Spriggs, *Chem. Eng. Sci.*, **20**, 931–940 (1965).

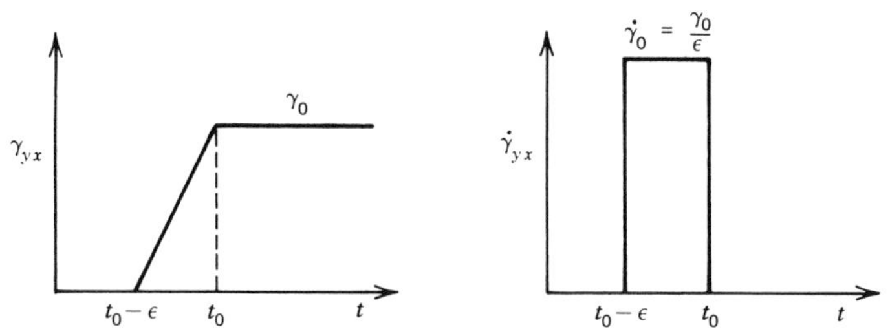

FIGURE 5.3-1. Time-dependent behavior of γ_{yx} and $\dot{\gamma}_{yx}$ in the sudden shear strain stress-relaxation experiment.

since the velocity gradient is zero except during the middle time interval. Now we take the limit as ε approaches zero using L'Hôpital's rule:

$$\tau_{yx}(t) = \lim_{\varepsilon \to 0}(-\gamma_0)\left[\frac{\dfrac{d}{d\varepsilon}\displaystyle\int_{t_0-\varepsilon}^{t_0} G(t-t')dt'}{\dfrac{d}{d\varepsilon}\varepsilon}\right]$$

$$= -\gamma_0 G(t-t_0) \tag{5.3-17}$$

Thus the function $G(t - t_0)$ describes the way in which the shear stress relaxes after the sudden shearing displacement. The top curve in either Fig. 3.4-15 or Fig. 3.4-16a gives a direct experimental measurement of G. If we insert the expression for the relaxation modulus for the generalized Maxwell fluid, we see that the stress dies out as a sum of exponentials. Keep in mind that for a Newtonian fluid there is no delayed stress relaxation—the stress drops instantly to zero as soon as the motion stops.

EXAMPLE 5.3-3 Stress Relaxation after Cessation of Steady Shear Flow

Next we turn our attention to the stress relaxation that occurs in a different type of experiment. Here we envisage a steady shear flow with shear rate $\dot{\gamma}_0 \ll 1/\lambda_{max}$, for time $t < 0$ (see Fig. 3.4-1d). Then at time $t = 0$ the flow is stopped suddenly. It is desired to describe the way in which the stress decays with time after the cessation of the steady shear flow. It is also desired to find the area under the $\eta^-(t)$ curve.

SOLUTION For this experiment Eq. 5.2-18 gives for $t < 0$ and $t > 0$

$$\tau_{yx}(t<0) = -\eta_0\dot{\gamma}_0 = -\dot{\gamma}_0\int_{-\infty}^{t} G(t-t')dt' \tag{5.3-18}$$

$$\tau_{yx}(t>0) = -\eta^-\dot{\gamma}_0 = -\dot{\gamma}_0\int_{-\infty}^{0} G(t-t')dt' \tag{5.3-19}$$

We now take the ratio of the above two expressions and then integrate it from $t = 0$ to $t = \infty$. This gives for the area under the stress-relaxation curve

$$\int_0^\infty \frac{\eta^-(t)}{\eta_0}dt = \int_0^\infty \frac{\int_{-\infty}^0 G(t-t')dt'}{\int_{-\infty}^t G(t-t')dt'}dt$$

$$= \frac{\int_0^\infty \int_t^\infty G(s)ds\, dt}{\int_0^\infty G(s)ds}$$

$$= \frac{\int_0^\infty \int_0^s G(s)dt\, ds}{\int_0^\infty G(s)ds}$$

$$= \frac{\int_0^\infty sG(s)ds}{\int_0^\infty G(s)ds} \tag{5.3-20}$$

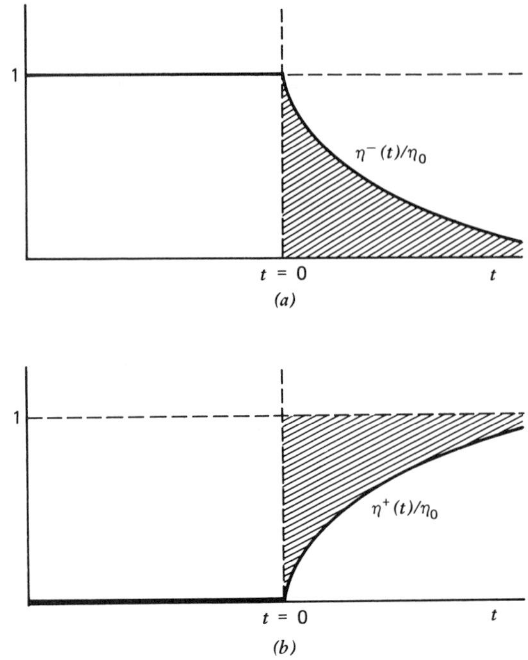

FIGURE 5.3-2. Integrals evaluated in Examples 5.3-3 and 4. (*a*) Integral in Eq. 5.3-20 for stress relaxation is given by the shaded area; (*b*) integral in Eq. 5.3-24 for stress growth is given by the shaded area.

In going from the second to the third line, we interchanged the order of integration so that one integration could be performed. See Fig. 5.3-2*a* for the graphical interpretation of this integral.

Note further that if we use the relaxation modulus for the generalized Maxwell model, then we find

$$\frac{\eta^-}{\eta_0} = \frac{\sum_k \eta_k e^{-t/\lambda_k}}{\sum_k \eta_k} \tag{5.3-21}$$

This result describes the top curve of the set of curves in Fig. 3.4-11, that is, the limiting curve for vanishingly small $\dot{\gamma}_0$. The linear theory cannot describe the dependence of the lower curves on $\dot{\gamma}_0$. To describe those curves we need a nonlinear viscoelastic theory, such as one of those described in Chapters 7 and 8.

EXAMPLE 5.3-4 Stress Growth at Inception of Steady Shear Flow

The next flow situation we examine is that shown in Fig. 3.4-1c where a fluid at rest is suddenly made to undergo steady-state shear flow after $t = 0$. Let the velocity gradient for $t > 0$ be $\dot{\gamma}_0$, where $\lambda_{\max}\dot{\gamma}_0 \ll 1$. Find $\eta^+(t)$ and the area under the curve of $1 - (\eta^+/\eta_0)$.

SOLUTION For this experiment Eq. 5.2-18 gives for $t < 0$ and $t > 0$

$$\tau_{yx}(t < 0) = 0 \tag{5.3-22}$$

$$\begin{aligned}\tau_{yx}(t > 0) &= -\eta^+\dot{\gamma}_0 \\ &= -\dot{\gamma}_0 \int_0^t G(t - t')dt' \end{aligned} \tag{5.3-23}$$

The area between the stress growth curve and its asymptote (see Fig. 5.3-2*b*) is given by

$$\int_0^\infty \left(1 - \frac{\eta^+(t)}{\eta_0}\right)dt = \int_0^\infty \frac{\int_{-\infty}^0 G(t-t')dt'}{\int_{-\infty}^t G(t-t')dt'}\,dt$$
$$= \frac{\int_0^\infty \int_t^\infty G(s)ds\,dt}{\int_0^\infty G(s)ds}$$
$$= \frac{\int_0^\infty sG(s)ds}{\int_0^\infty G(s)ds} \tag{5.3-24}$$

Thus the area between the stress-growth curve and its asymptote is the same as the area between the stress-relaxation curve and its asymptote.

When the relaxation modulus is specified as that for the generalized Maxwell model, then the stress-growth function is

$$\frac{\eta^+}{\eta^0} = \frac{\sum_k \eta_k(1 - e^{-t/\lambda_k})}{\sum_k \eta_k} \tag{5.3-25}$$

Note that this curve is monotonically increasing with t; it gives the upper "envelope" in Fig. 3.4-7. We cannot, by means of the linear viscoelastic theory, describe the "overshoot effect" shown in Figs. 3.4-7 and 8.

EXAMPLE 5.3-5 Constrained Recoil after Cessation of Steady Shear Flow

Next we investigate the system depicted in Fig. 3.4-18. Prior to $t = 0$ the fluid between the two parallel plates is undergoing steady shear flow with velocity gradient $\dot{\gamma}_0 \ll 1/\lambda_{max}$. After $t = 0$, the shear stress is removed so that $\tau_{yx} = 0$. The fluid then recoils and the strain $\gamma_{yx}(0, t)$ can be followed as a function of time. It is presumed that the plate spacing is small enough and the fluid viscous enough that a linear velocity profile is maintained throughout the experiment; that is, inertial effects can be neglected. We wish to find the "ultimate recoil," $\gamma_\infty = \gamma_{yx}(0, t)|_{t=\infty} = \int_0^\infty \dot{\gamma}_{yx}(t)\,dt$ (see Fig. 3.4-18). (*Note*: In this problem it is convenient to take $t = 0$ to be the reference time for the measurement of strain!)

SOLUTION Application of Eq. 5.2-18 to this problem for $t > 0$ gives

$$0 = -\int_{-\infty}^0 G(t-t')\dot{\gamma}_0\,dt' - \int_0^t G(t-t')\dot{\gamma}_{yx}(t')dt' \tag{5.3-26}$$

or, in terms of the variable $s = t - t'$

$$0 = -\dot{\gamma}_0 \int_t^\infty G(s)ds - \int_0^t G(s)\dot{\gamma}_{yx}(t-s)ds \tag{5.3-27}$$

Next we integrate over the time from $t = 0$ to $t = \infty$, and this gives

$$0 = -\dot{\gamma}_0 \int_0^\infty \int_t^\infty G(s)ds\,dt - \int_0^\infty \int_0^t G(s)\dot{\gamma}_{yx}(t-s)ds\,dt \tag{5.3-28}$$

Next interchange the order of integration to get

$$0 = -\dot{\gamma}_0 \int_0^\infty \left[\int_0^s dt\right] G(s)ds - \int_0^\infty \left[\int_s^\infty \dot{\gamma}_{yx}(t-s)dt\right] G(s)ds \tag{5.3-29}$$

or

$$0 = -\dot{\gamma}_0 \int_0^\infty sG(s)ds - \int_0^\infty \left[\int_0^\infty \dot{\gamma}_{yx}(t')dt'\right] G(s)ds \tag{5.3-30}$$

The quantity in brackets in Eq. 5.3-30 is just the ultimate recoil γ_∞, for which we then have the final result

$$\frac{-\gamma_\infty}{\dot{\gamma}_0} = \frac{\int_0^\infty sG(s)ds}{\int_0^\infty G(s)ds} \tag{5.3-31}$$

Notice that we did not actually solve the integral equation for $\dot{\gamma}_{yx}(t)$ in Eq. 5.3-26, but we were able to extract an integral of the solution, namely the ultimate recoil. This problem can also be solved by using Laplace transforms, as indicated in Problem 5D.1.

A quantity J_e^0, the steady-state compliance, is defined by Eq. 3.4-11

$$\gamma_\infty = J_e^0 \tau_0 \tag{5.3-32}$$

where $\tau_0 = -\eta_0 \dot{\gamma}_0$ is the shear stress prior to recoil. According to the linear viscoelasticity theory, this quantity is given by

$$J_e^0 = \frac{\int_0^\infty sG(s)ds}{[\int_0^\infty G(s)ds]^2} \tag{5.3-33}$$

This is obtained by combining the results in Eqs. 5.3-31 and 32.

EXAMPLE 5.3-6 Creep after Imposition of Constant Shear Stress

The next experiment we consider is that shown in Fig. 3.4-1f. Prior to $t = 0$ the fluid contained between two parallel plates is at rest. After $t = 0$ the fluid sample is subjected to a constant applied shear stress, so that the strain has a response similar to that shown in Fig. 5.3-3. We want to find an expression for the intercept γ_0. Designate the applied shear stress by τ_0 and the ultimate velocity gradient by $\dot{\gamma}_\infty$.

SOLUTION During the creep process, the rate of strain (or velocity gradient) will be (see Eq. 5.1-8a)

$$\dot{\gamma}_{yx} = \frac{d}{dt}\gamma_{yx}(0, t) \tag{5.3-34}$$

and therefore at any time t the strain will be

$$\gamma_{yx}(0, t) = \int_0^t \dot{\gamma}_{yx}(t')dt' \tag{5.3-35}$$

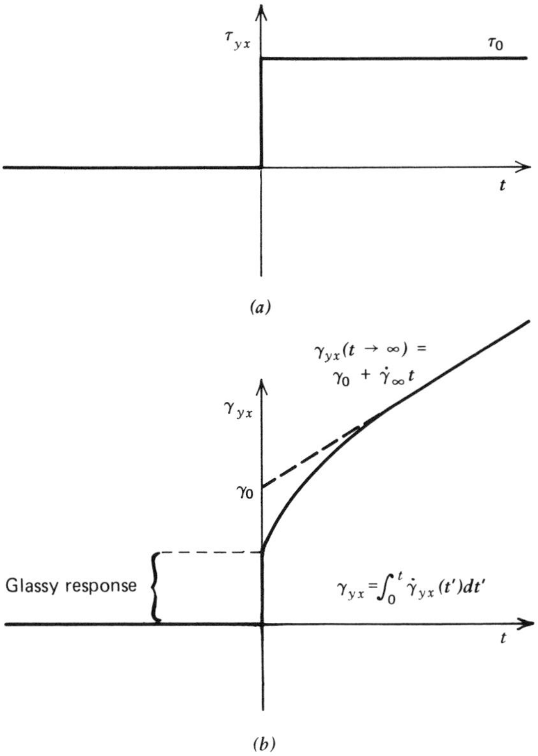

FIGURE 5.3-3. Creep experiment. (*a*) A constant shear stress τ_0 is applied at $t = 0$ and maintained for all times greater than $t = 0$; (*b*) the shear strain measured relative to $t = 0$ in the creep experiment increases with time and approaches an asymptote with slope $\dot{\gamma}_\infty$ for long times.

As in the preceding illustrative example we take $t = 0$ to be the reference time. For very large time, the strain curve will approach the dashed-line asymptote in Fig. 5.3-3, which is given by the equation

$$\lim_{t \to \infty} \gamma_{yx}(0, t) = \gamma_0 + \dot{\gamma}_\infty t \tag{5.3-36}$$

If we write Eq. 5.3-35 for $t \to \infty$ and equate the result to Eq. 5.3-36, then we get for the intercept γ_0

$$\gamma_0 = \int_0^\infty (\dot{\gamma}_{yx}(t') - \dot{\gamma}_\infty)dt' \tag{5.3-37}$$

Now we get an expression for this quantity by using the constitutive equation for the linear viscoelastic fluid.

If we write Eq. 5.2-18 for the creep experiment, we have

$$\tau_0 = -\int_0^t G(t - t')\dot{\gamma}_{yx}(t')dt' \tag{5.3-38}$$

But we can also write Eq. 5.2-18 for the steady-state shear flow with $\tau_{yx} = \tau_0$ and $\dot{\gamma}_{yx} = \dot{\gamma}_\infty$

$$\tau_0 = -\int_{-\infty}^t G(t - t')\dot{\gamma}_\infty \, dt' \tag{5.3-39}$$

We now equate these two expressions for τ_0 to obtain

$$\dot{\gamma}_\infty \int_0^\infty G(s)ds = \int_0^t G(s)\dot{\gamma}_{yx}(t-s)ds \tag{5.3-40}$$

This may be rewritten as

$$\dot{\gamma}_\infty \int_t^\infty G(s)ds = \int_0^t G(s)[\dot{\gamma}_{yx}(t-s) - \dot{\gamma}_\infty]ds \tag{5.3-41}$$

Next we integrate both sides from $t = 0$ to $t = \infty$ and then interchange the order of integration of s and t; this gives finally:

$$\frac{\gamma_0}{\dot{\gamma}_\infty} = \frac{\int_0^\infty sG(s)ds}{\int_0^\infty G(s)ds} \tag{5.3-42}$$

Here again we did not solve the integral equation for the velocity gradient, but extracted from the problem only the integral defined in Eq. 5.3-37.

From the above examples we find that certain quantities measured in different shearing experiments are given by the same ratio of integrals: $\int_0^\infty sG(s)ds/\int_0^\infty G(s)ds$. For future reference we list all of these results in Table 5.3-1. We also include one additional

TABLE 5.3-1

Shear Flow Experiments and "Analogous Quantities"

Experiment	Meaning of Symbols	Measurable Quantity[a] Equal to $\int_0^\infty sG(s)ds/\int_0^\infty G(s)ds$
Small-amplitude oscillatory motion (Eq. 5.3-7)	ω = frequency of oscillation	$\lim\limits_{\omega\to 0} \dfrac{\eta''/\omega}{\eta'}$
Stress relaxation after cessation of steady shear flow (Eq. 5.3-20)	$\dot{\gamma}_0$ = shear rate before stress relaxation	$\lim\limits_{\dot{\gamma}_0\to 0} \int_0^\infty \dfrac{\eta^-}{\eta_0} dt$
Stress growth after inception of steady shear flow (Eq. 5.3-24)	$\dot{\gamma}_0$ = shear rate during stress growth	$\lim\limits_{\dot{\gamma}_0\to 0} \int_0^\infty \left(1 - \dfrac{\eta^+}{\eta_0}\right)dt$
Constrained recoil after cessation of steady shear flow (Eq. 5.3-31)	$\dot{\gamma}_0$ = shear rate before recoil begins	$\lim\limits_{\dot{\gamma}_0\to 0} \dfrac{-\gamma_\infty}{\dot{\gamma}_0}$
Creep after application of steady shear stress (Eq. 5.3-42)	$\dot{\gamma}_\infty$ = shear rate when steady state is attained	$\lim\limits_{\tau_0\to 0} \dfrac{\gamma_0}{\dot{\gamma}_\infty}$
Steady-state shear flow (Problem 6B.3)	$\dot{\gamma}$ = shear rate at steady state	$\lim\limits_{\dot{\gamma}\to 0} \dfrac{\Psi_1}{2\eta}$

[a] Note that this ratio of integrals has the dimensions of time and is often used as the "characteristic time of the fluid" to construct the Deborah number. For the generalized Maxwell model the ratio of integrals is approximately equal to the longest relaxation time λ_1.

entry—that for the normal stress coefficient in steady-state shear flow—for which nonlinear viscoelasticity theory is required; this result will be obtained in Problem 6B.3, but we include it in the table for completeness.[4]

These relations among various experiments have been the subject of considerable experimentation, and they are useful in providing cross-checks of experimental techniques. The establishment of these interrelations requires no assumptions regarding the relaxation modulus. It must be kept in mind that all of the quantities listed in the right column of Table 5.3-1 are limiting values, and for some fluids these may have to be obtained experimentally by tedious extrapolation processes.

EXAMPLE 5.3-7 Determination of the Relaxation Spectrum from Linear Viscoelastic Data

In the preceding six examples we have shown how the response of a viscoelastic fluid to small deformation gradient experiments is related to the relaxation modulus $G(t)$. Here we want to show how this information can be used to find G from linear data. Take the relaxation modulus to be given by the generalized Maxwell model expression in Eq. 5.2-14 and use the storage and loss moduli for low-density polyethylene presented in Fig. 3.4-3 to illustrate the procedure for fitting G. Because time-temperature superposition has been used in obtaining these data, an extremely wide range of frequencies has been covered.

SOLUTION We begin by selecting a set of relaxation times λ_j for the spectrum. The spacing of these is conveniently taken to be decade intervals in order to reduce the amount of computation, but smaller spacings can be taken if more accurate fitting is required. The longest relaxation time λ_j is chosen so that $\lambda_1\omega_{\min} > 1$, where $\omega_{\min}$ is the lowest frequency for which data are available, unless the zero-frequency region is reached in the experiments. In the latter case take $\lambda_1\omega_0 \doteq 0.1$ where ω_0 is the critical frequency that marks the end of the zero-frequency regime. Similarly the smallest relaxation time $\lambda_{\min}$ is chosen so that $\lambda_{\min}\omega_{\max} < 1$, where $\omega_{\max}$ is the highest frequency for which data are available. For the low-density polyethylene melt we choose the λ_j to be $10^3, 10^2, \ldots, 10^{-3}, 10^{-4}$ s.

It remains to fit the viscosities η_j for each relaxation time. This is done by minimizing the difference between the measured and predicted moduli at N frequencies ω_j. If we denote the measured properties by G'_j and G''_j and the predicted properties by $G'(\omega_j)$ and $G''(\omega_j)$, then the quantity to be minimized is

$$\sum_{j=1}^{N}\left\{\left[\frac{G'(\omega_j)}{G'_j}-1\right]^2+\left[\frac{G''(\omega_j)}{G''_j}-1\right]^2\right\}$$

The calculated moduli are obtained from the following truncated forms of Eqs. 5.3-8 and 9:

$$G'(\omega_j) = \sum_{k=1}^{N}\frac{\eta_k\lambda_k\omega_j^2}{1+(\lambda_k\omega_j)^2} \tag{5.3-43}$$

$$G''(\omega_j) = \sum_{k=1}^{N}\frac{\eta_k\omega_j}{1+(\lambda_k\omega_j)^2} \tag{5.3-44}$$

For the low-density polyethylene this minimization[5] leads to the values of η_k shown in Table 5.3-2.

It is instructive to look at the spectral decomposition of the dynamic moduli that are predicted by the relaxation spectrum parameters in Table 5.3-2. Figure 5.3-4 shows the contribution of the

[4] The equality of the entries involving $\eta''/\omega\eta'$ and $\Psi_1/2\eta$ has been verified experimentally by T. Kotaka and K. Osaki, *J. Polym. Sci.*, **C15**, 453–479 (1966).

[5] A. C. Papanastasiou, L. E. Scriven, and C. W. Macosko, *J. Rheol.*, **27**, 387–410 (1983), discuss the use of nonlinear regression methods to determine simultaneously the best set of λ_k and η_k to fit $G(t)$.

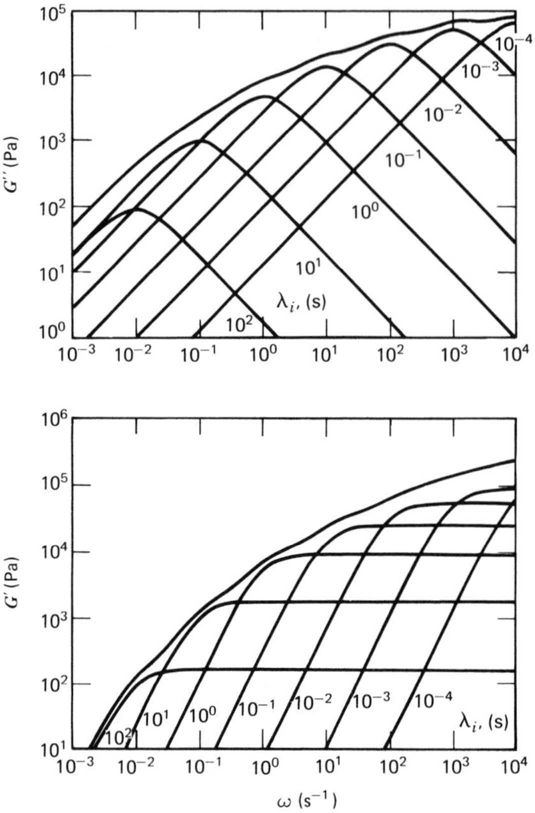

FIGURE 5.3-4. Spectral decomposition of the storage and loss moduli for the low-density polyethylene melt shown in Fig. 3.4-3. The moduli are calculated by Eqs. 5.3-43 and 44 with the λ_k and η_k given in Table 5.3-2. The upper, composite curves are also shown in Fig. 3.4-3 for comparison with the experimental data. Note that the contribution from the longest relaxation time is not shown. [H. M. Laun, *Rheol. Acta*, **17**, 1–15 (1978).]

TABLE 5.3-2

Constants in the Linear Viscoelastic Spectrum for the Low-Density Polyethylene Melt of Fig. 3.4-3

i	$\lambda_i(s)$	η_i (Pa · s)
1	10^3	1.00×10^3
2	10^2	1.80×10^4
3	10^1	1.89×10^4
4	10^0	9.80×10^3
5	10^{-1}	2.67×10^3
6	10^{-2}	5.86×10^2
7	10^{-3}	9.48×10^1
8	10^{-4}	1.29×10^1

Maxwell elements to the moduli. From this figure it is clear that the low-frequency behavior is dominated by the long relaxation times and that the high-frequency response is controlled by the short relaxation times. Once the spectrum is available it is straightforward to calculate other linear viscoelastic material functions by means of the formulas developed in the preceding examples. Predictions made in this way are included in the graphs of low-density polyethylene material functions presented in §§3.4 and 3.5.

§5.4 LINEAR VISCOELASTIC FLOW PROBLEMS

In determining the linear viscoelastic material functions in §5.3, the equations of continuity and motion were not needed because it was assumed there that the velocity profiles are, to a good approximation, linear in properly designed linear viscoelasticity experiments. In this section we turn our attention to the solution of flow problems, in which we are given only the boundary and initial conditions, and we have to solve the equations of continuity and motion along with the constitutive equation.

EXAMPLE 5.4-1 Wave Transmission in a Semi-Infinite Viscoelastic Liquid

A viscoelastic liquid is located in the region $0 \le y < \infty$. The velocity v_x at the surface $y = 0$ is maintained at $v_x = V \cos \omega t$, where V is the amplitude of the velocity and ω is the frequency. Find the velocity distribution $v_x(y, t)$ throughout the medium, after the initial transients have died out. For the analogous Newtonian fluid problem see Problem 1D.1.

SOLUTION We postulate that the velocity profile has the form $v_x = v_x(y, t)$, $v_y = 0$, $v_z = 0$. Then the equation of continuity is exactly satisfied, and the equation of motion and the constitutive equation give

$$\rho \frac{\partial v_x}{\partial t} = -\frac{\partial}{\partial y} \tau_{yx} \tag{5.4-1}$$

$$\tau_{yx}(y, t) = -\int_{-\infty}^{t} G(t - t') \frac{\partial v_x(y, t')}{\partial y} dt' \tag{5.4-2}$$

Combining these results in an integrodifferential equation for $v_x(y, t)$:

$$\rho \frac{\partial v_x}{\partial t} = \int_{-\infty}^{t} G(t - t') \frac{\partial^2 v_x(y, t')}{\partial y^2} dt' \tag{5.4-3}$$

Inasmuch as we are not interested in the fluid response immediately after the commencement of the wall oscillation, it is appropriate to make a further postulate for $v_x(y, t)$; since the system is linear, it is anticipated that the fluid will execute sinusoidal motion throughout, so that

$$v_x(y, t) = \mathscr{Re}\,\{v_x^0(y) e^{i\omega t}\} \tag{5.4-4}$$

Here $v_x^0(y)$ is a complex amplitude, with $v_x^0(0) = V$ and $v_x^0(\infty) = 0$. When this velocity expression is substituted into the previous equation we get

$$\begin{aligned} \rho\, \mathscr{Re}\,\{i\omega v_x^0(y) e^{i\omega t}\} &= \int_{-\infty}^{t} G(t - t')\, \mathscr{Re} \left\{ \frac{d^2 v_x^0}{dy^2} e^{i\omega t'} \right\} dt' \\ &= \mathscr{Re} \left\{ \frac{d^2 v_x^0}{dy^2} e^{i\omega t} \int_0^{\infty} G(s) e^{-i\omega s}\, ds \right\} \end{aligned} \tag{5.4-5}$$

The integral over $s = t - t'$ in the last line is just η^* according to Eq. 5.3-6. When the $\mathscr{R}e$-operator is removed and the factors $e^{i\omega t}$ eliminated, we get the ordinary differential equation for v_x^0

$$\frac{d^2 v_x^0}{dy^2} - \frac{\rho i \omega}{\eta^*} v_x^0 = 0 \tag{5.4-6}$$

If we let

$$\frac{\rho i \omega}{\eta^*} = (\alpha + i\beta)^2 \tag{5.4-7}$$

where α and β are real functions of ω, then the solution to Eq. 5.4-6 is

$$v_x^0(y) = Ve^{-(\alpha + i\beta)y} \tag{5.4-8}$$

and then from Eq. 5.4-4 we get

$$v_x(y, t) = Ve^{-\alpha y} \cos(\omega t - \beta y) \tag{5.4-9}$$

It is now clear that α is the attenuation of the velocity wave, and βy is the phase shift at a distance y from the wall; these frequency-dependent quantities are obtained by solving Eq. 5.4-7. After some tedious algebra we get

$$\alpha(\omega) = \frac{1}{|\eta^*|}\sqrt{\frac{\rho\omega}{2}(|\eta^*| - \eta'')} \tag{5.4-10}$$

$$\beta(\omega) = \frac{\eta'}{|\eta^*|}\sqrt{\frac{(\rho\omega/2)}{|\eta^*| - \eta''}} \tag{5.4-11}$$

where $|\eta^*| = \sqrt{\eta'^2 + \eta''^2}$. For the Newtonian fluid $\eta'' = 0$, $\eta' = |\eta^*| = \mu$, and $\alpha = \beta = \sqrt{\rho\omega/2\mu}$.

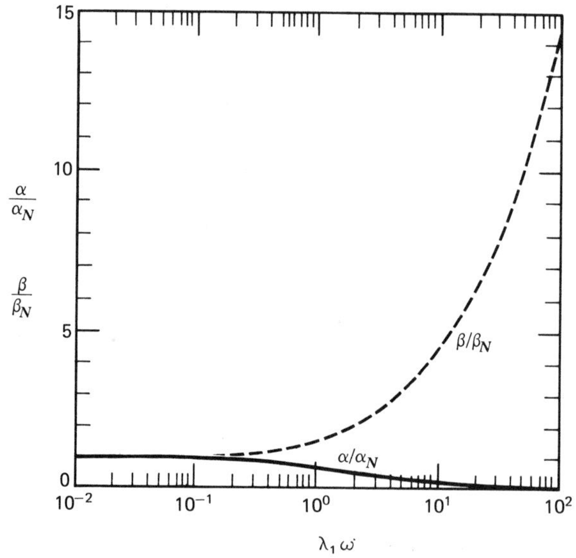

FIGURE 5.4-1. The functions $\alpha(\omega)$ and $\beta(\omega)$, from Eqs. 5.4-10 and 11, for the Maxwell model of Eq. 5.2-2. The functions have been normalized by dividing by the corresponding quantities for a Newtonian fluid with viscosity η_0, namely $\alpha_N = \beta_N = \sqrt{\rho\omega/2\eta_0}$.

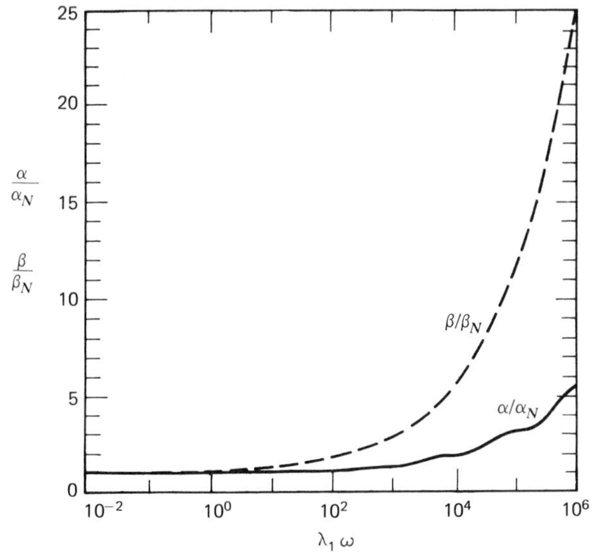

FIGURE 5.4-2. The functions $\alpha(\omega)$ and $\beta(\omega)$ of Eqs. 5.4-10 and 11, obtained from the experimental data on an LDPE melt, as represented by a finite sum of exponentials (see Example 5.3-7). The time constant λ_1 is the longest time constant in the relaxation spectrum; $T = 150°C$.

The functions $\alpha(\omega)$ and $\beta(\omega)$ are plotted in Fig. 5.4-1 for a Maxwell model; for this model $\alpha(\omega)$ is a monotone decreasing function. For the generalized Maxwell model, Eqs. 15.3-14 and 15 suggest that $\alpha(\omega)$ should increase somewhat with ω for large values of the frequency. This seems to be borne out by the function $\alpha(\omega)$ determined from experimental data on $\eta^*(\omega)$ for a polyethylene melt, as shown in Fig. 5.4-2. Both figures show that $\beta(\omega)$ is a rather rapidly increasing function of ω.

Finally it must be kept in mind that the analysis given here, leading to Eqs. 5.4-9 to 11, is valid only for linear viscoelasticity. This means that the final results can be used only if $V\sqrt{\rho/\omega\eta_0} \ll 1$.

EXAMPLE 5.4-2 Motion of a Viscoelastic Fluid Pulsating in a Tube[1]

A polymeric liquid in a circular tube of radius R is made to oscillate by means of a sinusoidally varying pressure gradient

$$-\frac{\partial p}{\partial z} = \mathcal{R}e\,\{P^0 e^{i\omega t}\} \tag{5.4-12}$$

where the magnitude of the complex amplitude P^0 is very small. Obtain an expression for the oscillatory volume flow rate when the "oscillatory steady state" has been reached.

SOLUTION We postulate that the velocity and shear-stress distributions are of the form

$$v_z = \mathcal{R}e\,\{v_z^0(r)e^{i\omega t}\}, \qquad v_r = 0, \qquad v_\theta = 0 \tag{5.4-13}$$

$$\tau_{rz} = \mathcal{R}e\,\{\tau_{rz}^0(r)e^{i\omega t}\} \tag{5.4-14}$$

[1] This problem was solved for a Maxwell fluid by L. J. F. Broer, *Appl. Sci. Res.*, **A6**, 226–236 (1957), and for the general linear viscoelastic fluid by A. G. Fredrickson, *Principles and Applications of Rheology*, Prentice-Hall, Englewood Cliffs, NJ (1964), pp. 133 *et seq.*

Here v_z^0 and τ_{rz}^0 are complex functions. The equation of continuity is automatically satisfied, and the equation of motion and constitutive equation are

$$\rho \frac{\partial v_z}{\partial t} = -\frac{\partial p}{\partial z} - \frac{1}{r}\frac{\partial}{\partial r}(r\tau_{rz}) \tag{5.4-15}$$

$$\tau_{rz} = -\int_{-\infty}^{t} G(t-t')\frac{\partial v_z(r,t')}{\partial r}\,dt' \tag{5.4-16}$$

Insertion of the expressions from Eqs. 5.4-12 to 14 into these last two equations gives

$$\rho i\omega v_z^0 = P^0 - \frac{1}{r}\frac{d}{dr}r\tau_{rz}^0 \tag{5.4-17}$$

$$\tau_{rz}^0 = -\eta^* \frac{dv_z^0}{dr} \tag{5.4-18}$$

where $\eta^* = \int_0^\infty G(s)\exp(-i\omega s)ds$. Combination of the last two equations gives

$$\eta^* \frac{1}{r}\frac{d}{dr}\left(r\frac{dv_z^0}{dr}\right) - \rho i\omega v_z^0 = -P^0 \tag{5.4-19}$$

which has to be solved with the boundary conditions that $v_z^0 = 0$ at $r = R$ and $v_z^0 =$ finite at $r = 0$. This equation is readily solved to give

$$v_z^0(r) = \frac{P^0}{\rho i\omega}\left(1 - \frac{J_0(\alpha r)}{J_0(\alpha R)}\right) \tag{5.4-20}$$

in which $\alpha^2 = -\rho i\omega/\eta^*$.

The volume flow rate

$$Q = 2\pi\int_0^R v_z r\,dr \tag{5.4-21}$$

will be expected to be of the form $Q = \mathscr{R}e\{Q^0\exp(i\omega t)\}$. When $v_z^0(r)$ from Eq. 5.4-20 is put into the expression for Q, the complex amplitude Q^0 can be found

$$\begin{aligned} Q^0 &= \frac{\pi R^2 P^0}{\rho i\omega}\left(1 - \frac{2J_1(\alpha R)}{\alpha R J_0(\alpha R)}\right) \\ &= \frac{\pi R^4 P^0}{8\eta^*}\left(1 - \frac{\rho i\omega R^2}{6\eta^*} + \cdots\right) \end{aligned} \tag{5.4-22}$$

In the second line, the power-series expansions of the Bessel functions have been used.

Let us now specialize to the Maxwell model for which

$$\eta^* = \frac{\eta_0}{1 + i\lambda_1\omega} \tag{5.4-23}$$

If we take P^0 to be real, then from Eq. 5.4-22 we get

$$Q = \frac{\pi R^4 P^0}{8\eta_0}\left[\left(1 + \frac{\rho R^2}{3\eta_0 \lambda_1}\lambda_1^2\omega^2 + \cdots\right)\cos\omega t - \left(1 - \frac{\rho R^2}{6\eta_0 \lambda_1}(1 - \lambda_1^2\omega^2) + \cdots\right)\lambda_1\omega \sin\omega t\right] \quad (5.4\text{-}24)$$

For the Newtonian liquid ($\lambda_1 = 0$), the volume flow rate Q is not in phase with the pressure gradient because of inertial effects. For the viscoelastic fluid, there is an additional phase shift associated with the time constant λ_1.

EXAMPLE 5.4-3 Analysis of a Torsional Oscillatory Viscometer with Small Slit Width[2]

In the torsional oscillatory viscometer the fluid is placed between a "cup" of radius aR and a "bob" of radius R as shown in Fig. 5.4-3. The cup is made to undergo very small sinusoidal oscillations in the tangential direction. This motion causes the bob, suspended by a torsion wire, to oscillate with the same frequency but with a different amplitude and a different phase. Show how measurement of the amplitude ratio and phase shift between the input (cup) oscillation and the output (bob) oscillation leads to the determination of the complex viscosity. Denote by θ_R and θ_{aR} the time-dependent angular displacements of the bob and cup, respectively.

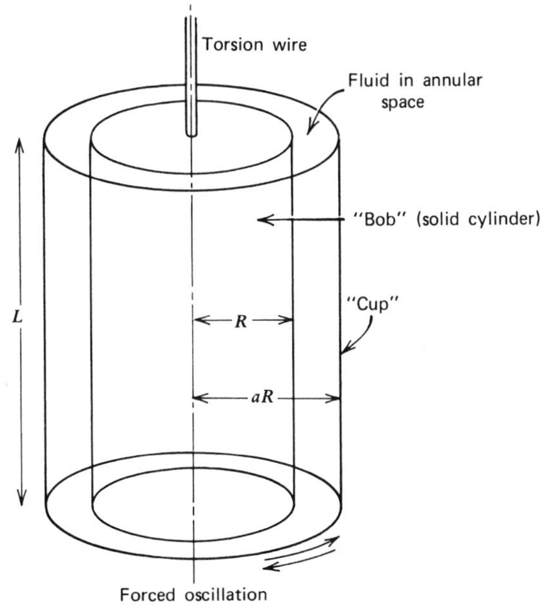

FIGURE 5.4-3. Torsional oscillatory viscometer.

[2] H. Markovitz, *J. Appl. Phys.*, **23**, 1070–1077 (1952) analyzed this problem without neglecting curvature, that is, without the narrow gap assumption. See also S. Oka, in *Rheology*, Vol. 3, F. R. Eirich, ed., Academic Press, New York (1960), Chap. 2.

SOLUTION We begin by applying Newton's second law of motion to the bob. The rate of change in its angular momentum, $I d^2\theta_R/dt^2$, is set equal to the sum of the torques acting on it; here I is the moment of inertia of the bob. When the bob has rotated through an angle θ_R, the torsion wire exerts a restoring torque $-K\theta_R$ on the bob, where K is the torsion constant of the wire; in addition the fluid exerts a torque on the bob through the shear stress at the bob surface $2\pi RL \cdot R \cdot (-\tau_{r\theta})|_R$. Combining these contributions gives an equation of motion for the bob

$$I \frac{d^2\theta_R}{dt^2} = -K\theta_R + 2\pi R^2 L(-\tau_{r\theta})|_R \tag{5.4-25}$$

subject to the initial condition that $\theta_R = 0$ at $t = 0$.

For the fluid in the gap, we assume that $v_\theta = v_\theta(r, t)$, $v_r = v_z = 0$ so that the θ-component of the equation of motion (Eq. B.1-5) becomes

$$\rho \frac{\partial v_\theta}{\partial t} = -\frac{1}{r^2}\frac{\partial}{\partial r}(r^2 \tau_{r\theta}) \tag{5.4-26}$$

The no-slip boundary conditions on the bob and cup surfaces are

$$\text{B.C. 1:} \quad \text{at } r = R \quad v_\theta = R\frac{d\theta_R}{dt} \tag{5.4-27}$$

$$\text{B.C. 2:} \quad \text{at } r = aR \quad v_\theta = aR\frac{d\theta_{aR}}{dt} \tag{5.4-28}$$

The initial condition is that $v_\theta = 0$ at $t = 0$; the initial conditions are not needed here since we shall be interested in the bob motion after all start-up transients have died out.

Next we use the natural frequency of the system $\sqrt{K/I}$ to define a dimensionless time $\tau = t\sqrt{K/I}$; a dimensionless driving frequency $\tilde{\omega}$ is similarly defined. The driving function, that is, the forced angular oscillations of the cup, may then be written

$$\theta_{aR} = \theta_{aR}^0 \, \mathscr{Re}\,\{e^{i\tilde{\omega}\tau}\} \tag{5.4-29}$$

The amplitude of the cup oscillations, which is small, is taken here to be real, so that the cup motion is described by a cosine function. After the initial transients have died out, the shear stress in the fluid, the fluid velocity, and the bob are all expected to oscillate at the same frequency ω but with different amplitudes and phase angles than the cup. We thus postulate

$$\theta_R = \mathscr{Re}\,\{\theta_R^0 e^{i\tilde{\omega}\tau}\} \tag{5.4-30}$$

$$v_\theta = \mathscr{Re}\,\{v_\theta^0(r) e^{i\tilde{\omega}\tau}\} \tag{5.4-31}$$

$$\begin{aligned} \tau_{r\theta} &= \mathscr{Re}\,\{\tau_{r\theta}^0(r) e^{i\tilde{\omega}\tau}\} \\ &= \mathscr{Re}\left\{-\eta^* \frac{\partial v_\theta^0}{\partial r} e^{i\tilde{\omega}\tau}\right\} \end{aligned} \tag{5.4-32}$$

where θ_R^0, v_θ^0, and $\tau_{r\theta}^0$ are the complex amplitudes of the bob angle, fluid velocity and shear stress, respectively. In the last line of Eq. 5.4-32 we have inserted the complex viscosity using its definition in Eq. 3.4-3b′.

When Eqs. 5.4-30 through 32 are substituted into Eqs. 5.4-25 and 26, the latter equations become

$$(1 - \tilde{\omega}^2)K \,\mathscr{R}e\, \{\theta_R^0 e^{i\tilde{\omega}\tau}\} = 2\pi R^2 L \,\mathscr{R}e \left\{ \eta^* \left. \frac{\partial v_\theta^0}{\partial r} \right|_{r=R} e^{i\tilde{\omega}\tau} \right\} \tag{5.4-33}$$

$$\rho\tilde{\omega} \sqrt{\frac{K}{I}} \,\mathscr{R}e\, \{i v_\theta^0 e^{i\tilde{\omega}\tau}\} = \mathscr{R}e \left\{ \frac{\eta^*}{r^2} \frac{\partial}{\partial r} \left(r^2 \frac{\partial}{\partial r} v_\theta^0 \right) e^{i\tilde{\omega}\tau} \right\} \tag{5.4-34}$$

We now remove the $\mathscr{R}e$-operators from these equations to get

$$\theta_R^0 = \frac{2\pi R^2 L \eta^*}{(1 - \tilde{\omega}^2)K} \left. \left(\frac{\partial v_\theta^0}{\partial r} \right) \right|_{r=R} \tag{5.4-35}$$

$$v_\theta^0 = -\frac{i\eta^*}{\rho\tilde{\omega}} \sqrt{\frac{I}{K}} \frac{1}{r^2} \frac{\partial}{\partial r} \left(r^2 \frac{\partial}{\partial r} v_\theta^0 \right) \tag{5.4-36}$$

Equations 5.4-35 and 36 can be simplified by assuming that $a \doteq 1$, so that curvature effects may be neglected. To specify positions between the cup and bob we introduce a dimensionless length variable $x = (r - R)/(a - 1)R$. The last two equations may thus be written as

$$\theta_R^0 = \frac{M}{(1 - \tilde{\omega}^2)} \left. \frac{d\phi^0}{dx} \right|_{x=0} \tag{5.4-37}$$

$$\frac{d^2\phi^0}{dx^2} - \frac{i\tilde{\omega}}{M} \phi^0 = 0 \tag{5.4-38}$$

where Eq. 5.4-38 must be solved with the boundary conditions

$$B.C.\ 1': \quad \text{at } x = 0 \quad \phi^0 = iA\tilde{\omega}\theta_R^0 \tag{5.4-39}$$

$$B.C.\ 2': \quad \text{at } x = 1 \quad \phi^0 = iA\tilde{\omega}\theta_{aR}^0 \tag{5.4-40}$$

In Equations 5.4-37 to 40 we have used the dimensionless groups

$$\phi^0 = \frac{2\pi R^3 L\rho(a - 1)}{\sqrt{KI}} v_\theta^0; \quad M = \frac{\eta^*/\rho}{(a - 1)^2 R^2} \sqrt{\frac{I}{K}}; \quad A = 2\pi R^4 \frac{L\rho(a - 1)}{I} \tag{5.4-41}$$

One can now determine θ_{aR}^0/θ_R^0 by first solving for ϕ^0 and then using that solution in Eq. 5.4-37. The result is conveniently expressed as a power series in $1/M$

$$\frac{\theta_{aR}^0}{\theta_R^0} = 1 + \frac{i}{M} \left(\frac{\tilde{\omega}^2 - 1}{A\tilde{\omega}} + \frac{\tilde{\omega}}{2} \right) - \frac{1}{M^2} \left(\frac{\tilde{\omega}^2 - 1}{6A} + \frac{\tilde{\omega}^2}{24} \right) + O\left(\frac{1}{M^3} \right) \tag{5.4-42}$$

The experimentally measurable quantities are the amplitude ratio $\mathscr{A} = |\theta_R^0|/\theta_{aR}^0$ and the phase angle δ. In terms of these quantities we may write the left-hand side of Eq. 5.4-42 as

$$\frac{\theta_{aR}^0}{\theta_R^0} = \frac{1}{\mathscr{A} e^{i\delta}} = \mathscr{A}^{-1}(\cos\delta - i \sin\delta) \tag{5.4-43}$$

To illustrate how η' and η'' are found from $\mathscr{A}$ and δ, we assume terms of $O(1/M^2)$ in Eq. 5.4-42 are negligible. That equation may then be easily solved for M

$$M = \frac{iW}{\mathscr{A}^{-1}(\cos\delta - i\sin\delta) - 1}$$

$$= -\frac{W\mathscr{A}\sin\delta}{1 + \mathscr{A}^2 - 2\mathscr{A}\cos\delta} - \frac{iW\mathscr{A}(\mathscr{A} - \cos\delta)}{1 + \mathscr{A}^2 - 2\mathscr{A}\cos\delta} \tag{5.4-44}$$

where W is given by

$$W = \frac{\tilde{\omega}^2 - 1}{A\tilde{\omega}} + \frac{\tilde{\omega}}{2} \tag{5.4-45}$$

Then from the definition of M in Eq. 5.4-41 and from the definition $\eta^* = \eta' - i\eta''$ we find

$$\eta' = \frac{-\rho BW\mathscr{A}\sin\delta}{1 + \mathscr{A}^2 - 2\mathscr{A}\cos\delta} \tag{5.4-46}$$

$$\eta'' = \frac{\rho BW\mathscr{A}(\mathscr{A} - \cos\delta)}{1 + \mathscr{A}^2 - 2\mathscr{A}\cos\delta} \tag{5.4-47}$$

in which B is an instrument parameter defined by

$$B = (a - 1)^2R^2\sqrt{\frac{K}{I}} \tag{5.4-48}$$

Equations 5.4-46 and 47 are the desired expressions for the complex viscosity functions. Here W, $\mathscr{A}$, and δ are all functions of the dimensionless frequency $\tilde{\omega}$, which is related to ω in Eq. 3.4-3 by $\tilde{\omega} = \omega/\sqrt{K/I}$.

§5.5 LIMITATIONS OF LINEAR VISCOELASTICITY AND RECOMMENDATIONS FOR ITS USE

It was emphasized that the general integral expressions for linear viscoelasticity in Eqs. 5.2-18 and 19 are recommended for the description of the stresses in fluid motions that involve infinitesimally small displacement gradients. The general linear viscoelastic model has the following limitations:

- **a.** It cannot describe shear-rate dependence of viscosity, inasmuch as that effect occurs for flows with $\lambda_{max}\dot{\gamma} \geq 1$.
- **b.** It cannot describe normal-stress phenomena, since they are nonlinear effects.
- **c.** It cannot describe small-strain phenomena if these involve large displacement gradients due to superposed rigid rotations (see Example 5.5-1).

The problem mentioned in (**c**) results from the fact that, in Eqs. 5.2-18 and 19, the stress tensor has an unwanted dependence on the orientation of a fluid element. This defect is remedied in Chapter 8 by replacing the infinitesimal strain tensor $\boldsymbol{\gamma}(t, t')$ by a "finite strain tensor" $\boldsymbol{\gamma}_{[0]}(t, t')$.

EXAMPLE 5.5-1 Illustration that the General Linear Viscoelastic Model Is Restricted to Motions with Small Displacement Gradients

Consider the flow system shown in Fig. 5.5-1, in which a steady-state shear flow experiment with $v_{\bar{x}} = \dot{\gamma}\bar{y}$ is being performed on a rotating turntable. The product $\lambda_{\max}\dot{\gamma}$ is presumed to be much less than unity. The time variable is chosen in such a way that the $\bar{x}$-axis is lined up with the x-axis at time $t = 0$.

Use Eq. 5.2-18 to find the components of the stress tensor in the xyz-system. Show that one is not led to the relation $\eta_0 = \int_0^\infty G(s)ds$, but that the viscosity is found to depend on the angular velocity of the turntable! This illustrates that there is a fundamental flaw in the formulation of the equation for linear viscoelasticity as given in Eq. 5.2-18.

SOLUTION First we have to get a relation between the coordinates $\bar{x}$, $\bar{y}$ referred to the turntable axes and the coordinates x, y referred to the axes fixed in space. These are

$$\begin{cases} \bar{x} = (x - x_0)\cos Wt + (y - y_0)\sin Wt & (5.5\text{-}1) \\ \bar{y} = -(x - x_0)\sin Wt + (y - y_0)\cos Wt & (5.5\text{-}2) \end{cases}$$

Then the velocity distribution $\mathbf{v}(x, y, t)$ in the flow system as seen by an observer in the xyz-system is

$$\begin{cases} v_x = \dot{\gamma}[-(x - x_0)\sin Wt\cos Wt + (y - y_0)\cos^2 Wt] - W(y - y_0) & (5.5\text{-}3) \\ v_y = \dot{\gamma}[-(x - x_0)\sin^2 Wt + (y - y_0)\sin Wt\cos Wt] + W(x - x_0) & (5.5\text{-}4) \end{cases}$$

The rate-of-strain tensor $\dot{\boldsymbol{\gamma}}$ then has the form

$$\dot{\boldsymbol{\gamma}}(t) = \begin{pmatrix} -\sin 2Wt & \cos 2Wt & 0 \\ \cos 2Wt & \sin 2Wt & 0 \\ 0 & 0 & 0 \end{pmatrix} \dot{\gamma} \tag{5.5-5}$$

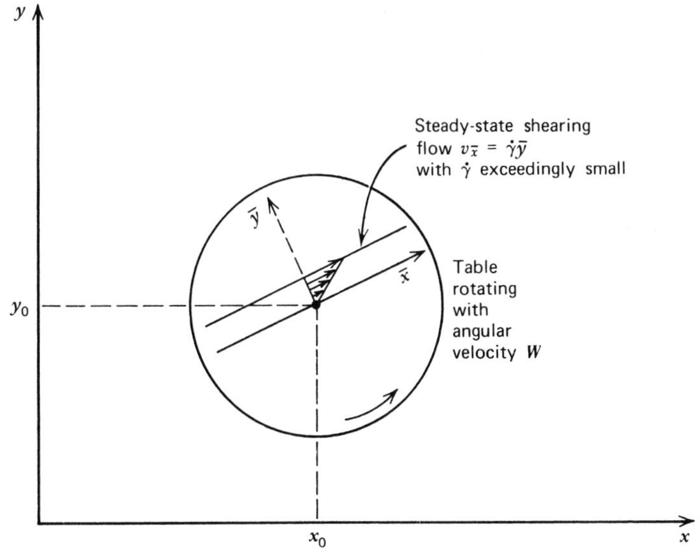

FIGURE 5.5-1. A steady-state shear flow experiment on a rotating turntable. The flow appears to be steady state to an observer standing on the turntable (i.e., $v_{\bar{x}} = \dot{\gamma}\bar{y}$), but the flow appears to be time-dependent to an observer who is not on the turntable (see Eqs. 5.5-3 and 4).

Then, according to Eq. 5.2-18 the stresses are given by

$$\boldsymbol{\tau} = -\dot{\gamma} \int_0^\infty G(s) \begin{pmatrix} -\sin 2W(t-s) & \cos 2W(t-s) & 0 \\ \cos 2W(t-s) & \sin 2W(t-s) & 0 \\ 0 & 0 & 0 \end{pmatrix} ds \tag{5.5-6}$$

At time $t = 0$, when the parallel-plate system is lined up with the x-axis, we get

$$\tau_{yx} = -\eta_0 \dot{\gamma} = -\dot{\gamma} \int_0^\infty G(s) \cos 2Ws \, ds \tag{5.5-7}$$

This tells us that the zero-shear-rate viscosity depends on W, which of course we find impossible to believe. This result serves to emphasize the fact that Eq. 5.2-18 carries with it the restriction of small displacement gradients (see Problem 5C.2); when viewed from the xyz-coordinate system the displacement gradients are not small. Note that Eq. 5.5-7 does give an acceptable result if the fluid memory is of short duration relative to the reciprocal of the angular velocity—that is, $\lambda_{max} W \ll 1$.

PROBLEMS

5A.1 Determination of Relaxation Modulus from Dynamic Viscosity

Compare the following two procedures for obtaining the relaxation modulus $G(s)$ from the dynamic viscosity:

a. Equation 5.3-4 can be inverted to give

$$G(s) = \frac{2}{\pi} \int_0^\infty \eta'(\omega) \cos \omega s \, d\omega \tag{5A.1-1}$$

Insert the dynamic viscosity data for 2% polyisobutylene in Primol in Fig. 3.4-4 into the integral and determine $G(s)$.

b. Assume that $G(s)$ has the form given by the expression in braces in Eq. 5.2-14, with $s = t - t'$, and use the relations in Eqs. 5.2-16 and 17 to reduce the number of parameters to three. Determine η_0, λ, and α by fitting Eqs. 5.3-12 and 14 to the 2% polyisobutylene in Primol data of Fig. 3.4-4. Then plot the curve for $G(s)$.

c. Follow the procedure in Example 5.3-7 to fit the same set of data.

5A.2 Experimental Test of the Kramers–Kronig Relations

There is a pair of relations (cf. Problem 5D.2) that enables one to compute η'' from η' data or vice versa; these are known as the Kramers–Kronig relations. For example,

$$\frac{\eta''}{\omega} = \frac{2}{\pi} \int_0^\infty \frac{\eta'(\omega) - \eta'(\bar{\omega})}{\bar{\omega}^2 - \omega^2} \, d\bar{\omega} \tag{5A.2-1}$$

Use the η' data from Fig. 3.4-4 for 2% polyisobutylene in Primol to see how well you are able to duplicate the η'' data for the same solution given in Fig. 3.4-5.

5B.1 Complex Viscosity for Elementary Linear Models

a. Obtain η' and η'' for the Maxwell model by using each of Eqs. 5.2-2, 6, and 8. Which equation is the easiest to use?

b. Repeat (a) for the Jeffreys model, using Eqs. 5.2-9, 11, and 12. From the experimental data, show that $\lambda_1 > \lambda_2$.

5B.2 Special Form for the Relaxation Modulus

It has been suggested[1] that

$$G(s) = \frac{\eta_0/\lambda}{\Gamma(1-\nu)}\left(\frac{\lambda}{s}\right)^{\nu} e^{-s/\lambda} \tag{5B.2-1}$$

may be a useful three-parameter empiricism for the relaxation modulus; η_0 is the zero-shear-rate viscosity, λ is a time constant, and ν is a dimensionless parameter ($0 < \nu \leq 1$). Verify that this $G(s)$ leads to

$$\eta^* = \eta_0 \frac{1}{(1 + i\lambda\omega)^{1-\nu}} \tag{5B.2-2}$$

Obtain η' and η'' for $\nu = 1/2$. What is the meaning of Eq. 5B.2-1 when $\nu = 0$?

5B.3 The Relaxation Spectrum[2]

Sometimes polymer chemists have chosen to use, instead of the relaxation modulus for the generalized Maxwell model (quantity in braces in Eq. 5.2-14), an expression that implies a "continuum" of exponentials. That is, they write

$$G(s) = \int_0^{\infty} \frac{H(\lambda)}{\lambda} e^{-s/\lambda}\, d\lambda = \int_{-\infty}^{+\infty} H(\lambda) e^{-s/\lambda}\, d\ln\lambda \tag{5B.3-1}$$

in which $H(\lambda)$ is called the relaxation spectrum. Specifying $H(\lambda)$ is equivalent to specifying $G(s)$. Empirical formulas for $H(\lambda)$ are available, containing two or three constants.

All of the linear viscoelasticity problems worked in this chapter can be developed in terms of $H(\lambda)$; for example, show that

$$\eta'(\omega) = \int_0^{\infty} \frac{H(\lambda)\, d\lambda}{1 + (\lambda\omega)^2}; \qquad \frac{\eta''(\omega)}{\omega} = \int_0^{\infty} \frac{\lambda H(\lambda)\, d\lambda}{1 + (\lambda\omega)^2} \tag{5B.3-2}$$

Interpret the results by comparing them with the expressions for η' and η'' for the Maxwell model.

5B.4 Relaxation Modulus for a Suspension of Rigid Dumbbells

It is shown in Chapter 14 that kinetic theory gives for the relaxation modulus for a dilute suspension of rigid dumbbells

$$G(t-t') = 2\eta_s \delta(t-t') + nkT[\tfrac{4}{5}\lambda\delta(t-t') + \tfrac{3}{5} e^{-(t-t')/\lambda}] \tag{5B.4-1}$$

[1] D. J. Segalman, private communication.

[2] J. D. Ferry, *Viscoelastic Properties of Polymers*, 3rd ed., Wiley, New York (1980), pp. 60, 343.

Here η_s is the viscosity of the solvent, n is the number of dumbbells per unit volume of solution, and $\lambda = \zeta L^2/12kT$ is a time constant; in the latter, L is the distance between the centers of the two "beads" making up the dumbbell, and ζ is the friction coefficient that gives the Stokes' law resistance of one bead of the dumbbell as it moves through the solvent ($\zeta = 6\pi\eta_s a$, where a is the bead radius).

a. How are the constants in the Jeffreys model of Eq. 5.2-11 related to the kinetic theory parameters which describe a rigid, rodlike macromolecule, idealized as a rigid dumbbell?

b. What are $\eta'(\omega)$ and $\eta''(\omega)$ for a dilute suspension of rigid dumbbells?

5B.5 High-Frequency Expressions for η' and η'' for the Generalized Maxwell Model

a. Show how to go from Eq. 5.3-10 to Eq. 5.3-14 for $\alpha = 2$ (which corresponds to the Rouse molecular theory of Chapter 15). In doing this make use of the Euler-Maclaurin expansion for converting a sum into an integral

$$\sum_{k=0}^{m} f(k) = \int_0^m f(t)dt + \frac{1}{2}[f(0) + f(m)] + \cdots \tag{5B.5-1}$$

Apply this to the sum in Eq. 5.3-10 to get

$$\begin{aligned}\sum_{k=1}^{\infty} \frac{k^2}{k^4 + (\lambda\omega)^2} &= \sum_{k=0}^{\infty} \frac{k^2}{k^4 + (\lambda\omega)^2} \\ &= \int_0^{\infty} \frac{k^2}{k^4 + (\lambda\omega)^2} dk + 0 + \cdots \\ &= \frac{1}{2\sqrt{\lambda\omega}} \int_0^{\infty} \frac{\sqrt{t}\, dt}{t^2 + 1} + \cdots \\ &= \frac{1}{2\sqrt{\lambda\omega}} \frac{\pi}{2\cos(\pi/4)} + \cdots\end{aligned} \tag{5B.5-2}$$

Then the dynamic viscosity becomes

$$\eta'(\omega) \sim \eta_0 \frac{3\sqrt{2}}{2\pi} \frac{1}{\sqrt{\lambda\omega}} \qquad (\lambda\omega \to \infty) \tag{5B.5-3}$$

b. Derive the results given in Eqs. 5.3-14 and 15 for arbitrary α.

5B.6 Step Strain Experiment

Derive Eq. 5.3-17 by two other methods:

a. Recognize that the sudden displacement is described by $\gamma_{yx}(-\infty, t) = \gamma_0 H(t)$, where $H(t)$ is the Heaviside unit step function ($H = 0$ for $t < 0$, and $H = 1$ for $t > 0$). Then

$$\dot{\gamma}_{yx}(t) = \frac{d}{dt}\gamma_{yx}(-\infty, t) = \gamma_0 \delta(t) \tag{5B.6-1}$$

where $\delta(t)$ is the Dirac delta function. Then obtain $\tau_{yx}(t)$ from the constitutive equation, Eq. 5.2-18.

b. Use the constitutive equation in Eq. 5.2-19 and let (using Eq. 5.1-9)

$$\begin{aligned}\gamma_{yx}(t, t') &= \gamma_{yx}(-\infty, t') + \gamma_{yx}(t, -\infty)\\ &= \gamma_0 H(t') - \gamma_0\end{aligned} \tag{5B.6-2}$$

Explain carefully how the second line is obtained.

5B.7 Complex Viscosity in Terms of the Memory Function

a. Start with Eq. 5.2-19 and obtain

$$\eta'(\omega) = \frac{1}{\omega}\int_0^\infty M(s) \sin \omega s \, ds \tag{5B.7-1}$$

$$\eta''(\omega) = \frac{1}{\omega}\int_0^\infty M(s)(1 - \cos \omega s) ds \tag{5B.7-2}$$

b. Show how to go from the expressions in (a) to Eqs. 5.3-4 and 5.

c. Use the expressions in (a) to get $\eta'(\omega)$ and $\eta''(\omega)$ for the Maxwell model.

5B.8 Linear Viscoelasticity from the Doi-Edwards Kinetic Theory for Polymer Melts

According to the Doi-Edwards theory (see Eq. 19.6-24, with $\varepsilon = 0$ and $\varepsilon' = 0$)

$$G(s) = \frac{96\eta_0}{\pi^2 \lambda} \sum_{k,\,\text{odd}} \frac{1}{k^2} e^{-\pi^2 k^2 s/\lambda} \tag{5B.8-1}$$

Show that this is equivalent to using the $G(s)$ that can be inferred from Eq. 5.2-14 for the generalized Maxwell model with the following expressions for η_k and λ_k

$$\eta_k = \eta_0 \frac{\lambda_k^2}{\sum_{k,\,\text{odd}} \lambda_k^2}; \qquad \lambda_k = \frac{\lambda}{\pi^2 k^2} \tag{5B.8-2}$$

where k is allowed to take on odd values only.

5C.1 Complex Viscosity for General Linear Viscoelastic Fluid

a. Use the linear viscoelastic constitutive equation:

$$\left(1 + \sum_{n=1}^{\infty} a_n \frac{\partial^n}{\partial t^n}\right)\tau_{ij} = -\eta_0\left(1 + \sum_{n=1}^{\infty} b_n \frac{\partial^n}{\partial t^n}\right)\dot{\gamma}_{ij} \tag{5C.1-1}$$

and show that $\eta^*/\eta_0 = (P + iQ)/(R + iS)$, or

$$\frac{\eta'}{\eta_0} = \frac{PR + QS}{R^2 + S^2}; \qquad \frac{\eta''}{\eta_0} = \frac{PS - QR}{R^2 + S^2} \tag{5C.1-2}$$

where

$$P(\omega) = 1 - b_2\omega^2 + b_4\omega^4 - \cdots \tag{5C.1-3}$$

$$Q(\omega) = b_1\omega - b_3\omega^3 + b_5\omega^5 - \cdots \tag{5C.1-4}$$

$$R(\omega) = 1 - a_2\omega^2 + a_4\omega^4 - \cdots \tag{5C.1-5}$$

$$S(\omega) = a_1\omega - a_3\omega^3 + a_5\omega^5 - \cdots \tag{5C.1-6}$$

b. It has been shown[3] that if all of the parameters a_n and b_n are expressed in terms of a single time constant λ:

$$a_n = \frac{\pi^{2n}\lambda^n}{(2n+1)!} = \left(\frac{2n+3}{3}\right)b_n \tag{5C.1-7}$$

then we get the results in Eqs. 5.3-10 and 11 when $\alpha = 2$. Show that these expressions for a_n and b_n lead to

$$\eta^* = \frac{3\eta_0}{\pi^2}\left[\frac{\pi\sqrt{i\lambda\omega}\cosh\pi\sqrt{i\lambda\omega} - \sinh\pi\sqrt{i\lambda\omega}}{i\lambda\omega\sinh\pi\sqrt{i\lambda\omega}}\right] \tag{5C.1-8^4}$$

or

$$\frac{\eta'}{\eta_0} = \frac{3}{\pi\sqrt{2\lambda\omega}}\left[\frac{\sinh\pi\sqrt{2\lambda\omega} - \sin\pi\sqrt{2\lambda\omega}}{\cosh\pi\sqrt{2\lambda\omega} - \cos\pi\sqrt{2\lambda\omega}}\right] \tag{5C.1-9}$$

$$\frac{\eta''}{\eta_0} = \frac{3}{\pi\sqrt{2\lambda\omega}}\left[\frac{\sinh\pi\sqrt{2\lambda\omega} + \sin\pi\sqrt{2\lambda\omega}}{\cosh\pi\sqrt{2\lambda\omega} - \cos\pi\sqrt{2\lambda\omega}} - \frac{\sqrt{2}}{\pi\sqrt{\lambda\omega}}\right] \tag{5C.1-10}$$

These expressions are then very nearly the same as those obtained from the Rouse theory (see Eqs. 15.3-29 and 30).

5C.2 Displacement Gradients in the Turntable Problem

In Example 5.5-1 we obtained the stress tensor starting with Eq. 5.2-18. Here we want to do the same using Eq. 5.2-19.

a. First we have to determine the location x', y' of a particle of fluid at time t'; the same fluid particle has a location x, y at time t. Show that

$$\begin{aligned} x' - x_0 = (x - x_0)[CC' + SS' - \dot{\gamma}(t' - t)C'S] \\ + (y - y_0)[C'S - S'C + \dot{\gamma}(t' - t)CC'] \end{aligned} \tag{5C.2-1}$$

$$\begin{aligned} y' - y_0 = (x - x_0)[S'C - C'S - \dot{\gamma}(t' - t)S'S] \\ + (y - y_0)[S'S + C'C + \dot{\gamma}(t' - t)S'C] \end{aligned} \tag{5C.2-2}$$

in which $S = \sin Wt$, $C = \cos Wt$, $S' = \sin Wt'$, $C' = \cos Wt'$.

[3] T. W. Spriggs and R. B. Bird, *Ind. Eng. Chem. Fundam.* **4**, 182–186 (1965).

[4] To get this result use L. B. W. Jolley, *Summation of Series*, Dover, New York (1961), Series No. 124, p. 22.

b. The displacements $u_x(t') = x' - x$ and $u_y(t') = y' - y$ can be obtained from (a). Show that the components of $\nabla \boldsymbol{u}$ are

$$(\nabla \boldsymbol{u})_{xx} = C'C + S'S - \dot{\gamma}(t' - t)C'S - 1 \tag{5C.2-3}$$

$$(\nabla \boldsymbol{u})_{xy} = S'C - C'S - \dot{\gamma}(t' - t)S'S \tag{5C.2-4}$$

$$(\nabla \boldsymbol{u})_{yx} = C'S - S'C + \dot{\gamma}(t' - t)C'C \tag{5C.2-5}$$

$$(\nabla \boldsymbol{u})_{yy} = S'S + C'C + \dot{\gamma}(t' - t)S'C - 1 \tag{5C.2-6}$$

Then obtain the components of the infinitesimal strain tensor.

c. Use Eq. 5.2-19 to obtain

$$\tau_{yx} = -\int_0^\infty M(s)s \cos W(2t - s)ds\, \dot{\gamma} \tag{5C.2-7}$$

Integrate by parts to obtain an integral involving $G(s)$ in the integrand. Is the result consistent with Eqs. 5.5-6 and 5.5-7? Explain.

d. Show that as W becomes quite small the displacement gradients become small, and that Eq. 5C.2-7 with $t = 0$ and Eq. 5.5-7 gives $\eta_0 = \int_0^\infty G(s)s\, ds$.

e. Show that Eq. 5.2-7 is true only in the limit of vanishingly small W, and that in this same limit the displacement gradients are vanishingly small.

5D.1 Use of Laplace Transform to Solve Constrained Recoil Problem[5]

Rework the recoil problem of Example 5.3-5 using Laplace transforms. Use the notation $G_n = \int_0^\infty G(s)s^n\, ds$. Begin by rewriting Eq. 5.3-27 as

$$-\dot{\gamma}_0 G_0 + \dot{\gamma}_0 \int_0^t G(s)ds = \int_0^t G(s)\dot{\gamma}_{yx}(t - s)ds \tag{5D.1-1}$$

Then take the Laplace transform of the entire equation, using the convolution theorem, to get

$$-\frac{\dot{\gamma}_0 G_0}{p} + \frac{\dot{\gamma}_0 \bar{G}(p)}{p} = \bar{G}(p)\bar{\dot{\gamma}}_{yx}(p) \tag{5D.1-2}$$

where $\bar{G} = \mathscr{L}\{G\}$ and $\bar{\dot{\gamma}}_{yx} = \mathscr{L}\{\dot{\gamma}_{yx}\}$. Next write $\bar{G}(p)$ as a Taylor expansion in the transform variable p:

$$\begin{aligned}\bar{G}(p) &= \int_0^\infty G(s)e^{-ps}\, ds \\ &= \int_0^\infty G(s)\left[1 - ps + \frac{1}{2!}(ps)^2 - \cdots\right]ds \\ &= G_0 - G_1 p + G_2 p^2 - \cdots \end{aligned} \tag{5D.1-3}$$

[5] A. S. Lodge, *Elastic Liquids*, Academic Press, New York (1964), pp. 144–147.

Substitute this into Eq. 5D.1-2 and solve for $\bar{\gamma}_{yx}$:

$$\bar{\gamma}_{yx} = \frac{\dot{\gamma}_0}{p} - \frac{\dot{\gamma}_0 G_0}{p\bar{G}(p)}$$

$$= -\dot{\gamma}_0 \frac{G_1}{G_0} + O(p) \qquad (5D.1\text{-}4)$$

To find the ultimate recoil use the "final value theorem" in the following way:

$$\gamma_\infty = \lim_{t\to\infty} \int_0^t \dot{\gamma}_{yx}(t')dt'$$

$$= \lim_{p\to 0} p\mathscr{L}\left\{ \int_0^t \dot{\gamma}_{yx}(t')dt' \right\}$$

$$= \lim_{p\to 0} p\left[\frac{1}{p}\bar{\gamma}_{yx}\right]$$

$$= \lim_{p\to 0}\left[-\dot{\gamma}_0 \frac{G_1}{G_0} + O(p)\right] \qquad (5D.1\text{-}5)$$

which is the same as the result in Eq. 5.3-31.

5D.2 Kramers-Kronig Relations[6]

a. By Fourier transformation show that Eqs. 5.3-4 and 5 may be inverted to give

$$G(s) = \frac{2}{\pi}\int_0^\infty \eta'(\omega)\cos\omega s\, d\omega \qquad (s \geq 0) \qquad (5D.2\text{-}1)$$

$$G(s) = \frac{2}{\pi}\int_0^\infty \eta''(\omega)\sin\omega s\, d\omega \qquad (s > 0) \qquad (5D.2\text{-}2)$$

b. Insert Eqs. 5D.2-1 and 2 for $G(s)$ in Eqs. 5.3-5 and 4, respectively. Show that the results may be written

$$\eta''(\omega) = \frac{2\omega}{\pi}\int_0^\infty \frac{\eta'(x)}{\omega^2 - x^2}\,dx \qquad (5D.2\text{-}3)$$

$$\eta'(\omega) - \eta'(\infty) = \frac{2}{\pi}\int_0^\infty \frac{x\eta''(x)}{x^2 - \omega^2}\,dx \qquad (5D.2\text{-}4)$$

This may be done by use of the methods of generalized functions.[7] Be careful in treating the behavior of $G(s)$ at $s = 0$.

[6] H. A. Kramers, *Atti Congr. Int. Fisici Como*, **2**, 545–557 (1927); R. de L. Kronig, *J. Opt. Soc. Am.*, **12**, 547–557 (1926).

[7] M. J. Lighthill, *Introduction to Fourier Analysis and Generalised Functions*, Cambridge University Press (1964).

c. Show that Eqs. 5D.2-3 and 4 may be written in the following forms that are more suited for computations:

$$\eta''(\omega) = \frac{2\omega}{\pi} \int_0^\infty \frac{\eta'(x) - \eta'(\omega)}{\omega^2 - x^2} dx \tag{5D.2-5}$$

$$\eta'(\omega) - \eta'(\infty) = \frac{2}{\pi} \int_0^\infty \frac{x\eta''(x) - \omega\eta''(\omega)}{x^2 - \omega^2} dx \tag{5D.2-6}$$

These are known as the *Kramers–Kronig relations*; similar relations for the amplitude and phase shift are also available.[8]

[8] H. C. Booij and G. P. J. M. Thoone, *Rheol. Acta*, **21**, 15–24 (1982).

PART III

NONLINEAR VISCOELASTIC CONSTITUTIVE EQUATIONS AND THEIR USE IN SOLVING FLUID DYNAMICS PROBLEMS

In Chapters 4 and 5 we studied two classes of constitutive equations and learned how to solve flow problems with them. In the generalized Newtonian fluid the stress tensor at some point in space is given in terms of the rate-of-strain tensor at that same point in space: both the stress tensor and the rate-of-strain tensor are evaluated at the same time t. In the general linear viscoelastic fluid, given by an integral containing the relaxation modulus, the stress tensor at time t at some point in space is given in terms of the rate-of-strain tensor at that same point in space but for all previous times t'. In both of these constitutive equations we are concerned with what is happening at one particular position in three-dimensional space.

In an illustrative example at the end of Chapter 5, we obtained a curious and disquieting result. We found that the general linear viscoelastic fluid constitutive equation led, in a steady shear flow experiment, to a dependence of the shear stress on the angular velocity of the turntable on which the experiment is being performed. Intuition suggests that this is an incorrect result, and that the stresses in a fluid should not depend on the local, instantaneous rate of rotation of the fluid element. If one believes this, then there is a need for a complete reexamination of the procedure for formulating constitutive equations in continuum mechanics.

In Chapter 9 we discuss the rules for the formulation of "admissible" constitutive equations, that is, relations that among other things are independent of the local rate of rotation of the fluid in space. It will be shown that any of the constitutive equations of linear viscoelasticity can be made admissible by replacing the various tensors in them by other tensors as follows:

Chapter	Physical Meaning	Tensors Appearing in Linear Viscoelasticity (Chapter 5)	Tensors Appearing in Nonlinear Viscoelasticity (Chapters 6, 7, 8)
6	Rate-of-strain tensor and its time derivatives	$\partial^n \boldsymbol{\gamma}/\partial t^n$	$\boldsymbol{\gamma}_{(n)}$ $(n = 1, 2, \ldots)$
7	Time derivative of the stress tensor	$\partial \boldsymbol{\tau}/\partial t$	$\boldsymbol{\tau}_{(1)}$
8	Strain tensor at t' referred to state at t	$\boldsymbol{\gamma}(t, t')$	$\boldsymbol{\gamma}^{[0]}(t, t')$, $\boldsymbol{\gamma}_{[0]}(t, t')$

In the limit of infinitesimally small displacement gradients the quantities in the rightmost column simplify to those in the linear viscoelastic column. The main thing to remember about the newly introduced tensors is that they correspond to tensors that are already familiar from Chapter 5 on linear viscoelasticity; a detailed discussion of the origin of these new tensors is postponed until Chapter 9.

In Chapter 6 we take as the object of study the "retarded-motion expansion." This is an attempt to generalize the Newtonian constitutive equation in a systematic way by expanding the stress tensor in a Taylor series about the Newtonian fluid. In this way we introduce the higher-order rate-of-strain tensors. The retarded-motion expansion allows us to study normal stress effects and to get some idea about the deviations from Newtonian behavior associated with second-order, third-order, etc., effects. Unfortunately one cannot proceed to very high shear rates or very rapid time changes in this way, so that the final results are of necessity of rather limited value as far as industrial problem solving is concerned.

In Chapter 7 we start with some differential models from linear viscoelasticity and generalize them by replacing the simple time derivatives by the more general kinematic tensors. In addition nonlinear terms can be added—empirically—in order to generate constitutive equations that can describe non-Newtonian viscosity on the one hand and linear viscoelasticity on the other. Hence, in this chapter we are able for the first time to reproduce some of the results of Chapter 4 (The Generalized Newtonian Fluid) and Chapter 5 (The General Linear Viscoelastic Fluid) and describe still other nonlinear effects as well.

In Chapter 8 we start with integral models from linear viscoelasticity and generalize them in order to obtain nonlinear models. Here again we are able to describe a wide variety of rheological phenomena—both linear and nonlinear—in terms of a single constitutive equation.

For the most part the constitutive equations given in Chapters 7 and 8 are presented as empirical relations that are so designed that a wide variety of experimental data can be described. It should be pointed out, however, that some of these constitutive equations have a molecular basis, and the kinetic-theory derivations are discussed at great length in Volume 2. As a matter of fact the constitutive equations that seem most satisfactory—both differential and integral—are those that are derived by, or at least inspired by, molecular theory.

It will be evident that these nonlinear viscoelastic constitutive equations obtained in Chapters 7 and 8 are sufficiently complex that very few flow problems can be solved analytically. Even numerical solutions of flow problems present serious difficulties. We try in both of these chapters to present material that will illustrate the methods and the problems encountered. In future years new constitutive equations will undoubtedly be developed, but it is hoped that the methodology taught in these chapters will prevail.

CHAPTER 6

THE RETARDED-MOTION EXPANSION

In this chapter the object of study is the "retarded-motion expansion," which is an expansion about the Newtonian fluid. Successive terms in the expansion account systematically for the deviations from Newtonian behavior because of "elastic effects." When the expansion is truncated after the second-order terms, the *second-order fluid* results; truncation after the third-order terms yields the *third-order fluid*, and so forth. Whereas the generalized Newtonian fluid models have been used mainly by engineers and the linear viscoelastic models have been used mainly by chemists, the second- and third-order fluid models have been studied primarily by mathematicians and continuum physicists. Because it is generally agreed that the retarded-motion expansion is the correct constitutive equation for flows in which the rate-of-strain tensor and its time derivatives are small, considerable effort has been expended to solve flow problems and to develop general theorems for this constitutive equation. These results obtained for the second-, third-,... order fluids (the "ordered fluids") have an aura of permanence, since no empiricism is involved in the development of the retarded-motion expansion.

A great deal of physical insight about "elastic effects" has been obtained by solving flow problems with the ordered-fluid models. One learns about the sign attached to the elastic effects and how elasticity results in secondary flows. One develops some qualitative feelings about how elasticity affects bubble shapes and particle orientations. In addition, one begins to develop an understanding of the mathematical procedures needed for coping with nonlinear viscoelastic problems and the role of boundary and initial conditions.

However, it must be kept in mind that the retarded-motion expansion is of little use to the engineer, since in industrial problems the non-Newtonian viscosity often plays the dominant role, and, as we shall see, the ordered fluids cannot describe the η vs. $\dot{\gamma}$ relation faithfully. Also, the retarded-motion expansion is of little use to the chemist doing linear viscoelasticity experiments, since the ordered fluids cannot describe the full range of time-dependent behavior. The retarded-motion expansion has a niche of its own and it provides valuable insight into some problems that cannot be studied by the generalized Newtonian fluids or by linear viscoelastic models. Keep in mind that all three of these models, covered in Chapters 4, 5, and 6, can be obtained as special cases of a much more general constitutive equation, as discussed in Chapter 9.

The chapter begins in §6.1 with the definition of the "nth rate-of-strain tensors". A full discussion of their derivation and geometrical significance is postponed until Chapter 9. In §6.2 we give the retarded-motion expansion and define a Deborah number to describe the importance of the elastic effects. Then in §6.3 some general theorems are given for the second-order fluid, which facilitate problem solving for certain important classes of flows. Examples in §§6.4 and 6.5 illustrate the fluid dynamical behavior of viscoelastic liquids in the limit of slow flows. Finally in §6.6 the use and limitations of the retarded-motion expansion are discussed.

§6.1 CONVECTED DERIVATIVES OF THE RATE-OF-STRAIN TENSOR

In this section we define the kinematic tensors to be used in the chapter. Imagine a velocity field specified by the function $\boldsymbol{v}(x_1, x_2, x_3, t)$ where (x_1, x_2, x_3) is a set of Cartesian coordinates. With respect to this coordinate system we define the following three kinematic tensors:[1]

		ijth Cartesian Component	
Velocity gradient tensor:	$\nabla \boldsymbol{v}$	$\dfrac{\partial}{\partial x_i} v_j$	(6.1-1)
Rate-of-strain tensor:	$\dot{\boldsymbol{\gamma}} = \nabla \boldsymbol{v} + (\nabla \boldsymbol{v})^\dagger$	$\dfrac{\partial}{\partial x_i} v_j + \dfrac{\partial}{\partial x_j} v_i$	(6.1-2)
Vorticity tensor:	$\boldsymbol{\omega} = \nabla \boldsymbol{v} - (\nabla \boldsymbol{v})^\dagger$	$\dfrac{\partial}{\partial x_i} v_j - \dfrac{\partial}{\partial x_j} v_i$	(6.1-3)

We have already encountered the rate-of-strain tensor $\dot{\boldsymbol{\gamma}}$ in earlier chapters. We see from the definitions that $\dot{\boldsymbol{\gamma}}$ is a symmetric tensor, whereas the vorticity tensor $\boldsymbol{\omega}$ is an antisymmetric tensor. The velocity gradient tensor is equal to one half the sum of the rate-of-strain tensor and the vorticity tensor:

$$\nabla \boldsymbol{v} = \tfrac{1}{2}(\dot{\boldsymbol{\gamma}} + \boldsymbol{\omega}) \tag{6.1-4}$$

That is, the velocity gradient tensor can be decomposed into its symmetric and antisymmetric parts. The symmetric tensor $\dot{\boldsymbol{\gamma}}$ has six independent Cartesian components, whereas the antisymmetric tensor $\boldsymbol{\omega}$ has three independent components to give a total of nine independent components, corresponding to the number of components of $\nabla \boldsymbol{v}$. For convenience the components of $\nabla \boldsymbol{v}$ in four coordinate systems are given in Tables A.7-1 to 4, and the components of $\dot{\boldsymbol{\gamma}}$ and $\boldsymbol{\omega}$ are presented in Tables B.3 and B.4.

The rate-of-strain tensor $\dot{\boldsymbol{\gamma}}$ describes the rate at which neighboring particles move with respect to each other independently of superposed rigid rotations. The vorticity tensor is a measure of the local rate of rotation of the fluid. For this reason the vorticity tensor does not appear in the constitutive equation for the Newtonian liquid. Thus the use of the rate-of-strain tensor rather than the velocity gradient tensor in the Newtonian fluid relation ensures that the stresses depend only on the rate at which neighboring particles move relative to each other, and not on superposed rigid rotations.

We now introduce the fundamental kinematic tensors[2] of this chapter, $\boldsymbol{\gamma}_{(1)}$, $\boldsymbol{\gamma}_{(2)} \cdots \boldsymbol{\gamma}_{(n)}$, called the *first, second, ⋯, nth rate-of-strain tensors*. The first of these is defined to be identical to the rate-of-strain tensor and the remaining are defined through a recurrence relation:

$$\boldsymbol{\gamma}_{(1)} \equiv \dot{\boldsymbol{\gamma}} \tag{6.1-5}$$

$$\boldsymbol{\gamma}_{(n+1)} = \frac{D}{Dt}\boldsymbol{\gamma}_{(n)} - \{(\nabla \boldsymbol{v})^\dagger \cdot \boldsymbol{\gamma}_{(n)} + \boldsymbol{\gamma}_{(n)} \cdot (\nabla \boldsymbol{v})\} \tag{6.1-6}$$

[1] *Caution*: There is considerable diversity in the literature regarding the definitions of these quantities. Some authors define the *ij*th component of $\nabla \boldsymbol{v}$ to be $\partial v_i/\partial x_j$. Factors of $\frac{1}{2}$ are commonly included on the right side of Eqs. 6.1-2 and 3, and some authors use a vorticity tensor that is the negative of, or $\frac{1}{2}$ the negative of, the one defined here.
[2] J. G. Oldroyd, *Proc. Roy. Soc.*, **A200**, 523–541 (1950).

The tensor $\boldsymbol{\gamma}_{(n+1)}$ is the nth *convected derivative*[3] of the rate-of-strain tensor $\boldsymbol{\gamma}_{(1)}$. These derivatives are defined in a manner such that they have a significance independent of superposed rigid rotations. The detailed explanation of the origin of these tensors is postponed until Chapter 9, and in this chapter we shall be concerned with the consequences of using the tensors in constitutive equations. Explicit expressions for $\boldsymbol{\gamma}_{(1)}$ and $\boldsymbol{\gamma}_{(2)}$ for homogeneous shear flows and shearfree flows are given as matrix displays in Appendix C. The following example illustrates how $\boldsymbol{\gamma}_{(1)}$, $\boldsymbol{\gamma}_{(2)}$, and $\boldsymbol{\gamma}_{(3)}$ can be written explicitly for homogeneous shear flow.

EXAMPLE 6.1-1 The Rate-of-Strain Tensors in Simple Shear Flow

Consider the simple shear flow $v_x = \dot{\gamma}_{yx}(t)y$, and display the first three rate-of-strain tensors in matrix form.

SOLUTION The standard procedure is to represent all tensors by their Cartesian components arranged in matrices. Tensor multiplications may then be performed as matrix multiplications. First we obtain for the velocity gradient tensor, its transpose, and the (first) rate-of-strain tensor:

$$\nabla \boldsymbol{v} = \begin{pmatrix} 0 & 0 & 0 \\ 1 & 0 & 0 \\ 0 & 0 & 0 \end{pmatrix} \dot{\gamma}_{yx}(t); \qquad (\nabla \boldsymbol{v})^{\dagger} = \begin{pmatrix} 0 & 1 & 0 \\ 0 & 0 & 0 \\ 0 & 0 & 0 \end{pmatrix} \dot{\gamma}_{yx}(t) \tag{6.1-7}$$

$$\boldsymbol{\gamma}_{(1)} \equiv \dot{\boldsymbol{\gamma}} = \begin{pmatrix} 0 & 1 & 0 \\ 1 & 0 & 0 \\ 0 & 0 & 0 \end{pmatrix} \dot{\gamma}_{yx}(t) \tag{6.1-8}$$

Then we use the definition in Eq. 6.1-6 written for $n = 1$. Since the flow is homogeneous the term $\{\boldsymbol{v} \cdot \nabla \boldsymbol{\gamma}_{(1)}\}$ is identically zero and we obtain

$$\begin{aligned} \boldsymbol{\gamma}_{(2)} &= \frac{\partial}{\partial t} \begin{pmatrix} 0 & 1 & 0 \\ 1 & 0 & 0 \\ 0 & 0 & 0 \end{pmatrix} \dot{\gamma}_{yx} - \begin{pmatrix} 0 & 1 & 0 \\ 0 & 0 & 0 \\ 0 & 0 & 0 \end{pmatrix} \begin{pmatrix} 0 & 1 & 0 \\ 1 & 0 & 0 \\ 0 & 0 & 0 \end{pmatrix} \dot{\gamma}_{yx}^2 - \begin{pmatrix} 0 & 1 & 0 \\ 1 & 0 & 0 \\ 0 & 0 & 0 \end{pmatrix} \begin{pmatrix} 0 & 0 & 0 \\ 1 & 0 & 0 \\ 0 & 0 & 0 \end{pmatrix} \dot{\gamma}_{yx}^2 \\ &= \begin{pmatrix} 0 & 1 & 0 \\ 1 & 0 & 0 \\ 0 & 0 & 0 \end{pmatrix} \frac{\partial \dot{\gamma}_{yx}}{\partial t} - 2 \begin{pmatrix} 1 & 0 & 0 \\ 0 & 0 & 0 \\ 0 & 0 & 0 \end{pmatrix} \dot{\gamma}_{yx}^2 \end{aligned} \tag{6.1-9}$$

For $\boldsymbol{\gamma}_{(3)}$ we obtain similarly

$$\boldsymbol{\gamma}_{(3)} = \begin{pmatrix} 0 & 1 & 0 \\ 1 & 0 & 0 \\ 0 & 0 & 0 \end{pmatrix} \frac{\partial^2 \dot{\gamma}_{yx}}{\partial t^2} - 6 \begin{pmatrix} 1 & 0 & 0 \\ 0 & 0 & 0 \\ 0 & 0 & 0 \end{pmatrix} \dot{\gamma}_{yx} \frac{\partial \dot{\gamma}_{yx}}{\partial t} \tag{6.1-10}$$

Notice that $\boldsymbol{\gamma}_{(n)} = 0$ for $n \geq 3$ if the flow is steady.

Before leaving this section we define the "order" of an expression in the rate-of-strain tensors. Notice from Example 6.1-1 that the tensor $\boldsymbol{\gamma}_{(n)}$ may contain terms up through

[3] In Chapter 9 this is called a "contravariant convected derivative."

nth degree in the components of the velocity gradient and their time derivatives taken together. The tensor $\boldsymbol{\gamma}_{(n)}$ is then defined[4] to be "of order n." We also define the order of any product involving rate-of-strain tensors as the sum of the orders of the factors. Such products could be scalar products, tensor products, or multiplication of a tensor by a scalar invariant. For example,

$$\boldsymbol{\gamma}_{(3)}, \quad \{\boldsymbol{\gamma}_{(2)} \cdot \boldsymbol{\gamma}_{(1)}\}, \quad \{\boldsymbol{\gamma}_{(1)} \cdot \boldsymbol{\gamma}_{(1)} \cdot \boldsymbol{\gamma}_{(1)}\}, \quad \text{and } (\boldsymbol{\gamma}_{(1)} : \boldsymbol{\gamma}_{(1)})\boldsymbol{\gamma}_{(1)}$$

are all terms of order three. Note that terms of order one have dimensions of reciprocal time; terms of order two, reciprocal time squared; terms of order three, reciprocal time cubed; and so on.

§6.2 THE RETARDED-MOTION EXPANSION

We wish to construct a constitutive equation that describes small departures from Newtonian behavior. To account systematically for such departures one could try developing a series expression involving increasing powers of the rate-of-strain tensor as well as powers of the first-, second-, and higher-order partial time derivatives. This idea is explored in Problem 6B.12, and it is shown that such an expression for $\boldsymbol{\tau}$ fails to give reasonable results in the turntable experiment discussed at the end of Chapter 5. The continuum-mechanics development in Chapter 9 shows that partial time derivatives of the rate-of-strain tensor should not appear in constitutive equations, but rather convected time derivatives.

We are then led to construct a constitutive equation as follows. First we assume[1] that the fluid is incompressible and that the stress tensor is symmetric and can be expressed as a polynomial in the rate-of-strain tensors $\boldsymbol{\gamma}_{(n)}$. Second we arrange the terms in the polynomial to be of increasing order, and terms of equal order are collected together. Then we arrive at the *retarded-motion expansion*, given here through terms of third order:[2]

$$\begin{aligned}\boldsymbol{\tau} = -[&b_1\boldsymbol{\gamma}_{(1)} + b_2\boldsymbol{\gamma}_{(2)} + b_{11}\{\boldsymbol{\gamma}_{(1)} \cdot \boldsymbol{\gamma}_{(1)}\} \\ &+ b_3\boldsymbol{\gamma}_{(3)} + b_{12}\{\boldsymbol{\gamma}_{(1)} \cdot \boldsymbol{\gamma}_{(2)} + \boldsymbol{\gamma}_{(2)} \cdot \boldsymbol{\gamma}_{(1)}\} + b_{1:11}(\boldsymbol{\gamma}_{(1)} : \boldsymbol{\gamma}_{(1)})\boldsymbol{\gamma}_{(1)} + \cdots]\end{aligned} \tag{6.2-1}$$

The scalars b_1, b_2, b_{11}, etc. are material parameters, and we shall call them *retarded-motion constants*. If we retain only the first-order term, this expansion reduces to the constitutive equation for the *incompressible Newtonian fluid*, and b_1 is then the viscosity. If all terms

[4] H. Giesekus, *Z. Angew. Math. Mech.*, **42**, 32–61 (1962).

[1] This assumption is inherent in R. S. Rivlin and J. L. Ericksen, *J. Rat. Mech. Anal.*, **4**, 323–425 (1955), for unsteady flow and in H. Giesekus, *Kolloid Z.*, **147**, 29–45 (1956), for steady flow.

[2] Among the third-order terms there is one containing $(\dot{\boldsymbol{\gamma}} : \dot{\boldsymbol{\gamma}})\dot{\boldsymbol{\gamma}}$ and another containing $\{\dot{\boldsymbol{\gamma}} \cdot \dot{\boldsymbol{\gamma}} \cdot \dot{\boldsymbol{\gamma}}\}$. The latter may be eliminated in favor of an additional term of the form $(\dot{\boldsymbol{\gamma}} : \dot{\boldsymbol{\gamma}})\dot{\boldsymbol{\gamma}}$ by using the Cayley–Hamilton theorem (Eq. A.3-28)

$$\dot{\boldsymbol{\gamma}}^3 - \dot{\boldsymbol{\gamma}}^2 \operatorname{tr} \dot{\boldsymbol{\gamma}} + \tfrac{1}{2}\dot{\boldsymbol{\gamma}}[(\operatorname{tr} \dot{\boldsymbol{\gamma}})^2 - \operatorname{tr} \dot{\boldsymbol{\gamma}}^2] - \boldsymbol{\delta}[\tfrac{1}{6}(\operatorname{tr} \dot{\boldsymbol{\gamma}})^3 - \tfrac{1}{2} \operatorname{tr} \dot{\boldsymbol{\gamma}} \operatorname{tr} \dot{\boldsymbol{\gamma}}^2 + \tfrac{1}{3} \operatorname{tr} \dot{\boldsymbol{\gamma}}^3] = \mathbf{0} \tag{6.2-1a}$$

and the fact that $\operatorname{tr} \dot{\boldsymbol{\gamma}} = 0$ for an incompressible fluid. In addition the term containing the unit tensor $\boldsymbol{\delta}$ can be discarded, inasmuch as the normal components of $\boldsymbol{\tau}$ can be determined only to within an additive isotropic function. H. Giesekus, *Z. Angew. Math. Mech.*, **42**, 32–61 (1962), developed the expansion complete through sixth order. When the fluid is compressible, additional terms have to be included in the retarded-motion expansion [see R. K. Prud'homme and R. B. Bird, *J. Non-Newtonian Fluid Mech.*, **3**, 261–279 (1978)].

through second order (i.e., the dashed underlined terms) are included, we obtain the *incompressible second-order fluid.*[3] Finally if all terms shown explicitly are included, we have the *incompressible third-order fluid.* That is, the "ordered fluids" are represented by constitutive equations obtained by truncating the retarded-motion expansion. Most of this chapter is concerned with second- and third-order fluids. In the following illustrative example we work out the expressions for the viscosity and normal stress coefficients for the third-order fluid.

EXAMPLE 6.2-1 Viscometric Functions for the Third-Order Fluid

Consider the steady shear flow $v_x = \dot{\gamma} y$ to find the viscometric functions η, Ψ_1, and Ψ_2 for the third-order fluid. Discuss conditions under which the shear stress is a monotone increasing function of shear rate.

SOLUTION The rate-of-strain tensors $\boldsymbol{\gamma}_{(1)}$, $\boldsymbol{\gamma}_{(2)}$ and $\boldsymbol{\gamma}_{(3)}$ may be obtained by simplifying the expressions found in Example 6.1-1 to steady state. The other combinations are found by matrix multiplication

$$\{\boldsymbol{\gamma}_{(1)} \cdot \boldsymbol{\gamma}_{(1)}\} = \begin{pmatrix} 1 & 0 & 0 \\ 0 & 1 & 0 \\ 0 & 0 & 0 \end{pmatrix} \dot{\gamma}^2, \qquad (\boldsymbol{\gamma}_{(1)} : \boldsymbol{\gamma}_{(1)}) = \text{tr}\{\boldsymbol{\gamma}_{(1)} \cdot \boldsymbol{\gamma}_{(1)}\} = 2\dot{\gamma}^2 \tag{6.2-2}$$

$$\{\boldsymbol{\gamma}_{(1)} \cdot \boldsymbol{\gamma}_{(2)} + \boldsymbol{\gamma}_{(2)} \cdot \boldsymbol{\gamma}_{(1)}\} = \begin{pmatrix} 0 & 1 & 0 \\ 1 & 0 & 0 \\ 0 & 0 & 0 \end{pmatrix} (-2\dot{\gamma}^3) \tag{6.2-3}$$

Up through terms of third order, Eq. 6.2-1 may now be written explicitly for steady simple shear flow as

$$\begin{pmatrix} \tau_{xx} & \tau_{yx} & 0 \\ \tau_{yx} & \tau_{yy} & 0 \\ 0 & 0 & \tau_{zz} \end{pmatrix} = - \left[\begin{pmatrix} 0 & 1 & 0 \\ 1 & 0 & 0 \\ 0 & 0 & 0 \end{pmatrix} (b_1 \dot{\gamma} - 2(b_{12} - b_{1:11})\dot{\gamma}^3) + \begin{pmatrix} 1 & 0 & 0 \\ 0 & 0 & 0 \\ 0 & 0 & 0 \end{pmatrix} (-2b_2 \dot{\gamma}^2) \right.$$
$$\left. + \begin{pmatrix} 1 & 0 & 0 \\ 0 & 1 & 0 \\ 0 & 0 & 0 \end{pmatrix} b_{11} \dot{\gamma}^2 \right] \tag{6.2-4}$$

so that the viscometric functions for the third-order fluid are

$$\eta = b_1 - (b_{12} - b_{1:11})2\dot{\gamma}^2 \tag{6.2-5}$$

$$\Psi_1 = -2b_2 \tag{6.2-6}$$

$$\Psi_2 = b_{11} \tag{6.2-7}$$

[3] The second-order fluid is usually given in the literature in terms of the kinematic tensors $\boldsymbol{\gamma}^{(n)}$ (commonly denoted $\boldsymbol{A}_n$ and called the "Rivlin–Ericksen tensors"; see Chapter 9) thus,

$$\boldsymbol{\tau} = -[a_1 \boldsymbol{\gamma}^{(1)} + a_2 \boldsymbol{\gamma}^{(2)} + a_{11}\{\boldsymbol{\gamma}^{(1)} \cdot \boldsymbol{\gamma}^{(1)}\}] \tag{6.2-1b}$$

The a's and the b's are interrelated by

$$b_1 = a_1; \quad b_2 = a_2; \quad b_{11} = a_{11} + 2a_2 \tag{6.2-1c}$$

From these results it is clear that the second-order fluid has a constant viscosity and constant first and second normal stress coefficients. In particular we see that $\Psi_{1,0}/2\eta_0 = -b_2/b_1$; furthermore from Table 5.3-1 it can be seen that $-b_2/b_1$ is equal to $\int_0^\infty sG(s)\,ds/\int_0^\infty G(s)\,ds$. In addition we note that the third-order fluid exhibits departure from a constant viscosity if $(b_{12} - b_{1:11}) \neq 0$. If $(b_{12} - b_{1:11}) < 0$ the fluid is shear thickening, and the negative of the shear stress is continually increasing with shear rate. If $(b_{12} - b_{1:11}) > 0$ the derivative of the negative of the shear stress with respect to shear rate is

$$d(-\tau_{yx})/d\dot{\gamma} = b_1[1 - 6\alpha^2\dot{\gamma}^2] \tag{6.2-8}$$

where $\alpha^2 = (b_{12} - b_{1:11})/b_1$. In order that this derivative be positive, we must require that $\dot{\gamma}^2 < 1/6\alpha^2$. Notice also that when $\dot{\gamma}^2 = 1/6\alpha^2$, we have $\eta = (2/3)b_1$, so that the third-order fluid can describe a shear-thinning viscosity that drops only by one third of its zero-shear-rate value (see Fig. 6.2-1).

In Example 6.2-1 we demonstrated that there must be a restriction on the shear rate in steady simple shear flow of a third-order fluid with coefficients chosen such that the fluid shows shear thinning. For the second-order fluid no such limitation appeared. In other kinds of flow, however, there will also be a restriction on the strain rates for the second-order fluid, as we may see, for example, from the elongational viscosity of a second-order fluid (Problem 6B.4):

$$\bar{\eta}(\dot{\varepsilon}) = 3b_1 - 3(b_2 - b_{11})\dot{\varepsilon} \tag{6.2-9}$$

We see that unless $b_2 = b_{11}$ we must require $\dot{\varepsilon} < b_1/(b_2 - b_{11})$ to ensure that $\bar{\eta}(\dot{\varepsilon})$ is positive. Since $\dot{\varepsilon}$ may be both positive (uniaxial elongation) or negative (biaxial extension) the above inequality is a real restriction irrespective of the sign of $(b_2 - b_{11})$. Such restrictions serve to illustrate that the retarded-motion expansion is indeed a constitutive equation designed for systematic investigations of small-velocity-gradient departure from Newtonian behavior.

We now show that the restriction of the retarded-motion expansion to "small velocity gradients" is really a restriction to small Deborah number. To do this we let κ denote a characteristic rate of strain for the flow. For example in tube flow we could take κ to be the average velocity divided by the tube radius, and for flow around a submerged object we could choose the velocity of the fluid far from the object (measured relative to the object) divided by a dimension of the object transverse to the flow (cf. Fig. 2.8-1). It is then reasonable to define a dimensionless stress $\boldsymbol{\tau}^*$ to be $\boldsymbol{\tau}/b_1\kappa = \boldsymbol{\tau}/\eta_0\kappa$ and a dimensionless nth rate-of-strain tensor $\boldsymbol{\gamma}^*_{(n)}$ to be $\boldsymbol{\gamma}_{(n)}/\kappa^n$. From Example 6.2-1 a good choice for a characteristic

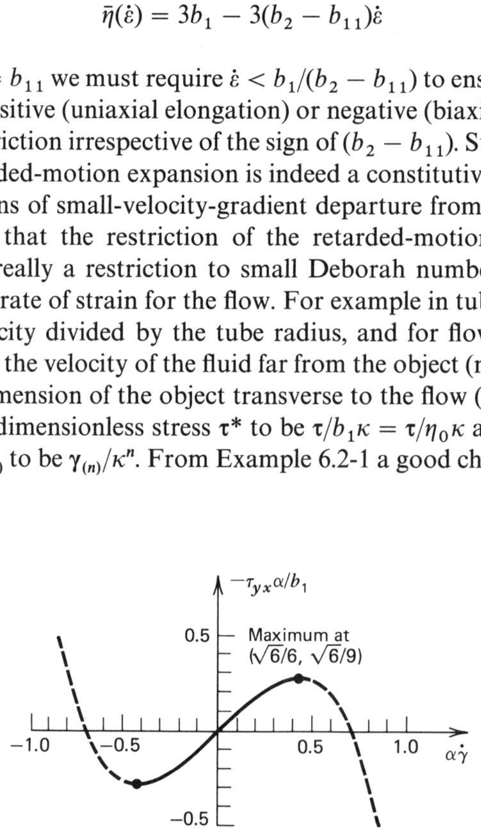

FIGURE 6.2-1. Illustration of the maximum in the shear stress for the third-order fluid. The maximum is located at $(\alpha\dot{\gamma}, -\tau_{yx}\alpha/b_1) = (\sqrt{6}/6, \sqrt{6}/9)$. If $\alpha\dot{\gamma}$ is increased above $\sqrt{6}/6$ the flow is unattainable. The model predicts an unacceptable negative viscosity at shear stresses $|\tau_{yx}\alpha/b_1| > \sqrt{6}/9$. The quantity α is defined below Eq. 6.2-8.

time for the fluid is $-b_2/b_1$, since this ratio is equal to the characteristic time found in linear viscoelasticity in Chapter 5. Thus we take the *Deborah number*[4] to be

$$\text{De} = \lambda\kappa = (-b_2/b_1)\kappa = (\Psi_{1,0}/2\eta_0)\kappa \tag{6.2-10}$$

Finally we denote dimensionless ratios of the retarded-motion expansion coefficients as follows:

$$B_{11} = \frac{b_{11}}{b_2}; \quad B_3 = \frac{b_3 b_1}{b_2^2}; \quad B_{12} = \frac{b_{12} b_1}{b_2^2}; \quad B_{1:11} = \frac{b_{1:11} b_1}{b_2^2} \tag{6.2-11}$$

With these definitions the dimensionless form of the retarded-motion expansion can be written:

$$\boxed{\begin{aligned} \tau^* = & -\gamma^*_{(1)} + \text{De}[\gamma^*_{(2)} + B_{11}\gamma^{*2}_{(1)}] \\ & -\text{De}^2[B_3\gamma^*_{(3)} + B_{12}\{\gamma^*_{(1)}\cdot\gamma^*_{(2)} + \gamma^*_{(2)}\cdot\gamma^*_{(1)}\} \\ & +B_{1:11}(\gamma^*_{(1)}:\gamma^*_{(1)})\gamma^*_{(1)}] + O(\text{De}^3) \end{aligned}} \tag{6.2-12}$$

The Deborah number is seen to be a measure of the importance of nonlinear terms on the flow. It is important to note, in addition, that the absolute values of the B's are expected to be less than unity, according to the kinetic-theory results described below.

The retarded-motion constants should ideally be determined from experiments. We note for instance from Example 6.2-1 that a knowledge of the viscometric functions at low shear rates determines the three constants in the second-order fluid. Rheogoniometer measurements may, however, not be sufficiently sensitive to determine these parameters for all fluids, and in §6.5 we illustrate how other flow situations may be used to obtain information about the retarded-motion constants. There is currently no available systematic tabulation of retarded-motion constants for real polymeric fluids.

Because of the limited experimental information available about the retarded-motion constants, it is important to utilize the results of molecular theories. Constitutive equations obtained from molecular theories can be cast in the form of a retarded-motion expansion. In so doing one obtains expressions for the retarded-motion constants in terms of the parameters appearing in specific molecular models. Table 6.2-1 gives the constants for two molecular theories for dilute polymer solutions and for one molecular theory for polymer melts. From the table we note that these theories, involving various structural models, lead to

1. The b_n alternate in sign with $b_1 > 0$.
2. At second order, $|b_2| > |b_{11}|$.
3. At third order, $b_3 > b_{12} > b_{1:11}$.

The positive value of b_1 corresponds, of course, to the fact that η_0 is positive, and it is shown in Chapter 9 that the alternation in the sign of the b_n can be deduced from continuum-mechanical arguments. However, the inequalities above, suggested by several molecular theories, cannot be obtained from continuum mechanics.

[4] For a discussion of the Deborah number see §2.8.

TABLE 6.2-1

Constants in the Retarded-Motion Expansion (Eq. 6.2-1) from Kinetic-Theory Models (Made Dimensionless with a_i)[a]

Order k	Constants	FENE Dumbbells (Dilute Solution)[b]	Multibead Rods (Dilute Solution)[c]	Freely Jointed Bead-Rod Chain (Melt)[d]
				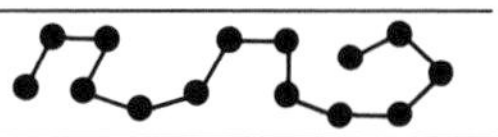
1	b_1/a_1	$\frac{\eta_s}{a_1}+\frac{b}{b+5}$	$\frac{\eta_s}{a_1}+\left[1-\frac{2}{5}\left(1-\frac{\lambda_N^{(2)}}{\lambda_N^{(1)}}\right)\right]$	$\frac{1}{60}+\frac{\varepsilon}{90}$
2	b_2/a_2	$-\frac{b^2}{(b+5)(b+7)}$	$-\frac{3}{5}$	$-\frac{1}{600}$
	b_{11}/a_2	0	$-\frac{12}{35}\left(1-\frac{\lambda_N^{(2)}}{\lambda_N^{(1)}}\right)$	$-\frac{1}{1050}+\frac{\varepsilon}{1050}$
3	b_3/a_3	$\frac{b^3(2b+11)/(2b+7)}{(b+5)(b+7)(b+9)}$	$\frac{3}{5}$	$\frac{17}{100{,}800}$
	b_{12}/a_3	$\frac{4b^3/(2b+7)}{(b+5)(b+7)(b+9)}$	$\frac{12}{35}\left[1-\frac{1}{2}\left(1-\frac{\lambda_N^{(2)}}{\lambda_N^{(1)}}\right)\right]$	$\frac{17}{117{,}600}-\frac{17\varepsilon}{352{,}800}$
	$b_{1:11}/a_3$	$\frac{3b^3/(2b+7)}{(b+5)^2(b+7)(b+9)}$	$\frac{3}{35}\left[1+\frac{6}{5}\left(1-\frac{\lambda_N^{(2)}}{\lambda_N^{(1)}}\right)\right]$	$\frac{17}{705{,}600}$
	a_k	$nkT\lambda_H^k$	$nkT\lambda_N^{(1)k}$	$NnkT\lambda^k$

[a] The second-order fluid constants are related to the zero-shear-rate values of the viscometric functions as follows: $b_1 = \eta_0$; $b_2 = -\frac{1}{2}\Psi_{1,0}$; $b_{11} = \Psi_{2,0}$. In the molecular-theory formulas n is the number density of polymer molecules, k is the Boltzmann constant, T is the absolute temperature, and η_s is the solvent viscosity. The lambdas are all time constants; b and ε are dimensionless parameters.

[b] The constants for the FENE (Finitely Extensible Nonlinear Elastic) dumbbells with no hydrodynamic interaction were derived by R. C. Armstrong, *J. Chem. Phys.*, **60**, 724–728 (1974); the parameter b is found to be in the range of about 10 to 300, and $b = \infty$ corresponds to dumbbells with linear (Hookean) springs. See §13.5.

[c] The constants for the multibead rod with hydrodynamic interaction were obtained by X. J. Fan, using R. B. Bird and C. F. Curtiss, *J. Non-Newtonian Fluid Mech.*, **14**, 85–101 (1984). For osculating beads the quantity $[1-(\lambda_N^{(2)}/\lambda_N^{(1)})]$ varies with the number of beads, N, from -0.5000 for $N = 2$ to -0.0284 for $N = 6$ and from $+0.0102$ for $N = 7$ to $+0.3082$ for $N = 70$. See also §§14.6 and 16.1–16.4.

[d] The constants for an undiluted system of interacting, freely jointed bead-rod chains are taken from C. F. Curtiss and R. B. Bird, *J. Chem. Phys.*, **74**, 2016–2025, 2026–2033 (1981). N is the number of beads in the chain, and ε is a parameter which is found to be in the range of 0.3 to 0.5 for some polymer melts. See also Chapter 19.

TABLE 6.2-2

Retarded-Motion Constants for Selected Empirical Constitutive Equations[a]

Order	Retarded-Motion Constants	Oldroyd Six-Constant[b] Model (Eq. 7.3-2)	Giesekus Model (Eq. 7.3-3)	Factorized K-BKZ Equation (Eq. 8.3-11)[c]	Factorized Rivlin-Sawyers Equation (Eq. 8.3-12)[c]
1	b_1	η_0	η_0	M_1	M_1
2	b_2	$-\eta_0(\lambda_1 - \lambda_2)$	$-\eta_0(\lambda_1 - \lambda_2)$	$-\frac{1}{2}M_2$	$-\frac{1}{2}M_2$
	b_{11}	$-\eta_0(\lambda_3 - \lambda_4)$	$-\alpha\eta_0(\lambda_1 - \lambda_2)$	$-W^{(01)}M_2$	$-\phi_2^{(00)}M_2$
3	b_3	$\eta_0\lambda_1(\lambda_1 - \lambda_2)$	$\eta_0\lambda_1(\lambda_1 - \lambda_2)$	$\frac{1}{6}M_3$	$\frac{1}{6}M_3$
	b_{12}	$\eta_0[\lambda_1(\lambda_3 - \lambda_4) + \frac{1}{2}\lambda_3(\lambda_1 - \lambda_2)]$	$2\alpha\eta_0\lambda_1(\lambda_1 - \lambda_2)$	$\frac{1}{2}W^{(01)}M_3$	$\frac{1}{2}\phi_2^{(00)}M_3$
	$b_{1:11}$	$\frac{1}{2}\eta_0[(\lambda_1 + \lambda_3)(\lambda_3 - \lambda_4) - \lambda_5(\lambda_1 - \lambda_2 - \lambda_3 + \lambda_4)]$	$\alpha(\alpha + \frac{1}{2})\eta_0\lambda_1(\lambda_1 - \lambda_2)$	$\frac{1}{2}(W^{(01)} + W^{(20)} + 2W^{(11)} + W^{(02)})M_3$	$\frac{1}{2}[\phi_2^{(00)} + \phi_1^{(10)} + (\phi_1^{(01)} + \phi_2^{(10)}) + \phi_2^{(01)}]M_3$

[a] See Problems 7B.6 and 9B.8 for the method involved in obtaining these entries.

[b] λ_6 and λ_7 have been set equal to zero in Eq. 7.3-2.

[c] The $W^{(mn)}$ and $\phi_j^{(mn)}$ are defined in Eq. 8.3-17; the M_j are given in Eq. 8.2-12.

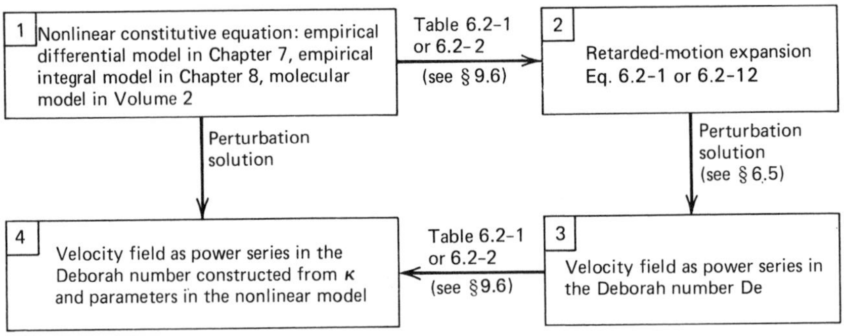

FIGURE 6.2-2. Use of the retarded-motion expansion to obtain perturbation solutions to flow problems for nonlinear constitutive equations.

In Chapters 7 and 8 a number of empirical nonlinear constitutive equations will be introduced. These empirical models will not have the restrictions inherent in the retarded-motion expansion, and are proposed with the hope of describing experimental data over a wider range of strain rates than is possible with the retarded-motion expansion. In the limit of small Deborah number, however, all these empirical models are equivalent[5] to a retarded-motion expansion with retarded-motion constants given in terms of the parameters of the empirical model. Values of such retarded-motion constants corresponding to a number of empirical constitutive equations are given in Table 6.2-2. Therefore any material function for a homogeneous flow field that has been calculated for the retarded-motion expansion may be immediately taken over for the empirical models in the limit of small Deborah number. For example, to obtain the lowest order terms in a power-series expansion of the viscometric functions of any of the nonlinear models in Table 6.2-2, all that is needed is to replace the retarded-motion constants in Eqs. 6.2-5 through 7 with the appropriate entries in Table 6.2-2. A more important application of Table 6.2-2, however, arises in connection with nonhomogeneous flow situations. In §6.5 it will be shown that the retarded-motion expansion is particularly well suited for developing perturbation solutions in which the velocity field is given as a power series in the Deborah number. Then Table 6.2-2 may be used to find the velocity field that would be obtained if the perturbation solution were performed with any of the empirical models in the table. This procedure, shown in the block diagram in Fig. 6.2-2, is important since the route from box 1 to box 4 via boxes 2 and 3 in the figure is usually both simpler and more systematic than the direct route from box 1 to box 4.

§6.3 USEFUL THEOREMS FOR THE SECOND-ORDER FLUID

In this section we consider three special theorems for the flow of second-order fluids: (a) the three-dimensional flow theorem of Giesekus, (b) the plane flow theorem of Tanner and Pipkin, and (c) the rectilinear flow theorem of Langlois, Rivlin, and Pipkin. These theorems are useful in reducing the total amount of work needed to obtain solutions of flow problems with the second-order fluid. The first two apply to creeping flow only.

We begin by giving the equations that one has to solve for an incompressible second-order fluid when the inertial terms can be neglected, (i.e., in the creeping-flow limit) or are

[5] K. Walters, *Z. Angew. Math. Phys.*, **21**, 592–600 (1970); see §9.6.

identically equal to zero. The incompressibility condition is Eq. 1.1-5, and the equation of motion is Eq. 1.1-8 with $\rho D\boldsymbol{v}/Dt = \mathbf{0}$. One then arrives at the following partial differential equations:

$$(\nabla \cdot \boldsymbol{v}) = 0 \tag{6.3-1}$$

$$[\nabla \cdot \{b_1 \boldsymbol{\gamma}_{(1)} + b_2 \boldsymbol{\gamma}_{(2)} + b_{11} \boldsymbol{\gamma}_{(1)} \cdot \boldsymbol{\gamma}_{(1)}\}] = \nabla \mathscr{P} \tag{6.3-2}$$

Here $\boldsymbol{\pi}$ in Eq. 1.1-8 has been decomposed into $\boldsymbol{\tau} + p\boldsymbol{\delta}$, and the dashed-underlined terms of Eq. 6.2-1 have been substituted for $\boldsymbol{\tau}$. In addition we have included the gravitational acceleration term in the modified pressure $\mathscr{P}$ so that $\nabla \mathscr{P} = \nabla p - \rho \boldsymbol{g}$.

In all three theorems it is assumed that we already know a solution for the creeping flow of an incompressible Newtonian fluid. That is, we know a velocity field $\boldsymbol{v}$ and a modified pressure field $\mathscr{P}_N$ that satisfy the boundary conditions, the incompressibility condition (Eq. 6.3-1), and the creeping-flow equation of motion:

$$[\nabla \cdot b_1 \boldsymbol{\gamma}_{(1)}] = b_1 \nabla^2 \boldsymbol{v} = \nabla \mathscr{P}_N \tag{6.3-3}$$

where b_1 is the viscosity. The modified pressure has the subscript "N" to indicate that it is associated with the first-order (i.e., Newtonian) fluid; the reason for doing this will be clear later.

For the Newtonian fluid it is possible to get a very useful expression involving the following combination of kinematic tensors:

$$\boldsymbol{\gamma}_{(2)} + \{\boldsymbol{\gamma}_{(1)} \cdot \boldsymbol{\gamma}_{(1)}\} = \frac{D}{Dt} \boldsymbol{\gamma}_{(1)} + \{(\nabla \boldsymbol{v}) \cdot (\nabla \boldsymbol{v})^\dagger - (\nabla \boldsymbol{v})^\dagger \cdot (\nabla \boldsymbol{v})\} \tag{6.3-4}$$

where the definition of $\boldsymbol{\gamma}_{(n)}$ in Eq. 6.1-6 has been used, and the rate-of-strain tensor has been split into the sum of the velocity gradient and its transpose. Next we form the divergence of Eq. 6.3-4, and in so doing we make use of the fact that the velocity field satisfies the incompressibility relation, Eq. 6.3-1, to obtain

$$\begin{aligned}[\nabla \cdot \{\boldsymbol{\gamma}_{(2)} + \boldsymbol{\gamma}_{(1)} \cdot \boldsymbol{\gamma}_{(1)}\}] &= \frac{D}{Dt} \nabla^2 \boldsymbol{v} + [(\nabla \boldsymbol{v}) \cdot \nabla^2 \boldsymbol{v}] + \frac{1}{2} \nabla((\nabla \boldsymbol{v}):(\nabla \boldsymbol{v}) + (\nabla \boldsymbol{v}):(\nabla \boldsymbol{v})^\dagger) \\ &= \frac{D}{Dt} \nabla^2 \boldsymbol{v} + [(\nabla \boldsymbol{v}) \cdot \nabla^2 \boldsymbol{v}] + \frac{1}{4} \nabla(\boldsymbol{\gamma}_{(1)} : \boldsymbol{\gamma}_{(1)}) \end{aligned} \tag{6.3-5}$$

Up to this point we have used only definitions and the incompressibility condition for the velocity field. If we now make use of the assumption that $\boldsymbol{v}$ satisfies Eq. 6.3-3 we find the *Giesekus equation*:

$$[\nabla \cdot \{\boldsymbol{\gamma}_{(2)} + \boldsymbol{\gamma}_{(1)} \cdot \boldsymbol{\gamma}_{(1)}\}] = \nabla \left[\frac{1}{b_1} \frac{D}{Dt} \mathscr{P}_N + \frac{1}{4} (\boldsymbol{\gamma}_{(1)} : \boldsymbol{\gamma}_{(1)}) \right] \tag{6.3-6}$$

This equation will be used several times in the remainder of this section for obtaining the pressure field in simple flows of the second-order fluid, and again in §6.5 for simplifying the procedure for obtaining perturbation solutions for the retarded-motion expansion.

a. The Three-Dimensional Flow Theorem[1] of Giesekus

Consider a three-dimensional creeping flow for a second-order fluid which has the restriction that:

$$b_{11} = b_2 \tag{6.3-7}$$

Note that this corresponds to $-\Psi_{2,0}/\Psi_{1,0} = \frac{1}{2}$, and that experimental values of this ratio are generally $\frac{1}{4}$ or smaller.

If Eq. 6.3-6 is multiplied by b_2 and added to Eq. 6.3-3, the result takes precisely the form of Eq. 6.3-2, with $\mathscr{P}$ now given by

$$\mathscr{P} = \mathscr{P}_N + \frac{b_2}{b_1}\frac{D}{Dt}\mathscr{P}_N + \frac{b_2}{4}(\boldsymbol{\gamma}_{(1)}:\boldsymbol{\gamma}_{(1)}) \tag{6.3-8}$$

Keep in mind that by the definition of the modified pressure $\mathscr{P} - \mathscr{P}_N = p - p_N$, so that a formula explicit in p is

$$p = p_N + \frac{b_2}{b_1}\frac{D}{Dt}\mathscr{P}_N + \frac{b_2}{4}(\boldsymbol{\gamma}_{(1)}:\boldsymbol{\gamma}_{(1)}) \tag{6.3-9}$$

The $\mathscr{P}_N$ operated on by the substantial derivative may be replaced by p_N only if gravity is negligible. This gives rise to the following theorem which we shall refer to as the *three-dimensional flow theorem of Giesekus*:

> Given a velocity field $\boldsymbol{v}$ and a pressure field $\mathscr{P}_N$ that satisfy the equations for creeping flow of an incompressible Newtonian fluid, then the same velocity field $\boldsymbol{v}$ and the pressure field $\mathscr{P}$ given by Eq. 6.3-8 satisfy the equations for creeping flow of an incompressible second-order fluid with $b_{11} = b_2$.

Note that the next two theorems do *not* have the restriction $b_{11} = b_2$.

b. The Plane Flow Theorem[2] of Tanner and Pipkin

This theorem is limited to *plane flow* defined by

$$v_1 = v_1(x_1, x_2); \quad v_2 = v_2(x_1, x_2); \quad v_3 = 0 \tag{6.3-10}$$

in which the x_i are Cartesian coordinates. For this flow the rate-of-strain tensor may be

[1] The treatment here closely follows H. Giesekus, *Rheol. Acta*, **3**, 59–71 (1963); see "Anhang II," p. 70. The theorem has been rederived by Fosdick and Rajagopal, who in addition gave a condition for uniqueness of the velocity field in the second-order fluid [R. L. Fosdick and K. R. Rajagopal, *Int. J. Non-Linear Mech.*, **13**, 131–137 (1978)].
[2] This theorem was developed by R. I. Tanner and A. C. Pipkin, *Trans. Soc. Rheol.*, **13**, 471–484 (1969), and R. I. Tanner, *Phys. Fluids*, **9**, 1246–1247 (1968), who were apparently unaware of Giesekus' earlier work. A condition for uniqueness of the flow is given by R. R. Huilgol, *SIAM J. Appl. Math.*, **24**, 226–233 (1973).

represented in matrix form by

$$\gamma_{(1)} = \begin{pmatrix} 2(\partial v_1/\partial x_1) & (\partial v_1/\partial x_2) + (\partial v_2/\partial x_1) & 0 \\ (\partial v_1/\partial x_2) + (\partial v_2/\partial x_1) & 2(\partial v_2/\partial x_2) & 0 \\ 0 & 0 & 0 \end{pmatrix} \tag{6.3-11}$$

If we take the dot product of this tensor with itself and use the incompressibility condition it follows that

$$[\nabla \cdot \{\gamma_{(1)} \cdot \gamma_{(1)}\}] = \tfrac{1}{2}\nabla(\gamma_{(1)} : \gamma_{(1)}) \tag{6.3-12}$$

This simple equation, along with the Giesekus equation (Eq. 6.3-6), is all that is needed to derive the plane flow theorem. Indeed if we multiply Eq. 6.3-12 by $(b_{11} - b_2)$, multiply Eq. 6.3-6 by b_2, and add both results to Eq. 6.3-3, the final equation takes the form of Eq. 6.3-2 with $\mathscr{P}$ now given by

$$\mathscr{P} = \mathscr{P}_N + \frac{b_2}{b_1}\frac{D}{Dt}\mathscr{P}_N + \left(\frac{b_{11}}{2} - \frac{b_2}{4}\right)(\gamma_{(1)} : \gamma_{(1)}) \tag{6.3-13}$$

Here again note that a formula explicit in p is obtained just by replacing the two $\mathscr{P}$'s on either side of the equality sign by p's. This then gives the *plane flow theorem of Tanner and Pipkin*:

> Given a plane velocity field $\boldsymbol{v}$ and a pressure field $\mathscr{P}_N$ that satisfy the equations for creeping plane flow of an incompressible Newtonian fluid, then the same velocity field and the pressure field $\mathscr{P}$ given by Eq. 6.3-13 satisfy the equations for creeping plane flow of an incompressible second-order fluid.

EXAMPLE 6.3-1 Force Exerted on a Solid Surface by a Second-Order Fluid in Steady Plane Flow

Use Eq. 6.3-13 to obtain an expression for the force per unit area exerted locally on an arbitrary, curved solid surface by an incompressible second-order fluid in steady plane flow.

SOLUTION With origin at the wall we introduce a local Cartesian coordinate system x_1, x_2, x_3 with $\boldsymbol{\delta}_1$ tangent to the wall and parallel to the direction of the fluid flow, $\boldsymbol{\delta}_2$ normal to the wall and pointing into the fluid, and $\boldsymbol{\delta}_3$ in the direction of no flow (see Fig. 6.3-1). In this coordinate system the velocity field is that of Eq. 6.3-10. In addition, we assume that the wall may slide relative to the axes in its own plane which means that at $x_2 = 0$ we have $v_2 = 0$ and $v_1 =$ constant. It follows that $\partial v_2/\partial x_1 = 0$ and $\partial v_1/\partial x_1 = 0$ at the origin, and therefore on account of incompressibility we also have $\partial v_2/\partial x_2 = 0$ at the origin. The velocity gradient and the rate-of-strain tensors therefore take the following simple forms at $x_1 = 0$ and $x_2 = 0$:

$$\nabla \boldsymbol{v} = \begin{pmatrix} 0 & 0 & 0 \\ 1 & 0 & 0 \\ 0 & 0 & 0 \end{pmatrix}[\dot{\gamma}_{12}]_w; \quad \dot{\boldsymbol{\gamma}} = \begin{pmatrix} 0 & 1 & 0 \\ 1 & 0 & 0 \\ 0 & 0 & 0 \end{pmatrix}[\dot{\gamma}_{12}]_w \tag{6.3-14}$$

where $\dot{\gamma}_{12} = (\partial v_1/\partial x_2)$, and the subscript "$w$" on a bracket indicates that the quantity inside the bracket is to be evaluated "at the origin," that is, at $x_1 = 0$ and $x_2 = 0$. The quantity $[\dot{\gamma}_{12}]_w$ may be

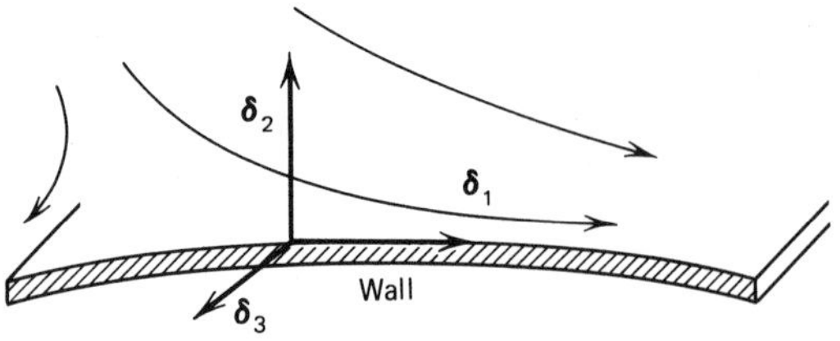

FIGURE 6.3-1. Steady plane flow of a second-order fluid near a curved wall. Unit vectors $\boldsymbol{\delta}_1, \boldsymbol{\delta}_2, \boldsymbol{\delta}_3$ define a local coordinate system with origin at the wall. At $x_2 = 0$, the velocity component $v_2 = 0$, but v_1 may be nonzero if the wall is moving.

either positive or negative, and its magnitude is equal to the shear rate at the wall. The second-order fluid pressure p given by Eq. 6.3-13 then becomes at the wall

$$p_w = \left[p_N + \frac{b_2}{b_1} v_1\left(\frac{\partial \mathscr{P}_N}{\partial x_1}\right) + \left(b_{11} - \frac{b_2}{2}\right)\dot{\gamma}_{12}^2\right]_w \tag{6.3-15}$$

Here we have also used the fact that $v_2 = 0$ at $x_2 = 0$. We now need to obtain the stress tensor components from the dashed-underlined terms in Eq. 6.2-1. When the expression is written out in matrix notation, and account is again taken of the simplifications at the origin of the coordinate system one obtains for the 22-component

$$[\tau_{22}]_w = -[b_{11}\dot{\gamma}_{12}^2]_w \tag{6.3-16}$$

Consequently we obtain the following expression for the total normal stress exerted by the second-order fluid on the wall

$$\begin{aligned} [-\pi_{22}]_w &= [-(p + \tau_{22})]_w \\ &= \left[-p_N - \frac{b_2}{b_1} v_1\left(\frac{\partial \mathscr{P}_N}{\partial x_1}\right) + \frac{b_2}{2}\dot{\gamma}_{12}^2\right]_w \end{aligned} \tag{6.3-17}$$

A similar development for the shear stress gives

$$[-\tau_{21}]_w = \left[b_1\dot{\gamma}_{12} + b_2 v_1 \frac{\partial}{\partial x_1}\dot{\gamma}_{12}\right]_w \tag{6.3-18}$$

These results may be combined to give the following expression for the force per unit area exerted by the fluid on the wall (cf. Example 1.2-1)

$$\begin{aligned} [-\boldsymbol{\delta}_2 \cdot \boldsymbol{\pi}]_w = &\left[p_N + \tfrac{1}{2}\Psi_{1,0}\left(-\frac{1}{\eta_0}(\boldsymbol{v}\cdot\nabla\mathscr{P}_N) + \tfrac{1}{2}\dot{\gamma}_{12}^2\right)\right]_w(-\boldsymbol{\delta}_2) \\ &+ [\eta_0\dot{\gamma}_{12} - \tfrac{1}{2}\Psi_{1,0}(\boldsymbol{v}\cdot\nabla\dot{\gamma}_{12})]_w\boldsymbol{\delta}_1 \end{aligned} \tag{6.3-19}$$

where the magnitude of $[\dot{\gamma}_{12}]_w = [\partial v_1/\partial x_2]_w$ is equal to the shear rate at the wall.

In connection with Eqs. 6.3-16 to 18 keep in mind that $\boldsymbol{\delta}_1$ must be in the direction of flow and $\boldsymbol{\delta}_2$ must be normal to the surface and pointing into the fluid. Each of these expressions may be used at several different locations; however, it is important that each set of local axes $\boldsymbol{\delta}_1, \boldsymbol{\delta}_2, \boldsymbol{\delta}_3$ have the same orientation with respect to the same global Cartesian coordinate system in which the flow is steady.

c. The Rectilinear Flow Theorem[3] of Langlois, Rivlin, and Pipkin

This theorem is limited to *rectilinear* flow defined by

$$v_1 = 0, \quad v_2 = 0, \quad v_3 = v_3(x_1, x_2) \tag{6.3-20}$$

in which the x_i are Cartesian coordinates. For rectilinear flow[4] it may be shown (Problem 6B.10) that if $\boldsymbol{v}$ satisfies Eq. 6.3-3 and if the pressure gradient $\nabla\mathscr{P}_N$ does not change in the direction of flow, then,

$$[\nabla\cdot\{\boldsymbol{\gamma}_{(1)}\cdot\boldsymbol{\gamma}_{(1)}\}] = \nabla\left(\frac{1}{b_1}(\boldsymbol{v}\cdot\nabla\mathscr{P}_N) + \tfrac{1}{2}((\nabla\boldsymbol{v}):(\nabla\boldsymbol{v})^\dagger)\right) \tag{6.3-21}$$

This equation along with the Giesekus equation may now be used to derive the rectilinear flow theorem in a manner analogous to the derivation of the plane flow theorem. This time we multiply Eq. 6.3-21 by $(b_{11} - b_2)$, multiply Eq. 6.3-6 by b_2, and add the results to Eq. 6.3-3. Then the result may be written in the form of Eq. 6.3-2, with $\mathscr{P}$ now given by

$$\mathscr{P} = \mathscr{P}_N + \frac{b_2}{b_1}\frac{\partial}{\partial t}\mathscr{P}_N + \frac{b_{11}}{b_1}(\boldsymbol{v}\cdot\nabla\mathscr{P}_N) + \frac{b_2}{4}(\dot{\boldsymbol{\gamma}}:\dot{\boldsymbol{\gamma}}) + \frac{(b_{11}-b_2)}{2}((\nabla\boldsymbol{v}):(\nabla\boldsymbol{v})^\dagger) \tag{6.3-22}$$

This gives rise to the *rectilinear flow theorem of Langlois, Rivlin, and Pipkin*:

> Given a rectilinear velocity field $\boldsymbol{v}$ and a pressure field $\mathscr{P}_N$ that satisfy the equation of motion for a Newtonian fluid and such that $\nabla\mathscr{P}_N$ does not change with distance in the flow direction, then the same velocity field and the pressure field $\mathscr{P}$ given by Eq. 6.3-22 satisfy the equations of continuity and motion for an incompressible second-order fluid.

Notice that the definition of rectilinear flow implies that the continuity equation is automatically satisfied and also that inertial terms do not enter into the equation of motion. This theorem, unlike the two preceding theorems, is not limited to creeping flow.

EXAMPLE 6.3-2 Force Exerted on a Solid Surface by a Second-Order Fluid in Steady Rectilinear Flow

Use Eq. 6.3-22 to obtain an expression for the total force per unit area exerted locally on a solid surface by an incompressible second-order fluid in steady rectilinear flow.

SOLUTION Arrange the coordinate system x_1, x_2, x_3 with origin at the wall and with $\boldsymbol{\delta}_1$ tangent to the wall, $\boldsymbol{\delta}_2$ locally normal to the wall and pointing into the fluid, and $\boldsymbol{\delta}_3$ tangent to the wall

[3] W. E. Langlois and R. S. Rivlin, *Rend. Mat.*, **22**, 169–185 (1963); A. C. Pipkin and R. S. Rivlin, *Z. Angew. Math. Phys.*, **14**, 738–742 (1963). See also E. A. Kearsley, *Trans. Soc. Rheol.*, **14**, 419–424 (1970); A. C. Pipkin, *Lectures on Viscoelasticity Theory*, Springer Verlag, Berlin (1972), Chapt. VIII, Sec. 11.

[4] Note that rectilinear flow is possible for Newtonian, second-order, and third-order fluids, but not in general for higher-order fluids (see Fig. 6.6-1).

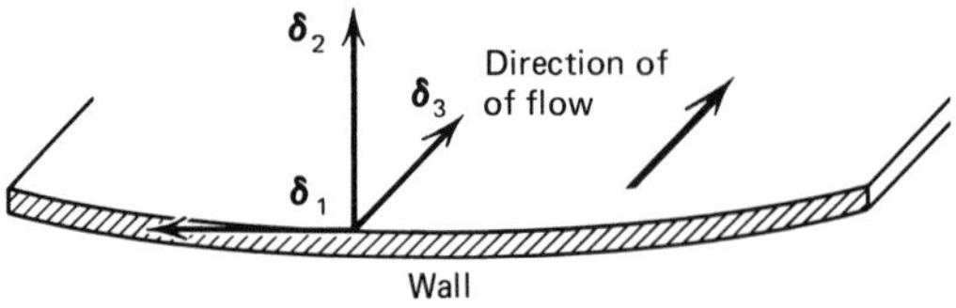

FIGURE 6.3-2. Steady rectilinear flow of a second-order fluid along a solid surface. Unit vectors $\boldsymbol{\delta}_1$ $\boldsymbol{\delta}_2$, $\boldsymbol{\delta}_3$ define a local coordinate system with origin at the wall. At $x_2 = 0$ the velocity component v_3 may be nonzero if the wall is moving, but $\partial v_3/\partial x_1$ and $\partial v_3/\partial x_3$ are zero.

and in the direction of flow (see Fig. 6.3-2). The velocity gradient and the rate-of-strain tensors become:

$$\nabla \mathbf{v} = \begin{pmatrix} 0 & 0 & \partial v_3/\partial x_1 \\ 0 & 0 & \partial v_3/\partial x_2 \\ 0 & 0 & 0 \end{pmatrix}; \quad \dot{\boldsymbol{\gamma}} = \begin{pmatrix} 0 & 0 & \partial v_3/\partial x_1 \\ 0 & 0 & \partial v_3/\partial x_2 \\ \partial v_3/\partial x_1 & \partial v_3/\partial x_2 & 0 \end{pmatrix} \tag{6.3-23}$$

By writing out all terms in Eq. 6.3-22 for steady flow one obtains

$$p = p_N + \frac{b_{11}}{b_1} v_3 \frac{\partial}{\partial x_3} \mathscr{P}_N + \frac{b_{11}}{2} \left[\left(\frac{\partial v_3}{\partial x_1} \right)^2 + \left(\frac{\partial v_3}{\partial x_2} \right)^2 \right] \tag{6.3-24}$$

This expression can then be simplified further at $x_2 = 0$ by noting that $\partial v_3/\partial x_1 = 0$ at the wall.

Similarly the expression for the stress tensor, the dotted-underlined terms in Eq. 6.2-1, may be written out in matrix form based on the velocity gradient in Eq. 6.3-23. When account is again taken of the fact that $\partial v_3/\partial x_1 = 0$ at the wall one obtains for τ_{22} at the wall,

$$[\tau_{22}]_w = -b_{11}[\dot{\gamma}_{23}^2]_w \tag{6.3-25}$$

Here $\dot{\gamma}_{23} = (\partial v_3/\partial x_2)$, and the subscript "$w$" on a bracket indicates that the quantity inside the bracket is to be evaluated "at the wall," that is, at $x_2 = 0$. Consequently we obtain the following simple expression for the total normal stress exerted by the fluid on the wall:

$$[\pi_{22}]_w = \left[p_N + \frac{b_{11}}{b_1} v_3 \frac{\partial}{\partial x_3} \mathscr{P}_N - \frac{b_{11}}{2} \dot{\gamma}_{23}^2 \right]_w \tag{6.3-26}$$

Similarly we find also the only non-zero shear stress component, which is identical to that for a Newtonian fluid:

$$[\tau_{23}]_w = -b_1[\dot{\gamma}_{23}]_w \tag{6.3-27}$$

Finally a compact expression for the force per unit area exerted by the fluid on the solid wall is

$$[-\boldsymbol{\delta}_2 \cdot \boldsymbol{\pi}]_w = \left[p_N - \Psi_{2,0} \left(\left(-\frac{v_3}{\eta_0} \frac{\partial \mathscr{P}_N}{\partial x_3} \right) + \tfrac{1}{2} \dot{\gamma}_{23}^2 \right) \right]_w (-\boldsymbol{\delta}_2) + [\eta_0 \dot{\gamma}_{23}]_w \boldsymbol{\delta}_3 \tag{6.3-28}$$

where the magnitude of $[\dot{\gamma}_{23}]_w = [\partial v_3/\partial x_2]_w$ is equal to the shear rate at the wall.

Notice that the first normal stress coefficient does not enter in the resultant normal force. In addition keep in mind that all available data show that $\Psi_{2,0}$ is negative or zero (Chapter 3).

§6.4 TWO-DIMENSIONAL AND RECTILINEAR FLOW PROBLEMS FOR THE SECOND-ORDER FLUID

In this section we turn to the application of the two-dimensional flow theorem and the rectilinear flow theorem, derived in the previous section. These theorems state that for two-dimensional creeping flows and rectilinear flows the velocity field for creeping motion of an incompressible Newtonian fluid will also be possible for a second-order fluid. This means that there is *no secondary motion due to fluid elasticity* in the examples considered here. The stresses and pressures are not, however, the same as in a Newtonian fluid, and therefore *the forces exerted on the boundaries are influenced by fluid elasticity*. Both examples considered in this section are concerned with investigations of such extra forces due to fluid elasticity. The reader may find that even though the derivations of the theorems were a bit tedious, their application to problem solving is straightforward.

EXAMPLE 6.4-1 Axial Eccentric Annular Flow[1]

Here we consider the axial motion of a second-order fluid in an eccentric annular region; a rod moves with speed V inside a circular cylindrical cavity as shown in Fig. 6.4-1. The rod has radius R_1,

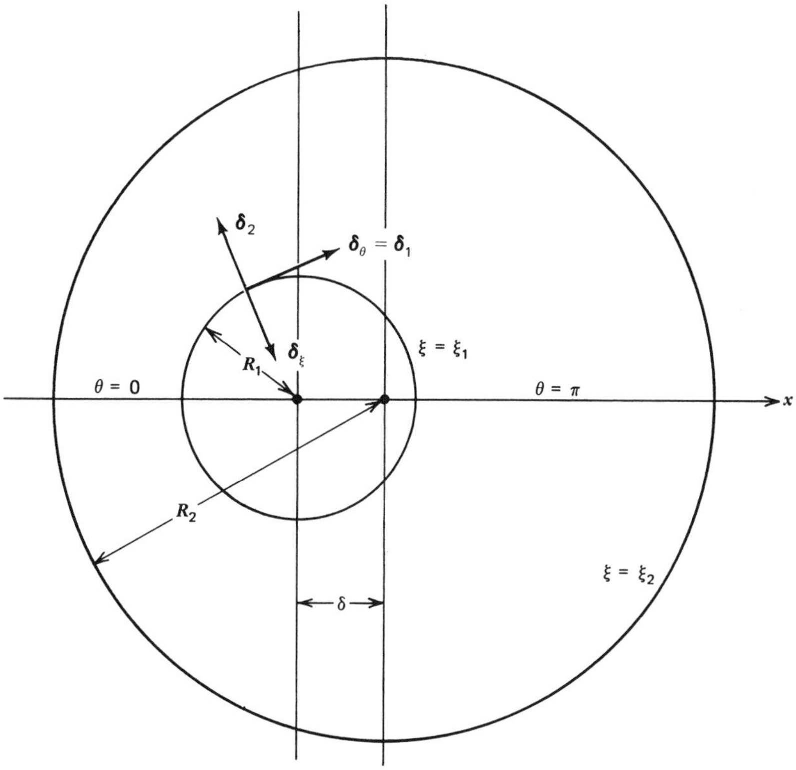

FIGURE 6.4-1. An off-centered circular rod of radius R_1 moving axially with speed V in a cylindrical cavity of radius R_2 filled with a viscoelastic coating material. The coordinates ξ and θ are coordinates in the bipolar coordinate system (see Fig. A.7-1).

[1] This problem was solved for an Oldroyd model by J. R. Jones, *J. Méch.*, **3**, 79–99 (1964); **4**, 121–132 (1965). For an alternative treatment based on the CEF equation of §9.6 see Z. Tadmor and R. B. Bird, *Polym. Eng. Sci.*, **14**, 124–136 (1974); see also Example 4.3-2.

and the radius of the cavity is R_2. The rod axis and cavity axis are not coincident, but separated by a distance δ. Because of the eccentric arrangement, the flow is most easily described in terms of bipolar coordinates (see Fig. A.7-1). It is desired to find the lateral force acting by the second-order fluid on the off-centered cylindrical rod.

This arrangement is of interest in connection with the stabilizing forces acting on an off-center wire in a wire-coating operation, where speeds of up to 25 m/s are encountered. Since wire-coating materials in their molten form are usually viscoelastic liquids, it is important to know whether the normal stresses in these liquids will help or hinder in the stabilizing of the wire-coating operation. Therefore it is of interest to find out if there is a force on an off-center cylinder that tends to push the cylinder toward the center of the cavity or in the opposite direction.

SOLUTION We anticipate that the flow will be a rectilinear flow, which in bipolar coordinates means that we look for a solution of the form

$$v_\xi = 0, \quad v_\theta = 0, \quad v_z = v_z(\xi, \theta) \tag{6.4-1}$$

We further imagine that $p = p(\xi, \theta)$ so that $\partial p/\partial z = 0$. The situation with a constant pressure gradient in the z-direction still results in a rectilinear flow. The equations of motion of a Newtonian fluid are

$$0 = \frac{X}{a}\frac{\partial p_N}{\partial \xi} \tag{6.4-2}$$

$$0 = \frac{X}{a}\frac{\partial p_N}{\partial \theta} \tag{6.4-3}$$

$$\left(\frac{X}{a}\right)^2 \frac{\partial^2 v_z}{\partial \xi^2} + \left(\frac{X}{a}\right)^2 \frac{\partial^2 v_z}{\partial \theta^2} = 0 \tag{6.4-4}$$

in which $X = \cosh \xi + \cos \theta$, and a is defined in Fig. A.7-1. Equations 6.4-2 through 4 are to be solved with the boundary conditions

$$v_z(\xi, \theta) = V, \qquad at \quad \xi = \xi_1, \qquad \text{for all } \theta \tag{6.4-5}$$

$$v_z(\xi, \theta) = 0, \qquad \text{at} \quad \xi = \xi_2, \qquad \text{for all } \theta \tag{6.4-6}$$

$$v_z(\xi, \theta) = v_z(\xi, \theta + 2\pi), \qquad \text{for all } \xi, \theta \tag{6.4-7}$$

The boundary conditions suggest that we look for a solution independent of θ, and we find then the Newtonian solution

$$\frac{v_z}{V} = \frac{\xi - \xi_2}{\xi_1 - \xi_2} \tag{6.4-8}$$

and

$$p_N = p_0 \tag{6.4-9}$$

where p_0 is a constant. This constant hydrostatic pressure does not produce any net lateral force on the inner cylinder.

Now that we have solved the Newtonian flow problem, we may use the rectilinear flow theorem of Langlois, Rivlin, and Pipkin to conclude that for the second-order fluid the velocity field is still given by Eq. 6.4-8 with the pressure field given by Eq. 6.3-22. Our interest is, however, not in the complete pressure field but only in the lateral force exerted on the moving cylinder, given by Eq. 6.3-28. In order to use this equation we need to "translate" it from the local rectangular coordinate system to the bipolar coordinate system used here. Since the bipolar coordinate system is an orthogonal coordinate system, and since $\boldsymbol{\delta}_\xi$ points into the inner cylinder and $\boldsymbol{\delta}_z$ is in the direction of flow we see that $\boldsymbol{\delta}_2 = -\boldsymbol{\delta}_\xi$ and $\boldsymbol{\delta}_3 = \boldsymbol{\delta}_z$. In addition we calculate the rate-of-strain tensor as follows:

$$\dot{\boldsymbol{\gamma}} = \dot{\gamma}_{\xi z}(\boldsymbol{\delta}_\xi \boldsymbol{\delta}_z + \boldsymbol{\delta}_z \boldsymbol{\delta}_\xi) = \dot{\gamma}_{23}(\boldsymbol{\delta}_2 \boldsymbol{\delta}_3 + \boldsymbol{\delta}_3 \boldsymbol{\delta}_2) \tag{6.4-10}$$

where $\dot{\gamma}_{\xi z}$ can be obtained from Eq. U of Table A.7-4:

$$\dot{\gamma}_{\xi z} = \frac{X}{a}\frac{\partial v_z}{\partial \xi} = \frac{XV}{a(\xi_2 - \xi_1)} \tag{6.4-11}$$

With the above relations between the unit base vectors we see that $\dot{\gamma}_{23} = -\dot{\gamma}_{\xi z}$, and therefore that Eq. 6.3-28 in the bipolar coordinate system reads

$$\boldsymbol{\pi}_\xi|_{\xi=\xi_1} = [\boldsymbol{\delta}_\xi \cdot \boldsymbol{\pi}]_{\xi=\xi_1} = [p_0 - \Psi_{2,0}\tfrac{1}{2}(-\dot{\gamma}_{\xi z})^2]_{\xi=\xi_1}\boldsymbol{\delta}_\xi + [\eta_0(-\dot{\gamma}_{\xi z})]_{\xi=\xi_1}\boldsymbol{\delta}_z \tag{6.4-12}$$

To get the lateral component of the net transmitted force we form the dot product with $\boldsymbol{\delta}_x$:

$$(\boldsymbol{\delta}_x \cdot \boldsymbol{\pi}_\xi)_{\xi=\xi_1} = \left[(p_0 - \tfrac{1}{2}\Psi_{2,0}\dot{\gamma}_{\xi z}^2)\frac{1 + \cosh\xi\cos\theta}{X(\xi, \theta)}\right]_{\xi=\xi_1} \tag{6.4-13}$$

where we have used the result for $(\boldsymbol{\delta}_\xi \cdot \boldsymbol{\delta}_x)$, which can be obtained from §A.7 (see Eq. A.7-11 and Exercise 5). The net force exerted in the x-direction on a length L of the cylinder, F_x, is then

$$\begin{aligned} F_x &= \int_S (\boldsymbol{\delta}_x \cdot \boldsymbol{\pi}_\xi)_{\xi=\xi_1}\, dS \\ &= \int_0^L \int_0^{2\pi} (\boldsymbol{\delta}_x \cdot \boldsymbol{\pi}_\xi)_{\xi=\xi_1} \frac{a}{X(\xi_1, \theta)}\, d\theta\, dz \end{aligned} \tag{6.4-14}$$

where we have used the fact that the element of area of a surface ξ = constant in bipolar coordinates is $(a/X)\, d\theta\, dz$. We now substitute $(\boldsymbol{\delta}_x \cdot \boldsymbol{\pi}_\xi)$ from Eq. 6.4-13 into Eq. 6.4-14 to find

$$F_x = La\int_0^{2\pi}\left[\frac{p_0}{X^2(\xi_1, \theta)} - \frac{\Psi_{2,0}V^2}{2a^2(\xi_1 - \xi_2)^2}\right](1 + \cosh\xi_1\cos\theta)\, d\theta \tag{6.4-15}$$

It is now a bit tedious to show that the contribution from the p_0 term is zero as expected. To do this one makes use of the integrals in part (c) of Problem 1C.4. It is considerably easier to find the contribution from the $\Psi_{2,0}$ term to get, finally,

$$F_x = -\Psi_{2,0}\frac{\pi V^2}{a(\xi_1 - \xi_2)^2}L \tag{6.4-16}$$

where $a(R_1, R_2, \delta)$ is given in the caption of Fig. A.7-1. We see that if the second normal stress coefficient $\Psi_{2,0}$ is negative, the force F_x will be positive and the wire self-centering.

The use of the second-order fluid as the constitutive equation means that the analysis is limited to small Deborah number. This means that $\text{De}(b_{11}/b_2) = -\Psi_{2,0}\dot{\gamma}/\eta_0 \ll 1$ where $\dot{\gamma}$ is a typical shear rate in the annular region. If $\text{De}(b_{11}/b_2) \geq 1$ there will in addition to the axial velocity component be a secondary tangential velocity component. Tadmor and Bird[1] have made an analysis based on the CEF equation of §9.6 and the assumption that the secondary flow is negligible. Their result for F_x is identical to Eq. 6.4-16 but with $\Psi_{2,0}$ replaced by $\Psi_2(\dot{\gamma})$, where $\dot{\gamma}$ is the shear rate in the annular region.

Many other physical phenomena may be important in wire coating, such as (a) phenomena associated with the wetting of the wire by the polymer, for instance entrainment of gas bubbles; and (b) phenomena associated with unsteady flow that could result from high coating speeds for which the flow is inherently unstable, or for which the lateral vibrations of the wire are significant.

EXAMPLE 6.4-2 Tangential Eccentric Annular Flow (Journal-Bearing Flow)[2]

In Problem 1C.4 the flow of a Newtonian fluid in a simplified journal-bearing system is discussed. Rework the same problem for the second-order fluid in order to assess the role of the normal stresses.

SOLUTION This is an example of a plane flow, and we may then use the plane flow theorem of Tanner and Pipkin to conclude that the flow field in the second-order fluid is the same as that found in Problem 1C.4 for the Newtonian fluid, and that the pressure field is given by Eq. 6.3-13. Again our interest is just in the forces exerted on the inner cylinder given by Eq. 6.3-19. In the use of this equation we may identify $\boldsymbol{\delta}_1$ with $\boldsymbol{\delta}_X$ and $\boldsymbol{\delta}_2$ with $\boldsymbol{\delta}_Y$ of Fig. 1C.4. Equation 6.3-19 then gives the following expression for the local force per unit area exerted by the fluid on the inner cylinder $\boldsymbol{\pi}_{-Y} = [-\boldsymbol{\delta}_Y \cdot \boldsymbol{\pi}]_w$:

$$\boldsymbol{\pi}_{-Y} = \left[p_N + \tfrac{1}{2}\Psi_{1,0}\left(-\frac{1}{\eta_0} W_1 r_1 \frac{d}{dX} p_N + \frac{1}{2}\left(\frac{\partial v_X}{\partial Y}\right)^2 \right) \right]_w (-\boldsymbol{\delta}_Y)$$
$$+ \left[\eta_0 \left(\frac{\partial v_X}{\partial Y}\right) - \tfrac{1}{2}\Psi_{1,0} W_1 r_1 \frac{\partial}{\partial X}\left(\frac{\partial v_X}{\partial Y}\right) \right]_w \boldsymbol{\delta}_X \tag{6.4-17}$$

where p_N, dp_N/dX, and $\partial v_X/\partial Y$ are given in Eqs. 1C.4-21, 1C.4-3, and 1C.4-5. When these expressions are used in Eq. 6.4-17 we may write the result in the notation of Problem 1C.4 as

$$\boldsymbol{\pi}_{-Y} = -\pi_{YX}\boldsymbol{\delta}_X - \pi_{YY}\boldsymbol{\delta}_Y \tag{6.4-18}$$

where

$$\pi_{YX} = \left[\eta_0 W_1 r_1 \left(\frac{4}{B} - \frac{3B_0}{B^2}\right) - \tfrac{1}{2}\Psi_{1,0} W_1^2 r_1 \frac{d}{d\theta}\left(\frac{4}{B} - \frac{3B_0}{B^2}\right) \right]_w \tag{6.4-19}$$

$$\pi_{YY} = \left[p_N + \tfrac{1}{2}\Psi_{1,0} W_1^2 r_1^2 \left(\frac{1}{B} - \frac{3B_0}{B^2}\right) \right]_w \tag{6.4-20}$$

and B_0 is defined below Eq. 1C.4-2.

Equations 6.4-18 to 20 are the result of the application of the Tanner–Pipkin theorem locally. The force expression may now be regarded as a function of θ to find the total torque and force on the inner cylinder. The developments given below are straightforward but somewhat tedious. They involve the use of the integrals in part **(c)** of Problem 1C.4.

[2] This example is based on the publication of J. M. Davies and K. Walters, *Rheology of Lubricants*, Elsevier Applied Science, London (1973). In this paper results are also given for the third-order fluid and for the Oldroyd model. See also R. I. Tanner, *Aust. J. Appl. Sci.*, **14**, 129–136 (1963); A. B. Metzner, *Rheol. Acta*, **10**, 434–444 (1971); R. I. Tanner, *Engineering Rheology*, Oxford University Press (1985), §6.3; and Example 7.5-1.

We now calculate the torque $\mathscr{T}$ on the inner cylinder which is defined to be positive in the direction of increasing θ:

$$
\begin{aligned}
\mathscr{T} &= L \int_0^{2\pi} (\boldsymbol{\delta}_X \cdot \boldsymbol{\pi}_{-Y}) r_1^2 \, d\theta \\
&= -L \int_0^{2\pi} \pi_{YX} r_1^2 \, d\theta \\
&= -\frac{2\pi\eta_0 L W_1 r_1^3 (c^2 + 2a^2)}{\sqrt{c^2 - a^2}\,(c^2 + \frac{1}{2}a^2)}
\end{aligned}
\tag{6.4-21}
$$

Notice that the term involving $\Psi_{1,0}$ vanishes when the integration is performed. Hence the expression for the torque is the same found for a Newtonian fluid.

The force in the y-direction is defined by

$$
\begin{aligned}
F_y &= L \int_0^{2\pi} (\boldsymbol{\delta}_y \cdot \boldsymbol{\pi}_{-Y}) r_1 \, d\theta = -L \int_0^{2\pi} (\pi_{YX} \cos\theta + \pi_{YY} \sin\theta) r_1 \, d\theta \\
&= L\eta_0 W_1 r_1^3 \int_0^{2\pi} \left[\frac{1}{r_1}\left(-\frac{4}{B} + \frac{6B_0}{B^2} \right) - \underline{6\left(\frac{1}{B^2} - \frac{B_0}{B^3} \right)} \right] \cos\theta \, d\theta \\
&\doteq -\frac{6\pi\eta_0 L W_1 r_1^3 a}{\sqrt{c^2 - a^2}\,(c^2 + \frac{1}{2}a^2)}
\end{aligned}
\tag{6.4-22}
$$

In going from the first to the second line of this equation, products that are odd on the interval $[0, 2\pi]$—and therefore vanish in the integration—have been omitted. In addition the term involving p_N has been integrated by parts, for simplicity. In going from the second to the third line, only the dashed underlined term has been retained since it will be the leading term for small eccentricities.

The force in the x-direction is given by

$$
\begin{aligned}
F_x &= L \int_0^{2\pi} (\boldsymbol{\delta}_x \cdot \boldsymbol{\pi}_{-Y}) r_1 \, d\theta = L \int_0^{2\pi} (\sin\theta \, \pi_{YX} - \cos\theta \, \pi_{YY}) r_1 \, d\theta \\
&\doteq -L\Psi_{1,0} W_1^2 r_1^3 \int_0^{2\pi} \left(\frac{1}{B} - \frac{3B_0}{B^2} \right)^2 \cos\theta \, d\theta \\
&= \frac{2\pi\Psi_{1,0} L W_1^2 r_1^3 a (c^4 - \frac{1}{8}a^2c^2 + \frac{1}{4}a^4)}{(c^2 - a^2)^{3/2}(c^2 + \frac{1}{2}a^2)^2}
\end{aligned}
\tag{6.4-23}
$$

Here again, in going from the first to the second line terms that are odd on the interval $[0, 2\pi]$ have been omitted. As before, only the leading term in the eccentricity has been retained.

The results of this example show that the magnitude of the force $F = (F_x^2 + F_y^2)^{1/2}$ for the second-order fluid is greater than the resultant force for the corresponding Newtonian fluid. This has been interpreted by Davies and Walters[2] to mean that lubricants with polymeric additives will support greater loads and therefore result in reduced wear. Keep in mind that this conclusion can be challenged, inasmuch as the analysis was done for a second-order fluid and is hence restricted to small values of $\text{De} = \Psi_{10} W_1 r_1 / 2\eta_0 B_0$; furthermore viscous heating effects and "shear thinning" of the viscosity may be far more imporant than the normal stress effects.

§6.5 PERTURBATION TECHNIQUE FOR CREEPING FLOWS

The governing partial differential equations for the flow of fluids described by the retarded-motion expansion are nonlinear in the velocity field. This may be seen specifically for the second-order fluid from Eqs. 6.3-1 and 2. In the previous two sections we presented some techniques that may be used for constructing exact analytical flow solutions for second-order fluids for certain special classes of flows. For most problems we cannot find exact analytical solutions, so we turn in this section to a perturbation technique that can be used to develop solutions to flow problems for the retarded-motion expansion for small Deborah numbers. Inasmuch as the retarded-motion expansion is itself restricted to small De, no significant new limitations are imposed by the use of the perturbation procedure.

To introduce the technique we use the Deborah number $\text{De} = -(b_2/b_1)\kappa = (\Psi_{1,0}/2\eta_0)\kappa$ introduced in connection with Eq. 6.2-12. Recall that κ is a characteristic strain rate in the flow, and that the Deborah number is a dimensionless strain rate that determines the importance of the various nonlinear terms in the retarded-motion expansion. Since the retarded-motion expansion is designed specifically to describe fluids in the limit of small Deborah number, it seems natural to expand the velocity and pressure fields for creeping motion as follows:

$$\boldsymbol{v} = \boldsymbol{v}_0 + \text{De}\, \boldsymbol{v}_1 + \cdots \quad \text{or} \quad \boldsymbol{v}^* = \boldsymbol{v}_0^* + \text{De}\, \boldsymbol{v}_1^* + \cdots \tag{6.5-1}$$

$$p = p_0 + \text{De}\, p_1 + \cdots \quad \text{or} \quad p^* = p_0^* + \text{De}\, p_1^* + \cdots \tag{6.5-2}$$

in which $\boldsymbol{v}^* = \boldsymbol{v}/V$ and $p^* = p/b_1\kappa$; here V is a characteristic velocity for the flow system. For illustrative purposes the expansion is carried only to first order in De here, but the technique may in principle be extended to any order. We then insert the expansion in Eq. 6.5-1 into Eq. 6.2-12 for the stress tensor, which gives

$$\begin{aligned} \boldsymbol{\tau}^* = & -[\boldsymbol{\gamma}_{(1)0}^* + \text{De}\, \boldsymbol{\gamma}_{(1)1}^* - \cdots] \\ & + \text{De}\, [\boldsymbol{\gamma}_{(2)0}^* + \cdots] \\ & + \text{De}\, B_{11}[\{\boldsymbol{\gamma}_{(1)0}^* \cdot \boldsymbol{\gamma}_{(1)0}^*\} + \cdots] \end{aligned} \tag{6.5-3}$$

in which $\boldsymbol{\gamma}_{(n)k}^* = \boldsymbol{\gamma}_{(n)k}/\kappa^n$, where $\boldsymbol{\gamma}_{(n)k}$ is the nth rate-of-strain tensor evaluated using $\boldsymbol{v}_k$ of Eq. 6.5-1.

Next we have to write the equations of continuity and motion in dimensionless form. The expression for $\boldsymbol{v}^*$ in Eq. 6.5-1 must be inserted into the equation $(\nabla \cdot \boldsymbol{v}) = 0$ in appropriate dimensionless form; and the above expression for $\boldsymbol{\tau}^*$ must be substituted into the equation of motion for creeping flow, which in dimensionless form is

$$\mathbf{0} = -\nabla^* p^* - [\nabla^* \cdot \boldsymbol{\tau}^*] + \boldsymbol{g}^* \tag{6.5-4}$$

in which $\nabla^* = (V/\kappa)\nabla$ and $\boldsymbol{g}^* = (V/b_1\kappa^2)\rho\boldsymbol{g}$. These substitutions lead to sets of equations of change obtained by equating the coefficients of De^0, De^1, De^2, and so on:

Zero Order:

$$(\nabla^* \cdot \boldsymbol{v}_0^*) = 0 \tag{6.5-5}$$

$$\nabla^{*2}\boldsymbol{v}_0^* - \nabla^*\mathscr{P}_0^* = 0 \tag{6.5-6}$$

First Order:

$$(\nabla^* \cdot \boldsymbol{v}_1^*) = 0 \tag{6.5-7}$$

$$\nabla^{*2}\boldsymbol{v}_1^* - \nabla^* p_1^* = [\nabla^* \cdot \boldsymbol{\gamma}_{(2)0}^*] + B_{11}[\nabla^* \cdot \boldsymbol{\gamma}_{(1)0}^{*2}] \tag{6.5-8}$$

in which $B_{11} = b_{11}/b_2$ and $\mathscr{P}^* = \mathscr{P}/b_1\kappa$. Second- and higher-order sets of equations can be written down by the same procedure. Each set has the same form, but the equation of motion for $\boldsymbol{v}_k^*$ (with $k \geq 1$) has nonhomogeneous terms on the right side, and these are known from the solution of the $(k-1)$th-order problem. The right side of Eq. 6.5-8 can be put into an alternative form by using the Giesekus equation, Eq. 6.3-6:

$$[\nabla^* \cdot \boldsymbol{\gamma}^*_{(2)0} + B_{11}\nabla^* \cdot \boldsymbol{\gamma}^{*2}_{(1)0}] = \nabla^*\left(\frac{D}{Dt^*}\mathscr{P}_0^* + \tfrac{1}{4}\boldsymbol{\gamma}^*_{(1)0}:\boldsymbol{\gamma}^*_{(1)0}\right) + (B_{11} - 1)[\nabla^* \cdot \boldsymbol{\gamma}^{*2}_{(1)0}] \tag{6.5-9}$$

The velocity field that enters into the substantial derivative D/Dt^* is $\boldsymbol{v}_0^*$, and $D/Dt^* = (D/Dt)/\kappa$.

It is now possible to put Eq. 6.5-8 into a slightly different form,

$$\nabla^{*2}\boldsymbol{v}_1^* - \nabla^*\tilde{p}_1^* + \boldsymbol{f}^* = \mathbf{0} \tag{6.5-10}$$

in which $\boldsymbol{f}^* = (V/b_1\kappa^2)\boldsymbol{f}$ (see Table 1.4-1) is given by

$$\boldsymbol{f}^* = (1 - B_{11})(\nabla^* \cdot \boldsymbol{\gamma}^{*2}_{(1)0}) \tag{6.5-11}$$

and

$$\tilde{p}_1^* = p_1^* + \frac{D}{Dt^*}\mathscr{P}_0^* + \tfrac{1}{4}(\boldsymbol{\gamma}^*_{(1)0}:\boldsymbol{\gamma}^*_{(1)0}) \tag{6.5-12}$$

Then Eq. 6.5-7 (continuity) and Eq. 6.5-10 (motion) are solved to get $\boldsymbol{v}_1^*$ and $\tilde{p}_1^*$, so that the final expressions for $\boldsymbol{v}^*$ and $\mathscr{P}^*$ become

$$\boldsymbol{v}^* = \boldsymbol{v}_0^* + \text{De}\,\boldsymbol{v}_1^* + \cdots \tag{6.5-13}$$

$$\mathscr{P}^* = \mathscr{P}_0^* + \text{De}\left[\tilde{p}_1^* - \frac{D}{Dt^*}\mathscr{P}_0^* - \tfrac{1}{4}(\boldsymbol{\gamma}^*_{(1)0}:\boldsymbol{\gamma}^*_{(1)0})\right] + \cdots \tag{6.5-14}$$

The zero-order functions $\boldsymbol{v}_0^*$ and p_0^* must satisfy the physical boundary conditions prescribed for the given problem. The perturbation fields $\boldsymbol{v}_k^*$ and p_k^* $(k \geq 1)$ must be subject to boundary conditions such that the functions $\boldsymbol{v}$ and p satisfy the boundary conditions of the stated problem for any value of De.

The simplest type of boundary condition arises when the velocity is specified on the entire boundary. Then the perturbation fields $\boldsymbol{v}_k$ for $k \geq 1$ must be zero on the entire boundary; an example of this situation is illustrated in Example 6.5-1. A different situation arises when a stress component is specified on part of the boundary. Then it is in fact necessary to evaluate $\boldsymbol{\gamma}_{(2)0}$ in connection with the formulation of the boundary condition for the perturbation fields, as illustrated in Example 6.5-2. Note that in both examples the perturbation fields satisfy the Newtonian fluid creeping flow equations augmented by a nonhomogeneous term. Hence we expect that the physical boundary conditions will be sufficient to determine all $\boldsymbol{v}_k^*$ and p_k^*.

Similar perturbation techniques have been developed for other types of models such as the differential models considered in Chapter 7. However, when these other models are

TABLE 6.5-1

Some Flow Systems That May be Used to Measure Second-Order Retarded-Motion Constants[a]

Flow System	Combination Evaluated	Comments
Plane flow		
(e.g., transverse slot)	Ψ_1	§6.4[b]
Translating bubble	$\Psi_1 + \frac{4}{7}\Psi_2$	Example 6.5-2
Rotating sphere	$\Psi_1 + 2\Psi_2$	Example 6.5-1
Rod-climbing	$\Psi_1 + 4\Psi_2$	Problem 6B.6
Rectilinear flow		
(e.g., tilted trough)	Ψ_2	§6.4[c]
Radial flow	$\Psi_1 - 3\Psi_2$	[a]

[a] A. Co and R. B. Bird, *Appl. Sci. Res.*, **33**, 385–404 (1977).
[b] R. I. Tanner and A. C. Pipkin, *Trans. Soc. Rheol.*, **13**, 471–484 (1964).
[c] A. C. Pipkin and R. I. Tanner, *Mech. Today*, **1**, 262–321 (1972); R. I. Tanner, *Trans. Soc. Rheol.*, **14**, 483–507 (1970); L. Sturges and D. D. Joseph, *Arch. Rat. Mech. Anal.*, **59**, 359–387 (1976).

used, the perturbation results are equivalent to those obtained from the retarded-motion expansion, as explained in connection with Fig. 6.2-2. In addition, although the models described in Chapters 7 and 8 are really designed to give a reasonable description over a wide range of strain rates, some of them are not complete at third order, and sometimes not even at second order. Therefore the perturbation technique based on the retarded-motion expansion is the simplest and safest method to obtain flow solutions at low Deborah number.

Table 6.5-1 lists some of the flow systems for which perturbation solutions are available. The remainder of this section is devoted to two examples that serve to illustrate the technique. The first concerns the determination of the first-order perturbation of a velocity field, and it is illustrated how this perturbation may be obtained in an efficient manner with the use of a stream function. The second example is somewhat more lengthy since the first-order perturbations in the pressure and stress fields are also required.

EXAMPLE 6.5-1 Flow Near a Rotating Sphere[1]

In Example 1.4-3 we considered the flow around a sphere rotating with angular velocity W in a Newtonian liquid of infinite extent. It was shown that inertial effects produce a secondary flow away from the sphere in the equatorial plane and towards the sphere along the axis of rotation. By contrast, the secondary flow around a sphere rotating slowly in a polymeric fluid goes in the opposite direction, as shown in Fig. 2.4-1; the fluid is drawn inward toward the sphere in the equatorial plane and then ejected along the axis of rotation. It is desired to obtain the secondary velocity field predicted by the second-order fluid by a perturbation analysis.

[1] H. Giesekus, *Rheol. Acta*, **3**, 59–71 (1963). In this landmark publication results are given complete through third order for the flow near a simultaneously rotating and translating sphere. Results for a rotating but not translating sphere are given complete through fourth order by R. L. Fosdick and B. G. Kao, *Rheol. Acta*, **19**, 675–697 (1980).

SOLUTION We take the origin of the coordinate system to be at the center of the sphere, with the z-axis along the axis of rotation of the sphere. We work in spherical coordinates, r, θ, ϕ, and we anticipate that the velocity and pressure fields will be independent of ϕ. We let the characteristic velocity V be WR so that $\mathbf{v}^* = \mathbf{v}/WR$, and we take the characteristic shear rate κ to be W so that $\boldsymbol{\gamma}^*_{(n)} = \boldsymbol{\gamma}_{(n)}/W^n$, $p^* = p/b_1W$, $\mathscr{P}^* = \mathscr{P}/b_1W$, $\mathbf{f}^* = \mathbf{f}R/b_1W^2$, and De $= -(b_2/b_1)W$. These definitions enable us to take over the general analysis in the introduction to this section.

The boundary conditions for this problem are

$$\text{At } r = R, \qquad \mathbf{v} = WR \sin\theta\, \boldsymbol{\delta}_\phi \tag{6.5-15}$$

$$\text{As } r \to \infty, \qquad \mathbf{v} \to \mathbf{0} \tag{6.5-16}$$

$$\text{As } r \to \infty, \qquad \mathscr{P} \to \text{constant} \tag{6.5-17}$$

As we know from Example 1.4-3, the solution to the *zero-order* equations of Eqs. 6.5-5 and 6 satisfying the above boundary conditions is

$$\mathbf{v}_0 = WR\,(R/r)^2 \sin\theta\, \boldsymbol{\delta}_\phi \tag{6.5-18}$$

$$\mathscr{P}_0 = \text{constant} \tag{6.5-19}$$

Next we have to solve the *first-order* equations with the boundary conditions

$$\text{At } r = R, \qquad \mathbf{v}_1 = \mathbf{0} \tag{6.5-20}$$

$$\text{As } r \to \infty, \qquad \mathbf{v}_1 \to \mathbf{0} \tag{6.5-21}$$

$$\text{As } r \to \infty, \qquad p_1 \to 0 \text{ at } \theta = \pi/2 \tag{6.5-22}$$

Obtaining the solution of the first-order equations in Eqs. 6.5-7 and 10 is a lengthy task, which is the main business of the remainder of this example.

Our first task is to work out the expression for $\mathbf{f}$ in Eq. 6.5-10. From the expression for $\mathbf{v}_0$ in Eq. 6.5-18 we first get

$$\boldsymbol{\gamma}_{(1)0} = -3W(R/r)^3 \sin\theta\, \{\boldsymbol{\delta}_r\boldsymbol{\delta}_\phi + \boldsymbol{\delta}_\phi\boldsymbol{\delta}_r\} \tag{6.5-23}$$

From this we obtain

$$\boldsymbol{\gamma}^2_{(1)0} = \{\boldsymbol{\gamma}_{(1)0}\cdot\boldsymbol{\gamma}_{(1)0}\} = 9W^2(R/r)^6 \sin^2\theta\, \{\boldsymbol{\delta}_r\boldsymbol{\delta}_r + \boldsymbol{\delta}_\phi\boldsymbol{\delta}_\phi\} \tag{6.5-24}$$

$$[\boldsymbol{\nabla}\cdot\boldsymbol{\gamma}^2_{(1)0}] = -9(W^2/R)(R/r)^7 \sin\theta\, (5\sin\theta\, \boldsymbol{\delta}_r + \cos\theta\, \boldsymbol{\delta}_\theta) \tag{6.5-25}$$

Therefore Eq. 6.5-10 becomes for the rotating-sphere problem:

$$b_1\nabla^2\mathbf{v}_1 - \boldsymbol{\nabla}\tilde{p}_1 = -\left(\frac{b_{11}}{b_2} - 1\right)9\sin\theta\,(W/R)b_1(R/r)^7\,(5\sin\theta\,\boldsymbol{\delta}_r + \cos\theta\,\boldsymbol{\delta}_\theta) \tag{6.5-26}$$

This has to be solved with the boundary conditions mentioned above (note that at $\theta = \pi/2$ and as $r \to \infty$, we have $\tilde{p}_1 \to 0$, since $\mathscr{P}_0 = \text{constant}$ and $\boldsymbol{\gamma}_{(1)0} \to \mathbf{0}$).

Because the flow is axisymmetric and because $\boldsymbol{v}$ and $\mathscr{P}$ do not depend on ϕ, it is expedient to use the stream-function method of §1.4. Use of Eq. E of Table 1.4-1 gives (with $-\boldsymbol{f}$ designating the right side of Eq. 6.5-26)

$$E^4\psi = E^2(E^2\psi) = -(r/b_1)\sin\theta\,[\nabla \times \boldsymbol{f}]_\phi$$

$$= \left(\frac{b_{11}}{b_2} - 1\right)144\,(W/R)(R/r)^7 \sin^2\theta\cos\theta \tag{6.5-27}$$

in which

$$E^2 = \frac{\partial^2}{\partial r^2} + \frac{\sin\theta}{r^2}\frac{\partial}{\partial\theta}\left(\frac{1}{\sin\theta}\frac{\partial}{\partial\theta}\right) \tag{6.5-28}$$

If we now postulate a solution to Eq. 6.5-27 of the form $\psi(r, \theta) = WR^3\, g(r)\sin^2\theta\cos\theta$, we get

$$\left(\frac{d^2}{dr^2} - \frac{6}{r^2}\right)\left(\frac{d^2}{dr^2} - \frac{6}{r^2}\right)g = \frac{144}{R^2}\left(\frac{b_{11}}{b_2} - 1\right)\left(\frac{R}{r}\right)^7 \tag{6.5-29}$$

This is a fourth-order, nonhomogeneous ordinary differential equation, for which we have to find the "complementary function," containing four constants of integration and a "particular integral." The *complementary function* is obtained by trying a solution of the form $g(r) = (r/R)^n$; substitution of this trial function into Eq. 6.5-29 with the right side set equal to zero leads to a quartic equation for n, and the latter has roots -2, 0, 3, and 5. Therefore, the complementary function is

$$[g(r)]_{\text{C.F.}} = c_1(R/r)^2 + c_2 + c_3(r/R)^3 + c_4(r/R)^5 \tag{6.5-30}$$

As $r \to \infty$, the function g must not be allowed to become infinite, and therefore c_3 and c_4 must both be zero. The *particular integral* is obtained by trying a function of the form $g(r) = c_0(r/R)^m$; insertion of this into the nonhomogeneous equation in Eq. 6.5-29 gives the values of c_0 and m. The particular integral is thus found to be

$$[g(r)]_{\text{P.I.}} = \left(\frac{b_{11}}{b_2} - 1\right)\left(\frac{R}{r}\right)^3 \tag{6.5-31}$$

The complete solution is given by the sum of these last two expressions. The two constants c_1 and c_2 are determined from

At $r = R$, $\quad v_{1r} = -(1/r^2\sin\theta)\,\partial\psi/\partial\theta = 0, \quad$ which implies that $g(R) = 0$ (6.5-32)

At $r = R$, $\quad v_{1\theta} = +(1/r\sin\theta)\,\partial\psi/\partial r = 0 \quad$ which implies that $g'(R) = 0$ (6.5-33)

These conditions give

$$c_1 = -\frac{3}{2}\left(\frac{b_{11}}{b_2} - 1\right); \quad c_2 = +\frac{1}{2}\left(\frac{b_{11}}{b_2} - 1\right) \tag{6.5-34}$$

and with these results $\psi(r, \theta)$ is now known completely. From $\psi(r, \theta)$ we can now obtain v_{1r} and $v_{1\theta}$, and hence $\boldsymbol{v}_1$. The velocity is then

$$\begin{aligned} \boldsymbol{v} &= WR(R/r)^2 \sin\theta\, \boldsymbol{\delta}_\phi \\ &\quad - \mathrm{De}\, WR\left(\frac{b_{11}}{b_2} - 1\right)\left[\left(\frac{1}{2}\left(\frac{R}{r}\right)^2 - \frac{3}{2}\left(\frac{R}{r}\right)^4 + \left(\frac{R}{r}\right)^5\right)(3\cos^2\theta - 1)\,\boldsymbol{\delta}_r \right.\\ &\quad \left. - 3\left(\left(\frac{R}{r}\right)^4 - \left(\frac{R}{r}\right)^5\right)\sin\theta\cos\theta\,\boldsymbol{\delta}_\theta\right] \\ &\quad + O(\mathrm{De}^2) \end{aligned} \tag{6.5-35}$$

or in terms of the zero-shear-rate viscometric functions

$$\begin{aligned} \boldsymbol{v} &= WR(R/r)^2 \sin\theta\, \boldsymbol{\delta}_\phi \\ &\quad + W^2 R\left(\frac{\Psi_{1,0} + 2\Psi_{2,0}}{2\eta_0}\right)\left[\left(\frac{1}{2}\left(\frac{R}{r}\right)^2 - \frac{3}{2}\left(\frac{R}{r}\right)^4 + \left(\frac{R}{r}\right)^5\right)(3\cos^2\theta - 1)\,\boldsymbol{\delta}_r \right.\\ &\quad \left. - 3\left(\left(\frac{R}{r}\right)^4 - \left(\frac{R}{r}\right)^5\right)\sin\theta\cos\theta\,\boldsymbol{\delta}_\theta\right] + \cdots \end{aligned} \tag{6.5-36}$$

Note that the primary velocity field has only a ϕ-component and is proportional to W, whereas the secondary flow field has only r- and θ-components and is proportional to W^2. If $\Psi_{1,0} + 2\Psi_{2,0} > 0$ the secondary flow far from the sphere will be toward the sphere near the equator and away from the sphere near the axis, with the two zones being separated by conical surfaces given by $3\cos^2\theta - 1 = 0$, or $\theta = \arccos(1/\sqrt{3}) \doteq 54°44'$ or $125°16'$. This behavior is in qualitative agreement with the flow pattern observed experimentally.[2]

The velocity field in Eq. 6.5-35 may be used to calculate the torque needed to maintain the angular velocity W. The result, $\mathscr{T} = 8\pi b_1 R^3 W$, is identical to that for a Newtonian fluid. Walters and Waters[3] have performed a perturbation calculation for third-order fluids, and they have included the inertial terms as well (i.e., the $[\boldsymbol{v}\cdot\nabla\boldsymbol{v}]$-term in the equation of motion). Their result for the torque is

$$\begin{aligned} \mathscr{T} = 8\pi b_1 R^3 W\left[1 + \frac{\mathrm{Re}^2}{1200} + \frac{\mathrm{ReDe}}{140}\left(1 + \frac{b_{11}}{b_2}\right) \right.\\ \left. - \frac{2\mathrm{De}^2}{15}\left(1 + \frac{b_{11}}{b_2}\right)^2 - \frac{24\mathrm{De}^2}{5}\frac{b_1(b_{12} - b_{1:11})}{b_2^2}\right] + \cdots \end{aligned} \tag{6.5-37}$$

in which $\mathrm{Re} = R^2 W\rho/b_1$ and $\mathrm{De} = -(b_2/b_1)W$. Note that the two dimensionless groups, Re and De, can be varied independently of one another. Note further that the third-order constant b_3 does not appear in this expression.

EXAMPLE 6.5-2 Flow Near a Translating Bubble[4]

In Example 1.4-2 we considered the flow around a spherical bubble rising slowly in an incompressible Newtonian fluid of density ρ. It is desired to extend this analysis to a bubble rising slowly in a second-order fluid as follows:

[2] K. Walters and J. G. Savins, *Trans. Soc. Rheol.*, **9**:1, 407–416 (1965); Y. Ide and J. L. White, *J. Appl. Polym. Sci.*, **18**, 2997–3018 (1974).

[3] K. Walters and N. D. Waters, *Br. J. Appl. Phys.*, **14**, 667–671 (1963); **15**, 989–991 (1964); *Rheol. Acta*, **3**, 312–315 (1964).

[4] O. Hassager in K. Østergaard and Aa. Fredenslund, eds., *Chemical Engineering with Per Søltoft*, Teknisk Forlag, Copenhagen (1977). The slow motion of a bubble in an Oldroyd fluid was studied by O. O. Ajayi, *J. Eng. Math.*, **9**, 273–280 (1975), through terms of first order, and by G. F. Tiefenbruck and L. G. Leal, *J. Non-Newtonian Fluid Mech.*, **7**, 257–264 (1980), through terms of second order in the Deborah number. The corresponding drop motion was first studied by M. G. Wagner and J. L. Slattery, *AIChE J.*, **17**, 1198–1207 (1971).

a. First, assume that the bubble remains spherical, and find the velocity perturbation in the fluid, the pressure and stress distributions on the surface of the bubble, and the velocity of rise of the bubble to first order in the Deborah number.

b. Second, use the pressure and stress distributions calculated above to estimate the shape of the bubble in the limit that it is only slightly deformed from spherical shape.

SOLUTION **(a)** This example differs from the preceding example in that the velocity is not completely specified on the entire boundary. On the surface of the bubble we specify that the normal component of velocity must be zero and that the shear stress must be zero. That is, if we describe the flow in the spherical coordinate system shown in Fig. 1.4-1 with origin at the center of the bubble we have the conditions

$$\text{At } r = R_0, \qquad v_r = 0; \qquad \tau_{r\theta} = 0 \tag{6.5-38}$$

$$\text{At } r = \infty \text{ and } \theta = \pi/2, \qquad p = p_\infty \tag{6.5-39}$$

Here R_0 is the radius of the undeformed sphere. We then expand the velocity and pressure according to Eqs. 6.5-1 and 2. For this problem the characteristic strain rate is $\kappa = V/R_0$, where V is the rise velocity of the bubble; this means that the dimensionless quantities used in the introduction to this section are: $\text{De} = -(b_2/b_1)(V/R_0)$, $\boldsymbol{\gamma}^*_{(n)} = \boldsymbol{\gamma}_{(n)}(R_0/V)^n$, $p^* = pR_0/b_1V$, $\boldsymbol{\tau}^* = \boldsymbol{\tau}R_0/b_1V$, and $\boldsymbol{f}^* = \boldsymbol{f}R_0^2/b_1V$. When Eq. 6.5-1 is inserted into Eq. 6.5-38 we find

$$\text{At } r = R_0, \qquad v_{0r} = 0 \quad v_{1r} = 0 \tag{6.5-40}$$

We also use Eq. 6.5-3 to reformulate the shear-stress condition in Eq. 6.5-38 to

$$\text{At } r = R_0, \qquad \gamma_{(1)0r\theta} = 0; \quad \gamma_{(1)1r\theta} = \frac{R_0}{V}(\gamma_{(2)0r\theta} + (b_{11}/b_2)\{\boldsymbol{\gamma}_{(1)0}\cdot\boldsymbol{\gamma}_{(1)0}\}_{r\theta}) \tag{6.5-41}$$

Notice that the boundary conditions specify values of the $r\theta$-component of the $\boldsymbol{\gamma}_{(1)}$ tensor constructed from $\boldsymbol{v}_0$ and $\boldsymbol{v}_1$, respectively. The $\boldsymbol{\gamma}_{(2)}$ tensor enters the boundary conditions for $\boldsymbol{v}_1$ only in $\boldsymbol{\gamma}_{(2)0}$, with the $\boldsymbol{v}_0$ field known from the solution of the unperturbed flow problem.

The perturbation solution now proceeds in the same way as in the preceding example and some detail will be omitted in the following. First we need the solution for the unperturbed velocity field and pressure distribution from Example 1.4-2:

$$\boldsymbol{v}_0 = V[-(1 - (R_0/r))\cos\theta\,\boldsymbol{\delta}_r + (1 - \tfrac{1}{2}(R_0/r))\sin\theta\,\boldsymbol{\delta}_\theta] \tag{6.5-42}$$

$$p_0 = (b_1VR_0r^{-2} - \rho g r)\cos\theta + p_\infty \tag{6.5-43}$$

From Eq. 6.5-42 we find

$$\boldsymbol{\gamma}_{(1)0} = VR_0r^{-2}\cos\theta\,(-2\boldsymbol{\delta}_r\boldsymbol{\delta}_r + \boldsymbol{\delta}_\theta\boldsymbol{\delta}_\theta + \boldsymbol{\delta}_\phi\boldsymbol{\delta}_\phi) \tag{6.5-44}$$

Then we calculate $\boldsymbol{\gamma}_{(2)0}$ and $\{\boldsymbol{\gamma}_{(1)0}\cdot\boldsymbol{\gamma}_{(1)0}\}$. These tensors turn out to have a vanishing $r\theta$-component at $r = R_0$, so that the boundary condition for $\boldsymbol{v}_1$ in Eq. 6.5-41 reduces to

$$\text{At } r = R_0, \qquad \gamma_{(1)1r\theta} = 0 \tag{6.5-45}$$

Next we need to calculate the expression for $\boldsymbol{f}$ in Eq. 6.5-11, which gives

$$\boldsymbol{f} = 2((b_{11}/b_2) - 1)\,b_1VR_0^3r^{-5}(5\cos^2\theta\boldsymbol{\delta}_r + \sin\theta\cos\theta\,\boldsymbol{\delta}_\theta) \tag{6.5-46}$$

We now may solve Eqs. 6.5-7, 10, and 46 with the use of a stream function with the help of Table 1.4-1. The results are

$$v_{1r} = [1 - (b_{11}/b_2)](VR_0^2/10)[-3r^{-2} + 5R_0r^{-3} - 2R_0^2r^{-4}](3\cos^2\theta - 1) \tag{6.5-47}$$

$$v_{1\theta} = [1 - (b_{11}/b_2)](VR_0^2/10)[+5R_0r^{-3} - 4R_0^2r^{-4}]\sin\theta\cos\theta \tag{6.5-48}$$

$$\tilde{p}_1 = b_1[1 - (b_{11}/b_2)]\,(VR_0^2/10)[-6r^{-3}(3\cos^2\theta - 1) + 5R_0r^{-4}(8\cos^2\theta - 1)] \tag{6.5-49}$$

The solution of the equation for the stream function provides just $\boldsymbol{v}_1$. Subsequently the known expressions for $\boldsymbol{v}_1$ and f must be introduced in the equation of motion in Eq. 6.5-10 and the resulting equation is then solved for $\tilde{p}_1$. In this last step we use the condition that $\tilde{p}_1 \to 0$ for $r \to \infty$ and $\theta = \pi/2$.

The complete solution for the velocity and pressure fields for the flow around a spherical bubble through terms of first order in the Deborah number may now be obtained from Eqs. 6.5-13 and 14 as

$$\boldsymbol{v} = \boldsymbol{v}_0 - (b_2/b_1)(V/R_0)\boldsymbol{v}_1 + \cdots \tag{6.5-50}$$

$$p = p_0 - (b_2/b_1)(V/R_0)\tilde{p}_1 - b_2V^2R_0[(3\cos^2\theta - 1)r^{-3} - \tfrac{1}{2}(2\cos^2\theta - 1)R_0r^{-4}] + \cdots \tag{6.5-51}$$

where $\boldsymbol{v}_0$, $\boldsymbol{v}_1$, p_0, and $\tilde{p}_1$ are given above.

The normal stress distribution at the surface of the bubble may now be obtained through terms of first order in De from Eq. 6.5-3 as

$$\begin{aligned}[\pi_{rr}]_{r=R_0} &= \left[p - b_1\left\{\gamma_{(1)0} + \text{De}\left(\gamma_{(1)1} - \frac{R_0}{V}\gamma_{(2)0}\right.\right.\right.\\ &\qquad \left.\left.\left.-(b_{11}R_0/b_2V)\{\gamma_{(1)0}\cdot\gamma_{(1)0}\}\right)\right\}_{rr}\right]_{r=R_0}\\ &= p_\infty + [3b_1(V/R_0) - \rho gR_0]\cos\theta\\ &\qquad + \frac{b_2}{10}(V/R_0)^2[(42 - 12(b_{11}/b_2))\cos^2\theta - 14 - (b_{11}/b_2)]\end{aligned} \tag{6.5-52}$$

Now we may parallel the treatment in Example 1.4-2 to find the following expression for the total force in the z-direction on the bubble:

$$\begin{aligned}F_z &= \int_0^{2\pi}\int_0^{\pi}[-\pi_{rr}\cos\theta]_{r=R_0}R_0^2\sin\theta\,d\theta\,d\phi\\ &= -4\pi b_1R_0V + \frac{4}{3}\pi R_0^3\rho g\end{aligned} \tag{6.5-53}$$

We see that the bubble will rise with the steady-state velocity

$$V = \tfrac{1}{3}\rho gR_0^2/b_1 \tag{6.5-54}$$

and that the normal stresses developed in the fluid do not affect the rise velocity at first order in the Deborah number. When Eq. 6.5-54 is introduced into the expression for the normal stress in Eq. 6.5-52, we see that the coefficient in front of $\cos\theta$ vanishes and we may rewrite the equation as

$$[\pi_{rr}]_{r=R_0} = p_m - \frac{2b_1V}{5R_0}\left(7 - 2\frac{b_{11}}{b_2}\right)\text{De}\,P_2(C) \tag{6.5-55}$$

with

$$p_m = p_\infty - \tfrac{1}{2}b_{11}(V/R_0)^2 \tag{6.5-56}$$

Here $P_2(C) = \frac{1}{2}(3C^2 - 1)$ is the second Legendre polynomial and $C = \cos\theta$. If one integrates Eq. 6.5-55 over the entire surface then the term involving $P_2(C)$ vanishes, so that p_m represents a mean pressure on the bubble.

(b) To describe the deformation of the bubble we need to know the velocity field and pressure distribution around the deformed bubble. We assume here that the bubble is deformed only so slightly from the spherical shape that the velocity field and pressure distribution for flow around the spherical bubble calculated in part **(a)** remain valid in a first approximation.

We now need to introduce the surface tension σ. This property will appear in the dimensionless group $\text{Ca} = b_1V/\sigma$ called the *capillary number*. Small values of the capillary number correspond to situations with dominant surface tension. In particular at zero capillary number there can be no deformation of the bubble from the spherical shape for any Deborah number. Conversely at zero Deborah number we see from Eq. 6.5-55 that only a normal stress independent of θ remains, and there will be no deformation of the spherical bubble independent of the capillary number. Consequently the leading term in a series approximation for the deformation of the bubble in powers of De and Ca will be linear in both De and Ca. That is, we consider that the bubble is deformed from a sphere of radius R_0 to a shape described by $R(\theta)$ as

$$R(\theta)/R_0 = 1 + \text{Ca De}\,\zeta(\theta) + \cdots \tag{6.5-57}$$

where we wish to determine $\zeta(\theta)$ subject to the conditions that $\text{De} < O(1)$ and $\text{Ca} < O(1)$. Now that we have the additional unknown function $\zeta(\theta)$ we need one more equation. This extra equation is furnished by the normal stress boundary condition at the surface of the bubble, which has so far been ignored. The condition states that at the surface of the bubble the gas pressure inside the bubble p_g equals the sum of the normal stress exerted by the fluid π_{rr} and the surface tension force:

$$p_g = \pi_{rr} + \sigma\left(\frac{1}{R_1} + \frac{1}{R_2}\right) \tag{6.5-58}$$

Here R_1 and R_2 are the principal radii of curvature of the surface. This boundary condition should really be applied at $r = R(\theta)$. However, in agreement with our assumption that the deformation is small, we now substitute $[\pi_{rr}]_{r=R_0}$ from Eq. 6.5-55 for π_{rr} into Eq. 6.5-58. Also for the small deformations from the spherical shape considered in Eq. 6.5-57 the surface tension terms may be linearized as follows:[5]

$$\frac{1}{R_1} + \frac{1}{R_2} = \frac{2}{R_0} - \frac{\text{Ca De}}{R_0}\left\{2\zeta + \frac{d}{dC}\left[(1 - C^2)\frac{d\zeta}{dC}\right]\right\} + O((\text{Ca De})^2) \tag{6.5-59}$$

where $C = \cos\theta$. When Eqs. 6.5-55 and 59 are substituted into Eq. 6.5-58 one obtains the following differential equation for $\zeta(\theta)$:

$$(\text{Ca De})\left\{\frac{d}{dC}\left[(1 - C^2)\frac{d\zeta}{dC}\right] + 2\zeta\right\} = \left[2 + \frac{R_0}{\sigma}(p_m - p_g)\right] - \tfrac{2}{5}\left(7 - 2\frac{b_{11}}{b_2}\right)\text{Ca De}\,P_2(C) \tag{6.5-60}$$

The function ζ has to satisfy the two conditions

$$\int_{-1}^{1} \zeta dC = 0; \quad \int_{-1}^{1} \zeta C dC = 0 \tag{6.5-61}$$

These equations are linearizations of the conditions that the volume of the bubble must remain equal to $(4\pi/3)R_0^3$ and that the "center of mass" of the bubble must remain at the origin of coordinates,

[5] L. D. Landau and E. M. Lifshitz, *Fluid Mechanics*, Addison-Wesley, Reading, MA (1959), pp. 238–239.

respectively. The condition that there is no kink in the surface at the poles, that is, that $d\zeta/d\theta$ is zero for $\theta = 0$ and π, is replaced by a condition that $d\zeta/dC$ is finite at $\theta = 0$ and π.

From a solution of Eq. 6.5-60 subject to the stated conditions one finds that $R(\theta)$ is

$$R/R_0 = 1 + \text{Ca}\,(V/20R_0)(7\Psi_{1,0} + 4\Psi_{2,0})P_2(\cos\theta) + \cdots \tag{6.5-62}$$

where the Deborah number has been rewritten in terms of the zero-shear-rate normal stress coefficients. Note that the second normal stress coefficient plays a relatively minor role in this expression, so that the deformation is primarily determined by the first normal stress coefficient. Since $\Psi_{1,0} \geq 0$ the calculation shows that the bubble is deformed into the shape of an ellipsoid elongated at the poles (a "prolate" ellipsoid). In the process of obtaining this solution we have applied the boundary conditions not at $R(\theta)$ but at R_0. To see that this approximation is consistent with the solution obtained let us consider the velocity perturbation to first order in the product of the Deborah number and capillary number. We call the velocity solution obtained in part **(a)** $\boldsymbol{v}^{(0)}$ and consider then a disturbance Ca De $\boldsymbol{v}^{(1)}$ due to the deformation in Eq. 6.5-57 such that the altered velocity field is

$$\boldsymbol{v} = \boldsymbol{v}^{(0)} + \text{Ca De}\,\boldsymbol{v}^{(1)} + \cdots \tag{6.5-63}$$

Now we expand this velocity field in a Taylor series in R around R_0 to obtain

$$\boldsymbol{v}(R, \theta) = \boldsymbol{v}^{(0)}(R_0, \theta) + \text{Ca De}\,[\zeta R_0(\partial \boldsymbol{v}^{(0)}/\partial R) + \boldsymbol{v}^{(1)}]_{R_0,\theta} + \cdots \tag{6.5-64}$$

Now we see that it is indeed consistent to apply the boundary conditions for $\boldsymbol{v}^{(0)}$ at R_0 rather than at R. Furthermore Eq. 6.5-64 gives a procedure for the construction of boundary conditions for $\boldsymbol{v}^{(1)}$ given the solution $\boldsymbol{v}^{(0)}$. However we are interested in the deformation just to first order in the capillary number, and for this purpose $\boldsymbol{v}^{(0)}$ is all that is needed.

§6.6 LIMITATIONS OF THE RETARDED-MOTION EXPANSION AND RECOMMENDATIONS FOR ITS USE

When the retarded-motion expansion was introduced in §6.2 it was described as a constitutive equation restricted to small Deborah number. An appreciation of the origin of this restriction may be obtained in Chapter 9 where a formal derivation of the expansion is given. At this point we merely summarize some of the difficulties that one may encounter if one tries to apply the retarded-motion expansion outside its range of validity:

- **i.** One may predict a shear stress vs. shear rate curve with a maximum, or even negative, viscosity or elongational viscosity (as discussed in Example 6.2-1 and in the text immediately following the example).
- **ii.** One may predict that the second-order fluid is unstable in the rest state unless one chooses some unrealistic values of the retarded-motion constants (see Problem 6C.1).
- **iii.** One may need additional boundary conditions due to the appearance of higher-order derivatives in terms such as $\{\boldsymbol{v}\cdot\nabla\boldsymbol{\gamma}_{(1)}\}$. However if the retarded-motion expansion is used in a perturbation expansion then the higher-order derivatives always appear in the nonhomogeneous terms (as explained in §6.5) and no additional boundary conditions are needed.
- **iv.** One cannot describe even qualitatively the stress relaxation experiments discussed in Chapter 3.

There may be more difficulties, but this list is sufficient to emphasize that the restriction to small De should be taken seriously. Any solution to a flow problem with a retarded-motion expansion where the second-order terms are not small corrections to the Newtonian terms (and the third-order terms small corrections to the second-order terms, and so on) must be regarded with suspicion and is likely to be a mathematical curiosity with no physical relevance.

Unfortunately, for flows in some geometries it is not possible to construct a solution where the second-order terms are everywhere small corrections to the Newtonian terms. Examples of such problems are flow around a sharp corner (where the second-order terms dominate the Newtonian terms close to the corner; see Problem 6C.2) and flow into a line sink (where the second-order terms dominate the Newtonian terms close to the sink; see Problem 6B.5). Consequently for flows that involve sharp corners, sinks, or sources the retarded-motion expansion should never be used.

Even when the retarded-motion expansion is used correctly, however, it is important to note that the use of more than the second-order terms in perturbation solutions often results in series with "diminishing returns." By this we mean that while retention of second-order terms gives a qualitative and quantitative description of the departure from Newtonian behavior, the inclusion of third- and higher-order terms provides only very minor additional improvements to the solution. This should be compared with the fact that the second-order terms may usually be obtained at a moderate analytical effort, but the higher-order terms require increasingly tedious and lengthy algebraic developments.

With the above restrictions in mind, then, we recommend that the retarded motion expansion be used primarily just through second-order terms for analytical investigations of the departure from Newtonian behavior. The purpose of such investigations could be:

i. To investigate qualitative behavior of polymer flow phenomena, such as the direction of secondary flows.
ii. To compare with numerical (or approximate) solutions of more realistic constitutive equations in limiting situations.
iii. To compare with experimental data in order to characterize polymeric fluids by their values of η_0, $\Psi_{1,0}$, and $\Psi_{2,0}$.
iv. To explore the nature of the motions of orientable or deformable particles in suspensions with a polymeric suspending fluid.

When experiments are performed to determine η_0, $\Psi_{1,0}$, and $\Psi_{2,0}$ from comparison with retarded-motion solutions it is often difficult to obtain data at sufficiently slow deformation rates that the fluid behaves as a second-order fluid and still retain sufficient sensitivity to enable accurate determinations of $\Psi_{1,0}$ and $\Psi_{2,0}$. Consequently for this purpose we would presently recommend the use of geometries in which no pressure transducers are required and the fluid acts as its own "manometer" as in the development of secondary flow or in deformation of a free surface (the rotating sphere, the rod climbing effect, the tilted trough, or the translating bubble). Keep in mind that these kinds of experimentally observed phenomena cannot be described by using the generalized Newtonian models of Chapter 4 or the linear viscoelastic models of Chapter 5.

The second-, third-, and higher-order fluid models have been used to examine flows in a wide variety of systems; we mention some of them here: secondary flow in curved tubes[1] and in conduits of noncircular cross section;[2–4] flow in eccentric annuli, both

[1] H. A. Barnes and K. Walters, *Proc. Roy. Soc. Ser. A*, **314**, 85–109 (1969).
[2] R. S. Rivlin in M. Reiner and D. Abir, eds., *Second Order Effects in Elasticity, Plasticity, and Fluid Dynamics*, Macmillan, New York (1964), pp. 668–677.
[3] R. S. Rivlin and W. E. Langlois, *Rend. Mat.*, **22**, 169–185 (1963).
[4] A. C. Pipkin, *Proceedings of the 4th International Congress on Rheology*, Vol. 1, Wiley-Interscience, New York (1965), pp. 213–222.

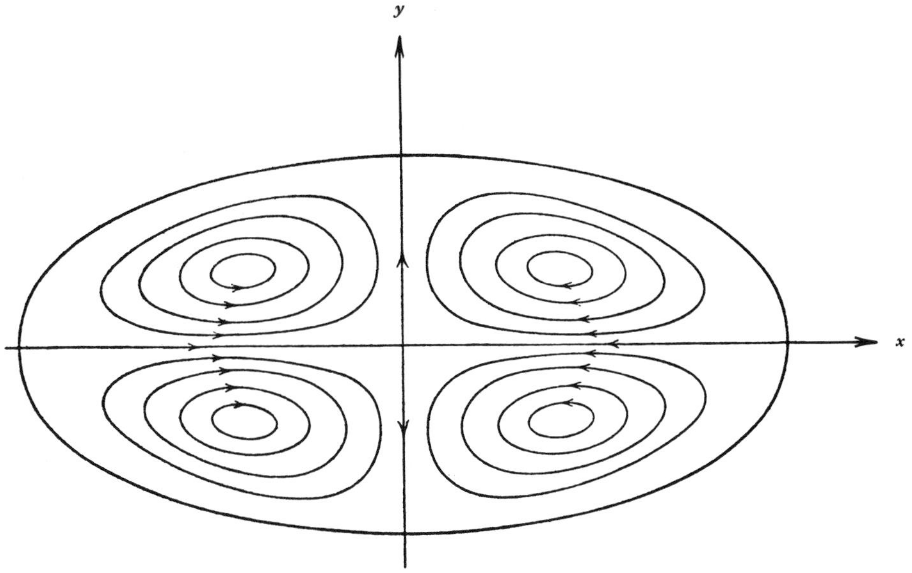

FIGURE 6.6-1. Secondary flow of a fourth-order fluid flowing in the z-direction through a straight pipe of elliptical cross section. [R. S. Rivlin in R. J. Seeger and G. Temple, eds., *Research Frontiers in Fluid Dynamics*, Wiley, New York (1965), Chapt. 5.]

tangential[5] and axial;[6] secondary flow in the disk and cylinder system[7] and in the cone-and-plate viscometer;[8–10] flow near a translating sphere;[11–13] flow in a converging tube;[14,15] radial flow between parallel disks;[16–18] climbing of a fluid near a rotating rod (Weissenberg effect);[19–21] boundary layer flows;[22–24] extrudate swell;[25] the hole pressure

[5] J. M. Davies and K. Walters in T. C. Davenport, ed., *Rheology of Lubricants*, Wiley, New York (1973), pp. 65–80.
[6] J. Y. Kazakia and R. S. Rivlin, *J. Non-Newtonian Fluid Mech.*, **8**, 311–317 (1981).
[7] D. F. Griffiths, D. T. Jones, and K. Walters, *J. Fluid Mech.*, **36**, 161–175 (1969).
[8] H. Giesekus, *Rheol. Acta*, **6**, 339–353 (1967).
[9] D. F. Griffiths and K. Walters, *J. Fluid Mech.*, **42**, 379–399 (1970).
[10] K. Walters and N. D. Waters in R. E. Wetton and R. W. Whorlow, eds., *Polymer Systems, Deformation and Flow*, Macmillan, London (1968), Chapt. 17, pp. 211–235.
[11] B. Caswell and W. H. Schwarz, *J. Fluid Mech.*, **13**, 417–426 (1962).
[12] H. Giesekus, *Rheol. Acta*, **3**, 59–71 (1963).
[13] M. G. Wagner and J. C. Slattery, *AIChE J.*, **17**, 1198–1207 (1971).
[14] E. B. Adams, J. C. Whitehead, and D. C. Bogue, *AIChE J.*, **11**, 1026–1032 (1965).
[15] P. Schümmer, *Rheol. Acta*, **6**, 192–200 (1967).
[16] W. H. Schwarz and C. Bruce, *Chem. Eng. Sci.*, **24**, 399–413 (1969).
[17] J. M. Piau and M. Piau, *C. R. Acad. Sci. Ser. A*, **270**, 159–161 (1970).
[18] A. Co and R. B. Bird, *Appl. Sci. Res.*, **33**, 385–404 (1977).
[19] A. Kaye, *Rheol. Acta*, **12**, 207–211 (1973).
[20] D. D. Joseph, G. S. Beavers, and R. L. Fosdick, *Arch. Rat. Mech. Anal.*, **49**, 381–401 (1973).
[21] D. D. Joseph and R. L. Fosdick, *Arch. Rat. Mech. Anal.*, **49**, 321–380 (1973).
[22] D. W. Beard and K. Walters, *Proc. Camb. Phil. Soc.*, **60**, 667–674 (1964).
[23] M. H. Davies, *Z. Angew. Math. Phys.*, **17**, 189–191 (1966).
[24] M. M. Denn, *Chem. Eng. Sci.*, **22**, 394–405 (1967).
[25] M. Zidan, *Rheol. Acta*, **8**, 89–123 (1969).

effect;[26–28] and motions of suspended, orientable particles.[29, 30] One of the oldest of the ordered-fluid calculations is that for noncircular tubes,[3] in which purely axial flow is in general not possible if terms of fourth order are retained; the axial flow must then be accompanied by secondary flows. An illustration of this is given in Fig. 6.6-1.

In some of the flow systems listed above the retarded-motion expansion has been used outside its range of applicability. An example of this is the extrudate swell problem, where the retarded-motion expansion should not be used because of the presence of the sharp corner.

PROBLEMS

6B.1 Kinematic Tensors for Shear and Shearfree Flows

a. Verify the matrix displays for $\boldsymbol{\nabla v}$, $\boldsymbol{\gamma}_{(1)}$, and $\boldsymbol{\gamma}_{(2)}$ for shear flows and for shearfree flows given in Table C.1.

b. Show that $\boldsymbol{\gamma}_{(n)} = \mathbf{0}$ for $n \geqslant 3$ for *steady-state* shear flow.

c. Show that

$$\boldsymbol{\gamma}_{(n)} = (-1)^{n+1}\boldsymbol{\gamma}_{(1)}^{n} \tag{6B.1-1}$$

for *steady-state* shearfree flows.

6B.2 Eccentric-Disk Rheometer Flow

In the eccentric-disk rheometer shown in Fig. 3B.1, the velocity distribution is given by

$$v_x = -W(y - Az); \quad v_y = Wx; \quad v_z = 0 \tag{6B.2-1}$$

in which W is the angular velocity of both disks (rotating in the same direction), and $A = a/b$, where a is the distance between the axes of rotation, and b is the spacing between the disks.

a. For this flow display the following tensors in matrix form: $\boldsymbol{\nabla v}$, $(\boldsymbol{\nabla v})^{\dagger}$, $\boldsymbol{\gamma}_{(1)}$, $\boldsymbol{\gamma}_{(2)}$, $\boldsymbol{\gamma}_{(3)}$, $\{\boldsymbol{\gamma}_{(1)} \cdot \boldsymbol{\gamma}_{(1)}\}$, and $\{\boldsymbol{\gamma}_{(1)} \cdot \boldsymbol{\gamma}_{(2)} + \boldsymbol{\gamma}_{(2)} \cdot \boldsymbol{\gamma}_{(1)}\}$.

b. Use the third-order fluid to find τ_{xz} and τ_{yz}, which can be measured in the eccentric-disk rheometer.

c. Next show that

$$\lim_{A \to 0}\left(-\frac{\tau_{xz}}{AW}\right) = b_1 - b_3 W^2 \tag{6B.2-2}$$

$$\lim_{A \to 0}\left(-\frac{\tau_{yz}}{AW}\right) = -b_2 W \tag{6B.2-3}$$

Compare these results with Eqs. 6B.3-2.

[26] E. A. Kearsley, *Trans. Soc. Rheol.*, **14**, 419–424 (1970).
[27] R. I. Tanner and A. C. Pipkin, *Trans. Soc. Rheol.*, **13**, 471–484 (1969).
[28] S. A. Trogdon and D. D. Joseph, *J. Non-Newtonian Fluid Mech.*, **10**, 185–213 (1982).
[29] L. G. Leal, *J. Fluid Mech.*, **69**, 305–338 (1975); *Ann. Rev. Fluid Mech.*, **12**, 435–476 (1976); P. C. Chan and L. G. Leal, *J. Fluid Mech.*, **82**, 549–559 (1977).
[30] P. Brunn, *J. Non-Newtonian Fluid Mech.*, 7 271–288 (1980); *Rheol. Acta*, **18**, 229–243 (1979); *J. Fluid Mech.*, **82**, 529–547 (1977); S. Kim, *J. Non-Newtonian Fluid Mech.*, **21**, 255–269 (1986).

6B.3 Complex Viscosity for Third-Order Fluid

a. Show that for the third-order fluid

$$\eta^*(\omega) = b_1 + b_2 i\omega - b_3\omega^2 \tag{6B.3-1}$$

b. Then obtain

$$\eta'(\omega) = b_1 - b_3\omega^2; \quad \eta''(\omega) = -b_2\omega \tag{6B.3-2}$$

Compare these equations with some data of Chapter 3 to assess the usefulness of these results.

c. From the results in **(b)** and the results in Eqs. 6.2-5 and 6, show that

$$\lim_{\omega \to 0} \frac{\eta''/\omega}{\eta'} = \lim_{\dot{\gamma} \to 0} \frac{\Psi_1}{2\eta} \tag{6B.3-3}$$

See Table 5.3-1, where this relation is given.

d. Can the relation in **(c)** be obtained from *linear* viscoelasticity?

6B.4 Shearfree Flows for Third-Order Fluid

a. Show that the functions $\bar{\eta}_1$ and $\bar{\eta}_2$ of Eqs. 3.5-1 and 2 are for the third-order fluid

$$\begin{aligned}\bar{\eta}_1 &= b_1(3 + b) + (b_{11} - b_2)(3 - 2b - b^2)\dot{\varepsilon} \\ &\quad + [(8 + (1 + b)^3)(b_3 - 2b_{12}) + (3 + b)(6 + 2b^2)b_{1:11}]\dot{\varepsilon}^2\end{aligned} \tag{6B.4-1}$$

$$\begin{aligned}\bar{\eta}_2 &= 2b_1 b - 4(b_{11} - b_2)b\dot{\varepsilon} \\ &\quad + [2(b_3 - 2b_{12})(3b + b^3) + 4b_{1:11}b(3 + b^2)]\dot{\varepsilon}^2\end{aligned} \tag{6B.4-2}$$

b. For elongational flow, with $b = 0$, it is known experimentally that $\bar{\eta}$ increases with $\dot{\varepsilon}$. What does this tell us about the quantity $(b_{11} - b_2)$? Is this consistent with the molecular-theory results given in Table 6.2-1?

6B.5 Flow into a Line Sink (or out of a Line Source) for a Third-Order Fluid

A third-order fluid is flowing into a line sink (or from a line source) whose strength is given by a constant $C = Q/2\pi L$where Q is the volume flow rate out of the sink and L is its length. For sink flow $Q < 0$ and for source flow $Q > 0$. Assume that the flow is purely radial and that there is no θ-dependence of the velocity v_r.

a. Obtain an expression for $v_r(r)$ in terms of the sink (or source) strength C by using the equation of continuity.

b. Show that the normal stresses are given by

$$\tau_{rr} = b_1 \frac{2C}{r^2} - b_{11}\frac{4C^2}{r^4} + b_{1:11}\frac{16C^3}{r^6} \tag{6B.5-1}$$

$$\tau_{\theta\theta} = -b_1 \frac{2C}{r^2} + (2b_2 - b_{11})\frac{4C^2}{r^4} - (3b_3 - 2b_{12} + b_{1:11})\frac{16C^3}{r^6} \tag{6B.5-2}$$

c. Obtain the pressure distribution in the fluid relative to the pressure far from the sink (or source), p_∞:

$$p_\infty - p(r) = \frac{\rho C^2}{2r^2} - (2b_{11} - b_2)\frac{2C^2}{r^4} - (3b_3 - 2b_{12} - 4b_{1:11})\frac{8C^3}{3r^6} \tag{6B.5-3}$$

Sketch the pressure distribution qualitatively for a typical polymer melt or solution. *Note*: Sufficiently near the line sink the third-order terms will dominate the second-order terms, which in turn will dominate the Newtonian terms. Hence in this problem the retarded-motion expansion is used outside of its range of validity.

6B.6 Rod-Climbing of Second-Order Fluids[1]

A vertical rod of radius R is rotating with angular velocity W in a fluid with a free surface. At small values of W the height of the free surface $h(r, W)$ (relative to the level far from the rod) as function of the radial coordinate r may be expanded as

$$h(r, W) = \frac{R^4}{gr^2}\left[\frac{2\beta}{\rho r^2} - \frac{1}{2}\right]W^2 + O(W^4) \tag{6B.6-1}$$

where ρ is the density of the fluid, g the gravitational acceleration, and $\beta = (2b_{11} - b_2)$ the "climbing constant." Surface tension has been neglected in this expression.

a. Show that when W is so small that terms of $O(W^4)$ can be neglected the fluid will climb the rod if and only if

$$\beta = \frac{1}{2}(\Psi_{1,0} + 4\Psi_{2,0}) > 0 \quad \text{and} \quad R^2 < 4\beta/\rho \tag{6B.6-2}$$

b. Show that for the Curtiss–Bird model of Table 6.2-1 we must require that the parameter ε be greater than 1/8 to describe "rod-climbing" at second order. In particular deduce that the Doi–Edwards model ($\varepsilon = 0$) incorrectly shows "rod-dipping" at second order.

6B.7 Radial Flow between Parallel Disks

The flow of Problem 4B.3 has been studied for the third-order fluid. It has been found[2] that the stress $p + \tau_{zz}$ exerted on the disks is

$$\frac{\Delta(p + \tau_{zz})}{\Delta(p + \tau_{zz})_N} = 1 + \frac{3}{10}\frac{(R_2/r)^2 - 1}{(R_2/B)\ln(R_2/r)}\left[2\left(1 + \frac{3}{2}\frac{b_{11}}{b_2}\right)\text{De} - \frac{6}{7}\text{Re} - 9\left(\frac{R_2}{B}\right)\left(\frac{b_1(b_{12} - b_{1:11})}{b_2^2}\right)\text{De}^2 + \cdots\right] \tag{6B.7-1}$$

Here $\text{Re} = \rho VB/b_1$, $\text{De} = (-b_2/b_1)(V/B)$, $V = Q/4\pi R_2 B$. In addition, we let $\Delta(p + \tau_{zz}) = (p + \tau_{zz})|_r - (p + \tau_{zz})|_{R_2}$, and $\Delta(p + \tau_{zz})_N = 3(R_2/B)(b_1 V/B)\ln(R_2/r)$ is the Newtonian creeping flow solution.

a. Interpret the solution.

b. Obtain the coefficient of the De term in Eq. 6B.7-1.

[1] This problem is based on G. S. Beavers and D. D. Joseph, *J. Fluid Mech.*, **69**, 475–511 (1975), who included the effect of surface tension in the special situation where the contact angle at the rod is 90°. The first analysis of rod-climbing in second-order fluids is that of H. Giesekus, *Rheol. Acta*, **1**, 403–413 (1961), who neglected both inertia and surface tension. See also A. S. Lodge, *Elastic Liquids*, Academic Press, New York (1964), pp. 193–194, 231–236, and O. Hassager, *J. Rheol.*, **29**, 361–364 (1985).

[2] A. Co and R. B. Bird, *Appl. Sci. Res.*, **33**, 385–404 (1977).

6B.8 Total Normal Stress on Spherical Bubble

Verify the following intermediate results needed to obtain Eq. 6.5-52:

a. $$[p]_{r=R_0} = p_\infty + [b_1(V/R_0) - \rho g R_0] \cos\theta + \frac{b_2}{10}(V/R_0)^2[(-2 + 22B_{11})\cos^2\theta - 6 + B_{11}] \qquad \text{(6B.8-1)}$$

b. $$[-b_1\gamma_{(1)0rr}]_{r=R_0} = 2b_1(V/R_0)\cos\theta \qquad \text{(6B.8-2)}$$

c. $$[-b_1 \text{De}\, \gamma_{(1)1rr}]_{r=R_0} = \frac{b_2}{10}(V/R_0)^2[(-6 + 6B_{11})\cos^2\theta + 2 - 2B_{11}] \qquad \text{(6B.8-3)}$$

d. $$[b_1 \text{De}(R_0/V)\gamma_{(2)0rr}]_{r=R_0} = \frac{b_2}{10}(V/R_0)^2[50\cos^2\theta - 10] \qquad \text{(6B.8-4)}$$

e. $$[b_1 \text{De}(B_{11}R_0/V)\{\boldsymbol{\gamma}_{(1)0}\cdot\boldsymbol{\gamma}_{(1)0}\}_{rr}]_{r=R_0} = \frac{b_2}{10}(V/R_0)^2[-40B_{11}\cos^2\theta] \qquad \text{(6B.8-5)}$$

where $B_{11} = b_{11}/b_2$.

6B.9 Bubble Deformation Equation

a. Derive Eqs. 6.5-61 as boundary conditions for the bubble deformation equation in Eq. 6.5-60.

b. Solve Eq. 6.5-60 to obtain

$$p_g = p_m + 2\frac{\sigma}{R_0} \qquad \text{(6B.9-1)}$$

$$\zeta = -\frac{1}{4}P_2 \qquad \text{(6B.9-2)}$$

Hint: The Legendre polynomials $P_n(x)$ are solutions of the equation:

$$\frac{d}{dx}\left[(1 - x^2)\frac{dP_n}{dx}\right] + n(n+1)P_n = 0 \qquad \text{(6B.9-3)}$$

6B.10 Rectilinear Flow

a. Use the definition in Eq. 6.3-20 to show that for rectilinear flow:

$$\{(\nabla \mathbf{v})\cdot(\nabla \mathbf{v})\} = \mathbf{0}, \qquad \{(\nabla \mathbf{v})^\dagger\cdot(\nabla \mathbf{v})^\dagger\} = \mathbf{0} \qquad \text{(6B.10-1)}$$

b. Also for rectilinear flow show that:

$$[\nabla\cdot\{(\nabla \mathbf{v})^\dagger\cdot(\nabla \mathbf{v})\}] = \mathbf{0} \qquad \text{(6B.10-2)}$$

c. Then use the above results to show that

$$[\nabla\cdot\{\boldsymbol{\gamma}_{(1)}\cdot\boldsymbol{\gamma}_{(1)}\}] = [(\nabla \mathbf{v})\cdot\nabla^2 \mathbf{v}] + \tfrac{1}{2}\nabla((\nabla \mathbf{v}):(\nabla \mathbf{v})^\dagger) \qquad \text{(6B.10-3)}$$

d. Finally show that if $\boldsymbol{v}$ satisfies Eq. 6.3-3 of the text, then

$$[\nabla \cdot \{\gamma_{(1)} \cdot \gamma_{(1)}\}] = \nabla\left(\frac{1}{b_1}(\boldsymbol{v} \cdot \nabla \mathscr{P}_N) + \tfrac{1}{2}((\nabla \boldsymbol{v}):(\nabla \boldsymbol{v})^\dagger)\right) \qquad (6B.10\text{-}4)$$

6B.11 Flow of a Second-Order Fluid into a Slit Located in a Wedge[3]

A polymeric fluid is flowing into a thin slit located at the bottom of a wedge, in steady, creeping flow. In the cylindrical coordinate system shown in Fig. 6B.11, the wedge is described by the planes $\theta = \pm\theta_0$. Let Q be the volume flow rate in the positive r-direction through a section of the slit extending a distance W in the z-direction. In the following the slit is replaced by a line sink of strength Q/W located at the intersection of the planes $\theta = \pm\theta_0$. For a Newtonian fluid the velocity distribution is known to be

$$v_r(r, \theta) = \frac{\alpha}{r}(\cos 2\theta - \cos 2\theta_0); \quad v_\theta = 0; \quad v_z = 0 \qquad (6B.11\text{-}1)$$

where

$$\alpha = \frac{-Q/W}{\sin 2\theta_0 - 2\theta_0 \cos 2\theta_0} \qquad (6B.11\text{-}2)$$

a. For a Newtonian fluid with viscosity η_0, show that the equation of motion for creeping flow gives

$$p(r, \theta) = p_\infty + \frac{2\alpha\eta_0}{r^2}\cos 2\theta \qquad (6B.11\text{-}3)$$

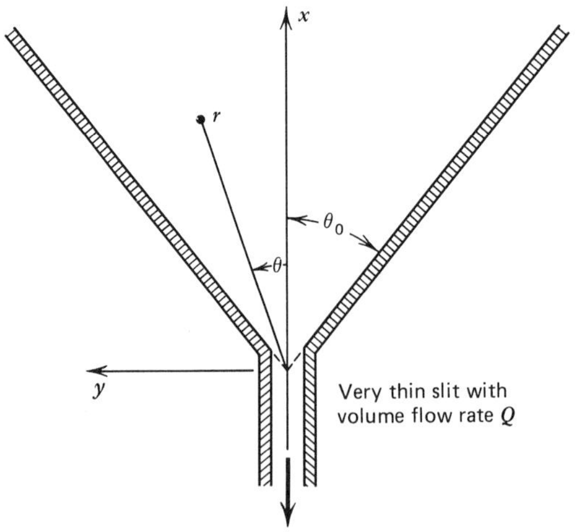

FIGURE 6B.11. Flow from a wedge into a thin slit. The flow is described in a cylindrical (r, θ, z) coordinate system. The z-axis is pointing outward from the plane of the paper.

[3] R. I. Tanner, *Phys. Fluids*, **9**, 1246–1247 (1968).

b. Now use Eq. 6.3-19 to show that the force per unit area exerted by a second-order fluid on the wedge at $\theta = \theta_0$ is

$$[\boldsymbol{\delta}_\theta \cdot \boldsymbol{\pi}]_{\theta=\theta_0} = \left(p_\infty + \frac{2\alpha}{r^2}\eta_0 \cos 2\theta_0 + \frac{\alpha^2}{r^4}\Psi_{1,0} \sin^2 2\theta_0\right)\boldsymbol{\delta}_\theta + \frac{2\alpha}{r^2}\eta_0 \sin 2\theta_0 \boldsymbol{\delta}_r \qquad (6B.11\text{-}4)$$

where $\boldsymbol{\delta}_\theta$ and $\boldsymbol{\delta}_r$ are evaluated at $\theta = \theta_0$.

c. Would it be possible to distinguish between a Newtonian fluid and a second-order fluid by measuring total normal thrust on the surface $\theta = \theta_0$ when $\theta_0 = \pi/2$ with flush-mounted pressure transducers?

Note: In the vicinity of the slit, the second-order terms associated with $\Psi_{1,0}$ are large compared to the Newtonian terms. Hence the second-order fluid is used outside its range of applicability, and the expression in Eq. 6B.11-4 may be devoid of physical significance.

6B.12 The Second-Order Fluid and the "Turntable Experiment"

Suppose that we decided to construct a constitutive equation as follows: We let $\boldsymbol{\tau}$ be a function of $\boldsymbol{\gamma}_{(1)}$ and all of its partial time derivatives $\partial^n\boldsymbol{\gamma}_{(1)}/\partial t^n$. A possibly useful expansion would then be

$$\begin{aligned}\boldsymbol{\tau} = &-[b_1\boldsymbol{\gamma}_{(1)} + b_2\partial\boldsymbol{\gamma}_{(1)}/\partial t + b_{11}\{\boldsymbol{\gamma}_{(1)}\cdot\boldsymbol{\gamma}_{(1)}\}\\ &+ b_3\partial^2\boldsymbol{\gamma}_{(1)}/\partial t^2 + b_{12}\{\boldsymbol{\gamma}_{(1)}\cdot\partial\boldsymbol{\gamma}_{(1)}/\partial t\\ &+ (\partial\boldsymbol{\gamma}_{(1)}/\partial t)\cdot\boldsymbol{\gamma}_{(1)}\} + b_{1:11}(\boldsymbol{\gamma}_{(1)}:\boldsymbol{\gamma}_{(1)})\boldsymbol{\gamma}_{(1)} + \cdots]\end{aligned} \qquad (6B.12\text{-}1)$$

This has the same form as Eq. 6.2-1; in the latter $\boldsymbol{\gamma}_{(n)}$ appears whereas $\partial^{n-1}\boldsymbol{\gamma}_{(1)}/\partial t^{n-1}$ appears in Eq. 6B.12-1. We now study both of these equations in terms of the turntable experiment in Example 5.5-1. We make this comparison using only the terms containing b_1, b_2, and b_{11} (i.e., the second-order fluid).

a. Use the velocity distribution in Eqs. 5.5-3 and 4 to verify the matrix display of $\boldsymbol{\gamma}_{(1)}$ given in Eq. 5.5-5, and then obtain the matrix display for $\{\boldsymbol{\gamma}_{(1)}\cdot\boldsymbol{\gamma}_{(1)}\}$. Next write both of these matrices for $t = 0$ (the time at which the parallel-plate system is lined up with the x-axis).

b. Evaluate $\partial\boldsymbol{\gamma}_{(1)}/\partial t$ and $\boldsymbol{\omega}$ for the flow field in Eqs. 5.5-3 and 4; display in matrix form.

c. Next show that the kinematic tensor $\boldsymbol{\gamma}_{(2)}$ can be written as

$$\boldsymbol{\gamma}_{(2)} = \frac{D}{Dt}\boldsymbol{\gamma}_{(1)} - \{\boldsymbol{\gamma}_{(1)}\cdot\boldsymbol{\gamma}_{(1)}\} + \tfrac{1}{2}\{\boldsymbol{\omega}\cdot\boldsymbol{\gamma}_{(1)} - \boldsymbol{\gamma}_{(1)}\cdot\boldsymbol{\omega}\} \qquad (6B.12\text{-}2)$$

where $\boldsymbol{\omega}$ is the vorticity tensor of Eq. 6.1-3.

d. Use the results of (**b**) and (**c**) to get $\boldsymbol{\gamma}_{(2)}$ at $t = 0$.

e. Display the expression for the stress tensor of a second-order fluid in matrix form for the flow field in Eqs. 5.5-3 and 4 at time $t = 0$. Compare with the second-order fluid terms in Eq. 6.2-4.

f. Do the same for the corresponding terms in Eq. 6B.12-1. Compare the result with that in (**e**). Discuss.

6B.13 Tangential Annular Flow Device for Determining the First Normal Stress Coefficient

Figure 6B.13 shows a flow device that has been suggested for measuring Ψ_1, the first normal stress coefficient. Geiger and Winter[4] list these assumptions in their analysis of the device:

[4] H. R. Osmers and P. F. Lobo, *Trans. Soc. Rheol.*, **20**, 239–252 (1976); K. Geiger and H. H. Winter, *Rheol. Acta*, **17**, 264–273 (1978).

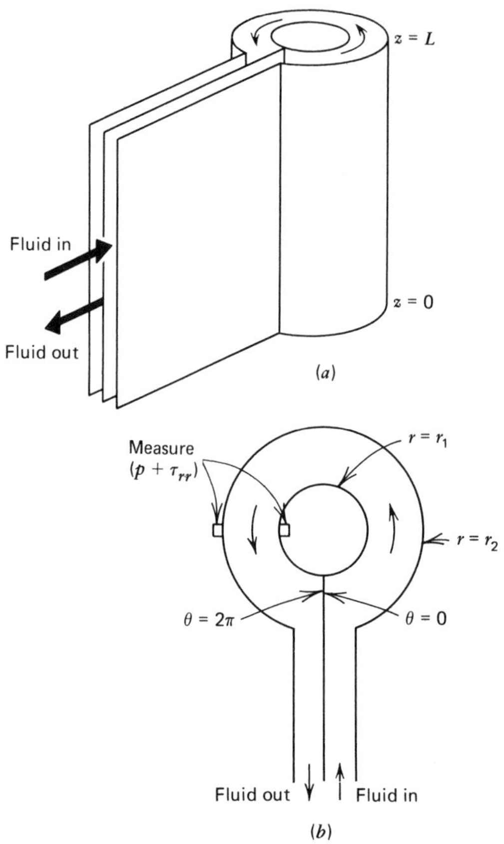

FIGURE 6B.13. Tangential annular flow device for measuring Ψ_1: (*a*) view of exterior of apparatus; (*b*) top view.

- **i.** The centrifugal force in the r-component of the equation of motion can be neglected.
- **ii.** Entrance and exit effects are neglected (i.e., near $\theta = 0$ and $\theta = 2\pi$).
- **iii.** Top and bottom effects are neglected (i.e., near $z = 0, L$).
- **iv.** $\partial p/\partial\theta$ is independent of r, and can be set equal to $-(p_0 - p_{2\pi})/2\pi \equiv -\Delta p/2\pi$.
- **v.** The system is isothermal.
- **vi.** The rheological properties are pressure-independent.
- **vii.** The fluid is incompressible.
- **viii.** There is no fluid slip at the solid surfaces.

It is postulated that $v_\theta = v_\theta(r)$, $v_r = 0$, $v_z = 0$, and $p = p(r, \theta)$. In the flow experiment one measures the tangential pressure drop Δp, the volume rate of flow Q, and the difference in "pressure readings" across the gap $\Delta\pi_{rr} = (p + \tau_{rr})|_{r=r_1} - (p + \tau_{rr})|_{r=r_2}$.

Geiger and Winter give an elaborate analysis for any kind of viscoelastic fluid. Here for the sake of simplicity, we analyze the apparatus for a second-order fluid.

a. For the postulated velocity profile write out the second-order fluid in matrix form (with components in the order r, θ, z).

b. From (**a**) obtain the expressions for the shear stress $\tau_{r\theta}$ and the first normal stress difference $\tau_{\theta\theta} - \tau_{rr}$.

c. Integrate the appropriately simplified r-component of the equation of motion from $r = r_1$ (inner cylinder) to $r = r_2$ (outer cylinder). Show that this leads to an expression for $\Delta\pi_{rr}$ as an integral involving $\tau_{\theta\theta} - \tau_{rr}$.

d. Substitute $\tau_{\theta\theta} - \tau_{rr}$ from (**b**) into the integral in (**c**) and obtain

$$\Delta\pi_{rr} = 2b_2 \int_{r_1}^{r_2} \left[r \frac{\partial}{\partial r}\left(\frac{v_\theta}{r}\right)\right]^2 \frac{dr}{r} \tag{6B.13-1}$$

e. Combine $\tau_{r\theta}$ from (**b**) and the θ-component of the equation of motion to get an ordinary differential equation for $v_\theta(r)$. Show that this can be integrated to give

$$\frac{v_\theta}{r} = \frac{\Delta p}{4\pi b_1}\left[\frac{\kappa^2 \ln \kappa}{\kappa^2 - 1}\left[1 - \left(\frac{r_2}{r}\right)^2\right] - \ln \frac{r}{r_2}\right] \tag{6B.13-2}$$

where $\kappa = r_1/r_2$.

f. By combining the results of (**d**) and (**e**) and performing the integration, obtain an expression for b_2 in terms of the measurable quantities $\Delta\pi_{rr}$ and Δp, the cylinder radii r_1 and r_2, and the zero-shear-rate viscosity b_1.

g. Finally, derive an equation for the volume flow rate from which b_1 may be determined.

6B.14 Centripetal Pumping between Parallel Disks[5]

Two disks are separated by a fixed distance H, and the upper disk is joined to a circular tube (see Fig. 6B.14). The upper disk is rotated with a constant angular velocity $W_0\boldsymbol{\delta}_z$. For a Newtonian fluid the rotation of the upper disk would cause a centrifugal force that would tend to suck the fluid downward out of the tube. For viscoelastic fluids, on the other hand, the normal stresses may cause the fluid to flow inward through the space between the disks and upward through the tube.

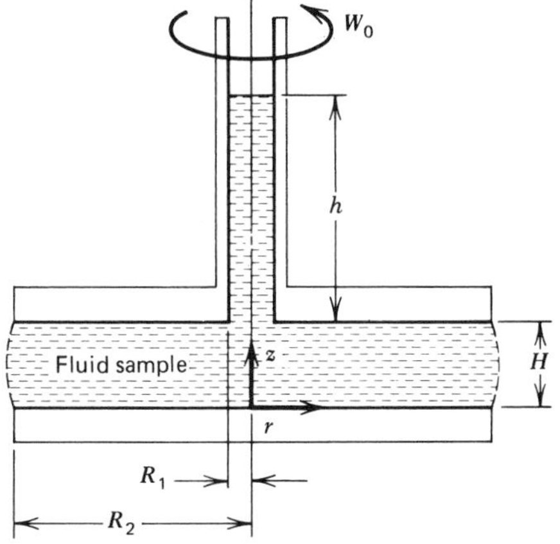

FIGURE 6B.14. Centripetal pumping in a disk-tube assembly. The viscoelastic fluid is pumped radially inward and up into the tube because of the action of normal stresses.

[5] Y. Tomita and H. Katō, *Trans. Jpn. Soc. Mech. Eng.* (*Nippon Kikai Gakkai Ronbunshū*), **32**, 241, 1399–1408 (Sept. 1966); B. Maxwell and A. J. Scalora, *Mod. Plast.*, **37**(2), 107 (1959); P. A. Good, A. J. Schwartz, and C. W. Macosko, *AIChE J.*, **20**, 67–73 (1974). Note that where we have 1/6 in Eq. 6B.14-7, a more detailed analysis gives 3/20 [K. Stewartson, *Proc. Camb. Phil. Soc.*, **49**, 333–341 (1953)]. For a discussion of tangential flow with slow superimposed radial flow see D. F. James, *Trans. Soc. Rheol.*, **19**, 67–80 (1975).

The complete calculation of the centripetal pumping is quite difficult, inasmuch as it is not a viscometric flow. However, it is relatively easy to use Eq. 6.2-1 to find out what pressure difference must be maintained between "1" and "2" in order to prevent flow. Find the height to which the fluid will rise in the tube to balance the normal stresses tending to drive the fluid up the tube.

a. Before doing the viscoelastic fluid calculations, it is instructive to solve the problem for the Newtonian fluid, to find the pressure difference required to prevent radially outward flow because of centrifugal effects. Make the postulate that $v_\theta = zf(r)$, $v_r = v_z = 0$; then show that the r-, θ-, and z-components of the equation of motion become

$$-\rho\frac{z^2f^2}{r} = -\frac{\partial p}{\partial r} \tag{6B.14-1}$$

$$0 = \mu\frac{d}{dr}\left(\frac{1}{r}\frac{d}{dr}(rf)\right) \tag{6B.14-2}$$

$$0 = -\frac{\partial p}{\partial z} - \rho g \tag{6B.14-3}$$

The first of these describes the balance between the centrifugal force and the radial pressure gradient. Solve the second equation to get the velocity distribution

$$v_\theta = \frac{W_0 rz}{H} \tag{6B.14-4}$$

Insert this into Eq. 6B.14-1 to get

$$-\rho W_0^2 r\left(\frac{z}{H}\right)^2 = -\frac{\partial p}{\partial r} \tag{6B.14-5}$$

It can now be seen that Eqs. 6B.14-3 and 5 are inconsistent, since the two equations give different results for the mixed second derivative $\partial^2 p/\partial r\partial z$. This means that the assumption made prior to Eq. 6B.14-1 regarding the velocity field is incorrect, and that strictly speaking secondary flows must be allowed for. To get an approximate expression for the radial pressure gradient, however, take an average of Eq. 6B.14-5 in the z-direction to get

$$-\frac{\rho W_0^2 r}{3} = -\frac{d\bar{p}}{dr} \tag{6B.14-6}$$

where $\bar{p} = (\int_0^H p\,dz)/H$. Then integrate this equation to get

$$\bar{p}_2 - \bar{p}_1 = \tfrac{1}{6}\rho W_0^2(R_2^2 - R_1^2) \tag{6B.14-7}$$

We see that we have to push harder at "2" than at "1" to keep the fluid from being thrown out of the gap.

b. Next rework the problem for the second-order fluid. For the velocity distribution in Eq. 6B.14-4 show that the stress tensor may be displayed in matrix form:

$$\tau = -b_1\begin{pmatrix} 0 & 0 & 0 \\ 0 & 0 & 1 \\ 0 & 1 & 0 \end{pmatrix}\frac{W_0 r}{H} + 2b_2\begin{pmatrix} 0 & 0 & 0 \\ 0 & 1 & 0 \\ 0 & 0 & 0 \end{pmatrix}\frac{W_0^2 r^2}{H^2} - b_{11}\begin{pmatrix} 0 & 0 & 0 \\ 0 & 1 & 0 \\ 0 & 0 & 1 \end{pmatrix}\frac{W_0^2 r^2}{H^2} \tag{6B.14-8}$$

Show that the second-order fluid analog to Eq. 6B.14-7 is

$$\bar{p}_2 - \bar{p}_1 = \frac{1}{6}\rho W_0^2(R_2^2 - R_1^2) + (2b_2 - b_{11})\frac{W_0^2}{2H^2}(R_2^2 - R_1^2) \tag{6B.14-9}$$

and finally that the height of rise of the fluid in the tube is

$$h = -\frac{W_0^2}{6g}(R_2^2 - R_1^2) - \frac{W_0^2}{2\rho g H^2}(2b_2 - b_{11})(R_2^2 - R_1^2) \tag{6B.14-10}$$

Interpret the result. The centripetal pumping has been observed experimentally and it has been used in plastics processing.[6]

6B.15 The Hole Pressure for Creeping Flow of a Second-Order Fluid Across a Transverse Slot[7]

Consider a plane flow across the deep, transverse slot shown in Fig. 2.3-4. Sufficiently far upstream and downstream from the slot, the fluid is flowing in the x-direction. It is desired to obtain an expression for the hole pressure p^* defined in §2.3c, for the pressure-driven, creeping flow of an incompressible, second-order fluid. Assume that the slot is so deep that the shear rate is zero at the bottom of the slot. Assume also that the presence of the slot does not significantly alter the flow near the plane $y = B$ so that the shear rate at the wall $y = B$, denoted by $\dot{\gamma}_w$, is a constant, independent of x.

a. Show that the plane $x = 0$ is a plane of symmetry for creeping flow of an incompressible Newtonian fluid, and that consequently $\partial\tau_{yx}/\partial x = 0$ along the line $x = 0$.

b. Then show from the y-component of the equation of motion that $\partial(\mathscr{P} + \tau_{yy})/\partial y = 0$ along the line $x = 0$, and that consequently $p^* = 0$ for plane creeping flow of an incompressible Newtonian fluid.

c. Finally use the expression in Eq. 6.3-19 to show that:

$$p^* = (1/4)\Psi_{1,0}\dot{\gamma}_w^2 \tag{6B.15-1}$$

for plane, creeping flow of an incompressible second-order fluid across a deep transverse slot.

6C.1 Stability of Second-Order Fluids[8]

Consider a second-order fluid of density ρ that fills the gap between two fixed plates of separation H as shown in Fig. 6C.1. Prior to time $t = 0$ the fluid is at rest. Then at $t = 0$ the fluid is subjected to a small velocity disturbance of the form $v_x = U(y)$, $v_y = v_z = 0$.

a. Assume that the velocity field for $t \geq 0$ is of the form $v_x = v_x(y, t)$, $v_y = v_z = 0$ with the conditions that $v_x(0, t) = v_x(H, t) = 0$. Show that v_x must satisfy

$$\rho\frac{\partial v_x}{\partial t} = b_1\Bigg[\underset{(1)}{\frac{\partial^2 v_x}{\partial y^2}} + \underset{(2)}{\frac{b_2}{b_1}\frac{\partial}{\partial t}\frac{\partial^2 v_x}{\partial y^2}}\Bigg] \tag{6C.1-1}$$

where term (2) on the right-hand side must be a small correction to term (1) for the constitutive equation to apply.

[6] Z. Tadmor and C. G. Gogos, *Principles of Polymer Processing*, Wiley, New York, (1979), p. 372.

[7] R. I. Tanner and A. C. Pipkin, *Trans. Soc. Rheol.*, **13**, 471–484 (1969).

[8] A. C. Pipkin, *Lectures in Viscoelasticity Theory*, Springer Verlag, New York (1972).

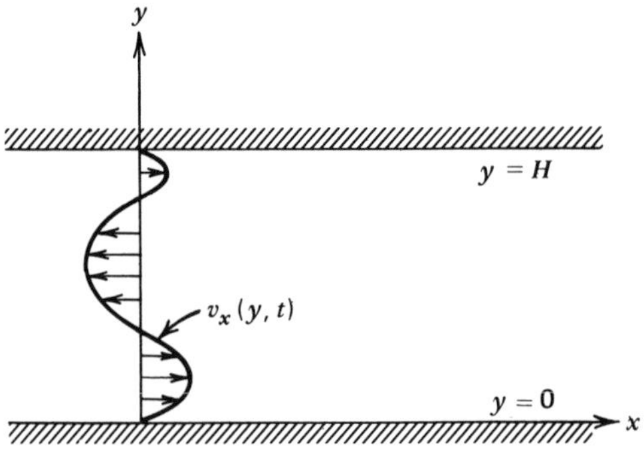

FIGURE 6C.1. A second-order fluid is contained in the gap between two parallel plates. At time $t = 0$ the velocity field is of the form $v_x = U(y)$, $v_y = v_z = 0$ with the conditions that $U(0) = U(H) = 0$. The plates are fixed at all times $t \geq 0$, and there is no pressure gradient other than that due to gravity.

b. Solve this equation by separation of variables to obtain formally

$$v_x(y, t) = \sum_{n=1}^{\infty} A_n \sin(n\pi y/H) \exp(\alpha_n t) \tag{6C.1-2}$$

where

$$A_n = \frac{2}{H} \int_0^H U(y) \sin(n\pi y/H)\, dy \tag{6C.1-3}$$

and

$$\alpha_n = -\left(\frac{b_2}{b_1} + \frac{\rho H^2}{b_1 n^2 \pi^2}\right)^{-1} \tag{6C.1-4}$$

where the α_n must be negative for the disturbances to die out.

c. Assume that the above solution must remain valid for all Fourier components, that is all values of n, and show that this results in the following necessary condition for the second-order fluid to be stable at rest

$$b_2/b_1 \geq 0 \quad \text{or} \quad \Psi_{1,0} \leq 0 \tag{6C.1-5}$$

d. Show that the condition for term (2) in Eq. 6C.1-1 to be small compared to term (1) may be formulated as

$$|\alpha_n(b_2/b_1)| \ll 1 \tag{6C.1-6}$$

for each Fourier mode n of the initial disturbance.

e. Finally show that if the initial disturbance is sufficiently smooth so that condition 6C.1-6 holds for all n with non-zero A_n then the corresponding α_n are all negative. Conclude that the condition in Eq. 6C.1-5 is then no longer necessary and that, within the range of validity of the solution and the assumed form of the velocity field, the disturbance will die out irrespective of the sign of $\Psi_{1,0}$.

6C.2 Flow of a Second-Order Fluid near a Corner

a. Review Problem 1C.3 for the plane flow of a Newtonian fluid near a sharp corner. Show that the stresses near the corner have the following dependence on the radial coordinate

$$\tau_{ij} \sim A_{ij} r^{\lambda_1 - 2} \qquad (r \to 0) \tag{6C.2-1}$$

where the A_{ij} are bounded functions of θ.

b. Use the plane-flow theorem of Tanner and Pipkin to show that the equations for plane flow of an incompressible second-order fluid are satisfied by the same velocity field, and that the stresses near the corner are

$$\tau_{ij} \sim B_{ij} r^{2(\lambda_1 - 2)} \qquad (r \to 0) \tag{6C.2-2}$$

where the B_{ij} are bounded functions of θ.

c. Compare the Newtonian and second-order contributions to the stresses, and conclude that the retarded-motion expansion should not be used unless $\lambda_1 \geq 2$.

d. Compare this value with those in Table 1C.3, and give examples of flow situations where the retarded-motion expansion can be used, and examples where it definitely should not be used.

CHAPTER 7

DIFFERENTIAL CONSTITUTIVE EQUATIONS

In Chapters 4, 5, and 6 we have encountered constitutive equations that are useful in three different flow regimes and for three different groups of people. The generalized Newtonian fluid model of Chapter 4 is restricted to steady-state shear flows and has been primarily used by engineers; the material information in the generalized Newtonian fluid model is contained in the viscosity $\eta(\dot{\gamma})$. In Chapter 5 we presented the linear viscoelastic fluid model which is restricted to small-displacement-gradient flows and has been primarily useful to chemists; the material information in this model is given by the relaxation modulus $G(t)$. Then in Chapter 6, the retarded-motion expansion, valid for slow, slowly varying flows, was presented. The retarded-motion expansion is of interest mostly to applied mathematicians and physicists; the material information is contained in the constants $b_1, b_2, b_{11}, \ldots$. Up to this point we have made no connection between $\eta(\dot{\gamma})$, $G(t)$, and the b's, except for $b_1 = \eta(0) = \int_0^\infty G(t)dt$.

In this chapter we first seek more general constitutive equations that can be used for arbitrary flows. It should be possible, of course, to simplify these more general constitutive equations to include the special models of Chapters 4, 5, and 6, but we also want to describe, at least qualitatively, as much additional nonlinear rheological material behavior in Chapter 3 as possible. Second, we want to focus attention on applying these constitutive equations to solving polymer fluid flow problems.

We begin the task of constructing more general, "admissible" constitutive equations by modifying the linear viscoelastic models of Chapter 5, which are valid for any time-dependent flow with small displacement gradients. In this chapter we modify these in their differential forms by replacing the time derivatives by convected derivatives as introduced in Chapter 6. In the next chapter we show how to get admissible equations by starting with the integral forms of the models. Constitutive equations obtained by reformulating the linear viscoelastic models are called *quasi-linear models.*

Once we see the kinds of rheological properties that can be predicted by the quasi-linear models, we will explore several ways in which they can be altered in order to obtain better agreement with the material functions presented in Chapter 3. This leads to the discussion of *nonlinear models* in §§7.3 and 8.3. Several of these nonlinear models describe, at least qualitatively, a wide variety of the material functions of Chapter 3; in particular, they show the well-known decrease of η and Ψ_1 with increasing $\dot{\gamma}$.

Once the rheological models have been presented, we turn to the primary task of solving fluid flow problems. For the most part these are solved here with the quasi-linear models so that we can illustrate the methods without becoming overcome with tedious algebra or numerical analysis; however, we do show some results for nonlinear models to emphasize the influence of differences in choice of constitutive equations on the calculated flow fields.

The illustration of problem solving is broken down into two sections: the first for problems involving one spatial variable (and possibly time) and the second for problems involving two or three spatial variables (and possibly time). As will be clear from the discussion and examples in these sections, success in obtaining analytical solutions to viscoelastic flow problems has been limited to situations in which the problem is either one-dimensional or can be reduced to one dimension by means of similarity or periodic solutions.

Finally in the last section we point out the limitations of the differential models and give recommendations for situations in which the different constitutive equations should be used.

§7.1 THE CONVECTED DERIVATIVE OF THE STRESS TENSOR

In Chapter 6 we introduced the convected time derivatives of the rate-of-strain tensor. In this chapter we introduce the *convected time derivative of the stress tensor*[1,2] $\boldsymbol{\tau}_{(1)}$ defined by (cf. Eq. 6.1-6)

$$\begin{aligned}\boldsymbol{\tau}_{(1)} &= \frac{D}{Dt}\boldsymbol{\tau} - \{(\nabla \boldsymbol{v})^{\dagger}\cdot\boldsymbol{\tau} + \boldsymbol{\tau}\cdot(\nabla \boldsymbol{v})\}\\ &= \frac{D}{Dt}\boldsymbol{\tau} - \{\boldsymbol{\tau}\cdot\nabla \boldsymbol{v}\}^{\dagger} - \{\boldsymbol{\tau}\cdot\nabla \boldsymbol{v}\} \quad \text{(for symmetric } \boldsymbol{\tau}) \qquad (7.1\text{-}1)\end{aligned}$$

In Chapter 9 we will show that $\boldsymbol{\tau}_{(1)}$ is a time derivative of stress that has a significance independent of superposed rotations. We can thus construct admissible constitutive equations from the differential linear viscoelastic models of Chapter 5 by replacing $\partial\boldsymbol{\tau}/\partial t$ by $\boldsymbol{\tau}_{(1)}$ and $\partial^n\dot{\boldsymbol{\gamma}}/\partial t^n$ by $\boldsymbol{\gamma}_{(n+1)}$.

In evaluating the quantities $\boldsymbol{\tau}_{(1)}$ and $\boldsymbol{\gamma}_{(n)}$ in differential constitutive equations it is convenient to use expressions for these that are tabulated in Appendix C for time-dependent

[1] In Chapter 9 this is referred to as the *contravariant convected time derivative* to distinguish it from the covariant convected time derivative denoted by a superscript (1). In this chapter we use only the former, and "contravariant" is implied in the phrase "convected time derivative."

[2] The method we use in this chapter for generating admissible equations is not unique. Instead of using convected time derivatives of the stress and rate-of-strain tensors in constitutive equations we could use the corotational or Jaumann derivative $\mathscr{D}/\mathscr{D}t$ defined by

$$\begin{aligned}\frac{\mathscr{D}}{\mathscr{D}t}\boldsymbol{\tau} &= \frac{D}{Dt}\boldsymbol{\tau} + \tfrac{1}{2}\{\boldsymbol{\omega}\cdot\boldsymbol{\tau} - \boldsymbol{\tau}\cdot\boldsymbol{\omega}\}\\ &= \boldsymbol{\tau}_{(1)} + \tfrac{1}{2}\{\dot{\boldsymbol{\gamma}}\cdot\boldsymbol{\tau} + \boldsymbol{\tau}\cdot\dot{\boldsymbol{\gamma}}\} \qquad (7.1\text{-}1a)\end{aligned}$$

The second line of Eq. 7.1-1a makes it clear that constitutive equations written with the corotational derivative can easily be rewritten in terms of the convected derivative and vice versa. The idea of the corotational derivative is quite old. See, for example, S. Zaremba, *Bull. Int. Acad. Sci. Cracovie*, 594–614, 614–621 (1903); G. Jaumann, *Grundlagen der Bewegungslehre*, Leipzig (1905), *Sitzungsberichte Akad. Wiss. Wien IIa*, **120**, 385–530 (1911); H. Fromm, *Z. Angew. Math. Mech.*, **25/27**, 146–150 (1947); **28**, 43–54 (1948); J. G. Oldroyd, *Proc. Roy. Soc.*, **A245**, 278–297 (1958); and J. D. Goddard and C. Miller, *Rheol. Acta*, **5**, 177–184 (1966). The first edition of *Dynamics of Polymeric Liquids* (Volume 1) contains an extensive discussion of the corotational derivative and corotational models. In addition, see H. Giesekus, *Rheol. Acta*, **3**, 59–71 (1963); *Zeits. f. angew. Math. u. Mech.*, **42**, 32–61 (1962).

shear and shearfree flows. In addition the contributions needed to evaluate $\boldsymbol{\tau}_{(1)}$ and $\boldsymbol{\gamma}_{(n)}$ are given in Appendix A for several orthogonal, curvilinear coordinate systems.

EXAMPLE 7.1-1 The Convected Derivative of the Stress Tensor for Simple Shear Flow

Show how the entry for $\boldsymbol{\tau}_{(1)}$ in Appendix C is obtained for a time-dependent simple shear flow $v_x = \dot{\gamma}_{yx}(t)y$, $v_y = v_z = 0$.

SOLUTION For simple shear flow the stress tensor is given in Eq. 3.2-1 and the velocity gradient and its transpose are shown in Eq. 6.1-7. To evaluate $\boldsymbol{\tau}_{(1)}$ we insert these into Eq. 7.1-1 and perform the indicated matrix multiplications. In this way we obtain

$$\boldsymbol{\tau}_{(1)} = \frac{\partial}{\partial t}\begin{pmatrix} \tau_{xx} & \tau_{yx} & 0 \\ \tau_{yx} & \tau_{yy} & 0 \\ 0 & 0 & \tau_{zz} \end{pmatrix} + \underline{v_x \frac{\partial}{\partial x}\begin{pmatrix} \tau_{xx} & \tau_{yx} & 0 \\ \tau_{yx} & \tau_{yy} & 0 \\ 0 & 0 & \tau_{zz} \end{pmatrix}}$$

$$- \begin{pmatrix} 0 & \dot{\gamma}_{yx} & 0 \\ 0 & 0 & 0 \\ 0 & 0 & 0 \end{pmatrix}\begin{pmatrix} \tau_{xx} & \tau_{yx} & 0 \\ \tau_{yx} & \tau_{yy} & 0 \\ 0 & 0 & \tau_{zz} \end{pmatrix} - \begin{pmatrix} \tau_{xx} & \tau_{yx} & 0 \\ \tau_{yx} & \tau_{yy} & 0 \\ 0 & 0 & \tau_{zz} \end{pmatrix}\begin{pmatrix} 0 & 0 & 0 \\ \dot{\gamma}_{yx} & 0 & 0 \\ 0 & 0 & 0 \end{pmatrix}$$

$$= \frac{\partial}{\partial t}\begin{pmatrix} \tau_{xx} & \tau_{yx} & 0 \\ \tau_{yx} & \tau_{yy} & 0 \\ 0 & 0 & \tau_{zz} \end{pmatrix} - \dot{\gamma}_{yx}\begin{pmatrix} 2\tau_{yx} & \tau_{yy} & 0 \\ \tau_{yy} & 0 & 0 \\ 0 & 0 & 0 \end{pmatrix} \qquad (7.1\text{-}2)$$

Note that the dashed-underline term arising in the substantial derivative is zero because the flow is homogeneous (i.e., the shear rate is independent of position) so that the stress tensor components do not depend on x, y, and z.

EXAMPLE 7.1-2 Calculation of $\boldsymbol{\tau}_{(1)}$ for Flow between Parallel Disks

A viscoelastic fluid is undergoing time-dependent torsional flow between parallel disks; the velocity field is assumed to be axisymmetric so that we write $v_r = v_r(r, z, t)$, $v_\theta = v_\theta(r, z, t)$, and $v_z = v_z(r, z, t)$. Determine $\boldsymbol{\tau}_{(1)}$ for this flow.

SOLUTION From the definition of the convected derivative in Eq. 7.1-1 (second line) we see that it is necessary to evaluate $\{\mathbf{v}\cdot\nabla\boldsymbol{\tau}\}$ and $\{\boldsymbol{\tau}\cdot\nabla\mathbf{v}\}$. The components of $\{\mathbf{v}\cdot\nabla\boldsymbol{\tau}\}$ are listed in cylindrical coordinates in Table A.7-2, Eqs. BB through JJ; we take those results over here by discarding terms involving $\partial/\partial\theta$ and by using the symmetry of $\boldsymbol{\tau}$. Similarly, we take the components of $\nabla\mathbf{v}$ from Eqs. S through AA of Table A.7-2, with $\partial(\cdot)/\partial\theta = 0$. When the matrix formed from these components is premultiplied by the matrix for $\boldsymbol{\tau}$ in cylindrical coordinates we obtain a matrix representation of $\{\boldsymbol{\tau}\cdot\nabla\mathbf{v}\}$. Adding this to its transpose gives, for example

$$[\{\boldsymbol{\tau}\cdot\nabla\mathbf{v}\}^\dagger + \{\boldsymbol{\tau}\cdot\nabla\mathbf{v}\}]_{rr} = 2\left[\tau_{rr}\frac{\partial v_r}{\partial r} + \tau_{r\theta}\left(-\frac{v_\theta}{r}\right) + \tau_{rz}\frac{\partial v_r}{\partial z}\right] \qquad (7.1\text{-}3)$$

$$[\{\boldsymbol{\tau}\cdot\nabla\mathbf{v}\}^\dagger + \{\boldsymbol{\tau}\cdot\nabla\mathbf{v}\}]_{r\theta} = \left[\tau_{rr}\frac{\partial v_\theta}{\partial r} + \tau_{r\theta}\left(\frac{v_r}{r} + \frac{\partial v_r}{\partial r}\right) + \tau_{rz}\frac{\partial v_\theta}{\partial z} - \tau_{\theta\theta}\frac{v_\theta}{r} + \tau_{\theta z}\frac{\partial v_r}{\partial z}\right] \qquad (7.1\text{-}4)$$

In a similar way the rz-, $\theta\theta$-, θz-, and zz-components can be obtained. When the various contributions to $\boldsymbol{\tau}_{(1)}$ are put together we obtain:

$$\tau_{(1)rr} = \frac{\partial \tau_{rr}}{\partial t} + v_r \frac{\partial}{\partial r} \tau_{rr} + v_z \frac{\partial}{\partial z} \tau_{rr} - 2\tau_{rr} \frac{\partial v_r}{\partial r} - 2\tau_{rz} \frac{\partial v_r}{\partial z} \tag{7.1-5}$$

$$\tau_{(1)r\theta} = \frac{\partial \tau_{r\theta}}{\partial t} + v_r \frac{\partial}{\partial r} \tau_{r\theta} + v_z \frac{\partial}{\partial z} \tau_{r\theta} + \tau_{rr}\left(\frac{v_\theta}{r} - \frac{\partial v_\theta}{\partial r}\right)$$
$$- \tau_{r\theta}\left(\frac{v_r}{r} + \frac{\partial v_r}{\partial r}\right) - \tau_{rz} \frac{\partial v_\theta}{\partial z} - \tau_{\theta z} \frac{\partial v_r}{\partial z} \tag{7.1-6}$$

$$\tau_{(1)rz} = \frac{\partial \tau_{rz}}{\partial t} + v_r \frac{\partial}{\partial r} \tau_{rz} + v_z \frac{\partial}{\partial z} \tau_{rz} - \tau_{rr} \frac{\partial v_z}{\partial r}$$
$$- \tau_{rz}\left(\frac{\partial v_z}{\partial z} + \frac{\partial v_r}{\partial r}\right) - \tau_{zz} \frac{\partial v_r}{\partial z} \tag{7.1-7}$$

$$\tau_{(1)\theta\theta} = \frac{\partial \tau_{\theta\theta}}{\partial t} + v_r \frac{\partial}{\partial r} \tau_{\theta\theta} + v_z \frac{\partial}{\partial z} \tau_{\theta\theta} - 2\tau_{\theta\theta} \frac{v_r}{r}$$
$$+ 2\tau_{r\theta}\left(\frac{v_\theta}{r} - \frac{\partial v_\theta}{\partial r}\right) - 2\tau_{\theta z} \frac{\partial v_\theta}{\partial z} \tag{7.1-8}$$

$$\tau_{(1)\theta z} = \frac{\partial \tau_{\theta z}}{\partial t} + v_r \frac{\partial}{\partial r} \tau_{\theta z} + v_z \frac{\partial}{\partial z} \tau_{\theta z} + \tau_{rz}\left(\frac{v_\theta}{r} - \frac{\partial v_\theta}{\partial r}\right)$$
$$- \tau_{\theta z}\left(\frac{v_r}{r} + \frac{\partial v_z}{\partial z}\right) - \tau_{r\theta} \frac{\partial v_z}{\partial r} - \tau_{zz} \frac{\partial v_\theta}{\partial z} \tag{7.1-9}$$

$$\tau_{(1)zz} = \frac{\partial \tau_{zz}}{\partial t} + v_r \frac{\partial}{\partial r} \tau_{zz} + v_z \frac{\partial}{\partial z} \tau_{zz} - 2\tau_{rz} \frac{\partial v_z}{\partial r} - 2\tau_{zz} \frac{\partial v_z}{\partial z} \tag{7.1-10}$$

We will need Eqs. 7.1-5 through 10 in working Example 7.5-2.

EXAMPLE 7.1-3 Evaluation of $\boldsymbol{\tau}_{(1)}$ for Steady Shear Flow Plus Superposed Rotation

Find the components of $\boldsymbol{\tau}_{(1)}$ in the fixed coordinate system (i.e., $\tau_{(1)xx}$, $\tau_{(1)yy}$, and $\tau_{(1)yx}$) for the "turntable flow" described in Example 5.5-1. These components are to be evaluated at the instant $t = 0$, when the $\bar{x}$- and $\bar{y}$-axes coincide with the x- and y-axes.

SOLUTION At any time t the components of the tensor $\boldsymbol{\tau}$ in the xyz-system are related to those in the $\bar{x}\bar{y}\bar{z}$-system by a coordinate transformation

$$\tau_{mn} = \sum_{\bar{m}} \sum_{\bar{n}} \Omega_{m\bar{m}} \Omega_{n\bar{n}} \tau_{\bar{m}\bar{n}} \tag{7.1-11}$$

in which the $\Omega_{m\bar{m}}$ are the elements of a rotation matrix that accounts for the fact that the $\bar{x}\bar{y}\bar{z}$-system is "tilted" with respect to the xyz-system in Fig. 5.5-1

$$(\Omega_{m\bar{m}}) = \begin{pmatrix} \cos Wt & -\sin Wt & 0 \\ \sin Wt & \cos Wt & 0 \\ 0 & 0 & 1 \end{pmatrix} \tag{7.1-12}$$

Let us now obtain expressions for the various terms that make up $\boldsymbol{\tau}_{(1)}$ in Eq. 7.1-1.

First we get $\partial\boldsymbol{\tau}/\partial t$ at $t = 0$. This is done by using Eq. 7.1-12 in Eq. 7.1-11, differentiating the latter with respect to t, and then setting t equal to zero. This gives

$$\frac{\partial}{\partial t}\begin{pmatrix} \tau_{xx} & \tau_{xy} & \tau_{xz} \\ \tau_{yx} & \tau_{yy} & \tau_{yz} \\ \tau_{zx} & \tau_{zy} & \tau_{zz} \end{pmatrix} = \begin{pmatrix} -2\tau_{\bar{y}\bar{x}} & \tau_{\bar{x}\bar{x}} - \tau_{\bar{y}\bar{y}} & 0 \\ \tau_{\bar{x}\bar{x}} - \tau_{\bar{y}\bar{y}} & 2\tau_{\bar{y}\bar{x}} & 0 \\ 0 & 0 & 0 \end{pmatrix}_{t=0} W \tag{7.1-13}$$

Note that $\partial\boldsymbol{\tau}/\partial t$ depends on the rate of rotation of the turntable, and therefore $\partial\boldsymbol{\tau}/\partial t$ is not admissible in a constitutive equation. The term $\{\boldsymbol{v}\cdot\nabla\boldsymbol{\tau}\}$ is zero, because the flow is homogeneous.

Next the $\nabla\boldsymbol{v}$ tensor must be calculated using the velocity profile in Eqs. 5.5-3 and 4. This gives

$$\nabla\boldsymbol{v} = \begin{pmatrix} -\dot{\gamma}\sin Wt\cos Wt & -\dot{\gamma}\sin^2 Wt + W & 0 \\ \dot{\gamma}\cos^2 Wt - W & \dot{\gamma}\sin Wt\cos Wt & 0 \\ 0 & 0 & 0 \end{pmatrix} \xrightarrow{t=0} \begin{pmatrix} 0 & W & 0 \\ \dot{\gamma} - W & 0 & 0 \\ 0 & 0 & 0 \end{pmatrix} \tag{7.1-14}$$

Therefore, at $t = 0$

$$-\{\boldsymbol{\tau}\cdot\nabla\boldsymbol{v}\} = \begin{pmatrix} (W - \dot{\gamma})\tau_{\bar{y}\bar{x}} & -W\tau_{\bar{x}\bar{x}} & 0 \\ (W - \dot{\gamma})\tau_{\bar{y}\bar{y}} & -W\tau_{\bar{y}\bar{x}} & 0 \\ 0 & 0 & 0 \end{pmatrix} \tag{7.1-15}$$

and $-\{(\nabla\boldsymbol{v})^\dagger\cdot\boldsymbol{\tau}\}$ is given by the transpose of this matrix.

If we now combine the various contributions to $\boldsymbol{\tau}_{(1)}$ above, we get at $t = 0$

$$\begin{pmatrix} \tau_{(1)xx} & \tau_{(1)xy} & \tau_{(1)xz} \\ \tau_{(1)yx} & \tau_{(1)yy} & \tau_{(1)yz} \\ \tau_{(1)zx} & \tau_{(1)zy} & \tau_{(1)zz} \end{pmatrix} = -\begin{pmatrix} 2\tau_{yx} & \tau_{yy} & 0 \\ \tau_{yy} & 0 & 0 \\ 0 & 0 & 0 \end{pmatrix}\dot{\gamma} \tag{7.1-16}$$

Here we have made use of the fact that $\tau_{\bar{m}\bar{n}} = \tau_{mn}$ at $t = 0$. It is thus seen that $\boldsymbol{\tau}_{(1)}$ does not contain W, the angular velocity of the turntable, and this is consistent with our assertion that it is admissible in a constitutive equation. It should be noted that the matrix in Eq. 7.1-16 is identical to that appearing in Eq. 7.1-2 for steady shear flow. We can think of $\boldsymbol{\tau}_{(1)}$ as being a special kind of time derivative which, unlike $\partial\boldsymbol{\tau}/\partial t$, "filters out" the effect of local rotation of the fluid.

If we calculate τ_{yx} at $t = 0$ in the system in Fig. 5.5-1 using the *linear Maxwell model* of Eq. 5.2-2, we find that $\tau_{yx}(t = 0) = -\eta_0\dot{\gamma}/(1 + 4\lambda_1^2 W^2)$; this result shows (incorrectly) that the shear stress depends on the rate of rotation of the turntable. If, on the other hand, we replace $\partial\boldsymbol{\tau}/\partial t$ by $\boldsymbol{\tau}_{(1)}$ to get the *convected Maxwell model*, as we do in the next section, we will get $\tau_{yx}(t = 0) = -\eta_0\dot{\gamma}$, with no dependence on W.

§7.2 QUASI-LINEAR DIFFERENTIAL MODELS

In this section we illustrate the construction of admissible constitutive equations by the procedure described in the preceding section. Because the Jeffreys model of Eq. 5.2-9 is known to describe linear viscoelastic behavior qualitatively, we use it to generate a quasi-linear model. This is done by replacing the partial time derivatives in the differential form of the Jeffreys model with the convected time derivatives to obtain the *convected Jeffreys model*[1] or *Oldroyd's fluid B*:[2]

$$\boxed{\boldsymbol{\tau} + \lambda_1\boldsymbol{\tau}_{(1)} = -\eta_0(\boldsymbol{\gamma}_{(1)} + \lambda_2\boldsymbol{\gamma}_{(2)})} \tag{7.2-1}$$

[1] In Chapter 9 this will be referred to as the "contravariant convected Jeffreys model" to distinguish it from the "covariant convected Jeffreys model" or Oldroyd's fluid A.

[2] J. G. Oldroyd, *Proc. Roy. Soc.*, **A200**, 523–541 (1950).

This constitutive equation contains three parameters: η_0, the zero-shear-rate viscosity; λ_1, the relaxation time; and λ_2, the retardation time. The kinematic tensors are those defined in §6.1 and used in the retarded-motion expansion. The convected Jeffreys model contains several other models as special cases:

a. If $\lambda_2 = 0$, the model reduces to the "convected Maxwell model."[3] The convected Maxwell model has been widely used for viscoelastic flow calculations, because of its simplicity.

b. If $\lambda_1 = 0$, the model simplifies to a second-order fluid with a vanishing second normal stress coefficient (see Eq. 6.2-1 with only the constants b_1 and b_2 nonzero).

c. If $\lambda_1 = \lambda_2$, the model reduces to a Newtonian fluid with viscosity η_0.

Although we know that the convected Jeffreys model is admissible, we do not know a priori whether or not it describes the kinds of material functions shown in Chapter 3 for typical polymeric liquids.[4] Hence it is very important (as it is for any model) to test it in as many different rheometric experiments as possible.

EXAMPLE 7.2-1 Time-Dependent Shearing Flows of the Convected Jeffreys Model

(a) Show how the convected Jeffreys model simplifies for a general time-dependent, simple shearing flow with velocity gradient $\partial v_x/\partial y = \dot{\gamma}_{yx}(t)$. Then use this result to evaluate the material functions for **(b)** steady shear flow, **(c)** small-amplitude oscillatory shearing flow, **(d)** start-up of steady shear flow, and **(e)** stress relaxation following steady shear flow.

SOLUTION **(a)** We start by writing out the various tensors that enter into the constitutive equation. Appendix C gives the matrix representations for $\boldsymbol{\tau}_{(1)}$, $\boldsymbol{\gamma}_{(1)}$, and $\boldsymbol{\gamma}_{(2)}$ for an unsteady shearing flow. When these are substituted into the constitutive equation for the convected Jeffreys model we obtain the following matrix equation:

$$\begin{pmatrix} \tau_{xx} & \tau_{yx} & 0 \\ \tau_{yx} & \tau_{yy} & 0 \\ 0 & 0 & \tau_{zz} \end{pmatrix} + \lambda_1 \frac{d}{dt}\begin{pmatrix} \tau_{xx} & \tau_{yx} & 0 \\ \tau_{yx} & \tau_{yy} & 0 \\ 0 & 0 & \tau_{zz} \end{pmatrix} - \begin{pmatrix} 2\tau_{yx} & \tau_{yy} & 0 \\ \tau_{yy} & 0 & 0 \\ 0 & 0 & 0 \end{pmatrix}\lambda_1\dot{\gamma}_{yx}$$

$$= -\eta_0\left[\begin{pmatrix} 0 & 1 & 0 \\ 1 & 0 & 0 \\ 0 & 0 & 0 \end{pmatrix}\dot{\gamma}_{yx} + \lambda_2\begin{pmatrix} 0 & 1 & 0 \\ 1 & 0 & 0 \\ 0 & 0 & 0 \end{pmatrix}\frac{d\dot{\gamma}_{yx}}{dt} - 2\begin{pmatrix} 1 & 0 & 0 \\ 0 & 0 & 0 \\ 0 & 0 & 0 \end{pmatrix}\lambda_2\dot{\gamma}_{yx}^2\right] \tag{7.2-2}$$

From this matrix equation we can now obtain a set of coupled differential equations for the stress tensor components:

$$\left(1 + \lambda_1 \frac{d}{dt}\right)\tau_{xx} - 2\tau_{yx}\lambda_1\dot{\gamma}_{yx}(t) = 2\eta_0\lambda_2\dot{\gamma}_{yx}^2(t) \tag{7.2-3}$$

$$\left(1 + \lambda_1 \frac{d}{dt}\right)\tau_{yy} = 0 \tag{7.2-4}$$

[3] The convected Maxwell model was first obtained from a molecular model by M. S. Green and A. V. Tobolsky, *J. Chem. Phys.*, **14**, 80–92 (1946).

[4] The convected Jeffreys model is derived in Chapter 13 from the kinetic theory of dilute solutions of elastic dumbbells.

$$\left(1 + \lambda_1 \frac{d}{dt}\right)\tau_{zz} = 0 \tag{7.2-5}$$

$$\left(1 + \lambda_1 \frac{d}{dt}\right)\tau_{yx} - \underline{\tau_{yy}\lambda_1\dot{\gamma}_{yx}(t)} = -\eta_0\left(1 + \lambda_2 \frac{d}{dt}\right)\dot{\gamma}_{yx}(t) \tag{7.2-6}$$

From these equations it can be seen that the normal stresses τ_{yy} and τ_{zz} are zero for all simple, time-dependent shearing flows. Thus we can drop the dashed underlined term in Eq. 7.2-6. We now use Eqs. 7.2-3 through 6 to calculate specific material functions.

(b) For steady shear flow the differential equations in Eqs. 7.2-3 and 6 simplify to algebraic equations. The two equations for τ_{xx} and τ_{yx} are easily solved to give the viscometric functions:

$$\begin{aligned} \tau_{yx} &= -\eta_0\dot{\gamma}_{yx} && \text{or} \quad \eta = \eta_0 \\ \tau_{xx} - \tau_{yy} &= -2\eta_0(\lambda_1 - \lambda_2)\dot{\gamma}_{yx}^2 && \text{or} \quad \Psi_1 = 2\eta_0(\lambda_1 - \lambda_2) \\ \tau_{yy} - \tau_{zz} &= 0 && \text{or} \quad \Psi_2 = 0 \end{aligned} \tag{7.2-7}$$

The convected Jeffreys model thus gives a constant viscosity and first normal stress coefficient. The second normal stress coefficient is zero.

(c) To get the small amplitude oscillatory shearing properties, we take the shear strain to be $\gamma_{yx}(0, t) = \int_0^t \dot{\gamma}_0 \cos \omega t' dt' = \gamma_0 \sin \omega t$, in which $\gamma_0 = \dot{\gamma}_0/\omega$ is the amplitude of the shear strain, and seek solutions to Eqs. 7.2-3 and 6 in the limit of small γ_0. For this flow the differential equation for the shear stress is

$$\left(1 + \lambda_1 \frac{d}{dt}\right)\tau_{yx} = -\eta_0\gamma_0\omega(\cos \omega t - \lambda_2\omega \sin \omega t) \tag{7.2-8}$$

Since we seek a steady periodic solution, the nonhomogeneous part of the first-order, linear, ordinary differential equation suggests we try a solution for τ_{yx} of the form

$$\tau_{yx} = A \cos \omega t + B \sin \omega t \tag{7.2-9}$$

By substituting Eq. 7.2-9 into Eq. 7.2-8 we can see that A and B must be given by (cf. Eqs. 3.4-3b)

$$A = -\eta_0\left(\frac{1 + \lambda_1\lambda_2\omega^2}{1 + \lambda_1^2\omega^2}\right)\gamma_0\omega = -\eta'(\omega)\gamma_0\omega \tag{7.2-10}$$

$$B = -\eta_0 \frac{(\lambda_1 - \lambda_2)\gamma_0\omega^2}{1 + \lambda_1^2\omega^2} = -\eta''(\omega)\gamma_0\omega \tag{7.2-11}$$

Note that η' and η'' are the same as for the linear viscoelastic Jeffreys model.

To show that the convected Jeffreys model gives identical predictions to the Jeffreys model in the linear viscoelastic limit is a little more complicated than the calculation of η' and η''. It is also necessary to demonstrate that as $\gamma_0 \to 0$ the normal stresses become negligible compared with the shear stress (see Problem 7B.1). Notice that $\dot{\gamma}_0$ can be large in the linear viscoelastic limit so that the demand above on $\gamma_0 \to 0$ is more restrictive than simply requiring τ_{xx} to be small compared with τ_{yx} as $\dot{\gamma}_0 \to 0$. The latter is easily seen to hold by inspection of the governing differential equations, Eqs. 7.2-3 and 6.

(d) For start-up of steady shear flow we let $\dot{\gamma}_{yx} = \dot{\gamma}_0 H(t)$, where H is the Heaviside unit step function. The stresses τ_{xx} and τ_{yx} are then given by the differential equations

$$\left(1 + \lambda_1 \frac{d}{dt}\right)\tau_{xx} - 2\lambda_1\dot{\gamma}_0\tau_{yx} = +2\eta_0\lambda_2\dot{\gamma}_0^2 \tag{7.2-12}$$

$$\left(1 + \lambda_1 \frac{d}{dt}\right)\tau_{yx} = -\eta_0\dot{\gamma}_0(1 + \lambda_2\delta(t)) \tag{7.2-13}$$

where the Dirac delta function has been introduced as the derivative of the step function $dH/dt = \delta(t)$ (see Eqs. 5.2-10a through c). Equation 7.2-13 is then multiplied by the integrating factor e^{t/λ_1} and integrated from $t = 0^-$ (where $\tau_{yx} = 0$) to an arbitrary time $t > 0$ to give (cf. Table 3.4-1)

$$\begin{aligned}\tau_{yx} &= -\eta_0\dot{\gamma}_0\frac{\lambda_2}{\lambda_1} - \eta_0\dot{\gamma}_0\left(1 - \frac{\lambda_2}{\lambda_1}\right)(1 - e^{-t/\lambda_1}) \\ &= -\eta^+(t, \dot{\gamma}_0)\dot{\gamma}_0\end{aligned} \tag{7.2-14}$$

When this last result is combined with Eq. 7.2-12 we obtain for τ_{xx}

$$\begin{aligned}\tau_{xx} &= -2\eta_0(\lambda_1 - \lambda_2)\dot{\gamma}_0^2\left[1 - \left(1 + \frac{t}{\lambda_1}\right)e^{-t/\lambda_1}\right] \\ &= -\Psi_1^+(t, \dot{\gamma}_0)\dot{\gamma}_0^2\end{aligned} \tag{7.2-15}$$

From Eqs. 7.2-14 and 15, it can be seen that the shear stress and first normal stress growth functions do not depend on shear rate as is found for the polymer solution and melt data in Chapter 3. Moreover, the stresses grow monotonically to the steady-state values, so that the convected Jeffreys model does not predict the stress overshoot observed experimentally in polymeric liquids. Finally, there is a jump in the shear stress at $t = 0$, following which the shear stress grows monotonically to steady state. This sudden jump is associated with the retardation time λ_2 in the model, but does not seem to be observed experimentally. However, as a result of the jump, the shear stress is predicted to grow more rapidly than the normal stress, in qualitative agreement with data.

(e) For stress relaxation following steady shear flow we take the shear rate to be $\dot{\gamma}(t) = \dot{\gamma}_0\{1 - H(t)\}$. When this is inserted into Eq. 7.2-6 we obtain the following differential equation for the shear stress:

$$\left(1 + \lambda_1 \frac{d}{dt}\right)\tau_{yx} = \eta_0\lambda_2\dot{\gamma}_0\delta(t) \tag{7.2-16}$$

This equation is integrated from $t = 0^-$ (where the shear stress is the steady-state value $-\eta_0\dot{\gamma}_0$) to time $t > 0$ to give the shear stress relaxation function (cf. Table 3.4-1)

$$\begin{aligned}\tau_{yx} &= -\eta_0\dot{\gamma}_0\left(1 - \frac{\lambda_2}{\lambda_1}\right)e^{-t/\lambda_1} \\ &= -\eta^-(t, \dot{\gamma}_0)\dot{\gamma}_0\end{aligned} \tag{7.2-17}$$

To find the normal stress relaxation we note from Eq. 7.2-3 that for $t \geq 0$ when the shear rate is zero, the normal stress τ_{xx} relaxes like e^{-t/λ_1} from its steady-state shear flow value. Thus,

$$\begin{aligned}\tau_{xx} &= -2\eta_0(\lambda_1 - \lambda_2)\dot{\gamma}_0^2 e^{-t/\lambda_1} \\ &= -\Psi_1^-(t, \dot{\gamma}_0)\dot{\gamma}_0^2\end{aligned} \tag{7.2-18}$$

As we found for the stress growth material functions, the rate of stress relaxation does not depend on the shear rate. In addition, there is again a discontinuity in the shear stress at $t = 0$, whereas the normal stresses are continuous. The jump in τ_{yx} results in the fact that the shear stress relaxes to zero more rapidly than the normal stress, in qualitative agreement with experiments.

EXAMPLE 7.2-2 Time-Dependent Shearfree Flows of the Convected Jeffreys Model

(a) Show how the convected Jeffreys model simplifies for an arbitrary, time-dependent shearfree flow. **(b)** Then use this result to find the model response to start-up of steady planar and elongational flow. **(c)** What are the steady-state material functions for the convected Jeffreys model in these flows?

SOLUTION **(a)** We again turn to Appendix C where the various tensors in the convected Jeffreys model are displayed in matrix form for shearfree flows. When these are inserted into the constitutive equation, Eq. 7.2-1, we obtain the following uncoupled, ordinary differential equations for the stress components:

$$\tau_{xx} + \lambda_1 \frac{d\tau_{xx}}{dt} + \lambda_1(1+b)\tau_{xx}\dot{\varepsilon}(t) = +\eta_0\left[(1+b)\dot{\varepsilon}[1+(1+b)\lambda_2\dot{\varepsilon}] + \lambda_2(1+b)\frac{d\dot{\varepsilon}}{dt}\right] \quad (7.2\text{-}19)$$

$$\tau_{zz} + \lambda_1 \frac{d\tau_{zz}}{dt} - 2\lambda_1\tau_{zz}\dot{\varepsilon}(t) = -\eta_0\left[2\dot{\varepsilon}[1-2\lambda_2\dot{\varepsilon}] + 2\lambda_2\frac{d\dot{\varepsilon}}{dt}\right] \quad (7.2\text{-}20)$$

The equation for τ_{yy} is the same as for τ_{xx} except that $+b$ is replaced by $-b$.

(b) In start-up of steady shearfree flow, the elongation rate is given by the Heaviside unit step function $\dot{\varepsilon}(t) = \dot{\varepsilon}_0 H(t)$ where $\dot{\varepsilon}_0$ is constant. When the differential equation for τ_{xx}, for example, is specialized to this strain rate function we obtain

$$\left(1 + (1+b)\lambda_1\dot{\varepsilon}_0 + \lambda_1\frac{d}{dt}\right)\tau_{xx} = \eta_0(1+b)\dot{\varepsilon}_0[1+(1+b)\lambda_2\dot{\varepsilon}_0 + \lambda_2\delta(t)] \quad (7.2\text{-}21)$$

which is to be solved subject to the initial condition that $\tau_{xx}(0^-) = 0$. Integrating Eq. 7.2-21 from $t = 0^-$ to t gives

$$\tau_{xx} = \eta_0(1+b)\dot{\varepsilon}_0 \frac{1+(1+b)\lambda_2\dot{\varepsilon}_0}{1+(1+b)\lambda_1\dot{\varepsilon}_0} - \eta_0\left(1 - \frac{\lambda_2}{\lambda_1}\right)\frac{(1+b)\dot{\varepsilon}_0 e^{-t(1+(1+b)\lambda_1\dot{\varepsilon}_0)/\lambda_1}}{1+(1+b)\lambda_1\dot{\varepsilon}_0} \quad (7.2\text{-}22)$$

Similarly we find that τ_{yy} is identical to τ_{xx} with b replaced by $-b$ and τ_{zz} is given by

$$\tau_{zz} = -2\eta_0\dot{\varepsilon}_0\frac{1-2\lambda_2\dot{\varepsilon}_0}{1-2\lambda_1\dot{\varepsilon}_0} + 2\eta_0\left(1-\frac{\lambda_2}{\lambda_1}\right)\frac{\dot{\varepsilon}_0 e^{-t(1-2\lambda_1\dot{\varepsilon}_0)/\lambda_1}}{1-2\lambda_1\dot{\varepsilon}_0} \quad (7.2\text{-}23)$$

An interesting result of the above formulas is that no steady state is attained if

$$\lambda_1\dot{\varepsilon}_0 > \tfrac{1}{2} \quad \text{or} \quad \lambda_1\dot{\varepsilon}_0 < \frac{-1}{(1+b)} \quad \text{or} \quad \lambda_1\dot{\varepsilon}_0 < \frac{-1}{(1-b)} \quad (7.2\text{-}24)$$

Since we specify shearfree flows with b in the range $0 \le b \le 1$, the first two of these conditions determine which shearfree flows of the convected Jeffreys model will approach a steady state in the start-up experiment.

For $-1/(1+b) < \lambda_1\dot{\varepsilon}_0 < 1/2$ we find the stress growth material functions (Table 3.5-1):

$$\begin{aligned}\bar{\eta}_1^+ &= -\frac{\tau_{zz}-\tau_{xx}}{\dot{\varepsilon}_0}\\ &= (3+b)\eta_0\frac{\lambda_2}{\lambda_1} + \frac{(3+b)\eta_0(1-(\lambda_2/\lambda_1))}{(1+(1+b)\lambda_1\dot{\varepsilon}_0)(1-2\lambda_1\dot{\varepsilon}_0)}\\ &\quad - \eta_0(1+b)\left(1-\frac{\lambda_2}{\lambda_1}\right)\frac{e^{-t(1+(1+b)\lambda_1\dot{\varepsilon}_0)/\lambda_1}}{(1+(1+b)\lambda_1\dot{\varepsilon}_0)}\\ &\quad - 2\eta_0\left(1-\frac{\lambda_2}{\lambda_1}\right)\frac{e^{-t(1-2\lambda_1\dot{\varepsilon}_0)/\lambda_1}}{(1-2\lambda_1\dot{\varepsilon}_0)}\end{aligned} \tag{7.2-25}$$

$$\begin{aligned}\bar{\eta}_2^+ &= -\frac{\tau_{yy}-\tau_{xx}}{\dot{\varepsilon}_0}\\ &= 2b\eta_0\frac{\lambda_2}{\lambda_1} + \frac{2b\eta_0(1-(\lambda_2/\lambda_1))}{(1+(1+b)\lambda_1\dot{\varepsilon}_0)(1+(1-b)\lambda_1\dot{\varepsilon}_0)}\\ &\quad - \eta_0(1+b)\left(1-\frac{\lambda_2}{\lambda_1}\right)\frac{e^{-t(1+(1+b)\lambda_1\dot{\varepsilon}_0)/\lambda_1}}{(1+(1+b)\lambda_1\dot{\varepsilon}_0)}\\ &\quad + \eta_0(1-b)\left(1-\frac{\lambda_2}{\lambda_1}\right)\frac{e^{-t(1+(1-b)\lambda_1\dot{\varepsilon}_0)/\lambda_1}}{(1+(1-b)\lambda_1\dot{\varepsilon}_0)}\end{aligned} \tag{7.2-26}$$

(c) For steady-state shearfree flows, the governing differential equations, Eqs. 7.2-19 and 20, reduce to algebraic equations that are easily solved for the stresses. These are then combined with the material function definitions to give for any $\dot{\varepsilon}_0$:

$$\bar{\eta}_1 = (3+b)\eta_0\frac{\lambda_2}{\lambda_1} + \frac{(3+b)\eta_0(1-(\lambda_2/\lambda_1))}{(1+(1+b)\lambda_1\dot{\varepsilon}_0)(1-2\lambda_1\dot{\varepsilon}_0)} \tag{7.2-27}$$

$$\bar{\eta}_2 = 2b\eta_0\frac{\lambda_2}{\lambda_1} + \frac{2b\eta_0(1-(\lambda_2/\lambda_1))}{(1+(1+b)\lambda_1\dot{\varepsilon}_0)(1+(1-b)\lambda_1\dot{\varepsilon}_0)} \tag{7.2-28}$$

Note that the steady material functions become infinite at the critical strain rates given by Eq. 7.2-24.

§7.3 NONLINEAR DIFFERENTIAL MODELS

In the preceding section we wrote down an admissible constitutive equation that is capable of describing time-dependent flows. However, the model is not able to portray well the rheological properties that are observed in typical polymer solutions and melts. For example, we noted previously the deficiencies of constant viscosity and normal stress coefficients in steady shear flow and the infinite elongational viscosity at finite elongation rates. In this section we illustrate some of the methods that have been used to generate constitutive equations that more accurately represent real material data. In each of the approaches, it is necessary to give up the starting point of a linear model that we used in §7.2.

TABLE 7.3-1

Material Functions for the White–Metzner Model (Eq. 7.3-1)[a]

Steady shear flow	$\eta = \eta(\dot{\gamma})$ $\Psi_1 = 2\eta(\dot{\gamma})\lambda(\dot{\gamma})$ $\Psi_2 = 0$	(A)
Small-amplitude oscillatory shearing	η', η'' not defined[b]	
Start-up of steady shear flow	$\eta^+(t, \dot{\gamma}) = \eta(\dot{\gamma})[1 - e^{-t/\lambda(\dot{\gamma})}]$ $\Psi_1^+(t, \dot{\gamma}) = 2\eta(\dot{\gamma})\lambda(\dot{\gamma})\left[1 - \left(1 + \frac{t}{\lambda(\dot{\gamma})}\right)e^{-t/\lambda(\dot{\gamma})}\right]$	(B)
Steady shearfree flow $(\dot{\gamma} \equiv \sqrt{\frac{1}{2}II}$ $= \sqrt{(3 + b^2)}\,\lvert\dot{\varepsilon}\rvert)$	$\bar{\eta}_1 = \frac{(3 + b)\eta(\dot{\gamma})}{(1 + (1 + b)\lambda\dot{\varepsilon})(1 - 2\lambda\dot{\varepsilon})}$ $\bar{\eta}_2 = \frac{2b\eta(\dot{\gamma})}{(1 + (1 - b)\lambda\dot{\varepsilon})(1 + (1 + b)\lambda\dot{\varepsilon})}$	(C)

[a] The time "constant" $\lambda(\dot{\gamma}) = \eta(\dot{\gamma})/G$.

[b] The White–Metzner model has no unique limit as strain amplitude approaches zero. For small shear rate amplitudes, the oscillatory shearing behavior is identical to that of the convected Maxwell model.

a. Inclusion of Invariants

We saw in Chapter 4 that for the generalized Newtonian fluid the viscosity was allowed to depend on the second invariant of the rate-of-strain tensor and that this led to a useful constitutive equation. It seems reasonable then to modify Eq. 7.2-1 to include dependence on $\dot{\gamma} = \sqrt{\frac{1}{2}(\boldsymbol{\gamma}_{(1)}:\boldsymbol{\gamma}_{(1)})}$. An example of this change is the *White–Metzner model*[1] which is the following modification of the convected Maxwell model:

$$\boxed{\boldsymbol{\tau} + \frac{\eta(\dot{\gamma})}{G}\boldsymbol{\tau}_{(1)} = -\eta(\dot{\gamma})\boldsymbol{\gamma}_{(1)}} \tag{7.3-1}$$

where G is a constant modulus. A few material functions for this model are summarized in Table 7.3-1. This model has the advantage of being relatively simple and yet still giving reasonable shapes for the shear-rate dependent viscosity and first normal stress coefficient. It can also be used in fast time-dependent motions, although its predictions are not completely realistic in these problems. This stems from its lack of a linear viscoelastic limit for small displacement gradients. In steady shearfree flows the model gives infinite elongational viscosities $\bar{\eta}_1$ and $\bar{\eta}_2$ in the same way as the convected Maxwell model; the exact value of elongation rate at which these viscosities become infinite depends on the particular form of $\eta(\dot{\gamma})$. The model has been found useful in exploratory hydrodynamic calculations aimed at assessing the interaction of shear thinning and memory on flow fields.[2]

There are of course other ways in which invariants could be introduced into the model. There is no reason, other than for preserving simplicity, to include the invariant II of

[1] J. L. White and A. B. Metzner, *J. Appl. Polym. Sci.*, **7**, 1867–1889 (1963).

[2] A. Beris, R. C. Armstrong, and R. A. Brown, *J. Non-Newtonian Fluid Mech.*, **13**, 109–148 (1983).

$\dot{\gamma}$ but not III in the model. Similarly, we could include invariants of stress. The molecularly based Phan-Thien–Tanner and the FENE-P models both include tr $\boldsymbol{\tau}$, with a certain amount of success. These are listed in §7.6 and discussed in Chapters 20 and 13, respectively.

b. Inclusion of Quadratic Terms in Velocity Gradient

Oldroyd realized that models like the convected Jeffreys model have included terms that are quadratic in velocity gradient, but that their inclusion has not been systematic. He thus suggested that a possibly useful generalization of the convected Jeffreys model could be obtained by adding to the latter all possible quadratic terms involving products of $\boldsymbol{\tau}$ with $\boldsymbol{\gamma}_{(1)}$ and of $\boldsymbol{\gamma}_{(1)}$ with itself. Thus Oldroyd suggested[3]

$$\begin{aligned} &\underline{\boldsymbol{\tau} + \lambda_1 \boldsymbol{\tau}_{(1)}} + \tfrac{1}{2}\lambda_3\{\boldsymbol{\gamma}_{(1)}\cdot\boldsymbol{\tau} + \boldsymbol{\tau}\cdot\boldsymbol{\gamma}_{(1)}\} + \tfrac{1}{2}\lambda_5(\mathrm{tr}\,\boldsymbol{\tau})\boldsymbol{\gamma}_{(1)} + \tfrac{1}{2}\lambda_6(\boldsymbol{\tau}:\boldsymbol{\gamma}_{(1)})\boldsymbol{\delta} \\ &\quad = \underline{-\eta_0[\boldsymbol{\gamma}_{(1)} + \lambda_2 \boldsymbol{\gamma}_{(2)}} + \lambda_4\{\boldsymbol{\gamma}_{(1)}\cdot\boldsymbol{\gamma}_{(1)}\} + \tfrac{1}{2}\lambda_7(\boldsymbol{\gamma}_{(1)}:\boldsymbol{\gamma}_{(1)})\boldsymbol{\delta}] \end{aligned} \tag{7.3-2}$$

The dashed-underlined terms belong to the convected Jeffreys model. Several simplified versions of this model, in which special values are assigned to some of the constants, have been used extensively in the literature; these are summarized in Table 7.3-2.

Some sample material functions for the Oldroyd 8-constant model are given in Table 7.3-3. In order that these material functions agree at least qualitatively with experimental data, there are some restrictions on the choice of the constants that have to be imposed:

1. Since η' is known to decrease with increasing ω, we must impose the requirement that $0 < \lambda_2 < \lambda_1$.
2. Since viscosity is generally a monotone decreasing function of $\dot{\gamma}$, we must require that $0 < \sigma_2 < \sigma_1$ [where $\sigma_i = \lambda_i(\lambda_3 + \lambda_5) + \lambda_{i+2}(\lambda_1 - \lambda_3 - \lambda_5) + \lambda_{i+5}(\lambda_1 - \lambda_3 - \tfrac{3}{2}\lambda_5)$, with $i = 1, 2$].
3. Since $|\tau_{xy}|$ is to be a monotone increasing function of $\dot{\gamma}$ for steady shear flow, we have to require that $\sigma_2 \geq \tfrac{1}{9}\sigma_1$.
4. When $\eta(\dot{\gamma})$ and $\eta'(\omega)$ are plotted on the same graph with $\dot{\gamma} = \omega$, the η-curve generally lies above the η'-curve. For this to be true in the region of moderate $\dot{\gamma}$ and ω, we must require that $\sigma_1 - \sigma_2 < \lambda_1(\lambda_1 - \lambda_2)$.
5. For the elongational viscosity to be bounded for positive and negative $\dot{\varepsilon}$ it is necessary for $\lambda_1 - \lambda_3$ to be between $\tfrac{2}{3}(\lambda_5 + \lambda_6) \pm \tfrac{1}{3}[4\lambda_6^2 - 11\lambda_5\lambda_6 + 4\lambda_5^2]^{1/2}$.

Consideration of additional material functions and additional flow patterns will in general provide further inequalities.

With eight constants and all the extra terms, considerably more variety in rheological response is possible than for the convected Jeffreys equation. Although we cannot obtain

[3] J. G. Oldroyd, *Proc. Roy. Soc.*, **A245**, 278–297 (1958); *Rheol. Acta*, **1**, 337–344 (1961). The parameters in Eq. 7.3-2 are related to the original constants used by Oldroyd as follows:

Eq. 7.3-2:	η_0	λ_1	λ_2	λ_3	λ_4	λ_5	λ_6	λ_7
Oldroyd:	η_0	λ_1	λ_2	$\lambda_1 - \mu_1$	$\lambda_2 - \mu_2$	μ_0	ν_1	ν_2

the kind of quantitative fit of the viscosity and first normal stress coefficient that is possible with the White–Metzner model, a wider range of properties can be correctly described qualitatively. For example, by suitable choice of the constants, stress overshoot in start-up of steady shear flow and a bounded elongational viscosity can be obtained. Also the algebraic form in which nonlinear terms have been included makes it generally easier to obtain analytical solutions than for the White–Metzner model. Thus, the Oldroyd 8-constant model has been found to be a useful and relatively simple constitutive equation for making exploratory fluid flow calculations.

The calculation of the material functions for the Oldroyd 8-constant model follows the same pattern that was illustrated in §7.2 for the convected Jeffreys model. For convenience we have tabulated in Table C.2 of Appendix C, the governing differential equations for the stresses in time-dependent shear and shearfree flows. The shapes of the material functions are illustrated in Figs. 7.3-1 through 4 for the special case of the Oldroyd 4-constant model. The model describes many of the rheological properties at least qualitatively correctly. In order to obtain more quantitative fits with data it is probably necessary to include a spectrum of relaxation times instead of the single λ_1. This is similar to the improvement found by use of the generalized Maxwell model in place of the Maxwell model in linear viscoelasticity. A defect in the Oldroyd 8-constant model that is not so easily removed is the presence of singularities in $\bar{\eta}_1$ and $\bar{\eta}_2$ for $\dot{\varepsilon} < 0$ and $0 < b < 1$. These arise from the polynomial nature of the Oldroyd 8-constant model[4] and are illustrated in Fig. 7.3-4.

c. Inclusion of Nonlinear Terms in Stress

There appear to be no rheological reasons why a constitutive equation must be restricted to terms linear in stress as Oldroyd has required. In this subsection we use the *Giesekus model*[5] as an example of a constitutive equation with nonlinear stress terms:

$$\boxed{\begin{aligned} \boldsymbol{\tau} &= \boldsymbol{\tau}_s + \boldsymbol{\tau}_p \\ \boldsymbol{\tau}_s &= -\eta_s \dot{\boldsymbol{\gamma}} \\ \boldsymbol{\tau}_p + \lambda_1 \boldsymbol{\tau}_{p(1)} - \alpha \frac{\lambda_1}{\eta_p} \{\boldsymbol{\tau}_p \cdot \boldsymbol{\tau}_p\} &= -\eta_p \dot{\boldsymbol{\gamma}} \end{aligned}} \quad (7.3\text{-}3)$$

Here the model is written as a superposition of solvent and polymer contributions, $\boldsymbol{\tau}_s$ and $\boldsymbol{\tau}_p$, to the stress tensor, which is the form in which constitutive equations derived by kinetic theory arise naturally for polymer solutions (see for example, Eqs. 13.3-1 and 13.4-4).

The Giesekus model contains four parameters: a relaxation time λ_1; the solvent and polymer contributions to the zero-shear-rate viscosity, η_s and η_p; and the dimensionless "mobility factor" α. The origin of the term involving α can be associated with anisotropic Brownian motion and/or anisotropic hydrodynamic drag on the constituent polymer molecules[6] (see §13.7).

[4] The shearfree flow properties of the Oldroyd 8-constant model are discussed in C. J. S. Petrie, *J. Non-Newtonian Fluid Mech.*, **14**, 189–202 (1984).

[5] H. Giesekus, *J. Non-Newtonian Fluid Mech.*, **11**, 69–109 (1982); **12**, 367–374 (1983); *Rheol. Acta*, **21**, 366–375 (1982).

[6] R. B. Bird and J. M. Wiest, *J. Rheol.*, **29**, 519–532 (1985).

TABLE 7.3-2

Models Included in the Oldroyd 8-Constant Model (Eq. 7.3-2)

Name of Model	Number of Constants	Values of Time Constants λ_1	λ_2	λ_3	λ_4	λ_5	λ_6	λ_7	Steady-State Shear Flow Material Functions[a]	Elongational Viscosity[b]	References
Oldroyd 6-constant model	6						0	0	η depends on $\dot{\gamma}$ Ψ_1 depends on $\dot{\gamma}$ Ψ_2 not simply related to Ψ_1	See Table 7.3-3 Eq. F	Problem 14B.4 Table 6.2-2 See footnote[f]
Oldroyd 4-constant model	4			0	0		0	0	η depends on $\dot{\gamma}$ Ψ_1 depends on $\dot{\gamma}$ $\Psi_2 = 0$	See Table 7.3-3 Eq. F	
Oldroyd fluid A	3			$2\lambda_1$	$2\lambda_2$	0	0	0	$\eta = \eta_0$ $\Psi_1 = 2\eta_0(\lambda_1 - \lambda_2)$ $\Psi_2 = -\Psi_1$	$3\eta_0 \dfrac{1 + \lambda_2\dot{\varepsilon} - 2\lambda_1\lambda_2\dot{\varepsilon}^2}{1 + \lambda_1\dot{\varepsilon} - 2\lambda_1^2\dot{\varepsilon}^2}$	Problem 7B.4
Oldroyd fluid B (Convected Jeffreys)	3			0	0	0	0	0	$\eta = \eta_0$ $\Psi_1 = 2\eta_0(\lambda_1 - \lambda_2)$ $\Psi_2 = 0$	$3\eta_0 \dfrac{1 - \lambda_2\dot{\varepsilon} - 2\lambda_1\lambda_2\dot{\varepsilon}^2}{1 - \lambda_1\dot{\varepsilon} - 2\lambda_1^2\dot{\varepsilon}^2}$	§7.2; Eq. 13.4-5

Corotational Jeffreys model	3			λ_1	λ_2	0	0	0	η depends on $\dot{\gamma}$ Ψ_1 depends on $\dot{\gamma}$ $\Psi_2 = -\frac{1}{2}\Psi_1$	$3\eta_0$	
Second-order fluid[c]	3	0		0		0	0	0	$\eta = \eta_0$ $\Psi_1 = -2\eta_0\lambda_2$ $\Psi_2 = \eta_0\lambda_4$	$3\eta_0(1 - (\lambda_2 - \lambda_4)\dot{\varepsilon})$	Chapter 6; Table 6.2-1
Convected Maxwell model	2		0	0	0	0	0	0	$\eta = \eta_0$ $\Psi_1 = 2\eta_0\lambda_1$ $\Psi_2 = 0$	$3\eta_0 \dfrac{1}{(1 + \lambda_1\dot{\varepsilon})(1 - 2\lambda_1\dot{\varepsilon})}$	
Gordon–Schowalter	3	$\dfrac{\eta_s\lambda_1}{\eta_0}$	$\xi\lambda_1$	$\dfrac{\xi\eta_s\lambda_1}{\eta_0}$	0	0	0	0	η depends on $\dot{\gamma}$ Ψ_1 depends on $\dot{\gamma}$ $\Psi_2 < 0$	Becomes infinite at finite $\dot{\varepsilon}$	See footnote[d] $\eta_0 = (1 - \xi)\eta_p + \eta_s$ Problem 13B.2
Johnson–Segalman	3	$\dfrac{\eta_s\lambda_1}{\eta_0}$	$\xi\lambda_1$	$\dfrac{\xi\eta_s\lambda_1}{\eta_0}$	0	0	0	0	η depends on $\dot{\gamma}$ Ψ_1 depends on $\dot{\gamma}$ $\Psi_2 < 0$	Becomes infinite at finite $\dot{\varepsilon}$	See footnote[e] $\eta_0 = \eta_p + \eta_s$

[a] See Table 7.3-3, Eqs. A, B, and C for general expressions.
[b] See Table 7.3-3, Eq. F for general expression.
[c] For the relation of the Oldroyd model to the third-order fluid see Problem 7B.6.
[d] R. J. Gordon and W. R. Schowalter, *Trans. Soc. Rheol.*, **16**, 79–97 (1972). This equation is derived from a kinetic theory of dilute polymer solutions in which η_s is the solvent viscosity; $(1 - \xi)\eta_p$ is the polymer contribution to the viscosity, and ξ is a slip parameter. The zero-shear-rate viscosity is given by $\eta_0 = (1 - \xi)\eta_p + \eta_s$.
[e] M. W. Johnson, Jr. and D. Segalman, *J. Non-Newtonian Fluid Mech.*, **2**, 255–270 (1977); **9**, 33–56 (1981); *Mech. Today*, **5**, 129–137 (1980).
[f] R. B. Bird and J. M. Wiest, *J. Rheol.*, **29**, 519–532 (1985), Table 1; J. M. Wiest and R. B. Bird, *J. Non-Newtonian Fluid Mech.*, **22**, 115–119 (1986).

TABLE 7.3-3

Material Functions for the Oldroyd 8-Constant Model (Eq. 7.3-2)

Steady shear flow:

$$\frac{\eta}{\eta_0} = \frac{1 + \sigma_2 \dot{\gamma}^2}{1 + \sigma_1 \dot{\gamma}^2} \tag{A}$$

$$\frac{\Psi_1}{2\eta_0 \lambda_1} = \frac{\eta(\dot{\gamma})}{\eta_0} - \frac{\lambda_2}{\lambda_1} \tag{B}$$

$$\frac{\Psi_2}{\eta_0 \lambda_1} = -\frac{\Psi_1}{2\eta_0 \lambda_1} + \frac{(\lambda_1 - \lambda_3)}{\lambda_1}\frac{\eta}{\eta_0} - \frac{(\lambda_2 - \lambda_4)}{\lambda_1} \tag{C}$$

where $\sigma_i = \lambda_i(\lambda_3 + \lambda_5) + \lambda_{i+2}(\lambda_1 - \lambda_3 - \lambda_5) + \lambda_{i+5}(\lambda_1 - \lambda_3 - \frac{3}{2}\lambda_5)$

Small-amplitude oscillatory shearing:

$$\frac{\eta'}{\eta_0} = \frac{1 + \lambda_1 \lambda_2 \omega^2}{1 + \lambda_1^2 \omega^2} \tag{D}$$

$$\frac{\eta''}{\omega \eta_0} = \frac{(\lambda_1 - \lambda_2)}{1 + \lambda_1^2 \omega^2} \tag{E}$$

Steady elongational flow:

$$\frac{\bar{\eta}}{3\eta_0} = \frac{1 - (\lambda_2 - \lambda_4)\dot{\varepsilon} + (\frac{3}{2}\lambda_5 - \lambda_1 + \lambda_3)(2\lambda_2 - 2\lambda_4 - 3\lambda_7)\dot{\varepsilon}^2}{1 - (\lambda_1 - \lambda_3)\dot{\varepsilon} + (\frac{3}{2}\lambda_5 - \lambda_1 + \lambda_3)(2\lambda_1 - 2\lambda_3 - 3\lambda_6)\dot{\varepsilon}^2} \tag{F}$$

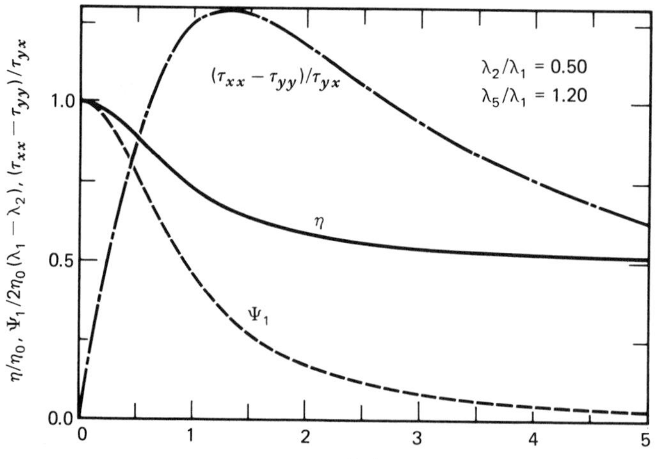

FIGURE 7.3-1. Dimensionless viscosity (——), first normal stress coefficient (----), and stress ratio (— — —) as functions of dimensionless shear rate for the Oldroyd 4-constant model defined in Table 7.3-2. The second normal stress coefficient is zero. The shear thinning in η and Ψ_1 is in qualitative agreement with experimental observations, but experimental data show a monotone increasing stress ratio (cf. Fig. 3.3-8) in disagreement with the model predictions.

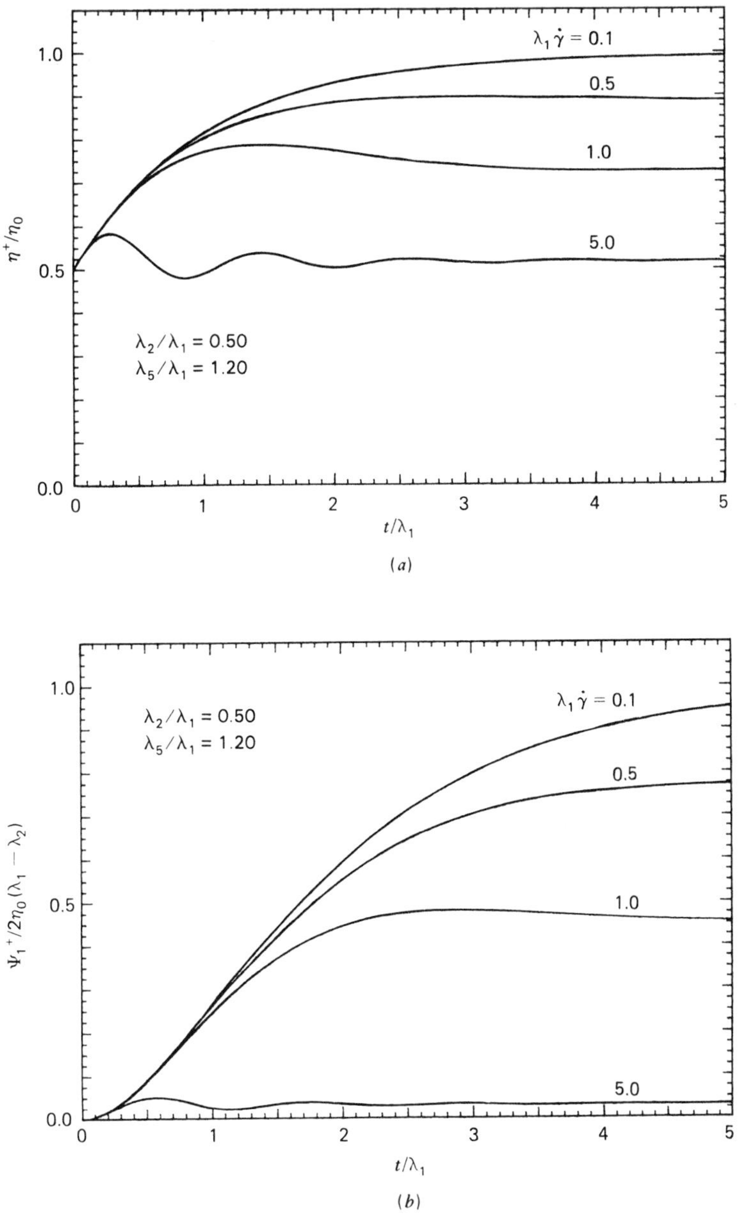

FIGURE 7.3-2. Stress growth material functions for start-up of steady shear flow for the Oldroyd 4-constant model defined in Table 7.3-2. These curves should be compared with the data in Figs. 3.4-7 through 3.4-10.

Note regarding Figures 7.3-1 through 4: Although the choice $\lambda_5/\lambda_1 = 1.20$ violates restriction 4 on p. 352, the set of parameters used here seems to be the best compromise for giving reasonably shaped curves for a wide variety of material functions.

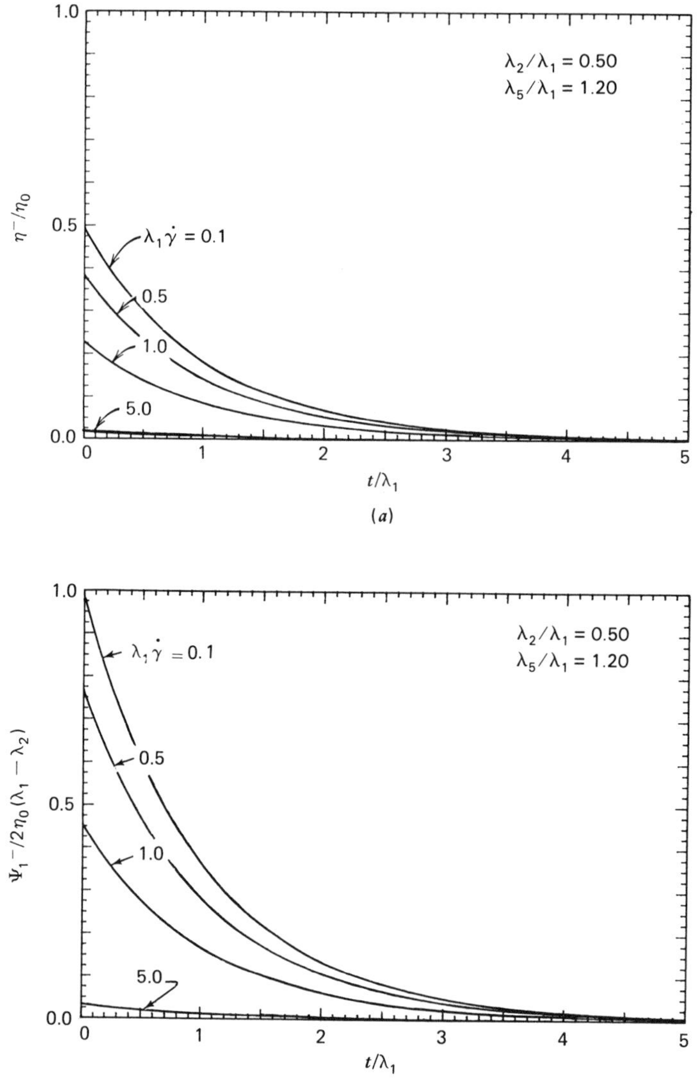

FIGURE 7.3-3. Stress relaxation material functions for cessation of steady shear flow of the Oldroyd 4-constant model defined in Table 7.3-2. These curves should be compared with the data in Figs. 3.4-11 through 3.4-13.

Equations 7.3-3 can be rewritten as a single constitutive equation by replacing $\boldsymbol{\tau}_p$ in the last equation with $\boldsymbol{\tau} - \boldsymbol{\tau}_s = \boldsymbol{\tau} + \eta_s \dot{\boldsymbol{\gamma}}$. This leads to

$$\begin{aligned} \boldsymbol{\tau} + \lambda_1 \boldsymbol{\tau}_{(1)} - a\frac{\lambda_1}{\eta_0}\{\boldsymbol{\tau}\cdot\boldsymbol{\tau}\} - a\lambda_2\{\boldsymbol{\gamma}_{(1)}\cdot\boldsymbol{\tau} + \boldsymbol{\tau}\cdot\boldsymbol{\gamma}_{(1)}\} \\ = -\eta_0\left[\boldsymbol{\gamma}_{(1)} + \lambda_2\boldsymbol{\gamma}_{(2)} - a\frac{\lambda_2^2}{\lambda_1}\{\boldsymbol{\gamma}_{(1)}\cdot\boldsymbol{\gamma}_{(1)}\}\right] \end{aligned} \tag{7.3-4}$$

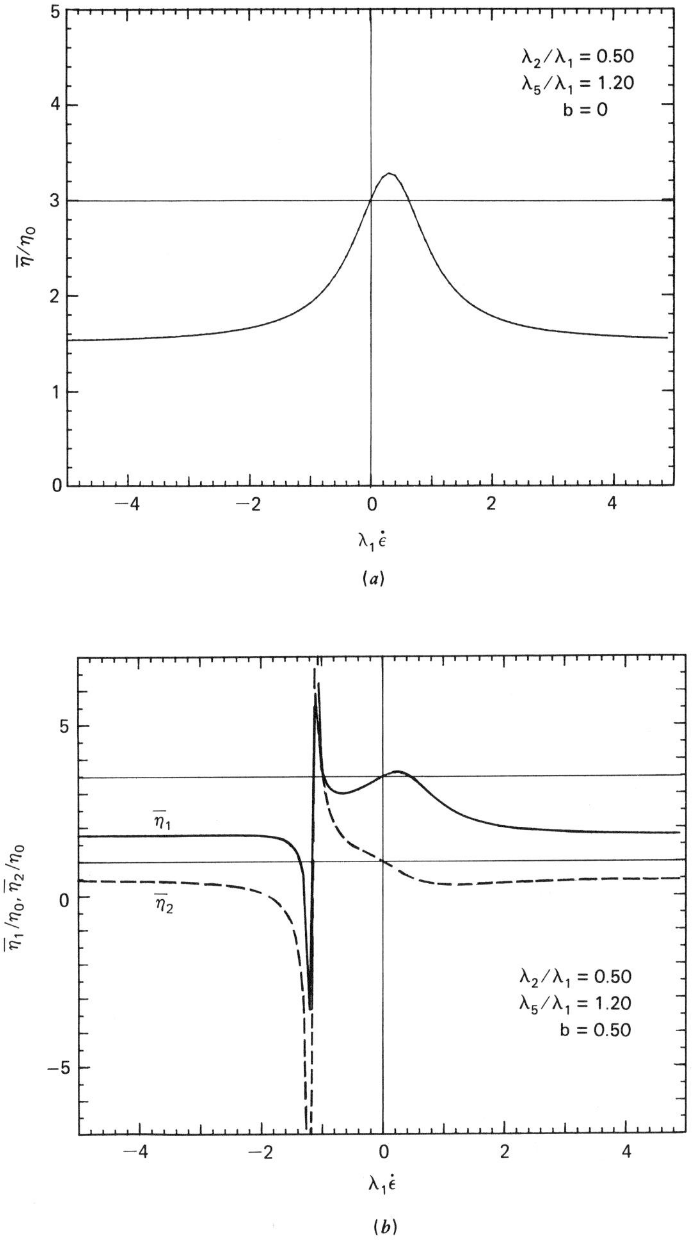

FIGURE 7.3-4. Steady shearfree flow material functions for the Oldroyd 4-constant model defined in Table 7.3-2 with $\lambda_2/\lambda_1 = 0.5$ and $\lambda_5/\lambda_1 = 1.2$ for several choices of the kinematic constant b: (*a*) $\bar{\eta}$ for elongational ($\dot{\varepsilon} > 0$) and biaxial stretching ($\dot{\varepsilon} < 0$) flow ($b = 0$); (*b*) $\bar{\eta}_1$ (——) and $\bar{\eta}_2$ (---) for $b = 0.5$. Note in part (*b*) that $\bar{\eta}_1$ and $\bar{\eta}_2$ are singular at a small negative value of $\dot{\varepsilon}$. Although there are no steady-state experimental data for $b = 0.5$, this singular behavior does not seem realistic.

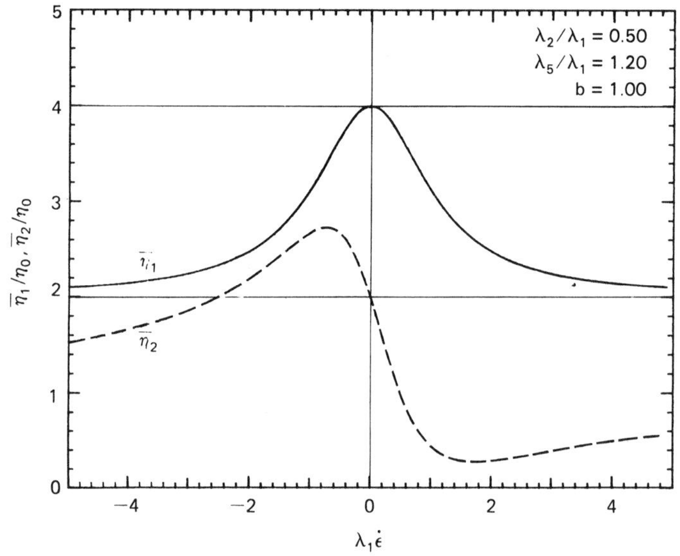

(c)

FIGURE 7.3-4. (c) $\bar{\eta}_1$ (——) and $\bar{\eta}_2$ (---) for planar elongational flow ($b = 1$).

where the zero-shear-rate viscosity η_0, the retardation time λ_2, and the modified mobility parameter a, are given in terms of η_s, η_p, and α as follows:

$$\eta_0 = \eta_s + \eta_p; \quad \lambda_2 = \lambda_1 \frac{\eta_s}{\eta_s + \eta_p}; \quad a = \alpha \frac{(\eta_s + \eta_p)}{\eta_p} \tag{7.3-5}$$

As given in Eq. 7.3-4, the Giesekus model is a special case of the Oldroyd 8-constant model to which a term involving $\{\boldsymbol{\tau}\cdot\boldsymbol{\tau}\}$ is added. Note that if $a = 0$ the convected Jeffreys model is recovered; thus the convected Jeffreys model can be written as the superposition in Eq. 7.3-3 with $\alpha = 0$ (cf. Eqs. 7.5-60). A number of equations used in the literature to which the Giesekus model can be reduced are listed in Table 7.3-4. Thus considerable diversity in the rheological predictions of the model is possible.

The inclusion of the $\{\boldsymbol{\tau}\cdot\boldsymbol{\tau}\}$ term in Eq. 7.3-3 gives material functions that are much more realistic than those obtained for the Oldroyd 8-constant model. For example, large decreases in the viscosity and normal stress coefficients with increasing shear rate are possible. For all $\alpha \neq 0$ or 1 the power-law slope of the viscosity is -1 when $\lambda_2 = 0$; this is unrealistically steep. However, by adding a small retardation term (e.g., $\lambda_2/\lambda_1 = 10^{-3}$), the value of $d \log \eta / d \log \dot{\gamma}$ can be kept larger than -1 so that the magnitude of the shear stress is always increasing with increasing shear rate. The second normal stress coefficient is nonzero and can be varied in size relative to the first normal stress coefficient; for example $\Psi_{2,0} = -(\alpha/2)\Psi_{1,0}$. Provided $\alpha \neq 0$ the elongational viscosity is bounded and reaches a constant value at large strain rates. Graphs of several steady and transient, shear and shearfree flow material functions are shown in Figs. 7.3-5 through 8. In general we must require $0 < \alpha < 1/2$ for realistic properties.[7]

[7] In §7.6 we present a multimode Giesekus model which is a superposition of contributions to the stress tensor, each given by an equation like Eq. 7.3-3 with $\eta_s = 0$. There it is seen that the multimode model is capable of nearly quantitative fits of η, Ψ_1, η^+, and $\bar{\eta}^+$ for low-density polyethylene (cf. Figs. 7.6-1 through 3).

TABLE 7.3-4

Models Included in the Giesekus Model (Eq. 7.3-3 or 4)
(see also Eqs. 13.7-10 and 11)

Name of Model	Values of Constants $\lambda_1 \geq 0$	λ_2	$0 \leq \alpha \leq 1$	Remarks
Newtonian	0, λ	0, λ	0	Eq. 1.2-2
Second-order fluid	0	$\lambda_2 < 0$		Eq. 6.2-1
Convected Maxwell		0	0	Eq. 7.2-1 (with $\lambda_2 = 0$)
Convected Jeffreys		$\lambda_2 > 0$	0	Eq. 7.2-1
"Leonov-like"		0	1/2	Reduces to Leonov[a] model for steady shear and shearfree flows
"Corotational Maxwell-like"		0	1	Reduces to corotational Maxwell model[b] for shearfree flow

[a] A. I. Leonov, *Rheol. Acta*, **15**, 85–98 (1976); A. I. Leonov, E. H. Lipkina, E. D. Paskhin, and A. N. Prokunin, *Rheol. Acta*, **15**, 411–426 (1976).
[b] S. Zaremba, *Bull. Int. Acad. Sci. Cracovie*, 594–614 (1903); 614–621 (1903); H. Fromm, *Z. Angew. Math. Mech.*, **25/27**, 146–150 (1947); **28**, 43–54 (1948); and T. W. DeWitt, *J. Appl. Phys.*, **26**, 889–894 (1955). The corotational Maxwell model is sometimes called the ZFD model.

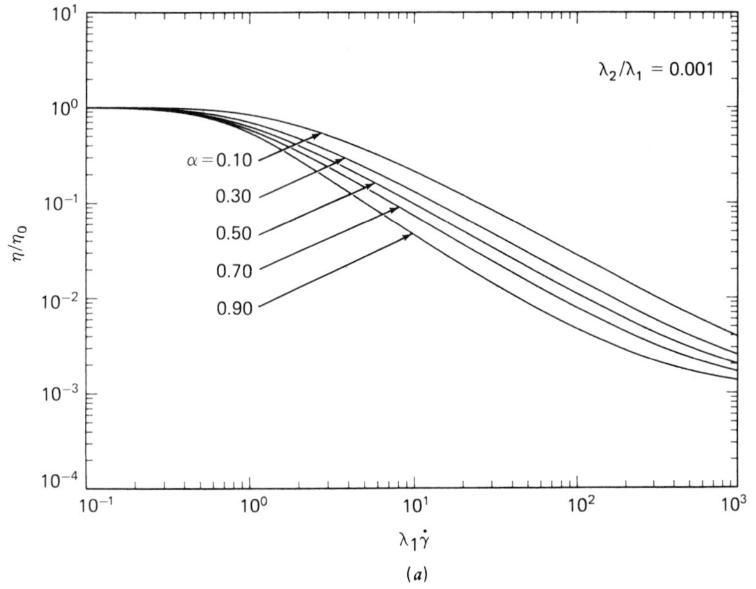

FIGURE 7.3-5. Dimensionless steady shear flow material functions for the Giesekus model with $\lambda_2/\lambda_1 = 10^{-3}$ and various values of α: (*a*) viscosity.

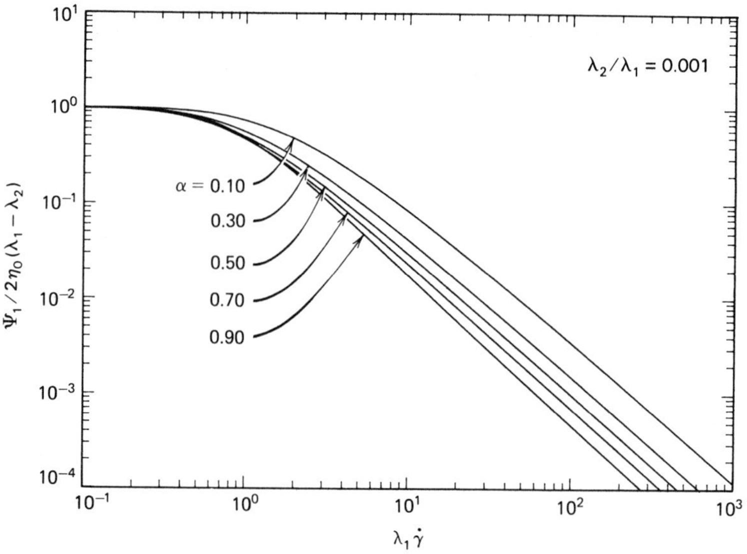

(*b*)

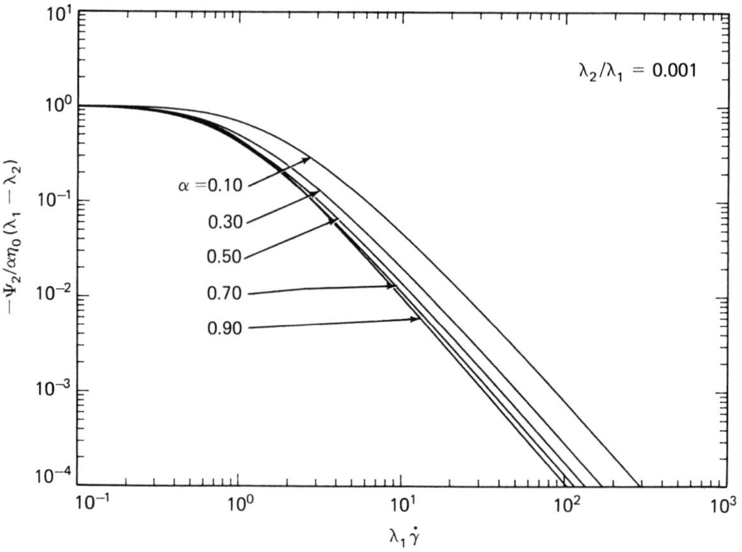

(*c*)

FIGURE 7.3-5. (*b*) first normal stress coefficient, (*c*) second normal stress coefficient.

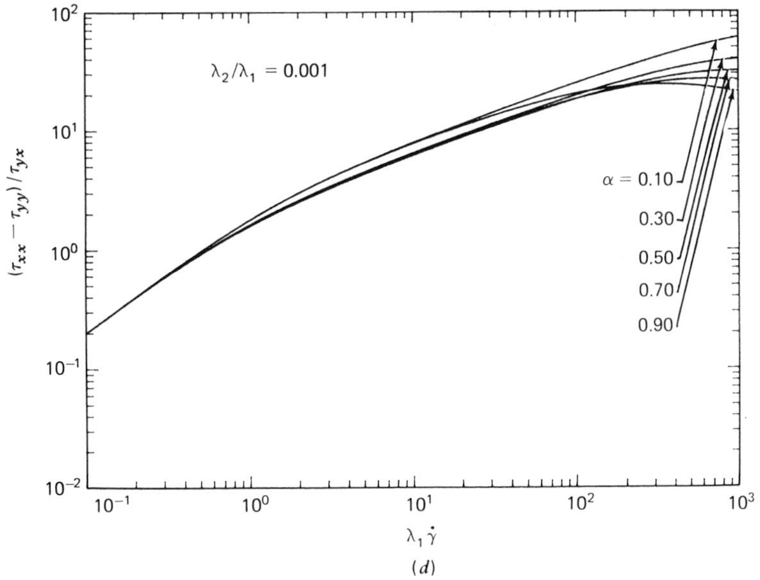

(*d*)

FIGURE 7.3-5. (*d*) stress ratio. The shear thinning of the viscometric functions is in good qualitative agreement with experimental data on polymer melts and solutions. The maximum in the stress ratio at high shear rates is probably unrealistic (cf. Fig. 3.3-8).

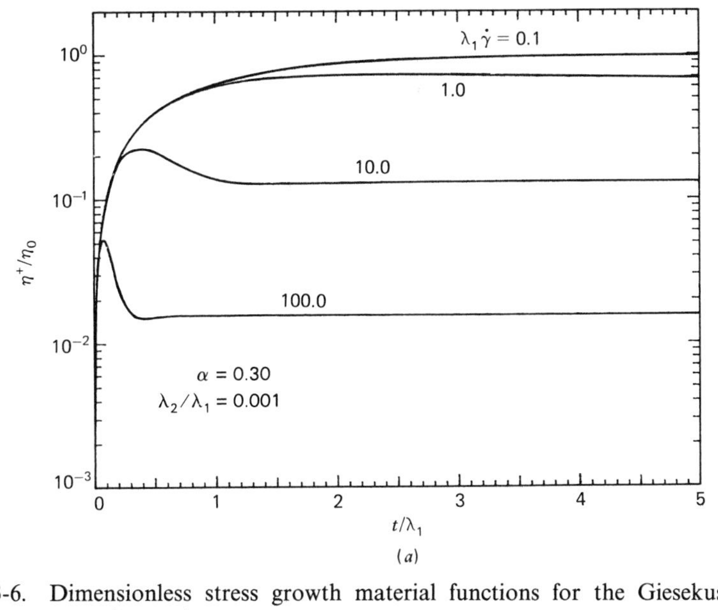

(*a*)

FIGURE 7.3-6. Dimensionless stress growth material functions for the Giesekus model with $\alpha = 0.3$ and $\lambda_2/\lambda_1 = 10^{-3}$. (*a*) η^+.

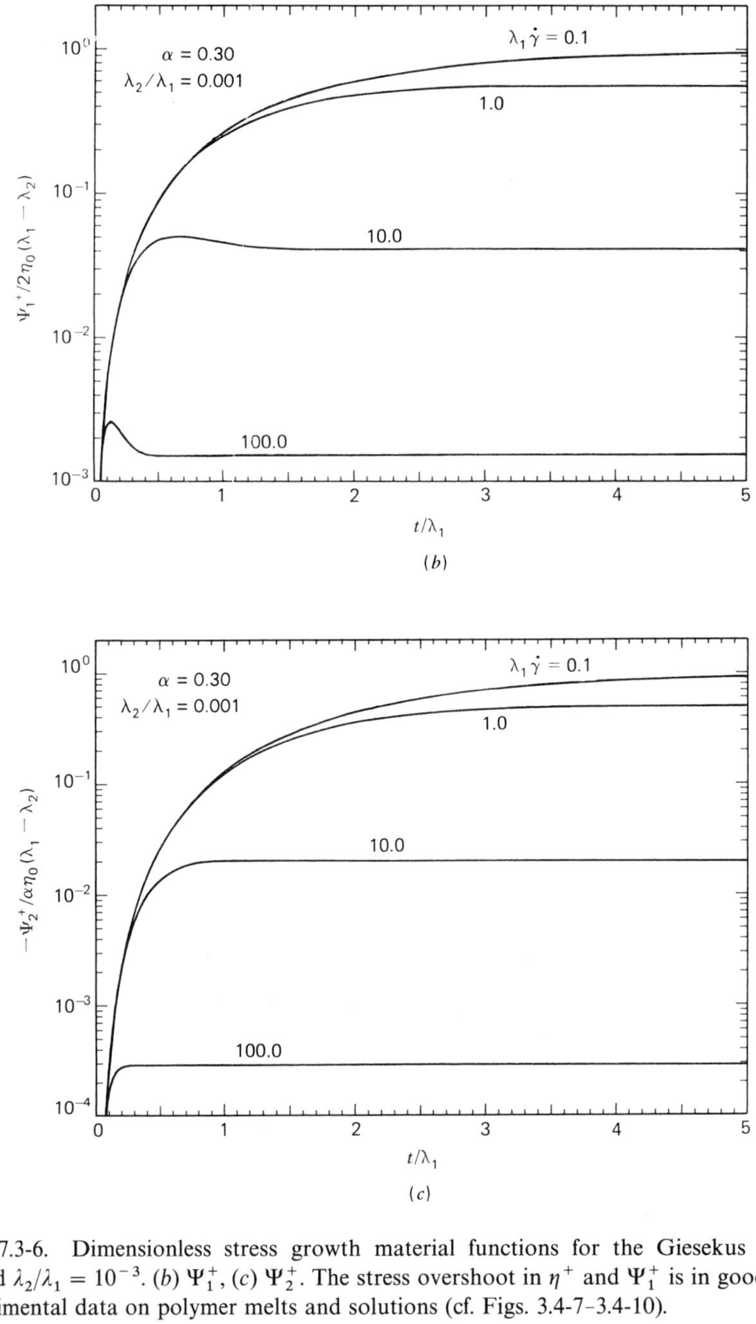

FIGURE 7.3-6. Dimensionless stress growth material functions for the Giesekus model with $\alpha = 0.3$ and $\lambda_2/\lambda_1 = 10^{-3}$. (*b*) Ψ_1^+, (*c*) Ψ_2^+. The stress overshoot in η^+ and Ψ_1^+ is in good agreement with experimental data on polymer melts and solutions (cf. Figs. 3.4-7–3.4-10).

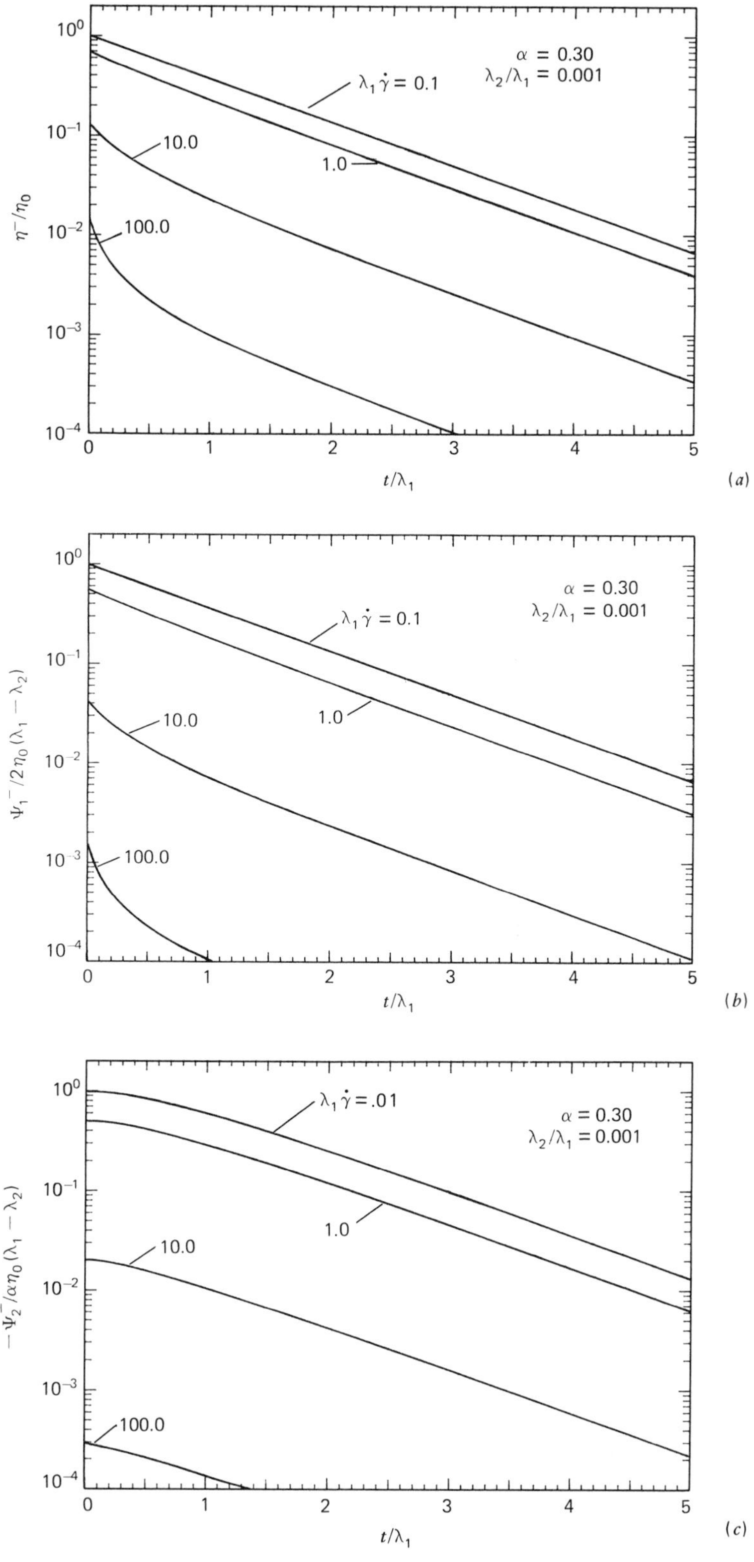

FIGURE 7.3-7. Dimensionless stress relaxation material functions for the Giesekus model with $\alpha = 0.3$ and $\lambda_2/\lambda_1 = 10^{-3}$. The increase in rate of relaxation with increasing shear rate for η^- and Ψ_1^- is in agreement with experimental data for polymer melts and solutions (cf. Figs. 3.4-11–3.4-13).

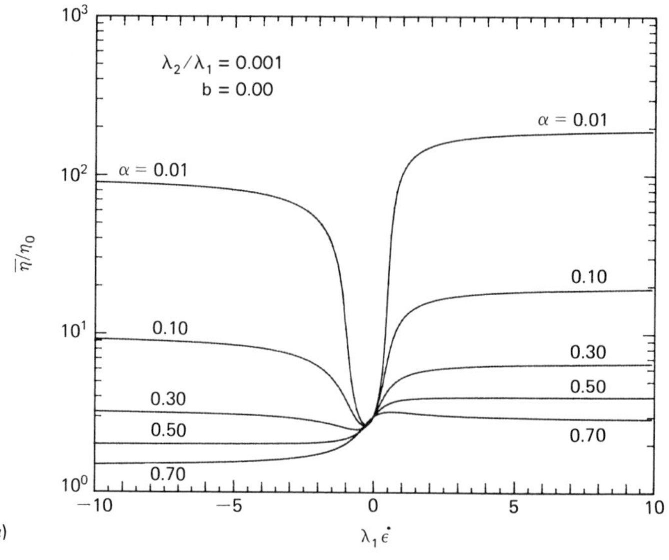

$\lambda_2/\lambda_1 = 0.001$
b = 0.00
α = 0.01
0.10
0.30
0.50
0.70
$\overline{\eta}/\eta_0$
$\lambda_1\dot{\epsilon}$

(a)

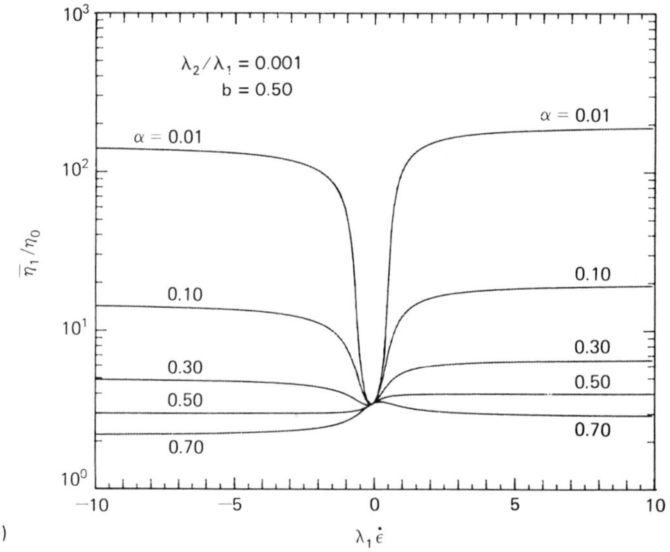

$\lambda_2/\lambda_1 = 0.001$
b = 0.50
α = 0.01
0.10
0.30
0.50
0.70
$\overline{\eta}_1/\eta_0$
$\lambda_1\dot{\epsilon}$

(b)

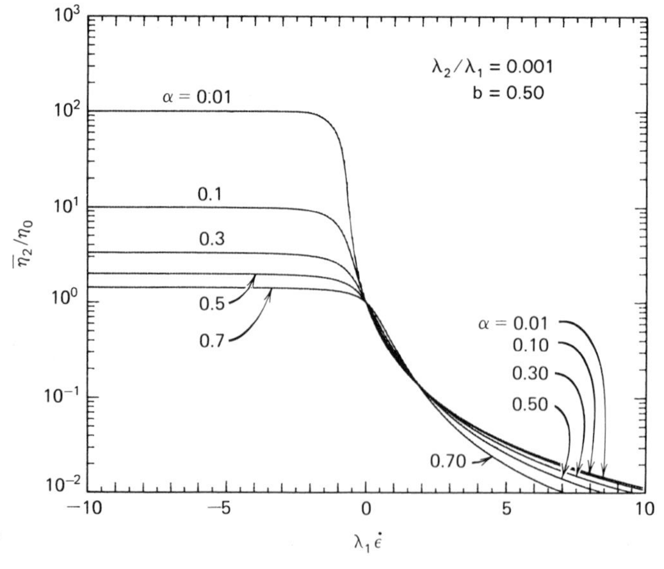

$\lambda_2/\lambda_1 = 0.001$
b = 0.50
α = 0.01
0.1
0.3
0.5
0.7
α = 0.01
0.10
0.30
0.50
0.70
$\overline{\eta}_2/\eta_0$
$\lambda_1\dot{\epsilon}$

(b)

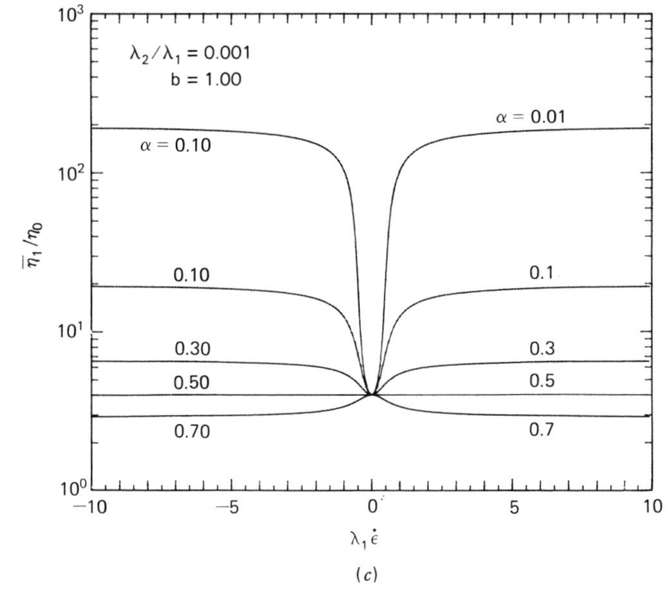

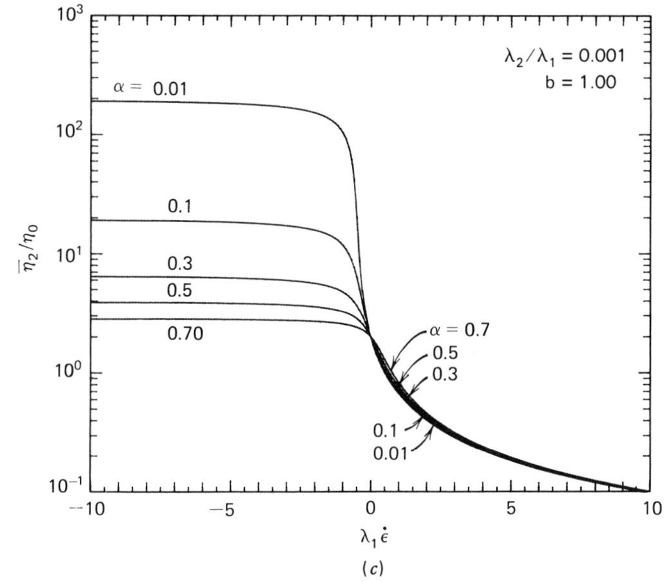

FIGURE 7.3-8. Dimensionless shearfree flow material functions for the Giesekus model with $\lambda_2/\lambda_1 = 10^{-3}$ and various values of α for several choices of the kinematic parameter b: (*a*) $\bar{\eta}$ for elongational ($\dot{\varepsilon} > 0$) and biaxial elongational ($\dot{\varepsilon} < 0$) flow for $b = 0$; (*b*) $\bar{\eta}_1$ and $\bar{\eta}_2$ for $b = 0.5$; and (*c*) $\bar{\eta}_1$ and $\bar{\eta}_2$ for planar elongational flow where $b = 1$.

TABLE 7.3-5

Material Functions for the Giesekus Model (Eq. 7.3-3)

Steady shear flow:

$$\frac{\eta}{\eta_0} = \frac{\lambda_2}{\lambda_1} + \left(1 - \frac{\lambda_2}{\lambda_1}\right)\frac{(1-f)^2}{1+(1-2\alpha)f} \tag{A}$$

$$\frac{\Psi_1}{2\eta_0(\lambda_1 - \lambda_2)} = \frac{f(1-\alpha f)}{(\lambda_1\dot{\gamma})^2\alpha(1-f)} \tag{B}$$

$$\frac{\Psi_2}{\eta_0(\lambda_1 - \lambda_2)} = \frac{-f}{(\lambda_1\dot{\gamma})^2} \tag{C}$$

where

$$f = \frac{1-\chi}{1+(1-2\alpha)\chi}; \quad \chi^2 = \frac{(1+16\alpha(1-\alpha)(\lambda_1\dot{\gamma})^2)^{1/2} - 1}{8\alpha(1-\alpha)(\lambda_1\dot{\gamma})^2}$$

For $\lambda_2 = 0$ and $\alpha \neq 0, 1$ the following asymptotic formulas are available:

$$\eta \sim \sqrt{(1-\alpha)/\alpha}\,\frac{\eta_0}{\lambda_1\dot{\gamma}} \qquad (\dot{\gamma} \to \infty) \tag{D}$$

$$\Psi_1 \sim \frac{\sqrt{2}}{\alpha}(\alpha(1-\alpha))^{1/4}\frac{\eta_0\lambda_1}{(\lambda_1\dot{\gamma})^{3/2}} \qquad (\dot{\gamma} \to \infty) \tag{E}$$

$$\Psi_2 \sim -\frac{\eta_0\lambda_1}{(\lambda_1\dot{\gamma})^2} \qquad (\dot{\gamma} \to \infty) \tag{F}$$

Small-amplitude oscillatory shear flow:

$$\frac{\eta'}{\eta_0} = \frac{1+\lambda_1\lambda_2\omega^2}{1+\lambda_1^2\omega^2} \tag{G}$$

$$\frac{\eta''}{\eta_0\omega} = \frac{(\lambda_1 - \lambda_2)}{1+\lambda_1^2\omega^2} \tag{H}$$

Steady elongational flow ($\alpha \neq 0$):

$$\frac{\bar{\eta}}{3\eta_0} = \frac{\lambda_2}{\lambda_1} + \left(1 - \frac{\lambda_2}{\lambda_1}\right)\frac{1}{6\alpha}\left[3 + \frac{1}{\lambda_1\dot{\varepsilon}}\{[1 - 4(1-2\alpha)\lambda_1\dot{\varepsilon} + 4\lambda_1^2\dot{\varepsilon}^2]^{1/2} - [1 + 2(1-2\alpha)\lambda_1\dot{\varepsilon} + \lambda_1^2\dot{\varepsilon}^2]^{1/2}\}\right] \tag{I}$$

For $\lambda_2 = 0$ and $\alpha \neq 0$ the following asymptotic formula holds:

$$\bar{\eta} \sim \frac{2\eta_0}{\alpha}, \quad (\dot{\varepsilon} \to \infty); \quad \bar{\eta} \sim \frac{\eta_0}{\alpha}(\dot{\varepsilon} \to -\infty) \tag{J}$$

The nonlinear stress term does make it more difficult to obtain analytical solutions for the material functions; a few of these are given in Table 7.3-5. For the transient material functions, numerical methods must be used. As a convenience, we provide in Table C.2 of Appendix C the governing equations to be used in finding these material functions.

§7.4 FLOW PROBLEMS IN ONE SPATIAL VARIABLE

In this section we illustrate the solution of one dimensional flow problems with the differential constitutive equations described in the previous two sections. The first two examples illustrate time-dependent effects in unsteady shearing flows, whereas the third example deals with elongational flow behavior.

EXAMPLE 7.4-1 The Rayleigh Problem for a Convected Jeffreys Fluid[1]

A semi-infinite body of viscoelastic liquid with constant density ρ is bounded at $y = 0$ by a flat solid surface (the xz-plane). Before $t = 0$ the fluid and the solid surface are at rest; after $t = 0$ the surface moves with constant velocity V in the positive x-direction. Find the velocity distribution if the fluid is described by a convected Jeffreys model. (See Problem 1B.6 for the analogous Newtonian fluid problem, and Problem 4C.3 for the power-law model.)

SOLUTION We postulate a solution of the form $v_x = v_x(y, t)$, that is, the flow is a nonhomogeneous shear flow with the shear rate a function of y and t. Then the equation of continuity is satisfied identically and the three components of the equation of motion are

$$\rho \frac{\partial v_x}{\partial t} = -\frac{\partial \tau_{yx}}{\partial y} \tag{7.4-1}$$

$$0 = -\frac{\partial p}{\partial y} - \frac{\partial \tau_{yy}}{\partial y} \tag{7.4-2}$$

$$0 = -\frac{\partial p}{\partial z} + \rho g \tag{7.4-3}$$

Since the shear rate depends on y and t we also expect $\tau_{ij} = \tau_{ij}(y, t)$. The $\{\boldsymbol{v}\cdot\nabla\boldsymbol{\tau}\}$ and $\{\boldsymbol{v}\cdot\nabla\boldsymbol{\gamma}_{(1)}\}$ contributions to $\boldsymbol{\tau}_{(1)}$ and $\boldsymbol{\gamma}_{(2)}$ are both zero since the velocity is perpendicular to the direction of the gradients. This means that the convected Jeffreys model will be the same as for homogeneous shear flow, which is given in Eqs. 7.2-3 through 6, with $\dot{\gamma}_{yx}(y, t) = \partial v_x/\partial y$ and with all d/dt replaced by $\partial/\partial t$.

From Eq. 7.2-4 we obtain immediately

$$\tau_{yy} = A(y)e^{-t/\lambda_1} \tag{7.4-4}$$

where $A(y)$ is an arbitrary function of y. But τ_{yy} is zero for $t \leq 0$, so $A(y)$ must be zero. Then the τ_{yy} term in Eq. 7.2-6 can be omitted (as it was for homogeneous flows) and Eqs. 7.4-1 and 7.2-6 can be combined to give:

$$\rho\left(\frac{\partial v_x}{\partial t} + \lambda_1 \frac{\partial^2 v_x}{\partial t^2}\right) = \eta_0\left(\frac{\partial^2 v_x}{\partial y^2} + \lambda_2 \frac{\partial}{\partial t}\frac{\partial^2 v_x}{\partial y^2}\right) \tag{7.4-5}$$

[1] This example is patterned after R. I. Tanner, *Z. Angew. Math. Phys.*, **13**, 573–580 (1962). Y. Mochimaru, *J. Non-Newtonian Fluid Mech.*, **12**, 135–152 (1983), has done start-up of steady shear flow between parallel plates.

It is convenient to introduce the following dimensionless quantities:

$$\phi = \frac{v_x}{V} \qquad T = \frac{t}{\lambda_1}$$
$$Y = \left(\frac{\rho}{\eta_0 \lambda_1}\right)^{1/2} y \qquad \Lambda = \frac{\lambda_2}{\lambda_1} \tag{7.4-6}$$

Note that y is scaled by the viscous diffusion penetration length for time λ_1. The mathematical problem that we have to solve then consists of the linear partial differential equation

$$\frac{\partial \phi}{\partial T} + \frac{\partial^2 \phi}{\partial T^2} = \frac{\partial^2 \phi}{\partial Y^2} + \Lambda \frac{\partial^3 \phi}{\partial T \partial Y^2} \tag{7.4-7}$$

along with the following boundary and initial conditions:

$$\phi = H(T), \qquad \text{at } Y = 0 \tag{7.4-8}$$

$$\phi \to 0, \qquad \text{as } Y \to \infty \tag{7.4-9}$$

$$\frac{\partial \phi}{\partial T} = \frac{\partial^2 \phi}{\partial T^2} = 0, \qquad \text{for } T \leq 0 \tag{7.4-10}$$

where $H(T)$ is the Heaviside unit step function.

Tanner solved the problem by taking the Laplace transform of Eq. 7.4-7 and using the initial condition; this gives

$$(s + s^2)\bar{\phi} = (1 + \Lambda s) \frac{d^2 \bar{\phi}}{dY^2} \tag{7.4-11}$$

where s is the Laplace transform variable. This ordinary differential equation can be solved with the boundary conditions that $\bar{\phi}$ approaches zero as $Y \to \infty$, and $\bar{\phi} = 1/s$ at $Y = 0$. The result is

$$\bar{\phi} = \frac{1}{s} \exp\left(-\left(\frac{s(1+s)}{1 + \Lambda s}\right)^{1/2} Y\right) \tag{7.4-12}$$

Tanner inverted Eq. 7.4-12 by performing a contour integral in the complex plane to obtain the solution

$$\phi(Y, T) = \frac{1}{2} + \frac{1}{\pi} \int_0^\infty \exp\left[-\left(\frac{u}{2}\right)^{1/2} M(\cos\theta - \sin\theta) Y\right]$$
$$\cdot \sin\left[uT - \left(\frac{u}{2}\right)^{1/2} M(\cos\theta + \sin\theta) Y\right] \frac{du}{u} \tag{7.4-13}$$

in which $M(u)$ and $\theta(u)$ are defined by:

$$M(u) = \left(\frac{1 + u^2}{1 + \Lambda^2 u^2}\right)^{1/4}; \quad \theta(u) = \frac{1}{2}(\arctan u - \arctan \Lambda u) \tag{7.4-14}$$

We see that for this very idealized problem and simplified constitutive equation, an analytical solution can be obtained. For the special case of $\Lambda = 1$ (both time constants equal), in which case the convected

Jeffreys model reduces to the Newtonian fluid, Tanner showed that the solution simplifies correctly to the error function expression given in Problem 1B.6 for the Newtonian fluid. For the limit $\Lambda = 0$ (the convected Maxwell model), Tanner showed that the solution reduced to

$$\phi(Y, T) = e^{-Y/2}H(T - Y) + \frac{Y}{2}\int_0^T \frac{e^{-u/2}}{\sqrt{u^2 - Y^2}} I_1\left(\tfrac{1}{2}\sqrt{u^2 - Y^2}\right)H(u - Y)du \qquad (7.4\text{-}15)$$

where I_1 is the modified Bessel function of the first kind. Thus the convected Maxwell model is seen to give a wave that propagates away from the plate at constant speed $\sqrt{\eta_0/\rho\lambda_1}$.

To show how the character of the solution changes as Λ goes from 0 (the "damped wave equation") to 1 (the "diffusion equation"), we have performed the integral in Eq. 7.4-13 numerically. Figure 7.4-1 illustrates the results of these calculations; the velocity is shown there as a function of position for different times and Λ. Note that for a given time t, small velocity disturbances are felt at larger distances y from the moving plate for larger values of Λ. That is, the retardation term which adds Newtonian character to the convected Jeffreys fluid produces more diffusivelike behavior, as expected. For the smallest value of $\Lambda = 0.01$ in Fig. 7.4-1, the wave front associated with elastic wave propagation is seen clearly in the rapid change in velocity over small distances (the flat regions). Comparison of the results for different values of Λ in this figure shows how the wave front is diffused as Λ increases.

EXAMPLE 7.4-2 Tube Flow of a White–Metzner Fluid with a Pulsatile Pressure Gradient[2,3]

Consider the unsteady flow of a viscoelastic fluid in a horizontal circular tube of radius R. Let the pressure gradient vary sinusoidally about some mean value $\Delta p_0/L$

$$\frac{\Delta p}{L} = \frac{\Delta p_0}{L}(1 + \varepsilon\,\mathcal{R}e\{e^{i\omega t}\}) \qquad (7.4\text{-}16)$$

where Δp_0 is the time-averaged pressure drop over a length L of tube, ε is the *small* amplitude of the pressure gradient variation, and ω is the frequency of the fluctuations in the pressure drop. Assume that inertial effects are negligible. We wish to find the mean volume flow rate $\langle Q \rangle$ as a function of the mean pressure gradient and frequency of fluctuation. Of particular interest is the flow enhancement relative to the steady-state volume flow rate Q_0 given by

$$I = \frac{\langle Q \rangle - Q_0}{Q_0} \qquad (7.4\text{-}17)$$

Since the volume flow rate for the base problem of steady tube flow can be given exactly by the White–Metzner model, we will use this model to describe the viscoelastic flow. Of course, for steady tube flow the White–Metzner model gives the same velocity field as the generalized Newtonian fluid.

[2] This example is patterned after J. M. Davies, S. Bhumiratana, and R. B. Bird, *J. Non-Newtonian Fluid Mech.*, **3**, 237–259 (1977/78) and N. Phan-Thien, *J. Non-Newtonian Fluid Mech.*, **4**, 167–176 (1978). Other discussions of this problem have been given by K. Walters and P. Townsend in S. Onogi, ed., *Proceedings of the Fifth International Congress on Rheology*, Vol. 4, University of Tokyo Press (1970), pp. 471–483, and H. A. Barnes, P. Townsend, and K. Walters, *Nature*, **224**, 585–587 (1969); *Rheol. Acta*, **10**, 517–527 (1971). A review article including the pulsatile flow problem as well as flow in a corrugated pipe, flow in a curved pipe, and flow in a noncircular pipe has been prepared by K. Walters in W. R. Schowalter, ed., *Progress in Heat and Mass Transfer*, Vol. 5, Pergamon Press, Elmsford, NY (1972), pp. 217–231.

[3] See Example 4.2-6 for pulsatile tube flow of a power-law fluid.

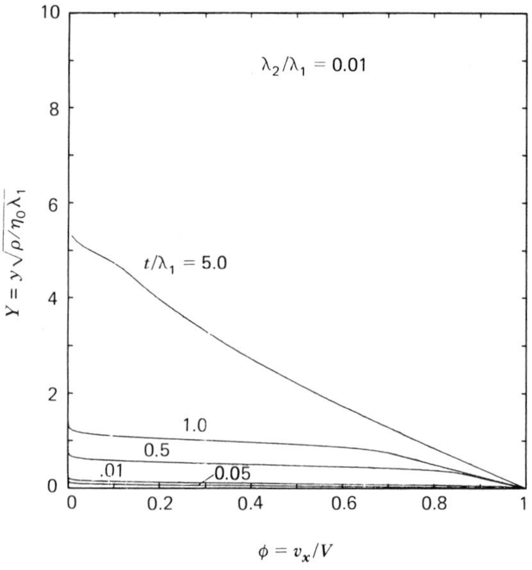

(a)

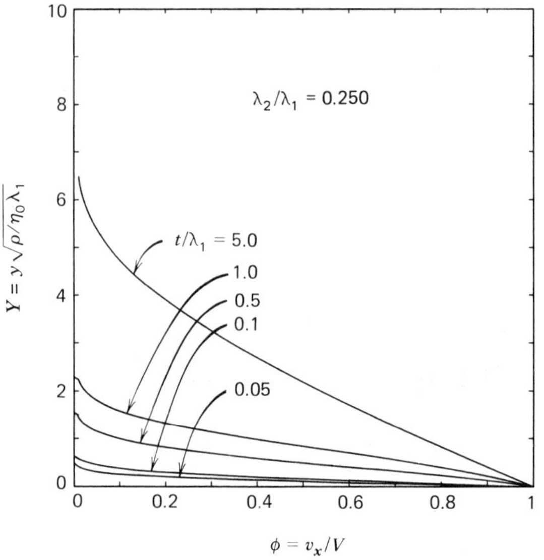

(b)

FIGURE 7.4-1. Calculated velocity profiles for the Rayleigh problem for a convected Jeffreys fluid. The curves show the velocity variation with position for a series of times $T = t/\lambda_1$. The influence of $\Lambda = \lambda_2/\lambda_1$ is illustrated by comparison of parts (*a*) through (*d*). As Λ decreases, the fluid becomes more

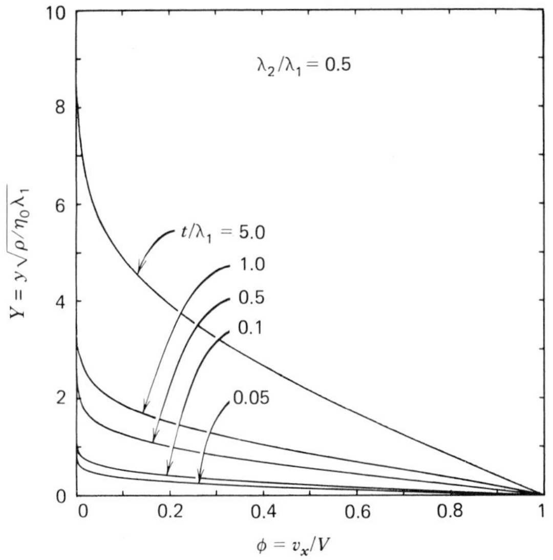

(c)

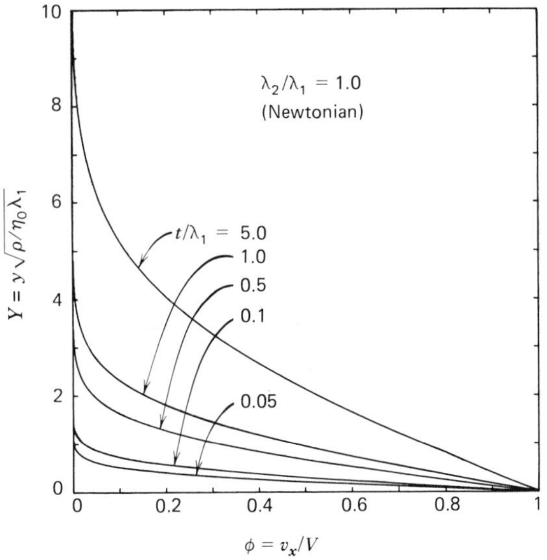

(d)

and more elastic (i.e., more solidlike). In the Newtonian limit ($\Lambda = 1$), the parameter λ_1 is an arbitrary constant.

SOLUTION It seems reasonable to assume that $v_z = v_z(r, t)$ and $v_r = v_\theta = 0$. For most polymer solutions and melts, the viscosity is so large that the inertial terms in the equation of motion can be neglected. The axial component of the equation of motion is then

$$0 = \frac{\Delta p_0}{L}(1 + \varepsilon\, \mathcal{R}e\{e^{i\omega t}\}) - \frac{1}{r}\frac{\partial}{\partial r}(r\tau_{rz}) \tag{7.4-18}$$

The shear stress is obtained by integrating with respect to r

$$\tau_{rz} = \frac{\Delta p_0 r}{2L}(1 + \varepsilon\, \mathcal{R}e\{e^{i\omega t}\}) \tag{7.4-19}$$

where we have required τ_{rz} to be finite at $r = 0$. This result must be combined with the constitutive equation to obtain an equation for the velocity field. The rz-component of the White–Metzner model gives (cf Eqs. 7.2-3 through 6)

$$\left(1 + \frac{\eta(\dot{\gamma})}{G}\frac{\partial}{\partial t}\right)\tau_{rz} = -\eta(\dot{\gamma})\dot{\gamma}_{rz} = \eta(\dot{\gamma})\dot{\gamma} \tag{7.4-20}$$

In the second equality of Eq. 7.4-20, we have used the fact that since the amplitude of the pressure gradient oscillation ε is small and inertial effects are negligible, the axial velocity is expected to oscillate as shown in Fig. 7.4-2. This means that everywhere in the tube $\dot{\gamma}_{rz}$ will be negative and equal to $-\dot{\gamma}$. When Eqs. 7.4-19 and 20 are combined we obtain

$$\frac{\Delta p_0 r}{2L}\left[1 + \varepsilon\,(\mathcal{R}e\{e^{i\omega t}\} + \frac{\eta(\dot{\gamma})}{G}\mathcal{R}e\{i\omega e^{i\omega t}\})\right] = \eta(\dot{\gamma})\dot{\gamma} \tag{7.4-21}$$

Solution of this algebraic equation for $\dot{\gamma}(r, t)$ is sufficient for obtaining the flow enhancement I, which is shown next.

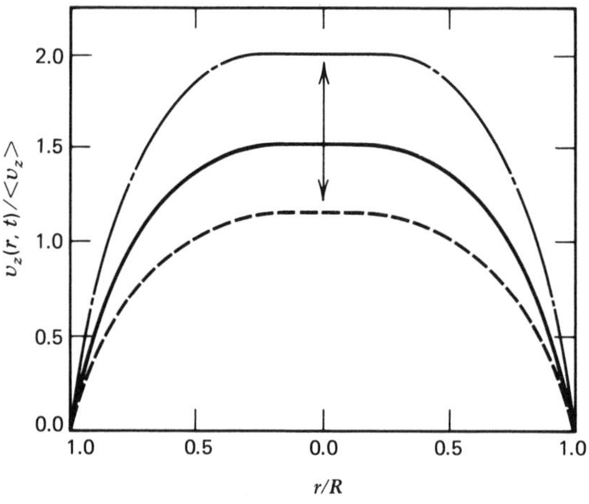

FIGURE 7.4-2. Schematic diagram illustrating the anticipated velocity profile $v_z(r, t)$ in pulsatile tube flow. The axial velocity at each time t is normalized by the average velocity $\langle v_z \rangle$ at that time. The solid curve (——) is the velocity at the average pressure drop Δp_0. The other two curves give the velocity for $\varepsilon = 0.1$: (—·—) corresponds to the maximum pressure drop, 1.1 Δp_0 and (- - -) corresponds to the minimum pressure drop 0.9 Δp_0.

The presence of the small parameter ε suggests that we develop a series solution for $\dot{\gamma}(r, t)$ of the form

$$\dot{\gamma}(r, t) = \dot{\gamma}_0 + \varepsilon[\dot{\gamma}_{10} + \mathscr{Re}\{\dot{\gamma}_{11}e^{i\omega t}\}] + \varepsilon^2[\dot{\gamma}_{20} + \mathscr{Re}\{\dot{\gamma}_{21}e^{i\omega t}\} + \mathscr{Re}\{\dot{\gamma}_{22}e^{2i\omega t}\}] + \cdots \quad (7.4\text{-}22)$$

in which the $\dot{\gamma}_{m0}(r)$ are real, the other $\dot{\gamma}_{mn}(r)$ are complex, and $\dot{\gamma}_0(r)$ is the shear rate distribution in steady tube flow. We now show that only $\dot{\gamma}_0$ and $\dot{\gamma}_{20}$ need to be found.

The time-averaged volume flow rate is given by

$$\langle Q \rangle = \frac{\omega}{2\pi}\int_0^{2\pi/\omega} 2\pi \int_0^R v_z(r, t)rdrdt \quad (7.4\text{-}23)$$

If this expression is integrated by parts and the no-slip boundary condition at the tube wall applied, then the result is

$$\langle Q \rangle = \pi \int_0^R \langle \dot{\gamma} \rangle r^2 dr \quad (7.4\text{-}24)$$

in which $\langle \cdot \rangle = (\omega/2\pi)\int_0^{2\pi/\omega}(\cdot)dt$. The steady-state flow rate corresponding to $\varepsilon = 0$ is

$$Q_0 = \pi \int_0^R \dot{\gamma}_0 r^2 dr \quad (7.4\text{-}25)$$

The enhancement is obtained by combining Eqs. 7.4-17, 22, 24, and 25:

$$\begin{aligned} I &= \frac{\int_0^R[\varepsilon\dot{\gamma}_{10} + \varepsilon^2\dot{\gamma}_{20} + \cdots]r^2dr}{\int_0^R\dot{\gamma}_0 r^2dr} \\ &= \frac{\varepsilon^2 \int_0^R\dot{\gamma}_{20}r^2dr}{\int_0^R\dot{\gamma}_0 r^2dr} + O(\varepsilon^4) \end{aligned} \quad (7.4\text{-}26)$$

In writing down the last line of Eq. 7.4-26 we have set $\dot{\gamma}_{10} = 0$, since the volume flow rate must be an even function of ε. This will be verified in the subsequent development. Thus finding I requires only that we determine $\dot{\gamma}_0$ and $\dot{\gamma}_{20}$ from Eqs. 7.4-21 and 22.

We now return to the problem of solving Eq. 7.4-21. To develop the series solution to this equation, we need in addition to Eq. 7.4-22 an expansion for η in powers of ε. A Taylor expansion in ε gives

$$\begin{aligned} \eta(\dot{\gamma}) = {} & \eta(\dot{\gamma}_0) + \varepsilon\eta_1(\dot{\gamma}_0)[\dot{\gamma}_{10} + \mathscr{Re}\{\dot{\gamma}_{11}e^{i\omega t}\}] \\ & + \frac{\varepsilon^2}{2}\Big[\eta_2(\dot{\gamma}_0)[\dot{\gamma}_{10} + \mathscr{Re}\{\dot{\gamma}_{11}e^{i\omega t}\}]^2 + 2\eta_1(\dot{\gamma}_0)[\dot{\gamma}_{20} \\ & + \mathscr{Re}\{\dot{\gamma}_{21}e^{i\omega t}\} + \mathscr{Re}\{\dot{\gamma}_{22}e^{2i\omega t}\}]\Big] + \cdots \end{aligned} \quad (7.4\text{-}27)$$

in which $\eta_j = d^j\eta/d\dot{\gamma}^j$. Also, by multiplying the expansions in Eqs. 7.4-22 and 27 we have

$$\begin{aligned} \eta(\dot{\gamma})\dot{\gamma} = {} & \eta(\dot{\gamma}_0)\dot{\gamma}_0 + \varepsilon[\eta_1(\dot{\gamma}_0)\dot{\gamma}_0 + \eta(\dot{\gamma}_0)][\dot{\gamma}_{10} + \mathscr{Re}\{\dot{\gamma}_{11}e^{i\omega t}\}] \\ & + \varepsilon^2\{(\eta(\dot{\gamma}_0)\dot{\gamma}_{20} + [\eta_1(\dot{\gamma}_0) + \tfrac{1}{2}\eta_2(\dot{\gamma}_0)\dot{\gamma}_0][\dot{\gamma}_{10}^2 + \tfrac{1}{2}\dot{\gamma}_{11}\bar{\dot{\gamma}}_{11}] \\ & + \eta_1(\dot{\gamma}_0)\dot{\gamma}_0\dot{\gamma}_{20}) \\ & + \text{terms involving } e^{i\omega t} \text{ and } e^{2i\omega t}\} + \cdots \end{aligned} \quad (7.4\text{-}28)$$

Since we seek only $\dot{\gamma}_0$ and $\dot{\gamma}_{20}$ we do not need the oscillatory terms at order ε^2. In obtaining the above we have made use of Eq. 4.2-56a, and we denote the complex conjugate by an overbar.

We now combine Eqs. 7.4-21, 27, and 28. Equating terms of equal order in ε and of the same frequency gives the following results:

$O(\varepsilon^0)$:
$$\frac{\Delta p_0 r}{2L} = \eta(\dot{\gamma}_0)\dot{\gamma}_0 \tag{7.4-29}$$

$O(\varepsilon)$:

Constant terms
$$0 = [\eta_1(\dot{\gamma}_0)\dot{\gamma}_0 + \eta(\dot{\gamma}_0)]\dot{\gamma}_{10} \tag{7.4-30}$$

$e^{i\omega t}$ *terms*
$$\frac{\Delta p_0 r}{2L}\left[1 + \frac{i\omega\eta(\dot{\gamma}_0)}{G}\right] = [\eta_1(\dot{\gamma}_0)\dot{\gamma}_0 + \eta(\dot{\gamma}_0)]\dot{\gamma}_{11} \tag{7.4-31}$$

$O(\varepsilon^2)$:

Constant terms
$$\eta(\dot{\gamma}_0)\dot{\gamma}_0\frac{\eta_1(\dot{\gamma}_0)}{2G}\mathscr{Re}\{-i\omega\dot{\gamma}_{11}\} = \eta(\dot{\gamma}_0)\dot{\gamma}_{20} + \eta_1(\dot{\gamma}_0)\dot{\gamma}_0\dot{\gamma}_{20} + [\eta_1(\dot{\gamma}_0) + \tfrac{1}{2}\eta_2(\dot{\gamma}_0)\dot{\gamma}_0][\dot{\gamma}_{10}^2 + \tfrac{1}{2}\dot{\gamma}_{11}\bar{\dot{\gamma}}_{11}] \tag{7.4-32}$$

There is no need to consider the other terms at $O(\varepsilon^2)$, since the constant terms are sufficient to give $\dot{\gamma}_{20}$, and this is all that affects the enhancement I. To find $\dot{\gamma}_{20}$ we note that in general the viscosity function will not be such that the [ ] term in Eq. 7.4-30 is zero. This means $\dot{\gamma}_{10}$ must be zero. Equation 7.4-29 can be combined with Eq. 7.4-31, and $\dot{\gamma}_{11}$ can be solved for explicitly. Then combination of these results for $\dot{\gamma}_{10}$ and $\dot{\gamma}_{11}$ with Eq. 7.4-32 leads to the final result for $\dot{\gamma}_{20}$

$$\dot{\gamma}_{20} = -\frac{(\eta\dot{\gamma}_0)^2}{2(\eta + \eta_1\dot{\gamma}_0)^3}\left\{(\eta_1 + \tfrac{1}{2}\eta_2\dot{\gamma}_0)\left(1 + \left(\frac{\eta}{G}\right)^2\omega^2\right) - \frac{\eta\eta_1\omega^2}{G^2}(\eta + \eta_1\dot{\gamma}_0)\right\} \tag{7.4-33}$$

where it is understood that η, η_1, and η_2 are all evaluated at $\dot{\gamma}_0$ (this same convention is used below in Eq. 7.4-34).

To evaluate the enhancement it is convenient to change the integration variable in Eq. 7.4-26 from r to $\dot{\gamma}_0$ by using Eq. 7.4-29. This results in:

$$\frac{I}{\varepsilon^2} = \frac{\int_0^{\dot{\gamma}_{0w}}\dot{\gamma}_{20}\eta^2[\eta + \eta_1\dot{\gamma}_0]\dot{\gamma}_0^2\,d\dot{\gamma}_0}{\int_0^{\dot{\gamma}_{0w}}\dot{\gamma}_0^3\eta^2[\eta + \eta_1\dot{\gamma}_0]d\dot{\gamma}_0} + \cdots \tag{7.4-34}$$

in which $\dot{\gamma}_{0w}$ is the wall shear rate at steady state which is given by Eq. 7.4-29 written for $r = R$. To examine I/ε^2 we have chosen a Carreau viscosity equation for η given by Eq. 4.1-9 with $a = 2$. With this choice of viscosity expression, I/ε^2 is a function of the following dimensionless quantities:

n	power-law index
$\dfrac{\eta_\infty}{\eta_0 - \eta_\infty}$	dimensionless infinite-shear-rate viscosity
$\Omega = \lambda\omega$	dimensionless frequency
$\dfrac{R\langle\Delta p\rangle\lambda}{L(\eta_0 - \eta_\infty)}$	dimensionless mean pressure gradient

Equation 7.4-34 was then integrated numerically by Romberg's method combined with Neville's algorithm for extrapolating to zero mesh size.[4] In these calculations it is reasonable to take the Carreau time constant λ to be equal to $(\eta_0 - \eta_\infty)/G$. Figure 7.4-3 shows some sample calculations. These suggest that the pulsatile pressure gradient may either increase or decrease the volume rate of flow. They further suggest that there will be a "resonance" effect, with appreciable increase or decrease near a particular pressure gradient.

Figure 7.4-4 contains some sample experimental data of Barnes, Townsend, and Walters for aqueous polyacrylamide solutions. The shapes of the curves are similar to the low-pressure-gradient results for the White–Metzner model. Note that the experimentally observed maximum enhancement occurs at approximately a constant mean pressure gradient, independent of the frequency of oscillation; this observation agrees with the White–Metzner model predictions at low $\Delta p_0/L$. Moreover, the measured values of the enhancement reported in Fig. 7.4-4 fall in the range $1.4 < I/\varepsilon^2 < 6$, so that the magnitude of I is also correctly given by the White–Metzner model over the same frequency range. However, the value of $\Delta p_0/L$ at which the maximum occurs is underestimated somewhat by the White–Metzner model calculations. If we assume the viscosity of the polyacrylamide solution used in the experiments is the same as that in Fig. 3.3-3, then the calculated values for $\Delta p_0/L$ at the maximum are too low by a factor of about four. At high pressure gradients where shear thinning in the viscosity and time constants is important, the White–Metzner model predicts negative flow enhancement, whereas only positive values are seen in the experimental results. Other data by Phan-Thien and Dudek[5] also show strictly positive I values, with I sometimes increasing and sometimes decreasing with frequency.

Comparison of the results presented here for the White–Metzner model with literature results for other constitutive equations shows that the enhancement predictions are very sensitive to the model used. This is illustrated in Fig. 7.4-5 where the enhancement results[6] are given for the Tanner model.[7] The Tanner model is the same as the convected Maxwell model except that the zero-shear-rate viscosity η_0 is replaced by the shear-rate-dependent viscosity $\eta(\dot{\gamma})$. The shapes of the enhancement versus mean pressure gradient curves are more realistic for the Tanner model than for the White–Metzner model; notice that no negative enhancement is predicted at large pressure gradients. Also the $\Delta p_0/L$ at which the maximum I occurs is only underestimated by a factor of two for the Tanner model. However, the magnitude of the enhancement is overestimated by an order of magnitude or more. Definitive experimental data and calculations are both needed to resolve the issue of the mean pressure drop and frequency dependence of the flow enhancement.

EXAMPLE 7.4-3 Fiber Spinning of a White–Metzner Fluid[8]

A viscoelastic fluid is formed into a fiber by extruding it through a small, circular orifice known as a spinneret and winding the resulting filament on a spool (see Fig. 7.4-6). The velocity of the fiber at the wind-up spool is greater than the extrusion velocity, so that the fiber is under tension. In addition to being stretched during the travel between the spinneret and the take-up spool, the polymer may pass through a solidification bath, be cooled, experience solvent evaporation, or be coagulated. Here we want to model the stretching of the polymeric fluid before it is solidified. To keep the analysis

[4] W. H. Press, B. P. Flannery, S. Teukolsky, and W. T. Vetterling, *Numerical Recipes*, Cambridge University Press (1985).

[5] N. Phan-Thien and J. Dudek, *J. Non-Newtonian Fluid Mech.*, **11**, 147–161 (1982).

[6] N. Phan-Thien, *J. Non-Newtonian Fluid Mech.*, **4**, 167–176 (1978).

[7] R. I. Tanner, *ASLE Trans.*, **8**, 179–183 (1965); see Table 7.6-1.

[8] This example is based on R. J. Fisher and M. M. Denn, *AIChE J.*, **22**, 236–246 (1976). In addition to the steady-state analysis presented here, Fisher and Denn also analyze the stability of the fiber spinning. For more information on fiber spinning see A. Ziabicki, *Fundamentals of Fibre Formation*, Wiley, New York (1976); J. R. A. Pearson, *Mechanics of Polymer Processing*, Elsevier, London (1985), Chapt. 15; Z. Tadmor and C. G. Gogos, *Principles of Polymer Processing*, Wiley, New York (1979), Sect. 15.1; M. M. Denn, *Process Modeling*, Longman, New York (1986), Chap. 12, pp. 230–269; M. M. Denn in J. R. A. Pearson and S. M. Richardson, eds., *Computational Analysis of Polymer Processing*, Applied Science, London (1983), Chapter 6, pp. 179–216.

simple, we will ignore the temperature and/or concentration dependence of the rheological properties in the region prior to solidification and assume the process is steady state. We further assume that the viscoelastic fluid can be described by the White–Metzner model with $\eta = m\dot{\gamma}^{n-1}$.

Let us choose a cylindrical coordinate system with origin at the position where the axial velocity profile can first be assumed independent of radial position and where the radius R of the fiber is not changing rapidly, that is, $|dR/dz| \ll 1$ (Fig. 7.4-6). This point should occur just downstream from the point of maximum extrudate swell. At $z = 0$ the velocity and radius of the filament are v_0 and R_0, and the axial stress τ_{zz} is taken to be τ_0. It is, of course, not possible to measure τ_0, and its introduction is a consequence of not treating the flow problem all the way back into the spinneret where fully developed tube flow boundary conditions on velocity and stress could be specified. Fortuitously, it turns out that the downstream solution is not very sensitive to the value of τ_0 chosen. A distance L downstream from the origin, the polymer solidifies. For $z > L$ the velocity of the filament is assumed

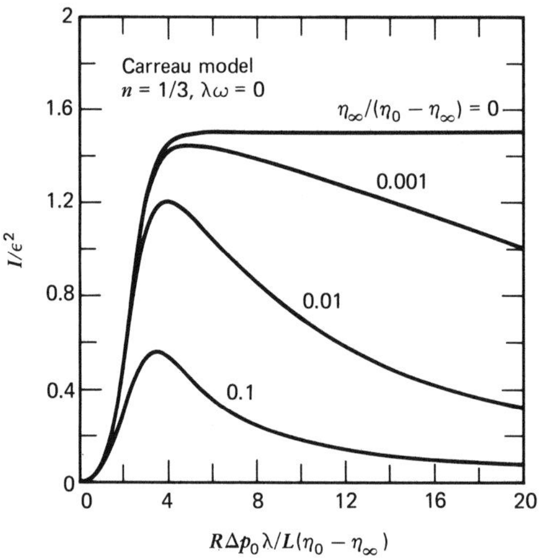

(a)

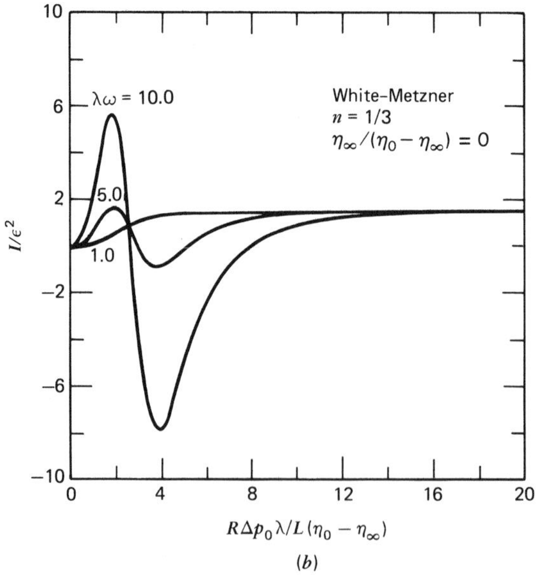

(b)

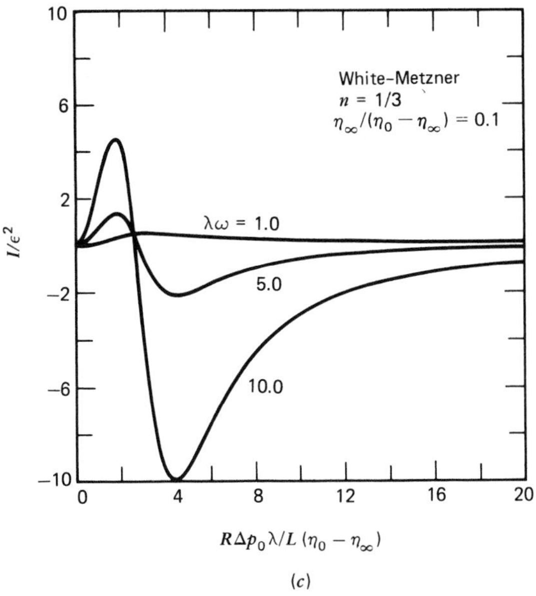

FIGURE 7.4-3. Flow enhancement I/ε^2 as a function of dimensionless pressure gradient $R\Delta p_0 \lambda/L(\eta_0 - \eta_\infty)$ for the White–Metzner model (Eq. 7.3-1) in which the viscosity is described by the Carreau equation (Eq. 4.1-9). All calculations assume a power-law index $n = \frac{1}{3}$. (*a*) $\lambda\omega = 0$ gives the generalized Newtonian fluid result; (*b*) and (*c*) show the influence of varying the dimensionless frequency $\lambda\omega$ and the dimensionless infinite-shear-rate viscosity given as $\eta_\infty/(\eta_0 - \eta_\infty)$. Here λ is the time constant in the Carreau equation.

constant (no further stretching occurs), so that the axial velocity v_L at $z = L$ is fixed by the speed at the wind-up spool. Find the velocity distribution in the filament and the tension F as functions of the "draw ratio" $D_R = v_L/v_0$.

SOLUTION In the region of interest it is reasonable to take $v_z = v_z(z)$ and $\tau_{ij} = \tau_{ij}(z)$. The continuity equation gives

$$\frac{v_z}{v_0} = \left(\frac{R_0}{R(z)}\right)^2 \tag{7.4-35}$$

where $R(z)$ is the local radius of the thread. In the thin filament[9] limit ($|dR/dz| \ll 1$), consideration of conservation of mass and momentum leads to (cf. Eq. 1C.5-2)

$$\frac{d}{dz}(\tau_{zz} - \tau_{rr}) = \frac{v_z'}{v_z}(\tau_{zz} - \tau_{rr}) \tag{7.4-36}$$

[9] The use of the thin filament approximation has been challenged by W. W. Schultz, "Slender Viscoelastic Fiber Flow," submitted to *J. Rheol.* A numerical simulation of the full fiber-spinning problem including flow in the capillary and the extrudate swell region by R. Keunings, M. J. Crochet, and M. M. Denn, *Ind. Eng. Chem. Fundam.*, **22**, 347–355 (1983) has been carried out for several viscoelastic fluid models including the convected Maxwell model. The results of this analysis support the use of the thin filament approximation for the region more than two spinneret diameters downstream from the spinneret and also downstream from the point of maximum extrudate swell. The results also indicate that the choice of the initial condition $T_0 = -1$ used in conjunction with Eq. 7.4-55 (see discussion after Eq. 7.4-64) is appropriate. Numerical results for the convected Maxwell model were obtained up to De $= \lambda_1 v_0/L = 0.025$ and for values of the tension F such that elastic effects were deemed to be important. The validity of the thin filament approximation at higher De is not known.

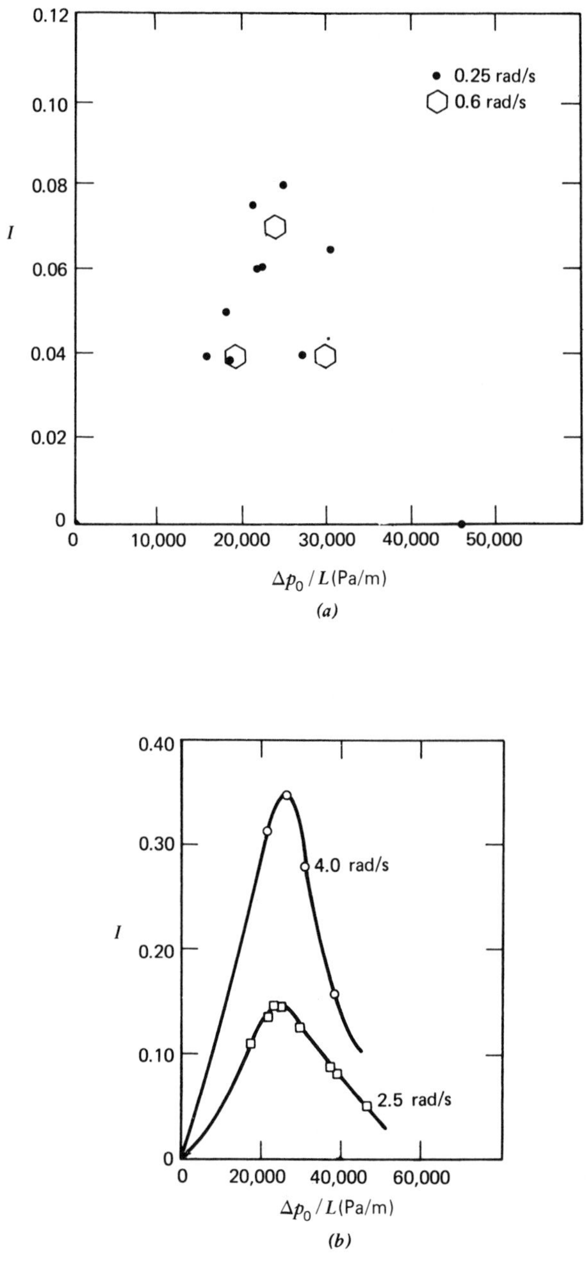

FIGURE 7.4-4. Experimental data on volume flow rate enhancement versus mean pressure gradient for a 1.75% aqueous polyacrylamide solution. In these experiments $\varepsilon = 0.25$ and $R = 0.16$ cm. (*a*) At low frequencies the flow enhancement is independent of frequency; (*b*) at higher frequencies the flow enhancement is seen to increase with increasing frequency. [H. A. Barnes, P. Townsend, and K. Walters, *Rheol. Acta*, **10**, 517–527 (1971).]

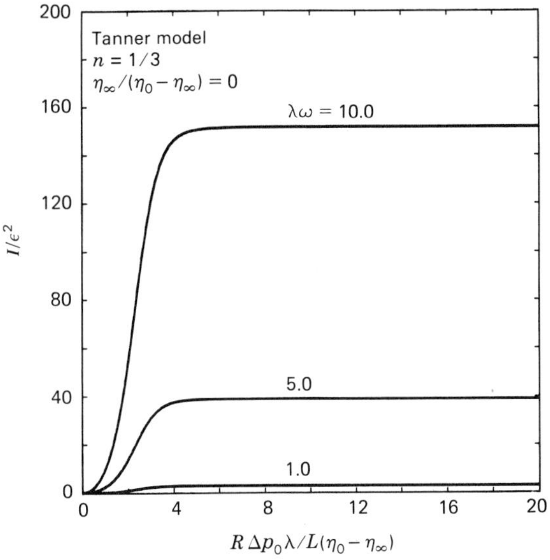

(*a*)

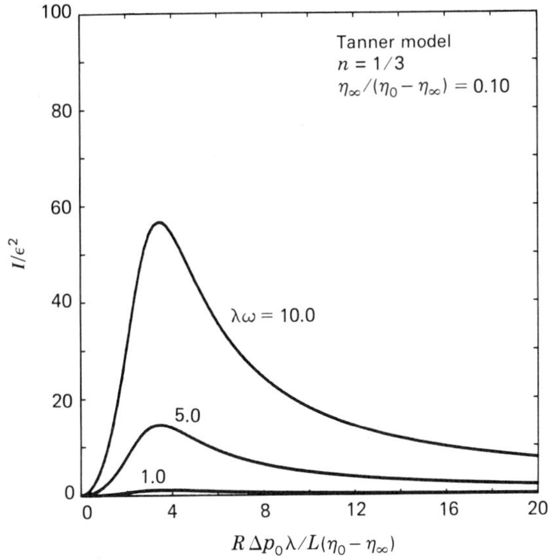

(*b*)

FIGURE 7.4-5. Flow enhancement I/ε^2 as a function of dimensionless pressure gradient $R\Delta p_0\lambda/L(\eta_0 - \eta_\infty)$ for the Tanner model in which the viscosity is described by the Carreau viscosity function (Eq. 4.1-9). The power-law index is $\frac{1}{3}$, and the dimensionless frequency is $\lambda\omega$ where λ is the time constant from the Carreau equation.

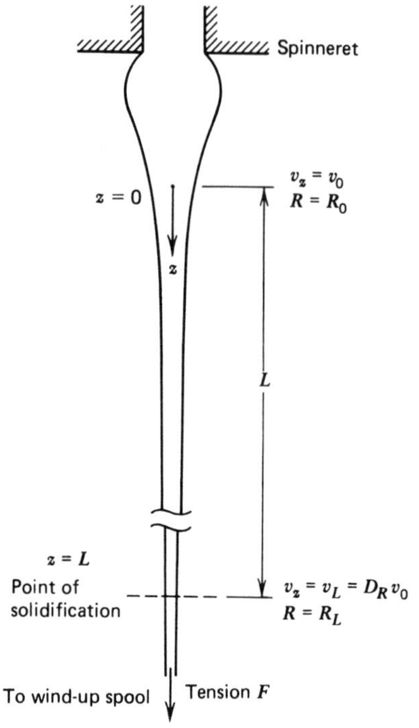

FIGURE 7.4-6. Schematic diagram of the fiber-spinning flow. The stretching of the polymer is modeled from a point $z = 0$ far enough from the spinneret for the axial velocity profile to be assumed flat to the point at $z = L$ where the polymer solidifies. From $z = L$ to the take-up spool no further stretching is assumed to occur. The draw ratio D_R is defined as v_L/v_0.

when we neglect inertia, gravitational forces, and surface tension; v_z' denotes dv_z/dz. Next, since $v_z = v_z(z)$, the continuity equation gives $v_r = -\frac{1}{2}r(dv_z/dz)$, and this leads to the following velocity gradient tensor:

$$\nabla \mathbf{v} = \begin{pmatrix} -\frac{1}{2}\frac{dv_z}{dz} & 0 & 0 \\ 0 & -\frac{1}{2}\frac{dv_z}{dz} & 0 \\ \underline{-\frac{1}{2}r\frac{d^2v_z}{dz^2}} & 0 & \frac{dv_z}{dz} \end{pmatrix} \tag{7.4-37}$$

The three diagonal terms in $\nabla \mathbf{v}$ are all of order v_L/L, whereas the dashed-underlined term is of order $(v_L/L)(R_0/L)$. Since $R_0/L \ll 1$, we neglect this latter term. It is then straightforward to write the rr- and zz-components of the constitutive equation, Eq. 7.3-1, as

$$\tau_{rr} + \frac{\eta(\dot{\gamma})}{G}\left[v_z\frac{d\tau_{rr}}{dz} + \frac{dv_z}{dz}\tau_{rr}\right] = \eta(\dot{\gamma})\frac{dv_z}{dz} \tag{7.4-38}$$

$$\tau_{zz} + \frac{\eta(\dot{\gamma})}{G}\left[v_z\frac{d\tau_{zz}}{dz} - 2\frac{dv_z}{dz}\tau_{zz}\right] = -2\eta(\dot{\gamma})\frac{dv_z}{dz} \tag{7.4-39}$$

where $\dot{\gamma} = \sqrt{\frac{1}{2}(\boldsymbol{\gamma}_{(1)}:\boldsymbol{\gamma}_{(1)})} = \sqrt{3}\, dv_z/dz$. Equations 7.4-36 through 39 are solved subject to the boundary conditions

$$\text{At } z = 0, \qquad v_z = v_0 \tag{7.4-40}$$

$$\tau_{zz} - \tau_{rr} = -F/\pi R_0^2 \tag{7.4-41}$$

$$\text{At } z = L, \qquad v_z = v_L = D_R v_0 \tag{7.4-42}$$

The boundary condition on $\tau_{zz} - \tau_{rr}$ is written in terms of the tension F in the filament and the radius of the thread by means of a total force balance in the z-direction from the take-up reel to any position z. This balance requires the tension and atmospheric pressure acting along the axis of the fiber to be balanced by π_{zz}

$$\pi_{zz} \cdot \pi R^2 = -F + p_a \pi R^2 \tag{7.4-43}$$

where p_a is ambient pressure. Since a radial force balance gives $\pi_{rr} = p_a$ we get Eq. 7.4-41.

Next we recast the problem in terms of the following dimensionless variables:

$$\zeta = \frac{z}{L}; \quad \phi = \frac{v_z}{v_0}; \quad T_{ij} = \tau_{ij}\frac{\pi R_0^2}{F} \tag{7.4-44}$$

When these definitions are introduced, Eq. 7.4-36 and the two contributions from the constitutive equation become

$$\frac{d}{d\zeta}(T_{zz} - T_{rr}) = \frac{1}{\phi}\phi'\,(T_{zz} - T_{rr}) \tag{7.4-45}$$

$$T_{rr} + \text{De}\,\phi'^{(n-1)}\left[\phi\frac{dT_{rr}}{d\zeta} + \phi' T_{rr}\right] = N\phi'^{n} \tag{7.4-46}$$

$$T_{zz} + \text{De}\,\phi'^{(n-1)}\left[\phi\frac{dT_{zz}}{d\zeta} - 2\phi' T_{zz}\right] = -2N\phi'^{n} \tag{7.4-47}$$

where $\phi' = d\phi/d\zeta$. In writing the constitutive equations we have introduced two dimensionless groups: the Deborah number

$$\text{De} = \frac{\lambda_0 v_0}{L} = 3^{(n-1)/2}\left(\frac{m}{G}\right)\left(\frac{v_0}{L}\right)^n \tag{7.4-48}$$

in which the time constant $\lambda_0 = (m/G)(\sqrt{3}\,v_0/L)^{n-1}$ is given as a characteristic value for η/G, and a dimensionless group N defined by

$$N = \frac{m(\sqrt{3}v_0/L)^{n-1}(v_0/L)}{F/\pi R_0^2} \tag{7.4-49}$$

Thus N gives a ratio of the viscous stress in the threadline to the applied tensile stress. The dimensionless boundary conditions are

$$\text{At } \zeta = 0, \qquad \phi = 1 \tag{7.4-50}$$

$$T_{zz} - T_{rr} = -1 \tag{7.4-51}$$

$$\text{At } \zeta = 1, \qquad \phi = D_R \tag{7.4-52}$$

We now integrate Eq. 7.4-45 to get:

$$T_{zz} - T_{rr} = C\phi \tag{7.4-53}$$

The integration constant C is -1 from Eqs. 7.4-50 and 51. Next we subtract Eq. 7.4-46 from 47 and combine with Eq. 7.4-53 to find

$$T_{zz} = -\frac{\phi}{3\ \mathrm{De}\ \phi'^n} - \tfrac{2}{3}\phi + \frac{N}{\mathrm{De}} \tag{7.4-54}$$

The group N/De appearing in the above is equal to $G/(F/\pi R_0^2)$ so that it gives the ratio of a characteristic elastic stress in the fluid to the stress imposed at $z = 0$. Finally we can eliminate T_{zz} in Eq. 7.4-47 by means of Eq. 7.4-54 to find a single, nonlinear, ordinary differential equation for ϕ

$$\phi + (\mathrm{De}\ \phi - 3N)\phi'^n - 2\mathrm{De}^2\ \phi\phi'^{2n} - n\mathrm{De}\ \phi^2\phi''\phi'^{n-2} = 0 \tag{7.4-55}$$

with

$$\begin{aligned} &\text{At } \zeta = 0, \qquad \phi = 1 \\ &\text{At } \zeta = 1, \qquad \phi = D_R \end{aligned} \tag{7.4-56}$$

Equations 7.4-55 and 56 are easy to solve in the two limits $\mathrm{De} = 0$ and $N/\mathrm{De} = 0$.

In the first limit, namely $\mathrm{De} = 0$, the constitutive equation reduces to that for a generalized Newtonian fluid with a power-law viscosity function and we find[10]

$$\phi - 3N\phi'^n = 0 \tag{7.4-57}$$

which has as its solution

$$\phi = [1 + (3N)^{-s}(1 - s)\zeta]^{1/(1-s)} \tag{7.4-58}$$

where $s = 1/n$ and we have applied the boundary condition at $\zeta = 0$. Then we apply the second boundary condition on ϕ at $\zeta = 1$ and solve the result for N in order to have the dimensionless reciprocal tension in the filament as a function of the draw ratio

$$N = \frac{1}{3}\left[\frac{D_R^{1-s} - 1}{1 - s}\right]^{-1/s} \tag{7.4-59}$$

Finally, to get the stress in the filament we insert the solution for the velocity into Eq. 7.4-54:

$$T_{zz} = -\tfrac{2}{3}\phi \tag{7.4-60}$$

This result follows at once, since the first and last terms on the right side of Eq. 7.4-54 cancel for a velocity field that satisfies Eq. 7.4-57. For the generalized Newtonian fluid it was not necessary to specify the upstream axial stress $T_0 = \tau_0\pi R_0^2/F$ to solve for ϕ, and T_0 is found to be $-2/3$ for all draw ratios.

The second limit $N/\mathrm{De} = 0$ is of more interest in this chapter, since it corresponds to high Deborah number and a large amount of tension in the fiber. Here the $-3N\phi'^n$ term in Eq. 7.4-55 can be neglected and it is easy to verify that the resulting equation has the solution

$$\phi = 1 + \zeta/\mathrm{De}^s \tag{7.4-61}$$

[10] J. R. A. Pearson and Y. T. Shah, *Ind. Eng. Chem. Fundam.*, **13**, 134–138 (1974).

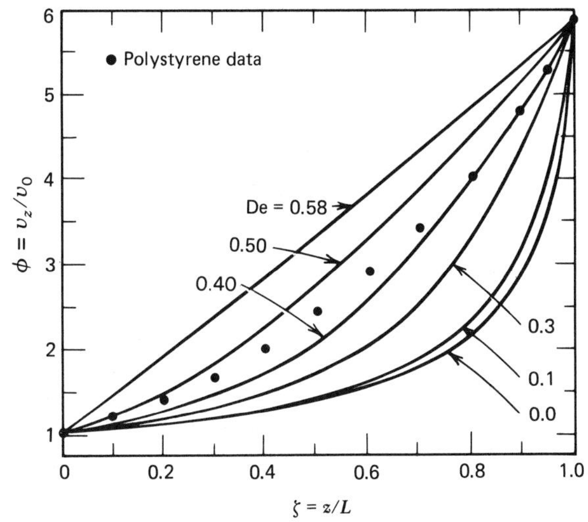

FIGURE 7.4-7. Comparison of calculated and measured dimensionless axial velocity profiles for different values of $\mathrm{De} = 3^{(n-1)/2}(m/G)(v_0/L)^n$. The calculations are based on Eqs. 7.4-55 and 56 with $n = 1/3$ and $D_R = 5.85$. Data are for Dow Styron 666 at 443 K. [R. J. Fisher and M. M. Denn, *AIChE J.*, **22**, 236–246 (1976); data are from G. R. Zeichner, M.Ch.E. Thesis, University of Delaware, Newark (1973). Reproduced by permission of the American Institute of Chemical Engineers.]

Thus the velocity profile is linear and the strain rate in the fiber is uniform. Applying the second boundary condition at $\zeta = 1$ gives

$$D_R = 1 + \mathrm{De}^{-s} \tag{7.4-62}$$

Note that for arbitrary De, the limit being considered here corresponds to infinite tension, so that in order to have bounded stresses for finite De we require

$$D_R < 1 + \mathrm{De}^{-s} \tag{7.4-63}$$

This provides an upper limit on the operating range for a fiber-spinning process as modeled here. Finally we calculate the stress distribution from Eq. 7.4-54:

$$T_{zz} = -(1 + \zeta/\mathrm{De}^s) \tag{7.4-64}$$

so that the stress also grows linearly with distance down the threadline. As in the previous limit, it is not necessary here to specify an upstream stress T_0 in this case because the coefficient N involving the tension in the differential equation is dropped. As before T_0 is found to be independent of draw ratio, and in this limit it is equal to -1.

For other choices of De, Fisher and Denn have solved Eqs. 7.4-55 and 56 numerically. Calculations were done for a series of De; in all cases $n = 1/3$, $D_R = 5.85$, and $T_0 = -1$. The choices of n and D_R were made to correspond as closely as possible to the experiments.[11] The choice of T_0 is reasonable, since for the limiting cases of $\mathrm{De} = 0$ and $N/\mathrm{De} = 0$, we have seen that T_0 is $-2/3$ and -1, respectively. In any case, Fisher and Denn report that the velocity field is insensitive to the choice of T_0. Solution of Eq. 7.4-55 requires that a value of N be given. However, specifying T_0 does not give a value of N directly but instead gives a combination of N and $\phi'(0)$ from Eq. 7.4-54. Thus the problem must be solved iteratively for different values of N until the solution for ϕ is found which gives $T_0 = -1$ in Eq. 7.4-54.

Results of the Fisher and Denn calculations for the velocity are compared in Fig. 7.4-7 to experimental data[11] on a polystyrene solution. The experimental data give $0.21 \le \mathrm{De} \le 0.31$ and

[11] G. R. Zeichner, M.Ch.E. Thesis, University of Delaware, Newark (1973).

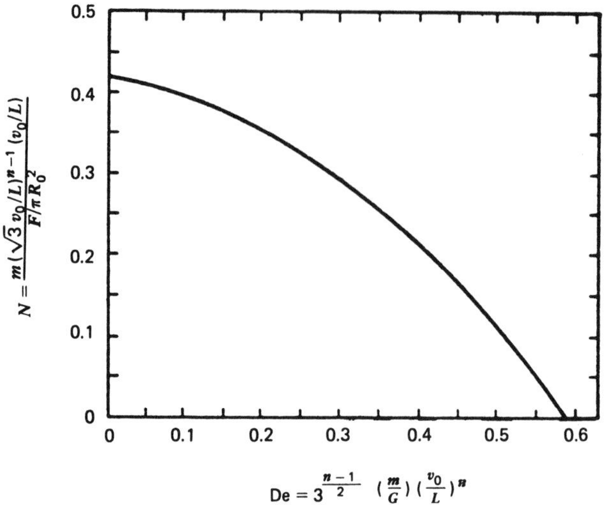

FIGURE 7.4-8. Dimensionless reciprocal tension N (Eq. 7.4-49) as a function of De (Eq. 7.4-48). The calculations are based on Eqs. 7.4-54, 55, and 56 with $n = 1/3$, $D_R = 5.85$, and $T_0 = -1$. [R. J. Fisher and M. M. Denn, *AIChE J.*, **22**, 236–246 (1976). Reproduced by permission of the American Institute of Chemical Engineers.]

$N \doteq 0.08$. Figure 7.4-7 shows that the inclusion of "elasticity" in the White–Metzner model improves the prediction of the velocity field relative to the power-law model. However, the model underestimates the velocity when the correct value of De is used. A good fit of the data is obtained with De between 0.4 and 0.5.

The solution for dimensionless reciprocal force as a function of De is shown in Fig. 7.4-8. There the force is seen to increase with increasing De as would be expected intuitively. Again the inclusion of elasticity improves the predictive ability of the model, but the calculated force is too small. To fit the measured value of $N = 0.08$, a value of De approximately equal to 0.52 must be chosen. This is somewhat higher than the De that best fits the velocity profiles.

§7.5 FLOW PROBLEMS IN TWO OR THREE SPATIAL VARIABLES

In this section we present two fairly lengthy examples illustrating how complex flows involving more than one spatial variable can be solved with differential constitutive equations. The first example, flow in a journal bearing, is important because it demonstrates how sensitive the calculated velocity, stress, and pressure fields are to the constitutive equation chosen to represent the fluid. We will see that constitutive equations which describe the memory of the fluid (e.g., the convected Maxwell and White–Metzner models) and a constitutive equation that describes only the shear thinning of the viscometric functions (the CEF model) differ markedly in predictions of (i) locations and presence of boundary layers in both the velocity and stress fields, (ii) direction of the force exerted by the fluid on the solid surfaces, and (iii) presence and location of separated flow regions.

The second example, flow of a convected Jeffreys model between parallel disks, is important because it shows the influence of the viscoelasticity on flow stability. We will see that even in creeping flow, the two-dimensional flow of the convected Jeffreys model is unstable above a critical Deborah number that is of order unity. Stability problems are important not only for viscometry but also for numerical simulations of viscoelastic flows.

EXAMPLE 7.5-1 Viscoelastic Flow in a Journal Bearing[1]

In Problem 1C.4 the flow of a Newtonian fluid in a simplified journal-bearing system (see Fig. 1C.4) is solved for small eccentricity, and in Example 6.4-2 the loads on the inner cylinder are calculated for the second-order fluid. Because of the Giesekus–Tanner–Pipkin theorem the velocity fields of the second-order and Newtonian fluids are identical. Here we examine the influence of viscoelasticity on both the velocity field and loads by considering the flow of a convected Maxwell model in the journal bearing.

As in the Newtonian and second-order fluid problems, we assume that both the gap and eccentricity are small; the small gap assumption is quite realistic, whereas small eccentricity is assumed simply to obtain analytical solutions. We will again approximate the local gap width $B(\theta)$ between the cylinders by

$$B(\theta) = c(1 + \varepsilon \cos \theta) \tag{7.5-1}$$

in which $c = r_2 - r_1$ is the average gap and $\varepsilon = a/c$ is the dimensionless eccentricity. Here, r_1 and r_2 are the radii of the inner and outer cylinders, respectively, and a is the distance between their axes. Whereas it is possible to solve exactly the lubrication equations for the Newtonian fluid with the approximate gap width given above, for more complicated constitutive equations such as the convected Maxwell model it is necessary to expand the velocity, stress, and pressure variables in powers of the small eccentricity ε. In order to develop consistent approximations in the description of the gap width, the equations of motion, and constitutive equations, and in order to organize the ordering of terms in the expansions, we use the domain perturbation approach of Joseph and Sturges.[2] Assume the flow is steady and two-dimensional.

SOLUTION **(a)** Equations to be Solved

If we neglect inertial and gravitational forces, then the continuity, momentum, and constitutive equations that govern the problem are

$$\begin{aligned} (\boldsymbol{\nabla} \cdot \boldsymbol{v}) &= 0 \\ \boldsymbol{\nabla} p + [\boldsymbol{\nabla} \cdot \boldsymbol{\tau}] &= 0 \\ \boldsymbol{\tau} + \lambda_1(\{\boldsymbol{v} \cdot \boldsymbol{\nabla} \boldsymbol{\tau}\} - \{(\boldsymbol{\nabla} \boldsymbol{v})^\dagger \cdot \boldsymbol{\tau}\}_s) &= -\eta_0 \dot{\boldsymbol{\gamma}} \end{aligned} \tag{7.5-2}$$

where the $\{\ \}_s$ symbol denotes $\{\ \} + \{\ \}^\dagger$. Since we take the flow to be two-dimensional, that is, $v_r = v_r(r, \theta)$, $v_\theta = v_\theta(r, \theta)$, and $v_z = 0$, then the necessary components of these equations are, in cylindrical coordinates

$$\frac{1}{r}\frac{\partial}{\partial r}(r v_r) + \frac{1}{r}\frac{\partial v_\theta}{\partial \theta} = 0 \tag{7.5-3}$$

$$\frac{\partial p}{\partial r} + \frac{\partial \tau_{rr}}{\partial r} + \left(\frac{\tau_{rr} - \tau_{\theta\theta}}{r}\right) + \frac{1}{r}\frac{\partial}{\partial \theta}\tau_{\theta r} = 0 \tag{7.5-4}$$

$$\frac{1}{r}\frac{\partial p}{\partial \theta} + \frac{1}{r}\frac{\partial}{\partial \theta}\tau_{\theta\theta} + \frac{\partial \tau_{r\theta}}{\partial r} + \frac{2}{r}\tau_{r\theta} = 0 \tag{7.5-5}$$

[1] This example is based on A. N. Beris, R. C. Armstrong, and R. A. Brown, *J. Non-Newtonian Fluid Mech.*, **13**, 109–148 (1983). In this paper solutions are presented for the second-order fluid, the CEF equation, the convected Maxwell model, and the White–Metzner model. For other viscoelastic calculations for this geometry see M. J. Davies and K. Walters in T. C. Davenport, ed., *Rheology of Lubricants*, Halsted Press, New York (1973); N. Phan-Thien and R. I. Tanner, *J. Non-Newtonian Fluid Mech.*, **9**, 107–117 (1981); B. Y. Ballal and R. S. Rivlin, *Trans. Soc. Rheol.*, **20**, 65–101 (1976); R. S. Rivlin, *J. Non-Newtonian Fluid Mech.*, **5**, 79–101 (1979); M. Reiner, M. Hanin, and A. Harnoy, *Israel J. Tech.*, **7**, 273–279 (1969).

[2] D. D. Joseph and L. Sturges, *J. Fluid Mech.*, **69**, 565–589 (1975).

$$\tau_{rr} + \lambda_1\left[v_r \frac{\partial}{\partial r}\tau_{rr} + \frac{v_\theta}{r}\frac{\partial}{\partial \theta}\tau_{rr} - 2\frac{v_\theta}{r}\tau_{r\theta} - 2(\kappa_{rr}\tau_{rr} + \kappa_{r\theta}\tau_{\theta r})\right] = -2\eta_0 \frac{\partial v_r}{\partial r} \tag{7.5-6}$$

$$\tau_{r\theta} + \lambda_1\left[v_r \frac{\partial}{\partial r}\tau_{r\theta} + \frac{v_\theta}{r}\frac{\partial}{\partial \theta}\tau_{r\theta} + \frac{v_\theta}{r}(\tau_{rr} - \tau_{\theta\theta}) - (\kappa_{rr}\tau_{r\theta} + \kappa_{r\theta}\tau_{\theta\theta}) - (\kappa_{\theta r}\tau_{rr} + \kappa_{\theta\theta}\tau_{\theta r})\right]$$
$$= -\eta_0\left(\frac{\partial v_\theta}{\partial r} + \frac{1}{r}\frac{\partial v_r}{\partial \theta} - \frac{v_\theta}{r}\right) \tag{7.5-7}$$

$$\tau_{\theta\theta} + \lambda_1\left[v_r \frac{\partial}{\partial r}\tau_{\theta\theta} + \frac{v_\theta}{r}\frac{\partial}{\partial \theta}\tau_{\theta\theta} - 2\frac{v_\theta}{r}\tau_{r\theta} - 2(\kappa_{\theta r}\tau_{r\theta} + \kappa_{\theta\theta}\tau_{\theta\theta})\right] = -2\eta_0\left(\frac{1}{r}\frac{\partial v_\theta}{\partial \theta} + \frac{v_r}{r}\right) \tag{7.5-8}$$

where we have used the abbreviation $\boldsymbol{\kappa} = (\nabla \mathbf{v})^\dagger$. For the plane flow under consideration we have in cylindrical coordinates

$$\nabla \boldsymbol{v} = \boldsymbol{\kappa}^\dagger = \begin{pmatrix} \dfrac{\partial v_r}{\partial r} & \dfrac{\partial v_\theta}{\partial r} & 0 \\ \left(\dfrac{1}{r}\dfrac{\partial v_r}{\partial \theta} - \dfrac{v_\theta}{r}\right) & \left(\dfrac{1}{r}\dfrac{\partial v_\theta}{\partial \theta} + \dfrac{v_r}{r}\right) & 0 \\ 0 & 0 & 0 \end{pmatrix} \tag{7.5-9}$$

Because of the invariance of the flow with respect to z, we expect no momentum flux in the z-direction, so we have set τ_{rz}, $\tau_{\theta z}$, and τ_{zz} to zero and have eliminated the corresponding components of the constitutive equation. Also in writing down the components of the constitutive equation we have taken advantage of the expressions for $\{\boldsymbol{v} \cdot \nabla \boldsymbol{\tau}\}_{ij}$ in cylindrical coordinates given in Table A.7-2.

Equations 7.5-3 through 8 completely determine the velocity and stress fields in this problem when taken together with the boundary conditions that

$$\text{On } r = r_1: \quad v_r = 0, \quad v_\theta = V \tag{7.5-10}$$

$$\text{On } r = r_1 + B(\theta): \quad v_r = 0, \quad v_\theta = 0 \tag{7.5-11}$$

The linear velocity V of the inner cylinder is equal to $r_1 W_1$, where W_1 is the angular velocity of the inner cylinder. The only conditions on stress and pressure are those of periodicity:

$$\tau_{ij}(r, \theta) = \tau_{ij}(r, \theta + 2\pi)$$
$$p(r, \theta) = p(r, \theta + 2\pi) \tag{7.5-12}$$

We now solve these equations by the domain perturbation technique.

(b) The Domain Perturbation Method

The key idea in the domain perturbation solution is the mapping of the original domain onto a simpler, reference domain. In the present problem, it is convenient to choose the reference domain to be a concentric cylinder geometry with coordinates $(\bar{r}, \bar{\theta})$ (see Fig. 7.5-1). The mapping between eccentric and concentric cylinders is done by the linear transformation $(r, \theta) \to (\bar{r}, \bar{\theta})$:

$$r = r^{[0]}(\bar{r}, \bar{\theta}; \varepsilon) = r_1 + (\bar{r} - r_1)(1 + \varepsilon \cos \bar{\theta}) \qquad (r_1 \le \bar{r} \le r_2)$$
$$\theta = \theta^{[0]}(\bar{r}, \bar{\theta}; \varepsilon) = \bar{\theta} \qquad (0 \le \bar{\theta} < 2\pi) \tag{7.5-13}$$

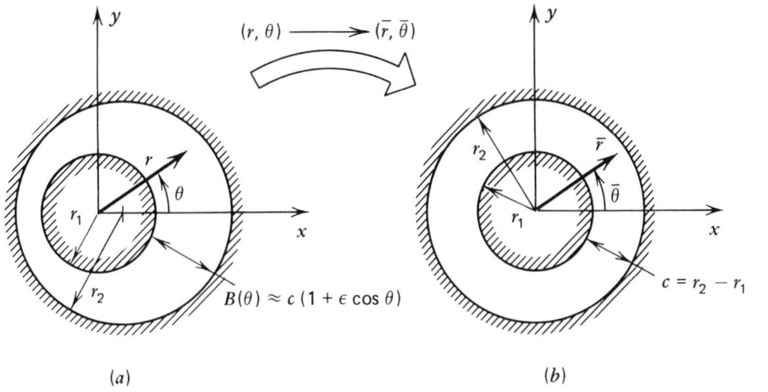

FIGURE 7.5-1. Mapping defined by Eq. 7.5-13 from the (*a*) eccentric cylinder geometry in which the coordinates are (r, θ) to a (*b*) reference (concentric cylinder) geometry with coordinates $(\bar{r}, \bar{\theta})$. The gap in the original problem is $c(1 + \varepsilon \cos \theta)$, whereas in the mapped problem it is a constant c.

where we have used a superscript square bracket $^{[\ \ ]}$ to denote functions of the transformed, "concentric" coordinates $(\bar{r}, \bar{\theta})$. We will also use a superscript angular bracket $^{\langle\ \ \rangle}$ to denote functions of the original coordinates (r, θ). A perturbation expansion in eccentricity for the original eccentric problem can be carried over to the concentric cylinder shape by a change of variables from (r, θ) to $(\bar{r}, \bar{\theta})$. Developing a perturbation expansion in either coordinate system involves taking partial derivatives of the dependent variables with respect to the small parameter ε. To get from the solution in the reference domain back to the original eccentric geometry requires knowing how to relate these two sets of partial derivatives. To illustrate these definitions and interrelations consider a function $u(r, \theta; \varepsilon) \equiv u^{\langle 0 \rangle}(r, \theta; \varepsilon) = u^{[0]}(\bar{r}, \bar{\theta}; \varepsilon)$. The two sets of partial derivatives are given by

$$u^{\langle n \rangle}(r, \theta; \varepsilon) \equiv \frac{\partial^n u^{\langle 0 \rangle}}{\partial \varepsilon^n}; \quad u^{[n]}(\bar{r}, \bar{\theta}; \varepsilon) \equiv \frac{\partial^n u^{[0]}}{\partial \varepsilon^n} \tag{7.5-14,15}$$

and they can be related through the chain rule, for example,

$$u^{[1]} = \frac{\partial u^{\langle 0 \rangle}}{\partial \varepsilon} + \frac{\partial r^{[0]}}{\partial \varepsilon}\frac{\partial u^{\langle 0 \rangle}}{\partial r} = u^{\langle 1 \rangle} + r^{[1]}\frac{\partial u^{\langle 0 \rangle}}{\partial r} \tag{7.5-16}$$

Equation 7.5-16 shows that $u^{[1]}$ can be regarded as a "substantial" derivative following the mapping. The factor $r^{[1]}$ in Eq. 7.5-16 arising from application of the chain rule can be evaluated from the mapping Eq. 7.5-13 as follows

$$r^{[1]} = \frac{\partial r^{[0]}}{\partial \varepsilon} = (\bar{r} - r_1) \cos \bar{\theta} \tag{7.5-17a}$$

$$r^{[1]}(\bar{r}, \bar{\theta}; 0) = (r - r_1) \cos \theta \tag{7.5-17b}$$

In Eq. 7.5-17b we have evaluated $r^{[1]}$ at $\varepsilon = 0$ and used the fact that $r = \bar{r}$ and $\theta = \bar{\theta}$ when the eccentricity is zero. Equation 7.5-17b is used in the subsequent expansions about $\varepsilon = 0$.

We now expand the dependent variables—velocity, stress, and pressure—in series in ε calculated in the reference domain, that is, with $\bar{r}$, $\bar{\theta}$ held constant:

$$\begin{pmatrix} \mathbf{v}(r, \theta; \varepsilon) \\ \boldsymbol{\tau}(r, \theta; \varepsilon) \\ p(r, \theta; \varepsilon) \end{pmatrix} = \sum_{n=0}^{\infty} \frac{\varepsilon^n}{n!} \begin{pmatrix} \mathbf{v}^{[n]}(\bar{r}, \bar{\theta}; \varepsilon) \\ \boldsymbol{\tau}^{[n]}(\bar{r}, \bar{\theta}; \varepsilon) \\ p^{[n]}(\bar{r}, \bar{\theta}; \varepsilon) \end{pmatrix}_{\varepsilon=0} \tag{7.5-18}$$

The reason for choosing this expansion rather than one in the original eccentric geometry is that the boundary conditions are simpler. The price that is paid for simplicity in the boundary location is that equations for the "substantial" derivatives, $\mathbf{v}^{[1]}$, etc., are not easily obtained. This difficulty is overcome by expressing the $\mathbf{v}^{[n]}(\bar{r}, \bar{\theta}; 0)\ldots$ in terms of the $\mathbf{v}^{\langle n\rangle}(r, \theta; 0)\ldots$ by means of the chain rule as shown in Eqs. 7.5-15 through 17.

The equations for the $\mathbf{v}^{\langle n\rangle}(r, \theta; 0)$ are easily obtained by taking $\partial^n/\partial\varepsilon^n$ of the governing equation set, Eqs. 7.5-2, and then evaluating these at $\varepsilon = 0$. Keep in mind that in Eqs. 7.5-2 the $\mathbf{v}(r, \theta; \varepsilon)$, etc., are identical to the $\mathbf{v}^{\langle 0\rangle}(r, \theta; \varepsilon)$, etc. In this way we obtain

$$\begin{gathered}(\nabla \cdot \mathbf{v}^{\langle n\rangle}) = 0 \\ \nabla p^{\langle n\rangle} + [\nabla \cdot \boldsymbol{\tau}^{\langle n\rangle}] = \mathbf{0} \\ \boldsymbol{\tau}^{\langle n\rangle} + \lambda_1(\{\mathbf{v} \cdot \nabla\boldsymbol{\tau}\}^{\langle n\rangle} - \{(\nabla\mathbf{v})^\dagger \cdot \boldsymbol{\tau}\}_s^{\langle n\rangle}) = -\eta_0\dot{\boldsymbol{\gamma}}^{\langle n\rangle}\end{gathered} \tag{7.5-19}$$

in which all dependent variables are evaluated at $(r, \theta; 0)$. The specific forms of these equations for given values of n are conveniently obtained from Eqs. 7.5-2 by substituting power series in ε for the velocity, stress, and pressure with coefficients $\mathbf{v}^{\langle n\rangle}$, $p^{\langle n\rangle}$, and $\boldsymbol{\tau}^{\langle n\rangle}$ and equating equal orders in ε.

To obtain the boundary conditions for the $\mathbf{v}^{\langle n\rangle}(r, \theta; 0)$ we insert the expansion for $\mathbf{v}$ in Eqs. 7.5-18 into Eqs. 7.5-10 and 11. On $r = \bar{r} = r_1$ we have

$$\begin{aligned} v_r &= \sum_{n=0}^{\infty} \frac{1}{n!} v_r^{[n]}(\bar{r}, \bar{\theta}; 0)\varepsilon^n \Bigg|_{\bar{r}=r_1} \\ &= v_r^{\langle 0\rangle}(r_1, \theta; 0) + \varepsilon\left[v_r^{\langle 1\rangle}(r, \theta; 0) + (r - r_1)\cos\theta \frac{\partial v_r^{\langle 0\rangle}(r, \theta; 0)}{\partial r}\right]_{r=r_1} + \cdots \\ &= v_r^{\langle 0\rangle}(r_1, \theta; 0) + \varepsilon v_r^{\langle 1\rangle}(r_1, \theta; 0) + \cdots \\ v_\theta &= v_\theta^{\langle 0\rangle}(r_1, \theta; 0) + \varepsilon v_\theta^{\langle 1\rangle}(r_1, \theta; 0) + \cdots \end{aligned} \tag{7.5-20}$$

Since from here on all of the $v_i^{\langle n\rangle}$ are evaluated at $\varepsilon = 0$, we will drop this dependence from the arguments. For example, we abbreviate $v_\theta^{\langle 1\rangle}(r_1, \theta; 0)$ as $v_\theta^{\langle 1\rangle}(r_1, \theta)$. Comparing these equations with Eq. 7.5-10 gives

$$\begin{array}{lll} \text{On } r = r_1 & v_r^{\langle 0\rangle}(r_1, \theta) = 0 & v_\theta^{\langle 0\rangle}(r_1, \theta) = V \\ & v_r^{\langle 1\rangle}(r_1, \theta) = 0 & v_\theta^{\langle 1\rangle}(r_1, \theta) = 0 \\ & \vdots & \vdots \end{array} \tag{7.5-21}$$

Similarly for the outer boundary at $r = r_1 + B(\theta)$ or $\bar{r} = r_2$ we find

$$\begin{aligned} v_r(r_2, \theta) &= v_r^{\langle 0\rangle}(r_2, \theta) + \varepsilon\Big[v_r^{\langle 1\rangle}(r, \theta) \\ &\qquad + (r - r_1)\cos\theta \frac{\partial v_r^{\langle 0\rangle}(r, \theta)}{\partial r}\Big]_{r=r_2} + \cdots \\ &= v_r^{\langle 0\rangle}(r_2, \theta) + \varepsilon\left[v_r^{\langle 1\rangle}(r_2, \theta) + c\cos\theta \frac{\partial v_r^{\langle 0\rangle}(r_2, \theta)}{\partial r}\right] + \cdots \\ v_\theta(r_2, \theta) &= v_\theta^{\langle 0\rangle}(r_2, \theta) + \varepsilon\left[v_\theta^{\langle 1\rangle}(r_2, \theta) + c\cos\theta \frac{\partial v_\theta^{\langle 0\rangle}(r_2, \theta)}{\partial r}\right] + \cdots \end{aligned} \tag{7.5-22}$$

Comparing Eqs. 7.5-22 and 11 gives the conditions at $r = r_2$,

$$
\begin{aligned}
r = r_2 \qquad & v_r^{\langle 0\rangle}(r_2, \theta) = 0; \quad v_\theta^{\langle 0\rangle}(r_2, \theta) = 0 \\
& v_r^{\langle 1\rangle}(r_2, \theta) = -c \cos\theta \frac{\partial v_r^{\langle 0\rangle}(r_2, \theta)}{\partial r} \\
& v_\theta^{\langle 1\rangle}(r_2, \theta) = -c \cos\theta \frac{\partial v_\theta^{\langle 0\rangle}(r_2, \theta)}{\partial r} \\
& \qquad \vdots
\end{aligned}
\tag{7.5-23}
$$

(c) Solution of the Zeroth-Order Domain Perturbation Problem

We now solve the lowest-order problem given by Eqs. 7.5-19 with $n = 0$. Specifically,

$$
\begin{aligned}
(\nabla \cdot \boldsymbol{v}^{\langle 0\rangle}) &= 0 \\
\nabla p^{\langle 0\rangle} + [\nabla \cdot \boldsymbol{\tau}^{\langle 0\rangle}] &= \mathbf{0} \\
\boldsymbol{\tau}^{\langle 0\rangle} + \lambda_1(\{\boldsymbol{v}^{\langle 0\rangle} \cdot \nabla \boldsymbol{\tau}^{\langle 0\rangle}\} - \{(\nabla \boldsymbol{v}^{\langle 0\rangle})^\dagger \cdot \boldsymbol{\tau}^{\langle 0\rangle}\}_s) &= -\eta_0 \dot{\boldsymbol{\gamma}}^{\langle 0\rangle}
\end{aligned}
\tag{7.5-24}
$$

The component forms of these equations are given by Eqs. 7.5-3 through 8. The boundary conditions on $v_r^{\langle 0\rangle}$ and $v_\theta^{\langle 0\rangle}$ in Eqs. 7.5-21 and 23 suggest that we try $v_r^{\langle 0\rangle} = 0$, $v_\theta^{\langle 0\rangle} = v_\theta^{\langle 0\rangle}(r)$, $\tau_{ij}^{\langle 0\rangle} = \tau_{ij}^{\langle 0\rangle}(r)$, and $p^{\langle 0\rangle} = p^{\langle 0\rangle}(r)$. With this assumed solution the continuity equation is automatically satisfied.

The velocity gradient tensor is

$$
\nabla \boldsymbol{v}^{\langle 0\rangle} = \begin{pmatrix} 0 & \dfrac{\partial v_\theta^{\langle 0\rangle}}{\partial r} & 0 \\ -\dfrac{v_\theta^{\langle 0\rangle}}{r} & 0 & 0 \\ 0 & 0 & 0 \end{pmatrix} = \boldsymbol{\kappa}^{\langle 0\rangle\dagger}
\tag{7.5-25}
$$

When this is combined with the constitutive equations for $\tau_{rr}^{\langle 0\rangle}$, $\tau_{r\theta}^{\langle 0\rangle}$, and $\tau_{\theta\theta}^{\langle 0\rangle}$ we find

$$
\tau_{rr}^{\langle 0\rangle} = 0
\tag{7.5-26}
$$

$$
\tau_{r\theta}^{\langle 0\rangle} = -\eta_0 \left[\frac{\partial v_\theta^{\langle 0\rangle}}{\partial r} - \frac{v_\theta^{\langle 0\rangle}}{r}\right] = -\eta_0 \dot{\gamma}_{r\theta}^{\langle 0\rangle}
\tag{7.5-27}
$$

$$
\tau_{\theta\theta}^{\langle 0\rangle} = 2\lambda_1 \left(\frac{\partial v_\theta^{\langle 0\rangle}}{\partial r} - \frac{v_\theta^{\langle 0\rangle}}{r}\right) \tau_{r\theta}^{\langle 0\rangle} = -2\eta_0 \lambda_1 \dot{\gamma}_{r\theta}^{\langle 0\rangle 2}
\tag{7.5-28}
$$

Note that these results just give the viscometric properties of the convected Maxwell model at the shear rate $\dot{\gamma}_{r\theta}^{\langle 0\rangle}$.

The equations of motion now can be written as

r-component:
$$
\frac{dp^{\langle 0\rangle}}{dr} = \frac{\tau_{\theta\theta}^{\langle 0\rangle}}{r}
\tag{7.5-29}
$$

θ-component:
$$
\frac{d}{dr}\tau_{r\theta}^{\langle 0\rangle} + \frac{2}{r}\tau_{r\theta}^{\langle 0\rangle} = 0
\tag{7.5-30}
$$

For use in the next order problem in the ε-series we develop solutions to the above equations as series in the small dimensionless gap μ defined as

$$\mu = c/r_1 \tag{7.5-31}$$

In this development it is convenient to use the following dimensionless variables and groups:

$$\begin{aligned}
\zeta &= \frac{(r - r_1)}{c} \\
T_{ij}^{\langle 0 \rangle}(\zeta) &= \frac{\tau_{ij}^{\langle 0 \rangle}(r)}{(\eta_0 V/c)} \\
P^{\langle 0 \rangle}(\zeta) &= \frac{p^{\langle 0 \rangle}(r)}{(\eta_0 V/c)} \\
\hat{v}_i^{\langle 0 \rangle}(\zeta) &= \frac{v_i^{\langle 0 \rangle}(r)}{V} \\
\text{De} &= \text{We}\mu = \lambda_1 V/r_1 \\
\text{We} &= \lambda_1 V/c
\end{aligned} \tag{7.5-32}$$

The Weissenberg number We is a dimensionless shear rate, and for the small gap, concentric cylinder limit of the convected Maxwell model it is equal to $\Psi_1\dot{\gamma}/2\eta_0$. The Deborah number De is a ratio of the time constant for the convected Maxwell model to the time required for the fluid to move once around the journal bearing. Note that when μ is small, We $\gg$ De. We wish to develop solutions valid for De of order unity, so that memory of events in one part of the journal bearing is felt in other parts.

In terms of the dimensionless variables the θ-component of the equation of motion can be written for small μ as

$$\frac{d}{d\zeta} T_{r\theta}^{\langle 0 \rangle} = -2\mu(1 - \mu\zeta + \cdots)T_{r\theta}^{\langle 0 \rangle} \tag{7.5-33}$$

which has the solution

$$T_{r\theta}^{\langle 0 \rangle} = c_0 + \mu(c_1 - 2c_0\zeta) + \cdots \tag{7.5-34}$$

where c_0 and c_1 are constants. When this is combined with the constitutive equation for $T_{r\theta}^{\langle 0 \rangle}$, an ordinary differential equation for $\hat{v}_\theta^{\langle 0 \rangle}$ is obtained. Solving this equation together with the boundary conditions that $\hat{v}_\theta^{\langle 0 \rangle}(0) = 1$ and $\hat{v}_\theta^{\langle 0 \rangle}(1) = 0$ leads to

$$\hat{v}_\theta^{\langle 0 \rangle} = (1 - \zeta) + \frac{\mu}{2}(\zeta^2 - \zeta) + O(\mu^2) \tag{7.5-35}$$

In obtaining Eq. 7.5-35 we find the integration constants in Eq. 7.5-34 to be $c_0 = 1$ and $c_1 = 3/2$. It is then straightforward to develop the remaining series solutions, which have the leading behavior

$$\begin{aligned}
\hat{v}_r^{\langle 0 \rangle} &= 0 \\
T_{rr}^{\langle 0 \rangle} &= 0 \\
T_{r\theta}^{\langle 0 \rangle} &= 1 + \mu(\tfrac{3}{2} - 2\zeta) + O(\mu^2) \\
T_{\theta\theta}^{\langle 0 \rangle} &= -\frac{2\text{De}}{\mu}(1 + O(\mu)) \\
P^{\langle 0 \rangle} &= P_0 - 2\text{De}\,\zeta + O(\mu)
\end{aligned} \tag{7.5-36}$$

The constant P_0 is an arbitrary reference pressure. We note in Eq. 7.5-36 that the normal stress $T_{\theta\theta}^{\langle 0 \rangle}$ is singular and is thus the dominant term for $\mu \to 0$.

(d) Solution to the First-Order Domain Perturbation Problem

We now move to the next order in ε and solve Eqs. 7.5-19 for $n = 1$. The governing equations are

$$(\nabla \cdot \boldsymbol{v}^{\langle 1 \rangle}) = 0$$

$$\nabla p^{\langle 1 \rangle} + [\nabla \cdot \boldsymbol{\tau}^{\langle 1 \rangle}] = \mathbf{0} \tag{7.5-37}$$

$$\boldsymbol{\tau}^{\langle 1 \rangle} + \lambda_1[\{\boldsymbol{v}^{\langle 0 \rangle} \cdot \nabla \boldsymbol{\tau}^{\langle 1 \rangle}\} + \{\boldsymbol{v}^{\langle 1 \rangle} \cdot \nabla \boldsymbol{\tau}^{\langle 0 \rangle}\} - \{(\nabla \boldsymbol{v}^{\langle 0 \rangle})^\dagger \cdot \boldsymbol{\tau}^{\langle 1 \rangle}\}_s - \{(\nabla \boldsymbol{v}^{\langle 1 \rangle})^\dagger \cdot \boldsymbol{\tau}^{\langle 0 \rangle}\}_s] = -\eta_0 \dot{\boldsymbol{\gamma}}^{\langle 1 \rangle}$$

The boundary conditions on $v_r^{\langle 1 \rangle}$ and $v_\theta^{\langle 1 \rangle}$ are given by Eqs. 7.5-21 and 23 with Eq. 7.5-35 used to evaluate $\partial v_i^{\langle 0 \rangle}/\partial r$ in Eq. 7.5-23. Including terms through order μ gives

$$\begin{aligned} \text{At } r = r_1, \quad & v_r^{\langle 1 \rangle} = 0 \\ & v_\theta^{\langle 1 \rangle} = 0 \\ \text{At } r = r_2, \quad & v_r^{\langle 1 \rangle} = 0 \\ & v_\theta^{\langle 1 \rangle} = V \cos\theta + O(\mu^2) \end{aligned} \tag{7.5-38}$$

The periodic nature of the boundary conditions suggests the use of a complex stream function $F(r)$:

$$\begin{aligned} v_r^{\langle 1 \rangle} &= \mathcal{R}e\{iF(r)e^{i\theta}c/r\} \\ v_\theta^{\langle 1 \rangle} &= \mathcal{R}e\{-F'(r)e^{i\theta}c\} \end{aligned} \tag{7.5-39}$$

The continuity equation is thus automatically satisfied. We note for later use the sizes of the following terms: $v_r^{\langle 1 \rangle} \sim \mu V$, $v_\theta^{\langle 1 \rangle} \sim V$, and $F \sim V$.

The rr-component of the constitutive equation given in Eq. 7.5-37 can be written down with the aid of Eq. 7.5-6

$$\tau_{rr}^{\langle 1 \rangle} + \lambda_1\left[\frac{V(1-\zeta)}{r}\frac{\partial}{\partial\theta}\tau_{rr}^{\langle 1 \rangle} - \frac{2}{r}\frac{\partial v_r^{\langle 1 \rangle}}{\partial\theta}\eta_0\frac{V}{c}(1 + \mu(1-\zeta) + \cdots)\right] = -2\eta_0\frac{\partial v_r^{\langle 1 \rangle}}{\partial r} \tag{7.5-40}$$

We next introduce dimensionless variables as defined in Eq. 7.5-32, and also a dimensionless stream function f defined by

$$f(\zeta) = F(r)/V \tag{7.5-41}$$

Also we assume periodic solutions for the stresses and pressure

$$\begin{aligned} \tau_{ij}^{\langle 1 \rangle} &= \mathcal{R}e\{T_{ij}^{\langle 1 \rangle}(\zeta)e^{i\theta}\}(\eta_0 V/c) \\ p^{\langle 1 \rangle} &= \mathcal{R}e\{P^{\langle 1 \rangle}(\zeta)e^{i\theta}\}(\eta_0 V/c) \end{aligned} \tag{7.5-42}$$

so that the stress and pressure, $\boldsymbol{T}^{\langle 1 \rangle}$ and $P^{\langle 1 \rangle}$, we seek at this order are functions of ζ alone. When Eq. 7.5-40 is rewritten with these new variables we obtain

$$T_{rr}^{\langle 1 \rangle} + i\mathrm{De}(1-\zeta)(1 - \tfrac{3}{2}\mu\zeta + \cdots)T_{rr}^{\langle 1 \rangle} = -2\mu(if' + \mathrm{De}f) + O(\mu^2) \tag{7.5-43}$$

where the prime denotes $d/d\zeta$. Clearly $T_{rr}^{\langle 1\rangle} \sim \mu$ so we can develop $T_{rr}^{\langle 1\rangle}$ as a series in μ with leading term

$$T_{rr}^{\langle 1\rangle} = -\frac{2(if' + \text{De}f)}{[1 + i\text{De}(1 - \zeta)]}\mu + O(\mu^2) \tag{7.5-44}$$

Similarly we find

$$\begin{aligned} T_{r\theta}^{\langle 1\rangle}[1 + i\text{De}(1 - \zeta)] &= f'' + 2\text{De}^2 f - \frac{\text{De}}{\mu}T_{rr}^{\langle 1\rangle} \\ &= f'' + 2\text{De}^2 f + \frac{2\text{De}(if' + \text{De}f)}{[1 + i\text{De}(1 - \zeta)]} + O(\mu) \end{aligned} \tag{7.5-45}$$

$$T_{\theta\theta}^{\langle 1\rangle}[1 + i\text{De}(1 - \zeta)] = \frac{2\text{De}}{\mu}[2i\text{De}f' - f'' - T_{r\theta}^{\langle 1\rangle}] + O(1) \tag{7.5-46}$$

We note that $T_{r\theta}^{\langle 1\rangle} \sim 1$ and $T_{\theta\theta}^{\langle 1\rangle} \sim 1/\mu$. Thus $T_{\theta\theta}^{\langle 1\rangle}$ is singular and dominant as $T_{\theta\theta}^{\langle 0\rangle}$ was at the previous order.

The constitutive equation must now be combined with the equations of motion. This leads to

r-component:
$$P^{\langle 1\rangle\prime} + T_{rr}^{\langle 1\rangle\prime} + \frac{\mu(T_{rr}^{\langle 1\rangle} - T_{\theta\theta}^{\langle 1\rangle})}{(1 + \mu\zeta)} + \frac{\mu}{(1 + \mu\zeta)}iT_{r\theta}^{\langle 1\rangle} = 0 \tag{7.5-47}$$

θ-component:
$$i\mu P^{\langle 1\rangle} + i\mu T_{\theta\theta}^{\langle 1\rangle} + (1 + \mu\zeta)T_{r\theta}^{\langle 1\rangle\prime} + 2\mu T_{r\theta}^{\langle 1\rangle} = 0 \tag{7.5-48}$$

From the θ-component of the equation of motion we can see that $P^{\langle 1\rangle}$ is of the same size as $T_{\theta\theta}^{\langle 1\rangle}$, that is $P^{\langle 1\rangle} \sim 1/\mu$, and from the radial component of the equation of motion we see that $P^{\langle 1\rangle\prime} \sim 1$. Thus the series for $P^{\langle 1\rangle}$ is

$$P^{\langle 1\rangle} = \frac{P_0^{\langle 1\rangle}}{\mu} + O(1) \tag{7.5-49}$$

so that the pressure is constant across the gap at lowest order in μ.

An equation for the stream function f can be obtained by first eliminating the pressure in Eq. 7.5-48 by taking $d/d\zeta$ of this equation. The leading behavior of the result is given by the dominant balance

$$i\mu T_{\theta\theta}^{\langle 1\rangle\prime} + T_{r\theta}^{\langle 1\rangle\prime\prime} = 0 \tag{7.5-50}$$

The stress coefficients $T_{r\theta}^{\langle 1\rangle}$ and $T_{\theta\theta}^{\langle 1\rangle}$ are then eliminated in favor of the stream function by means of Eqs. 7.5-45 and 46. After much tedious algebra this results in

$$(-2\xi + i(1 - \xi^2))g^{(iv)} - (-2\xi^2 + 2i\xi)g''' + 2i\xi^2 g'' - 4i\xi g' + 4ig = 0 \tag{7.5-51}$$

where

$$\xi = \text{De}(1 - \zeta); \quad g(\xi) = f(\zeta); \quad g'(\xi) = -(1/\text{De})f'(\zeta) \tag{7.5-52}$$

The general solution to Eq. 7.5-51 is

$$g(\xi) = Ae^{(1-i)\xi} + Be^{-(1+i)\xi} + C\xi^2 + D\xi \tag{7.5-53}$$

The constants A, B, C, and D are determined from the boundary conditions. The relevant boundary conditions are obtained by rewriting Eqs. 7.5-38 in terms of the stream function g and new independent variable ξ:

$$g(0) = g'(\text{De}) = g(\text{De}) = 0; \quad g'(0) = 1/\text{De} \tag{7.5-54}$$

These conditions lead to

$$\begin{aligned} \frac{1}{A} &= \text{De}\left\{\left(1 - i - \frac{2}{\text{De}}\right)e^{(1-i)\text{De}} + \left(1 + i + \frac{2}{\text{De}}\right)e^{-(1+i)\text{De}} + 2\right\} \\ B &= -A \\ C &= \text{De}^{-2}\{A(2\text{De} + e^{-(1+i)\text{De}} - e^{(1-i)\text{De}}) - 1\} \\ D &= -2A + \text{De}^{-1} \end{aligned} \tag{7.5-55}$$

Once the stream function is known the stresses are readily calculated from Eqs. 7.5-44 through 46 and the pressure from Eq. 7.5-48.

(e) Discussion of Results for Velocity and Stress Fields

Solutions for the velocity and stress fields for the White–Metzner and CEF[3] models can be obtained in a similar fashion, although the differential equation for the stream function for the White–Metzner model is most conveniently solved numerically.[4] In the following we compare results for these two constitutive equations with those for the convected Maxwell model. In the White–Metzner and CEF models the viscosity is described by a power-law expression, the first normal stress coefficient is given by $2\eta^2(\dot{\gamma})/G$, and $\Psi_2 = 0$. We do this to facilitate comparison between the models. The use of a power-law for η amounts to a Taylor expansion for the viscosity about the mean shear rate $\dot{\gamma}_0 = V/c$ in the small gap limit. The constant modulus G is related to λ_1 in the convected Maxwell model by $\lambda_1 = \eta(\dot{\gamma}_0)/G$. In scaling the stresses and pressure for the shear-thinning models the zero-shear-rate viscosity of the convected Maxwell model is replaced by $\eta^{\langle 0 \rangle} = \eta(\dot{\gamma}_0)$.

Figure 7.5-2 shows the radial variation in the correction to the radial velocity $v_r^* = v_r^{\langle 1 \rangle}/V\mu$ at an angular position of $\theta = 3\pi/2$. From Eq. 7.5-39 we see that at this position $v_r^{\langle 1 \rangle}$ is proportional to the real part of the stream function F. Note that since $v_r^{\langle 0 \rangle} = 0$, v_r^* gives the total radial velocity up through order ε in the expansion in Eq. 7.5-18. Differences in the various model predictions are dramatic. Whereas the inclusion of shear thinning does not greatly affect the De dependence of v_r^*, the absence of memory in the CEF equation allows the development of a boundary layer at the outer cylinder as De is increased. This is in contrast to the boundary layer in $dv_r^*/d\zeta$ that develops with increasing De at the inner cylinder for the convected Maxwell and White–Metzner models. This latter boundary layer results from the no-slip condition on v_θ and the interrelation of $v_r^{\langle 1 \rangle}$ and $v_\theta^{\langle 1 \rangle}$ through the stream function in Eq. 7.5-39.

The tangential velocity expansion in Eq. 7.5-18 has contributions from both $v_\theta^{[0]}$ and[5] $v_\theta^{[1]} = v_\theta^{\langle 1 \rangle} - \zeta V \cos\theta$ and thus we plot the dimensionless total tangential velocity $v_\theta^* = v_\theta/V$ in Fig. 7.5-3. These contours of v_θ^* are shown for the three constitutive equations on the reference domain which has been cut at $\theta = 0$ and unwrapped. An eccentricity of $\varepsilon = 0.4$ has been used to emphasize

[3] The CEF model, mentioned briefly in §4.6, is developed in Example 9.6-3 (see Eq. 9.6-18). This model describes the stresses in viscometric flows exactly and thus is sufficient for the solution of the concentric cylinder problem. However, the CEF model, like the generalized Newtonian fluid, predicts that stress depends on the instantaneous value of the velocity gradient tensor. Thus the CEF equation is incapable of describing memory effects, and comparison of its predictions with those for the White–Metzner and convected Maxwell models allows us to assess the importance of fluid memory, as given by the Deborah number, in the journal-bearing problem.

[4] A. N. Beris, R. C. Armstrong, and R. A. Brown, *op cit.*

[5] The term $v_\theta^{[1]}$ is obtained from Eq. 7.5-16. It is given evaluated at $\bar{r} = r_2$ as the [ ] term in the last line of Eq. 7.5-22. Note also $\partial v_\theta^{\langle 0 \rangle}/\partial r = -V/c$ at lowest order.

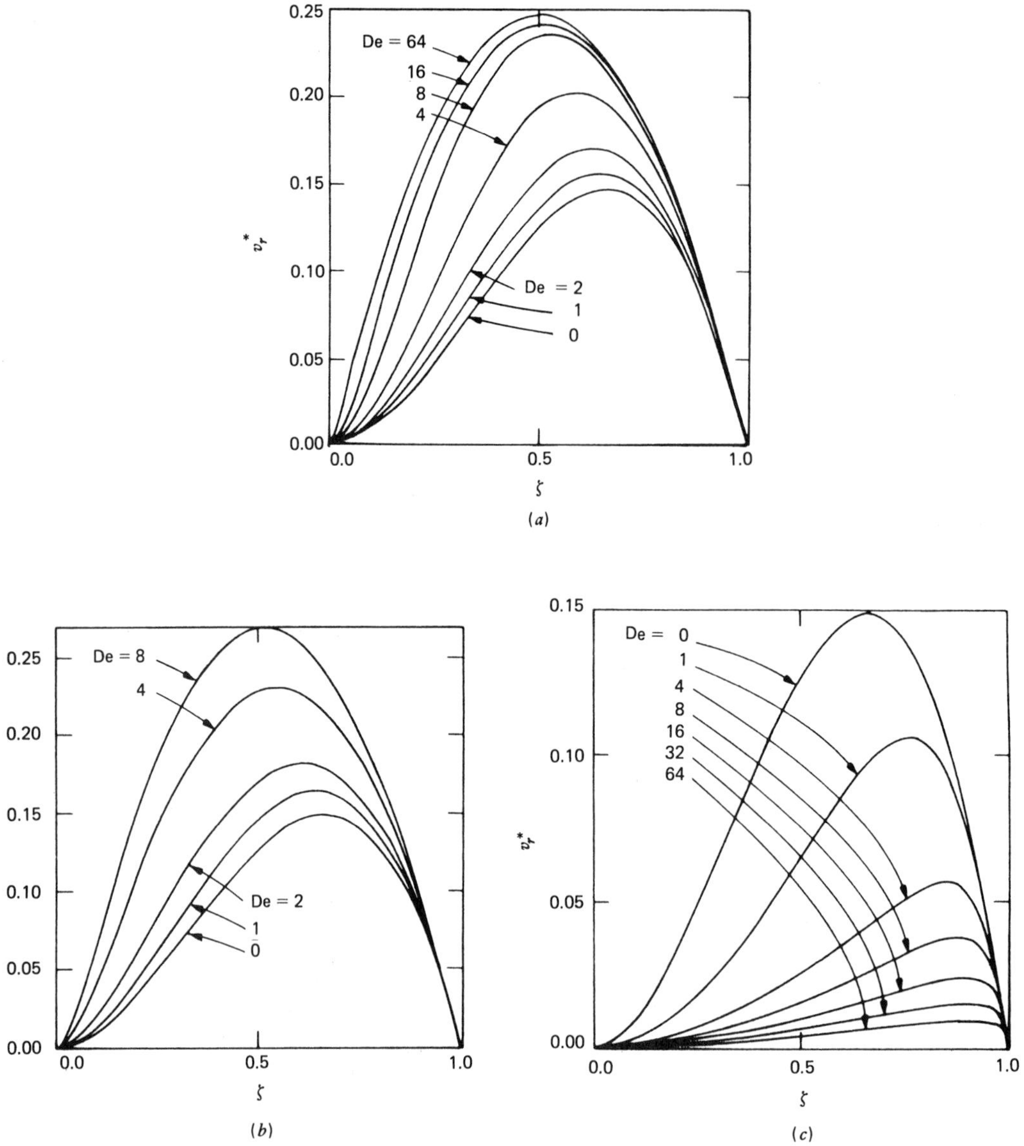

FIGURE 7.5-2. Dependence of the correction to the radial velocity $v_r^* = v_r^{\langle 1 \rangle}/V\mu$ on the dimensionless radial coordinate $\zeta = (r - r_1)/c(1 + \varepsilon \cos \theta)$ at $\theta = 3\pi/2$ for (*a*) the convected Maxwell model, (*b*) the White–Metzner model with power-law index $n = 0.3$, and (*c*) the CEF model with $n = 0.3$. [A. N. Beris, R. C. Armstrong, and R. A. Brown, *J. Non-Newtonian Fluid Mech.*, **13**, 109–148 (1983).]

differences in the models. Note that at De = 0.01 there is flow separation at the outer cylinder at $\theta = 0$ as is predicted from the exact Newtonian solution. Increasing De in the convected Maxwell model causes this secondary flow to disappear, whereas in the White–Metzner model it shrinks slightly and is convected downstream. The CEF model again predicts very different results; for this constitutive equation the secondary flow regime grows with increasing De and stays centered on $\theta = 0$.

The tangential normal stress is shown in Fig. 7.5-4. Since the $\tau_{\theta\theta}^{[0]}$ term is constant at lowest order in μ, we show only the correction term in the dimensionless form $\tau_{\theta\theta}^* = \tau_{\theta\theta}^{\langle 1 \rangle}\,\mu/(\eta^{\langle 0 \rangle}V/c)\text{De}$. Note that $\tau_{\theta\theta}^*$ contains all of the spatial dependence of $\tau_{\theta\theta}$ in the distinguished limit $0 < \mu \ll \varepsilon \ll 1$. A boundary layer near the outer wall is seen to develop in $\tau_{\theta\theta}^*$ in all of the models at large De. In the

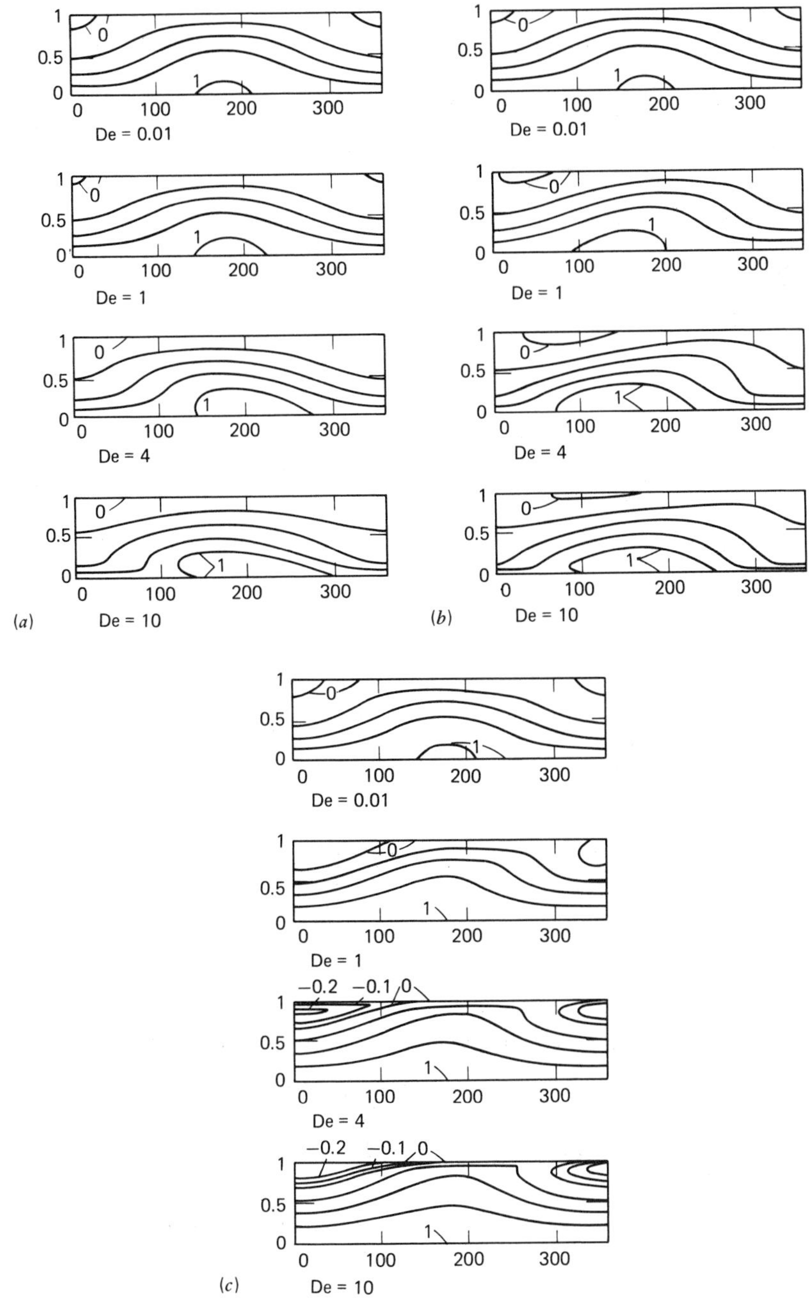

FIGURE 7.5-3. Contours of the tangential velocity $v_\theta^*(\zeta, \theta) = v_\theta(\zeta, \theta)/V$ at Deborah numbers ranging from 0.01 to 10 for the same models as in Fig. 7.5-2: (*a*) convected Maxwell; (*b*) White–Metzner ($n = 0.3$); (*c*) CEF ($n = 0.3$). The total tangential velocity is calculated from the perturbation results as $v_\theta = v_\theta^{\langle 0 \rangle} + \varepsilon(v_\theta^{\langle 1 \rangle} - \zeta\, V \cos\, \theta)$, and an eccentricity of $\varepsilon = 0.4$ was used. The contours are equally spaced between the minimum and maximum values. [A. N. Beris, R. C. Armstrong, and R. A. Brown, *J. Non-Newtonian Fluid Mech.*, **13**, 109–148 (1983).]

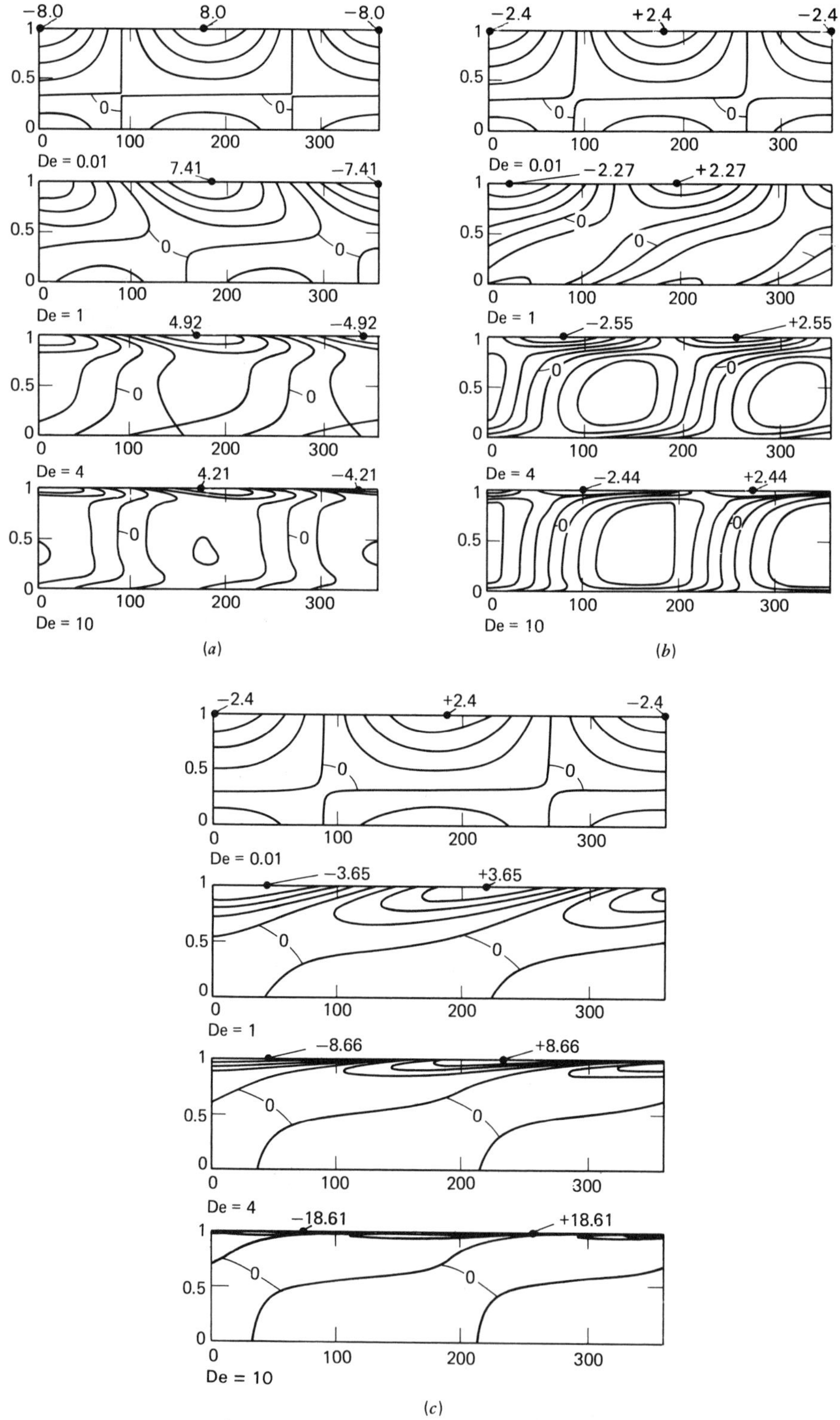

FIGURE 7.5-4. Contours of the correction to the tangential normal stress $\tau^*_{\theta\theta} = \tau^{\langle 1 \rangle}_{\theta\theta}\mu/(\eta^{\langle 0 \rangle}V/c)\mathrm{De}$ for the same three models as in Fig. 7.5-2: (*a*) convected Maxwell; (*b*) White–Metzner ($n = 0.3$); (*c*) CEF ($n = 0.3$). Contours are evenly spaced between the minimum and maximum values. [A. N. Beris, R. C. Armstrong, and R. A. Brown, *J. Non-Newtonian Fluid Mech.*, **13**, 109–148 (1983).]

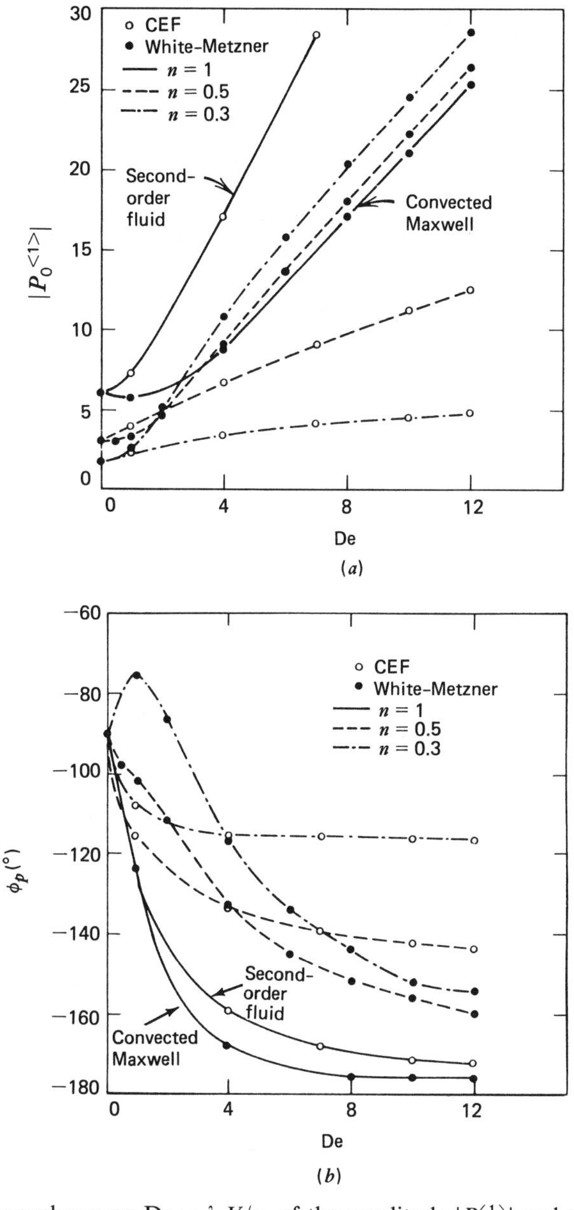

FIGURE 7.5-5. Dependence on De $= \lambda_1 V/r_1$ of the amplitude $|P_0^{\langle 1 \rangle}|$ and phase angle ϕ_p of the dimensionless correction to the pressure $p^* = p^{\langle 1 \rangle}\mu/(\eta_0 V/c) = |P_0^{\langle 1 \rangle}| \cos(\theta + \phi_p)$. Results are shown for the White-Metzner and CEF constitutive equations for several power-law indices. Note that the $n = 1$ curves for White-Metzner and CEF correspond to the convected Maxwell and second-order fluid results, respectively. [A. N. Beris, R. C. Armstrong, and R. A. Brown, *J. Non-Newtonian Fluid Mech.*, **13**, 109–148 (1983).]

models with memory, away from the outer wall $\tau_{\theta\theta}^*$ is nearly independent of ζ. For the CEF equation, however, the tangential normal stress is almost uniformly zero away from the outer wall; moreover, the stress gradient in the boundary layer is much more intense for this constitutive equation.

Finally in Fig. 7.5-5 we depict the dimensionless pressure. When the solutions to $p^{\langle 0 \rangle}$ and $p^{\langle 1 \rangle}$ are combined in the expansion for pressure we find that $p/(\eta_0 V/c) = (\varepsilon/\mu)|P_0^{\langle 1 \rangle}| \cos(\theta + \phi_p) + O(1)$, that is, the largest contribution to p comes from the $P^{\langle 1 \rangle}$ term. The dimensionless pressure is thus defined as $p^* = p^{\langle 1 \rangle}\mu/(\eta_0 V/c) = |P_0^{\langle 1 \rangle}| \cos(\theta + \phi_p)$ where ϕ_p is the phase angle between the pressure and shape variations. The magnitude of the pressure oscillation governs the load-carrying capability of the journal bearing. In this regard it is interesting that at high De, increasing shear thinning increases $|P_0^{\langle 1 \rangle}|$ for the models with memory and decreases it for the model without memory. The phase angle determines the steady-state position of the journal relative to the outer cylinder. For a Newtonian fluid $\phi_p = -90°$ and the offset direction is perpendicular to the load. The White–Metzner model with power-law index of 0.3 shows that the fluid pushes the inner cylinder toward the direction of the narrow part of the gap at moderate De. All other models show that the viscoelastic nature of the fluid causes the direction of the pressure force to shift toward the wide part of the gap.

Experimental observations[6] of a 1.5% polyacrylamide solution in a journal bearing with dimensionless gap μ of approximately 0.1 and $\varepsilon = 0.6$ show that the amplitude of the pressure oscillation decreases rapidly with increasing De and that the minimum in the pressure field shifts upstream, that is, the phase angle ϕ_p becomes less negative, as De increases. None of the models correctly predicts the pressure amplitude dependence and only the White–Metzner model gives the correct trend in ϕ_p (and then only for moderate De). The experiments also show that the separated flow region rapidly disappears with increasing De. This observation is consistent with the predictions of the convected Maxwell model, although the experiments show the recirculation to vanish at much lower De than those calculated here. The CEF model predicts the opposite trend of growth of the separated flow cell, whereas the White–Metzner model predicts first a shift downstream and then a small shrinkage at very high De (~ 10). Thus none of these models is completely satisfactory for this flow but the White–Metzner model seems to give the best overall, qualitative agreement with data. Numerical calculations with the Giesekus model[7] show better agreement than any of the models used here.

EXAMPLE 7.5-2. Torsional Flow of a Convected Jeffreys Fluid between Parallel Disks.[8]

A problem of considerable importance in the characterization of viscoelastic fluids is flow between parallel, coaxial disks, one of which is rotating. This problem was treated for Newtonian fluids in Problem 1B.5, and the geometry is shown in Fig. 1B.5. The disks are separated by a gap B; the bottom plate is fixed, and the upper one rotates with constant angular velocity W for $t \geq 0$. For $t < 0$ the fluid is at rest. Determine the time dependent velocity profiles in the gap for a convected Jeffreys fluid. Is the flow stable? Assume for this analysis that the disks are of infinite radius.

SOLUTION **(a)** Governing Equations

The fluid is taken to be incompressible and the flow is assumed to be axisymmetric: $v_i = v_i(r, z, t)$, $i = r, \theta, z$. The equation of continuity and equations of motion then give

$$\frac{1}{r}\frac{\partial}{\partial r}(rv_r) + \frac{\partial v_z}{\partial z} = 0 \tag{7.5-56}$$

$$\rho\left(\frac{\partial v_r}{\partial t} + v_r\frac{\partial v_r}{\partial r} - \frac{v_\theta^2}{r} + v_z\frac{\partial v_r}{\partial z}\right) = -\frac{\partial p}{\partial r} - \left[\frac{1}{r}\frac{\partial}{\partial r}(r\tau_{rr}) + \frac{\partial}{\partial z}\tau_{zr} - \frac{\tau_{\theta\theta}}{r}\right] \tag{7.5-57}$$

[6] J. V. Lawler, S. J. Muller, R. A. Brown, and R. C. Armstrong, *J. Non-Newtonian Fluid Mech.*, **20**, 51–92 (1986); R. C. Armstrong, R. A. Brown, J. V. Lawler, and S. J. Muller in *Proceedings of the Symposium on Viscoelasticity* (Mathematics Research Center, Madison, WI), Elsevier, New York (1985).

[7] A. N. Beris, R. C. Armstrong, and R. A. Brown, *J. Non-Newtonian Fluid Mech.*, **16**, 141-172 (1984); **19**, 323-347 (1986).

[8] This example is based on N. Phan-Thien, *J. Non-Newtonian Fluid Mech.*, **13**, 325–340 (1983). See also N. Phan-Thien, *J. Fluid Mech.*, **125**, 427–442 (1983); D. F. Griffiths, D. T. Jones, and K. Walters, *J. Fluid Mech.*, **36**, 161–176 (1969); R. K. Bhatnagar and M. G. N. Perera, *J. Rheol.*, **26**, 19–41 (1982).

$$\rho\left(\frac{\partial v_\theta}{\partial t} + v_r \frac{\partial v_\theta}{\partial r} - \frac{v_\theta v_r}{r} + v_z \frac{\partial v_\theta}{\partial z}\right) = -\left[\frac{1}{r^2}\frac{\partial}{\partial r}(r^2 \tau_{r\theta}) + \frac{\partial \tau_{z\theta}}{\partial z}\right] \tag{7.5-58}$$

$$\rho\left(\frac{\partial v_z}{\partial t} + v_r \frac{\partial v_z}{\partial r} + v_z \frac{\partial v_z}{\partial z}\right) = -\frac{\partial p}{\partial z} - \left[\frac{1}{r}\frac{\partial}{\partial r}(r\tau_{rz}) + \frac{\partial \tau_{zz}}{\partial z}\right] \tag{7.5-59}$$

These equations must be solved together with the constitutive equation for the convected Jeffreys model, Eq. 7.2-1. The constitutive equation is somewhat simplified if we split the stress tensor into two parts $\boldsymbol{\tau}_s$ and $\boldsymbol{\tau}_p$ where $\boldsymbol{\tau}_s$ is a "Newtonian solvent" contribution and $\boldsymbol{\tau}_p$ is given by the convected Maxwell model:[9]

$$\boldsymbol{\tau} = \boldsymbol{\tau}_s + \boldsymbol{\tau}_p \tag{7.5-60a}$$

$$\boldsymbol{\tau}_s = -\eta_s \dot{\boldsymbol{\gamma}} \tag{7.5-60b}$$

$$\boldsymbol{\tau}_p + \lambda_1 \boldsymbol{\tau}_{p(1)} = -\eta_p \dot{\boldsymbol{\gamma}} \tag{7.5-60c}$$

If we add Eq. 7.5-60b to λ_1 times the convected derivative of Eq. 7.5-60b and then add this sum to Eq. 7.5-60c, we see that the result is identical to the convected Jeffreys model of Eq. 7.2-1 if we require

$$\eta_0 = \eta_s + \eta_p; \quad \lambda_2 = (\eta_s/\eta_0)\lambda_1 \tag{7.5-61}$$

For the axisymmetric flow, the Newtonian part of the constitutive equation is readily evaluated (see Eqs. B.3-7 through 12 in Appendix B):

$$\boldsymbol{\tau}_s = -\eta_s \begin{pmatrix} 2\dfrac{\partial v_r}{\partial r} & r\dfrac{\partial}{\partial r}\left(\dfrac{v_\theta}{r}\right) & \dfrac{\partial v_r}{\partial z} + \dfrac{\partial v_z}{\partial r} \\ r\dfrac{\partial}{\partial r}\left(\dfrac{v_\theta}{r}\right) & 2\dfrac{v_r}{r} & \dfrac{\partial v_\theta}{\partial z} \\ \dfrac{\partial v_r}{\partial z} + \dfrac{\partial v_z}{\partial r} & \dfrac{\partial v_\theta}{\partial z} & 2\dfrac{\partial v_z}{\partial z} \end{pmatrix} \tag{7.5-62}$$

The six constitutive equations for $\boldsymbol{\tau}_p$ are written by combining the results for $\boldsymbol{\tau}_{p(1)}$ in Example 7.1-2, the rate-of-strain tensor on the right side of Eq. 7.5-62, and Eq. 7.5-60c.

The boundary conditions require no slip on the solid plates:

$$\text{At } z = 0, \quad v_r = v_\theta = v_z = 0$$

$$\text{At } z = B, \quad v_r = v_z = 0; \quad v_\theta = Wr \tag{7.5-63}$$

These boundary conditions suggest we try[10]

$$v_\theta = G(z, t)r \tag{7.5-64}$$

If we let v_r and v_z depend on z through a second function $H(z, t)$ then the continuity equation requires that

$$v_z = -2H; \quad v_r = r\frac{\partial H}{\partial z} \tag{7.5-65}$$

We must then find G and H from the equations of motion and constitutive equations.

[9] This splitting of $\boldsymbol{\tau}$ was used in Eq. 7.3-3 for the Giesekus model.

[10] The velocity field in Eqs. 7.5-64 and 65 is the form of the classical solution due to von Karman for Newtonian fluids; T. von Kármán, *Z. Angew. Math. Mech.*, **1**, 233–252 (1921).

It is convenient to introduce dimensionless variables as follows:

$$\begin{gathered} \zeta = z/B; \quad \xi = r/B; \quad \tau = Wt \\ g(\zeta, \tau) = G/W; \quad h(\zeta, \tau) = H/WB \\ \mathbf{T} = \boldsymbol{\tau}/\eta_0 W = -\Lambda\dot{\boldsymbol{\Gamma}} + \boldsymbol{t} \\ \dot{\boldsymbol{\Gamma}} = \dot{\boldsymbol{\gamma}}/W; \quad \boldsymbol{t} = \boldsymbol{\tau}_p/\eta_0 W \\ P = p/\eta_0 W; \quad \Lambda = \eta_s/\eta_0 = \lambda_2/\lambda_1; \\ \text{De} = \lambda_1 W; \quad \text{Re} = \rho W B^2/\eta_0 \end{gathered} \tag{7.5-66}$$

Note that De is the Deborah number for the problem and gives the relative importance of elasticity in the problem, and $\boldsymbol{t}$ is the dimensionless "polymer" contribution to the stress tensor. The dimensionless velocity field is given by

$$v_r/WB = \xi h_\zeta; \quad v_\theta/WB = \xi g; \quad v_z/WB = -2h \tag{7.5-67}$$

in which h_ζ denotes $\partial h(\zeta, \tau)/\partial \zeta$.

In terms of the functions g and h the dimensionless rate-of-strain tensor is

$$\dot{\boldsymbol{\Gamma}} = \begin{pmatrix} 2h_\zeta & 0 & \xi h_{\zeta\zeta} \\ 0 & 2h_\zeta & \xi g_\zeta \\ \xi h_{\zeta\zeta} & \xi g_\zeta & -4h_\zeta \end{pmatrix} \tag{7.5-68}$$

and the dimensionless components of the "polymer" contribution to the stress tensor are

$$t_{rr} + \text{De}\left[\frac{\partial t_{rr}}{\partial \tau} + \xi h_\zeta \frac{\partial}{\partial \xi} t_{rr} - 2h \frac{\partial}{\partial \zeta} t_{rr} - 2h_\zeta t_{rr} - 2\xi h_{\zeta\zeta} t_{rz}\right] = -2(1 - \Lambda)h_\zeta \tag{7.5-69}$$

$$t_{r\theta} + \text{De}\left[\frac{\partial t_{r\theta}}{\partial \tau} + \xi h_\zeta \frac{\partial}{\partial \xi} t_{r\theta} - 2h \frac{\partial}{\partial \zeta} t_{r\theta} - 2h_\zeta t_{r\theta} - \xi g_\zeta t_{rz} - \xi h_{\zeta\zeta} t_{\theta z}\right] = 0 \tag{7.5-70}$$

$$t_{rz} + \text{De}\left[\frac{\partial t_{rz}}{\partial \tau} + \xi h_\zeta \frac{\partial}{\partial \xi} t_{rz} - 2h \frac{\partial}{\partial \zeta} t_{rz} + h_\zeta t_{rz} - \xi h_{\zeta\zeta} t_{zz}\right] = -(1 - \Lambda)\xi h_{\zeta\zeta} \tag{7.5-71}$$

$$t_{\theta\theta} + \text{De}\left[\frac{\partial t_{\theta\theta}}{\partial \tau} + \xi h_\zeta \frac{\partial}{\partial \xi} t_{\theta\theta} - 2h \frac{\partial}{\partial \zeta} t_{\theta\theta} - 2h_\zeta t_{\theta\theta} - 2\xi g_\zeta t_{\theta z}\right] = -2(1 - \Lambda)h_\zeta \tag{7.5-72}$$

$$t_{\theta z} + \text{De}\left[\frac{\partial t_{\theta z}}{\partial \tau} + \xi h_\zeta \frac{\partial}{\partial \xi} t_{\theta z} - 2h \frac{\partial}{\partial \zeta} t_{\theta z} + h_\zeta t_{\theta z} - \xi g_\zeta t_{zz}\right] = -(1 - \Lambda)\xi g_\zeta \tag{7.5-73}$$

$$t_{zz} + \text{De}\left[\frac{\partial t_{zz}}{\partial \tau} + \xi h_\zeta \frac{\partial}{\partial \xi} t_{zz} - 2h \frac{\partial}{\partial \zeta} t_{zz} + 4h_\zeta t_{zz}\right] = 4(1 - \Lambda)h_\zeta \tag{7.5-74}$$

The structure of the constitutive equations suggests we try power series solutions for the stress components:

$$t_{ij} = t_{ij}^0 + \xi t_{ij}^1 + \xi^2 t_{ij}^2 + \cdots = \sum_{k=0}^{\infty} \xi^k t_{ij}^k(\zeta, \tau) \tag{7.5-75}$$

Inserting these expansions into Eqs. 7.5-69 through 74 gives constitutive equations for the following nonzero t_{ij}^k:

$$t_{\theta z}^1 + \mathrm{De}\left[\frac{\partial t_{\theta z}^1}{\partial \tau} + 2h_\zeta t_{\theta z}^1 - 2h\frac{\partial}{\partial \zeta}t_{\theta z}^1 - g_\zeta t_{zz}^0\right] = -(1-\Lambda)g_\zeta \tag{7.5-76}$$

$$t_{zz}^0 + \mathrm{De}\left[\frac{\partial t_{zz}^0}{\partial \tau} - 2h\frac{\partial}{\partial \zeta}t_{zz}^0 + 4h_\zeta t_{zz}^0\right] = 4(1-\Lambda)h_\zeta \tag{7.5-77}$$

$$t_{rz}^1 + \mathrm{De}\left[\frac{\partial t_{rz}^1}{\partial \tau} - 2h\frac{\partial}{\partial \zeta}t_{rz}^1 + 2h_\zeta t_{rz}^1 - h_{\zeta\zeta}t_{zz}^0\right] = -(1-\Lambda)h_{\zeta\zeta} \tag{7.5-78}$$

$$t_{r\theta}^2 + \mathrm{De}\left[\frac{\partial t_{r\theta}^2}{\partial \tau} - 2h\frac{\partial t_{r\theta}^2}{\partial \zeta} - g_\zeta t_{rz}^1 - h_{\zeta\zeta}t_{\theta z}^1\right] = 0 \tag{7.5-79}$$

$$t_{rr}^0 + \mathrm{De}\left[\frac{\partial t_{rr}^0}{\partial \tau} - 2h\frac{\partial t_{rr}^0}{\partial \zeta} - 2h_\zeta t_{rr}^0\right] = -2(1-\Lambda)h_\zeta \tag{7.5-80a}$$

$$t_{rr}^2 + \mathrm{De}\left[\frac{\partial t_{rr}^2}{\partial \tau} - 2h\frac{\partial t_{rr}^2}{\partial \zeta} - 2h_{\zeta\zeta}t_{rr}^1\right] = 0 \tag{7.5-80b}$$

$$t_{\theta\theta}^0 + \mathrm{De}\left[\frac{\partial t_{\theta\theta}^0}{\partial \tau} - 2h\frac{\partial}{\partial \zeta}t_{\theta\theta}^0 - 2h_\zeta t_{\theta\theta}^0\right] = -2(1-\Lambda)h_\zeta \tag{7.5-81a}$$

$$t_{\theta\theta}^2 + \mathrm{De}\left[\frac{\partial t_{\theta\theta}^2}{\partial \tau} - 2h\frac{\partial}{\partial \zeta}t_{\theta\theta}^2 - 2g_\zeta t_{\theta z}^1\right] = 0 \tag{7.5-81b}$$

All other t_{ij}^k are zero. Note also that $t_{\theta\theta}^0 = t_{rr}^0$.

We can now return to the equation of motion and rewrite its components in terms of the assumed velocity field and the expansion coefficients for the stress components. At the same time we expand the pressure in a series in ξ:

$$P = P_0(\zeta, \tau) + \xi P_1(\zeta, \tau) + \xi^2 P_2(\zeta, \tau) + \cdots \tag{7.5-82}$$

From the r-component of the equation of motion we find that $P_1 = P_3 = P_4 \ldots = 0$ and that only P_0 and P_2 are nonzero. The r-, θ-, and z-components of the equation of motion then give, respectively,

$$\mathrm{Re}(h_{\zeta\tau} + h_\zeta^2 - g^2 - 2hh_{\zeta\zeta}) = 2P_2 + \Lambda h_{\zeta\zeta\zeta} - 3t_{rr}^2 + t_{\theta\theta}^2 - \frac{\partial t_{rz}^1}{\partial \zeta} \tag{7.5-83}$$

$$\mathrm{Re}(g_\tau + 2h_\zeta g - 2hg_\zeta) = \Lambda g_{\zeta\zeta} - 4t_{r\theta}^2 - \frac{\partial}{\partial \zeta}t_{\theta z}^1 \tag{7.5-84}$$

$$2\mathrm{Re}(h_\tau - 2hh_\zeta) = \frac{\partial}{\partial \zeta}P_0 + 2\Lambda h_{\zeta\zeta} + 2t_{rz}^1 + \frac{\partial}{\partial \zeta}t_{zz}^0 \tag{7.5-85a}$$

$$0 = \frac{\partial}{\partial \zeta}P_2 \tag{7.5-85b}$$

From the last equation we see that $P_2 = P_2(\tau)$. We can thus differentiate the r-component of the equation of motion with respect to ζ and thereby eliminate P_2:

$$\Lambda h_{\zeta\zeta\zeta\zeta} - 3\frac{\partial}{\partial\zeta}t_{rr}^2 + \frac{\partial}{\partial\zeta}t_{\theta\theta}^2 - \frac{\partial^2}{\partial\zeta^2}t_{rz}^1 - \text{Re}(h_{\zeta\zeta\tau} - 2gg_\zeta - 2hh_{\zeta\zeta\zeta}) = 0 \tag{7.5-86}$$

In order to solve for h, g, and the t_{ij}^k we use the above ζ-derivative of the r-component of the equation of motion together with the θ-component of the equation of motion and the constitutive equations, Eqs. 7.5-76 through 81. Once these are known we can find P_0 from the z-component of the equation of motion and P_2 from the r-component of the equation of motion. These equations are to be solved with the boundary conditions

$$\begin{aligned} &\text{At } \zeta = 0, \quad && h = \frac{\partial h}{\partial\zeta} = g = 0 \\ &\text{At } \zeta = 1, \quad && g = 1,\ h = \frac{\partial h}{\partial\zeta} = 0 \\ &\text{At } \tau = 0, \quad && g = h = \frac{\partial h}{\partial\zeta} = t_{ij}^k = 0 \end{aligned} \tag{7.5-87}$$

(b) Small De, Re Expansion

We illustrate the solution to these equations for small Deborah and Reynolds numbers

$$0 < \text{De} \ll \text{Re} \ll 1 \tag{7.5-88}$$

The dependent variables are expanded as power series in De and Re. For example, for h we write

$$h = h_0 + \text{Re}\, h_{01} + \text{De}\, h_{10} + \text{Re}^2\, h_{02} + \text{Re De}\, h_{11} + \cdots \tag{7.5-89}$$

At lowest order the constitutive equations give

$$\begin{aligned} t_{\theta z0}^1 &= -(1-\Lambda)g_{0\zeta} \\ t_{zz0}^0 &= 4(1-\Lambda)h_{0\zeta} \\ t_{rz0}^1 &= -(1-\Lambda)h_{0\zeta\zeta} \\ t_{r\theta 0}^2 &= 0 \\ t_{rr0}^0 &= -2(1-\Lambda)h_{0\zeta}; \qquad t_{rr0}^2 = 0 \\ t_{\theta\theta 0}^k &= t_{rr0}^k \qquad (k = 0, 2) \end{aligned} \tag{7.5-90}$$

Combining these with the equations of motion gives

$$h_{0\zeta\zeta\zeta\zeta} = 0; \quad g_{0\zeta\zeta} = 0 \tag{7.5-91}$$

which when solved with the boundary conditions leads directly to

$$h_0 = 0; \quad g_0 = \zeta \tag{7.5-92}$$

At lowest order the θ-component of velocity varies linearly from zero at the bottom disk to the rigid rotation at the top plate.

Continuing to higher order gives

$$
\begin{aligned}
h &= -\tfrac{1}{60}\,\mathrm{Re}(2\zeta^2 - 3\zeta^3 + \zeta^5) + \cdots \\
g &= \zeta + \tfrac{1}{30}(1-\Lambda)\mathrm{ReDe}(2\zeta^2 - 3\zeta^3 + \zeta^5) \\
&\quad + \tfrac{1}{30}\,\mathrm{Re}^2(-\tfrac{4}{105}\zeta - \tfrac{1}{6}\zeta^4 + \tfrac{3}{10}\zeta^5 - \tfrac{2}{21}\zeta^7) + \cdots
\end{aligned}
\tag{7.5-93}
$$

These results can be used to show that the torques on the upper and lower disks are the same until order Re^2.

(c) Stability

Here we perform a linear stability analysis for the limit of creeping flow, Re = 0, which is the limit of interest in most rheometer applications. The steady state solution in this limit is easily found to be

$$
\begin{aligned}
g &= \zeta \\
h &= 0 \\
t^1_{\theta z} &= -(1-\Lambda) \\
t^2_{\theta\theta} &= -2(1-\Lambda)\mathrm{De} \\
&\text{all other } t^k_{ij} = 0
\end{aligned}
\tag{7.5-94}
$$

This solution is valid for all De.

Now consider small disturbances about this flow. Let

$$
\begin{aligned}
g &= \zeta + g_1 \\
h &= h_1 \\
t^1_{\theta z} &= -(1-\Lambda) + \sigma_{\theta z} \\
t^2_{\theta\theta} &= -2(1-\Lambda)\mathrm{De} + \sigma_{\theta\theta} \\
t^2_{rr} &= \sigma_{rr}; \quad t^0_{\theta\theta} = t^0_{rr} = \sigma_{rr,0} \\
t^1_{rz} &= \sigma_{rz}; \quad t^0_{zz} = \sigma_{zz}; \quad t^2_{r\theta} = \sigma_{r\theta}
\end{aligned}
\tag{7.5-95}
$$

where h_1 and g_1 are the velocity disturbances and the σ_{ij} are the stress disturbances. When Eqs. 7.5-95 are substituted into the constitutive equations and the creeping flow equations of motion and the Laplace transforms of these equations taken, we obtain the following transformed constitutive equation and equations of motion:

$$\bar{\sigma}_{rr} + s\mathrm{De}\,\bar{\sigma}_{rr} = \mathrm{De}\,\sigma^0_{rr} \tag{7.5-96}$$

$$\bar{\sigma}_{zz} + s\mathrm{De}\,\bar{\sigma}_{zz} = \mathrm{De}\,\sigma^0_{zz} + 4(1-\Lambda)\bar{h}'_1 \tag{7.5-97}$$

$$\bar{\sigma}_{rz} + s\mathrm{De}\,\bar{\sigma}_{rz} = \mathrm{De}\,\sigma^0_{rz} - (1-\Lambda)\bar{h}''_1 \tag{7.5-98}$$

$$\bar{\sigma}_{r\theta} + s\mathrm{De}\,\bar{\sigma}_{r\theta} = \mathrm{De}\,(\sigma^0_{r\theta} + \bar{\sigma}_{rz} - (1-\Lambda)\bar{h}''_1) \tag{7.5-99}$$

$$\bar{\sigma}_{\theta\theta} + s\mathrm{De}\,\bar{\sigma}_{\theta\theta} = \mathrm{De}\,(\sigma^0_{\theta\theta} + 2\bar{\sigma}_{\theta z} - 2(1-\Lambda)\bar{g}'_1) \tag{7.5-100}$$

$$\bar{\sigma}_{\theta z} + s\mathrm{De}\,\bar{\sigma}_{\theta z} = \mathrm{De}\,(\sigma^0_{\theta z} + \bar{\sigma}_{zz} + 2(1-\Lambda)\bar{h}'_1) - (1-\Lambda)\bar{g}'_1 \tag{7.5-101}$$

$$\Lambda\bar{g}''_1 - 4\bar{\sigma}_{r\theta} - \bar{\sigma}'_{\theta z} = 0 \tag{7.5-102}$$

$$\Lambda\bar{h}^{(iv)}_1 - 3\bar{\sigma}'_{rr} + \bar{\sigma}'_{\theta\theta} - \bar{\sigma}''_{rz} = 0 \tag{7.5-103}$$

In these equations the overbar denotes the transformed variable, (e.g., $\mathscr{L}\{\sigma_{rr}\} = \bar{\sigma}_{rr}$), the initial conditions are denoted by the superscript zero (e.g., $\sigma_{rr}(\zeta, 0) = \sigma_{rr}^0(\zeta)$), the primes denote differentiation with respect to ζ (e.g., $d\bar{h}_1/d\zeta = \bar{h}_1'$), and s is the Laplace transform variable. We have not included a constitutive equation for $\sigma_{rr,0}$ inasmuch as this does not enter the equations of motion.

We now proceed to obtain a single ordinary differential equation for $\bar{h}$. This is done by first substituting the constitutive equations into the equations of motion, Eqs. 7.5-102 and 103. The equation that results from Eq. 7.5-102 is

$$\bar{g}_1'' - \frac{2(1-\Lambda)\,\mathrm{De}\,\bar{h}_1''}{(1+\Lambda s\mathrm{De})} = \frac{\mathrm{De}}{(1+\Lambda s\mathrm{De})}\left[4\sigma_{r\theta}^0 + \sigma_{\theta z}^{0\prime} + \frac{\mathrm{De}(4\sigma_{rz}^0 + \sigma_{zz}^{0\prime})}{(1+s\mathrm{De})}\right] \tag{7.5-104}$$

This result can be used to eliminate $\bar{g}_1$ from the equation obtained from Eq. 7.5-103 and the constitutive equations. The resulting equation for $\bar{h}$ is (after some lengthy algebra)

$$\begin{aligned}\bar{h}_1^{(\mathrm{iv})} &+ \frac{4(1-\Lambda)\mathrm{De}^2(s^2\mathrm{De}^2 + 4s\mathrm{De} - 2\Lambda + 5)}{(1+\Lambda s\mathrm{De})^2(1+s\mathrm{De})^2}\bar{h}_1'' \\ &= \frac{\mathrm{De}}{(1+\Lambda s\mathrm{De})}\left[3\sigma_{rr}^{0\prime} - \sigma_{\theta\theta}^{0\prime} + \sigma_{rz}^{0\prime\prime} - \frac{2\,\mathrm{De}}{(1+s\mathrm{De})}\left(\frac{\mathrm{De}}{(1+s\mathrm{De})}\sigma_{zz}^{0\prime} + \sigma_{\theta z}^{0\prime}\right)\right. \\ &\quad \left. + \frac{2(1-\Lambda)(2+s\mathrm{De})\mathrm{De}}{(1+s\mathrm{De})(1+\Lambda s\mathrm{De})}\left(\frac{\mathrm{De}}{1+s\mathrm{De}}(4\sigma_{rz}^{0\prime} + \sigma_{zz}^{0\prime}) + 4\sigma_{r\theta}^0 + \sigma_{\theta z}^{0\prime}\right)\right]\end{aligned} \tag{7.5-105}$$

For simplicity let us assume an initial disturbance in only one of the stress components, say $\sigma_{\theta\theta}^0$. That is, we take

$$\sigma_{\theta\theta}^0 = \alpha_k \mathscr{Re}\{ie^{2\pi ik\zeta}\} \qquad (k = 1, 2\ldots) \tag{7.5-106}$$

where the amplitudes α_k are real and all other $\sigma_{ij}^0 = 0$. Then,

$$\sigma_{\theta\theta}^{0\,\prime} = -2\pi k\alpha_k \mathscr{Re}\{e^{2\pi ik\zeta}\} \tag{7.5-107}$$

When these disturbances are substituted into Eq. 7.5-105, we see that $\bar{h}_1$ is of the form

$$\bar{h}_1 = \bar{\hbar}_1(s)e^{2\pi ik\zeta} \tag{7.5-108}$$

in which $\bar{\hbar}_1$ is given by

$$\begin{aligned}\bar{\hbar}_1 &= \frac{\alpha_k \mathrm{De}\,(1+\Lambda s\mathrm{De})(1+s\mathrm{De})^2}{8\pi k[\pi^2k^2(1+\Lambda s\mathrm{De})^2(1+s\mathrm{De})^2 - (1-\Lambda)\mathrm{De}^2(s^2\mathrm{De}^2 + 4s\mathrm{De} - 2\Lambda + 5)]} \\ &= \frac{\alpha_k \mathrm{De}(1+\Lambda s\mathrm{De})(1+s\mathrm{De})^2}{8\pi k[a_0(s\mathrm{De})^4 + a_1(s\mathrm{De})^3 + a_2(s\mathrm{De})^2 + a_3 s\mathrm{De} + a_4]}\end{aligned} \tag{7.5-109}$$

where the a_i's are

$$\begin{aligned}a_0 &= \pi^2k^2\Lambda^2 \\ a_1 &= 2\pi^2k^2\Lambda(1+\Lambda) \\ a_2 &= \pi^2k^2(\Lambda^2 + 4\Lambda - 11) - (1-\Lambda)\mathrm{De}^2 \\ a_3 &= 2\pi^2k^2(1+\Lambda) - 4(1-\Lambda)\mathrm{De}^2 \\ a_4 &= \pi^2k^2 + (1-\Lambda)(2\Lambda - 5)\mathrm{De}^2\end{aligned} \tag{7.5-110}$$

To investigate the stability for $0 < \Lambda < 1$ we notice[11] that a *necessary* condition for stability is that all coefficients of the characteristic polynomial, $a_0, a_1, \ldots a_4$, be positive. This leads directly to the condition

$$\mathrm{De} < \mathrm{De}_c = \frac{\pi}{\sqrt{(1-\Lambda)(5-2\Lambda)}} \tag{7.5-111}$$

for stability, where De_c denotes the critical Deborah number. Note that the kth mode becomes unstable at a critical value $\mathrm{De}_{c,k} = \pi k/\sqrt{(1-\Lambda)(5-2\Lambda)}$. The stability of the flow is dictated by the smallest such number.

To show that Eq. 7.5-111 is also a sufficient condition for stability we must construct the Routh array:

$$\begin{matrix} a_0 & a_2 & a_4 \\ a_1 & a_3 & \\ b_1 & b_2 & \\ c_1 & & \\ d_1 & & \end{matrix} \tag{7.5-112}$$

where

$$\begin{aligned} b_1 &= (a_1 a_2 - a_0 a_3)/a_1; \qquad & b_2 &= a_4 \\ c_1 &= (b_1 a_3 - a_1 b_2)/b_1; \qquad & d_1 &= a_4 \end{aligned} \tag{7.5-113}$$

The necessary and sufficient condition for stability is that all coefficients in the first column of the Routh array are positive for $0 < \Lambda < 1$. Clearly a_0 and a_1 are positive. Also the condition that $d_1 > 0$ is given by Eq. 7.5-111. It is straightforward to show that b_1 and c_1 are also positive provided that Eq. 7.5-111 is satisfied. Therefore this equation is necessary and sufficient for $0 < \Lambda < 1$. Note that as $\Lambda \to 1$ the critical Deborah number in Eq. 7.5-111 goes to infinity, which is consistent with the fact that the creeping flow of a Newtonian fluid in this geometry is stable.

When $\Lambda = 0$ (the convected Maxwell model) the characteristic polynomial is only of second order, that is, the denominator of Eq. 7.5-109 is

$$8\pi k[a_0(s\mathrm{De})^2 + a_1 s\mathrm{De} + a_2] \tag{7.5-114}$$

in which

$$\begin{aligned} a_0 &= \pi^2 k^2 - \mathrm{De}^2 \\ a_1 &= 2\pi^2 k^2 - 4\mathrm{De}^2 \\ a_2 &= \pi^2 k^2 - 5\mathrm{De}^2 \end{aligned} \tag{7.5-115}$$

[11] D. R. Coughanowr and L. B. Koppel, *Process Systems Analysis and Control*, McGraw-Hill, New York (1965).

The Routh test results in the following possibilities:

De	a_0	a_1	a_2	Stability
$\mathrm{De}^2 < \dfrac{\pi^2 k^2}{5}$	+	+	+	Stable
$\dfrac{\pi^2 k^2}{5} < \mathrm{De}^2 < \dfrac{\pi^2 k^2}{2}$	+	+	−	Unstable
$\dfrac{\pi^2 k^2}{2} < \mathrm{De}^2 < \pi^2 k^2$	+	−	−	Unstable
$\pi^2 k^2 < \mathrm{De}^2$	−	−	−	Unstable

The critical Deborah number is then $\mathrm{De}_c = \pi/\sqrt{5}$. The fact that the kth mode becomes stable as De is increased beyond $\pi^2 k^2$ does not affect the overall stability of the flow, since as soon as $\mathrm{De} > \mathrm{De}_c$ there will always be a value of k large enough so that the mode is unstable.

§7.6 LIMITATIONS OF DIFFERENTIAL MODELS AND RECOMMENDATIONS FOR THEIR USE

The differential constitutive equations presented in this chapter are not limited to small deformation gradients as are the linear viscoelastic models of Chapter 5 or to small, slowly changing deformation rates as is the retarded-motion expansion of Chapter 6. Moreover, the quasi-linear and nonlinear models introduced in §§7.2 and 7.3 are simple enough to allow (barely!) analytical solutions to be obtained for some interesting flows. For this reason these constitutive equations have proven very useful in making exploratory fluid dynamical calculations to gain insight into the qualitative effects of viscoelasticity on complex flow fields. The price of simplicity in the constitutive equation is paid in the poor approximation to some of the material functions. The utility of one of these models in a specific flow problem depends on which material functions are most closely related to the flow and on how well the model describes those properties. Often, making this assessment must be based on experience. For example, in calculating flows where there is expected to be a strong elongational flow component, the Oldroyd models are not good choices, unless care is taken to ensure that the parameters are chosen to avoid an infinite elongational viscosity. Similarly, in flows that are primarily shearing and involve large shear rates, the Giesekus model should be used only if a retardation time is included so that the power-law region in which $\eta \propto \dot{\gamma}^{-1}$ is avoided.

There are many nonlinear differential models other than the ones we have chosen in §7.2 and 3 which can be found in the literature. We list some of these in Table 7.6-1. Note that these models contain no new types of terms.

All of the models presented here can be viewed as nonlinear generalizations of the linear viscoelastic Jeffreys model; all[1] have the same linear viscoelastic material functions as the Jeffreys model. The inability to describe linear viscoelasticity correctly is the result of having a single relaxation time and is the most serious defect of these models. There are two ways that we can remedy this problem. The first method, which can be used for only some of the nonlinear differential models, is to reformulate the model in integral form where it is

[1] Except for the White–Metzner model, which does not have a linear viscoelastic limit.

straightforward to include as many time constants as desired. The use of integral models with multiple time constants is discussed in Chapter 8. The connection between the differential and integral forms of the same model, for example, the convected Jeffreys model, is described in Chapter 9.

A second approach, which can be used for any of the nonlinear differential constitutive equations, is to superpose a series of models, each with a different time constant. This was done in Chapter 5 to construct the generalized Maxwell model of linear viscoelasticity. The corresponding generalized convected Maxwell model is

$$\boldsymbol{\tau} = \sum_i \boldsymbol{\tau}_i \tag{7.6-1}$$

$$\boldsymbol{\tau}_i + \lambda_i \boldsymbol{\tau}_{i(1)} = -\eta_i \boldsymbol{\gamma}_{(1)} \tag{7.6-2}$$

Here the constants η_i and λ_i are determined from linear viscoelasticity (see Example 5.3-7). Equations 7.6-1 and 2 are also known as the Rouse–Zimm model (see Chapter 15), which is obtained from the kinetic theory of a dilute solution of macromolecules modeled as freely jointed bead-spring units. The generalized convected Maxwell model is not any better than the convected Maxwell model at describing nonlinear material functions such as viscosity and normal stress coefficients. More successful data fitting can be achieved with superposition of nonlinear models, such as the Giesekus model, Eq. 7.3-4. To illustrate the improvement obtained by including additional modes we show a comparison in Figs. 7.6-1 through 3 of an 8-mode Giesekus model with experimental data in shear and elongational flow for a low-density polyethylene melt (see §§3.3–3.5). The time constants and viscosities, λ_i and η_i, are the same as those listed in Table 5.3-2. The α_i for each mode were chosen empirically to give a reasonable fit of the data. Clearly the fit can be made quite good. The drawback to using this, or any other multimode differential model, is that it is computationally impractical. In numerical simulations, the number of unknowns in a problem increases nearly linearly with the number of modes; with present computers, using more than two or possibly three modes is not reasonable.

Two other successful examples of multimode models are those of Acierno, La Mantia, Marrucci, and Titomanlio[2] and of Phan-Thien and Tanner;[3] these models are given in Table 7.6-1. Both models were derived from network theories[4] of concentrated polymer solutions and melts, and both have been shown to describe data on low-density polyethylene melts well.[3,5] Fitting data with the Phan-Thien–Tanner model is particularly straightforward, since the term involving ξ is most important in shear flows, and the parameter ε is most important in shearfree flows.

Still other superpositions are possible. In particular, Giesekus has pointed out that there is no reason a priori to assume the different time scale events are uncoupled.[6] This leads, however, to computationally cumbersome equations and is not recommended for fluid flow problem solving.

[2] D. Acierno, F. P. La Mantia, G. Marrucci, and G. Titomanlio, *J. Non-Newtonian Fluid Mech.*, **1**, 125–146 (1976); see also H. Janeschitz-Kriegl, *Polymer Melt Rheology and Flow Birefringence*, Springer, Berlin (1983), §2.5.5.

[3] N. Phan-Thien, *J. Rheol.*, **22**, 259–283 (1977); an alternative formulation of the nonlinear stress term involving ε is given by N. Phan-Thien and R. I. Tanner, *J. Non-Newtonian Fluid Mech.*, **2**, 353–365 (1977). A very similar model has been proposed by M. W. Johnson and D. Segalman, *J. Non-Newtonian Fluid Mech.*, **2**, 255–270 (1977).

[4] R. J. J. Jongschaap, *J. Non-Newtonian Fluid Mech.*, **8**, 183–190 (1981), has a clear discussion of the connection between the Acierno, La Mantia, Marrucci, and Titomanlio model and network theories. See also Chapter 20.

[5] D. Acierno, F. P. La Mantia, G. Marrucci, G. Rizzio, and G. Titomanlio, *J. Non-Newtonian Fluid Mech.*, **1**, 147–157 (1976); D. Acierno, F. P. La Mantia, and G. Marrucci, *J. Non-Newtonian Fluid Mech.*, **2**, 271–280 (1977).

[6] H. Giesekus, *Rheol. Acta*, **11**, 69–109 (1982); **12**, 367–374 (1983).

TABLE 7.6-1

Differential Constitutive Equations

Name	Reference	Constants or Functions	Equation for Model		Comments
Tanner	[a]	$\eta(\dot{\gamma})$, λ	$\boldsymbol{\tau} + \lambda \boldsymbol{\tau}_{(1)} = -\eta \boldsymbol{\gamma}_{(1)}$	(A)	Modification of Oldroyd B model (cf. Eq. 7.2-1) with $\lambda_2 = 0$.
FENE-P	[b]	$\eta_s, \eta_p, \lambda_1, b$ $\varepsilon = 2/(b(b+2))$	$\boldsymbol{\tau} = -\eta_s \dot{\boldsymbol{\gamma}} + \boldsymbol{\tau}_p$ $Z\boldsymbol{\tau}_p + \lambda_1 \boldsymbol{\tau}_{p(1)} - \lambda_1 \left(\boldsymbol{\tau}_p - (1 - \varepsilon b) \frac{\eta_p \boldsymbol{\delta}}{\lambda_1} \right) \frac{D \ln Z}{Dt}$ $= -(1 - \varepsilon b)\eta_p \dot{\boldsymbol{\gamma}}$ $Z = 1 + \left(\frac{3}{b}\right)\left(1 - \lambda_1 \frac{\operatorname{tr} \boldsymbol{\tau}_p}{3\eta_p}\right)$	(B)	Approximate constitutive equation for FENE dumbbells (cf. Eq. 13.5-56). The splitting of the stress tensor is similar to Eq. 7.3-3.
Nearly Hookean dumbbell	[c]	$\eta_s, \eta_p, \lambda_1, b$	$\boldsymbol{\tau} = -\eta_s \dot{\boldsymbol{\gamma}} + \boldsymbol{\tau}_p$ $\boldsymbol{\tau}_p = \eta_p \boldsymbol{\alpha}_{(1)}$ $\boldsymbol{\alpha} + \lambda_1 \boldsymbol{\alpha}_{(1)} + \frac{1}{b}\{(\operatorname{tr} \boldsymbol{\alpha})\boldsymbol{\alpha} + 2\boldsymbol{\alpha} \cdot \boldsymbol{\alpha}\} = \boldsymbol{\delta}$	(C)	Derived for dilute suspension of nearly-Hookean dumbbells. (cf. Eqs. 13.5-56a–c)

Acierno, La Mantia, Marrucci, and Titomanlio [d]	$\eta_{i0}(=G_{i0}\lambda_{i0})$, λ_{i0}, a	$\boldsymbol{\tau} = \sum_i \boldsymbol{\tau}_i$ $\left(1 - \lambda_i \frac{D \ln G_i}{Dt}\right)\boldsymbol{\tau}_i + \lambda_i \boldsymbol{\tau}_{i(1)} = -\eta_i \dot{\boldsymbol{\gamma}}$ $G_i = G_{i0} x_i;\ \lambda_i = \lambda_{i0} x_i^{1.4};\ \eta_i = G_i \lambda_i = \eta_{i0} x_i^{2.4}$ $\lambda_i \frac{dx_i}{dt} = (1 - x_i) - a x_i \sqrt{-(\text{tr}\ \boldsymbol{\tau}_i)/2G_i}$ (D)	Derived from a network theory (see §20.5) The subscript '0' on the a_i, λ_i, and η_i denote values fit from linear viscoelastic data (see Example 5.3-7)
Phan-Thien and Tanner [e]	η_i, λ_i, ξ, ε	$\boldsymbol{\tau} = \sum_i \boldsymbol{\tau}_i$ $Z(\text{tr}\ \boldsymbol{\tau}_i)\boldsymbol{\tau}_i + \lambda_i \boldsymbol{\tau}_{i(1)} + \frac{\xi}{2} \lambda_i \{\dot{\boldsymbol{\gamma}} \cdot \boldsymbol{\tau}_i + \boldsymbol{\tau}_i \cdot \dot{\boldsymbol{\gamma}}\} = -\eta_i \dot{\boldsymbol{\gamma}}$ $Z = \begin{cases} 1 - \varepsilon\lambda_i\ \text{tr}\ \boldsymbol{\tau}_i/\eta_i \\ \exp(-\varepsilon\lambda_i\ \text{tr}\ \boldsymbol{\tau}_i/\eta_i) \end{cases}$ (E)	Derived from a network theory (Chapter 20). Two forms of $Z(\text{tr}\ \boldsymbol{\tau}_i)$ have been used; the exponential form causes $\bar{\eta}$ to have a maximum.

[a] R. I. Tanner, *ASLE Trans.*, **8**, 179–183 (1965).

[b] R. I. Tanner, *Trans. Soc. Rheol.*, **19**, 37–65 (1975); R. B. Bird, *J. Non-Newtonian Fluid Mech.*, **5**, 1–12 (1979); R. B. Bird, P. J. Dotson, and N. L. Johnson, *J. Non-Newtonian Fluid Mech.*, **7**, 213–235 (1980); **8**, 193 (1981); X.-J. Fan, *J. Non-Newtonian Fluid Mech.*, **17**, 125–144 (1985).

[c] R. C. Armstrong and S. Ishikawa, *J. Rheol.*, **24**, 143–165 (1980); R. C. Armstrong, S. K. Gupta, and O. Basaran, *Polym. Eng. Sci.*, **20**, 466–472 (1980).

[d] D. Acierno, F. P. La Mantia, G. Marrucci, and G. Titomanlio, *J. Non-Newtonian Fluid Mech.*, **1**, 125–146 (1976).

[e] N. Phan-Thien, *J. Rheol.*, **22**, 259–283 (1977); an alternative formulation of the nonlinear stress term involving ε is given by N. Phan-Thien and R. I. Tanner, *J. Non-Newtonian Fluid Mech.*, **2**, 353–365 (1977). A very similar model has been proposed by M. W. Johnson and D. Segalman, *J. Non-Newtonian Fluid Mech.*, **2**, 255–270 (1977).

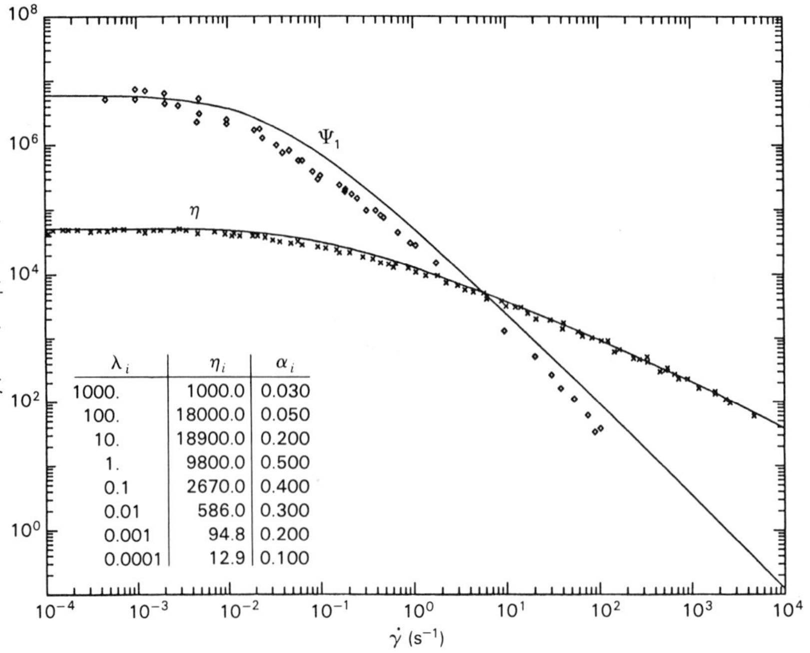

FIGURE 7.6-1. Viscosity and first normal stress coefficient for an 8-mode Giesekus model as compared to experimental data for a low-density polyethylene melt (cf. Fig. 3.3-2). The λ_i and η_i for the eight modes are listed in Table 5.3-2; the α_i were found empirically.

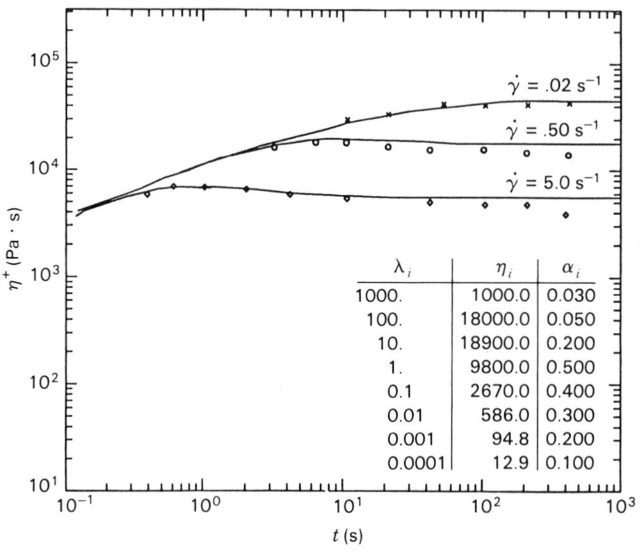

FIGURE 7.6-2. Comparison of η^+ calculated by an 8-mode Giesekus model with experimental data for a low-density polyethylene melt (cf. Fig. 3.4-7). The model parameters are given in Table 5.3-2 and in the figure.

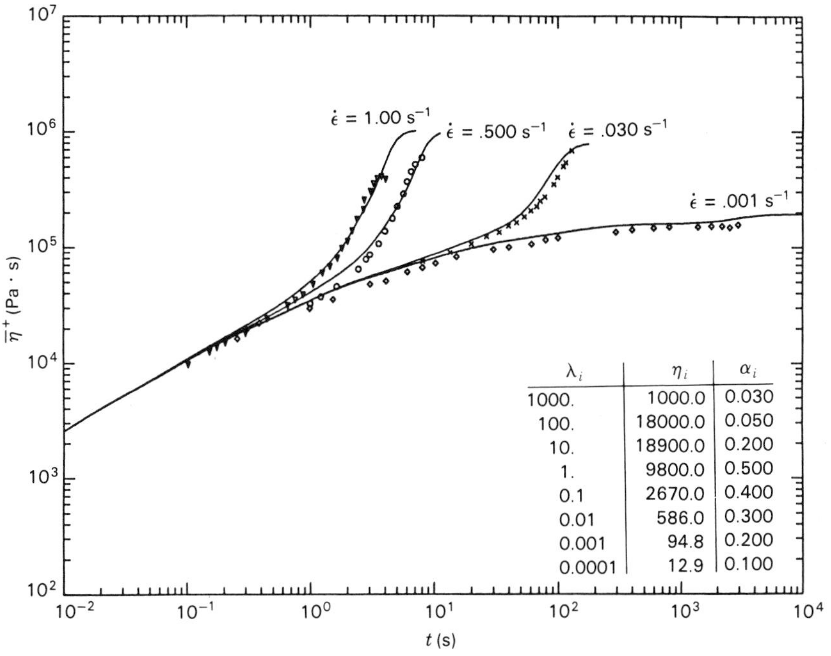

FIGURE 7.6-3. Comparison of $\bar{\eta}^+$ calculated by an 8-mode Giesekus model with experimental data for a low-density polyethylene melt (cf. Fig. 3.5-3). The model parameters are given in Table 5.3-2 and in the figure.

Finally, Leonov and co-workers have proposed a differential model based on nonequilibrium thermodynamics.[7] In this model the stress tensor $\boldsymbol{\tau}$ is given in terms of an elastic potential function, an elastic strain tensor, and an irreversible rate-of-strain tensor. The constitutive equation is very closely related to nonlinear elasticity; at least for simple shear flow it is identical to the Giesekus model discussed in §7.3.

Clearly still other nonlinear differential models can be constructed by using the methods of §§7.1-7.3. It is, however, very difficult to decide what new kinds of terms should be added and time consuming to evaluate the resulting constitutive equations by comparison with experimental data. Helpful guidance in the selection of new terms can be obtained from molecular theory, which is the subject of Volume 2. It is interesting that many of the most successful models presented in this chapter have been obtained by molecular theory.

PROBLEMS

7A.1 Estimation of Oldroyd Model Parameters

Consider the 4-constant Oldroyd model obtained by setting $\lambda_3 = \lambda_4 = \lambda_6 = \lambda_7 = 0$ in Eq. 7.3-2. Obtain the numerical values of η_0, λ_1, λ_2, and λ_5 by using the data on $\eta(\dot{\gamma})$ and $\eta'(\omega)$ on 2% polyisobutylene in Primol (see Figs. 3.3-3 and 3.4-4). Having obtained the four parameters, plot the curve of Ψ_1 given by Eq. B of Table 7.3-3 and compare with the experimental data in Fig. 3.3-5. What conclusions do you draw from this comparison?

[7] A. I. Leonov, *Rheol. Acta*, **15**, 85–93 (1976); A. I. Leonov, E. H. Lipkina, E. D. Paskhin, and A. N. Prokunin, *Rheol. Acta*, **15**, 411–426 (1976). See also P. A. Dashner and W. E. Van Arsdale, *J. Non-Newtonian Fluid Mech.*, **8**, 59–67 (1981); **12**, 375–382 (1983).

7A.2 Deviation from Stokes' Law Because of Viscoelasticity of a Macromolecular Fluid.

Stokes' law for steady-state, creeping flow around a sphere was given in Example 1.4-1. The analogous problem of solving the equations of continuity and motion (with the inertial terms omitted) from the Oldroyd model of Eq. 7.3-2 was first done by Leslie.[1] In this work λ_3 and λ_4 were set equal to 0 and the abbreviations $\sigma_i = \lambda_i\lambda_5 + \lambda_{i+5}(\lambda_1 - \frac{3}{2}\lambda_5)$ were used (see the definition σ_i in Table 7.3-3). Leslie performed a perturbation analysis about the known Newtonian fluid solution, expanding in powers of the Deborah number $\text{De} = v_\infty \lambda_1/R$. His solution is valid for

$$\text{Re} \ll \text{De} \ll 1 \tag{7A.2-1}$$

where $\text{Re} = Rv_\infty\rho/\eta_0$ is the Reynolds number. Although the details of the derivation are tedious and lengthy, the final result for the drag force up through $O(\text{De}^2)$ is rather simple. The result, as corrected by Giesekus[2] in a later publication, is

$$F_D = 6\pi\eta_0 R v_\infty\left\{1 - \frac{1}{2275}\left[\left(\frac{401}{11} - 39\frac{\lambda_2}{\lambda_1}\right)\left(1 - \frac{\lambda_2}{\lambda_1}\right) - 471\left(\frac{\sigma_2 - \sigma_1}{\lambda_1^2}\right)\right]\text{De}^2 + \cdots\right\} \tag{7A.2-2}$$

a. From Eq. 7A.2-1 show that in applying the above result for F_D to spheres falling in viscoelastic fluids, there is an upper limit to the radius of the sphere that can be used and an upper limit to the speed of descent of the sphere.

b. Design an experiment for testing Eq. 7A.2-2 for a fluid with the following properties:

$$\eta_0 = 1.05 \text{ Pa·s} \qquad \lambda_5 = 0.05 \text{ s}$$
$$\lambda_1 = 0.1 \text{ s} \qquad \lambda_6 = \lambda_7 = 0$$
$$\lambda_2 = 0.02 \text{ s}$$

c. Find an expression for the terminal velocity of the falling sphere.

7B.1 Normal Stresses in Small Amplitude Oscillatory Shearing of the Convected Jeffreys Model

In Example 7.2-1 it was shown that η' and η'' for the convected Jeffreys model were identical to those for the Jeffreys model of linear viscoelasticity. Here we finish the task of showing that the two models also agree in predicting zero normal stresses in the limit of small strains.

a. Show that the normal stress τ_{xx} in small-amplitude oscillatory shear flow is the solution to

$$\left(1 + \lambda_1 \frac{d}{dt}\right)\tau_{xx} = T(1 + \cos 2\omega t + \lambda_1\omega \sin 2\omega t) \tag{7B.1-1}$$

in which

$$T = -\frac{\eta_0\gamma_0^2\omega^2(\lambda_1 - \lambda_2)}{(1 + \lambda_1^2\omega^2)} \tag{7B.1-2}$$

[1] F. M. Leslie, *Q. J. Mech. Appl. Math.*, **14**, 36–48 (1961), with an appendix by R. I. Tanner.

[2] H. Giesekus, *Rheol. Acta*, **3**, 59–71 (1963), see footnote 16 on p. 69. In this paper the problem of simultaneous translation and rotation of a sphere in a viscoelastic liquid is studied. The effect of the container boundaries on the movement of a particle through a non-Newtonian fluid has been studied by B. Caswell, *Chem. Eng. Sci.*, **25**, 1167–1176 (1970).

b. Obtain a periodic solution to τ_{xx} of the form

$$\frac{\tau_{xx}}{T} = A \cos 2\omega t + B \sin 2\omega t + C \tag{7B.1-3}$$

where A, B, and C are functions of ω. Notice that the normal stress oscillates at a frequency 2ω about a non-zero mean value.

c. Use the results of part **(b)** together with Eqs. 7.2-9 through 11 to show that the ratio of the magnitude of τ_{xx} to the magnitude of τ_{xy} goes to zero as γ_0 approaches zero for all ω.

7B.2 Shearfree Flow for the White–Metzner Model

a. Show that the second invariant of the rate-of-deformation tensor is

$$\dot{\gamma} = \sqrt{(3 + b^2)}\,|\dot{\varepsilon}_0| \tag{7B.2-1}$$

where $\dot{\varepsilon}_0$ is the constant strain rate.

b. Derive Eqs. C in Table 7.3-1.

c. Suppose $\eta(\dot{\gamma})$ is given by the power-law function (Eq. 4.1-10). What is the maximum value of $\dot{\varepsilon}_0$ for which the results in part **(b)** are physically realistic. Is this an improvement on the convected Maxwell model?

7B.3 Step Strain Stress Relaxation for the Oldroyd 8-Constant Model

It is desired to determine the linear viscoelastic properties of the Oldroyd 8-constant model by finding its response to a step strain experiment, $\dot{\gamma}_{yx} = \gamma_0 \delta(t)$ where the shear strain amplitude γ_0 is small.

a. Show that in the limit $\gamma_0 \to 0$ if we retain only the terms first order in γ_0, all normal stresses are zero and the shear stress is given by

$$\tau_{yx} = \tau_{yx,1}\gamma_0 + O(\gamma_0^3) \tag{7B.3-1}$$

$$\left[1 + \lambda_1 \frac{d}{dt}\right]\tau_{yx,1} = -\eta_0\left[1 + \lambda_2 \frac{d}{dt}\right]\delta(t) \tag{7B.3-2}$$

where $\tau_{yx}(-\infty) = 0$.

b. Integrate Eq. 7B.3-2 to obtain

$$\tau_{yx,1} = 0 \qquad t < 0 \tag{7B.3-3a}$$

$$\tau_{yx,1} = \tau_0 e^{-t/\lambda_1} \qquad t > 0 \tag{7B.3-3b}$$

in which $\tau_0 = \tau_{yx,1}(0^+)$.

c. In order for the solution to be nontrivial it is clear that τ_0 must be singular. What form for τ_0 does the differential equation suggest?

d. Combine the results of **(b)** and **(c)** to obtain

$$\begin{aligned}\tau_{yx,1} &= -\frac{\eta_0\gamma_0}{\lambda_1}\left[\left(1 - \frac{\lambda_2}{\lambda_1}\right)H(t)e^{-t/\lambda_1} + \lambda_2\delta(t)\right] \\ &= -\frac{\eta_0\gamma_0}{\lambda_1}H(t)\left[\left(1 - \frac{\lambda_2}{\lambda_1}\right)e^{-t/\lambda_1} + 2\lambda_2\delta(t)\right]\end{aligned} \tag{7B.3-4}$$

What is the linear relaxation modulus $G(t)$?

7B.4 Elongational Viscosity for 8-Constant Oldroyd Fluid

a. Write out Eq. 7.3-2 in matrix form for steady-state elongational flow $v_z = \dot{\varepsilon}z$, $v_x = -(1/2)\dot{\varepsilon}x$, and $v_y = -(1/2)\dot{\varepsilon}y$, where $\dot{\varepsilon}$ is a constant.

b. From the matrix equation obtain a pair of equations for $\tau_{xx} - \tau_{zz}$ and $\text{tr}\,\boldsymbol{\tau} = \tau_{xx} + \tau_{yy} + \tau_{zz}$

$$(1 - (\lambda_1 - \lambda_3)\dot{\varepsilon})(\tau_{xx} - \tau_{zz}) - (\tfrac{3}{2}\lambda_5 - \lambda_1 + \lambda_3)\dot{\varepsilon}\,\text{tr}\,\boldsymbol{\tau} = 3\eta_0\dot{\varepsilon}(1 - (\lambda_2 - \lambda_4)\dot{\varepsilon}) \tag{7B.4-1}$$

$$(2(\lambda_1 - \lambda_3) - 3\lambda_6)\dot{\varepsilon}(\tau_{xx} - \tau_{zz}) + \text{tr}\,\boldsymbol{\tau} = 3\eta_0\dot{\varepsilon}^2(2(\lambda_2 - \lambda_4) - 3\lambda_7) \tag{7B.4-2}$$

c. Solve the equations in **(b)** for $\tau_{xx} - \tau_{zz}$ and find that the elongational viscosity $\bar{\eta}$ is given by Eq. F in Table 7.3-3. To what extent is this result in accord with known experimental facts?

7B.5 A 3-Constant Oldroyd Model[3]

The Oldroyd 8-constant model in §7.3 has often been used in simplified form by reducing the number of parameters. One simplification is to require that Ψ_2 be zero (this leads to $\lambda_3 = \lambda_4 = 0$) and to require arbitrarily that $\boldsymbol{\tau}$ be traceless (this leads to $\lambda_6 = \tfrac{2}{3}\lambda_1$, $\lambda_7 = \tfrac{2}{3}\lambda_2$).

a. Show that this leads to the following expressions for the steady shear flow functions:

$$\frac{\eta}{\eta_0} = \frac{1 + \tfrac{2}{3}\lambda_1\lambda_2\dot{\gamma}^2}{1 + \tfrac{2}{3}(\lambda_1\dot{\gamma})^2} \tag{7B.5-1}$$

$$\frac{\Psi_1}{2\eta_0\lambda_1} = \frac{1 - (\lambda_2/\lambda_1)}{1 + \tfrac{2}{3}(\lambda_1\dot{\gamma})^2} \tag{7B.5-2}$$

and that the small-amplitude oscillatory functions are given by Eqs. D and E in Table 7.3-3. What restrictions have to be placed on the time constants λ_1 and λ_2?

b. Show that $\eta(\dot{\gamma})$ is the same function as $\eta'(c\omega)$, and that $\Psi_1\dot{\gamma}/\sqrt{6}$ vs. $\dot{\gamma}$ is the same function as η'' vs. $c\omega$, with $c = \sqrt{2/3} = 0.82$. The constant c is called the "shift factor," and experimental values[4] lie between about 0.4 and 0.7.

c. An oft-quoted empiricism[3] known as the Cox–Merz rule (cf. §3.6) states that $\eta(\dot{\gamma})$ should be the same function as $|\eta^*|$ vs. ω. To what extent is the 3-constant Oldroyd model in agreement with this rule?

7B.6 Third-Order Retarded Motion Constants Corresponding to the Oldroyd 6-Constant Model

The Oldroyd and Giesekus models presented in §7.3 can easily be developed into a retarded-motion expansion by successive substitution. We illustrate the procedure here for the Oldroyd 6-constant model with $\lambda_6 = \lambda_7 = 0$.

a. Recognize that the first-order approximation to the stress in Eq. 7.3-2 is $\boldsymbol{\tau}_1 = -\eta_0\boldsymbol{\gamma}_{(1)}$ so that successive approximations to $\boldsymbol{\tau}$ are conveniently given by

$$\begin{aligned}\boldsymbol{\tau}_k = {} & -\eta_0[\boldsymbol{\gamma}_{(1)} + \lambda_2\boldsymbol{\gamma}_{(2)} + \lambda_4\boldsymbol{\gamma}_{(1)}^2] - \lambda_1\boldsymbol{\tau}_{k-1(1)} \\ & -\tfrac{1}{2}\lambda_3\{\boldsymbol{\tau}_{k-1}\cdot\boldsymbol{\gamma}_{(1)} + \boldsymbol{\gamma}_{(1)}\cdot\boldsymbol{\tau}_{k-1}\} - \tfrac{1}{2}\lambda_5(\text{tr}\,\boldsymbol{\tau}_{k-1})\boldsymbol{\gamma}_{(1)}\end{aligned} \tag{7B.6-1}$$

[3] M. C. Williams and R. B. Bird, *Phys. Fluids*, **5**, 1126–1127 (1962); the oscillatory normal stresses were given incorrectly in that publication, but were corrected in *Ind. Eng. Chem. Fundam.*, **3**, 42–49 (1964).

[4] See M. C. Williams, *Chem. Eng. Sci.*, **20**, 693–702 (1965), Figs. 4a and b.

b. Show then that τ_2 is

$$\tau_2 = -\eta_0[\gamma_{(1)} - (\lambda_1 - \lambda_2)\gamma_{(2)} - (\lambda_3 - \lambda_4)\gamma_{(1)}^2] \tag{7B.6-2}$$

c. Repeat the iteration to obtain the third-order expansion coefficients given in Table 6.2-2. In obtaining these coefficients it is necessary to use the Cayley–Hamilton theorem.

7B.7 Relation between Ψ_1 and η^-

The following equation has been found to be valid for a dilute suspension of rigid dumbbells as well as for dilute solutions of chainlike bead-spring models (see Example 15.4-1):

$$\Psi_1(\dot{\gamma}) = 2\int_0^\infty \eta^-(t, \dot{\gamma})dt \tag{7B.7-1}$$

Does this result hold for

a. The convected Jeffreys model?
b. The White–Metzner model?
c. The Oldroyd 4-constant model?

For part **(c)** you will need η^- from Problem 7D.1. What conclusions can you draw from these results?

7B.8 Steady Shearfree Flow of the Giesekus Model

a. Show that for the Giesekus model in a steady shearfree flow the three normal stresses are given by

$$aT_{pxx}^2 - (1 + (1 + b)\chi)T_{pxx} + (1 - \Lambda)(1 + b)\chi = 0 \tag{7B.8-1}$$

$$aT_{pyy}^2 - (1 + (1 - b)\chi)T_{pyy} + (1 - \Lambda)(1 - b)\chi = 0 \tag{7B.8-2}$$

$$aT_{pzz}^2 - (1 - 2\chi)T_{pzz} - 2(1 - \Lambda)\chi = 0 \tag{7B.8-3}$$

in which the following dimensionless quantities have been used: $\mathbf{T}_p = \boldsymbol{\tau}_p/(\eta_0/\lambda_1)$, $\chi = \lambda_1\dot{\varepsilon}$, $\Lambda = \lambda_2/\lambda_1$, and $a = \alpha/(1 - \Lambda)$.

b. Solve these equations and obtain results for $\bar{\eta}_1$ and $\bar{\eta}_2$. In order to choose the correct roots from Eqs. 7B.8-1 through 3, require that the results simplify properly in the limit as $\alpha \to 0$.

c. Show that for steady elongational flow ($b = 0$) the following asymptotes hold when $\Lambda = 0$:

$$\frac{\bar{\eta}}{\eta_0} \sim \frac{2}{\alpha} \quad \text{as } \lambda_1\dot{\varepsilon} \to \infty \tag{7B.8-4}$$

$$\frac{\bar{\eta}}{\eta_0} \sim \frac{1}{\alpha} \quad \text{as } \lambda_1\dot{\varepsilon} \to -\infty \tag{7B.8-5}$$

7B.9 Eccentric-Disk Rheometer Flow

The eccentric-disk rheometer consists of two parallel disks, both of which rotate with a constant angular velocity W. The axes of rotation of the disks are separated by a distance a, and the gap between the disks is b (see Fig. 3B.1). The flow in this system may be reasonably well described by the velocity distribution

$$v_x = -W(y - Az); \quad v_y = Wx; \quad v_z = 0 \tag{7B.9-1}$$

in which $A = a/b$.

a. Find the tensors $\nabla \boldsymbol{v}$, $(\nabla \boldsymbol{v})^\dagger$, and $\boldsymbol{\gamma}_{(1)}$ for this flow. Note that $\boldsymbol{\gamma}_{(1)}$ is independent of position and time, so that it can be concluded that $\boldsymbol{\tau}$ is constant throughout the gap; therefore $D\boldsymbol{\tau}/Dt$ vanishes.

b. Display the convected Maxwell model in matrix form for this flow.

c. Show that this model gives

$$\left(-\frac{\tau_{xz}}{AW}\right) = \eta_0 \frac{1}{1 + \lambda_1^2 W^2} \tag{7B.9-2}$$

$$\left(-\frac{\tau_{yz}}{AW}\right) = \eta_0 \frac{\lambda_1 W}{1 + \lambda_1^2 W^2} \tag{7B.9-3}$$

d. Compare the results of **(c)** with Eqs. 7.2-10 and 11 with $\lambda_2 = 0$ (or Eqs. 5.3-8 and 9 with one term in the sum). What do you conclude? It is shown in Example 10.1-2 that a model-independent result of this sort can be obtained in the limit that $A \to 0$—that is, in the limit of small deformation gradients.

7B.10 Radial Flow between Two Lubricated Disks[5]

A high-viscosity polymer flows radially inward in the space between two parallel porous circular disks as shown in Fig. 7B.10. The disk surfaces are lubricated by injection of a small amount of low-viscosity liquid through the disks, so that the polymer fluid velocity profiles are roughly as shown in the figure. Find the relation between the volume flow rate and the pressure drop between the outer radius R_o and the inner radius R_i. Use the convected Maxwell model.

a. Use the continuity equation to get the velocity distribution $v_r = -C/r$, where $C = Q/2\pi B$ is a constant of integration.

b. Show next that the convected Maxwell model gives the following expressions for τ_{rr} and $\tau_{\theta\theta}$:

$$r\frac{d\tau_{rr}}{dr} - \left(\frac{r^2}{C\lambda_1} - 2\right)\tau_{rr} = \frac{2\eta_0}{\lambda_1} \tag{7B.10-1}$$

$$r\frac{d\tau_{\theta\theta}}{dr} - \left(\frac{r^2}{C\lambda_1} + 2\right)\tau_{\theta\theta} = -\frac{2\eta_0}{\lambda_1} \tag{7B.10-2}$$

Then rewrite these equations in terms of the dimensionless variables: $x = r^2/2\lambda_1 C$, $T_{rr} = \tau_{rr}\lambda_1/\eta_0$, $T_{\theta\theta} = \tau_{\theta\theta}\lambda_1/\eta_0$.

c. Solve the differential equations in **(b)**, with the boundary conditions $T_{rr} = T_{rr,o}$ and $T_{\theta\theta} = T_{\theta\theta,o}$ at $r = R_o$ (or $x = x_o$) to get

$$T_{rr} = -x^{-1} + C_1 x^{-1} e^x \tag{7B.10-3}$$

$$C_1 = x_o e^{-x_o}(T_{rr,o} + x_o^{-1}) \tag{7B.10-4}$$

$$T_{\theta\theta} = 1 - xe^x E_1(x) + C_2 x e^x \tag{7B.10-5}$$

$$C_2 = x_o^{-1} e^{-x_o}[T_{\theta\theta,o} - 1 + x_o e^{x_o} E_1(x_o)] \tag{7B.10-6}$$

in which C_1 and C_2 are constants of integration and $E_1(x) = -Ei(-x) = \int_x^\infty e^{-x} x^{-1}\,dx$.

d. Next use the r-component of the equation of motion in the form $r\,d(p + \tau_{rr})/dr = -(\tau_{rr} - \tau_{\theta\theta})$ to get

$$P + T_{rr} = \tfrac{1}{2}\{-x^{-1} + C_1[x^{-1}e^x - Ei(x)] - e^x E_1(x) + C_2 e^x\} + C_3 \tag{7B.10-7}$$

$$C_3 = P_o + \tfrac{1}{2}T_{rr,o}[1 + x_o e^{-x_o} Ei(x_o)] + \tfrac{1}{2}e^{-x_o}Ei(x_o) - \tfrac{1}{2}x_o^{-1}T_{\theta\theta,o} + \tfrac{1}{2}x_o^{-1} \tag{7B.10-8}$$

[5] D. K. Gartling, *J. Non-Newtonian Fluid Mech.*, **17**, 203–231 (1985).

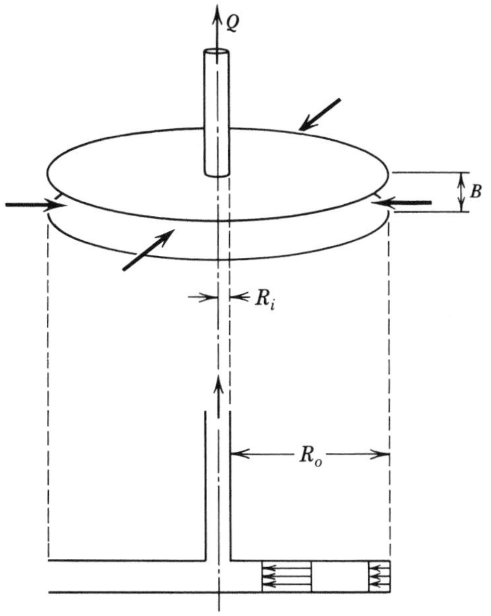

FIGURE 7B.10. Inward flow between two parallel porous circular disks through which a low-viscosity liquid lubricant is injected.

in which $Ei(x) = \int_{-\infty}^{x} e^x x^{-1}\, dx$ is the standard "exponential integral" function, $P = p\lambda_1/\eta_0$, and P_o is the dimensionless pressure at $r = R_o$.

e. The "pressure drop" desired now is $(P + T_{rr})_o - (P + T_{rr})_i$. However, when we obtain this expression, we find that it contains the unmeasurable quantities $T_{rr,o}$ and $T_{\theta\theta,o}$. Since, at the outer rim the flow is very slow, these quantities may at a lowest level of approximation be set equal to zero, and then P_o can be equated to $p_{atm}\lambda_1/\eta_o$. An improved approximation can be obtained by using the retarded-motion expansion equivalent to the Maxwell model (see Eq. 7B.6-2 with $\lambda_2 = \lambda_3 = \lambda_4 = 0$); since the flow field is known at $r = R_o$, the $\boldsymbol{\gamma}_{(1)}$ and $\boldsymbol{\gamma}_{(2)}$ in Eq. 7B.6-2 can be evaluated in terms of known quantities.

7C.1 Arbitrary Shearing and Shearfree Flows of the Convected Jeffreys Model

a. Show that the differential equations, Eqs. 7.2-3 and 6, for the stresses in a convected Jeffreys model undergoing arbitrary time-dependent shear flow can be integrated to give

$$\begin{aligned}\tau_{yx} &= -\eta_0 \frac{\lambda_2}{\lambda_1}\dot{\gamma}_{yx}(t) - \frac{\eta_0}{\lambda_1}\int_{-\infty}^{t}\left[\left(1 - \frac{\lambda_2}{\lambda_1}\right)e^{-(t-t')/\lambda_1}\dot{\gamma}_{yx}(t')\right]dt' \\ &= -\eta_0 \frac{\lambda_2}{\lambda_1}\dot{\gamma}_{yx}(t) + \frac{\eta_0}{\lambda_1^2}\int_{-\infty}^{t}\left[\left(1 - \frac{\lambda_2}{\lambda_1}\right)e^{-(t-t')/\lambda_1}\gamma_{yx}(t, t')\right]dt' \qquad (7C.1\text{-}1)\end{aligned}$$

$$\begin{aligned}\tau_{xx} &= -2\frac{\eta_0}{\lambda_1}\int_{-\infty}^{t}\int_{-\infty}^{t'}\left[\left(1 - \frac{\lambda_2}{\lambda_1}\right)e^{-(t-t'')/\lambda_1}\dot{\gamma}_{yx}(t'')\dot{\gamma}_{yx}(t')\right]dt''\, dt' \\ &= -\frac{\eta_0}{\lambda_1^2}\int_{-\infty}^{t}\left[\left(1 - \frac{\lambda_2}{\lambda_1}\right)e^{-(t-t')/\lambda_1}\gamma_{yx}^2(t, t')\right]dt' \qquad (7C.1\text{-}2)\end{aligned}$$

where $\gamma_{yx}(t, t') = \int_t^{t'} \dot{\gamma}_{yx}(t'')\, dt''$ is the shear strain and the stress tensor is taken to be zero at $t = -\infty$.

b. Use the integral results of part **(a)** to obtain the material functions η', η'', η^+, Ψ_1^+, η^-, and Ψ_1^- that were given in §7.2.

c. Assume that a convected Jeffreys fluid is at equilibrium prior to time $t = 0$. For $t \geq 0$ the fluid is subjected to a shearfree flow with arbitrary $\dot{\varepsilon}$ and b. Show that the differential equations that describe this problem, Eqs. 7.2-19 and 20 can be integrated to give

$$\tau_{xx} = \eta_0(1+b)\frac{\lambda_2}{\lambda_1}\dot{\varepsilon}(t) + \frac{\eta_0(1+b)}{\lambda_1}\left(1 - \frac{\lambda_2}{\lambda_1}\right)\int_0^t e^{-(t-t')/\lambda_1}e^{+(1+b)\varepsilon(t,t')}\dot{\varepsilon}(t')dt' \quad (7C.1\text{-}3)$$

$$\tau_{zz} = -2\eta_0\frac{\lambda_2}{\lambda_1}\dot{\varepsilon}(t) - \frac{2\eta_0}{\lambda_1}\left(1 - \frac{\lambda_2}{\lambda_1}\right)\int_0^t e^{-(t-t')/\lambda_1}e^{-2\varepsilon(t,t')}\dot{\varepsilon}(t')dt' \quad (7C.1\text{-}4)$$

in which $\varepsilon(t, t') = \int_t^{t'} \dot{\varepsilon}(t'')\,dt''$ is the Hencky strain between times t and t'. The normal stress τ_{yy} is given by Eq. 7C.1-3 with $(1 + b)$ replaced everywhere by $(1 - b)$.

d. Show how the results from part **(c)** can be used to obtain the shearfree flow material functions presented in Example 7.2-2.

7C.2 Tube Flow of an Oldroyd 8-Constant Fluid[6,7]

Obtain the expression for the volume flow rate vs. pressure drop for the flow of the Oldroyd 8-constant fluid through a circular pipe. Use the abbreviation σ_i for the groups of time constants as given in Table 7.3-3:

$$\sigma_i = \lambda_i(\lambda_3 + \lambda_5) + \lambda_{i+2}(\lambda_1 - \lambda_3 - \lambda_5) + \lambda_{i+5}(\lambda_1 - \lambda_3 - \tfrac{3}{2}\lambda_5) \quad (7C.2\text{-}1)$$

Also use the following dimensionless quantities:

$$\Omega = \frac{3\sqrt{\sigma_1}Q}{\pi R^3} \quad (7C.2\text{-}2)$$

$$X = \sigma_1\dot{\gamma}_R^2 \qquad \left(\dot{\gamma}_R = -\left.\frac{dv_z}{dr}\right|_{r=R}\right) \quad (7C.2\text{-}3)$$

$$n = \frac{\sigma_2}{\sigma_1} \quad (7C.2\text{-}4)$$

Show that

$$\Omega = \left[1 - \frac{1}{2X^2}\left(\frac{1+X}{1+nX}\right)^3 f\right]\sqrt{X} \quad (7C.2\text{-}5)$$

in which

$$f = \tfrac{1}{2}n^3X^2 - 3n^2(n-1)X + 3n(n-1)(2n-1)\ln(1+X)$$
$$-\tfrac{1}{2}X\left(\frac{n-1}{1+X}\right)^2[6n + (7n-1)X] \quad (7C.2\text{-}6)$$

[6] M. C. Williams and R. B. Bird, *AIChE J.*, **8**, 378–382 (1962).
[7] K. Walters, *Arch. Rat. Mech. Anal.*, **9**, 411–414 (1962).

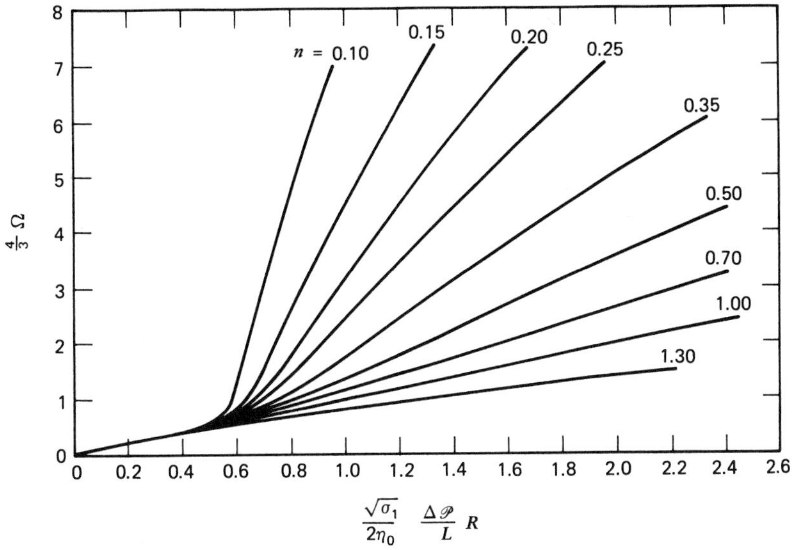

FIGURE 7C.2. Volumetric flow rate versus pressure difference for an 8-constant Oldroyd liquid in a cylindrical pipe, for various values of $n = \sigma_2/\sigma_1$. The quantities σ_i and Ω are defined in Eqs. 7C.2-1 and 2. [M. C. Williams and R. B. Bird, *AIChE J.*, **8**, 378–382 (1962).]

This gives us, in dimensionless form, the volume rate of flow in terms of the wall shear rate. Show further that the shear rate may be eliminated in favor of the pressure drop through the system by means of the relation

$$\frac{\sqrt{\sigma_1}}{2\eta_0}\frac{\Delta\mathscr{P}}{L}R = \left(\frac{1+nX}{1+X}\right)\sqrt{X} \tag{7C.2-7}$$

To get Ω in terms of $\Delta\mathscr{P}/L$, Eqs. 7C.2-5, 6, and 7 must be combined. When this is done numerically the plot in Fig. 7C.2 is obtained. To what extent does the result appear to be realistic?

7C.3 Spriggs-Bird Two Constant Model

Spriggs and Bird[8] considered the following nonlinear generalization of Eq. 5C.1-1:

$$\left(1 + \sum_{n=1}^{\infty} a_n \mathscr{F}^n\right)\boldsymbol{\tau} = -\eta_0\left(1 + \sum_{n=1}^{\infty} b_n \mathscr{F}^n\right)\dot{\boldsymbol{\gamma}} \tag{7C.3-1}$$

in which

$$\mathscr{F}\boldsymbol{\tau} = \boldsymbol{\tau}_{(1)} + \tfrac{1}{3}(\boldsymbol{\tau}:\dot{\boldsymbol{\gamma}})\boldsymbol{\delta} \tag{7C.3-2}$$

By choosing a_n and b_n in a certain way (see Eq. 5C.1-7), the linear viscoelastic material functions of the Rouse theory can be very nearly reproduced.

[8] T. W. Spriggs and R. B. Bird, *Ind. Eng. Chem. Fundam.*, **4**, 182–186 (1965). See also R. Roscoe, *Br. J. Appl. Phys.*, **15**, 1095–1101 (1964), for a somewhat more general model.

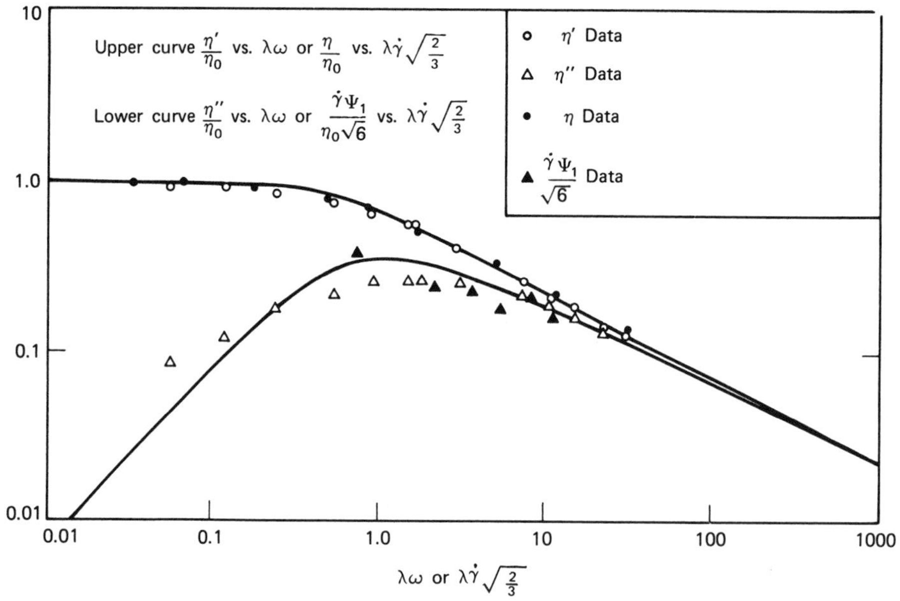

FIGURE 7C.3. Solid curves are functions obtained from the Spriggs–Bird model. The experimental data for η, η', and η'' for 5% polyisobutylene in decalin are taken from T. W. Dewitt, H. Markovitz, F. J. Padden, and L. J. Zapas, *J. Colloid Sci.*, **10**, 174–188 (1955), and the normal stress data are those of H. Markovitz and R. B. Williamson, *Trans. Soc. Rheol.*, **1**, 25–36 (1957). [Reprinted with permission from T. W. Spriggs and R. B. Bird, *Ind. Eng. Chem. Fundam.*, **4**, 182–186 (1965). Copyright by the American Chemical Society.]

Show that Eq. 7C.3-1 leads to the following expressions for η and Ψ_1

$$\frac{\eta}{\eta_0} = \frac{PR + QS}{R^2 + S^2} \tag{7C.3-3}$$

$$\frac{\Psi_1 \dot{\gamma}}{\sqrt{6}\eta_0} = \frac{PS - QR}{R^2 + S^2} \tag{7C.3-4}$$

in which P, Q, R, and S are functions given in Eqs. 5C.1-3 to 6, except that the argument ω is replaced by $\sqrt{\frac{2}{3}}\dot{\gamma}$. This suggests that η and η' should superpose if a "shift factor" of $\sqrt{2/3}$ is applied, and that a similar superposition of η'' and $\Psi_1\dot{\gamma}/\sqrt{6}$ should be possible. One comparison with experimental data is given in Fig. 7C.3; there Eqs. 5C.1-9 and 10 are used for η' and η'', and the same equations with ω replaced by $\sqrt{2/3}\dot{\gamma}$ are used for η and $\Psi_1\dot{\gamma}\sqrt{6}$, respectively. The only parameters are the zero-shear-rate viscosity η_0 and the time constant λ.

7C.4 A Constitutive Equation Suggested from a Theory of Suspensions

From a kinetic theory study of a dilute suspension of nearly spherical rigid particles, Leal and Hinch[9] have deduced a constitutive equation of the form

$$\boldsymbol{\tau} = -c_1\dot{\boldsymbol{\gamma}} - c_2\boldsymbol{\alpha} - c_3\{\dot{\boldsymbol{\gamma}}\cdot\boldsymbol{\alpha} + \boldsymbol{\alpha}\cdot\dot{\boldsymbol{\gamma}} - \tfrac{2}{3}(\boldsymbol{\alpha}:\dot{\boldsymbol{\gamma}})\boldsymbol{\delta}\} \tag{7C.4-1}$$

[9] L. G. Leal and E. J. Hinch, *J. Fluid Mech.*, **55**, 745–765 (1972); Eqs. 7C.4-1 and 2 are a special case of a model by G. L. Hand, *J. Fluid Mech.*, **13**, 33–46 (1962). For an extended discussion of dilute suspension rheology and the Hand equation see D. Barthès-Biesel and A. Acrivos, *Int. J. Multiphase Flow*, **1**, 1–24 (1973).

in which $\boldsymbol{\alpha}$ is a "structure tensor" (symmetric and traceless) that obeys the relation

$$\boldsymbol{\alpha}_{(1)} + c_4\boldsymbol{\alpha} + \tfrac{1}{2}\{\dot{\boldsymbol{\gamma}}\cdot\boldsymbol{\alpha} + \boldsymbol{\alpha}\cdot\dot{\boldsymbol{\gamma}}\} = c_5\dot{\boldsymbol{\gamma}} \tag{7C.4-2}$$

The c_i are positive constants related to the particle shape and concentration. What properties does this constitutive equation have?

7D.1 Stress Growth and Stress Relaxation for the 4-Constant Oldroyd Model

a. For the stress relaxation after cessation of a steady shear flow with shear rate $\dot{\gamma}_0$, show that the constitutive equation, for $t > 0$, simplifies to

$$\tau_{xx} + \lambda_1 \frac{d}{dt}\tau_{xx} = 0 \tag{7D.1-1}$$

$$\tau_{yx} + \lambda_1 \frac{d}{dt}\tau_{yx} = -\eta_0\lambda_2 \frac{d}{dt}\dot{\gamma}_0[1 - H(t)] \tag{7D.1-2}$$

in which $H(t)$ is the Heaviside unit step function, and $dH/dt = \delta(t)$. Solve these equations using Laplace transforms; make use of the viscometric functions tabulated in Table 7.3-3 to supply the initial conditions for τ_{xx} and τ_{yx}. Obtain, finally,

$$\eta^- = \eta_0\left(\frac{1 + \lambda_2\lambda_5\dot{\gamma}_0^2}{1 + \lambda_1\lambda_5\dot{\gamma}_0^2} - \frac{\lambda_2}{\lambda_1}\right)e^{-t/\lambda_1} \tag{7D.1-3}$$

$$\Psi_1^- = 2\eta_0\left(\frac{\lambda_1 - \lambda_2}{1 + \lambda_1\lambda_5\dot{\gamma}_0^2}\right)e^{-t/\lambda_1} \tag{7D.1-4}$$

To what extent do these material functions describe the experimental data in Chapter 3?

b. Use Laplace transforms to do a similar analysis for stress growth, when a shear rate $\dot{\gamma}_0$ is suddenly applied at time $t = 0$. Show that

$$\eta^+ = \eta_0\left[\left(\frac{\lambda_2}{\lambda_1} - F\right)e^{-t/\lambda_1}\cos\Gamma t/\lambda_1 + (1 - F)\Gamma^{-1}e^{-t/\lambda_1}\sin\Gamma t/\lambda_1 + F\right] \tag{7D.1-5}$$

where $\Gamma = \sqrt{\lambda_1\lambda_5}\dot{\gamma}_0$ and $F = (1 + \lambda_2\lambda_5\dot{\gamma}_0^2)/(1 + \lambda_1\lambda_5\dot{\gamma}_0^2)$. How does η^+ behave as $t \to \infty$ and as $t \to 0$? Does the function η^+ as a function of t go through a maximum similar to that seen in the experimental data? Does the maximum shift to shorter times as $\dot{\gamma}_0$ increases?

CHAPTER 8

SINGLE-INTEGRAL CONSTITUTIVE EQUATIONS

In this chapter we turn our attention to single-integral constitutive equations. The starting point is the general linear viscoelastic model. Since this model is constructed specifically for flows with small displacement gradients, we need to generalize it to describe flows with large displacement gradients. Hence we start in §8.1 with a description of fluid particle paths and define the displacement functions that describe the motion of fluid particles in flow fields. Using the displacement functions we then define a set of finite strain tensors that can be used to formulate integral constitutive equations. Then in §8.2 we construct the general quasi-linear integral model with an arbitrary memory function. For a specific choice of the memory function the model is identical to the convected Jeffreys model (or Oldroyd-B model) presented in §7.2, but the form is such that the dependence of the stress at the present time on past configurations is illustrated more clearly. The proof of the equivalence of the two forms is postponed until Example 9.4-1. In §8.3 we modify the general quasi-linear model to obtain a nonlinear model providing an improved description of measured material functions In parallel with the previous chapter we consider in §8.4 flow problems in one spatial variable, and in §8.5 flow problems in two or three spatial variables. Finally in §8.6 we discuss the limitations of single-integral models and give recommendations for their use.

§8.1 THE FINITE STRAIN TENSORS

In §§6.1 and 7.1 where we introduced the tensors $\boldsymbol{\gamma}_{(n)}$ and $\boldsymbol{\tau}_{(1)}$, we described the fluid motion by means of the velocity field $\boldsymbol{v}(\boldsymbol{r}, t)$. In this section, where we introduce the finite strain tensors, we make use of an alternative description of the fluid motion in which the motions of fluid particles are given; that is, we select a fluid particle and then describe its trajectory through the three-dimensional space occupied by the fluid.

Consider a fluid particle moving as part of a fluid continuum. At some past time t' the particle has position $\boldsymbol{r}'$, and at the present time t the particle has position $\boldsymbol{r}$. The designation $\boldsymbol{r}, t$ (or $\boldsymbol{r}', t'$) then uniquely specifies the particle and is referred to as a "particle label." The motion of the particle may then be described by the functions

$$\boldsymbol{r}' = \boldsymbol{r}'(\boldsymbol{r}, t, t') \quad \text{or} \quad x_i' = x_i'(x, t, t') \tag{8.1-1}$$

which give the past position $\boldsymbol{r}'$ of the particle $\boldsymbol{r}, t$ as a function of time t', or by the functions

$$\boldsymbol{r} = \boldsymbol{r}(\boldsymbol{r}', t', t) \quad \text{or} \quad x_i = x_i(x', t', t) \tag{8.1-2}$$

which give the present position $\boldsymbol{r}$ of the particle $\boldsymbol{r}'$, t' as a function of time t. The equations have been given also in terms of the Cartesian coordinates (x_1, x_2, x_3) of $\boldsymbol{r}$ and (x_1', x_2', x_3') of $\boldsymbol{r}'$; for the sake of brevity the sets of coordinates are denoted simply by x and x' respectively. The functions on the right side of Eqs. 8.1-1 and 2 are called the *displacement functions.* In these functions the first two arguments identify the particle and the last argument indicates the time dependence. The displacement functions for all particles and all times completely specify the motion of the fluid. Notice that in Eq. 8.1-1 the symbol $\boldsymbol{r}'$(or x_i') appears with two different meanings: on the right the symbol denotes a function and on the left the symbol denotes the value of the function. Similar remarks apply to $\boldsymbol{r}$ (or x_i) in Eq. 8.1-2.

From a mathematical point of view $\boldsymbol{r}(\boldsymbol{r}', t', t)$ and $\boldsymbol{r}'(\boldsymbol{r}, t, t')$ are really the same function of the arguments in the indicated order. For this reason there is strictly speaking no reason to use separate symbols for the functions as long as the arguments are always included. However we shall keep the prime on the function $\boldsymbol{r}'(\boldsymbol{r}, t, t')$. This allows us to drop the list of arguments when no confusion can arise.

From the displacement functions we now define two *displacement gradient tensors* $\boldsymbol{\Delta}(\boldsymbol{r}, t, t')$ and $\mathbf{E}(\boldsymbol{r}, t, t')$. These are defined in terms of their Cartesian components $\Delta_{ij}(\boldsymbol{r}, t, t')$ and $\mathrm{E}_{ij}(\boldsymbol{r}, t, t')$ by

$$\Delta_{ij}(\boldsymbol{r}, t, t') = \frac{\partial x_i'(x, t, t')}{\partial x_j} \tag{8.1-3}$$

$$\mathrm{E}_{ij}(\boldsymbol{r}, t, t') = \frac{\partial x_i(x', t', t)}{\partial x_j'} \tag{8.1-4}$$

The Δ_{ij} are measures of the displacements at time t' relative to the positions at time t, whereas E_{ij} are measures of the displacements at time t relative to the positions at time t'. The tensors $\boldsymbol{\Delta}(\boldsymbol{r}, t, t')$ and $\mathbf{E}(\boldsymbol{r}, t, t')$ are inverse to each other; that is, in Cartesian coordinates[1]

$$\sum_m \Delta_{im}\mathrm{E}_{mj} = \delta_{ij} \tag{8.1-5}$$

as may be seen from the definitions. In addition it may be shown (see Problem 8D.2) that for an incompressible fluid $\det(\Delta_{ij}) = \det(\mathrm{E}_{ij}) = 1$. The displacement gradient tensors are generally not symmetrical and contain information about both deformation and rotation in the neighborhood of a given particle. They are therefore not immediately suitable for use in constitutive equations. In terms of the displacement gradient tensors, however, we can

[1]From the definitions in Eqs. 8.1-3 and 4 we could write $\mathrm{E}_{ij}(\boldsymbol{r}, t, t') = \Delta_{ij}(\boldsymbol{r}', t', t)$ and $\Delta_{ij}(\boldsymbol{r}, t, t') = \mathrm{E}_{ij}(\boldsymbol{r}', t', t)$ so that instead of having two separate symbols $\boldsymbol{\Delta}$ and $\mathbf{E}$ we could use just one symbol and then keep careful track of the order of the arguments. However, whenever we use $\boldsymbol{\Delta}$ and $\mathbf{E}$ we assume that the arguments are in the order given in the definitions in Eqs. 8.1-3 and 4 with t indicating the present state and t' a past state, and never in the reverse order. The use of two separate symbols enables us then to omit the arguments in situations when no confusion can occur, as we have done in Eq. 8.1-5; in other situations we retain the arguments t and t' but omit the $\boldsymbol{r}$.

We have used the term "displacement gradient tensors" here for both $\boldsymbol{\Delta}$ and $\mathbf{E}$ and, in Chapter 5, for $\nabla\boldsymbol{u}$. These three tensors are very closely related: $\boldsymbol{\Delta} = \mathbf{E}^{-1} = (\nabla\boldsymbol{u})^{\dagger} + \boldsymbol{\delta}$. Calling all three tensors by the same name will not cause confusion, since generally a specific symbol will be used. However, when we say that linear viscoelastic formulas are applicable "in the limit of vanishingly small displacement gradients," we are referring to $\nabla\boldsymbol{u}$, rather than $\boldsymbol{\Delta}$ or $\mathbf{E}$.

define two *finite strain tensors* suitable for this purpose. These are the *Cauchy strain tensor*, $\mathbf{B}^{-1}(\boldsymbol{r}, t, t')$:

$$\mathbf{B}^{-1}(\boldsymbol{r}, t, t') = \{\boldsymbol{\Delta}^{\dagger} \cdot \boldsymbol{\Delta}\} \quad \text{or} \quad \mathrm{B}_{ij}^{-1}(\boldsymbol{r}, t, t') = \sum_m \frac{\partial x'_m}{\partial x_i} \frac{\partial x'_m}{\partial x_j} \tag{8.1-6}$$

and the *Finger strain tensor*, $\mathbf{B}(\boldsymbol{r}, t, t')$:

$$\mathbf{B}(\boldsymbol{r}, t, t') = \{\mathbf{E} \cdot \mathbf{E}^{\dagger}\} \quad \text{or} \quad \mathrm{B}_{ij}(\boldsymbol{r}, t, t') = \sum_m \frac{\partial x_i}{\partial x'_m} \frac{\partial x_j}{\partial x'_m} \tag{8.1-7}$$

From the properties of the displacement gradient tensors, we see that $\mathbf{B}$ and $\mathbf{B}^{-1}$ are indeed inverse to each other as implied by the notation, and further that for incompressible fluids their determinants are unity. Additional properties of $\mathbf{B}$ and $\mathbf{B}^{-1}$ are:

- **i.** They are symmetric (as seen from their definitions).
- **ii.** They are positive definite (Problem 8B.3).
- **iii.** They describe the deformation, from time t' to time t, of the material in the neighborhood of the given particle with a significance independent of superposed rigid rotations (Example 8.1-2 and §9.3).
- **iv.** They reduce to the unit tensor if the displacement from t' to t involves no deformation, that is, if it is a translation plus a rotation (Example 8.1-2).

The last property listed above leads us to define two closely related finite strain tensors $\boldsymbol{\gamma}^{[0]}$ and $\boldsymbol{\gamma}_{[0]}$, which are zero for a rigid body motion:

$$\boldsymbol{\gamma}^{[0]}(\boldsymbol{r}, t, t') = \{\boldsymbol{\Delta}^{\dagger} \cdot \boldsymbol{\Delta}\} - \boldsymbol{\delta}; \quad \gamma_{ij}^{[0]}(\boldsymbol{r}, t, t') = \sum_m \frac{\partial x'_m}{\partial x_i} \frac{\partial x'_m}{\partial x_j} - \delta_{ij} \tag{8.1-8}$$

$$\boldsymbol{\gamma}_{[0]}(\boldsymbol{r}, t, t') = \boldsymbol{\delta} - \{\mathbf{E} \cdot \mathbf{E}^{\dagger}\}; \quad \gamma_{[0]ij}(\boldsymbol{r}, t, t') = \delta_{ij} - \sum_m \frac{\partial x_i}{\partial x'_m} \frac{\partial x_j}{\partial x'_m} \tag{8.1-9}$$

To distinguish these tensors from those defined in Eqs. 8.1-6 and 7, we call $\boldsymbol{\gamma}^{[0]}$ and $\boldsymbol{\gamma}_{[0]}$ the *relative (finite) strain tensors.*[2] For small displacement gradients both of these tensors reduce to the infinitesimal strain tensor $\boldsymbol{\gamma}$ encountered in Chapter 5 (see Example 9.3-1).

The above component forms of the displacement gradient tensors and the strain tensors are valid in rectangular (Cartesian) coordinates only. In Table B.5 through B.7 of Appendix B, we give component forms of $\boldsymbol{\Delta}$, $\mathbf{E}$, $\boldsymbol{\gamma}^{[0]}$, $\boldsymbol{\gamma}_{[0]}$ for general deformations in rectangular, cylindrical, and spherical coordinate systems.[3] In addition in Appendix C we give expressions for $\boldsymbol{\Delta}$, $\mathbf{E}$, $\boldsymbol{\gamma}^{[0]}$ and $\boldsymbol{\gamma}_{[0]}$ in rectangular coordinates worked out specifically for simple shear flows and for shearfree flows. In Example 8.1-1 at the end of this section we illustrate the construction of some of the entries in Appendix C. Whenever components of $\boldsymbol{\Delta}$, $\mathbf{E}$, $\boldsymbol{\gamma}^{[0]}$, or $\boldsymbol{\gamma}_{[0]}$ are needed, the reader will find the tabulations in Appendices B and C helpful.

[2] J. G. Oldroyd, *J. Non-Newtonian Fluid Mech.*, **14**, 9–46 (1984), see Eq. 55.

[3] For the corresponding formulas for nonorthogonal coordinate systems, see footnote f of Table 9.3-1.

This completes the definition of the kinematic quantities needed in the chapter. Note that the finite strain tensors are defined in terms of the displacement functions and not in terms of the velocity field. Indeed the nonlinear integral models to be introduced in this chapter give explicit expressions for the stress tensor in terms of the displacement functions. Fluid mechanical problems with integral constitutive models may be formulated and solved just in terms of the displacement functions. One may then subsequently construct the velocity field if so desired. This is in contrast with all nonlinear models of Chapters 4, 6, and 7, in which the stresses are related directly to the velocity field, and for which one most naturally will formulate and solve flow problems for the velocity field. For completeness we now proceed to relate the displacement functions to (the history of) the velocity field.

Starting from the displacement functions in Eqs. 8.1-1 and 2 we define the velocity of a fluid particle at time t' and at time t by

$$\boldsymbol{w}'(\boldsymbol{r}, t, t') = \frac{\partial}{\partial t'} \boldsymbol{r}'(\boldsymbol{r}, t, t') \tag{8.1-10}$$

$$\boldsymbol{w}(\boldsymbol{r}', t', t) = \frac{\partial}{\partial t} \boldsymbol{r}(\boldsymbol{r}', t', t) \tag{8.1-11}$$

The spatial velocity field $\boldsymbol{v}$ at times t' and t is equal to the particle velocities at those instants, that is

$$\boldsymbol{v}'(\boldsymbol{r}', t') = \boldsymbol{w}'(\boldsymbol{r}, t, t') \tag{8.1-12}$$

where $\boldsymbol{r}'$ is given by Eq. 8.1-1 and

$$\boldsymbol{v}(\boldsymbol{r}, t) = \boldsymbol{w}(\boldsymbol{r}', t', t) \tag{8.1-13}$$

where $\boldsymbol{r}$ is given by Eq. 8.1-2. When use is made of the displacement function $\boldsymbol{r}'(\boldsymbol{r}, t, t')$ in Eq. 8.1-1, then Eqs. 8.1-10 and 12 give explicit recipes for the construction of the particle velocity field $\boldsymbol{w}'(\boldsymbol{r}, t, t')$ and the spatial velocity field $\boldsymbol{v}(\boldsymbol{r}', t')$. Similarly given the displacement function $\boldsymbol{r}(\boldsymbol{r}', t', t)$, then $\boldsymbol{w}(\boldsymbol{r}', t', t)$ and $\boldsymbol{v}(\boldsymbol{r}, t)$ are constructed from Eqs. 8.1-11 and 13. Here again note that there is only one particle velocity function, with the first two arguments denoting the particle dependence and the last argument denoting the time dependence, so that the prime on $\boldsymbol{w}'$ in Eq. 8.1-12 merely indicates the time dependence. Similar remarks hold for the prime on $\boldsymbol{v}'$.

We now turn the problem around to show how the displacement functions may be determined from a given spatial velocity field. For this purpose insert $\boldsymbol{v}(\boldsymbol{r}, t)$ into Eq. 8.1-11 to obtain the initial value problem

$$\frac{\partial}{\partial t} \boldsymbol{r}(\boldsymbol{r}', t', t) = \boldsymbol{v}(\boldsymbol{r}, t) \tag{8.1-14}$$

to be solved with the initial condition that $\boldsymbol{r} = \boldsymbol{r}'$ at $t = t'$. The initial value problem for $\boldsymbol{r}'$ is obtained by interchanging primed and unprimed quantities in Eq. 8.1-14 and in the initial condition. For simple flow fields these equations can sometimes be solved analytically (see Examples 8.1-1 and 10.1-2 as well as Problems 8B.5 and 9D.1). In general, however, numerical methods must be used.

EXAMPLE 8.1-1 Evaluation of Kinematic Tensors

It is desired to calculate the kinematic tensors $\mathbf{\Delta}$, $\mathbf{E}$, $\mathbf{B}^{-1}$, $\mathbf{B}$, $\boldsymbol{\gamma}^{[0]}$ and $\boldsymbol{\gamma}_{[0]}$ for **(a)** simple shear flow (Eq. 3.1-1) and **(b)** shearfree flows (Eq. 3.1-3).

SOLUTION **(a)** For the simple shear flow $v_x = \dot{\gamma}_{yx}(t)y$, Eq. 8.1-14 for the displacement functions reads

$$\frac{\partial x}{\partial t} = \dot{\gamma}_{yx}(t)y; \quad \frac{\partial y}{\partial t} = 0; \quad \frac{\partial z}{\partial t} = 0 \tag{8.1-15}$$

The solution to these three equations subject to the condition that $(x, y, z) = (x', y', z')$ at $t = t'$ is

$$x = x' - \gamma_{yx}(t, t')y'; \quad y = y'; \quad z = z' \tag{8.1-16}$$

where $\gamma_{yx}(t, t') = \int_t^{t'} \dot{\gamma}_{yx}(t'')dt''$, and $\dot{\gamma}_{yx}$ is the velocity gradient. Then the matrix representations are easily found for the following tensors

$$\mathbf{E} = \begin{pmatrix} 1 & -\gamma_{yx} & 0 \\ 0 & 1 & 0 \\ 0 & 0 & 1 \end{pmatrix}; \quad \mathbf{\Delta} = \begin{pmatrix} 1 & \gamma_{yx} & 0 \\ 0 & 1 & 0 \\ 0 & 0 & 1 \end{pmatrix} \tag{8.1-17}$$

As a check on the calculations at this point we may verify that $\mathbf{E}$ and $\mathbf{\Delta}$ are inverse to one another by performing a matrix multiplication. By matrix multiplication we also find

$$\mathbf{B} = \begin{pmatrix} 1 + \gamma_{yx}^2 & -\gamma_{yx} & 0 \\ -\gamma_{yx} & 1 & 0 \\ 0 & 0 & 1 \end{pmatrix}; \quad \mathbf{B}^{-1} = \begin{pmatrix} 1 & \gamma_{yx} & 0 \\ \gamma_{yx} & 1 + \gamma_{yx}^2 & 0 \\ 0 & 0 & 1 \end{pmatrix} \tag{8.1-18}$$

Again at this point we may check the calculations by performing a matrix multiplication to verify that $\mathbf{B}$ and $\mathbf{B}^{-1}$ are inverse to one another, and by showing that they have determinants equal to unity. Finally we may form $\boldsymbol{\gamma}^{[0]} = \mathbf{B}^{-1} - \boldsymbol{\delta}$ and $\boldsymbol{\gamma}_{[0]} = \boldsymbol{\delta} - \mathbf{B}$ by using Eqs. 8.1-8 and 9 to obtain the entries in Appendix C. Note that the finite strain tensors contain components that are nonlinear in the displacement gradients.

(b) For shearfree flows, $v_x = -\frac{1}{2}\dot{\varepsilon}(1 + b)x$, $v_y = -\frac{1}{2}\dot{\varepsilon}(1 - b)y$, $v_z = \dot{\varepsilon}z$, Eq. 8.1-14 for the displacement functions reads

$$\frac{\partial x}{\partial t} = -\tfrac{1}{2}(1 + b)\dot{\varepsilon}(t)x; \quad \frac{\partial y}{\partial t} = -\tfrac{1}{2}(1 - b)\dot{\varepsilon}(t)y; \quad \frac{\partial z}{\partial t} = \dot{\varepsilon}(t)z \tag{8.1-19}$$

The solution subject to the condition that $(x, y, z) = (x', y', z')$ at $t = t'$ may be written

$$x = x'\lambda_x; \quad y = y'\lambda_y; \quad z = z'\lambda_z \tag{8.1-20}$$

where we have defined

$$\lambda_x = \exp\left[\tfrac{1}{2}(1 + b)\varepsilon(t, t')\right] \tag{8.1-21}$$

$$\lambda_y = \exp\left[\tfrac{1}{2}(1 - b)\varepsilon(t, t')\right] \tag{8.1-22}$$

$$\lambda_z = \exp\left[-\varepsilon(t, t')\right] \tag{8.1-23}$$

and $\varepsilon(t, t') = \int_t^{t'} \dot{\varepsilon}(t'')dt''$; this quantity is called the *Hencky strain.* The quantities $\lambda_x, \lambda_y, \lambda_z$ are called the *principal elongation ratios.* If we imagine for example, that the fluid at the past time t' has the shape of a cube with unit dimensions and faces perpendicular to the coordinate axes, then this cube becomes distorted into the shape of a rectangular box with dimensions given by the principal elongation ratios at the present time t (see Fig. 3.1-3). The displacement gradient tensors become

$$\mathbf{E} = \begin{pmatrix} \lambda_x & 0 & 0 \\ 0 & \lambda_y & 0 \\ 0 & 0 & \lambda_z \end{pmatrix}; \quad \boldsymbol{\Delta} = \begin{pmatrix} \lambda_x^{-1} & 0 & 0 \\ 0 & \lambda_y^{-1} & 0 \\ 0 & 0 & \lambda_z^{-1} \end{pmatrix} \tag{8.1-24}$$

We see that the condition that $\mathbf{E}$ and $\boldsymbol{\Delta}$ have determinants equal to unity requires that the product of the principal elongation ratios be unity, in agreement with Eqs. 8.1-21 through 23. This condition corresponds to the fact that the above-mentioned cube is distorted into a box of the same volume. The finite strain tensors become

$$\mathbf{B} = \begin{pmatrix} \lambda_x^2 & 0 & 0 \\ 0 & \lambda_y^2 & 0 \\ 0 & 0 & \lambda_z^2 \end{pmatrix}; \quad \mathbf{B}^{-1} = \begin{pmatrix} \lambda_x^{-2} & 0 & 0 \\ 0 & \lambda_y^{-2} & 0 \\ 0 & 0 & \lambda_z^{-2} \end{pmatrix} \tag{8.1-25}$$

Again from these we may construct $\boldsymbol{\gamma}^{[0]} = \mathbf{B}^{-1} - \boldsymbol{\delta}$ and $\boldsymbol{\gamma}_{[0]} = \boldsymbol{\delta} - \mathbf{B}$ by using Eqs. 8.1-8 and 9 to obtain the entries in Appendix C.

EXAMPLE 8.1-2 Evaluation of Strain Tensors for a Steady Shear Flow with a Superposed Rotation

It is desired to evaluate the strain tensors $\mathbf{B}(\boldsymbol{r}, t, t')$ and $\mathbf{B}^{-1}(\boldsymbol{r}, t, t')$ at time $t = 0$ for a typical particle in the shear flow with superposed rotation described in Example 5.5-1 and Problem 5C.2.

SOLUTION We use the same notation as in Example 5.5-1. Then the coordinates $\bar{x}, \bar{y}$ refer to the turntable on which we have a steady shear flow, $v_{\bar{x}} = \dot{\gamma}\bar{y}$, $v_{\bar{y}} = 0$, so that the displacement functions are

$$\begin{cases} \bar{x} = \bar{x}' - \dot{\gamma}\bar{y}'t' \\ \bar{y} = \bar{y}' \end{cases} \tag{8.1-26}$$

where the unprimed coordinates refer to the position of the particle at time $t = 0$, and the primed coordinates refer to the position at time t'. In addition we need transformation rules between the turntable coordinates and the space fixed coordinates. These are given by Eqs. 5.5-1 and 2 repeated here for $t = 0$ and $t = t'$ separately

At time 0:

$$\begin{cases} \bar{x} = x - x_0 \\ \bar{y} = y - y_0 \end{cases} \quad \text{or} \quad \begin{cases} x = \bar{x} + x_0 \\ y = \bar{y} + y_0 \end{cases} \tag{8.1-27a,b}$$

At time t':

$$\begin{cases} \bar{x}' = (x' - x_0)C + (y' - y_0)S \\ \bar{y}' = -(x' - x_0)S + (y' - y_0)C \end{cases} \quad \text{or} \quad \begin{cases} x' = \bar{x}'C - \bar{y}'S + x_0 \\ y' = \bar{x}'S + \bar{y}'C + y_0 \end{cases} \tag{8.1-28a,b}$$

where $C = \cos Wt'$ and $S = \sin Wt'$. The displacement functions referred to the space fixed axes are then obtained by inserting Eqs. 8.1-28a into Eqs. 8.1-26, which then in turn are inserted into Eqs. 8.1-27b to get

$$\begin{cases} x = (x' - x_0)(C + S\dot{\gamma}t') + (y' - y_0)(S - C\dot{\gamma}t') + x_0 \\ y = -(x' - x_0)S + (y' - y_0)C + y_0 \end{cases} \tag{8.1-29}$$

Then by differentiation we obtain

$$\mathbf{E}(0, t') = \begin{pmatrix} C + S\dot{\gamma}t' & S - C\dot{\gamma}t' & 0 \\ -S & C & 0 \\ 0 & 0 & 1 \end{pmatrix} \tag{8.1-30}$$

Notice that the displacement gradient tensor depends on the angular orientation of the turntable at the past time t'. However when we perform the matrix multiplication to form $\mathbf{B} = \{\mathbf{E} \cdot \mathbf{E}^{\dagger}\}$ this dependence drops out

$$\mathbf{B}(0, t') = \begin{pmatrix} 1 + (\dot{\gamma}t')^2 & -\dot{\gamma}t' & 0 \\ -\dot{\gamma}t' & 1 & 0 \\ 0 & 0 & 1 \end{pmatrix}; \quad \mathbf{B}^{-1}(0, t') = \begin{pmatrix} 1 & \dot{\gamma}t' & 0 \\ \dot{\gamma}t' & 1 + (\dot{\gamma}t')^2 & 0 \\ 0 & 0 & 1 \end{pmatrix} \tag{8.1-31}$$

In addition we have given $\mathbf{B}^{-1}$, which we have obtained simply as the inverse of $\mathbf{B}$; in this derivation we used the fact that det $\mathbf{B} = 1$. The important point about Eqs. 8.1-31 is that the expressions for $\mathbf{B}$ and $\mathbf{B}^{-1}$ are those given in Eqs. 8.1-18 when $\gamma_{yx}(0, t') = \dot{\gamma}t'$. This means that the influence of the rotation clearly present in the displacement gradient tensors has been "filtered out" in the finite strain tensors.

§8.2 QUASI-LINEAR INTEGRAL MODELS

We are now in a position to modify the general linear viscoelastic model to obtain an admissible constitutive equation. We may do this simply by replacing the infinitesimal strain tensor $\boldsymbol{\gamma}(t, t')$ of Eq. 5.2-19 by either of the relative strain tensors $\boldsymbol{\gamma}_{[0]}(t, t')$ or $\boldsymbol{\gamma}^{[0]}(t, t')$ of the previous section.[1] Constitutive equations obtained by using $\boldsymbol{\gamma}_{[0]}(t, t')$ and $\boldsymbol{\gamma}^{[0]}(t, t')$ give rather different predictions outside the linear regime, but several different molecular theories and comparison with experiments (see Problem 8B.4) suggest very strongly that it is more appropriate to replace $\boldsymbol{\gamma}(t, t')$ by $\boldsymbol{\gamma}_{[0]}(t, t')$. Therefore we use the following quasi-linear integral model:

$$\boxed{\boldsymbol{\tau}(t) = \int_{-\infty}^{t} M(t - t')\boldsymbol{\gamma}_{[0]}(t, t')dt'} \tag{8.2-1}$$

where $M(t - t')$ is the *memory function*. We refer to this model as the *Lodge rubberlike liquid*.[2] Since $\boldsymbol{\gamma}_{[0]}$ reduces to $\boldsymbol{\gamma}$ in the limit of small displacement gradients, the model

[1] In place of $\boldsymbol{\gamma}_{[0]}$ we could also use $-\mathbf{B}$, and in place of $\boldsymbol{\gamma}^{[0]}$ we could also use $\mathbf{B}^{-1}$. These replacements do not alter the properties of the constitutive equations, since the materials are incompressible.

[2] A. S. Lodge, *Elastic Liquids*, Academic Press, New York (1964); many rheological properties of this model are worked out in detail in Chapts. 6 and 7. The equation of the form of Eq. 8.2-1 but with $\boldsymbol{\gamma}^{[0]}$ in place of $\boldsymbol{\gamma}_{[0]}$ is sometimes called the equation of *finite linear viscoelasticity* [see W. R. Schowalter, *Mechanics of Non-Newtonian Fluids*, Pergamon, New York (1978), §11.3; and R. Huilgol, *Continuum Mechanics of Viscoelastic Liquids*, Wiley, New York (1975), p. 164].

becomes identical to the general linear viscoelastic model (Eq. 5.2-19) in that limit. Hence the memory function is the same as that of linear viscoelasticity and is related to the relaxation modulus $G(t - t')$ by $M(t - t') = \partial G(t - t')/\partial t'$. The model retains the feature of the general linear viscoelastic model that the integrand is factorized into a product of two functions, one depending on the nature of the specific fluid and the other on the deformation history. We call the model "quasi-linear" because the stress depends linearly on the history of a strain tensor. Keep in mind, however, that $\boldsymbol{\gamma}_{[0]}(t, t')$ is itself nonlinear in the displacement gradients, and that Eq. 8.2-1 therefore really represents a nonlinear model for the stress in terms of the displacements.

The following specific choices for $M(t - t')$ lead to commonly used constitutive equations:

a.
$$M(t - t') = \frac{\eta_0}{\lambda_1^2} \exp\left[-(t - t')/\lambda_1\right] \tag{8.2-2}$$

This gives the *convected Maxwell model* introduced in Chapter 7. It is recognized as a nonlinear version of the Maxwell model in Eq. 5.2-8.

b.
$$M(t - t') = \frac{\eta_0}{\lambda_1^2}\left[\left(1 - \frac{\lambda_2}{\lambda_1}\right) \exp\left[-(t - t')/\lambda_1\right] + 2\lambda_1\lambda_2 \frac{\partial}{\partial t'}\delta(t - t')\right] \tag{8.2-3}$$

This gives the *convected Jeffreys model* (or Oldroyd fluid B) introduced in Chapter 7. It is recognized as a nonlinear version of the Jeffreys model in Eq. 5.2-12.

c.
$$M(t - t') = \sum_{i=1}^{\infty} \frac{\eta_i}{\lambda_i^2} \exp\left[-(t - t')/\lambda_i\right] \tag{8.2-4}$$

This gives the *convected generalized Maxwell model*, which is recognized as a nonlinear modification of Eq. 5.2-15. This model, also called the *Lodge network model*,[3] has been specifically derived from a molecular theory of polymer melts (see Chapter 20). The Rouse and Zimm theories of dilute polymer solutions (see Chapter 15) also give a constitutive equation of the form of Eq. 8.2-1 with the memory function of Eq. 8.2-4; these theories, in contrast to the network theories, give explicit expressions for the parameters η_i and λ_i in terms of the constants describing the mechanical model. In these dilute solution theories, and in the network model of Green and Tobolsky,[4] it is the finite strain tensor $\boldsymbol{\gamma}_{[0]}(t, t')$ that enters naturally in the theory and not the tensor $\boldsymbol{\gamma}^{[0]}(t, t')$. This is the reason for our choice in Eq. 8.2-1 of $\boldsymbol{\gamma}_{[0]}(t, t')$ as the strain tensor used to replace $\boldsymbol{\gamma}(t, t')$ in the quasi-linear models. In the following example we investigate some of the properties of the Lodge rubberlike liquid. A rather complete and extensive investigation of the rheological properties of this model is given by Lodge.[2]

EXAMPLE 8.2-1 Simple Shear Flow of the Lodge Rubberlike Liquid

Consider a time-dependent simple shear flow of a Lodge rubberlike liquid with the flow field $v_x = \dot{\gamma}_{yx}(t)y$ and $v_y = v_z = 0$. Obtain the general expressions for the stress tensor components, and use

[3] A. S. Lodge, *Elastic Liquids*, Academic Press, New York (1964), Chapts. 6 and 7; A. S. Lodge, *Trans. Faraday Soc.*, **52**, 120–130 (1956); A. S. Lodge, R. C. Armstrong, M. H. Wagner, and H. H. Winter, *Pure Appl. Chem.*, **54**, 1349–1359 (1983); see also §20.4.

[4] M. S. Green and A. V. Tobolsky, *J. Chem. Phys.*, **14**, 80–92 (1946).

these to find the material functions for **(a)** steady-state shear flow, **(b)** start-up of shear flow, and **(c)** stress relaxation upon cessation of steady shear flow.

SOLUTION We use the entry for $\gamma_{[0]}(t, t')$ in homogeneous shear flow in Appendix C. When the expressions for the components of $\gamma_{[0]}$ are inserted into Eq. 8.2-1 we obtain for the stress components of interest

$$\tau_{yx}(t) = \int_{-\infty}^{t} M(t - t')\gamma_{yx}(t, t')dt' \tag{8.2-5}$$

$$\tau_{xx}(t) - \tau_{yy}(t) = -\int_{-\infty}^{t} M(t - t')\gamma_{yx}^2(t, t')dt' \tag{8.2-6}$$

$$\tau_{yy}(t) - \tau_{zz}(t) = 0 \tag{8.2-7}$$

where

$$\gamma_{yx}(t, t') = \int_{t}^{t'} \dot{\gamma}_{yx}(t'')dt'' \tag{8.2-8}$$

is the shear strain from t to t'. Note that the shear stress is precisely the same as that predicted from the general linear viscoelastic model in Eq. 5.2-19.

(a) For *steady-state shear flow* $\gamma_{yx}(t, t') = -\dot{\gamma}(t - t')$, where $\dot{\gamma}$ is the shear rate. We then obtain

$$\tau_{yx} = -M_1\dot{\gamma}; \qquad \eta = M_1 \overset{\text{(LNM)}}{=} \sum_i \eta_i \tag{8.2-9}$$

$$\tau_{xx} - \tau_{yy} = -M_2\dot{\gamma}^2; \quad \Psi_1 = M_2 \overset{\text{(LNM)}}{=} 2\sum_i \eta_i\lambda_i \tag{8.2-10}$$

$$\tau_{yy} - \tau_{zz} = 0; \qquad \Psi_2 = 0 \tag{8.2-11}$$

where

$$M_n = \int_0^{\infty} M(s)s^n\, ds = n\int_0^{\infty} G(s)s^{n-1}\, ds \tag{8.2-12}$$

and $\overset{\text{(LNM)}}{=}$ means that the particular memory function for the *L*odge *N*etwork *M*odel has been assumed (see Eq. 8.2-4).

Note that η and Ψ_1 are constants for any choice of the memory function. This prediction is not in agreement with most experimental data over a wide range of shear rates. The model also predicts incorrectly that $\Psi_1/2\eta = \int_0^\infty G(s)s\, ds/\int_0^\infty G(s)ds$ is independent of the shear rate. Moreover since the shear stress predicted is identical to that of the linear viscoelastic model, none of the nonlinear effects observed for the shear stresses in these experiments (see Chapter 3) can be described by this model. Also the equalities between the measurable shear quantities in Table 5.3-1 are predicted to hold for stress relaxation, stress growth, constrained recoil, creep and steady-state flow without the limiting process on shear rate (or shear stress).

(b) For *start-up of shear flow* with shear rate $\dot{\gamma}_0$ after time $t = 0$ we have from Eq. 8.2-8

$$\gamma_{yx}(t, t') = -\dot{\gamma}_0 t, \qquad t > 0,\ t' < 0 \tag{8.2-13}$$

$$\gamma_{yx}(t, t') = -\dot{\gamma}_0(t - t'), \qquad t > 0,\ t' > 0 \tag{8.2-14}$$

The resulting stresses are

$$\tau_{yx} = -\eta^+\dot{\gamma}_0 = -[M_1 - J_1(t)]\dot{\gamma}_0 \overset{\text{(LNM)}}{=} -\left[\sum_i \eta_i(1 - e^{-t/\lambda_i})\right]\dot{\gamma}_0 \tag{8.2-15}$$

$$\tau_{xx} - \tau_{yy} = -\Psi_1^+\dot{\gamma}_0^2 = -[M_2 - 2tJ_1(t) - J_2(t)]\dot{\gamma}_0^2 \overset{\text{(LNM)}}{=} -\left[2\sum_i \eta_i\lambda_i(1 - (1 + t/\lambda_i)e^{-t/\lambda_i})\right]\dot{\gamma}_0^2 \tag{8.2-16}$$

where

$$J_n(t) = \int_t^\infty M(s)(s - t)^n\, ds \overset{\text{(LNM)}}{=} n! \sum_i \eta_i\lambda_i^{n-1}e^{-t/\lambda_i} \tag{8.2-17}$$

and $\tau_{yy} - \tau_{zz} = 0$. Note that for the Lodge network model $(\eta^+)_{t=0} = (\Psi_1^+)_{t=0} = 0$ and that $(\partial\Psi_1^+/\partial t)_{t=0} = 0$, but $(\partial\eta^+/\partial t)_{t=0} > 0$. The network model does predict correctly that the normal stresses rise more slowly than the shear stresses if there are at least two distinct time constants in the memory function.

(c) For *stress relaxation* upon cessation of a steady shear flow with shear rate $\dot{\gamma}_0$ at time $t = 0$ we have

$$\gamma_{yx}(t, t') = \dot{\gamma}_0 t' \qquad t > 0,\ t' < 0 \tag{8.2-18}$$

$$\gamma_{yx}(t, t') = 0 \qquad t > 0,\ t' > 0 \tag{8.2-19}$$

The resulting stresses are

$$\tau_{yx} = -\eta^-\dot{\gamma}_0 = -J_1(t)\dot{\gamma}_0 \overset{\text{(LNM)}}{=} -\left[\sum_i \eta_i e^{-t/\lambda_i}\right]\dot{\gamma}_0 \tag{8.2-20}$$

$$\tau_{xx} - \tau_{yy} = -\Psi_1^-\dot{\gamma}_0^2 = -J_2(t)\dot{\gamma}_0^2 \overset{\text{(LNM)}}{=} -\left[2\sum_i \eta_i\lambda_i e^{-t/\lambda_i}\right]\dot{\gamma}_0^2 \tag{8.2-21}$$

and $\tau_{yy} - \tau_{zz} = 0$. For the Lodge network model it may be shown that the normal stresses relax more slowly than the shear stresses provided there are at least two distinct exponentials in the memory function; this is in qualitative agreement with experimental observations.

§8.3 NONLINEAR INTEGRAL CONSTITUTIVE EQUATIONS

It was demonstrated in Example 8.2-1 that the Lodge rubberlike liquid has a constant viscosity equal to the first moment M_1 of the memory function $M(s)$ (see Eq. 8.2-9). The fact that the viscosity function calculated from that model does not depend on shear rate may not be altered by the choice of different memory functions, but is a direct result of the use of an expression for the stress tensor that is linear in $\boldsymbol{\gamma}_{[0]}(t, t')$ or $\mathbf{B}(t, t')$. In this section we show how the quasi-linear constitutive equation may be modified to give a nonlinear equation capable of describing a shear-rate-dependent viscosity and the shapes of other nonlinear material functions as well. There are several ways to introduce additional nonlinearities into Eq. 8.2-1:

a. Include invariants of $\mathbf{B}$ in the memory function.

b. Allow $\boldsymbol{\gamma}_{[0]}$ to appear nonlinearly in the integrand, for example, by including $\{\boldsymbol{\gamma}_{[0]} \cdot \boldsymbol{\gamma}_{[0]}\}$.

c. Add additional terms with second-, third-, and higher-order integrals containing double, triple, and higher products of $\boldsymbol{\gamma}_{[0]}$ evaluated at different past times $t', t'', t'''\ldots$

In this section we discuss several constitutive equations that incorporate the notions listed under **(a)** and **(b)**. The matter of multiple integrals is postponed until Chapter 9, inasmuch as such "memory-integral expansions" have not been used for fluid dynamics calculations.

Recall that in §4.1 we modified the Newtonian fluid constitutive equation (linear in $\dot{\boldsymbol{\gamma}}$) to obtain a nonlinear equation capable of describing shear-rate dependent viscosity. The process here is in a way similar to that in §4.1. Just as in §4.1 where we started with a discussion of the scalar invariant $\dot{\gamma} = \sqrt{\frac{1}{2}(\boldsymbol{\gamma}_{(1)}:\boldsymbol{\gamma}_{(1)})}$ of the rate-of-strain tensor, we start this section with a discussion of the scalar invariants of the strain tensors.

We begin by defining the scalar invariants I_1 and I_2 (Eqs. A.3-23 and 24) of the Finger strain tensor $\mathbf{B}(t, t')$ by

$$I_1 = \text{tr}\,\mathbf{B} \tag{8.3-1}$$

$$I_2 = \tfrac{1}{2}[(\text{tr}\,\mathbf{B})^2 - \text{tr}\,(\mathbf{B}^2)] \tag{8.3-2}$$

The third scalar invariant, the determinant of $\mathbf{B}$, need not be considered since for incompressible fluids det $\mathbf{B} = 1$. Indeed the incompressibility condition allows us to write the Cayley–Hamilton theorem for $\mathbf{B}$ (Eq. A.3-28) in the simple form

$$\mathbf{B}^2 - I_1\mathbf{B} + I_2\boldsymbol{\delta} = \mathbf{B}^{-1} \tag{8.3-3}$$

If we take the trace of Eq. 8.3-3 and use the definitions in Eqs. 8.3-1 and 8.3-2 we may write the invariants as follows

$$I_1 = \text{tr}\,\mathbf{B} = \text{tr}(\boldsymbol{\delta} - \boldsymbol{\gamma}_{[0]}); \quad I_2 = \text{tr}\,\mathbf{B}^{-1} = \text{tr}(\boldsymbol{\delta} + \boldsymbol{\gamma}^{[0]}) \tag{8.3-4}$$

These invariants will appear in constitutive equations in this section.

The invariant $\dot{\gamma} = \sqrt{\frac{1}{2}(\boldsymbol{\gamma}_{(1)}:\boldsymbol{\gamma}_{(1)})}$ defined in §4.1 is never negative, that is, for all possible $\boldsymbol{\gamma}_{(1)}$ tensors $\dot{\gamma} \geq 0$. We used this fact for example in the definition of the power-law viscosity function, in which $\dot{\gamma}$ is raised to some power n, which could be a fractional number. In the same way it is useful to know the ranges of the values taken by I_1 and I_2 for real deformations. To understand this range we recall from §8.1 that $\mathbf{B}$ is symmetric and positive definite. This means that the eigenvalues of $\mathbf{B}$ are real and positive and may consequently be written as $\lambda_x^2, \lambda_y^2, \lambda_z^2$ where the λ's are positive. Then we have

$$I_1 = \lambda_x^2 + \lambda_y^2 + \lambda_z^2; \quad I_2 = \lambda_x^{-2} + \lambda_y^{-2} + \lambda_z^{-2} \tag{8.3-5}$$

The possible combinations of I_1 and I_2 are those that can be generated by the above equations for all combinations of positive $\lambda_x, \lambda_y, \lambda_z$ compatible with the incompressibility condition det $\mathbf{B} = 1$ which is

$$\lambda_x\lambda_y\lambda_z = 1 \tag{8.3-6}$$

The values of the λ's corresponding to deformations generated by shearfree motions were given in Eqs. 8.1-21 to 23 and may also be found in Appendix C. Without proof we now state that the region of possible I_1 and I_2 values is that bounded on one side by deformations generated in uniaxial elongation ($b = 0$, $\varepsilon < 0$) and on the other side by deformations generated in biaxial stretching ($b = 0, \varepsilon > 0$); here ε is the Hencky strain. The resulting range is shown in Fig. 8.3-1.

We are now in a position to introduce the two nonlinear single-integral models that are the primary objects of study in this chapter. These are

The K-BKZ Equation[1]

$$\boldsymbol{\tau}(t) = \int_{-\infty}^{t} \left[\frac{\partial V(t-t', I_1, I_2)}{\partial I_1} \boldsymbol{\gamma}_{[0]} + \frac{\partial V(t-t', I_1, I_2)}{\partial I_2} \boldsymbol{\gamma}^{[0]} \right] dt' \tag{8.3-7}$$

and

The Rivlin-Sawyers Equation[2]

$$\boldsymbol{\tau}(t) = \int_{-\infty}^{t} [\psi_1(t-t', I_1, I_2)\boldsymbol{\gamma}_{[0]} + \psi_2(t-t', I_1, I_2)\boldsymbol{\gamma}^{[0]}] dt' \tag{8.3-8}$$

In these equations V and the ψ_i are scalar functions of the arguments indicated for $-\infty < t' \leq t$ and for I_1 and I_2 in the shaded areas shown in Fig. 8.3-1. Certainly the Rivlin-Sawyers model includes the K-BKZ equation but we find it convenient to be able to refer to the equations separately. Note that these constitutive equations include both $\boldsymbol{\gamma}_{[0]}$ and $\boldsymbol{\gamma}^{[0]}$. According to Eq. 8B.3-2 this is tantamount to including $\boldsymbol{\gamma}_{[0]}$ and $\{\boldsymbol{\gamma}_{[0]} \cdot \boldsymbol{\gamma}_{[0]}\}$, that is, to allowing the strain tensor to enter the integrand nonlinearly. Hence Eqs. 8.3-7 and 8 could be formulated entirely in terms of the relative strain tensor $\boldsymbol{\gamma}_{[0]}$. However, in most published works it has been common practice to use $\mathbf{B}$ and $\mathbf{B}^{-1}$, or alternatively $\boldsymbol{\gamma}_{[0]}$ and $\boldsymbol{\gamma}^{[0]}$, and we therefore follow the customary usage. Another point to note is that if the $\boldsymbol{\gamma}^{[0]}$ term is omitted, the second normal-stress coefficient is always zero.

The arguments leading to the K-BKZ equation go back to the ideas proposed in Chapter 5 that polymer melts behave for sudden changes in shape as an elastic solid and for very slow changes in shape as a Newtonian fluid. The K-BKZ equation arises from an ad hoc transformation of a general nonlinear expression for the stress tensor of an ideal elastic solid in large deformations to a constitutive equation for a fluid that combines elastic and viscous ideas.

As an alternative to the K-BKZ equation, the Rivlin-Sawyers constitutive equation is based on the physical assumption that the effects on the stress at time t of the deformations at different past times t' are independent of each other. With this assumption it may be shown from some rather formal arguments[2] that Eq. 8.3-8 represents the most general constitutive equation for isotropic fluids.

Very little work on describing material functions or solving flow problems has been done with the K-BKZ and the Rivlin-Sawyers equations in their general forms shown above. Instead it has been customary to introduce the additional assumptions that the scalar functions V and ψ_i may be written as a product of time-dependent and

[1] Proposed independently by A. Kaye, College of Aeronautics, Cranfield, *Note No. 134* (1962), and B. Bernstein, E. Kearsley, and L. Zapas, *Trans. Soc. Rheol.*, **7**, 391-410 (1963); B. Bernstein, E. A. Kearsley, and L. J. Zapas, *J. Res. Natl. Bur. Stand.*, **68B**, 103-113 (1964). See also L. J. Zapas and T. Craft, *J. Res. Natl. Bur. Stand.*, **69A**, 541-546 (1965), and L. J. Zapas, *J. Res. Natl. Bur. Stand.*, **70A**, 525-532 (1966); E. A. Kearsley and L. J. Zapas, *Trans. Soc. Rheol.*, **20**, 623-637 (1976); L. J. Zapas and J. C. Phillips, *J. Rheol.*, **25**, 405-420 (1981) for data comparisons and B. Bernstein, *Acta Mech.*, **2**, 329-354 (1966), and *Int. J. Nonlinear Mech.*, **4**, 183-200 (1969), for a rather complete compilation of material function expressions.

[2] Proposed by R. S. Rivlin and K. N. Sawyers, *Ann. Rev. Fluid Mech.*, **3**, 117-146 (1971). See also R. V. Chacon and N. Friedman, *Arch. Rat. Mech. Anal.*, **18**, 230-240 (1965); A. D. Martin and V. J. Mizel, *Arch. Rat. Mech. Anal.*, **15**, 353-367 (1964); and R. Huilgol, *Continuum Mechanics of Viscoelastic Liquids*, Wiley, New York (1975).

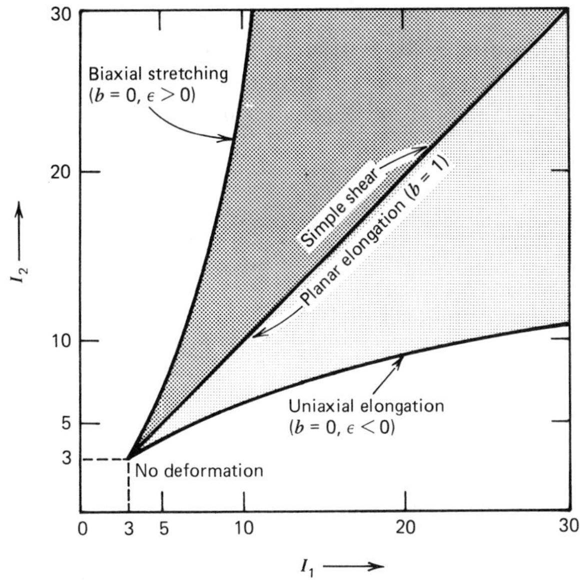

FIGURE 8.3-1. Range of allowable combinations of I_1, I_2 for incompressible materials according to Eqs. 8.3-5 and 6 (shown here for $I_1 < 30$ and $I_2 < 30$). All combinations within the shaded zones are accessible in shearfree flows (see Eqs. 8.1-21 to 23) with combinations in the light gray sector corresponding to $\varepsilon < 0$, and combinations in the dark gray sector corresponding to $\varepsilon > 0$. The line $I_1 = I_2$ that separates the two regions corresponds to both planar elongation and simple shear. The asymptotes of the boundaries are (for large I_1, I_2): $I_1 \sim I_2^2/4$ (for uniaxial elongation) and $I_2 \sim I_1^2/4$ (for biaxial stretching). The point $I_1 = 3$, $I_2 = 3$ corresponds to no deformation gradients. [This kind of sketch was first set forth by K. N. Sawyers, *J. Elast.*, **7**, 99–102 (1977); see also P. K. Currie, *J. Non-Newtonian Fluid Mech.*, **11**, 53–68 (1982); V. K. Stokes, *Trans. ASME*, **48**, 664–666 (1981); L. R. G. Treloar, *The Physics of Rubber Elasticity*, 3rd ed., Oxford University Press, London, (1975), Chap. 10.]

strain-dependent factors as follows:[3]

$$V(t - t', I_1, I_2) = M(t - t')W(I_1, I_2) \tag{8.3-9}$$

$$\psi_i(t - t', I_1, I_2) = M(t - t')\phi_i(I_1, I_2) \tag{8.3-10}$$

In these equations $M(t - t')$ is the linear viscoelastic memory function that appears in Eq. 5.2-19, and W and the ϕ_i are functions defined within the shaded regions shown in Fig. 8.3-1. With these assumptions one arrives at the following constitutive equations, which are still quite flexible since they contain unspecified functions of time and the strain invariants:

The Factorized K–BKZ Equation:

$$\boldsymbol{\tau}(t) = \int_{-\infty}^{t} M(t - t')\left[\frac{\partial W(I_1, I_2)}{\partial I_1}\boldsymbol{\gamma}_{[0]} + \frac{\partial W(I_1, I_2)}{\partial I_2}\boldsymbol{\gamma}^{[0]}\right]dt' \tag{8.3-11}$$

The Factorized Rivlin–Sawyers Equation:

$$\boldsymbol{\tau}(t) = \int_{-\infty}^{t} M(t - t')[\phi_1(I_1, I_2)\boldsymbol{\gamma}_{[0]} + \phi_2(I_1, I_2)\boldsymbol{\gamma}^{[0]}]dt' \tag{8.3-12}$$

[3] Such a factorization was first suggested by J. L. White and N. Tokita, *J. Phys. Soc. Jpn.*, **22**, 719–724 (1967).

The function W in Eq. 8.3-11 is called the *potential function.* In the limit of small displacement gradients, I_1 and I_2 both approach three, and both nonlinear strain tensors simplify to the infinitesimal strain tensor $\boldsymbol{\gamma}$ introduced in Chapter 5. Consequently, since we require that $M(t - t')$ be the linear viscoelastic memory function, we must require that W and the ϕ_i satisfy the following relations:

$$\left(\frac{\partial W}{\partial I_1}\right)_{3,3} + \left(\frac{\partial W}{\partial I_2}\right)_{3,3} = 1 \tag{8.3-13}$$

$$\phi_1(3, 3) + \phi_2(3, 3) = 1 \tag{8.3-14}$$

Equations 8.3-11 and 12 will be featured in the remainder of this section. These constitutive equations are sufficiently flexible that they can describe many rheological phenomena, and it is anticipated that equations of this form will be increasingly used for numerical flow calculations. Expressions for several material functions derivable from Eqs. 8.3-11 and 12 are displayed in Table 8.3-1.

Several molecular theories for polymeric systems result in constitutive equations in the form of the factorized K–BKZ equation: the Rouse–Zimm theory for dilute polymer solutions (Chapter 15), the Curtiss–Bird ($\varepsilon = 0$) theory for polymer melts (Chapter 19), and the Lodge network theory (Chapter 20). The memory functions and W functions for these models are shown in Table 8.3-2; the Curtiss–Bird ($\varepsilon = 0$) model is further considered in Examples 8.3-3 and 4. In addition several single-integral equations have been proposed that may be considered as empirical modifications of molecularly derived equations or as completely empirical equations. A number of these are shown in Tables 8.3-2 and 3. Some of the proposed empirical equations are of the Rivlin–Sawyers form and may not be written in the K–BKZ form.

In connection with Eqs. 8.3-11 and 12 we note also that:

i. If Eqs. 8.3-11 and 12 are to have retarded-motion expansions, then $W(I_1, I_2)$ and $\phi_j(I_1, I_2)$ must have Taylor expansions[4] about $I_1 = 3$, $I_2 = 3$

$$\begin{aligned} W = W^{(00)} &+ W^{(10)}(I_1 - 3) + W^{(01)}(I_2 - 3) \\ &+ \tfrac{1}{2}W^{(20)}(I_1 - 3)^2 + W^{(11)}(I_1 - 3)(I_2 - 3) \\ &+ \tfrac{1}{2}W^{(02)}(I_2 - 3)^2 + \cdots \end{aligned} \tag{8.3-15}$$

$$\phi_j = \phi_j^{(00)} + \phi_j^{(10)}(I_1 - 3) + \phi_j^{(01)}(I_2 - 3) + \cdots \tag{8.3-16}$$

where

$$W^{(mn)} = \left.\frac{\partial^{m+n} W}{\partial I_1^m \partial I_2^n}\right|_{3,3} \tag{8.3-17a}$$

$$\phi_j^{(mn)} = \left.\frac{\partial^{m+n} \phi_j}{\partial I_1^m \partial I_2^n}\right|_{3,3} \tag{8.3-17b}$$

When these Taylor expansions are combined with Eqs. 8.3-11 and 12, the retarded-motion expansion coefficients listed in Table 6.2-2 are obtained (see

[4] O. Hassager, unpublished.

TABLE 8.3-1

Material Functions from the Factorized K-BKZ Equation (Eq. 8.3-11)[a] and the Factorized Rivlin-Sawyers Equation (Eq. 8.3-12)

$$\eta(\dot{\gamma}) = \int_0^\infty M(s)s(\phi_1 + \phi_2)ds \tag{A}$$

$$\Psi_1(\dot{\gamma}) = \int_0^\infty M(s)s^2(\phi_1 + \phi_2)ds \tag{B}$$

$$\Psi_2(\dot{\gamma}) = -\int_0^\infty M(s)s^2\phi_2\, ds \tag{C}$$

$$\eta'(\omega) = \frac{1}{\omega}\int_0^\infty M(s) \sin \omega s\, ds \qquad \text{(D)}^b$$

$$\eta''(\omega) = \frac{1}{\omega}\int_0^\infty M(s)(1 - \cos \omega s)ds \qquad \text{(E)}^b$$

$$\bar{\eta}_1(\dot{\varepsilon}, b) = \frac{1}{\dot{\varepsilon}}\int_0^\infty M(s)[\phi_1(e^{2\dot{\varepsilon}s} - e^{-(1+b)\dot{\varepsilon}s}) + \phi_2(e^{(1+b)\dot{\varepsilon}s} - e^{-2\dot{\varepsilon}s})]ds \qquad \text{(F)}^c$$

$$\bar{\eta}_2(\dot{\varepsilon}, b) = \frac{1}{\dot{\varepsilon}}\int_0^\infty M(s)[(\phi_1 e^{-\dot{\varepsilon}s} + \phi_2 e^{\dot{\varepsilon}s})(e^{b\dot{\varepsilon}s} - e^{-b\dot{\varepsilon}s})]ds \qquad \text{(G)}^c$$

[a] To obtain the entry for the factorized K-BKZ model replace ϕ_j by $\partial W/\partial I_j$; note that $\phi_j = \phi_j(I_1, I_2)$ and $W = W(I_1, I_2)$, where $I_j = I_j(t, t-s)$.
[b] See Problem 5B.7.
[c] For $0 \le b \le 1$ and $-\infty < \dot{\varepsilon} < \infty$. When $b = 0$, the function $\bar{\eta}_1$ becomes $\bar{\eta}$, and $\bar{\eta}_2 = 0$.

Problem 9B.8). It is interesting to note that for the K-BKZ fluid the four coefficients $W^{(01)}$, $W^{(20)}$, $W^{(11)}$, and $W^{(02)}$ appear in the retarded-motion coefficients up through third order; the corresponding quantities for the Rivlin-Sawyers fluid are $\phi_2^{(00)}$, $\phi_1^{(10)}$, $(\phi_1^{(01)} + \phi_2^{(10)})$, and $\phi_2^{(01)}$. Because the coefficients $\phi_1^{(01)}$ and $\phi_2^{(10)}$ appear only as a sum, there is at third order no additional flexibility provided by the Rivlin-Sawyers model—that is, the models give equivalent results at the third-order fluid level.

An example of a model not possessing a retarded-motion expansion is the Wagner model of Table 8.3-3. Consequently for this model one cannot use the procedure outlined in Fig. 6.2-2 to obtain perturbation solutions in the limit of slow flow. However, for the purpose of making fluid mechanical calculations outside the slow flow limit, the lack of a retarded-motion expansion is probably not important.

ii. The factorized K-BKZ equation may be written in an alternative form by the use of the chain rule for partial differentiation[5]

$$\frac{1}{2}\sum_k \mathrm{E}_{ik}\frac{\partial W}{\partial \mathrm{E}_{jk}} = -\frac{\partial W}{\partial I_1}\gamma_{[0]ij} - \frac{\partial W}{\partial I_2}\gamma_{ij}^{[0]} + \left(I_2 + \frac{\partial W}{\partial I_1} - \frac{\partial W}{\partial I_2}\right)\delta_{ij} \tag{8.3-18}$$

[5] One use of this expression is illustrated in Example 8.3-3. See also O. Hassager, *J. Non-Newtonian Fluid Mech.*, **9**, 321–328 (1981), in which a variational principle for factorized K-BKZ fluids is given for unsteady-state flows.

TABLE 8.3-2

Examples of Molecular Models Included in the Factorized K-BKZ Model (Eq. 8.3-11)

Model	$M(s)$	$W(I_1, I_2)$	Adjustable Parameters
Rouse-Zimm[a] (Chapter 15)	$-2\eta_s\delta'(s) + \sum_{j=1}^{N} \frac{\eta_j}{\lambda_j^2} e^{-s/\lambda_j}$	I_1	$\eta_s, \eta_0, \lambda_1, \sigma$ in $\lambda_j \doteq \lambda_1/j^\sigma$; $\eta_j = (\eta_0 - \eta_s)\lambda_j/\sum_j \lambda_j$ ($\sigma = 2$ in the Rouse model)
Lodge network[b] (Chapter 20)	$\sum_{j=1}^{N} \frac{\eta_j}{\lambda_j^2} e^{-s/\lambda_j}$	I_1	$n_j, \lambda_j,$ for $j = 1, 2\ldots$
Tanner-Simmons network rupture[c] (Problem 8B.7)	$\sum_{j=1}^{N} \frac{\eta_j}{\lambda_j^2} e^{-s/\lambda_j}$	I_1, for $I_1 \le I_0$; I_0, for $I_1 > I_0$	$\eta_j, \lambda_j,$ for $j = 1, 2\ldots$; I_0
Curtiss-Bird ($\varepsilon = 0$) or Doi-Edwards[d] (Chapter 19)	$\frac{96\eta_0}{\lambda^2} \sum_{\alpha\,\text{odd}} e^{-\pi^2\alpha^2 s/\lambda}$	$\frac{5}{4\pi}\int \ln\,(\mathbf{B}:\boldsymbol{uu})d\boldsymbol{u}$	η_0 and λ

[a] P. E. Rouse, Jr., *J. Chem. Phys.*, **21**, 1272–1280 (1953); B. H. Zimm, *J. Chem. Phys.*, **24**, 269–278 (1956); the solvent contribution to the stress tensor has been included in Eq. 8.3-11 by the use of the derivative of the Dirac delta function $\delta'(s)$ (see Eq. 5.2-10c).
[b] A. S. Lodge, *Trans. Faraday Soc.*, **52**, 120–130 (1956).
[c] R. I. Tanner and J. M. Simmons, *Chem. Eng. Sci.*, **22**, 1803–1815 (1967).
[d] C. F. Curtiss and R. B. Bird, *J. Chem. Phys.*, **74**, 2016–2025 (1981); **74**, 2026–2033 (1981). In this model $\boldsymbol{u}$ is a unit vector, and $\int \ldots d\boldsymbol{u}$ is an integral over the surface of a unit sphere. See also P. K. Currie, *J. Non-Newtonian Fluid Mech.*, **11**, 53–68 (1982); O. Hassager, *J. Non-Newtonian Fluid Mech.*, **9**, 321–328 (1981). Note that $\int \ln\,(\mathbf{B}:\boldsymbol{uu})d\boldsymbol{u} = \int \ln\, \Sigma_r\Sigma_s\Sigma_q\, \mathrm{E}_{rs}\mathrm{E}_{rq}u_s u_q d\boldsymbol{u}$. This follows from the fact that the eigenvalues of XY and YX are identical for all square matrices X and Y. Both expressions are therefore the same functions of the scalar invariants I_1 and I_2.

where $W = W(I_1, I_2)$. For incompressible fluids the isotropic terms may be omitted in constitutive equations.

iii. Initial value problems with the factorized K-BKZ model will be well posed for finite positive memory functions $M(s)$ when W is monotone in each argument, strictly monotone in I_1 or I_2, and a convex function of $\sqrt{I_1}$ and $\sqrt{I_2}$. We refer to this as the "Renardy condition."[6] An application of this condition is illustrated in the following example.

iv. Standard techniques exist for the determination of M from linear viscoelastic measurements.[7] There is currently no standard technique for the determination of $W(I_1, I_2)$, but beginnings have been made for tackling this difficult problem.[8]

[6] M. Renardy, *Arch. Rat. Mech. Anal.*, **88**, 83–94 (1985). A function $f(x)$ is convex at a point a if in a neighborhood around a we have $f(x) \ge f(a) + (x - a)f'(a)$. If $f''(a) > 0$, then $f(x)$ is strictly convex at a.
[7] J. D. Ferry, *Viscoelastic Properties of Polymers*, 3rd ed., Wiley, New York (1980). See also Example 5.3-7.
[8] J. Meissner, S. E. Stephenson, A. Demarmels, and P. Portmann, *J. Non-Newtonian Fluid Mech.*, **11**, 221–237 (1981); P. Bach and O. Hassager, *Mathematics Research Center Technical Summary Report No. 2755*, University of Wisconsin, Madison (Sept. 1984).

TABLE 8.3-3

Examples of Models Included in the Factorized Rivlin–Sawyers Model (Eq. 8.3-12)[a]

Model	Flow	ϕ_1	ϕ_2	Adjustable Parameters	Comments
Phillips[b]	Simple shear[c]	$(1-q)\exp(-\beta\lvert\gamma_{yx}\rvert)$	$q\exp(-\beta\lvert\gamma_{yx}\rvert)$	q, β	$q = -\Psi_2/\Psi_1$
	General	Not specified	Not specified		This model has no retarded-motion expansion
Wagner[d]	Simple shear[c]	$\exp(-\beta\lvert\gamma_{yx}\rvert)$	0	α, β	$\Psi_2 = 0$
	General	$\exp(-\beta\sqrt{\alpha I_1 + (1-\alpha)I_2 - 3})$	0		This model has no retarded-motion expansion
Papanastasiou, Scriven, and Macosko[e]	Simple shear[c]	$\dfrac{\alpha}{\alpha + \gamma_{yx}^2}$	0	α, β	$\Psi_2 = 0$
	General	$\dfrac{\alpha}{(\alpha-3) + \beta I_1 + (1-\beta)I_2}$	0		

[a] For a criticism of the Rivlin–Sawyers model see R. G. Larson and K. Monroe, *Rheol. Acta*, **23**, 10–13 (1984).
[b] M. C. Phillips, *J. Non-Newtonian Fluid Mech.*, **2**, 109–121, 123–137, 139–149 (1977).
[c] $v_x = \dot{\gamma}_{yx}(t)y$, $v_y = v_z = 0$; $\gamma_{yx}(t, t') = \int_t^{t'} \dot{\gamma}_{yx}(t'')dt''$.
[d] M. H. Wagner, *Rheol. Acta*, **18**, 33–50 (1979). See also M. H. Wagner and J. Meissner, *Macromol. Chem.*, **181**, 1533–1550 (1980). For further discussion of the Wagner model and its molecular basis see §20.5. In §20.5 the function ϕ_1 is designated by h.
[e] A. C. Papanastasiou, L. E. Scriven, and C. W. Macosko, *J. Rheol.*, **27**, 387–410 (1983).

EXAMPLE 8.3-1 A Nonlinear Single-Integral Constitutive Equation

As an example of a potential function for a factorized K–BKZ constitutive equation consider the following function:[9]

$$W = \begin{cases} K[(1-q)I_1^p + qI_2^p], & \text{for } p > 0 \\ 3[(1-q)\ln I_1 + q \ln I_2], & \text{for } p = 0 \end{cases} \tag{8.3-19}$$

where $K = (1/p)3^{1-p}$ is a factor determined by Eq. 8.3-13, and $p \geq 0$ so that W is a monotone increasing function of I_1 and I_2. For the constitutive equation in Eq. 8.3-11 with the above W function find: **(a)** limits on the two parameters q and p suggested by the Renardy condition, **(b)** expressions for the second-order fluid constants, **(c)** expressions for the viscometric functions. Find also the asymptotic forms for the viscometric functions in the limit of high shear rates for the following memory functions: **(d)** the Maxwell model, **(e)** the Jeffreys model, and **(f)** the Segalman model.

SOLUTION **(a)** If we denote $I_1^{1/2} = x_1$ and $I_2^{1/2} = x_2$ then (for $p > 0$)

$$\frac{\partial W}{\partial x_1} = 2K(1-q)px_1^{2p-1}; \quad \frac{\partial W}{\partial x_2} = 2Kqpx_2^{2p-1} \tag{8.3-20}$$

and

$$\frac{\partial^2 W}{\partial x_1^2} = 2K(1-q)p(2p-1)x_1^{2p-2}; \quad \frac{\partial^2 W}{\partial x_2^2} = 2Kqp(2p-1)x_2^{2p-2} \tag{8.3-21}$$

and $\partial^2 W/\partial x_1 \partial x_2 = 0$. We see that the Renardy conditions are satisfied when

$$0 \leq q \leq 1 \quad \text{and} \quad p \geq 1/2 \tag{8.3-22}$$

which ensures that both second partial derivatives are positive.

(b) To find the second-order fluid constants we expand the potential in I_1 and I_2 around $I_1 = I_2 = 3$

$$\begin{aligned} W &= W(3, 3) + \left(\frac{\partial W}{\partial I_1}\right)_{3,3}(I_1 - 3) + \left(\frac{\partial W}{\partial I_2}\right)_{3,3}(I_2 - 3) + \cdots \\ &= (3/p) + (1-q)(I_1 - 3) + q(I_2 - 3) + \cdots \end{aligned} \tag{8.3-23}$$

where we have used $K = (1/p)3^{1-p}$. Then from Table 6.2-2 we see that

$$b_1 = M_1; \quad b_2 = -\tfrac{1}{2}M_2; \quad b_{11} = -qM_2 \tag{8.3-24}$$

and consequently

$$\eta_0 = M_1; \quad \Psi_{1,0} = M_2; \quad \Psi_{2,0} = -qM_2 \tag{8.3-25}$$

where M_n is the nth moment of the memory function as defined in Eq. 8.2-12. We see that $q = -\Psi_{2,0}/\Psi_{1,0}$. This enables us to narrow the useful range of q-values further than given in Eq. 8.3-22. For example if we wish to describe a bulge in Tanners tilted trough (§2.3b) as well as

[9] O. Hassager, unpublished. The separate expression given for $W(p = 0)$ is chosen so that:

$$\frac{\partial W(p=0)}{\partial I_j} = \left.\frac{\partial W(p>0)}{\partial I_j}\right|_{p=0} \quad \text{for } j = 1, 2.$$

rod-climbing (§2.3a) at second order we must have $0 < q < \frac{1}{4}$. This process of using experimental facts to narrow the useful range of parameters has been encountered earlier in §§6.2 and 7.3.

(c) To find the rheological properties in steady simple shear flow we first write the constitutive equation of Eq. 8.3-11:

$$\tau = 3^{1-p} \int_{-\infty}^{t} M(t-t')[(1-q)I_1^{p-1}\gamma_{[0]} + qI_2^{p-1}\gamma^{[0]}]dt' \tag{8.3-26}$$

in which we have used the potential of Eq. 8.3-19 for $p \geq 0$. For steady shear flow with shear rate $\dot{\gamma}$ we obtain the finite strain tensors from Appendix C with $\gamma_{yx} = -\dot{\gamma}s$, where $s = t - t'$; we also find $I_1 = I_2 = 3 + (\dot{\gamma}s)^2$. Thus the relevant stress components become

$$\tau_{yx} = -\left[3^{1-p} \int_0^\infty M(s)[3 + (\dot{\gamma}s)^2]^{p-1} s\, ds\right]\dot{\gamma} = -\eta(\dot{\gamma})\dot{\gamma} \tag{8.3-27}$$

$$\tau_{xx} - \tau_{yy} = -\left[3^{1-p} \int_0^\infty M(s)[3 + (\dot{\gamma}s)^2]^{p-1} s^2\, ds\right]\dot{\gamma}^2 = -\Psi_1(\dot{\gamma})\dot{\gamma}^2 \tag{8.3-28}$$

and $\Psi_2(\dot{\gamma}) = -q\Psi_1(\dot{\gamma})$. These expressions may be used to compute $\eta(\dot{\gamma})$ or $\Psi_1(\dot{\gamma})$ for any memory function. The limiting slopes of $\eta(\dot{\gamma})$ and $\Psi_1(\dot{\gamma})$ as $\dot{\gamma} \to \infty$ depend on the memory function, as illustrated in the following.

(d) For the single-relaxation-time memory function, corresponding to the Maxwell model in Eq. 8.2-2, we obtain

$$\begin{aligned}\eta &= \left[3^{1-p} \frac{\eta_0}{\lambda_1^2} \int_0^\infty e^{-s/\lambda_1}[3(\lambda_1\dot{\gamma})^{-2} + (s/\lambda_1)^2]^{p-1}\, s\, ds\right](\lambda_1\dot{\gamma})^{2p-2} \\ &\sim \left[3^{1-p}\eta_0 \int_0^\infty e^{-x}x^{2p-1}\, dx\right](\lambda_1\dot{\gamma})^{2p-2} = 3^{1-p}\eta_0\Gamma(2p)(\lambda_1\dot{\gamma})^{2p-2}\end{aligned} \tag{8.3-29}$$

for $\lambda_1\dot{\gamma} \to \infty$. Similarly for Ψ_1 we find the asymptote

$$\Psi_1 \sim \left[3^{1-p}\eta_0\lambda_1 \int_0^\infty e^{-x}x^{2p}\, dx\right](\lambda_1\dot{\gamma})^{2p-2} = 3^{1-p}\eta_0\lambda_1\Gamma(2p+1)(\lambda_1\dot{\gamma})^{2p-2} \tag{8.3-30}$$

for $\lambda_1\dot{\gamma} \to \infty$ and in this limit we still have $\Psi_2(\dot{\gamma}) = -q\Psi_1(\dot{\gamma})$. We see that we have achieved partial success in that the viscosity function has a power-law shape for large $\dot{\gamma}$. We see that we must require $p > \frac{1}{2}$ to ensure that the shear stress vs. shear rate curve does not show a maximum. This condition is the same as that derived in **(a)**. We also note that the stress ratio

$$\frac{\tau_{xx} - \tau_{yy}}{\tau_{yx}} \sim \frac{\Gamma(2p+1)}{\Gamma(2p)}\lambda_1\dot{\gamma} = 2p\,\lambda_1\dot{\gamma}, \qquad \text{for } \lambda_1\dot{\gamma} \to \infty \tag{8.3-31}$$

Inasmuch as this result is in conflict with the data in Chapter 3 (see particularly Fig. 3.3-8), we try using several other memory functions.

(e) For the memory function in Eq. 8.2-3 corresponding to the Jeffreys model we obtain the asymptotes[10]

$$\eta \sim \eta_0 \frac{\lambda_2}{\lambda_1}, \qquad \text{for } \lambda_1\dot{\gamma} \to \infty \tag{8.3-32}$$

[10] The next term in the asymptotic expression for $\eta(\dot{\gamma})$ is proportional to $(\lambda_1\dot{\gamma})^{2p-2}$. This term must be retained to get Eq. 8.3-29 from Eq. 8.3-32 by setting λ_2 equal to zero.

and

$$\Psi_1 \sim 3^{1-p}\eta_0(\lambda_1 - \lambda_2)\Gamma(2p+1)(\lambda_1\dot{\gamma})^{2p-2}, \qquad \text{for } \lambda_1\dot{\gamma} \to \infty \tag{8.3-33}$$

This model may be used with any parameter $p > 0$, but when $p < \frac{1}{2}$ we must choose λ_2/λ_1 so that the shear stress vs. shear rate curve is monotone increasing. The constraint that $p \geq \frac{1}{2}$ obtained in **(a)** is not applicable here, inasmuch as the Renardy condition requires that $M(s)$ be bounded. We also find that

$$\frac{\tau_{xx} - \tau_{yy}}{\tau_{yx}} \sim \left(\frac{\lambda_1}{\lambda_2} - 1\right)3^{1-p}\Gamma(2p+1)(\lambda_1\dot{\gamma})^{2p-1}, \qquad \text{for } \lambda_1\dot{\gamma} \to \infty \tag{8.3-34}$$

We see that it is possible to adjust this ratio to increase with $\lambda_1\dot{\gamma}$, become constant, or eventually decrease with $\lambda_1\dot{\gamma}$ depending on the value of p.

(f) Next we could consider the memory function of Eq. 8.2-4. If we use a finite number of relaxation times, then when the shear rate or frequency exceeds the reciprocal of the smallest relaxation time, the model will fail to give a physically reasonable response. If one uses an infinite number of relaxation times, then one has the problem that these are not experimentally determinable. Therefore we use the Segalman memory function[11] corresponding to the $G(s)$ given in Eq. 5B.2-1 (which is a three-constant empiricism that accounts for a complete spectrum of relaxation times)

$$M(s) = \frac{\eta_0}{\lambda^2\Gamma(1-\nu)}\left(\frac{\lambda}{s}\right)^{\nu+1} e^{-s/\lambda}\left(\nu + \frac{s}{\lambda}\right) \tag{8.3-35}$$

containing three constants: η_0 (zero-shear-rate viscosity); λ (time constant); and ν (a dimensionless parameter between 0 and 1). Substitution of this memory function into Eqs. 8.3-27 and 28 gives $\eta = \eta_0$ and $\Psi_1 = 2(1-\nu)\eta_0\lambda$ in the limit as $\lambda\dot{\gamma} \to 0$.

Now we return to the problem of getting the asymptotic expressions for $\lambda\dot{\gamma} \to \infty$. These turn out to depend on combinations of the p and ν values. For the viscosity we find from Eqs. 8.3-27 and 25

$$\begin{aligned}
\eta &= \left[3^{1-p}\int_0^\infty M(s)[3\dot{\gamma}^{-2} + s^2]^{p-1}\, s\, ds\right]\dot{\gamma}^{2p-2} \\
&\sim \left[3^{1-p}\int_0^\infty M(s)s^{2p-1}\, ds\right]\dot{\gamma}^{2p-2} \\
&= \left[\frac{3^{1-p}\eta_0}{\Gamma(1-\nu)}\int_0^\infty e^{-x}(\nu + x)x^{2p-2-\nu}\, dx\right](\lambda\dot{\gamma})^{2p-2} \\
&= \frac{3^{1-p}\eta_0(2p-1)\Gamma(2p-\nu-1)}{\Gamma(1-\nu)}(\lambda\dot{\gamma})^{2p-2} \qquad \begin{pmatrix}\lambda\dot{\gamma} \to \infty \\ p > (\nu+1)/2\end{pmatrix}
\end{aligned} \tag{8.3-36}$$

The restriction $p > (\nu + 1)/2$ must be placed on this result, since it is needed to ensure that the integral does not diverge at $x = s/\lambda = 0$. In the limit as $\nu \to 0$, the Maxwell model result in Eq. 8.3-29 is obtained. Alternatively if we make the change of variable $x = \dot{\gamma}s$ we are led to

$$\begin{aligned}
\eta &= \left[\frac{3^{1-p}\eta_0}{\Gamma(1-\nu)}\int_0^\infty e^{-x/\lambda\dot{\gamma}}x^{-\nu}\left(\nu + \frac{x}{\lambda\dot{\gamma}}\right)(3 + x^2)^{p-1}\, dx\right](\lambda\dot{\gamma})^{\nu-1} \\
&\sim \left[\frac{3^{1-p}\eta_0\nu}{\Gamma(1-\nu)}\int_0^\infty \frac{(3+x^2)^{p-1}}{x^\nu}\, dx\right](\lambda\dot{\gamma})^{\nu-1} \qquad \begin{pmatrix}\lambda\dot{\gamma} \to \infty \\ p < (\nu+1)/2\end{pmatrix}
\end{aligned} \tag{8.3-37}$$

which carries with it the restriction that $p < (\nu + 1)/2$.

[11] This part of the example was worked out by S. R. Burdette.

The normal stress coefficient Ψ_1 also depends on the p and ν values:

$$\Psi_1 \sim \left[3^{1-p}\int_0^\infty M(s)s^{2p}\,ds\right]\dot{\gamma}^{2p-2}$$
$$= \frac{3^{1-p}2\eta_0\lambda p\Gamma(2p-\nu)}{\Gamma(1-\nu)}(\lambda\dot{\gamma})^{2p-2} \qquad \begin{pmatrix}\lambda\dot{\gamma}\to\infty \\ p>\nu/2\end{pmatrix} \tag{8.3-38}$$

which simplifies to the Maxwell result in Eq. 8.3-30 when $\nu \to 0$; Eq. 8.3-38 is valid only when $p > \nu/2$. An alternative expression is obtained by using the change of variable $x = \dot{\gamma}s$:

$$\Psi_1 \sim \left[\frac{3^{1-p}\eta_0\lambda\nu}{\Gamma(1-\nu)}\int_0^\infty \frac{(3+x^2)^{p-1}}{x^{\nu-1}}\,dx\right](\lambda\dot{\gamma})^{\nu-2} \qquad \begin{pmatrix}\lambda\dot{\gamma}\to\infty \\ p<\nu/2\end{pmatrix} \tag{8.3-39}$$

and this is valid for $p < \nu/2$.

The asymptotic behavior of the stress ratio as $\lambda\dot{\gamma} \to \infty$ is as follows:

$$\frac{\tau_{xx}-\tau_{yy}}{\tau_{yx}} \propto \begin{cases} (\lambda\dot{\gamma})^0 & p < \dfrac{\nu}{2} \\ (\lambda\dot{\gamma})^{2p-\nu} & \dfrac{\nu}{2} < p < \dfrac{\nu+1}{2} \\ (\lambda\dot{\gamma}) & p > \dfrac{\nu+1}{2} \end{cases} \tag{8.3-40}$$

Therefore the stress ratio will be constant or increase with some power of $\lambda\dot{\gamma}$ between 0 and 1, the behavior depending on the choice of the constants p and ν. The various regions of parameter space are depicted in Fig. 8.3-2. An examination of the various values of the power-law parameters shown in Table 8.3-4 suggests that the model considered here has sufficient capability to describe the limiting behavior of the viscometric functions. From the values of m, n, m', and n', the model parameters η_0, λ, ν, and p can be determined; the parameter q can be obtained from Ψ_2 or $\bar{\eta}$ data.

The curves of the various material function^c ^re shown in Fig. 8.3-3. It can be seen that the shapes of the curves are reasonably satisfactory. With a total of five constants (η_0, λ, ν, p, and q) it is thus possible to describe a wide variety of rheological phenomena, at least qualitatively.

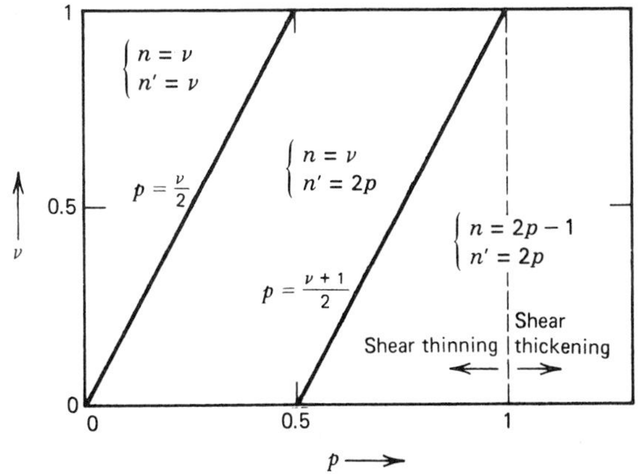

FIGURE 8.3-2. Sketch showing the values of n and n', the "power-law indices" for the high-shear-rate asymptotes of the viscosity and first normal stress coefficient: $\eta \sim m\dot{\gamma}^{n-1}$ and $\Psi_1 \sim m'\dot{\gamma}^{n'-2}$. There are three regions corresponding to different combinations of ν and p.

TABLE 8.3-4

Power-Law Function Parameters for Describing the High Shear Rate Asymptotes of the Viscosity and First Normal Stress Coefficient

	$\eta = m\dot{\gamma}^{n-1}$		$\Psi_1 = m'\dot{\gamma}^{n'-2}$	
Fluid	m Pa · s^n	n	m' Pa · s$^{n'}$	n'
Solutions:[a]				
Hydroxyethylcellulose	21	0.400	30	0.567
Separan	25	0.333	410	0.830
Polyisobutylene	140	0.350	1700	0.677
Melts:[b]				
High-density polyethylene	1.89×10^4	0.40	4.28×10^4	0.41
Low-density polyethylene	0.48×10^4	0.45	1.61×10^4	0.51
Polystyrene	1.56×10^4	0.32	3.31×10^4	0.31
Polypropylene	0.44×10^4	0.50	1.02×10^4	0.51

[a] P. J. Leider, *Ind. Eng. Chem. Fundam.*, **13**, 342–346 (1974).
[b] C. D. Han, K. U. Kim, N. Siskovic, and C. R. Huang, *Rheol. Acta*, **14**, 533–549 (1975), as cited by M. H. Wagner, *Rheol. Acta*, **18**, 33–50 (1979).

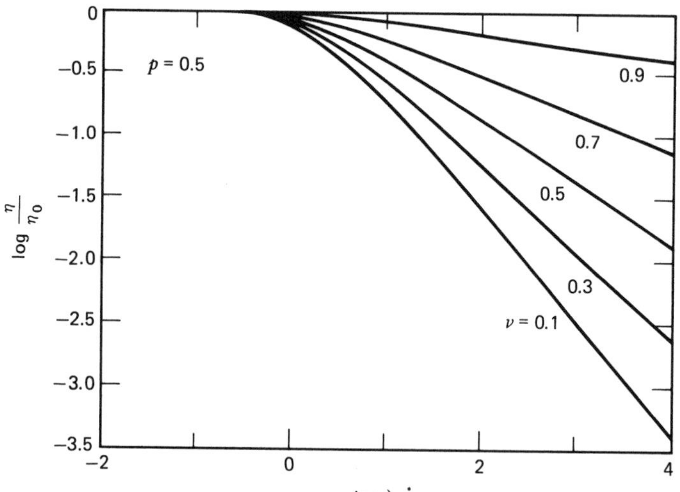

(*a*)

FIGURE 8.3-3. Material functions computed for a factorized K–BKZ fluid with potential function $W(I_1, I_2)$ given by Eq. 8.3-19 and the memory function $M(s)$ by Eq. 8.3-35: (*a*) viscosity.

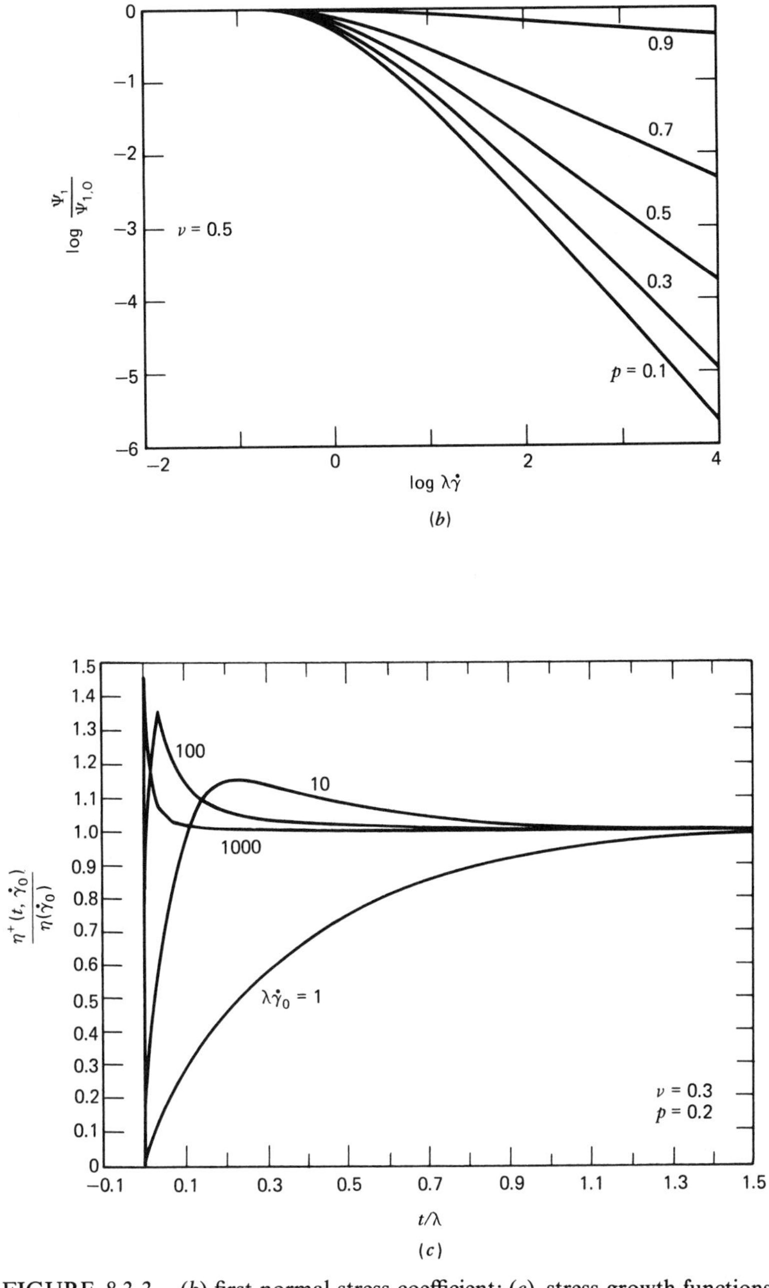

FIGURE 8.3-3. (*b*) first normal stress coefficient; (*c*), stress growth functions.

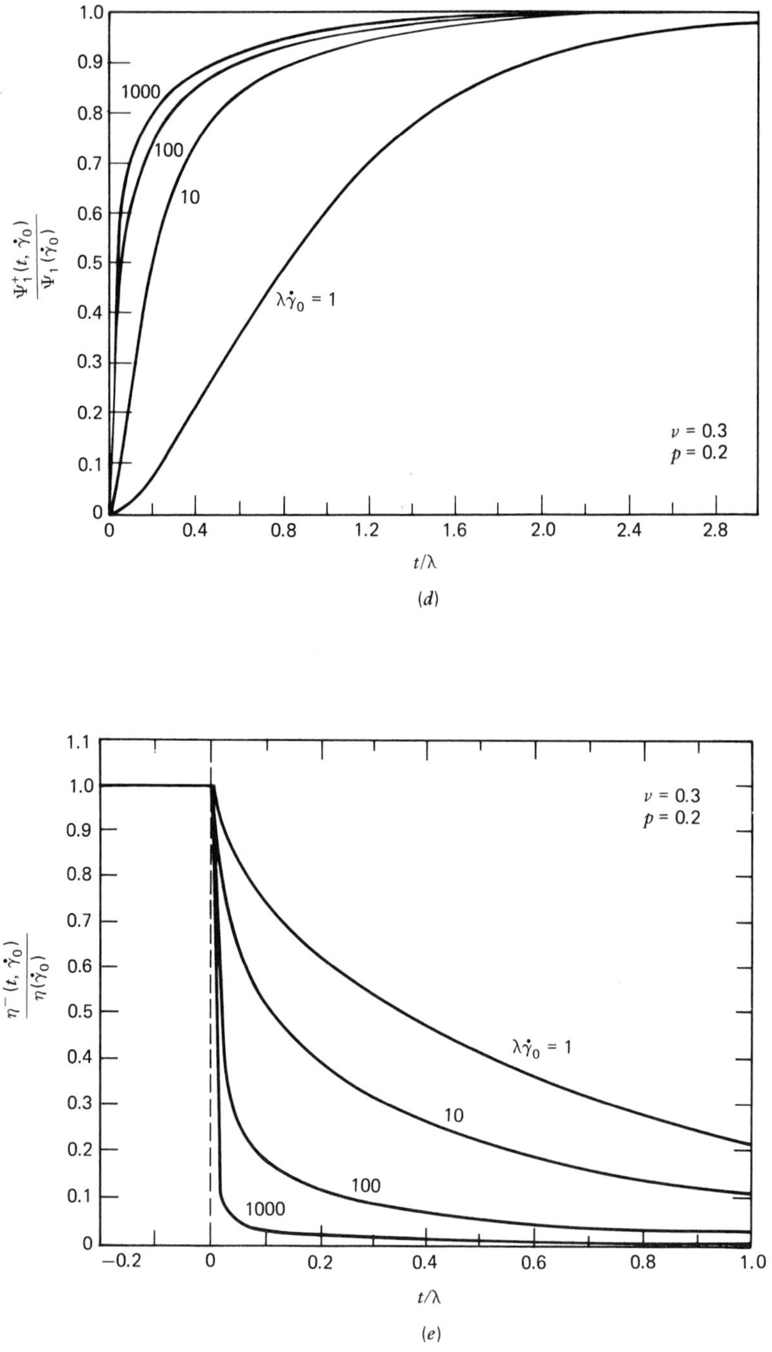

FIGURE 8.3-3. (*d*) stress growth functions; (*e*) stress relaxation functions.

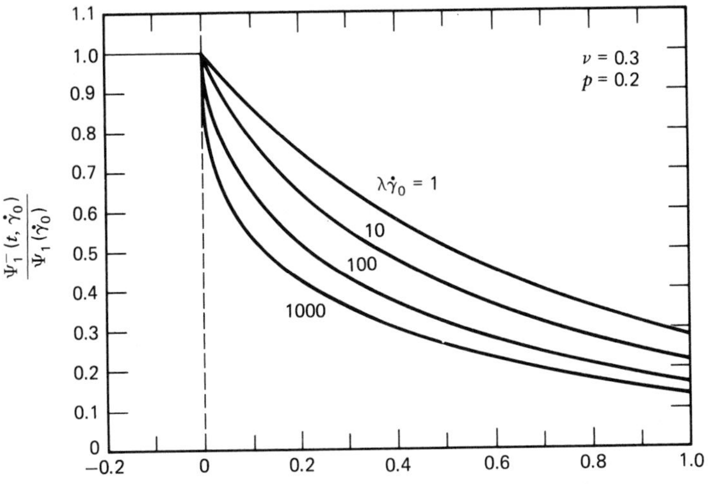

(*f*)

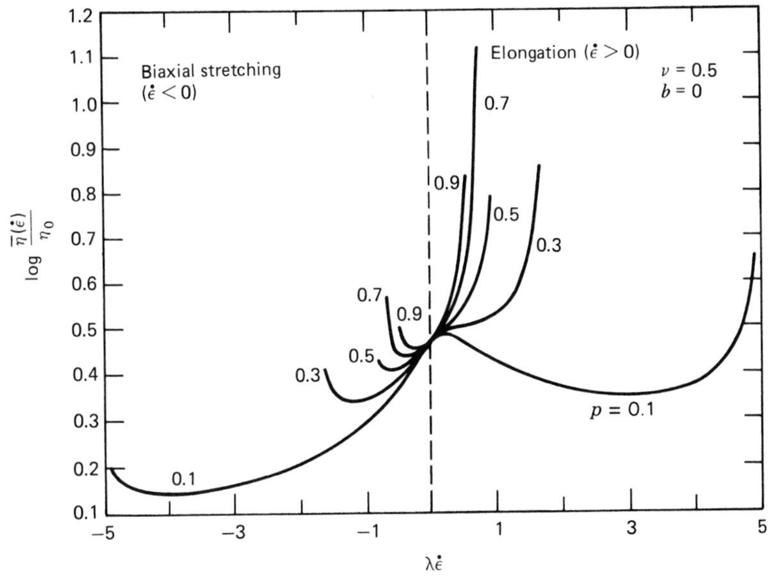

(*g*)

FIGURE 8.3-3. (*f*) stress relaxation functions; (*g*) elongational viscosity ($b = 0$).

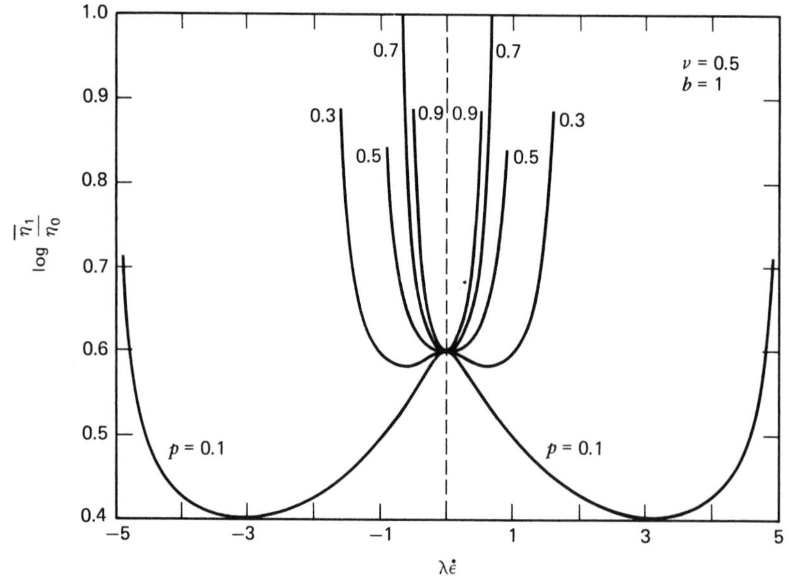

(*h*)

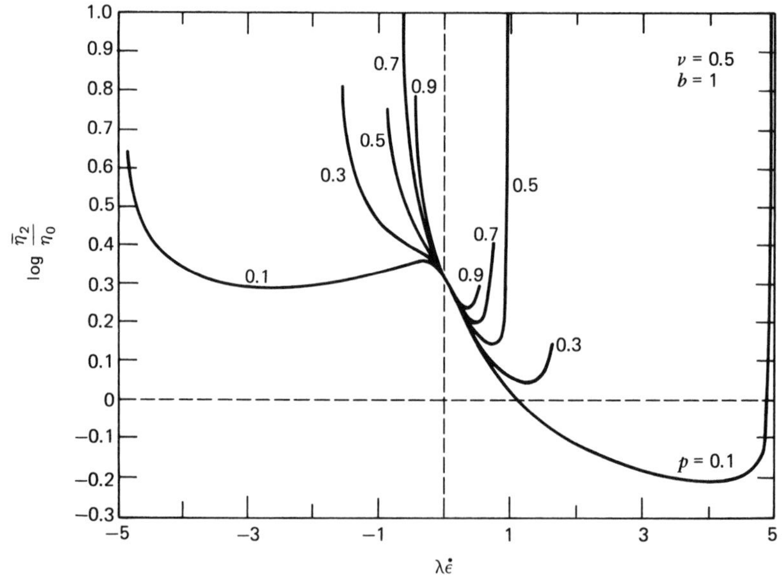

(*i*)

FIGURE 8.3-3. (*h*), (*i*) planar elongational viscosities ($b = 1$). Note in (*h*) that $\bar{\eta}_1(-\dot{\varepsilon}) = \bar{\eta}_1(\dot{\varepsilon})$, and in (*i*) that $\bar{\eta}_2(-\dot{\varepsilon}) = -\bar{\eta}_2(\dot{\varepsilon}) + \bar{\eta}_1(\dot{\varepsilon})$. [Curves calculated by S. R. Burdette.]

EXAMPLE 8.3-2 The Wagner Model

The most carefully tested constitutive equation of the Rivlin–Sawyers type is the Wagner model given in Table 8.3-3. Wagner called $\phi_1(I_1, I_2)$ the "damping function," since it describes the diminishing of the fluid memory by the various kinematic events of the past. Obtain the shear flow and steady elongational flow material functions for this model if $M(t - t')$ is given as a sum of exponentials as in Eq. 8.2-4. Show how the model parameters can be obtained from shear flow experiments.

SOLUTION **(a)** Shear Flow ($v_x = \dot{\gamma}_{yx}(t)y$, $v_y = 0$, $v_z = 0$)
For the particular choice of ϕ_1 in Table 8.3-3 we find for shearing flows

$$\phi_1(I_1, I_2) = e^{-\beta|\gamma_{yx}|} \tag{8.3-41}$$

since $I_1 = I_2 = 3 + \gamma^2{}_{yx}$ for these flows. A convenient method for determining ϕ_1 is from stress relaxation following a step shear strain of magnitude γ_0 (cf. Table 3.4-1e and §3.4, Experiment e). For the step strain experiment $\gamma_{yx}(t, t') = -\gamma_0[1 - H(t')]$ if the step strain is applied at time $t = 0$; here $H(t)$ is the Heaviside unit step function. By inserting this expression for γ_{yx} into the Wagner model we find

$$\begin{aligned}\tau_{yx} &= \int_{-\infty}^{t} M(t - t')\phi_1(|\gamma_{yx}|)\gamma_{yx}(t, t')dt' \\ &= -\gamma_0 \int_{-\infty}^{0} M(t - t')dt'\phi_1(\gamma_0) \\ &= -\gamma_0 G(t)\phi_1(\gamma_0)\end{aligned} \tag{8.3-42}$$

where $\gamma_{[0]}$ is obtained from Appendix C. Here we have used the facts that $M(t - t') = \partial G(t - t')/\partial t'$ and that $G(t) \to 0$ as $t \to \infty$. Equation 8.3-42 shows that the nonlinear relaxation modulus (Table 3.4-1e) can be factored into a function of time alone and a function of strain alone

$$G(t, \gamma_0) = G(t)\phi_1(\gamma_0) \tag{8.3-43}$$

Experimental data (cf. Figs. 3.4-15 and 16) show that this factorization[12] is indeed observed, except perhaps at very short times, and this provides a justification for the factorization of the K–BKZ and Rivlin–Sawyers models introduced in Eqs. 8.3-11 and 12. The function $G(t)$, which is the linear viscoelastic relaxation modulus, can be determined from the top curve ($\gamma_0 \to 0$) in Fig. 3.4-15 (or from other linear viscoelastic experiments). The strain-dependent part describes the vertical displacement of the relaxation curves in Figs. 3.4-15 and 16. By shifting the large γ_0 curves in Fig. 3.4-15 so as to make them overlap with the top, linear viscoelastic curve, we obtain $\phi_1(\gamma_0)$ as shown in Fig. 8.3-4.

When ϕ_1 is given by the single exponential in Eq. 8.3-41, the data in Fig. 8.3-4 are well described[13] for $\gamma_0 \leq 10$ by $\beta = 0.18$ (dashed curve). In order to improve the description of ϕ_1 for shear flows, Wagner has also used a two-exponential function

$$\phi_1 = ce^{-\beta_1|\gamma_{yx}|} + (1 - c)e^{-\beta_2|\gamma_{yx}|} \tag{8.3-44}$$

where c, β_1, and β_2 are constants. This functional form does not apply to flows other than shearing flow. Fitting Eq. 8.3-44 to the data in Fig. 8.3-4 results in $c = 0.57$, $\beta_1 = 0.31$, and $\beta_2 = 0.106$, which is seen to give a good description of the data for shear strains up to 30. To finish determining the model,

[12] "Strain-time" factorization is also observed for step strain biaxial stretching flows. See P. R. Soskey and H. H. Winter, *J. Rheol.*, **29**, 493–517 (1985), and A. C. Papanastasiou, L. E. Scriven, and C. W. Macosko, *J. Rheol.*, **27**, 387–410 (1983).

[13] M. H. Wagner, *Rheol. Acta*, **15**, 136–142 (1976); **16**, 43–50 (1977); H. M. Laun, *Rheol. Acta*, **17**, 1–15 (1978).

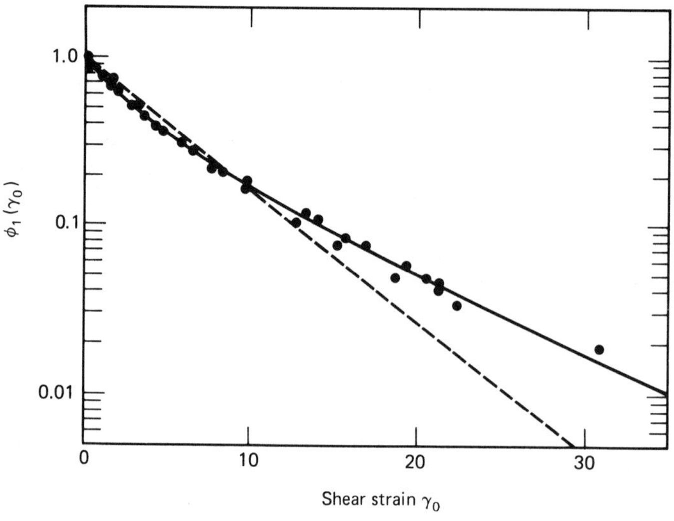

FIGURE 8.3-4. The damping function ϕ_1 as a function of shear strain γ_0 obtained by vertical shifting of the relaxation curves in Fig. 3.4-15. The dashed line corresponds to Eq. 8.3-41 and the solid line to Eq. 8.3-44. [H. M. Laun, *Rheol. Acta*, **17**, 1-15 (1978).]

we must fit $M(t - t')$. This can be done from linear viscoelastic experiments as shown in Example 5.3-7.

It is now of interest to compare the Wagner model predictions with other nonlinear shear flow material functions. We use Eq. 8.3-41 for ϕ_1 to illustrate the calculation of material functions. For steady shear flow with shear rate $\dot{\gamma}$, we have $|\gamma_{yx}| = \dot{\gamma}(t - t')$ so that (with $s = t - t'$)

$$\eta(\dot{\gamma}) = \int_0^\infty M(s)e^{-\beta\dot{\gamma}s}\, s\, ds \tag{8.3-45}$$

$$\Psi_1(\dot{\gamma}) = \int_0^\infty M(s)e^{-\beta\dot{\gamma}s}s^2\, ds \tag{8.3-46}$$

$$\Psi_2(\dot{\gamma}) = 0 \tag{8.3-47}$$

Note that these equations suggest that $\Psi_1(\dot{\gamma}) = -(1/\beta)(d\eta/d\dot{\gamma})$. When $M(s)$ is given as a sum of exponentials the integrals can be performed to give

$$\eta(\dot{\gamma}) = \sum_{p=1}^\infty \frac{\eta_p}{(1 + \beta\lambda_p\dot{\gamma})^2} \tag{8.3-48}$$

$$\Psi_1(\dot{\gamma}) = 2\sum_{p=1}^\infty \frac{\eta_p\lambda_p}{(1 + \beta\lambda_p\dot{\gamma})^3} \tag{8.3-49}$$

for the shear rate dependence of η and Ψ_1. These predictions with η_p and λ_p given by Table 5.3-2, and $\beta = 0.18$ are compared with experimental data in Fig. 3.3-2 (dashed curves). For comparisen, the corresponding predictions for the two-exponential ϕ_1 are shown in Fig. 3.3-2 as the solid curve. Only a minor improvement at large shear rates in the description of Ψ_1 is obtained from the two-exponential form. Similarly, other nonlinear shear flow material functions can be calculated with Eqs. 8.3-41 and 44, and comparisons[13] of some of these with experimental data are given in Figs. 3.4-17 and 19 to 21.

(b) Steady Elongational Flow ($v_x = -\frac{1}{2}\dot{\varepsilon}x$, $v_y = -\frac{1}{2}\dot{\varepsilon}y$, $v_z = \dot{\varepsilon}z$)
For this flow (with $I_1 = 2e^{-\dot{\varepsilon}s} + e^{2\dot{\varepsilon}s}$ and $I_2 = 2e^{\dot{\varepsilon}s} + e^{-2\dot{\varepsilon}s}$) the damping function is

$$\phi_1 = \exp\left[-\beta\sqrt{\alpha(2e^{-\dot{\varepsilon}s} + e^{2\dot{\varepsilon}s}) + (1-\alpha)(2e^{\dot{\varepsilon}s} + e^{-2\dot{\varepsilon}s}) - 3}\right] \tag{8.3-50}$$

The steady elongational viscosity $\bar{\eta}(\dot{\varepsilon})$ is then obtained as

$$\bar{\eta}(\dot{\varepsilon}) = \frac{1}{\dot{\varepsilon}}\int_0^\infty \left\{\sum_{p=1}^{\infty} \frac{\eta_0}{\lambda_p^2} e^{-s/\lambda_p}\right\}\phi_1(e^{2\dot{\varepsilon}s} - e^{-\dot{\varepsilon}s})ds \tag{8.3-51}$$

This integral must be done numerically.

Wagner has shown how a combination of shear flow data and transient elongational flow data can be used to obtain the parameter α in the damping function.[14] The ability of the Wagner model to fit a wide variety of shear and elongational flow material functions with only two parameters, α and β, to describe the nonlinear part of the memory function, is impressive.

EXAMPLE 8.3-3 The Doi–Edwards Constitutive Equation

In Volume 2 we consider constitutive equations derived from molecular theory. In Chapter 19 the Curtiss–Bird theory of polymer melts is given. The single-integral constitutive equation obtained in this theory does not fall within the categories listed in Tables 8.3-2 and 3. However a special case of the equation, namely with $\varepsilon = 0$ in Eq. 19.6-9, known as the Doi–Edwards equation, is of the factorized K–BKZ form; see Table 8.3-2 for the potential function for this model. Insert the potential for the Doi–Edwards model from Table 8.3-2 into the K–BKZ model and rewrite the resulting model in terms of $\mathbf{B}^{-1}$.

SOLUTION We use the form of the potential W with the E_{ij} shown explicitly as in footnote (d) of Table 8.3-2. This form is then differentiated with respect to E_{jk} and we form the combination

$$\frac{1}{2}\sum_k \mathrm{E}_{ik}\frac{\partial W}{\partial \mathrm{E}_{jk}} = \frac{5}{4\pi}\int \frac{\sum_k \sum_m \mathrm{E}_{ik}u_k \mathrm{E}_{jm}u_m}{\sum_k \sum_m \sum_n \mathrm{E}_{nk}u_k \mathrm{E}_{nm}u_m}\,d\boldsymbol{u} \tag{8.3-52}$$

Keep in mind that $\boldsymbol{u}$ is a unit vector and $\int \ldots d\boldsymbol{u}$ is shorthand for an integration over the surface of a unit sphere. Note that the expression on the right is symmetric in "i" and "j" as it should be. We then insert this expression into Eq. 8.3-18, which is in turn inserted into the K–BKZ equation to obtain the constitutive equation:[15]

$$\boldsymbol{\tau}(t) = -\frac{5}{4\pi}\int_{-\infty}^{t} M(t-t')\int \hat{\boldsymbol{u}}\hat{\boldsymbol{u}}\,d\boldsymbol{u}\,dt' \tag{8.3-53}$$

where

$$\hat{\boldsymbol{u}} = [\mathbf{E}\cdot\boldsymbol{u}]/|\mathbf{E}\cdot\boldsymbol{u}| \tag{8.3-54}$$

is a unit vector. The two vertical bars in the denominator of Eq. 8.3-54 indicate the magnitude of the vector (see Eq. A.2-17). Note that the strain dependence enters through the displacement gradient tensor $\mathbf{E}(t, t')$. In order to rewrite the constitutive equation in a form containing the strain tensors, we must perform a transformation of variables in the integration over the unit sphere. Specifically we

[14] M. H. Wagner, *Rheol. Acta*, **18**, 33–50, 427–428 (1979); *J. Non-Newtonian Fluid Mech.*, **4**, 39–55 (1978).
[15] R. B. Bird, H. H. Saab, and C. F. Curtiss, *J. Chem. Phys.*, **77**, 4747–4757 (1982); see also Eq. 19.6-9.

wish to transform from the spherical coordinates θ, ϕ of the unit vector $\boldsymbol{u} = \boldsymbol{u}(\theta, \phi)$ to the spherical coordinates $\hat{\theta}$, $\hat{\phi}$ of the unit vector $\hat{\boldsymbol{u}} = \hat{\boldsymbol{u}}(\hat{\theta}, \hat{\phi})$, related to $\boldsymbol{u}$ through Eq. 8.3-54. To do this we need the Jacobian of the transformation (see Problem 8D.1), and we then obtain

$$d\boldsymbol{u} = \frac{\det \boldsymbol{\Delta}}{|\boldsymbol{\Delta} \cdot \hat{\boldsymbol{u}}|^3} \, d\hat{\boldsymbol{u}} \tag{8.3-55}$$

where $d\boldsymbol{u} = \sin\theta \, d\theta \, d\phi$ and $d\hat{\boldsymbol{u}} = \sin\hat{\theta} \, d\hat{\theta} \, d\hat{\phi}$, and $\boldsymbol{\Delta}$ is the tensor inverse to $\mathbf{E}$. We recall from §8.1 that the determinants of $\mathbf{E}$ and $\boldsymbol{\Delta}$ are equal to unity for incompressible fluids. When this is taken into account and Eq. 8.3-55 is inserted into Eq. 8.3-53 we obtain

$$\boldsymbol{\tau} = -\frac{5}{4\pi} \int_{-\infty}^{t} M(t - t') \int \frac{\hat{\boldsymbol{u}}\hat{\boldsymbol{u}}}{|\boldsymbol{\Delta} \cdot \hat{\boldsymbol{u}}|^3} \, d\hat{\boldsymbol{u}} \, dt' \tag{8.3-56}$$

If we introduce the expression for the magnitude of the vector in the denominator as the square root of the scalar product of the vector with itself we find that the equation may finally be rewritten as[15]

$$\boldsymbol{\tau} = -\frac{5}{4\pi} \int_{-\infty}^{t} M(t - t') \int \frac{\boldsymbol{u}\boldsymbol{u}}{(\mathbf{B}^{-1} : \boldsymbol{u}\boldsymbol{u})^{3/2}} \, d\boldsymbol{u} \, dt' \tag{8.3-57}$$

in which we have dropped the circumflex on the integration variable. In this form the deformation history enters through the finite strain tensor $\mathbf{B}^{-1}(t, t')$. In this example we have shown how the constitutive equation for the Doi–Edwards model may be written in three different ways. The form in Eq. 8.3-53 has been used in calculations of shear flow properties,[16] and the form in Eq. 8.3-57 has been used in calculations of elongational flow properties.[17]

EXAMPLE 8.3-4 The Approximate Currie Potential for the Doi–Edwards Model

The following approximate potential function[18] for the Curtiss–Bird ($\varepsilon = 0$) or Doi–Edwards constitutive equation has been used in numerical simulations[19] of viscoelastic flows:

$$W(J) = 5 \ln (J - 1) \tag{8.3-58}$$

where

$$J = I_1 + 2(I_2 + \tfrac{13}{4})^{1/2} \tag{8.3-59}$$

Derive the corresponding approximate constitutive equation and investigate its properties in steady simple shear flow.

SOLUTION The approximate constitutive equation is obtained by substituting Eqs. 8.3-58 and 8.3-59 into Eq. 8.3-11,

$$\boldsymbol{\tau} = 5 \int_{-\infty}^{t} \frac{M(t - t')}{(J - 1)} \left[\boldsymbol{\gamma}_{[0]} + (I_2 + \tfrac{13}{4})^{-1/2} \boldsymbol{\gamma}^{[0]} \right] dt' \tag{8.3-60}$$

[16] H. H. Saab, R. B. Bird, and C. F. Curtiss, *op. cit.*, pp. 4758–4766.
[17] R. B. Bird, H. H. Saab, and C. F. Curtiss, *J. Phys. Chem.*, **86**, 1102–1106 (1982); see also Eq. 19.6-9.
[18] This example is based on P. K. Currie, *J. Non-Newtonian Fluid Mech.*, **11**, 53–68 (1982).
[19] B. Bernstein, D. S. Malkus, and E. T. Olsen, *Int. J. Num. Meth. Fluids*, **5**, 43–70 (1985); D. S. Malkus and B. Bernstein, *J. Non-Newtonian Fluid Mech.*, **16**, 77–116 (1984); S. Dupont, J. M. Marchal, and M. J. Crochet, *J. Non-Newtonian Fluid Mech.*, **17**, 157–183 (1985).

Now for steady shear flow with shear rate $\dot{\gamma}$ we find

$$J = 3 + (\dot{\gamma}s)^2 + (25 + 4(\dot{\gamma}s)^2)^{1/2} \tag{8.3-61}$$

The expressions for the shear stress and normal stress differences then become

$$\tau_{yx} = -\left[5\int_0^\infty \frac{M(s)s}{(J-1)}[1 + (I_2 + \tfrac{13}{4})^{-1/2}]ds\right]\dot{\gamma} = -\eta(\dot{\gamma})\dot{\gamma} \tag{8.3-62}$$

$$\tau_{xx} - \tau_{yy} = -\left[5\int_0^\infty \frac{M(s)s^2}{(J-1)}[1 + (I_2 + \tfrac{13}{4})^{-1/2}]ds\right]\dot{\gamma}^2 = -\Psi_1(\dot{\gamma})\dot{\gamma}^2 \tag{8.3-63}$$

$$\tau_{yy} - \tau_{zz} = \left[5\int_0^\infty \frac{M(s)s^2}{(J-1)}(I_2 + \tfrac{13}{4})^{-1/2}\,ds\right]\dot{\gamma}^2 = -\Psi_2(\dot{\gamma})\dot{\gamma}^2 \tag{8.3-64}$$

These expressions plus the memory function $M(s)$ from Table 8.3-2 determine the viscometric functions over the entire range of shear rates. It is rather simple to obtain the viscometric functions in the limit of small shear rates. To do so we let $\dot{\gamma} = 0$ in the integrals in Eqs. 8.3-62, 63, and 64 so that $I_1 = I_2 = 3$ and $J = 8$. We then find

$$\eta_0 = \int_0^\infty M(s)s\,ds = \eta_0 \tag{8.3-65}$$

$$\Psi_{1,0} = \int_0^\infty M(s)s^2\,ds = \tfrac{1}{5}\eta_0\lambda \tag{8.3-66}$$

$$\Psi_{2,0} = -\tfrac{2}{7}\int_0^\infty M(s)s^2\,ds = -\tfrac{2}{35}\eta_0\lambda \tag{8.3-67}$$

In the limit of high shear rates it may be shown that the viscometric functions exhibit power-law behavior corresponding to $\eta \propto \dot{\gamma}^{-3/2}$, $\Psi_1 \propto \dot{\gamma}^{-2}$, and $\Psi_2 \propto \dot{\gamma}^{-5/2}$.

Even though we have used an approximate expression for the potential, the zero-shear-rate values in Eqs. 8.3-65, 66, and 67 are exact. The ratio of $\Psi_{2,0}/\Psi_{1,0} = -\frac{2}{7}$ means that the Curtiss–Bird ($\varepsilon = 0$) or Doi–Edwards equation incorrectly predicts a depression in the free surface of a fluid near a rotating rod (see Problem 6B.6).[20] Also since the viscosity drops faster than $(\lambda\dot{\gamma})^{-1}$ with shear rate, the equation cannot describe shear flow above a certain critical shear rate.

§8.4 FLOW PROBLEMS IN ONE SPATIAL VARIABLE

In this section we illustrate the solution of flow problems with two examples. In Example 8.4-1 we consider the steady flow of a viscoelastic fluid into a line sink, and in Example 8.4-2 we consider the transient inflation of a spherical viscoelastic film. The particular situations have been chosen for their simplicity and for illustrating two main techniques for the solution of flow problems with integral constitutive equations. We shall refer to these as the Eulerian technique and the Lagrangian technique, respectively.

In the *Eulerian technique*, illustrated in the first example, the dependent variables are the velocity, pressure, and stress. Since the velocity does not enter explicitly into the integral constitutive equation, we must guess a form for the velocity field and then use this guess to

[20] O. Hassager, *J. Rheol.*, **29**, 361–364 (1985).

evaluate the displacement functions and subsequently the stress components. Finally we combine the stress tensor obtained in this way with the equation of motion and attempt to satisfy the resulting equation by suitable choice of adjustable constants or functions in the assumed velocity field.

By contrast in the *Lagrangian technique*, illustrated in the second example, the dependent variables are the particle positions (displacement functions), pressure, and stress. Since an integral constitutive equation can be written explicitly in terms of the displacement functions, these can be combined with the equations of motion to give one or more integro-differential equations for the particle positions as functions of time. The result is an initial value problem in which the particle motion can be determined by forward integration from some time at which the entire past flow history is known.

In certain simple situations analytical solutions may be obtained from the steady Eulerian technique. By contrast the integro-differential equations obtained in the Lagrangian technique must almost always be solved numerically. Note also that whereas applying the Eulerian technique to steady and unsteady problems results in time-independent and time-dependent differential equations, respectively, the Lagrangian technique always results in an unsteady problem. More will be said about these techniques in §8.5; we turn now to the examples. These examples are chosen so that inversion of the displacement functions to calculate $\gamma_{[0]}$ is trivial; a problem in which this inversion is nontrivial is given in Example 10.1-2.

EXAMPLE 8.4-1 Flow of a Viscoelastic Fluid into a Line Sink

Consider the steady two-dimensional plane flow of a Lodge rubberlike liquid (Eq. 8.2-1) moving radially towards the z-axis in a cylindrical coordinate system. Find the velocity field and find the distribution of stresses and pressure in the fluid. Neglect gravitational effects.

SOLUTION We begin by postulating a form for the velocity field: $v_r = v_r(r)$, $v_\theta = 0$, $v_z = 0$. In the cylindrical coordinate system the equation of continuity reduces to

$$\frac{d}{dr}(rv_r) = 0 \tag{8.4-1}$$

This equation may immediately be integrated once to give

$$v_r = -\frac{K}{r} \tag{8.4-2}$$

where K is a constant of integration, and $K \geq 0$ since we assume that the velocity is towards the axis. From an integration over a cylindrical volume of length L and radius r (Fig. 8.4-1) it may be seen that $K = Q/2\pi L$ where Q is the volume rate of flow into the sink. The r-component of Eq. 8.1-14 for the displacement function now reads

$$\frac{dr}{dt} = -\frac{K}{r} \tag{8.4-3}$$

which is to be solved subject to the condition that $r = r'$ at $t = t'$. The final results for the displacement functions then become

$$r = \sqrt{r'^2 - 2K(t - t')}; \quad \theta = \theta'; \quad z = z' \tag{8.4-4}$$

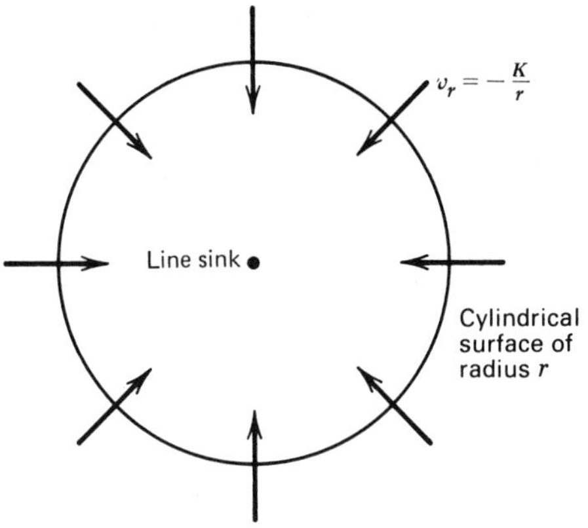

FIGURE 8.4-1. Axisymmetrical flow into a line sink. The volume rate of flow through a cylindrical surface of radius r and length L is $Q = 2\pi KL$.

With these displacement functions we see from Eqs. B.6-7 through 12 that the only nonzero components of $\gamma_{[0]}$ are

$$\gamma_{[0]rr} = 1 - \left(\frac{\partial r}{\partial r'}\right)^2 = -\frac{2K(t-t')}{r^2} \tag{8.4-5a}$$

$$\gamma_{[0]\theta\theta} = 1 - \left(\frac{r}{r'}\frac{\partial \theta}{\partial \theta'}\right)^2 = \frac{2K(t-t')}{r^2 + 2K(t-t')} \tag{8.4-5b}$$

Notice that after the differentiations have been performed, the strain tensor components must be rewritten so that only r, t, and t' appear in the expressions. In order to do this it is necessary to invert the displacement functions in Eq. 8.4-4. When these quantities are substituted into the constitutive equation (Eq. 8.2-1) and when the variable $s = t - t'$ is used, the only nonzero stress tensor components are

$$\tau_{rr}(r) = -\frac{2K}{r^2}\int_0^\infty M(s)\, s\, ds \tag{8.4-6}$$

$$\tau_{\theta\theta}(r) = 2K\int_0^\infty M(s)\frac{s}{r^2 + 2Ks}\, ds \tag{8.4-7}$$

where $s = t - t'$. In the expression for $\tau_{\theta\theta}$ we see why we need to have $K \geq 0$, and that radial flow out from a line source (with $K < 0$) may not be described.

With the above components of the stress tensor, the θ- and z-components of the equation of motion may be satisfied identically by $p = p(r)$, and the radial component of the equation of motion then gives

$$\begin{aligned}\frac{dp}{dr} &= -\rho v_r \frac{dv_r}{dr} - \left(\frac{1}{r}\frac{d}{dr}(r\tau_{rr})\right) + \frac{\tau_{\theta\theta}}{r} \\ &= +\frac{\rho K^2}{r^3} + 2K\int_0^\infty M(s)s\left[-\frac{1}{r^3} + \frac{1}{r(r^2 + 2Ks)}\right]ds\end{aligned} \tag{8.4-8}$$

This is a first-order, separable differential equation for $p(r)$, which may be integrated with the boundary condition that $p = p_\infty$ at $r = \infty$, to give

$$p(r) = p_\infty - \frac{\rho K^2}{2r^2} + \int_0^\infty M(s)\left[\frac{Ks}{r^2} - \tfrac{1}{2}\ln\left(1 + \frac{2Ks}{r^2}\right)\right]ds \qquad (8.4\text{-}9)$$

and then combination of Eqs. 8.4-6 and 9 gives for π_{rr}

$$\pi_{rr}(r) = p_\infty - \frac{\rho K^2}{2r^2} - \int_0^\infty M(s)\left[\frac{Ks}{r^2} + \tfrac{1}{2}\ln\left(1 + \frac{2Ks}{r^2}\right)\right]ds \qquad (8.4\text{-}10)$$

This is the general solution to the line-sink flow problem with arbitrary memory function $M(s)$. It is interesting to specialize the above solution to the expression given in Eq. 8.2-3 corresponding to the convected Jeffreys model. We then find

$$\pi_{rr} \sim p_\infty - [\text{Re} + 2 - (1 - (\lambda_2/\lambda_1))](\eta_0 K/r^2) \qquad (r \to 0) \qquad (8.4\text{-}11)$$

$$\pi_{rr} \sim p_\infty - (\text{Re} + 2)(\eta_0 K/r^2) \qquad (r \to \infty) \qquad (8.4\text{-}12)$$

where $\text{Re} = \rho K/2\eta_0$ is the Reynolds number. Recall that when $\lambda_2/\lambda_1 = 1$ the convected Jeffreys model becomes a Newtonian fluid, and when $\lambda_2/\lambda_1 = 0$ it becomes a convected Maxwell fluid. We see that $d\pi_{rr}/dr$ near the sink is predicted to be somewhat smaller for an elastic fluid than for a viscous fluid. Moreover we see in Eq. 8.4-11 that inertial (Re), viscous (2), and elastic $(1 - (\lambda_2/\lambda_1))$ terms all have the same r-dependence close to the sink. This latter conclusion is in remarkable contrast with that obtained from the solution of the same flow problem with the third-order fluid (Problem 6B.5) where terms associated with elasticity always dominate inertial and viscous terms close to the sink, and the solution becomes physically unacceptable.

See Problem 8C.1 for the flow from the infinite half space into a line sink located in a plane.

EXAMPLE 8.4-2 Transient Inflation of a Spherical Viscoelastic Film[1]

Consider a thin spherical viscoelastic film forming a bubble initially of radius r_0 and thickness Δr_0 such that $\Delta r_0/r_0 \ll 1$ (Fig. 8.4-2). The bubble has been at rest for a long time with its interior filled with a gas of pressure $p_{g,0}$ equal to the surrounding pressure. At time $t = 0$ the external pressure is suddenly removed, and the bubble begins to expand. It is desired to formulate an integro-differential equation from which the radius of the film may be calculated as function of time. For illustrative purposes use a Lodge rubberlike liquid with the memory function corresponding to the convected Jeffreys model in Eq. 8.2-3. Neglect inertial effects.

SOLUTION We first write the convected Jeffreys model in the form

$$\boldsymbol{\tau}(t) = -\eta_0 \frac{\lambda_2}{\lambda_1}\boldsymbol{\gamma}_{(1)} + \frac{\eta_0}{\lambda_1^2}\left(1 - \frac{\lambda_2}{\lambda_1}\right)\int_{-\infty}^{t} e^{-(t-t')/\lambda_1}\boldsymbol{\gamma}_{[0]}(t, t')dt' \qquad (8.4\text{-}13)$$

We then introduce a spherical coordinate system and label the coordinates of a typical fluid particle P by r, θ, ϕ at the present time t and by r', θ', ϕ' at a past time t' (see Fig. 8.4-2). From symmetry

[1] For discussions of the related problems of bubble growth and collapse in an infinite sea of liquid see H. Lamb, *Hydrodynamics*, 6th ed., Cambridge University Press (1932), p. 122, and G. K. Batchelor, *An Introduction to Fluid Dynamics*, Cambridge University Press (1967), pp. 486–490. A treatment of the mass-transfer process coupled with the rheological problem has been given in a paper on the dissolution of a gas bubble in a viscoelastic fluid by E. Zana and L. G. Leal, *Ind. Eng. Chem. Fundam.*, **14**, 175–182 (1975). See also J. R. Street, *Trans. Soc. Rheol.*, **12**, 103–131 (1968), and A. C. Papanastasiou, L. E. Scriven, and C. W. Macosko, *J. Non-Newtonian Fluid Mech.*, **16**, 53–75 (1984).

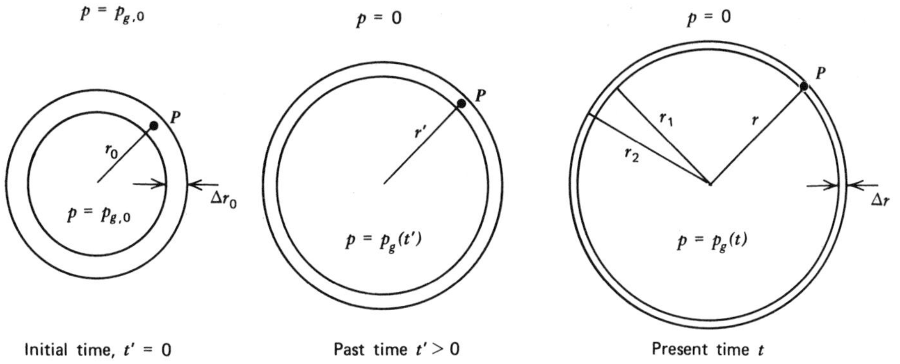

FIGURE 8.4-2. Cross section of a thin viscoelastic film forming a bubble filled with a gas. The bubble is initially at equilibrium with the inside gas pressure equal to the surrounding pressure. At time $t' = 0$ the external pressure is set equal to zero and the bubble expands.

considerations we see that the r-component of the equation of motion in spherical coordinates simplifies for creeping flow to

$$0 = -\frac{\partial p}{\partial r} - \left(\frac{1}{r^2}\frac{\partial}{\partial r}(r^2\tau_{rr}) - \frac{\tau_{\theta\theta} + \tau_{\phi\phi}}{r}\right) \tag{8.4-14}$$

Hence the pressure drop across the film is

$$p_g = -\int_{r_1}^{r_2} \frac{\partial}{\partial r}(p + \tau_{rr})dr = \int_{r_1}^{r_2} \frac{2\tau_{rr} - \tau_{\theta\theta} - \tau_{\phi\phi}}{r}\,dr \doteq (2\tau_{rr} - \tau_{\theta\theta} - \tau_{\phi\phi})\frac{\Delta r}{r} \tag{8.4-15}$$

In the last expression we have introduced the approximation that the stress tensor is constant across the thin film. Let us now calculate the components of the kinematic tensors $\boldsymbol{\gamma}_{(1)}$ and $\boldsymbol{\gamma}_{[0]}$. From Table B.3 we find

$$\boldsymbol{\gamma}_{(1)} = 2\begin{pmatrix} \dfrac{\partial v_r}{\partial r} & 0 & 0 \\ 0 & \dfrac{v_r}{r} & 0 \\ 0 & 0 & \dfrac{v_r}{r} \end{pmatrix} = 2\begin{pmatrix} -2 & 0 & 0 \\ 0 & 1 & 0 \\ 0 & 0 & 1 \end{pmatrix}\frac{1}{r}\frac{dr}{dt} \tag{8.4-16}$$

In the last step of Eq. 8.4-16 we have used the incompressibility condition for the viscoelastic liquid in the film in the form $\operatorname{tr}\boldsymbol{\gamma}_{(1)} = 0$. Note also that we have expressed the velocity component as the time derivative of the displacement function. Similarly from Table B.6 we find

$$\boldsymbol{\gamma}_{[0]} = \begin{pmatrix} 1 - \left(\dfrac{\partial r}{\partial r'}\right)^2 & 0 & 0 \\ 0 & 1 - \left(\dfrac{r}{r'}\right)^2 & 0 \\ 0 & 0 & 1 - \left(\dfrac{r}{r'}\right)^2 \end{pmatrix}$$

$$= \begin{pmatrix} 1 - \left(\dfrac{r'}{r}\right)^4 & 0 & 0 \\ 0 & 1 - \left(\dfrac{r}{r'}\right)^2 & 0 \\ 0 & 0 & 1 - \left(\dfrac{r}{r'}\right)^2 \end{pmatrix} \tag{8.4-17}$$

In the last step we have again made use of the incompressibility condition for the viscoelastic liquid, this time in the form $\det(\boldsymbol{\delta} - \boldsymbol{\gamma}_{[0]}) = \det \mathbf{B} = 1$.

Before we combine Eqs. 8.4-16 and 17 with Eq. 8.4-15 we need expressions for Δr and $p_g(t)$ as functions of r. First Δr is obtained from the conservation of liquid in the film

$$r^2 \Delta r = r_0^2 \Delta r_0 \tag{8.4-18}$$

We now approximate the thermodynamic equation of state for the gas inside the bubble by an ideal gas law and assume adiabatic expansion so that

$$p_g(t)V^{\gamma} = p_{g,0}V_0^{\gamma} \quad \text{or} \quad p_g(t)r^{3\gamma} = p_{g,0}r_0^{3\gamma} \tag{8.4-19}$$

in which $\gamma = \hat{C}_p/\hat{C}_v$. When Eqs. 8.4-13, 16, 17, 18, and 19 are combined with Eq. 8.4-15 we obtain the desired integro-differential equation for $r(t)$. In dimensionless form this equation may be written

$$\frac{dx}{d\tau} = \frac{1}{(\lambda_2/\lambda_1)\beta}\frac{1}{x^{\alpha}} - \frac{(1-(\lambda_2/\lambda_1))x}{6(\lambda_2/\lambda_1)\text{De}^2}\int_{-\infty}^{\tau}\left[\left(\frac{x(\tau)}{x(\tau')}\right)^2 - \left(\frac{x(\tau)}{x(\tau')}\right)^{-4}\right]e^{-(\tau-\tau')/\text{De}}d\tau' \tag{8.4-20}$$

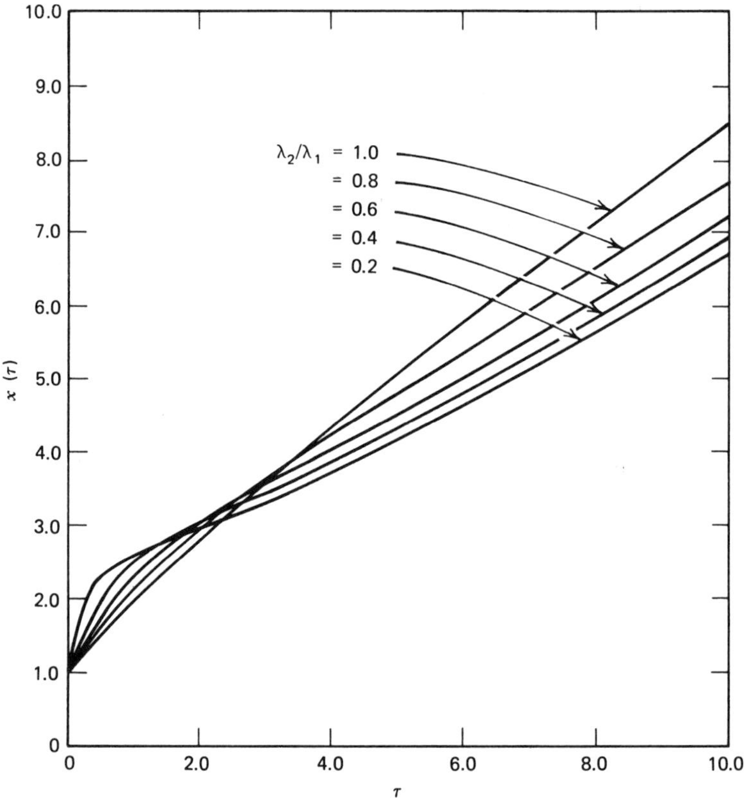

FIGURE 8.4-3. The reduced radius $x(\tau) = r(\tau)/r_0$ versus reduced time $\tau = tp_{g,0}/\eta_0$ as given by the solution of Eq. 8.4-20 for the inflation of a spherical film of a convected Jeffreys model. The parameters are $\alpha = 0.2$, $\beta = 1.0$, and $\text{De} = 2.0$.

where we have introduced the dimensionless quantities

$$x(\tau') = r(t')/r_0 \tag{8.4-21}$$

$$\mathrm{De} = \lambda_1 p_{g,0} \eta_0^{-1}; \quad \tau' = t' p_{g,0} \eta_0^{-1} \tag{8.4-22}$$

$$\alpha = 3\gamma - 4; \quad \beta = 12\Delta r_0/r_0 \tag{8.4-23}$$

Here the Deborah number De is the ratio of the time constant for the fluid (λ_1) to a characteristic time for the process ($\eta_0/p_{g,0}$). Equation 8.4-20 with the condition that $x(\tau') = 1$ for $\tau' \leq 0$ is an initial value problem. For given values of α, β, De, and λ_2/λ_1 we may therefore solve numerically for $x(\tau)$ by a forward integration technique. Figure 8.4-3 shows the reduced radius $x(\tau)$ as a function of the reduced time τ. We have used the values $\alpha = 0.2$ (corresponding to an ideal diatomic gas), $\beta = 1$, De $= 2$, and $\lambda_2/\lambda_1 = 0.2, 0.4, 0.6, 0.8$, and 1.0. We see that a viscoelastic film is predicted to expand somewhat faster than a purely viscous film for small times τ, but that for larger times the viscoelastic film is predicted to expand considerably more slowly than a Newtonian film of the same zero-shear-rate viscosity.

§8.5 FLOW PROBLEMS IN TWO OR THREE SPATIAL VARIABLES

The solution of flow problems in two or three spatial variables for integral models is an area of very active research. The two main techniques, the Eulerian technique and the Lagrangian technique, described in §8.4 are both used in these more complex flow situations. Most research has focused on finite element discretizations to represent velocity, pressure, and displacement fields.

In the *Eulerian technique* a fixed region of space is discretized by a mesh of finite elements. Approximations to the velocity and pressure fields are then constructed from low-order polynomial trial functions on the mesh. The velocity and pressure fields contain the values of velocity and pressure at specified points or "nodes" in the mesh as parameters. Based on the velocity field, the displacement functions and resulting stresses are found, and residuals of the equation of continuity and the equations of motion are computed. Methods are then developed to adjust the velocity and pressure field to minimize a weighted sum of the residuals (cf. §4.3). Some methods[1] have the built-in feature that the equation of continuity is satisfied almost exactly in the final converged solution.

In the *Lagrangian technique*[2] a given amount of fluid is discretized by a mesh of finite elements. Each element node corresponds to a fluid particle, and all fluid particles inside the elements are denoted by unique sets of element coordinates. The motion of the fluid is described by the motion of the element nodes in space. Particles inside the element move with constant element coordinates in a manner determined by the motion of the element nodes. The dependent variables in this Lagrangian discretization then become the coordinates of the element nodes as functions of time and values of the pressure at pressure nodes associated with each element. The object here again is to construct a set of integro-differential equations from which the motion of the nodes may be found by forward integration in time.

[1] B. Bernstein, D. S. Malkus, and E. T. Olsen, *Int. J. Num. Mech. Fluids*, **5**, 43–70 (1985); D. S. Malkus and B. Bernstein, *J. Non-Newtonian Fluid Mech.*, **16**, 77–116 (1984). For other methods see S. Dupont, J. M. Marchal, and M. J. Crochet, *J. Non-Newtonian Fluid Mech.*, **17**, 157–183 (1985); B. Caswell and M. Viryayuthakorn, *J. Non-Newtonian Fluid Mech.*, **12**, 13–29 (1983); H. Court, A. R. Davies, and K. Walters, *J. Non-Newtonian Fluid Mech.*, **8**, 95–117 (1981); M. J. Crochet, A. R. Davies, and K. Walters, *Numerical Simulation of Non-Newtonian Flow*, Elsevier, Amsterdam (1984).

[2] O. Hassager and C. Bisgaard, *J. Non-Newtonian Fluid Mech.*, **12**, 153–164 (1983).

The Eulerian technique seems to have attracted most interest in research developments, possibly because it is somewhat more directly implemented; it has the disadvantage that tracking particles is difficult, since the velocity components rather than the particle positions are used as the dependent variables. The Lagrangian technique, on the other hand, has the advantage that particle tracking is easily and naturally handled.

A detailed discussion of any of the numerical techniques is outside the scope of this textbook. Instead we just give two examples of predictions obtained by the methods. The examples are not problems to be worked step by step, but are intended to illustrate the results obtained.

EXAMPLE 8.5-1 The Hole-Pressure Effect for Flow Across a Transverse Slot

It is desired to illustrate the calculations of the hole-pressure effect for flow across a transverse slot by Malkus and Bernstein,[3] who used the Curtiss–Bird constitutive equation, Eq. 19.6-9, with parameters $\eta_0 = 410$ Pa · s, $\lambda = 9.54$ s, $\varepsilon = 0.375$, and density $\rho = 1363$ kg/m^3. These fluid parameters have been selected to fit measurements of the viscosity function and first normal stress coefficient of a nearly monodisperse solution of polystyrene in a chlorinated biphenyl.[4]

SOLUTION The geometry is shown in Fig. 8.5-1. A fluid is driven by a pressure gradient in the channel, from left to right, and the desired quantity is the "hole pressure"

$$p^* = (\mathscr{P} + \tau_{yy})_1 - (\mathscr{P} + \tau_{yy})_2 \tag{8.5-1}$$

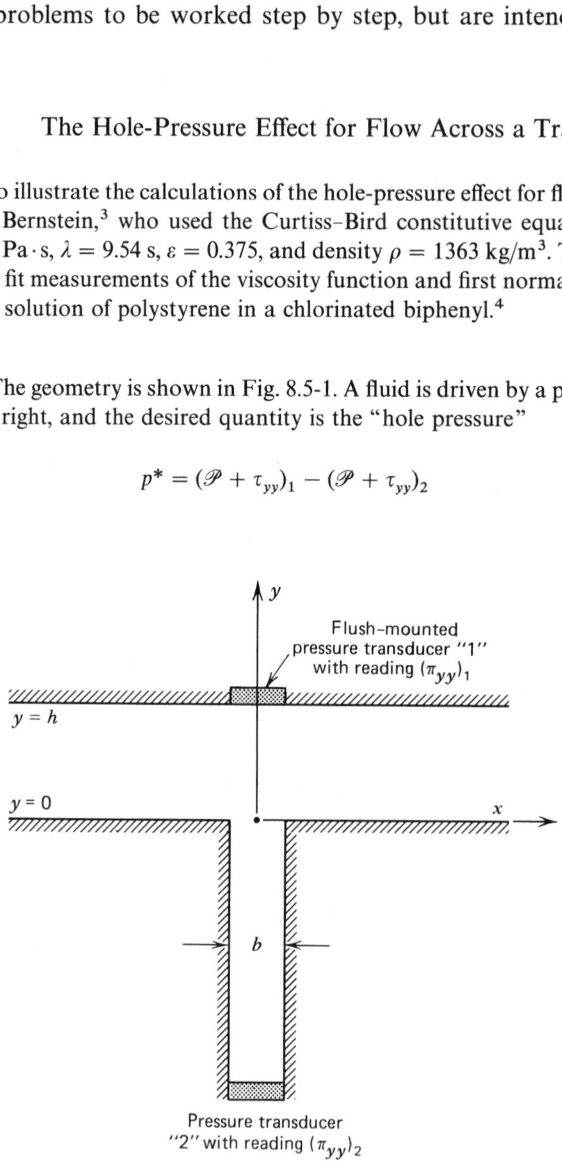

FIGURE 8.5-1. Geometry for plane flow across a transverse slot. Fluid is driven by a pressure gradient from left to right and the velocity field is of the form $v_x = v_x(x, y)$, $v_y = v_y(x, y)$, $v_z = 0$.

[3] D. S. Malkus and B. Bernstein, *J. Non-Newtonian Fluid Mech.*, **13**, 77–116 (1984); corrected calculations have been made by D. S. Malkus, personal communication (1986).

[4] H. H. Saab, R. B. Bird, and C. F. Curtiss, *J. Chem. Phys.*, **77**, 4758–4766 (1982).

Here $(\mathscr{P} + \tau_{yy})_1$ and $(\mathscr{P} + \tau_{yy})_2$ are the values of the total normal stress π_{yy} corrected for gravitational head that are measured by flush-mounted pressure transducers "1" and "2" respectively, located as shown in Fig. 8.5-1. In Fig. 8.5-2 we show the Malkus–Bernstein prediction for p^* as a function of the first normal stress difference $(\tau_{xx} - \tau_{yy})_1$ at the position of pressure transducer "1" based on their numerical solution of the complete flow problem.

In order to facilitate the comparison of the Malkus–Bernstein prediction for p^* with approximate analytical formulas we assume in the following that the slot is sufficiently narrow compared to the channel height ($h \gg b$ in Fig. 8.5-1) that the flow in the neighborhood of pressure transducer "1" is undisturbed by the presence of the slot. Then in the limit of very slow flow, p^* is given by the Tanner and Pipkin analytical relation for second order fluids ($p^* = -(\tau_{xx} - \tau_{yy})/4$; see Eq. 6B.15-1), as shown also in Fig. 8.5-2. The calculations by Malkus and Bernstein[3] cover stress ratios $[(\tau_{xx} - \tau_{yy})/\tau_{yx}]_1$ at pressure transducer "1" up to a maximum of two, and clearly show significant deviations from second-order behavior as evidenced by the discrepancy between the numerical calculations and the Tanner–Pipkin relation in Fig. 8.5-2. Also shown in Fig. 8.5-2 is the prediction for p^* from the Higashitani–Pritchard relation[5] computed by Malkus and Bernstein using the viscometric functions for the Curtiss–Bird model. For this particular constitutive equation, the Higashitani–Pritchard relation apparently differs only slightly from the Tanner–Pipkin relation in the region where both differ significantly from the p^* result obtained from the full numerical solution to the flow problem using the Curtiss-Bird constitutive equation. Experimental measurements,[6] on the other hand, support the validity of the Higashitani–Pritchard relation. Thus here is a situation where numerical simulations and experiments disagree. The reason for this discrepancy is not known.

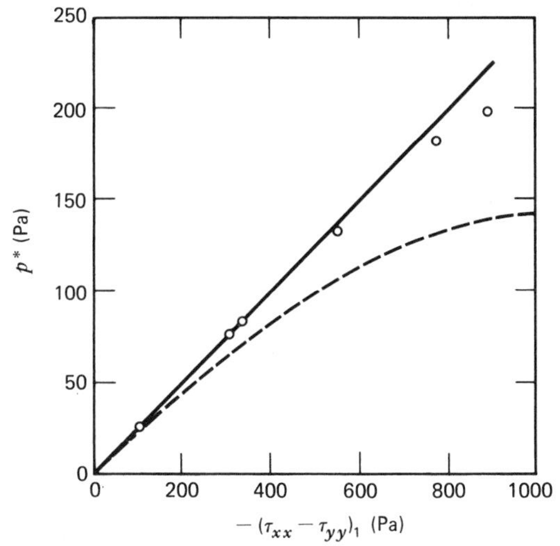

FIGURE 8.5-2. Hole pressure p^* defined by Eq. 8.5-1 as function of the first normal stress difference at pressure transducer "1" in Fig. 8.5-1 for the Curtiss-Bird model, Eq. 19.6-9, with parameters $\eta_0 = 410$ Pa·s, $\lambda = 9.54$ s, $\varepsilon = 0.375$, and density $\rho = 1363$ kg/m^3:

- \- - - Values computed by Malkus and Bernstein by solving the complete flow problem using the Curtiss–Bird model.[3]
- ○ Values based on the Higashitani–Pritchard relation (Eq. 2.3-15) with $(\tau_{xx} - \tau_{yy})$ and τ_{yx} obtained from the Curtiss–Bird model as computed by Bernstein and Malkus.[3]
- — Values from the second-order fluid as given by the Tanner and Pipkin relation ($p^* = -(\tau_{xx} - \tau_{yy})/4$) (Eq. 6B.15-1).

[5] K. Higashitani and W. G. Pritchard, *Trans. Soc. Rheol.*, **16**, 687–696 (1972).

[6] D. Pike and D. G. Baird, *J. Rheol.*, **28**, 439–447 (1984); A. S. Lodge and L. de Vargas, *Rheol. Acta*, **22**, 151–170 (1983); A. S. Lodge, *Polym. News*, **9**, 242–246 (1984).

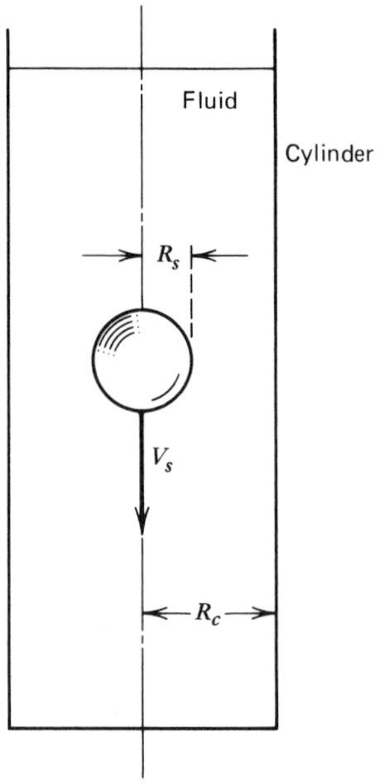

FIGURE 8.5-3. Sphere of radius R_s moving axially with velocity V_s on the centerline of a cylinder of radius R_c.

EXAMPLE 8.5-2 Flow Around a Rigid Sphere Moving Along the Axis of a Cylindrical Tube[2]

A sphere of radius R_s is moving with velocity V_s on the axis of a cylinder of radius R_c as shown in Fig. 8.5-3. The motion of the sphere is caused by a net force F_s acting on the sphere in the direction of V_s. The cylinder is filled with a viscoelastic liquid modeled by a convected Maxwell model with zero-shear-rate viscosity η_0 and time constant λ_1. The dimensionless groups for this problem are the geometric ratio R_s/R_c, the Deborah number De, and a dimensionless force K where

$$\text{De} = \frac{V_s \lambda_1}{R_s} \tag{8.5-2}$$

$$K = \frac{F_s}{6\pi\eta_0 R_s V_s} \tag{8.5-3}$$

It is desired to illustrate the calculations of Hassager and Bisgaard of the relation $K = K(R_s/R_c, \text{De})$.

SOLUTION Hassager and Bisgaard formulate the flow problem as an initial value problem, in which the sphere is at rest in a quiescent fluid for time $t < 0$. Then the force of magnitude F_s is exerted on the sphere for $t \geq 0$, and the resulting transient motion of the sphere and fluid is calculated. Hassager and Bisgaard use a Lagrangian technique, and for moderately small Deborah numbers observe an approach to a steady velocity V_s. The resulting values of De and K as calculated by Hassager and Bisgaard[7] are shown in Fig. 8.5-4.

[7] These results agree closely with results by F. Sugeng and R. I. Tanner, *J. Non-Newtoniaan Fluid Mech.*, **20** 281–292 (1986), who used an Eulerian technique to solve the steady flow problem for the convected Maxwell model in differential form with both a finite element method and a boundary element method; see also M. J. Crochet in *Finite Elements in Fluids*, Vol. 4, ed. R. H. Gallagher, Wiley, New York (1982), pp. 573–579.

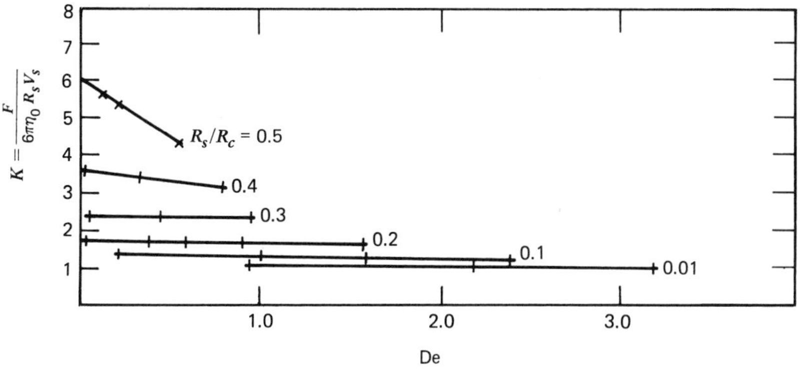

FIGURE 8.5-4. Calculated results[2] for K as a function of De and R_s/R_c for the motion of a sphere in a cylinder as shown in Fig. 8.5-3. + Calculated points. — Interpolation of the calculated points. [O. Hassager and C. Bisgaard, *J. Non-Newtonian Fluid Mech.*, **12**, 153–164 (1983).]

The simulated K-values were compared with experiments for $(R_s/R_c) = 0.3$, 0.4, and 0.5 and a 1% solution of polyacrylamide in glycerine at 293 K, as shown in Fig. 8.5-5. The measurements have been reduced to dimensionless form with the parameters $\eta_0 = 21.0$ Pa·s and $\lambda_1 = 12.0$ s. We see that with this choice of parameters there is good agreement between the measurements and the simulations.

The significance of these values of the parameters $\eta_0 = 21.0$ Pa·s and $\lambda_1 = 12.0$ s might be questioned, since they are determined in an inhomogeneous flow experiment. Certainly from a conceptual point of view it would be simpler to determine η_0 and λ_1 in steady shear flow, for which the Maxwell model has a constant viscosity $\eta(\dot{\gamma}) = \eta_0$ and a first normal stress difference $-(\tau_{xx} - \tau_{yy}) = 2\eta_0\lambda_1\dot{\gamma}^2$. We compare therefore in Fig. 8.5-6 the convected Maxwell model values for the viscosity and first normal stress difference (based on the parameters $\eta_0 = 21.0$ Pa·s and $\lambda_1 = 12.0$ s determined in the sphere experiment) with independent measurements by Prud'homme.[8] The comparison indicates that at least for this particular fluid and model the parameters may be determined equally well in either the sphere experiment or in a steady shear flow experiment. Thus here is a situation where numerical simulations and experiments do agree.

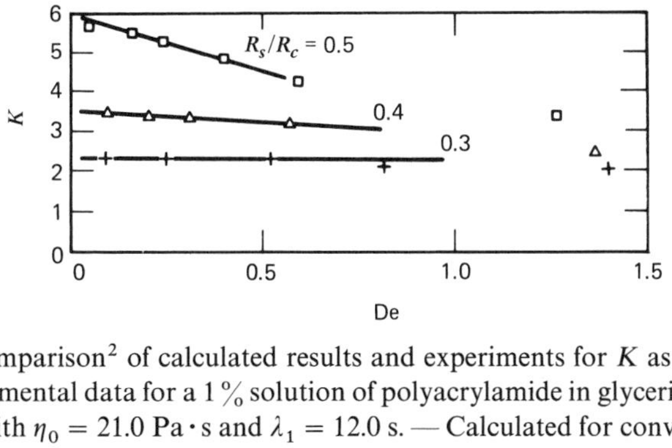

FIGURE 8.5-5. Comparison[2] of calculated results and experiments for K as a function of De and R_s/R_c: □, △, + Experimental data for a 1% solution of polyacrylamide in glycerine at 293 K plotted in dimensionless form with $\eta_0 = 21.0$ Pa·s and $\lambda_1 = 12.0$ s. — Calculated for convected Maxwell model based on Fig. 8.5-4. [O. Hassager and C. Bisgaard, *J. Non-Newtonian Fluid Mech.*, **12**, 153–164 (1983).]

[8] R. K. Prud'homme, private communication.

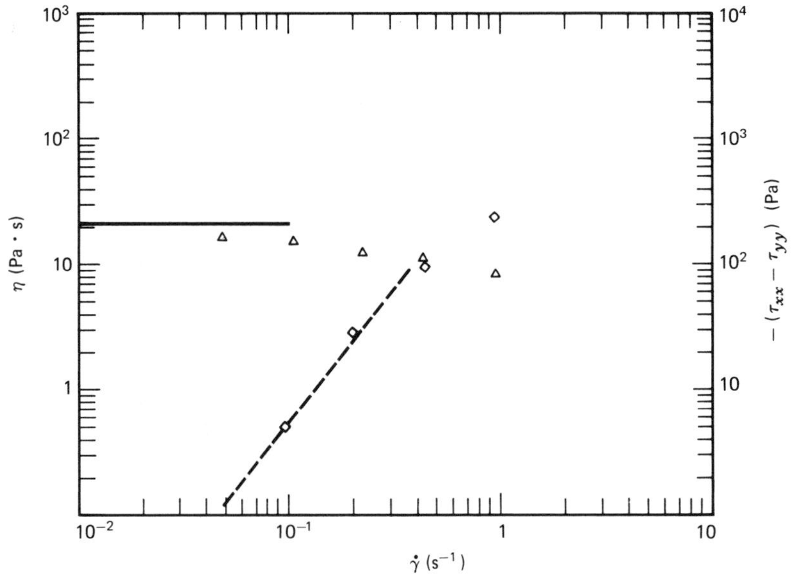

FIGURE 8.5-6. Viscosity η and first normal stress difference $-(\tau_{xx} - \tau_{yy})$ in steady shear flow. Predictions for the convected Maxwell model with parameters $\eta_0 = 21.0$ Pa · s and $\lambda_1 = 12.0$ s are — (for $\eta = \eta_0$) and ----- (for $-(\tau_{xx} - \tau_{yy}) = 2\eta_0\lambda_1\dot{\gamma}^2$). Experimental measurements in cone-and-plate geometry (by R. K. Prud'homme[8]) △ for η, and ◇ for $-(\tau_{xx} - \tau_{yy})$.

§8.6 LIMITATIONS OF SINGLE-INTEGRAL MODELS AND RECOMMENDATIONS FOR THEIR USE

The most general constitutive equations considered in this chapter are the K-BKZ and the Rivlin-Sawyers models in Eqs. 8.3-7 and 8.3-8, respectively. These models are too general to be useful for numerical flow calculations, and we have concentrated our discussion on the factorized versions of the two models in Eqs. 8.3-11 and 8.3-12. The factorized K-BKZ equation and the factorized Rivlin-Sawyers model have the following positive features:

i. They include the general linear viscoelastic fluid completely. This means that all of the material presented in Chapter 5 is included in the Rivlin-Sawyers and the K-BKZ equations.

ii. They provide a framework that includes a rather large number of nonlinear constitutive equations both of molecular and empirical origin. It is possible to choose simple empirical functions for $M(s)$ and $W(I_1, I_2)$ (or $\phi_j(I_1, I_2)$) to obtain reasonably satisfactory constitutive equations containing only four or five constants; these constants usually have simple physical meaning and are easy to determine from rheometric data.

iii. They provide a basis for the characterization of polymers, a subject important in connection with many problems relating to quality control. It is possible to use these constitutive equations to interrelate material functions (see, for example, Problem 8C.2).

No constitutive equation will ever be perfect, and the factorized Rivlin-Sawyers and K-BKZ equations also have their limitations:

i. The models generally predict too much recoil in elastic recoil experiments[1] (see, however, comment under **(c)** below).

ii. In certain fast strain experiments, the Rivlin–Sawyers model may form the basis[2] for a work-producing perpetual motion machine which is physically unreasonable.

Since the second limitation does not apply to the K–BKZ equation, we favor it somewhat.

Several considerations must be taken into account in the choice of a specific factorized K–BKZ equation. For example the potential function should be analytic at $(I_1, I_2) = (3, 3)$ in order that the model simplify properly to the retarded-motion expansion. Also the potential function and the memory function should be chosen so that shear stress vs. shear rate curve is monotone increasing. Considerations of this kind were illustrated in connection with Example 8.3-1.

There are some integral models that we have omitted because they cannot be formulated as either Rivlin–Sawyers or K–BKZ equations. In particular we have omitted:

a. *Integral models in which the memory function depends not only on the elapsed time $t - t'$ but also on the strain rate at time t'.*

Examples of such models are the Bird–Carreau[3] and Bogue–Chen[4] models. Models of this type have in the past attracted considerable attention but are no longer used extensively. One serious defect of models that include the strain rate in the memory function is that they do not simplify correctly to a linear viscoelastic model in the limit of small displacement gradients.[5]

b. *Integral models in which the memory function depends on the stress at time t'.*

Examples of such models are the Kaye and Phan-Thien–Tanner network models.[6] This type of model, however, may not be convenient from a computational point of view. Indeed one of the advantages of the integral models of this chapter over differential models is that they are explicit in the stress. In flow simulations this means that the stress components can be eliminated, and the total number of dependent variables may be reduced. However if the stress is included in the memory function this elimination is no longer possible.

c. *Models in which the effect on the stress at time t of the deformations at different past times, t' and t'' say, couple with one another.*

One example of this type of model is the Wagner damping functional model.[1] In this model the function ϕ_1 in Eq. 8.3-12 is replaced by a functional that selects the minimum value of ϕ_1 between times t' and t. This functional is introduced to model "irreversible loss

[1] M. H. Wagner, *Rheol. Acta*, **18**, 681–692 (1979).
[2] R. G. Larson and K. Monroe, *Rheol. Acta*, **23**, 10–13 (1984).
[3] R. B. Bird and P. J. Carreau, *Chem. Eng. Sci.*, **23**, 427–434 (1968); P. J. Carreau, I. F. Macdonald, and R. B. Bird, *ibid.*, 901–911 (1968).
[4] I. Chen and D. C. Bogue, *Trans. Soc. Rheol.*, **16**, 59–78 (1972).
[5] For still other objections, see H. E. van Es and R. M. Christensen, *Trans. Soc. Rheol.*, **17**, 325–330 (1973).
[6] A. Kaye, *Br. J. Appl. Phys.*, **17**, 803–806 (1966); A. S. Lodge, *Rheol. Acta*, **7**, 379–392 (1968); N. Phan-Thien and R. I. Tanner, *J. Non-Newtonian Fluid Mech.*, **2**, 353–365 (1977); N. Phan-Thien, *J. Rheol.*, **22**, 259–283 (1978); R. I. Tanner, *J. Non-Newtonian Fluid Mech.*, **5**, 103–112 (1979). See also §20.5.

of network junctions" (see Chapter 20), and Wagner has shown that this modification significantly improves the description of elastic recovery. Another example of this type of model is the Curtiss–Bird model[7] with $\varepsilon \neq 0$ (Eq. 19.6-9). In this model the contribution to the stress at time t of the strain rate at time t depends upon past states of deformation. We shall see in §9.6 that from a continuum mechanical point of view it is also natural to include the possibility of coupling between deformations at different times in the past and that such coupling contributes to the stress at the present time.

This chapter is the last in a sequence of five chapters, each devoted to a specific type of constitutive equation. It is not possible to recommend any one of these types of constitutive equations to be used for the modeling of the flow properties of polymeric liquids in all possible situations. Rather the choice of constitutive equation depends not only on the material but also on the particular application and the accuracy required by the user; the amount of time available and the computing techniques available may also play important roles in determining the choice of constitutive relation. At this point we wish to formulate some recommendations for a number of different constitutive-equation users:

1. Some rheologists are interested primarily in the experimental characterization of polymeric liquids. This can be done by the determination of the parameters in a model so that the model agrees with the rheometric experiments to the extent possible. For this purpose we recommend the Wagner model or the factorized K–BKZ model. As previously mentioned these automatically include the general linear viscoelastic model of Chapter 5, which has customarily been used by polymer chemists as a basis for the characterization of polymeric liquids in the linear regime.

2. Some fluid mechanicians are interested primarily in the low Reynolds number hydrodynamics of dilute suspensions or emulsions near equilibrium. Questions of interest here may be particle rotation, sedimentation, or deformation and breakup of particle clusters. For this purpose we recommend the retarded-motion expansion of Chapter 6.

3. Some engineers may deal with flow in complex geometric configurations, but may be interested only in rough exploratory calculations. In this situation it must first be determined whether the flow is primarily a shearing flow or an extensional flow. This determination will be based on a physical understanding of the flow situation, and the process will usually also involve some idealization of the geometry. Then for flows that are primarily shearing flows (including those involving heat and mass transfer) we recommend the use of the generalized Newtonian fluid, if necessary in connection with the lubrication approximation or the quasi-steady-state approximation as described in Chapter 4. For flows that involve a mixture of shear and shearfree kinematics or for shearing flows in which normal stress effects are of interest, we recommend the use of single-mode differential models, such as an Oldroyd model or the Giesekus model of Chapter 7.

4. Finally some engineers and fluid dynamicists are interested in precise numerical simulations of flow in complex geometries. For this purpose we recommend the nonlinear models of Chapters 7 and 8. The development of reliable routines for numerical simulations has been the object of much research effort, but is still a largely unresolved problem.[8] It is not possible, therefore, to make more specific recommendations for this last group of users, since any such recommendations will inevitably be altered by future research.

[7] C. F. Curtiss and R. B. Bird, *J. Chem. Phys.*, **74**, 2016–2025, 2026–2033 (1981); *Phys. Today*, **37**(1), 36–43 (1984). R. B. Bird, H. H. Saab, and C. F. Curtiss, *J. Phys. Chem.*, **86**, 1102–1105 (1982); R. B. Bird, H. H. Saab, and C. F. Curtiss, *J. Chem. Phys.*, **77**, 4747–4757 (1982); H. H. Saab, R. B. Bird, and C. F. Curtiss, *ibid.*, 4758–4766 (1982).

[8] See M. J. Crochet and K. Walters, *Ann. Rev. Fluid Mech.*, **15**, 241–260 (1983). For two status reports on numerical methods, see *J. Non-Newtonian Fluid Mech.*, **16**, 1–209 (1984) and **20**, 1–339 (1986).

Gradually there has been a shift in emphasis away from the models in Chapters 4, 5, and 6 and toward those of Chapters 7 and 8, and we expect this trend to continue. Also we expect that molecular-theory-based constitutive equations will replace empirical equations because they have the added advantage of providing extra insight into molecular orientation and stretching. A crucial need is the accumulation of experimental data on material functions, particularly for shearfree flows, and flow-field mappings (by laser Doppler velocimetry, for example), for well characterized fluids over wide ranges of system parameters.

PROBLEMS

8B.1 Potential Functions for Special Models

a. It is desired to demonstrate that two simple integral models in Table 8.3-3 may be formulated as factorized K-BKZ equations for special choices of the parameters. To do this find the potential functions (constructed to be zero at $I_1, I_2 = 3,3$) for the models of Wagner[1] for $\alpha = 1$ and of Papanastasiou et al.[2] for $\beta = 1$.

b. Do the above models obey the Renardy conditions?

Answer: **a.** Wagner: $W = \dfrac{2}{\beta^2}[1 - (\beta\sqrt{I_1 - 3} + 1)\exp(-\beta\sqrt{I_1 - 3})]$

Papanastasiou–Scriven–Macosko: $W = \alpha \ln [(\alpha + I_1 - 3)/\alpha]$

8B.2 Transient Stretching of a Thin Liquid Filament[3]

Consider a thin liquid filament with one end fixed and one end free. For time $t < t_0$ the filament is at rest with uniform radius R_0, and for $t > t_0$ the filament is being stretched by means of a gravitational acceleration g in the direction of the filament axis and also because of an imposed force on the free end. The initial locations of the fluid particles are described by the coordinates r_0, z_0 for time $t \leq t_0$ (see Fig. 8B.2); the particle locations are given by r, z for time $t > t_0$. The radius of the filament is $R(z_0, t_0, t)$, and for $t > t_0$ it is assumed that $z = z(z_0, t_0, t)$ and $r = r_0 R(z_0, t_0, t)/R_0$.

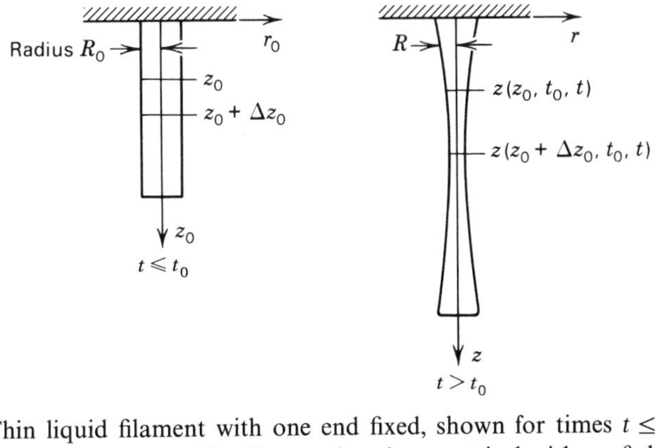

FIGURE 8B.2. Thin liquid filament with one end fixed, shown for times $t \leq t_0$ and $t > t_0$ with cylindrical coordinate r_0, z_0 and r, z. The filament is axisymmetrical with no θ-dependence.

[1] M. H. Wagner, *Rheol. Acta*, **18**, 33–50 (1979).
[2] A. C. Papanastasiou, L. E. Scriven, and C. W. Macosko, *J. Rheol.*, **27**, 387–410 (1983).
[3] P. Markovich and M. Renardy, *J. Non-Newtonian Fluid Mech.*, **17**, 13–22 (1985).

a. Make a mass balance on the material located between z_0 and $z_0 + \Delta z_0$ for $t \leq t_0$ to obtain

$$R(z_0, t_0, t) = \frac{R_0}{\sqrt{\partial z/\partial z_0}} \tag{8B.2-1}$$

where $z = z(z_0, t_0, t)$.

b. Make a momentum balance on the same material to obtain

$$\rho \frac{\partial^2 z}{\partial t^2} = \frac{\partial}{\partial z_0}\left(- \frac{1}{\partial z/\partial z_0}(\tau_{zz} - \tau_{rr}) + \frac{\sigma}{R_0\sqrt{\partial z/\partial z_0}}\right) + \rho g \tag{8B.2-2}$$

where σ is the surface tension. All the constitutive assumptions enter in the expression for the normal stress difference $\tau_{zz} - \tau_{rr}$.

c. Show that for the Newtonian fluid of Eq. 1.2-2

$$\tau_{zz} - \tau_{rr} = - \frac{3\mu}{\partial z/\partial z_0}\frac{\partial^2 z}{\partial t\, \partial z_0} \tag{8B.2-3}$$

d. Show that for the Lodge rubberlike liquid of Eq. 8.2-1

$$\tau_{zz} - \tau_{rr} = - \int_{-\infty}^{t} M(t - t')\left[\left(\frac{\partial z/\partial z_0}{\partial z'/\partial z_0}\right)^2 - \left(\frac{\partial z'/\partial z_0}{\partial z/\partial z_0}\right)\right] dt' \tag{8B.2-4}$$

where $z' = z'(z_0, t_0, t')$.

e. Look up the article by Markovich and Renardy[3] and discuss their numerical solutions of the above evolution equation for $z(z_0, t_0, t)$ for a number of initial and boundary conditions.

8B.3 Properties of Finite Strain Tensors

Consider a particle with coordinates x_i and a neighboring particle with coordinates $x_i + dx_i$ both at time t.

a. Show that the separation, $(ds')^2$, of the two particles at some other time t' is given by

$$(ds')^2 = \sum_i \sum_j \mathrm{B}_{ij}^{-1}(t, t')\, dx_i\, dx_j \tag{8B.3-1}$$

and conclude from this that $\mathbf{B}$ and $\mathbf{B}^{-1}$ are positive definite.

b. By using the Cayley–Hamilton theorem, verify that $\boldsymbol{\gamma}^{[0]}$ and $\boldsymbol{\gamma}_{[0]}$ are related by

$$\boldsymbol{\gamma}^{[0]} = (I_2 - I_1)\boldsymbol{\delta} + (I_1 - 2)\boldsymbol{\gamma}_{[0]} + \{\boldsymbol{\gamma}_{[0]} \cdot \boldsymbol{\gamma}_{[0]}\} \tag{8B.3-2}$$

for incompressible fluids.

c. How are the invariants of $\mathbf{B}$ and those of $\boldsymbol{\gamma}_{[0]}$ related?

8B.4 An Alternative to the Lodge Rubberlike Liquid

a. Show that if $\boldsymbol{\gamma}_{[0]}(t, t')$ is replaced by $\boldsymbol{\gamma}^{[0]}(t, t')$ in Eq. 8.2-1, one obtains for steady-state shear flow

$$\tau_{yx} = -M_1\dot{\gamma}; \quad \tau_{xx} - \tau_{yy} = -M_2\dot{\gamma}^2; \quad \tau_{yy} - \tau_{zz} = +M_2\dot{\gamma}^2 \tag{8B.4-1}$$

instead of the results shown in Eqs. 8.2-9 to 11. What experimental data in Chapter 3 show that $\boldsymbol{\gamma}_{[0]}$ is to be preferred?

b. What happens if $\boldsymbol{\gamma}_{[0]}$ in Eq. 8.2-1 is replaced by $(1 - q)\boldsymbol{\gamma}_{[0]} + q\boldsymbol{\gamma}^{[0]}$, where $0 \leq q \leq 1$?

8B.5 Eccentric-Disk Rheometer

The eccentric-disk rheometer of Fig. 3B.1 has been examined in Problem 6B.2 (retarded-motion expansion) and in Problem 7B.9 (convected Maxwell model). Here we analyze the eccentric-disk rheometer by using the factorized Rivlin–Sawyers model.

a. Show that the displacement functions corresponding to the velocity distribution in Eq. 6B.2-1 are

$$x(x', y', z', t', t) = x' \cos W(t - t') - (y' - Az') \sin W(t - t') \tag{8B.5-1}$$

$$y(x', y', z', t', t) = x' \sin W(t - t') + (y' - Az') \cos W(t - t') + Az' \tag{8B.5-2}$$

$$z(x', y', z', t', t) = z' \tag{8B.5-3}$$

Invert the displacement functions by solving for x', y', z'.

b. Find the displacement gradient tensors $\mathbf{\Delta}$ and $\mathbf{E}$ for this flow. Check your results by verifying that $\{\mathbf{\Delta} \cdot \mathbf{E}\} = \boldsymbol{\delta}$.

c. Then find the Finger and Cauchy strain tensors

$$\mathbf{B} = \begin{pmatrix} 1 + A^2S^2 & A^2S(1 - C) & AS \\ A^2S(1 - C) & 1 + A^2(1 - C)^2 & A(1 - C) \\ AS & A(1 - C) & 1 \end{pmatrix} \tag{8B.5-4}$$

$$\mathbf{B}^{-1} = \begin{pmatrix} 1 & 0 & -AS \\ 0 & 1 & -A(1 - C) \\ -AS & -A(1 - C) & 2A^2(1 - C) + 1 \end{pmatrix} \tag{8B.5-5}$$

in which $S = \sin W(t - t')$ and $C = \cos W(t - t')$

d. Obtain the expressions for τ_{xz} and τ_{yz} using the factorized Rivlin–Sawyers model.

e. Next obtain expressions for $I_1 - 3$ and $I_2 - 3$. Note that, according to Eqs. 8.3-16 and 17b, ϕ_1 and ϕ_2 are given as power series in A^2.

f. Finally obtain the result

$$\lim_{A \to 0} \left(-\frac{\tau_{xz}}{AW} \right) = \eta'(W) \tag{8B.5-6}$$

$$\lim_{A \to 0} \left(-\frac{\tau_{yz}}{AW} \right) = \eta''(W) \tag{8B.5-7}$$

by using Eqs. 5B.7-1 and 2.

8B.6 Planar Elongational Flow

Consider the flow of Eq. 3.1-3 with $b = 1$. Such a flow can be realized approximately for slow flow in the central portion of the crossed-slit apparatus in Fig. 8B.6:

a. Find the displacement functions.

b. Obtain the Finger and Cauchy tensors.

c. Verify the statement in Fig. 8.3-1 that $I_1 = I_2$ for planar elongational flow.

d. For the Lodge elastic liquid, with $M(s) = (\eta_0/\lambda_1^2) \exp(-s/\lambda_1)$, obtain $\bar{\eta}_1(\dot{\varepsilon})$ defined in Eq. 3.5-1. What restrictions must be placed on the result?

e. Show that $\bar{\eta}_1(0) = 4\eta_0$ (cf. Fig. 7.3-8c). Verify that this result is true for the Lodge elastic liquid with any memory function $M(s)$.

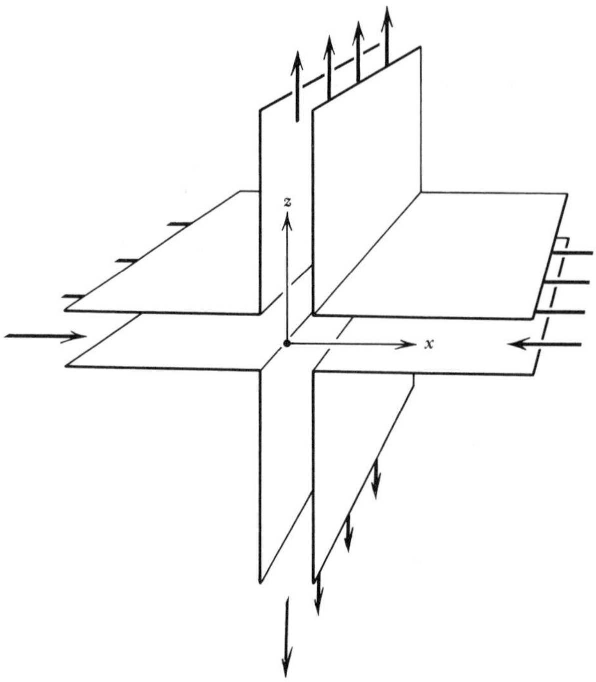

FIGURE 8B.6. Crossed-slit apparatus.

8B.7 The "Network-Rupture Model"

The Lodge rubberlike liquid model (Eq. 8.2-1) includes the idea that the stress in a fluid element at the present time t depends on the strain undergone by the fluid element during all past times t' going from $t' = -\infty$ to $t' = t$. Tanner and Simmons[4] suggest that, as a concentrated polymer solution or polymer melt flows, the "network" of polymer chains is gradually destroyed so that the fluid loses its "memory" of everything that happened t_R time units before t; t_R is called the "rupture time." This is an appealing and simple idea worth investigating. Tanner and Simmons give a general formula for calculating t_R, but for shearing flows t_R is given by the relation $t_R = B/\dot{\gamma}$, where B is a positive constant (characteristic of each fluid) and $\dot{\gamma} = \sqrt{\frac{1}{2}(\boldsymbol{\gamma}_{(1)}:\boldsymbol{\gamma}_{(1)})}$ is the shear rate. Hence in the Lodge rubberlike liquid they suggest replacing $\int_{-\infty}^{t}$ by $\int_{t-t_R}^{t}$.

They also suggest that, instead of using $\boldsymbol{\gamma}_{[0]}(t, t')$ in the integral, one should use a linear combination of $\boldsymbol{\gamma}_{[0]}$ and $\boldsymbol{\gamma}^{[0]}$. Then the Tanner–Simmons model is

$$\boldsymbol{\tau} = \int_{t-t_R}^{t} M(t - t')[(1 - q)\boldsymbol{\gamma}_{[0]}(t, t') + q\boldsymbol{\gamma}^{[0]}(t, t')]dt' \tag{8B.7-1}$$

where q is a small, positive constant characteristic of each polymer.

a. Show that for steady-state shear flow

$$\left\{\begin{aligned} \eta &= \int_0^{t_R} M(s)s\, ds && \text{(8B.7-2)} \\ \Psi_1 &= \int_0^{t_R} M(s)s^2\, ds && \text{(8B.7-3)} \\ \Psi_2 &= -q\int_0^{t_R} M(s)s^2\, ds && \text{(8B.7-4)} \end{aligned}\right.$$

[4] R. I. Tanner and J. M. Simmons, *Chem. Eng. Sci.*, **22**, 1803–1815 (1967); R. I. Tanner, *AIChE J.*, **15**, 177–183 (1969).

b. Note that $\Psi_2/\Psi_1 = -q$. What figure in Chapter 3 can you use to evaluate this result, and what is the approximate value of q suggested by the data?

c. Next make a specific choice of the memory function $M(s)$:

$$M(s) = \frac{\eta_0}{\lambda^2} e^{-s/\lambda} \tag{8B.7-5}$$

where η_0 is the zero-shear-rate viscosity and λ is a time constant; this means that the model now contains *four* adjustable parameters: η_0, λ, B, and q. Obtain an expression for the viscosity η as a function of the shear rate $\dot{\gamma}$.

d. In the viscosity found above what are the limiting values of η/η_0 as $\dot{\gamma} \to 0$ and $\dot{\gamma} \to \infty$?

Note: Tanner and Simmons were able to fit experimental data quite well when $M(s)$ was taken to be a sum of exponentials. They found that $\bar{\eta}(\dot{\varepsilon})$ remained finite, and their $\bar{\eta}(\dot{\varepsilon})$ had qualitatively the same shape as that of polymer melt data.

8B.8 Material Functions for the Wagner Model (Segalman Relaxation Modulus)

Use Segalman's memory function $M(s)$ of Eq. 8.3-35. Insert this into Eq. 8.3-12 along with Wagner's ϕ_1 and ϕ_2 functions in Table 8.3-3.

a. Obtain the expressions for $\eta(\dot{\gamma})$ and $\Psi_1(\dot{\gamma})$ for this model.

b. Show that in the limit of zero shear rate

$$\eta = \eta_0; \quad \Psi_1 = 2\eta_0\lambda(1 - \nu) \tag{8B.8-1}$$

c. Show that in the limit of high shear rates

$$\eta = \eta_0 \nu(\beta\lambda\dot{\gamma})^{\nu-1}; \quad \Psi_1 = \eta_0\lambda\nu(1 - \nu)(\beta\lambda\dot{\gamma})^{\nu-2} \tag{8B.8-2}$$

Are these results satisfactory?

8C.1 Flow of a Viscoelastic Fluid into a Line Sink[5]

It is desired to investigate possible generalizations of the plane radial flow problem in Example 8.4-1 to situations with angular dependence of the velocity. That is when Eq. 8.4-2 is replaced by

$$v_r = -\frac{f(\theta)}{r} \tag{8C.1-1}$$

Use the Lodge rubberlike liquid model.

a. Show that the only nonzero components of the $\gamma_{[0]}$ tensor are

$$\gamma_{[0]rr} = -\frac{2fs}{r^2} - \frac{f'^2 s^2}{r^2(r^2 + 2fs)} \tag{8C.1-2}$$

$$\gamma_{[0]\theta\theta} = \frac{2fs}{r^2 + 2fs} \tag{8C.1-3}$$

$$\gamma_{[0]r\theta} = \gamma_{[0]\theta r} = \frac{f's}{r^2 + 2fs} \tag{8C.1-4}$$

[5] A. M. Hull, *J. Non-Newtonian Fluid Mech.*, **8**, 327–336 (1981).

b. Show then that the equation of motion is satisfied if it is possible to choose a pressure field p such that

$$\frac{\partial p}{\partial r} = -\int_0^\infty M(s)\,\frac{4f^2s^2 + f'^2s^2 + r^2f''s}{r^3(r^2 + 2fs)}\,ds \qquad \text{(8C.1-5}a\text{)}$$

$$\frac{\partial p}{\partial \theta} = -\int_0^\infty M(s)\,\frac{2f's}{r^2 + 2fs}\,ds \qquad \text{(8C.1-5b)}$$

c. Eliminate p by cross differentiation between Eqs. 8C.1-5a and 5b and show that the resulting equation requires that $f(\theta)$ satisfy

$$f''' + 4f' = 0 \qquad \text{(8C.1-6)}$$

$$f'(4f^2 + 2ff'' - f'^2) = 0 \qquad \text{(8C.1-7)}$$

d. Show that the only solutions for $f(\theta)$ that satisfy both equations are

$$f = K \qquad \text{(8C.1-8)}$$

and

$$f = K\cos^2(\theta - \theta_0) \qquad \text{(8C.1-9)}$$

Equation 8C.1-8 is the solution in Example 8.4-1, whereas Eq. 8C.1-9 is the Hull solution for flow into a line sink located in an infinite plane, that is, $v_r = 0$ at $\theta = \theta_0 \pm \pi/2$. Verify that θ_0 may be taken to be zero without loss of generality.

e. Show that for the Hull solution (with $\theta_0 = 0$)

$$p = \int_0^\infty M(s)\left[\frac{Ks}{r^2} - \ln\left(1 + \frac{2Ks\cos^2\theta}{r^2}\right)\right]ds + p_\infty \qquad \text{(8C.1-10)}$$

and

$$\tau_{rr} = -\left[\int_0^\infty F\left(1 + \frac{2Ks}{r^2}\right)ds\right]\cos^2\theta \qquad \text{(8C.1-11)}$$

$$\tau_{r\theta} = \tau_{\theta r} = -\left[\int_0^\infty F\,ds\right]\sin\theta\cos\theta \qquad \text{(8C.1-12)}$$

$$\tau_{\theta\theta} = \left[\int_0^\infty F\,ds\right]\cos^2\theta \qquad \text{(8C.1-13)}$$

where

$$F(r, \theta, s) = \frac{2KM(s)s}{r^2 + 2Ks\cos^2\theta} \qquad \text{(8C.1-14)}$$

8C.2 Interrelation between Material Functions for the Factorized K–BKZ Model

Consider the stress relaxation after cessation, at time $t = 0$, of the steady shear flow with shear rate $\dot{\gamma}_0$. Show that for $t \geq 0$ the K–BKZ model gives

$$\tau_{yx}|_{t\geq 0} = \dot{\gamma}_0 \int_t^{\infty} M(s)(t-s)\left(\frac{\partial W}{\partial I_1} + \frac{\partial W}{\partial I_2}\right)_{t-t'=s} ds \qquad (8C.2\text{-}1)$$

From Eq. B of Table 8.3-1 we see that for $t \leq 0$,

$$(\tau_{xx} - \tau_{yy})|_{t\leq 0} = -\dot{\gamma}_0^2 \int_0^{\infty} M(s)s^2\left(\frac{\partial W}{\partial I_1} + \frac{\partial W}{\partial I_2}\right)_{t-t'=s} ds \qquad (8C.2\text{-}2)$$

From these two equations does the K–BKZ model predict the following relation?

$$(\tau_{xx} - \tau_{yy})|_{t\leq 0} = 2\dot{\gamma}_0 \int_0^{\infty} \tau_{yx}\, dt \qquad (8C.2\text{-}3)$$

Equation 8C.2-3 suggests that measuring the area under the stress relaxation curve gives information equivalent to that obtained in a steady-state normal stress measurement.[6]

8D.1 Unit Vector Transformation in Deforming Continua[7]

It is desired to develop the determinant J of the Jacobian matrix of the transformation from spherical coordinates θ, ϕ of the radial unit vector $\boldsymbol{u}(\theta, \phi)$ to spherical coordinates $\hat{\theta}$, $\hat{\phi}$ of the radial unit vector $\hat{\boldsymbol{u}}(\hat{\theta}, \hat{\phi})$, where $\boldsymbol{u}$ and $\hat{\boldsymbol{u}}$ are related by Eq. 8.3-54. Use the following steps:

a. First show that Eq. 8.3-54 can be inverted to yield

$$\boldsymbol{u} = \frac{[\boldsymbol{\Delta}\cdot\hat{\boldsymbol{u}}]}{|\boldsymbol{\Delta}\cdot\hat{\boldsymbol{u}}|} \qquad (8D.1\text{-}1)$$

b. Instead of the desired transformation consider the transformation from the spherical coordinates $(\hat{r}, \hat{\theta}, \hat{\phi})$ of the position vector $\hat{\boldsymbol{r}}(\hat{r}, \hat{\theta}, \hat{\phi})$ to the spherical coordinates (r, θ, ϕ) of the position vector $\boldsymbol{r}(r, \theta, \phi)$, where $\hat{\boldsymbol{r}}$ and $\boldsymbol{r}$ are related by

$$\boldsymbol{r} = [\boldsymbol{\Delta}\cdot\hat{\boldsymbol{r}}] \qquad (8D.1\text{-}2)$$

Denote the elements of the Jacobian matrix of the transformation $\hat{r}, \hat{\theta}, \hat{\phi} \to r, \theta, \phi$ by A_{ij} and show

$$\begin{aligned} J &= (A_{22}A_{33} - A_{23}A_{32})|_{r=|\boldsymbol{\Delta}\cdot\hat{\boldsymbol{r}}|} \\ &= (AA^{11})|_{r=|\boldsymbol{\Delta}\cdot\hat{\boldsymbol{r}}|} \end{aligned} \qquad (8D.1\text{-}3)$$

where A is the determinant of (A_{ij}), and A^{ij} is an element of the matrix inverse to (A_{ij}).

[6] Equation 8C.2-3 has been obtained from the kinetic theory of bead-spring chains (see Eq. 15.4-35). Several experiments tend to substantiate this relation: P. Attane, P. LeRoy, J. M. Pierrard, and G. Turrel, *J. Non-Newtonian Fluid Mech.*, **3**, 1–12 (1977), for polystyrene melts and concentrated solutions of polyisobutylene in mineral oil; H. W. Gao, S. Ramachandran, and E. B. Christiansen, *J. Rheol.*, **25**, 213–235 (1981), for concentrated solutions of polystyrene in *n*-butylbenzene.

[7] The development in this problem was supplied to us by A. S. Lodge. A similar derivation is given by S. M. Dinh, Sc.D. Thesis, Massachussetts Institute of Technology, Cambridge (1983).

c. Introduce the notation that the spherical coordinates of $\boldsymbol{r}$ are denoted by ρ_i, with $\rho_1, \rho_2, \rho_3 = r, \theta, \phi$, and the rectangular coordinates of $\boldsymbol{r}$ are x_i. Similarly denote the spherical coordinates of $\hat{\boldsymbol{r}}$ by $\hat{\rho}_i$ and its rectangular coordinates by $\hat{x}_i$. Then show that

$$A_{ij} = \frac{\partial \rho_i}{\partial \hat{\rho}_j} = \sum_m \sum_n \frac{\partial \rho_i}{\partial x_m} \frac{\partial x_m}{\partial \hat{x}_n} \frac{\partial \hat{x}_n}{\partial \hat{\rho}_j} = \sum_m \sum_n \frac{\partial \rho_i}{\partial x_m} \Delta_{mn} \frac{\partial \hat{x}_n}{\partial \hat{\rho}_j} \tag{8D.1-4}$$

$$A^{ij} = \frac{\partial \hat{\rho}_i}{\partial \rho_j} = \sum_m \sum_n \frac{\partial \hat{\rho}_i}{\partial \hat{x}_m} \mathrm{E}_{mn} \frac{\partial x_n}{\partial \rho_j} \tag{8D.1-5}$$

d. Then show that

$$A = (\det \boldsymbol{\Delta}) \frac{\hat{r}^2 \sin \hat{\theta}}{r^2 \sin \theta} \tag{8D.1-6}$$

$$A^{11} = (\hat{\boldsymbol{u}} \cdot [\mathbf{E} \cdot \boldsymbol{u}]) \tag{8D.1-7}$$

e. Combine the above to show that

$$J = \frac{\sin \hat{\theta}\, (\det \boldsymbol{\Delta})}{\sin \theta\, |\boldsymbol{\Delta} \cdot \hat{\boldsymbol{u}}|^3} \tag{8D.1-8}$$

8D.2 Incompressibility Conditions

Consider an arbitrary deformation that is described by the displacement functions Eqs. 8.1-2. It is desired to establish incompressibility conditions by comparing the mass of material contained in a volume $V(t)$, the boundaries of which are fluid particles, at the present time t and at an arbitrary past time t'. By a change of variable in the expression for the mass at t', show that

$$\frac{\rho}{\rho'} = \det \boldsymbol{\Delta} \tag{8D.2-1}$$

where ρ and ρ' denote the fluid density at t and t'. From this deduce

$$\det \boldsymbol{\Delta} = \det \mathbf{E} = 1 \tag{8D.2-2}$$

$$\det \mathbf{B} = \det \mathbf{B}^{-1} = 1 \tag{8D.2-3}$$

for incompressible fluids.

PART IV
CONTINUUM MECHANICS AND ITS USE IN SOLVING FLUID DYNAMICS PROBLEMS

Chapters 4 through 8 have dealt with the problems of solving fluid dynamics problems once a constitutive equation has been specified. These constitutive equations contained various tensors that were introduced so that the stress in a fluid element would not depend on the instantaneous rate of rotation of the fluid element in space, but the detailed origins of these tensors were not discussed.

In Chapter 9 we do give the derivations for these various tensors starting from the notion of a "convected coordinate system" that is embedded in the fluid and moves along with it. In this way we show how the kinematic tensors of Chapters 6, 7, and 8 are generated, and we also derive expressions for a number of additional kinematic tensors that we did not need for these chapters. In this way we build up a rather complete notational scheme that includes many more kinematic tensors, some of which are necessary in connection with reading the research literature. In Chapter 9 we also show the relations among the various types of constitutive equations discussed so far in terms of a "family tree"; this discussion should be helpful to the reader who feels the need for a summary of the equations, their limitations, and interrelations.

Finally in Chapter 10 we turn to the subject of "rheometry," the measurement of material functions by means of various laboratory devices—some standard, and some not so standard. In order for a rheometric device to be of use, we have to be able to describe the flow in the system with no (or at least minimal) information about the constitutive equation for the material. This challenging subject then involves the solution of fluid dynamics problems in rather simple flow systems, but with minimal use of specific constitutive equations. It also requires a careful exploration of some aspects of the kinematics of the flow patterns that occur in the commonly used rheometric devices.

After the subject of rheometry has been covered, we are finally in a position to understand the entire program of activity that needs to be carried out in the area of polymer fluid dynamics:

a. Given a small sample of a fluid, one measures some of the material functions for the fluid. There is a general feeling among rheologists that both shear and shearfree experimental measurements need to be made. Unsteady-state experiments give information not supplied by the steady-state experiments.

b. With the experimental data obtained in (**a**) one determines the constants or functions that appear in the constitutive equation that is to be used. Clearly the more experimental material-function data that are available, the more carefully the constants can be determined.

c. Then the fluid dynamics problem is solved for some given geometrical arrangement. Usually the flow problems must be solved by numerical methods, and large-scale computing devices must be available.

d. Finally, the solution in (**c**) should be compared with experimental data on the flow system; this could be data on pressure drops, flow rates, forces, and torques, or, even better, flow visualization data or else a mapping of the flow field by laser Doppler velocimetry.

The importance of experimental data on material functions and the experimental data on the flow system cannot be overemphasized. The subject of polymer fluid dynamics is still largely an experimental one.

The field of polymer fluid dynamics does not end with Chapter 10, however. Another viewpoint of the subject is provided by taking the molecular approach. Molecular theories have been very helpful in suggesting forms for constitutive equations; they also give additional information about the constants and functions that occur in various ordered expansions. Furthermore, if one wishes to know about the stretching and orientation of polymer molecules in various flow patterns, a molecular approach is mandatory. Readers interested in learning about this approach to polymer fluid dynamics can find an extensive discussion in Volume 2.

CHAPTER 9

CONTINUUM-MECHANICS CONCEPTS[1]

In Chapters 6, 7, and 8 we discussed some nonlinear viscoelastic constitutive equations and showed how to use them to solve flow problems. These constitutive equations contained finite strain tensors, rate-of-strain tensors, and the convected derivative of the stress tensor. These various quantities were defined and it was shown how to evaluate them for problem-solving purposes. In this chapter we show how these quantities arise naturally by considering the transformation rules between fixed and convected coordinates; we also discuss why they are needed in order to produce constitutive equations that are "admissible."

The establishment of criteria for admissibility is the subject of §9.1. This subject has been approached in a number of ways in the literature, and the interrelation of the various approaches is by no means easy to understand.[2] We have chosen here to present the criteria for admissibility according to the convected-component method of Oldroyd.[3] Having done this we then turn to the problems of interrelating convected coordinates and fixed coordinates and the tensor components in the two coordinate systems. Thus §9.2 is devoted to the relation between convected coordinates and fixed coordinates, and the relation between the base vectors in the two coordinate systems. Then §9.3 describes the transformation rules for the kinematic tensors used in continuum mechanics; this section culminates in the presentation of a table that is helpful for understanding the notation and the relations among the various kinematic tensors. In §9.4 a discussion of the transformation rules for the stress tensor and its time derivatives is given, and this, too, ends with a table summarizing the notation. Then in §9.5 we return to the subject addressed in §9.1—that of constructing admissible constitutive equations—this time showing how the tables of the two preceding sections can be used to write down constitutive equations in a coordinate system fixed in space. It is at this point that we can understand how the equations given in Chapters 6, 7, and 8 satisfy the criteria of admissibility. In the final section, §9.6, we introduce the very general "memory-integral expansion" and show how it provides important perspectives about constitutive equations, and in particular how the key equations of Chapters 4 through 8 are interrelated.

[1] The authors wish to thank Professor A. S. Lodge for many helpful discussions in connection with this chapter.

[2] For some perspectives on admissibility criteria see A. S. Lodge, *Body Tensor Fields in Continuum Mechanics*, Academic Press, New York (1974), pp. 254–256; this book is recommended for those interested in a more advanced and comprehensive study of the fundamental basis of continuum mechanics.

[3] J. G. Oldroyd, *J. Non-Newtonian Fluid Mech.*, **14**, 9–46 (1984).

This chapter is intended to be only a brief introduction to the subject of continuum mechanics, which is a very large branch of classical mechanics. Those wishing more on this subject should consult one of the many available textbooks or reference works.[4]

§9.1 OLDROYD'S CRITERIA FOR ADMISSIBILITY OF CONSTITUTIVE EQUATIONS[1]

The aim of this section is to obtain the general form of the equations that interrelate the components of the stress and strain tensors (and their time derivatives or integrals) in a material element moving in an arbitrary way as a part of a flowing continuum. In general, this constitutive equation may involve the stress components at time t' for $-\infty < t' \leq t$ and the strain components at time t' for $-\infty < t' \leq t$. Oldroyd proposed that the constitutive equation thus expressed should be independent of (a) any frame of reference, (b) the position in space, the translational motion, and the rotational motion of the fluid element, and (c) the stress and strain in the neighboring fluid elements (except for the continuity requirements on velocity and stress across any surface). In order to construct an equation satisfying these requirements it is most convenient to use a "convected coordinate system," with coordinate surfaces $\hat{x}^i = \text{constant}$ $(i = 1, 2, 3)$, embedded in the fluid and deforming with it; this kind of coordinate system was first proposed by Hencky.[2] In this kind of coordinate system a fluid particle has by definition the same coordinates $\hat{x}^1$, $\hat{x}^2$, $\hat{x}^3$ for all time. In Fig. 9.1-1 we show how a convected coordinate system moves through space as it is swept along with the moving fluid; for simplicity a two-dimensional flow is depicted.

Next we want to describe the fundamental kinematic and dynamic quantities that should appear in the constitutive equation. The kinematic quantities pertain to the shape and change of shape of fluid elements, and the dynamic quantities have to do with the forces transmitted across fluid surfaces. The next two paragraphs are concerned with the kinematic and dynamic variables central to the formulation of constitutive equations.

We first examine the kinematic description of the flow. At any point in the convected coordinate system we can construct a set of three *convected base vectors* $\hat{\boldsymbol{g}}_i = (\partial/\partial\hat{x}^i)\boldsymbol{r}$, where $\boldsymbol{r}$ is the position vector (see §A.8). A base vector $\hat{\boldsymbol{g}}_i$ is tangent to the $\hat{x}^i$-coordinate curve, and its change in length indicates the extent to which the fluid is stretched in the $\hat{x}^i$-direction. The base vectors depend on the convected coordinates $\hat{x} \equiv \hat{x}^1, \hat{x}^2, \hat{x}^3$ and on the time. The vector between two neighboring fluid particles $\hat{x}$ and $\hat{x} + d\hat{x}$ at time t is given by

$$d\boldsymbol{r} = \sum_i \hat{\boldsymbol{g}}_i(\hat{x}, t)d\hat{x}^i \tag{9.1-1}$$

[4] Traditional presentations of continuum mechanics have been given by W. Prager, *Introduction to Mechanics of Continua*, Ginn, Boston (1961), and L. E. Malvern, *Introduction to the Mechanics of a Continuous Medium*, Prentice-Hall, Englewood Cliffs, NJ, (1969). An introductory text with a less traditional flavor and with a number of novel viewpoints is A. S. Lodge, *Elastic Liquids*, Academic Press, New York (1964); this book has been particularly influential in the field of rheological measurements, inasmuch as the opening chapters are devoted to homogeneous flows. For extensive bibliographies consult C. Truesdell and R. Toupin, *Encylopedia of Physics*, Vol. III/1, Springer-Verlag, Berlin (1960), pp. 226–793 (with an appendix on tensor fields by J. L. Ericksen, pp. 794–858), and C. Truesdell and W. Noll, *Encyclopedia of Physics*, Vol. III/3, Springer-Verlag, Berlin (1965), pp. 1–579.

[1] J. G. Oldroyd, *Proc. Roy. Soc.*, **A200**, 523–541 (1950); **A202**, 345–358 (1950); **A245**, 278–297 (1958); **A283**, 115–133 (1965); J. G. Oldroyd and B. R. Duffy, *J. Non-Newtonian Fluid Mech.*, **5**, 141–145 (1979). A particularly fine summary of Oldroyd's work is that portion of an essay submitted in competition for the Adams Prize in 1964, which was published posthumously as J. G. Oldroyd, *J. Non-Newtonian Fluid Mech.*, **14**, 9–46 (1984); those wishing to study Oldroyd's approach to continuum mechanics should begin by studying this very readable publication. Much of §9.1 is a paraphrasing of certain portions of pp. 17–24 of this paper.

[2] H. Hencky, *Z. Angew. Math. Mech.*, **5**, 144–146 (1925).

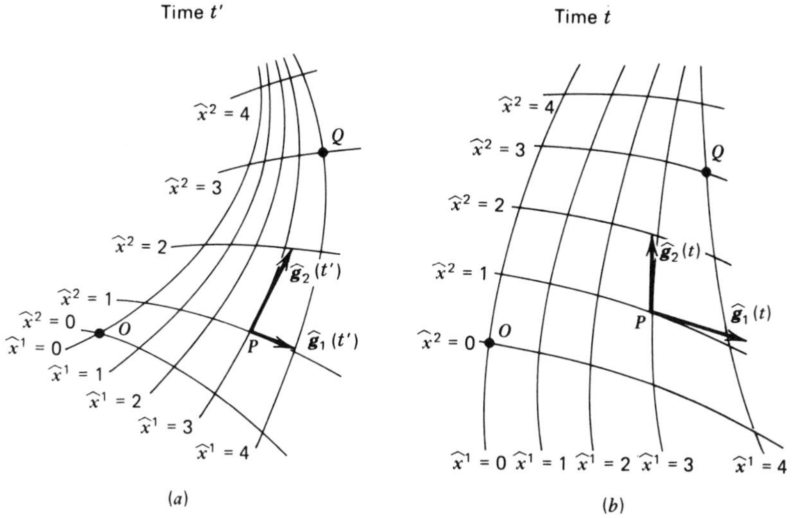

FIGURE 9.1-1. An arbitrarily chosen coordinate system, embedded in a flowing fluid, at two different times (*a*) t' and (*b*) t. Fluid particle P is located at $\hat{x}^1 = 3$, $\hat{x}^2 = 1$ at all times; fluid particle Q is at $\hat{x}^1 = 4$, $\hat{x}^2 = 3$ at all times. The base vectors $\hat{\boldsymbol{g}}_1$ and $\hat{\boldsymbol{g}}_2$ at fluid particle P are also shown.

so that the square of the separation between the particles is given by

$$\begin{aligned}(\boldsymbol{dr}\cdot\boldsymbol{dr}) &= \sum_i \sum_j (\hat{\boldsymbol{g}}_i \cdot \hat{\boldsymbol{g}}_j) d\hat{x}^i \, d\hat{x}^j \\ &= \sum_i \sum_j \hat{g}_{ij}(\hat{x}, t') d\hat{x}^i \, d\hat{x}^j \end{aligned} \tag{9.1-2}$$

in which the $\hat{g}_{ij}(\hat{x}, t)$ are the covariant metric coefficients. These quantities, which describe the relative distances between arbitrary pairs of neighboring particles, contain complete information about the shape of a fluid element at time t'. All the kinematic quantities that may appear in a constitutive equation must be derivable from the metric coefficients $\hat{g}_{ij}(\hat{x}, t)$ for $-\infty < t' \leq t$ which is sometimes called the *deformation history* at the fluid particle $\hat{x}$. Since the $\hat{g}_{ij}$ are obtained by forming scalar products, they do not depend on the instantaneous position or orientation of the fluid element in space.

Next we turn our attention to the dynamical quantities. Let $\hat{\boldsymbol{f}}_i$ be the instantaneous force (per unit area) transmitted across a fluid surface element perpendicular to $\hat{\boldsymbol{g}}_i$. Then according to §1.1b

$$\hat{\boldsymbol{f}}_i = [(\hat{\boldsymbol{g}}_i/|\hat{\boldsymbol{g}}_i|)\cdot\boldsymbol{\pi}] \tag{9.1-3}$$

where $\boldsymbol{\pi}$ is the total stress tensor and $\hat{\boldsymbol{g}}_i/|\hat{\boldsymbol{g}}_i|$ is the unit vector in the $\hat{x}^i$ direction. We now form the scalar product of Eq. 9.1-3 with $\hat{\boldsymbol{g}}_j$

$$(\hat{\boldsymbol{f}}_i \cdot \hat{\boldsymbol{g}}_j) = \frac{1}{|\hat{\boldsymbol{g}}_i|}(\boldsymbol{\pi} : \hat{\boldsymbol{g}}_j \hat{\boldsymbol{g}}_i) \tag{9.1-4}$$

But the double dot product on the right side is just $\hat{\pi}_{ij}$, the ijth component of $\boldsymbol{\pi}$ in the convected coordinate system. Therefore

$$\hat{\pi}_{ij} = \sqrt{(\hat{\boldsymbol{g}}_i \cdot \hat{\boldsymbol{g}}_i)}(\hat{\boldsymbol{f}}_i \cdot \hat{\boldsymbol{g}}_j) \tag{9.1-5}$$

The quantities $\hat{\pi}_{ij}(\hat{x}, t)$ then describe completely the state of stress at a fluid particle at time t, and $\hat{\pi}_{ij}(\hat{x}, t')$ for $-\infty < t' \leq t$ is called the *stress history* at the fluid particle $\hat{x}$. Since the $\hat{\pi}_{ij}$ are obtained by forming scalar products, they are independent of the instantaneous position and orientation of the fluid element in space.

The constitutive equation will generally be some kind of relation containing the stress history and the strain history. If one desires to take into account nonisothermal effects, then the *temperature history*, $T(\hat{x}, t')$ for $-\infty < t' \leq t$, must be included as well. In addition the time difference $t - t'$ may be expected to appear, as well as some material constants (or even material constant tensors whose components are $\hat{\kappa}_{N\,kl\ldots}^{\;ij\cdots}(\hat{x})$, and whose time derivatives with respect to t' are zero). The constitutive equation might then contain time derivatives such as $\partial\hat{g}_{ij}/\partial t$, $\partial^2\hat{g}_{ij}/\partial t^2$, $\partial\hat{\pi}_{ij}/\partial t, \ldots$ or quantities such as $\hat{g}_{ij}(t')$, $\partial\hat{g}_{ij}/\partial t'$, $\partial\hat{\pi}_{ij}/\partial t', \ldots$ in the integrands of time integrals $\int_{-\infty}^{t} \ldots dt'$. So far we have written only the covariant convected components $\hat{g}_{ij}$ and $\hat{\pi}_{ij}$, but contravariant and mixed components may also be used. For incompressible fluids Oldroyd summarized all of the above in the following way:

> The constitutive equation for a flowing continuum can be written as an invariant set of integro-differential equations relating
>
> $$\hat{g}_{ij}(\hat{x}, t'), \quad \hat{\tau}_{ij}(\hat{x}, t'), \quad T(\hat{x}, t'), \quad \hat{\kappa}_{N\,pqr\ldots}^{\;ijk\cdots}(\hat{x}) \qquad (N = 1, 2, 3\ldots) \tag{9.1-6}$$
>
> for $-\infty < t' \leq t$. Because of the incompressibility constraint it is necessary that
>
> $$\sqrt{\det \hat{g}_{ij}(\hat{x}, t')} = 1 \qquad (-\infty < t' \leq t) \tag{9.1-7}$$
>
> which is equivalent to the statement that the volume of a fluid element is a constant for all t' (see §A.8). The auxiliary relation
>
> $$\hat{\pi}_{ij}(\hat{x}, t) = p(\hat{x}, t)\hat{g}_{ij}(\hat{x}, t) + \hat{\tau}_{ij}(\hat{x}, t) \tag{9.1-8}$$
>
> then gives the total stress tensor components in terms of the $\hat{\tau}_{ij}$ and an isotropic pressure[3] p.

The term "invariant" used above refers to the fact that the equation must have a form independent of the coordinate system; this is accomplished by writing the equations in tensor component form, with the two sides of each equation being tensor components of the same kind (e.g., covariant or contravariant) and order.

Oldroyd pointed out that equations of the above type are *admissible* in the sense that they are (a) form invariant under a change of coordinate system, (b) value invariant under a change of translational or rotational motion of the fluid element as it goes through space, and (c) value invariant under a change of rheological history of neighboring fluid elements. Oldroyd further offered the term *rheological invariance* to describe these three invariances. In this book we use the term "admissible" to indicate that a constitutive equation is rheologically invariant.

The second invariance requirement listed above (labeled (b)) has met with some resistance. It is, however, probably very good for polymer solutions and polymer melts, and indeed almost all molecular theories for polymeric liquids give constitutive equations that

[3] In this text we use the convention that $\hat{\tau}_{ij} = 0$ at equilibrium; this serves to define p.

are consistent with this invariance requirement;[4] this is presumably a consequence of the fact that centrifugal and Coriolis forces can be neglected at the molecular level.[5] It may well be, however, that for a two-phase fluid the invariance requirement (b) will not be appropriate if one wishes to develop a constitutive equation for the equivalent one-phase continuum.

We conclude by writing down several constitutive equations from previous chapters in convected components

Second Order Fluid: (Eq. 6.2-1)

$$\hat{\tau}^{ij} = -b_1 \frac{\partial \hat{g}^{ij}}{\partial t} - b_2 \frac{\partial^2 \hat{g}^{ij}}{\partial t^2} - b_{11} \sum_k \sum_l \frac{\partial \hat{g}^{ik}}{\partial t} \hat{g}_{kl} \frac{\partial \hat{g}^{lj}}{\partial t} \tag{9.1-9}$$

Contravariant Convected Jeffreys Model: (Eq. 7.2-1)

$$\hat{\tau}^{ij} + \lambda_1 \frac{\partial}{\partial t} \hat{\tau}^{ij} = -\eta_0 \left(\frac{\partial \hat{g}^{ij}}{\partial t} + \lambda_2 \frac{\partial^2 \hat{g}^{ij}}{\partial t^2} \right) \tag{9.1-10}$$

Lodge Rubberlike Liquid: (Eq. 8.2-1)

$$\hat{\tau}^{ij} = -\int_{-\infty}^{t} M(t - t')[\hat{g}^{ij}(t') - \hat{g}^{ij}(t)]\, dt' \tag{9.1-11}$$

These equations are all "admissible," in the sense that they satisfy Oldroyd's rules of rheological invariance. In the next several sections we develop the connection between the convected and fixed coordinates and the convected and fixed components of the kinematic and dynamic tensors. Then we will have all the necessary transformation rules to enable us to go back and forth between constitutive equations written in the convected-component form and the fixed-component form. Also we will be able to restate Oldroyd's rules so that we can construct admissible constitutive equations directly using fixed-component notation.

§9.2 CONVECTED COORDINATES AND CONVECTED BASE VECTORS[1]

In §9.1 we introduced the idea of a convected coordinate system embedded in a flowing fluid and deforming with it; in that section no reference was made to any coordinate system fixed in space. In this section we discuss the relation between the convected coordinate system and a specific Cartesian coordinate system fixed in space.[2] We also want to relate the convected base vectors to the unit vectors of the fixed coordinate system.

[4] As one example of a kinetic theory that leads to terms not satisfying this invariance we can cite R. B. Bird, X. J. Fan, and C. F. Curtiss, *J. Non-Newtonian Fluid Mech.*, **15**, 85–92 (1984); even in this theory, however, it was proven that the terms accounting for the deviations from the invariance requirement are negligibly small.

[5] P.-G. de Gennes, *Physica*, **118A**, 44–45 (1983); G. Ryskin, *Phys. Rev.*, **A32**, 1239–1240 (1985).

[1] The presentation in this and the following section has been influenced by L. I. Sedov, *Introduction to the Mechanics of a Continuous Medium*, Addison-Wesley, Reading, MA (1965), and by the first three chapters of A. S. Lodge, *Elastic Liquids*, Academic Press, New York (1964).

[2] One can, of course, choose an orthogonal (or even nonorthogonal) curvilinear coordinate system for the fluid at time t; by using Cartesian coordinates the discussion is simplified somewhat.

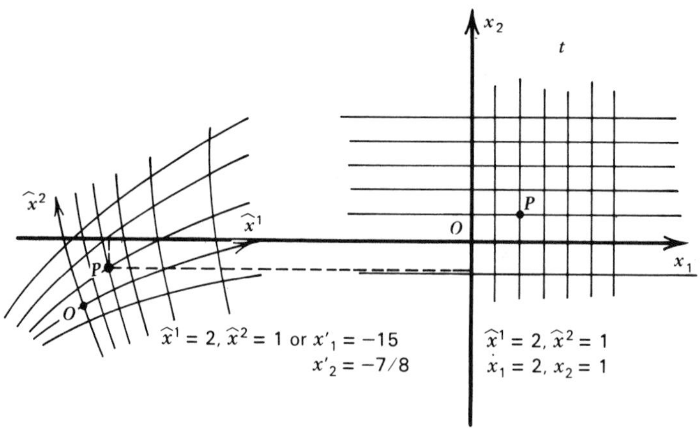

FIGURE 9.2-1. Convected coordinates; the convected coordinates $\hat{x}^i$ exactly coincide with the fixed coordinates x_i at time t.

We consider now the special convected coordinate system $\hat{x}^1$, $\hat{x}^2$, $\hat{x}^3$ in Fig. 9.2-1, which at time t exactly coincides with a Cartesian coordinate system x_1, x_2, x_3. A fluid particle P can be designated by giving its convected coordinates $\hat{x}^i$, which remain the same for all values of the time. The fluid particle P can also be identified by giving its Cartesian coordinates x_i' at time t'; these coordinates are given with respect to the frame $Ox_1x_2x_3$. At time t particle P has coordinates x_i in the frame $Ox_1x_2x_3$. Note that the coordinates x_i at time t have exactly the same numerical values as the particle coordinates $\hat{x}^i$ in the $\hat{O}\hat{x}^1\hat{x}^2\hat{x}^3$ system.[3]

The motion of the fluid is described by giving the location of all fluid particles for all past times t':

$$x_i' = x_i'(\hat{x}^1, \hat{x}^2, \hat{x}^3, t') \qquad i = 1, 2, 3 \tag{9.2-1}$$

The fluid particle given by the convected coordinates $\hat{x}^1$, $\hat{x}^2$, $\hat{x}^3$ is located at x_1, x_2, x_3, at time t. That is we can use either $\hat{x}^1$, $\hat{x}^2$, $\hat{x}^3$ as the "particle label" or alternatively the variables x_1, x_2, x_3, and t. Hence we may rewrite Eq. 9.2-1 as

$$x_i' = x_i'(x_1, x_2, x_3, t, t') \quad \text{or} \quad \boldsymbol{r}' = \boldsymbol{r}'(\boldsymbol{r}, t, t') \tag{9.2-2}$$

These relations may be inverted to give

$$x_i = x_i(x_1', x_2', x_3', t', t) \quad \text{or} \quad \boldsymbol{r} = \boldsymbol{r}(\boldsymbol{r}', t', t) \tag{9.2-3}$$

The functions on the right sides of these equations are the *displacement functions* (cf. Eqs. 8.1-1 and 2).

As pointed out in §9.1 we may define *convected base vectors* $\hat{\boldsymbol{g}}_i$ at each fluid particle. They are functions of the location in the convected coordinate system and also depend on the time t' since the coordinate system itself is changing with time as the fluid moves along.

[3] In comparing the discussion here with that of §A.8 note that the Cartesian coordinates x_i of Appendix A correspond to the x_i' here, and the nonorthogonal, curvilinear coordinates q^i of Appendix A correspond to $\hat{x}^i$ here.

Hence we indicate the dependence on fluid particle and time by writing $\hat{\boldsymbol{g}}_i(\hat{x}^1, \hat{x}^2, \hat{x}^3, t')$ or alternatively $\hat{\boldsymbol{g}}_i(\boldsymbol{r}, t, t')$—that is, we can use the particle label $\hat{x}^i(i = 1, 2, 3)$ or $(\boldsymbol{r}, t)$. Then,

$$\hat{\boldsymbol{g}}_i(\boldsymbol{r}, t, t') = \frac{\partial}{\partial \hat{x}^i}\boldsymbol{r}' \qquad \text{(Definition of } \hat{\boldsymbol{g}}_i\text{)}$$

$$= \frac{\partial}{\partial x_i}\left(\sum_j \boldsymbol{\delta}_j x'_j\right) \qquad \text{(Expand position vector } \boldsymbol{r}' \text{ in its Cartesian components)}$$

$$= \sum_j \boldsymbol{\delta}_j \Delta_{ji} \qquad \text{(Use definition of Eq. 8.1-3)}$$

$$= [\boldsymbol{\Delta} \cdot \boldsymbol{\delta}_i] \tag{9.2-4}$$

Note that the tensor $\boldsymbol{\Delta}(\boldsymbol{r}, t, t') = \sum_i \sum_j \boldsymbol{\delta}_i \boldsymbol{\delta}_j \Delta_{ij}$ operates on the ith unit vector $\boldsymbol{\delta}_i$ to give the ith convected base vector $\hat{\boldsymbol{g}}_i$; we can also regard Δ_{ji} as the jth Cartesian component of $\hat{\boldsymbol{g}}_i$. By virtue of the way we define the convected coordinate system, at $t' = t$ we must have $\hat{\boldsymbol{g}}_i(\boldsymbol{r}, t, t) = \boldsymbol{\delta}_i$ and $\boldsymbol{\Delta}(\boldsymbol{r}, t, t) = \boldsymbol{\delta}$. The base vector $\hat{\boldsymbol{g}}_i$ is tangent to the $\hat{x}^i$ coordinate curve; the set of three base vectors $\hat{\boldsymbol{g}}_i(\boldsymbol{r}, t, t')$ describes how the fluid in the neighborhood of the particle $\boldsymbol{r}, t$ is oriented and distorted as it moves through space.

At every fluid particle we may also define a set of *convected reciprocal base vectors*

$$\hat{\boldsymbol{g}}^i(\boldsymbol{r}, t, t') = \frac{\partial}{\partial \boldsymbol{r}'}\hat{x}^i \qquad \text{(Definition of } \hat{\boldsymbol{g}}^i\text{)}$$

$$= \left(\sum_j \boldsymbol{\delta}_j \frac{\partial}{\partial x'_j}\right) x_i \qquad \text{(Expand } \partial/\partial \boldsymbol{r}' \text{ in its Cartesian components)}$$

$$= \sum_j \mathrm{E}_{ij} \boldsymbol{\delta}_j \qquad \text{(Use definition of Eq. 8.1-4)}$$

$$= [\boldsymbol{\delta}_i \cdot \mathbf{E}] \tag{9.2-5}$$

Note that the tensor $\mathbf{E}(\boldsymbol{r}, t, t') = \sum_i \sum_j \boldsymbol{\delta}_i \boldsymbol{\delta}_j \mathrm{E}_{ij}$ operates on the ith unit vector to give the ith convected reciprocal base vector $\hat{\boldsymbol{g}}^i$; we can also regard E_{ij} as the jth Cartesian component of $\hat{\boldsymbol{g}}^i$. It should be pointed out that to calculate the components of $\mathbf{E}$, we first differentiate the inverse displacement functions and then eliminate the x'_j in favor of the x_j so that $\mathbf{E}$ finally depends on $\boldsymbol{r}$, t and t'. At $t' = t$, we have $\hat{\boldsymbol{g}}^i = \boldsymbol{\delta}_i$ and $\mathbf{E} = \boldsymbol{\delta}$. The reciprocal base vector $\hat{\boldsymbol{g}}^i$ is normal to the surface $\hat{x}^i$ = constant; the set of $\hat{\boldsymbol{g}}^i$ gives information about how the fluid is being deformed. We also mention that for an incompressible fluid $\hat{\boldsymbol{g}}^i = [\hat{\boldsymbol{g}}_j \times \hat{\boldsymbol{g}}_k]$ (for ijk = 123, 231, and 312). Keep in mind that $(\hat{\boldsymbol{g}}_i \cdot \hat{\boldsymbol{g}}^j) = (\hat{\boldsymbol{g}}^i \cdot \hat{\boldsymbol{g}}_j) = \delta_{ij}$ (see §A.8).

In Fig. 9.2-2 we show the convected base vectors and reciprocal base vectors in a homogeneous flow, and in Fig. 9.2-3 the convected base vectors for four different times in a nonhomogeneous flow. These figures should be helpful in showing how the convected coordinate system and the $\hat{\boldsymbol{g}}_i$ and $\hat{\boldsymbol{g}}^i$ change with time as the fluid flows along. In addition Problems 9B.3 and 4 provide an opportunity for calculating the convected base vectors and various kinematic tensors for the flows depicted in Figs. 9.2-2 and 3.

In the following development we need to know how the convected base vectors and the displacement gradient tensors $\boldsymbol{\Delta}$ and $\mathbf{E}$ change with t' as we follow a fluid particle along

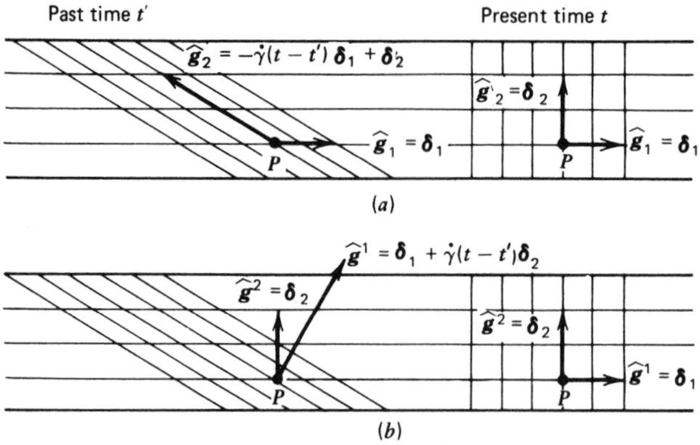

FIGURE 9.2-2. Steady shear flow $v_1 = \dot{\gamma}x_2, v_2 = 0, v_3 = 0$ showing the convected base vectors and reciprocal base vectors associated with a fluid particle P as it moves along from some past time t' to the present time t. The vectors $\hat{\boldsymbol{g}}_3$ and $\hat{\boldsymbol{g}}^3$ are perpendicular to the plane of the paper and both equal to $\boldsymbol{\delta}_3$. In this *homogeneous* flow the convected base vectors $\hat{\boldsymbol{g}}_i$ are coincident with material lines; that is, they are "embedded" in the fluid. The convected reciprocal base vectors $\hat{\boldsymbol{g}}^i$ are perpendicular to material surfaces.

its trajectory. As we show in Example 9.2-1 the time derivatives of the convected base vectors are

$$\frac{\partial}{\partial t'}\hat{\boldsymbol{g}}_i = +[\hat{\boldsymbol{g}}_i \cdot \nabla \boldsymbol{v}] \tag{9.2-6}$$

$$\frac{\partial}{\partial t'}\hat{\boldsymbol{g}}^i = -[(\nabla \boldsymbol{v}) \cdot \hat{\boldsymbol{g}}^i] \tag{9.2-7}$$

Here it is understood that the base vectors and $\nabla \boldsymbol{v}$ are all evaluated at t' for constant $\boldsymbol{r}$ and t (that is, following the fluid particle). From these relations and Eqs. 9.2-4 and 5 it follows that the time derivatives of the displacement gradient tensors are given by

$$\frac{\partial}{\partial t'}\boldsymbol{\Delta} = +\{(\nabla \boldsymbol{v})^\dagger \cdot \boldsymbol{\Delta}\} \tag{9.2-8}$$

$$\frac{\partial}{\partial t'}\mathbf{E} = -\{\mathbf{E} \cdot (\nabla \boldsymbol{v})^\dagger\} \tag{9.2-9}$$

These expressions are needed in §9.3 to get the rate-of-strain tensors.

EXAMPLE 9.2-1 Change of Convected Base Vectors with Time

Derive Eqs. 9.2-6 to 9.

SOLUTION **(a)** First make use of the definition of the $\hat{\boldsymbol{g}}_i$ and write

$$\frac{\partial}{\partial t'}\hat{\boldsymbol{g}}_i(\hat{x}, t') = \frac{\partial}{\partial t'}\left(\frac{\partial}{\partial \hat{x}^i}\boldsymbol{r}'\right) \tag{9.2-10}$$

Next we interchange the order of differentiation:

$$\frac{\partial}{\partial t'}\hat{\boldsymbol{g}}_i(\hat{x}, t') = \frac{\partial}{\partial \hat{x}^i}\left(\frac{\partial}{\partial t'}\boldsymbol{r}'\right) \tag{9.2-11}$$

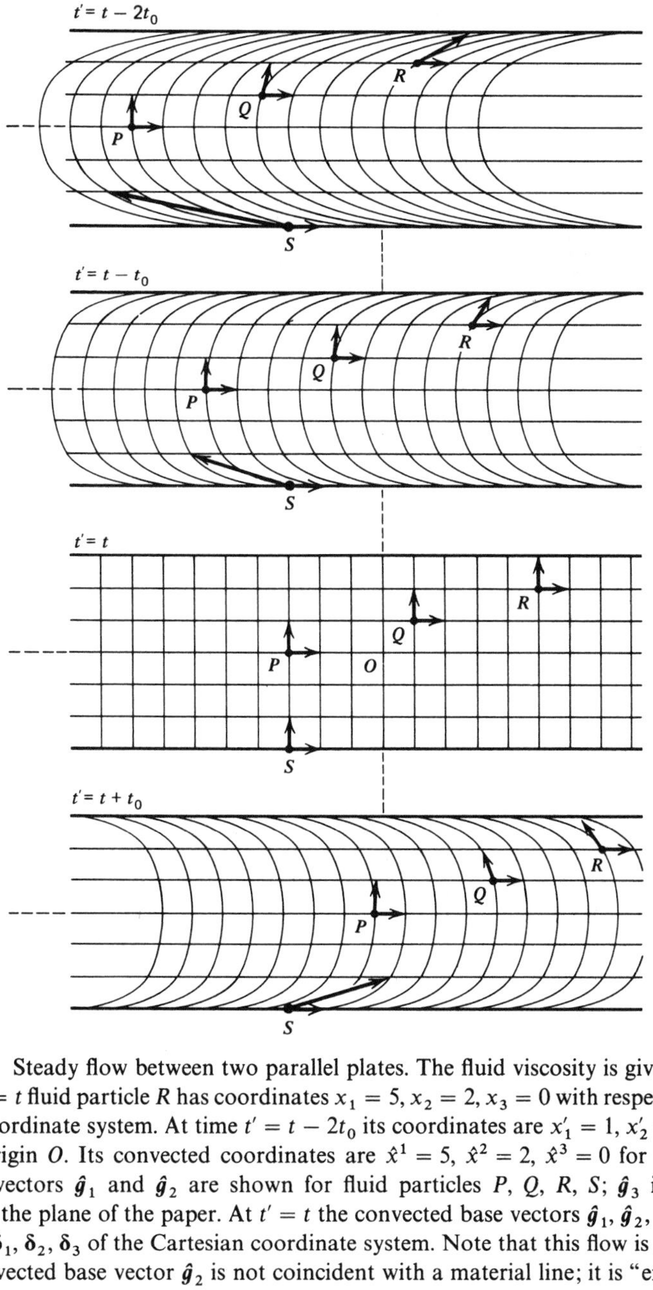

FIGURE 9.2-3. Steady flow between two parallel plates. The fluid viscosity is given by $\eta = m\dot{\gamma}^{n-1}$ with $n = \frac{1}{3}$. At $t' = t$ fluid particle R has coordinates $x_1 = 5, x_2 = 2, x_3 = 0$ with respect to the origin O of a Cartesian coordinate system. At time $t' = t - 2t_0$ its coordinates are $x_1' = 1, x_2' = 2, x_3' = 0$ with respect to the origin O. Its convected coordinates are $\hat{x}^1 = 5$, $\hat{x}^2 = 2$, $\hat{x}^3 = 0$ for all times t'. The convected base vectors $\hat{\boldsymbol{g}}_1$ and $\hat{\boldsymbol{g}}_2$ are shown for fluid particles P, Q, R, S; $\hat{\boldsymbol{g}}_3$ is of unit length perpendicular to the plane of the paper. At $t' = t$ the convected base vectors $\hat{\boldsymbol{g}}_1, \hat{\boldsymbol{g}}_2, \hat{\boldsymbol{g}}_3$ coincide with the unit vectors $\boldsymbol{\delta}_1, \boldsymbol{\delta}_2, \boldsymbol{\delta}_3$ of the Cartesian coordinate system. Note that this flow is *nonhomogeneous* and that the convected base vector $\hat{\boldsymbol{g}}_2$ is not coincident with a material line; it is "embedded" in the fluid in the sense that it is always tangent to the same material curve.

But the change of $\boldsymbol{r}'$ with respect to t' at constant fluid particle label $\hat{x}^i$ is the definition of the fluid velocity $\boldsymbol{v}'$ at the time t', so that

$$\frac{\partial}{\partial t'} \hat{\boldsymbol{g}}_i(\hat{x}, t') = \frac{\partial}{\partial \hat{x}^i} \boldsymbol{v}'$$

$$= \sum_j \frac{\partial x_j'}{\partial \hat{x}^i} \left(\frac{\partial}{\partial x_j'} \boldsymbol{v}' \right) \tag{9.2-12}$$

However $\partial x_j'/\partial \hat{x}^i$ is the same as $\partial x_j'/\partial x_i$, and therefore, according to Eq. 9.2-4, is also the same as the jth component of $\hat{\boldsymbol{g}}_i$. Hence Eq. 9.2-6 follows directly.

(b) To get Eq. 9.2-7 we start by differentiating the relation $(\hat{\boldsymbol{g}}_i \cdot \hat{\boldsymbol{g}}^j) = \delta_{ij}$ with respect to t'

$$\left(\left(\frac{\partial}{\partial t'}\hat{\boldsymbol{g}}_i\right)\cdot\hat{\boldsymbol{g}}^j\right) + \left(\hat{\boldsymbol{g}}_i\cdot\left(\frac{\partial}{\partial t'}\hat{\boldsymbol{g}}^j\right)\right) = 0 \tag{9.2-13}$$

Then we use Eq. 9.2-6 in the first term to get

$$([\hat{\boldsymbol{g}}_i\cdot\nabla\boldsymbol{v}]\cdot\hat{\boldsymbol{g}}^j) = -\left(\hat{\boldsymbol{g}}_i\cdot\frac{\partial}{\partial t'}\hat{\boldsymbol{g}}^j\right) \tag{9.2-14}$$

The dot products on the left side may be regrouped:

$$(\hat{\boldsymbol{g}}_i\cdot[(\nabla\boldsymbol{v})\cdot\hat{\boldsymbol{g}}^j]) = -\left(\hat{\boldsymbol{g}}_i\cdot\frac{\partial}{\partial t'}\hat{\boldsymbol{g}}^j\right) \tag{9.2-15}$$

Since $\hat{\boldsymbol{g}}_i$ is arbitrary (i.e., i can be 1, 2, or 3), the quantities dotted into the $\hat{\boldsymbol{g}}_i$ may be equated and Eq. 9.2-7 is obtained.

(c) To get Eq. 9.2-8 we substitute $\hat{\boldsymbol{g}}_i = [\boldsymbol{\Delta}\cdot\boldsymbol{\delta}_i]$ into both sides of Eq. 9.2-6:

$$\left[\frac{\partial\boldsymbol{\Delta}}{\partial t'}\cdot\boldsymbol{\delta}_i\right] = [[\boldsymbol{\Delta}\cdot\boldsymbol{\delta}_i]\cdot\nabla\boldsymbol{v}] \tag{9.2-16}$$

When the right side is rearranged we get:

$$\left[\frac{\partial\boldsymbol{\Delta}}{\partial t'}\cdot\boldsymbol{\delta}_i\right] = [\{(\nabla\boldsymbol{v})^\dagger\cdot\boldsymbol{\Delta}\}\cdot\boldsymbol{\delta}_i] \tag{9.2-17}$$

When the quantities dotted into $\boldsymbol{\delta}_i$ are equated, Eq. 9.2-8 results. Equation 9.2-9 may be derived in a similar way.

§9.3 TRANSFORMATION RULES FOR THE STRAIN TENSOR AND ITS TIME DERIVATIVES

It was pointed out in §9.1 that the quantities $\hat{g}_{ij} = (\hat{\boldsymbol{g}}_i\cdot\hat{\boldsymbol{g}}_j)$ contain complete information about the shape of a fluid element. The same can be said about the quantities $\hat{g}^{ij} = (\hat{\boldsymbol{g}}^i\cdot\hat{\boldsymbol{g}}^j)$. This section is devoted to a study of these quantities and their time derivatives. In particular we want to provide the transformation rules that allow us to write the kinematic quantities, defined in the convected coordinate system, in terms of the Cartesian components in a fixed coordinate system.

In Chapter 5, in the discussion of linear viscoelasticity, we introduced infinitesimal strain tensors describing the strain at time t', relative to the fluid configuration at time t. This suggests that we ought to use the quantity $[\hat{g}_{ij}(\boldsymbol{r}, t, t') - \hat{g}_{ij}(\boldsymbol{r}, t, t)]$ as a measure of the relative strain, where $\boldsymbol{r}$, t is the label for a particular fluid particle. This quantity can be expressed in terms of Cartesian components in the fixed frame by using Eq. 9.2-4:

$$\begin{aligned}[\hat{g}_{ij}(\boldsymbol{r}, t, t') - \hat{g}_{ij}(\boldsymbol{r}, t, t)] &= \left(\left(\sum_m \boldsymbol{\delta}_m\Delta_{mi}\right)\cdot\left(\sum_n \boldsymbol{\delta}_n\Delta_{nj}\right)\right) - (\boldsymbol{\delta}_i\cdot\boldsymbol{\delta}_j)\\ &= \sum_n \Delta^\dagger_{in}(\boldsymbol{r}, t, t')\Delta_{nj}(\boldsymbol{r}, t, t') - \delta_{ij}\\ &= \gamma^{[0]}_{ij}(\boldsymbol{r}, t, t')\end{aligned} \tag{9.3-1}$$

These are the Cartesian components of the *relative (finite) strain tensor* $\boldsymbol{\gamma}^{[0]} = \sum_i \sum_j \boldsymbol{\delta}_i \boldsymbol{\delta}_j \gamma_{ij}^{[0]}$ given earlier in Eq. 8.1-8. In a similar way we can use Eq. 9.2-5 to show that

$$\begin{aligned} -[\hat{g}^{ij}(\boldsymbol{r}, t, t') - \hat{g}^{ij}(\boldsymbol{r}, t, t)] &= (\boldsymbol{\delta}_i \cdot \boldsymbol{\delta}_j) - \left(\left(\sum_m \mathrm{E}_{im}\boldsymbol{\delta}_m\right)\cdot\left(\sum_n \mathrm{E}_{jn}\boldsymbol{\delta}_n\right)\right) \\ &= \delta_{ij} - \sum_n \mathrm{E}_{in}(\boldsymbol{r}, t, t')\mathrm{E}_{nj}^{\dagger}(\boldsymbol{r}, t, t') \\ &= \gamma_{[0]ij}(\boldsymbol{r}, t, t') \end{aligned} \tag{9.3-2}$$

These are the Cartesian components of the *relative (finite) strain tensor* $\boldsymbol{\gamma}_{[0]} = \sum_i \sum_j \boldsymbol{\delta}_i \boldsymbol{\delta}_j \gamma_{[0]ij}$ given in Eq. 8.1-9.

Several comments deserve to be made regarding the relative strain tensors:

a. They are both obtainable from the displacement functions (as shown in §8.1).

b. They both reduce to the infinitesimal strain tensor $\boldsymbol{\gamma}$ for small strains (see Example 9.3-1).

c. They both depend on the particle label $\boldsymbol{r}$, t and on the time t'; when $t' = t$, the relative finite strain tensors vanish.

d. They both arise naturally in molecular theories, and both are used in empirical constitutive equations.

The main point of the above discussion is that we have seen how the finite strain tensors introduced in Chapter 8 are related to the quantities $\hat{g}_{ij}$ and $\hat{g}^{ij}$ that arise in the convected-coordinate formulation of constitutive equations.

EXAMPLE 9.3-1 The Relation of the Relative Strain Tensors to the Infinitesimal Strain Tensor

Let $\boldsymbol{u} = \boldsymbol{r}' - \boldsymbol{r}$ be the location of a fluid particle at time t' relative to its location at time t. Show that both $\boldsymbol{\gamma}^{[0]}$ and $\boldsymbol{\gamma}_{[0]}$ become $\boldsymbol{\gamma} = \nabla\boldsymbol{u} + (\nabla\boldsymbol{u})^{\dagger}$ in the limit of very small displacement gradients.

SOLUTION **(a)** First we substitute $x_k' = x_k + u_k$ into the second line of Eq. 9.3-1 to get

$$\begin{aligned} \gamma_{ij}^{[0]} &= \sum_k \left(\delta_{ik} + \frac{\partial u_k}{\partial x_i}\right)\left(\delta_{jk} + \frac{\partial u_k}{\partial x_j}\right) - \delta_{ij} \\ &= \left(\frac{\partial}{\partial x_i} u_j + \frac{\partial}{\partial x_j} u_i\right) + \sum_k \left(\frac{\partial}{\partial x_i} u_k\right)\left(\frac{\partial}{\partial x_j} u_k\right) \end{aligned} \tag{9.3-3}$$

In the limit of very small displacement gradients, the quadratic terms can be omitted so that

$$\boldsymbol{\gamma}^{[0]} \doteq \nabla\boldsymbol{u} + (\nabla\boldsymbol{u})^{\dagger} \tag{9.3-4}$$

This is the infinitesimal strain tensor $\boldsymbol{\gamma}$ used in Chapter 5.

(b) Similarly if we substitute $x_k = x_k' - u_k$ into Eq. 9.3-2 we get

$$\begin{aligned} \gamma_{[0]ij} &= \delta_{ij} - \sum_k \left(\delta_{ik} - \frac{\partial u_i}{\partial x_k'}\right)\left(\delta_{jk} - \frac{\partial u_j}{\partial x_k'}\right) \\ &= \left(\frac{\partial}{\partial x_j'} u_i + \frac{\partial}{\partial x_i'} u_j\right) + \text{quadratic terms} \end{aligned} \tag{9.3-5}$$

But for very small displacement gradients, the quadratic terms can be omitted; also x'_j and x_j will be indistinguishable in the derivatives. Hence,

$$\gamma_{[0]} \doteq \nabla \boldsymbol{u} + (\nabla \boldsymbol{u})^{\dagger} \tag{9.3-6}$$

for small displacement gradients.

In linear viscoelasticity we used not only the infinitesimal strain tensor $\boldsymbol{\gamma}$ but also the rate of strain tensor $\dot{\boldsymbol{\gamma}} = \partial\boldsymbol{\gamma}/\partial t$ and higher time derivatives, $\ddot{\boldsymbol{\gamma}} = \partial\dot{\boldsymbol{\gamma}}/\partial t = \partial^2\boldsymbol{\gamma}/\partial t^2$, etc. It is not surprising, then, that in nonlinear viscoelasticity time derivatives of the relative strain tensors have been defined

$$\boldsymbol{\gamma}^{[n]} = \partial^n \boldsymbol{\gamma}^{[0]}(\boldsymbol{r}, t, t')/\partial t'^n \quad \text{and} \quad \boldsymbol{\gamma}_{[n]} = \partial^n \boldsymbol{\gamma}_{[0]}(\boldsymbol{r}, t, t')/\partial t'^n \tag{9.3-7}$$

By way of illustration we show how to get $\boldsymbol{\gamma}^{[1]}$ and $\boldsymbol{\gamma}^{[2]}$.

To get $\boldsymbol{\gamma}^{[1]}$ we take the time derivative of Eq. 9.3-1 and use Eq. 9.2-8 to evaluate $(\partial/\partial t')\boldsymbol{\Delta}$:

$$\begin{aligned}
\boldsymbol{\gamma}^{[1]}(\boldsymbol{r}, t, t') &= \frac{\partial}{\partial t'}\{\boldsymbol{\Delta}^{\dagger}\cdot\boldsymbol{\Delta} - \boldsymbol{\delta}\} \\
&= \left\{\frac{\partial\boldsymbol{\Delta}^{\dagger}}{\partial t'}\cdot\boldsymbol{\Delta} + \boldsymbol{\Delta}^{\dagger}\cdot\frac{\partial\boldsymbol{\Delta}}{\partial t'}\right\} \\
&= \left\{\{\boldsymbol{\Delta}^{\dagger}\cdot\nabla\boldsymbol{v}\}\cdot\boldsymbol{\Delta} + \boldsymbol{\Delta}^{\dagger}\cdot\{(\nabla\boldsymbol{v})^{\dagger}\cdot\boldsymbol{\Delta}\}\right\} \\
&= \{\boldsymbol{\Delta}^{\dagger}\cdot(\nabla\boldsymbol{v} + (\nabla\boldsymbol{v})^{\dagger})\cdot\boldsymbol{\Delta}\} \\
&= \{\boldsymbol{\Delta}^{\dagger}\cdot\boldsymbol{\gamma}^{(1)}\cdot\boldsymbol{\Delta}\}
\end{aligned} \tag{9.3-8}$$

in which $\boldsymbol{\Delta}$, $\nabla\boldsymbol{v}$, and $\boldsymbol{\gamma}^{(1)}$ are understood to be functions of $\boldsymbol{r}$, t, and t'. The last equality serves to define the tensor $\boldsymbol{\gamma}^{(1)}(\boldsymbol{r}, t, t')$. Note that when $t' = t$, the displacement gradient tensor $\boldsymbol{\Delta}$ becomes the unit tensor and hence $\boldsymbol{\gamma}^{[1]}(\boldsymbol{r}, t, t) = \boldsymbol{\gamma}^{(1)}(\boldsymbol{r}, t)$. The tensor $\boldsymbol{\gamma}^{(1)}(\boldsymbol{r}, t)$ is exactly the same as the rate-of-strain tensor $\dot{\boldsymbol{\gamma}}(\boldsymbol{r}, t) = \nabla\boldsymbol{v} + (\nabla\boldsymbol{v})^{\dagger}$.

To get $\boldsymbol{\gamma}^{[2]}$ we take the time derivative of Eq. 9.3-8:

$$\begin{aligned}
\boldsymbol{\gamma}^{[2]}(\boldsymbol{r}, t, t') &= \frac{\partial}{\partial t'}\{\boldsymbol{\Delta}^{\dagger}\cdot\boldsymbol{\gamma}^{(1)}\cdot\boldsymbol{\Delta}\} \\
&= \left\{\frac{\partial\boldsymbol{\Delta}^{\dagger}}{\partial t'}\cdot\boldsymbol{\gamma}^{(1)}\cdot\boldsymbol{\Delta} + \boldsymbol{\Delta}^{\dagger}\cdot\frac{\partial\boldsymbol{\gamma}^{(1)}}{\partial t'}\cdot\boldsymbol{\Delta} + \boldsymbol{\Delta}^{\dagger}\cdot\boldsymbol{\gamma}^{(1)}\cdot\frac{\partial\boldsymbol{\Delta}}{\partial t'}\right\} \\
&= \left\{\{\boldsymbol{\Delta}^{\dagger}\cdot\nabla\boldsymbol{v}\}\cdot\boldsymbol{\gamma}^{(1)}\cdot\boldsymbol{\Delta} + \boldsymbol{\Delta}^{\dagger}\cdot\frac{\partial\boldsymbol{\gamma}^{(1)}}{\partial t'}\cdot\boldsymbol{\Delta} + \boldsymbol{\Delta}^{\dagger}\cdot\boldsymbol{\gamma}^{(1)}\cdot\{(\nabla\boldsymbol{v})^{\dagger}\cdot\boldsymbol{\Delta}\}\right\} \\
&= \left\{\boldsymbol{\Delta}^{\dagger}\cdot\left\{\frac{\partial\boldsymbol{\gamma}^{(1)}}{\partial t'} + (\nabla\boldsymbol{v})\cdot\boldsymbol{\gamma}^{(1)} + \boldsymbol{\gamma}^{(1)}\cdot(\nabla\boldsymbol{v})^{\dagger}\right\}\cdot\boldsymbol{\Delta}\right\} \\
&= \{\boldsymbol{\Delta}^{\dagger}\cdot\boldsymbol{\gamma}^{(2)}\cdot\boldsymbol{\Delta}\}
\end{aligned} \tag{9.3-9}$$

Here also $\boldsymbol{\Delta}$, $\nabla\boldsymbol{v}$, and $\boldsymbol{\gamma}^{(2)}$ are understood to be functions of $\boldsymbol{r}$, t, t'—that is, they are evaluated at time t' at the fluid particle $\boldsymbol{r}$, t. The tensor $\boldsymbol{\gamma}^{(2)}(\boldsymbol{r}, t\ t')$ is defined by the last equality above. When $t' = t$, the tensor $\boldsymbol{\Delta}$ becomes the unit tensor, and the derivative $(\partial\boldsymbol{\gamma}^{(1)}/\partial t')_{\boldsymbol{r},t}$ following the particle $\boldsymbol{r}, t$ becomes identical to the substantial derivative[1] $D\boldsymbol{\gamma}^{(1)}/Dt$. As a result of Eqs. 9.3-8 and 9 and straightforward extensions we have

$$\boldsymbol{\gamma}^{[1]}(\boldsymbol{r}, t, t) = \boldsymbol{\gamma}^{(1)}(\boldsymbol{r}, t) = \nabla\boldsymbol{v} + (\nabla\boldsymbol{v})^{\dagger} \tag{9.3-10}$$

$$\boldsymbol{\gamma}^{[2]}(\boldsymbol{r}, t, t) = \boldsymbol{\gamma}^{(2)}(\boldsymbol{r}, t) = \frac{D\boldsymbol{\gamma}^{(1)}}{Dt} + \{(\nabla\boldsymbol{v})\cdot\boldsymbol{\gamma}^{(1)} + \boldsymbol{\gamma}^{(1)}\cdot(\nabla\boldsymbol{v})^{\dagger}\} \tag{9.3-11}$$

$$\vdots \qquad\qquad \vdots \qquad\qquad \vdots$$

$$\boldsymbol{\gamma}^{[n]}(\boldsymbol{r}, t, t) = \boldsymbol{\gamma}^{(n)}(\boldsymbol{r}, t) = \frac{D\boldsymbol{\gamma}^{(n-1)}}{Dt} + \{(\nabla\boldsymbol{v})\cdot\boldsymbol{\gamma}^{(n-1)} + \boldsymbol{\gamma}^{(n-1)}\cdot(\nabla\boldsymbol{v})^{\dagger}\} \tag{9.3-12}$$

In this way the higher rate-of-strain tensors are defined. The tensor $\boldsymbol{\gamma}^{(n)}$ is said to be the *covariant convected derivative* of $\boldsymbol{\gamma}^{(n-1)}$; it is also the $(n-1)$th covariant convected derivative of $\boldsymbol{\gamma}^{(1)}$.

The tensors $\boldsymbol{\gamma}_{(n)}$, first introduced in Eqs. 6.1-5 and 6, are obtained by a procedure analogous to that described above. The results may be summarized as follows

$$\boldsymbol{\gamma}_{[1]}(\boldsymbol{r}, t, t) = \boldsymbol{\gamma}_{(1)}(\boldsymbol{r}, t) = \nabla\boldsymbol{v} + (\nabla\boldsymbol{v})^{\dagger} \tag{9.3-13}$$

$$\boldsymbol{\gamma}_{[2]}(\boldsymbol{r}, t, t) = \boldsymbol{\gamma}_{(2)}(\boldsymbol{r}, t) = \frac{D\boldsymbol{\gamma}_{(1)}}{Dt} - \{(\nabla\boldsymbol{v})^{\dagger}\cdot\boldsymbol{\gamma}_{(1)} + \boldsymbol{\gamma}_{(1)}\cdot\nabla\boldsymbol{v}\} \tag{9.3-14}$$

$$\vdots \qquad\qquad \vdots \qquad\qquad \vdots$$

$$\boldsymbol{\gamma}_{[n]}(\boldsymbol{r}, t, t) = \boldsymbol{\gamma}_{(n)}(\boldsymbol{r}, t) = \frac{D\boldsymbol{\gamma}_{(n-1)}}{Dt} - \{(\nabla\boldsymbol{v})^{\dagger}\cdot\boldsymbol{\gamma}_{(n-1)} + \boldsymbol{\gamma}_{(n-1)}\cdot\nabla\boldsymbol{v}\} \tag{9.3-15}$$

The tensor $\boldsymbol{\gamma}_{(n)}$ is the *contravariant convected derivative* of $\boldsymbol{\gamma}_{(n-1)}$ and the $(n-1)$th contravariant convected derivative of $\boldsymbol{\gamma}_{(1)}$.

The interrelation of the tensors $\boldsymbol{\gamma}^{[n]}$ (with $n \geq 0$) and $\boldsymbol{\gamma}^{(n)}$ with $(n \geq 1)$ is summarized in Table 9.3-1, and also the relation of these quantities to $\hat{g}_{ij}$ and its time derivatives. In addition the same information is shown for the tensors $\boldsymbol{\gamma}_{[n]}$ and $\boldsymbol{\gamma}_{(n)}$. The table is useful for showing how the notation for the kinematic tensors is systematized:

a. The superscript [ ] and ( ) quantities arise from the behavior of the dot products of the convected base vectors $\hat{\boldsymbol{g}}_i$ whereas the subscript [ ] and ( ) quantities result from the dot products of the convected reciprocal base vectors $\hat{\boldsymbol{g}}^i$. The subscript quantities seem to

[1] The tensor $\boldsymbol{\gamma}^{(1)}$ can be regarded either as being a function of the fluid particle label $\boldsymbol{r}, t$ and the time t', or else as a function of the spatial coordinates $\boldsymbol{r}'$ and the time t'. Then by the chain rule of partial differention,

$$\frac{\partial}{\partial t'}\boldsymbol{\gamma}^{(1)}(\boldsymbol{r}, t, t') = \frac{\partial}{\partial t'}\boldsymbol{\gamma}^{(1)}(\boldsymbol{r}', t') + \left\{\left(\frac{\partial}{\partial t'}\boldsymbol{r}'\right)_{\boldsymbol{r},t}\cdot\left(\frac{\partial}{\partial \boldsymbol{r}'}\boldsymbol{\gamma}^{(1)}(\boldsymbol{r}', t')\right)\right\} \tag{9.3-9a}$$

But $(\partial\boldsymbol{r}'/\partial t')$ is just the velocity $\boldsymbol{v}'$ at $\boldsymbol{r}', t'$. When $t' = t$, the right side of Eq. 9.3-9a becomes

$$\frac{D}{Dt}\boldsymbol{\gamma}^{(1)}(\boldsymbol{r}, t) = \frac{\partial\boldsymbol{\gamma}^{(1)}}{\partial t} + \{\boldsymbol{v}\cdot\nabla\boldsymbol{\gamma}^{(1)}\} \tag{9.3-9b}$$

which is the substantial derivative of $\boldsymbol{\gamma}^{(1)}$.

TABLE 9.3-1

Kinematic Tensors and Their Relation to One Another[a–f]

		Displacement functions $\boldsymbol{r}'(\boldsymbol{r}, t, t')$	$\overset{\S 8.1}{\longleftrightarrow}$	Velocity field $\boldsymbol{v}(\boldsymbol{r}, t)$		
$+[\hat{g}_{ij}(\hat{x}, t') - \hat{g}_{ij}(\hat{x}, t)]$	$=$	$\gamma_{ij}^{[0]} = +[\Delta_{ik}^{\dagger}\Delta_{kj} - \delta_{ij}]$	$\overset{t'=t}{\longrightarrow}$	0		
$-[\hat{g}^{ij}(\hat{x}, t') - \hat{g}^{ij}(\hat{x}, t)]$	$=$	$\gamma_{[0]ij} = -[\mathrm{E}_{ik}\mathrm{E}_{kj}^{\dagger} - \delta_{ij}]$				
$\downarrow \partial/\partial t'$		$\downarrow \partial/\partial t'$				
$+\dfrac{\partial \hat{g}_{ij}}{\partial t'}$	$=$	$\gamma_{ij}^{[1]} = \sum_k \sum_l \Delta_{ik}^{\dagger}\gamma_{kl}^{(1)}\Delta_{lj}$	$\overset{t'=t}{\longrightarrow}$	$\gamma_{ij}^{(1)} = \dfrac{\partial v_j}{\partial x_i} + \dfrac{\partial v_i}{\partial x_j} \equiv \dot{\gamma}_{ij}$	$=$	$+\dfrac{\partial \hat{g}_{ij}}{\partial t}$
$-\dfrac{\partial \hat{g}^{ij}}{\partial t'}$	$=$	$\gamma_{[1]ij} = \sum_k \sum_l \mathrm{E}_{ik}\gamma_{(1)kl}\mathrm{E}_{lj}^{\dagger}$		$\gamma_{(1)ij} = \dfrac{\partial v_j}{\partial x_i} + \dfrac{\partial v_i}{\partial x_j} \equiv \dot{\gamma}_{ij}$	$=$	$-\dfrac{\partial \hat{g}^{ij}}{\partial t}$
$\downarrow \partial/\partial t'$		$\downarrow \partial/\partial t'$		$\downarrow \mathfrak{d}/\mathfrak{d}t$		$\downarrow \partial/\partial t$
$+\dfrac{\partial^2 \hat{g}_{ij}}{\partial t'^2}$	$=$	$\gamma_{ij}^{[2]} = \sum_k \sum_l \Delta_{ik}^{\dagger}\gamma_{kl}^{(2)}\Delta_{lj}$	$\overset{t'=t}{\longrightarrow}$	$\gamma_{ij}^{(2)} = \dfrac{D}{Dt}\gamma_{ij}^{(1)} + \sum_k \left(\dfrac{\partial v_k}{\partial x_i}\gamma_{kj}^{(1)} + \gamma_{ik}^{(1)}\dfrac{\partial v_k}{\partial x_j}\right)$	$=$	$+\dfrac{\partial^2 \hat{g}_{ij}}{\partial t^2}$
$-\dfrac{\partial^2 \hat{g}^{ij}}{\partial t'^2}$	$=$	$\gamma_{[2]ij} = \sum_k \sum_l \mathrm{E}_{ik}\gamma_{(2)kl}\mathrm{E}_{lj}^{\dagger}$		$\gamma_{(2)ij} = \dfrac{D}{Dt}\gamma_{(1)ij} - \sum_k \left(\dfrac{\partial v_i}{\partial x_k}\gamma_{(1)kj} + \gamma_{(1)ik}\dfrac{\partial v_j}{\partial x_k}\right)$	$=$	$-\dfrac{\partial^2 \hat{g}^{ij}}{\partial t^2}$
$\downarrow \partial/\partial t'$		$\downarrow \partial/\partial t'$		$\downarrow \mathfrak{d}/\mathfrak{d}t$		$\downarrow \partial/\partial t$

[a] This table is based on the discussion of kinematic tensors given by J. G. Oldroyd, *Proc. Roy. Soc.*, **A200**, 45–63 (1950); *J. Non-Newtonian Fluid Mech.*, **14**, 9–46 (1984).

[b] The symbol $\mathfrak{d}/\mathfrak{d}t$ in the table is Oldroyd's symbol for convected differentiation.

$$\boldsymbol{\gamma}^{(2)} = (\boldsymbol{\gamma}^{(1)})^{(1)} = (\mathfrak{d}/\mathfrak{d}t)\boldsymbol{\gamma}^{(1)} \text{ and } \boldsymbol{\gamma}_{(2)} = (\boldsymbol{\gamma}_{(1)})_{(1)} = (\mathfrak{d}/\mathfrak{d}t)\boldsymbol{\gamma}_{(1)}.$$

Since the meaning of the operator $\mathfrak{d}/\mathfrak{d}t$ depends on whether (1) appears as a superscript or a subscript, the notation $\mathfrak{d}/\mathfrak{d}t$ is not used elsewhere in this book. The $\boldsymbol{\gamma}^{(n)}$ are exactly the same as the Rivlin–Ericksen tensors, and twice Oldroyd's $e^{(n)}$; the tensors $\boldsymbol{\gamma}^{(1)}$ and $\boldsymbol{\gamma}^{(2)}$ were obtained earlier for steady-state flows by Y. Dupont, *Bull. Sci. Acad. Belg.*, **17**, 441–459 (1931), Eq. 27.

[c] The first and fourth columns show the kinematic tensors in terms of covariant and contravariant convected components. The second and third columns show the equivalent tensors given as Cartesian components in a fixed coordinate system.

[d] The partial derivative $\partial/\partial t'$ means $(\partial/\partial t')_{\hat{x}}$ in the first column and $(\partial/\partial t')_{r,t}$ in the second; that is, both represent a differentiation with respect to time following a fluid particle. In the fourth column $\partial\hat{g}_{ij}/\partial t$ means $(\partial\hat{g}_{ij}/\partial t')_{\hat{x}}$ evaluated at $t' = t$.

[e] All tensor components in the second column ($\gamma_{ij}^{[0]}$, $\gamma_{ij}^{(1)}$, Δ_{ij}, etc.) are understood to be functions of $\boldsymbol{r}$, t, and t'. All tensor components in the third column are understood to be functions of $\boldsymbol{r}$ and t.

[f] For nonorthogonal coordinate systems, the Δ_{ij} and E_{ij} are replaced by $\Delta_j^i = x'^i{}_{,j}$ and $\mathrm{E}_j^i = x^i{}_{,j'}$, where ",$j$" denotes covariant differentiation with respect to x^j. In Cartesian coordinates these reduce to partial derivatives: $\Delta_{ij} = \partial x_i'/\partial x_j$ and $\mathrm{E}_{ij} = \partial x_i/\partial x_j'$. In generalized coordinates the entries in column two may be calculated by replacing Δ_j^i with $\partial x'^i/\partial x^j$ and E_j^i with $\partial x^i/\partial x'^j$. These latter forms are the most convenient for calculations in geometries where $\boldsymbol{\Delta}$ and $\mathbf{E}$ are not already tabulated in Appendix B.

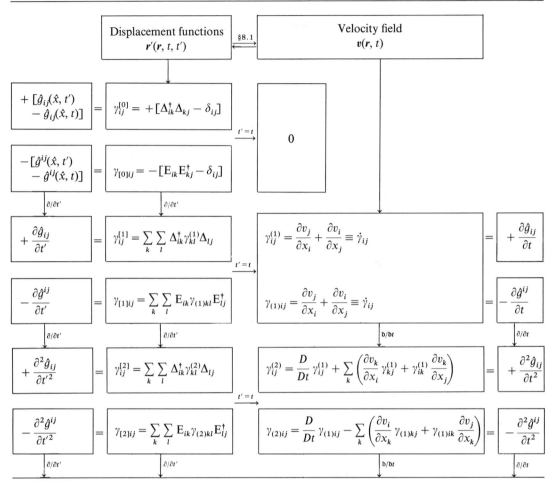

arise more naturally in the molecular theories and have been used in most of the empirical constitutive equations.

b. The [ ] quantities, which depend on t and t', appear in integral constitutive equations, in integrals over t'. The ( ) quantities appear as functions of t in differential constitutive equations.

c. The [ ] quantities are identical to the corresponding ( ) quantities when $t' = t$; there are, however, no tensors $\gamma^{(0)}$ and $\gamma_{(0)}$ since both $\gamma^{[0]}$ and $\gamma_{[0]}$ go to $\mathbf{0}$ when $t' = t$.

d. Higher order [ ] quantities are obtained by successive *partial* differentiation with respect to t' with $\boldsymbol{r}$, t being held constant. Higher order ( ) quantities are obtained by successive *convected* differentiation.

e. The ( ) quantities are easily obtained from the velocity field since the convected differentiation operators involve only $(\boldsymbol{v} \cdot \nabla)$, $\nabla \boldsymbol{v}$, and $(\nabla \boldsymbol{v})^{\dagger}$. The [ ] quantities require a knowledge of the displacement functions, which are needed to get the displacement gradient tensors $\boldsymbol{\Delta}$ and $\mathbf{E}$.

f. In the limit of very small deformations (i.e., linear viscoelasticity), the relative strain tensors, $\gamma^{[0]}$ and $\gamma_{[0]}$, both simplify to the infinitesimal strain tensor γ as pointed out earlier. In addition in this limit the tensors $\gamma^{[n]}$ and $\gamma_{[n]}$ both simplify to $\partial^n \gamma / \partial t'^n$, and similarly the tensors $\gamma^{(n)}$ and $\gamma_{(n)}$ both simplify to $\partial^n \gamma / \partial t^n$.

g. In Appendix C the most important kinematic tensors are tabulated for homogeneous shear flows and shearfree flows, that is, for those flows encountered in material-function measurements. Table C.1 is particularly useful for evaluating constitutive equations. Appendix B contains tables of the components of various kinematic tensors in cylindrical and spherical coordinates.

The systematic notation shown here should make it easy to remember the meaning of the various kinematic quantities as well as their interrelation. In Chapters 6 and 7 we use the nth rate of strain tensors $\gamma_{(n)}$, and in Chapter 8 we use the relative strain tensors $\gamma_{[0]}$ and $\gamma^{[0]}$. In research publications many of the other tensors in this table (e.g., $\gamma^{(n)}$, $\gamma^{[1]}$, $\gamma_{[1]}$) have been used; unfortunately there seems to be no standard systematic set of symbols for these kinematic tensors in the literature.

§9.4 TRANSFORMATION RULES FOR THE STRESS TENSOR AND ITS TIME DERIVATIVES

Next we turn to the stress tensor $\boldsymbol{\tau}$ at time t' evaluated at the fluid particle $\boldsymbol{r}, t$. This tensor may be expanded in terms of unit vectors and Cartesian components as follows:

$$\boldsymbol{\tau}(\boldsymbol{r}, t, t') = \sum_i \sum_j \boldsymbol{\delta}_i \boldsymbol{\delta}_j \tau_{ij}(\boldsymbol{r}, t, t') \tag{9.4-1}$$

It may also be expanded in terms of the base vectors and contravariant components (or alternatively, in terms of the reciprocal base vectors and the covariant components)

$$\boldsymbol{\tau}(\boldsymbol{r}, t, t') = \sum_i \sum_j \hat{\boldsymbol{g}}_i \hat{\boldsymbol{g}}_j \hat{\tau}^{ij}(\boldsymbol{r}, t, t') \tag{9.4-2}$$

$$\boldsymbol{\tau}(\boldsymbol{r}, t, t') = \sum_i \sum_j \hat{\boldsymbol{g}}^i \hat{\boldsymbol{g}}^j \hat{\tau}_{ij}(\boldsymbol{r}, t, t') \tag{9.4-3}$$

where the $\hat{\boldsymbol{g}}_i$ and $\hat{\boldsymbol{g}}^i$ are functions of $\boldsymbol{r}$, t and t'. But the base vectors and the reciprocal base vectors can be written in terms of the unit vectors (as in Eqs. 9.2-4 and 5), and therefore the covariant and contravariant components can be expressed in terms of the Cartesian components as follows:

$$\hat{\tau}_{ij}(\boldsymbol{r}, t, t') = \sum_m \sum_n \tau_{mn}(\boldsymbol{r}, t, t')\Delta_{mi}\Delta_{nj} \equiv \tau_{ij}^{[0]}(\boldsymbol{r}, t, t') \tag{9.4-4}$$

$$\hat{\tau}^{ij}(\boldsymbol{r}, t, t') = \sum_m \sum_n \mathrm{E}_{im}\mathrm{E}_{jn}\tau_{mn}(\boldsymbol{r}, t, t') \equiv \tau_{[0]ij}(\boldsymbol{r}, t, t') \tag{9.4-5}$$

These are the Cartesian components of the tensors $\boldsymbol{\tau}^{[0]} = \sum_i \sum_j \boldsymbol{\delta}_i\boldsymbol{\delta}_j\tau_{ij}^{[0]}$ and $\boldsymbol{\tau}_{[0]} = \sum_i \sum_j \boldsymbol{\delta}_i\boldsymbol{\delta}_j\tau_{[0]ij}$. Note that when $t' = t$, both $\boldsymbol{\tau}^{[0]}$ and $\boldsymbol{\tau}_{[0]}$ become equal to $\boldsymbol{\tau}(\boldsymbol{r}, t)$.

We can now find the derivatives with respect to t', using the procedure of §9.3, to get the quantities $\boldsymbol{\tau}^{[n]} = \partial^n\boldsymbol{\tau}^{[0]}/\partial t'^n$ and $\boldsymbol{\tau}_{[n]} = \partial^n\boldsymbol{\tau}_{[0]}/\partial t'^n$. We illustrate the method by getting $\boldsymbol{\tau}^{[1]}$ and $\boldsymbol{\tau}_{[1]}$:

$$\begin{aligned}\boldsymbol{\tau}^{[1]}(\boldsymbol{r}, t, t') &= \frac{\partial}{\partial t'}\{\boldsymbol{\Delta}^\dagger \cdot \boldsymbol{\tau} \cdot \boldsymbol{\Delta}\} \\ &= \left\{\boldsymbol{\Delta}^\dagger \cdot \left\{\frac{\partial \boldsymbol{\tau}}{\partial t'} + (\nabla \boldsymbol{v}) \cdot \boldsymbol{\tau} + \boldsymbol{\tau} \cdot (\nabla \boldsymbol{v})^\dagger\right\} \cdot \boldsymbol{\Delta}\right\} \\ &\equiv \{\boldsymbol{\Delta}^\dagger \cdot \boldsymbol{\tau}^{(1)} \cdot \boldsymbol{\Delta}\}\end{aligned} \tag{9.4-6}$$

$$\begin{aligned}\boldsymbol{\tau}_{[1]}(\boldsymbol{r}, t, t') &= \frac{\partial}{\partial t'}\{\mathbf{E} \cdot \boldsymbol{\tau} \cdot \mathbf{E}^\dagger\} \\ &= \left\{\mathbf{E} \cdot \left\{\frac{\partial \boldsymbol{\tau}}{\partial t'} - (\nabla \boldsymbol{v})^\dagger \cdot \boldsymbol{\tau} - \boldsymbol{\tau} \cdot (\nabla \boldsymbol{v})\right\} \cdot \mathbf{E}^\dagger\right\} \\ &\equiv \{\mathbf{E} \cdot \boldsymbol{\tau}_{(1)} \cdot \mathbf{E}^\dagger\}\end{aligned} \tag{9.4-7}$$

In these relations it is understood that $\boldsymbol{\Delta}$, $\mathbf{E}$, $\nabla \boldsymbol{v}$, $\boldsymbol{\tau}^{(1)}$, and $\boldsymbol{\tau}_{(1)}$ are all functions of t' and the particle label $\boldsymbol{r}$, t. When t' becomes equal to t we get

$$\boldsymbol{\tau}^{[1]}(\boldsymbol{r}, t, t) = \boldsymbol{\tau}^{(1)}(\boldsymbol{r}, t) = \frac{D\boldsymbol{\tau}}{Dt} + \{(\nabla \boldsymbol{v}) \cdot \boldsymbol{\tau} + \boldsymbol{\tau} \cdot (\nabla \boldsymbol{v})^\dagger\} \tag{9.4-8}$$

$$\boldsymbol{\tau}_{[1]}(\boldsymbol{r}, t, t) = \boldsymbol{\tau}_{(1)}(\boldsymbol{r}, t) = \frac{D\boldsymbol{\tau}}{Dt} - \{(\nabla \boldsymbol{v})^\dagger \cdot \boldsymbol{\tau} + \boldsymbol{\tau} \cdot (\nabla \boldsymbol{v})\} \tag{9.4-9}$$

The quantities $\boldsymbol{\tau}^{(1)}$ and $\boldsymbol{\tau}_{(1)}$ are the *covariant and contravariant convected derivatives* of the stress tensor $\boldsymbol{\tau}$. The convected derivative $\boldsymbol{\tau}_{(1)}$ has already been encountered in Eq. 7.1-1. In Table 9.4-1 we summarize the notation developed in this section and show how the various quantities are interrelated. We also give the equivalent quantities in the convected-component notation. In the table we also introduce the symbols $\boldsymbol{\tau}^{(0)}$ and $\boldsymbol{\tau}_{(0)}$ (both identical to $\boldsymbol{\tau}$) just to preserve the notational rule that [ ]-suffixed quantities become ( )-suffixed quantities when t' goes to t.

A few words are in order regarding the operation of *convected differentiation* used in both §9.3 and §9.4.

TABLE 9.4-1

The Stress Tensor and Its Time Derivatives[a–f]

$\hat{\tau}_{ij}(\hat{x}, t')$	=	$\tau_{ij}^{[0]} = \sum_k \sum_l \Delta_{ik}^{\dagger} \tau_{kl}^{(0)} \Delta_{lj}$	$\xrightarrow{t'=t}$	$\tau_{ij}^{(0)} \equiv \tau_{ij}$	=	$\hat{\tau}_{ij}(\hat{x}, t)$
$\hat{\tau}^{ij}(\hat{x}, t')$	=	$\tau_{[0]ij} = \sum_k \sum_l \mathrm{E}_{ik} \tau_{(0)kl} \mathrm{E}_{lj}^{\dagger}$		$\tau_{(0)ij} \equiv \tau_{ij}$	=	$\hat{\tau}^{ij}(\hat{x}, t)$
$\downarrow \partial/\partial t'$		$\downarrow \partial/\partial t'$		$\downarrow \mathfrak{D}/\mathfrak{D}t$		$\downarrow \partial/\partial t$
$\dfrac{\partial \hat{\tau}_{ij}}{\partial t'}$	=	$\tau_{ij}^{[1]} = \sum_k \sum_l \Delta_{ik}^{\dagger} \tau_{kl}^{(1)} \Delta_{lj}$	$\xrightarrow{t'=t}$	$\tau_{ij}^{(1)} = \dfrac{D\tau_{ij}}{Dt} + \sum_k \left(\dfrac{\partial v_k}{\partial x_i} \tau_{kj} + \tau_{ik} \dfrac{\partial v_k}{\partial x_j} \right)$	=	$\dfrac{\partial \hat{\tau}_{ij}}{\partial t}$
$\dfrac{\partial \hat{\tau}^{ij}}{\partial t'}$	=	$\tau_{[1]ij} = \sum_k \sum_l \mathrm{E}_{ik} \tau_{(1)kl} \mathrm{E}_{lj}^{\dagger}$		$\tau_{(1)ij} = \dfrac{D\tau_{ij}}{Dt} - \sum_k \left(\dfrac{\partial v_i}{\partial x_k} \tau_{kj} + \tau_{ik} \dfrac{\partial v_j}{\partial x_k} \right)$	=	$\dfrac{\partial \hat{\tau}^{ij}}{\partial t}$
$\downarrow \partial/\partial t'$		$\downarrow \partial/\partial t'$		$\downarrow \mathfrak{D}/\mathfrak{D}t$		$\partial/\partial t$

[a] This table is based on J. G. Oldroyd, *Proc. Roy. Soc.*, **A200**, 45–63 (1950); *J. Non-Newtonian Fluid Mech.*, **14**, 9–46 (1984).
[b] The symbol $\mathfrak{D}/\mathfrak{D}t$ is Oldroyd's convected derivative operator: $\boldsymbol{\tau}^{(2)} = (\boldsymbol{\tau}^{(1)})^{(1)} = (\mathfrak{D}/\mathfrak{D}t)\boldsymbol{\tau}^{(1)}$ and $\boldsymbol{\tau}_{(2)} = (\boldsymbol{\tau}_{(1)})_{(1)} = (\mathfrak{D}/\mathfrak{D}t)\boldsymbol{\tau}_{(1)}$. Since the meaning of the operator $\mathfrak{D}/\mathfrak{D}t$ depends on whether (1) appears as a superscript or a subscript, we do not use it elsewhere in this book.
[c] The first and fourth columns are written in terms of convected components. The second and third columns give the tensors in fixed Cartesian components.
[d] The derivative $\partial/\partial t'$ means $(\partial/\partial t')_{\hat{x}}$ in the first column and $(\partial/\partial t')_{r,t}$ in the second.
[e] All tensor components in the second column are understood to be functions of $\boldsymbol{r}$, t, and t'; in the third column they are functions of $\boldsymbol{r}$ and t.
[f] See Note f of Table 9.3-1.

1. Higher convected derivatives can be obtained by successive application of the convected derivative operator

$$\boldsymbol{\Lambda}^{(n+1)} = (\boldsymbol{\Lambda}^{(n)})^{(1)} \qquad (\boldsymbol{\Lambda} = \text{any second-order tensor}) \tag{9.4-10}$$

2. The convected derivatives of the unit tensor are

$$\boldsymbol{\delta}^{(1)} = \boldsymbol{\gamma}^{(1)}; \quad \boldsymbol{\delta}_{(1)} = -\boldsymbol{\gamma}_{(1)} \tag{9.4-11}$$

It must be emphasized that $\boldsymbol{\gamma}^{(1)}$ and $\boldsymbol{\gamma}_{(1)}$ are *not* the convected derivatives of the infinitesimal strain tensor $\boldsymbol{\gamma}$; see also Eq. 9B.5-5.

3. The convected derivative of the product of a scalar function $a(\boldsymbol{r}, t)$ and a tensor $\boldsymbol{\Lambda}(\boldsymbol{r}, t)$ is

$$(a\boldsymbol{\Lambda})_{(1)} = a\boldsymbol{\Lambda}_{(1)} + (Da/Dt)\boldsymbol{\Lambda} \tag{9.4-12}$$

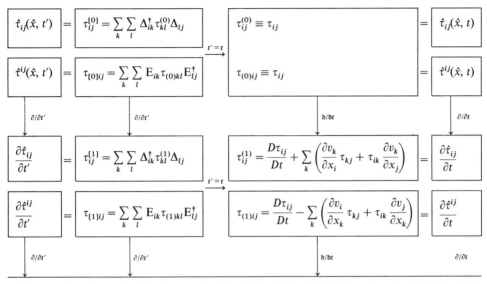

4. The convected derivative of the tensor product of two tensors, $\mathbf{\Lambda}$ and $\mathbf{\Theta}$, *does not* obey the simple rule for differentiating a product used in ordinary differential calculus; instead it may be shown that

$$\{\mathbf{\Lambda}\cdot\mathbf{\Theta}\}_{(1)} = \{\mathbf{\Lambda}_{(1)}\cdot\mathbf{\Theta}\} + \{\mathbf{\Lambda}\cdot\mathbf{\Theta}_{(1)}\} + \{\mathbf{\Lambda}\cdot\dot{\boldsymbol{\gamma}}\cdot\mathbf{\Theta}\} \tag{9.4-13a}$$

$$\{\mathbf{\Lambda}\cdot\mathbf{\Theta}\}^{(1)} = \{\mathbf{\Lambda}^{(1)}\cdot\mathbf{\Theta}\} + \{\mathbf{\Lambda}\cdot\mathbf{\Theta}^{(1)}\} - \{\mathbf{\Lambda}\cdot\dot{\boldsymbol{\gamma}}\cdot\mathbf{\Theta}\} \tag{9.4-13b}$$

5. The Jaumann (or corotational derivative) is simply related to the convected derivatives as follows:

$$\frac{\mathscr{D}}{\mathscr{D}t}\mathbf{\Lambda} = \frac{1}{2}(\mathbf{\Lambda}_{(1)} + \mathbf{\Lambda}^{(1)}) \tag{9.4-14}$$

This derivative *does* obey the usual differential-calculus rule for the differentiation of products:

$$\frac{\mathscr{D}}{\mathscr{D}t}\{\mathbf{\Lambda}\cdot\mathbf{\Theta}\} = \left\{\frac{\mathscr{D}\mathbf{\Lambda}}{\mathscr{D}t}\cdot\mathbf{\Theta}\right\} + \left\{\mathbf{\Lambda}\cdot\frac{\mathscr{D}\mathbf{\Theta}}{\mathscr{D}t}\right\} \tag{9.4-15}$$

The Jaumann derivative gives the time rate of change in a coordinate system that goes along with a fluid particle and rotates with the instantaneous fluid angular velocity. This derivative can be used for constructing constitutive equations. All of the models discussed in Chapters 6 and 7 can be written in terms of the Jaumann derivatives $\mathscr{D}^n\dot{\boldsymbol{\gamma}}/\mathscr{D}t^n$ and $\mathscr{D}\boldsymbol{\tau}/\mathscr{D}t$ just as conveniently[1] as in terms of the quantities $\boldsymbol{\gamma}_{(n)}$ and $\boldsymbol{\tau}_{(1)}$ (see Problem 9B.9).

EXAMPLE 9.4-1 The Integral Form of the Convected Jeffreys Model (Oldroyd-B Model)

It was pointed out in Chapter 8 that the convected Jeffreys model, introduced in Chapter 7, can be put into integral form. We are now in a position to show how this is done, using Tables 9.3-1 and 9.4-1. First rewrite the differential form of the model (see Eq. 7.2-1) for some past time t' and then integrate the resulting equation.

SOLUTION The convected Jeffreys model in differential form is

$$\boldsymbol{\tau}_{(0)} + \lambda_1\boldsymbol{\tau}_{(1)} = -\eta_0[\boldsymbol{\gamma}_{(1)} + \lambda_2\boldsymbol{\gamma}_{(2)}] \tag{9.4-16}$$

where all tensor quantities are evaluated at the present time t. The equation may also be written for the fluid particle $\mathbf{r}, t$ at some past time t' thus,

$$\boldsymbol{\tau}_{[0]} + \lambda_1\boldsymbol{\tau}_{[1]} = -\eta_0[\boldsymbol{\gamma}_{[1]} + \lambda_2\boldsymbol{\gamma}_{[2]}] \tag{9.4-17}$$

But, according to the table this is the same as

$$\boldsymbol{\tau}_{[0]} + \lambda_1\frac{\partial}{\partial t'}\boldsymbol{\tau}_{[0]} = -\eta_0\left[\frac{\partial}{\partial t'}\boldsymbol{\gamma}_{[0]} + \lambda_2\frac{\partial^2}{\partial t'^2}\boldsymbol{\gamma}_{[0]}\right] \tag{9.4-18}$$

[1] In Chapters 7 and 8 of the 1977 edition of this book we based our presentation on the corotating reference frame. See also J. G. Oldroyd, *Proc. Roy. Soc.*, **A245**, 278–297 (1958); J. D. Goddard and C. Miller, *Rheol. Acta*, **5**, 177–184 (1966); J. D. Goddard, *Trans. Soc. Rheol.*, **11**, 381–399 (1967).

This is a first-order differential equation for $\tau_{[0]}$ as a function of t'; it may be integrated to give

$$\tau_{[0]}(\boldsymbol{r}, t, t') = e^{-t'/\lambda_1}\left[\int_{-\infty}^{t'} -\frac{\eta_0}{\lambda_1}\left(\frac{\partial}{\partial t''}\gamma_{0]} + \lambda_2 \frac{\partial^2}{\partial t''^2}\gamma_{[0]}\right)e^{t''/\lambda_1}\,dt'' + \mathbf{K}\right] \tag{9.4-19}$$

where $\gamma_{[0]}$ is now a function of $\boldsymbol{r}$, t, and t'', and $\mathbf{K}$ is an arbitrary tensor function of t. If we require that the stresses in the fluid be finite at $t' = -\infty$, then $\mathbf{K} = \mathbf{0}$. Subsequently integration by parts gives

$$\tau_{[0]}(\boldsymbol{r}, t, t') = +\int_{-\infty}^{t'}\left\{\frac{\eta_0}{\lambda_1^2}\left[\left(1 - \frac{\lambda_2}{\lambda_1}\right)e^{-(t'-t'')/\lambda_1} + 2\lambda_1\lambda_2\frac{\partial}{\partial t''}\delta(t'' - t')\right]\right\}\gamma_{[0]}(\boldsymbol{r}, t, t'')dt'' + \left(\frac{\eta_0}{\lambda_1^2}\right)(\lambda_2 - \lambda_1)\gamma_{[0]}(\boldsymbol{r}, t, \tag{9.4-20}$$

This gives the stress at particle $\boldsymbol{r}$, t at some past time t' in terms of the deformation history for all times $t'' \leq t'$. If now we let t' become equal to t, we get

$$\tau_{(0)}(\boldsymbol{r}, t) = +\int_{-\infty}^{t}\left\{\frac{\eta_0}{\lambda_1^2}\left[\left(1 - \frac{\lambda_2}{\lambda_1}\right)e^{-(t-t')/\lambda_1} + 2\lambda_1\lambda_2\frac{\partial}{\partial t'}\delta(t' - t)\right]\right\}\gamma_{[0]}(\boldsymbol{r}, t, t')dt' \tag{9.4-21}$$

where the dummy variable of integration has been changed from t'' to t'. Equation 9.4-21 is the same as the integral form of the convected Jeffreys model given by Eqs. 8.2-1 and 3.

§9.5 CONSTRUCTION OF ADMISSIBLE CONSTITUTIVE EQUATIONS IN TERMS OF FIXED COMPONENTS

In §9.1 Oldroyd's criteria for admissibility for constitutive equations were discussed in terms of the convected components of the stress tensor and the metric coefficients of the convected coordinate system. In the intervening sections we showed how to relate convected coordinates to fixed coordinates, and also how to transform convected components to fixed components (including time derivatives and quantities appearing in time integrals). We are now in a position to construct constitutive equations in terms of the tensors given in Tables 9.3-1 and 9.4-1, these equations being admissible in the sense defined in §9.1.

Admissible equations may be created by writing down an algebraic relation among the ( ) quantities and single or multiple time integrals of algebraic relations of the [ ] quantities which may also involve time differences, such as $t - t'$. The unit tensor $\boldsymbol{\delta}$ may also be used. In combining these ( ) and [ ] quantities dot products and multiple dot products may be formed, but the final equation must be such that each term is a second-order tensor. Material constants (or even material constant tensors, whose components are $\kappa_{N\,kl\ldots}^{\;ij\ldots}(\boldsymbol{r}, t)$, whose convected time derivatives are zero) may appear in the constitutive equation; the temperature history may be included if it is desired to account for nonisothermal effects. No operations may be performed that would violate the admissibility criteria laid down in §9.1; in case of doubt one can transform the proposed constitutive equation back into the convected coordinate form. Inclusion of tensor constants would lead to constitutive equations for materials such as liquid crystals. Since we are interested only in polymers that do not form liquid crystals we include only scalar material constants in constitutive equations in this book.

Constitutive equations have been generated in a number of ways:

1. One can replace the tensors occurring in linear viscoelasticity by the appropriate [ ] and ( ) quantities defined in this chapter to generate "quasi-linear constitutive

equations." There is, however, no unique way to do this. For example, the linear Jeffreys model of Eq. 5.2-9 could be transformed into a nonlinear viscoelastic equation in a number of ways; two possibilities are:

Oldroyd fluid A:
$$\boldsymbol{\tau} + \lambda_1 \boldsymbol{\tau}^{(1)} = -\eta_0(\boldsymbol{\gamma}^{(1)} + \lambda_2 \boldsymbol{\gamma}^{(2)}) \tag{9.5-1}$$

Oldroyd fluid B:
$$\boldsymbol{\tau} + \lambda_1 \boldsymbol{\tau}_{(1)} = -\eta_0(\boldsymbol{\gamma}_{(1)} + \lambda_2 \boldsymbol{\gamma}_{(2)}) \tag{9.5-2}$$

Only comparison with experimental data or molecular theories can enable one to choose between these two models. Neither model gives a non-Newtonian viscosity or shear-rate dependent normal stress coefficients in steady-state shear flow. Equation 9.5-1 gives $\Psi_2 = -\Psi_1$, whereas Eq. 9.5-2 gives $\Psi_2 = 0$; the latter is somewhat closer to the experimental facts, and therefore Eq. 9.5-2 is preferred.

2. One can put together empirical expressions, using the [ ] and ( ) quantities, in a wide variety of combinations, linear or nonlinear, differential or integral. The models of Oldroyd and Giesekus (Chapter 7) and those of K-BKZ and Rivlin and Sawyers (Chapter 8) are examples of this. Here one proceeds on a try-it-and-see basis until combinations are obtained that are capable of describing experimental data on material functions to the degree desired.

3. One can proceed in a mathematical fashion and make various kinds of ordered expansions. One such attempt leads to the "retarded-motion expansion" discussed in Chapter 6.

4. One can attempt to see how the ordered expansions in (3) simplify for various special categories of flows. An example of this is the CEF equation, for steady shear flows, discussed later in this chapter.

5. One can make use of molecular theories; these theories, of course, involve making some kind of model for the macromolecules in the fluid and some kind of assumptions as to how these molecules interact with one another. If the molecular modeling is sufficiently simple, a complete constitutive equation can be obtained. However, because of the crudeness of the modeling, the final constitutive equation may not be very realistic. If more faithful modeling is done at the molecular level, then it may not even be possible to work all the way through to a constitutive equation, or if it is possible, the resulting constitutive equation may be too complicated for general use in hydrodynamic calculations. In any case, the molecular theories have provided some very useful ideas as to what kinds of terms ought to be included in constitutive equations. Molecular theories are discussed extensively in Volume 2.

§9.6 MEMORY-INTEGRAL EXPANSIONS[1,2]

In Chapters 4 through 8 a number of apparently unrelated constitutive equations were proposed and studied. Some assistance in interrelating these models and understanding their limits of applicability can be obtained by looking at a much more general

[1] The presentation in this section has been strongly influenced by K. Walters, *Z. Angew. Math. Phys.*, **21**, 592–600 (1970), and R. S. Rivlin and K. N. Sawyers, *Ann. Rev. Fluid Mech.*, **3**, 117–146 ((1971).

[2] The basis for this section is the fundamental development on materials with memory by A. E. Green and R. S. Rivlin, *Arch. Rat. Mech. Anal.*, **1**, 1–21 (1957), with further developments by A. E. Green, R. S. Rivlin, and A. J. M. Spencer, *Arch. Rat. Mech. Anal.*, **3**, 82–92 (1959); A. E. Green and R. S. Rivlin, *Arch. Rat. Mech. Anal.*, **4**, 387–404 (1960); B. D. Coleman and W. Noll. *Rev. Mod. Phys.*, **33**, 239–249 (1961), errata: *ibid.*, **36**, 1103 (1964); A. C. Pipkin, *Rev. Mod. Phys.*, **36**, 1034–1041 (1964).

constitutive equation, namely the "memory-integral expansion." We postulate that the stress tensor $\boldsymbol{\tau}(\boldsymbol{r}, t)$ is a functional of the strain history $\boldsymbol{\gamma}_{[0]}(\boldsymbol{r}, t, t')$, where $-\infty < t' \leq t$ (this is less general than Oldroyd's constitutive equation in Eqs. 9.1-6 to 8). Then the functional is expanded a Fréchet series (roughly speaking, a Fréchet series is to a functional as a Taylor series is to a function). This gives the stress tensor as a series of integrals of ever-increasing dimensionality:

$$\begin{aligned}\boldsymbol{\tau}(\boldsymbol{r}, t) = &\int_{-\infty}^{t} M_I(t-t')\boldsymbol{\gamma}'_{[0]}\,dt' \\ &+ \int_{-\infty}^{t}\int_{-\infty}^{t} M_{II}(t-t', t-t'')\{\boldsymbol{\gamma}'_{[0]}\cdot\boldsymbol{\gamma}''_{[0]} + \boldsymbol{\gamma}''_{[0]}\cdot\boldsymbol{\gamma}'_{[0]}\}dt''\,dt' \\ &+ \int_{-\infty}^{t}\int_{-\infty}^{t}\int_{-\infty}^{t} [M_{III}(t-t', t-t'', t-t''')\boldsymbol{\gamma}'_{[0]}(\boldsymbol{\gamma}''_{[0]}:\boldsymbol{\gamma}'''_{[0]}) \\ &+ M_{IV}(t-t', t-t'', t-t''')\{\boldsymbol{\gamma}'_{[0]}\cdot\boldsymbol{\gamma}''_{[0]}\cdot\boldsymbol{\gamma}'''_{[0]} + \boldsymbol{\gamma}'''_{[0]}\cdot\boldsymbol{\gamma}''_{[0]}\cdot\boldsymbol{\gamma}'_{[0]}\}]dt'''\,dt''\,dt' + \cdots\end{aligned} \qquad (9.6\text{-}1)$$

in which $\boldsymbol{\gamma}'_{[0]} \equiv \boldsymbol{\gamma}_{[0]}(\boldsymbol{r}, t, t')$. The expansion is given here in terms of the relative strain tensor $\boldsymbol{\gamma}_{[0]}$, but it could just as well have been given in terms of the other relative strain tensor $\boldsymbol{\gamma}^{[0]}$; or still other kinematic tensors could have been used.[1,3]

A discussion of the convergence of the series in Eq. 9.6-1, the usefulness of a truncated series, and other mathematical questions are outside the scope of this text, inasmuch as we do not use Eq. 9.6-1 for solving fluid dynamics problems. Our main interest in the memory-integral expansion is as a vehicle for explaining how various constitutive equations are related. We do this by means of Fig. 9.6-1 and the illustrative examples that follow.

In Fig. 9.6-1 we see that the memory-integral expansion contains four important constitutive equations, which in turn include other less general equations:

a. If only the first term is retained we get the *Lodge rubberlike liquid*, which has been studied in Chapter 8. This constitutive equation includes all of linear viscoelasticity but is not capable of describing the shear-rate dependence of the viscometric functions, and it gives $\Psi_2 = 0$.

b. If we specialize to a fluid in which the effects on $\boldsymbol{\tau}$ at time t of the deformations experienced by a fluid element at different past times t' and t'' are independent of one another, then the *Rivlin–Sawyers fluid* is obtained. Several widely studied integral constitutive equations (including the Lodge rubberlike liquid) fall into this category. The Rivlin-Sawyers fluid has the capability of giving shear-rate-dependent viscometric functions and also $\Psi_2 \neq 0$.

c. If in the memory-integral expansion we expand $\boldsymbol{\gamma}_{[0]}(\boldsymbol{r}, t, t')$ about $t' = t$, then the *retarded-motion expansion* is obtained. Example 9.6-1 shows how the retarded-motion expansion systematically builds up a general description of the stress tensor for flows with small, slowly changing velocity gradients.

[3] A. C. Pipkin, *op cit.*, has shown that a memory-integral expansion equivalent to Eq. 9.6-1 may be obtained by replacing $\boldsymbol{\gamma}_{[0]}$ everywhere by $\boldsymbol{\gamma}_{[1]}$. J. D. Goddard, in *Trans. Soc. Rheol.*, **11**, 381–399 (1967), gave a memory-integral expansion in terms of a corotational rate-of-strain tensor; this expansion was further explored by R. B. Bird, O. Hassager, and S. I. Abdel-Khalik, *AIChE J.*, **20**, 1041–1066 (1975). See also Chapters 8 and 9 of the first edition of this book.

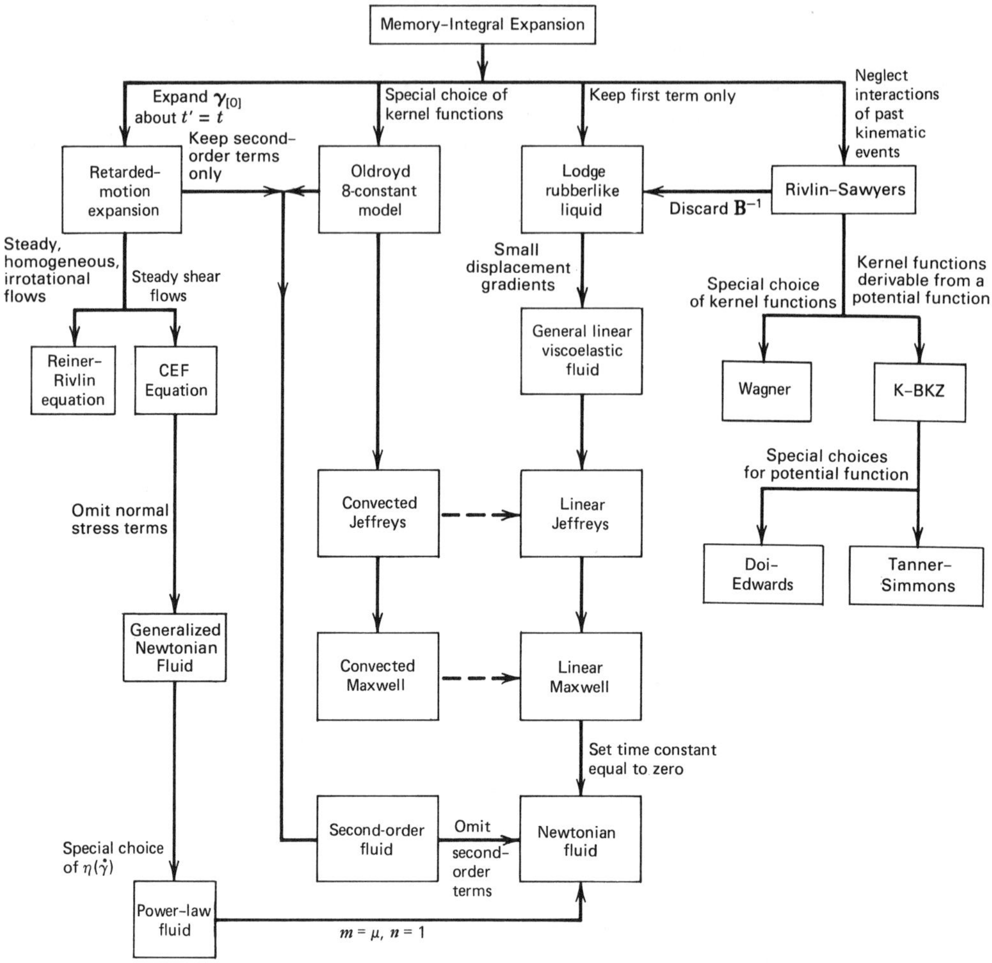

FIGURE 9.6-1. Relations among constitutive equations.

d. For certain classes of flows the memory-integral expansion can be collapsed into rather compact form. For example, for steady shear flows the expansion reduces to the *Criminale–Ericksen–Filbey* (*CEF*) *equation*; for steady, homogeneous, irrotational flows, the *Reiner–Rivlin* equation is obtained. These results are particularly convenient, since we have relatively simple constitutive equations that are exact for flows within these two categories; they are useful in analyzing the flows in rheometric devices (see Chapter 10).

e. Finally, by making very special choices for the kernels M_I, M_{II}, etc., in the memory-integral expansion, one can in principle generate constitutive equations with small numbers of constants. In Example 9.6-2 we show this by taking a differential model and showing how to put it into the form of a memory-integral expansion; in this way we find what M_I, M_{II}, etc. must be.

In addition, constitutive equations generated by molecular theories have been found to fit into the form of the memory-integral expansion or one of the special constitutive equations derived from it. This point is pursued much further in Volume 2.

EXAMPLE 9.6-1 Relation of the Memory-Integral Expansion to the Retarded-Motion Expansion[4]

Expand $\boldsymbol{\gamma}_{[0]}(\boldsymbol{r}, t, t')$ in the memory-integral expansion in a Taylor series about $t' = t$ as is done in Eq. 9B.2-2. Show that this leads to the retarded-motion expansion in Eq 6.2-1.

SOLUTION Let us write the memory-integral expansion as $\boldsymbol{\tau} = \sum_n \boldsymbol{\tau}_n$, where $\boldsymbol{\tau}_n$ is the term containing an n-fold integral. When the expansion for $\boldsymbol{\gamma}_{[0]}$ is substituted into $\boldsymbol{\tau}_1$ and the variable of integration is changed from t' to $s = t - t'$, we get

$$\boldsymbol{\tau}_1 = \sum_{p=1}^{\infty} (p!)^{-1}(-1)^p \boldsymbol{\gamma}_{(p)}(\boldsymbol{r}, t) \int_0^\infty M_I(s) s^p \, ds \tag{9.6-2}$$

Similarly, for $\boldsymbol{\tau}_2$, with $s' = t - t''$, we have

$$\begin{aligned} \boldsymbol{\tau}_2 &= \sum_{p=1}^{\infty} \sum_{q=1}^{\infty} (p!q!)^{-1}(-1)^{p+q}\{\boldsymbol{\gamma}_{(p)} \cdot \boldsymbol{\gamma}_{(q)}\} \int_0^\infty \int_0^\infty M_{II}(s, s')(s^p s'^q + s^q s'^p) ds\, ds' \\ &= 2\{\boldsymbol{\gamma}_{(1)} \cdot \boldsymbol{\gamma}_{(1)}\} \int_0^\infty \int_0^\infty M_{II}(s, s') s s' \, ds'\, ds \\ &\quad - \tfrac{1}{2}\{\boldsymbol{\gamma}_{(1)} \cdot \boldsymbol{\gamma}_{(2)} + \boldsymbol{\gamma}_{(2)} \cdot \boldsymbol{\gamma}_{(1)}\} \int_0^\infty \int_0^\infty M_{II}(s, s')(ss'^2 + s^2 s') ds'\, ds + \cdots \end{aligned} \tag{9.6-3}$$

where we have displayed explicitly terms through third order in velocity gradients. Then for $\boldsymbol{\tau}_3$ we can write down a triple sum, which starts out as follows (with $s'' = t - t'''$)

$$\begin{aligned} \boldsymbol{\tau}_3 &= -(\boldsymbol{\gamma}_{(1)} : \boldsymbol{\gamma}_{(1)})\boldsymbol{\gamma}_{(1)} \int_0^\infty \int_0^\infty \int_0^\infty M_{III}(s, s', s'') s s' s'' \, ds''\, ds'\, ds \\ &\quad - 2\{\boldsymbol{\gamma}_{(1)} \cdot \boldsymbol{\gamma}_{(1)} \cdot \boldsymbol{\gamma}_{(1)}\} \int_0^\infty \int_0^\infty \int_0^\infty M_{IV}(s, s', s'') s s' s'' \, ds''\, ds'\, ds + \cdots \end{aligned} \tag{9.6-4}$$

The Cayley–Hamilton theorem (Eq. A.3-28) enables us to replace $\boldsymbol{\gamma}_{(1)}^3$ in the second term of Eq. 9.6-4 by $\frac{1}{2}(\boldsymbol{\gamma}_{(1)} : \boldsymbol{\gamma}_{(1)})\boldsymbol{\gamma}_{(1)}$; in so doing we make use of the assumption of incompressibility ($\text{tr}\, \boldsymbol{\gamma}_{(1)} = 0$) and an isotropic term (a scalar multiplied by $\boldsymbol{\delta}$) is discarded.

We can now compare the terms generated by the above procedure with those in the retarded-motion expansion in Eq. 6.2-1. This gives expressions for the constants in the expansion:

$$b_p = (p!)^{-1}(-1)^{p+1} \int_0^\infty M_I(s) s^p \, ds, \qquad p = 1, 2, 3, \ldots \tag{9.6-5}$$

$$b_{11} = -2 \int_0^\infty \int_0^\infty M_{II}(s, s') s s' \, ds'\, ds \tag{9.6-6}$$

$$b_{12} = \tfrac{1}{2} \int_0^\infty \int_0^\infty M_{II}(s, s')(ss'^2 + s^2 s') ds'\, ds \tag{9.6-7}$$

$$b_{1:11} = \int_0^\infty \int_0^\infty \int_0^\infty [M_{III}(s, s', s'') + M_{IV}(s, s', s'')] s s' s'' ds''\, ds'\, ds \tag{9.6-8}$$

$$\vdots$$

[4] This example is taken from K. Walters, *op cit*.

Higher-order coefficients may be obtained by continuing the process. Since $M_I(s)$ is positive for all s, it is clear that the b_p alternate in sign (see Tables 6.2-1 and 6.2-2).

EXAMPLE 9.6-2 Development of the Oldroyd 6-Constant Model in a Memory-Integral Expansion[4]

Follow the procedure of Example 9.4-1 to put the Oldroyd 6-constant model (see Eq. 7.3-2 with $\lambda_6 = \lambda_7 = 0$) into integral form.

SOLUTION First we rewrite the Oldroyd model at a past time t' thus,

$$\underline{\boldsymbol{\tau}'_{[0]} + \lambda_1 \frac{\partial}{\partial t'} \boldsymbol{\tau}'_{[0]}} + \tfrac{1}{2}\lambda_3\{\boldsymbol{\tau}'_{[0]} \cdot \boldsymbol{\gamma}'_{[1]} + \boldsymbol{\gamma}'_{[1]} \cdot \boldsymbol{\tau}'_{[0]}\} + \tfrac{1}{2}\lambda_5(\text{tr}\,\boldsymbol{\tau}'_{[0]})\boldsymbol{\gamma}'_{[1]}$$

$$= \underline{-\eta_0(\boldsymbol{\gamma}'_{[1]} + \lambda_2 \frac{\partial}{\partial t'} \boldsymbol{\gamma}'_{[1]}} + \lambda_4\{\boldsymbol{\gamma}'_{[1]} \cdot \boldsymbol{\gamma}'_{[1]}\}) \tag{9.6-9}$$

The single accent is a reminder that these quantities are functions of the particle label $\boldsymbol{r}$, t and the past time t'. The dashed-underlined terms are those linear in stresses and velocity gradients. We start by considering the differential equation formed by these terms only; this is a first-order differential equation in t' and can be integrated by the method of Example 9.4-1 to give

$$\boldsymbol{\tau}'_{[0]} = -\frac{\eta_0 \lambda_2}{\lambda_1} \boldsymbol{\gamma}'_{[1]} - \frac{\eta_0(\lambda_1 - \lambda_2)}{\lambda_1^2} \int_{-\infty}^{t'} e^{-(t'-t'')/\lambda_1} \boldsymbol{\gamma}''_{[1]} dt'' \tag{9.6-10}$$

where $\boldsymbol{\gamma}''_{[1]}$ is a function of $\boldsymbol{r}$, t, and t''.

Next this solution is inserted for $\boldsymbol{\tau}'_{[0]}$ in the nonunderlined (i.e., nonlinear) terms in Eq. 9.6-9. This gives a more complicated first-order differential equation for $\boldsymbol{\tau}'_{[0]}$, which nonetheless can be integrated to give:

$$\begin{aligned}\boldsymbol{\tau}'_{[0]} &= (\text{two terms identical to right side of Eq. 9.6-10}) \\ &\quad - \frac{\eta_0(\lambda_2\lambda_3 - \lambda_1\lambda_4)}{\lambda_1^2} \int_{-\infty}^{t'} e^{-(t'-t'')/\lambda_1} \{\boldsymbol{\gamma}''_{[1]} \cdot \boldsymbol{\gamma}''_{[1]}\} dt'' \\ &\quad - \frac{\eta_0(\lambda_1 - \lambda_2)\lambda_3}{2\lambda_1^3} \int_{-\infty}^{t'} \int_{-\infty}^{t''} e^{-(t'-t''')/\lambda_1} \{\boldsymbol{\gamma}'''_{[1]} \cdot \boldsymbol{\gamma}''_{[1]} + \boldsymbol{\gamma}''_{[1]} \cdot \boldsymbol{\gamma}'''_{[1]}\} dt'''\, dt''\end{aligned} \tag{9.6-11}$$

Note that the term containing λ_5 does not appear at second order inasmuch as $\text{tr}\,\boldsymbol{\tau}'_{[0]} = 0$ according to Eq. 9.6-10. This procedure can be continued to third and higher order with triple and higher multiple integrals being generated.

If we stop at second order, Eq. 9.6-11 can be reorganized into the form of a memory-integral expansion; this is done by writing Eq. 9.6-11 for $t' = t$ and then forcing all terms linear in $\boldsymbol{\gamma}'_{[1]}$ to appear in a single-integral term, and all quadratic terms in a double-integral term (by introducing the appropriate δ-functions):

$$\begin{aligned}\boldsymbol{\tau}(t) &= -\int_{-\infty}^{t} G_I(t - t')\boldsymbol{\gamma}'_{[1]} dt' \\ &\quad - \int_{-\infty}^{t} \int_{-\infty}^{t} G_{II}(t - t', t - t'')\{\boldsymbol{\gamma}'_{[1]} \cdot \boldsymbol{\gamma}''_{[1]} + \boldsymbol{\gamma}''_{[1]} \cdot \boldsymbol{\gamma}'_{[1]}\} dt''\, dt'\end{aligned} \tag{9.6-12}$$

in which

$$G_I = \frac{\eta_0(\lambda_1 - \lambda_2)}{\lambda_1^2} e^{-(t-t')/\lambda_1} + \frac{2\eta_0\lambda_2}{\lambda_1} \delta(t - t') \tag{9.6-13}$$

$$G_{II} = \begin{cases} 0 \quad \text{if } t' < t'' \leq t \\ \dfrac{\eta_0(\lambda_1 - \lambda_2)\lambda_3}{2\lambda_1^3} e^{-(t-t'')/\lambda_1} + \dfrac{\eta_0(\lambda_2\lambda_3 - \lambda_1\lambda_4)}{2\lambda_1^2} e^{-(t-t')/\lambda_1}\delta(t' - t'') \quad \text{if } -\infty < t'' < t' \end{cases} \tag{9.6-14}$$

We are thus led to a memory-integral expansion in $\gamma_{[1]}$, of which we have worked out the first two terms. The function $G_I(t - t')$ is identical to the relaxation modulus $G(t - t')$ of linear viscoelasticity (see Eq. 5.2-18). The final results may be put in the form of Eq. 9.6-1 (an expansion in $\gamma_{[0]}$) as may be seen in problem 9B.6.

EXAMPLE 9.6-3 The Criminale–Ericksen–Filbey (CEF) Equation and the Reiner–Rivlin Equation

Show how the memory-integral expansion simplifies for **(a)** steady shear flows, and **(b)** steady, homogeneous, irrotational flows.

SOLUTION **(a)** In Example 9.6-1 it is shown how the memory-integral expansion is related to the retarded-motion expansion. If the latter is applied to the steady shear flow $v_x = \dot{\gamma}y$, $v_y = 0$, $v_z = 0$, then from Appendix C

$$\gamma_{(1)} = \begin{pmatrix} 0 & 1 & 0 \\ 1 & 0 & 0 \\ 0 & 0 & 0 \end{pmatrix} \dot{\gamma}, \quad \gamma_{(2)} = -2\begin{pmatrix} 1 & 0 & 0 \\ 0 & 0 & 0 \\ 0 & 0 & 0 \end{pmatrix} \dot{\gamma}^2 \tag{9.6-15}$$

and $\gamma_{(n)} = 0$ for $n \geq 3$. Furthermore terms of third order involve the following products:[5]

$$\{\gamma_{(1)} \cdot \gamma_{(2)} + \gamma_{(2)} \cdot \gamma_{(1)}\} = -2\dot{\gamma}^2\gamma_{(1)}$$

$$\{\gamma_{(1)} \cdot \gamma_{(1)} \cdot \gamma_{(1)}\} = \dot{\gamma}^2\gamma_{(1)} \tag{9.6-16}$$

When these are substituted into the retarded-motion expansion (Eq 6.2-1), we get

$$\begin{aligned} \tau = &-(b_1 - 2b_{12}\dot{\gamma}^2 + 2b_{1:11}\dot{\gamma}^2 + \cdots)\gamma_{(1)} \\ &-(b_2 + \cdots)\gamma_{(2)} \\ &-(b_{11} + \cdots)\{\gamma_{(1)} \cdot \gamma_{(1)}\} \end{aligned} \tag{9.6-17}$$

Here the dots indicate additional terms that arise when one includes terms of fourth, fifth, and higher orders. Thus τ depends only on the three kinematic tensors $\gamma_{(1)}$, $\gamma_{(2)}$, and $\{\gamma_{(1)} \cdot \gamma_{(1)}\}$, and the coefficients of these tensors are functions of the shear rate $\dot{\gamma}$ alone. We now assign symbols to the scalar coefficients thus,

$$\boxed{\tau = -\eta\gamma_{(1)} + \tfrac{1}{2}\Psi_1\gamma_{(2)} - \Psi_2\{\gamma_{(1)} \cdot \gamma_{(1)}\}} \tag{9.6-18}$$

where $\eta(\dot{\gamma})$, $\Psi_1(\dot{\gamma})$, and $\Psi_2(\dot{\gamma})$ are the three (experimentally measurable) "viscometric functions" defined in §3.3. Equation 9.6-18 is the *Criminale-Ericksen-Filbey ("CEF") equation.*[6]

[5] At fourth and higher order the following two relations are needed:

$$\{\gamma_{(2)} \cdot \gamma_{(2)}\} = -2\dot{\gamma}^2\gamma_{(2)}; \quad \{\gamma_{(1)}^2 \cdot \gamma_{(2)} + \gamma_{(2)} \cdot \gamma_{(1)}^2\} = 2\dot{\gamma}^2\gamma_{(2)} \tag{9.6-16a}$$

[6] W. O. Criminale, Jr., J. L. Ericksen, and G. L. Filbey, Jr., *Arch. Rat. Mech. Anal.*, **1**, 410–417 (1958); see also J. L. Ericksen in, J. T. Bergen ed., *Viscoelasticity: Phenomenological Aspects*, Academic Press, New York (1960). The unsteady-state analog of the CEF equation has been derived, but it contains scalar functionals (A. S. Lodge, *Body Tensor Fields in Continuum Mechanics*, Academic Press, New York (1974), Eq. 7.1-11).

Note that the first term in the CEF equation is the generalized Newtonian fluid of Chapter 4; the second and third terms account for the normal-stress effects. The CEF equation is very helpful in the fluid mechanical analysis of standard rheometers.[7]

(b) For flows that are time independent and also homogeneous $D\boldsymbol{\gamma}_{(n)}/Dt$ is exactly zero. For flows that are irrotational $\nabla\boldsymbol{v}$ is the same as $(\nabla\boldsymbol{v})^\dagger$. Therefore, for flows that are steady, homogeneous, and irrotational, it can be shown that

$$\boldsymbol{\gamma}_{(n+1)} = -\tfrac{1}{2}\{\boldsymbol{\gamma}_{(1)}\cdot\boldsymbol{\gamma}_{(n)} + \boldsymbol{\gamma}_{(n)}\cdot\boldsymbol{\gamma}_{(1)}\} \qquad (9.6\text{-}19)$$

From this recursion relation we can start with $n = 1$ and derive expressions for all the $\boldsymbol{\gamma}_{(n)}$; doing this we find

$$\boldsymbol{\gamma}_{(n)} = (-1)^{n+1}\boldsymbol{\gamma}_{(1)}^n \qquad (9.6\text{-}20)$$

When this is substituted into the retarded-motion expansion, we find that $\boldsymbol{\tau}$ is a sum of terms containing multiple products of $\boldsymbol{\gamma}_{(1)}$. By repeated application of the Cayley–Hamilton theorem (Eq. A.3-28) these products ultimately yield terms containing $\boldsymbol{\delta}$, $\boldsymbol{\gamma}_{(1)}$, and $\boldsymbol{\gamma}_{(1)}^2$, with scalar coefficients depending only on the invariants of $\boldsymbol{\gamma}_{(1)} \equiv \dot{\boldsymbol{\gamma}}$ (which we call I, II, III). For incompressible fluids we omit the isotropic contributions containing $\boldsymbol{\delta}$ and use the fact that $I = 0$. Then the retarded-motion expansion collapses to

$$\boxed{\boldsymbol{\tau} = -f_1(II, III)\dot{\boldsymbol{\gamma}} - f_2(II, III)\{\dot{\boldsymbol{\gamma}}\cdot\dot{\boldsymbol{\gamma}}\}} \qquad (9.6\text{-}21)$$

This is the *Reiner–Rivlin* equation[8–10]. When originally proposed this equation was believed to be more generally applicable than it is. Hence there are unfortunately many publications in which the Reiner–Rivlin equation has been applied inappropriately to flows that are not steady, homogeneous, and irrotational.

PROBLEMS

9B.1 Relations Among Kinematic Tensors

a. The tensors $\boldsymbol{\gamma}^{(1)}$ and $\boldsymbol{\gamma}_{(1)}$ are both defined to be identical to $\dot{\boldsymbol{\gamma}} = \nabla\boldsymbol{v} + (\nabla\boldsymbol{v})^\dagger$. Show from the definitions that the higher-order rate-of-strain tensors are related thus

$$\boldsymbol{\gamma}^{(2)} = \boldsymbol{\gamma}_{(2)} + 2\{\boldsymbol{\gamma}_{(1)}\cdot\boldsymbol{\gamma}_{(1)}\} \qquad (9B.1\text{-}1)$$

$$\boldsymbol{\gamma}_{(2)} = \boldsymbol{\gamma}^{(2)} - 2\{\boldsymbol{\gamma}^{(1)}\cdot\boldsymbol{\gamma}^{(1)}\} \qquad (9B.1\text{-}2)$$

$$\boldsymbol{\gamma}^{(3)} = \boldsymbol{\gamma}_{(3)} + 3\{\boldsymbol{\gamma}_{(1)}\cdot\boldsymbol{\gamma}_{(2)} + \boldsymbol{\gamma}_{(2)}\cdot\boldsymbol{\gamma}_{(1)}\} + 6\{\boldsymbol{\gamma}_{(1)}\cdot\boldsymbol{\gamma}_{(1)}\cdot\boldsymbol{\gamma}_{(1)}\} \qquad (9B.1\text{-}3)$$

$$\boldsymbol{\gamma}_{(3)} = \boldsymbol{\gamma}^{(3)} - 3\{\boldsymbol{\gamma}^{(1)}\cdot\boldsymbol{\gamma}^{(2)} + \boldsymbol{\gamma}^{(2)}\cdot\boldsymbol{\gamma}^{(1)}\} + 6\{\boldsymbol{\gamma}^{(1)}\cdot\boldsymbol{\gamma}^{(1)}\cdot\boldsymbol{\gamma}^{(1)}\} \qquad (9B.1\text{-}4)$$

[7] All of the viscometric flow problems solved in B. D. Coleman, H. Markovitz, and W. Noll, *Viscometric Flows of Non-Newtonian Fluids*, Springer, New York (1966), can be solved using Eq. 9.6-18.
[8] M. Reiner, *Am. J. Math.*, **67**, 350–362 (1945).
[9] R. S. Rivlin, *Proc. Roy. Soc.* (*London*), **A193**, 260–281 (1948); *Proc. Camb. Phil. Soc.*, **45**, 88–91 (1949).
[10] K. Weissenberg, *Arch. Sci. Phys. Nat.* (5), **140**, 44–106, 130–171 (1935), made an earlier attempt to construct a constitutive equation starting with the premise that $\boldsymbol{\tau}$ must be some general function of $\dot{\boldsymbol{\gamma}}$. The unsteady state analog of the Reiner–Rivlin equation has been derived [A. S. Lodge, *Body Tensor Fields in Continuum Mechanics*, Academic Press, New York (1974), Eq. 7.4-5], but it contains scalar functionals.

b. Show that Eqs. 9B.1-2 and 4 can also be obtained from the Walters–Waterhouse relation:[1]

$$\boldsymbol{\gamma}_{(n)} = (-1)^{n+1} n! \begin{vmatrix} \boldsymbol{\gamma}^{(1)} & \boldsymbol{\delta} & \mathbf{0} & \mathbf{0} & \cdots \\ \frac{1}{2!}\boldsymbol{\gamma}^{(2)} & \boldsymbol{\gamma}^{(1)} & \boldsymbol{\delta} & \mathbf{0} & \cdots \\ \frac{1}{3!}\boldsymbol{\gamma}^{(3)} & \frac{1}{2!}\boldsymbol{\gamma}^{(2)} & \boldsymbol{\gamma}^{(1)} & \boldsymbol{\delta} & \cdots \\ \frac{1}{4!}\boldsymbol{\gamma}^{(4)} & \frac{1}{3!}\boldsymbol{\gamma}^{(3)} & \frac{1}{2!}\boldsymbol{\gamma}^{(2)} & \boldsymbol{\gamma}^{(1)} & \cdots \\ \vdots & \vdots & \vdots & \vdots & \end{vmatrix} \tag{9B.1-5}$$

Then show that Eqs. 9B.1-1 and 3 can be obtained from the inverse relation: change all $\boldsymbol{\gamma}^{(n)}$ to $\boldsymbol{\gamma}_{(n)}$ and vice versa, omit the factor $(-1)^{n+1}$, and replace the unit tensor $\boldsymbol{\delta}$ by $-\boldsymbol{\delta}$. In multiplying out the determinant left-to-right multiplication must be preserved.

9B.2 Expansion of the Relative Strain Tensors in Taylor Series

Verify that the relative strain tensors, considered as functions of t', can be expanded in Taylor series about $t' = t$ to give

$$\boldsymbol{\gamma}^{[0]}(\boldsymbol{r}, t, t') = \sum_{k=1}^{\infty} (k!)^{-1} \boldsymbol{\gamma}^{(k)}(\boldsymbol{r}, t)(t' - t)^k \tag{9B.2-1}$$

$$\boldsymbol{\gamma}_{[0]}(\boldsymbol{r}, t, t') = \sum_{k=1}^{\infty} (k!)^{-1} \boldsymbol{\gamma}_{(k)}(\boldsymbol{r}, t)(t' - t)^k \tag{9B.2-2}$$

Table 9.3-1 contains all the information needed to obtain these expansions.

9B.3 Base Vectors and Kinematic Tensors for Steady Shear Flow

For the flow depicted in Fig. 9.2-2 work out the following:

a. From the velocity field find the displacement functions.

b. Display the components of the displacement gradient tensors $\boldsymbol{\Delta}$ and $\mathbf{E}$ in matrix form.

c. Verify that the expressions for $\hat{\boldsymbol{g}}_i$ and $\hat{\boldsymbol{g}}^i$ shown in Fig. 9.2-2 are correct.

d. Next work out the matrix displays for the relative strain tensors $\boldsymbol{\gamma}^{[0]}$ and $\boldsymbol{\gamma}_{[0]}$, and check them against the entries in Table C.1.

e. From **(d)** get $\boldsymbol{\gamma}^{[n]}$ and $\boldsymbol{\gamma}_{[n]}$ for $n = 1, 2$ by differentiating with respect to t'; then set t' equal to t to get the corresponding $\boldsymbol{\gamma}^{(n)}$ and $\boldsymbol{\gamma}_{(n)}$ quantities.

9B.4 Base Vectors and Kinematic Tensors for Flow of a Non-Newtonian Fluid in a Thin Slit

Repeat Problem 9B.3 for the flow shown in Fig. 9.2-3. Verify that the base vectors have been drawn properly for fluid particles *P*, *Q*, *R*, and *S*.

[1] K. Walters and W. M. Waterhouse, *J. Non-Newtonian Fluid Mech.*, **3**, 293–296 (1977/78).

9B.5 Operations with Kinematic Tensors

Which of the following relations are true?

a. $\text{tr}\,\boldsymbol{\gamma}_{[1]}(\boldsymbol{r}, t, t') = 0$ for incompressible fluids (9B.5-1)
b. $\{\boldsymbol{\gamma}_{(1)} \cdot \boldsymbol{\gamma}_{(1)}\}_{(1)} = \{\boldsymbol{\gamma}_{(2)} \cdot \boldsymbol{\gamma}_{(1)} + \boldsymbol{\gamma}_{(1)} \cdot \boldsymbol{\gamma}_{(2)} + \boldsymbol{\gamma}_{(1)} \cdot \boldsymbol{\gamma}_{(1)} \cdot \boldsymbol{\gamma}_{(1)}\}$ (9B.5-2)
c. $(s\boldsymbol{\delta})_{(1)} = \boldsymbol{\delta}\, Ds/Dt - s\boldsymbol{\gamma}_{(1)}$ where s is a scalar (9B.5-3)
d. $(\boldsymbol{\delta} - \boldsymbol{\gamma}_{[0]})_{(1)} = \boldsymbol{0}$ (9B.5-4)
e. $(\boldsymbol{\gamma}^{[0]})^{(1)} = -\boldsymbol{\gamma}^{(1)}$ (9B.5-5)

9B.6 Interrelation of Two Memory-Integral Expansions

Verify that Eq. 9.6-12 can be transformed into the form of Eq. 9.6-1 and that $M_I(t - t') = \partial G_I(t - t')/\partial t'$ and $M_{II}(t - t', t - t'') = -\partial^2 G_{II}(t - t', t - t'')/\partial t' \partial t''$.

9B.7 A Convected Maxwell Model

a. What are the good and bad features of the following constitutive equation?

$$\boldsymbol{\tau} + \lambda_1[(1 - \varepsilon)\boldsymbol{\tau}_{(1)} + \varepsilon\boldsymbol{\tau}^{(1)}] = -\eta_0\dot{\boldsymbol{\gamma}} \tag{9B.7-1}$$

Here η_0, λ_1, and ε are the model parameters. What can be said about the model when $\varepsilon = 0, \frac{1}{2}$, and 1?

b. Is the model in Eq. 9B.7-1 the same as the following integral model?

$$\boldsymbol{\tau} = +\int_{-\infty}^{t} \left\{\frac{\eta_0}{\lambda_1^2} e^{-(t-t')/\lambda_1}\right\}((1 - \varepsilon)\boldsymbol{\gamma}'_{[0]} + \varepsilon\boldsymbol{\gamma}^{[0]\prime})dt' \tag{9B.7-2}$$

Here $\boldsymbol{\gamma}'_{[0]} \equiv \boldsymbol{\gamma}_{[0]}(\boldsymbol{r}, t, t')$.

9B.8 The Retarded Motion Expansion for the Factorized Rivlin–Sawyers Fluid

a. Substitute Eq. 8.3-16 for $\phi_1(I_1, I_2)$ and $\phi_2(I_1, I_2)$ into the constitutive equation in Eq. 8.3-12. Then use the expansions in Problem 9B.2 to eliminate the strain tensors in favor of the series involving the nth-order rate-of-strain tensors.

b. Show that for incompressible fluids $\text{tr}\,\boldsymbol{\gamma}^{(2)} = -\text{tr}\,\boldsymbol{\gamma}_{(2)} = \text{tr}\,\{\boldsymbol{\gamma}^{(1)} \cdot \boldsymbol{\gamma}^{(1)}\}$. Next expand the invariants of the Finger tensor as follows:

$$I_1 - 3 = \tfrac{1}{2}(\text{tr}\{\boldsymbol{\gamma}_{(1)} \cdot \boldsymbol{\gamma}_{(1)}\})s^2 + \tfrac{1}{6}(\text{tr}\,\boldsymbol{\gamma}_{(3)})s^3 - \cdots \tag{9B.8-1}$$

$$I_2 - 3 = \tfrac{1}{2}(\text{tr}\{\boldsymbol{\gamma}^{(1)} \cdot \boldsymbol{\gamma}^{(1)}\})s^2 - \tfrac{1}{6}(\text{tr}\,\boldsymbol{\gamma}^{(3)})s^3 + \cdots \tag{9B.8-2}$$

in which $s = t - t'$. It is thus seen that I_1 and I_2 are identical through terms of second order in s.

c. Then use the condition in Eq. 8.3-14 and the definition of the M_n in Eq. 8.2-12 to get the retarded-motion coefficients in Table 6.2-2.

9B.9 The Jaumann ("Corotational") Derivative[2]

The *Jaumann derivative* of the tensor $\boldsymbol{\tau}$ is defined as

$$\frac{\mathscr{D}}{\mathscr{D}t}\boldsymbol{\tau} = \frac{\partial}{\partial t}\boldsymbol{\tau} + \{\boldsymbol{v}\cdot\nabla\boldsymbol{\tau}\} + \tfrac{1}{2}\{\boldsymbol{\omega}\cdot\boldsymbol{\tau} - \boldsymbol{\tau}\cdot\boldsymbol{\omega}\} \qquad (9B.9\text{-}1)$$

in which $\boldsymbol{\omega}$ is the vorticity tensor.[2] This derivative has often been used in the rheology and continuum-mechanics literature. Show that

$$\frac{\mathscr{D}}{\mathscr{D}t}\dot{\boldsymbol{\gamma}} = \boldsymbol{\gamma}^{(2)} - \{\boldsymbol{\gamma}^{(1)}\cdot\boldsymbol{\gamma}^{(1)}\}$$

$$= \boldsymbol{\gamma}_{(2)} + \{\boldsymbol{\gamma}_{(1)}\cdot\boldsymbol{\gamma}_{(1)}\} \qquad (9B.9\text{-}2)$$

$$\frac{\mathscr{D}^2\dot{\boldsymbol{\gamma}}}{\mathscr{D}t^2} = \boldsymbol{\gamma}^{(3)} - \tfrac{3}{2}\{\boldsymbol{\gamma}^{(1)}\cdot\boldsymbol{\gamma}^{(2)} + \boldsymbol{\gamma}^{(2)}\cdot\boldsymbol{\gamma}^{(1)}\} + 2\{\boldsymbol{\gamma}^{(1)}\cdot\boldsymbol{\gamma}^{(1)}\cdot\boldsymbol{\gamma}^{(1)}\}$$

$$= \boldsymbol{\gamma}_{(3)} + \tfrac{3}{2}\{\boldsymbol{\gamma}_{(1)}\cdot\boldsymbol{\gamma}_{(2)} + \boldsymbol{\gamma}_{(2)}\cdot\boldsymbol{\gamma}_{(1)}\} + 2\{\boldsymbol{\gamma}_{(1)}\cdot\boldsymbol{\gamma}_{(1)}\cdot\boldsymbol{\gamma}_{(1)}\} \qquad (9B.9\text{-}3)$$

9C.1 Time Derivatives of the Displacement Gradient Tensors

a. Start with the definition of E_{ij} in Eq. 9.2-5 and show that

$$\left(\frac{\partial}{\partial t}\mathbf{E}\right)_{\boldsymbol{r}',t'} = \{(\nabla\boldsymbol{v})^\dagger\cdot\mathbf{E}\} \qquad (9C.1\text{-}1)$$

where $\nabla\boldsymbol{v}$ is evaluated at $\boldsymbol{r}$ and t.

b. Knowing that $\{\mathbf{E}\cdot\boldsymbol{\Delta}\} = \boldsymbol{\delta}$, show from Eq. 9C.1-1 that

$$\left(\frac{\partial}{\partial t}\boldsymbol{\Delta}\right)_{\boldsymbol{r}',t'} = -\{\boldsymbol{\Delta}\cdot(\nabla\boldsymbol{v})^\dagger\} \qquad (9C.1\text{-}2)$$

where $\nabla\boldsymbol{v}$ is evaluated at $\boldsymbol{r}$ and t.

9D.1 Relations Among the Tensors $\boldsymbol{\gamma}^{[1]}$, $\boldsymbol{\gamma}_{[1]}$, and $\dot{\boldsymbol{\gamma}}$

a. Consider a fluid particle located at position $\boldsymbol{r}$ at the present time t, and at $\boldsymbol{r}'$ at some past time t'. Show that for homogeneous flows the function $\boldsymbol{r}' = \boldsymbol{r}'(\boldsymbol{r}, t, t')$ is obtained by solving the equation

$$\frac{\partial}{\partial t'}\boldsymbol{r}' = [\boldsymbol{\kappa}'\cdot\boldsymbol{r}'] \qquad (9D.1\text{-}1)$$

along with the condition that $\boldsymbol{r}'(t) = \boldsymbol{r}$. Here $\boldsymbol{\kappa}' = \boldsymbol{\kappa}(\boldsymbol{r}, t, t')$ is $(\nabla\boldsymbol{v})^\dagger$ evaluated at t'.

[2] For an extensive discussion of the corotational formulation of continuum mechanics, the Jaumann derivative, and corotational integration, see J. D. Goddard and C. Miller, *Rheol. Acta*, **5**, 177–184 (1966), and J. D. Goddard, *Trans. Soc. Rheol.*, **11**, 381–399 (1967). See also R. B. Bird, O. Hassager, and S. I. Abdel-Khalik, *AIChE J.*, **20**, 1041–1066 (1975), and J. G. Oldroyd, *Proc. Roy. Soc.*, **A245**, 278–297 (1958).

b. Show that the solution to Eq. 9D.1-1 for $\boldsymbol{\kappa} = \boldsymbol{\kappa}(t)$ is

$$\boldsymbol{r}'(t') = [\boldsymbol{\Omega}_t^{t'}(\boldsymbol{\kappa}) \cdot \boldsymbol{r}(t)] \tag{9D.1-2}$$

where the matrizant[3] $\boldsymbol{\Omega}_t^{t'}(\boldsymbol{\kappa})$ is given by

$$\begin{aligned}\boldsymbol{\Omega}_t^{t'}(\boldsymbol{\kappa}) &= \boldsymbol{\delta} + \int_t^{t'} \boldsymbol{\kappa}''\, dt'' + \int_t^{t'} \int_t^{t''} \{\boldsymbol{\kappa}'' \cdot \boldsymbol{\kappa}'''\} dt'''\, dt'' \\ &\quad + \int_t^{t'} \int_t^{t''} \int_t^{t'''} \{\boldsymbol{\kappa}'' \cdot \boldsymbol{\kappa}''' \cdot \boldsymbol{\kappa}''''\} dt''''\, dt'''\, dt'' + \cdots\end{aligned} \tag{9D.1-3}$$

in which $\boldsymbol{\kappa}'' \equiv \boldsymbol{\kappa}(\boldsymbol{r}, t, t'')$.

c. Verify that

$$\boldsymbol{\Delta} = \boldsymbol{\delta} - \int_{t'}^{t} \boldsymbol{\kappa}''\, dt'' + \int_{t'}^{t} \int_{t''}^{t} \{\boldsymbol{\kappa}'' \cdot \boldsymbol{\kappa}'''\} dt'''\, dt'' + \cdots \tag{9D.1-4}$$

$$\mathbf{E} = \boldsymbol{\delta} + \int_{t'}^{t} \boldsymbol{\kappa}''\, dt'' + \int_{t'}^{t} \int_{t''}^{t} \{\boldsymbol{\kappa}'' \cdot \boldsymbol{\kappa}'''\} dt'''\, dt'' + \cdots \tag{9D.1-5}$$

d. Then show that:[4]

$$\begin{aligned}\dot{\boldsymbol{\gamma}}' &= \boldsymbol{\gamma}'_{[1]} - \frac{1}{2}\int_{t'}^{t} (\{\boldsymbol{\gamma}'_{[1]} \cdot \boldsymbol{\gamma}''_{[1]} + \boldsymbol{\gamma}''_{[1]} \cdot \boldsymbol{\gamma}'_{[1]}\} \\ &\quad - \{\boldsymbol{\omega}'' \cdot \boldsymbol{\gamma}'_{[1]} - \boldsymbol{\gamma}'_{[1]} \cdot \boldsymbol{\omega}''\}) dt'' + \cdots\end{aligned} \tag{9D.1-6}$$

and obtain a similar relation giving $\dot{\boldsymbol{\gamma}}'$ in terms of integrals over $\boldsymbol{\gamma}^{[1]}$ and $\boldsymbol{\omega}$ at past times; here $\boldsymbol{\omega} = \nabla \boldsymbol{v} - (\nabla \boldsymbol{v})^{\dagger}$ is the vorticity tensor.

[3] N. R. Amundson, *Mathematical Methods in Chemical Engineering*, Prentice-Hall, Englewood Cliffs, NJ (1966), p. 199.

[4] See R. C. Armstrong and R. B. Bird, *J. Chem. Phys.*, **58**, 2715–2723 (1973); erratum: in Eq. 22, replace 27/280 by 27/980.

CHAPTER 10

FLUID DYNAMICS OF RHEOMETRY

In Chapter 3 we described the material functions that can be used to characterize non-Newtonian fluids. There we introduced the two main types of flows, shear and shearfree, used for this purpose. For each of these flows we considered only homogeneous flow, and we gave definitions for the material functions in terms of the uniform stresses generated in the fluid. In the laboratory we do not usually generate homogeneous flows nor directly measure the material functions. Instead we measure forces and torques exerted on pieces of equipment, rates of rotation in an apparatus, volume flow rates, and so on. It is necessary to be able to relate these measurable quantities to the desired material functions.

We consider here a number of experimental arrangements used for studying shear and shearfree flows. We begin in §10.1 with a discussion of linear viscoelastic property measurements, since these provide the common time-dependent basis to which large deformation shear and shearfree flow results must simplify in the limit of small displacement gradients. Then in §10.2 we consider steady-state shear flows. These are the most commonly performed experiments in rheometry; many standard instruments are available commercially to do these. Next in §10.3 we describe shearfree flow experiments. These methods have been under active development over the past decade and there remain large classes of materials and material functions for which measurements cannot yet be made. Finally, in §10.4 we present several examples of experiments that use complex flows (in the sense of being neither simple shear flow nor shearfree flow) to learn about the material functions.

Although this chapter focuses on mechanical means for measuring material functions, rheo-optical techniques may ultimately prove more useful for certain types of properties, including those involving short-time, transient shear flows and all types of shearfree flows.[1] We do not discuss laboratory techniques or experimental difficulties in any depth here, as these can be found in other references.[2]

[1] A general discussion of birefringence methods is given by H. Janeschitz-Kriegl, *Polymer Melt Rheology and Flow Birefringence*, Elsevier, Amsterdam (1983). Measurement of shearfree flow properties is described by J. A. van Aken and H. Janeschitz-Kriegl, *Rheol. Acta*, **19**, 744–752 (1980); **20**, 419–432 (1981). Transient measurements of the shear stress and first normal stress difference made by birefringence and by standard mechanical methods for shear flow of a polystyrene melt have been shown to agree by F. H. Gortemaker, M. G. Hansen, B. deCindio, H. M. Laun, and H. Janeschitz-Kriegl, *Rheol. Acta*, **15**, 256–267 (1976). The start-up of steady shear flow has been studied by birefringence by A. W. Chow and G. G. Fuller, *J. Non-Newtonian Fluid Mech.*, **17**, 233–243 (1985).

[2] See J. Meissner, *Ann. Rev. Fluid Mech.*, **17**, 45–64 (1985); H. Janeschitz-Kriegl, *Polymer Melt Rheology and Flow Birefringence*, Springer-Verlag, Berlin (1983); J. M. Dealy, *Rheometers for Molten Plastics*, Van Nostrand, New York (1982); R. W. Whorlow, *Rheological Techniques*, Ellis Horwood, Chichester, UK (1980); K. Walters, *Rheometry*, Chapman and Hall, London (1975); S. Oka in F. R. Eirich, ed., *Rheology*, Vol. 3, Academic Press, New York (1960), Chapt. 2.

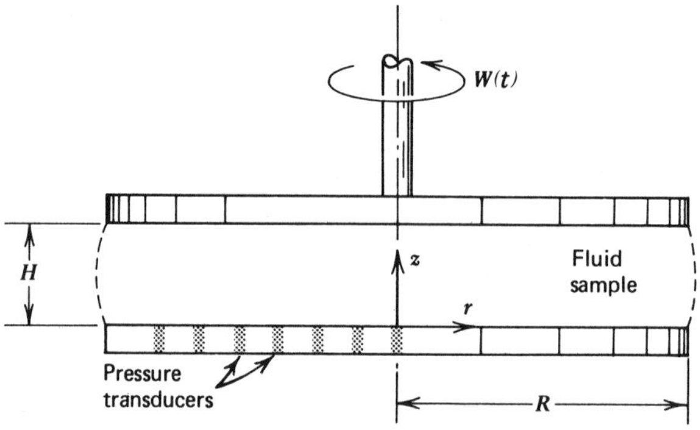

FIGURE 10.1-1. Parallel-disk instrument. In Example 10.1-1, $W(t) = \theta_0\, \mathscr{Re}\{i\omega e^{i\omega t}\}$, and in Example 10.2-2, W is a constant.

§10.1 LINEAR VISCOELASTIC MEASUREMENTS[1]

As has been described in Chapter 5, the linear viscoelastic properties of a material can be obtained from a wide variety of experiments. Chapter 5 has further shown how the results of the various experimental measurements can be interrelated. In Example 5.4-3, we worked out the relationships for determining η' and η'' from an oscillating, concentric cylinder device. In this section, we analyze two additional instrumental configurations frequently used to measure these linear viscoelastic material functions. The second example is particularly interesting because we must use a general, nonlinear model, namely the first term of the memory-integral expansion (Eq. 9.6-1), to analyze the flow.

EXAMPLE 10.1-1 The Parallel-Disk Viscometer

A viscoelastic fluid is contained between two parallel disks of radius R separated by a gap H (see Fig. 10.1-1). The upper plate is caused to oscillate in the θ-direction sinusoidally with frequency ω and angular amplitude θ_0, and the time-dependent torque on the fixed, lower plate is measured. Show how the torque measurement can be used to determine η' and η'' from measurements with $\theta_0 \ll 1$. Assume that inertial forces are negligible.

SOLUTION The motion of the top plate is given by the angular displacement θ as a function of the time t

$$\theta = \theta_0\, \mathscr{Re}\{e^{i\omega t}\} \tag{10.1-1}$$

where θ_0 is real. The boundary conditions on the fluid velocity are

$$\text{At } z = H, \qquad v_\theta = \theta_0 r\, \mathscr{Re}\{i\omega e^{i\omega t}\}$$

$$v_r = v_z = 0 \tag{10.1-2}$$

$$\text{At } z = 0, \qquad v_\theta = v_r = v_z = 0 \tag{10.1-3}$$

[1] The standard reference on linear viscoelastic measurement methods is J. D. Ferry, *Viscoelastic Properties of Polymers*, 3rd ed., Wiley, New York (1980). We have also drawn extensively in this section on K. Walters, *Rheometry*, Chapman and Hall, London (1975), Chapts. 6 and 8.

The form of the boundary conditions strongly suggests that we assume $v_r = v_z = 0$ and

$$v_\theta = r\,\mathcal{R}e\{f(z)e^{i\omega t}\} \tag{10.1-4}$$

where $f(z)$ is complex. This form for v_θ satisfies the continuity equation automatically. The boundary conditions on f are

$$f(0) = 0; \quad f(H) = i\omega\theta_0 \tag{10.1-5}$$

The constitutive equation appropriate for small displacement gradients ($\theta_0 \ll 1$) is the general linear viscoelastic model, Eq. 5.2-18. When we combine Eq. 5.2-18 with Eq. 5.3-6, which gives the complex viscosity in terms of $G(s)$, and use the assumed velocity field together with Eqs. B.3-7 through 12 for the rate of strain components, we find that

$$\tau_{\theta z} = \tau_{z\theta} = -r\,\mathcal{R}e\{\eta^* f'(z)e^{i\omega t}\} \tag{10.1-6}$$

and that all other $\tau_{ij} = 0$.

Next we insert the stress tensor into the equation of motion in which the inertial terms are neglected. The r- and z-components show that the pressure is everywhere constant between the plates. The θ-component gives

$$f'' = 0 \tag{10.1-7}$$

The solution to Eq. 10.1-7 that satisfies the boundary conditions is

$$f = i\omega\theta_0 z/H \tag{10.1-8}$$

so that

$$\tau_{\theta z} = -(\theta_0 r/H)\,\mathcal{R}e\{\eta^* i\omega e^{i\omega t}\} \tag{10.1-9}$$

We can now calculate the torque $\mathcal{T}$ that the lower shaft must exert on the bottom plate in order to keep it stationary. This is given by

$$\begin{aligned} \mathcal{T} &= \mathcal{T}_0\,\mathcal{R}e\{e^{i(\omega t + \delta - \pi)}\} \\ &= -\mathcal{T}_0\,\mathcal{R}e\{(\cos\delta + i\sin\delta)e^{i\omega t}\} \\ &= \int_0^{2\pi}\int_0^R r\,\mathcal{R}e\{\tau_{\theta z}\}\cdot r\,dr\,d\theta \\ &= -(\pi R^4\omega\theta_0/2H)\,\mathcal{R}e\{(i\eta' + \eta'')e^{i\omega t}\} \end{aligned} \tag{10.1-10}$$

where $\mathcal{T}_0$ is real. By comparing the second and last lines of Eq. 10.1-10 we find the components of the complex viscosity in terms of the measured torque amplitude $\mathcal{T}_0$ and phase shift δ as follows:

$$\eta' = \frac{2H\mathcal{T}_0 \sin\delta}{\pi R^4\omega\theta_0} \tag{10.1-11}$$

$$\eta'' = \frac{2H\mathcal{T}_0 \cos\delta}{\pi R^4\omega\theta_0} \tag{10.1-12}$$

The phase angle δ must vary between 0 for a purely elastic solid and $\pi/2$ for a purely viscous fluid.

The most serious assumption involved in the derivation of Eqs. 10.1-11 and 12 is the neglect of inertia. It can be shown that this is a good approximation provided that (see Problem 10B.2)

$$H\left|\frac{i\rho\omega}{\eta^*}\right|^{1/2} \ll 1 \tag{10.1-13}$$

An advantage of the parallel-disk geometry for measuring η^* for low-viscosity liquids is that at fixed frequency ω the gap H can be reduced in order to minimize inertial effects.

EXAMPLE 10.1-2 The Eccentric Rotating Disk Rheometer[2]

The eccentric disk or orthogonal rheometer flow[3] is shown in Fig. 3B.1. The material to be tested is contained between parallel disks of radius R, each of which rotates with angular velocity W. The resulting motion would be a rigid rotation except for the fact that the axes of rotation of the disks are displaced from each other by an amount a. The gap between the disks is b. Show how the linear viscoelastic material functions η' and η'' can be determined from the steady-state forces F_x and F_y exerted by the fluid in the x- and y-directions on the lower plate. Assume that inertial forces are negligible, and take the origin of the xyz-coordinate system to be at the midpoint of the line connecting the centers of the two disks.

SOLUTION **(a)** Boundary Conditions and Assumed Solution

The boundary conditions for the flow in Fig. 3B.1 (with the coordinate system described above) are

$$\text{On } z = b/2, \quad \boldsymbol{v} = [W\boldsymbol{\delta}_z \times [\boldsymbol{r} - \tfrac{1}{2}a\boldsymbol{\delta}_y]]$$
$$= [W\boldsymbol{\delta}_z \times [r\boldsymbol{\delta}_r - \tfrac{1}{2}a(\sin\theta\,\boldsymbol{\delta}_r + \cos\theta\,\boldsymbol{\delta}_\theta)]] \tag{10.1-14a}$$
$$= \tfrac{1}{2}Wa\cos\theta\,\boldsymbol{\delta}_r + W(r - \tfrac{1}{2}a\sin\theta)\boldsymbol{\delta}_\theta$$

$$\text{On } z = -b/2, \quad \boldsymbol{v} = [W\boldsymbol{\delta}_z \times [\boldsymbol{r} + \tfrac{1}{2}a\boldsymbol{\delta}_y]]$$
$$= -\tfrac{1}{2}Wa\cos\theta\,\boldsymbol{\delta}_r + W(r + \tfrac{1}{2}a\sin\theta)\boldsymbol{\delta}_\theta \tag{10.1-14b}$$

where $\boldsymbol{\delta}_i$ is the unit vector in the ith coordinate direction. Equations 10.1-14a and b are of the form

$$v_r = aW\,\mathcal{Re}\{v_r^0 e^{i\theta}\} \tag{10.1-15a}$$

$$v_\theta = rW + aW\,\mathcal{Re}\{v_\theta^0 e^{i\theta}\} \tag{10.1-15b}$$

$$v_z = aW\,\mathcal{Re}\{v_z^0 e^{i\theta}\} \tag{10.1-15c}$$

in which the dimensionless complex amplitude functions v_i^0 are given by

$$\text{On } z = b/2, \quad v_r^0 = 1/2, \quad v_\theta^0 = i/2, \quad v_z^0 = 0 \tag{10.1-16a}$$

$$\text{On } z = -b/2, \quad v_r^0 = -1/2, \quad v_\theta^0 = -i/2, \quad v_z^0 = 0 \tag{10.1-16b}$$

The factor aW is included in the velocity perturbation definitions in Eq. 10.1-15 because we expect the flow will reduce to a rigid rotation in the limit as $a \to 0$.

[2] This analysis is based on K. Walters, *Rheometry*, Chapman and Hall, London (1975), pp. 164–172.

[3] The kinematics of flow between eccentric rotating disks was discussed in Problems 3B.1 and 3C.2. This flow has also been analyzed in Problems 6B.2, 7B.9, and 8B.5.

The boundary conditions suggest a solution of the form given by Eq. 10.1-15 with $v_r^0 = v_r^0(z)$, $v_\theta^0 = v_\theta^0(z)$, $v_z^0 = 0$. If these are inserted into the continuity equation we obtain the following relation between v_r^0 and v_θ^0:

$$v_r^0 + iv_\theta^0 = 0 \tag{10.1-17}$$

(b) Constitutive Equation

Because the material particles experience rotations through large angles during this experiment, the general linear viscoelastic model cannot be used (see Example 8.1-2). However, the first term of the memory-integral expansion, Eq. 9.6-1 (namely, the Lodge rubberlike liquid constitutive equation), can be used for this analysis if the strain (a/b) imposed on the material is sufficiently small; we have seen in §9.6 that this equation is valid in the limit of small strains for a very broad class of fluids.

To calculate the stress at position $\boldsymbol{r}$ and time t, the Lodge rubberlike liquid requires that we integrate the relative strain tensor $\boldsymbol{\gamma}_{[0]}(\boldsymbol{r}, t, t')$ following a particle that is at position $\boldsymbol{r}$ at time t. This is most easily done if we take the position $\boldsymbol{r}'$ of the particle at past times t' to be the dependent variable and the present location $\boldsymbol{r}$ at which we want to evaluate stress to be the independent variable. Evaluation of $\boldsymbol{\gamma}_{[0]}$ is then accomplished by using the displacement functions $\boldsymbol{r}' = \boldsymbol{r}'(\boldsymbol{r}, t, t')$ to calculate the displacement gradient tensor $\boldsymbol{\Delta}$, by inverting $\boldsymbol{\Delta}$ to obtain $\mathbf{E}$, and then finally by using $\mathbf{E}$ to calculate $\boldsymbol{\gamma}_{[0]}$. The displacement functions are given by the velocities as follows:

$$\begin{aligned} \frac{\partial r'}{\partial t'} &= v_r = aW\, \mathcal{R}e\{v_r^0(z')e^{i\theta'}\} \\ r'\frac{\partial \theta'}{\partial t'} &= v_\theta = r'W + aW\, \mathcal{R}e\{v_\theta^0(z')e^{i\theta'}\} \\ \frac{\partial z'}{\partial t'} &= v_z = 0 \end{aligned} \tag{10.1-18}$$

subject to the condition

$$\text{At } t' = t \qquad r' = r; \quad \theta' = \theta; \quad z' = z \tag{10.1-19}$$

where the velocities $v_i = v_i(r', \theta', z', t')$ are functions of the space coordinates r', θ', z' and we have taken $v_z = 0$. The solution to Eqs. 10.1-18 is made easy by the fact that the amplitudes aWv_i^0 of the perturbations to the rigid rotation are small. Thus we can construct a successive substitution scheme to obtain a solution good to terms linear in the small offset a.

At lowest order we neglect all terms containing a. Thus Eqs. 10.1-18 reduce to

$$\frac{\partial r'}{\partial t'} = 0; \quad \frac{\partial \theta'}{\partial t'} = W; \quad \frac{\partial z'}{\partial t'} = 0 \tag{10.1-20}$$

which have the solutions (when condition Eq. 10.1-19 is applied)

$$r' = r; \quad \theta' = \theta - W(t - t'); \quad z' = z \tag{10.1-21}$$

The terms that were neglected in obtaining Eqs. 10.1-20 are now evaluated by using Eqs. 10.1-21 and the following equations for the next approximation to the displacement functions are obtained:

$$\begin{aligned} \frac{\partial r'}{\partial t'} &= aW\, \mathcal{R}e\{v_r^0(z)e^{i\theta}e^{-iW(t-t')}\} \\ \frac{\partial \theta'}{\partial t'} &= W + aW\, \mathcal{R}e\left\{\frac{v_\theta^0(z)}{r}e^{i\theta}e^{-iW(t-t')}\right\} \\ \frac{\partial z'}{\partial t'} &= 0 \end{aligned} \tag{10.1-22}$$

The solutions to Eqs. 10.1-22 are

$$r' = r + a\,\mathcal{Re}\{iv_r^0(z)e^{i\theta}[1 - e^{-iW(t-t')}]\}$$

$$\theta' = \theta - W(t-t') + a\,\mathcal{Re}\left\{i\,\frac{v_\theta^0(z)}{r}\,e^{i\theta}[1 - e^{-iW(t-t')}]\right\} \qquad (10.1\text{-}23)$$

$$z' = z$$

Equations 10.1-23 are the desired solution for the displacement functions since the quadratic terms in a obtained in the next iteration are negligibly small.

To calculate $\mathbf{\Delta}$ we need the following derivatives which are all evaluated to order a

$$\frac{\partial r'}{\partial r} = \frac{\partial z'}{\partial z} = 1 \qquad \frac{\partial \theta'}{\partial r} = -a\,\mathcal{Re}\left\{i\,\frac{v_\theta^0}{r^2}\,\Omega\right\}$$

$$\frac{\partial r'}{\partial \theta} = -a\,\mathcal{Re}\{v_r^0\Omega\} \qquad \frac{\partial \theta'}{\partial \theta} = 1 - a\,\mathcal{Re}\left\{\frac{v_\theta^0}{r}\,\Omega\right\} \qquad (10.1\text{-}24)$$

$$\frac{\partial r'}{\partial z} = a\,\mathcal{Re}\left\{i\,\frac{dv_r^0}{dz}\,\Omega\right\} \qquad \frac{\partial \theta'}{\partial z} = a\,\mathcal{Re}\left\{\frac{i}{r}\,\frac{dv_\theta^0}{dz}\,\Omega\right\}$$

where we have abbreviated

$$\Omega = e^{i\theta}[1 - e^{-iW(t-t')}] \qquad (10.1\text{-}25)$$

The displacement gradient tensor $\mathbf{\Delta}$ is given in Appendix B, Eqs. B.7-3 and 5 as

$$\Delta_{ij} = \sum_k T_{ik} D_{kj} \qquad (i, j = r, \theta, z) \qquad (10.1\text{-}26)$$

in which (T_{ij}) is the orthogonal matrix in Eq. B.7-5 which rotates $\boldsymbol{r}$ into $\boldsymbol{r}'$, and (D_{ij}) is given in the present problem (according to Eq. B.7-3) by

$$(D_{ij}) = \begin{pmatrix} 1 & \dfrac{1}{r}\dfrac{\partial r'}{\partial \theta} & \dfrac{\partial r'}{\partial z} \\ r'\dfrac{\partial \theta'}{\partial r} & \dfrac{r'}{r}\dfrac{\partial \theta'}{\partial \theta} & r'\dfrac{\partial \theta'}{\partial z} \\ 0 & 0 & 1 \end{pmatrix} \qquad (10.1\text{-}27)$$

The displacement gradient tensor $\mathbf{E}$ then has components

$$\mathrm{E}_{ij} = \Delta_{ij}^{-1} = \sum_k D_{ik}^{-1} T_{jk} \qquad (10.1\text{-}28)$$

since (T_{ij}) is orthogonal. However, to calculate $\gamma_{[0]}$ we need only the components of the matrix inverse to (D_{ij}), since

$$\gamma_{[0]ij} = \delta_{ij} - \sum_k \mathrm{E}_{ik}\mathrm{E}_{jk} = \delta_{ij} - \sum_k D_{ik}^{-1} D_{jk}^{-1} \qquad (10.1\text{-}29)$$

It is straightforward to invert (D_{ij}), and in doing so we use the fact that det $(D_{ij}) = 1$ because the fluid is assumed incompressible. In this way we obtain

$$(D_{ij}^{-1}) = \begin{pmatrix} \dfrac{r'}{r}\dfrac{\partial\theta'}{\partial\theta} & -\dfrac{1}{r}\dfrac{\partial r'}{\partial\theta} & \dfrac{r'}{r}\left(\dfrac{\partial r'}{\partial\theta}\dfrac{\partial\theta'}{\partial z} - \dfrac{\partial\theta'}{\partial\theta}\dfrac{\partial r'}{\partial z}\right) \\ -r'\dfrac{\partial\theta'}{\partial r} & 1 & r'\left(\dfrac{\partial\theta'}{\partial r}\dfrac{\partial r'}{\partial z} - \dfrac{\partial\theta'}{\partial z}\right) \\ 0 & 0 & 1 \end{pmatrix} \tag{10.1-30}$$

Equations 10.1-24, 29, and 30 can now be combined to give $\gamma_{[0]}$. If we retain only terms through first order in a, then the only nonzero components of $\gamma_{[0]}$ are

$$\gamma_{[0]rz} = \gamma_{[0]zr} = a\,\mathscr{R}e\left\{i\Omega\frac{dv_r^0}{dz}\right\} \tag{10.1-31}$$

$$\gamma_{[0]\theta z} = \gamma_{[0]z\theta} = a\,\mathscr{R}e\left\{i\Omega\frac{dv_\theta^0}{dz}\right\} \tag{10.1-32}$$

where we have made use of the continuity equation, Eq. 10.1-17, in showing that $\gamma_{[0]rr}$ and $\gamma_{[0]r\theta}$ are zero.

When combined with the Lodge rubberlike liquid, Eqs. 10.1-31 and 32 give the stress components up to and including linear terms in a:

$$\tau_{rz} = \tau_{zr} = \mathscr{R}e\left\{\left[-\int_0^\infty \frac{iM(s)}{W}(1 - e^{-iWs})ds\right]\left[-aWe^{i\theta}\frac{dv_r^0}{dz}\right]\right\} = -aW\mathscr{R}e\left\{\eta^*(W)e^{i\theta}\frac{dv_r^0}{dz}\right\} \tag{10.1-33}$$

$$\tau_{\theta z} = \tau_{z\theta} = -aW\,\mathscr{R}e\left\{\eta^*(W)e^{i\theta}\frac{dv_\theta^0}{dz}\right\} \tag{10.1-34}$$

where we have used Eq. 5.3-6, after integration by parts, to replace the Fourier transform of the relaxation modulus with the complex viscosity (see also Problem 5B.7). All other stress components are zero to the present order of approximation.

(c) Equation of Motion

Since the stresses are periodic in θ, it is reasonable also to assume that the pressure varies periodically in θ so we write

$$p = p_0 + \mathscr{R}e\{\bar{p}(r, z)e^{i\theta}\} \tag{10.1-35}$$

where p_0 is a constant. When Eqs. 10.1-15 and 33 through 35 are put into the equation of motion written in cylindrical coordinates we find to first order in the offset a:

r-component: $$0 = -\mathscr{R}e\left\{\frac{\partial\bar{p}}{\partial r}\right\} + aW\,\mathscr{R}e\left\{\eta^*\frac{d^2v_r^0}{dz^2}\right\} \tag{10.1-36}$$

θ-component: $$0 = -\mathscr{R}e\left\{i\frac{\bar{p}}{r}\right\} + aW\,\mathscr{R}e\left\{\eta^*\frac{d^2v_\theta^0}{dz^2}\right\} \tag{10.1-37}$$

z-component: $$0 = -\mathscr{R}e\left\{\frac{\partial\bar{p}}{\partial z}\right\} \tag{10.1-38}$$

where we have made use of the continuity equation and the fact that inertial terms are negligible. Equation 10.1-38 shows that $\bar{p} = \bar{p}(r)$. Note that Eqs. 10.1-36 through 38 are the same as the creeping flow Navier–Stokes equations for the assumed velocity and pressure fields except that the constant Newtonian viscosity μ is replaced by the complex viscosity η^*. This means, of course, that *the material behavior in the eccentric rotating disk flow is completely determined by the complex viscosity* η^* *for small eccentricity and that no other material functions enter.*[4] To use this experiment to measure η^* we must complete finding the details of the velocity field.

(d) Velocity Field and Forces on the Lower Plate

It is easy to see that the equations of continuity and motion and the boundary conditons are satisfied by

$$v_r^0 = z/b \tag{10.1-39}$$

$$v_\theta^0 = iz/b \tag{10.1-40}$$

$$\bar{p} = 0 \tag{10.1-41}$$

In the eccentric rotating disk experiment, two orthogonal components of the force in the xy-plane on one of the plates are measured. If we let F_x and F_y be the components of this force in the x- and y-directions, respectively, then we have for the bottom plate ($z = -b/2$)

$$F_x = -\int_0^{2\pi}\int_0^R [\tau_{zr}\cos\theta - \tau_{z\theta}\sin\theta]_{z=-b/2}\, r\, dr\, d\theta \tag{10.1-42}$$

$$F_y = -\int_0^{2\pi}\int_0^R [\tau_{zr}\sin\theta + \tau_{z\theta}\cos\theta]_{z=-b/2}\, r\, dr\, d\theta \tag{10.1-43}$$

The necessary stresses are found by substituting Eqs. 10.1-39 and 40 into Eqs. 10.1-33 and 34, and these can then be put back into the force expressions. When we replace η^* by $\eta' - i\eta''$ we obtain the rheological properties:

$$\boxed{\eta'(W) = \left[\frac{b}{\pi W a R^2}\right] F_x} \tag{10.1-44}$$

$$\boxed{\eta''(W) = \left[\frac{b}{\pi W a R^2}\right] F_y} \tag{10.1-45}$$

Equations 10.1-44 and 45 show how measurements of the (steady) forces F_x and F_y can be converted immediately into the linear viscoelastic material functions.

Throughout this analysis we have used the fact that the disturbance velocities v_r^0 and v_θ^0 were small. To be certain that this condition is satisfied, it is necessary to vary a in the experiments. In the limit as $a \to 0$, the quantity Wa can be made small enough for the theory to be satisfied, in which case F_x/a and F_y/a approach constant values which yield the complex viscosity according to Eqs. 10.1-44 and 45.

[4] K. Walters, *loc cit.*, gives a more general result valid for other eccentric geometries.

§10.2 STEADY-STATE SHEAR FLOWS

In this section we focus our attention on the determination of the viscometric functions. The usual geometries for achieving steady shear flow were introduced in §3.7, where viscometric flows were defined. The most important arrangements are the cone-and-plate, parallel-disk, circular tube, rectangular slit, and concentric cylinder apparatuses. For each, the equations of motion must be used together with the experimentally applied boundary conditions in order to relate the steady shear flow material functions to the measured forces and velocities. Several relations obtainable in this way are tabulated in Table 10.2-1 for reference purposes along with a listing of the quantities measured for each type of experiment.[1] The derivations of these relations are treated in detail in the examples at the end of this section and in problems at the end of the chapter.

The most widely used geometry for complete characterization of the viscometric functions is the cone-and-plate configuration; the angle between the cone and plate is very small, typically less than 4°. It is available in a variety of commercial rheological instruments. At high rates of rotation in the cone-and-plate apparatus, a steady shear flow is not possible because of inertial, viscoelastic, and possibly other effects. A particularly severe limitation results from the tendency of viscoelastic samples to fracture[2] at the free surface in cone-and-plate flow at moderate to high rotation rates. Thus, this geometry cannot be used to obtain rheological data at very high shear rates. The cone-and-plate instrument is also convenient for measuring unsteady unidirectional shear flow material functions.

Because of its freedom from inertial effects (see Example 10.2-3) and edge fracture, the capillary viscometer is used to determine viscosity at very high shear rates ($\dot{\gamma} > 10^2\ s^{-1}$). However, flow in the capillary viscometer yields no information on the normal stress coefficients.[3] The rectangular slit is similarly useful for high shear rate measurements of the viscosity, and by using a combination of flush-mounted and hole-mounted pressure transducers it is possible to measure the first normal stress coefficient with this instrument at high shear rates.[4]

A parallel-disk configuration can also be used to obtain all of the viscometric functions. This arrangement has the same drawback as the cone-and-plate that at high rotation rates edge fracture and/or inertial forces become important. It has the advantage, however, that the fluid motion is viscometric at all low rates of rotation, whereas the cone-and-plate flow is steady shear only in the limit of small cone angles. Moreover, high shear rates can be obtained in the parallel-plate device at moderate rotation rates by using very small gaps. From Table 10.2-1, we see that to find either of the normal stress coefficients separately from torsional flow measurements, we must know values of the pressure exerted against one of the disks at specific positions. Many early local pressure measurements were

[1] A review of experimental geometries and appropriate equations for determining the viscometric functions is given by A. C. Pipkin and R. I. Tanner, *Mech. Today*, **1**, 262–321 (1972). H. Markovitz in F. R. Eirich, ed., *Rheology*, Vol. 4, Academic Press, New York (1967), Chapt. 6, pp. 347–410. R. W. Whorlow, *Rheological Techniques*, Ellis Horwood, Chichister, UK (1980). See also R. I. Tanner, *Engineering Rheology*, Oxford University Press (1985), Chapter 3.

[2] J. F. Hutton, *Rheol. Acta*, **8**, 54–59 (1969); R. I. Tanner and M. Keentok, *J. Rheol.*, **27**, 47–57 (1983).

[3] Measurements have been made of the thrust of the fluid exiting from the capillary tube on a wall, and attempts have been made to relate these measurements to the normal stress coefficients. For a summary of errors in the literature associated with the measurement of normal stresses in capillaries and slits, see J. M. Davies, J. F. Hutton, and K. Walters, *J. Phys. D.: Appl. Phys.*, **6**, 2259–2266 (1973), and K. Walters, *Rheometry*, Chapman and Hall, London (1975), pp. 96–105. For a discussion of the use of the exit pressure in capillaries and slits to determine normal stresses, see D. V. Boger and M. M. Denn, *J. Non-Newtonian Fluid Mech.*, **6**, 163–185 (1980).

[4] A. S. Lodge and L. deVargas, *Rheol. Acta*, **22**, 151–170 (1983); A. S. Lodge, *Polym. News*, **9**, 242–246 (1984); A. S. Lodge, *Chem. Eng. Commun.*, **32**, 1–60 (1985).

TABLE 10.2-1

Examples of Relations for Determining the Viscometric Functions (η, Ψ_1, Ψ_2) in Standard Experimental Arrangements

A. *Capillary Viscometer* (Table 3.7-2*a* and Example 10.2-3)

Q = Volume rate of flow
$\Delta\mathscr{P}$ = Pressure drop through tube
R = Tube radius
L = Tube length
$\dot{\gamma}_R$ = Shear rate at tube wall
τ_R = Shear stress at tube wall

$$\eta(\dot{\gamma}_R) = \frac{\tau_R}{(Q/\pi R^3)}\left[3 + \frac{d \ln (Q/\pi R^3)}{d \ln \tau_R}\right]^{-1} \tag{A-1}$$

$$\dot{\gamma}_R = \frac{1}{\tau_R^2}\frac{d}{d\tau_R}(\tau_R^3 Q/\pi R^3) \tag{A-2}$$

$$\tau_R = \Delta\mathscr{P}R/2L \tag{A-3}$$

B. *Cone-and-Plate Instrument* (Table 3.7-2*d* and Example 10.2-1)

R = Radius of circular plate
ϑ_0 = Angle between cone and plate (usually less than 4°)
W_0 = Angular velocity of cone
$\mathscr{T}$ = Torque on plate
$\mathscr{F}$ = Force required to keep tip of cone in contact with circular plate
$\pi_{\theta\theta}(r)$ = Pressure measured by flush-mounted pressure transducers located on plate
p_a = Atmospheric pressure

$$\eta(\dot{\gamma}) = \frac{3\mathscr{T}}{2\pi R^3 \dot{\gamma}} \tag{B-1}$$

$$\Psi_1(\dot{\gamma}) = \frac{2\mathscr{F}}{\pi R^2 \dot{\gamma}^2} \tag{B-2}$$

$$\Psi_1(\dot{\gamma}) + 2\Psi_2(\dot{\gamma}) = -\frac{1}{\dot{\gamma}^2}\frac{\partial \pi_{\theta\theta}}{\partial \ln r} \tag{B-3}$$

$$\Psi_2(\dot{\gamma}) = \frac{p_a - \pi_{\theta\theta}(R)}{\dot{\gamma}^2} \tag{B-4}$$

$$\dot{\gamma} = W_0/\vartheta_0 \tag{B-5}$$

C. *Parallel-Disk Instrument* (Table 3.7-2*c* and Example 10.2-2)

R = Radius of disks
H = Separation of disks
W_0 = Angular velocity of upper disk

$$\eta(\dot{\gamma}_R) = \frac{(\mathscr{T}/2\pi R^3)}{\dot{\gamma}_R}\left[3 + \frac{d \ln (\mathscr{T}/2\pi R^3)}{d \ln \dot{\gamma}_R}\right] \tag{C-1}$$

$\mathscr{T}$ = Torque required to rotate upper disk $\mathscr{F}$ = Force required to keep separation of two disks constant $\dot{\gamma}_R$ = Shear rate at edge of system $\pi_{zz}(0)$, $\pi_{zz}(R)$ = Normal pressure measured on disk at center and at rim p_a = Atmospheric pressure	$$\Psi_1(\dot{\gamma}_R) - \Psi_2(\dot{\gamma}_R) = \frac{(\mathscr{F}/\pi R^2)}{\dot{\gamma}_R^2}\left[2 + \frac{d\ln(\mathscr{F}/\pi R^2)}{d\ln\dot{\gamma}_R}\right] \quad \text{(C-2)}$$ $$\Psi_1(\dot{\gamma}_R) + \Psi_2(\dot{\gamma}_R) = \frac{1}{\dot{\gamma}_R^2}\frac{d\pi_{zz}(0)}{d\ln\dot{\gamma}_R} \quad \text{(C-3)}$$ $$\Psi_2(\dot{\gamma}_R) = \frac{p_a - \pi_{zz}(R)}{\dot{\gamma}_R^2} \quad \text{(C-4)}$$ $$\dot{\gamma}_R = \frac{W_0 R}{H} \quad \text{(C-5)}$$
D. *Couette Viscometer* (Table 3.7-2*b*) *Narrow Gap* R_1, R_2 = Radii of inner and outer cylinders H = Height of cylinders W_1, W_2 = Angular velocities of inner and outer cylinders $\mathscr{T}$ = Torque on inner cylinder $\pi_{rr}(R_1)$, $\pi_{rr}(R_2)$ = Normal pressures measured on inner and outer cylinders	$$\eta(\dot{\gamma}) = \frac{\mathscr{T}(R_2 - R_1)}{2\pi R_1^3 H \lvert W_2 - W_1\rvert} \quad \text{(D-1)}$$ $$\Psi_1(\dot{\gamma}) = \frac{-[\pi_{rr}(R_2) - \pi_{rr}(R_1)]R_1}{(R_2 - R_1)\dot{\gamma}^2} + \frac{\rho R_1^2}{3\dot{\gamma}^2}(W_1^2 + W_2^2 + W_1 W_2) \quad \text{(D-2)}$$ $$\dot{\gamma} = \frac{\lvert W_2 - W_1\rvert R_1}{R_2 - R_1} \quad \text{(D-3)}$$
E. *Axial Annular Flow* (Problem 2B.5) R_1, R_2 = Radii of inner and outer cylinders $(\pi_{rr})_i$ = Reading of a flush-mounted pressure transducer at R_i	$$(\pi_{rr})_1 - (\pi_{rr})_2 = -\int_{R_1}^{R_2} \frac{\Psi_2\dot{\gamma}^2}{r}\,dr \quad \text{(E-1)}$$

(Continued)

TABLE 10.2-1 (*Continued*)

F. *Torsional Flow between Two Disks, the Upper One of Which is Rotating and Attached to a Vertical Tube* (Problem 10B.6)

h = Height of rise of fluid in tube
R_1 = Radius of tube
R_2 = Radius of disks
W_0 = Angular velocity of tube-disk assembly
H = Gap between disks
g = Gravitational acceleration

$$h = \frac{W_0^2}{\rho g H^2} \int_{R_1}^{R_2} (\Psi_1 + \Psi_2) r\, dr - \frac{W_0^2}{6g}(R_2^2 - R_1^2) \qquad \text{(F-1)}$$

G. *Truncated Cone-and-Plate Instrument* (Problem 10B.7)

R = Radius of circular plate
R_0 = Radius of truncated section of cone
ϑ_0 = Angle between cone and plate (usually less than 4°)
H = Gap between truncated section of cone and circular plate
$\pi'_{\theta\theta}(r)$ = Pressure measured by transducers mounted at the bottom of holes along bottom disk
$\pi_{\theta\theta}(r)$ = Normal stress on bottom circular disk
W = Angular velocity of truncated cone
$\dot{\gamma}_0$ = Shear rate for $R_0 \le r \le R$
p^* = Hole pressure
p_a = Atmospheric pressure

$$\pi_{\theta\theta}(0) = \pi'_{\theta\theta}(0) \qquad \text{(G-1)}$$

$$\Psi_1(\dot{\gamma}_0) + 2\Psi_2(\dot{\gamma}_0) = -\frac{1}{\dot{\gamma}_0^2}\frac{\partial \pi'_{\theta\theta}}{\partial \ln r} \quad (R_0 \le r \le R) \qquad \text{(G-2)}$$

$$\Psi_1(\dot{\gamma}_0) + \Psi_2(\dot{\gamma}_0) = \frac{1}{\dot{\gamma}_0}\frac{d}{d\dot{\gamma}_0}\left\{(\pi_{\theta\theta}(0) - p_a) - [\Psi_1(\dot{\gamma}_0) + 2\Psi_2(\dot{\gamma}_0)]\dot{\gamma}_0^2 \ln \frac{R}{R_0}\right\} \qquad \text{(G-3)}$$

$$p^*(\dot{\gamma}_0) = p_a - \pi'_{\theta\theta}(R) - \Psi_2(\dot{\gamma}_0)\dot{\gamma}_0^2 \qquad \text{(G-4)}$$

$$\dot{\gamma}_0 = \frac{WR_0}{H} \qquad \text{(G-5)}$$

misinterpreted because of the hole pressure effect. Note, on the other hand, that with a cone-and-plate instrument, the first normal stress coefficient can be determined by measuring the total force required to keep the cone and plate together; this gives Ψ_1 directly without interference from the presence of holes.

The Couette viscometer, which utilizes the concentric cylinder geometry, allows a straightforward determination of viscosity when the gap between the cylinders is small. For large gaps the analysis is complicated because the velocity field depends on the viscosity function. Finally, the hole pressure can cause significant errors in this instrument, making measurement of the normal stresses difficult.

With most of the above devices it is difficult to make measurements in the zero-shear-rate region because the instrument forces become very small. To get the zero-shear-rate viscosity, a parallel-plate "sandwich" viscometer[5] or a falling sphere experiment (see §10.4 and Example 8.5-2) is useful. If the linear viscoelastic properties can be measured to lower frequencies, then the equality between η' and η at low ω and $\dot{\gamma}$ can be used to get η_0. Similarly, $\Psi_{1,0}$ can be evaluated if low-frequency η'' data are available (cf. Eq. 3.6-21).

In the following examples, we show how a few of the systems mentioned here are analyzed. Some of the operational difficulties encountered with each are also discussed. Analysis of these flow problems is sometimes simplified by the fact that for steady-state shear flow, the constitutive equation is known to be the CEF equation, Eq. 9.6-18.

EXAMPLE 10.2-1 Measurement of Viscosity and Normal Stress Coefficients in the Cone-and-Plate Instrument

A cone-and-plate geometry is shown in Fig. 10.2-1. The fluid to be tested is placed in the gap between the cone and plate; the cone is rotated with an angular velocity W. Three measurements can be made: the torque $\mathscr{T}$ on the plate, the total normal thrust $\mathscr{F}$ on the plate, and the "pressure distribution," $(p + \tau_{\theta\theta})|_{\theta=\pi/2}$ across the plate. From these measurements we want to get the material

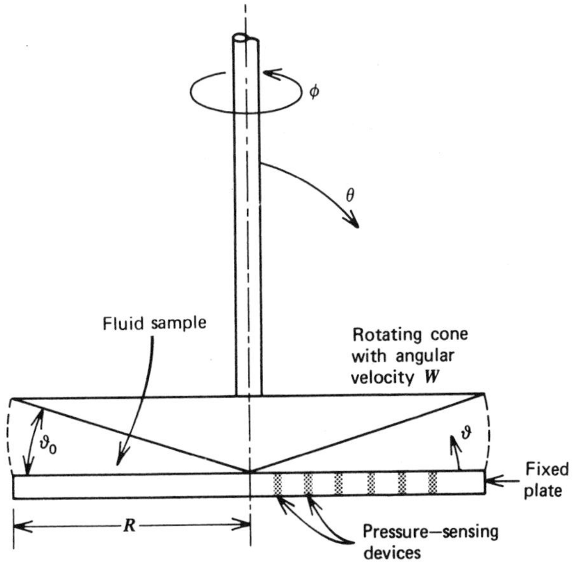

FIGURE 10.2-1. Cone-and-plate device; ϑ_0 is quite small, of the order of several degrees.

[5] H. M. Laun and J. Meissner, *Rheol. Acta*, **19**, 60–67 (1980).

functions, η, Ψ_1, and Ψ_2. Assume that inertial forces can be neglected. Recall that in Example 1.3-4 it was shown how to find the viscosity of a Newtonian fluid from cone-and-plate viscometer data.

SOLUTION **(a)** The Shear Rate

The small cone angle analysis of Example 1.3-4 can be taken over directly here to show that the shear rate $\dot{\gamma}$ is uniform throughout the gap and is equal to

$$\dot{\gamma} = -\dot{\gamma}_{\theta\phi} = \frac{W}{\vartheta_0} \tag{10.2-1}$$

Since η, Ψ_1, and Ψ_2 are functions of $\dot{\gamma}$ alone, the CEF equation tells us that the stress tensor $\boldsymbol{\tau}$ is also uniform throughout the gap.

(b) The Shear Stress

Since the shear stress is constant in the fluid, we can immediately take over Eqs. 1.3-36 and 37 to give the shear stress and viscosity in terms of the measured torque and rotation rate of the cone

$$\boxed{\eta(\dot{\gamma}) = \frac{3\mathscr{T}\vartheta_0}{2\pi R^3 W}} \tag{10.2-2}$$

(c) The Radial Pressure Distribution

From the radial component of the equation of motion (neglecting the centrifugal force term $-\rho v_\phi^2/r$) we get

$$0 = -\frac{\partial p}{\partial r} - \frac{1}{r^2}\frac{\partial}{\partial r}(r^2\tau_{rr}) + \frac{\tau_{\theta\theta} + \tau_{\phi\phi}}{r} \tag{10.2-3}$$

Since $\tau_{\theta\theta}$ is constant, we can replace $\partial p/\partial r$ by $\partial \pi_{\theta\theta}/\partial r$; we do this because pressure transducers will measure $\pi_{\theta\theta}$ rather than p. The normal stress terms can be rewritten in terms of the material functions to give

$$\frac{\partial}{\partial \ln r}\pi_{\theta\theta} = -(\Psi_1 + 2\Psi_2)\dot{\gamma}^2 \tag{10.2-4}$$

Hence, if $\pi_{\theta\theta}$ is measured on the plate (where $\theta = \pi/2$) and plotted against ln r, the slope will yield the indicated combination of normal stress coefficients. Since data in §3.3 indicate that Ψ_1 is positive and Ψ_2 is roughly $-0.1\Psi_1$, we would expect the above slope to be negative; this is illustrated in Fig. 10A.2.

The pressure at the edge of the plate $\pi_{\theta\theta}(R)$ can be related to the second normal stress coefficient by means of a normal force balance across the fluid–air interface at the rim. If we assume the liquid surface at $r = R$ to be spherical, then $\pi_{rr}(R) = p_a$ where p_a is atmospheric pressure. When this is combined with the definition of the second normal stress coefficient we find

$$\Psi_2 = (p_a - \pi_{\theta\theta}(R))/\dot{\gamma}^2 \tag{10.2-5}$$

Equation 10.2-4 may be integrated over the plate from R to r and combined with Eq. 10.2-5 to give

$$\begin{aligned} \pi_{\theta\theta} &= \pi_{\theta\theta}(R) - (\Psi_1 + 2\Psi_2)\dot{\gamma}^2 \ln\left(\frac{r}{R}\right) \\ &= p_a - \Psi_2\dot{\gamma}^2 - (\Psi_1 + 2\Psi_2)\dot{\gamma}^2 \ln\left(\frac{r}{R}\right) \end{aligned} \tag{10.2-6}$$

This result gives the radial distribution of the force per unit area measured by pressure transducers on the bottom plate.

(d) Total Thrust on the Lower Plate

The total thrust of the fluid on the plate minus the thrust associated with atmospheric pressure p_a is $\mathscr{F} = 2\pi \int_0^R \pi_{\theta\theta}\, r dr - \pi R^2 p_a$. Insertion of $\pi_{\theta\theta}$ from Eq. 10.2-6 then gives after integration

$$\boxed{\Psi_1 = \frac{2\mathscr{F}}{\pi R^2 \dot{\gamma}^2}} \qquad (10.2\text{-}7)$$

Many commercially available cone-and-plate instruments are equipped to measure the total thrust $\mathscr{F}$, and thus it is possible to determine the first normal stress difference by means of Eq. 10.2-7. On a few instruments it is also possible to measure the local pressure distribution along the plate; for these devices both Ψ_1 and Ψ_2 can be measured.[6]

There are several assumptions that have been made in the derivations of the material functions given above that can introduce error into interpretation of experimental measurements.[1,6] First, the fluid inertia,[7] which tends to throw the fluid out of the gap, has been neglected. Consequently, when inertial effects are important the total thrust measured will be less than that given by Eq. 10.2-7 for the true first normal stress difference, and Ψ_1 will be underestimated. In practice, the best way to correct for the centrifugal force is to calibrate the instrument using Newtonian liquids. Inertial effects can also be accounted for approximately by theoretical means (see Problem 10B.3).

Second, the assumed rim condition is that the fluid–air interface is spherical, and experimentally this is not found exactly. The nonspherical free surface implies that the flow is not viscometric all the way to the edge of the instrument. However, varying the free surface shape does not seem to affect the total thrust measurements,[8] and errors associated with the nonspherical interface appear to be no more than 5 %. To achieve reproducibility the cone-and-plate device is sometimes operated in a sea of liquid to avoid a free surface.

Third, for large cone angles and high speeds of rotation, secondary flows are observed in the gap as a result of the competing centrifugal and normal stress effects.[9] Secondary flows may occur for cone angles as small as 10°. If secondary flows are present, the overall fluid motion will not be viscometric and the analysis presented here is invalid.

Fourth, the temperature of the fluid in the gap may not be uniform due to viscous heating. An approximate analysis of nonisothermal effects in the cone-and-plate viscometer gives a maximum temperature rise within the gap[10]

$$(T - T_0)_{\max} = \frac{3\mathscr{T} W \vartheta_0}{16\pi k R} \qquad (10.2\text{-}8)$$

[6] See, for example, S. Ramachandran, H. W. Gao, and E. B. Christiansen, *Macromolecules*, **18**, 695–699 (1985); H. W. Gao, S. Ramachandran, and E. B. Christiansen, *J. Rheol.*, **25**, 213-235 (1981); E. B. Christiansen and W. R. Leppard, *Trans. Soc. Rheol.*, **18**, 65–86 (1974).

A method that has been suggested for measuring Ψ_2 in a cone-and-plate geometry without small pressure transducers is to vary the gap between the cone and plate. See R. Jackson and A. Kaye, *Br. J. Appl. Phys.*, **17**, 1355-1360 (1966); B. D. Marsh and J. R. A. Pearson, *Rheol. Acta*, **7**, 326–331 (1968); K. Walters, *Rheometry*, Chapman and Hall, London (1975), pp. 56–58.

[7] Neglect of fluid inertia can also lead to errors in the measurement of small-amplitude oscillatory shear properties in the cone-and-plate geometry. The errors can be serious at high frequencies, particularly for η''. See K. Walters and R. A. Kemp in R. E. Wetton and R. W. Whorlow, eds., *Polymer Systems*, Macmillan, London (1968), pp. 237–250.

[8] E. Ashare, Ph.D. Thesis, University of Wisconsin, Madison (1968), pp. 50–52, 67–68, 105.

[9] H. Giesekus, *Rheol. Acta*, **4**, 85–101 (1965); W. H. Hoppmann II and C. E. Miller, *Trans. Soc. Rheol.*, **7**, 181–193 (1963); K. Walters and N. D. Waters in R. E. Wetton and R. W. Whorlow, eds., *Polymer Systems*, Macmillan, London (1968), pp. 211–235; R. M. Turian, *Ind. Eng. Chem. Fundam.*, **11**, 361–368 (1972).

[10] This result was obtained by R. B. Bird and R. M. Turian, *Chem. Eng. Sci.*, **17**, 331–334 (1961), by using a variational method.

where T_0 is the temperature at which the cone-and-plate surfaces are held constant, and k is the thermal conductivity of the fluid. Equation 10.2-8 can be used to estimate the importance of viscous heating; if a temperature rise of 1 or 2 K is predicted, then serious errors can result from use of the isothermal analysis presented here. Viscous heating for some simple non-Newtonian fluid models is discussed in §4.4.

Finally, there are other problems associated with polymer degradation,[11] melt fracture, bubble inclusion, solvent evaporation, possible slip at the wall, balling of the material, and alignment of the instrument that add uncertainty to the reported values.

EXAMPLE 10.2-2 Measurement of the Viscometric Functions in the Parallel-Disk Instrument

The parallel-disk system (Fig. 10.1-1) is very similar in operation to the cone-and-plate device discussed in the preceding example. As before, when the upper disk is rotated with constant angular velocity W, the torque $\mathscr{T}$ required to achieve this rotation as well as the total force $\mathscr{F}$ required to maintain the disks at a separation H, and the pressure distribution $(p + \tau_{zz})|_{z=0}$ across one of the plates can be measured. Show how to obtain the viscometric functions in terms of these measurements when inertial effects are unimportant. (See Problem 1B.5 for Newtonian flow in this system.)

SOLUTION (a) Governing Equations

We begin by postulating that a fluid plane of constant z rotates with an angular velocity $w(z)$ that depends on its position between the two disks. The velocity field is then $v_\theta = rw(z)$, $v_r = 0$, and $v_z = 0$; accordingly, the shear rate is $\dot{\gamma} = r\,dw/dz = rw'$. Unlike the cone-and-plate rheometer, the shear rate is not constant throughout the fluid. Since the flow is steady-state shearing, the CEF equation, Eq. 9.6-18, can be used as the constitutive equation.

To evaluate the stress tensor with the CEF equation, we need the following kinematic tensors derived from the postulated velocity field

$$\nabla \boldsymbol{v} = \begin{pmatrix} 0 & w & 0 \\ -w & 0 & 0 \\ 0 & rw' & 0 \end{pmatrix}; \qquad \boldsymbol{\gamma}_{(1)} = \dot{\gamma}\begin{pmatrix} 0 & 0 & 0 \\ 0 & 0 & 1 \\ 0 & 1 & 0 \end{pmatrix}; \qquad \dot{\gamma} = rw' \tag{10.2-9}$$

$$\boldsymbol{\gamma}_{(1)}^2 = \dot{\gamma}^2\begin{pmatrix} 0 & 0 & 0 \\ 0 & 1 & 0 \\ 0 & 0 & 1 \end{pmatrix} \tag{10.2-10}$$

$$\boldsymbol{\gamma}_{(2)} = -2\dot{\gamma}^2\begin{pmatrix} 0 & 0 & 0 \\ 0 & 1 & 0 \\ 0 & 0 & 0 \end{pmatrix} \tag{10.2-11}$$

When these contributions are assembled into the CEF equation the stress tensor is

$$\boldsymbol{\tau} = -\begin{pmatrix} 0 & 0 & 0 \\ 0 & (\Psi_1 + \Psi_2)\dot{\gamma}^2 & \eta\dot{\gamma} \\ 0 & \eta\dot{\gamma} & \Psi_2\dot{\gamma}^2 \end{pmatrix} \tag{10.2-12}$$

[11] J. M. Lipson and A. S. Lodge, *Rheol. Acta*, **7**, 364–379 (1968); W. G. Pritchard, *Phil. Trans. Roy. Soc. London*, **A270**, 507–556 (1971).

If fluid inertia and gravity are neglected, the equation of motion in cylindrical coordinates takes the form

$$r\text{-Component:} \qquad 0 = -\frac{\partial p}{\partial r} - \frac{(\Psi_1 + \Psi_2)\dot{\gamma}^2}{r} \tag{10.2-13}$$

$$\theta\text{-Component:} \qquad 0 = \frac{\partial}{\partial z}\,\eta\dot{\gamma} \tag{10.2-14}$$

$$z\text{-Component:} \qquad 0 = -\frac{\partial p}{\partial z} + \frac{\partial}{\partial z}\Psi_2\dot{\gamma}^2 \tag{10.2-15}$$

Equation 10.2-14 says that the shear stress, and hence the shear rate, is independent of z. Thus the expression for the shear rate given in Eq. 10.2-9 can be integrated to yield w, which in turn gives

$$v_\theta = \frac{rWz}{H} \tag{10.2-16}$$

Thus the shear rate is a function of r alone:

$$\dot{\gamma} = \frac{rW}{H} = \dot{\gamma}_R r/R \tag{10.2-17}$$

where $\dot{\gamma}_R = RW/H$ is the shear rate at the rim.

(b) The Viscosity

In order to find the viscosity of the sample, we consider the total torque $\mathcal{T}$ required to rotate the upper disk:

$$\begin{aligned} \mathcal{T} &= 2\pi \int_0^R (-r\tau_{z\theta})r\,dr \\ &= 2\pi \int_0^R \eta\dot{\gamma}r^2\,dr \\ &= \frac{2\pi R^3}{\dot{\gamma}_R^3} \int_0^{\dot{\gamma}_R} \eta(\dot{\gamma})\dot{\gamma}^3\,d\dot{\gamma} \end{aligned} \tag{10.2-18}$$

In going from the second to third line of Eq. 10.2-18 we have made the change of variable indicated by Eq. 10.2-17. Because of the inhomogeneity of the shear rate in torsional flow, the viscosity is not an explicit function of the applied torque. However by differentiating this last result with respect to $\dot{\gamma}_R$ we find[12]

$$\boxed{\eta(\dot{\gamma}_R) = \frac{(\mathcal{T}/2\pi R^3)}{\dot{\gamma}_R}\left[3 + \frac{d\ln(\mathcal{T}/2\pi R^3)}{d\ln\dot{\gamma}_R}\right]} \tag{10.2-19}$$

Thus by varying $\dot{\gamma}_R$ and computing the change in torque as indicated above, the viscosity may be determined explicitly.

[12] Use the "Leibniz formula" for differentiating an integral:

$$\frac{d}{dt}\int_{a_1(t)}^{a_2(t)} f(x,t)dx = \int_{a_1(t)}^{a_2(t)} \frac{\partial f}{\partial t}dx + f(a_2,t)\frac{da_2}{dt} - f(a_1,t)\frac{da_1}{dt} \tag{10.2-19a}$$

(c) The Radial Pressure Distribution

The r-component of the equation of motion suggests that we can get the sum of the first and second normal stress coefficients from the radial pressure distribution. To see this we integrate this equation from a position r to the rim $r = R$ to obtain

$$p(r) - p_a = \int_{\dot{\gamma}}^{\dot{\gamma}_R} (\Psi_1 + \Psi_2)\dot{\gamma}\, d\dot{\gamma} \tag{10.2-20}$$

where we have set $p(R) = p_a$, since $\tau_{rr} = 0$, and we have changed the integration variable from r to $\dot{\gamma}$. The quantity measured by pressure transducers along the bottom plate is π_{zz} rather than p, so we add τ_{zz} as given by the CEF equation to both sides of Eq. 10.2-20. The result is

$$\pi_{zz}(\dot{\gamma}) - p_a = \int_{\dot{\gamma}}^{\dot{\gamma}_R} (\Psi_1 + \Psi_2)\dot{\gamma}\, d\dot{\gamma} - \Psi_2\dot{\gamma}^2 \tag{10.2-21}$$

To obtain an explicit expression for the normal stress coefficients we evaluate this last result at the center of the disk ($\dot{\gamma} = 0$), and then differentiate the result with respect to $\dot{\gamma}_R$ as we did in Eq. 10.2-19. In this way we obtain

$$\frac{d}{d\dot{\gamma}_R}(\pi_{zz}(0) - p_a) = (\Psi_1(\dot{\gamma}_R) + \Psi_2(\dot{\gamma}_R))\dot{\gamma}_R \tag{10.2-22}$$

This shows how one combination of normal stresses can be obtained from measurements of π_{zz} at the center of the disk at different shear rates. Note that since the shear rate is zero at $r = 0$, then $\pi_{zz}(0)$ can be measured with a hole mounted pressure transducer without having to corrrect for the hole pressure.

(d) The Second Normal Stress Coefficient

The second normal stress coefficient can easily be found by evaluating Eq. 10.2-21 at the rim:

$$\pi_{zz}(R) - p_a = -\Psi_2(\dot{\gamma}_R)\dot{\gamma}_R^2 \tag{10.2-23}$$

Accurate measurement of the rim pressure $\pi_{zz}(R)$ requires use of flush-mounted pressure transducers or careful correction for the hole pressure if hole-mounted transducers are used.

(e) The Total Thrust on the Lower Plate

The total force $\mathcal{F}$ that must be applied to keep the disks from separating is found by integrating $\pi_{zz}(r) - p_a$ over the area of one of the disks. Using π_{zz} from Eq. 10.2-21 gives

$$\begin{aligned}
\mathcal{F} &= \frac{2\pi R^2}{\dot{\gamma}_R^2}\int_0^{\dot{\gamma}_R}\left[\int_{\dot{\gamma}}^{\dot{\gamma}_R}[\Psi_1(x) + \Psi_2(x)]x\,dx - \Psi_2(\dot{\gamma})\dot{\gamma}^2\right]\dot{\gamma}\, d\dot{\gamma} \\
&= \frac{2\pi R^2}{\dot{\gamma}_R^2}\left[\int_0^{\dot{\gamma}_R}\int_0^{x}[\Psi_1(x) + \Psi_2(x)]\dot{\gamma}\, d\dot{\gamma}\, x\,dx - \int_0^{\dot{\gamma}_R}\Psi_2(\dot{\gamma})\dot{\gamma}^3\, d\dot{\gamma}\right] \\
&= \frac{\pi R^2}{\dot{\gamma}_R^2}\int_0^{\dot{\gamma}_R}(\Psi_1 - \Psi_2)\dot{\gamma}^3\, d\dot{\gamma}
\end{aligned} \tag{10.2-24}$$

Solving Eq. 10.2-24 for the normal stress coefficients gives

$$\boxed{\Psi_1 - \Psi_2 = \frac{1}{\dot{\gamma}_R^2}\left(\frac{\mathcal{F}}{\pi R^2}\right)\left[2 + \frac{d\ln(\mathcal{F}/\pi R^2)}{d\ln\dot{\gamma}_R}\right]} \tag{10.2-25}$$

Thus in a parallel-disk rheometer, one can determine the non-Newtonian viscosity and both normal stress coefficients by varying the angular velocity of the upper disk and the separation of the disks.

EXAMPLE 10.2-3 Obtaining the Non-Newtonian Viscosity from the Capillary Rheometer

Show how the viscosity as a function of shear rate can be determined from measurements of volume flow rate and pressure drop in a capillary tube.

SOLUTION **(a)** The Shear Stress

It is not difficult to show from the equation of motion that for steady tube flow the shear stress varies linearly with distance from the center of the tube (see Example 4.2-1)

$$\tau_{rz} = \tau_R r/R \tag{10.2-26}$$

where τ_R is the shear stress at the tube wall and is given by

$$\tau_R = (\mathscr{P}_0 - \mathscr{P}_L)R/2L \tag{10.2-27}$$

Note that, unlike the situation for the rotational instruments discussed above, the inertial terms in the equation of motion are identically zero for the capillary viscometer. This means that the results from this example can be applied to obtain the viscosity irrespective of the flow rate, provided only that the flow is laminar.

(b) The Shear Rate[13]

The volume rate of flow is given by

$$Q = 2\pi \int_0^R v_z r \, dr \tag{10.2-28}$$

Integration by parts gives

$$Q = -\pi \int_0^R \left(\frac{dv_z}{dr}\right) r^2 \, dr = \pi \int_0^R \dot{\gamma} r^2 \, dr \tag{10.2-29}$$

In Eq. 10.2-29 we have introduced the shear rate $\dot{\gamma} = -dv_z/dr$. Next we change the variable of integration from r to τ_{rz} according to Eq. 10.2-26:

$$\left(\frac{Q}{\pi R^3}\right) = \frac{1}{\tau_R^3} \int_0^{\tau_R} \dot{\gamma} \tau_{rz}^2 \, d\tau_{rz} \tag{10.2-30}$$

The shear rate is to be regarded as a function of shear stress in this last equation. Equation 10.2-30 indicates that data taken in tubes of different lengths and radii should collapse onto a single curve when plotted as $(Q/\pi R^3)$ vs. τ_R. To obtain the desired expression for the shear rate at the wall, we differentiate Eq. 10.2-30 with respect to τ_R

$$\dot{\gamma}_R = \frac{1}{\tau_R^2} \frac{d}{d\tau_R} \left[\tau_R^3 \frac{Q}{\pi R^3} \right] \tag{10.2-31}$$

[13] K. Weissenberg as cited by B. Rabinowitsch, *Z. Phys. Chem.*, **A145**, 1–26 (1929), and R. Eisenschitz, *Kolloid-Z.*, **64**, 184–195 (1933).

This well-known result is sometimes called the "Weissenberg–Rabinowitsch equation." It tells how the wall shear rate $\dot{\gamma}_R$ can be obtained by differentiating pressure drop-flow rate data. Once $\dot{\gamma}_R$ and τ_R are known the viscosity is easily found to be

$$\boxed{\eta(\dot{\gamma}_R) = \frac{\tau_R}{\dot{\gamma}_R} = \frac{\tau_R}{(Q/\pi R^3)}\left[3 + \frac{d\ln(Q/\pi R^3)}{d\ln \tau_R}\right]^{-1}} \tag{10.2-32}$$

Equation 10.2-32 gives the viscosity as a function of shear rate at the wall. If we in addition assume that $\eta(\dot{\gamma}_R)$ obtained at the wall is the same as $\eta(\dot{\gamma})$ throughout the tube, then Eq. 10.2-32 can be used to plot $\eta(\dot{\gamma})$. This assumption appears to be valid for typical polymeric fluids. However, for fluids where the wall may cause changes in the microstructure, for example, preferential orientation of fibers in suspensions of long fibers, Eq. 10.2-32 may not yield information about the viscosity away from the wall.

(c) End Effects

We now know how to calculate the viscosity, given data on the volume flow rate and pressure gradient for fully developed steady shear flow in a tube. The volume rate of flow is obtained in standard capillary viscometer measurements; however, the steady shear flow pressure gradient needed to compute the wall shear stress is not. Figure 10.2-2 illustrates a typical experimental arrangement for this instrument. The test fluid is pushed from a reservoir (barrel) through the tube by a ram which is driven at a specified constant velocity. The load on the ram, and thus the pressure p in the fluid

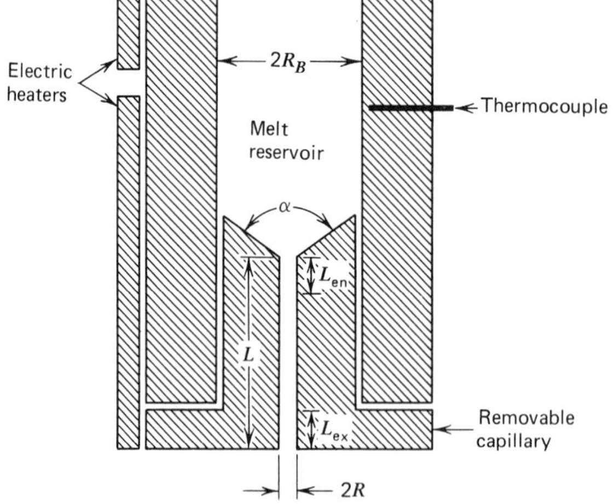

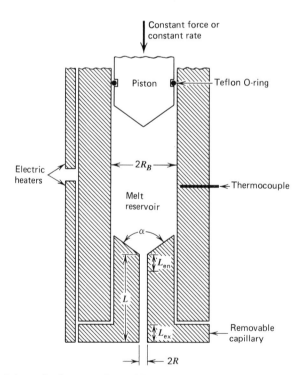

FIGURE 10.2-2. Schematic diagram of a capillary viscometer. The capillary test section is a tube of length L and radius R. Polymer is maintained in a fluid state in an upstream barrel of radius R_B; the converging, transition region between the barrel and capillary has an included angle α. The steady shear flow velocity profile develops over an entrance length L_{en}; the exit may disturb the fully developed velocity profile over a distance L_{ex} from the capillary end.

adjacent to the ram, is the measured pressure. Thus, in order to use the formulas just developed we must know how to compute the steady shear flow pressure gradient (or equivalently τ_R) from p.

The desired wall shear stress can be obtained by performing two experiments,[14] one with a capillary of length L_A attached to the barrel and the other with a capillary of length L_B. Both capillaries have radius R and are longer than the sum of the entrance length L_{en} required for the establishment of fully developed velocity profiles and the exit length L_{ex} over which the velocity profiles may be influenced by the end of the tube. This requirement ensures that there will be steady shear flow in some section of each capillary. The ram pressures p_A and p_B are then measured for the two capillaries for the same volume flow rate Q. The pressure gradient that exists in the steady shear flow portions of the two capillaries may then be computed by applying the macroscopic mechanical energy balance[15] to the two systems (see Problem 10C.1). By subtracting these two results, one can show that

$$\tau_R = \left[\frac{p_B - p_A}{L_B - L_A} + \underline{\rho g \cos \beta}\right]\frac{R}{2} = -\left(\frac{d\mathscr{P}}{dz}\right)\frac{R}{2} \qquad \begin{matrix} L_A > L_{\text{en}} + L_{\text{ex}} \\ L_B > L_{\text{en}} + L_{\text{ex}} \end{matrix} \tag{10.2-33}$$

where β is the angle between the tube axis and the vertical. This is then the corrected wall shear stress to be used with a flow rate Q through a capillary tube of radius R. On the right-hand side of Eq. 10.2-33, $d\mathscr{P}/dz$ is the constant pressure gradient corresponding to τ_R. Under ordinary operating conditions, the gravitational (dashed-underlined) term is negligible and $d\mathscr{P}/dz$ is very nearly equal to dp/dz. It is customary to measure the ram pressure for a series of tubes of radius R and different lengths L. The steady shear flow pressure gradient is then found from the slope of the linear region of a plot of p vs. L by replacing $(p_B - p_A)/(L_B - L_A)$ by dp/dL. The linear region is usually reached by L/R of about 10.

In addition to end effects, viscous dissipation heating, fluid compressibility, change of viscosity with pressure, and flow instabilities can introduce errors into capillary viscometer measurements. The temperature rise associated with viscous dissipation can be reduced by using a smaller diameter capillary, since for shear-thinning fluids the rate of heat conduction to the wall increases more rapidly than viscous heat generation with decreasing tube radius. The influence of pressure on viscosity can be seen as a concave-up shape to the pressure drop versus L/D curve at large L/D. If this upward curvature is observed it is necessary to use shorter or wider capillaries. At present, the best practice is to design and operate the viscometer so that these effects can be minimized.

§10.3 SHEARFREE FLOWS[1]

Methods for measuring material functions are not nearly so well developed for shearfree flows as for shearing flows. The primary difficulty is that of mechanically generating the flow because of the fact that in steady shearfree flow, neighboring particles move apart exponentially in time (cf. Eq. 3.1-4). This difficulty has led to many approximate methods for measuring shearfree flow material functions.

It is convenient to divide the measurement techniques for shearfree flows into those methods that yield material functions defined in Chapter 3 and those that do not.[2] We

[14] B. A. Toms in F. R. Eirich, ed., *Rheology*, Vol. 2, Academic Press (1958), pp. 475–501. The correction given in Eq. 10.2-33 was first suggested by Couette.

[15] A. G. Fredrickson, *Principles and Applications of Rheology*, Prentice-Hall, Englewood Cliffs, NJ (1964), pp. 196–200.

[1] For recent reviews of experimental methods for measuring shearfree flow material functions, see J. Meissner, *Chem. Eng. Commun.*, **33**, 159–180 (1985); *Ann. Rev. Fluid Mech.*, **17**, 45–64 (1985). We have drawn heavily on these reviews in this section.

[2] The terms *controllable* and *noncontrollable* have also been used to categorize shearfree flow experiments by F. Nazem and C. T. Hill, *Trans. Soc. Rheol.*, **18**, 87–101 (1974); see also K. Walters, *Rheometry*, Chapman and Hall, London (1975). These terms have also been applied with a different meaning to the description of shear flows by A. C. Pipkin, *Quart. Appl. Math.*, **26**, 87–100 (1968) and A. C. Pipkin and R. I. Tanner, *Mech. Today*, **1**, 262–321 (1972).

introduce this distinction here because, unlike the situation for shear flows where there are numerous methods available for measuring material functions of a wide variety of polymeric liquids, techniques for measuring the kinds of shearfree flow material functions defined in Chapter 3 have been developed only for polymer melts and high-viscosity solutions. Thus complex flows that are predominantly shearfree flows have been used to obtain qualitative information about the shearfree flow properties of low-viscosity polymer solutions.

We first consider uniaxial elongational flows given by Eq. 3.1-3 with $b = 0$ and $\dot{\varepsilon} > 0$. Then in the following two subsections, we look at methods for generating biaxial stretching flow (Eq. 3.1-3 with $b = 0$ and $\dot{\varepsilon} < 0$) and general shearfree flows (Eq. 3.1-3 with arbitrary b).

a. Uniaxial Elongational Flow ($b = 0$, $\dot{\varepsilon} > 0$)

We begin by discussing methods that yield material functions for uniaxial elongational flows. Two basic methods have been used for this purpose: one in which the ends of the sample are clamped and the clamps separated so as to generate either a known elongation rate or a known tensile stress in the material; and the second in which known velocities are applied to two places along the sample so as to achieve a controlled elongation rate or tensile stress.

The first method in which fixed ends are clamped is similar to the operation of a tensile testing apparatus. It is necessary, of course, that the cross head that moves the clamp(s) be programmable. In order to perform start-up of steady-state elongational flow the clamps must be separated exponentially in time. The most serious restriction resulting from this exponential motion is that only modest (Hencky) strains $\varepsilon_{max} = \dot{\varepsilon}t_{max} = \ln(L_{max}/L_0)$ can be achieved because of the size limitations of the instrument. Here L_0 and L_{max} are the initial and maximum sample lengths. For vertical configurations the sample must be surrounded by a density matching medium which also serves to maintain temperature uniformity. It is necessary to be sure that this surrounding fluid does not interact chemically with the test specimen over the time scale of the experiments. Figure 10.3-1 shows two examples of this type of elongational viscometer.[3,4] In one, a cam is used to vary the force on the sample in such a way that the tensile stress remains constant; the other uses a servocontrolled motor to achieve either fixed kinematics or, through feedback control, a constant stress. Total extension ratios L_{max}/L_0 of 50, corresponding to $\varepsilon_{max} = 3.9$, have been achieved with this latter device.[5] A variation on the configurations shown in Fig. 10.3-1 has been used in which the sample is floated horizontally on a supporting liquid.[6] The horizontal configuration does not require the density of the supporting liquid to be matched as closely to that of the sample as does the vertical arrangement.

The second arrangement for generating a uniaxial elongational flow is illustrated in Fig. 10.3-2. The sample is floated horizontally on a supporting liquid that does not interact chemically with it. The elongation is produced by one or two sets of counter-rotating gears through which the sample is pulled.[7] Since the gears do not move relative to one another,

[3] F. N. Cogswell, *Plast. Polym.*, **36**, 109–111 (1968).

[4] H. Münstedt, *Rheol. Acta*, **14**, 1077–1088 (1975); *J. Rheol.*, **23**, 421–436 (1979).

[5] H. M. Laun and H. Münstedt, *Rheol. Acta*, **15**, 517–524 (1976).

[6] G. V. Vinogradov, B. V. Radushkevich, and V. D. Fikhman, *J. Polym. Sci. Part A-2*, **8**, 1–17 (1970); G. V. Vinogradov, V. D. Fikhman, and B. V. Radushkevich, *Rheol. Acta*, **11**, 286–291 (1972); A. Franck and J. Meissner, *Rheol. Acta*, **23**, 117–123 (1984).

[7] J. Meissner, *Rheol. Acta*, **10**, 230–242 (1971); *Trans. Soc. Rheol.*, **16**, 405–420 (1972); J. Meissner, T. Raible, and S. E. Stephenson, *J. Rheol.*, **25**, 1–28 (1981).

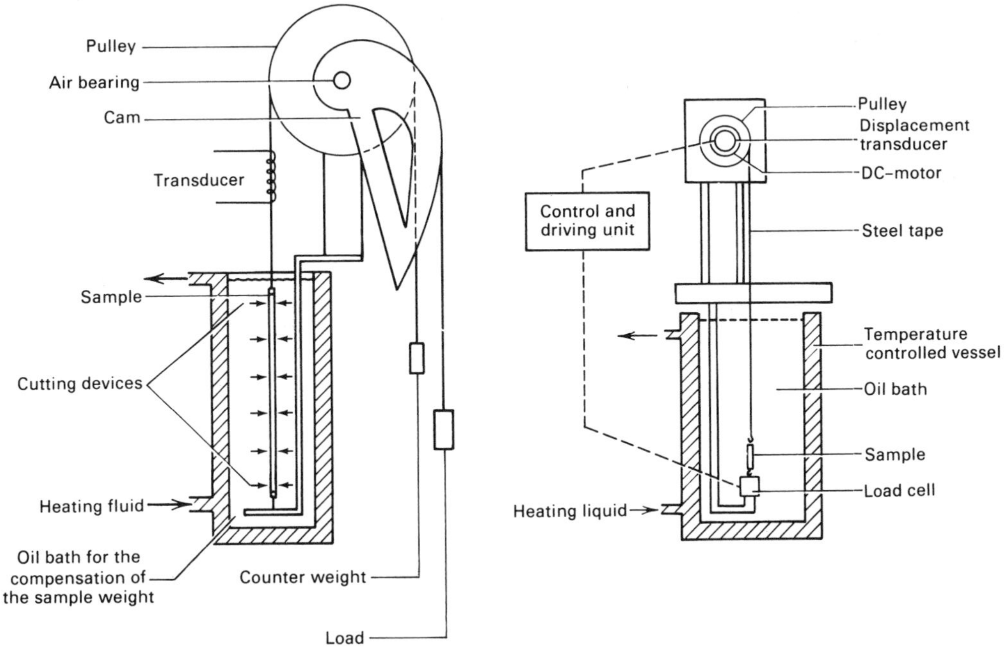

FIGURE 10.3-1. Schematic diagrams for two devices used to generate uniaxial elongational flows by separating clamped ends of the sample. The device on the left uses a cam to maintain mechanically a constant tensile stress on the sample. [H. Münstedt, *Rheol. Acta*, **14**, 1077–1088 (1975).] The device on the right uses a servocontrol system to maintain either a specified (positive) elongation rate or a specified tensile stress. [H. Münstedt, *J. Rheol.*, **23**, 421–436 (1979).]

this method has the advantage over the geometries in Fig. 10.3-1 that the maximum strain $\varepsilon_{\max}$ is not device-limited. Hencky strains $\varepsilon_{\max}$ as large as seven (corresponding to an extension ratio $L_{\max}/L_0$ of 1100) have been obtained.[8] The disadvantage of the rotating gear method relative to the moving clamps is that larger samples are required. This is a serious limitation inasmuch as the samples must be very nearly free of inhomogeneities.

With either of the two methods, analysis of the force-extension data to obtain the elongational material functions is straightforward. We assume that the surroundings are at

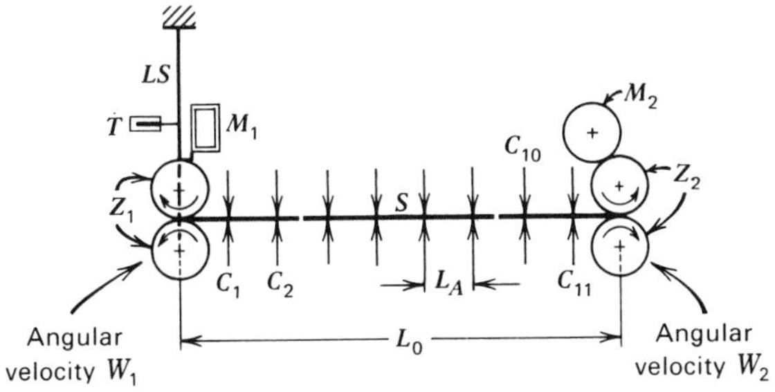

FIGURE 10.3-2. Schematic diagram of an elongational viscometer which uses two sets of rotary clamps (Z_1 and Z_2) to elongate the sample (S). The clamps are driven by two independent motors (M_1 and M_2) at constant rotation rates W_1 and W_2. The force in the sample is measured by the transducer (T) mounted on a leaf spring (LS). At the end of the test the sample can be cut into small segments by the scissors C_i in order to test for uniformity of the elongation throughout the sample and to obtain the true strain rate for the test. [J. Meissner, *Rheol. Acta*, **10**, 230–242 (1971).]

[8] J. Meissner, T. Raible, and S. E. Stephenson, *op. cit.*

atmospheric pressure p_a. The total force per unit area exerted by the load cell and atmospheric pressure on the sample must be balanced by π_{zz}. Thus if the load cell indicates a tensile force F:

$$\pi_{zz} = -(F(t)/A(t)) + p_a \tag{10.3-1}$$

where A is the instantaneous cross-sectional area of the sample. A force balance in the radial direction gives

$$\pi_{rr} = p_a \tag{10.3-2}$$

Subtracting these last two results gives the normal stress difference as a function of time

$$\boxed{\tau_{zz} - \tau_{rr} = -F(t)/A(t)} \tag{10.3-3}$$

The material functions can then be calculated from the normal stress difference. For example, in start-up of steady elongational flow with elongation rate $\dot{\varepsilon}_0$, the cross-sectional area varies with time according to $A(t) = A_0 e^{-\dot{\varepsilon}_0 t}$, where A_0 is the initial cross-sectional area of the sample, and we have

$$\bar{\eta}^+ = \frac{F(t)e^{\dot{\varepsilon}_0 t}}{A_0 \dot{\varepsilon}_0} \tag{10.3-4}$$

This expression makes it clear that when steady state is reached, F will be an exponentially decreasing function of time and will thus become very difficult to measure at large strains.

Particularly for large strains ($\varepsilon > 4$), extreme care must be taken to be sure that the sample is deformed homogeneously.[8] This can be checked by cutting the sample into small sections of uniform length at the end of the test, and then weighing them to check for uniformity. Small errors in strain rate are found to be amplified in the stress as strain is increased. It is also found that temperature uniformity must be maintained to within 0.1 °C throughout the rheometer to achieve uniform elongations. In addition interfacial tension between the sample and supporting fluid can be important for low-strain-rate experiments[9] and for recovery experiments.[8]

Both of the test methods described above are limited to polymer melts. In order to investigate the elongational properties of polymer solutions, a variety of experiments have been used that do not yield the material functions defined in Chapter 3. These include fiber spinning,[10] the tubeless siphon,[11] converging flow,[12] and the triple-jet method.[13]

[9] H. M. Laun and H. Münstedt, *Rheol. Acta*, **17**, 415–425 (1978).

[10] J. Ferguson and K. Missaghi, *J. Non-Newtonian Fluid Mech.*, **11**, 269–281 (1982); N. E. Hudson, J. Ferguson, and P. Mackie, *Trans. Soc. Rheol.*, **18**, 541–562 (1974); J. Mewis and A. B. Metzner, *J. Fluid Mech.*, **62**, 593–600 (1974); J. A. Spearot and A. B. Metzner, *Trans. Soc. Rheol.*, **16**, 495–518 (1972); D. Acierno, J. N. Dalton, J. M. Rodriguez, and J. L. White, *J. Appl. Polym. Sci.*, **15**, 2395–2415 (1971). See also Example 7.4-3.

[11] S. T. J. Peng and R. F. Landel in G. Astarita, G. Marrucci, and L. Nicolais, eds., *Rheology*, Vol. 2, Plenum Press, New York (1980), pp. 385–391; S. T. J. Peng and R. F. Landel, *J. Appl. Phys.*, **47**, 4255–4260 (1976); J. R. A. Pearson and T. J. F. Pickup, *Polymer*, **14**, 209–214 (1973); F. A. Kanel, Ph.D Thesis, University of Delaware, Newark (1972); G. Astarita and L. Nicodemo, *Chem. Eng. J.*, **1**, 57–00 (1970). See also §2.5b.

[12] T. Kizior and F. A. Seyer, *Trans. Soc. Rheol.*, **18**, 271–285 (1974); D. R. Oliver and R. Bragg, *Can. J. Chem. Eng.*, **51**, 287–290 (1973); D. R. Oliver, *Chem. Eng. J.*, **6**, 265–271 (1973); F. N. Cogswell, *Polym. Eng. Sci.*, **12**, 64–73 (1972); A. P. Metzner and A. B. Metzner, *Rheol. Acta*, **9**, 174–181 (1970); G. Marrucci and R. E. Murch, *Ind. Eng. Chem. Fundam.*, **9**, 498–499 (1970); H. Giesekus, *Rheol. Acta*, **7**, 127–138 (1968); **8**, 411–421 (1969); A. B. Metzner, E. A. Uebler, and C. F. Chan Man Fong, *AIChE J.*, **15**, 750–758 (1969).

[13] R. Bragg and D. R. Oliver, *Nature*, **241**, 371 (1973); D. R. Oliver and R. Bragg, *Rheol. Acta*, **13**, 830–835 (1974).

b. Biaxial Stretching Flow ($b = 0$, $\dot{\varepsilon} < 0$)

Several methods have been used to obtain biaxial stretching flow for polymer melts: sheet inflation, axisymmetric stagnation flow, lubricated squeezing flow, and sheet stretching. All of these experiments yield shearfree flow material functions.

In the sheet inflation method, a circular sample is clamped around its perimeter and then inflated by applying a differential pressure to the two sides. Both inert gas[14] and silicone oil[15] have been used to inflate the bubble. A biaxial stretching flow occurs in the region near the pole of the bubble, and the elongation rate is determined by measuring the deformation of a grid printed on the sample. By controlling the applied pressure, either a constant stress or constant elongation rate can be achieved.

An axisymmetric stagnation flow, obtained by impinging two polystyrene melt streams through lubricated hyperboloid-shaped walls, has been used to obtain constant elongation rates in biaxial stretching.[16] Birefringence was used to determine the stresses.

The two most promising techniques at this time for biaxial stretching measurements appear to be lubricated squeezing flow[17] and film stretching with rotary clamps. These are illustrated in Figs. 10.3-3 and 4. The rotary clamp method will be discussed in the next subsection on general shearfree flows. The lubricated squeeze flow is accomplished by placing a disk of polymer sample of radius R_0 in the center of two lubricated plates of radius $R(> R_0)$. The lubricant is chosen so that its viscosity is between 500 and 1000 times smaller than the zero-shear-rate viscosity of the polymer melt at the test temperature. After the

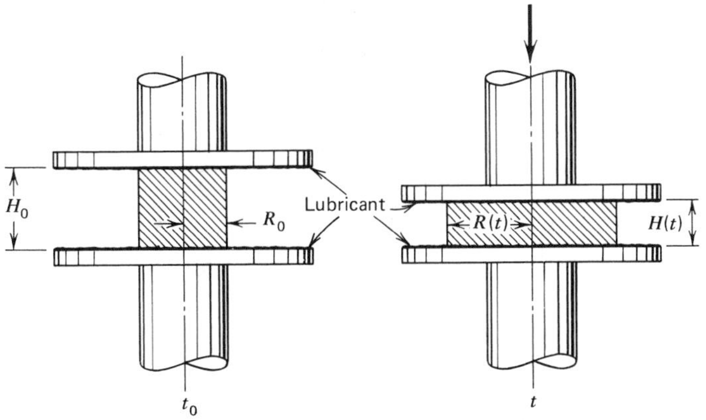

FIGURE 10.3-3. Lubricated squeeze flow device used to generate biaxial stretching flow. The left side of the figure (t_0) shows the initial configuration of the sample; the right side shows the sample at some other time $t > t_0$. The top plate can be driven downward at a programmable rate and the total normal force on the bottom plate measured. [P. R. Soskey and H. H. Winter, *J. Rheol.*, **29**, 493–517 (1985).]

[14] C. D. Denson and R. J. Gallo, *Polym. Eng. Sci.*, **11**, 174–176 (1971); D. D. Joye, G. W. Poehlein, and C. D. Denson, *Trans. Soc. Rheol.*, **16**, 421–449 (1972); **17**, 287–302 (1973); E. D. Baily, *Trans. Soc. Rheol.*, **18**, 635–640 (1974); J. M. Maerker and W. R. Schowalter, *Rheol. Acta*, **13**, 627–638 (1974); A. J. DeVries, C. Bonnebat, and J. Beautemps, *J. Polym. Sci. Polym. Symp.*, **58** 109–156 (1977).

[15] C. D. Denson and D. C. Hylton, *Polym. Eng. Sci.*, **20**, 535–539 (1980); J. Rhi-Sausi and J. M. Dealy, *Polym. Eng. Sci.*, **21**, 227–232 (1981).

[16] J. A. van Aken and H. Janeschitz-Kriegl, *Rheol. Acta*, **19**, 744–752 (1980); **20**, 419–432 (1981); H. H. Winter, C. W. Macosko, and K. E. Bennet, *Rheol. Acta.*, **18**, 323–334 (1979).

[17] S. Chatraei, C. W. Macosko, and H. H. Winter, *J. Rheol.*, **25**, 433–443 (1981).

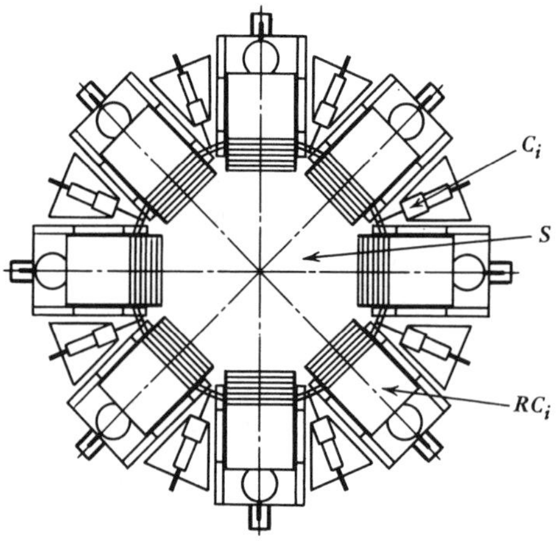

(a)

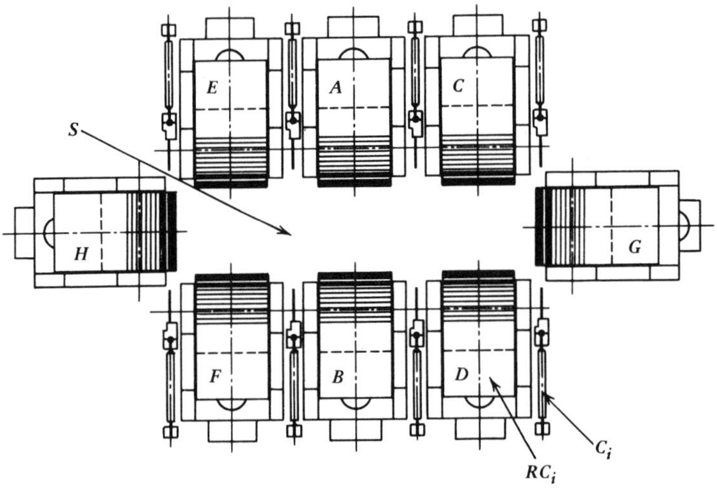

(b)

FIGURE 10.3-4. Two arrangements of the rotary clamp device which can be used to generate a wide variety of shearfree flows. The rotary clamps RC_i are capable of pulling the sample S at a programmed rate and of measuring the force in the direction of pulling. Scissors C_i located between the clamps are used to cut sample after it has been pulled outside the central region outlined by the clamps. (*a*) The rotary clamps are located in a circular arrangement and all driven at the same rate to produce biaxial stretching flow ($b = 0$, $\dot{\varepsilon} < 0$). (*b*) The clamps are located in a rectangular arrangement. By rotating clamps A–F and keeping G and H fixed, planar elongational flow ($b = 1$) can be achieved. [J. Meissner, *Chem. Eng. Commun.*, **33**, 159–180 (1985).]

sample is melted, squeezing is produced by moving the top plate downward in a programmable manner. The elongation rate is given by

$$\dot{\varepsilon} = \frac{1}{H}\frac{dH}{dt} \tag{10.3-5}$$

(note that $\dot{\varepsilon}$ is negative) so that to achieve a constant elongation rate, the gap H between the two plates must decrease exponentially in time. Since the sample is surrounded by atmospheric pressure, the normal stress difference $\tau_{zz} - \tau_{rr}$ is given by the total downward force F on the bottom plate by

$$\boxed{\tau_{zz} - \tau_{rr} = \frac{F(t)}{\pi[R(t)]^2}} \tag{10.3-6}$$

provided that the radius $R(t)$ of the deforming sample is less than the radius of the plates. Once the sample fills the region between the two plates, the plate radius is substituted for $R(t)$ in Eq. 10.3-6. Equations 10.3-5 and 6 can be combined to calculate biaxial stretching material functions (cf. Table 3.5-1) for specific $\dot{\varepsilon}(t)$. This geometry has proven particularly convenient for step strain stress relaxation experiments[18] after a suddenly applied strain ε:

$$\varepsilon = \ln\left(\frac{H}{H_0}\right) \tag{10.3-7}$$

where H_0 and H are the gaps before and after the strain is applied. The lubricated squeeze flow is a particularly simple design and has the advantages of small sample size and ability to cover a wide range of temperatures for polymer melts.

c. General Shearfree Flow (Arbitrary b)

The only currently available method for generating well defined, variable shearfree flows is one in which a sheet of polymer is stretched along its periphery by means of specially designed rotary clamps.[19] This device is illustrated in Fig. 10.3-4 in two configurations capable of generating biaxial stretching and planar elongational flow; other configurations have also been used to produce different values of b. It is also possible to pivot the clamps during the experiment and thus generate a shearfree flow with b and $\dot{\varepsilon}$ both functions of time.

§10.4 COMPLEX FLOWS

It is not always possible to measure material functions in one of the well-defined geometries described in the previous sections. For example, we have pointed out that it is presently impossible to measure elongational flow properties of polymer solutions in geometries where either the stresses or the velocity field can be precisely controlled.

[18] P. R. Soskey and H. H. Winter, *J. Rheol.*, **29**, 493–517 (1985); A. C. Papanastasiou, L. E. Scriven, and C. W. Macosko, *J. Rheol.*, **27**, 387–410 (1983).

[19] J. Meissner, *Chem. Eng. Commun.*, **33**, 159–180 (1985); J. Meissner, *Ann. Rev. Fluid Mech.*, **17**, 45–64 (1985); J. Meissner, S. E. Stephenson, A. Demarmels, and P. Portmann, *J. Non-Newtonian Fluid Mech.*, **11**, 221–237 (1982).

Similarly sample fracture often prohibits the use of the cone-and-plate or parallel-plate instruments at shear rates high enough for normal stresses to be appreciable. The capillary viscometer fails to provide data on the low-shear-rate viscosity behavior or to give any normal stress information at all.

Shortcomings such as these of standard instruments coupled with the practical necessity of having some material measurements (even though perhaps qualitative in nature), particularly on the "elastic" properties of polymeric fluids, have fostered the use of a variety of "complex" flows for rheological characterization. By complex we mean that the measureable features of the flow and stress fields cannot be related to the material functions defined in Chapter 3 without an unjustified, a priori assumption about the constitutive equation. This is in contrast to the situation we have dealt with in the previous sections where either no constitutive equation had to be assumed (e.g., the capillary viscometer) or the flow was simple enough so that an exact constitutive equation was known (e.g., the CEF equation for the parallel-plate instrument in steady shear flow). Generally, the complexity in the flow results from its being nonhomogeneous and from its involving some mixture of shear and shearfree flow.

As one example we cite the falling-ball viscometer which is a standard device for measuring the viscosity of Newtonian fluids (by means of Eq. 1.4-31).[1] Creeping flow around a sphere is very complicated, inasmuch as it is nonhomogeneous, is unsteady from the material point of view, and contains elements of both shear and shearfree flows. However, for very small Deborah numbers ($\mathrm{De} = \lambda v_\infty / R$, where λ is a characteristic time for the fluid; v_∞, the terminal velocity of the sphere; and R, the sphere radius), it is reasonable to expect that we might be able to extract information on the zero-shear-rate properties from this experiment. This is because for very small De we can use the retarded-motion expansion. In the limit $\mathrm{De} = 0$, the Newtonian fluid is recovered and when the first-order terms in De are retained, the second-order fluid, whose constants are simply related to the zero-shear-rate viscometric functions, is obtained (see §6.2).

When wall effects can be neglected, the perturbation analysis of Leslie[2] can be used to calculate the zero-shear-rate viscosity in terms of the terminal velocity of the sphere (see Problem 7A.2). The result of Leslie can also be used to estimate a characteristic time for the fluid if the liquid is assumed to be described by a convected Maxwell model. An empirical approach to estimating a time constant from falling-ball viscometry is also available.[3]

In typical laboratory implementations of the falling-ball viscometer, the walls and bottom of the fluid container can substantialy influence the terminal velocity of the sphere. Wall corrections to the equations for determining the viscosity can be appreciably larger in non-Newtonian fluids than in Newtonian fluids.[4] As was shown in Example 8.5-2, numerical simulations for a sphere falling through a viscoelastic fluid in a cylinder can be very useful in obtaining the zero-shear-rate viscometric functions. There comparison of numerical simulations[5] with experimentally measured terminal velocities allowed quantitative determination of η_0 and $\Psi_{1,0}$. Present methods are not capable of yielding information

[1] A description of many of the standard techniques for measuring μ for Newtonian fluids is contained in J. R. Van Wazer, J. W. Lyons, K. Y. Kim, and R. E. Colwell, *Viscosity and Flow Measurement*, Wiley, New York (1963). See also R. W. Whorlow, *Rheological Techniques*, Ellis Horwood, Chichester, UK (1980).

[2] F. M. Leslie, *Q. J. Mech. Appl. Math.*, **14**, 36–48 (1961); erratum: H. Giesekus, *Rheol. Acta*, **3**, 59–71 (1963), p. 69, footnote 16. See also K. Adachi, N. Yoshioka, and K. Sakai, *J. Non-Newtonian Fluid Mech.*, **3**, 107–125 (1977/78).

[3] Y. I. Cho and J. P. Hartnett, *Lett. Heat Mass Transfer*, **6**, 335–342 (1979); Y. I. Cho, J. P. Hartnett, and W. Y. Lee, *J. Non-Newtonian Fluid Mech.*, **15**, 61–74 (1984).

[4] Y. I. Cho, J. P. Hartnett, and E. Y. Kwack, *Chem. Eng. Commun.*, **6**, 141–149 (1980). A theoretical analysis of wall effects is given by B. Caswell, *Chem. Eng. Sci.*, **25**, 1167–1176 (1970). For an experimental study see M. Gottlieb, *J. Non-Newtonian Fluid Mech.*, **6**, 97–109 (1979).

[5] O. Hassager and C. Bisgaard, *J. Non-Newtonian Fluid Mech.*, **12**, 153–164 (1983).

outside of the zero-shear-rate regime.[6] Numerical simulation techniques have also been developed for flow of a Bingham plastic around a sphere,[7] and these methods could be used to determine the viscosity and yield stress of a Bingham plastic by means of falling-ball viscometry.

A closely related geometry is the rolling-ball viscometer.[8] A rolling-ball viscometer consists of an inclined tube containing a sphere whose diameter is slightly smaller than the diameter of the tube (see Fig. 10.4-1). For a power-law fluid the following formula[9] has been derived to allow the power-law constants m and n to be determined from the speed V with which the ball rolls down the tube inclined at an angle β with respect to the horizontal:

$$\tfrac{1}{3}(2R)^{n+1}(\rho_s - \rho)g \sin\beta = m\left[\pi V\left(\frac{2n+1}{n}\right)\right]^n \left(\frac{R}{R-r}\right)^{2n+(1/2)} J_n \tag{10.4-1}$$

where R and r are the tube and sphere radii, respectively, and ρ_s and ρ are the densities of the sphere and fluid. The quantities J_n are

$$J_n = 2\int_0^\infty \frac{d\xi}{[I_n(\xi^2)]^n} \tag{10.4-2}$$

where

$$I_n(\alpha) = \int_{-\pi}^{\pi} (\cos^2 \tfrac{1}{2}\theta + \alpha)^{2+(1/n)}\, d\theta \tag{10.4-3}$$

An extensive table of J_n has been prepared by Sestak and Ambros.[10] A few sample values are given in Table 10.4-1. Equation 10.4-1 has been found to be satisfactory for aqueous solutions of carboxymethylcellulose and polyacrylamide.[9]

Still other geometries have been used. Several additional systems suggested for determining the second-order fluid constants (and thus η_0, $\Psi_{1,0}$, and $\Psi_{2,0}$) are listed in Table 6.5-1. These include rod-climbing, radial flow between parallel disks, and flow around a rotating sphere. The falling cylinder viscometer is discussed in Problem 1C.2.

Finally we mention the parallel-plate plastometer which involves squeezing flow between parallel disks. This system has been shown in Examples 1.3-5 and 4.2-7 to be useful in obtaining the viscosity of a Newtonian fluid and the power-law constants m and n. In Example 10.4-1 we show that this system can also be used to obtain information about the normal stresses under fast squeezing conditions.

[6] R. P. Chhabra, P. H. T. Uhleherr, and D. V. Boger, *J. Non-Newtonian Fluid Mech.*, **6**, 187–199 (1980).

[7] A. N. Beris, J. A. Tsamopoulos, R. C. Armstrong, and R. A. Brown, *J. Fluid Mech.*, **158**, 219–244 (1985).

[8] J. R. Van Wazer, J. W. Lyons, K. Y. Kim, and R. E. Colwell, *op. cit.*, pp. 276–281.

[9] R. B. Bird and R. M. Turian, *Ind. Eng. Chem. Fundam.*, **3**, 87 (1964). The corresponding formula for a Newtonian fluid with viscosity μ was given by H. W. Lewis, *Anal. Chem.*, **25**, 507–508 (1953)

$$\mu = \frac{4}{9\pi J}\frac{R^2(\rho_s - \rho)g\sin\beta}{V}\left(\frac{R-r}{R}\right)^{5/2} \tag{10.4-1a}$$

where J is a constant:

$$J = \frac{4}{3}\left[\sqrt{2} - \frac{1}{\sqrt{5}}(\sqrt{10}+2)^{1/2}\right] = 0.531 \tag{10.4-2a}$$

[10] J. Šesták and F. Ambros, *Rheol. Acta*, **12**, 70–76 (1973).

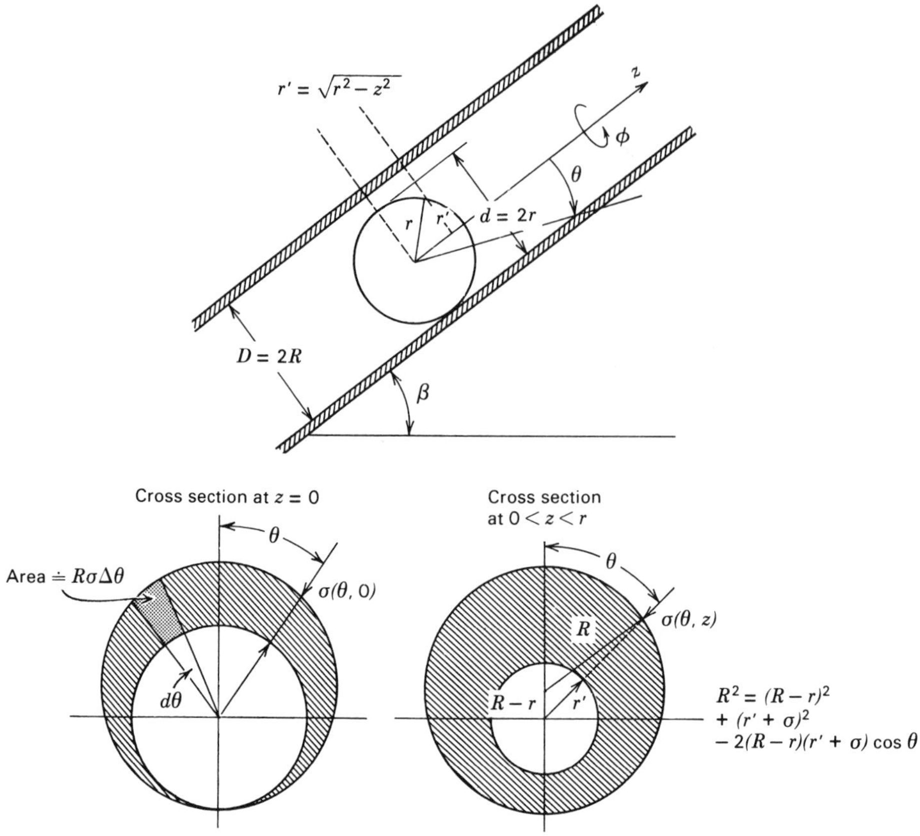

FIGURE 10.4-1 The rolling-ball viscometer. A tightly fitting ball of radius r rolls inside of a tube of radius R filled with fluid. The tube makes an angle β with the horizontal. The local gap $\sigma(\theta, z)$ between the sphere surface and tube wall is given approximately by

$$\sigma \doteq 2(R - r)\left[\cos^2 \tfrac{1}{2}\theta + \frac{R - \sqrt{R^2 - z^2}}{2(R - r)}\right]$$

TABLE 10.4-1

Values of the Quantity J_n Defined by Eq. 10.4-2 as a Function of the Power-Law Index n

n	J_n
0.20	1.8698
0.30	1.5317
0.40	1.2778
0.50	1.0812
0.60	0.9249
0.70	0.7979
0.80	0.6930
0.90	0.6052
1.00	0.5308

EXAMPLE 10.4-1. Squeezing Flow of Viscoelastic Fluids between Parallel Disks[11]

A viscoelastic liquid is contained in the gap between two parallel disks that are initially separated by a gap $2h_0$ (see Fig. 1.3-5). A constant normal force F is applied to the plates to cause the fluid in the gap to be squeezed out. For very small gaps ($h/R \ll 1$) it is reasonable to assume that the flow is sufficiently close to steady shear flow that the CEF equation can be used for the constitutive equation. With the additional assumption that the quasi-steady-state approximation can be used, obtain a relation between the force F applied to the plates, the instantaneous rate of movement of the plates $\dot{h}$, and the viscometric properties of the fluid. Assume gravitational and inertial effects are negligible.

SOLUTION It is reasonable to take $v_r = v_r(r, z)$, $v_z = v_z(r, z)$, $v_\theta = 0$, and $p = p(r, z)$. The continuity equation and the r- and z-components of the equation of motion are then

$$\frac{1}{r}\frac{\partial}{\partial r}(rv_r) + \frac{\partial}{\partial z}v_z = 0 \tag{10.4-4}$$

$$0 = -\frac{\partial p}{\partial r} - \left(\frac{1}{r}\frac{\partial}{\partial r}(r\tau_{rr}) - \frac{\tau_{\theta\theta}}{r} + \frac{\partial \tau_{rz}}{\partial z}\right) \tag{10.4-5}$$

$$0 = -\frac{\partial p}{\partial z} - \left(\frac{1}{r}\frac{\partial}{\partial r}(r\tau_{rz}) + \frac{\partial \tau_{zz}}{\partial z}\right) \tag{10.4-6}$$

These equations are to be solved together with the constitutive equation, Eq. 9.6-18. To evaluate the three terms that contribute to the CEF equation we need first

$$\nabla \boldsymbol{v} = \begin{pmatrix} \dfrac{\partial v_r}{\partial r} & 0 & \dfrac{\partial v_z}{\partial r} \\ 0 & \dfrac{v_r}{r} & 0 \\ \dfrac{\partial v_r}{\partial z} & 0 & \dfrac{\partial v_z}{\partial z} \end{pmatrix} \tag{10.4-7}$$

where matrix entries are written in the order r, θ, z. The rate-of-strain tensor is then

$$\boldsymbol{\gamma}_{(1)} = \begin{pmatrix} 2\dfrac{\partial v_r}{\partial r} & 0 & \dfrac{\partial v_z}{\partial r} + \dfrac{\partial v_r}{\partial z} \\ 0 & 2\dfrac{v_r}{r} & 0 \\ \dfrac{\partial v_z}{\partial r} + \dfrac{\partial v_r}{\partial z} & 0 & 2\dfrac{\partial v_z}{\partial z} \end{pmatrix} \tag{10.4-8}$$

[11] This example is motivated by M. A. McClelland and B. A. Finlayson, *J. Non-Newtonian Fluid Mech.*, **13**, 181–201 (1983); *J. Rheol.*, **32**, 101–133 (1988). The authors are indebted to L. M. Quinzani and L. E. Wedgewood for valuable contributions to this example.

The Ψ_2 term in Eq. 9.6-18 involves $\{\gamma_{(1)} \cdot \gamma_{(1)}\}$, the components of which are

$$\{\gamma_{(1)} \cdot \gamma_{(1)}\}_{rr} = 4\left(\frac{\partial v_r}{\partial r}\right)^2 + \left(\frac{\partial v_r}{\partial z} + \frac{\partial v_z}{\partial r}\right)^2 \tag{10.4-9a}$$

$$\{\gamma_{(1)} \cdot \gamma_{(1)}\}_{\theta\theta} = 4\left(\frac{v_r}{r}\right)^2 \tag{10.4-9b}$$

$$\{\gamma_{(1)} \cdot \gamma_{(1)}\}_{zz} = 4\left(\frac{\partial v_z}{\partial z}\right)^2 + \left(\frac{\partial v_r}{\partial z} + \frac{\partial v_z}{\partial r}\right)^2 \tag{10.4-9c}$$

$$\{\gamma_{(1)} \cdot \gamma_{(1)}\}_{rz} = \{\gamma_{(1)} \cdot \gamma_{(1)}\}_{zr} = 2\left(\frac{\partial v_r}{\partial z} + \frac{\partial v_z}{\partial r}\right)\left(\frac{\partial v_r}{\partial r} + \frac{\partial v_z}{\partial z}\right) \tag{10.4-9d}$$

and all other components are zero. Similarly for the Ψ_1-term in the CEF equation we need to know the components of $\gamma_{(2)}$:

$$\gamma_{(2)rr} = 2\left(v_r \frac{\partial^2 v_r}{\partial r^2} + v_z \frac{\partial^2 v_r}{\partial r \partial z}\right) - 4\left(\frac{\partial v_r}{\partial r}\right)^2 - 2\frac{\partial v_r}{\partial z}\left(\frac{\partial v_r}{\partial z} + \frac{\partial v_z}{\partial r}\right) \tag{10.4-10a}$$

$$\gamma_{(2)\theta\theta} = 2v_r \frac{\partial}{\partial r}\left(\frac{v_r}{r}\right) + 2v_z \frac{\partial}{\partial z}\left(\frac{v_r}{r}\right) - 4\left(\frac{v_r}{r}\right)^2 \tag{10.4-10b}$$

$$\gamma_{(2)zz} = 2v_r \frac{\partial^2 v_z}{\partial r \partial z} + 2v_z \frac{\partial^2 v_z}{\partial z^2} - 2\frac{\partial v_z}{\partial r}\left(\frac{\partial v_r}{\partial z} + \frac{\partial v_z}{\partial r}\right) - 4\left(\frac{\partial v_z}{\partial z}\right)^2 \tag{10.4-10c}$$

$$\gamma_{(2)rz} = \gamma_{(2)zr} = v_r\left(\frac{\partial^2 v_r}{\partial r \partial z} + \frac{\partial^2 v_z}{\partial r^2}\right) + v_z\left(\frac{\partial^2 v_r}{\partial z^2} + \frac{\partial^2 v_z}{\partial r \partial z}\right)$$
$$- \left(\frac{\partial v_r}{\partial z} + \frac{\partial v_z}{\partial r}\right)\left(\frac{\partial v_r}{\partial r} + \frac{\partial v_z}{\partial z}\right) - 2\frac{\partial v_r}{\partial r}\frac{\partial v_z}{\partial r} - 2\frac{\partial v_z}{\partial z}\frac{\partial v_r}{\partial z} \tag{10.4-10d}$$

with all other components being zero. Finally for the CEF equation we need the shear rate

$$\dot{\gamma} = \sqrt{\tfrac{1}{2}(\gamma_{(1)} : \gamma_{(1)})} = \sqrt{\tfrac{1}{2}\,\mathrm{tr}\{\gamma_{(1)} \cdot \gamma_{(1)}\}}$$
$$= \sqrt{\tfrac{1}{2}\left(2\left(\frac{\partial v_r}{\partial z} + \frac{\partial v_z}{\partial r}\right)^2 + 4\left(\frac{v_r}{r}\right)^2 + 4\left(\frac{\partial v_r}{\partial r}\right)^2 + 4\left(\frac{\partial v_z}{\partial z}\right)^2\right)} \tag{10.4-11}$$

No approximations have been made in writing down Eqs. 10.4-4 through 11 except for the form of the velocity field given above Eq. 10.4-4 and the quasi-steady-state, creeping flow, and negligible gravity assumptions listed in the problem statement.

We now seek a perturbation solution to these equations, in which we take advantage of the fact that $\varepsilon = h/R \ll 1$. We begin by estimating the orders of magnitude of the velocity components. Since the plates move with speed $-\dot{h}$, we have

$$v_z \sim O(-\dot{h}) \tag{10.4-12}$$

Then from the continuity equation it follows that

$$v_r \sim O(-\dot{h}R/h) \tag{10.4-13}$$

so that v_r will be larger than v_z by a factor of $\varepsilon^{-1} \gg 1$. As a convenient shorthand let us denote the order of magnitude of v_r by $V = -\dot{h}R/h$. Then Eqs. 10.4-12 and 13 suggest that we expand v_r and v_z thus

$$v_r = V[\hat{v}_{r0} + \varepsilon\hat{v}_{r1} + \cdots] \tag{10.4-14}$$

$$v_z = V[\varepsilon\hat{v}_{z1} + \cdots] \tag{10.4-15}$$

where the ^ denotes dimensionless quantities. In addition we define dimensionless radial and axial variables as

$$\xi = r/R; \qquad \zeta = z/h \tag{10.4-16}$$

Then by combining Eqs. 10.4-11, 14 and 15 we obtain the dimensionless shear rate $\dot{\Gamma}$

$$\dot{\Gamma} = \frac{\dot{\gamma}h}{V} = -\frac{\partial \hat{v}_{r0}}{\partial \zeta}\left[1 + \varepsilon\frac{(\partial \hat{v}_{r1}/\partial \zeta)}{(\partial \hat{v}_{r0}/\partial \zeta)} + \cdots\right] \qquad (0 \leq \zeta \leq 1) \tag{10.4-17}$$

so that at lowest order the shear rate is $(-\partial v_r/\partial z)$. The restriction to $0 \leq \zeta \leq 1$ in Eq. 10.4-17 arises from the fact that the shear rate must be positive. It is also assumed that the ε-term in Eq. 10.4-17 is sufficiently small that the term in brackets is positive. For the remainder of this example we will consider only the region $0 \leq \zeta \leq 1$; the flow for $-1 \leq \zeta \leq 0$ can be obtained from symmetry.

We can now obtain expansions for the components of the stress tensor. In doing this it is convenient to make the stress tensor, viscosity, and normal stress coefficients dimensionless as follows:

$$\mathbf{T} = \frac{\tau}{\eta^0 V/h}; \qquad \hat{\eta} = \frac{\eta}{\eta^0}; \qquad \hat{\Psi}_1 = \frac{\Psi_1}{\eta^0 h/V}; \qquad \hat{\Psi}_2 = -q\hat{\Psi}_1 \tag{10.4-18}$$

where η^0 is a characteristic viscosity and q is a constant. The relation between Ψ_2 and Ψ_1 is supported by data in Fig. 3.7-7. To get expansions for T_{ij} it is necessary to expand $\hat{\eta}$ and $\hat{\Psi}_1$ thus

$$\begin{aligned}\hat{\eta} &= \hat{\eta}(\dot{\Gamma}_0) + \varepsilon\left(\frac{d\hat{\eta}}{d\dot{\Gamma}}\frac{d\dot{\Gamma}}{d\varepsilon}\right)_{\varepsilon=0} + \cdots \\ &= \hat{\eta}(\dot{\Gamma}_0) - \varepsilon\left(\frac{d\hat{\eta}}{d\dot{\Gamma}}\right)_{\dot{\Gamma}=\dot{\Gamma}_0}\frac{\partial \hat{v}_{r1}}{\partial \zeta} + O(\varepsilon^2)\end{aligned} \tag{10.4-19}$$

$$\hat{\Psi}_1 = \hat{\Psi}_1(\dot{\Gamma}_0) - \varepsilon\left(\frac{d\hat{\Psi}_1}{d\dot{\Gamma}}\right)_{\dot{\Gamma}=\dot{\Gamma}_0}\frac{\partial \hat{v}_{r1}}{\partial \zeta} + O(\varepsilon^2) \tag{10.4-20}$$

where $\dot{\Gamma}_0 = |\partial \hat{v}_{r0}/\partial \zeta| = -\partial \hat{v}_{r0}/\partial \zeta$. When Eqs. 10.4-14 through 20 are combined with the shear stress expression from the CEF equation we obtain

$$\begin{aligned}\mathrm{T}_{rz} = -\hat{\eta}(\dot{\Gamma}_0)\frac{\partial \hat{v}_{r0}}{\partial \zeta} &+ \varepsilon\left\{-\left[\hat{\eta}(\dot{\Gamma}_0)\frac{\partial \hat{v}_{r1}}{\partial \zeta} - \left(\frac{d\hat{\eta}}{d\dot{\Gamma}}\right)_{\dot{\Gamma}=\dot{\Gamma}_0}\frac{\partial \hat{v}_{r1}}{\partial \zeta}\frac{\partial \hat{v}_{r0}}{\partial \zeta}\right]\right. \\ &+ \tfrac{1}{2}\hat{\Psi}_1(\dot{\Gamma}_0)\left[\hat{v}_{r0}\frac{\partial^2 \hat{v}_{r0}}{\partial \zeta \partial \xi} + \hat{v}_{z1}\frac{\partial^2 \hat{v}_{r0}}{\partial \zeta^2} - (3 - 4q)\frac{\partial \hat{v}_{r0}}{\partial \zeta}\frac{\partial \hat{v}_{z1}}{\partial \zeta}\right. \\ &\left.\left. - (1 - 4q)\frac{\partial \hat{v}_{r0}}{\partial \zeta}\frac{\partial \hat{v}_{r0}}{\partial \xi}\right]\right\} + O(\varepsilon^2)\end{aligned} \tag{10.4-21}$$

For convenience and to facilitate comparison with the results of Example 4.2-7, let us assume that the viscosity and first normal stress coefficients are given by power-law expressions (see text below Eq. 4.2-68). Then we take the characteristic viscosity η^0 to be

$$\eta^0 = m(V/h)^{n-1} \tag{10.4-22}$$

That is, we will take η^0 to be the power-law viscosity evaluated at the characteristic shear rate V/h. It then follows that

$$\hat{\eta} = \dot{\Gamma}^{n-1}; \qquad \hat{\Psi}_1 = M\dot{\Gamma}^{n'-2} \tag{10.4-23}$$

where M is a dimensionless parameter

$$M = \left(\frac{m'}{m}\right)\left(\frac{V}{h}\right)^{n'-n} \tag{10.4-24}$$

In the expansion for $\hat{\eta}$ we then find

$$\left.\frac{d\hat{\eta}}{d\dot{\Gamma}}\right|_{\dot{\Gamma}=\dot{\Gamma}_0} = (n-1)\dot{\Gamma}_0^{n-2} \tag{10.4-25}$$

Combining Eqs. 10.4-21 through 25 gives

$$\begin{aligned} \mathrm{T}_{rz} = \left(-\frac{\partial \hat{v}_{r0}}{\partial \zeta}\right)^n + \varepsilon\Bigg[-n\left(-\frac{\partial \hat{v}_{r0}}{\partial \zeta}\right)^{n-1}\frac{\partial \hat{v}_{r1}}{\partial \zeta} \\ + \tfrac{1}{2}M\left(-\frac{\partial \hat{v}_{r0}}{\partial \zeta}\right)^{n'-2}\left(\hat{v}_{r0}\frac{\partial^2 \hat{v}_{r0}}{\partial \zeta \partial \xi} + \hat{v}_{z1}\frac{\partial^2 \hat{v}_{r0}}{\partial \zeta^2} - (3-4q)\frac{\partial \hat{v}_{r0}}{\partial \zeta}\frac{\partial \hat{v}_{z1}}{\partial \zeta} \\ - (1-4q)\frac{\partial \hat{v}_{r0}}{\partial \zeta}\frac{\partial \hat{v}_{r0}}{\partial \xi}\right)\Bigg] + O(\varepsilon^2) \end{aligned} \tag{10.4-26}$$

By a similar procedure we find expansions for the other dimensionless stress components:

$$\mathrm{T}_{rr} = -(1-q)M\left(-\frac{\partial \hat{v}_{r0}}{\partial \zeta}\right)^{n'} + O(\varepsilon) \tag{10.4-27}$$

$$\mathrm{T}_{\theta\theta} = -2\varepsilon\left(\frac{\hat{v}_{r0}}{\xi}\right)\left(-\frac{\partial \hat{v}_{r0}}{\partial \zeta}\right)^{n-1} + O(\varepsilon^2) \tag{10.4-28}$$

$$\mathrm{T}_{zz} = qM\left(-\frac{\partial \hat{v}_{r0}}{\partial \zeta}\right)^{n'} - 2\varepsilon\left[\left(-\frac{\partial \hat{v}_{r0}}{\partial \zeta}\right)^{n-1}\frac{\partial \hat{v}_{z1}}{\partial \zeta}\right] + O(\varepsilon^2) \tag{10.4-29}$$

In these expansions we assume that the stress ratio is of order unity, and this implies that M is also of order unity.

The last variable that needs to be scaled and expanded is the pressure. From an order of magnitude analysis of the r-component of the equation of motion we see that at lowest order in ε the $\partial p/\partial r$ and $\partial \tau_{rz}/\partial z$ terms must balance. Hence the pressure must be of order $(\eta^0 V/h)\varepsilon^{-1}$, which suggests an expansion of the form

$$P = \frac{p}{(\eta^0 V/h)} = P_{-1}\varepsilon^{-1} + P_0 + O(\varepsilon) \tag{10.4-30}$$

Note that the pressure is singular and will be much larger than any of the stresses as $\varepsilon \to 0$.

Dimensionless forms of the equation of motion are obtained by multiplying Eqs. 10.4-5 and 6 by $h/(\eta^0 V/h)$:

$$r\text{-component} \quad 0 = -\varepsilon\frac{\partial P}{\partial \xi} - \left(\varepsilon\frac{1}{\xi}\frac{\partial}{\partial \xi}(\xi \mathrm{T}_{rr}) - \varepsilon\frac{\mathrm{T}_{\theta\theta}}{\xi} + \frac{\partial \mathrm{T}_{rz}}{\partial \zeta}\right) \tag{10.4-31}$$

$$z\text{-component} \quad 0 = -\frac{\partial P}{\partial \zeta} - \left(\varepsilon\frac{1}{\xi}\frac{\partial}{\partial \xi}(\xi \mathrm{T}_{rz}) + \frac{\partial \mathrm{T}_{zz}}{\partial \zeta}\right) \tag{10.4-32}$$

When the expansions for P and the T_{ij} are inserted into these equations and the coefficients of equal orders in ε equated, we obtain the following set of differential equations:

$$O(\varepsilon^{-1}): \qquad \frac{\partial P_{-1}}{\partial \zeta} = 0 \tag{10.4-33}$$

$$O(\varepsilon^{0}): \qquad \frac{\partial P_{-1}}{\partial \xi} + \frac{\partial}{\partial \zeta}\left(-\frac{\partial \hat{v}_{r0}}{\partial \zeta}\right)^{n} = 0 \tag{10.4-34}$$

$$\frac{\partial P_{0}}{\partial \zeta} + qM\frac{\partial}{\partial \zeta}\left(-\frac{\partial \hat{v}_{r0}}{\partial \zeta}\right)^{n'} = 0 \tag{10.4-35}$$

$$O(\varepsilon): \qquad \frac{\partial P_{0}}{\partial \xi} - (1-q)\frac{M}{\xi}\frac{\partial}{\partial \xi}\left[\xi\left(-\frac{\partial \hat{v}_{r0}}{\partial \zeta}\right)^{n'}\right] + \frac{\partial}{\partial \zeta}\left[-n\left(-\frac{\partial \hat{v}_{r0}}{\partial \zeta}\right)^{n-1}\frac{\partial \hat{v}_{r1}}{\partial \zeta}\right.$$
$$+ \tfrac{1}{2}M\left(-\frac{\partial \hat{v}_{r0}}{\partial \zeta}\right)^{n'-2}\left(\hat{v}_{r0}\frac{\partial^2 \hat{v}_{r0}}{\partial \zeta \partial \xi} + \hat{v}_{z1}\frac{\partial^2 \hat{v}_{r0}}{\partial \zeta^2}\right.$$
$$\left.\left. - (3-4q)\frac{\partial \hat{v}_{r0}}{\partial \zeta}\frac{\partial \hat{v}_{z1}}{\partial \zeta} - (1-4q)\frac{\partial \hat{v}_{r0}}{\partial \zeta}\frac{\partial \hat{v}_{r0}}{\partial \xi}\right)\right] = 0 \tag{10.4-36}$$

$$\frac{\partial P_{1}}{\partial \zeta} + \frac{1}{\xi}\frac{\partial}{\partial \xi}\left[\xi\left(-\frac{\partial \hat{v}_{r0}}{\partial \zeta}\right)^{n}\right] - 2\frac{\partial}{\partial \zeta}\left[\left(-\frac{\partial \hat{v}_{r0}}{\partial \zeta}\right)^{n-1}\frac{\partial \hat{v}_{z1}}{\partial \zeta}\right] = 0 \tag{10.4-37}$$

The forms of Eqs. 10.4-33 and 34 suggest that the lowest-order contribution to the pressure, P_{-1}, will be independent of the normal stresses in the fluid, inasmuch as M and q do not appear in these equations.

Before Eqs. 10.4-33 through 37 can be solved, we need to specify boundary conditions. These are

$$\text{At } z = h \qquad v_z = \dot{h}; \qquad v_r = 0 \tag{10.4-38}$$

$$\text{At } z = 0 \qquad \frac{\partial v_z}{\partial z} = 0; \qquad \frac{\partial v_r}{\partial z} = 0 \tag{10.4-39}$$

$$\text{At } r = R \text{ and } z = h \qquad \pi_{rr} = p_a \tag{10.4-40}$$

where p_a is atmospheric pressure. This last boundary condition is clearly an approximation since there is no fluid-air interface at $r = R$ for $t > 0$. We feel nonetheless that this is a reasonable choice at least for the early part of the experiment; this point is discussed further by McClelland and Finlayson[11] and at the end of this example. When the dimensionless expansions for velocity, stress, and pressure are combined with Eqs. 10.4-38 through 40, we obtain

$$\text{At } \zeta = 0 \qquad \frac{\partial \hat{v}_{zn}}{\partial \zeta} = \frac{\partial \hat{v}_{rn}}{\partial \zeta} = 0 \qquad (n \geq 0) \tag{10.4-41}$$

$$\text{At } \zeta = 1 \qquad \hat{v}_{z1} = -1; \qquad \hat{v}_{rn} = 0 \qquad (n \geq 0)$$
$$\hat{v}_{zn} = 0 \qquad (n > 1) \tag{10.4-42}$$

$$\text{At } \xi = 1, \zeta = 1 \qquad P_{-1} = 0$$
$$P_0 = P_a + (1-q)M\left(-\frac{\partial \hat{v}_{r0}}{\partial \zeta}\right)^{n'}_{\xi=1,\zeta=1} \tag{10.4-43}$$
$$P_n + T_{rr,n} = 0 \qquad (n \geq 1)$$

in which $T_{rr,n}$ is the nth order term in the expansion for T_{rr}.

We are now in a position to solve the perturbation equations. Before doing so we note that the objective of the calculation is to obtain the force on the upper plate. This is given by Eq. 4.2-64, which in dimensionless form is

$$\hat{F} = \frac{F}{\pi R^2 \eta^0 V/h} = 2\int_0^1 (P - P_a + \mathrm{T}_{zz})_{\zeta=1}\xi d\xi$$

$$= 2\int_0^1 \left[\varepsilon^{-1}P_{-1} + (P_0 - P_a) + qM\left(-\frac{\partial \hat{v}_{r0}}{\partial \zeta}\right)^{n'}\right]_{\zeta=1} \xi d\xi + O(\varepsilon) \tag{10.4-44}$$

$$= \frac{1}{\varepsilon}\hat{F}_{-1} + \hat{F}_0 + \cdots$$

From this we see that in order to get the first correction to the force, $\hat{F}_0$, we need to obtain only P_0 and $\hat{v}_{r0}$.

At lowest order we see immediately that $P_{-1} = P_{-1}(\xi)$. We next find at $O(\varepsilon^0)$ that

$$\hat{v}_{r0} = \left(-\frac{dP_{-1}}{d\xi}\right)^{1/n}\left(\frac{n}{n+1}\right)(1 - \zeta^{(1/n)+1}) \tag{10.4-45}$$

To obtain an equation for the pressure gradient we apply overall conservation of mass (cf. Eq. 4.2-61), which, when combined with the expansion for the radial velocity, gives

$$\int_0^1 \hat{v}_{r0}\, d\zeta = \tfrac{1}{2}\xi \tag{10.4-46}$$

$$\int_0^1 \hat{v}_{rn}\, d\zeta = 0 \qquad (n \geq 1) \tag{10.4-47}$$

Combining Eqs. 10.4-45 and 46 gives an ordinary differential equation for P_{-1}, which is easily solved together with the boundary condition in Eq. 10.4-43. Finally $\hat{v}_{z1}$ can be obtained from $\hat{v}_{r0}$ by means of the continuity equation. In this way we obtain

$$\hat{v}_{r0} = \tfrac{1}{2}\xi\left(\frac{2n+1}{n+1}\right)(1 - \zeta^{(1/n)+1}) \tag{10.4-48}$$

$$\hat{v}_{z1} = -\left(\frac{2n+1}{n+1}\right)\left(\zeta - \frac{n}{2n+1}\zeta^{(1/n)+2}\right) \tag{10.4-49}$$

$$P_{-1} = \left(\frac{2n+1}{2n}\right)^n \left(\frac{1}{n+1}\right)(1 - \xi^{n+1}) \tag{10.4-50}$$

Equations 10.4-48 to 50 are identical to the results obtained in Example 4.2-7 except for the p_a contribution to the pressure (Eq. 4.2-63), which does not enter in this analysis until the next order. Thus normal stress effects are not important at the lowest order in the force expression.

We now move to $O(\varepsilon)$ where we must solve the r-component of the equation of motion. All terms in Eq. 10.4-36 are now known except P_0 and $\hat{v}_{r1}$. These are found from Eqs. 10.4-35, 36, 41, 42, 43, and 47 in the same way as $\hat{v}_{r0}$ and P_{-1} were determined. The final result for P_0 is

$$P_0 = P_a - M\left(\frac{2n+1}{2n}\right)^{n'}\left\{1 - q\xi^{n'}\zeta^{n'/n} + \frac{K}{n'}(1 - \xi^{n'})\right\} \tag{10.4-51}$$

where

$$K = \frac{(3n + n' + 1) - n(2n + 1)(n' + 4 - 3q)}{(n + n')(2n + n' + 1)} \tag{10.4-52}$$

The first normal stress coefficient affects P_0 through M and n' in this result, and the influence of the second normal stress coefficient arises from the q-terms. Note that if $\Psi_2 = 0$ then $P_0 = P_a$, and the pressure distribution for the power-law fluid given in Eq. 4.2-63 is recovered.

Finally we develop the solution for the force by combining Eqs. 10.4-44, 48, 50, and 51. This leads to

$$\hat{F}_{-1} = \frac{1}{n + 3}\left(\frac{2n + 1}{2n}\right)^n \tag{10.4-53}$$

$$\hat{F}_0 = \frac{M}{(n' + 2)}\left(\frac{2n + 1}{2n}\right)^{n'}[(n' + 2) + K] \tag{10.4-54}$$

To find the separation of the plates as a function of time, we use the quasi-steady-state approximation, which allows us to obtain at each instant dh/dt from the force expression in Eqs. 10.4-44, 53, and 54. It is necessary to remove the h and $\dot{h}$ from the scaling of $\hat{F}$ in order to do this, and for that purpose we introduce the following additional dimensionless variables and groups:

$$\hat{t} = \frac{t}{\lambda}; \qquad \hat{h} = \frac{h}{h_0}; \qquad \varepsilon_0 = \frac{h_0}{R}$$
$$F^* = \frac{F\lambda^n}{\pi R^2 m} = \hat{F}\left(\frac{\lambda V}{h}\right)^n = \hat{F}\left(-\frac{d\hat{h}/d\hat{t}}{\varepsilon_0 \hat{h}^2}\right)^n \tag{10.4-55}$$

By combining these definitions with the force expression we get

$$F^* = \frac{\varepsilon_0^{-(n+1)}}{(n + 3)}\left(\frac{2n + 1}{2n}\right)^n \frac{(-d\hat{h}/d\hat{t})^n}{\hat{h}^{2n+1}} + \frac{2\varepsilon_0^{-n'}}{(n' + 2)}\left(\frac{2n + 1}{2n}\right)^{n'}[(n' + 2) + K]\frac{(-d\hat{h}/d\hat{t})^{n'}}{\hat{h}^{2n'}} \tag{10.4-56}$$

which is the desired relation between F^*, $d\hat{h}/d\hat{t}$, and $\hat{h}$.

If the plate speed $\dot{h}$ were held constant during the squeeze flow experiment, then Eq. 10.4-56 could be used in a straightforward way to calculate the force as a function of time. On the other hand, when the force is held constant, as we consider it to be in this example, then Eq. 10.4-56 must be solved numerically for $\hat{h}(\hat{t})$. This is easily done following the method suggested by McClelland and Finlayson[11]. Let us interchange the dependent and independent variables in Eq. 10.4-56; that is, we consider $\hat{t}$ to be a function of $\hat{h}$. We then find $\hat{t}(\hat{h})$ by forward integration of

$$\frac{d\hat{t}}{d\hat{h}} = f(\hat{h}) \tag{10.4-57}$$

subject to the initial condition that

$$\text{At } \hat{h} = 1, \qquad \hat{t} = 0 \tag{10.4-58}$$

The value of $f(\hat{h})$ is found at each $\hat{h}$ from Eq. 10.4-56 by means of Newton's method.

The results of this integration are shown in Fig. 10.4-2 where they are compared with Leider's data[12] for a polyisobutylene solution. The value of q is arbitrarily chosen to be 0.1, since there are no

[12] P. J. Leider and R. B. Bird, University of Wisconsin Rheology Research Center Report No. 22 (1973).

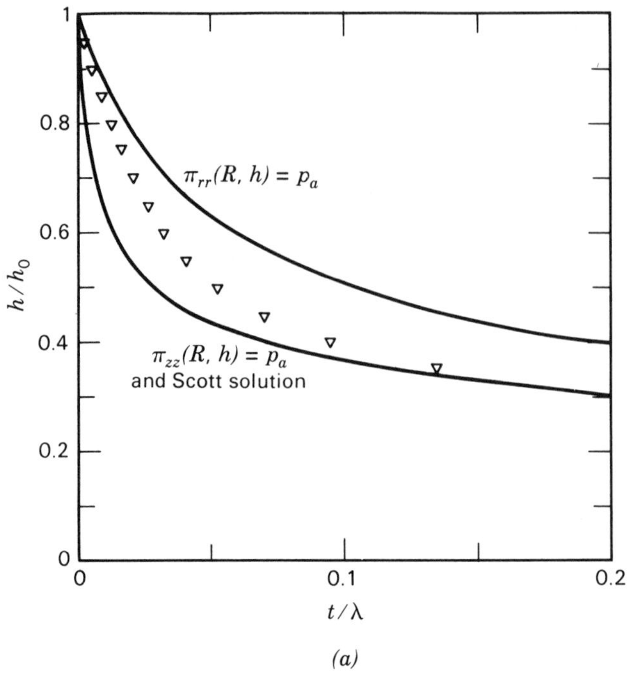

(a)

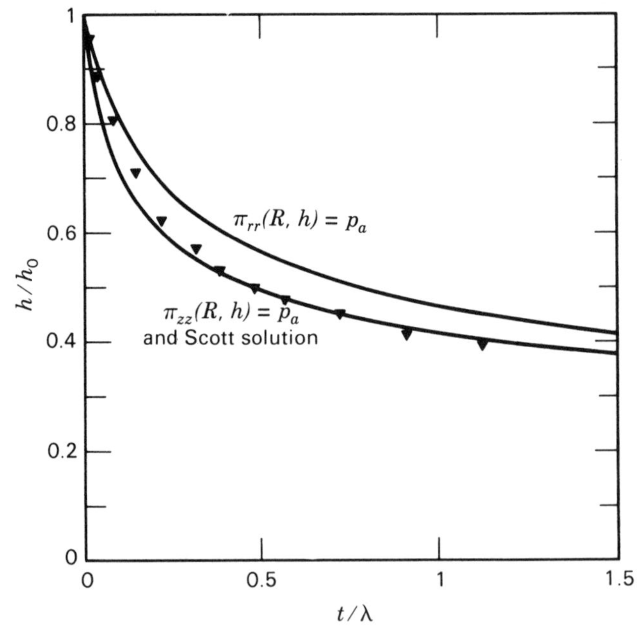

(b)

FIGURE 10.4-2. Comparison of calculated and experimental results for squeezing flow of a polyisobutylene solution with $q = 0.1$ and (a) $F^* = F\lambda^n/\pi R^2 m = 970$ and $R/h_0 = 91.7$; (b) $F^* = 0.265$ and $R/h_0 = 189.6$. The curve labeled $\pi_{rr}(R, h) = p_a$ is calculated from Eq. 10.4-56, which uses $\pi_{rr} = p_a$ at $r = R$, $z = h$; the curve labeled $\pi_{zz}(R, h) = p_a$ was obtained with $\pi_{zz} = p_a$ at $r = R$, $z = h$. The Scott solution from Eq. 4.2-65 is also shown for comparison. Data are from P. J. Leider and R. B. Bird, University of Wisconsin Rheology Research Center Report No. 22 (1973).

Ψ_2 data available on this particular solution; changing q by a factor of two in either direction does not affect $h(t)$ much. The calculated results are seen to describe the data accurately only for very short times. At long times, it is interesting that the data are well described by the Scott equation (Eq. 4.2-65).

We believe that the discrepancy between the results obtained from Eq. 10.4-56 and the data at moderate and long times may be due to the inappropriateness of the boundary condition, Eq. 10.4-40. When the experiment is first started, there is a free surface at $r = R$, so that requiring $\pi_{rr} = p_a$ is reasonable at $t = 0$. As fluid is squeezed out of the gap, the actual free surface will be beyond $r = R$, and it is questionable whether or not $\pi_{rr} = p_a$ at a position inside the fluid. To see how quickly deviations from this boundary condition might be expected, note that when the free surface has moved a distance h beyond the plate edge, the dimensionless gap will be roughly $\hat{h} \doteq (1 + \varepsilon_0)^{-1/2}$. For $\varepsilon_0 = 0.01$ this gives $\hat{h} \doteq 0.995$. Furthermore, the sensitivity of the solution to this boundary condition can be seen by replacing Eq. 10.4-40 by

$$\text{At } r = R \text{ and } z = h \qquad \pi_{zz} = p_a \tag{10.4-59}$$

This boundary condition might be reasonable if the free surface of the extruded material were nearly parallel to the plate surfaces and if velocity rearrangements near the plate rim could be ignored.[13] With this boundary condition, $[(n' + 2) + K]$ in Eq. 10.4-56 is replaced by K and the resulting solution for $\hat{h}(\hat{t})$ is seen in Fig. 10.4-2 to be very close to the Scott solution. The choice of an appropriate rim boundary condition is an unresolved issue in this problem. Experimentally, it might be feasible to circumvent this difficulty by using a partially filled gap.

Other possible sources of the discrepancy between theory and experiment are the quasi-steady-state approximation and the CEF constitutive equation. The latter should be used only for steady-state shear flows.

PROBLEMS

10A.1 Analysis of Capillary Viscometer Data[1]

The volumetric flow rate through a capillary tube has been measured at a series of imposed pressure drops for a 3.5% (weight) solution of carboxymethylcellulose in water at 303 K. The data have been corrected for end effects using the procedure of Example 10.2-3.

$4Q/\pi R^3$ (s^{-1})	τ_R (Pa)
250	220
350	255
500	298
700	341
900	382
1250	441
1750	509
2500	584
3500	670
5000	751
7000	825
9000	887
12500	1000
17500	1070
25000	1200

[13] D. V. Boger and M. M. Denn, *J. Non-Newtonian Fluid Mech.*, **6**, 163–185 (1980).

[1] Tabular data from A. G. Fredrickson, *Principles and Applications of Rheology*, 1964. Reprinted by permission of Prentice-Hall, Inc., Englewood Cliffs, NJ, pp. 309–310.

Construct a graph of log η vs. log $\dot{\gamma}$ for this fluid. What is the slope of this curve in the power-law region?

10A.2 Normal Stress Measurements in the Cone-and-Plate Instrument

In Fig. 10A.2 data are shown on the radial variation of the total normal stress $\pi_{\theta\theta}$ exerted on the plate in the cone-and-plate device. The data are for a 2.5% polyacrylamide solution and were taken using flush-mounted pressure transducers at the positions indicated. The radius of the plate is 5 cm. From these data determine Ψ_1 and Ψ_2 for the polyacrylamide solution at the four shear rates shown.

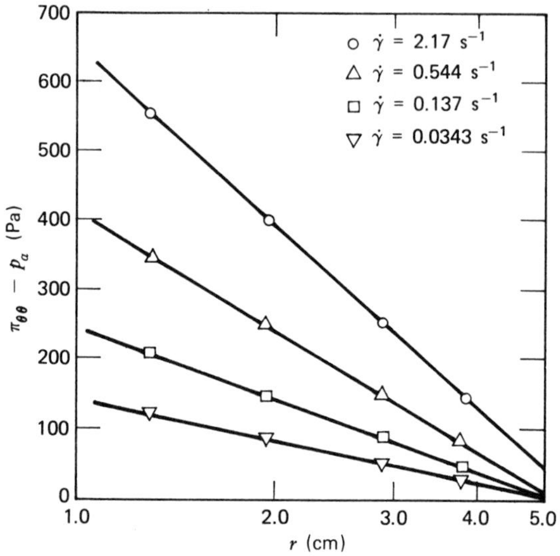

FIGURE 10A.2. Total normal stress on the plate relative to atmospheric pressure $\pi_{\theta\theta} - p_a$ versus distance from the center of the plate for a 2.5% polyacrylamide solution. [E. B. Christiansen and W. R. Leppard, *Trans. Soc. Rheol.*, **18**, 65–86 (1974).]

10A.3 Applicability of Concentric Cylinder Instrument Results

In order to use the formulas presented in Table 10.2-1D, the gap between the two cylinders of this instrument must be small. How must the ratio R_1/R_2 be restricted if the shear stress is to vary no more than 2% within the gap?

10A.4 Viscosity Measurements with an Epprecht Viscometer

An Epprecht viscometer utilizes a concentric cylinder geometry in which the inner cylinder is rotated at angular velocity W_1 and the outer cylinder is held in a fixed position. The shear stress at the inner cylinder τ_1 is measured for a series of rotation rates. Sample data[2] taken on a 1% solution of carboxymethylcellulose are given below. For the particular instrument used, the inner cylinder had a

[2] Tabular data from S. Middleman, *The Flow of High Polymers*, Wiley-Interscience, New York (1968), p. 23.

radius $R_1 = 1.54$ cm and the outer cylinder had a radius $R_2 = 1.94$ cm. Use the results of Problem 10C.2 to construct a plot of $\eta(\dot{\gamma})$ for this polymer solution.

τ_1(Pa)	W_1(rad/s)
4.1	2.63
5.2	3.53
6.5	4.64
8.1	6.19
10.1	8.17
13.6	11.8
17.0	15.9
20.7	20.9
25.4	28.0
30.9	36.9

10B.1 Forced Oscillation Measurements with the Cone-and-Plate Geometry

In Example 10.1-1 we showed how the complex viscosity could be determined from a parallel-plate system in which the upper plate was executing small-amplitude oscillations, and the torque $\mathscr{T}$ on the bottom, fixed plate was monitored. Rework that example for a cone-and-plate system (see Fig. 10.2-1) in which the upper cone is made to oscillate according to $\phi = \phi_0 \mathscr{Re}\{e^{i\omega t}\}$, in which $\phi_0 \ll 1$ is the small, real amplitude of oscillation. Show that the formulas analogous to Eqs. 10.1-11 and 12 are

$$\eta' = \frac{3\vartheta_0 \mathscr{T}_0 \sin \delta}{2\pi R^3 \omega\phi_0} \tag{10B.1-1}$$

$$\eta'' = \frac{3\vartheta_0 \mathscr{T}_0 \cos \delta}{2\pi R^3 \omega\phi_0} \tag{10B.1-2}$$

in which $\mathscr{T}_0$ is the amplitude of the oscillating torque defined in the first line of Eq. 10.1-10. Note that the phase angle δ is in the range $0 \leq \delta < \pi/2$.

10B.2 Inertial Effects in Oscillatory Parallel-Plate Flow

If inertial effects are allowed for in the oscillatory parallel-plate flow considered in Example 10.1-1, show that the torque is given by

$$\mathscr{T}_0(\cos \delta + i \sin \delta) \sin \alpha H = -\frac{\pi R^4}{2} i\omega\theta_0 \alpha\eta^* \tag{10B.2-1}$$

where $\alpha = \sqrt{(i\omega\rho/\eta^*)}$ and $\mathscr{T}_0$ is the real torque amplitude. Suggest an iterative scheme for solving this equation.

10B.3 Inertial Corrections in the Cone-and-Plate Instrument

In obtaining Eq. 10.2-7 for the total normal thrust in the cone-and-plate system, inertial forces were assumed to be negligible. The size of the error introduced in this way can be estimated theoretically in the following manner by considering the equation of motion for a Newtonian fluid:

a. Show that for a Newtonian fluid, the pressure distribution between the cone and plate (for small cone angles) is

$$p - p_a = \tfrac{1}{2}\rho W^2 R^2 \left(\frac{\vartheta}{\vartheta_0}\right)^2 \left(\frac{r^2}{R^2} - 1\right) \tag{10B.3-1}$$

b. Determine the average pressure in the gap[3] at any radial position $\bar{p}(r) = (1/\vartheta_0) \int_0^{\vartheta_0} p \, d\theta$ and use this result to compute the total normal thrust (over that due to atmospheric pressure) that a Newtonian fluid exerts on the plate

$$\mathscr{F}_{\text{Newtonian}} = -\frac{\pi}{12} \rho R^4 W^2 \tag{10B.3-2}$$

c. Interpret the result in **(b)**. The averaged form of Eq. 10B.3-1 has been compared with experimental data by Adams and Lodge,[4] who find a reasonable fit. It appears that this method slightly overestimates the inertial terms.

10B.4 End Correction for the Couette Viscometer

The assumption of tangential annular flow used in obtaining the results in Table 10.2-1D does not hold near the bottom of the viscometer. In order to compensate for this end effect, the bottom of the inner cylinder is actually constructed with a conical shape (see Fig. 10B.4); the flow at the bottom is thus a cone-and-plate flow. The gap is assumed to be narrow, that is, $(R_2 - R_1)/R_1 \ll 1$.

a. For what choice of the cone angle ϑ_0 is the shear rate uniform throughout the fluid?

b. Obtain an expression for η for this geometry that is analogous to Eq. D-1 in Table 10.2-1.

Answer: b. $\eta = \dfrac{\mathscr{T}(R_2 - R_1)}{2\pi R_1^3 H[1 + (R_1/3H)]W_1}$

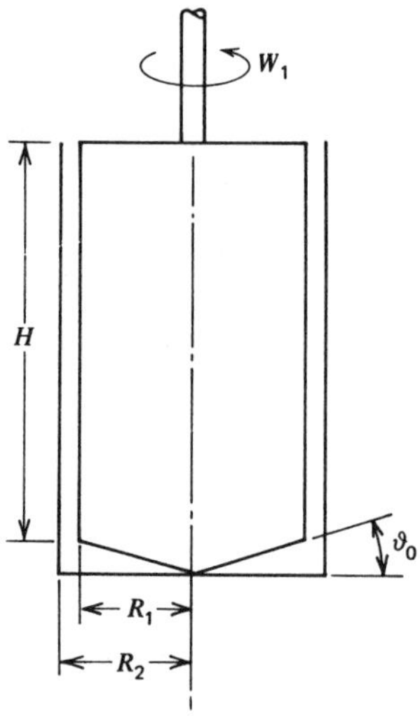

FIGURE 10B.4. Concentric cylinder viscometer with conical bottom. The inner cylinder rotates and the outer cylinder is fixed.

[3] This averaging procedure was first used by H. W. Greensmith and R. S. Rivlin, *Phil. Trans. Roy. Soc. London*, **A245**, 399–428 (1953), to fit experimental data for a torsional flow system. See also K. Walters, *Rheometry*, Chapman and Hall, London (1975), pp. 61–66.

[4] N. Adams and A. S. Lodge, *Phil. Trans. Roy. Soc. London*, **A256**, 149–184 (1964).

10B.5 Viscous Heating in a Concentric Cylinder Viscometer

Estimate the temperature rise caused by viscous heating in a concentric cylinder viscometer by considering steady tangential annular flow between two cylinders. The inner cylinder has radius κR and rotates with constant angular velocity W; the outer cylinder has radius R and is held fixed. In addition, the temperature of the outer cylinder is kept at T_0. It is assumed that the temperature rise is small enough so that the viscosity is not substantially altered by it. What is the maximum temperature rise in the fluid if $\kappa \doteq 1$ and no heat is transferred to the inner cylinder?

Answer: $T_{\max} - T_0 = \dfrac{\eta R^2 W^2}{2k}$

10B.6 Centripetal Pumping between Parallel Disks[5]

Rework Problem 6B.14 by using the CEF constitutive equation. In this way verify Eq. F-1 in Table 10.2-1. Note that since the flow in this problem is viscometric, the CEF equation will be valid for a wide class of non-Newtonian fluids.

10B.7 The Truncated Cone-and-Plate Instrument[6]

The truncated cone-and-plate instrument shown in Fig. 10B.7 is a hybrid of the cone-and-plate and parallel-plate instruments. The normal force $\pi_{\theta\theta}$ is measured along the bottom plate by means of hole-mounted pressure transducers. Thus these readings are affected by the hole pressure, which depends on the shear rate. Denote the measured values by $\pi'_{\theta\theta}$; these are related to $\pi_{\theta\theta}$ by

$$p^* = \pi_{\theta\theta} - \pi'_{\theta\theta} \tag{10B.7-1}$$

where p^* is the hole pressure (see §2.3c). By paralleling the analyses in Examples 10.2-1 and 2, verify Eqs. G-1 through G-5 in Table 10.2-1. These formulas permit the truncated cone-and-plate device to be used to determine the first and second normal stress coefficients and the hole pressure as functions of the shear rate.

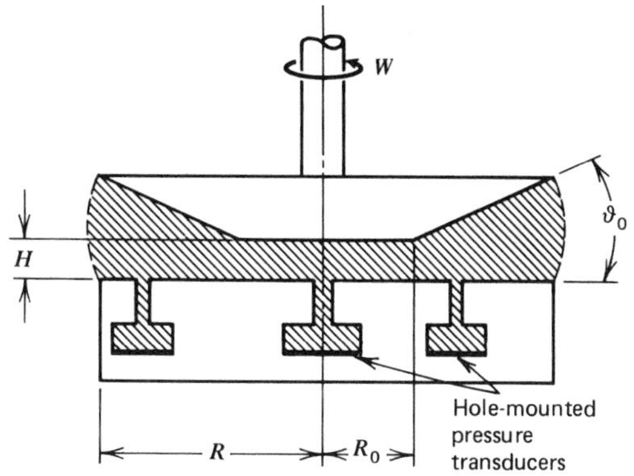

FIGURE 10B.7. The truncated cone-and-plate instrument.

10C.1 End Correction for a Capillary Viscometer[7]

Consider the operation of two capillary viscometers A and B (see Fig. 10.2-2). The reservoirs are identical and are located at the same elevation, but one is equipped with a tube of length L_A and the other a tube of length L_B. The radius of each tube is R, and the entrance angles are the same for the

[5] Y. Tomita and H. Katō, *Trans. Jpn. Soc. Mech. Eng.* (*Nippon Kikai Gakkai Ronbunshū*), **32**, 241, 1399–1408 (Sept. 1966); B. Maxwell and A. J. Scalora, *Mod. Plast.*, **37**(2), 107 (1959); P. A. Good, A. J. Schwartz, and C. W. Macosko, *AIChE J.*, **20**, 67–73 (1974).

[6] A. S. Lodge, *Rheol. Acta*, **10**, 554–556 (1971); A. S. Lodge and T. H. Hou, *Rheol. Acta*, **20**, 247–260 (1981).

[7] A. G. Fredrickson, *Principles and Applications of Rheology*, Prentice-Hall, Englewood Cliffs, NJ (1964), pp. 196–200.

two capillaries. Denote, respectively, by h_A and p_A the height of fluid in the reservoir and driving pressure (at the ram) for viscometer A. Use similar definitions for B. The ram speeds in the two viscometers are the same so that the flow rate Q through each of the two tubes is the same.

a. Apply the macroscopic mechanical energy balance[8] to the fluid in the viscometer, taking plane number 1 to be the tip of the ram and plane number 2 to be at the exit of the tube. For the purposes of this balance, take h to be constant. Show that for viscometer A

$$(\hat{\Phi}_{A_2} - \hat{\Phi}_{A_1}) + \frac{1}{2}\left(\frac{\langle v_z^3\rangle}{\langle v_z\rangle}\bigg|_{A_2} - \frac{\langle v_z^3\rangle}{\langle v_z\rangle}\bigg|_{A_1}\right) + \frac{1}{\rho Q}\int_{A_2} p v_z dS - \frac{p_A}{\rho} = -(\hat{E}_v)_A \tag{10C.1-1}$$

and that a similar relation holds for B. Here $\hat{\Phi}$ is the potential energy per unit mass, v_z is the axial velocity, $\langle v_z^3\rangle$ and $\langle v_z\rangle$ are cross-sectional averages of v_z^3 and v_z, and $\hat{E}_v$ is defined by

$$\hat{E}_v = -\frac{1}{\rho Q}\int_V (\boldsymbol{\tau}:\nabla \boldsymbol{v})dV \tag{10C.1-2}$$

b. Subtract the two mechanical energy balances to obtain

$$(\hat{\Phi}_{B_2} - \hat{\Phi}_{B_1}) - (\hat{\Phi}_{A_2} - \hat{\Phi}_{A_1}) - \frac{1}{\rho}(p_B - p_A) = +(\hat{E}_v)_A - (\hat{E}_v)_B \tag{10C.1-3}$$

c. Next, by splitting the total volume of each system into a volume in which there is steady shear flow plus volumes associated with the flow in the reservoir and in the entrance and exit regions, show that

$$(\hat{E}_v)_A - (\hat{E}_v)_B = \frac{2\pi(L_B - L_A)}{\rho Q}\int_0^R \tau_{rz}\left(\frac{dv_z}{dr}\right) r\, dr = -\frac{2}{\rho}\frac{\tau_R}{R}(L_B - L_A) \tag{10C.1-4}$$

Assume that $h_A = h_B$ in deriving this result. What other assumptions are needed?

d. Choose the reference potential to be at the bottom of the tube of length B so that $\hat{\Phi}_{B_2} = 0$. Use the resulting expressions for the potentials together with the results from **(b)** and **(c)** above to show

$$\tau_R = \left(\frac{p_B - p_A}{L_B - L_A} + \rho g \cos\beta\right)\frac{R}{2} \tag{10C.1-5}$$

where β is the angle between the tube axis and the vertical.

10C.2 Measurements in a Wide-Gap Couette Viscometer[9]

a. Show that for tangential annular flow between concentric cylinders with a wide gap, the appropriate generalizations of Eqs. D-1 and D-2 in Table 10.2-1 are

$$\Delta W = W_2 - W_1 = \frac{1}{2}\int_{\tau_1}^{\tau_2} \eta^{-1}(\tau)d\tau \tag{10C.2-1}$$

$$\Delta\pi_{rr} = \pi_{rr}(R_2) - \pi_{rr}(R_1) = \int_{R_1}^{R_2}\left(\rho\frac{v_\theta^2}{r} + \frac{\tau_{\theta\theta} - \tau_{rr}}{r}\right)dr \tag{10C.2-2}$$

In Eq. 10C.2-1, $\tau \equiv \tau_{r\theta} = \mathscr{T}/2\pi r^2 H$.

[8] See, for example, R. B. Bird, W. E. Stewart, and E. N. Lightfoot, *Transport Phenomena*, Wiley, New York (1960), pp. 211–214.

[9] B. D. Coleman, H. Markovitz, and W. Noll, *Viscometric Flows of Non-Newtonian Fluids*, Springer, New York (1966), pp. 42–44; S. Middleman, *The Flow of High Polymers*, Wiley-Interscience, New York (1968), pp. 19–25.

b. For wide gaps the above formulas may be conveniently inverted by differentiating with respect to the torque $\mathscr{T}$. The resulting expressions, valid for any value of $\mathscr{T}$, must also hold for a torque equal to $\beta\mathscr{T}$, where $\beta \equiv R_1^2/R_2^2$. Use this fact to obtain

$$\dot{\gamma}_{r\theta}(\tau_1) = \sum_{j=0}^{\infty} \left(2\mathscr{T}' \frac{\partial \Delta W}{\partial \mathscr{T}'} \right) \Bigg|_{\mathscr{T}' = \mathscr{T}\beta^j} \tag{10C.2-3}$$

$$(\tau_{\theta\theta} - \tau_{rr})|_{\tau_1} = \sum_{j=0}^{\infty} \left(2\mathscr{T}' \frac{\partial \Delta \pi_{rr}^{(c)}}{\partial \mathscr{T}'} \right) \Bigg|_{\mathscr{T}' = \mathscr{T}\beta^j} \tag{10C.2-4}$$

where

$$\Delta \pi_{rr}^{(c)} = \Delta \pi_{rr} - \int_{R_1}^{R_2} \rho \frac{v_\theta^2}{r} \, dr \tag{10C.2-5}$$

c. Explain how Eqs. 10C.2-3 and 4 can be used to interpret experimental data. Why are they useful only for wide gaps?

APPENDIX A

VECTOR AND TENSOR NOTATION[1]

The physical quantities encountered in the dynamics of polymeric liquids can be placed in the following categories: *scalars* such as shear rate, temperature, energy, volume, and time; *vectors*, such as velocity, momentum, acceleration, and force; and second order *tensors*, such as the stress, rate-of-strain, and vorticity tensors. We distinguish among these quantities by the following notation:[2]

$$s = \text{scalar (lightface italic)}$$

$$\boldsymbol{v} = \text{vector (boldface italic)}$$

$$\boldsymbol{\tau} = \text{second-order tensor (boldface Greek)}$$

$$\mathsf{\mathbf{B}} = \text{tensor of arbitrary order (boldface sans serif)}$$

In addition, boldface Greek symbols with one subscript (such as $\boldsymbol{\delta}_1$) are vectors. For vectors and tensors, several different kinds of multiplication are possible. Some of the operations require the use of special multiplication signs to be defined later: the "single dot" $\cdot$, the "double dot" $:$, and the "cross" $\times$. The parentheses enclosing these special multiplications, or sums of dot or cross multiplications, indicate the type of quantity produced by the multiplication:[3]

$$(\ ) = \text{scalar}; \qquad [\ ] = \text{vector}; \qquad \{\ \} = \text{second-order tensor}$$

No special significance is attached to the kind of parentheses if the operation enclosed is addition or subtraction, or a multiplication in which $\cdot$, $:$, and $\times$ do not appear. Hence $(\boldsymbol{v} \cdot \boldsymbol{w})$ and $(\boldsymbol{\sigma} : \boldsymbol{\tau})$ are scalars, $[\boldsymbol{v} \times \boldsymbol{w}]$ and $[\boldsymbol{\tau} \cdot \boldsymbol{v}]$ are vectors, and $\{\boldsymbol{\sigma} \cdot \boldsymbol{\tau} + \boldsymbol{\tau} \cdot \boldsymbol{\sigma}\}$ is a tensor. On the other hand, $(\boldsymbol{v} - \boldsymbol{w})$ may be written $[\boldsymbol{v} - \boldsymbol{w}]$ or $\{\boldsymbol{v} - \boldsymbol{w}\}$ when convenient to do so; similarly $\boldsymbol{vw}$ may be written $(\boldsymbol{vw})$ or $[\boldsymbol{vw}]$. Actually, scalars can be regarded as zero-order tensors, vectors as first-order tensors. The multiplication signs may be interpreted thus:

Multiplication Sign	Order of Result
None	$\sum$
$\times$	$\sum - 1$
$\cdot$	$\sum - 2$
$:$	$\sum - 4$

[1] This appendix is a modification of Appendix A of *Transport Phenomena*, by R. B. Bird, W. E. Stewart, and E. N. Lightfoot, Wiley, New York (1960). See also P. M. Morse and H. Feshbach, *Methods of Theoretical Physics*, McGraw-Hill (1953), Chap. 1, and H. D. Block, *Introduction to Tensor Analysis*, Merrill, Columbus, OH (1962).

[2] A convenient notation for blackboard use is: scalar (no underline), vector (single underline), tensor (double underline).

[3] The parentheses rules are sometimes relaxed in Volume 2, where statistical averages are indicated by $\langle\ \rangle$ and $[\![\ ]\!]$.

in which $\sum$ represents the sum of the orders of the quantities being multiplied. For example $s\boldsymbol{\tau}$ is of the order $0 + 2 = 2$, $\boldsymbol{vw}$ is of the order $1 + 1 = 2$, $\boldsymbol{\delta}_1\boldsymbol{\delta}_2$ is of the order $1 + 1 = 2$, $[\boldsymbol{v} \times \boldsymbol{w}]$ is of the order $1 + 1 - 1 = 1$, $(\boldsymbol{\sigma}:\boldsymbol{\tau})$ is of the order $2 + 2 - 4 = 0$, and $\{\boldsymbol{\sigma}\cdot\boldsymbol{\tau}\}$ is of the order $2 + 2 - 2 = 2$.

The basic operations that can be performed on scalar quantities need not be elaborated on here. However, the laws for the algebra of scalars may be used to illustrate three terms that arise in the subsequent discussion of vector operations:

a. For the multiplication of two scalars, r and s, the order of multiplication is immaterial so that the *commutative* law is valid: $rs = sr$.

b. For the successive multiplication of three scalars, q, r, and s, the order in which the multiplications are performed is immaterial, so that the *associative* law is valid: $(qr)s = q(rs)$.

c. For the multiplication of a scalar s by the sum of scalars p, q, and r, it is immaterial whether the addition or multiplication is performed first, so that the *distributive* law is valid: $s(p + q + r) = sp + sq + sr$.

These laws are not generally valid for the analogous vector and tensor operations described in the paragraphs to follow.

§A.1 VECTOR OPERATIONS FROM A GEOMETRICAL VIEWPOINT

In elementary physics courses one is introduced to vectors from a geometrical standpoint. In this section we extend this approach to include the operations of vector multiplication. In §A.2 a parallel analytic treatment is given.

Definition of a Vector and Its Magnitude

A vector $\boldsymbol{v}$ is defined as a quantity of a given magnitude and direction. The magnitude of the vector is designated by $|\boldsymbol{v}|$ or simply by the corresponding lightface symbol v. Two vectors $\boldsymbol{v}$ and $\boldsymbol{w}$ are equal when their magnitudes are equal and when they point in the same direction; they do not have to be collinear or have the same point of origin. If $\boldsymbol{v}$ and $\boldsymbol{w}$ have the same magnitude but point in opposite directions, then $\boldsymbol{v} = -\boldsymbol{w}$.

Addition and Subtraction of Vectors

The addition of two vectors can be accomplished by the familiar parallelogram construction, as indicated by Fig. A.1-1a. Vector addition obeys the following laws:

Commutative: $$(\boldsymbol{v} + \boldsymbol{w}) = (\boldsymbol{w} + \boldsymbol{v}) \qquad \text{(A.1-1)}$$

Associative: $$(\boldsymbol{v} + \boldsymbol{w}) + \boldsymbol{u} = \boldsymbol{v} + (\boldsymbol{w} + \boldsymbol{u}) \qquad \text{(A.1-2)}$$

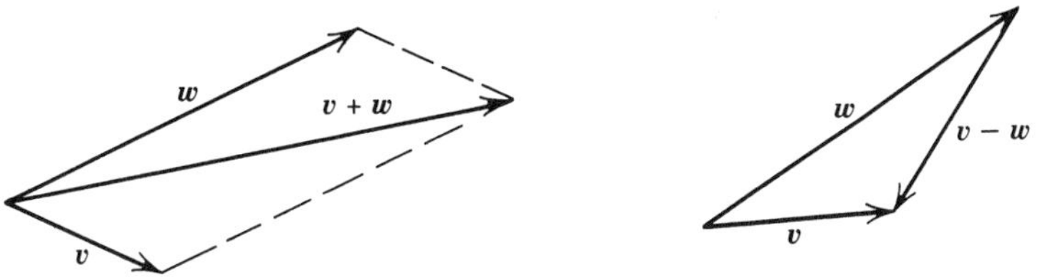

FIGURE A.1-1. Addition and subtraction of vectors.

Vector subtraction is performed by reversing the sign of one vector and adding; thus $\boldsymbol{v} - \boldsymbol{w} = \boldsymbol{v} + (-\boldsymbol{w})$. The geometrical construction for this is shown in Fig. A.1-1b.

Multiplication of a Vector by a Scalar

When a vector is multiplied by a scalar, the magnitude of the vector is altered but its direction is not. The following laws are applicable

Commutative: $$s\boldsymbol{v} = \boldsymbol{v}s \tag{A.1-3}$$

Associative: $$r(s\boldsymbol{v}) = (rs)\boldsymbol{v} \tag{A.1-4}$$

Distributive: $$(q + r + s)\boldsymbol{v} = q\boldsymbol{v} + r\boldsymbol{v} + s\boldsymbol{v} \tag{A.1-5}$$

Scalar Product (or Dot Product) of Two Vectors

The scalar product of two vectors $\boldsymbol{v}$ and $\boldsymbol{w}$ is a scalar quantity defined by

$$(\boldsymbol{v} \cdot \boldsymbol{w}) = vw \cos \phi_{vw} \tag{A.1-6}$$

in which ϕ_{vw} is the angle between the vectors $\boldsymbol{v}$ and $\boldsymbol{w}$. The scalar product is then the magnitude of $\boldsymbol{w}$ multiplied by the projection of $\boldsymbol{v}$ on $\boldsymbol{w}$, or vice versa (Fig. A.1-2a). Note that the scalar product of a vector with itself is just the square of the magnitude of the vector

$$(\boldsymbol{v} \cdot \boldsymbol{v}) = |\boldsymbol{v}|^2 = v^2 \tag{A.1-7}$$

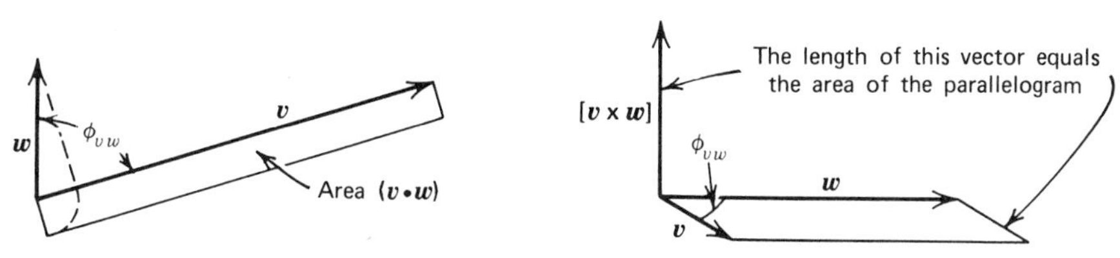

FIGURE A.1-2. Products of two vectors: (*a*) the scalar product; and (*b*) the vector product.

The rules governing scalar products are as follows

Commutative: $$(\boldsymbol{u}\cdot\boldsymbol{v}) = (\boldsymbol{v}\cdot\boldsymbol{u}) \tag{A.1-8}$$

Not Associative: $$(\boldsymbol{u}\cdot\boldsymbol{v})\boldsymbol{w} \neq \boldsymbol{u}(\boldsymbol{v}\cdot\boldsymbol{w}) \tag{A.1-9}$$

Distributive: $$(\boldsymbol{u}\cdot\{\boldsymbol{v}+\boldsymbol{w}\}) = (\boldsymbol{u}\cdot\boldsymbol{v}) + (\boldsymbol{u}\cdot\boldsymbol{w}) \tag{A.1-10}$$

Vector Product (or Cross Product) of Two Vectors

The vector product of two vectors $\boldsymbol{v}$ and $\boldsymbol{w}$ is a vector defined by

$$[\boldsymbol{v}\times\boldsymbol{w}] = \{vw\sin\phi_{vw}\}\boldsymbol{n}_{vw} \tag{A.1-11}$$

in which $\boldsymbol{n}_{vw}$ is a vector of unit length (a "unit vector") normal to the plane containing $\boldsymbol{v}$ and $\boldsymbol{w}$ and pointing in the direction that a right-handed screw will move if turned from $\boldsymbol{v}$ toward $\boldsymbol{w}$ through the angle ϕ_{vw}. The vector product is illustrated in Fig. A.1-2b. The magnitude of the vector product is just the area of the parallelogram defined by the vectors $\boldsymbol{v}$ and $\boldsymbol{w}$. It follows from the definition of the vector product that

$$[\boldsymbol{v}\times\boldsymbol{v}] = \boldsymbol{0} \tag{A.1-12}$$

Note the following summary of laws governing the vector product operation:

Not Commutative: $$[\boldsymbol{v}\times\boldsymbol{w}] = -[\boldsymbol{w}\times\boldsymbol{v}] \tag{A.1-13}$$

Not Associative: $$[\boldsymbol{u}\times[\boldsymbol{v}\times\boldsymbol{w}]] \neq [[\boldsymbol{u}\times\boldsymbol{v}]\times\boldsymbol{w}] \tag{A.1-14}$$

Distributive: $$[\{\boldsymbol{u}+\boldsymbol{v}\}\times\boldsymbol{w}] = [\boldsymbol{u}\times\boldsymbol{w}] + [\boldsymbol{v}\times\boldsymbol{w}] \tag{A.1-15}$$

Multiple Products of Vectors

Somewhat more complicated are multiple products formed by combinations of the multiplication processes just described

(a) $rs\boldsymbol{v}$ (b) $s(\boldsymbol{v}\cdot\boldsymbol{w})$ (c) $s[\boldsymbol{v}\times\boldsymbol{w}]$
(d) $(\boldsymbol{u}\cdot[\boldsymbol{v}\times\boldsymbol{w}])$ (e) $[\boldsymbol{u}\times[\boldsymbol{v}\times\boldsymbol{w}]]$ (f) $([\boldsymbol{u}\times\boldsymbol{v}]\cdot[\boldsymbol{w}\times\boldsymbol{z}])$
(g) $[[\boldsymbol{u}\times\boldsymbol{v}]\times[\boldsymbol{w}\times\boldsymbol{z}]]$

The geometrical interpretations of the first three of these are straightforward. The magnitude of $(\boldsymbol{u}\cdot[\boldsymbol{v}\times\boldsymbol{w}])$ can easily be shown to represent the volume of a parallelepiped defined by the vectors $\boldsymbol{u}$, $\boldsymbol{v}$, and $\boldsymbol{w}$.

EXERCISES

1. What are the "orders" of the following quantities: $(\boldsymbol{v}\cdot\boldsymbol{w})$, $(\boldsymbol{v}-\boldsymbol{u})\boldsymbol{w}$, $(\boldsymbol{ab}:\boldsymbol{cd})$, $[\boldsymbol{v}\cdot\rho\boldsymbol{wu}]$, $[[\boldsymbol{a}\times\boldsymbol{f}]\times[\boldsymbol{b}\times\boldsymbol{g}]]$?

2. Draw a sketch to illustrate the inequality in Eq. A.1-9. Are there any special cases for which it becomes an equality?
3. A mathematical surface of area S has an orientation given by a unit normal vector $\boldsymbol{n}$, pointing downstream of the surface. A fluid of density ρ flows through this surface with a velocity $\boldsymbol{v}$. Show that the mass rate of flow through the surface is $w = \rho(\boldsymbol{n} \cdot \boldsymbol{v})S$.
4. The angular velocity $\boldsymbol{W}$ of a rotating solid body is a vector whose magnitude is the rate of angular displacement (radians per second) and whose direction is that in which a right-handed screw would advance if turned in the same direction. The position vector $\boldsymbol{r}$ of a point is that vector from the origin of coordinates to the point. Show that the velocity of any point in a rotating solid body is $\boldsymbol{v} = [\boldsymbol{W} \times \boldsymbol{r}]$. The origin is located on the axis of rotation.
5. A constant force $\boldsymbol{F}$ acts on a body moving with a velocity $\boldsymbol{v}$, which is not necessarily collinear with $\boldsymbol{F}$. Show that the rate at which $\boldsymbol{F}$ does work on the body is $W = (\boldsymbol{F} \cdot \boldsymbol{v})$.

§A.2 VECTOR OPERATIONS FROM AN ANALYTICAL VIEWPOINT

In this section a parallel analytical treatment is given to each of the topics presented geometrically in §A.1. In the discussion here we restrict ourselves to rectangular coordinates and label the axes as 1, 2, 3 corresponding to the usual notation of x, y, z; only right-handed coordinates are used.

Many formulas can be expressed compactly in terms of the *Kronecker delta* δ_{ij} and the *permutation symbol* ε_{ijk}. These quantities are defined thus,

$$\begin{cases} \delta_{ij} = +1, & \text{if } i = j \\ \delta_{ij} = 0, & \text{if } i \neq j \end{cases} \qquad \begin{matrix} \text{(A.2-1)} \\ \text{(A.2-2)} \end{matrix}$$

$$\begin{cases} \varepsilon_{ijk} = +1, & \text{if } ijk = 123, 231, \text{ or } 312 \\ \varepsilon_{ijk} = -1, & \text{if } ijk = 321, 132, \text{ or } 213 \\ \varepsilon_{ijk} = 0, & \text{if any two indices are alike} \end{cases} \qquad \begin{matrix} \text{(A.2-3)} \\ \text{(A.2-4)} \\ \text{(A.2-5)} \end{matrix}$$

Note also that $\varepsilon_{ijk} = (1/2)(i - j)(j - k)(k - i)$.

Several relations involving these quantities are useful in proving some vector and tensor identities

$$\sum_{j=1}^{3} \sum_{k=1}^{3} \varepsilon_{ijk} \varepsilon_{hjk} = 2\delta_{ih} \qquad \text{(A.2-6)}$$

$$\sum_{k=1}^{3} \varepsilon_{ijk} \varepsilon_{mnk} = \delta_{im}\delta_{jn} - \delta_{in}\delta_{jm} \qquad \text{(A.2-7)}$$

Note that a three-by-three determinant may be written in terms of the ε_{ijk}

$$\begin{vmatrix} a_{11} & a_{12} & a_{13} \\ a_{21} & a_{22} & a_{23} \\ a_{31} & a_{32} & a_{33} \end{vmatrix} = \sum_{i=1}^{3} \sum_{j=1}^{3} \sum_{k=1}^{3} \varepsilon_{ijk} a_{1i} a_{2j} a_{3k} \qquad \text{(A.2-8)}$$

The quantity ε_{ijk} thus selects the necessary terms that appear in the determinant and affixes the proper sign to each term.

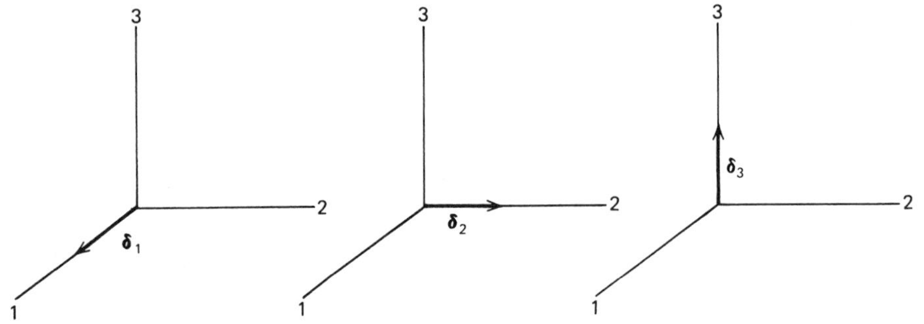

FIGURE A.2-1. The unit vectors $\boldsymbol{\delta}_i$; each vector is of unit magnitude and points in one of the coordinate directions.

The Unit Vectors

Let $\boldsymbol{\delta}_1, \boldsymbol{\delta}_2, \boldsymbol{\delta}_3$, be the "unit vectors" (i.e., vectors of unit magnitude) in the direction of the 1, 2, 3 axes[1] (Fig. A.2-1). We can use the definitions of the scalar and vector products to tabulate all possible products of each type

$$\begin{cases} (\boldsymbol{\delta}_1 \cdot \boldsymbol{\delta}_1) = (\boldsymbol{\delta}_2 \cdot \boldsymbol{\delta}_2) = (\boldsymbol{\delta}_3 \cdot \boldsymbol{\delta}_3) = 1 & \text{(A.2-9)} \\ (\boldsymbol{\delta}_1 \cdot \boldsymbol{\delta}_2) = (\boldsymbol{\delta}_2 \cdot \boldsymbol{\delta}_3) = (\boldsymbol{\delta}_3 \cdot \boldsymbol{\delta}_1) = 0 & \text{(A.2-10)} \end{cases}$$

$$\begin{cases} [\boldsymbol{\delta}_1 \times \boldsymbol{\delta}_1] = [\boldsymbol{\delta}_2 \times \boldsymbol{\delta}_2] = [\boldsymbol{\delta}_3 \times \boldsymbol{\delta}_3] = 0 & & & \text{(A.2-11)} \\ [\boldsymbol{\delta}_1 \times \boldsymbol{\delta}_2] = \boldsymbol{\delta}_3; & [\boldsymbol{\delta}_2 \times \boldsymbol{\delta}_3] = \boldsymbol{\delta}_1; & [\boldsymbol{\delta}_3 \times \boldsymbol{\delta}_1] = \boldsymbol{\delta}_2 & \text{(A.2-12)} \\ [\boldsymbol{\delta}_2 \times \boldsymbol{\delta}_1] = -\boldsymbol{\delta}_3; & [\boldsymbol{\delta}_3 \times \boldsymbol{\delta}_2] = -\boldsymbol{\delta}_1; & [\boldsymbol{\delta}_1 \times \boldsymbol{\delta}_3] = -\boldsymbol{\delta}_2 & \text{(A.2-13)} \end{cases}$$

All of these relations may be summarized by the following two relations:

$$\boxed{(\boldsymbol{\delta}_i \cdot \boldsymbol{\delta}_j) = \delta_{ij}} \tag{A.2-14}$$

$$\boxed{[\boldsymbol{\delta}_i \times \boldsymbol{\delta}_j] = \sum_{k=1}^{3} \varepsilon_{ijk} \boldsymbol{\delta}_k} \tag{A.2-15}$$

in which δ_{ij} is the Kronecker delta and ε_{ijk} is the permutation symbol defined in the introduction to this section. These two relations enable us to develop analytic expressions for all the common "dot" and "cross" operations. In the remainder of this section, and in the next section, in developing expressions for vector and tensor operations all we do is to break all vectors up into components and then apply Eqs. A.2-14 and 15.

[1] In most elementary texts the unit vectors are called ***i, j, k***. We prefer to use $\boldsymbol{\delta}_1, \boldsymbol{\delta}_2, \boldsymbol{\delta}_3$ because the components of these vectors are given by the Kronecker delta. That is, the component of $\boldsymbol{\delta}_1$ in the 1-direction is δ_{11} or unity; the component of $\boldsymbol{\delta}_1$ in the 2-direction is δ_{12} or zero.

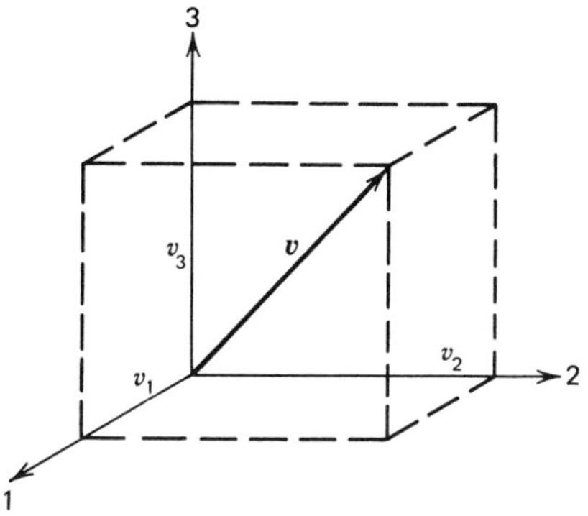

FIGURE A.2-2. The projections of a vector on the coordinate axes 1, 2, and 3.

Expansion of a Vector in Terms of Its Components

Any vector $\boldsymbol{v}$ can be completely specified by giving the values of its projections v_1, v_2, v_3, on the coordinate axes 1, 2, 3 (Fig. A.2-2). The vector can be constructed by adding vectorially the components multiplied by their corresponding unit vectors:

$$\boldsymbol{v} = \boldsymbol{\delta}_1 v_1 + \boldsymbol{\delta}_2 v_2 + \boldsymbol{\delta}_3 v_3 = \sum_{i=1}^{3} \boldsymbol{\delta}_i v_i \tag{A.2-16}$$

Note that *a vector associates a scalar with each coordinate direction.*[2] The v_i are called the "components of the vector $\boldsymbol{v}$" and they are scalars, whereas the $\boldsymbol{\delta}_i v_i$ are vectors, which when added together vectorially give $\boldsymbol{v}$.

The magnitude of a vector is given by

$$|\boldsymbol{v}| = v = \sqrt{v_1^2 + v_2^2 + v_3^2} = \sqrt{\sum_i v_i^2} \tag{A.2-17}$$

Two vectors $\boldsymbol{v}$ and $\boldsymbol{w}$ are equal if their components are equal: $v_1 = w_1$, $v_2 = w_2$, and $v_3 = w_3$. Also $\boldsymbol{v} = -\boldsymbol{w}$, if $v_1 = -w_1$, etc.,

Addition and Subtraction of Vectors

The addition or subtraction of vectors $\boldsymbol{v}$ and $\boldsymbol{w}$ may be written in terms of components

$$\boldsymbol{v} + \boldsymbol{w} = \sum_i \boldsymbol{\delta}_i v_i + \sum_i \boldsymbol{\delta}_i w_i = \sum_i \boldsymbol{\delta}_i (v_i + w_i) \tag{A.2-18}$$

Geometrically, this corresponds to adding up the projections of $\boldsymbol{v}$ and $\boldsymbol{w}$ on each individual axis and then constructing a vector with these new components. Three or more vectors may be added in exactly the same fashion.

[2] For a discussion of the relation of this definition of a vector to the definition in terms of the rules for transformation of coordinates, see W. Prager, *Mechanics of Continua*, Ginn, Boston (1961).

Multiplication of a Vector by a Scalar

Multiplication of a vector by a scalar corresponds to multiplying each component of the vector by the scalar

$$s\boldsymbol{v} = s\left\{\sum_i \boldsymbol{\delta}_i v_i\right\} = \sum_i \boldsymbol{\delta}_i\{sv_i\} \tag{A.2-19}$$

Scalar Product (or Dot Product) of Two Vectors

The scalar product of two vectors $\boldsymbol{v}$ and $\boldsymbol{w}$ is obtained by writing each vector in terms of components according to Eq. A.2-16 and then performing the scalar-product operations on the unit vectors, using Eq. A.2-14

$$\begin{aligned}(\boldsymbol{v}\cdot\boldsymbol{w}) &= \left(\left\{\sum_i \boldsymbol{\delta}_i v_i\right\}\cdot\left\{\sum_j \boldsymbol{\delta}_j w_j\right\}\right) = \sum_i\sum_j (\boldsymbol{\delta}_i\cdot\boldsymbol{\delta}_j)v_i w_j \\ &= \sum_i\sum_j \delta_{ij} v_i w_j = \sum_i v_i w_i \end{aligned} \tag{A.2-20}$$

Hence the scalar product of two vectors is obtained by summing the products of the corresponding components of the two vectors. Note that $(\boldsymbol{v}\cdot\boldsymbol{v})$ (sometimes written as $\boldsymbol{v}^2$ or as v^2) is a scalar representing the square of the magnitude of $\boldsymbol{v}$.

Vector Product (or Cross Product) of Two Vectors

The vector product of two vectors $\boldsymbol{v}$ and $\boldsymbol{w}$ may be worked out by using Eqs. A.2-16 and 15:

$$\begin{aligned}[\boldsymbol{v}\times\boldsymbol{w}] &= \left[\left\{\sum_j \boldsymbol{\delta}_j v_j\right\}\times\left\{\sum_k \boldsymbol{\delta}_k w_k\right\}\right] \\ &= \sum_j\sum_k [\boldsymbol{\delta}_j\times\boldsymbol{\delta}_k] v_j w_k = \sum_i\sum_j\sum_k \varepsilon_{ijk}\boldsymbol{\delta}_i v_j w_k \\ &= \begin{vmatrix} \boldsymbol{\delta}_1 & \boldsymbol{\delta}_2 & \boldsymbol{\delta}_3 \\ v_1 & v_2 & v_3 \\ w_1 & w_2 & w_3 \end{vmatrix}\end{aligned} \tag{A.2-21}$$

Here we have made use of Eq. A.2-8. Note that the ith-component of $[\boldsymbol{v}\times\boldsymbol{w}]$ is given by $\sum_j\sum_k \varepsilon_{ijk} v_j w_k$; this result is often used in proving vector identities.

Multiple Vector Products

Expressions for the multiple products mentioned in §A.1 can be obtained by using the foregoing analytical expressions for the scalar and vector products. For example, the product $(\boldsymbol{u}\cdot[\boldsymbol{v}\times\boldsymbol{w}])$ may be written

$$(\boldsymbol{u}\cdot[\boldsymbol{v}\times\boldsymbol{w}]) = \sum_i u_i[\boldsymbol{v}\times\boldsymbol{w}]_i = \sum_i\sum_j\sum_k \varepsilon_{ijk} u_i v_j w_k \tag{A.2-22}$$

Then, from Eq. A.2-8, we obtain

$$(\boldsymbol{u}\cdot[\boldsymbol{v}\times\boldsymbol{w}]) = \begin{vmatrix} u_1 & u_2 & u_3 \\ v_1 & v_2 & v_3 \\ w_1 & w_2 & w_3 \end{vmatrix} \tag{A.2-23}$$

The magnitude of $(\boldsymbol{u}\cdot[\boldsymbol{v}\times\boldsymbol{w}])$ is the volume of a parallelepiped defined by the vectors $\boldsymbol{u}$, $\boldsymbol{v}$, $\boldsymbol{w}$. Furthermore, the vanishing of the determinant is a necessary and sufficient condition that the vectors $\boldsymbol{u}$, $\boldsymbol{v}$, and $\boldsymbol{w}$ be coplanar.

The Position Vector

The usual symbol for the position vector—that is, the vector specifying the location of a point in space—is $\boldsymbol{r}$. The components of $\boldsymbol{r}$ are then x_1, x_2, and x_3, so that

$$\boldsymbol{r} = \sum_i \boldsymbol{\delta}_i x_i \tag{A.2-24}$$

This is an irregularity in the notation, since the components have a symbol different from that for the vector. The magnitude of $\boldsymbol{r}$ is usually called $r = \sqrt{x_1^2 + x_2^2 + x_3^2}$, and this r is the radial coordinate in spherical coordinates (see Fig. A.6-1).

EXAMPLE A.2-1 Proof of a Vector Identity

The analytical expressions for dot and cross products may be used to prove vector identities; for example, verify the relation

$$[\boldsymbol{u}\times[\boldsymbol{v}\times\boldsymbol{w}]] = \boldsymbol{v}(\boldsymbol{u}\cdot\boldsymbol{w}) - \boldsymbol{w}(\boldsymbol{u}\cdot\boldsymbol{v}) \tag{A.2-25}$$

SOLUTION The i-component of the expression on the left side can be expanded as

$$\begin{aligned}
[\boldsymbol{u}\times[\boldsymbol{v}\times\boldsymbol{w}]]_i &= \sum_j\sum_k \varepsilon_{ijk}u_j[\boldsymbol{v}\times\boldsymbol{w}]_k = \sum_j\sum_k \varepsilon_{ijk}u_j\left\{\sum_l\sum_m \varepsilon_{klm}v_l w_m\right\} \\
&= \sum_j\sum_k\sum_l\sum_m \varepsilon_{ijk}\varepsilon_{klm}u_j v_l w_m = \sum_j\sum_k\sum_l\sum_m \varepsilon_{ijk}\varepsilon_{lmk}u_j v_l w_m
\end{aligned} \tag{A.2-26}$$

Use may now be made of Eq. A.2-7 to complete the proof

$$\begin{aligned}
[\boldsymbol{u}\times[\boldsymbol{v}\times\boldsymbol{w}]]_i &= \sum_j\sum_l\sum_m (\delta_{il}\delta_{jm} - \delta_{im}\delta_{jl})u_j v_l w_m = v_i\sum_j\sum_m \delta_{jm}u_j w_m - w_i\sum_j\sum_l \delta_{jl}u_j v_l \\
&= v_i\sum_j u_j w_j - w_i\sum_j u_j v_j = v_i(\boldsymbol{u}\cdot\boldsymbol{w}) - w_i(\boldsymbol{u}\cdot\boldsymbol{v})
\end{aligned} \tag{A.2-27}$$

which is just the i-component of the right side of Eq. A.2-25. In a similar way one may verify such identities as

$$(\boldsymbol{u}\cdot[\boldsymbol{v}\times\boldsymbol{w}]) = (\boldsymbol{v}\cdot[\boldsymbol{w}\times\boldsymbol{u}]) \tag{A.2-28}$$

$$([\boldsymbol{u}\times\boldsymbol{v}]\cdot[\boldsymbol{w}\times\boldsymbol{z}]) = (\boldsymbol{u}\cdot\boldsymbol{w})(\boldsymbol{v}\cdot\boldsymbol{z}) - (\boldsymbol{u}\cdot\boldsymbol{z})(\boldsymbol{v}\cdot\boldsymbol{w}) \tag{A.2-29}$$

$$[[\boldsymbol{u}\times\boldsymbol{v}]\times[\boldsymbol{w}\times\boldsymbol{z}]] = ([\boldsymbol{u}\times\boldsymbol{v}]\cdot\boldsymbol{z})\boldsymbol{w} - ([\boldsymbol{u}\times\boldsymbol{v}]\cdot\boldsymbol{w})\boldsymbol{z} \tag{A.2-30}$$

EXERCISES

1. Write out the following summations:

$$\text{(a)}\ \sum_{k=1}^{3} k^2 \qquad \text{(b)}\ \sum_{k=1}^{3} a_k^2 \qquad \text{(c)}\ \sum_{j=1}^{3}\sum_{k=1}^{3} a_{jk}b_{kj} \qquad \text{(d)}\ \left(\sum_{j=1}^{3} a_j\right)^2 = \sum_{j=1}^{3}\sum_{k=1}^{3} a_j a_k$$

2. A vector $\boldsymbol{v}$ has components $v_x = 1$, $v_y = 2$, $v_z = -5$. A vector $\boldsymbol{w}$ has components $w_x = 3$, $w_y = -1$, $w_z = 1$. Evaluate:
   (a) $(\boldsymbol{v} \cdot \boldsymbol{w})$ (c) The length of $\boldsymbol{v}$ (e) $[\boldsymbol{\delta}_1 \times \boldsymbol{w}]$
   (b) $[\boldsymbol{v} \times \boldsymbol{w}]$ (d) $(\boldsymbol{\delta}_1 \cdot \boldsymbol{v})$ (f) ϕ_{vw}
   (g) $[\boldsymbol{r} \times \boldsymbol{v}]$, where $\boldsymbol{r}$ is the position vector.
3. Evaluate: (a) $([\boldsymbol{\delta}_1 \times \boldsymbol{\delta}_2] \cdot \boldsymbol{\delta}_3)$ (b) $[[\boldsymbol{\delta}_2 \times \boldsymbol{\delta}_3] \times [\boldsymbol{\delta}_1 \times \boldsymbol{\delta}_3]]$.
4. Show that Eq. A.2-6 is valid for the particular case $i = 1$, $h = 2$.
   Show that Eq. A.2-7 is valid for the particular case $i = j = m = 1$, $n = 2$.
5. Verify that $\sum_{j=1}^{3} \sum_{k=1}^{3} \varepsilon_{ijk}\alpha_{jk} = 0$ if $\alpha_{jk} = \alpha_{kj}$.
6. Explain carefully the statement after Eq. A.2-21 that the ith component of $[\boldsymbol{v} \times \boldsymbol{w}]$ is $\sum_j \sum_k \varepsilon_{ijk} v_j w_k$.
7. Verify that $([\boldsymbol{v} \times \boldsymbol{w}] \cdot [\boldsymbol{v} \times \boldsymbol{w}]) + (\boldsymbol{v} \cdot \boldsymbol{w})^2 = v^2 w^2$ (the "identity of Lagrange").

§A.3 TENSOR OPERATIONS

In the last section it was shown that expressions could be developed for all common "dot" and "cross" operations for vectors by knowing how to write a vector $\boldsymbol{v}$ as a sum $\sum_i \boldsymbol{\delta}_i v_i$, and by knowing how to manipulate the unit vectors $\boldsymbol{\delta}_i$. In this section we follow a parallel procedure. We write a tensor $\boldsymbol{\tau}$ as a sum $\sum_i \sum_j \boldsymbol{\delta}_i \boldsymbol{\delta}_j \tau_{ij}$, and give formulas for the manipulation of the unit dyads $\boldsymbol{\delta}_i \boldsymbol{\delta}_j$; in this way expressions are developed for the commonly occurring "dot" and "cross" operations for tensors.

The Unit Dyads

The unit vectors $\boldsymbol{\delta}_i$ were defined in the foregoing discussion and then the *scalar products* $(\boldsymbol{\delta}_i \cdot \boldsymbol{\delta}_j)$ and *vector products* $[\boldsymbol{\delta}_i \times \boldsymbol{\delta}_j]$ were given. There is a third kind of product that can be formed with the unit vectors, namely, the *dyadic products* $\boldsymbol{\delta}_i \boldsymbol{\delta}_j$ (written without parentheses, brackets, or multiplication symbols). According to the rules of notation given in the introduction to Appendix A, the products $\boldsymbol{\delta}_i \boldsymbol{\delta}_j$ are tensors of the second order. Since $\boldsymbol{\delta}_i$ and $\boldsymbol{\delta}_j$ are of unit magnitude, we will refer to the products $\boldsymbol{\delta}_i \boldsymbol{\delta}_j$ as *unit dyads*. Whereas each unit vector in Fig. A.2-1 represents a single coordinate direction, the unit dyads in Fig. A.3-1 represent *ordered* pairs of coordinate directions.

(In physical problems we often work with quantities that require the simultaneous specification of two directions. For example, the flux of x-momentum across a unit area of surface perpendicular to the y-direction is a quantity of this type. Since this quantity is sometimes not the same as the flux of y-momentum perpendicular to the x-direction, it is evident that specifying the two directions is not sufficient; we must also agree upon the order in which the directions are given.)

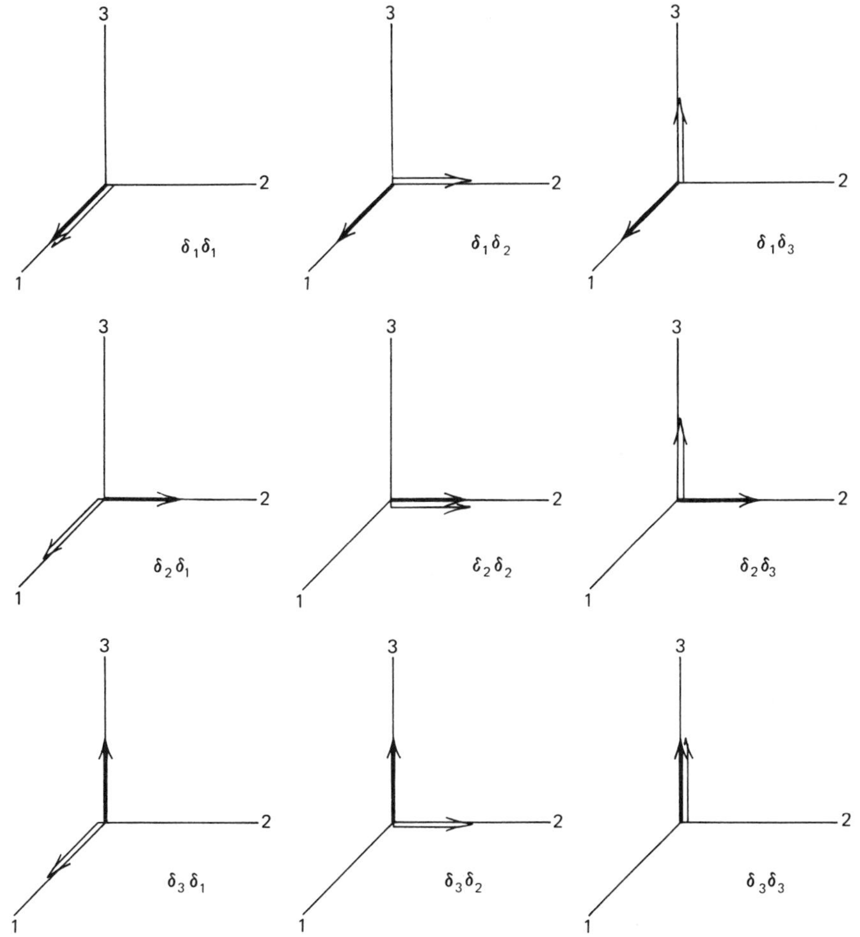

FIGURE A.3-1. The unit dyads $\boldsymbol{\delta}_i\boldsymbol{\delta}_j$. The solid arrows give the first unit vector in the dyadic product and the hollow arrows the second. Note that $\boldsymbol{\delta}_1\boldsymbol{\delta}_2$ is not the same as $\boldsymbol{\delta}_2\boldsymbol{\delta}_1$.

The dot and cross products of unit vectors were introduced by means of the geometrical definitions of these operations. The analogous operations for the unit dyads are introduced formally by relating them to the operations for unit vectors

$$(\boldsymbol{\delta}_i\boldsymbol{\delta}_j:\boldsymbol{\delta}_k\boldsymbol{\delta}_l) = (\boldsymbol{\delta}_j\cdot\boldsymbol{\delta}_k)(\boldsymbol{\delta}_i\cdot\boldsymbol{\delta}_l) = \delta_{jk}\delta_{il} \tag{A.3-1}$$

$$[\boldsymbol{\delta}_i\boldsymbol{\delta}_j\cdot\boldsymbol{\delta}_k] = \boldsymbol{\delta}_i(\boldsymbol{\delta}_j\cdot\boldsymbol{\delta}_k) = \boldsymbol{\delta}_i\delta_{jk} \tag{A.3-2}$$

$$[\boldsymbol{\delta}_i\cdot\boldsymbol{\delta}_j\boldsymbol{\delta}_k] = (\boldsymbol{\delta}_i\cdot\boldsymbol{\delta}_j)\boldsymbol{\delta}_k = \delta_{ij}\boldsymbol{\delta}_k \tag{A.3-3}$$

$$\{\boldsymbol{\delta}_i\boldsymbol{\delta}_j\cdot\boldsymbol{\delta}_k\boldsymbol{\delta}_l\} = \boldsymbol{\delta}_i(\boldsymbol{\delta}_j\cdot\boldsymbol{\delta}_k)\boldsymbol{\delta}_l = \delta_{jk}\boldsymbol{\delta}_i\boldsymbol{\delta}_l \tag{A.3-4}$$

$$\{\boldsymbol{\delta}_i\boldsymbol{\delta}_j\times\boldsymbol{\delta}_k\} = \boldsymbol{\delta}_i[\boldsymbol{\delta}_j\times\boldsymbol{\delta}_k] = \sum_{l=1}^{3}\varepsilon_{jkl}\boldsymbol{\delta}_i\boldsymbol{\delta}_l \tag{A.3-5}$$

$$\{\boldsymbol{\delta}_i\times\boldsymbol{\delta}_j\boldsymbol{\delta}_k\} = [\boldsymbol{\delta}_i\times\boldsymbol{\delta}_j]\boldsymbol{\delta}_k = \sum_{l=1}^{3}\varepsilon_{ijl}\boldsymbol{\delta}_l\boldsymbol{\delta}_k \tag{A.3-6}$$

These results are easy to remember: one simply takes the dot (or cross) product of the unit vectors on either side of the dot (or cross); in Eq. A.3-1 two such operations are performed.

Expansion of a Tensor in Terms of Its Components

In Eq. A.2-16 we expanded a vector in terms of its components, each component being multiplied by the appropriate unit vector. Here we extend this idea and define[1] a (second-order) tensor as *a quantity that associates a scalar with each ordered pair of coordinate directions* in the following sense:

$$\begin{aligned}\boldsymbol{\tau} &= \boldsymbol{\delta}_1\boldsymbol{\delta}_1\tau_{11} + \boldsymbol{\delta}_1\boldsymbol{\delta}_2\tau_{12} + \boldsymbol{\delta}_1\boldsymbol{\delta}_3\tau_{13} \\ &\quad + \boldsymbol{\delta}_2\boldsymbol{\delta}_1\tau_{21} + \boldsymbol{\delta}_2\boldsymbol{\delta}_2\tau_{22} + \boldsymbol{\delta}_2\boldsymbol{\delta}_3\tau_{23} \\ &\quad + \boldsymbol{\delta}_3\boldsymbol{\delta}_1\tau_{31} + \boldsymbol{\delta}_3\boldsymbol{\delta}_2\tau_{32} + \boldsymbol{\delta}_3\boldsymbol{\delta}_3\tau_{33} \\ &= \sum_{i=1}^{3}\sum_{j=1}^{3} \boldsymbol{\delta}_i\boldsymbol{\delta}_j\tau_{ij} \end{aligned} \qquad \text{(A.3-7)}$$

The scalars τ_{ij} are referred to as the "components of the tensor $\boldsymbol{\tau}$."

There are several special kinds of second-order tensors worth noting:

1. If $\tau_{ij} = \tau_{ji}$, the tensor is said to be *symmetric.*
2. If $\tau_{ij} = -\tau_{ji}$, the tensor is said to be *antisymmetric.*
3. If the components of a tensor are taken to be the components of $\boldsymbol{\tau}$, but with the indices transposed, the resulting tensor is called the *transpose* of $\boldsymbol{\tau}$ and given the symbol $\boldsymbol{\tau}^\dagger$

$$\boldsymbol{\tau}^\dagger = \sum_i\sum_j \boldsymbol{\delta}_i\boldsymbol{\delta}_j\tau_{ji} \qquad \text{(A.3-8)}$$

4. If the components of the tensor are formed by ordered pairs of the components of two vectors $\boldsymbol{v}$ and $\boldsymbol{w}$, the resulting tensor is called the *dyadic product of* $\boldsymbol{v}$ *and* $\boldsymbol{w}$ and given the symbol $\boldsymbol{vw}$:

$$\boldsymbol{vw} = \sum_i\sum_j \boldsymbol{\delta}_i\boldsymbol{\delta}_j v_i w_j \qquad \text{(A.3-9)}$$

Note that $\boldsymbol{vw} \neq \boldsymbol{wv}$, but that $(\boldsymbol{vw})^\dagger = \boldsymbol{wv}$.

5. If the components of the tensor are given by the Kronecker delta δ_{ij}, the resulting tensor is called the *unit tensor* and given the symbol $\boldsymbol{\delta}$:

$$\boldsymbol{\delta} = \sum_i\sum_j \boldsymbol{\delta}_i\boldsymbol{\delta}_j\delta_{ij} \qquad \text{(A.3-10)}$$

The magnitude of a tensor is defined by

$$\begin{aligned}|\boldsymbol{\tau}| = \tau &= \sqrt{\tfrac{1}{2}(\boldsymbol{\tau}:\boldsymbol{\tau}^\dagger)} \\ &= \sqrt{\tfrac{1}{2}\sum_i\sum_j \tau_{ij}^2}\end{aligned} \qquad \text{(A.3-11)}$$

[1] Tensors are often defined in terms of the transformation rules; the connections between such a definition and that given above is discussed by W. Prager, *Mechanics of Continua*, Ginn, Boston (1961). See also p. 606.

Addition of Tensors and Dyadic Products

Two tensors are added thus,

$$\boldsymbol{\sigma} + \boldsymbol{\tau} = \sum_i \sum_j \boldsymbol{\delta}_i \boldsymbol{\delta}_j \sigma_{ij} + \sum_i \sum_j \boldsymbol{\delta}_i \boldsymbol{\delta}_j \tau_{ij} = \sum_i \sum_j \boldsymbol{\delta}_i \boldsymbol{\delta}_j (\sigma_{ij} + \tau_{ij}) \tag{A.3-12}$$

That is, the sum of two tensors is that tensor whose components are the sums of the corresponding components of the two tensors. The same is true for dyadic products.

Multiplication of a Tensor by a Scalar

Multiplication of a tensor by a scalar corresponds to multiplying each component of the tensor by the scalar

$$s\boldsymbol{\tau} = s\left\{\sum_i \sum_j \boldsymbol{\delta}_i \boldsymbol{\delta}_j \tau_{ij}\right\} = \sum_i \sum_j \boldsymbol{\delta}_i \boldsymbol{\delta}_j \{s\tau_{ij}\} \tag{A.3-13}$$

The same is true for dyadic products.

The Scalar Product (or Double Dot Product) of Two Tensors

Two tensors may be multiplied according to the double dot operation

$$(\boldsymbol{\sigma} : \boldsymbol{\tau}) = \left(\left\{\sum_i \sum_j \boldsymbol{\delta}_i \boldsymbol{\delta}_j \sigma_{ij}\right\} : \left\{\sum_k \sum_l \boldsymbol{\delta}_k \boldsymbol{\delta}_l \tau_{kl}\right\}\right) = \sum_i \sum_j \sum_k \sum_l (\boldsymbol{\delta}_i \boldsymbol{\delta}_j : \boldsymbol{\delta}_k \boldsymbol{\delta}_l) \sigma_{ij} \tau_{kl}$$

$$= \sum_i \sum_j \sum_k \sum_l \delta_{il} \delta_{jk} \sigma_{ij} \tau_{kl} = \sum_i \sum_j \sigma_{ij} \tau_{ji} \tag{A.3-14}$$

in which Eq. A.3-1 has been used. Similarly, we may show that

$$(\boldsymbol{\tau} : \boldsymbol{v}\boldsymbol{w}) = \sum_i \sum_j \tau_{ij} v_j w_i \tag{A.3-15}$$

$$(\boldsymbol{u}\boldsymbol{v} : \boldsymbol{w}\boldsymbol{z}) = \sum_i \sum_j u_i v_j w_j z_i \tag{A.3-16}$$

The Tensor Product (the Single Dot Product) of Two Tensors

Two tensors may also be multiplied according to the single dot operation

$$\{\boldsymbol{\sigma} \cdot \boldsymbol{\tau}\} = \left\{\left(\sum_i \sum_j \boldsymbol{\delta}_i \boldsymbol{\delta}_j \sigma_{ij}\right) \cdot \left(\sum_k \sum_l \boldsymbol{\delta}_k \boldsymbol{\delta}_l \tau_{kl}\right)\right\} = \sum_i \sum_j \sum_k \sum_l \{\boldsymbol{\delta}_i \boldsymbol{\delta}_j \cdot \boldsymbol{\delta}_k \boldsymbol{\delta}_l\} \sigma_{ij} \tau_{kl}$$

$$= \sum_i \sum_j \sum_k \sum_l \delta_{jk} \boldsymbol{\delta}_i \boldsymbol{\delta}_l \sigma_{ij} \tau_{kl} = \sum_i \sum_l \boldsymbol{\delta}_i \boldsymbol{\delta}_l \left(\sum_j \sigma_{ij} \tau_{jl}\right) \tag{A.3-17}$$

That is, the *il*-component of $\{\boldsymbol{\sigma} \cdot \boldsymbol{\tau}\}$ is $\sum_j \sigma_{ij} \tau_{jl}$. Similar operations may be performed with dyadic products. It is common practice to write $\{\boldsymbol{\sigma} \cdot \boldsymbol{\sigma}\}$ as $\boldsymbol{\sigma}^2$, $\{\boldsymbol{\sigma} \cdot \boldsymbol{\sigma}^2\}$ as $\boldsymbol{\sigma}^3$, and so on.

The Vector Product (or Dot Product) of a Tensor with a Vector

When a tensor is dotted into a vector, we get a vector

$$[\boldsymbol{\tau}\cdot\boldsymbol{v}] = \left[\left\{\sum_i\sum_j \boldsymbol{\delta}_i\boldsymbol{\delta}_j\tau_{ij}\right\}\cdot\left\{\sum_k \boldsymbol{\delta}_k v_k\right\}\right] = \sum_i\sum_j\sum_k [\boldsymbol{\delta}_i\boldsymbol{\delta}_j\cdot\boldsymbol{\delta}_k]\tau_{ij}v_k$$

$$= \sum_i\sum_j\sum_k \boldsymbol{\delta}_i\delta_{jk}\tau_{ij}v_k = \sum_i \boldsymbol{\delta}_i\left\{\sum_j \tau_{ij}v_j\right\} \quad \text{(A.3-18)}$$

That is, the ith component of $[\boldsymbol{\tau}\cdot\boldsymbol{v}]$ is $\sum_j \tau_{ij}v_j$. Similarly, the ith component of $[\boldsymbol{v}\cdot\boldsymbol{\tau}]$ is $\sum_j v_j\tau_{ji}$. Clearly, $[\boldsymbol{\tau}\cdot\boldsymbol{v}] \neq [\boldsymbol{v}\cdot\boldsymbol{\tau}]$ unless $\boldsymbol{\tau}$ is symmetric.

Recall that when a vector $\boldsymbol{v}$ is multiplied by a scalar s the resultant vector $s\boldsymbol{v}$ points in the same direction as $\boldsymbol{v}$ but has a different length. But, when $\boldsymbol{\tau}$ is dotted into $\boldsymbol{v}$, the resultant vector $[\boldsymbol{\tau}\cdot\boldsymbol{v}]$ differs from $\boldsymbol{v}$ in *both* length and direction; that is, the tensor $\boldsymbol{\tau}$ "deflects" or "twists" the vector $\boldsymbol{v}$ to form a new vector pointing in a different direction.

The Tensor Product (or Cross Product) of a Tensor with a Vector

When a tensor is crossed with a vector, we get a tensor:

$$\{\boldsymbol{\tau}\times\boldsymbol{v}\} = \left\{\left(\sum_i\sum_j \boldsymbol{\delta}_i\boldsymbol{\delta}_j\tau_{ij}\right)\times\left(\sum_k \boldsymbol{\delta}_k v_k\right)\right\} = \sum_i\sum_j\sum_k [\boldsymbol{\delta}_i\boldsymbol{\delta}_j\times\boldsymbol{\delta}_k]\tau_{ij}v_k$$

$$= \sum_i\sum_j\sum_k\sum_l \varepsilon_{jkl}\boldsymbol{\delta}_i\boldsymbol{\delta}_l\tau_{ij}v_k = \sum_i\sum_l \boldsymbol{\delta}_i\boldsymbol{\delta}_l\left\{\sum_j\sum_k \varepsilon_{jkl}\tau_{ij}v_k\right\} \quad \text{(A.3-19)}$$

Hence, the il-component of $\{\boldsymbol{\tau}\times\boldsymbol{v}\}$ is $\sum_j\sum_k \varepsilon_{jkl}\tau_{ij}v_k$. Similarly the lk-component of $\{\boldsymbol{v}\times\boldsymbol{\tau}\}$ is $\sum_i\sum_j \varepsilon_{ijl}v_i\tau_{jk}$.

The Invariants of a Tensor

In §A.2 it was pointed out that a scalar may be formed from a single vector $\boldsymbol{v}$ by forming the product $(\boldsymbol{v}\cdot\boldsymbol{v}) = \sum_i v_iv_i$. This is the square of the magnitude of the vector $\boldsymbol{v}$, and is known as a *scalar invariant* of $\boldsymbol{v}$, because its value is independent of the coordinate system to which the components of $\boldsymbol{v}$ are referred. Only a single, independent scalar invariant can be constructed from a vector. From a tensor $\boldsymbol{\tau}$, three independent scalars can be formed by taking the *trace* of (i.e., summing the diagonal elements of) $\boldsymbol{\tau}$, $\boldsymbol{\tau}^2$, and $\boldsymbol{\tau}^3$

$$I = \text{tr}\,\boldsymbol{\tau} = \sum_i \tau_{ii} \quad \text{(A.3-20)}$$

$$II = \text{tr}\,\boldsymbol{\tau}^2 = \sum_i\sum_j \tau_{ij}\tau_{ji} \quad \text{(A.3-21)}$$

$$III = \text{tr}\,\boldsymbol{\tau}^3 = \sum_i\sum_j\sum_k \tau_{ij}\tau_{jk}\tau_{ki} \quad \text{(A.3-22)}$$

These are called *invariants* of the tensor $\boldsymbol{\tau}$, because their values are independent of the choice of coordinate system to which the components of $\boldsymbol{\tau}$ are referred. Other scalars can of course be formed, but they will be combinations of these three. For example, one often encounters the invariants I_1, I_2, and I_3, defined as

$$I_1 = I \tag{A.3-23}$$

$$I_2 = \tfrac{1}{2}(I^2 - II) \tag{A.3-24}$$

$$I_3 = \tfrac{1}{6}(I^3 - 3I \cdot II + 2III) = \det \boldsymbol{\tau} \tag{A.3-25}$$

It is also possible to form *joint invariants* of two tensors[2] $\boldsymbol{\sigma}$ and $\boldsymbol{\tau}$; these will be tr $\boldsymbol{\sigma}$, tr $\boldsymbol{\tau}$, tr $\boldsymbol{\sigma}^2$, tr $\boldsymbol{\tau}^2$, tr $\boldsymbol{\sigma}^3$, tr $\boldsymbol{\tau}^3$, tr $\{\boldsymbol{\sigma} \cdot \boldsymbol{\tau}\}$, tr $\{\boldsymbol{\sigma}^2 \cdot \boldsymbol{\tau}\}$, tr $\{\boldsymbol{\sigma} \cdot \boldsymbol{\tau}^2\}$, and tr $\{\boldsymbol{\sigma}^2 \cdot \boldsymbol{\tau}^2\}$.

The Eigenvalues of a Tensor

From a tensor $\boldsymbol{\tau}$ we may form the "characteristic equation":

$$\det(\lambda \boldsymbol{\delta} - \boldsymbol{\tau}) = 0 \tag{A.3-26}$$

where λ is a scalar. When the determinant on the left side is multiplied out, we get

$$\lambda^3 - I_1 \lambda^2 + I_2 \lambda - I_3 = 0 \tag{A.3-27}$$

from which three roots λ_1, λ_2, λ_3 may be obtained. These are the *eigenvalues* of the tensor $\boldsymbol{\tau}$.

An important theorem is that of Cayley and Hamilton which states that a tensor satisfies its own characteristic equation:

$$\boldsymbol{\tau}^3 - I_1 \boldsymbol{\tau}^2 + I_2 \boldsymbol{\tau} - I_3 \boldsymbol{\delta} = \mathbf{0} \tag{A.3-28}$$

This allows $\boldsymbol{\tau}^3$ to be expressed in terms of $\boldsymbol{\delta}$, $\boldsymbol{\tau}$, and $\boldsymbol{\tau}^2 = \{\boldsymbol{\tau} \cdot \boldsymbol{\tau}\}$. Extensions of the Cayley–Hamilton theorem have been worked out by Rivlin.[2]

Other Operations

From the foregoing results, it is not difficult to prove the following identities:

$$[\boldsymbol{\delta} \cdot \boldsymbol{v}] = [\boldsymbol{v} \cdot \boldsymbol{\delta}] = \boldsymbol{v} \tag{A.3-29}$$

$$[\boldsymbol{uv} \cdot \boldsymbol{w}] = \boldsymbol{u}(\boldsymbol{v} \cdot \boldsymbol{w}) \tag{A.3-30}$$

$$[\boldsymbol{w} \cdot \boldsymbol{uv}] = (\boldsymbol{w} \cdot \boldsymbol{u})\boldsymbol{v} \tag{A.3-31}$$

$$(\boldsymbol{uv} : \boldsymbol{wz}) = (\boldsymbol{uw} : \boldsymbol{vz}) = (\boldsymbol{u} \cdot \boldsymbol{z})(\boldsymbol{v} \cdot \boldsymbol{w}) \tag{A.3-32}$$

$$(\boldsymbol{\tau} : \boldsymbol{uv}) = ([\boldsymbol{\tau} \cdot \boldsymbol{u}] \cdot \boldsymbol{v}) \tag{A.3-33}$$

$$(\boldsymbol{uv} : \boldsymbol{\tau}) = (\boldsymbol{u} \cdot [\boldsymbol{v} \cdot \boldsymbol{\tau}]) \tag{A.3-34}$$

[2] R. S. Rivlin, *J. Rat. Mech. Anal.*, **4**, 681–702 (1955). See also R. S. Rivlin in R. J. Seeger and G. Temple, eds., *Research Frontiers in Fluid Dynamics*, Wiley-Interscience, New York (1965), Chapt. 5.

EXERCISES

1. The components of a symmetrical tensor $\boldsymbol{\tau}$ are

$$\tau_{xx} = 3 \quad \tau_{xy} = 2 \quad \tau_{xz} = -1$$

$$\tau_{yx} = 2 \quad \tau_{yy} = 2 \quad \tau_{yz} = 1$$

$$\tau_{zx} = -1 \quad \tau_{zy} = 1 \quad \tau_{zz} = 4$$

The components of a vector $\boldsymbol{v}$ are

$$v_x = 5 \quad v_y = 3 \quad v_z = -2$$

Evaluate
(a) $[\boldsymbol{\tau}\cdot\boldsymbol{v}]$ (b) $[\boldsymbol{v}\cdot\boldsymbol{\tau}]$ (c) $(\boldsymbol{\tau}:\boldsymbol{\tau})$
(d) $(\boldsymbol{v}\cdot[\boldsymbol{\tau}\cdot\boldsymbol{v}])$ (e) $\boldsymbol{v}\boldsymbol{v}$ (f) $[\boldsymbol{\tau}\cdot\boldsymbol{\delta}_1]$
(g) $\{\boldsymbol{\tau}\cdot\boldsymbol{\delta}\}$

2. Evaluate
(a) $[[\boldsymbol{\delta}_1\boldsymbol{\delta}_2\cdot\boldsymbol{\delta}_2]\times\boldsymbol{\delta}_1]$ (c) $(\boldsymbol{\delta}:\boldsymbol{\delta})$
(b) $(\boldsymbol{\delta}:\boldsymbol{\delta}_1\boldsymbol{\delta}_2)$ (d) $\{\boldsymbol{\delta}\cdot\boldsymbol{\delta}\}$

3. Find the eigenvalues of the tensor $\boldsymbol{\tau}$ given in Exercise 1.
4. If $\boldsymbol{\alpha}$ is symmetrical and $\boldsymbol{\beta}$ is antisymmetrical, show that $(\boldsymbol{\alpha}:\boldsymbol{\beta}) = 0$.
5. Explain carefully the statement after Eq. A.3-17 that the *il*-component of $\{\boldsymbol{\sigma}\cdot\boldsymbol{\tau}\}$ is $\sum_j \sigma_{ij}\tau_{jl}$.
6. Consider a rigid structure composed of point particles joined by massless rods. The particles are numbered 1, 2, 3, ... N, and the particle masses are m_ν ($\nu = 1, 2, \ldots N$). The locations of the particles with respect to the center of mass are $\boldsymbol{R}_\nu$. The entire structure rotates on an axis passing through the center of mass with an angular velocity $\boldsymbol{W}$. Show that the angular momentum with respect to the center of mass is

$$\boldsymbol{L} = \sum_\nu m_\nu [\boldsymbol{R}_\nu \times [\boldsymbol{W} \times \boldsymbol{R}_\nu]]$$

Then show that the latter expression may be rewritten as

$$\boldsymbol{L} = [\boldsymbol{\Phi}\cdot\boldsymbol{W}]$$

where

$$\boldsymbol{\Phi} = \sum_\nu m_\nu \{(\boldsymbol{R}_\nu\cdot\boldsymbol{R}_\nu)\boldsymbol{\delta} - \boldsymbol{R}_\nu\boldsymbol{R}_\nu\}$$

is the *moment of inertia tensor.*

7. The kinetic energy of rotation of the rigid structure in Exercise 6 is

$$K = \sum_\nu \tfrac{1}{2} m_\nu (\dot{\boldsymbol{R}}_\nu\cdot\dot{\boldsymbol{R}}_\nu)$$

where $\dot{\boldsymbol{R}}_\nu = [\boldsymbol{W}\times\boldsymbol{R}_\nu]$ is the velocity of the νth particle. Show that

$$K = \tfrac{1}{2}(\boldsymbol{\Phi}:\boldsymbol{W}\boldsymbol{W})$$

§A.4 THE VECTOR AND TENSOR DIFFERENTIAL OPERATIONS

The vector differential operator $\boldsymbol{\nabla}$, known as "nabla" or "del," is defined in rectangular coordinates as

$$\boldsymbol{\nabla} = \boldsymbol{\delta}_1 \frac{\partial}{\partial x_1} + \boldsymbol{\delta}_2 \frac{\partial}{\partial x_2} + \boldsymbol{\delta}_3 \frac{\partial}{\partial x_3} = \sum_i \boldsymbol{\delta}_i \frac{\partial}{\partial x_i} \tag{A.4-1}$$

in which the $\boldsymbol{\delta}_i$ are the unit vectors and the x_i are the variables associated with the 1, 2, 3 axes (i.e., the x_1, x_2, x_3 are the Cartesian coordinates normally referred to as x, y, z). The symbol $\boldsymbol{\nabla}$ is a vector-operator—it has components like a vector but it cannot stand alone; it must operate on a scalar, vector, or tensor function. In this section we summarize the various operations of $\boldsymbol{\nabla}$ on scalars, vectors, and tensors. Just as in §§A.2 and A.3, we decompose vectors and tensors into their components and then use Eqs. A.2-14 and 15, and Eqs. A.3-1 to 6. Keep in mind that in this section equations written out in component form are valid only for rectangular coordinates, for which the unit vectors $\boldsymbol{\delta}_i$ are constants; curvilinear coordinates are discussed in §§A.6 and 7. (*Note*: In the kinetic theory discussion in Volume 2 we often use the symbol $\partial/\partial \boldsymbol{r}$ in lieu of $\boldsymbol{\nabla}$. This notation is particularly convenient in multidimensional spaces.)

The Gradient of a Scalar Field

If s is a scalar function of the variables x_1, x_2, x_3, then the operation of $\boldsymbol{\nabla}$ on s is:

$$\boldsymbol{\nabla} s = \boldsymbol{\delta}_1 \frac{\partial s}{\partial x_1} + \boldsymbol{\delta}_2 \frac{\partial s}{\partial x_2} + \boldsymbol{\delta}_3 \frac{\partial s}{\partial x_3} = \sum_i \boldsymbol{\delta}_i \frac{\partial s}{\partial x_i} \tag{A.4-2}$$

The vector thus constructed from the derivatives of s is designated by $\boldsymbol{\nabla} s$ (or grad s) and is called the *gradient* of the scalar field s. The following properties of the gradient operation should be noted

Not Commutative: $$\boldsymbol{\nabla} s \neq s\boldsymbol{\nabla} \tag{A.4-3}$$

Not Associative: $$(\boldsymbol{\nabla} r)s \neq \boldsymbol{\nabla}(rs) \tag{A.4-4}$$

Distributive: $$\boldsymbol{\nabla}(r + s) = \boldsymbol{\nabla} r + \boldsymbol{\nabla} s \tag{A.4-5}$$

The Divergence of a Vector Field

If the vector $\boldsymbol{v}$ is a function of the space variables x_1, x_2, x_3, then a scalar product may be formed with the operator $\boldsymbol{\nabla}$; in obtaining the final form, we use Eq. A.2-14:

$$\begin{aligned}(\boldsymbol{\nabla} \cdot \boldsymbol{v}) &= \left(\left\{\sum_i \boldsymbol{\delta}_i \frac{\partial}{\partial x_i}\right\} \cdot \left\{\sum_j \boldsymbol{\delta}_j v_j\right\}\right) = \sum_i \sum_j (\boldsymbol{\delta}_i \cdot \boldsymbol{\delta}_j) \frac{\partial}{\partial x_i} v_j \\ &= \sum_i \sum_j \delta_{ij} \frac{\partial}{\partial x_i} v_j = \sum_i \frac{\partial v_i}{\partial x_i}\end{aligned} \tag{A.4-6}$$

This collection of derivatives of the components of the vector $\boldsymbol{v}$ is called the *divergence* of $\boldsymbol{v}$ (sometimes abbreviated div $\boldsymbol{v}$). Some properties of the divergence operator should be noted

Not Commutative: $$(\nabla \cdot \boldsymbol{v}) \neq (\boldsymbol{v} \cdot \nabla) \tag{A.4-7}$$

Not Associative: $$(\nabla \cdot s\boldsymbol{v}) \neq (\nabla s \cdot \boldsymbol{v}) \tag{A.4-8}$$

Distributive: $$(\nabla \cdot \{\boldsymbol{v} + \boldsymbol{w}\}) = (\nabla \cdot \boldsymbol{v}) + (\nabla \cdot \boldsymbol{w}) \tag{A.4-9}$$

The Curl of a Vector Field

A cross product may also be formed between the ∇ operator and the vector $\boldsymbol{v}$, which is a function of the three space variables. This cross product may be simplified by using Eq. A.2-15 and written in a variety of forms

$$\begin{aligned}
[\nabla \times \boldsymbol{v}] &= \left[\left\{\sum_j \boldsymbol{\delta}_j \frac{\partial}{\partial x_j}\right\} \times \left\{\sum_k \boldsymbol{\delta}_k v_k\right\}\right] \\
&= \sum_j \sum_k [\boldsymbol{\delta}_j \times \boldsymbol{\delta}_k] \frac{\partial}{\partial x_j} v_k = \sum_i \sum_j \sum_k \varepsilon_{ijk} \boldsymbol{\delta}_i \frac{\partial}{\partial x_j} v_k \\
&= \begin{vmatrix} \boldsymbol{\delta}_1 & \boldsymbol{\delta}_2 & \boldsymbol{\delta}_3 \\ \dfrac{\partial}{\partial x_1} & \dfrac{\partial}{\partial x_2} & \dfrac{\partial}{\partial x_3} \\ v_1 & v_2 & v_3 \end{vmatrix} \\
&= \boldsymbol{\delta}_1 \left\{\frac{\partial v_3}{\partial x_2} - \frac{\partial v_2}{\partial x_3}\right\} + \boldsymbol{\delta}_2 \left\{\frac{\partial v_1}{\partial x_3} - \frac{\partial v_3}{\partial x_1}\right\} + \boldsymbol{\delta}_3 \left\{\frac{\partial v_2}{\partial x_1} - \frac{\partial v_1}{\partial x_2}\right\}
\end{aligned} \tag{A.4-10}$$

The vector thus constructed is called the *curl* of $\boldsymbol{v}$. Other notations for $[\nabla \times \boldsymbol{v}]$ are curl $\boldsymbol{v}$ and rot $\boldsymbol{v}$, the latter being common in the German literature. The curl operation, like the divergence, is distributive but not commutative or associative. Note that the ith component of $[\nabla \times \boldsymbol{v}]$ is $\sum_j \sum_k \varepsilon_{ijk}(\partial v_k/\partial x_j)$.

The Gradient of a Vector Field

In addition to the scalar product $(\nabla \cdot \boldsymbol{v})$ and the vector product $[\nabla \times \boldsymbol{v}]$ one may also form the dyadic product $\nabla \boldsymbol{v}$:

$$\nabla \boldsymbol{v} = \left\{\sum_i \boldsymbol{\delta}_i \frac{\partial}{\partial x_i}\right\}\left\{\sum_j \boldsymbol{\delta}_j v_j\right\} = \sum_i \sum_j \boldsymbol{\delta}_i \boldsymbol{\delta}_j \frac{\partial}{\partial x_i} v_j \tag{A.4-11}$$

This is called the *gradient* of the vector $\boldsymbol{v}$ and is sometimes written grad $\boldsymbol{v}$. It is a second-order tensor whose ij-component[1] is $(\partial/\partial x_i)v_j$. Its transpose is

$$(\nabla \boldsymbol{v})^\dagger = \sum_i \sum_j \boldsymbol{\delta}_i \boldsymbol{\delta}_j \frac{\partial}{\partial x_j} v_i \tag{A.4-12}$$

whose ij-component is $\partial v_i/\partial x_j$. Note that $\nabla \boldsymbol{v} \neq \boldsymbol{v}\nabla$ and $(\nabla \boldsymbol{v})^\dagger \neq \boldsymbol{v}\nabla$.

[1] *Caution*: Some authors define the ij-component of $\nabla \boldsymbol{v}$ to be $\partial v_i/\partial x_j$.

The Divergence of a Tensor Field

If the tensor $\boldsymbol{\tau}$ is a function of the space variables x_1, x_2, x_3, then a vector product may be formed with operator $\boldsymbol{\nabla}$; in obtaining the final form we use Eq. A.3-3:

$$[\boldsymbol{\nabla}\cdot\boldsymbol{\tau}] = \left[\left\{\sum_i \boldsymbol{\delta}_i \frac{\partial}{\partial x_i}\right\}\cdot\left\{\sum_j\sum_k \boldsymbol{\delta}_j\boldsymbol{\delta}_k\tau_{jk}\right\}\right] = \sum_i\sum_j\sum_k [\boldsymbol{\delta}_i\cdot\boldsymbol{\delta}_j\boldsymbol{\delta}_k]\frac{\partial}{\partial x_i}\tau_{jk}$$

$$= \sum_i\sum_j\sum_k \delta_{ij}\boldsymbol{\delta}_k\frac{\partial}{\partial x_i}\tau_{jk} = \sum_k \boldsymbol{\delta}_k\left\{\sum_i \frac{\partial}{\partial x_i}\tau_{ik}\right\} \tag{A.4-13}$$

This is called the *divergence* of the tensor $\boldsymbol{\tau}$, and is sometimes written div $\boldsymbol{\tau}$. The kth component of $[\boldsymbol{\nabla}\cdot\boldsymbol{\tau}]$ is $\sum_i (\partial\tau_{ik}/\partial x_i)$. If $\boldsymbol{\tau}$ is the product $s\boldsymbol{v}\boldsymbol{w}$, then

$$[\boldsymbol{\nabla}\cdot s\boldsymbol{v}\boldsymbol{w}] = \sum_k \boldsymbol{\delta}_k\left\{\sum_i \frac{\partial}{\partial x_i}(sv_iw_k)\right\} \tag{A.4-14}$$

The Laplacian of a Scalar Field

If we take the divergence of the gradient of the scalar function s, we obtain

$$(\boldsymbol{\nabla}\cdot\boldsymbol{\nabla}s) = \left(\left\{\sum_i \boldsymbol{\delta}_i \frac{\partial}{\partial x_i}\right\}\cdot\left\{\sum_j \boldsymbol{\delta}_j\frac{\partial s}{\partial x_j}\right\}\right)$$

$$= \sum_i\sum_j \delta_{ij}\frac{\partial}{\partial x_i}\frac{\partial s}{\partial x_j} = \left\{\sum_i \frac{\partial^2}{\partial x_i^2}s\right\} \tag{A.4-15}$$

The collection of differential operators operating on s in the last line is given the symbol $\boldsymbol{\nabla}^2$; hence in rectangular coordinates

$$(\boldsymbol{\nabla}\cdot\boldsymbol{\nabla}) = \boldsymbol{\nabla}^2 = \frac{\partial^2}{\partial x_1^2} + \frac{\partial^2}{\partial x_2^2} + \frac{\partial^2}{\partial x_3^2} \tag{A.4-16}$$

This is called the *Laplacian* operator. (Some authors use the symbol $\boldsymbol{\Delta}$ for the Laplacian operator—this is particularly true in the older German literature; hence $(\boldsymbol{\nabla}\cdot\boldsymbol{\nabla}s)$, $(\boldsymbol{\nabla}\cdot\boldsymbol{\nabla})s$, $\boldsymbol{\nabla}^2 s$, and Δs are all equivalent forms of notation.) The Laplacian operator has only the distributive property, just as the gradient, divergence, and curl.

The Laplacian of a Vector Field

If we take the divergence of the gradient of the vector function $\boldsymbol{v}$, we obtain

$$[\boldsymbol{\nabla}\cdot\boldsymbol{\nabla}\boldsymbol{v}] = \left[\left\{\sum_i \boldsymbol{\delta}_i \frac{\partial}{\partial x_i}\right\}\cdot\left\{\sum_j\sum_k \boldsymbol{\delta}_j\boldsymbol{\delta}_k\frac{\partial}{\partial x_j}v_k\right\}\right]$$

$$= \sum_i\sum_j\sum_k [\boldsymbol{\delta}_i\cdot\boldsymbol{\delta}_j\boldsymbol{\delta}_k]\frac{\partial}{\partial x_i}\frac{\partial}{\partial x_j}v_k$$

$$= \sum_i\sum_j\sum_k \delta_{ij}\boldsymbol{\delta}_k\frac{\partial}{\partial x_i}\frac{\partial}{\partial x_j}v_k = \sum_k \boldsymbol{\delta}_k\left(\sum_i \frac{\partial^2}{\partial x_i^2}v_k\right) \tag{A.4-17}$$

That is, the kth component of $[\nabla \cdot \nabla \boldsymbol{v}]$ is, in rectangular coordinates, just $\nabla^2 v_k$. Alternative notations for $[\nabla \cdot \nabla \boldsymbol{v}]$ are $(\nabla \cdot \nabla)\boldsymbol{v}$ and $\nabla^2 \boldsymbol{v}$.

Other Differential Relations

Numerous identities can be proved using the definitions just given:

$$\nabla rs = r\nabla s + s\nabla r \tag{A.4-18}$$

$$(\nabla \cdot s\boldsymbol{v}) = (\nabla s \cdot \boldsymbol{v}) + s(\nabla \cdot \boldsymbol{v}) \tag{A.4-19}$$

$$(\nabla \cdot [\boldsymbol{v} \times \boldsymbol{w}]) = (\boldsymbol{w} \cdot [\nabla \times \boldsymbol{v}]) - (\boldsymbol{v} \cdot [\nabla \times \boldsymbol{w}]) \tag{A.4-20}$$

$$[\nabla \times s\boldsymbol{v}] = [\nabla s \times \boldsymbol{v}] + s[\nabla \times \boldsymbol{v}] \tag{A.4-21}$$

$$[\nabla \cdot \nabla \boldsymbol{v}] = \nabla(\nabla \cdot \boldsymbol{v}) - [\nabla \times [\nabla \times \boldsymbol{v}]] \tag{A.4-22}$$

$$[\boldsymbol{v} \cdot \nabla \boldsymbol{v}] = \tfrac{1}{2}\nabla(\boldsymbol{v} \cdot \boldsymbol{v}) - [\boldsymbol{v} \times [\nabla \times \boldsymbol{v}]] \tag{A.4-23}$$

$$[\nabla \cdot \boldsymbol{v}\boldsymbol{w}] = [\boldsymbol{v} \cdot \nabla \boldsymbol{w}] + \boldsymbol{w}(\nabla \cdot \boldsymbol{v}) \tag{A.4-24}$$

$$(s\boldsymbol{\delta} : \nabla \boldsymbol{v}) = s(\nabla \cdot \boldsymbol{v}) \tag{A.4-25}$$

$$[\nabla \cdot s\boldsymbol{\delta}] = \nabla s \tag{A.4-26}$$

$$[\nabla \cdot s\boldsymbol{\tau}] = [\nabla s \cdot \boldsymbol{\tau}] + s[\nabla \cdot \boldsymbol{\tau}] \tag{A.4-27}$$

$$\nabla(\boldsymbol{v} \cdot \boldsymbol{w}) = [(\nabla \boldsymbol{v}) \cdot \boldsymbol{w}] + [(\nabla \boldsymbol{w}) \cdot \boldsymbol{v}] \tag{A.4-28}$$

EXAMPLE A.4-1 Proof of a Tensor Identity

Prove that for symmetrical $\boldsymbol{\tau}$:

$$(\boldsymbol{\tau} : \nabla \boldsymbol{v}) = (\nabla \cdot [\boldsymbol{\tau} \cdot \boldsymbol{v}]) - (\boldsymbol{v} \cdot [\nabla \cdot \boldsymbol{\tau}]) \tag{A.4-29}$$

SOLUTION First we write out the right side in terms of components:

$$(\nabla \cdot [\boldsymbol{\tau} \cdot \boldsymbol{v}]) = \sum_i \frac{\partial}{\partial x_i} [\boldsymbol{\tau} \cdot \boldsymbol{v}]_i = \sum_i \sum_j \frac{\partial}{\partial x_i} \tau_{ij} v_j \tag{A.4-30}$$

$$(\boldsymbol{v} \cdot [\nabla \cdot \boldsymbol{\tau}]) = \sum_j v_j [\nabla \cdot \boldsymbol{\tau}]_j = \sum_j \sum_i v_j \frac{\partial}{\partial x_i} \tau_{ij} \tag{A.4-31}$$

The left side may be written as

$$(\boldsymbol{\tau} : \nabla \boldsymbol{v}) = \sum_i \sum_j \tau_{ji} \frac{\partial}{\partial x_i} v_j = \sum_i \sum_j \tau_{ij} \frac{\partial}{\partial x_i} v_j \tag{A.4-32}$$

the second form resulting from the symmetry of $\boldsymbol{\tau}$. Subtraction of Eq. A.4-31 from Eq. A.4-30 will give Eq. A.4-32.

EXERCISES

1. Perform all the operations in Eq. A.4-6 by writing out all the summations instead of using the $\sum$ notation.
2. A field $\boldsymbol{v}(x, y, z)$ is said to be *irrotational* if $[\nabla \times \boldsymbol{v}] = \mathbf{0}$. Which of the following fields are irrotational:
   (a) $v_x = by \quad v_y = 0 \quad v_z = 0$
   (b) $v_x = bx \quad v_y = 0 \quad v_z = 0$
   (c) $v_x = by \quad v_y = bx \quad v_z = 0$
   (d) $v_x = -by \quad v_y = bx \quad v_z = 0$
   Here b is a constant.
3. Evaluate $(\nabla \cdot \boldsymbol{v})$, $\nabla \boldsymbol{v}$, and $[\nabla \cdot \boldsymbol{v}\boldsymbol{v}]$ for the four fields in Exercise 2.
4. A vector $\boldsymbol{v}$ has components:

$$v_i = \sum_{j=1}^{3} \alpha_{ij} x_j$$

   with $\alpha_{ij} = \alpha_{ji}$ and $\sum_{i=1}^{3} \alpha_{ii} = 0$; the α_{ij} are constants. Evaluate $(\nabla \cdot \boldsymbol{v})$, $[\nabla \times \boldsymbol{v}]$, $\nabla \boldsymbol{v}$, $(\nabla \boldsymbol{v})^{\dagger}$ and $[\nabla \cdot \boldsymbol{v}\boldsymbol{v}]$. (*Hint*: In connection with evaluating $[\nabla \times \boldsymbol{v}]$, see Exercise 5 in §A.2.)
5. Verify that $\nabla^2(\nabla \cdot \boldsymbol{v}) = (\nabla \cdot (\nabla^2 \boldsymbol{v}))$, and that $[\nabla \cdot (\nabla \boldsymbol{v})^{\dagger}] = \nabla(\nabla \cdot \boldsymbol{v})$.
6. Verify that $(\nabla \cdot [\nabla \times \boldsymbol{v}]) = 0$ and $[\nabla \times \nabla s] = \mathbf{0}$.
7. If $\boldsymbol{r}$ is the position vector (with components x_1, x_2, x_3) and $\boldsymbol{v}$ is any vector, show that
   (a) $(\nabla \cdot \boldsymbol{r}) = 3$
   (b) $[\nabla \times \boldsymbol{r}] = \mathbf{0}$
   (c) $[\boldsymbol{r} \times [\nabla \cdot \boldsymbol{v}\boldsymbol{v}]] = [\nabla \cdot \boldsymbol{v}[\boldsymbol{r} \times \boldsymbol{v}]]$ (where $\boldsymbol{v}$ is a function of position)
8. Develop an alternative expression for $[\nabla \times [\nabla \cdot s\boldsymbol{v}\boldsymbol{v}]]$.
9. Is the following relation valid?

$$\tfrac{1}{2}\rho(\nabla \cdot \boldsymbol{v}) = [\nabla \times \rho \boldsymbol{v}] + \tfrac{1}{2}(\nabla \cdot \rho \boldsymbol{v}) - (\boldsymbol{v} \cdot \nabla \rho)$$

10. Write out in full in rectangular coordinates

   (a) $\dfrac{\partial}{\partial t} \rho \boldsymbol{v} = -[\nabla \cdot \rho \boldsymbol{v}\boldsymbol{v}] - \nabla p - [\nabla \cdot \boldsymbol{\tau}] + \rho \boldsymbol{g}$

   (b) $\boldsymbol{\tau} = -\mu\{\nabla \boldsymbol{v} + (\nabla \boldsymbol{v})^{\dagger} - \tfrac{2}{3}(\nabla \cdot \boldsymbol{v})\boldsymbol{\delta}\}$
11. If $(\nabla \cdot \boldsymbol{v}) = 0$, then $\boldsymbol{v} = [\nabla \times \boldsymbol{A}]$ where $\boldsymbol{A}$ is some vector. If $[\nabla \cdot \boldsymbol{\tau}] = \mathbf{0}$ and $\boldsymbol{\tau}$ is symmetrical, then $\boldsymbol{\tau} = \{\nabla \times \{\nabla \times \boldsymbol{\Phi}\}^{\dagger}\}$ where $\boldsymbol{\Phi}$ is some symmetrical tensor. Prove these statements.
12. If $\boldsymbol{r}$ is the position vector and r is its magnitude, verify that

   (a) $\nabla \dfrac{1}{r} = -\dfrac{\boldsymbol{r}}{r^3}$ (c) $\nabla(\boldsymbol{a} \cdot \boldsymbol{r}) = \boldsymbol{a}$ if $\boldsymbol{a}$ is a constant vector

   (b) $\nabla f(r) = \dfrac{1}{r}\dfrac{df}{dr}\boldsymbol{r}$ (d) $\nabla\nabla r^n = n r^{n-2}[(n-2)r^{-2}\boldsymbol{r}\boldsymbol{r} + \boldsymbol{\delta}]$

§A.5 VECTOR AND TENSOR INTEGRAL THEOREMS

For performing general proofs in continuum physics, several integral theorems are extremely useful.

The Gauss-Ostrogradskii Divergence Theorem

If V is a closed region in space enclosed by a surface S, then

$$\int_V (\nabla \cdot \boldsymbol{v})\, dV = \int_S (\boldsymbol{n} \cdot \boldsymbol{v}) dS \tag{A.5-1}$$

in which $\boldsymbol{n}$ is the outwardly directed unit normal vector. This is known as the *divergence theorem* of Gauss and Ostrogradskii. Two closely allied theorems for scalars and tensors are

$$\int_V \nabla s \, dV = \int_S \boldsymbol{n} s \, dS \tag{A.5-2}$$

$$\int_V [\nabla \cdot \boldsymbol{\tau}]\, dV = \int_S [\boldsymbol{n} \cdot \boldsymbol{\tau}] dS \tag{A.5-3}$$ [1]

The last relation is also valid for dyadic products $\boldsymbol{vw}$. Note that in all three equations ∇ in the volume integral is just replaced by $\boldsymbol{n}$ in the surface integral.

The Stokes Curl Theorem

If S is a surface bounded by the closed curve C, then

$$\int_S (\boldsymbol{n} \cdot [\nabla \times \boldsymbol{v}])\, dS = \oint_C (\boldsymbol{t} \cdot \boldsymbol{v}) dC \tag{A.5-4}$$

in which $\boldsymbol{t}$ is a unit tangential vector in the direction of integration along C; $\boldsymbol{n}$ is the unit normal vector to S in the direction that a right-hand screw would move if its head were twisted in the direction of integration along C. A similar relation exists for tensors[1]

$$\int_S [\boldsymbol{n} \cdot \{\nabla \times \boldsymbol{\tau}\}]\, dS = \oint_C [\boldsymbol{t} \cdot \boldsymbol{\tau}] dC \tag{A.5-5}$$

The Leibniz Formula for Differentiating a Triple Integral

Let V be a closed moving region in space enclosed by a surface S; let the velocity of any surface element be $\boldsymbol{v}_S$. Then, if $s(x, y, z, t)$ is a scalar function of position and time,

$$\frac{d}{dt} \int_V s \, dV = \int_V \frac{\partial s}{\partial t} dV + \int_S s(\boldsymbol{v}_S \cdot \boldsymbol{n}) dS \tag{A.5-6}$$

[1] See P. M. Morse and H. Feshbach, *Methods of Theoretical Physics*, McGraw-Hill, New York (1953), p. 66.

This is an extension of the *Leibniz formula* for differentiating an integral; keep in mind that $V = V(t)$ and $S = S(t)$. Equation A.5-6 also applies to vectors and tensors.

EXERCISES

1. Consider the vector field

$$\boldsymbol{v} = \boldsymbol{\delta}_1 x_1 + \boldsymbol{\delta}_2 x_3 + \boldsymbol{\delta}_3 x_2$$

   Evaluate both sides of Eq. A.5-1 over the region bounded by the planes $x_1 = 0$, $x_1 = 1$; $x_2 = 0$, $x_2 = 2$; $x_3 = 0$, $x_3 = 4$.
2. Use the same vector field to evaluate both sides of Eq. A.5-4 for the face $x_1 = 1$ in Exercise 1.
3. Consider the time-dependent scalar function:

$$s = x + y + zt$$

   Evaluate both sides of Eq. A.5-6 over the volume bounded by the planes: $x = 0$, $x = t$; $y = 0$, $y = 2t$; $z = 0$, $z = 4t$. The quantities x, y, z, t are dimensionless.
4. Use Eq. A.5-5 to show that

$$2\int_S \boldsymbol{n}\, dS = \oint_C [\boldsymbol{r} \times \boldsymbol{t}]dC$$

   where $\boldsymbol{r}$ is the position vector locating a point on C with respect to the origin.
5. Evaluate both sides of Eq. A.5-2 for the function $s(x, y, z) = x^2 + y^2 + z^2$. The volume V is the triangular prism lying between the two triangles whose vertices are (2, 0, 0), (2, 1, 0), (2, 0, 3), and (−2, 0, 0), (−2, 1, 0), (−2, 0, 3).

§A.6 VECTOR AND TENSOR ALGEBRA IN CURVILINEAR COORDINATES

Thus far we have considered only rectangular coordinates x, y, and z. Although formal derivations are usually made in rectangular coordinates, for working problems it is often more natural to use curvilinear coordinates. The two most commonly occurring curvilinear coordinate systems are the *cylindrical* and the *spherical*. In the following we discuss only these two systems, but the method can also be applied to all *orthogonal* coordinate systems, that is, those in which the three families of coordinate surfaces are mutually perpendicular.

We are primarily interested in knowing how to write various differential operations, such as ∇s, $[\nabla \times \boldsymbol{v}]$, and $(\boldsymbol{\tau}:\nabla\boldsymbol{v})$ in curvilinear coordinates. It turns out that we can do this quite simply if we know, for the coordinate system being used, just two things: (a) the expression for ∇ in curvilinear coordinates; and (b) the spatial derivatives of the unit vectors in curvilinear coordinates. Hence, we want to focus our attention on these two points. However, in this section we discuss transformation of coordinates and transformation of unit vectors.

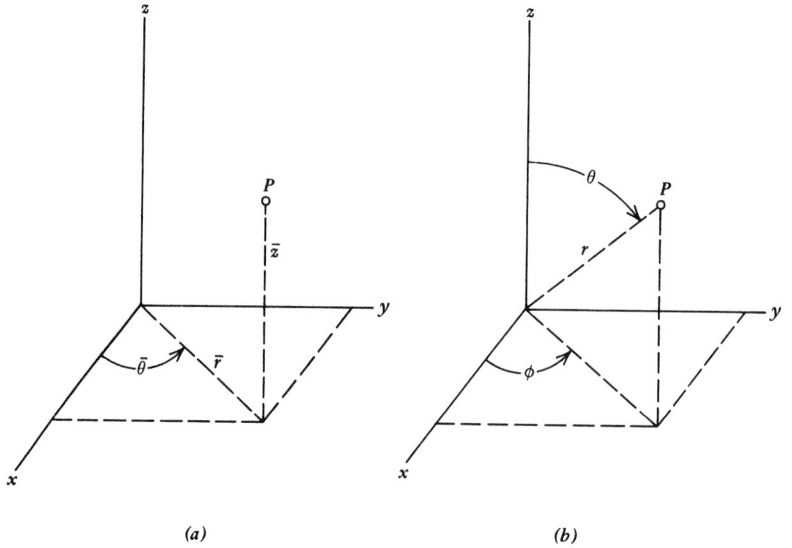

FIGURE A.6-1. (*a*) Cylindrical coordinates with $0 \leq \bar{r} < \infty, 0 \leq \bar{\theta} < 2\pi, -\infty < \bar{z} < \infty$. (*b*) Spherical coordinates with $0 \leq r < \infty, 0 \leq \theta \leq \pi, 0 \leq \phi < 2\pi$. Note that $\bar{r}$ and $\bar{\theta}$ in cylindrical coordinates are not the same as r and θ in spherical coordinates. Note also how the position vector is written in the three coordinate systems:

Rectangular: $\boldsymbol{r} = \boldsymbol{\delta}_x x + \boldsymbol{\delta}_y y + \boldsymbol{\delta}_z z$; $r = \sqrt{x^2 + y^2 + z^2}$

Cylindrical: $\boldsymbol{r} = \boldsymbol{\delta}_r \bar{r} + \boldsymbol{\delta}_z \bar{z}$; $r = \sqrt{\bar{r}^2 + \bar{z}^2}$

Spherical: $\boldsymbol{r} = \boldsymbol{\delta}_r r$; $r = r$

See footnote 1 of §A.6 regarding the omission of the bars in cylindrical coordinates.

Cylindrical Coordinates

In cylindrical coordinates, instead of designating the coordinates of a point by x, y, z, we locate the point by giving the values of r, θ, z. These coordinates[1] are shown in Fig. A.6-1*a*. They are related to the rectangular coordinates by

$$x = r\cos\theta \qquad \text{(A.6-1)} \qquad\qquad r = +\sqrt{x^2 + y^2} \qquad \text{(A.6-4)}$$

$$y = r\sin\theta \qquad \text{(A.6-2)} \qquad\qquad \theta = \arctan(y/x) \qquad \text{(A.6-5)}$$

$$z = z \qquad \text{(A.6-3)} \qquad\qquad z = z \qquad \text{(A.6-6)}$$

[1] *Caution*: We have chosen to use the familiar r, θ, z-notation for cylindrical coordinates rather than to switch to some less familiar symbols, even though there are two situations in which confusion can arise: (a) occasionally one has to use cylindrical and spherical coordinates in the same problem, and the symbols r and θ have different meanings in the two systems; (b) occasionally one deals with the position vector $\boldsymbol{r}$ in problems involving cylindrical coordinates, but then the magnitude of $\boldsymbol{r}$ is not the same as the coordinate r, but rather $\sqrt{r^2 + z^2}$. In such situations, as in Fig. A.6-1, we can use overbars for the cylindrical coordinates and write $\bar{r}, \bar{\theta}, \bar{z}$. For most discussions bars will not be needed.

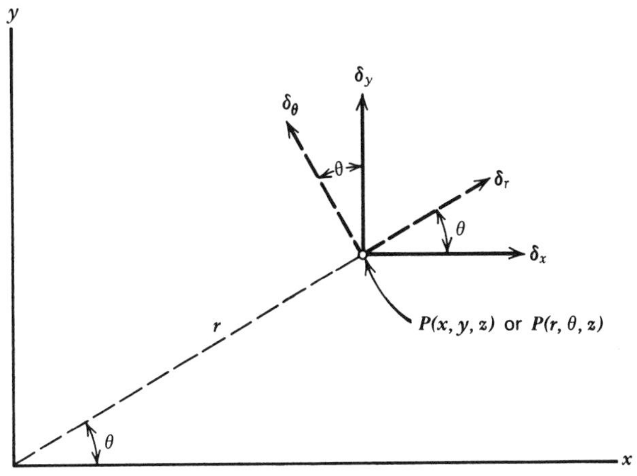

FIGURE A.6-2. Unit vectors in rectangular and cylindrical coordinates; the z-axis and unit vector $\boldsymbol{\delta}_z$ have been omitted for simplicity.

In order to convert derivatives of scalars with respect to x, y, z into derivatives with respect to r, θ, z, the "chain rule" of partial differentiation[2] is used. The derivative operators are readily found to be related thus:

$$\frac{\partial}{\partial x} = (\cos\theta)\frac{\partial}{\partial r} + \left(-\frac{\sin\theta}{r}\right)\frac{\partial}{\partial \theta} + (0)\frac{\partial}{\partial z} \tag{A.6-7}$$

$$\frac{\partial}{\partial y} = (\sin\theta)\frac{\partial}{\partial r} + \left(\frac{\cos\theta}{r}\right)\frac{\partial}{\partial \theta} + (0)\frac{\partial}{\partial z} \tag{A.6-8}$$

$$\frac{\partial}{\partial z} = (0)\frac{\partial}{\partial r} + (0)\frac{\partial}{\partial \theta} + (1)\frac{\partial}{\partial z} \tag{A.6-9}$$

With these relations, derivatives of any scalar functions (including, of course, components of vectors and tensors) with respect to x, y, and z can be expressed in terms of derivatives with respect to r, θ, and z.

Having discussed the interrelationship of the coordinates and derivatives in the two coordinate systems, we now turn to the relation between the unit vectors. We begin by noting that the unit vectors $\boldsymbol{\delta}_x$, $\boldsymbol{\delta}_y$, $\boldsymbol{\delta}_z$ (or $\boldsymbol{\delta}_1$, $\boldsymbol{\delta}_2$, $\boldsymbol{\delta}_3$ as we have been calling them) are independent of position—that is, independent of x, y, z. In cylindrical coordinates the unit vectors $\boldsymbol{\delta}_r$ and $\boldsymbol{\delta}_\theta$ will depend on position, as can be seen in Fig. A.6-2. The unit vector $\boldsymbol{\delta}_r$ is a vector of unit length in the direction of increasing r; the unit vector $\boldsymbol{\delta}_\theta$ is a vector of unit length in the direction of increasing θ. Clearly as the point P is moved around on the

[2] For example, for a scalar function $\chi(x, y, z) = \psi(r, \theta, z)$:

$$\left(\frac{\partial \chi}{\partial x}\right)_{y,z} = \left(\frac{\partial r}{\partial x}\right)_{y,z}\left(\frac{\partial \psi}{\partial r}\right)_{\theta,z} + \left(\frac{\partial \theta}{\partial x}\right)_{y,z}\left(\frac{\partial \psi}{\partial \theta}\right)_{r,z} + \left(\frac{\partial z}{\partial x}\right)_{y,z}\left(\frac{\partial \psi}{\partial z}\right)_{r,\theta}$$

Note that we are careful to use different symbols χ and ψ, since χ is a different function of x, y, z than ψ is of r, θ, and z!

xy-plane, the direction of $\boldsymbol{\delta}_r$ and $\boldsymbol{\delta}_\theta$ change. Elementary trigonometrical arguments lead to the following relations:

$$\boldsymbol{\delta}_r = (\cos\theta)\boldsymbol{\delta}_x + (\sin\theta)\boldsymbol{\delta}_y + (0)\boldsymbol{\delta}_z \tag{A.6-10}$$

$$\boldsymbol{\delta}_\theta = (-\sin\theta)\boldsymbol{\delta}_x + (\cos\theta)\boldsymbol{\delta}_y + (0)\boldsymbol{\delta}_z \tag{A.6-11}$$

$$\boldsymbol{\delta}_z = (0)\boldsymbol{\delta}_x + (0)\boldsymbol{\delta}_y + (1)\boldsymbol{\delta}_z \tag{A.6-12}$$

These may be solved for $\boldsymbol{\delta}_x$, $\boldsymbol{\delta}_y$, and $\boldsymbol{\delta}_z$ to give

$$\boldsymbol{\delta}_x = (\cos\theta)\boldsymbol{\delta}_r + (-\sin\theta)\boldsymbol{\delta}_\theta + (0)\boldsymbol{\delta}_z \tag{A.6-13}$$

$$\boldsymbol{\delta}_y = (\sin\theta)\boldsymbol{\delta}_r + (\cos\theta)\boldsymbol{\delta}_\theta + (0)\boldsymbol{\delta}_z \tag{A.6-14}$$

$$\boldsymbol{\delta}_z = (0)\boldsymbol{\delta}_r + (0)\boldsymbol{\delta}_\theta + (1)\boldsymbol{\delta}_z \tag{A.6-15}$$

The utility of these two sets of relations will be made clear in the next section.

Vectors and tensors can be decomposed into components with respect to cylindrical coordinates just as was done for rectangular coordinates in Eqs. A.2-16 and A.3-7 (i.e., $\boldsymbol{v} = \boldsymbol{\delta}_r v_r + \boldsymbol{\delta}_\theta v_\theta + \boldsymbol{\delta}_z v_z$). Also the multiplication rules for the unit vectors and unit dyads are the same as in Eqs. A.2-14 and 15 and A.3-1 to 6. Consequently the various dot and cross product operations (but *not* the differential operations!) are performed as described in §§A.2 and 3. For example,

$$(\boldsymbol{v}\cdot\boldsymbol{w}) = v_r w_r + v_\theta w_\theta + v_z w_z \tag{A.6-16}$$

$$\begin{aligned}[\boldsymbol{v}\times\boldsymbol{w}] = \boldsymbol{\delta}_r(v_\theta w_z - v_z w_\theta) + \boldsymbol{\delta}_\theta(v_z w_r - v_r w_z) \\ + \boldsymbol{\delta}_z(v_r w_\theta - v_\theta w_r)\end{aligned} \tag{A.6-17}$$

$$\begin{aligned}\{\boldsymbol{\sigma}\cdot\boldsymbol{\tau}\} = \boldsymbol{\delta}_r\boldsymbol{\delta}_r(\sigma_{rr}\tau_{rr} + \sigma_{r\theta}\tau_{\theta r} + \sigma_{rz}\tau_{zr}) \\ + \boldsymbol{\delta}_r\boldsymbol{\delta}_\theta(\sigma_{rr}\tau_{r\theta} + \sigma_{r\theta}\tau_{\theta\theta} + \sigma_{rz}\tau_{z\theta}) \\ + \boldsymbol{\delta}_r\boldsymbol{\delta}_z(\sigma_{rr}\tau_{rz} + \sigma_{r\theta}\tau_{\theta z} + \sigma_{rz}\tau_{zz}) \\ + \text{etc.}\end{aligned} \tag{A.6-18}$$

Spherical Coordinates

We now tabulate for reference the same kind of information for spherical coordinates r, θ, ϕ. These coordinates are shown in Figure A.6-1*b*. They are related to the rectangular coordinates by

$$x = r\sin\theta\cos\phi \tag{A.6-19}$$

$$y = r\sin\theta\sin\phi \tag{A.6-20}$$

$$z = r\cos\theta \tag{A.6-21}$$

$$r = +\sqrt{x^2 + y^2 + z^2} \tag{A.6-22}$$

$$\theta = \arctan\left(\sqrt{x^2 + y^2}/z\right) \tag{A.6-23}$$

$$\phi = \arctan(y/x) \tag{A.6-24}$$

For the spherical coordinates we have the following relations for the derivative operators:

$$\frac{\partial}{\partial x} = (\sin\theta\cos\phi)\frac{\partial}{\partial r} + \left(\frac{\cos\theta\cos\phi}{r}\right)\frac{\partial}{\partial\theta} + \left(-\frac{\sin\phi}{r\sin\theta}\right)\frac{\partial}{\partial\phi} \tag{A.6-25}$$

$$\frac{\partial}{\partial y} = (\sin\theta\sin\phi)\frac{\partial}{\partial r} + \left(\frac{\cos\theta\sin\phi}{r}\right)\frac{\partial}{\partial\theta} + \left(\frac{\cos\phi}{r\sin\theta}\right)\frac{\partial}{\partial\phi} \tag{A.6-26}$$

$$\frac{\partial}{\partial z} = (\cos\theta)\frac{\partial}{\partial r} + \left(-\frac{\sin\theta}{r}\right)\frac{\partial}{\partial\theta} + (0)\frac{\partial}{\partial\phi} \tag{A.6-27}$$

The relations between the unit vectors are

$$\boldsymbol{\delta}_r = (\sin\theta\cos\phi)\boldsymbol{\delta}_x + (\sin\theta\sin\phi)\boldsymbol{\delta}_y + (\cos\theta)\boldsymbol{\delta}_z \tag{A.6-28}$$

$$\boldsymbol{\delta}_\theta = (\cos\theta\cos\phi)\boldsymbol{\delta}_x + (\cos\theta\sin\phi)\boldsymbol{\delta}_y + (-\sin\theta)\boldsymbol{\delta}_z \tag{A.6-29}$$

$$\boldsymbol{\delta}_\phi = (-\sin\phi)\boldsymbol{\delta}_x + (\cos\phi)\boldsymbol{\delta}_y + (0)\boldsymbol{\delta}_z \tag{A.6-30}$$

and

$$\boldsymbol{\delta}_x = (\sin\theta\cos\phi)\boldsymbol{\delta}_r + (\cos\theta\cos\phi)\boldsymbol{\delta}_\theta + (-\sin\phi)\boldsymbol{\delta}_\phi \tag{A.6-31}$$

$$\boldsymbol{\delta}_y = (\sin\theta\sin\phi)\boldsymbol{\delta}_r + (\cos\theta\sin\phi)\boldsymbol{\delta}_\theta + (\cos\phi)\boldsymbol{\delta}_\phi \tag{A.6-32}$$

$$\boldsymbol{\delta}_z = (\cos\theta)\boldsymbol{\delta}_r + (-\sin\theta)\boldsymbol{\delta}_\theta + (0)\boldsymbol{\delta}_\phi \tag{A.6-33}$$

And, finally, some sample operations in spherical coordinates are

$$\begin{aligned}(\boldsymbol{\sigma}:\boldsymbol{\tau}) &= \sigma_{rr}\tau_{rr} + \sigma_{r\theta}\tau_{\theta r} + \sigma_{r\phi}\tau_{\phi r} \\ &\quad + \sigma_{\theta r}\tau_{r\theta} + \sigma_{\theta\theta}\tau_{\theta\theta} + \sigma_{\theta\phi}\tau_{\phi\theta} \\ &\quad + \sigma_{\phi r}\tau_{r\phi} + \sigma_{\phi\theta}\tau_{\theta\phi} + \sigma_{\phi\phi}\tau_{\phi\phi}\end{aligned} \tag{A.6-34}$$

$$(\boldsymbol{u}\cdot[\boldsymbol{v}\times\boldsymbol{w}]) = \begin{vmatrix} u_r & u_\theta & u_\phi \\ v_r & v_\theta & v_\phi \\ w_r & w_\theta & w_\phi \end{vmatrix} \tag{A.6-35}$$

That is, the relations (not involving $\boldsymbol{\nabla}$!) given in §§A.2 and 3 can be written directly in terms of spherical components.

EXERCISES

1. Show that

$$\int_0^{2\pi}\int_0^{\pi} \boldsymbol{\delta}_r \sin\theta \, d\theta \, d\phi = \mathbf{0}$$

$$\int_0^{2\pi}\int_0^{\pi} \boldsymbol{\delta}_r\boldsymbol{\delta}_r \sin\theta \, d\theta \, d\phi = (4\pi/3)\boldsymbol{\delta}$$

   where $\boldsymbol{\delta}_r$ is the unit vector in the r-direction in spherical coordinates.

2. Verify that in spherical coordinates $\boldsymbol{\delta} = \boldsymbol{\delta}_r\boldsymbol{\delta}_r + \boldsymbol{\delta}_\theta\boldsymbol{\delta}_\theta + \boldsymbol{\delta}_\phi\boldsymbol{\delta}_\phi$.

§A.7 DIFFERENTIAL OPERATIONS IN CURVILINEAR COORDINATES

We now turn to the use of the $\boldsymbol{\nabla}$-operator in curvilinear coordinates. As in the previous section, we work out in detail the results for cylindrical and spherical coordinates. Then we summarize the procedure for getting the $\boldsymbol{\nabla}$-operations for any orthogonal curvilinear coordinates.

Cylindrical Coordinates

From Eqs. A.6-10, 11, and 12 we can obtain expressions for the spatial derivatives of the unit vectors $\boldsymbol{\delta}_r$, $\boldsymbol{\delta}_\theta$, and $\boldsymbol{\delta}_z$:

$$\boxed{\frac{\partial}{\partial r}\boldsymbol{\delta}_r = \mathbf{0} \qquad \frac{\partial}{\partial r}\boldsymbol{\delta}_\theta = \mathbf{0} \qquad \frac{\partial}{\partial r}\boldsymbol{\delta}_z = \mathbf{0}} \tag{A.7-1}$$

$$\boxed{\frac{\partial}{\partial \theta}\boldsymbol{\delta}_r = \boldsymbol{\delta}_\theta \qquad \frac{\partial}{\partial \theta}\boldsymbol{\delta}_\theta = -\boldsymbol{\delta}_r \qquad \frac{\partial}{\partial \theta}\boldsymbol{\delta}_z = \mathbf{0}} \tag{A.7-2}$$

$$\boxed{\frac{\partial}{\partial z}\boldsymbol{\delta}_r = \mathbf{0} \qquad \frac{\partial}{\partial z}\boldsymbol{\delta}_\theta = \mathbf{0} \qquad \frac{\partial}{\partial z}\boldsymbol{\delta}_z = \mathbf{0}} \tag{A.7-3}$$

The reader would do well to interpret these derivatives geometrically by considering the way $\boldsymbol{\delta}_r$, $\boldsymbol{\delta}_\theta$, $\boldsymbol{\delta}_z$ change as the location of P is changed in Fig. A.6-2.

We now use the definition of the $\boldsymbol{\nabla}$-operator in Eq. A.4-1, the expressions in Eqs. A.6-13, 14, and 15, and the derivative operators in Eqs. A.6-7, 8, and 9 to obtain the formula for $\boldsymbol{\nabla}$ in cylindrical coordinates

$$\begin{aligned}\boldsymbol{\nabla} &= \boldsymbol{\delta}_x \frac{\partial}{\partial x} + \boldsymbol{\delta}_y \frac{\partial}{\partial y} + \boldsymbol{\delta}_z \frac{\partial}{\partial z} \\ &= (\boldsymbol{\delta}_r \cos\theta - \boldsymbol{\delta}_\theta \sin\theta)\left(\cos\theta \frac{\partial}{\partial r} - \frac{\sin\theta}{r}\frac{\partial}{\partial \theta}\right) \\ &\quad + (\boldsymbol{\delta}_r \sin\theta + \boldsymbol{\delta}_\theta \cos\theta)\left(\sin\theta \frac{\partial}{\partial r} + \frac{\cos\theta}{r}\frac{\partial}{\partial \theta}\right) + \boldsymbol{\delta}_z \frac{\partial}{\partial z}\end{aligned} \tag{A.7-4}$$

When this is multiplied out, there is considerable simplification, and we get

$$\boxed{\boldsymbol{\nabla} = \boldsymbol{\delta}_r \frac{\partial}{\partial r} + \boldsymbol{\delta}_\theta \frac{1}{r}\frac{\partial}{\partial \theta} + \boldsymbol{\delta}_z \frac{\partial}{\partial z}} \tag{A.7-5}$$

for *cylindrical* coordinates. This may be used for obtaining all differential operations in cylindrical coordinates, provided that Eqs. A.7-1, 2, and 3 are used to differentiate any unit vectors on which $\boldsymbol{\nabla}$ operates. This point will be made clear in the subsequent illustrative example.

Spherical Coordinates

The spatial derivatives of $\boldsymbol{\delta}_r$, $\boldsymbol{\delta}_\theta$, and $\boldsymbol{\delta}_\phi$ are given by differentiating Eqs. A.6-28, 29, and 30:

$$\frac{\partial}{\partial r}\boldsymbol{\delta}_r = \mathbf{0} \qquad \frac{\partial}{\partial r}\boldsymbol{\delta}_\theta = \mathbf{0} \qquad \frac{\partial}{\partial r}\boldsymbol{\delta}_\phi = \mathbf{0} \tag{A.7-6}$$

$$\frac{\partial}{\partial \theta}\boldsymbol{\delta}_r = \boldsymbol{\delta}_\theta \qquad \frac{\partial}{\partial \theta}\boldsymbol{\delta}_\theta = -\boldsymbol{\delta}_r \qquad \frac{\partial}{\partial \theta}\boldsymbol{\delta}_\phi = \mathbf{0} \tag{A.7-7}$$

$$\frac{\partial}{\partial \phi}\boldsymbol{\delta}_r = \boldsymbol{\delta}_\phi \sin\theta \qquad \frac{\partial}{\partial \phi}\boldsymbol{\delta}_\theta = \boldsymbol{\delta}_\phi \cos\theta \qquad \frac{\partial}{\partial \phi}\boldsymbol{\delta}_\phi = -\boldsymbol{\delta}_r \sin\theta - \boldsymbol{\delta}_\theta \cos\theta \tag{A.7-8}$$

The use of Eqs. A.6-31, 32, and 33 and Eqs. A.6-25, 26, and 27 gives the following expression for the ∇-operator:

$$\nabla = \boldsymbol{\delta}_r \frac{\partial}{\partial r} + \boldsymbol{\delta}_\theta \frac{1}{r}\frac{\partial}{\partial \theta} + \boldsymbol{\delta}_\phi \frac{1}{r\sin\theta}\frac{\partial}{\partial \phi} \tag{A.7-9}$$

in *spherical* coordinates. This expression may be used for obtaining differential operations in spherical coordinates, provided that Eqs. A.7-6, 7, and 8 are used concomitantly.

General Orthogonal Coordinates

Thus far we have discussed the two most-used curvilinear coordinates. We now present without proof the relations for any orthogonal curvilinear coordinates. Let the relation between Cartesian coordinates x_i and the curvilinear coordinates q_α be given by

$$\begin{aligned} x_1 &= x_1(q_1, q_2, q_3) \\ x_2 &= x_2(q_1, q_2, q_3) \qquad \text{or} \qquad x_i = x_i(q_\alpha) \\ x_3 &= x_3(q_1, q_2, q_3) \end{aligned} \tag{A.7-10}$$

These can be solved for the q_α to get the inverse relations $q_\alpha = q_\alpha(x_i)$. Then[1] the unit vectors in rectangular coordinates $\boldsymbol{\delta}_i$ and those in curvilinear coordinates $\boldsymbol{\delta}_\alpha$ are related thus:

$$\boldsymbol{\delta}_\alpha = \sum_i \frac{1}{h_\alpha}\left(\frac{\partial x_i}{\partial q_\alpha}\right)\boldsymbol{\delta}_i = \sum_i h_\alpha\left(\frac{\partial q_\alpha}{\partial x_i}\right)\boldsymbol{\delta}_i \tag{A.7-11}$$

$$\boldsymbol{\delta}_i = \sum_\alpha h_\alpha\left(\frac{\partial q_\alpha}{\partial x_i}\right)\boldsymbol{\delta}_\alpha = \sum_\alpha \frac{1}{h_\alpha}\left(\frac{\partial x_i}{\partial q_\alpha}\right)\boldsymbol{\delta}_\alpha \tag{A.7-12}$$

[1] P. Morse and H. Feshbach, *Methods of Theoretical Physics*, McGraw-Hill, New York (1953), p. 26.

in which the "scale factors" h_α are given by

$$h_\alpha^2 = \sum_i \left(\frac{\partial x_i}{\partial q_\alpha}\right)^2 = \left[\sum_i \left(\frac{\partial q_\alpha}{\partial x_i}\right)^2\right]^{-1} \tag{A.7-13}$$

The spatial derivatives of the unit vectors $\boldsymbol{\delta}_\alpha$ can then be found to be

$$\boxed{\frac{\partial \boldsymbol{\delta}_\alpha}{\partial q_\beta} = \frac{\boldsymbol{\delta}_\beta}{h_\alpha}\frac{\partial h_\beta}{\partial q_\alpha} - \delta_{\alpha\beta}\sum_{\gamma=1}^{3}\frac{\boldsymbol{\delta}_\gamma}{h_\gamma}\frac{\partial h_\alpha}{\partial q_\gamma}} \tag{A.7-14}$$

and the $\boldsymbol{\nabla}$-operator is

$$\boxed{\boldsymbol{\nabla} = \sum_\alpha \frac{\boldsymbol{\delta}_\alpha}{h_\alpha}\frac{\partial}{\partial q_\alpha}} \tag{A.7-15}$$

The reader should verify that Eqs. A.7-14 and 15 can be used to get Eqs. A.7-1 to 3, A.7-5, and A.7-6 to 9.

The scale factors introduced above also arise in the expressions for the volume and surface elements $dV = h_1 h_2 h_3 dq_1 dq_2 dq_3$ and $dS_{\alpha\beta} = h_\alpha h_\beta dq_\alpha dq_\beta (\alpha \neq \beta)$; here $dS_{\alpha\beta}$ is a surface element on a surface of constant γ, where $\gamma \neq \alpha$ and $\gamma \neq \beta$. The reader should verify that the volume elements and various surface elements in cylindrical and spherical coordinates can be found in this way.

In Tables A.7-1, 2, 3, and 4 we summarize the differential operations most commonly encountered in rectangular, cylindrical, spherical, and bipolar coordinates.[2] The curvilinear coordinate expressions given can be obtained by the method illustrated in the following two examples.

EXAMPLE A.7-1 Differential Operations in Cylindrical Coordinates

Derive expressions for $(\boldsymbol{\nabla}\cdot\boldsymbol{v})$ and $\boldsymbol{\nabla}\boldsymbol{v}$ in cylindrical coordinates.

SOLUTION **(a)** We begin by writing $\boldsymbol{\nabla}$ in cylindrical coordinates and decomposing $\boldsymbol{v}$ into its components

$$(\boldsymbol{\nabla}\cdot\boldsymbol{v}) = \left(\left\{\boldsymbol{\delta}_r \frac{\partial}{\partial r} + \boldsymbol{\delta}_\theta \frac{1}{r}\frac{\partial}{\partial \theta} + \boldsymbol{\delta}_z \frac{\partial}{\partial z}\right\}\cdot\left\{\boldsymbol{\delta}_r v_r + \boldsymbol{\delta}_\theta v_\theta + \boldsymbol{\delta}_z v_z\right\}\right) \tag{A.7-16}$$

Expanding, we get

$$\begin{aligned}(\boldsymbol{\nabla}\cdot\boldsymbol{v}) &= \left(\boldsymbol{\delta}_r\cdot\frac{\partial}{\partial r}\boldsymbol{\delta}_r v_r\right) + \left(\boldsymbol{\delta}_r\cdot\frac{\partial}{\partial r}\boldsymbol{\delta}_\theta v_\theta\right) + \left(\boldsymbol{\delta}_r\cdot\frac{\partial}{\partial r}\boldsymbol{\delta}_z v_z\right)\\ &+ \left(\boldsymbol{\delta}_\theta\cdot\frac{1}{r}\frac{\partial}{\partial \theta}\boldsymbol{\delta}_r v_r\right) + \left(\boldsymbol{\delta}_\theta\cdot\frac{1}{r}\frac{\partial}{\partial \theta}\boldsymbol{\delta}_\theta v_\theta\right) + \left(\boldsymbol{\delta}_\theta\cdot\frac{1}{r}\frac{\partial}{\partial \theta}\boldsymbol{\delta}_z v_z\right)\\ &+ \left(\boldsymbol{\delta}_z\cdot\frac{\partial}{\partial z}\boldsymbol{\delta}_r v_r\right) + \left(\boldsymbol{\delta}_z\cdot\frac{\partial}{\partial z}\boldsymbol{\delta}_\theta v_\theta\right) + \left(\boldsymbol{\delta}_z\cdot\frac{\partial}{\partial z}\boldsymbol{\delta}_z v_z\right)\end{aligned} \tag{A.7-17}$$

[2] For other coordinate systems see the extensive compilation of P. Moon and D. E. Spencer, *Field Theory Handbook*, Springer, Berlin (1961). In addition, an orthogonal coordinate system is available in which one of the three sets of coordinate surfaces is made up of coaxial cones (but with noncoincident apexes); all of the $\boldsymbol{\nabla}$-operations have been tabulated by the originators of this coordinate system, J. F. Dijksman and E. P. W. Savenije, *Rheol. Acta*, **24**, 105–118 (1985).

We now use the relations given in Eqs. A.7-1, 2, and 3 to evaluate the derivatives of the unit vectors. This gives

$$(\nabla \cdot \boldsymbol{v}) = (\boldsymbol{\delta}_r \cdot \boldsymbol{\delta}_r)\frac{\partial v_r}{\partial r} + (\boldsymbol{\delta}_r \cdot \boldsymbol{\delta}_\theta)\frac{\partial v_\theta}{\partial r} + (\boldsymbol{\delta}_r \cdot \boldsymbol{\delta}_z)\frac{\partial v_z}{\partial r} + (\boldsymbol{\delta}_\theta \cdot \boldsymbol{\delta}_r)\frac{1}{r}\frac{\partial v_r}{\partial \theta} + (\boldsymbol{\delta}_\theta \cdot \boldsymbol{\delta}_\theta)\frac{1}{r}\frac{\partial v_\theta}{\partial \theta} + (\boldsymbol{\delta}_\theta \cdot \boldsymbol{\delta}_z)\frac{1}{r}\frac{\partial v_z}{\partial \theta}$$

$$+ \frac{v_r}{r}(\boldsymbol{\delta}_\theta \cdot \boldsymbol{\delta}_\theta) + \frac{v_\theta}{r}(\boldsymbol{\delta}_\theta \cdot \{-\boldsymbol{\delta}_r\}) + (\boldsymbol{\delta}_z \cdot \boldsymbol{\delta}_r)\frac{\partial v_r}{\partial z} + (\boldsymbol{\delta}_z \cdot \boldsymbol{\delta}_\theta)\frac{\partial v_\theta}{\partial z} + (\boldsymbol{\delta}_z \cdot \boldsymbol{\delta}_z)\frac{\partial v_z}{\partial z} \tag{A.7-18}$$

Since $(\boldsymbol{\delta}_r \cdot \boldsymbol{\delta}_r) = 1$, $(\boldsymbol{\delta}_r \cdot \boldsymbol{\delta}_\theta) = 0$, etc., the latter simplifies to

$$(\nabla \cdot \boldsymbol{v}) = \frac{\partial v_r}{\partial r} + \frac{1}{r}\frac{\partial v_\theta}{\partial \theta} + \frac{v_r}{r} + \frac{\partial v_z}{\partial z} \tag{A.7-19}$$

which is the same as Eq. A of Table A.7-2. The procedure is a bit tedious, but it is straightforward.

(b) Next we examine the dyadic product $\nabla \boldsymbol{v}$:

$$\nabla \boldsymbol{v} = \left\{\boldsymbol{\delta}_r \frac{\partial}{\partial r} + \boldsymbol{\delta}_\theta \frac{1}{r}\frac{\partial}{\partial \theta} + \boldsymbol{\delta}_z \frac{\partial}{\partial z}\right\}\{\boldsymbol{\delta}_r v_r + \boldsymbol{\delta}_\theta v_\theta + \boldsymbol{\delta}_z v_z\}$$

$$= \boldsymbol{\delta}_r\boldsymbol{\delta}_r \frac{\partial v_r}{\partial r} + \boldsymbol{\delta}_r\boldsymbol{\delta}_\theta \frac{\partial v_\theta}{\partial r} + \boldsymbol{\delta}_r\boldsymbol{\delta}_z \frac{\partial v_z}{\partial r} + \boldsymbol{\delta}_\theta\boldsymbol{\delta}_r \frac{1}{r}\frac{\partial v_r}{\partial \theta} + \boldsymbol{\delta}_\theta\boldsymbol{\delta}_\theta \frac{1}{r}\frac{\partial v_\theta}{\partial \theta} + \boldsymbol{\delta}_\theta\boldsymbol{\delta}_z \frac{1}{r}\frac{\partial v_z}{\partial \theta}$$

$$+ \boldsymbol{\delta}_\theta\boldsymbol{\delta}_\theta \frac{v_r}{r} - \boldsymbol{\delta}_\theta\boldsymbol{\delta}_r \frac{v_\theta}{r} + \boldsymbol{\delta}_z\boldsymbol{\delta}_r \frac{\partial v_r}{\partial z} + \boldsymbol{\delta}_z\boldsymbol{\delta}_\theta \frac{\partial v_\theta}{\partial z} + \boldsymbol{\delta}_z\boldsymbol{\delta}_z \frac{\partial v_z}{\partial z}$$

$$= \boldsymbol{\delta}_r\boldsymbol{\delta}_r \frac{\partial v_r}{\partial r} + \boldsymbol{\delta}_r\boldsymbol{\delta}_\theta \frac{\partial v_\theta}{\partial r} + \boldsymbol{\delta}_r\boldsymbol{\delta}_z \frac{\partial v_z}{\partial r} + \boldsymbol{\delta}_\theta\boldsymbol{\delta}_r\left(\frac{1}{r}\frac{\partial v_r}{\partial \theta} - \frac{v_\theta}{r}\right) + \boldsymbol{\delta}_\theta\boldsymbol{\delta}_\theta\left(\frac{1}{r}\frac{\partial v_\theta}{\partial \theta} + \frac{v_r}{r}\right)$$

$$+ \boldsymbol{\delta}_\theta\boldsymbol{\delta}_z \frac{1}{r}\frac{\partial v_z}{\partial \theta} + \boldsymbol{\delta}_z\boldsymbol{\delta}_r \frac{\partial v_r}{\partial z} + \boldsymbol{\delta}_z\boldsymbol{\delta}_\theta \frac{\partial v_\theta}{\partial z} + \boldsymbol{\delta}_z\boldsymbol{\delta}_z \frac{\partial v_z}{\partial z} \tag{A.7-20}$$

Hence, the rr-component is $\partial v_r/\partial r$, the $r\theta$-component is $\partial v_\theta/\partial r$, etc. Note that the trace of this expression for $\nabla \boldsymbol{v}$ gives the expression for $(\nabla \cdot \boldsymbol{v})$ in Eq. A.7-19.

EXAMPLE A.7-2 Differential Operations in Spherical Coordinates

Find the r-component of $[\nabla \cdot \boldsymbol{\tau}]$ in spherical coordinates

SOLUTION Using Eq. A.7-9 we have

$$[\nabla \cdot \boldsymbol{\tau}]_r = \left[\left\{\boldsymbol{\delta}_r \frac{\partial}{\partial r} + \boldsymbol{\delta}_\theta \frac{1}{r}\frac{\partial}{\partial \theta} + \boldsymbol{\delta}_\phi \frac{1}{r \sin\theta}\frac{\partial}{\partial \phi}\right\} \cdot \{\boldsymbol{\delta}_r\boldsymbol{\delta}_r\tau_{rr} + \boldsymbol{\delta}_r\boldsymbol{\delta}_\theta\tau_{r\theta} + \boldsymbol{\delta}_r\boldsymbol{\delta}_\phi\tau_{r\phi}\right.$$

$$\left. + \boldsymbol{\delta}_\theta\boldsymbol{\delta}_r\tau_{\theta r} + \boldsymbol{\delta}_\theta\boldsymbol{\delta}_\theta\tau_{\theta\theta} + \boldsymbol{\delta}_\theta\boldsymbol{\delta}_\phi\tau_{\theta\phi} + \boldsymbol{\delta}_\phi\boldsymbol{\delta}_r\tau_{\phi r} + \boldsymbol{\delta}_\phi\boldsymbol{\delta}_\theta\tau_{\phi\theta} + \boldsymbol{\delta}_\phi\boldsymbol{\delta}_\phi\tau_{\phi\phi}\}\right]_r \tag{A.7-21}$$

TABLE A.7-1

Summary of Differential Operations Involving the ∇-Operator in Rectangular Coordinates (x, y, z)

$$(\nabla \cdot \boldsymbol{v}) = \frac{\partial v_x}{\partial x} + \frac{\partial v_y}{\partial y} + \frac{\partial v_z}{\partial z} \tag{A}$$

$$(\nabla^2 s) = \frac{\partial^2 s}{\partial x^2} + \frac{\partial^2 s}{\partial y^2} + \frac{\partial^2 s}{\partial z^2} \tag{B}$$

$$\begin{aligned}(\boldsymbol{\tau} : \nabla \boldsymbol{v}) = \tau_{xx}\left(\frac{\partial v_x}{\partial x}\right) + \tau_{xy}\left(\frac{\partial v_x}{\partial y}\right) + \tau_{xz}\left(\frac{\partial v_x}{\partial z}\right) \\ + \tau_{yx}\left(\frac{\partial v_y}{\partial x}\right) + \tau_{yy}\left(\frac{\partial v_y}{\partial y}\right) + \tau_{yz}\left(\frac{\partial v_y}{\partial z}\right) \\ + \tau_{zx}\left(\frac{\partial v_z}{\partial x}\right) + \tau_{zy}\left(\frac{\partial v_z}{\partial y}\right) + \tau_{zz}\left(\frac{\partial v_z}{\partial z}\right)\end{aligned} \tag{C}$$

$$[\nabla s]_x = \frac{\partial s}{\partial x} \tag{D}$$

$$[\nabla s]_y = \frac{\partial s}{\partial y} \tag{E}$$

$$[\nabla s]_z = \frac{\partial s}{\partial z} \tag{F}$$

$$[\nabla \times \boldsymbol{v}]_x = \frac{\partial v_z}{\partial y} - \frac{\partial v_y}{\partial z} \tag{G}$$

$$[\nabla \times \boldsymbol{v}]_y = \frac{\partial v_x}{\partial z} - \frac{\partial v_z}{\partial x} \tag{H}$$

$$[\nabla \times \boldsymbol{v}]_z = \frac{\partial v_y}{\partial x} - \frac{\partial v_x}{\partial y} \tag{I}$$

$$[\nabla \cdot \boldsymbol{\tau}]_x = \frac{\partial \tau_{xx}}{\partial x} + \frac{\partial \tau_{yx}}{\partial y} + \frac{\partial \tau_{zx}}{\partial z} \tag{J}$$

$$[\nabla \cdot \boldsymbol{\tau}]_y = \frac{\partial \tau_{xy}}{\partial x} + \frac{\partial \tau_{yy}}{\partial y} + \frac{\partial \tau_{zy}}{\partial z} \tag{K}$$

$$[\nabla \cdot \boldsymbol{\tau}]_z = \frac{\partial \tau_{xz}}{\partial x} + \frac{\partial \tau_{yz}}{\partial y} + \frac{\partial \tau_{zz}}{\partial z} \tag{L}$$

$$[\nabla^2 \boldsymbol{v}]_x = \frac{\partial^2 v_x}{\partial x^2} + \frac{\partial^2 v_x}{\partial y^2} + \frac{\partial^2 v_x}{\partial z^2} \tag{M}$$

$$[\nabla^2 \boldsymbol{v}]_y = \frac{\partial^2 v_y}{\partial x^2} + \frac{\partial^2 v_y}{\partial y^2} + \frac{\partial^2 v_y}{\partial z^2} \tag{N}$$

$$[\nabla^2 \boldsymbol{v}]_z = \frac{\partial^2 v_z}{\partial x^2} + \frac{\partial^2 v_z}{\partial y^2} + \frac{\partial^2 v_z}{\partial z^2} \tag{O}$$

$$[\boldsymbol{v} \cdot \nabla \boldsymbol{w}]_x = v_x\left(\frac{\partial w_x}{\partial x}\right) + v_y\left(\frac{\partial w_x}{\partial y}\right) + v_z\left(\frac{\partial w_x}{\partial z}\right) \tag{P}$$

$$[\boldsymbol{v} \cdot \nabla \boldsymbol{w}]_y = v_x\left(\frac{\partial w_y}{\partial x}\right) + v_y\left(\frac{\partial w_y}{\partial y}\right) + v_z\left(\frac{\partial w_y}{\partial z}\right) \tag{Q}$$

$$[\boldsymbol{v} \cdot \nabla \boldsymbol{w}]_z = v_x\left(\frac{\partial w_z}{\partial x}\right) + v_y\left(\frac{\partial w_z}{\partial y}\right) + v_z\left(\frac{\partial w_z}{\partial z}\right) \tag{R}$$

TABLE A.7-1 (*Continued*)

$$\{\nabla v\}_{xx} = \frac{\partial v_x}{\partial x} \quad \text{(S)}$$

$$\{\nabla v\}_{xy} = \frac{\partial v_y}{\partial x} \quad \text{(T)}$$

$$\{\nabla v\}_{xz} = \frac{\partial v_z}{\partial x} \quad \text{(U)}$$

$$\{\nabla v\}_{yx} = \frac{\partial v_x}{\partial y} \quad \text{(V)}$$

$$\{\nabla v\}_{yy} = \frac{\partial v_y}{\partial y} \quad \text{(W)}$$

$$\{\nabla v\}_{yz} = \frac{\partial v_z}{\partial y} \quad \text{(X)}$$

$$\{\nabla v\}_{zx} = \frac{\partial v_x}{\partial z} \quad \text{(Y)}$$

$$\{\nabla v\}_{zy} = \frac{\partial v_y}{\partial z} \quad \text{(Z)}$$

$$\{\nabla v\}_{zz} = \frac{\partial v_z}{\partial z} \quad \text{(AA)}$$

$$\{v \cdot \nabla \tau\}_{xx} = (v \cdot \nabla)\tau_{xx} \quad \text{(BB)}$$

$$\{v \cdot \nabla \tau\}_{xy} = (v \cdot \nabla)\tau_{xy} \quad \text{(CC)}$$

$$\{v \cdot \nabla \tau\}_{xz} = (v \cdot \nabla)\tau_{xz} \quad \text{(DD)}$$

$$\{v \cdot \nabla \tau\}_{yx} = (v \cdot \nabla)\tau_{yx} \quad \text{(EE)}$$

$$\{v \cdot \nabla \tau\}_{yy} = (v \cdot \nabla)\tau_{yy} \quad \text{(FF)}$$

$$\{v \cdot \nabla \tau\}_{yz} = (v \cdot \nabla)\tau_{yz} \quad \text{(GG)}$$

$$\{v \cdot \nabla \tau\}_{zx} = (v \cdot \nabla)\tau_{zx} \quad \text{(HH)}$$

$$\{v \cdot \nabla \tau\}_{zy} = (v \cdot \nabla)\tau_{zy} \quad \text{(II)}$$

$$\{v \cdot \nabla \tau\}_{zz} = (v \cdot \nabla)\tau_{zz} \quad \text{(JJ)}$$

where the operator $(v \cdot \nabla) = v_x \frac{\partial}{\partial x} + v_y \frac{\partial}{\partial y} + v_z \frac{\partial}{\partial z}$

TABLE A.7-2

Summary of Differential Operations Involving the ∇-Operator in Cylindrical Coordinates (r, θ, z)

$$(\nabla \cdot \boldsymbol{v}) = \frac{1}{r}\frac{\partial}{\partial r}(rv_r) + \frac{1}{r}\frac{\partial v_\theta}{\partial \theta} + \frac{\partial v_z}{\partial z} \tag{A}$$

$$(\nabla^2 s) = \frac{1}{r}\frac{\partial}{\partial r}\left(r\frac{\partial s}{\partial r}\right) + \frac{1}{r^2}\frac{\partial^2 s}{\partial \theta^2} + \frac{\partial^2 s}{\partial z^2} \tag{B}$$

$$\begin{aligned}(\boldsymbol{\tau}:\nabla \boldsymbol{v}) = {} & \tau_{rr}\left(\frac{\partial v_r}{\partial r}\right) + \tau_{r\theta}\left(\frac{1}{r}\frac{\partial v_r}{\partial \theta} - \frac{v_\theta}{r}\right) + \tau_{rz}\left(\frac{\partial v_r}{\partial z}\right) \\ & + \tau_{\theta r}\left(\frac{\partial v_\theta}{\partial r}\right) + \tau_{\theta\theta}\left(\frac{1}{r}\frac{\partial v_\theta}{\partial \theta} + \frac{v_r}{r}\right) + \tau_{\theta z}\left(\frac{\partial v_\theta}{\partial z}\right) \\ & + \tau_{zr}\left(\frac{\partial v_z}{\partial r}\right) + \tau_{z\theta}\left(\frac{1}{r}\frac{\partial v_z}{\partial \theta}\right) + \tau_{zz}\left(\frac{\partial v_z}{\partial z}\right)\end{aligned} \tag{C}$$

$$[\nabla s]_r = \frac{\partial s}{\partial r} \tag{D}$$

$$[\nabla s]_\theta = \frac{1}{r}\frac{\partial s}{\partial \theta} \tag{E}$$

$$[\nabla s]_z = \frac{\partial s}{\partial z} \tag{F}$$

$$[\nabla \times \boldsymbol{v}]_r = \frac{1}{r}\frac{\partial v_z}{\partial \theta} - \frac{\partial v_\theta}{\partial z} \tag{G}$$

$$[\nabla \times \boldsymbol{v}]_\theta = \frac{\partial v_r}{\partial z} - \frac{\partial v_z}{\partial r} \tag{H}$$

$$[\nabla \times \boldsymbol{v}]_z = \frac{1}{r}\frac{\partial}{\partial r}(rv_\theta) - \frac{1}{r}\frac{\partial v_r}{\partial \theta} \tag{I}$$

$$[\nabla \cdot \boldsymbol{\tau}]_r = \frac{1}{r}\frac{\partial}{\partial r}(r\tau_{rr}) + \frac{1}{r}\frac{\partial}{\partial \theta}\tau_{\theta r} + \frac{\partial}{\partial z}\tau_{zr} - \frac{\tau_{\theta\theta}}{r} \tag{J}$$

$$[\nabla \cdot \boldsymbol{\tau}]_\theta = \frac{1}{r^2}\frac{\partial}{\partial r}(r^2\tau_{r\theta}) + \frac{1}{r}\frac{\partial}{\partial \theta}\tau_{\theta\theta} + \frac{\partial}{\partial z}\tau_{z\theta} + \frac{\tau_{\theta r} - \tau_{r\theta}}{r} \tag{K}$$

$$[\nabla \cdot \boldsymbol{\tau}]_z = \frac{1}{r}\frac{\partial}{\partial r}(r\tau_{rz}) + \frac{1}{r}\frac{\partial}{\partial \theta}\tau_{\theta z} + \frac{\partial}{\partial z}\tau_{zz} \tag{L}$$

$$[\nabla^2 \boldsymbol{v}]_r = \frac{\partial}{\partial r}\left(\frac{1}{r}\frac{\partial}{\partial r}(rv_r)\right) + \frac{1}{r^2}\frac{\partial^2 v_r}{\partial \theta^2} + \frac{\partial^2 v_r}{\partial z^2} - \frac{2}{r^2}\frac{\partial v_\theta}{\partial \theta} \tag{M}$$

$$[\nabla^2 \boldsymbol{v}]_\theta = \frac{\partial}{\partial r}\left(\frac{1}{r}\frac{\partial}{\partial r}(rv_\theta)\right) + \frac{1}{r^2}\frac{\partial^2 v_\theta}{\partial \theta^2} + \frac{\partial^2 v_\theta}{\partial z^2} + \frac{2}{r^2}\frac{\partial v_r}{\partial \theta} \tag{N}$$

$$[\nabla^2 \boldsymbol{v}]_z = \frac{1}{r}\frac{\partial}{\partial r}\left(r\frac{\partial v_z}{\partial r}\right) + \frac{1}{r^2}\frac{\partial^2 v_z}{\partial \theta^2} + \frac{\partial^2 v_z}{\partial z^2} \tag{O}$$

$$[\boldsymbol{v} \cdot \nabla \boldsymbol{w}]_r = v_r\left(\frac{\partial w_r}{\partial r}\right) + v_\theta\left(\frac{1}{r}\frac{\partial w_r}{\partial \theta} - \frac{w_\theta}{r}\right) + v_z\left(\frac{\partial w_r}{\partial z}\right) \tag{P}$$

$$[\boldsymbol{v} \cdot \nabla \boldsymbol{w}]_\theta = v_r\left(\frac{\partial w_\theta}{\partial r}\right) + v_\theta\left(\frac{1}{r}\frac{\partial w_\theta}{\partial \theta} + \frac{w_r}{r}\right) + v_z\left(\frac{\partial w_\theta}{\partial z}\right) \tag{Q}$$

$$[\boldsymbol{v} \cdot \nabla \boldsymbol{w}]_z = v_r\left(\frac{\partial w_z}{\partial r}\right) + v_\theta\left(\frac{1}{r}\frac{\partial w_z}{\partial \theta}\right) + v_z\left(\frac{\partial w_z}{\partial z}\right) \tag{R}$$

TABLE A.7-2 *(Continued)*

$$\{\nabla \boldsymbol{v}\}_{rr} = \frac{\partial v_r}{\partial r} \tag{S}$$

$$\{\nabla \boldsymbol{v}\}_{r\theta} = \frac{\partial v_\theta}{\partial r} \tag{T}$$

$$\{\nabla \boldsymbol{v}\}_{rz} = \frac{\partial v_z}{\partial r} \tag{U}$$

$$\{\nabla \boldsymbol{v}\}_{\theta r} = \frac{1}{r}\frac{\partial v_r}{\partial \theta} - \frac{v_\theta}{r} \tag{V}$$

$$\{\nabla \boldsymbol{v}\}_{\theta\theta} = \frac{1}{r}\frac{\partial v_\theta}{\partial \theta} + \frac{v_r}{r} \tag{W}$$

$$\{\nabla \boldsymbol{v}\}_{\theta z} = \frac{1}{r}\frac{\partial v_z}{\partial \theta} \tag{X}$$

$$\{\nabla \boldsymbol{v}\}_{zr} = \frac{\partial v_r}{\partial z} \tag{Y}$$

$$\{\nabla \boldsymbol{v}\}_{z\theta} = \frac{\partial v_\theta}{\partial z} \tag{Z}$$

$$\{\nabla \boldsymbol{v}\}_{zz} = \frac{\partial v_z}{\partial z} \tag{AA}$$

$$\{\boldsymbol{v}\cdot\nabla\boldsymbol{\tau}\}_{rr} = (\boldsymbol{v}\cdot\nabla)\tau_{rr} - \frac{v_\theta}{r}(\tau_{r\theta} + \tau_{\theta r}) \tag{BB}$$

$$\{\boldsymbol{v}\cdot\nabla\boldsymbol{\tau}\}_{r\theta} = (\boldsymbol{v}\cdot\nabla)\tau_{r\theta} + \frac{v_\theta}{r}(\tau_{rr} - \tau_{\theta\theta}) \tag{CC}$$

$$\{\boldsymbol{v}\cdot\nabla\boldsymbol{\tau}\}_{rz} = (\boldsymbol{v}\cdot\nabla)\tau_{rz} - \frac{v_\theta}{r}\tau_{\theta z} \tag{DD}$$

$$\{\boldsymbol{v}\cdot\nabla\boldsymbol{\tau}\}_{\theta r} = (\boldsymbol{v}\cdot\nabla)\tau_{\theta r} + \frac{v_\theta}{r}(\tau_{rr} - \tau_{\theta\theta}) \tag{EE}$$

$$\{\boldsymbol{v}\cdot\nabla\boldsymbol{\tau}\}_{\theta\theta} = (\boldsymbol{v}\cdot\nabla)\tau_{\theta\theta} + \frac{v_\theta}{r}(\tau_{r\theta} + \tau_{\theta r}) \tag{FF}$$

$$\{\boldsymbol{v}\cdot\nabla\boldsymbol{\tau}\}_{\theta z} = (\boldsymbol{v}\cdot\nabla)\tau_{\theta z} + \frac{v_\theta}{r}\tau_{rz} \tag{GG}$$

$$\{\boldsymbol{v}\cdot\nabla\boldsymbol{\tau}\}_{zr} = (\boldsymbol{v}\cdot\nabla)\tau_{zr} - \frac{v_\theta}{r}\tau_{z\theta} \tag{HH}$$

$$\{\boldsymbol{v}\cdot\nabla\boldsymbol{\tau}\}_{z\theta} = (\boldsymbol{v}\cdot\nabla)\tau_{z\theta} + \frac{v_\theta}{r}\tau_{zr} \tag{II}$$

$$\{\boldsymbol{v}\cdot\nabla\boldsymbol{\tau}\}_{zz} = (\boldsymbol{v}\cdot\nabla)\tau_{zz} \tag{JJ}$$

where the operator $(\boldsymbol{v}\cdot\nabla) = v_r\frac{\partial}{\partial r} + \frac{v_\theta}{r}\frac{\partial}{\partial \theta} + v_z\frac{\partial}{\partial z}$

TABLE A.7-3

Summary of Differential Operations Involving the ∇-Operator in Spherical Coordinates (r, θ, ϕ)

$$(\nabla\cdot v) = \frac{1}{r^2}\frac{\partial}{\partial r}(r^2 v_r) + \frac{1}{r\sin\theta}\frac{\partial}{\partial\theta}(v_\theta \sin\theta) + \frac{1}{r\sin\theta}\frac{\partial v_\phi}{\partial\phi} \tag{A}$$

$$(\nabla^2 s) = \frac{1}{r^2}\frac{\partial}{\partial r}\left(r^2\frac{\partial s}{\partial r}\right) + \frac{1}{r^2\sin\theta}\frac{\partial}{\partial\theta}\left(\sin\theta\frac{\partial s}{\partial\theta}\right) + \frac{1}{r^2\sin^2\theta}\frac{\partial^2 s}{\partial\phi^2} \tag{B}$$

$$\begin{aligned}(\tau:\nabla v) = {} & \tau_{rr}\left(\frac{\partial v_r}{\partial r}\right) + \tau_{r\theta}\left(\frac{1}{r}\frac{\partial v_r}{\partial\theta} - \frac{v_\theta}{r}\right) + \tau_{r\phi}\left(\frac{1}{r\sin\theta}\frac{\partial v_r}{\partial\phi} - \frac{v_\phi}{r}\right)\\ & + \tau_{\theta r}\left(\frac{\partial v_\theta}{\partial r}\right) + \tau_{\theta\theta}\left(\frac{1}{r}\frac{\partial v_\theta}{\partial\theta} + \frac{v_r}{r}\right) + \tau_{\theta\phi}\left(\frac{1}{r\sin\theta}\frac{\partial v_\theta}{\partial\phi} - \frac{v_\phi}{r}\cot\theta\right)\\ & + \tau_{\phi r}\left(\frac{\partial v_\phi}{\partial r}\right) + \tau_{\phi\theta}\left(\frac{1}{r}\frac{\partial v_\phi}{\partial\theta}\right) + \tau_{\phi\phi}\left(\frac{1}{r\sin\theta}\frac{\partial v_\phi}{\partial\phi} + \frac{v_r}{r} + \frac{v_\theta}{r}\cot\theta\right)\end{aligned} \tag{C}$$

$$[\nabla s]_r = \frac{\partial s}{\partial r} \tag{D}$$

$$[\nabla s]_\theta = \frac{1}{r}\frac{\partial s}{\partial\theta} \tag{E}$$

$$[\nabla s]_\phi = \frac{1}{r\sin\theta}\frac{\partial s}{\partial\phi} \tag{F}$$

$$[\nabla\times v]_r = \frac{1}{r\sin\theta}\frac{\partial}{\partial\theta}(v_\phi\sin\theta) - \frac{1}{r\sin\theta}\frac{\partial v_\theta}{\partial\phi} \tag{G}$$

$$[\nabla\times v]_\theta = \frac{1}{r\sin\theta}\frac{\partial v_r}{\partial\phi} - \frac{1}{r}\frac{\partial}{\partial r}(r v_\phi) \tag{H}$$

$$[\nabla\times v]_\phi = \frac{1}{r}\frac{\partial}{\partial r}(r v_\theta) - \frac{1}{r}\frac{\partial v_r}{\partial\theta} \tag{I}$$

$$[\nabla\cdot\tau]_r = \frac{1}{r^2}\frac{\partial}{\partial r}(r^2\tau_{rr}) + \frac{1}{r\sin\theta}\frac{\partial}{\partial\theta}(\tau_{\theta r}\sin\theta) + \frac{1}{r\sin\theta}\frac{\partial}{\partial\phi}\tau_{\phi r} - \frac{\tau_{\theta\theta}+\tau_{\phi\phi}}{r} \tag{J}$$

$$[\nabla\cdot\tau]_\theta = \frac{1}{r^3}\frac{\partial}{\partial r}(r^3\tau_{r\theta}) + \frac{1}{r\sin\theta}\frac{\partial}{\partial\theta}(\tau_{\theta\theta}\sin\theta) + \frac{1}{r\sin\theta}\frac{\partial}{\partial\phi}\tau_{\phi\theta} + \frac{(\tau_{\theta r}-\tau_{r\theta}) - \tau_{\phi\phi}\cot\theta}{r} \tag{K}$$

$$[\nabla\cdot\tau]_\phi = \frac{1}{r^3}\frac{\partial}{\partial r}(r^3\tau_{r\phi}) + \frac{1}{r\sin\theta}\frac{\partial}{\partial\theta}(\tau_{\theta\phi}\sin\theta) + \frac{1}{r\sin\theta}\frac{\partial}{\partial\phi}\tau_{\phi\phi} + \frac{(\tau_{\phi r}-\tau_{r\phi}) + \tau_{\phi\theta}\cot\theta}{r} \tag{L}$$

$$\begin{aligned}[\nabla^2 v]_r = {} & \frac{\partial}{\partial r}\left(\frac{1}{r^2}\frac{\partial}{\partial r}(r^2 v_r)\right) + \frac{1}{r^2\sin\theta}\frac{\partial}{\partial\theta}\left(\sin\theta\frac{\partial v_r}{\partial\theta}\right) + \frac{1}{r^2\sin^2\theta}\frac{\partial^2 v_r}{\partial\phi^2}\\ & - \frac{2}{r^2\sin\theta}\frac{\partial}{\partial\theta}(v_\theta\sin\theta) - \frac{2}{r^2\sin\theta}\frac{\partial v_\phi}{\partial\phi}\end{aligned} \tag{M}$$

$$[\nabla^2 v]_\theta = \frac{1}{r^2}\frac{\partial}{\partial r}\left(r^2\frac{\partial v_\theta}{\partial r}\right) + \frac{1}{r^2}\frac{\partial}{\partial\theta}\left(\frac{1}{\sin\theta}\frac{\partial}{\partial\theta}(v_\theta\sin\theta)\right) + \frac{1}{r^2\sin^2\theta}\frac{\partial^2 v_\theta}{\partial\phi^2} + \frac{2}{r^2}\frac{\partial v_r}{\partial\theta} - \frac{2\cot\theta}{r^2\sin\theta}\frac{\partial v_\phi}{\partial\phi} \tag{N}$$

$$\begin{aligned}[\nabla^2 v]_\phi = {} & \frac{1}{r^2}\frac{\partial}{\partial r}\left(r^2\frac{\partial v_\phi}{\partial r}\right) + \frac{1}{r^2}\frac{\partial}{\partial\theta}\left(\frac{1}{\sin\theta}\frac{\partial}{\partial\theta}(v_\phi\sin\theta)\right) + \frac{1}{r^2\sin^2\theta}\frac{\partial^2 v_\phi}{\partial\phi^2}\\ & + \frac{2}{r^2\sin\theta}\frac{\partial v_r}{\partial\phi} + \frac{2\cot\theta}{r^2\sin\theta}\frac{\partial v_\theta}{\partial\phi}\end{aligned} \tag{O}$$

$$[v\cdot\nabla w]_r = v_r\left(\frac{\partial w_r}{\partial r}\right) + v_\theta\left(\frac{1}{r}\frac{\partial w_r}{\partial\theta} - \frac{w_\theta}{r}\right) + v_\phi\left(\frac{1}{r\sin\theta}\frac{\partial w_r}{\partial\phi} - \frac{w_\phi}{r}\right) \tag{P}$$

$$[v\cdot\nabla w]_\theta = v_r\left(\frac{\partial w_\theta}{\partial r}\right) + v_\theta\left(\frac{1}{r}\frac{\partial w_\theta}{\partial\theta} + \frac{w_r}{r}\right) + v_\phi\left(\frac{1}{r\sin\theta}\frac{\partial w_\theta}{\partial\phi} - \frac{w_\phi}{r}\cot\theta\right) \tag{Q}$$

TABLE A.7-3 (*Continued*)

$$[\boldsymbol{v}\cdot\nabla\boldsymbol{w}]_\phi = v_r\left(\frac{\partial w_\phi}{\partial r}\right) + v_\theta\left(\frac{1}{r}\frac{\partial w_\phi}{\partial \theta}\right) + v_\phi\left(\frac{1}{r\sin\theta}\frac{\partial w_\phi}{\partial \phi} + \frac{w_r}{r} + \frac{w_\theta}{r}\cot\theta\right) \quad \text{(R)}$$

$$\{\nabla\boldsymbol{v}\}_{rr} = \frac{\partial v_r}{\partial r} \quad \text{(S)}$$

$$\{\nabla\boldsymbol{v}\}_{r\theta} = \frac{\partial v_\theta}{\partial r} \quad \text{(T)}$$

$$\{\nabla\boldsymbol{v}\}_{r\phi} = \frac{\partial v_\phi}{\partial r} \quad \text{(U)}$$

$$\{\nabla\boldsymbol{v}\}_{\theta r} = \frac{1}{r}\frac{\partial v_r}{\partial \theta} - \frac{v_\theta}{r} \quad \text{(V)}$$

$$\{\nabla\boldsymbol{v}\}_{\theta\theta} = \frac{1}{r}\frac{\partial v_\theta}{\partial \theta} + \frac{v_r}{r} \quad \text{(W)}$$

$$\{\nabla\boldsymbol{v}\}_{\theta\phi} = \frac{1}{r}\frac{\partial v_\phi}{\partial \theta} \quad \text{(X)}$$

$$\{\nabla\boldsymbol{v}\}_{\phi r} = \frac{1}{r\sin\theta}\frac{\partial v_r}{\partial \phi} - \frac{v_\phi}{r} \quad \text{(Y)}$$

$$\{\nabla\boldsymbol{v}\}_{\phi\theta} = \frac{1}{r\sin\theta}\frac{\partial v_\theta}{\partial \phi} - \frac{v_\phi}{r}\cot\theta \quad \text{(Z)}$$

$$\{\nabla\boldsymbol{v}\}_{\phi\phi} = \frac{1}{r\sin\theta}\frac{\partial v_\phi}{\partial \phi} + \frac{v_r}{r} + \frac{v_\theta}{r}\cot\theta \quad \text{(AA)}$$

$$\{\boldsymbol{v}\cdot\nabla\boldsymbol{\tau}\}_{rr} = (\boldsymbol{v}\cdot\nabla)\tau_{rr} - \left(\frac{v_\theta}{r}\right)(\tau_{r\theta} + \tau_{\theta r}) - \left(\frac{v_\phi}{r}\right)(\tau_{r\phi} + \tau_{\phi r}) \quad \text{(BB)}$$

$$\{\boldsymbol{v}\cdot\nabla\boldsymbol{\tau}\}_{r\theta} = (\boldsymbol{v}\cdot\nabla)\tau_{r\theta} + \left(\frac{v_\theta}{r}\right)(\tau_{rr} - \tau_{\theta\theta}) - \left(\frac{v_\phi}{r}\right)(\tau_{\phi\theta} + \tau_{r\phi}\cot\theta) \quad \text{(CC)}$$

$$\{\boldsymbol{v}\cdot\nabla\boldsymbol{\tau}\}_{r\phi} = (\boldsymbol{v}\cdot\nabla)\tau_{r\phi} - \left(\frac{v_\theta}{r}\right)\tau_{\theta\phi} + \left(\frac{v_\phi}{r}\right)[(\tau_{rr} - \tau_{\phi\phi}) + \tau_{r\theta}\cot\theta] \quad \text{(DD)}$$

$$\{\boldsymbol{v}\cdot\nabla\boldsymbol{\tau}\}_{\theta r} = (\boldsymbol{v}\cdot\nabla)\tau_{\theta r} + \left(\frac{v_\theta}{r}\right)(\tau_{rr} - \tau_{\theta\theta}) - \left(\frac{v_\phi}{r}\right)(\tau_{\theta\phi} + \tau_{\phi r}\cot\theta) \quad \text{(EE)}$$

$$\{\boldsymbol{v}\cdot\nabla\boldsymbol{\tau}\}_{\theta\theta} = (\boldsymbol{v}\cdot\nabla)\tau_{\theta\theta} + \left(\frac{v_\theta}{r}\right)(\tau_{r\theta} + \tau_{\theta r}) - \left(\frac{v_\phi}{r}\right)(\tau_{\theta\phi} + \tau_{\phi\theta})\cot\theta \quad \text{(FF)}$$

$$\{\boldsymbol{v}\cdot\nabla\boldsymbol{\tau}\}_{\theta\phi} = (\boldsymbol{v}\cdot\nabla)\tau_{\theta\phi} + \left(\frac{v_\theta}{r}\right)\tau_{r\phi} + \left(\frac{v_\phi}{r}\right)[\tau_{\theta r} + (\tau_{\theta\theta} - \tau_{\phi\phi})\cot\theta] \quad \text{(GG)}$$

$$\{\boldsymbol{v}\cdot\nabla\boldsymbol{\tau}\}_{\phi r} = (\boldsymbol{v}\cdot\nabla)\tau_{\phi r} - \left(\frac{v_\theta}{r}\right)\tau_{\phi\theta} + \left(\frac{v_\phi}{r}\right)[(\tau_{rr} - \tau_{\phi\phi}) + \tau_{\theta r}\cot\theta] \quad \text{(HH)}$$

$$\{\boldsymbol{v}\cdot\nabla\boldsymbol{\tau}\}_{\phi\theta} = (\boldsymbol{v}\cdot\nabla)\tau_{\phi\theta} + \left(\frac{v_\theta}{r}\right)\tau_{\phi r} + \left(\frac{v_\phi}{r}\right)[\tau_{r\theta} + (\tau_{\theta\theta} - \tau_{\phi\phi})\cot\theta] \quad \text{(II)}$$

$$\{\boldsymbol{v}\cdot\nabla\boldsymbol{\tau}\}_{\phi\phi} = (\boldsymbol{v}\cdot\nabla)\tau_{\phi\phi} + \left(\frac{v_\phi}{r}\right)[(\tau_{r\phi} + \tau_{\phi r}) + (\tau_{\theta\phi} + \tau_{\phi\theta})\cot\theta] \quad \text{(JJ)}$$

where the operator $(\boldsymbol{v}\cdot\nabla) = v_r\dfrac{\partial}{\partial r} + \dfrac{v_\theta}{r}\dfrac{\partial}{\partial \theta} + \dfrac{v_\phi}{r\sin\theta}\dfrac{\partial}{\partial \phi}$

TABLE A.7-4

Summary of Differential Operations Involving the ∇-Operator in Bipolar Coordinates (ξ, θ, z)

In this table the abbreviation $X = \cosh\xi + \cos\theta$ is used.

$$(\nabla \cdot \boldsymbol{v}) = \frac{X^2}{a}\frac{\partial}{\partial \xi}\left(\frac{v_\xi}{X}\right) + \frac{X^2}{a}\frac{\partial}{\partial \theta}\left(\frac{v_\theta}{X}\right) + \frac{\partial v_z}{\partial z} \tag{A}$$

$$\nabla^2 s = \left(\frac{X}{a}\right)^2 \frac{\partial^2 s}{\partial \xi^2} + \left(\frac{X}{a}\right)^2 \frac{\partial^2 s}{\partial \theta^2} + \frac{\partial^2 s}{\partial z^2} \tag{B}$$

$$\begin{aligned}(\boldsymbol{\tau}:\nabla \boldsymbol{v}) = {} & \tau_{\xi\xi}\left(\frac{X}{a}\frac{\partial v_\xi}{\partial \xi} + \frac{v_\theta}{a}\sin\theta\right) + \tau_{\xi\theta}\left(\frac{X}{a}\frac{\partial v_\xi}{\partial \theta} + \frac{v_\theta}{a}\sinh\xi\right) + \tau_{\xi z}\left(\frac{\partial v_\xi}{\partial z}\right) + \tau_{\theta\xi}\left(\frac{X}{a}\frac{\partial v_\theta}{\partial \xi} - \frac{v_\xi}{a}\sin\theta\right) \\ & + \tau_{\theta\theta}\left(\frac{X}{a}\frac{\partial v_\theta}{\partial \theta} - \frac{v_\xi}{a}\sinh\xi\right) + \tau_{\theta z}\left(\frac{\partial v_\theta}{\partial z}\right) + \tau_{z\xi}\left(\frac{X}{a}\frac{\partial v_z}{\partial \xi}\right) + \tau_{z\theta}\left(\frac{X}{a}\frac{\partial v_z}{\partial \theta}\right) + \tau_{zz}\left(\frac{\partial v_z}{\partial z}\right)\end{aligned} \tag{C}$$

$$[\nabla s]_\xi = \frac{X}{a}\frac{\partial s}{\partial \xi} \tag{D}$$

$$[\nabla s]_\theta = \frac{X}{a}\frac{\partial s}{\partial \theta} \tag{E}$$

$$[\nabla s]_z = \frac{\partial s}{\partial z} \tag{F}$$

$$[\nabla \times \boldsymbol{v}]_\xi = \frac{X}{a}\frac{\partial v_z}{\partial \theta} - \frac{\partial v_\theta}{\partial z} \tag{G}$$

$$[\nabla \times \boldsymbol{v}]_\theta = \frac{\partial v_\xi}{\partial z} - \frac{X}{a}\frac{\partial v_z}{\partial \xi} \tag{H}$$

$$[\nabla \times \boldsymbol{v}]_z = \frac{X^2}{a}\frac{\partial}{\partial \xi}\left(\frac{v_\theta}{X}\right) - \frac{X^2}{a}\frac{\partial}{\partial \theta}\left(\frac{v_\xi}{X}\right) \tag{I}$$

$$[\nabla \cdot \boldsymbol{\tau}]_\xi = \frac{X}{a}\frac{\partial \tau_{\xi\xi}}{\partial \xi} + \frac{X}{a}\frac{\partial \tau_{\theta\xi}}{\partial \theta} + \frac{\partial \tau_{z\xi}}{\partial z} + \frac{\tau_{\theta\theta} - \tau_{\xi\xi}}{a}\sinh\xi + \frac{\tau_{\theta\xi} + \tau_{\xi\theta}}{a}\sin\theta \tag{J}$$

$$[\nabla \cdot \boldsymbol{\tau}]_\theta = \frac{X}{a}\frac{\partial \tau_{\xi\theta}}{\partial \xi} + \frac{X}{a}\frac{\partial \tau_{\theta\theta}}{\partial \theta} + \frac{\partial \tau_{z\theta}}{\partial z} + \frac{\tau_{\theta\theta} - \tau_{\xi\xi}}{a}\sin\theta - \frac{\tau_{\theta\xi} + \tau_{\xi\theta}}{a}\sinh\xi \tag{K}$$

$$[\nabla \cdot \boldsymbol{\tau}]_z = \frac{X}{a}\frac{\partial \tau_{\xi z}}{\partial \xi} + \frac{X}{a}\frac{\partial \tau_{\theta z}}{\partial \theta} + \frac{\partial \tau_{zz}}{\partial z} - \frac{\tau_{\xi z}}{a}\sinh\xi + \frac{\tau_{\theta z}}{a}\sin\theta \tag{L}$$

$$[\nabla^2 \boldsymbol{v}]_\xi = \left(\frac{X}{a}\right)^2 \frac{\partial^2 v_\xi}{\partial \xi^2} + \left(\frac{X}{a}\right)^2 \frac{\partial^2 v_\xi}{\partial \theta^2} + \frac{\partial^2 v_\xi}{\partial z^2} - \frac{v_\xi}{a^2}(\sinh^2\xi + \sin^2\theta) + 2\frac{X}{a^2}\left(\frac{\partial v_\theta}{\partial \theta}\sinh\xi + \frac{\partial v_\theta}{\partial \xi}\sin\theta\right) \tag{M}$$

$$[\nabla^2 \boldsymbol{v}]_\theta = \left(\frac{X}{a}\right)^2 \frac{\partial^2 v_\theta}{\partial \xi^2} + \left(\frac{X}{a}\right)^2 \frac{\partial^2 v_\theta}{\partial \theta^2} + \frac{\partial^2 v_\theta}{\partial z^2} - \frac{v_\theta}{a^2}(\sin^2\theta + \sinh^2\xi) - 2\frac{X}{a^2}\left(\frac{\partial v_\xi}{\partial \xi}\sin\theta + \frac{\partial v_\xi}{\partial \theta}\sinh\xi\right) \tag{N}$$

$$[\nabla^2 \boldsymbol{v}]_z = \left(\frac{X}{a}\right)^2 \frac{\partial^2 v_z}{\partial \xi^2} + \left(\frac{X}{a}\right)^2 \frac{\partial^2 v_z}{\partial \theta^2} + \frac{\partial^2 v_z}{\partial z^2} \tag{O}$$

$$[\boldsymbol{v} \cdot \nabla \boldsymbol{w}]_\xi = v_\xi\left(\frac{X}{a}\frac{\partial w_\xi}{\partial \xi} + \frac{w_\theta}{a}\sin\theta\right) + v_\theta\left(\frac{X}{a}\frac{\partial w_\xi}{\partial \theta} + \frac{w_\theta}{a}\sinh\xi\right) + v_z\frac{\partial w_\xi}{\partial z} \tag{P}$$

$$[\boldsymbol{v} \cdot \nabla \boldsymbol{w}]_\theta = v_\xi\left(\frac{X}{a}\frac{\partial w_\theta}{\partial \xi} - \frac{w_\xi}{a}\sin\theta\right) + v_\theta\left(\frac{X}{a}\frac{\partial w_\theta}{\partial \theta} - \frac{w_\xi}{a}\sinh\xi\right) + v_z\frac{\partial w_\theta}{\partial z} \tag{Q}$$

$$[\boldsymbol{v} \cdot \nabla \boldsymbol{w}]_z = v_\xi\left(\frac{X}{a}\frac{\partial w_z}{\partial \xi}\right) + v_\theta\left(\frac{X}{a}\frac{\partial w_z}{\partial \theta}\right) + v_z\frac{\partial w_z}{\partial z} \tag{R}$$

TABLE A.7-4 (*Continued*)

$$\{\nabla \boldsymbol{v}\}_{\xi\xi} = \frac{X}{a}\frac{\partial v_\xi}{\partial \xi} + \frac{v_\theta}{a}\sin\theta \tag{S}$$

$$\{\nabla \boldsymbol{v}\}_{\xi\theta} = \frac{X}{a}\frac{\partial v_\theta}{\partial \xi} - \frac{v_\xi}{a}\sin\theta \tag{T}$$

$$\{\nabla \boldsymbol{v}\}_{\xi z} = \frac{X}{a}\frac{\partial v_z}{\partial \xi} \tag{U}$$

$$\{\nabla \boldsymbol{v}\}_{\theta\xi} = \frac{X}{a}\frac{\partial v_\xi}{\partial \theta} + \frac{v_\theta}{a}\sinh\xi \tag{V}$$

$$\{\nabla \boldsymbol{v}\}_{\theta\theta} = \frac{X}{a}\frac{\partial v_\theta}{\partial \theta} - \frac{v_\xi}{a}\sinh\xi \tag{W}$$

$$\{\nabla \boldsymbol{v}\}_{\theta z} = \frac{X}{a}\frac{\partial v_z}{\partial \theta} \tag{X}$$

$$\{\nabla \boldsymbol{v}\}_{z\xi} = \frac{\partial v_\xi}{\partial z} \tag{Y}$$

$$\{\nabla \boldsymbol{v}\}_{z\theta} = \frac{\partial v_\theta}{\partial z} \tag{Z}$$

$$\{\nabla \boldsymbol{v}\}_{zz} = \frac{\partial v_z}{\partial z} \tag{AA}$$

$$\{\boldsymbol{v}\cdot\nabla\boldsymbol{\tau}\}_{\xi\xi} = (\boldsymbol{v}\cdot\nabla)\tau_{\xi\xi} + \left(\frac{\tau_{\xi\theta}+\tau_{\theta\xi}}{a}\right)(v_\xi\sin\theta + v_\theta\sinh\xi) \tag{BB}$$

$$\{\boldsymbol{v}\cdot\nabla\boldsymbol{\tau}\}_{\xi\theta} = (\boldsymbol{v}\cdot\nabla)\tau_{\xi\theta} + \left(\frac{\tau_{\theta\theta}-\tau_{\xi\xi}}{a}\right)(v_\xi\sin\theta + v_\theta\sinh\xi) \tag{CC}$$

$$\{\boldsymbol{v}\cdot\nabla\boldsymbol{\tau}\}_{\xi z} = (\boldsymbol{v}\cdot\nabla)\tau_{\xi z} + \frac{\tau_{\theta z}}{a}(v_\xi\sin\theta + v_\theta\sinh\xi) \tag{DD}$$

$$\{\boldsymbol{v}\cdot\nabla\boldsymbol{\tau}\}_{\theta\xi} = (\boldsymbol{v}\cdot\nabla)\tau_{\theta\xi} + \left(\frac{\tau_{\theta\theta}-\tau_{\xi\xi}}{a}\right)(v_\xi\sin\theta + v_\theta\sinh\xi) \tag{EE}$$

$$\{\boldsymbol{v}\cdot\nabla\boldsymbol{\tau}\}_{\theta\theta} = (\boldsymbol{v}\cdot\nabla)\tau_{\theta\theta} - \left(\frac{\tau_{\xi\theta}+\tau_{\theta\xi}}{a}\right)(v_\xi\sin\theta + v_\theta\sinh\xi) \tag{FF}$$

$$\{\boldsymbol{v}\cdot\nabla\boldsymbol{\tau}\}_{\theta z} = (\boldsymbol{v}\cdot\nabla)\tau_{\theta z} - \left(\frac{\tau_{\xi z}}{a}\right)(v_\xi\sin\theta + v_\theta\sinh\xi) \tag{GG}$$

$$\{\boldsymbol{v}\cdot\nabla\boldsymbol{\tau}\}_{z\xi} = (\boldsymbol{v}\cdot\nabla)\tau_{z\xi} + \left(\frac{\tau_{z\theta}}{a}\right)(v_\xi\sin\theta + v_\theta\sinh\xi) \tag{HH}$$

$$\{\boldsymbol{v}\cdot\nabla\boldsymbol{\tau}\}_{z\theta} = (\boldsymbol{v}\cdot\nabla)\tau_{z\theta} - \left(\frac{\tau_{z\xi}}{a}\right)(v_\xi\sin\theta + v_\theta\sinh\xi) \tag{II}$$

$$\{\boldsymbol{v}\cdot\nabla\boldsymbol{\tau}\}_{zz} = (\boldsymbol{v}\cdot\nabla)\tau_{zz} \tag{JJ}$$

where the operator $(\boldsymbol{v}\cdot\nabla) = v_\xi \frac{X}{a}\frac{\partial}{\partial\xi} + v_\theta \frac{X}{a}\frac{\partial}{\partial\theta} + v_z \frac{\partial}{\partial z}$

We now use Eqs. A.7-6, 7, 8 and Eq. A.3-3. Since we want only the r-component, we sort out only those terms which contribute to the coefficient of $\boldsymbol{\delta}_r$:

$$\left[\boldsymbol{\delta}_r \frac{\partial}{\partial r} \cdot \boldsymbol{\delta}_r \boldsymbol{\delta}_r \tau_{rr}\right] = [\boldsymbol{\delta}_r \cdot \boldsymbol{\delta}_r \boldsymbol{\delta}_r] \frac{\partial \tau_{rr}}{\partial r} = \boldsymbol{\delta}_r \frac{\partial \tau_{rr}}{\partial r} \tag{A.7-22}$$

$$\left[\boldsymbol{\delta}_\theta \frac{1}{r}\frac{\partial}{\partial \theta} \cdot \boldsymbol{\delta}_\theta \boldsymbol{\delta}_r \tau_{\theta r}\right] = [\boldsymbol{\delta}_\theta \cdot \boldsymbol{\delta}_\theta \boldsymbol{\delta}_r] \frac{1}{r}\frac{\partial}{\partial \theta} \tau_{\theta r} + \text{other term} \tag{A.7-23}$$

$$\left[\boldsymbol{\delta}_\phi \frac{1}{r \sin\theta}\frac{\partial}{\partial \phi} \cdot \boldsymbol{\delta}_\phi \boldsymbol{\delta}_r \tau_{\phi r}\right] = [\boldsymbol{\delta}_\phi \cdot \boldsymbol{\delta}_\phi \boldsymbol{\delta}_r] \frac{1}{r \sin\phi}\frac{\partial}{\partial \phi} \tau_{\phi r} + \text{other term} \tag{A.7-24}$$

$$\left[\boldsymbol{\delta}_\theta \frac{1}{r}\frac{\partial}{\partial \theta} \cdot \boldsymbol{\delta}_r \boldsymbol{\delta}_r \tau_{rr}\right] = \frac{\tau_{rr}}{r}\left[\boldsymbol{\delta}_\theta \cdot \left\{\frac{\partial}{\partial \theta}\boldsymbol{\delta}_r\right\}\boldsymbol{\delta}_r\right] + \frac{\tau_{rr}}{r}\left[\boldsymbol{\delta}_\theta \cdot \boldsymbol{\delta}_r \left\{\frac{\partial}{\partial \theta}\boldsymbol{\delta}_r\right\}\right]$$
$$= \frac{\tau_{rr}}{r}[\boldsymbol{\delta}_\theta \cdot \boldsymbol{\delta}_\theta \boldsymbol{\delta}_r] = \boldsymbol{\delta}_r \frac{\tau_{rr}}{r} \tag{A.7-25}$$

$$\left[\boldsymbol{\delta}_\phi \frac{1}{r \sin\theta}\frac{\partial}{\partial \phi} \cdot \boldsymbol{\delta}_r \boldsymbol{\delta}_r \tau_{rr}\right] = \frac{\tau_{rr}}{r \sin\theta}\left[\boldsymbol{\delta}_\phi \cdot \left\{\frac{\partial}{\partial \phi}\boldsymbol{\delta}_r\right\}\boldsymbol{\delta}_r\right] = \frac{\tau_{rr}}{r \sin\theta}[\boldsymbol{\delta}_\phi \cdot \boldsymbol{\delta}_\phi \sin\theta \boldsymbol{\delta}_r] = \boldsymbol{\delta}_r \frac{\tau_{rr}}{r} \tag{A.7-26}$$

$$\left[\boldsymbol{\delta}_\theta \frac{1}{r}\frac{\partial}{\partial \theta} \cdot \boldsymbol{\delta}_\theta \boldsymbol{\delta}_\theta \tau_{\theta\theta}\right] = \boldsymbol{\delta}_r\left(-\frac{\tau_{\theta\theta}}{r}\right) + \text{other term} \tag{A.7-27}$$

$$\left[\boldsymbol{\delta}_\phi \frac{1}{r \sin\theta}\frac{\partial}{\partial \phi} \cdot \boldsymbol{\delta}_\theta \boldsymbol{\delta}_r \tau_{\theta r}\right] = \boldsymbol{\delta}_r \frac{\tau_{\theta r}\cos\theta}{r \sin\theta} \tag{A.7-28}$$

$$\left[\boldsymbol{\delta}_\phi \frac{1}{r \sin\theta}\frac{\partial}{\partial \phi} \cdot \boldsymbol{\delta}_\phi \boldsymbol{\delta}_\phi \tau_{\phi\phi}\right] = \boldsymbol{\delta}_r\left(\frac{-\tau_{\phi\phi}}{r}\right) + \text{other terms} \tag{A.7-29}$$

Combining the above results we get

$$[\nabla \cdot \boldsymbol{\tau}]_r = \frac{1}{r^2}\frac{\partial}{\partial r}(r^2 \tau_{rr}) + \frac{\tau_{\theta r}}{r}\cot\theta + \frac{1}{r}\frac{\partial}{\partial \theta}\tau_{\theta r} + \frac{1}{r \sin\theta}\frac{\partial \tau_{\phi r}}{\partial \phi} - \frac{\tau_{\theta\theta} + \tau_{\phi\phi}}{r} \tag{A.7-30}$$

Note that this expression is correct whether or not $\boldsymbol{\tau}$ is symmetrical.

EXERCISES

1. If $\boldsymbol{r}$ is the instantaneous position vector for a particle, show that the velocity and acceleration of the particle are given by (use Eq. A.7-2):

$$\boldsymbol{v} = \frac{d}{dt}\boldsymbol{r} = \boldsymbol{\delta}_r \dot{r} + \boldsymbol{\delta}_\theta r\dot{\theta} + \boldsymbol{\delta}_z \dot{z}$$

$$\boldsymbol{a} = \boldsymbol{\delta}_r(\ddot{r} - r\dot{\theta}^2) + \boldsymbol{\delta}_\theta(r\ddot{\theta} + 2\dot{r}\dot{\theta}) + \boldsymbol{\delta}_z \ddot{z}$$

in cylindrical coordinates. The dots indicate time derivatives of the coordinates.

2. Obtain $(\nabla \cdot \boldsymbol{v})$, $[\nabla \times \boldsymbol{v}]$, and $\nabla \boldsymbol{v}$ in spherical coordinates, and $[\nabla \cdot \boldsymbol{\tau}]$ in cylindrical coordinates.

3. Use Table A.7-2 to write down directly the following quantities in cylindrical coordinates:

a. $(\nabla \cdot \rho \boldsymbol{v})$, where ρ is a scalar
b. $[\nabla \cdot \rho \boldsymbol{v}\boldsymbol{v}]_r$, where ρ is a scalar
c. $[\nabla \cdot p\boldsymbol{\delta}]_\theta$, where p is a scalar
d. $(\nabla \cdot [\boldsymbol{\tau} \cdot \boldsymbol{v}])$
e. $[\boldsymbol{v} \cdot \nabla \boldsymbol{v}]_\theta$
f. $\nabla \boldsymbol{v} + (\nabla \boldsymbol{v})^\dagger$

4. Show that in any orthogonal coordinate system q_1, q_2, q_3:

$$(\nabla \cdot \boldsymbol{v}) = \frac{1}{h_1 h_2 h_3}\left[\frac{\partial}{\partial q_1}(h_2 h_3 v_1) + \frac{\partial}{\partial q_2}(h_3 h_1 v_2) + \frac{\partial}{\partial q_3}(h_1 h_2 v_3)\right]$$

For a summary of vector and tensor operations in terms of the h_i, see the tabulation in Morse and Feshbach.[3]

5. For problems involving eccentric annuli *bipolar coordinates* ξ, θ, z, are used[4] (see Fig. A.7-1):

$$x = \frac{a \sinh \xi}{\cosh \xi + \cos \theta} \qquad \xi = \operatorname{arctanh} \frac{2ax}{a^2 + (x^2 + y^2)}$$

$$y = \frac{a \sin \theta}{\cosh \xi + \cos \theta} \qquad \theta = \arctan \frac{2ay}{a^2 - (x^2 + y^2)}$$

$$z = z \qquad z = z$$

where a is a constant. The coordinates have the following ranges: $-\infty < \xi < +\infty$; $0 \leq \theta < 2\pi$; $-\infty < z < +\infty$.

(a) Show that (if $X = \cosh \xi + \cos \theta$):

$$\nabla = \boldsymbol{\delta}_\xi \frac{X}{a}\frac{\partial}{\partial \xi} + \boldsymbol{\delta}_\theta \frac{X}{a}\frac{\partial}{\partial \theta} + \boldsymbol{\delta}_\theta \frac{\partial}{\partial z}$$

and

$$\frac{\partial}{\partial \xi}\boldsymbol{\delta}_\xi = -X^{-1} \sin \theta \boldsymbol{\delta}_\theta; \quad \frac{\partial}{\partial \xi}\boldsymbol{\delta}_\theta = +X^{-1} \sin \theta \boldsymbol{\delta}_\xi; \quad \frac{\partial}{\partial \xi}\boldsymbol{\delta}_z = \mathbf{0}$$

$$\frac{\partial}{\partial \theta}\boldsymbol{\delta}_\xi = -X^{-1} \sinh \xi \boldsymbol{\delta}_\theta; \quad \frac{\partial}{\partial \theta}\boldsymbol{\delta}_\theta = +X^{-1} \sinh \xi \boldsymbol{\delta}_\xi; \quad \frac{\partial}{\partial \theta}\boldsymbol{\delta}_z = \mathbf{0}$$

$$\frac{\partial}{\partial z}\boldsymbol{\delta}_\xi = \mathbf{0}; \quad \frac{\partial}{\partial z}\boldsymbol{\delta}_\theta = \mathbf{0}; \quad \frac{\partial}{\partial z}\boldsymbol{\delta}_z = \mathbf{0}$$

(b) Evaluate $(\nabla \cdot \boldsymbol{v})$ and $[\nabla \cdot \boldsymbol{\tau}]$ in bipolar cocordinates.
(c) Show that an element of volume is given by $dx\,dy\,dz = (a/X)^2\,d\xi\,d\theta\,dz$.
(d) Show that an element of area $dS_{\theta z}$ is given by $dS_{\theta z} = (a/X)\,d\theta\,dz$.

[3] P. M. Morse and H. Feshbach, *op. cit.*, p. 115.
[4] P. M. Morse and H. Feshbach, *op. cit.*, p. 1210.

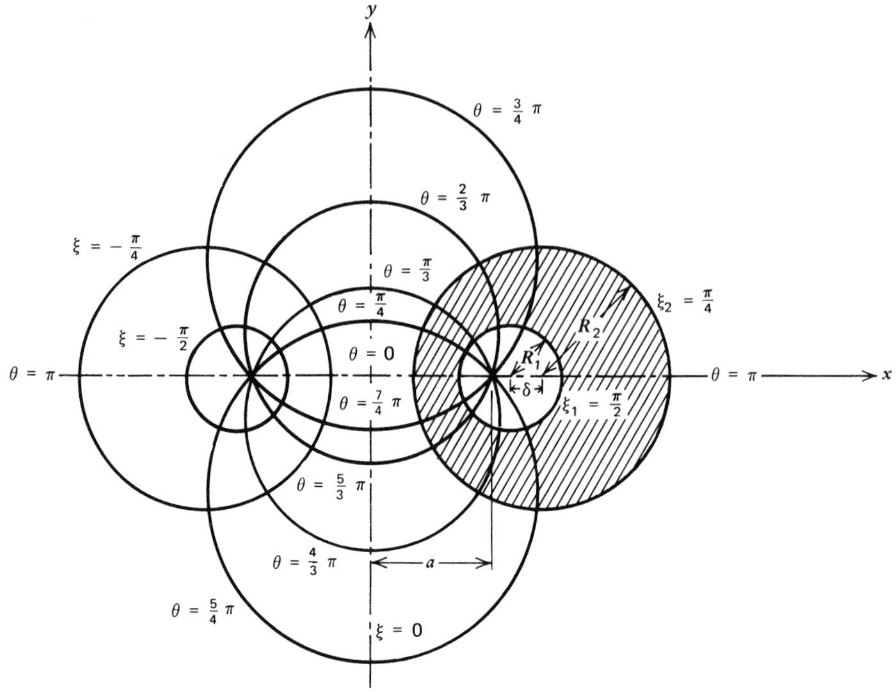

FIGURE A.7-1. Bipolar coordinate system, showing how these coordinates are useful for describing the eccentric annular region between two cylinders of radii R_1 and R_2 with the centers displaced by a distance δ. When R_1, R_2, and δ are given, the parameter a is obtained from[5]

$$a = \frac{1}{2\delta}\sqrt{(R_1^2 + R_2^2 - \delta^2)^2 - 4R_1^2R_2^2}$$

and the values ξ_1 and ξ_2 defining the boundaries of the eccentric annular region are given by $\xi_1 = \text{arcsinh}\ a/R_1$ and $\xi_2 = \text{arcsinh}\ a/R_2$.

6. Verify that the entries for $\nabla^2 \boldsymbol{v}$ in Table A.7-2 can be obtained by any one of the following methods:

(a) First verify that, in cylindrical coordinates the operator $(\nabla \cdot \nabla)$ is

$$(\nabla \cdot \nabla) = \frac{\partial^2}{\partial r^2} + \frac{1}{r}\frac{\partial}{\partial r} + \frac{1}{r^2}\frac{\partial^2}{\partial \theta^2} + \frac{\partial^2}{\partial z^2}$$

and then apply the operator to $\boldsymbol{v}$.

(b) Use the expression for $[\nabla \cdot \boldsymbol{\tau}]$ in Table A.7-2, but substitute the components for $\nabla \boldsymbol{v}$ in place of the components of $\boldsymbol{\tau}$, so as to obtain $[\nabla \cdot \nabla \boldsymbol{v}]$.

(c) Use Eq. A.4-22:

$$\nabla^2 \boldsymbol{v} = \nabla(\nabla \cdot \boldsymbol{v}) - [\nabla \times [\nabla \times \boldsymbol{v}]]$$

and use the gradient, divergence, and curl operations in Table A.7-2 to evaluate the operations on the right side.

[5] B. Y. Ballal and R. S. Rivlin, *Trans. Soc. Rheol.*, **20**, 65–101 (1976).

§A.8 NONORTHOGONAL CURVILINEAR COORDINATES[1]

In the foregoing sections it has been shown how the various algebraic and differential operations for vectors and tensors can be described in terms of unit vectors and the space derivatives of the unit vectors. Alternatively one can present the subject by using "base vectors." We illustrate this for spherical coordinates in which $\boldsymbol{r} = \boldsymbol{\delta}_r r$; the *base vectors* are defined by

$$\boldsymbol{g}_r = \frac{\partial}{\partial r}\boldsymbol{r} = \boldsymbol{\delta}_r; \quad \boldsymbol{g}_\theta = \frac{\partial}{\partial \theta}\boldsymbol{r} = r\boldsymbol{\delta}_\theta; \quad \boldsymbol{g}_\phi = \frac{\partial}{\partial \phi}\boldsymbol{r} = r \sin\theta\, \boldsymbol{\delta}_\phi \tag{A.8-1}$$

That is, the base vectors differ from the unit vectors only in the inclusion of the "scale factors" (these are the h_r, h_θ, and h_ϕ of Eq. A.7-13). We can also define *reciprocal base vectors* as follows:

$$\boldsymbol{g}^r = \nabla r = \boldsymbol{\delta}_r; \quad \boldsymbol{g}^\theta = \nabla\theta = \frac{1}{r}\boldsymbol{\delta}_\theta; \quad \boldsymbol{g}^\phi = \nabla\phi = \frac{1}{r\sin\theta}\boldsymbol{\delta}_\phi \tag{A.8-2}$$

These include the reciprocals of the scale factors. With these definitions we find that

$$(\boldsymbol{g}_\alpha \cdot \boldsymbol{g}^\beta) = \delta_{\alpha\beta} \tag{A.8-3}$$

where α and β can be r, θ, or ϕ. Vectors can then be written as

$$\boldsymbol{v} = \sum_\alpha \boldsymbol{g}_\alpha v^\alpha \quad \text{or} \quad \boldsymbol{v} = \sum_\alpha \boldsymbol{g}^\alpha v_\alpha \tag{A.8-4}$$

in which v_α and v^α are referred to respectively as the *covariant* and *contravariant* components of the vector $\boldsymbol{v}$. (The v_α and v^α are different from the *physical components* v_i of Eq. A.2-16; and indeed the v_α and v^α do not in general have the same dimensions as $\boldsymbol{v}$.) Similarly second-order tensors may be expanded as

$$\boldsymbol{\tau} = \sum_\alpha \sum_\beta \boldsymbol{g}_\alpha \boldsymbol{g}_\beta \tau^{\alpha\beta} \quad \text{or} \quad \boldsymbol{\tau} = \sum_\alpha \sum_\beta \boldsymbol{g}^\alpha \boldsymbol{g}^\beta \tau_{\alpha\beta} \tag{A.8-5a}$$

in which $\tau_{\alpha\beta}$ and $\tau^{\alpha\beta}$ are the covariant and contravariant components, respectively. One can also have "mixed components" if one writes

$$\boldsymbol{\tau} = \sum_\alpha \sum_\beta \boldsymbol{g}_\alpha \boldsymbol{g}^\beta \tau^{\alpha}_{\cdot\beta} \quad \text{or} \quad \boldsymbol{\tau} = \sum_\alpha \sum_\beta \boldsymbol{g}^\alpha \boldsymbol{g}_\beta \tau_{\alpha\cdot}^{\ \beta} \tag{A.8-5b}$$

the dots being inserted in order to make the order of the indices clear. Next expressions can be developed for the spatial derivatives of the base vectors and reciprocal base vectors (analogous to Eqs. A.7-6 to 8) and for the del-operator

$$\nabla = \boldsymbol{g}^r \frac{\partial}{\partial r} + \boldsymbol{g}^\theta \frac{\partial}{\partial \theta} + \boldsymbol{g}^\phi \frac{\partial}{\partial \phi} \tag{A.8-6}$$

[1] The treatment given here is similar to the development given by H. D. Block, *Introduction to Tensor Analysis*, Merrill, Columbus, OH (1962), and by A. P. Wills, *Vector Analysis*, Dover Reprint, New York (1958), Chapt. 10. See also L. I. Sedov, *Introduction to the Mechanics of a Continuous Medium*, Addison-Wesley, Reading, MA (1965), and A. J. McConnell, *Applications of Tensor Analysis*, Dover Reprint, New York (1957).

Then using these relations one can develop the expressions for ∇s, $(\nabla \cdot \boldsymbol{v})$, $[\nabla \cdot \boldsymbol{\tau}]$, $\nabla^2 s$, etc., just as before; the only difference is that one works in terms of the base vectors $\boldsymbol{g}_\alpha$ and the reciprocal base vectors $\boldsymbol{g}^\alpha$, instead of with the $\boldsymbol{\delta}_i$. Exactly the same results are obtained, of course. In fact there is no particular advantage to using base vectors and covariant and contravariant components, as long as one is dealing with orthogonal coordinates. In problem solving one almost always uses orthogonal coordinates, and hence the procedures of §§A.1 to A.7 are sufficient—that is, the method using unit vectors and physical components.

In developments using nonorthogonal coordinates, and in particular in the formal continuum mechanics derivations using "convected coordinates" in Chapter 9, it is very useful and also conventional to use the formulation of general tensor analysis in terms of covariant and contravariant components. In addition some authors feel that it is stylish to use general tensor notation even when it is not needed, and hence it is necessary to be familiar with the use of base vectors and covariant and contravariant components in order to study the literature of rheology and fluid mechanics.

We now consider a curvilinear coordinate system, which in general is nonorthogonal. The location of a point in space is given by coordinates q^i (with $i = 1, 2, 3$);[2] the curvilinear coordinates are related to the Cartesian coordinates by

$$x_i = x_i(q^1, q^2, q^3) \quad \text{and} \quad q^i = q^i(x_1, x_2, x_3) \qquad (i = 1, 2, 3) \tag{A.8-7}$$

The directed line segment $\boldsymbol{dr}$, joining two points a differential distance apart, may be written as[2] (see Fig. A.8-1):

$$\boldsymbol{dr} = \sum_i \boldsymbol{g}_i dq^i \tag{A.8-8}$$

This serves to define the *base vectors*[3] $\boldsymbol{g}_i$. Then

$$\boldsymbol{g}_i = \frac{\partial}{\partial q^i}\boldsymbol{r} = \sum_j \frac{\partial x_j}{\partial q^i}\frac{\partial}{\partial x_j}\boldsymbol{r} = \sum_j \frac{\partial x_j}{\partial q^i}\boldsymbol{\delta}_j \tag{A.8-9}$$

Hence the base vectors $\boldsymbol{g}_i$ can be written as a linear combination of the unit vectors $\boldsymbol{\delta}_j$. The base vectors are directed tangentially along the coordinate curves, but they are not of unit length.

Because of the nonorthogonality of the $\boldsymbol{g}_i$, the dot and cross product relations are not particularly simple. In order to develop more conveninent dot and cross product

[2] Note that the q^i are nonorthogonal coordinates; they are *not* the same as the x_i (rectangular coordinates) in §§A.1 to A.4. A superscript i is introduced because one customarily adopts the rule that summations will be performed on pairs of repeated indices—one a *super*script and the other a *sub*script. A second rule is that a nonsummed (and hence, nonrepeated) index must appear in every term in the equation in the same position, that is, as a superscript or as a subscript. In applying these rules we find that a superscript in the denominator is equivalent to a subscript in the numerator—that is, in $\partial s/\partial q^i$ the i is counted as a subscript. One might wonder why it is not usual to write $\boldsymbol{dr} = \sum_i \boldsymbol{g}^i dq_i$ using the reciprocal base vectors defined in Eq. A.8-10 or 11. The reason for this is that the dq_i are *not* exact differentials (see L. I. Sedov. *op. cit.*, p. 6).

[3] The use of a boldface roman symbol with one subscript as a vector is in violation of the rules given at the beginning of Appendix A. We do this because the closely related metric coefficients, defined presently, are always designated by g_{ij}. It would be unwise to tamper with a universally adopted notation such as g_{ij}. Note that for orthogonal coordinates the g_{ii} are just the same as h_i^2 ("scale factors") in §A.7, and $g_{ij} = 0$ for $i \neq j$.

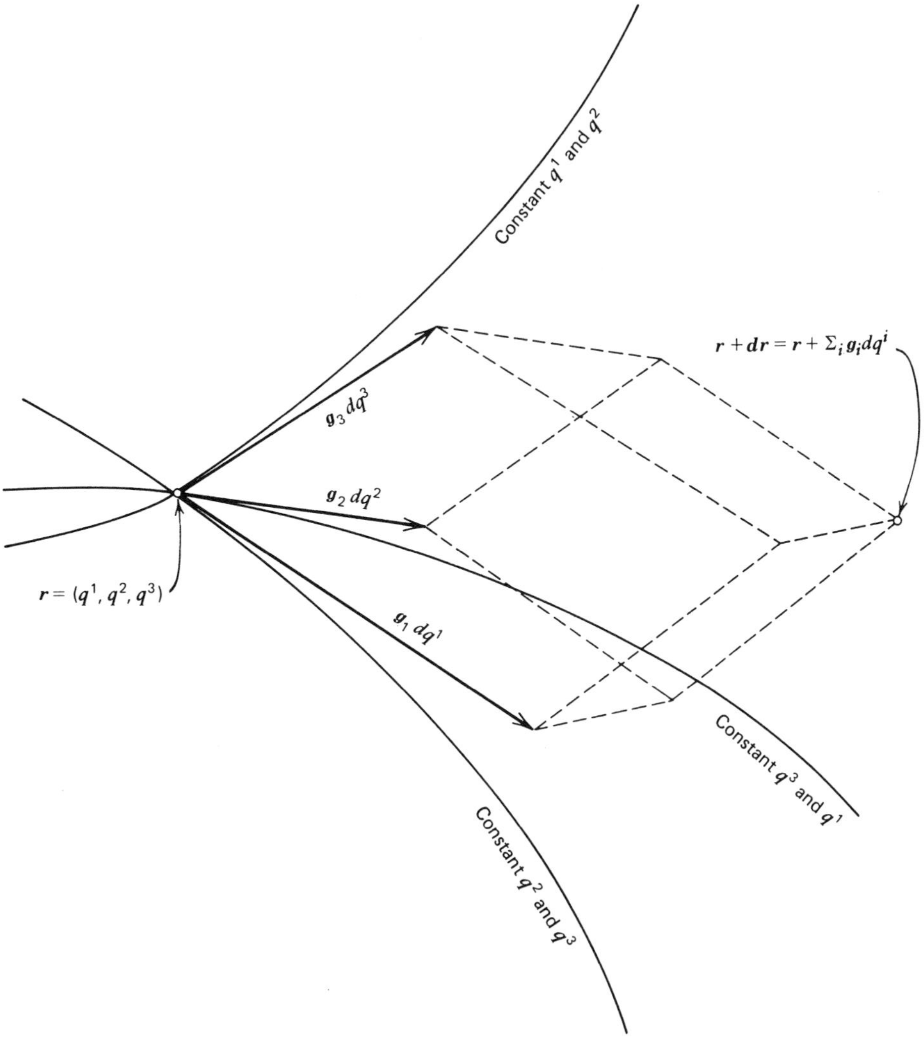

FIGURE A.8-1. The base vectors $\boldsymbol{g}_i$ in a nonorthogonal curvilinear coordinate system. The parallelepiped formed by the $\boldsymbol{g}_i$ has a volume $(\boldsymbol{g}_1 \cdot [\boldsymbol{g}_2 \times \boldsymbol{g}_3]) = \sqrt{g}$.

relations among the base vectors, it is customary to introduce a set of *reciprocal base vectors* $\boldsymbol{g}^i$ defined by

$$\boldsymbol{g}^1 = \frac{[\boldsymbol{g}_2 \times \boldsymbol{g}_3]}{(\boldsymbol{g}_1 \cdot [\boldsymbol{g}_2 \times \boldsymbol{g}_3])} \tag{A.8-10}$$

with $\boldsymbol{g}^2$ and $\boldsymbol{g}^3$ obtained by cyclic permutation; note that the denominator is just the volume of the parallelepiped formed by the vectors $\boldsymbol{g}_1$, $\boldsymbol{g}_2$, and $\boldsymbol{g}_3$ (cf. Eq. A.2-23). The reciprocal base vectors can also be obtained from

$$\boldsymbol{g}^i = \boldsymbol{\nabla} q^i = \sum_j \frac{\partial q^i}{\partial x_j} \boldsymbol{\delta}_j \tag{A.8-11}$$

The reciprocal base vector $\boldsymbol{g}^i$ is perpendicular to the surface q^i = constant.

From the definitions it is clear that

$$(\boldsymbol{g}^i \cdot \boldsymbol{g}_j) = \delta^i_j; \quad (\boldsymbol{g}_i \cdot \boldsymbol{g}^j) = \delta^j_i \tag{A.8-12}$$

where δ^i_j is the Kronecker delta, which is 0 if $i \neq j$ and 1 if $i = j$; this way of writing the Kronecker delta is required by the notation rules of footnote 2. The other dot products do not give 0 or 1, and we define the *metric coefficients*[4] g_{ij} and g^{ij} by

$$(\boldsymbol{g}_i \cdot \boldsymbol{g}_j) = g_{ij}; \quad (\boldsymbol{g}^i \cdot \boldsymbol{g}^j) = g^{ij} \tag{A.8-13}$$

The determinant of the g_{ij} is called g, and $\sqrt{g}$ is in turn equal to $(\boldsymbol{g}_1 \cdot [\boldsymbol{g}_2 \times \boldsymbol{g}_3])$. Note further that $\sum_l g_{kl} g^{lj} = \delta^j_k$. Equations A.8-12 and 13 are analogous to Eq. A.2-14 for orthogonal coordinates. Analogously to Eq. A.2-15 we have

$$[\boldsymbol{g}_i \times \boldsymbol{g}_j] = \sum_k \mathscr{E}_{ijk} \boldsymbol{g}^k \tag{A.8-14}$$

$$[\boldsymbol{g}^i \times \boldsymbol{g}^j] = \sum_k \mathscr{E}^{ijk} \boldsymbol{g}_k \tag{A.8-15}$$

Here $\mathscr{E}_{ijk} = \varepsilon_{ijk}\sqrt{g}$ and $\mathscr{E}^{ijk} = \varepsilon^{ijk}/\sqrt{g}$, where $\varepsilon_{ijk} = \varepsilon^{ijk}$ are defined in Eqs. A.2-3, 4, and 5. In addition to the operations in Eqs. A.8-12 to 15 there are dyadic operations corresponding to those in Eqs. A.3-1 to 6; for example,

$$[\boldsymbol{g}_i \cdot \boldsymbol{g}_j \boldsymbol{g}_k] = (\boldsymbol{g}_i \cdot \boldsymbol{g}_j)\boldsymbol{g}_k = g_{ij}\boldsymbol{g}_k, \quad \text{and} \quad [\boldsymbol{g}^i \cdot \boldsymbol{g}_j \boldsymbol{g}_k] = (\boldsymbol{g}^i \cdot \boldsymbol{g}_j)\boldsymbol{g}_k = \delta^i_j \boldsymbol{g}_k.$$

It is now possible to express vectors and tensors in terms of the base vectors, analogously to Eqs. A.8-4 and 5:

$$\boldsymbol{v} = \sum_i \boldsymbol{g}_i v^i = \sum_i \boldsymbol{g}^i v_i \tag{A.8-16}$$

$$\begin{aligned} \boldsymbol{\tau} &= \sum_i \sum_j \boldsymbol{g}_i \boldsymbol{g}_j \tau^{ij} = \sum_i \sum_j \boldsymbol{g}^i \boldsymbol{g}^j \tau_{ij} \\ &= \sum_i \sum_j \boldsymbol{g}_i \boldsymbol{g}^j \tau^i_{\cdot j} = \sum_i \sum_j \boldsymbol{g}^i \boldsymbol{g}_j \tau_i^{\cdot j} \end{aligned} \tag{A.8-17}$$

The components v_i and v^i are called *covariant* and *contravariant* components of the vector $\boldsymbol{v}$, respectively. Similarly τ_{ij}, τ^{ij}, and $\tau_i^{\cdot j}$ (or $\tau^i_{\cdot j}$) are called the *covariant*, *contravariant*, and *mixed* components of the tensor $\boldsymbol{\tau}$, respectively.

It should be added that the covariant and contravariant components are simply related

$$v_j = \sum_i g_{ij} v^i \tag{A.8-18}$$

[4] Note that from Eqs. A.8-8 and 13 the square of the separation of two adjacent points is

$$(d\boldsymbol{r} \cdot d\boldsymbol{r}) = \sum_i \sum_j g_{ij} dq^i dq^j \tag{A.8-12a}$$

The metric coefficients may also be regarded as the covariant and contravariant components of the unit tensor:

$$\boldsymbol{\delta} = \sum_i \sum_j \boldsymbol{g}_i \boldsymbol{g}_j g^{ij} = \sum_i \sum_j \boldsymbol{g}^i \boldsymbol{g}^j g_{ij} \tag{A.8-12b}$$

as may easily be shown by taking the dot product of $\boldsymbol{g}_j$ with Eq. A.8-16. Similarly,

$$\tau^{ij} = \sum_k \sum_l g^{ik} g^{jl} \tau_{kl} \tag{A.8-19}$$

Equations A.8-18 and 19 illustrate the use of the g_{ij} and g^{ij} as "raising and lowering operators."

The various kinds of vector and tensor multiplications can be performed as before; however, because of the several ways of expanding vectors and tensors in their components, these operations may be written in a number of ways. For example, consider the scalar product of two vectors:

$$\begin{aligned}(\boldsymbol{v} \cdot \boldsymbol{w}) &= \left(\left\{\sum_i \boldsymbol{g}_i v^i\right\} \cdot \left\{\sum_j \boldsymbol{g}^j w_j\right\}\right) \\ &= \sum_i \sum_j (\boldsymbol{g}_i \cdot \boldsymbol{g}^j) v^i w_j \\ &= \sum_i \sum_j \delta_i^j v^i w_j \\ &= \sum_i v^i w_i \end{aligned} \tag{A.8-20}$$

or

$$\begin{aligned}(\boldsymbol{v} \cdot \boldsymbol{w}) &= \left(\left\{\sum_i \boldsymbol{g}_i v^i\right\} \cdot \left\{\sum_j \boldsymbol{g}_j w^j\right\}\right) \\ &= \sum_i \sum_j (\boldsymbol{g}_i \cdot \boldsymbol{g}_j) v^i w^j \\ &= \sum_i \sum_j g_{ij} v^i w^j \end{aligned} \tag{A.8-21}$$

Clearly two other combinations may be used. Note that the results are all really the same. The result in Eq. A.8-21 may be written $\sum_i v^i \{\sum_j g_{ij} w^j\} = \sum_i v^i w_i$ using Eq. A.8-18, thereby obtaining the result in Eq. A.8-20.

Other operations proceed similarly:

$$\begin{aligned}[\boldsymbol{\tau} \cdot \boldsymbol{v}] &= \left[\left\{\sum_i \sum_j \boldsymbol{g}_i \boldsymbol{g}_j \tau^{ij}\right\} \cdot \left\{\sum_k \boldsymbol{g}^k v_k\right\}\right] \\ &= \sum_i \sum_j \sum_k [\boldsymbol{g}_i \boldsymbol{g}_j \cdot \boldsymbol{g}^k] \tau^{ij} v_k \\ &= \sum_i \sum_j \sum_k \boldsymbol{g}_i \delta_j^k \tau^{ij} v_k \\ &= \sum_i \sum_j \boldsymbol{g}_i \tau^{ij} v_j \\ &= \sum_i \boldsymbol{g}_i \left(\sum_j \tau^{ij} v_j\right) \end{aligned} \tag{A.8-22}$$

Hence, the ith contravariant component of $[\boldsymbol{\tau} \cdot \boldsymbol{v}]$ is $\sum_j \tau^{ij} v_j$.

Cross products may also be performed using the rules in Eqs. A.8-14 and 15:

$$\begin{aligned}[\boldsymbol{v} \times \boldsymbol{w}] &= \left[\left\{\sum_i \boldsymbol{g}_i v^i\right\} \times \left\{\sum_j \boldsymbol{g}_j w^j\right\}\right] \\ &= \sum_i \sum_j [\boldsymbol{g}_i \times \boldsymbol{g}_j] v^i w^j \\ &= \sum_i \sum_j \sum_k \mathscr{E}_{ijk} \boldsymbol{g}^k v^i w^j \end{aligned} \tag{A.8-23}$$

so that the kth covariant component of $[\boldsymbol{v} \times \boldsymbol{w}]$ is $\sum_i \sum_j \mathscr{E}_{ijk} v^i w^j$.

Differentiation operations are performed similarly to those described in §A.7. Recall that, in differentiating the unit vectors in curvilinear coordinates, one obtains results of the type shown in Eqs. A.7-1 to 3, 6 to 8, and 14. Note that the spatial derivative of one unit vector gives in general a linear combination of all the unit vectors, the coefficients of which are functions of position. Hence the derivatives of $\boldsymbol{g}_i$ and $\boldsymbol{g}^i$ will be expected to have the form

$$\boxed{\frac{\partial}{\partial q^j} \boldsymbol{g}_i = \sum_k \begin{Bmatrix} k \\ ij \end{Bmatrix} \boldsymbol{g}_k; \quad \frac{\partial}{\partial q^j} \boldsymbol{g}^i = -\sum_k \begin{Bmatrix} i \\ kj \end{Bmatrix} \boldsymbol{g}^k} \tag{A.8-24}$$

where the coefficients $\{^{\,k}_{ij}\}$ are called *Christoffel symbols* (some authors use the notation Γ^k_{ij}); they are *not* the components of a third-order tensor. By forming appropriate combinations of dot products of the base vectors with Eqs. A.8-24 it is possible[1] to get explicit expressions for the $\{^{\,k}_{ij}\}$ in terms of the g_{ij}:

$$\begin{Bmatrix} k \\ ij \end{Bmatrix} = \frac{1}{2} \sum_l g^{kl} \left(\frac{\partial g_{il}}{\partial q^j} + \frac{\partial g_{jl}}{\partial q^i} - \frac{\partial g_{ij}}{\partial q^l} \right) \tag{A.8-25}$$

Christoffel symbols for three orthogonal coordinate systems are given in Table A.8-1. The $\boldsymbol{\nabla}$ operator is

$$\boxed{\boldsymbol{\nabla} = \sum_i \boldsymbol{g}^i \frac{\partial}{\partial q^i}} \tag{A.8-26}$$

This is analogous to Eq. A.7-15 (note once again that the "scale factors" have been absorbed into the $\boldsymbol{g}^i$). As in §A.7, once we know the expressions for $\boldsymbol{\nabla}$ and the differentiation of the base vectors, any $\boldsymbol{\nabla}$-operations can be worked out. We begin by considering the gradient operations $\boldsymbol{\nabla} s$, $\boldsymbol{\nabla v}$, and $\boldsymbol{\nabla \tau}$. For a *scalar*:

$$\boldsymbol{\nabla} s = \sum_i \boldsymbol{g}^i \frac{\partial s}{\partial q^i} \equiv \sum_i \boldsymbol{g}^i s_{,i} \tag{A.8-27}$$

TABLE A.8-1

Christoffel Symbols and Metric Coefficients for Several Orthogonal Coordinate Systems

	Christoffel Symbols	Metric Coefficients[a]
Rectangular:	All $\{\ \} = 0$	$g_{xx} = g_{yy} = g_{zz} = 1$ All other g_{ij} are zero
Cylindrical:	$\begin{Bmatrix} r \\ \theta\theta \end{Bmatrix} = -r$ $\begin{Bmatrix} \theta \\ \theta r \end{Bmatrix} = \begin{Bmatrix} \theta \\ r\theta \end{Bmatrix} = \dfrac{1}{r}$ All other $\{\ \}$ are zero	$g_{rr} = 1$ $g_{\theta\theta} = r^2$ $g_{zz} = 1$ All other g_{ij} are zero
Spherical:	$\begin{Bmatrix} r \\ \theta\theta \end{Bmatrix} = -r$ $\begin{Bmatrix} r \\ \phi\phi \end{Bmatrix} = -r \sin^2 \theta$ $\begin{Bmatrix} \theta \\ \phi\phi \end{Bmatrix} = -\sin \theta \cos \theta$ $\begin{Bmatrix} \phi \\ r\phi \end{Bmatrix} = \begin{Bmatrix} \phi \\ \phi r \end{Bmatrix} = \dfrac{1}{r}$ $\begin{Bmatrix} \theta \\ r\theta \end{Bmatrix} = \begin{Bmatrix} \theta \\ \theta r \end{Bmatrix} = \dfrac{1}{r}$ $\begin{Bmatrix} \phi \\ \theta\phi \end{Bmatrix} = \begin{Bmatrix} \phi \\ \phi\theta \end{Bmatrix} = \cot \theta$ All other $\{\ \}$ are zero	$g_{rr} = 1$ $g_{\theta\theta} = r^2$ $g_{\phi\phi} = r^2 \sin^2 \theta$ All other g_{ij} are zero

[a] In orthogonal coordinates g_{ij} and g^{ij} are zero if $i \neq j$; furthermore, $g^{ii} = (g_{ii})^{-1}$. The components g_{ii} are the same as h_i^2 in §A.7.

For a *vector*:

$$\begin{aligned}
\nabla \boldsymbol{v} &= \left\{\sum_i \boldsymbol{g}^i \frac{\partial}{\partial q^i}\right\}\left\{\sum_j \boldsymbol{g}^j v_j\right\} \\
&= \sum_i \sum_j \left(\boldsymbol{g}^i \boldsymbol{g}^j \frac{\partial}{\partial q^i} v_j + \boldsymbol{g}^i v_j \frac{\partial}{\partial q^i} \boldsymbol{g}^j\right) \\
&= \sum_i \sum_j \left(\boldsymbol{g}^i \boldsymbol{g}^j \frac{\partial}{\partial q^i} v_j + \boldsymbol{g}^i v_j\left(-\sum_k \begin{Bmatrix} j \\ ki \end{Bmatrix} \boldsymbol{g}^k\right)\right) \\
&= \sum_i \sum_j \boldsymbol{g}^i \boldsymbol{g}^j\left(\frac{\partial}{\partial q^i} v_j - \sum_k \begin{Bmatrix} k \\ ji \end{Bmatrix} v_k\right) \\
&\equiv \sum_i \sum_j \boldsymbol{g}^i \boldsymbol{g}^j v_{j,i}
\end{aligned} \tag{A.8-28}$$

or alternatively,

$$\begin{aligned}\nabla \boldsymbol{v} &= \left\{\sum_i \boldsymbol{g}^i \frac{\partial}{\partial q^i}\right\}\left\{\sum_j \boldsymbol{g}_j v^j\right\} \\ &= \sum_i \sum_j \boldsymbol{g}^i \boldsymbol{g}_j \left(\frac{\partial}{\partial q^i} v^j + \sum_k \begin{Bmatrix} j \\ ki \end{Bmatrix} v^k\right). \\ &\equiv \sum_i \sum_j \boldsymbol{g}^i \boldsymbol{g}_j v^j_{,i} \end{aligned} \tag{A.8-29}$$

For a second-order *tensor* we have several choices

$$\begin{aligned}\nabla \boldsymbol{\tau} &= \sum_i \sum_j \sum_k \boldsymbol{g}^i \boldsymbol{g}^j \boldsymbol{g}^k \left(\frac{\partial}{\partial q^i} \tau_{jk} - \sum_l \begin{Bmatrix} l \\ ji \end{Bmatrix} \tau_{lk} - \sum_l \begin{Bmatrix} l \\ ki \end{Bmatrix} \tau_{jl}\right) \\ &\equiv \sum_i \sum_j \sum_k \boldsymbol{g}^i \boldsymbol{g}^j \boldsymbol{g}^k \tau_{jk,i} \end{aligned} \tag{A.8-30}$$

$$\begin{aligned}\nabla \boldsymbol{\tau} &= \sum_i \sum_j \sum_k \boldsymbol{g}^i \boldsymbol{g}_j \boldsymbol{g}_k \left(\frac{\partial}{\partial q^i} \tau^{jk} + \sum_l \begin{Bmatrix} j \\ li \end{Bmatrix} \tau^{lk} + \sum_l \begin{Bmatrix} k \\ li \end{Bmatrix} \tau^{jl}\right) \\ &\equiv \sum_i \sum_j \sum_k \boldsymbol{g}^i \boldsymbol{g}_j \boldsymbol{g}_k \tau^{jk}_{,i} \end{aligned} \tag{A.8-31}$$

There are two additional choices involving the mixed components.

In the above expressions we have introduced the comma notation for *covariant differentiation*; it is simply a convenient shorthand notation for the components of the gradient operations. The terms in the covariant derivatives containing the Christoffel symbols arise from the differentiation of the unit vectors. [Note that these correspond to the "extra terms" in $\nabla \boldsymbol{v}$ obtained in Eq. A.7-20, namely, $\boldsymbol{\delta}_\theta \boldsymbol{\delta}_\theta (v_r/r) - \boldsymbol{\delta}_\theta \boldsymbol{\delta}_r (v_\theta/r)$].

There are several important properties of covariant differentiation that deserve mention. One, known as *Ricci's lemma*, is that $g_{ij,k} = 0$ and $g^{ij}{}_{,k} = 0$. A second is that the usual differentiation rules of calculus apply: $(v_i w_j)_{,k} = v_i w_{j,k} + v_{i,k} w_j$.

Various other ∇ operations can be worked out by the foregoing procedures, for example,

$$(\nabla \cdot \boldsymbol{v}) = \sum_i v^i{}_{,i} \tag{A.8-32}$$

$$[\nabla \times \boldsymbol{v}] = \sum_i \sum_j \sum_k \mathscr{E}^{ijk} \boldsymbol{g}_i v_{k,j} \tag{A.8-33}$$

$$\nabla^2 s = \sum_i \sum_j g^{ij} s_{,ij} \tag{A.8-34}$$

$$[\nabla \cdot \boldsymbol{\tau}] = \sum_i \sum_j \boldsymbol{g}_i \tau^{ji}{}_{,j} \tag{A.8-35}$$

Many alternate forms are possible by "juggling" the indices using the raising and lowering operations with the g_{ij} and g^{ij}. For example, $[\nabla \cdot \boldsymbol{\tau}]$ can be written $\sum_i \sum_j \boldsymbol{g}^i \tau^{j}_{\cdot i,j}$ or $\sum_i \sum_j \sum_k \boldsymbol{g}^i g^{jk} \tau_{ji,k}$.

Thus far we have restricted ourselves to a single set of coordinates q^i. Let us now consider a second set of coordinates $\bar{q}^i$. These two sets of coordinates are related to one another by the coordinate transformations:

$$\bar{q}^i = \bar{q}^i(q^1, q^2, q^3); \quad q^i = q^i(\bar{q}^1, \bar{q}^2, \bar{q}^3) \tag{A.8-36}$$

which are assumed to be single valued and differentiable. We may then write:

$$d\bar{q}^i = \sum_j \left(\frac{\partial \bar{q}^i}{\partial q^j}\right) dq^j; \quad dq^i = \sum_j \left(\frac{\partial q^i}{\partial \bar{q}^j}\right) d\bar{q}^j \tag{A.8-37}$$

Now, if we call $\bar{\boldsymbol{g}}_i$ the base vectors in the barred coordinate system, we may write Eq. A.8-8 in terms of both sets of coordinates:

$$d\boldsymbol{r} = \sum_i \boldsymbol{g}_i dq^i; \quad d\boldsymbol{r} = \sum_i \bar{\boldsymbol{g}}_i d\bar{q}^i \tag{A.8-38}$$

Combining the last two sets of equations we get the transformation rules for the base vectors:

$$\boxed{\boldsymbol{g}_i = \sum_j \frac{\partial \bar{q}^j}{\partial q^i} \bar{\boldsymbol{g}}_j \quad \Bigg| \quad \bar{\boldsymbol{g}}_i = \sum_j \frac{\partial q^j}{\partial \bar{q}^i} \boldsymbol{g}_j} \tag{A.8-39}$$

It may further be shown that there are corresponding transformation rules for the reciprocal base vectors:

$$\boxed{\boldsymbol{g}^i = \sum_j \frac{\partial q^i}{\partial \bar{q}^j} \bar{\boldsymbol{g}}^j \quad \Bigg| \quad \bar{\boldsymbol{g}}^i = \sum_j \frac{\partial \bar{q}^i}{\partial q^j} \boldsymbol{g}^j} \tag{A.8-40}$$

Once the transformation rules for the base vectors are known the transformation rules for the components of any vector or tensor can easily be written down. Consider a vector $\boldsymbol{v}$, which may be written in either of two forms in both coordinate systems:

$$\boldsymbol{v} = \sum_i \boldsymbol{g}_i v^i = \sum_i \bar{\boldsymbol{g}}_i \bar{v}^i \qquad \boldsymbol{v} = \sum_i \boldsymbol{g}^i v_i = \sum_i \bar{\boldsymbol{g}}^i \bar{v}_i \tag{A.8-41, 42}$$

By combining Eq. A.8-41 with Eq. A.8-39, and by combining Eq. A.8-42 with Eq. A.8-40, we get

$$v^i = \sum_j \frac{\partial q^i}{\partial \bar{q}^j} \bar{v}^j \qquad \bar{v}^i = \sum_j \frac{\partial \bar{q}^i}{\partial q^j} v^j \tag{A.8-43}$$

$$v_i = \sum_j \frac{\partial \bar{q}^j}{\partial q^i} \bar{v}_j \qquad \bar{v}_i = \sum_j \frac{\partial q^j}{\partial \bar{q}^i} v_j \tag{A.8-44}$$

These equations are often taken to be the definitions of the contravariant and covariant components of a vector, respectively. Similar transformation rules may be found for second- and higher-order tensors, for example,

$$\bar{\tau}^{ij} = \sum_k \sum_l \frac{\partial \bar{q}^i}{\partial q^k} \frac{\partial \bar{q}^j}{\partial q^l} \tau^{kl} \tag{A.8-45}$$

$$\bar{\tau}_{ij} = \sum_k \sum_l \frac{\partial q^k}{\partial \bar{q}^i} \frac{\partial q^l}{\partial \bar{q}^j} \tau_{kl} \tag{A.8-46}$$

$$\bar{\tau}^{i}_{\cdot j} = \sum_k \sum_l \frac{\partial \bar{q}^i}{\partial q^k} \frac{\partial q^l}{\partial \bar{q}^j} \tau^{k}_{\cdot l} \tag{A.8-47}$$

Here, too, these relations are often taken to be the definitions of the contravariant, covariant, and mixed components of a tensor, respectively. We have chosen, however, to define vectors and tensors in terms of the base vectors as in Eqs. A.8-16 and 17 and to develop the rules for operating with the base vectors (Eqs. A.8-12 to 15 for the algebraic rules, Eq. A.8-24 for the differentiation rules, and Eqs. A.8-39 to 40 for the transformation rules). With these rules for the base vectors, the expressions for the components of the vectors and tensors can then be derived. This mode of presentation is that adopted by Wills[5] and Sedov[6], whereas McConnell[7] uses relations such as Eqs. A.8-43 to 45 as the basic starting point.

EXERCISES

1. Verify that[8]

$$(\nabla \cdot \boldsymbol{v}) = \frac{1}{\sqrt{g}} \sum_i \frac{\partial}{\partial q^i} (\sqrt{g}\, v^i)$$

$$\nabla^2 s = \frac{1}{\sqrt{g}} \sum_i \sum_j \frac{\partial}{\partial q^i} \left(\sqrt{g}\, g^{ij} \frac{\partial s}{\partial x^j} \right)$$

2. Use the result of (1) to derive the expressions for $(\nabla \cdot \boldsymbol{v})$ and $\nabla^2 s$ given in Tables A.7-2 and 3 for cylindrical and spherical coordinates, respectively. Note that the vector components v^i in (1) must be converted into components referred to the unit vectors.

3. Show that if a second-order tensor is symmetric (i.e., $\tau^{ij} = \tau^{ji}$) in one coordinate system, it is symmetric in all other coordinate systems.

[5] A. P. Wills, *op. cit.*
[6] L. I. Sedov, *op cit.*
[7] A. J. McConnell, *op. cit.*
[8] A. P. Wills, *op cit.*, p. 261.

§A.9 FURTHER COMMENTS ON VECTOR-TENSOR NOTATION

The bold-face notation used in this book is called *Gibbs notation.*[1] Also widely used is another notation referred to as Cartesian tensor notation.[2] As shown in Table A.9-1 a few examples suffice to compare the two systems. The two outer columns are just two different

TABLE A.9-1

Gibbs Notation	Expanded Notation in Terms of Unit Vectors and Unit Dyads	Cartesian Tensor Notation
$(\boldsymbol{v} \cdot \boldsymbol{w})$	$\sum_i v_i w_i$	$v_i w_i$
$[\boldsymbol{v} \times \boldsymbol{w}]$	$\sum_i \sum_j \sum_k \varepsilon_{ijk} \boldsymbol{\delta}_i v_j w_k$	$\varepsilon_{ijk} v_j w_k$
$[\nabla \cdot \boldsymbol{\tau}]$	$\sum_i \sum_j \boldsymbol{\delta}_i \frac{\partial}{\partial x_j} \tau_{ji}$	$\partial_j \tau_{ji}$
$\nabla^2 s$	$\sum_i \frac{\partial^2}{\partial x_i^2} s$	$\partial_i \partial_i s$
$[\nabla \times [\nabla \times \boldsymbol{v}]]$	$\sum_i \sum_j \sum_k \sum_m \sum_n \boldsymbol{\delta}_i \varepsilon_{ijk} \varepsilon_{kmn} \frac{\partial}{\partial x_j} \frac{\partial}{\partial x_m} v_n$	$\varepsilon_{ijk} \varepsilon_{kmn} \partial_j \partial_m v_n$
$\{\boldsymbol{\tau} \times \boldsymbol{v}\}$	$\sum_i \sum_j \sum_k \sum_l \varepsilon_{jkl} \boldsymbol{\delta}_i \boldsymbol{\delta}_l \tau_{ij} v_k$	$\varepsilon_{jkl} \tau_{ij} v_k$

ways of abbreviating the operations described explicitly in the middle column in rectangular coordinates. The rules for converting from one system to another are easy:

To convert from expanded notation to Cartesian tensor notation:

1. Omit all summation signs.
2. Omit all unit vectors and unit dyads.
3. Replace $\partial/\partial x_i$ by ∂_i.

To convert from Cartesian tensor notation to expanded notation:

1. Supply summation signs for all repeated indices.
2. Supply unit vectors and unit dyads for all nonrepeated indices; in each term of a tensor equation the unit vectors must appear in the same order in the unit dyads.
3. Replace ∂_i by $\partial/\partial x_i$.

The Gibbs notation is compact, easy to read, and devoid of any reference to a particular coordinate system; however, one has to know the meaning of the dot and cross operations

[1] J. W. Gibbs. *Vector Analysis*, Dover Reprint, New York (1960).
[2] W. Prager, *Mechanics of Continua*, Ginn, Boston (1961).

and the use of boldface symbols. The Cartesian tensor notation indicates the nature of the operations explicitly in rectangular coordinates, but errors in reading or writing subscripts can be most aggravating. People who know both systems equally well prefer the Gibbs notation for general discussions and for presenting results, but revert to Cartesian tensor notation for doing proofs of identities.

Similar index notation can also be used in lieu of the expanded notation in §A.8. One simply remembers that base vectors (or reciprocal base vectors) have to be supplied for every nonrepeated contravariant (or covariant) index, and that sums are then implied on all pairs of repeated indices (one subscript and one superscript). For example, the relation

$$[\boldsymbol{v} \times \boldsymbol{w}] = [\nabla \cdot \boldsymbol{\tau}] \tag{A.9-1}$$

in expanded form becomes

$$\sum_i \sum_j \sum_k \mathscr{E}_{ijk} \boldsymbol{g}^k v^i w^j = \sum_i \sum_k \boldsymbol{g}_k \tau^{ik}{}_{,i} \tag{A.9-2}$$

In abbreviated notation we may simply write down the statement of the equality of the kth components:

$$\mathscr{E}_{ijl} g^{lk} v^i w^j = \tau^{ik}{}_{,i} \tag{A.9-3}$$

or

$$\mathscr{E}_{ijk} v^i w^j = \tau^{i}{}_{.k,i} \tag{A.9-4}$$

depending on whether we choose to write the equation in contravariant or covariant components. It is in this notation that most tensor books and continuum mechanics research papers are written. Usually the scaffolding provided by the base vectors is carefully removed so that the reader sees nothing but the relations among the components.

Occasionally *matrix notation* is used to display the components of vectors and tensors with respect to designated coordinate systems. For example, when $v_x = \dot{\gamma} y$, $v_y = 0$, $v_z = 0$, $\nabla \boldsymbol{v}$ can be written in two ways:

$$\nabla \boldsymbol{v} = \boldsymbol{\delta}_y \boldsymbol{\delta}_x \dot{\gamma} = \begin{pmatrix} 0 & 0 & 0 \\ \dot{\gamma} & 0 & 0 \\ 0 & 0 & 0 \end{pmatrix} \tag{A.9-5}$$

The second " = " is not really an "equals" sign, but has to be interpreted as "may be displayed as." Note that this notation is somewhat dangerous since one has to infer the unit dyads that are to be multiplied by the matrix elements—in this case, $\boldsymbol{\delta}_x \boldsymbol{\delta}_x$, $\boldsymbol{\delta}_x \boldsymbol{\delta}_y$, etc. If we had used cylindrical coordinates, $\nabla \boldsymbol{v}$ would be represented by the matrix:

$$\nabla \boldsymbol{v} = \begin{pmatrix} \dot{\gamma} \sin\theta \cos\theta & -\dot{\gamma} \sin^2\theta & 0 \\ \dot{\gamma} \cos^2\theta & -\dot{\gamma} \sin\theta \cos\theta & 0 \\ 0 & 0 & 0 \end{pmatrix} \tag{A.9-6}$$

where the matrix elements are understood to be multiplied by $\boldsymbol{\delta}_r \boldsymbol{\delta}_r$, $\boldsymbol{\delta}_r \boldsymbol{\delta}_\theta$, etc, and then added together.

Despite the hazard of misinterpretation and the loose use of "=," the matrix notation enjoys widespread use, the main reason being that the "dot" operations correspond to standard matrix multiplication rules. For example,

$$(\boldsymbol{v} \cdot \boldsymbol{w}) = (v_1 \quad v_2 \quad v_3)\begin{pmatrix} w_1 \\ w_2 \\ w_3 \end{pmatrix} = v_1 w_1 + v_2 w_2 + v_3 w_3 \tag{A.9-7}$$

$$[\boldsymbol{\tau} \cdot \boldsymbol{v}] = \begin{pmatrix} \tau_{11} & \tau_{12} & \tau_{13} \\ \tau_{21} & \tau_{22} & \tau_{23} \\ \tau_{31} & \tau_{32} & \tau_{33} \end{pmatrix}\begin{pmatrix} v_1 \\ v_2 \\ v_3 \end{pmatrix} = \begin{pmatrix} \tau_{11}v_1 + \tau_{12}v_2 + \tau_{13}v_3 \\ \tau_{21}v_1 + \tau_{22}v_2 + \tau_{23}v_3 \\ \tau_{31}v_1 + \tau_{32}v_2 + \tau_{33}v_3 \end{pmatrix} \tag{A.9-8}$$

Of course such matrix multiplications can be performed only when the components are referred to the same unit vectors or base vectors.

APPENDIX B

COMPONENTS OF THE EQUATION OF MOTION AND KINEMATIC TENSORS

TABLE B.1

The Equation of Motion[a] in Terms of τ

Rectangular Coordinates (x, y, z):

$$\rho\left(\frac{\partial v_x}{\partial t} + v_x \frac{\partial}{\partial x} v_x + v_y \frac{\partial}{\partial y} v_x + v_z \frac{\partial}{\partial z} v_x\right) = -\left[\frac{\partial}{\partial x}\tau_{xx} + \frac{\partial}{\partial y}\tau_{yx} + \frac{\partial}{\partial z}\tau_{zx}\right] - \frac{\partial p}{\partial x} + \rho g_x \tag{B.1-1}$$

$$\rho\left(\frac{\partial v_y}{\partial t} + v_x \frac{\partial}{\partial x} v_y + v_y \frac{\partial}{\partial y} v_y + v_z \frac{\partial}{\partial z} v_y\right) = -\left[\frac{\partial}{\partial x}\tau_{xy} + \frac{\partial}{\partial y}\tau_{yy} + \frac{\partial}{\partial z}\tau_{zy}\right] - \frac{\partial p}{\partial y} + \rho g_y \tag{B.1-2}$$

$$\rho\left(\frac{\partial v_z}{\partial t} + v_x \frac{\partial}{\partial x} v_z + v_y \frac{\partial}{\partial y} v_z + v_z \frac{\partial}{\partial z} v_z\right) = -\left[\frac{\partial}{\partial x}\tau_{xz} + \frac{\partial}{\partial y}\tau_{yz} + \frac{\partial}{\partial z}\tau_{zz}\right] - \frac{\partial p}{\partial z} + \rho g_z \tag{B.1-3}$$

Cylindrical Coordinates (r, θ, z):

$$\rho\left(\frac{\partial v_r}{\partial t} + v_r \frac{\partial v_r}{\partial r} + \frac{v_\theta}{r}\frac{\partial v_r}{\partial \theta} - \frac{v_\theta^2}{r} + v_z \frac{\partial v_r}{\partial z}\right) = -\left[\frac{1}{r}\frac{\partial}{\partial r}(r\tau_{rr}) + \frac{1}{r}\frac{\partial}{\partial \theta}\tau_{\theta r} + \frac{\partial}{\partial z}\tau_{zr} - \frac{\tau_{\theta\theta}}{r}\right] - \frac{\partial p}{\partial r} + \rho g_r \tag{B.1-4}$$

$$\rho\left(\frac{\partial v_\theta}{\partial t} + v_r \frac{\partial v_\theta}{\partial r} + \frac{v_\theta}{r}\frac{\partial v_\theta}{\partial \theta} + \frac{v_r v_\theta}{r} + v_z \frac{\partial v_\theta}{\partial z}\right) = -\left[\frac{1}{r^2}\frac{\partial}{\partial r}(r^2\tau_{r\theta}) + \frac{1}{r}\frac{\partial}{\partial \theta}\tau_{\theta\theta} + \frac{\partial}{\partial z}\tau_{z\theta} + \frac{\tau_{\theta r} - \tau_{r\theta}}{r}\right] - \frac{1}{r}\frac{\partial p}{\partial \theta} + \rho g_\theta \tag{B.1-5}$$

$$\rho\left(\frac{\partial v_z}{\partial t} + v_r \frac{\partial v_z}{\partial r} + \frac{v_\theta}{r}\frac{\partial v_z}{\partial \theta} + v_z \frac{\partial v_z}{\partial z}\right) = -\left[\frac{1}{r}\frac{\partial}{\partial r}(r\tau_{rz}) + \frac{1}{r}\frac{\partial}{\partial \theta}\tau_{\theta z} + \frac{\partial}{\partial z}\tau_{zz}\right] - \frac{\partial p}{\partial z} + \rho g_z \tag{B.1-6}$$

Spherical Coordinates (r, θ, ϕ):

$$\rho\left(\frac{\partial v_r}{\partial t} + v_r \frac{\partial v_r}{\partial r} + \frac{v_\theta}{r}\frac{\partial v_r}{\partial \theta} + \frac{v_\phi}{r\sin\theta}\frac{\partial v_r}{\partial \phi} - \frac{v_\theta^2 + v_\phi^2}{r}\right)$$
$$= -\left[\frac{1}{r^2}\frac{\partial}{\partial r}(r^2\tau_{rr}) + \frac{1}{r\sin\theta}\frac{\partial}{\partial \theta}(\tau_{\theta r}\sin\theta) + \frac{1}{r\sin\theta}\frac{\partial}{\partial \phi}\tau_{\phi r} - \frac{\tau_{\theta\theta} + \tau_{\phi\phi}}{r}\right] - \frac{\partial p}{\partial r} + \rho g_r \tag{B.1-7}$$

$$\rho\left(\frac{\partial v_\theta}{\partial t} + v_r \frac{\partial v_\theta}{\partial r} + \frac{v_\theta}{r}\frac{\partial v_\theta}{\partial \theta} + \frac{v_\phi}{r\sin\theta}\frac{\partial v_\theta}{\partial \phi} + \frac{v_r v_\theta}{r} - \frac{v_\phi^2 \cot\theta}{r}\right)$$
$$= -\left[\frac{1}{r^3}\frac{\partial}{\partial r}(r^3\tau_{r\theta}) + \frac{1}{r\sin\theta}\frac{\partial}{\partial \theta}(\tau_{\theta\theta}\sin\theta) + \frac{1}{r\sin\theta}\frac{\partial}{\partial \phi}\tau_{\phi\theta} + \frac{(\tau_{\theta r} - \tau_{r\theta}) - \tau_{\phi\phi}\cot\theta}{r}\right] - \frac{1}{r}\frac{\partial p}{\partial \theta} + \rho g_\theta \tag{B.1-8}$$

$$\rho\left(\frac{\partial v_\phi}{\partial t} + v_r \frac{\partial v_\phi}{\partial r} + \frac{v_\theta}{r}\frac{\partial v_\phi}{\partial \theta} + \frac{v_\phi}{r\sin\theta}\frac{\partial v_\phi}{\partial \phi} + \frac{v_\phi v_r}{r} + \frac{v_\theta v_\phi}{r}\cot\theta\right)$$
$$= -\left[\frac{1}{r^3}\frac{\partial}{\partial r}(r^3\tau_{r\phi}) + \frac{1}{r\sin\theta}\frac{\partial}{\partial \theta}(\tau_{\theta\phi}\sin\theta) + \frac{1}{r\sin\theta}\frac{\partial}{\partial \phi}\tau_{\phi\phi} + \frac{(\tau_{\phi r} - \tau_{r\phi}) + \tau_{\phi\theta}\cot\theta}{r}\right]$$
$$- \frac{1}{r\sin\theta}\frac{\partial p}{\partial \phi} + \rho g_\phi \tag{B.1-9}$$

[a] In these equations no assumption is made regarding the symmetry of τ.

TABLE B.2

The Equation of Motion for a Newtonian Fluid with Constant Density (ρ) and Constant Viscosity (μ)

Rectangular Coordinates (x, y, z):

$$\rho\left(\frac{\partial v_x}{\partial t} + v_x \frac{\partial v_x}{\partial x} + v_y \frac{\partial v_x}{\partial y} + v_z \frac{\partial v_x}{\partial z}\right) = \mu\left[\frac{\partial^2}{\partial x^2} v_x + \frac{\partial^2}{\partial y^2} v_x + \frac{\partial^2}{\partial z^2} v_x\right] - \frac{\partial p}{\partial x} + \rho g_x \tag{B.2-1}$$

$$\rho\left(\frac{\partial v_y}{\partial t} + v_x \frac{\partial v_y}{\partial x} + v_y \frac{\partial v_y}{\partial y} + v_z \frac{\partial v_y}{\partial z}\right) = \mu\left[\frac{\partial^2}{\partial x^2} v_y + \frac{\partial^2}{\partial y^2} v_y + \frac{\partial^2}{\partial z^2} v_y\right] - \frac{\partial p}{\partial y} + \rho g_y \tag{B.2-2}$$

$$\rho\left(\frac{\partial v_z}{\partial t} + v_x \frac{\partial v_z}{\partial x} + v_y \frac{\partial v_z}{\partial y} + v_z \frac{\partial v_z}{\partial z}\right) = \mu\left[\frac{\partial^2}{\partial x^2} v_z + \frac{\partial^2}{\partial y^2} v_z + \frac{\partial^2}{\partial z^2} v_z\right] - \frac{\partial p}{\partial z} + \rho g_z \tag{B.2-3}$$

Cylindrical Coordinates (r, θ, z):

$$\rho\left(\frac{\partial v_r}{\partial t} + v_r \frac{\partial v_r}{\partial r} + \frac{v_\theta}{r}\frac{\partial v_r}{\partial \theta} - \frac{v_\theta^2}{r} + v_z \frac{\partial v_r}{\partial z}\right) = \mu\left[\frac{\partial}{\partial r}\left(\frac{1}{r}\frac{\partial}{\partial r}(r v_r)\right) + \frac{1}{r^2}\frac{\partial^2 v_r}{\partial \theta^2} + \frac{\partial^2 v_r}{\partial z^2} - \frac{2}{r^2}\frac{\partial v_\theta}{\partial \theta}\right] - \frac{\partial p}{\partial r} + \rho g_r \tag{B.2-4}$$

$$\rho\left(\frac{\partial v_\theta}{\partial t} + v_r \frac{\partial v_\theta}{\partial r} + \frac{v_\theta}{r}\frac{\partial v_\theta}{\partial \theta} + \frac{v_r v_\theta}{r} + v_z \frac{\partial v_\theta}{\partial z}\right) = \mu\left[\frac{\partial}{\partial r}\left(\frac{1}{r}\frac{\partial}{\partial r}(r v_\theta)\right) + \frac{1}{r^2}\frac{\partial^2 v_\theta}{\partial \theta^2} + \frac{\partial^2 v_\theta}{\partial z^2} + \frac{2}{r^2}\frac{\partial v_r}{\partial \theta}\right] - \frac{1}{r}\frac{\partial p}{\partial \theta} + \rho g_\theta \tag{B.2-5}$$

$$\rho\left(\frac{\partial v_z}{\partial t} + v_r \frac{\partial v_z}{\partial r} + \frac{v_\theta}{r}\frac{\partial v_z}{\partial \theta} + v_z \frac{\partial v_z}{\partial z}\right) = \mu\left[\frac{1}{r}\frac{\partial}{\partial r}\left(r\frac{\partial v_z}{\partial r}\right) + \frac{1}{r^2}\frac{\partial^2 v_z}{\partial \theta^2} + \frac{\partial^2 v_z}{\partial z^2}\right] - \frac{\partial p}{\partial z} + \rho g_z \tag{B.2-6}$$

Spherical Coordinates (r, θ, ϕ):

$$\begin{aligned}\rho&\left(\frac{\partial v_r}{\partial t} + v_r \frac{\partial v_r}{\partial r} + \frac{v_\theta}{r}\frac{\partial v_r}{\partial \theta} + \frac{v_\phi}{r \sin\theta}\frac{\partial v_r}{\partial \phi} - \frac{v_\theta^2 + v_\phi^2}{r}\right)\\ &= \mu\left[\frac{1}{r^2}\frac{\partial^2}{\partial r^2}(r^2 v_r) + \frac{1}{r^2 \sin\theta}\frac{\partial}{\partial \theta}\left(\sin\theta \frac{\partial v_r}{\partial \theta}\right) + \frac{1}{r^2 \sin^2\theta}\frac{\partial^2 v_r}{\partial \phi^2}\right] - \frac{\partial p}{\partial r} + \rho g_r\end{aligned} \tag{B.2-7}$$

$$\begin{aligned}\rho&\left(\frac{\partial v_\theta}{\partial t} + v_r \frac{\partial v_\theta}{\partial r} + \frac{v_\theta}{r}\frac{\partial v_\theta}{\partial \theta} + \frac{v_\phi}{r \sin\theta}\frac{\partial v_\theta}{\partial \phi} + \frac{v_r v_\theta}{r} - \frac{v_\phi^2 \cot\theta}{r}\right)\\ &= \mu\left[\frac{1}{r^2}\frac{\partial}{\partial r}\left(r^2\frac{\partial v_\theta}{\partial r}\right) + \frac{1}{r^2}\frac{\partial}{\partial \theta}\left(\frac{1}{\sin\theta}\frac{\partial}{\partial \theta}(v_\theta \sin\theta)\right) + \frac{1}{r^2 \sin^2\theta}\frac{\partial^2 v_\theta}{\partial \phi^2} + \frac{2}{r^2}\frac{\partial v_r}{\partial \theta} - \frac{2\cot\theta}{r^2 \sin\theta}\frac{\partial v_\phi}{\partial \phi}\right]\\ &\quad - \frac{1}{r}\frac{\partial p}{\partial \theta} + \rho g_\theta\end{aligned} \tag{B.2-8}$$

$$\begin{aligned}\rho&\left(\frac{\partial v_\phi}{\partial t} + v_r \frac{\partial v_\phi}{\partial r} + \frac{v_\theta}{r}\frac{\partial v_\phi}{\partial \theta} + \frac{v_\phi}{r \sin\theta}\frac{\partial v_\phi}{\partial \phi} + \frac{v_\phi v_r}{r} + \frac{v_\theta v_\phi}{r}\cot\theta\right)\\ &= \mu\left[\frac{1}{r^2}\frac{\partial}{\partial r}\left(r^2\frac{\partial v_\phi}{\partial r}\right) + \frac{1}{r^2}\frac{\partial}{\partial \theta}\left(\frac{1}{\sin\theta}\frac{\partial}{\partial \theta}(v_\phi \sin\theta)\right)\right.\\ &\quad \left. + \frac{1}{r^2 \sin^2\theta}\frac{\partial^2 v_\phi}{\partial \phi^2} + \frac{2}{r^2 \sin\theta}\frac{\partial v_r}{\partial \phi} + \frac{2\cot\theta}{r^2 \sin\theta}\frac{\partial v_\theta}{\partial \phi}\right] - \frac{1}{r \sin\theta}\frac{\partial p}{\partial \phi} + \rho g_\phi\end{aligned} \tag{B.2-9}$$

TABLE B.3

The Rate-of-Strain Tensor $\dot{\boldsymbol{\gamma}} = \nabla \boldsymbol{v} + (\nabla \boldsymbol{v})^\dagger$

Rectangular Coordinates (x, y, z):

$$\dot{\gamma}_{xx} = 2\frac{\partial v_x}{\partial x} \tag{B.3-1}$$

$$\dot{\gamma}_{yy} = 2\frac{\partial v_y}{\partial y} \tag{B.3-2}$$

$$\dot{\gamma}_{zz} = 2\frac{\partial v_z}{\partial z} \tag{B.3-3}$$

$$\dot{\gamma}_{xy} = \dot{\gamma}_{yx} = \frac{\partial v_y}{\partial x} + \frac{\partial v_x}{\partial y} \tag{B.3-4}$$

$$\dot{\gamma}_{yz} = \dot{\gamma}_{zy} = \frac{\partial v_z}{\partial y} + \frac{\partial v_y}{\partial z} \tag{B.3-5}$$

$$\dot{\gamma}_{zx} = \dot{\gamma}_{xz} = \frac{\partial v_x}{\partial z} + \frac{\partial v_z}{\partial x} \tag{B.3-6}$$

Cylindrical Coordinates (r, θ, z):

$$\dot{\gamma}_{rr} = 2\frac{\partial v_r}{\partial r} \tag{B.3-7}$$

$$\dot{\gamma}_{\theta\theta} = 2\left(\frac{1}{r}\frac{\partial v_\theta}{\partial \theta} + \frac{v_r}{r}\right) \tag{B.3-8}$$

$$\dot{\gamma}_{zz} = 2\frac{\partial v_z}{\partial z} \tag{B.3-9}$$

$$\dot{\gamma}_{r\theta} = \dot{\gamma}_{\theta r} = r\frac{\partial}{\partial r}\left(\frac{v_\theta}{r}\right) + \frac{1}{r}\frac{\partial v_r}{\partial \theta} \tag{B.3-10}$$

$$\dot{\gamma}_{\theta z} = \dot{\gamma}_{z\theta} = \frac{1}{r}\frac{\partial v_z}{\partial \theta} + \frac{\partial v_\theta}{\partial z} \tag{B.3-11}$$

$$\dot{\gamma}_{zr} = \dot{\gamma}_{rz} = \frac{\partial v_r}{\partial z} + \frac{\partial v_z}{\partial r} \tag{B.3-12}$$

Spherical Coordinates (r, θ, ϕ):

$$\dot{\gamma}_{rr} = 2\frac{\partial v_r}{\partial r} \tag{B.3-13}$$

$$\dot{\gamma}_{\theta\theta} = 2\left(\frac{1}{r}\frac{\partial v_\theta}{\partial \theta} + \frac{v_r}{r}\right) \tag{B.3-14}$$

$$\dot{\gamma}_{\phi\phi} = 2\left(\frac{1}{r\sin\theta}\frac{\partial v_\phi}{\partial \phi} + \frac{v_r}{r} + \frac{v_\theta \cot\theta}{r}\right) \tag{B.3-15}$$

TABLE B.3 (*Continued*)

$$\dot{\gamma}_{r\theta} = \dot{\gamma}_{\theta r} = r\frac{\partial}{\partial r}\left(\frac{v_\theta}{r}\right) + \frac{1}{r}\frac{\partial v_r}{\partial \theta} \tag{B.3-16}$$

$$\dot{\gamma}_{\theta\phi} = \dot{\gamma}_{\phi\theta} = \frac{\sin\theta}{r}\frac{\partial}{\partial \theta}\left(\frac{v_\phi}{\sin\theta}\right) + \frac{1}{r\sin\theta}\frac{\partial v_\theta}{\partial \phi} \tag{B.3-17}$$

$$\dot{\gamma}_{\phi r} = \dot{\gamma}_{r\phi} = \frac{1}{r\sin\theta}\frac{\partial v_r}{\partial \phi} + r\frac{\partial}{\partial r}\left(\frac{v_\phi}{r}\right) \tag{B.3-18}$$

TABLE B.4

The Vorticity Tensor $\boldsymbol{\omega} = \nabla \boldsymbol{v} - (\nabla \boldsymbol{v})^\dagger$

Rectangular Coordinates (x, y, z):

$$\omega_{xy} = -\omega_{yx} = \frac{\partial v_y}{\partial x} - \frac{\partial v_x}{\partial y} \tag{B.4-1}$$

$$\omega_{yz} = -\omega_{zy} = \frac{\partial v_z}{\partial y} - \frac{\partial v_y}{\partial z} \tag{B.4-2}$$

$$\omega_{zx} = -\omega_{xz} = \frac{\partial v_x}{\partial z} - \frac{\partial v_z}{\partial x} \tag{B.4-3}$$

Cylindrical Coordinates (r, θ, z):

$$\omega_{r\theta} = -\omega_{\theta r} = \frac{1}{r}\frac{\partial}{\partial r}(r v_\theta) - \frac{1}{r}\frac{\partial v_r}{\partial \theta} \tag{B.4-4}$$

$$\omega_{\theta z} = -\omega_{z\theta} = \frac{1}{r}\frac{\partial v_z}{\partial \theta} - \frac{\partial v_\theta}{\partial z} \tag{B.4-5}$$

$$\omega_{zr} = -\omega_{rz} = \frac{\partial v_r}{\partial z} - \frac{\partial v_z}{\partial r} \tag{B.4-6}$$

Spherical Coordinates (r, θ, ϕ):

$$\omega_{r\theta} = -\omega_{\theta r} = \frac{1}{r}\frac{\partial}{\partial r}(r v_\theta) - \frac{1}{r}\frac{\partial v_r}{\partial \theta} \tag{B.4-7}$$

$$\omega_{\theta\phi} = -\omega_{\phi\theta} = \frac{1}{r\sin\theta}\frac{\partial}{\partial \theta}(v_\phi \sin\theta) - \frac{1}{r\sin\theta}\frac{\partial v_\theta}{\partial \phi} \tag{B.4-8}$$

$$\omega_{\phi r} = -\omega_{r\phi} = \frac{1}{r\sin\theta}\frac{\partial v_r}{\partial \phi} - \frac{1}{r}\frac{\partial}{\partial r}(r v_\phi) \tag{B.4-9}$$

TABLE B.5

The Relative Strain Tensor $\gamma^{[0]}$

Rectangular Coordinates (x, y, z):

$$x' = x'(x, y, z, t, t')$$
$$y' = y'(x, y, z, t, t')$$
$$z' = z'(x, y, z, t, t')$$

$$\gamma_{xx}^{[0]} = \left(\frac{\partial x'}{\partial x}\right)^2 + \left(\frac{\partial y'}{\partial x}\right)^2 + \left(\frac{\partial z'}{\partial x}\right)^2 - 1 \tag{B.5-1}$$

$$\gamma_{yy}^{[0]} = \left(\frac{\partial x'}{\partial y}\right)^2 + \left(\frac{\partial y'}{\partial y}\right)^2 + \left(\frac{\partial z'}{\partial y}\right)^2 - 1 \tag{B.5-2}$$

$$\gamma_{zz}^{[0]} = \left(\frac{\partial x'}{\partial z}\right)^2 + \left(\frac{\partial y'}{\partial z}\right)^2 + \left(\frac{\partial z'}{\partial z}\right)^2 - 1 \tag{B.5-3}$$

$$\gamma_{xy}^{[0]} = \gamma_{yx}^{[0]} = \left(\frac{\partial x'}{\partial x}\right)\left(\frac{\partial x'}{\partial y}\right) + \left(\frac{\partial y'}{\partial x}\right)\left(\frac{\partial y'}{\partial y}\right) + \left(\frac{\partial z'}{\partial x}\right)\left(\frac{\partial z'}{\partial y}\right) \tag{B.5-4}$$

$$\gamma_{yz}^{[0]} = \gamma_{zy}^{[0]} = \left(\frac{\partial x'}{\partial y}\right)\left(\frac{\partial x'}{\partial z}\right) + \left(\frac{\partial y'}{\partial y}\right)\left(\frac{\partial y'}{\partial z}\right) + \left(\frac{\partial z'}{\partial y}\right)\left(\frac{\partial z'}{\partial z}\right) \tag{B.5-5}$$

$$\gamma_{zx}^{[0]} = \gamma_{xz}^{[0]} = \left(\frac{\partial x'}{\partial x}\right)\left(\frac{\partial x'}{\partial z}\right) + \left(\frac{\partial y'}{\partial x}\right)\left(\frac{\partial y'}{\partial z}\right) + \left(\frac{\partial z'}{\partial x}\right)\left(\frac{\partial z'}{\partial z}\right) \tag{B.5-6}$$

Cylindrical Coordinates (r, θ, z):

$$r' = r'(r, \theta, z, t, t')$$
$$\theta' = \theta'(r, \theta, z, t, t')$$
$$z' = z'(r, \theta, z, t, t')$$

$$\gamma_{rr}^{[0]} = \left(\frac{\partial r'}{\partial r}\right)^2 + \left(r'\frac{\partial \theta'}{\partial r}\right)^2 + \left(\frac{\partial z'}{\partial r}\right)^2 - 1 \tag{B.5-7}$$

$$\gamma_{\theta\theta}^{[0]} = \frac{1}{r^2}\left[\left(\frac{\partial r'}{\partial \theta}\right)^2 + \left(r'\frac{\partial \theta'}{\partial \theta}\right)^2 + \left(\frac{\partial z'}{\partial \theta}\right)^2\right] - 1 \tag{B.5-8}$$

$$\gamma_{zz}^{[0]} = \left(\frac{\partial r'}{\partial z}\right)^2 + \left(r'\frac{\partial \theta'}{\partial z}\right)^2 + \left(\frac{\partial z'}{\partial z}\right)^2 - 1 \tag{B.5-9}$$

$$\gamma_{r\theta}^{[0]} = \gamma_{\theta r}^{[0]} = \frac{1}{r}\left[\left(\frac{\partial r'}{\partial r}\right)\left(\frac{\partial r'}{\partial \theta}\right) + r'^2\left(\frac{\partial \theta'}{\partial r}\right)\left(\frac{\partial \theta'}{\partial \theta}\right) + \left(\frac{\partial z'}{\partial r}\right)\left(\frac{\partial z'}{\partial \theta}\right)\right] \tag{B.5-10}$$

$$\gamma_{\theta z}^{[0]} = \gamma_{z\theta}^{[0]} = \frac{1}{r}\left[\left(\frac{\partial r'}{\partial \theta}\right)\left(\frac{\partial r'}{\partial z}\right) + r'^2\left(\frac{\partial \theta'}{\partial \theta}\right)\left(\frac{\partial \theta'}{\partial z}\right) + \left(\frac{\partial z'}{\partial \theta}\right)\left(\frac{\partial z'}{\partial z}\right)\right] \tag{B.5-11}$$

$$\gamma_{zr}^{[0]} = \gamma_{rz}^{[0]} = \left(\frac{\partial r'}{\partial r}\right)\left(\frac{\partial r'}{\partial z}\right) + r'^2\left(\frac{\partial \theta'}{\partial r}\right)\left(\frac{\partial \theta'}{\partial z}\right) + \left(\frac{\partial z'}{\partial r}\right)\left(\frac{\partial z'}{\partial z}\right) \tag{B.5-12}$$

TABLE B.5 *(Continued)*

Spherical Coordinates (r, θ, ϕ):

$$r' = r'(r, \theta, \phi, t, t')$$

$$\theta' = \theta'(r, \theta, \phi, t, t')$$

$$\phi' = \phi'(r, \theta, \phi, t, t')$$

$$\gamma_{rr}^{[0]} = \left(\frac{\partial r'}{\partial r}\right)^2 + \left(r' \frac{\partial \theta'}{\partial r}\right)^2 + \left(r' \sin \theta' \frac{\partial \phi'}{\partial r}\right)^2 - 1 \tag{B.5-13}$$

$$\gamma_{\theta\theta}^{[0]} = \frac{1}{r^2}\left[\left(\frac{\partial r'}{\partial \theta}\right)^2 + \left(r' \frac{\partial \theta'}{\partial \theta}\right)^2 + \left(r' \sin \theta' \frac{\partial \phi'}{\partial \theta}\right)^2\right] - 1 \tag{B.5-14}$$

$$\gamma_{\phi\phi}^{[0]} = \frac{1}{(r \sin \theta)^2}\left[\left(\frac{\partial r'}{\partial \phi}\right)^2 + \left(r' \frac{\partial \theta'}{\partial \phi}\right)^2 + \left(r' \sin \theta' \frac{\partial \phi'}{\partial \phi}\right)^2\right] - 1 \tag{B.5-15}$$

$$\gamma_{r\theta}^{[0]} = \gamma_{\theta r}^{[0]} = \frac{1}{r}\left[\left(\frac{\partial r'}{\partial r}\right)\left(\frac{\partial r'}{\partial \theta}\right) + r'^2\left(\frac{\partial \theta'}{\partial r}\right)\left(\frac{\partial \theta'}{\partial \theta}\right) + (r' \sin \theta')^2\left(\frac{\partial \phi'}{\partial r}\right)\left(\frac{\partial \phi'}{\partial \theta}\right)\right] \tag{B.5-16}$$

$$\gamma_{\theta\phi}^{[0]} = \gamma_{\phi\theta}^{[0]} = \frac{1}{r^2 \sin \theta}\left[\left(\frac{\partial r'}{\partial \theta}\right)\left(\frac{\partial r'}{\partial \phi}\right) + r'^2\left(\frac{\partial \theta'}{\partial \theta}\right)\left(\frac{\partial \theta'}{\partial \phi}\right) + (r' \sin \theta')^2\left(\frac{\partial \phi'}{\partial \theta}\right)\left(\frac{\partial \phi'}{\partial \phi}\right)\right] \tag{B.5-17}$$

$$\gamma_{\phi r}^{[0]} = \gamma_{r\phi}^{[0]} = \frac{1}{r \sin \theta}\left[\left(\frac{\partial r'}{\partial r}\right)\left(\frac{\partial r'}{\partial \phi}\right) + r'^2\left(\frac{\partial \theta'}{\partial r}\right)\left(\frac{\partial \theta'}{\partial \phi}\right) + (r' \sin \theta')^2\left(\frac{\partial \phi'}{\partial r}\right)\left(\frac{\partial \phi'}{\partial \phi}\right)\right] \tag{B.5-18}$$

TABLE B.6

The Relative Strain Tensor $\gamma_{[0]}$

Rectangular Coordinates (x, y, z):

$$x = x(x', y', z', t', t)$$
$$y = y(x', y', z', t', t)$$
$$z = z(x', y', z', t', t)$$

$$\gamma_{[0]xx} = 1 - \left[\left(\frac{\partial x}{\partial x'}\right)^2 + \left(\frac{\partial x}{\partial y'}\right)^2 + \left(\frac{\partial x}{\partial z'}\right)^2\right] \quad \text{(B.6-1)}$$

$$\gamma_{[0]yy} = 1 - \left[\left(\frac{\partial y}{\partial x'}\right)^2 + \left(\frac{\partial y}{\partial y'}\right)^2 + \left(\frac{\partial y}{\partial z'}\right)^2\right] \quad \text{(B.6-2)}$$

$$\gamma_{[0]zz} = 1 - \left[\left(\frac{\partial z}{\partial x'}\right)^2 + \left(\frac{\partial z}{\partial y'}\right)^2 + \left(\frac{\partial z}{\partial z'}\right)^2\right] \quad \text{(B.6-3)}$$

$$\gamma_{[0]xy} = \gamma_{[0]yx} = -\left[\left(\frac{\partial x}{\partial x'}\right)\left(\frac{\partial y}{\partial x'}\right) + \left(\frac{\partial x}{\partial y'}\right)\left(\frac{\partial y}{\partial y'}\right) + \left(\frac{\partial x}{\partial z'}\right)\left(\frac{\partial y}{\partial z'}\right)\right] \quad \text{(B.6-4)}$$

$$\gamma_{[0]yz} = \gamma_{[0]zy} = -\left[\left(\frac{\partial y}{\partial x'}\right)\left(\frac{\partial z}{\partial x'}\right) + \left(\frac{\partial y}{\partial y'}\right)\left(\frac{\partial z}{\partial y'}\right) + \left(\frac{\partial y}{\partial z'}\right)\left(\frac{\partial z}{\partial z'}\right)\right] \quad \text{(B.6-5)}$$

$$\gamma_{[0]zx} = \gamma_{[0]xz} = -\left[\left(\frac{\partial x}{\partial x'}\right)\left(\frac{\partial z}{\partial x'}\right) + \left(\frac{\partial x}{\partial y'}\right)\left(\frac{\partial z}{\partial y'}\right) + \left(\frac{\partial x}{\partial z'}\right)\left(\frac{\partial z}{\partial z'}\right)\right] \quad \text{(B.6-6)}$$

Cylindrical Coordinates (r, θ, z):

$$r = r(r', \theta', z', t', t)$$
$$\theta = \theta(r', \theta', z', t', t)$$
$$z = z(r', \theta', z', t', t)$$

$$\gamma_{[0]rr} = 1 - \left[\left(\frac{\partial r}{\partial r'}\right)^2 + \left(\frac{1}{r'}\frac{\partial r}{\partial \theta'}\right)^2 + \left(\frac{\partial r}{\partial z'}\right)^2\right] \quad \text{(B.6-7)}$$

$$\gamma_{[0]\theta\theta} = 1 - r^2\left[\left(\frac{\partial \theta}{\partial r'}\right)^2 + \left(\frac{1}{r'}\frac{\partial \theta}{\partial \theta'}\right)^2 + \left(\frac{\partial \theta}{\partial z'}\right)^2\right] \quad \text{(B.6-8)}$$

$$\gamma_{[0]zz} = 1 - \left[\left(\frac{\partial z}{\partial r'}\right)^2 + \left(\frac{1}{r'}\frac{\partial z}{\partial \theta'}\right)^2 + \left(\frac{\partial z}{\partial z'}\right)^2\right] \quad \text{(B.6-9)}$$

$$\gamma_{[0]r\theta} = \gamma_{[0]\theta r} = -r\left[\left(\frac{\partial r}{\partial r'}\right)\left(\frac{\partial \theta}{\partial r'}\right) + \frac{1}{r'^2}\left(\frac{\partial r}{\partial \theta'}\right)\left(\frac{\partial \theta}{\partial \theta'}\right) + \left(\frac{\partial r}{\partial z'}\right)\left(\frac{\partial \theta}{\partial z'}\right)\right] \quad \text{(B.6-10)}$$

$$\gamma_{[0]\theta z} = \gamma_{[0]z\theta} = -r\left[\left(\frac{\partial \theta}{\partial r'}\right)\left(\frac{\partial z}{\partial r'}\right) + \frac{1}{r'^2}\left(\frac{\partial \theta}{\partial \theta'}\right)\left(\frac{\partial z}{\partial \theta'}\right) + \left(\frac{\partial \theta}{\partial z'}\right)\left(\frac{\partial z}{\partial z'}\right)\right] \quad \text{(B.6-11)}$$

$$\gamma_{[0]zr} = \gamma_{[0]rz} = -\left[\left(\frac{\partial r}{\partial r'}\right)\left(\frac{\partial z}{\partial r'}\right) + \frac{1}{r'^2}\left(\frac{\partial r}{\partial \theta'}\right)\left(\frac{\partial z}{\partial \theta'}\right) + \left(\frac{\partial r}{\partial z'}\right)\left(\frac{\partial z}{\partial z'}\right)\right] \quad \text{(B.6-12)}$$

TABLE B.6 *(Continued)*

Spherical Coordinates (r, θ, ϕ):

$$r = r(r', \theta', \phi', t', t)$$

$$\theta = \theta(r', \theta', \phi', t', t)$$

$$\phi = \phi(r', \theta', \phi', t', t)$$

$$\gamma_{[0]rr} = 1 - \left[\left(\frac{\partial r}{\partial r'}\right)^2 + \left(\frac{1}{r'}\frac{\partial r}{\partial \theta'}\right)^2 + \left(\frac{1}{r' \sin \theta'}\frac{\partial r}{\partial \phi'}\right)^2\right] \qquad \text{(B.6-13)}$$

$$\gamma_{[0]\theta\theta} = 1 - r^2\left[\left(\frac{\partial \theta}{\partial r'}\right)^2 + \left(\frac{1}{r'}\frac{\partial \theta}{\partial \theta'}\right)^2 + \left(\frac{1}{r' \sin \theta'}\frac{\partial \theta}{\partial \phi'}\right)^2\right] \qquad \text{(B.6-14)}$$

$$\gamma_{[0]\phi\phi} = 1 - r^2 \sin^2 \theta\left[\left(\frac{\partial \phi}{\partial r'}\right)^2 + \left(\frac{1}{r'}\frac{\partial \phi}{\partial \theta'}\right)^2 + \left(\frac{1}{r' \sin \theta'}\frac{\partial \phi}{\partial \phi'}\right)^2\right] \qquad \text{(B.6-15)}$$

$$\gamma_{[0]r\theta} = \gamma_{[0]\theta r} = -r\left[\left(\frac{\partial r}{\partial r'}\right)\left(\frac{\partial \theta}{\partial r'}\right) + \frac{1}{r'^2}\left(\frac{\partial r}{\partial \theta'}\right)\left(\frac{\partial \theta}{\partial \theta'}\right) + \frac{1}{(r' \sin \theta')^2}\left(\frac{\partial r}{\partial \phi'}\right)\left(\frac{\partial \theta}{\partial \phi'}\right)\right] \qquad \text{(B.6-16)}$$

$$\gamma_{[0]\theta\phi} = \gamma_{[0]\phi\theta} = -r^2 \sin \theta\left[\left(\frac{\partial \theta}{\partial r'}\right)\left(\frac{\partial \phi}{\partial r'}\right) + \frac{1}{r'^2}\left(\frac{\partial \theta}{\partial \theta'}\right)\left(\frac{\partial \phi}{\partial \theta'}\right) + \frac{1}{(r' \sin \theta')^2}\left(\frac{\partial \theta}{\partial \phi'}\right)\left(\frac{\partial \phi}{\partial \phi'}\right)\right] \qquad \text{(B.6-17)}$$

$$\gamma_{[0]\phi r} = \gamma_{[0]r\phi} = -r \sin \theta\left[\left(\frac{\partial r}{\partial r'}\right)\left(\frac{\partial \phi}{\partial r'}\right) + \frac{1}{r'^2}\left(\frac{\partial r}{\partial \theta'}\right)\left(\frac{\partial \phi}{\partial \theta'}\right) + \frac{1}{(r' \sin \theta')^2}\left(\frac{\partial r}{\partial \phi'}\right)\left(\frac{\partial \phi}{\partial \phi'}\right)\right] \qquad \text{(B.6-18)}$$

TABLE B.7

The Displacement Gradient Tensors[a] $\boldsymbol{\Delta}$ and $\mathbf{E}$

Rectangular Coordinates (x, y, z):

$$\begin{pmatrix} \Delta_{xx} & \Delta_{xy} & \Delta_{xz} \\ \Delta_{yx} & \Delta_{yy} & \Delta_{yz} \\ \Delta_{zx} & \Delta_{zy} & \Delta_{zz} \end{pmatrix} = \begin{pmatrix} \partial x'/\partial x & \partial x'/\partial y & \partial x'/\partial z \\ \partial y'/\partial x & \partial y'/\partial y & \partial y'/\partial z \\ \partial z'/\partial x & \partial z'/\partial y & \partial z'/\partial z \end{pmatrix} \tag{B.7-1}$$

$$\begin{pmatrix} \mathrm{E}_{xx} & \mathrm{E}_{xy} & \mathrm{E}_{xz} \\ \mathrm{E}_{yx} & \mathrm{E}_{yy} & \mathrm{E}_{yz} \\ \mathrm{E}_{zx} & \mathrm{E}_{zy} & \mathrm{E}_{zz} \end{pmatrix} = \begin{pmatrix} \partial x/\partial x' & \partial x/\partial y' & \partial x/\partial z' \\ \partial y/\partial x' & \partial y/\partial y' & \partial y/\partial z' \\ \partial z/\partial x' & \partial z/\partial y' & \partial z/\partial z' \end{pmatrix} \tag{B.7-2}$$

Cylindrical Coordinates (r, θ, z): $[C = \cos\theta,\ S = \sin\theta,\ C' = \cos\theta',\ S' = \sin\theta']$

$$\begin{pmatrix} \Delta_{rr} & \Delta_{r\theta} & \Delta_{rz} \\ \Delta_{\theta r} & \Delta_{\theta\theta} & \Delta_{\theta z} \\ \Delta_{zr} & \Delta_{z\theta} & \Delta_{zz} \end{pmatrix} = \begin{pmatrix} T_{rr} & T_{r\theta} & T_{rz} \\ T_{\theta r} & T_{\theta\theta} & T_{\theta z} \\ T_{zr} & T_{z\theta} & T_{zz} \end{pmatrix} \begin{pmatrix} \partial r'/\partial r & (1/r)\partial r'/\partial\theta & \partial r'/\partial z \\ r'\partial\theta'/\partial r & (r'/r)\partial\theta'/\partial\theta & r'\partial\theta'/\partial z \\ \partial z'/\partial r & (1/r)\partial z'/\partial\theta & \partial z'/\partial z \end{pmatrix} \tag{B.7-3}$$

$$\begin{pmatrix} \mathrm{E}_{rr} & \mathrm{E}_{r\theta} & \mathrm{E}_{rz} \\ \mathrm{E}_{\theta r} & \mathrm{E}_{\theta\theta} & \mathrm{E}_{\theta z} \\ \mathrm{E}_{zr} & \mathrm{E}_{z\theta} & \mathrm{E}_{zz} \end{pmatrix} = \begin{pmatrix} \partial r/\partial r' & (1/r')\partial r/\partial\theta' & \partial r/\partial z' \\ r\partial\theta/\partial r' & (r/r')\partial\theta/\partial\theta' & r\partial\theta/\partial z' \\ \partial z/\partial r' & (1/r')\partial z/\partial\theta' & \partial z/\partial z' \end{pmatrix} \begin{pmatrix} T_{rr} & T_{\theta r} & T_{zr} \\ T_{r\theta} & T_{\theta\theta} & T_{z\theta} \\ T_{rz} & T_{\theta z} & T_{zz} \end{pmatrix} \tag{B.7-4}$$

where the orthogonal matrix (T_{ij}) is given by

$$\begin{pmatrix} T_{rr} & T_{r\theta} & T_{rz} \\ T_{\theta r} & T_{\theta\theta} & T_{\theta z} \\ T_{zr} & T_{z\theta} & T_{zz} \end{pmatrix} = \begin{pmatrix} CC' + SS' & SC' - CS' & 0 \\ CS' - SC' & SS' + CC' & 0 \\ 0 & 0 & 1 \end{pmatrix} \tag{B.7-5}$$

Spherical Coordinates (r, θ, ϕ): $[C = \cos\theta,\ S = \sin\theta,\ C' = \cos\theta',\ S' = \sin\theta';\ c = \cos\phi,\ s = \sin\phi,$ $c' = \cos\phi',\ s' = \sin\phi']$

$$\begin{pmatrix} \Delta_{rr} & \Delta_{r\theta} & \Delta_{r\phi} \\ \Delta_{\theta r} & \Delta_{\theta\theta} & \Delta_{\theta\phi} \\ \Delta_{\phi r} & \Delta_{\phi\theta} & \Delta_{\phi\phi} \end{pmatrix} = \begin{pmatrix} T_{rr} & T_{r\theta} & T_{r\phi} \\ T_{\theta r} & T_{\theta\theta} & T_{\theta\phi} \\ T_{\phi r} & T_{\phi\theta} & T_{\phi\phi} \end{pmatrix} \begin{pmatrix} \partial r'/\partial r & (1/r)\partial r'/\partial\theta & (1/rS)\partial r'/\partial\phi \\ r'\partial\theta'/\partial r & (r'/r)\partial\theta'/\partial\theta & (r'/rS)\partial\theta'/\partial\phi \\ r'S'\partial\phi'/\partial r & (r'S'/r)\partial\phi'/\partial\theta & (r'S'/rS)\partial\phi'/\partial\phi \end{pmatrix} \tag{B.7-6}$$

$$\begin{pmatrix} \mathrm{E}_{rr} & \mathrm{E}_{r\theta} & \mathrm{E}_{r\phi} \\ \mathrm{E}_{\theta r} & \mathrm{E}_{\theta\theta} & \mathrm{E}_{\theta\phi} \\ \mathrm{E}_{\phi r} & \mathrm{E}_{\phi\theta} & \mathrm{E}_{\phi\phi} \end{pmatrix} = \begin{pmatrix} \partial r/\partial r' & (1/r')\partial r/\partial\theta' & (1/r'S')\partial r/\partial\phi' \\ r\partial\theta/\partial r' & (r/r')\partial\theta/\partial\theta' & (r/r'S')\partial\theta/\partial\phi' \\ rS\partial\phi/\partial r' & (rS/r')\partial\phi/\partial\theta' & (rS/r'S')\partial\phi/\partial\phi' \end{pmatrix} \begin{pmatrix} T_{rr} & T_{\theta r} & T_{\phi r} \\ T_{r\theta} & T_{\theta\theta} & T_{\phi\theta} \\ T_{r\phi} & T_{\theta\phi} & T_{\phi\phi} \end{pmatrix} \tag{B.7-7}$$

where the orthogonal matrix (T_{ij}) is given by

$$\begin{pmatrix} T_{rr} & T_{r\theta} & T_{r\phi} \\ T_{\theta r} & T_{\theta\theta} & T_{\theta\phi} \\ T_{\phi r} & T_{\phi\theta} & T_{\phi\phi} \end{pmatrix} = \begin{pmatrix} ScS'c' + SsS's' + CC' & ScC'c' + SsC's' - CS' & -Scs' + Ssc' \\ CcS'c' + CsS's' - SC' & CcC'c' + CsC's' + SS' & -Ccs' + Csc' \\ -sS'c' + cS's' & -sC'c' + cC's' & ss' + cc' \end{pmatrix} \tag{B.7-8}$$

[a] The tensors $\boldsymbol{\Delta}$ and $\mathbf{E}$ are examples of quantities that are sometimes called "double tensor fields"; for more on these quantities and their transformations see J. L. Erickson, "Tensor Fields," in *Handbuch der Physik*, Springer, Berlin (1960), pp. 794–858, see §§15 and 16.

APPENDIX C

EQUATIONS AND TENSORS SPECIALIZED FOR HOMOGENEOUS SHEAR AND SHEARFREE FLOWS

TABLE C.1

Continuum-Mechanics Tensors

	Shear Flows	Shearfree Flows
Velocity distributions ⟹	$v_x = \dot{\gamma}_{yx}(t)y$ $v_y = 0$ $v_z = 0$	$v_x = -\frac{1}{2}(1+b)\dot{\varepsilon}(t)x$ $v_y = -\frac{1}{2}(1-b)\dot{\varepsilon}(t)y$ $v_z = +\dot{\varepsilon}(t)z$
$\nabla \boldsymbol{v}$	$\begin{pmatrix} 0 & 0 & 0 \\ 1 & 0 & 0 \\ 0 & 0 & 0 \end{pmatrix} \dot{\gamma}_{yx}(t)$	$\begin{pmatrix} -\frac{1}{2}(1+b) & 0 & 0 \\ 0 & -\frac{1}{2}(1-b) & 0 \\ 0 & 0 & 1 \end{pmatrix} \dot{\varepsilon}(t)$
$\dot{\boldsymbol{\gamma}} = \boldsymbol{\gamma}^{(1)} = \boldsymbol{\gamma}_{(1)}$	$\begin{pmatrix} 0 & 1 & 0 \\ 1 & 0 & 0 \\ 0 & 0 & 0 \end{pmatrix} \dot{\gamma}_{yx}(t)$	$\begin{pmatrix} -(1+b) & 0 & 0 \\ 0 & -(1-b) & 0 \\ 0 & 0 & 2 \end{pmatrix} \dot{\varepsilon}(t)$
$\boldsymbol{\gamma}^{(2)}$ [f,g]	$\begin{pmatrix} 0 & 1 & 0 \\ 1 & 0 & 0 \\ 0 & 0 & 0 \end{pmatrix} \frac{\partial \dot{\gamma}_{yx}}{\partial t} + \begin{pmatrix} 0 & 0 & 0 \\ 0 & 2 & 0 \\ 0 & 0 & 0 \end{pmatrix} \dot{\gamma}_{yx}^2$	$\begin{pmatrix} -(1+b) & 0 & 0 \\ 0 & -(1-b) & 0 \\ 0 & 0 & 2 \end{pmatrix} \frac{\partial \dot{\varepsilon}}{\partial t} + \begin{pmatrix} (1+b)^2 & 0 & 0 \\ 0 & (1-b)^2 & 0 \\ 0 & 0 & 4 \end{pmatrix} \dot{\varepsilon}^2$
$\boldsymbol{\gamma}_{(2)}$ [f,g]	$\begin{pmatrix} 0 & 1 & 0 \\ 1 & 0 & 0 \\ 0 & 0 & 0 \end{pmatrix} \frac{\partial \dot{\gamma}_{yx}}{\partial t} - \begin{pmatrix} 2 & 0 & 0 \\ 0 & 0 & 0 \\ 0 & 0 & 0 \end{pmatrix} \dot{\gamma}_{yx}^2$	$\begin{pmatrix} -(1+b) & 0 & 0 \\ 0 & -(1-b) & 0 \\ 0 & 0 & 2 \end{pmatrix} \frac{\partial \dot{\varepsilon}}{\partial t} - \begin{pmatrix} (1+b)^2 & 0 & 0 \\ 0 & (1-b)^2 & 0 \\ 0 & 0 & 4 \end{pmatrix} \dot{\varepsilon}^2$
$\boldsymbol{\tau} = \boldsymbol{\tau}^{(0)} = \boldsymbol{\tau}_{(0)}$	$\begin{pmatrix} \tau_{xx} & \tau_{yx} & 0 \\ \tau_{yx} & \tau_{yy} & 0 \\ 0 & 0 & \tau_{zz} \end{pmatrix}$	$\begin{pmatrix} \tau_{xx} & 0 & 0 \\ 0 & \tau_{yy} & 0 \\ 0 & 0 & \tau_{zz} \end{pmatrix}$
$\boldsymbol{\tau}^{(1)}$	$\frac{\partial}{\partial t}\begin{pmatrix} \tau_{xx} & \tau_{yx} & 0 \\ \tau_{yx} & \tau_{yy} & 0 \\ 0 & 0 & \tau_{zz} \end{pmatrix} + \begin{pmatrix} 0 & \tau_{xx} & 0 \\ \tau_{xx} & 2\tau_{yx} & 0 \\ 0 & 0 & 0 \end{pmatrix} \dot{\gamma}_{yx}(t)$	$\frac{\partial}{\partial t}\begin{pmatrix} \tau_{xx} & 0 & 0 \\ 0 & \tau_{yy} & 0 \\ 0 & 0 & \tau_{zz} \end{pmatrix} + \begin{pmatrix} -(1+b)\tau_{xx} & 0 & 0 \\ 0 & -(1-b)\tau_{yy} & 0 \\ 0 & 0 & 2\tau_{zz} \end{pmatrix} \dot{\varepsilon}(t)$

$\boldsymbol{\tau}_{(1)}$	$\frac{\partial}{\partial t}\begin{pmatrix}\tau_{xx} & \tau_{yx} & 0\\ \tau_{yx} & \tau_{yy} & 0\\ 0 & 0 & \tau_{zz}\end{pmatrix} - \begin{pmatrix}2\tau_{yx} & \tau_{yy} & 0\\ \tau_{yy} & 0 & 0\\ 0 & 0 & 0\end{pmatrix}\dot{\gamma}_{yx}(t)$	$\frac{\partial}{\partial t}\begin{pmatrix}\tau_{xx} & 0 & 0\\ 0 & \tau_{yy} & 0\\ 0 & 0 & \tau_{zz}\end{pmatrix} - \begin{pmatrix}-(1+b)\tau_{xx} & 0 & 0\\ 0 & -(1-b)\tau_{yy} & 0\\ 0 & 0 & 2\tau_{zz}\end{pmatrix}\dot{\varepsilon}(t)$
Displacement functions ⟹	$x = x' - y'\gamma_{yx}(t, t')$ [b] $y = y'$ $z = z'$ where γ_{yx} is the shear strain $t \to t'$ $\gamma_{yx}(t, t') = \int_t^{t'} \dot{\gamma}_{yx}(t'')dt''$	$x = \lambda_x x';\ \lambda_x = \exp[\frac{1}{2}(1+b)\varepsilon(t, t')]$ [c,d] $y = \lambda_y y';\ \lambda_y = \exp[\frac{1}{2}(1-b)\varepsilon(t, t')]$ $z = \lambda_z z';\ \lambda_z = \exp[-\varepsilon(t, t')]$ where the $(\lambda_x, \lambda_y, \lambda_z)$ are the "principal elongation ratios" from $t' \to t$ and $\varepsilon(t, t') = \int_t^{t'} \dot{\varepsilon}(t'')dt''$
$\boldsymbol{\Delta}(t, t')$	$\begin{pmatrix}1 & \gamma_{yx} & 0\\ 0 & 1 & 0\\ 0 & 0 & 1\end{pmatrix}$	$\begin{pmatrix}\lambda_x^{-1} & 0 & 0\\ 0 & \lambda_y^{-1} & 0\\ 0 & 0 & \lambda_z^{-1}\end{pmatrix}$
$\mathbf{E}(t, t')$	$\begin{pmatrix}1 & -\gamma_{yx} & 0\\ 0 & 1 & 0\\ 0 & 0 & 1\end{pmatrix}$	$\begin{pmatrix}\lambda_x & 0 & 0\\ 0 & \lambda_y & 0\\ 0 & 0 & \lambda_z\end{pmatrix}$
$\boldsymbol{\gamma}^{[0]}(t, t') = \mathbf{B}^{-1}(t, t') - \boldsymbol{\delta}$	$\begin{pmatrix}0 & \gamma_{yx} & 0\\ \gamma_{yx} & \gamma_{yx}^2 & 0\\ 0 & 0 & 0\end{pmatrix}$	$\begin{pmatrix}\lambda_x^{-2}-1 & 0 & 0\\ 0 & \lambda_y^{-2}-1 & 0\\ 0 & 0 & \lambda_z^{-2}-1\end{pmatrix}$
$\boldsymbol{\gamma}_{[0]}(t, t') = \boldsymbol{\delta} - \mathbf{B}(t, t')$	$\begin{pmatrix}-\gamma_{yx}^2 & \gamma_{yx} & 0\\ \gamma_{yx} & 0 & 0\\ 0 & 0 & 0\end{pmatrix}$	$\begin{pmatrix}1-\lambda_x^2 & 0 & 0\\ 0 & 1-\lambda_y^2 & 0\\ 0 & 0 & 1-\lambda_z^2\end{pmatrix}$ [d]
$\boldsymbol{\gamma}^{[1]}(t, t') = \frac{\partial}{\partial t'}\boldsymbol{\gamma}^{[0]}(t, t')$	$\begin{pmatrix}0 & 1 & 0\\ 1 & 2\gamma_{yx} & 0\\ 0 & 0 & 0\end{pmatrix}\dot{\gamma}_{yx}(t')$	$\begin{pmatrix}-(1+b)\lambda_x^{-2} & 0 & 0\\ 0 & -(1-b)\lambda_y^{-2} & 0\\ 0 & 0 & 2\lambda_z^{-2}\end{pmatrix}\dot{\varepsilon}(t')$

(Continued)

TABLE C.1 *(Continued)*

$\boldsymbol{\gamma}_{[1]}(t,t') = \frac{\partial}{\partial t'}\boldsymbol{\gamma}_{[0]}(t,t')$		$\begin{pmatrix} -2\gamma_{yx} & 1 & 0 \\ 1 & 0 & 0 \\ 0 & 0 & 0 \end{pmatrix}\dot{\gamma}_{yx}(t')$	$\begin{pmatrix} -(1+b)\lambda_x^2 & 0 & 0 \\ 0 & -(1-b)\lambda_y^2 & 0 \\ 0 & 0 & 2\lambda_z^2 \end{pmatrix}\dot{\varepsilon}(t')$
$\boldsymbol{\tau}^{[0]}$	[e]	$\begin{pmatrix} \tau'_{xx} & \tau'_{yx}+\gamma_{yx}\tau'_{xx} & 0 \\ \tau'_{yx}+\gamma_{yx}\tau'_{xx} & \tau'_{yy}+2\gamma_{yx}\tau'_{yx}+\gamma_{yx}^2\tau'_{xx} & 0 \\ 0 & 0 & \tau'_{zz} \end{pmatrix}$	$\begin{pmatrix} \lambda_x^{-2}\tau'_{xx} & 0 & 0 \\ 0 & \lambda_y^{-2}\tau'_{yy} & 0 \\ 0 & 0 & \lambda_z^{-2}\tau'_{zz} \end{pmatrix}$
$\boldsymbol{\tau}_{[0]}$	[e]	$\begin{pmatrix} \tau'_{xx}-2\gamma_{yx}\tau'_{yx}+\gamma_{yx}^2\tau'_{xx} & \tau'_{yx}-\gamma_{yx}\tau'_{yy} & 0 \\ \tau'_{yx}-\gamma_{yx}\tau'_{yy} & \tau'_{yy} & 0 \\ 0 & 0 & \tau'_{zz} \end{pmatrix}$	$\begin{pmatrix} \lambda_x^2\tau'_{xx} & 0 & 0 \\ 0 & \lambda_y^2\tau'_{yy} & 0 \\ 0 & 0 & \lambda_z^2\tau'_{zz} \end{pmatrix}$
I_1	[h]	$3+\gamma_{yx}^2$	$\lambda_x^2+\lambda_y^2+\lambda_z^2$
I_2		$3+\gamma_{yx}^2$	$\lambda_x^{-2}+\lambda_y^{-2}+\lambda_z^{-2}$
I_3		1	$\lambda_x^2\lambda_y^2\lambda_z^2 = 1$

[a] $\dot{\gamma}_{yx}$ is the yx-component of the rate-of-strain tensor $\dot{\boldsymbol{\gamma}}$. For steady-state (shear) flow we replace $\dot{\gamma}_{yx}(t)$ by the shear rate $\dot{\gamma} = \sqrt{\frac{1}{2}(\boldsymbol{\gamma}_{(1)}:\boldsymbol{\gamma}_{(1)})}$; $\dot{\gamma}_{yx}(t)$ may be positive or negative, but $\dot{\gamma}$ is always positive.

[b] γ_{yx} is the yx-component of the infinitesimal strain tensor as defined in Eq. 5.2-7. For steady-state shear flow $\gamma_{yx}(t,t') = -\dot{\gamma}(t-t')$.

[c] When $b = 0$ (elongational flow), $\varepsilon(t,t')$ is the Hencky strain of the deformation from $t \to t'$.

[d] The principal elongation ratios are defined for the deformation from $t' \to t$ to be consistent with the usage of λ_x^2, λ_y^2, λ_z^2 as the eigenvalues of $\mathbf{B}(t,t')$.

[e] In the entries for $\boldsymbol{\tau}^{[0]}$ and $\boldsymbol{\tau}_{[0]}$ the quantity $\tau_{ij}(t')$ is abbreviated as τ'_{ij}.

[f] For steady shear flow, $\boldsymbol{\gamma}_{(n)}$ and $\boldsymbol{\gamma}^{(n)}$ are zero for $n \geq 3$.

[g] For steady shearfree flows, $\boldsymbol{\gamma}_{(n)} = -\{\boldsymbol{\gamma}_{(1)}\cdot\boldsymbol{\gamma}_{(n-1)}\} = (-1)^{n+1}\boldsymbol{\gamma}_{(1)}^n$ and $\boldsymbol{\gamma}^{(n)} = \{\boldsymbol{\gamma}^{(1)}\cdot\boldsymbol{\gamma}^{(n-1)}\} = \boldsymbol{\gamma}^{(1)n}$ for $n \geq 2$.

[h] $I_1 = \text{tr}\,\mathbf{B} = \text{tr}(\boldsymbol{\delta} - \boldsymbol{\gamma}_{[0]})$; $I_2 = \text{tr}\,\mathbf{B}^{-1} = \text{tr}(\boldsymbol{\delta} + \boldsymbol{\gamma}^{[0]})$; and $I_3 = \det\mathbf{B} = 1$ for incompressible materials.

TABLE C.2

Oldroyd 8-Constant and Giesekus Equations[a] (Eqs. 7.3-2 and 7.3-4)

Shear Flows: $(v_x = \dot{\gamma}_{yx}(t)y,\ v_y = 0,\ v_z = 0)$

$$(xx):\left[1 + \lambda_1 \frac{d}{dt}\right]\tau_{xx} - (2\lambda_1 - \lambda_3 - \lambda_6)\tau_{yx}\dot{\gamma}_{yx} - a\frac{\lambda_1}{\eta_0}(\tau_{xx}^2 + \tau_{yx}^2) = \eta_0(2\lambda_2 - \lambda_4 - \lambda_7)\dot{\gamma}_{yx}^2 \quad \text{(A)}$$

$$(yy):\left[1 + \lambda_1 \frac{d}{dt}\right]\tau_{yy} + (\lambda_3 + \lambda_6)\tau_{yx}\dot{\gamma}_{yx} - a\frac{\lambda_1}{\eta_0}(\tau_{yx}^2 + \tau_{yy}^2) = -\eta_0(\lambda_4 + \lambda_7)\dot{\gamma}_{yx}^2 \quad \text{(B)}$$

$$(zz):\left[1 + \lambda_1 \frac{d}{dt}\right]\tau_{zz} + \lambda_6\tau_{yx}\dot{\gamma}_{yx} - a\frac{\lambda_1}{\eta_0}\tau_{zz}^2 = -\eta_0\lambda_7\dot{\gamma}_{yx}^2 \quad \text{(C)}$$

$$(yx):\left[1 + \lambda_1 \frac{d}{dt}\right]\tau_{yx} + \tfrac{1}{2}(\lambda_3 + \lambda_5)\tau_{xx}\dot{\gamma}_{yx} - \tfrac{1}{2}(2\lambda_1 - \lambda_3 - \lambda_5)\tau_{yy}\dot{\gamma}_{yx} + \tfrac{1}{2}\lambda_5\tau_{zz}\dot{\gamma}_{yx} - a\frac{\lambda_1}{\eta_0}(\tau_{xx} + \tau_{yy})\tau_{yx} = -\eta_0\left[1 + \lambda_2\frac{d}{dt}\right]\dot{\gamma}_{yx} \quad \text{(D)}$$

$$\tau_{zx} = \tau_{yz} = 0$$

(Continued)

TABLE C.2 *(Continued)*

Shearfree Flows: $(v_x = -\frac{1}{2}(1+b)\dot{\varepsilon}(t)x,\ v_y = -\frac{1}{2}(1-b)\dot{\varepsilon}(t)y,\ v_z = +\dot{\varepsilon}(t)z)$

$$(xx){:}\ \lambda_1 \frac{d\tau_{xx}}{dt} + \tau_{xx}[1 + (1+b)(\lambda_1 - \lambda_3 - \tfrac{1}{2}\lambda_5 - \tfrac{1}{2}\lambda_6)\dot{\varepsilon}] - a\frac{\lambda_1}{\eta_0}\tau_{xx}^2 - \tfrac{1}{2}\tau_{yy}[(1+b)\lambda_5 + (1-b)\lambda_6]\dot{\varepsilon} + \tau_{zz}[\lambda_6 - \tfrac{1}{2}(1+b)\lambda_5]\dot{\varepsilon} \tag{E}$$

$$= \eta_0\left[(1+b)\dot{\varepsilon} + (1+b)\lambda_2\frac{d\dot{\varepsilon}}{dt} - \dot{\varepsilon}^2[(1+b)^2(\lambda_4 - \lambda_2) + (3+b^2)\lambda_7]\right]$$

$$(yy){:}\ \lambda_1 \frac{d\tau_{yy}}{dt} + \tau_{yy}[1 + (1-b)(\lambda_1 - \lambda_3 - \tfrac{1}{2}\lambda_5 - \tfrac{1}{2}\lambda_6)\dot{\varepsilon}] - a\frac{\lambda_1}{\eta_0}\tau_{yy}^2 - \tfrac{1}{2}\tau_{xx}[(1-b)\lambda_5 + (1+b)\lambda_6]\dot{\varepsilon} + \tau_{zz}[\lambda_6 - \tfrac{1}{2}(1-b)\lambda_5]\dot{\varepsilon} \tag{F}$$

$$= \eta_0\left[(1-b)\dot{\varepsilon} + (1-b)\lambda_2\frac{d\dot{\varepsilon}}{dt} - \dot{\varepsilon}^2[(1-b)^2(\lambda_4 - \lambda_2) + (3+b^2)\lambda_7]\right]$$

$$(zz){:}\ \lambda_1 \frac{d\tau_{zz}}{dt} + \tau_{zz}[1 - 2(\lambda_1 - \lambda_3 - \tfrac{1}{2}\lambda_5 - \tfrac{1}{2}\lambda_6)\dot{\varepsilon}] - a\frac{\lambda_1}{\eta_0}\tau_{zz}^2 + \tau_{xx}[\lambda_5 - \tfrac{1}{2}(1+b)\lambda_6]\dot{\varepsilon} + \tau_{yy}[\lambda_5 - \tfrac{1}{2}(1-b)\lambda_6]\dot{\varepsilon} \tag{G}$$

$$= -\eta_0\left[2\dot{\varepsilon} + 2\lambda_2\frac{d\dot{\varepsilon}}{dt} + \dot{\varepsilon}^2[4(\lambda_4 - \lambda_2) + (3+b^2)\lambda_7]\right]$$

[a] Equations for the Oldroyd 8-constant model are obtained by setting $a = 0$; equations for the Giesekus model are obtained by setting λ_5 through $\lambda_7 = 0$, $\lambda_3 = -2a\lambda_2$, and $\lambda_4 = -a\lambda_2^2/\lambda_1$. See §7.3.

AUTHOR INDEX

FOR VOLUME 1

SUBJECT INDEX

FOR VOLUMES 1 AND 2

Page numbers following (1) refer to Volume 1, "Fluid Mechanics."
Page numbers following (2) refer to Volume 2, "Kinetic Theory."